THE PERIODIC TABLE OF ELEMENTS (Long Period Form)

There are seven periods (horizontal rows) and eighteen groups (vertical columns) of elements. The number of electrons in filled shells is shown in the column at the extreme left; the remaining electrons for each element are shown immediately below the symbol for each element. Atomic numbers are enclosed by brackets. Atomic weights are shown above the symbols. Atomic weight values in parentheses are those of the isotopes of longest half-life for certain radioactive elements whose atomic weights cannot be quoted precisely without knowledge of origin of the element. Atomic weights are based on carbon-12. *(By permission of International Union of Pure and Applied Chemistry, 1975.)*

METALS · TRANSITION METALS · NONMETALS

PERIODS	I A	II A	III B	IV B	V B	VI B	VII B	VIII	VIII	VIII	I B	II B	III A	IV A	V A	VI A	VII A	0
1 (0)	1.0079 H[1] 1																	4.00260 He[2] 2
2 (2)	6.941 Li[3] 1	9.01218 Be[4] 2											10.81 B[5] 3	12.011 C[6] 4	14.0067 N[7] 5	15.9994 O[8] 6	18.99840 F[9] 7	20.179 Ne[10] 8
3 (2,8)	22.98977 Na[11] 1	24.305 Mg[12] 2											26.98154 Al[13] 3	28.086 Si[14] 4	30.97376 P[15] 5	32.06 S[16] 6	35.453 Cl[17] 7	39.948 Ar[18] 8
4 (2,8)	39.098 K[19] 8,1	40.08 Ca[20] 8,2	44.9559 Sc[21] 9,2	47.90 Ti[22] 10,2	50.9414 V[23] 11,2	51.996 Cr[24] 13,1	54.9380 Mn[25] 13,2	55.847 Fe[26] 14,2	58.9332 Co[27] 15,2	58.70 Ni[28] 16,2	63.546 Cu[29] 18,1	65.38 Zn[30] 18,2	69.72 Ga[31] 18,3	72.59 Ge[32] 18,4	74.9216 As[33] 18,5	78.96 Se[34] 18,6	79.904 Br[35] 18,7	83.80 Kr[36] 18,8
5 (2,8,18)	85.4678 Rb[37] 8,1	87.62 Sr[38] 8,2	88.9059 Y[39] 9,2	91.22 Zr[40] 10,2	92.9064 Nb[41] 12,1	95.94 Mo[42] 13,1	(97) Tc[43] 14,1	101.07 Ru[44] 15,1	102.9055 Rh[45] 16,1	106.4 Pd[46] 18	107.868 Ag[47] 18,1	112.41 Cd[48] 18,2	114.82 In[49] 18,3	118.69 Sn[50] 18,4	121.75 Sb[51] 18,5	127.60 Te[52] 18,6	126.9045 I[53] 18,7	131.30 Xe[54] 18,8
6 (2,8,18)	132.9054 Cs[55] 18,8,1	137.33 Ba[56] 18,8,2	[57-71] *	178.49 Hf[72] 32,10,2	180.9479 Ta[73] 32,11,2	183.85 W[74] 32,12,2	186.207 Re[75] 32,13,2	190.2 Os[76] 32,14,2	192.22 Ir[77] 32,15,2	195.09 Pt[78] 32,17,1	196.9665 Au[79] 32,18,1	200.59 Hg[80] 32,18,2	204.37 Tl[81] 32,18,3	207.2 Pb[82] 32,18,4	208.9804 Bi[83] 32,18,5	(209) Po[84] 32,18,6	(210) At[85] 32,18,7	(222) Rn[86] 32,18,8
7 (2,8,18,32)	(223) Fr[87] 18,8,1	226.0254 Ra[88] 18,8,2	[89-103] †	Rf[104] 32,10,2	Ha[105] 32,11,2	[106] 32,12,2	[107]	[108]										

*LANTHANIDE SERIES

138.9055 La[57] 18,9,2	140.12 Ce[58] 20,8,2	140.9077 Pr[59] 21,8,2	144.24 Nd[60] 22,8,2	(145) Pm[61] 23,8,2	150.4 Sm[62] 24,8,2	151.96 Eu[63] 25,8,2	157.25 Gd[64] 25,9,2	158.9254 Tb[65] 27,8,2	162.50 Dy[66] 28,8,2	164.9304 Ho[67] 29,8,2	167.26 Er[68] 30,8,2	168.9342 Tm[69] 31,8,2	173.04 Yb[70] 32,8,2	174.97 Lu[71] 32,9,2

†ACTINIDE SERIES

(227) Ac[89] 18,9,2	232.0381 Th[90] 18,10,2	231.0359 Pa[91] 20,9,2	238.029 U[92] 21,9,2	237.0482 Np[93] 23,8,2	238.029 Pu[94] 24,8,2	(243) Am[95] 25,8,2	(247) Cm[96] 25,9,2	(247) Bk[97] 26,9,2	(251) Cf[98] 28,8,2	(254) Es[99] 29,8,2	(257) Fm[100] 30,8,2	(258) Md[101] 31,8,2	(255) No[102] 32,8,2	(260) Lr[103] 32,9,2

Preface

The *Fifth Edition* is a thorough revision that the authors believe further improves the qualities which have contributed to the wide acceptance of this introductory chemistry text. Theory and principles have been increased with a small reduction in descriptive chemistry, particularly that of the lesser known elements, to provide an excellent balance between theory and reaction chemistry. The teaching philosophy of the authors is that a blend of the two constitutes the best method of teaching freshman chemistry and, in fact, is essential to providing a proper background for both science and nonscience students. All material is presented at a sound pedagogical level, with close attention given to achieving a clear presentation that students can understand and, hopefully, enjoy.

College Chemistry, Fifth Edition, as in previous editions, has the study of the metals according to the qualitative analysis scheme. Its companion edition, *General Chemistry, Fifth Edition,* (by the same authors) is identical in level and content, but differs in organization with the metals being studied according to the periodic groups. The content is the same in both texts for the first thirty-three chapters.

College Chemistry, Fifth Edition, is designed primarily for use in courses which integrate general chemistry and qualitative analysis and includes a supplement of analytical procedures. Ample attention is given throughout the book to the periodic relationships of the elements, however. The analytical procedures in the supplement are essentially those developed by A. A. Noyes. The principal objective of qualitative analysis in any elementary course is the chemistry of the common inorganic cations and anions. Hence, inorganic reagents are specified almost exclusively in the supplement. Emphasis is placed on understanding the reasons for the various steps utilized in the analytical procedures.

Chief among the many new features that have been incorporated in the Fifth Edition to improve its usefulness and to bring it completely up-to-date are two new chapters. In one of these chapters, *Molecular Structure* (Chapter 6), the valence shell electron-pair repulsion theory and the valence bond concept are discussed as methods for predicting the molecular structures of a wide variety of molecules and ions. In the other new chapter, *Spectroscopy and Chromatography* (Chapter 33), several methods of characterizing substances are covered, with particular emphasis on visible, ultraviolet, infrared, nuclear magnetic resonance, and electron paramagnetic resonance spectroscopy; mass spectrometry; and gas phase and liquid phase chromatography. Examples of these methods being used to predict structure and to characterize compounds are given in each of the new chapters, to aid the student in understanding the methods.

The chapter *Chemical Thermodynamics* maintains the substantial depth of treatment, without the use of calculus, which characterized it in the Fourth Edition. But there is now increased emphasis on entropy, as one of the most important concepts of science; an expanded discussion on predicting reaction spontaneity; and attitional worked-out examples of thermo problems.

Several changes in order of presentation of material have been made in the Fifth

Edition, both as a result of suggestions by users of previous editions and the experiences of the authors in their own teaching. Changes in order of presentation include (1) placing the chapter *The Relationship of the Periodic Classification to the Properties of the Elements* earlier in the text, as Chapter 7, to follow more closely the discussion of periodic classification in Chapter 3 and to utilize the background provided by the discussions of atomic structure, bonding, and molecular structure in Chapters 3, 4, 5, and 6; (2) making *Ionic Equilibria of Weak Electrolytes* and *The Solubility Product Principle,* old Chapters 30 and 31, new Chapters 18 and 19, so that problems involving equilibrium constants immediately follow discussions of chemical equilibrium (Chapter 17); (3) consolidation of material on the gaseous state and the kinetic-molecular theory, formerly Chapters 8 and 9, into a unified treatment in one chapter (Chapter 10); (4) incorporation of a somewhat shortened treatment of colloid chemistry, previously a separate chapter, with *Solutions of Electrolytes* to form a single unit (Chapter 14); and (5) placing the chapter *Acids, Bases, and Salts* earlier in the text, as Chapter 15, so as to follow immediately after the two chapters on solutions.

It is believed that the revised order of presentation in the *Fifth Edition* will please both old and new users of the text. It should be pointed out, however, that those who may prefer to use the order of presentation in the previous edition will have no difficulty in doing so.

Other special features of the *Fifth Edition* include:

(1) Expanded sections on atomic spectra and electronic structure that include a discussion of the Rydberg constant and several worked-out problems on the energy changes that occur during electron transitions.

(2) Discussions of pollution and the environment brought completely up-to-date and placed throughout the text where relevant.

(3) Additional worked-out problems using the General Gas Law Equation, but only after preliminary emphasis on working problems from basic principles of the gas laws and a careful derivation of the General Gas Law Equation from basic principles.

(4) In depth treatment of ionic radii in crystal structures, including several worked-out examples of calculations of ionic radii from unit cell dimensions.

(5) A discussion of the radius-ratio rule for crystalline ionic compounds.

(6) Increased emphasis on heterogenous and homogeneous catalysts.

(7) Increased emphasis on solvents other than water.

(8) Additional worked-out examples of problems involving equilibrium concepts, including gaseous-state equilibrium (Chapter 17).

(9) New material on order of reaction and half-lives of first and second order reactions, including application of the half-life concept to nuclear reactions; many worked-out examples of problems involving half-lives and time of reaction in the discussions of first and second order reactions in general (Chapter 17) and for nuclear reactions in particular (Chapter 30).

(10) Increased emphasis on calculations that involve equilibrium constants for polyprotic acids, including use of the principle of successive approximations (Chapter 18).

(11) A more in-depth discussion of metal-to-metal bonds, with diagrams of some typical compounds which have metal-to-metal bonds.

(12) Expanded treatment of the metallic bond and of intermetallic compounds, including the significance of the ratio of outer shell electrons to number of atoms.

(13) Increased attention to units and cancellation of units in working out answers to problems, with continued emphasis on significant figures in all problems.

(14) New appendix table listing the half-lives for many isotopes.

(15) Improved artwork and addition of several new figures and table to increase clarity of presentation.

(16) Updating of all topics and inclusion of new developments, for example, preparation of element 106, photosynthesis in plants, the 1975 recommendations of the International Union of Pure and Applied Chemistry for atomic weight values, catalytic converters, no-lead gasoline, effects of ozone on the environment, solar energy, and fuel cells.

(17) An unusually extensive index containing over 5,000 entries.

(18) Updated list of references at the end of each chapter for enrichment reading.

(19) Completely revised end-of-chapter problems totaling over 600 in number; the problems range in difficulty from easy straightforward ones that illustrate principles, to some that will challenge the best students. An increased number of worked-out problems within the chapters prepare the student for those at the end. More than 1000 questions round out an unusually complete set of end-of-chapter exercises.

The authors have meticulously examined the present edition word-by-word and have rewritten many phrases and sentences for added clarity in the new edition.

The number of supplemental materials especially designed for use with the text has been increased with the addition of a new manual, *Problems and Solutions for General and College Chemistry, Fifth Editions,* prepared by John H. Meiser, F. Keith Ault (both of Ball State University), and Henry F. Holtzclaw, Jr. The manual includes the worked-out solutions to approximately one-third of the text problems, those marked ⑤, at the ends of chapters. A study guide to assist the student, written by Norman E. Griswold of Nebraska Wesleyan University, and an instructor's guide by the same author are both available from the publisher of the text. Available also from the publisher is *Basic Laboratory Studies in College Chemistry, Fifth Edition,* by Grace Hered, William H. Nebergall, and William Hered. This manual complements the text but is also suitable for use with most other beginning chemistry texts.

We are indebted to the Sadtler Research Laboratories, Philadelphia, Pennsylvania, and to Varian Associates, Palo Alto, California, for the infrared and nuclear magnetic resonance spectra used as examples in the new Chapter 33. The authors express their gratitude to Professor John C. Bailar, Jr., of the University of Illinois, consulting editor for D. C. Heath and Company, for his detailed and thorough reading of the manuscript. His many excellent suggestions resulted in a significant improvement of the text. Professor Robert S. Marianelli, University of Nebraska, formulated the new problem sets and checked several sections of the manuscript for accuracy and clarity. Professor W. F. Couser, San Antonio College, read the entire manscript and made a number of helpful suggestions. Others who read sections of the manuscript and contributed much by their suggestions include Professors John H. Meiser and F. Keith Ault, both of Ball State University, and Professor Robin J. Hood, now of Marygrove College, Detroit. David Busby, University of Nebraska, independently checked the answers to all problems at the ends of chapters. Jean Holtzclaw typed the entire manuscript and reports that the increase in her knowledge of chemistry with each page typed helped to offset the tedium of the typing job! In addition, the authors wish to thank the many users of previous editions whose suggestions and constructive criticism brought about a number of improvements.

The authors also express their appreciation to Jeff Holtmeier and Cynthia Nickerson of the editorial staff of D. C. Heath and Company for their gracious help and willing cooperation. Above all, the authors owe a special debt of gratitude to Dr. Paul P. Bryant, Science Editor of D. C. Heath, College Division. We are convinced that there is no finer science editor anywhere. It has been a stimulating experience and true privilege to work with such an outstanding person.

Contents

Some Fundamental Concepts

1

Within the last two decades, science and technology have acquired what can be described as an ambivalent reputation. On the one side, many valuable contributions to our lives have been made; but on another side, the contributions have sometimes led to some circumstances that have not been for the good of all. For example, the same fertilizers and insecticides that made possible great increases in the production of food have polluted our air and water. The automobile and the jet airplane revolutionized travel and the movement of goods, but they also have contaminated much of the air we breathe. We worry justifiably about the toxicity of some of the very medicines that have saved countless lives. Energy is the key to our existence. Yet we deplore the destruction of natural beauty caused by strip mining for coal; the pollution of our oceans, lakes, and beaches by oil spills; the possibility of excess radioactive contamination from accidents in nuclear plants; and the alarmingly rapid depletion of our natural sources of energy. It is readily apparent that the interaction and interdependence of various aspects of our society bring about good results but also vexing and often unexpected problems; problems which we have sometimes been slow to recognize and almost always slow to act on.

Still, most of the problems created by science, in the development of new products and processes, can also be solved by science through a cooperative effort with many different occupations.

Today, chemists as member of the scientific community face great challenges. They can legitimately take well deserved credit for many scientific and technological advances, but they must also share in the responsibility for solving problems that are created by some of these same advances. The responsibility is not the chemists' alone, however. It is the *joint* responsibility of the physicists, the sociologists, the biologists, the theologians, the political scientists, *and* the chemists. Together they must work to solve the problems of a world which sometimes seems to be trying to destroy itself in the process of its own technological progress. No single discipline offers the total perspective or the expertise to accomplish this task. Each discipline has specialized knowledge which, together with that of other disciplines, can contribute to bringing about a better life for all peoples. We must draw upon all of our knowledge if we are to achieve that goal while at the same time preserving ethical and moral standards.

There is, therefore, a great need for chemists to know their field extremely well and to have the necessary understanding and perspective in chemistry and other fields to work effectively for the wise use of their discoveries. Equally important is the need for persons in the other fields to know enough chemistry, and the other sciences, to allow them to apply their specialized knowledge to problems involving science. The question as to why a person should today study chemistry or any other science, regardless of the intended main field of endeavor, has never had a more demanding or obvious answer.

Introduction

1.1 Chemistry

In our study of chemistry we shall be concerned with the composition and structure of matter. Such forms of matter as wood and glass, water and gasoline, salt and sugar, coal and granite, and iron and gold differ strikingly from each other in many ways. These differences result from differences in the composition and structure of the various substances. This is not the whole story of chemistry, however, for matter is not static. Our existence, in fact, depends upon changes which occur in matter. The changes that take place when food is digested and assimilated by the body are critical to the life process. The changes occurring during the combustion in a gasoline engine cause the formation of pollutant gases that can be extremely harmful to our health. Detergents in sewage-waste can produce changes in the composition of rivers that may ultimately cause huge fish-kills, thereby affecting our food supply.

As we proceed in our study of chemistry, we shall examine some of the changes in the composition and structure of matter, the causes which produce these changes, the changes in energy which accompany them, and the principles and laws involved in these changes. In brief, then, the science of **chemistry is the study of the composition, structure, and properties of matter, and the changes that matter undergoes.**

The amount of chemical knowledge has become so vast during the last century that chemists usually specialize in one of several principal branches. **Analytical chemistry** is concerned with the identification, separation, and quantitative determination of the composition of different substances. **Physical chemistry** is primarily concerned with the structure of matter, energy changes, and the laws, principles, and theories which explain the transformations of one form of matter into another. **Organic chemistry** is the branch dealing with the compounds of carbon. **Inorganic chemistry** is concerned with the chemistry of elements other than carbon and their compounds. **Biochemistry** is the chemistry of the substances comprising living organisms. A course in **general chemistry** is a survey of all of the branches of chemistry and introduces the student to the entire field of the science.

It should be mentioned that the boundaries between the branches of chemistry are arbitrarily defined. Much of the work in chemistry cuts across these defined boundaries.

1.2   Matter and Energy

Matter, by definition, is anything that occupies space and has mass. All the objects in the universe, since they occupy space and have mass, are composed of matter. The property of occupying space is often easily perceived by our senses of sight and touch. The property of **mass** of an object pertains to the quantity of matter that the object contains. Hence, the force required to give an object a given acceleration, or the resistance of the object to being moved (inertia), is a measure of its mass.

Energy can be defined as the capacity for doing work. We recognize heat, electricity, and light as forms of energy when we see that heat generates the steam that drives the steam turbine, electricity causes electric motors to turn, and light is consumed in the manufacture of foods by plants. Scientists recognize two kinds of energy: potential energy and kinetic energy. A piece of matter is said to possess **potential energy** by virtue of its position, condition, or composition. Water at the top of a waterfall possesses potential energy because of its position; if it falls in a hydroelectric plant it does work and produces electricity. A compressed spring, because of its compression, can do work such as making a clock run. Natural gas is recognized as having potential energy, because it will burn and produce heat that does work, for example, in a steam turbine.

When a body is in motion, it also has energy—the capacity for doing work. The energy which a body possesses because of its motion is called **kinetic energy.** Kinetic energy is the kind of energy which can be transferred most readily from one system to another.

1.3   Law of Conservation of Matter and Energy

When a piece of copper metal is burned in air it unites with oxygen in the air. If the product (copper oxide) is collected and weighed, it is found to weigh more than the original piece of metal. If, however, the weight of the oxygen of the air that combines with the burning metal is taken into consideration, it can be shown that the weight of the product (copper oxide) is, within the limits of accuracy of the weighing instrument, equal to the sum of the weights of the reactants (copper and oxygen). This behavior of matter is in accord with what is called the **Law of Conservation of Matter: There is no detectable increase or decrease in the quantity of matter during an ordinary chemical change.**

Chemical changes are always accompanied by either the conversion of chemical energy into other forms of energy, or of other forms of energy into chemical energy. Usually, heat energy is evolved or absorbed, but sometimes the conversion involves light or electrical energy instead of, or in addition to, heat energy. Many transformations of energy, of course, do not involve chemical changes. Electrical energy can be converted into mechanical, light, heat, or chemical energy without chemical changes. Mechanical energy is converted into electrical energy in the dynamo. Potential and kinetic energy can be converted into one another. Many other conversions are possible. All of the energy involved in any change appears in some form after the change is completed. This fact is expressed in the **Law of Conservation of Energy: Energy cannot be created or destroyed, although it can be changed in form.**

In more precise terms, it is necessary to regard matter and energy not as distinct entities, but as two forms of a single entity. In the decomposition of atoms, such as takes place in the production of nuclear energy, matter is actually converted into energy. The reverse conversion of energy into matter can also be demonstrated. This interconversion of matter and energy takes place in ordinary chemical reactions, but the total mass of the products differs so slightly from the mass of the reactants that it is impossible to measure the difference experimentally. Thus, for practical purposes we say that the law of conservation of matter holds in ordinary chemical reactions. However, to be more precise, we may combine the separate laws of conservation of matter and energy into one statement: **The total quantity of matter and energy available in the universe is fixed.**

1.4   States of Matter

Matter can exist in three different states, designated as solid, liquid, and gas (Fig. 1-1). Each state can be distinguished by certain characteristics.

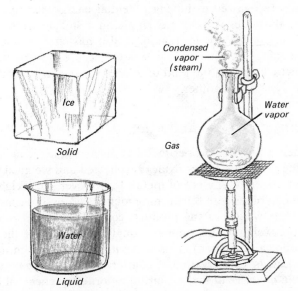

FIGURE 1-1

The three states of matter as illustrated by water.

A substance in the **solid state** is rigid, possessing a definite shape, and has a volume which is very nearly independent of changes in temperature and pressure.

A **liquid** flows and thus takes the shape of its container, except that it assumes a horizontal surface. Liquids are only slightly compressible and so for practical purposes have definite volumes.

A substance in the **gaseous (vapor) state** takes both the shape and volume of its container. Gases are readily compressible and capable of infinite expansion.

1.5 Chemical and Physical Properties

The characteristics which enable us to distinguish one substance from another are known as **properties.** Those characteristics involved in a transformation of one substance into another are known as **chemical properties.** Thus, a chemical property is exhibited when wood burns, for the constituents of the wood combine with oxygen to form substances different from the wood or the oxygen. A chemical property is also exhibited when iron combines with oxygen and water to form the very different reddish-brown iron oxide which we know as iron rust.

In addition to chemical properties, every substance also possesses definite **physical properties**—those which do not involve a change in composition of the material. Some familiar physical properties of matter are color, hardness, crystalline form, ductility, malleability, physical state, melting point, boiling point, density, electrical and thermal conductivity, and specific heat. Taste and odor are often classed as physical properties, but these sensations actually involve chemical changes. Changes in physical conditions such as temperature or pressure may modify the physical properties of a substance. For example, a substance in the gaseous state has certain physical properties such as density, specific heat, and thermal conductivity. By decreasing the temperature of the substance or compressing it to a smaller volume, we may change it from the gaseous to the liquid state, in which condition it has an entirely different density, specific heat, and thermal conductivity.

A substance can be identified by its chemical and physical properties and by its composition because no two substances are alike in all respects.

1.6 Chemical and Physical Changes

When charcoal, a solid black substance composed mostly of carbon, burns in air, an invisible gas consisting of both carbon and oxygen (carbon dioxide) is formed. When milk sours, the sugar in the milk is converted into an acid, and the composition and the properties of the acid differ greatly from those of the sugar. Iron rust formed by the corrosion of iron metal contains oxygen as well as iron, and it is therefore a different substance with different properties than those of iron metal. All such changes are called **chemical changes.** A chemical change always produces one or more substances entirely different in chemical composition and in properties from those that existed before the change occurred. In addition, all chemical changes are accompanied by either the production or absorption of some form of energy.

Changes that do not alter the composition of a substance are known as **physical changes.** The melting of ice, the freezing of water, the conversion of water to steam, the condensation of steam to water, and the magnetization of iron are all examples of physical change. In each of these there is a change in properties but there is no alteration of the chemical composition of the substances involved. Water in either the solid, liquid, or gaseous state has the same chemical composition. Iron is still the same substance whether magnetized or not.

1.7   Substances

A pure substance is defined as any matter, all specimens of which have identical chemical and physical properties and composition. Pure water is an example of a substance. All samples of pure water, regardless of their source, have exactly the same composition, 2.0158 parts by weight of hydrogen to 15.9994 parts of oxygen, and are identical in melting point, boiling point, and all other properties. Pure iron, pure aluminum, pure carbon, pure sugar, pure oxygen, and pure carbon dioxide are representative substances. Two substances can be distinguished from each other by a study of their characteristic properties. For example, sugar and salt may be distinguished by taste, iron and gold by their colors, and silver and mercury by their physical states at normal temperatures.

1.8   Mixtures

Can be separated physically

A mixture is composed of two or more substances each of which retains its identity and specific properties. The composition of a mixture can be varied continuously. Black gunpowder is a mixture of carbon, sulfur, and potassium nitrate; granite is a mixture of quartz, feldspar, and mica; a solution of sugar and water is a mixture; and air is a mixture of nitrogen, oxygen, carbon dioxide, water vapor, and other gases. Milk, butter, cement, flour, and gasoline are other examples of mixtures. Many naturally occurring materials are mixtures.

Because each component of a mixture possesses and retains its own set of characteristic properties, the various components can be separated by physical methods. The heterogeneous character of black gunpowder is readily detected by examining it under a microscope. Treatment of a sample of gunpowder with water causes the potassium nitrate to dissolve, leaving the sulfur and charcoal as solid particles; subsequent treatment of the residue with carbon disulfide causes the sulfur to dissolve, leaving only the carbon in the form of solid particles. The potassium nitrate and sulfur may be reclaimed as crystalline particles by evaporating their respective solutions to dryness. An intimate mixture of iron filings and sulfur may be separated either by dissolving the sulfur in carbon disulfide, leaving the iron, or by removing the iron with a magnet, leaving the sulfur.

1.9   Elements

There are two classes of substances: elements and compounds. **Elements are pure substances which cannot be decomposed by a chemical change.** Familiar examples are iron, silver, gold, aluminum, sulfur, oxygen, and carbon.

One hundred and six elements are known at the present time; a list of these is printed on the inside front cover of this book. Eighty-eight elements have been found in nature and the other eighteen have been synthesized. Eleven of the 88 elements make up about 99 per cent of the earth's crust and the atmosphere (Table 1-1). Oxygen constitutes nearly one-half and silicon nearly one-fourth of the total quantity of the elements in the atmosphere and the earth's crust. Only about one-fourth of the elements ever occur in nature in the free state; the others are found only in chemical combination with other elements.

Analyses of rock samples brought to earth by lunar astronauts show the elements of at least part of the moon's crust to be the same as those on earth. Since the analyses were only of minerals in rocks, the moon's elements identified first were almost all in combination with other elements. (One investigator reported iron in its elemental state.)

TABLE 1-1   Percentages of Elements in the Atmosphere and the Earth's Crust

Oxygen	49.20%	Chlorine	0.19%
Silicon	25.67	Phosphorus	0.11
Aluminum	7.50	Manganese	0.09
Iron	4.71	Carbon	0.08
Calcium	3.39	Sulfur	0.06
Sodium	2.63	Barium	0.04
Potassium	2.40	Nitrogen	0.03
Magnesium	1.93	Fluorine	0.03
Hydrogen	0.87	Strontium	0.02
Titanium	0.58	All others	0.47

1.10   Compounds

Can only be separated by chemical means

Compounds are substances which are composed of two or more different elements and can be decomposed by chemical changes. Elements in combination are different from elements in the free, or uncombined, state. However, the term *element* is used to designate an elemental substance whether free or in combination. For example, white crystalline sugar is a compound consisting of the element carbon, which is usually a black solid when free, and the two elements hydrogen and oxygen, which are colorless gases when uncombined. If heated sufficiently in the absence of air, sugar decomposes to carbon and water. Water, a compound, can be decomposed by an electric current into its two constituent elements, hydrogen and oxygen. Table salt, also a compound, can be broken down by an electric current into sodium and chlorine.

Whereas there are only 106 known elements, there are hundreds of thousands of chemical compounds representing different combinations of these elements. Each compound possesses definite chemical and physical properties by which chemists can distinguish it from all other compounds.

1.11 Molecules

Water has a definite composition and a set of chemical and physical properties which enable us to recognize it as a distinct substance. One might ask the question, "To what extent can a drop of water be subdivided with the smallest particle still retaining the chemical and physical properties of water?" The limit to which such a subdivision could be carried is a particle called a **molecule** of water. Subdivision of a molecule of water would result in the formation of two substances, hydrogen and oxygen, each with a set of properties quite different from water and from each other. **The smallest particle of an element or compound that can have a stable, independent existence is called a molecule** (Fig. 1–2). Molecules are too small to be seen even with very powerful optical microscopes. All but the very largest molecules cannot be seen even with an electron microscope. An idea of the minute size of molecules can be appreciated from the fact that if a drop of water were to be magnified to the size of the earth, its constituent molecules would appear to be about the size of baseballs. One hundred million molecules of water laid side by side would make a row about one inch long.

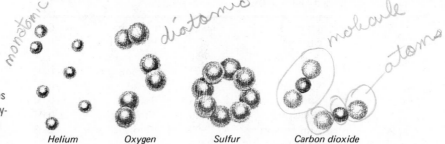

FIGURE 1–2

Representations of molecules for the elements helium, oxygen, and sulfur and for the compound carbon dioxide.

Helium   *Oxygen*   *Sulfur*   *Carbon dioxide*

1.12 Atoms

We define an atom as the smallest particle of an element which can enter into a chemical combination. For example, one atom of carbon can combine with two atoms of oxygen to form one molecule of carbon dioxide.

An atom of an element may or may not be capable of independent existence. In some cases, an atom of an element and a molecule of it are identical. When the molecule of an element contains only one atom, the molecule is said to be monatomic. Examples of elements which are composed of monatomic molecules are helium, neon, and xenon. On the other hand, the elements hydrogen, oxygen, nitrogen, fluorine, chlorine, bromine, and iodine consist of diatomic molecules (two atoms per molecule). Molecules of phosphorus and sulfur normally contain four and eight atoms, respectively. It follows, then, that the term *molecule* applies to small particles of either elements or compounds, whereas the term *atom* applies only to the smallest particle of an element.

The word *atom* comes from the Greek word *atomos,* which means indivisible. The early Greek philosophers were the originators of the concept of atoms, and they taught that matter, since it is composed of atoms, is therefore finitely divisible. Although generally credited as being the father of the atomic theory (1808), John

Dalton, an English chemist and physicist, revived the old Greek atomic hypothesis and put it on a quantitative basis. Like the Greeks, Dalton made no distinction between atoms and molecules as we do today, and he applied the word *atom* to particles of both elements and compounds. We shall see in Chapter 3 that even atoms are not the fundamental units of matter; atoms are made up of still smaller particles, comprising a very complex system.

1.13 The Scientific Method

Your study of chemistry will be concerned with the hypotheses, theories, and laws which give this science its foundation and framework; a framework into which bits of information fit to make an integrated area of knowledge.

The **scientific method** provides a logical way of finding the answer to the many questions which can be subjected to experimental inquiry and investigation. The first step in applying the scientific method to the solution of a problem involves the carrying out of carefully planned experiments to gather facts which give information about all phases of the problem. The second step consists of an attempt to formulate a simple generalization which will correlate these facts. If this attempt is successful, the simple generalization becomes a law. Usually, however, no general law can be formulated to correlate the facts, and a provisional conjecture, known as a **hypothesis,** is advanced to explain the data. A hypothesis is then tested by further experiments, and if it is capable of explaining a large body of facts in a given field it is dignified by the name of **theory.** Theories serve as guides for further work by serving as the bases for predicting new information or the direction in which additional information must be sought. Finally, a theory must be established or modified in such a manner that it can be accepted as a general truth, which then is often referred to as a **law.**

The theories and laws of chemistry are, in general, less precise than those of physics. This lack of perfection, however, tends to stimulate interest in chemistry and points up the fact that there are still great opportunities for discovery in chemistry. Theories and laws help to simplify the study of chemistry by systematizing and ordering the vast body of chemical knowledge.

Measurements in Chemistry

1.14 Units of Measurement

In any kind of quantitative work it is necessary to have a system of units of measurement. In scientific work the **metric system** of weights and measures, based upon the decimal system, is used throughout the world. A system of units known as the **International System of Units** (SI) was recommended officially by the General Conference on Weights and Measures in 1960 and adopted by the National Bureau of Standards in 1964. Although a complete change to SI units must of necessity take place over a lengthy period of time, the universal use of an international system will promote ease of communication between scientists and a better understanding by the layman of technical material written expressly for him.

The SI system of units begins with the following seven base units, from which other units of weights and measures can be derived.

Physical Property	Name of Unit	Symbol
Length	Meter	m
Mass	Kilogram	kg
Time	Second	s
Electric current	Ampere	A
Thermodynamic temperature	Kelvin	K
Luminous intensity	Candela	cd
Quantity of substance	Mole	mol

Appendix C lists several important units of measurement and their relationship to each other.

1.15   Mass and Weight

We saw in Section 1.2 that the **mass** is the quantity of matter which a body contains and that the force required to give the body a given acceleration is a measure of its mass. The mass of a body of matter is an invariable quantity. On the other hand, the **weight** of a body pertains to the force of attraction of the earth for the body and is variable, since the attraction is dependent upon the distance the body is from the earth's center. If this book were taken up in a plane it would weigh less it does at sea level. If it were to be taken far out into space its weight would become negligible. It is well known that astronauts experience weightlessness while in outer space. Scientists measure quantities of matter in terms of mass rather than weight because the mass of a body remains constant, whereas the weight of a body is an "accident of its environment." It should be noted, however, that in common terminology the term "weight" is often used for the mass of a substance.

An attraction (called gravitational attraction) between two bodies of matter depends upon their masses and the distance separating them. The instrument used in science for determining the mass of an object is called a **balance;** one type of balance is shown in Fig. 1–3. The body, or object, whose mass is to be determined is placed on the pan of the balance. "Weights" of known mass within the balance are adjusted by means of knobs, to balance the object on the pan. "Weighing" an object on a balance on the earth's surface makes use of the fact that the earth's gravitational attraction on objects of equal mass is the same; i.e., their weights are equal.

In the weighing process, the mass of an object is compared with a standard defined mass. The **unit of mass** in the International System of Units (SI) is the **standard kilogram,** which is a cylinder of platinum-iridium alloy kept at the

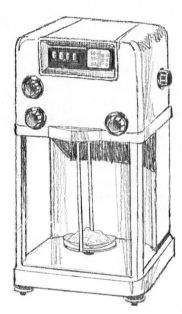

Reading scale of balance

Weight = 12.8505 grams

FIGURE 1-3
A modern balance with scale for direct reading.

International Bureau of Weights and Measures at Sèvres, France. The gram is equal to one one-thousandth of the mass of the standard kilogram and is very nearly equal to the weight of one cubic centimeter of water at 4° Celsius, the temperature of water at its maximum density (see Section 1.21 for a discussion of temperature scales).

1.16   Volume

It is oftentimes more convenient to measure the volume of a body of matter than to weigh it; this is particularly true of substances in the liquid or gaseous state. The **volume** of a body of matter is the space that it occupies. The unit of volume is the **liter,** which is defined in SI units as 10^{-3} cubic meters. Therefore, one thousandth of a liter, referred to as the **milliliter,** is equal to one cubic centimeter (Fig. 1-4).

FIGURE 1-4
Comparison between a cubic centimeter block and a dime. One cubic centimeter is equal to one milliliter. One milliliter of water at 4° C weighs one gram.

Dime

1 cubic centimeter
= 1 milliliter

1.17   Density

One of the physical properties of a solid, liquid, or gas is its density. **Density is defined as mass per unit volume.** This may be expressed mathematically as

$$\text{Density} = \frac{\text{mass}}{\text{volume}} \quad \text{or} \quad D = \frac{M}{V}$$

Substances can usually be distinguished by measuring their densities because it is rare that any two substances have identical densities. One milliliter of mercury (a liquid) at 25° C has a mass of 13.5339 grams. We say, therefore, that mercury has a density of 13.5339 g/ml. Because the milliliter and cubic centimeter are the same size (see previous definition of the liter), the density of mercury may also be designated as 13.5339 g/cm³. The density of water at 25° C is 0.99707 g/cm³ and at 4° C it is 1 g/cm³. At 0° C and 1 atmosphere of pressure one liter of hydrogen has a mass of 0.08987 gram; its density is 0.08987 g/liter. The density of air is 1.2929 g/liter at 0° C and 1 atmosphere of pressure. The following problems are based upon density.

Example 1. Calculate the density of a body that weighs (has a mass of) 320 g and has a volume of 45 cm³.

$$\text{Density} = \frac{\text{mass}}{\text{volume}} = \frac{320 \text{ g}}{45 \text{ cm}^3} = 7.1 \text{ g/cm}^3$$

Example 2. What volume will 4.00 g of air occupy? The density of air is 1.29 g/liter.

$$\text{Volume} = \frac{\text{mass}}{\text{density}} = \frac{4.00 \text{ g}}{1.29 \text{ g/liter}} = 3.10 \text{ liters}$$

Example 3. What is the mass of a piece of iron which has a volume of 120 cm³ and a density of 7.20 g/cm³?

$$\text{Mass} = \text{volume} \times \text{density} = 120 \text{ cm}^3 \times 7.20 \text{ g/cm}^3 = 864 \text{ g}$$

Example 4. The density of a solution of sulfuric acid which contains 38.0 per cent by weight of sulfuric acid (H_2SO_4) is 1.30 g/ml. How many grams of pure sulfuric acid are contained in 400 ml of this solution?

One ml of the solution of sulfuric acid has a mass of 1.30 g. Then 400 ml of acid will have the mass

$$\text{Mass} = \text{volume} \times \text{density} = 400 \text{ ml} \times 1.30 \text{ g/ml} = 520 \text{ g}$$

Because 38.0 per cent by weight of the acid is pure H_2SO_4, the number of grams of H_2SO_4 in 400 ml of the solution of the acid is

$$520 \text{ g} \times \frac{38.0}{100} = 198 \text{ g}$$

1.18   Specific Gravity

The term **specific gravity** denotes the ratio of the mass of a substance to the mass of an equal volume of a reference substance. The reference substance for solids and liquids is usually water.

$$\text{Specific gravity of a solid or liquid} = \frac{\text{mass of the solid or liquid}}{\text{mass of an equal volume of water}}$$

Note that specific gravity, being the ratio of two masses, has no units. The units in the numerator and denominator cancel each other.

Example 1. A pycnometer being used by a student (see Section 1.19) weighs 210 g when empty, 370 g when filled with water, and 412 g when filled with glycerine. Calculate the specific gravity of glycerine.

$$\text{Specific gravity of glycerine} = \frac{\text{mass of glycerine}}{\text{mass of an equal volume of water}}$$

$$= \frac{(412 - 210)\ \text{g}}{(370 - 210)\ \text{g}} = \frac{202\ \text{g}}{160\ \text{g}} = 1.26$$

Common reference substances used in specifying the specific gravities for gases are air and hydrogen.

When measured in the metric system of units, the density of any substance has practically the same numerical value as its specific gravity referred to water as the reference substance. For example, the density of glycerine is 1.26 g/cm^3 whereas its specific gravity referred to water as the reference substance is 1.26 (no units).

The following example is based upon specific gravity.

Example 2. Calculate the mass of 50 ml of kerosene, specific gravity 0.82.

1 ml of kerosene has a mass of 0.82 g. Then 50 ml has a mass of 50 ml × 0.82 g/ml = 41 g.

1.19 Measurement of the Density of a Liquid

FIGURE 1–5
A pycnometer.

Although the density of a liquid may be measured in many ways, the simplest method involves the use of a **pycnometer.** This piece of apparatus is a small bottle (10-ml to 250-ml capacity) with a ground-glass stopper which is bored with a moderately fine capillary (Fig. 1–5). In determining the density of a liquid the empty pycnometer is first weighed; then it is filled with pure water and weighed again. The difference in weight when the pycnometer is filled with water and when it is empty equals the weight of water. The volume of the pycnometer is obtained by dividing the weight of the water by the density of water at the temperature at which the determination was made, usually 25° C. The pycnometer is then filled with the liquid whose density is to be determined, and it is weighed again. This weight minus that of the empty container equals the weight of the liquid. The density of the liquid is obtained by dividing its weight by the volume of the pycnometer.

1.20   Temperature and Its Measurement

The word **temperature** refers to the "hotness" or "coldness" of a body of matter. In order to measure temperature some physical property of a substance which varies with temperature must be used. Practically all substances expand with an increase in temperature and contract when the temperature falls. It is this property which is the basis for the common thermometer. The mercury or alcohol in our common glass thermometers rises when the temperature increases because the volume of the mercury or alcohol expands more than does the column of the glass container. The mercury thermometer has a smaller bore than the alcohol thermometer; hence it is usually more difficult to read. The bore of the mercury thermometer is made smaller because mercury expands about one-sixth as much as alcohol for the same increase in temperature.

1.21   Standard Reference Temperatures

In order that we may agree on a set of temperature values, it is necessary to have **fixed temperatures** which can be readily determined. Two such fixed temperatures which are commonly used are the freezing and boiling points of water at an atmospheric pressure of 760 mm of mercury. On the Celsius (sometimes called centigrade) scale the freezing point of water is taken as $0°$ and the boiling point as $100°$. The space between these two fixed points is divided into 100 equal intervals, or degrees. On the Fahrenheit scale the freezing point of water is taken as $32°$ and the boiling point as $212°$. The space between these two fixed points is divided into 180 equal parts, or degrees. Thus, a degree Fahrenheit is 100/180, or 5/9, of a degree Celsius (Fig. 1–6). The relationships are shown by the equations

$$\frac{F - 32}{180} = \frac{C}{100} \qquad C = \frac{5}{9}(F - 32) \qquad F = \frac{9}{5}C + 32$$

The readings below $0°$ on either scale are treated as negative.

Another temperature scale, used mainly by scientists, is called the **Kelvin** scale, named after Lord Kelvin, a British physicist. On the Kelvin scale, the zero point on the scale is $-273° C$ (or more exactly, $-273.15° C$). The size of the degree on the Kelvin scale is the same as that on the Celsius scale. The freezing temperature of water on the Kelvin scale, therefore, is $273° K$ ($0° C$), and the boiling temperature is $373° K$ ($100° C$). A temperature on the Celsius scale is converted to the Kelvin scale by adding $273°$ to the Celsius reading; to convert from Kelvin to Celsius, $273°$ is subtracted from the Kelvin reading.

$$K = C + 273 \qquad C = K - 273$$

The Kelvin (K) is the basic unit of thermodynamic temperature recommended in the International System of Units (SI). The Kelvin temperature scale is discussed in greater detail in Section 10.5.

Figure 1–6 shows the relationships between the three temperature scales. Temperatures in this book are in Celsius (centigrade) unless otherwise specified. The following are examples of conversions between the three scales.

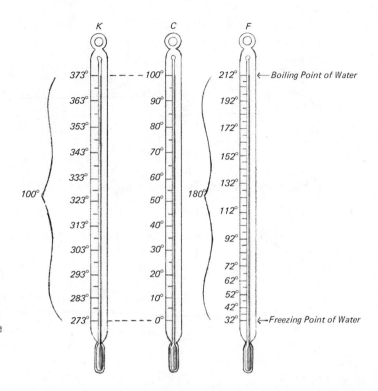

FIGURE 1-6

The relationships between the Kelvin, Celsius (centigrade), and Fahrenheit temperature scales.

Example 1.   Ethyl alcohol boils at 78.5° C at one atmosphere of pressure. What is its boiling point on the Fahrenheit scale and on the Kelvin scale?

$$F = \tfrac{9}{5}C + 32 = \tfrac{9}{5} \times 78.5 + 32 = 141 + 32 = 173°$$
$$K = C + 273 = 78.5 + 273 = 351.5°$$

Example 2.   Convert 50° F to the Celsius scale and the Kelvin scale.

$$C = \tfrac{5}{9}(F - 32) = \tfrac{5}{9}(50 - 32) = \tfrac{5}{9} \times 18 = 10°$$
$$K = C + 273 = 10 + 273 = 283°$$

1.22   The Measurement of Heat

Chemical reactions are accompanied by either the evolution or the absorption of energy, usually in the form of heat. The unit of measurement of heat is the **calorie,** defined in SI units as being equal to 4.184 joules. It is approximately equal to the quantity of heat which will raise the temperature of one gram of pure water one degree Celsius. The amount of heat necessary to raise the temperature of 1 gram of water 1 degree Celsius is not quite the same at all temperatures, but for our purposes it is sufficiently accurate to assume that it is the same. The **kilocalorie** (kcal) is equal to 1000 calories.

   The **heat capacity** of a body of matter is the number of calories necessary to raise its temperature 1 degree Celsius. The heat capacity of 1 gram of water is one calorie, so the heat capacity of any amount of water is numerically equal to its weight in grams. The greater the mass of a substance the greater its heat capacity.

The specific heat of a substance is a physical property that may be used in describing the substance. The **specific heat** of a substance is the number of calories of heat required to raise the temperature of one gram of the substance one degree Celsius. Every substance has its own specific heat. Metals have much lower specific heats than does water. Only a very few substances have higher specific heats than water. The specific heat of copper is 0.09 calorie per gram per degree, that of aluminum is 0.21 calorie per gram per degree, and that of zinc is 0.093 calorie per gram per degree. The following problems are based upon the measurement of heat.

Example 1. Calculate the quantity of heat required to raise the temperature of 200 g of water from 10° C to 40° C.

$$\text{Heat required} = \text{mass} \times \text{specific heat} \times \text{temperature change}$$
$$= 200 \text{ g} \times 1 \text{ cal/g °C} \times (40 - 10) \text{ °C} = 6000 \text{ cal}$$

Example 2. Calculate the quantity of heat required to raise the temperature of 300 g of mercury from 20° to 100° C. The specific heat of mercury is 0.033 calorie per gram per degree.

$$\text{Heat required} = \text{mass} \times \text{specific heat} \times \text{temperature change}$$
$$= 300 \text{ g} \times 0.033 \text{ cal/g °C} \times (100 - 20) \text{ °C} = 792 \text{ cal}$$

Example 3. Calculate the temperature rise which results when 400 calories are supplied to 100 g of ethyl alcohol, the specific heat of which is 0.58 cal/g °C.

$$\text{Heat required} = \text{mass} \times \text{specific heat} \times \text{temperature change}$$
$$\text{Temperature change} = \frac{\text{heat required}}{\text{mass} \times \text{specific heat}}$$
$$= \frac{400 \text{ cal}}{100 \text{ g} \times 0.58 \text{ cal/g °C}} = 6.9°$$

Example 4. The heat of combustion of carbon is 780 calories per gram. How much water could be heated from 22° C to 100° C using the heat from the burning of 100 g of carbon?

$$\text{Heat produced} = 100 \text{ g} \times 780 \text{ cal/g} = 78,000 \text{ cal}$$
$$\text{Heat required} = \text{mass} \times \text{specific heat} \times \text{temperature change}$$
$$\text{Mass} = \frac{\text{heat required}}{\text{specific heat} \times \text{temperature change}}$$
$$= \frac{78,000 \text{ cal}}{1 \text{ cal/g °C} \times (100 - 22) \text{ °C}} = 1000 \text{ g}$$

QUESTIONS

entity — something that exists independently

1. With what is the science of chemistry concerned?
2. List the five principal branches of chemistry and describe the area covered by each branch.
3. What is the meaning of the statement that matter and energy are not distinct entities but are different forms of a single entity?
4. Describe a chemical change that illustrates the Law of Conservation of Matter.
5. State the Law of Conservation of Matter, the Law of Conservation of Energy, and the combined Law of Conservation of Matter and Energy.
6. Classify each of the following as element, compound, or mixture: steel, toothpaste, sugar, air, aluminum, gold, bread, crude oil, alcohol, concrete, tincture of iodine, table salt.
7. Classify each of the following as a physical or chemical change: melting of ice; emission of light by a firefly; emission of light by an electric light bulb; fermenting of sugar; melting of gold; "turning" of leaves in color; magnetizing a screwdriver.
8. Name some properties that are useful in identifying specimens of pure substances.
9. Define the terms "atom" and "molecule."
10. Name three different compounds and list the elements of which each is composed.
11. How do molecules of elements and compounds differ?
12. How many atoms are there in a molecule of carbon dioxide; of helium; of sulfur?
13. How would you go about separating the following mixtures: sand and sugar; wax and lead shot; sulfur and iron filings; sugar and water?
14. Classify each of the following as possessing potential or kinetic energy: an automobile (a) just beginning to coast down from the top of a hill, (b) coasting halfway down the hill, (c) coasting on the level at the bottom of the hill; a book on a bookshelf; a stretched rubber band; a pendulum at the end of its swing; a pendulum at the middle of its swing; gasoline; falling water. Can any of these possess both potential and kinetic energy? Explain. *yes*
15. Give an example which illustrates the transformation of chemical potential energy into kinetic energy. Is it possible to convert kinetic energy into potential energy?
16. What properties distinguish each of the three states of matter?
17. Compare the three states of matter with regard to volume, shape, and compressibility.
18. Trace the progression of a concept from hypothesis to theory to law.
19. Distinguish between the terms "mass" and "weight."
20. Why is it unnecessary to consider the distance from the earth's surface when the mass of an object is determined by weighing it on an analytical balance as described in Section 1.15?
21. The specific gravity of a substance is sometimes defined as the ratio of the density of the substance to the density of a reference substance. Is this

definition compatible with the definition of specific gravity as given in Section 1.18? Explain your answer.

22. Show that the density of a substance, when measured in the metric system of units, has practically the same numerical value as the specific gravity referred to water as the reference substance.

23. The density of mercury is 13.5 g/cm^3 whereas its specific gravity is simply 13.5. Why is specific gravity a dimensionless number?

24. How do the densities of most substances change as a result of an increase in temperature?

PROBLEMS

NOTE: *In all problem sets at the ends of chapters the symbol* ⓢ *before the number of a problem indicates that the solution to that problem is worked out in the manual prepared by Meiser, Ault, and Holtzclaw titled* Problems and Solutions for General and College Chemistry, 5th editions, *by Nebergall, Schmidt, and Holtzclaw.*

The student should make a special effort to develop good habits with respect to expressing answers for problems to the correct number of significant figures. To this end, the authors have paid careful attention to significant figures in the answers provided for the problems in all problem sets. See Appendix A.2 at the back of the book for a discussion of significant figures.

ⓢ1. Express 4.71 kg in grams, milligrams, metric tons, and pounds. (See Appendix C.) *Ans. 4.71 × 10³ g; 4.71 × 10⁶ mg; 4.71 × 10⁻³ ton; 10.4 lb.*

2. A man weighs 163 pounds. What is his weight in kilograms? (See Appendix C.) *Ans. 73.9 kg*

ⓢ3. Calculate the number of liters in a gallon and in a cubic inch. (See Appendix C.) *Ans. 3.785 liters; 1.639 × 10⁻² liter*

ⓢ4. A rectangular tank which measures 500 cm in width and 5.00 m in length is filled with water to a depth of 100 mm. What volume of water is required? What mass of water is required? (See Appendix C.)
Ans. 2.50 × 10³ liters; 2.50 × 10³ kg

5. Express 15.1 ounces in centigrams, milligrams, kilograms, and metric tons. (See Appendix C.)
Ans. 4.28 × 10⁴ cg; 4.28 × 10⁵ mg; 0.428 kg; 4.28 × 10⁻⁴ ton

ⓢ6. How many millimeters are in one yard? How many yards are in one centimeter? How many Angstrom units are in one yard? (See Appendix C.)
Ans. 9.144 × 10² mm; 1.094 × 10⁻² yd; 9.144 × 10⁹ Å

7. Express 3.72 millimeters in centimeters, meters, and Angstrom units. (See Appendix C.) *Ans. 0.372 cm; 3.72 × 10⁻³ m; 3.72 × 10⁷ Å*

8. Calculate the length of the 220-yard dash in meters and centimeters. (See Appendix C.) *Ans. 201 m; 2.01 × 10⁴ cm*

ⓢ9. If milk is sold for 85 cents per half-gallon, what is its cost per liter? (See Appendix C.) *Ans. 45 cents/liter*

[S]10. What is the volume of 6.75 g of mercury; of 27.0 g of mercury; of 362 g of mercury? (The density of mercury is 13.5 g/cm³.)

Ans. 0.500 cm³; 2.00 cm³; 26.8 cm³

[S]11. (a) The world record for the 100-meter dash is 9.90 seconds. If a runner could maintain this pace for a mile, what would be his time in minutes for the mile? (See Appendix C.)   *Ans. 2.65 min*

(b) What would be the time in seconds for running the 100-meter dash if it were run at the average speed required for a 4.00-minute mile?

Ans. 15.0 sec

12. The speed of light is 2.998×10^{10} cm/sec. It takes light 2.564 seconds to get from a light source on earth to a reflector on the moon and back. What is the distance in miles from the light source to the moon? (See Appendix C.)

Ans. 238,900 miles

[S]13. If 5.84×10^{13} helium atoms (spherical) are laid in a line, each touching the next, they measure 6.75 miles. What is the diameter of a helium atom in Å? (See Appendix C.)   *Ans. 1.86 Å*

[S]14. A student has two beakers. The sum of their masses is 264.2 grams. The difference in their masses is 11.40 grams. What is the mass of each beaker?

Ans. 137.8 g; 126.4 g

15. Use the proper exponential form to express each of the following: (See Appendix A and Appendix C.)

(a) The number of kilometers in one foot.   *Ans. 3.048×10^{-4} km/ft*
(b) The number of meters in one Angstrom unit.   *Ans. 1.0×10^{-10} m/Å*
(c) The number of U.S. gallons in one liter.   *Ans. 2.642×10^{-1} gal/liter*
(d) The number of grams in one pound.   *Ans. 4.536×10^{2} g/lb*

[S]16. Calculate the density of a solid for which 40.0 g occupies a volume of 21.7 cm³.

Ans. 1.84 g/cm³

17. What is the density of a liquid if 90.0 cm³ weighs 75.2 g? *Ans. 0.836 g/cm³*

[S]18. What is the specific gravity of a liquid if 500 cm³ weighs as much as 850 cm³ of water?   *Ans. 1.70*

[S]19. What is the weight of each of the following?
[S](a) 33.3 cm³ of a liquid, density = 1.836 g/cm³   *Ans. 61.1 g*
(b) 50.0 cm³ of a liquid, specific gravity = 0.790   *Ans. 39.5 g*
(c) 4.00 cm³ of mercury, density = 13.5 g/cm³   *Ans. 54.0 g*

[S]20. Use the following data to calculate the specific gravity of an unknown liquid:
Weight of pycnometer + liquid: 16.7539 g
Weight of pycnometer empty: 10.6402 g
Weight of pycnometer + water: 14.7572 g   *Ans. 1.4850*

21. What mass of mercury will occupy 85.0 cm³ volume? (The density of mercury is 13.5 g/cm³.)   *Ans. 1.15 kg*

22. Find the volume of 536 g of copper (density = 8.94 g/cm³).

Ans. 60.0 cm³

[S]23. A piece of lead when put into water displaces 119.0 cubic centimeters. The specific gravity of pure lead is 11.3. Calculate the weight of the piece of lead.

Ans. 1.34 kg

⚲24. The density of uranium is 18.7 g/cm³. Which of the following contains the greatest mass of uranium: 1.0 lb; 0.50 kg; 0.050 liter? (See Appendix C.)
Ans. 0.050 liter

25. The density of aluminum is 2.70 g/cm³. Which of the following will occupy the greatest volume: 1.0 lb; 0.50 kg; 0.050 liter? (See Appendix C.)
Ans. 0.50 kg

⚲26. The density of a solution that is 96.0% sulfuric acid by weight is 1.84 g/ml. How many milliliters volume will be occupied by 25.0 g of the acid? How many grams of water will be present in this volume? *Ans. 13.6 ml; 1.00 g*

27. Use the following data to calculate the specific gravity of aluminum:
Weight of empty graduated cylinder: 24.5289 g
Weight of graduated cylinder + water: 28.2210 g
Volume of water: 3.69 ml
Weight of graduated cylinder + water + aluminum: 41.3320 g
Volume of water + aluminum: 8.54 ml *Ans. 2.70*

⚲28. Make the following temperature conversions:
(a) 0.00° F to degrees Celsius. *Ans. −17.8° C*
⚲(b) 37.0° C to degrees Fahrenheit. *Ans. 98.6° F*
⚲(c) 66.0° F to degrees Kelvin. *Ans. 292° K*
(d) −40° C to degrees Fahrenheit. *Ans. −40° F*
(e) 150.0° C to degrees Fahrenheit. *Ans. 302.0° F*
(f) 7.00° K to degrees Fahrenheit. *Ans. −447° F*

29. Derive an equation for the conversion of °K to °F, and convert −40° F to °K. Then, as a check on your work, convert −40° F to degrees Celsius and, using the conversion equation for Celsius to Kelvin, calculate degrees Kelvin.
Ans. $F = 9/5(K − 273) + 32$; 233° K

⚲30. How many calories of heat would be required to increase the temperature of 600 g of water from 80.6° F to 107.6° F? *Ans. 9,000 cal*

31. Calculate the quantity of heat required to raise the temperature of 0.840 kg of water from 35.0° C to 108.5° F. *Ans. 6.30×10^3 cal*

⚲32. If 360 calories are added to 30.0 g of water at 298° K, what is the resulting temperature? *Ans. 310° K*

⚲33. Calculate the heat capacity of the following:
⚲(a) 22.6 g of water (specific heat, 1.00 cal/g °C). *Ans. 22.6 cal/°C*
(b) 1.10 kg of aluminum (specific heat, 0.21 cal/g °C). *Ans. 230 cal/°C*
⚲(c) 50.0 ml of copper (density, 8.94 g/cm³, specific heat, 0.09 cal/g °C).
Ans. 40 cal/°C

34. Calculate the specific heat of water in units of cal/lb °F. The specific heat of water is 1.00 cal/g °C. (See Appendix C.) *Ans. 252 cal/lb °F*

⚲35. How many grams of water at 25° C must be mixed with 1.50 kilograms of water at 100° C so that the resulting combination will have a temperature of 70° C? *Ans. 1,000 g*

36. Calculate the temperature of a sample of water made by mixing 250 g of water at 25° C and 125 g of water at 61° C. *Ans. 37° C*

⚲37. When 18 g of water is formed from its elements, 68,315 cal of heat is evolved. What would be the change in temperature of a piece of zinc weighing 10.0

kilograms if heated by the quantity of heat liberated by the formation of 18.0 g of water? (The specific heat of zinc is 0.093 cal per gram per degree.)

Ans. 73° C

38. A 10.0-kg piece of iron at 50.0° C is placed in 1.00×10^3 g of water at 10.0° C. The iron and water come to the same temperature of 31.4° C. Assuming that no heat is lost to the outside, what is the specific heat of iron?

Ans. 0.115 cal/g °C

39. A cube of cork weighs 500 grams. The density of this cork is 0.250 g/cm³. Calculate the edge length of the cube.   *Ans. 12.6 cm*

⑤40. A spherical ball of iron weighs one metric ton. What would be its diameter? (The specific gravity of iron is 7.86.) (See Appendix C.)   *Ans. 62.4 cm*

REFERENCES

"A Program to Coordinate Environmental Research," R. Bowers (Solid State Physics), P. Hohenberg (Economics), G. Likens (Ecology), W. Lynn (Civil and Environmental Engineering), D. Nelkin (Senior Research Associate in Program), M. Nelkin (Applied Physics). (Prepared under auspices of Cornell University Interdisciplinary Program on Science, Technology, and Society.) *Amer. Scientist,* **59,** 183 (1971).

"Scientific Methods in Art and Archeology," A. E. Werner, *Chem. in Britain,* **6,** 55 (1970).

"The Chemist and Society," L. H. Williams, *Chem. in Britain,* **4,** 194 (1968).

"Conversion to the Metric System," Lord Ritchie-Calder, *Sci. American,* July, 1970; p. 17.

"Chemistry, Curiosity, and Culture," W. F. Kieffer, *J. Chem. Educ.,* **45,** 550 (1968).

"Description of Fahrenheit's Thermometer," (provided by D. B. Murphy from *Encyclopedia Britannica,* first American edition, Thomas Dobson, Philadelphia, 1798, pp. 496–7, Vol. XVIII), *J. Chem. Educ.,* **46,** 192 (1969).

"Cleaning Our Environment—The Chemical Basis for Action," (Staff), *Chem. & Eng. News,* Sept. 8, 1969; p. 58.

"Public Policy and the Environment," M. Harris (Introduction, p. 74), L. A. DuBridge (Government's Role, p. 75), H. P. Doan (Industry's Role, p. 77), B. Commoner (Ecological Problems, p. 78), (Panel Discussion, p. 80), *Chem. & Eng. News,* Feb. 9, 1970; p. 74.

"Truth and Aesthetics in Chemistry," M. H. Klapper, *J. Chem. Educ.,* **46,** 577 (1969).

"SI Units" (International System of Units), G. Socrates, *J. Chem. Educ.,* **46,** 710 (1969).

"Physical versus Chemical Change," W. J. Gensler, *J. Chem. Educ.,* **47,** 154 (1970).

"Differentiating Physical and Chemical Changes," L. E. Strong, *J. Chem. Educ.,* **47,** 689 (1970).

"Pollution Problems, Resource, Policy, and the Scientist," A. W. Eipper, *Science,* **169,** 11 (1970).

"Policy for NBS Usage of SI Units," National Bureau of Standards, *J. Chem. Educ.,* **48,** 569 (1971).

"Two Concepts of Matter," G. Gale, *Chemistry,* **46** (10), 13 (1973).

"Chemistry is Real and Relevant," E. Wolthuis, *J. Chem. Educ.,* **50,** 423 (1973).

"The Origin of Chemical Elements, 1," J. Selbin, *J. Chem. Educ.,* **50,** 306 (1973).

"The Origin of Chemical Elements, 2," J. Selbin, *J. Chem. Educ.,* **50,** 380 (1973).

"The Chemical Elements of Life," E. Frieden, *Sci. American,* July, 1972; p. 52.

"Chemistry and Science: The Next Hundred Years," A. L. Hammond, *Science,* **185,** 847 (1974).

"Precision Balances—A Chapter in their Development," J. T. Stock, *Chem. in Britain,* **7,** 385 (1971).

"Dalton Revisited," (Staff), *Chem. in Britain,* **9,** 228 (1973).

"The Scientist in Society: Inspiration and Obligation," M. Rees, *Amer. Scientist,* **63,** 144 (1975).

"A Metric (SI) Energy Scale: Conversions and Comparisons," L. Petrakis, *J. Chem. Educ.,* **51,** 459 (1974).

"The World of Science and the World of People," B. M. Steckler, *J. Chem. Educ.,* **50,** 46 (1973).

"The Mole and Avogadro's Number," R. M. Hawthorne, Jr., *J. Chem. Educ.,* **50,** 282 (1973).

Symbols, Formulas, and Equations

2

Chemists use chemical symbols, formulas, and equations when speaking and writing about matter and the changes it undergoes. It is essential, therefore, that students become skilled in the use of the symbolism of the field in order to study chemistry effectively.

2.1 Symbols

As a matter of convenience in indicating the elements, the chemist usually uses abbreviations, since they are more quickly written than names. These abbreviations are called **symbols.** For hydrogen, the symbol is H; for oxygen, O; and for carbon, C. Sometimes two letters are necessary to distinguish between two or more elements when the names of the elements begin with the same letter. For example, Ca is the symbol for calcium, Co for cobalt, Cr for chromium, Cl for chlorine, and Ni for nickel. No symbol contains more than two letters and the first letter is always capitalized. Some symbols are abbreviations of the Latin names of the elements, as Fe for iron (Latin, *ferrum*), Na for sodium (Latin, *natrium*), and Cu for copper (Latin, *cuprum*). Tungsten has the symbol W from the German word *wolfram*. The symbols for all known elements are given in the table of atomic weights on the inside front cover of this book.

2.2 Formulas

A **formula** is a single symbol or a group of symbols which represents the composition of a substance. The symbols in a formula identify the elements present in the substance. Thus, NaCl is the formula for sodium chloride (common table salt) and so identifies the elements sodium and chlorine. Notice that in this formula there are equal numbers of atoms of the elements sodium and chlorine. Subscripts are used in formulas to indicate the numbers of atoms in the compound, but only when more than one atom of a given element is present. For example, the formula for water, H_2O indicates that each molecule contains two atoms of hydrogen and one atom of oxygen. Sulfuric acid is represented by the formula H_2SO_4, which indicates two atoms of hydrogen, one of sulfur, and four of oxygen. The formula for aluminum sulfate, $Al_2(SO_4)_3$, specifies two atoms of aluminum and three sulfate groups. Each sulfate group contains one atom of sulfur and four atoms of oxygen. Hence, the formula shows a total of two atoms of aluminum, three atoms of sulfur, and twelve atoms of oxygen.

The formula for a molecule of the most common form of sulfur is S_8, showing that each molecule of this element consists of eight atoms (see Fig. 1–2 in Chapter 1). The molecular formulas for elemental hydrogen and oxygen (diatomic molecules) are H_2 and O_2, while the formula for helium (a monatomic molecule) is He. Note that H_2 and 2H do not mean the same thing. H_2 represents a molecule of hydrogen consisting of two atoms of the element, chemically combined. The expression 2H, on the other hand, indicates that the two hydrogen atoms are not in combination as a unit but are separate particles.

Additional information on writing formulas for compounds will be presented in Chapter 4.

2.3 Chemical Equations

Atoms are the fundamental particles of the elements that enter into chemical changes. Substances that take part in chemical changes are made up of atoms in the form of molecules or ions (ions are atoms or groups of atoms that are electrically charged). Chemical changes involve the regrouping of atoms or ions to form other substances. The **chemical equation** is the chemist's shorthand expression for describing a chemical change, and symbols and formulas are used to indicate the composition of the substances involved in the change. The writer of an equation must know what substances react and what substances are formed, as well as the correct formulas for these substances.

A statement of the decomposition of water by electrolysis, "When water is decomposed by an electric current, hydrogen and oxygen are formed," can be expressed as

$$H_2O \xrightarrow{\text{Elec.}} H_2 + O_2 \tag{1}$$

The formula for the reactant is written to the left of the arrow and the formulas for the products are written to the right. The arrow is read as either "gives," "produces," "yields," or "forms." The + sign on the right side of the expression is read as "and"; it does not imply mathematical addition. When a + sign appears between the formulas for two reactants on the left side of an equation, it implies "reacts with." Special conditions under which the reaction takes place are often indicated above and below the arrow. In equation **(1)**, for example, the abbreviation "Elec." over the arrow indicates that the reaction occurs by means of an electric current.

As it now stands, expression **(1)** does not conform with the law of conservation of matter. Two atoms of oxygen in the molecule O_2 could not be formed from one molecule of water containing but one oxygen atom. To conform with the law, the equation is *balanced* by introducing the proper number (coefficient) before each formula. Thus, two molecules of water will furnish two oxygen atoms for the production of one diatomic oxygen molecule. Two water molecules will then supply four hydrogen atoms for the production of two diatomic molecules of hydrogen. The balanced equation is

$$2H_2O \longrightarrow 2H_2 + O_2 \tag{2}$$

It is important to remember that the subscripts in a formula cannot be changed in order to make an equation balance, because substances have definite atomic compositions that cannot be changed by merely changing the subscripts in their formulas.

Sometimes the symbol $\triangle$ is used in an equation to represent heat that must be supplied in order to make a reaction proceed. For example, potassium chlorate decomposes upon heating to form potassium chloride and oxygen.

$$\underset{\text{Potassium chlorate}}{2KClO_3} + \text{Heat (or } \triangle) \longrightarrow \underset{\text{Potassium chloride}}{2KCl} + \underset{\text{Oxygen}}{3O_2} \tag{3}$$

or

$$2KClO_3 \xrightarrow{\triangle} 2KCl + 3O_2$$

When the fact that heat is produced during a chemical reaction is to be emphasized, the symbol $\triangle$ is placed to the right of the arrow.

$$\underset{\text{Carbon}}{C} + \underset{\text{Oxygen}}{O_2} \longrightarrow \underset{\text{Carbon dioxide}}{CO_2} + \text{Heat (or } \triangle) \tag{4}$$

A gaseous product formed by a reaction in a solution can be designated by an ascending arrow (↑).

$$\underset{\text{Sodium metal}}{2Na} + \underset{\text{Water}}{2H_2O} \longrightarrow \underset{\text{Sodium hydroxide}}{2NaOH} + \underset{\text{Hydrogen}}{H_2\uparrow} \tag{5}$$

If a particular substance is in solid (precipitated) form in a reaction mixture, either its formula is underlined or a descending arrow (↓) is used.

$$AgNO_3 + NaCl \longrightarrow \underline{AgCl} + NaNO_3 \tag{6}$$

or

$$\underset{\substack{\text{Silver}\\\text{nitrate}}}{AgNO_3} + \underset{\substack{\text{Sodium}\\\text{chloride}}}{NaCl} \longrightarrow \underset{\substack{\text{Silver}\\\text{chloride}}}{AgCl\downarrow} + \underset{\substack{\text{Sodium}\\\text{nitrate}}}{NaNO_3}$$

2.4 Atomic Weights

It has been determined by experiment that an atom of hydrogen (the lightest of all atoms) weighs 1.67×10^{-24} g (0.000 000 000 000 000 000 000 001 67 g). Obviously, expressing weights of atoms in this manner is somewhat inconvenient. Since the weights of the individual atoms of all the elements are of the same small order of magnitude, 10^{-24} g, it has been found more convenient to use *relative* weights, rather than actual weights measured in grams or other units. **The relative weights of the atoms of the different elements are known as atomic weights and are proportional to the actual weights of the atoms.** For example, the atomic weight of hydrogen is 1.0079 on this arbitrary scale, that of carbon is 12.011, and that of oxygen is 15.999. Hence, carbon atoms weigh about twelve times as much as hydrogen atoms, and oxygen atoms are approximately sixteen times heavier than hydrogen atoms. Sulfur atoms weigh about twice as much as oxygen atoms, or 32.06 on the relative scale. The standard used in expressing the atomic weights is a particular kind of carbon atom called the carbon-12 isotope. This will be discussed in Section 3.10. Refer to the inside front cover of this book for a complete list of the atomic weights of the elements.

In addition to identifying the element and representing one atom of that element, a chemical symbol can be used to represent one chemical unit of weight of an element. The symbol O can represent one atom of oxygen, or it can represent 15.999 units of weight of oxygen. The unit of weight may be a gram, an ounce, a pound, etc. In quantitative work, the chemist finds it convenient to use the atomic weight of an element expressed in grams and so calls it a **gram-atomic weight,** or **gram-atom.** Thus, one gram-atomic weight of oxygen is 15.999 g; that of hydrogen is 1.0079 g.

2.5 The Avogadro Number

It can easily be shown that an equal number of atoms, **6.022×10^{23},** is present in one gram-atomic weight of any element. This number is called **Avogadro's number** in honor of the Italian professor of physics Amedeo Avogadro (1776–1856). It is important to remember that one gram-atomic weight of *any* of the elements, e.g., 15.999 g of oxygen, 32.06 g of sulfur, or 1.0079 g of hydrogen, contains 6.022×10^{23} atoms.

Knowing the atomic weight of an element and Avogadro's number, one can easily calculate the weight in grams of one atom of the element. The weight of one oxygen atom would be, therefore,

$$\frac{15.999 \text{ g}}{6.022 \times 10^{23}} = 2.656 \times 10^{-23} \text{ g}$$

This extremely small weight suggests that atoms are very small particles.

2.6 Formula Weights, Molecular Weights, Moles

The sum of the atomic weights of all the atoms shown by the formula of a substance is its formula weight. Formula weights are relative, just as are the atomic weights upon which they are based. The formula weight of sulfuric acid (H_2SO_4) is $(2 \times 1.0) + (1 \times 32.1) + (4 \times 16.0) = 98.1$. For most of our work it will be

63.5
32.1
.64
159.6

permissible to round off the atomic weights of the elements at one digit after the decimal point, e.g., 1.0079 to 1.0 for hydrogen, 32.06 to 32.1 for sulfur, 26.98154 to 27.0 for aluminum, and 35.453 to 35.5 for chlorine.

The formula weight of a substance expressed in grams is a **gram-formula weight.** The gram-formula weight is also referred to as a **mole** of that substance. Thus the gram-formula weight (or a mole) of sulfuric acid is 98.1 grams, whereas its formula weight is simply 98.1.

When the true formula for a molecule of a substance is known, the molecular weight can be calculated by adding the atomic weights of the atoms given in the formula of the substance. It is quite correct to speak of the molecular weight (or formula weight) of substances which consist of discrete molecules, such as meth-ane, CH_4; carbon dioxide, CO_2; and water, H_2O. On the other hand, ionic com-pounds such as NaCl, $KClO_3$, NaOH, and $CuSO_4$ do not consist of physically distinct and electrically neutral molecules (see Section 4.1) either in the crystalline state or in solution. Hence it is customary to use formula weights to represent the total composition of such substances. **The use of the term "mole" in this book will be taken to mean a gram-formula weight of ionic compounds and a gram-molecular weight of nonionic (molecular) substances (either elements of compounds).**

It is often convenient in quantitative chemical work to use millimoles in place of moles of a substance. **A millimole of a substance is its formula weight expressed in milligrams.** For example, the formula weight of H_2SO_4 is 98.1. Then one mole of H_2SO_4 is 98.1 g, and one millimole of it is 98.1 mg. In other words, one millimole is one-thousandth of a mole.

The following examples are based upon gram-atomic weights, gram-atoms, gram-formula weights, moles, and millimoles.

Example 1. How many gram-atoms of sulfur are there in 80.3 g of sulfur?

$$\text{Gram-atoms of S} = \frac{\text{grams of S}}{\text{gram-atomic weight of S}} = \frac{80.3 \text{ g}}{32.1 \text{ g}} = 2.50$$

Example 2. How many moles of sulfur are there in 80.3 g of sulfur if the molecular formula of sulfur is S_8?

The gram-formula weight of S_8 is $8 \times 32.1 \text{ g} = 256.8 \text{ g}$.

$$\text{Moles of } S_8 = \frac{\text{grams of } S_8}{\text{gram-formula weight of } S_8} = \frac{80.3 \text{ g}}{256.8 \text{ g}} = 0.313$$

Example 3. How many moles of ethyl alcohol, C_2H_5OH, are contained in 138 g of that substance?

The formula weight of C_2H_5OH is calculated as follows:

$$2C = 2 \times 12.0 = 24.0$$
$$6H = 6 \times 1.0 = 6.0$$
$$1O = 1 \times 16.0 = 16.0$$
$$\overline{\text{Formula weight of } C_2H_5OH = 46.0}$$

The gram-formula weight of ethyl alcohol is 46.0 g and the number of moles contained in 138 g of the compound is

$$\text{Moles of } C_2H_5OH = \frac{\text{grams of } C_2H_5OH}{\text{gram-formula weight of } C_2H_5OH} = \frac{138 \text{ g}}{46.0 \text{ g}} = 3.00$$

Example 4.   **How many millimoles are contained in 0.160 gram of NaOH?**

The formula weight of NaOH is $(23.0 + 16.0 + 1.0) = 40.0$.

$$\text{Millimoles of NaOH} = \frac{\text{milligrams of NaOH}}{\text{formula weight of NaOH in mg}} = \frac{160 \text{ mg}}{40.0 \text{ mg}} = 4.00$$

2.7   Percentage Composition from Formulas

The percentage by weight of each element in a compound may be calculated easily from the formula of the compound. This percentage will be equal to the fraction of the total weight of the compound which is attributable to the element, multiplied by 100. For example, suppose we calculate the percentage of hydrogen and that of oxygen in the compound water, H_2O.

$$\text{Percentage of hydrogen by wt. in } H_2O = \frac{2 \times \text{gram-atomic weight of H}}{\text{gram-formula weight of } H_2O} \times 100$$

$$= \frac{2 \times 1.0}{(2 \times 1.0) + (1 \times 16.0)} \times 100 = 11\%$$

$$\text{Percentage of oxygen by wt. in } H_2O = \frac{1 \times \text{gram-atomic weight of O}}{\text{gram-formula weight of } H_2O}$$

$$= \frac{1 \times 16.0}{(2 \times 1.0) + (1 \times 16.0)} \times 100 = 89\%$$

All samples of pure water, H_2O, no matter what their source may be, contain the same elements, hydrogen and oxygen, united in the same proportion by weight, 11 parts of hydrogen to 89 parts of oxygen. A similar statement may be made about any pure compound. This is one of the fundamental concepts of chemistry and is known as the **Law of Definite Proportions,** or the **Law of Definite Composition.** This law may be stated as follows: **Different samples of a pure compound always contain the same elements in the same proportions by weight.** This law, among others, convinced the English chemist and physicist John Dalton (1766–1844) of the atomic nature of matter and led him to outline his atomic theory.

2.8   Derivation of Formulas

The first step in deriving the formula of a new compound is the determination of its percentage composition. The percentage composition of a compound can be determined experimentally either by analysis or synthesis. For example, by analysis, a weighed sample of the compound may be separated into its constituent elements and the weight of each element in the sample determined. Thus, a

1.0000-g sample of a certain compound is found to contain 0.2729 g of carbon and 0.7271 g of oxygen.

$$\text{Percentage of carbon in the compound} = \frac{0.2729 \text{ g}}{1.0000 \text{ g}} \times 100 = 27.29\%$$

$$\text{Percentage of oxygen in the compound} = \frac{0.7271 \text{ g}}{1.0000 \text{ g}} \times 100 = 72.71\%$$

The relative number of carbon and oxygen atoms in the compound can now be obtained by dividing the above percentages by the weights (which are relative) of the respective atoms.

Element	Percentage	Relative Number of Atoms	Divide by the Smaller Number	Smallest Integral Number of Atoms
Carbon	27.29	$\frac{27.29}{12.01} = 2.27$	$\frac{2.27}{2.27} = 1.00$	1
Oxygen	72.71	$\frac{72.71}{16.00} = 4.54$	$\frac{4.54}{2.27} = 2.00$	2

Thus for every 2.27 atoms of carbon there are 4.54 atoms of oxygen. Reducing to simple whole numbers, since only whole numbers of atoms can enter into chemical combination, makes it evident that there are twice as many oxygen atoms as carbon atoms in the compound. The simplest formula for the compound must therefore be CO_2. This **simplest formula** corresponds to a formula weight of 44. It can be shown by experiment that this compound has a molecular weight of 44. Thus in this case the simplest formula, that is, the formula which indicates the ratio of the numbers of different kinds of atoms present, and the **molecular formula,** the formula which indicates the kinds of atoms present and the actual number of each kind per molecule, are one and the same. Had the molecular weight for the compound been 88, then the molecular formula would have indicated twice as many atoms as the simplest formula, i.e., C_2O_4 rather than CO_2.

As another example of the determination of the formula of a compound, let us derive the formula for an oxide of iron which contains 69.94 per cent iron and 30.06 per cent oxygen.

Element	Percentage	Relative Number of Atoms	Divide by the Smaller Number	Smallest Integral Number of Atoms
Iron	69.94	$\frac{69.94}{55.85} = 1.25$	$\frac{1.25}{1.25} = 1.0$	$2 \times 1.0 = 2$
Oxygen	30.06	$\frac{30.06}{16.00} = 1.88$	$\frac{1.88}{1.25} = 1.5$	$2 \times 1.5 = 3$

Unlike the previous example, step 2 did not give two integers, so a third step was necessary. This step involves multiplying both 1.0 and 1.5 by the smallest whole number which will give whole numbers of atoms of each element. Thus, $2 \times 1.0 = 2$ and $2 \times 1.5 = 3$. Hence, the simplest formula for the oxide of iron is Fe_2O_3.

Let us derive the formula of a hydrated salt, $Na_2SO_4 \cdot xH_2O$, a 1.500-g sample of which was found to contain 0.705 g of water. The weight of Na_2SO_4 in the 1.500-g sample = $(1.500 - 0.705)$ g = 0.795 g. The relative number of moles (formula weights) of Na_2SO_4 and H_2O in the salt may be calculated by dividing the weight of each of the compounds by its formula weight.

Compound	Weight	Relative Number of Moles	Divide by the Smaller Number	Smallest Integral Number of Moles
Na_2SO_4	0.795 g	$\dfrac{0.795 \text{ g}}{142 \text{ g}} = 0.0056$	$\dfrac{0.0056}{0.0056} = 1.0$	1
H_2O	0.705 g	$\dfrac{0.705 \text{ g}}{18.0 \text{ g}} = 0.0392$	$\dfrac{0.0392}{0.0056} = 7.0$	7

Hence the simplest formula of the hydrated salt is $Na_2SO_4 \cdot 7H_2O$.

2.9   Calculations Based on Equations

There was little real progress in the development of chemistry before the French chemist Lavoisier (1743–1794) put this science on a quantitative basis. In this section we shall become familiar with the quantitative aspects of chemistry through the study of calculations based on chemical equations.

Since symbols and formulas represent atomic and formula weights of substances, respectively, then a *balanced* chemical equation represents quantitative relationships for a given chemical change. Thus, the balanced equation for the thermal decomposition of potassium chlorate to form potassium chloride and oxygen,

$$2KClO_3 + \triangle \longrightarrow 2KCl + 3O_2\uparrow$$

possesses the following quantitative significance: When decomposed by heating, two formula weights of potassium chlorate always yield two formula weights of potassium chloride and three formula weights of oxygen. The chemical units of formula weights may be grams, kilograms, ounces, pounds, tons, etc.

$$2KClO_3 \quad + \triangle \longrightarrow \quad 2KCl \quad + \quad 3O_2\uparrow$$
$$\begin{matrix} (2 \times 122.6) & & (2 \times 74.6) & (3 \times 32.0) \\ \text{or } 245.2 \text{ units} & & 149.2 \text{ units} & 96.0 \text{ units} \end{matrix}$$
$$245.2 \text{ units}$$

Note that the number of units of weight for the reactant ($KClO_3$) is equal to the total number of units of weight for the products (KCl and O_2), in accord with the Law of Conservation of Mass.

Example 1. **Calculate the weight of oxygen produced during the thermal decomposition of 100 g of potassium chlorate.**

a. *Unit Method*

If 245.2 g of $KClO_3$ will produce 96.0 g of O_2, then 1 g of $KClO_3$ will produce $(1/245.2)$ (96.0) g of oxygen. Hence, 100 g of $KClO_3$ will produce

$$100 \times \frac{1 \text{ g}}{245.2 \text{ g}} \times 96.0 \text{ g} = 39.2 \text{ g of oxygen}$$

This method is sometimes called the "unit method" because (in this case) the weight of oxygen produced from *one unit* (1 g) of $KClO_3$ is calculated.

b. *Molar Method*

$$2KClO_3 + \triangle \xrightarrow{\quad} 2KCl + 3O_2\uparrow$$
$$\text{2 moles} \qquad\qquad \text{2 moles} \quad \text{3 moles}$$

First, calculate the number of moles of $KClO_3$ in 100 g of the compound.

$$\frac{100 \text{ g}}{122.6 \text{ g/mole}} = 0.816 \text{ mole of } KClO_3$$

Since, according to the equation, two moles of $KClO_3$ yields 3 moles of O_2, then 1 mole of $KClO_3$ would yield $\frac{3}{2} = 1.5$ moles of O_2. Thus 0.816 mole of $KClO_3$ would form $1.5 \times 0.816 = 1.224$ moles of O_2. The grams of O_2 in 1.224 moles $= 1.224$ moles $\times$ 32.00 g/mole $= 39.2$ g of oxygen. A summary of the molar method of solving the problem is

$$\underbrace{\frac{100 \text{ g}}{122.6 \text{ g/mole}} \times \frac{3}{2} \times 32.00 \text{ g/mole}}_{} = 39.2 \text{ g of oxygen}$$

(moles of $KClO_3$ used)

(moles of O_2 produced)

(grams of O_2 produced)

c. *Factor Method*

$$2KClO_3 + \triangle \longrightarrow 2KCl + 3O_2\uparrow$$
$$\text{245.2 g} \qquad\qquad\qquad \text{96.0 g}$$

If 245.2 g of $KClO_3$ will produce 96.0 g of oxygen, then 100 g of $KClO_3$ will produce a smaller quantity of oxygen than will 245.2 g of $KClO_3$. Multiplying the 96.0 g of oxygen by a fraction made up of the two weights of $KClO_3$ and having a value less than unity (a proper fraction) will give the quantity of oxygen obtainable from 100 g of $KClO_3$.

$$96.0 \text{ g} \times \frac{100.0 \text{ g}}{245.2 \text{ g}} = 39.2 \text{ g of oxygen}$$

Example 2. Calculate the weight of limestone which is 95 per cent pure $CaCO_3$ that must be decomposed by heating to produce 500 g of lime, CaO.

The following calculation is by the molar method. First write the balanced equation for the reaction.

$$CaCO_3 + \triangle \longrightarrow CaO + CO_2\uparrow$$
$$\text{1 mole} \qquad\qquad \text{1 mole}$$
$$\text{100.1 g} \qquad\qquad \text{56.1 g}$$

Inasmuch as one mole of CaO is 56.1 g, the number of moles of CaO in 500 g is $^{500}/_{56.1}$. According to the equation, one mole of $CaCO_3$ is required for the production of each mole of CaO. Hence, $^{500}/_{56.1}$ moles of $CaCO_3$ is required to produce $^{500}/_{56.1}$ moles of CaO.

Multiplying the required number of moles of $CaCO_3$ by the weight of $CaCO_3$ in one mole (that is, by the gram-molecular weight of $CaCO_3$) will give the number of grams of *pure* $CaCO_3$ necessary to prepare 500 g of CaO.

$$\frac{500 \text{ g}}{56.1 \text{ g/mole}} \times 100.1 \text{ g/mole} = 892 \text{ g of pure } CaCO_3$$

Thus, 892 g of pure $CaCO_3$ is required. It follows that more than 892 g of impure limestone (95% pure $CaCO_3$) would be required. The weight of limestone required may be found by multiplying the 892 g of pure $CaCO_3$ by an improper fraction made up of 100 per cent and 95 per cent.

$$892 \text{ g} \times \frac{100 \text{ per cent}}{95 \text{ per cent}} = 939 \text{ g of limestone}$$

Example 3. Calculate the weight of carbon dioxide produced by the complete combustion of 10.0 lb of ethane, C_2H_6, in air.

First write the equation for the reaction.

$$\underset{\text{2 moles}}{2C_2H_6} + 7O_2 \longrightarrow \underset{\text{4 moles}}{4CO_2} + 6H_2O$$

The number of pound-molecular weights of C_2H_6 in 10.0 lb is $^{10.0}/_{30.0}$. According to the equation, four pound-molecular weights of CO_2 are produced from two pound-molecular weights of C_2H_6.

Hence, the number of pound-molecular weights of CO_2 produced from $^{10.0}/_{30.0}$ pound-molecular weight of C_2H_6 is $^{10.0}/_{30.0} \times ^4/_2$.

The number of pounds of CO_2 is then calculated by multiplying the number of pound-molecular weights of CO_2 by the number of pounds in each pound-molecular weight (that is, by 44.0 lb).

$$\frac{10.0 \text{ lb}}{30.0 \text{ lb}} \times \frac{4}{2} \times 44.0 \text{ lb} = 29.3 \text{ lb of carbon dioxide}$$

QUESTIONS

1. What are the names of the elements represented by the following symbols: P, Ge, In, Ti, Fr, Al, Cd, K, As, Ag, Zn, Mo, Ru, C, Xe, Na, Br?
2. Write the chemical formulas for molecules of the following:
   (a) Oxygen
   (b) Sulfur
   (c) Potassium chlorate
   (d) Ethyl alcohol
   (e) Carbon dioxide
3. Distinguish between the terms *symbol, formula,* and *equation.*

4. Contrast the meanings of the expressions 4P and P_4.
5. Define the following terms:
   (a) Formula weight
   (b) Gram-formula weight
   (c) Gram-molecular weight
   (d) Mole
6. What is meant by the statement that atomic weights are relative weights?
7. Write out in words the meaning of the following chemical equations:
   (a) $C + O_2 \longrightarrow CO_2$
   (b) $CaO + CO_2 \longrightarrow CaCO_3$
   (c) $2SO_2 + O_2 \longrightarrow 2SO_3$
   (d) $CH_4 + 2O_2 \longrightarrow CO_2 + 2H_2O$
8. Balance the following equations:
   (a) $Ag + H_2S + O_2 \longrightarrow Ag_2S + H_2O$
   (b) $P_4 + O_2 + H_2O \longrightarrow H_3PO_4$
   (c) $Pb + H_2O + O_2 \longrightarrow Pb(OH)_2$
   (d) $Sb + O_2 \longrightarrow Sb_4O_6$
   (e) $Cu + HNO_3 \longrightarrow Cu(NO_3)_2 + H_2O + NO$
   (f) $SbCl_5 + H_2O \longrightarrow SbOCl_3 + HCl$
   (g) $PtCl_4 + XeF_2 \longrightarrow PtF_6 + ClF + Xe$
   (h) $CO + H_2O \longrightarrow H_2 + CO_2$
9. Balance the following equations:
   (a) $H_2 + O_2 \longrightarrow H_2O$
   (b) $Al_2O_3 + F_2 \longrightarrow AlF_3 + O_2F_2$
   (c) $NH_4NO_2 \longrightarrow N_2 + H_2O$
   (d) $Cu(NO_3)_2 \longrightarrow CuO + NO_2 + O_2$
   (e) $H_3PO_3 \longrightarrow H_3PO_4 + PH_3$
   (f) $N_2 + H_2 \longrightarrow NH_3$
   (g) $(NH_4)_2Cr_2O_7 \longrightarrow Cr_2O_3 + H_2O + N_2$
   (h) $HI + Cl_2 \longrightarrow HCl + ICl_3$
   (i) $Fe_3O_4 + C \longrightarrow Fe + CO$
10. Balance the following equations:
    (a) $KClO_2 \longrightarrow KCl + KClO_3$
    (b) $PCl_5 + H_2O \longrightarrow H_3PO_4 + HCl$
    (c) $PCl_3 + Cl_2 \longrightarrow PCl_5$
    (d) $Al + H_2SO_4 \longrightarrow Al_2(SO_4)_3 + H_2$
    (e) $Ca_3(PO_4)_2 + C \longrightarrow Ca_3P_2 + CO$
    (f) $CaC_2 + H_2O \longrightarrow Ca(OH)_2 + C_2H_2$
    (g) $NH_3 + O_2 \longrightarrow H_2O + NO$
    (h) $Ca_3(PO_4)_2 + H_3PO_4 \longrightarrow Ca(H_2PO_4)_2$
11. Write balanced chemical equations which represent the reactions described as follows:
    (a) Hydrogen (H_2) and chlorine (Cl_2) combine to form hydrogen chloride (HCl).
    (b) Magnesium carbonate ($MgCO_3$) is heated to drive off carbon dioxide (CO_2), leaving behind a residue of magnesium oxide (MgO).

(c) Water and hot carbon react to form hydrogen and carbon monoxide (CO).

(d) Propane (C_3H_8) burns in oxygen to produce carbon dioxide and water.

(e) Carbon burns in oxygen to produce carbon monoxide (CO).

(f) Carbon burns in oxygen to produce carbon dioxide. Compare with (e). Would you expect the use of excess oxygen to favor the production of CO or CO_2? Would an excess of carbon favor the production of CO or CO_2?

(g) Water vapor reacts with sodium metal to produce hydrogen and sodium hydroxide (NaOH).

12. Explain why the following equation does not conform to the Law of Conservation of Matter and change the equation so that it does.

$$C_{12}H_{22}O_{11} + O_2 \longrightarrow 12CO_2 + 11H_2O + \triangle$$

13. Compare, by means of ratios, the relative weights of the following:

(a) An oxygen atom of weight 16 and a nitrogen atom of weight 14.

(b) A carbon atom of weight 12 and a sulfur atom of weight 32.

(c) A mole of oxygen atoms and a mole of nitrogen atoms.

(d) A mole of carbon atoms and a mole of sulfur atoms.

14. Show that one gram-atom of oxygen contains the same number of atoms as does one gram-atom of iron.

15. Extract the maximum amount of qualitative and quantitative information from the chemical formula for sulfuric acid (H_2SO_4).

16. (a) Why do chemists find it convenient to use the value 196.9665 for the atomic weight of gold and 15.999 for the atomic weight of oxygen instead of the actual weights of the atoms in grams?

(b) Calculate the actual weights of the atoms of gold and oxygen in grams.

PROBLEMS

$\boxed{s}$1. Calculate the formula weight of each of the following compounds:

$\boxed{s}$(a) $C_2H_4Cl_2$	Ans. 98.96
(b) KO_2	Ans. 71.10
(c) $OS(CH_3)_2$	Ans. 78.13
$\boxed{s}$(d) $[Co(NH_3)_6]Br_3$	Ans. 400.8
(e) CsF	Ans. 151.9
(f) $NaZn(UO_2)_3(C_2H_3O_2)_9 \cdot 6H_2O$	Ans. 1537.94
(g) S_4N_4	Ans. 184.3

2. Calculate the gram-formula weight of each of the following materials:

(a) Na_2O_2	Ans. 77.98 g
(b) $(NH_4)_2S_2O_8$	Ans. 228.2 g
(c) $C_7H_6O_3$	Ans. 138.1 g
(d) S_8	Ans. 256.5 g
(e) CF_3COOH	Ans. 114.0 g
(f) $MgI(C_2H_5)$	Ans. 180.3 g
(g) $Hg[Co(CO)_4]_2$	Ans. 542.5 g

⌜s⌝3. Calculate the formula weight of each of the following minerals:

⌜s⌝(a) Carnotite, $K_2(UO_2)_2(VO_4)_2 \cdot 3H_2O$ *Ans. 902.2*

(b) Cacoxenite, $Fe_4(PO_4)_3(OH)_3 \cdot 12H_2O$ *Ans. 775.5*

(c) Euclasite, $BeAlSiO_4(OH)$ *Ans. 145.1*

(d) Turquois, $CuAl_6(PO_4)_4(OH)_8 \cdot 4H_2O$ *Ans. 813.4*

(e) Stannite, Cu_2FeSn_4 *Ans. 657.7*

⌜s⌝4. What is the weight of the following in grams:

(a) Two gram-atoms of beryllium?   *Ans. 18.02 g*

⌜s⌝(b) Four gram-formula weights of magnesium oxide?   *Ans. 161.22 g*

(c) One-half gram-molecular weight of ethyl alcohol (C_2H_5OH)?

                    Ans. 23.03 g

⌜s⌝(d) One millimole of sulfur dioxide?   *Ans. 6.41×10^{-2} g*

⌜s⌝(e) 5.25×10^{23} molecules of hydrogen chloride?   *Ans. 31.8 g*

⌜s⌝5. Calculate the number of moles in

(a) 18 g of $Sr(OH)_2$ *Ans. 0.15 mole*

⌜s⌝(b) 315 g of $Mg(HCO_3)_2$ *Ans. 2.15 moles*

(c) 12 g of N_2F_4 *Ans. 0.12 mole*

⌜s⌝(d) 3.6 mg of NH_3 *Ans. 0.00021 mole*

6. Determine the number of

(a) moles of sulfur dioxide in 17.54 grams of sulfur dioxide.   *Ans. 0.2738*

(b) atoms of magnesium in 2.3 gram-atoms of magnesium. *Ans. 1.4×10^{24}*

(c) gram-atoms of fluorine in 76.0 grams of fluorine (F_2).   *Ans. 4.00*

(d) moles of fluorine in 76.0 grams of fluorine (F_2).   *Ans. 2.00*

(e) gram-atoms of selenium in 119 atoms of selenium.   *Ans. 1.98×10^{-22}*

7. Which contains the greatest mass of oxygen: one mole of hydrogen peroxide (H_2O_2), one mole of calcium carbonate ($CaCO_3$), or one mole of water? Show why.   *Ans. One mole of calcium carbonate*

⌜s⌝8. Which of the following amounts contains the greatest number of atoms of any type: 2.0 grams of lithium metal, 0.30 gram-atom of hydrogen, one-sixth mole of hydrogen molecules, 2.0×10^{23} atoms of mercury, or 150 millimoles of bromine molecules? Show why.

                 Ans. One-sixth mole of hydrogen molecules

9. (a) Which element contains the greatest mass in one gram-atom: helium, chlorine, carbon, or rubidium? Show why.   *Ans. Rubidium*

(b) Which contains the greatest mass in one mole: ammonia gas (NH_3), oxygen gas (O_2), xenon gas (Xe), or chlorine gas (Cl_2)? Show why.

                  Ans. Xenon gas

⌜s⌝10. (a) Which contains the greatest number of atoms: 71 g of hydrogen, 90 g of oxygen, 32 g of fluorine, or 1500 g of chlorine? Show why.

                 Ans. 71 g of hydrogen

(b) Which contains the greatest number of gram-atoms: 7 g of hydrogen, 100 g of oxygen, 43 g of fluorine, or 56 g of magnesium? Show why.

                 Ans. 7 g of hydrogen

11. Calculate the weight of

(a) 4.500 gram-atoms of tantalum   *Ans. 814.3 g*

(b) 1.464 moles of K_3PO_4 *Ans. 310.8 g*
(c) 3.491×10^{19} molecules of chlorine (Cl_2)   *Ans. 4.110 mg*
(d) 0.400 mole of $(NH_4)_2SO_4$ *Ans. 52.9 g*

12. Calculate the number of formula weights represented by 4.50 kg of TNT, $C_6H_2(CH_3)(NO_2)_3$ *Ans. 19.8*

s 13. Calculate the weight of one atom of gold from its gram-atomic weight and Avogadro's number.   *Ans. 3.271×10^{-22} g*

14. Calculate each of the following to three significant figures:
    (a) The percentage of sodium, of aluminum, and of hydrogen by weight in $NaAlH_4$.   *Ans. Na = 42.6%; Al = 50.0%; H = 7.47%*
    (b) The percentage of ammonia in $RhCl_3 \cdot 6NH_3$.   *Ans. 32.8%*
    (c) The percentage of water in $MgSO_4 \cdot 7H_2O$.   *Ans. 51.2%*

s 15. Calculate the percentage composition of each of the following to three significant figures:
    s (a) C_3H_8 *Ans. C = 81.7%; H = 18.3%*
    s (b) Na_3PO_4 *Ans. Na = 42.1%; P = 18.9%; O = 39.0%*
    (c) $Rb_2S_2O_3$ *Ans. Rb = 60.4%; S = 22.7%; O = 17.0%*

16. Calculate the percentage composition of each of the following to three significant figures:
    (a) $(VO_2)_2SO_4$ *Ans. V = 38.9%; S = 12.2%; O = 48.9%*
    (b) $Al_2(HPO_4)_3$ *Ans. Al = 15.8%; H = 0.884%; P = 27.2%; O = 56.2%*
    (c) CH_3COCl *Ans. C = 30.6%; H = 3.85%; O = 20.4%; Cl = 45.2%*
    (d) $C_6H_2(CH_3)(NO_2)_3$
                         Ans. C = 37.0%; H = 2.22%; N = 18.5%; O = 42.3%

17. Determine the simplest formulas for the compounds with the following compositions:
    (a) 46.7% silicon and 53.3% oxygen   *Ans. SiO_2*
    (b) 25.9% nitrogen and 74.1% oxygen   *Ans. N_2O_5*
    (c) 12.64% lithium, 0.917% hydrogen, 28.19% phosphorus, and 58.25% oxygen.   *Ans. Li_2HPO_4*
    (d) 37.49% carbon, 12.58% hydrogen, and 49.93% oxygen. *Ans. CH_4O*

s 18. (a) What is the simplest formula of the compound which has the following percentage composition: Carbon, 37.51%; hydrogen, 3.15%; fluorine, 59.34%?   *Ans. CHF*
    (b) The molecular weight of the compound is 64. What is its molecular formula?   *Ans. $C_2H_2F_2$*

s 19. A 5.00-gram sample of an oxide of lead contains 4.33 g of lead. Derive the simplest formula for the compound.   *Ans. PbO_2*

20. A compound consists of hydrogen, carbon, and nitrogen. It contains 51.83% nitrogen and 44.44% carbon. What is the simplest formula for the compound?   *Ans. HCN*

21. (a) A compound has the following percentage composition: boron, 78.14%; hydrogen, 21.86%. Calculate the simplest formula for the compound.   *Ans. BH_3*
    (b) The molecular weight of the compound is 27.7. What is its true molecular formula?   *Ans. B_2H_6*

22. (a) What is the simplest formula of the compound which has the following percentage composition? Carbon, 85.6%; hydrogen, 14.4%.   *Ans. CH_2*
    (b) The molecular weight of the compound is 28. What is its molecular formula?   *Ans. C_2H_4*

23. A 1.610-g sample of a chloride of iron contains 0.709 g of iron. Calculate the simplest formula of the compound.   *Ans. $FeCl_2$*

24. A 24.00-g sample of an oxide of osmium contains 6.04 g of oxygen. Calculate the simplest formula for the compound.   *Ans. OsO_4*

$\boxed{s}$25. Determine the simplest formula of a compound which has the following analysis: Na, 16.78%; NH_4, 13.16%; H (in addition to that in the NH_4), 0.74%; PO_4, 69.32%.   *Ans. $NaNH_4HPO_4$*

26. A certain compound contains 42.66% Ag, 12.25% P, and 45.09% F. Calculate the simplest formula of the compound.   *Ans. $AgPF_6$*

$\boxed{s}$27. Calculate the simplest formula for the compound which has the following composition: Cr, 20.78%; H, 5.64% C, 48.00%; and O, 25.58%.
    Ans. $CrC_{10}H_{14}O_4$ or, as commonly expressed, $Cr(C_5H_7O_2)_2$

28. Determine the formula of a crystalline salt which has the following percentage composition: N, 8.70%; C, 29.85%; H, 7.51%; B, 6.72%; F, 47.22%.
    Ans. $NC_4H_{12}BF_4$ or, as commonly expressed, $[N(CH_3)_4]BF_4$

29. A sample of an element weighs 1.1567 g and is known to be 0.010074 gram-atom of that element. Calculate the atomic weight of this element. Which element is this?   *Ans. 114.82; In*

$\boxed{s}$30. What weight of Al_2O_3 will contain 30.00 g of oxygen?   *Ans. 63.73 g*

31. What weight of oxygen is contained in 62 g of $Mg(C_2H_3O_2)_2$?   *Ans. 28 g*

32. Determine the simplest formula of a compound which contains 17.55% Na, 39.70% Cr, and 42.75% O.   *Ans. $Na_2Cr_2O_7$*

$\boxed{s}$33. A sample of a molecular compound weighs 3.45 g and is known to be 13.4 millimoles of that compound. Calculate the molecular weight of the compound.   *Ans. 257*

34. What weight of oxygen is contained in 200 g of $Na_2S_2O_3$?   *Ans. 60.7 g*

$\boxed{s}$35. What weight of NaN_3 will be needed to produce 100 g of nitrogen (N_2): ($3NaN_3 + \triangle \longrightarrow Na_3N + 4N_2$)?   *Ans. 174 g*

$\boxed{s}$36. Calculate the weight of oxygen produced by the thermal decomposition of 379 g of potassium nitrate: ($2KNO_3 + \triangle \longrightarrow 2KNO_2 + O_2$). *Ans. 60.0 g*

37. What weight of germanium dioxide will be formed by the combustion of 20.0 lb of germanium in an excess of oxygen: ($Ge + O_2 \longrightarrow GeO_2$)? What weight of impure oxygen (93.7% pure oxygen by weight) will be required for actual reaction?   *Ans. 28.8 lb; 9.41 lb*

38. Calculate the weight of hydrogen which would be formed by the reaction of 50 g of manganese with excess HCl. The equation for the reaction is:
    $Mn + 2HCl \longrightarrow MnCl_2 + H_2$ *Ans. 1.8 g*

39. A sample containing 0.1200 mole of a compound weighs 5.64 g. Calculate the molecular weight of the compound.   *Ans. 47.0*

$\boxed{s}$40. Write the balanced equation and calculate the number of moles of carbon monoxide (CO) required to react with 15.0 g of nickel metal (Ni) to produce tetracarbonyl nickel [$Ni(CO)_4$]. How many grams of CO would this be?
    Ans. 1.02 mole; 28.6 g

41. What weight of BeO will contain 128.0 g of beryllium? *Ans. 355.2 g*

S42. DDT, a chlorinated hydrocarbon containing only C, H, and Cl, was analyzed. When a 22.21 mg sample was burned (reaction of DDT with oxygen of the air), 38.60 mg of CO_2 and 5.079 mg of H_2O were obtained. What is the simplest formula of DDT? The molecular weight of DDT is approximately 350. What is the molecular formula of DDT? *Ans. $C_{14}H_9Cl_5$; $C_{14}H_9Cl_5$*

43. Urea (an organic compound composed of carbon, hydrogen, oxygen and nitrogen) has been used in the treatment of sickle cell anemia. When a 47.65-mg sample of urea was burned in air, 34.92 mg of CO_2 and 28.59 mg of H_2O were produced. When all the nitrogen in a 56.42-mg sample of urea was converted to ammonia, 32.00 mg of ammonia was produced. What is the simplest formula of urea? If the molecular weight of urea is approximately 60, what is its molecular formula? *Ans. CON_2H_4; CON_2H_4*

44. If a grain is equal to 64.8 milligrams, how many moles of aspirin, $C_8H_9O_4$, are in a 5.00-grain aspirin tablet? *Ans. 1.92×10^{-3} mole*

45. What is the maximum amount of calcium carbide (CaC_2) which can be prepared from 8.40 g of calcium and 4.80 g of carbon?
Ans. 0.200 gram-formula weight or 12.8 g

S46. (a) What weight of copper(II) nitrate, $Cu(NO_3)_2$, must be decomposed to produce 1500 g (assume the zeros are significant figures) of copper(II) oxide, CuO? $Cu(NO_3)_2 \longrightarrow CuO + NO_2 + O_2$ (unbalanced)
Ans. 3537 g

    (b) What weight of nitrogen dioxide, NO_2, would be produced by the decomposition of 75.0 g of copper(II) nitrate? *Ans. 36.8 g*

    (c) How many moles of oxygen would be produced by the decomposition of 75.0 g of copper(II) nitrate? *Ans. 0.200 mole*

47. How many tons of sulfur would have to be burned to produce 25.0 tons of sulfur dioxide: $(S + O_2 \longrightarrow SO_2)$? *Ans. 12.5 tons*

48. Calculate the weight of CH_3MgBr necessary to produce 25.0 g of $Sn(CH_3)_4$ by the reaction: $4CH_3MgBr + SnCl_4 \longrightarrow Sn(CH_3)_4 + 4MgClBr$.
Ans. 66.7 g

S49. What weight of chlorine would be required in the reaction with 10.0 metric tons of hydrogen: $(H_2 + Cl_2 \longrightarrow 2HCl)$? What weight of HCl would be produced, assuming the yield to be 98.0%? *Ans. 352 tons; 354 tons*

S50. A solution containing 50.0 g of LiOH is mixed with a solution containing 50.0 g of HCl: $(LiOH + HCl \longrightarrow LiCl + H_2O)$. Which compound is in excess and what weight of LiCl is produced? *Ans. LiOH; 58.1 g*

S51. A mixture containing 248.1 g of H_2O_2 and 32.00 g of N_2H_4 reacts as follows: $7H_2O_2 + N_2H_4 \longrightarrow 2HNO_3 + 8H_2O$. What is the composition of the system by weight after the reaction, assuming the reaction goes to completion?
Ans. 125.8 g of HNO_3, 143.9 g of water, and 10.3 g of H_2O_2

52. What weight of sulfur which is 98.0% pure by weight would be required in the production of 237 kilograms of H_2SO_4?
$$S + O_2 \longrightarrow SO_2$$
$$2SO_2 + O_2 \longrightarrow 2SO_3$$
$$SO_3 + H_2O \longrightarrow H_2SO_4$$
Ans. 79.1 kg

⑤53. What weight of $P(CH_3)_4I$ will be formed by the reaction of 16.0 lb of CH_3I with $P(CH_3)_3$: $[P(CH_3)_3 + CH_3I \longrightarrow P(CH_3)_4I]$? What weight of impure $P(CH_3)_3$ (90% pure by weight) will be required in this reaction?

Ans. 24.6 lb; 9.5 lb

54. A 12.0-g sample of sulfur(IV) fluoride, SF_4, is allowed to react with water: $(SF_4 + 3H_2O \longrightarrow H_2SO_3 + 4HF)$. What weight of HF is produced and what weight of water enters the reaction? *Ans. 8.89 g; 6.00 g*

55. (a) Balance the following equation and calculate the weight of oxygen which is required to react with 37.0 g of cyclopentene, C_5H_8:

$$C_5H_8 + O_2 \longrightarrow CO_2 + H_2O \text{ (unbalanced)}$$

Ans. 122 g

(b) How many moles of CO_2 would be produced?   *Ans. 2.72 moles*

56. For the following reaction, 110.0 g of aluminum oxide (Al_2O_3), 60.0 g of carbon, and 200.0 g of chlorine are present:

$$Al_2O_3 + C + Cl_2 \longrightarrow AlCl_3 + CO \text{ (unbalanced)}$$

(a) Which substance is in greatest excess in terms of the quantities required for reaction? Show why, by calculation.   *Ans. Carbon*

(b) Calculate the maximum amount of aluminum chloride $(AlCl_3)$ which could be produced.   *Ans. 250.7 g*

57. The following quantities are combined in a lead chamber: 1.2×10^{24} atoms of hydrogen, 1.0 gram-atom of sulfur, and 88 g of oxygen.

(a) What is the total weight in grams of all the atoms in the chamber?

Ans. 1.2×10^2 g

Note that only two significant figures are justified by the data (see Appendix A.2). Hence, the answer should be expressed as 1.2×10^2 g (or 120 g, in which case the zero is not a significant figure but is necessary to establish the decimal point). The answer 122 g is not in accord with the significant figures in the data and hence is incorrect for this problem.

(b) How many gram-atoms (of all kinds) are there in the collection?

Ans. 8.5 gram-atoms

(c) If the three elements are induced to react to form a compound which has two hydrogen atoms and four oxygen atoms for each sulfur atom, how many grams of material will remain unreacted?   *Ans. 24 g*

REFERENCES

"Avogadro's Number: Early Values of Loschmidt and Others," R. M. Hawthorne, Jr., *J. Chem. Educ.*, **47**, 751 (1970).

"History of the Chemical Sign Language," R. Winderlich (transl. by R. E. Oesper), *J. Chem. Educ.*, **30**, 58 (1953).

The Mole Concept in Chemistry, W. F. Kieffer, Reinhold Publishing Corp., New York, 1962 (Paperback).

"Size and Shape of A Molecule," M. J. Demchik and V. C. Demchik, *J. Chem. Educ.*, **48**, 770 (1971).

"Avogadro's Hypothesis and the Duhemian Pitfall," R. L. Causey, *J. Chem. Educ.,* **48,** 365 (1971).

"Atomic Weights of the Elements; Changes in Atomic Weight Values," (Report of the Inorganic Chemistry Division Commission on Atomic Weights, International Union of Pure and Applied Chemistry), *Pure and Applied Chemistry,* **37** (4), 591 (1974).

"Metrology: A More Accurate Value for Avogadro's Number," A. L. Robinson, *Science,* **185,** 1037 (1974).

Structure of the Atom and the Periodic Law

3

A series of discoveries during the latter part of the 19th century has modified the Daltonian concept of the atom. Atoms are no longer considered to be simple, compact bodies as supposed by Dalton; they are complex systems composed of a number of smaller particles. The only particles of the atoms we shall consider here, however, are electrons, protons, and neutrons.

Knowledge of the structure of atoms has made it possible to systematize the facts of chemistry in such a way that the subject is easier to understand and remember. Chemical reactions that occur between atoms, and the forces that hold atoms together in molecules, can be explained in terms of atomic structure.

3.1   Electrons

Much of the information concerning the structure of atoms resulted from experiments involving the conduction of electricity through gases exerting a very low pressure. The apparatus used for experiments of this type, called a **discharge tube,** consists of a sealed glass tube provided with two metal electrodes connected to a source of electric current (Fig. 3–1a). When the tube is highly evacuated and an electrical potential is applied which is sufficient to cause a current to pass between the electrodes, rays of light extend from the negative electrode **(cathode)** toward the positive electrode **(anode).** These rays are called **cathode rays.**

Cathode rays have the following characteristics: (1) they travel from the cathode to the anode in straight lines, as indicated by the fact that an object placed in their path casts a sharp shadow on the end of the tube (Fig. 3–1b), (2) they cause a piece of metal foil to become hot after striking it for awhile, and (3) they are

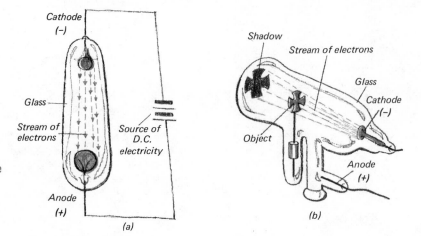

FIGURE 3-1

Two types of discharge tubes showing (a) streams of electrons flowing from the cathode to the anode, and (b) a shadow being cast by an object placed in a stream of electrons.

deflected by magnetic and electrical fields (Fig. 3–2). The character of the deflection is the same as that shown by negatively charged particles passing through such fields (a charged object will attract an object of opposite charge, whereas objects having charges of the same sign repel each other). These characteristics of cathode rays are best explained by assuming that they are streams of small particles of matter which are negatively charged. These particles are called **electrons.**

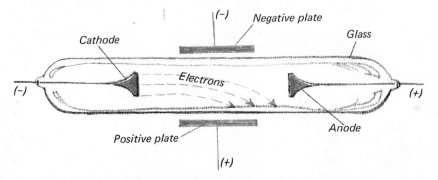

FIGURE 3-2

Electrons in a discharge tube are repelled by the negative plate and attracted by the positive plate.

In 1897, the English physicist Sir J. J. Thomson determined the ratio of the charge to mass, e/m, of electrons, and found e/m to be identical for all cathode ray particles, irrespective of the gas in the tube or the metal of which the electrodes were made. The charge on the electrons was measured by Millikan at the University of Chicago in 1909. From the values of e/m and e, the mass of the electron was calculated to be 1/1837 of that of the hydrogen atom. Thus, on the atomic weight scale the mass of the electron is 0.00055.

Electrons are emitted by certain metals when the metals are heated to high temperatures. This process is called **thermal emission.** Electrons are also emitted from very active metals such as cesium, sodium, and potassium when they are exposed to light (Fig. 3–3). This type of emission is known as the **photoelectric**

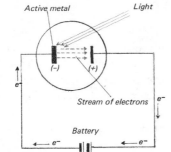

FIGURE 3-3
The photoelectric cell in an electric circuit. Light striking the active metal electrode causes electrons to be emitted and a current to flow.

effect and is the basis of the "electric eye," or photoelectric cell. X rays liberate electrons from all forms of matter.

The fact that all electrons are identical, irrespective of their source or method of liberation from matter, and can be liberated from any kind of atom proves quite conclusively that they are parts of all atoms.

3.2 Protons

In 1886, Goldstein, using a discharge tube with a perforated cathode, observed that another kind of ray was being emitted from the anode and passing through the holes in the cathode. Wien, in 1898, showed these rays to be positively charged particles; they are today called **positive rays.** The ratio of the charge on one of the particles to its mass was found to be much smaller than that for the electron and to vary with the kind of gas in the tube. Since the ratio varied, it follows that either the charge varied or the mass varied or both. Actually, it was found that both the charge and the mass varied. Measurements showed the charge on each particle to be either a positive unit charge, or some whole number multiple of this unit. The unit positive charge was shown to be equal to that on the electron but opposite in sign. The mass of the positive particle was found to be less when hydrogen was used in the discharge tube than when any other gas was employed. From the values of e and e/m for the positive particles when hydrogen is used, the value of m can be calculated; the mass is 1.0073 on the atomic weight scale. This particle is called the **proton** and is one of the basic units of structure of all atoms.

The formation of positively charged particles from neutral atoms or molecules of whatever gas is used in a discharge tube is caused by the loss of electrons by these neutral atoms or molecules when they are struck by high speed cathode rays. These charged atoms are known as **gaseous ions.** From the fact that any neutral atom can be made to form positive ions by the loss of one or more electrons, it follows that every atom must contain one or more positive units. The simplest atom is that of hydrogen. It contains one proton (hydrogen ion) and one electron.

3.3 Radioactivity and Atomic Structure

The first conclusive evidence that atoms are complex rather than "indivisible," as stated in the original atomic theory of matter, came with the discovery of radioactivity by Becquerel, a French physicist, in 1896. **Radioactivity** is the term applied to the spontaneous decomposition of atoms of certain elements, such as radium and uranium, into other elements with the simultaneous production of rays of particles. There are three types of rays, and these can be studied and characterized by placing samples of radioactive material in a narrow hole bored in a block of lead and allowing the emitted rays to pass through a strong electric field (Fig. 3–4). One type of ray is observed to curve toward the negative part of

FIGURE 3-4

The effect of an electric field on rays from the radioactive substance radium. Beta, β, rays are electrons. Alpha, α, rays are nuclei of helium atoms. Gamma, γ, rays are similar to x rays.

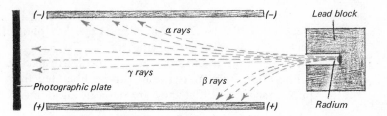

the electric field and so must consist of positively charged particles. These are called **alpha (α) particles.** A second type of ray is deflected toward the positive part of the field, with the path bent much more than that of the alpha particles. These facts indicate that these particles are negatively charged and of much less mass than alpha particles. They are called **beta (β) particles.** The third type of ray is not deflected when it passes through the electric field, indicating that the particles are neither negatively nor positively charged. These rays are similar to x rays but more penetrating than x rays, or either α particles or β particles (Fig. 3–5). They are called **gamma (γ) rays.** Several other kinds of rays, in addition to the three discussed here, are also known and will be discussed later in Chapter 30, Nuclear Chemistry.

Alpha particles have a mass of 4 on the atomic weight scale and a charge of +2. Experiment has shown that they are helium ions, i.e., helium atoms that have lost two electrons each. Beta particles are known to be electrons with velocities approaching that of the speed of light (about 186,000 miles/second).

Convincing proof that atoms contain protons was obtained by Rutherford in 1919. Using high velocity α particles from radium as projectiles, he bombarded atoms such as nitrogen and aluminum and found that protons were ejected from the atoms as a result of these collisions. These experiments indicated quite definitely that the proton is a unit of the atom.

3.4 Neutrons

In 1932, the English physicist Chadwick discovered a third basic unit of the atom. He observed that when atoms of beryllium (and other elements) are bombarded with high velocity α particles, uncharged particles are emitted. Chadwick called

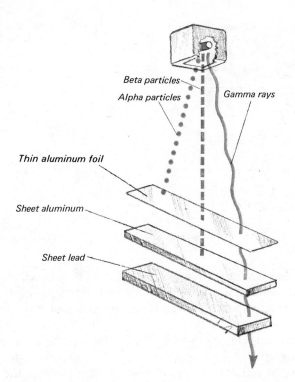

Beta particles

Alpha particles

Gamma rays

Thin aluminum foil

Sheet aluminum

Sheet lead

FIGURE 3-5
A diagram showing the relative penetrating power of the radiations from radium.

these neutral particles **neutrons**; they were calculated to have a mass of 1.0087 on the atomic weight scale. It appears that under certain conditions a neutron may disintegrate and form a proton and an electron. This may be the origin of the electrons composing the β rays, which are emitted by radioactive elements.

3.5 Summary of Three Basic Particles in an Atom

In the preceding sections we discussed three basic particles present in an atom—electrons, protons, and neutrons (the ordinary hydrogen atom, which contains one proton, one electron, and no neutrons is the only exception).

Particle	Mass (at. wt. scale)	Charge
Electron	0.00055	−1
Proton	1.0073	+1
Neutron	1.0087	0

3.6 The Nuclear Atom

Alpha particles from radioactive elements travel at very high speeds—on the average about 10,000 miles per second. Their paths may be photographed in an apparatus (cloud chamber, Fig. 3–6) devised by C. T. R. Wilson. As alpha particles

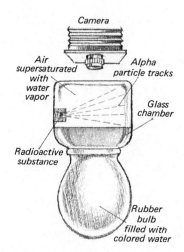

FIGURE 3-6

The Wilson cloud chamber. Pressure on the bulb compresses the air and water vapor to produce a supersaturated condition. Molecules of the gases in the air ionized by radiations serve as nuclei on which water vapor condenses when the pressure is reduced to produce a fog track, marking the path of the radiation particles.

pass through the air in the cloud chamber they knock electrons from the molecules of the gases that make up the air and thus form ions. The air is supersaturated with water vapor, and the ionized gaseous molecules serve as nuclei upon which droplets of water condense. Thus, trails of small droplets of water (fog tracks) appear, marking the paths of the alpha particles through the cloud chamber. Photographs of fog tracks show that the paths are usually straight, indicating that the alpha particles generally pass directly through the molecules of air without being deflected from their paths; however, a few paths show an abrupt change in direction near the end of the course, which means some alpha particles are deflected (Fig. 3–7).

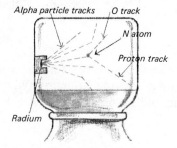

FIGURE 3-7

Fog tracks in a cloud chamber. Alpha particles colliding with molecules in the air are often merely deflected, but sometimes they disrupt a nitrogen nucleus to produce an oxygen nucleus and a proton.

When Rutherford projected a beam of α particles from a radioactive source upon very thin gold foil, he found that most of the particles passed through the solid foil without deflection; only a few of them were diverted from their paths. From the results of a series of such experiments Rutherford concluded that (1) the volume occupied by an atom must be largely empty space, as indicated by the fact that most of the α particles pass through the foil undeflected, and (2) there must be located within each atom a heavy, positively charged body (the nucleus). This follows because an abrupt change in path (as noted for a few α particles) of a relatively heavy and positively charged α particle can result only from its hitting or from its close approach to another particle (the nucleus) with a highly concentrated, positive charge. Rutherford determined the diameter of the nucleus to be about $\frac{1}{10,000}$ of the diameter of the atom (the diameter of an atom is of the order of 10^{-8} cm). Rutherford presumed the atom to consist of a very small, positively charged **nucleus,** in which most of the mass of the atom is concentrated, surrounded by a number of electrons necessary to produce an electrically neutral whole. This is Rutherford's nuclear theory of the atom, proposed in 1911. We still use this model today.

From this same series of experiments, Rutherford found that the number of the positive charges on the nucleus (and thus the number of protons in the nucleus) is, for many of the lighter elements, approximately one-half the atomic weight of the element. The number of positive charges on the nucleus of an atom of a given element is equal to the **atomic number** of that element. Long before Rutherford's experiments, however, the number representing the position of an element in the Periodic Table (first flyleaf of this book) had been called the atomic number of that element.

3.7   Moseley's Determination of Atomic Numbers

In 1914, Moseley worked out a method for determining the number of positive charges on the nucleus of the atom for all elements. This method involved the use of x rays. A modern x-ray tube (Fig. 3–8) is a modified cathode ray tube in which electrons are obtained by thermal emission from a filament heated by an electric current. When a solid target is placed in the path of the speeding electrons, very penetrating rays are produced. These penetrating rays are called **x rays.** In order to study the x rays produced when different elements (or their compounds) were made the targets in the x-ray tube, Moseley used certain salt crystals as diffraction gratings to spread the x rays into a **spectrum.** This effect

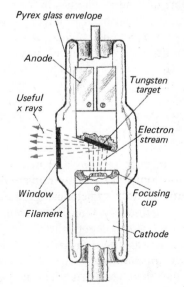

FIGURE 3-8

In the x-ray tube, electrons striking a metal target such as tungsten cause the production of x rays. In Moseley's experiment, various elements were used as targets.

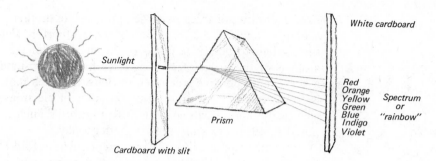

FIGURE 3-9
Sunlight passing through a prism is separated into its component colors by refraction.

Increasing wavelength

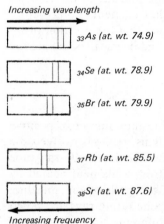

$_{33}$As (at. wt. 74.9)

$_{34}$Se (at. wt. 78.9)

$_{35}$Br (at. wt. 79.9)

$_{37}$Rb (at. wt. 85.5)

$_{38}$Sr (at. wt. 87.6)

Increasing frequency

FIGURE 3-10
Diagram representing several x-ray spectra as found by Moseley. Note the greater shift in wavelength between Br and Rb. The element having atomic number 36, krypton, was left out by Moseley.

is similar to that produced when a beam of sunlight is spread into its component colors (a spectrum) by passing it through a glass prism (Fig. 3-9). When the spectrum of x rays was recorded on a photographic film, it appeared as lines characteristic of the material of which the target was made. Moseley arranged the x-ray spectra of different elements in order of their increasing atomic weights and found that (with three exceptions) the heavier the element, the shorter the wavelengths of the principal lines making up the spectrum. He also found that the wavelength of an element differed from that of the element next to it by the same amount (Fig. 3-10). This result suggested that nuclei of atoms of elements adjacent in the series differ from each other by one proton. It then became possible to arrange all the elements in order of increasing positive charge on the nuclei of their atoms, beginning with hydrogen as 1. **The atomic number of an element, then, is the number of positive charges in the nucleus of each of its atoms.** Because an atom is electrically neutral, it follows that the atomic number represents the number of electrons outside its nucleus as well as the number of protons in its nucleus. The atomic numbers for the elements used in Fig. 3-10 are 33 (As), 34 (Se), 35 (Br), 37 (Rb), and 38 (Sr). The atomic number for krypton, an element unknown in Moseley's time, is 36.

3.8   The Composition of the Nucleus

The nuclei of atoms contain both protons and neutrons except for the nucleus of the ordinary hydrogen atom, which only contains a proton. The particles inside the nucleus are often referred to as **nucleons.** Because each proton and each neutron has a mass of approximately 1 on the atomic weight scale, the atomic weight is approximately equal to the sum of the number of protons and neutrons in the nucleus, or, in other words, the number of nucleons (neglecting the small mass of the electron). The composition of each of the atomic nuclei of the elements having atomic numbers one through ten is given in Table 3-1.

TABLE 3-1 Nuclear Compositions of the Very Light Elements

	Symbol	Atomic Number	Number of Protons	Number of Neutrons	Atomic Weight
Hydrogen	H	1	1	0	1.0079
Helium	He	2	2	2	4.00260
Lithium	Li	3	3	4	6.941
Beryllium	Be	4	4	5	9.01218
Boron	B	5	5	6	10.81
Carbon	C	6	6	6	12.011
Nitrogen	N	7	7	7	14.0067
Oxygen	O	8	8	8	15.9994
Fluorine	F	9	9	10	18.998403
Neon	Ne	10	10	10	20.179

3.9 Isotopes

When we note that each proton and neutron contributes approximately one unit of atomic mass to atoms, we expect to find the atomic weights of the elements to be approximately whole numbers. The fact that many atomic weights are far from being whole numbers led to the discovery that most elements exist as mixtures of two or more kinds of atoms of different atomic masses but of similar chemical properties. For example, chlorine, with an atomic weight of 35.453, exists as two kinds of chlorine atoms. These have masses very close to the whole numbers 35 and 37. Both kinds of chlorine atoms have an atomic number of 17, which means that they each have 17 protons in the nucleus. The difference must lie, then, in the number of neutrons in the nuclei of the two types of atoms; chlorine-35 has 18 neutrons and chlorine-37 has 20 neutrons. **Atoms of the same atomic number and different masses are called isotopes.** It is important to remember that the only difference in composition between isotopes of the same element is in the number of neutrons in the nucleus. The atomic weight of an element is, therefore, an average of the weights of the isotopes of the element in the proportions in which they normally occur in nature. The atomic weight of 35.453 for chlorine indicates that the atoms of weight 35 are almost three times as abundant as the atoms of weight 37.

A symbolism has been devised to distinguish between isotopes of an element. The two isotopes of chlorine are designated by the symbols $^{35}_{17}Cl$ and $^{37}_{17}Cl$. Note that the mass number is made the superscript and the atomic number the subscript.

3.10 Atomic Weight Scale

In Section 2.4, mention was made of a carbon isotope being used as the present standard in the atomic weight scale. Carbon has three principal isotopes, $^{12}_{6}C$, $^{13}_{6}C$, and $^{14}_{6}C$. Each has six protons in the nucleus and six electrons outside the nucleus. Carbon-12 has, in addition, six neutrons in the nucleus; carbon-13, seven neutrons; and carbon-14, eight neutrons. The atomic weight scale is based upon the arbitrarily assigned value of the exact number 12 for the mass of the lightest and most abundant carbon isotope, $^{12}_{6}C$. The three isotopes of carbon occur in such proportion in a natural mixture that the average weight of the mixture, on the atomic weight scale, is 12.011, indicating that carbon-12 is the most abundant. Magnesium atoms weigh on the average a little more than twice as much as carbon-12 atoms, or 24.305 units on the atomic weight scale; and hydrogen atoms, on the average, about one-twelfth as much as carbon-12 atoms, or 1.0079. The table on the inside front cover lists the atomic weights of the elements, based upon the arbitrarily assigned value of 12 for $^{12}_{6}C$.

3.11   Atomic Spectra and Electronic Structure

Sir Isaac Newton showed that sunlight is a mixture of different kinds of light. He separated sunlight into its component colors (its spectrum) by allowing it to pass through a refracting glass prism. Sunlight contains all wavelengths of visible light and hence gives the **continuous spectrum** (Fig. 3–9) so familiar in the rainbow. Sunlight also contains ultraviolet (very short wavelengths) and infrared light (very long wavelengths), both of which may be detected and recorded photographically but which are beyond the range of the human eye. Incandescent solids, liquids, and gases under great pressure give continuous spectra. When an electric current is passed through a gas contained in a vacuum tube at low pressure, the gas gives off light, which when passed through a prism shows a spectrum made up of a number of bright lines **(a bright line spectrum).** These lines can be recorded photographically and the wavelength of the light producing each line can be calculated from the positions of the lines on the photograph.

Monatomic gases produce bright line spectra that consist of sets of lines which are characteristic of the elements making up the gas. Because these spectra are produced by the excitation of atoms they are called **atomic spectra.** Many attempts have been made to explain how an excited atom radiates energy and why it radiates the particular frequencies that it does. The German physicist M. Planck concluded, in 1900, that an atom cannot absorb or emit radiation energy continuously but instead must do so in discontinuous quantities called **quanta.** Experiment has shown that these energy quanta are always multiples of the frequency, v. Mathematically,

$$E = hv$$

where E is energy; h is a constant, referred to as Planck's constant, equal to 6.626×10^{-27} erg-sec; v is the frequency. After careful measurement of the light frequencies of the lines in the visible spectrum of hydrogen, J. R. Rydberg found that they are related by the general equation

$$v = R\left(\frac{1}{n_1{}^2} - \frac{1}{n_2{}^2}\right)$$

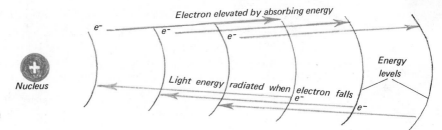

FIGURE 3-11
Electron transitions which give rise to atomic spectra.

where *R*, now called the **Rydberg constant,** has the value 3.289×10^{15} cycles per second; and n_1 and n_2 are integers, n_1 being smaller than n_2. Hence,

$$\nu = 3.289 \times 10^{15} \left(\frac{1}{n_1{}^2} - \frac{1}{n_2{}^2} \right)$$

The Rydberg equation is empirical; that is, it is derived from observation rather than theory.

In 1913, the Danish physicist Niels Bohr provided an explanation for the regularities observed by Rydberg. Bohr pictured the atom as a miniature solar system, with the nucleus corresponding to the sun and the electrons to the orbiting planets. While revolving in an orbit about the nucleus, the electron, according to Bohr, obeys the ordinary laws of mechanics. He suggested that an atom radiates energy as light only when the electron suddenly changes from its present orbit to one where it has less energy (Fig. 3–11). However, the planetary theory was discredited when it was pointed out that whereas planets attract each other because of gravitational forces, the electrons in an atom repel one another. furthermore, the planets revolve in the same direction and in nearly the same plane, while electrons move in all directions and in all planes.

Bohr found the quantity of energy radiated to be equal to the difference between the energies the atom possessed before and after the electron changed orbits. He found it necessary to assume, in agreement with Planck's ideas, that the electron does not gain or lose energy continuously in passing from a lower orbit to a higher one during excitation of the atom or in dropping to a lower orbit from a higher one when radiating energy. The gain or loss is only in discrete quantities (quanta). A quantum of energy is always directly proportional to the frequency of the radiation.

According to Bohr, the electron in the hydrogen atom can have only those energies, E_n, permitted by the equation

$$E_n = \frac{-21.79 \times 10^{-12}}{n^2} \text{ erg}$$

The integer *n* corresponds to the energy level, or orbit, in which the electron is located, and the value -21.79×10^{-12} erg is the energy required to remove the electron from the level closest to the nucleus to a point completely out of the atom.

Radiation is emitted when the electron (often designated by e⁻) "falls" from an outer energy level to one closer to the nucleus. Conversely, energy is required (absorbed) to move an electron from an energy level closer to the nucleus to one further away. The energy emitted or absorbed in such cases is equal to the energy difference between the two levels and can be calculated from the Bohr equation

$$E = E_{n_2} - E_{n_1} = 21.79 \times 10^{-12} \frac{\text{erg}}{\text{atom}} \left(\frac{1}{n_1{}^2} - \frac{1}{n_2{}^2} \right)$$

where n_1 and n_2 are integers representing the two energy levels, n_1 being the level closer to the nucleus and n_2 an energy level farther away from the nucleus than n_1.

Example 1. Calculate the amount of energy required (absorbed) to pull an electron in the hydrogen atom from the energy level closest to the nucleus ($n_1 = 1$) to the third energy level from the nucleus ($n_2 = 3$).

$$E = E_{n_2} - E_{n_1} = 21.79 \times 10^{-12} \left(\frac{1}{n_1{}^2} - \frac{1}{n_2{}^2} \right)$$

$$= 21.79 \times 10^{-12} \left(\frac{1}{1^2} - \frac{1}{3^2} \right)$$

$$= 21.79 \times 10^{-12} \left(\frac{1}{1} - \frac{1}{9} \right) = 1.937 \times 10^{-11} \text{ erg per atom}$$

Example 2. Calculate the amount of energy emitted when an electron in an excited hydrogen atom falls from the third energy level (counting from the nucleus) to the energy level closest to the nucleus.

The energy emitted is the same as that calculated in Example 1 to pull the electron from the energy level closest to the nucleus to the third energy level from the nucleus. Hence, the answer is 1.937×10^{-11} erg per atom.

Example 3. Calculate the frequency of radiation emitted when an electron in an excited hydrogen atom falls from the third energy level counting from the nucleus to the energy level closest to the nucleus.

$$\nu = 3.289 \times 10^{15} \left(\frac{1}{n_1{}^2} - \frac{1}{n_2{}^2} \right)$$

$$= 3.289 \times 10^{15} \left(\frac{1}{1^2} - \frac{1}{3^2} \right) = 2.924 \times 10^{15} \text{ cycles per second}$$

The Rydberg equation and the Bohr equation can be brought into excellent agreement by dividing the Bohr equation expression for E by Planck's constant, h, equal to 6.626×10^{-27} erg-sec/atom.

$$E = h\nu$$

Then,
$$\nu = \frac{E}{H} = \frac{21.79 \times 10^{-12} \left(\dfrac{1}{n_1{}^2} - \dfrac{1}{n_2{}^2} \right)}{6.626 \times 10^{-27}}$$

$$= 3.289 \times 10^{15} \left(\frac{1}{n_1{}^2} - \frac{1}{n_2{}^2} \right)$$

The agreement of the two equations provides powerful and indispensable evidence for the Bohr concept of atomic structure. A difficulty in the Bohr theory needs mention, however. The simple Bohr equation given above is derived for hydrogen, which has only one electron. If more than one electron is present in an atom, repulsive forces between the electrons complicate the calculations to the extent that predictions are virtually impossible. Even after modification and expansion the theory can account for the spectra of only the very simplest atomic systems, such as the hydrogen atom. Nevertheless, the Bohr theory remains a milestone in scientific progress, inasmuch as, on the basis of his theory, Bohr was able to account satisfactorily for the bright lines in several atomic spectral series.

It is now known that the laws of ordinary mechanics do not adequately explain the properties of such small particles as electrons. Bohr's theory of the electron as a compact particle of matter moving in a circular or elliptical orbit about a nucleus has been largely supplanted by the more mathematical theories of modern quantum mechanics. Unfortunately, the quantum mechanical concept of the atom does not lend itself readily to a mechanical model that can be visualized.

For our purposes, it is enough to consider that the Bohr theory and the quantum mechanical treatment indicate that the hydrogen atom consists of one proton (as a nucleus) and one electron moving about the proton. Larger atoms have several electrons moving about a positive nucleus, which consists of several protons and neutrons. Although the electrons are known to move about the nucleus, the exact path they take within a particular energy level cannot be determined. The German physicist Heisenberg expressed this in a form which has come to be known as the **Heisenberg Uncertainty Principle: It is impossible to determine accurately both the momentum and the position of an electron simultaneously.** (The momentum is the mass of the particle multiplied by its velocity.) The Heisenberg principle maintains that the more accurately we measure the momentum of a moving electron, the less accurately we can determine its position (and conversely). An experimental procedure which is designed to measure the position of an electron exactly will alter the momentum of the electron, and an experiment which seeks to measure accurately the momentum of an electron unavoidably changes its position. If measurements are obtained for both position and momentum at the same time, the values are inexact for one or the other or both. Therefore, since the exact position or path of an electron cannot be determined, the best that can be done is to speak of **the probability of finding an electron at a given location within the atom.**

The electrons, however, *are* located in definite *energy levels*. Inasmuch as the physical and chemical properties of an atom are primarily determined by the

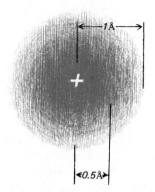

FIGURE 3-12

Electron cloud, or electron density distribution, corresponding to the 1s orbital around the nucleus. The + sign indicates the nucleus.

energies associated with the electrons in their movement within the atom, the energy of the electron is much more important to the chemist than the actual position of the electron.

The results of a quantum mechanical treatment of the problem show that the electron may be visualized as being in rapid motion within a relatively large region around the nucleus, but spending most of its time in certain high probability regions. For the hydrogen atom, the single electron effectively occupies all the space within about 1 Å (1×10^{-8} cm) of the nucleus, with the greatest probability of being at a distance of about 0.5 Å; this gives the hydrogen atom a spherical shape. Some chemists prefer to consider the electron in terms of a cloud of negative charge (**electron cloud**), with the cloud being dense in regions of high electron probability and more diffuse in regions of low probability. These electron clouds are spoken of as **atomic orbitals.**

Each atomic orbital can be described by a quantity called a **wave function,** which is given the symbol ψ. The square of the wave function, ψ^2, is a measure of the probability of finding the electron at a given point at a distance r from the nucleus; and $4\pi r^2 \psi^2$ (**the radial probability density**) is a measure of the probability of finding the electron within the volume of a thin spherical shell of thickness $r + dr$ (where dr is a very small increment of r).

Figure 3–12 illustrates the electron density distribution of the atomic orbital nearest to the positive nucleus. Most of the time, but not always, the electron will be located inside the spherical outline. Putting it another way, the probability of finding the electron inside the spherical boundary is high. For hydrogen, the probability (and hence the electron density) is practically zero right at the nucleus but increases rapidly just outside the nucleus and becomes highest at a distance of about 0.5 Å from the nucleus. This distance may be considered as the radius of the electron shell. The probability decreases rapidly as the distance from the nucleus increases and becomes exceedingly small at a distance of 1 Å. Figure 3–13 is a plot of the radial probability density, $4\pi r^2 \psi^2$, as a function of distance from the nucleus, r, for particular orbitals of the $n = 1$, $n = 2$, and $n = 3$ energy levels.

We have already noted that certain atomic orbitals are approximately spherical. Some orbitals have more complex shapes, as shown in Section 3.20. The nuclei of atoms which are more complex than that of hydrogen are surrounded by electrons which are arranged in a series of **shells,** or **energy levels.** The closer an energy level is to the nucleus, the more strongly the electrons in that level are attracted to the nucleus because the electrostatic force of attraction between particles of opposite electrical charge increases rapidly with decreased distance of separation. The orbital electrons are classified in terms of their energy levels, i.e., on the basis of the firmness with which they are bound to the nucleus in the atom. The energy levels, or shells, are designated by either the numbers 1, 2, 3, 4, 5, 6, and 7 or the letters K, L, M, N, O, P, and Q, starting with the one nearest the nucleus. The maximum number of electrons found to occupy each of the various electron shells is $2n^2$, where n is the number of the shell counting from the nucleus. Hence the shell nearest the nucleus, for which $n = 1$, can contain a maximum of $2(1)^2$ or 2 electrons; the second shell from the nucleus ($n = 2$), $2(2)^2$ or 8 electrons; and the third shell from the nucleus $2(3)^2$ or 18 electrons.

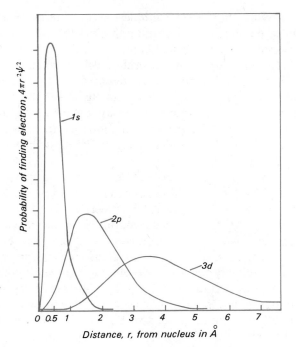

FIGURE 3-13
The probability of finding the electron at a given distance *r* from the nucleus for the 1*s*, 2*p*, and 3*d* orbitals. Note the overlap of orbitals.

Table 3–2 summarizes the maximum number of electrons possible for the first five major shells.

TABLE 3-2

Energy Levels, or Shells		Maximum Number of Electrons, $2n^2$
Letter Designation	n	
K	1	2
L	2	8
M	3	18
N	4	32
O	5	50*

*As can be seen in Tables 3–3 and 3–8, no atom of the known elements is large enough to contain more than 32 electrons in the O shell. The number 50, although hypothetical, is a logical prediction of the maximum number of electrons theoretically possible for the O shell.

3.12 Atomic Structures

A single hydrogen atom consists of one proton and one electron. Customarily, the proton is represented by p^+ and the electron by e^-. The single electron is in motion in the K shell, or first energy level, around the proton.

$1p^+$ | $\dfrac{\text{K}}{1e^-}$

Hydrogen atom

$\begin{matrix}2p^+\\2n\end{matrix}$ | $\dfrac{\text{K}}{2e^-}$

Helium atom

$\begin{matrix}3p^+\\4n\end{matrix}$ | $\dfrac{\text{K}\quad\text{L}}{2e^-\ \ 1e^-}$

Lithium atom

$\begin{matrix}4p^+\\5n\end{matrix}$ | $\dfrac{\text{K}\quad\text{L}}{2e^-\ \ 2e^-}$

Beryllium atom

$\begin{matrix}10p^+\\10n\end{matrix}$ | $\dfrac{\text{K}\quad\text{L}}{2e^-\ \ 8e^-}$

Neon atom

$\begin{matrix}11p^+\\12n\end{matrix}$ | $\dfrac{\text{K}\quad\text{L}\quad\text{M}}{2e^-\ \ 8e^-\ \ 1e^-}$

Sodium atom

$\begin{matrix}15p^+\\16n\end{matrix}$ | $\dfrac{\text{K}\quad\text{L}\quad\text{M}}{2e^-\ \ 8e^-\ \ 5e^-}$

Phosphorus atom

$\begin{matrix}19p^+\\20n\end{matrix}$ | $\dfrac{\text{K}\quad\text{L}\quad\text{M}\quad\text{N}}{2e^-\ \ 8e^-\ \ 8e^-\ \ 1e^-}$

Potassium atom

$\begin{matrix}21p^+\\24n\end{matrix}$ | $\dfrac{\text{K}\quad\text{L}\quad\text{M}\quad\text{N}}{2e^-\ \ 8e^-\ \ 9e^-\ \ 2e^-}$

Scandium atom

After hydrogen, the next simplest atom in terms of atomic structure is that of the **noble gas** helium, with an atomic number of two and an atomic weight of approximately four. The helium atom contains two protons and two neutrons in its nucleus, and two electrons in the K shell, which is completely filled by these two electrons. The neutron is represented by n.

The next atom in complexity after helium is that of the metal lithium, with an atomic number of three and an atomic weight of approximately seven. The nucleus contains three protons and four neutrons, and there are three orbital electrons. Two of the electrons make up the K shell; the third is in the next higher energy level, i.e., the L shell.

An atom of the metal beryllium, of atomic number four and atomic weight approximately nine, contains four protons and five neutrons in the nucleus, and four electrons, two each in the K and L shells.

Boron (At. No. 5) has three electrons in its L shell, carbon (At. No. 6) has four, nitrogen (At. No. 7) has five, oxygen (At. No. 8) has six, fluorine (At. No. 9) has seven, and neon (At. No. 10) has eight electrons in the L shell. Just as the K shell is completely filled by two electrons, so the L shell is filled by eight, and neon resembles helium in that both have their shells completely filled by electrons.

The sodium atom (At. No. 11) contains one more electron than the neon atom. This electron is found in the next higher energy level, i.e., the third, or M, shell. The sodium atom resembles the lithium atom in having one electron in its outermost shell.

The magnesium atom (At. No. 12) has two electrons in its outer shell (the M shell) and so is analogous to beryllium, which also has two electrons in its outer shell (the L shell). Aluminum (At. No. 13), with three electrons in the outer shell, silicon with four, phosphorus with five, sulfur with six, chlorine with seven, and argon with eight, correspond respectively to boron, carbon, nitrogen, oxygen, fluorine, and neon. Any shell, other than the K shell (which is completely filled with two electrons), is complete with eight electrons if it is the outermost shell. Hence, argon has a complete outer electron configuration, with eight electrons in the M shell. The M shell, when it becomes an inner shell in larger atoms, can hold up to eighteen electrons.

Potassium and calcium (At. Nos. 19 and 20) have one and two electrons, respectively, in the N shell. Hence, potassium corresponds to lithium and sodium in outer configuration, whereas calcium corresponds to beryllium and magnesium.

Beginning with scandium (At. No. 21), the M shell, *which is now the second shell from the outside,* starts building from eight toward eighteen electrons, scandium having nine electrons in the M shell. Its outer shell, N, has the same number of electrons as calcium with two.

Table 3–3 lists the atomic numbers and the numbers of electrons in the various electron shells of atoms of the known elements.

3.13 The Periodic Law

The known elements vary greatly in their physical and chemical properties and in the nature of the compounds which they form. The study of the individual

TABLE 3-3   Electron Distribution in the Atoms of the Elements

(handwritten annotation above headers: 1s² 2s² 3p⁶ 4)

Element	At. No.	K	L	M	N	O	P	Q
H	1	1						
He	2	2						
Li	3	2	1					
Be	4	2	2					
B	5	2	3					
C	6	2	4					
N	7	2	5					
O	8	2	6					
F	9	2	7					
Ne	10	2	8					
Na	11	2	8	1				
Mg	12	2	8	2				
Al	13	2	8	3				
Si	14	2	8	4				
P	15	2	8	5				
S	16	2	8	6				
Cl	17	2	8	7				
Ar	18	2	8	8				
K	19	2	8	8	1			
Ca	20	2	8	8	2			
Sc	21	2	8	9	2			
Ti	22	2	8	10	2			
V	23	2	8	11	2			
Cr	24	2	8	13	1			
Mn	25	2	8	13	2			
Fe	26	2	8	14	2			
Co	27	2	8	15	2			
Ni	28	2	8	16	2			
Cu	29	2	8	18	1			
Zn	30	2	8	18	2			
Ga	31	2	8	18	3			
Ge	32	2	8	18	4			
As	33	2	8	18	5			
Se	34	2	8	18	6			
Br	35	2	8	18	7			
Kr	36	2	8	18	8			
Rb	37	2	8	18	8	1		
Sr	38	2	8	18	8	2		
Y	39	2	8	18	9	2		
Zr	40	2	8	18	10	2		
Nb	41	2	8	18	12	1		
Mo	42	2	8	18	13	1		
Tc	43	2	8	18	14	1		
Ru	44	2	8	18	15	1		
Rh	45	2	8	18	16	1		
Pd	46	2	8	18	18	0		
Ag	47	2	8	18	18	1		
Cd	48	2	8	18	18	2		
In	49	2	8	18	18	3		
Sn	50	2	8	18	18	4		
Sb	51	2	8	18	18	5		
Te	52	2	8	18	18	6		
I	53	2	8	18	18	7		
Xe	54	2	8	18	18	8		
Cs	55	2	8	18	18	8	1	
Ba	56	2	8	18	18	8	2	
La	57	2	8	18	18	9	2	
Ce	58	2	8	18	20	8	2	
Pr	59	2	8	18	21	8	2	
Nd	60	2	8	18	22	8	2	
Pm	61	2	8	18	23	8	2	
Sm	62	2	8	18	24	8	2	
Eu	63	2	8	18	25	8	2	
Gd	64	2	8	18	25	9	2	
Tb	65	2	8	18	27	8	2	
Dy	66	2	8	18	28	8	2	
Ho	67	2	8	18	29	8	2	
Er	68	2	8	18	30	8	2	
Tm	69	2	8	18	31	8	2	
Yb	70	2	8	18	32	8	2	
Lu	71	2	8	18	32	9	2	
Hf	72	2	8	18	32	10	2	
Ta	73	2	8	18	32	11	2	
W	74	2	8	18	32	12	2	
Re	75	2	8	18	32	13	2	
Os	76	2	8	18	32	14	2	
Ir	77	2	8	18	32	15	2	
Pt	78	2	8	18	32	17	1	
Au	79	2	8	18	32	18	1	
Hg	80	2	8	18	32	18	2	
Tl	81	2	8	18	32	18	3	
Pb	82	2	8	18	32	18	4	
Bi	83	2	8	18	32	18	5	
Po	84	2	8	18	32	18	6	
At	85	2	8	18	32	18	7	
Rn	86	2	8	18	32	18	8	
Fr	87	2	8	18	32	18	8	1
Ra	88	2	8	18	32	18	8	2
Ac	89	2	8	18	32	18	9	2
Th	90	2	8	18	32	18	10	2
Pa	91	2	8	18	32	20	9	2
U	92	2	8	18	32	21	9	2
Np	93	2	8	18	32	23	8	2
Pu	94	2	8	18	32	24	8	2
Am	95	2	8	18	32	25	8	2
Cm	96	2	8	18	32	25	9	2
Bk	97	2	8	18	32	26	9	2
Cf	98	2	8	18	32	28	8	2
Es	99	2	8	18	32	29	8	2
Fm	100	2	8	18	32	30	8	2
Md	101	2	8	18	32	31	8	2
No	102	2	8	18	32	32	8	2
Lr	103	2	8	18	32	32	9	2
Rf	104	2	8	18	32	32	10	2
Ha	105	2	8	18	32	32	11	2
—	106	2	8	18	32	32	12	2

properties and the many compounds of each of the elements would prove to be extremely laborious and time consuming, but though every element is different from every other element, similarities make possible groupings that simplify the study. The most widely used grouping is that called the Periodic Table.

It became evident early in the development of chemistry that certain elements known at that time could be grouped together by reason of their similar properties. Lithium, sodium, and potassium (a portion of the group now referred to as the alkali metals), calcium, strontium, and barium (in the group called alkaline earth metals), and chlorine, bromine, and iodine (in the halogen group) are examples of groups of elements exhibiting similar properties.

The German chemist Döbereiner, in 1829, was the first to suggest the existence of a relationship between the atomic weights and the properties of the elements. He showed that the atomic weight of strontium lies about midway between the atomic weights of two chemically similar elements, calcium and barium, and that the properties of strontium are intermediate between those of calcium and barium. He later recognized the existence of other **triads** of similar elements. Among these are the triads chlorine, bromine, and iodine; and lithium, sodium, and potassium. By 1854, other chemists had shown that oxygen, sulfur, selenium, and tellurium could be classed as a **family** of elements, and that nitrogen, phosphorus, arsenic, antimony, and bismuth could make up another family.

In 1865, Newlands in England recognized a correlation between the magnitude of the atomic weights and the properties of the elements. By 1870, Mendeleev in Russia and Lothar Meyer in Germany, working independently and apparently unaware of the work of Newlands, outlined the nature of this relationship between properties and atomic weights in some detail. Their conclusions led to the statement that the properties of the elements are periodic functions of their atomic weights. The idea is illustrated in the following two series of elements. Note that the elements are arranged in order of increasing atomic weights from left to right.

Li	Be	B	C	N	O	F	Ne
6.941	9.01218	10.81	12.011	14.0067	15.9994	18.998403	20.179

Na	Mg	Al	Si	P	S	Cl	Ar
22.98977	24.305	26.98154	28.0855	30.97376	32.06	35.453	39.948

These two rows of elements comprise what are called the second and third **periods** of the modern Periodic Table (Table 3–4 and the front endleaf of this book). Elements in each horizontal row differ decidedly in both physical and chemical properties from one another. There is, however, an interesting and useful gradation in properties of the elements in a given row as the atomic weight increases. Furthermore, when the elements are arranged in order of increasing atomic weight, elements in the same vertical columns have similar properties. For example, lithium (Li) is found to be very similar physically and chemically to sodium (Na), and beryllium (Be) to be very similar to magnesium (Mg).

The next heavier element after argon is potassium, which is similar to lithium and sodium. Beginning with potassium, however, eighteen elements must be added to the series in order of increasing atomic weights before rubidium (which is similar to potassium) is reached. These eighteen elements constitute a long

THE PERIODIC TABLE OF ELEMENTS (Long Period Form)

There are seven periods (horizontal rows) and eighteen groups (vertical columns) of elements. The number of electrons in filled shells is shown in the column at the extreme left; the remaining electrons are shown immediately below the symbol for each element. Atomic numbers are enclosed by brackets. Atomic weights are shown above the symbols. Atomic weight values in parentheses are those of the isotopes of longest half-life for certain radioactive elements whose atomic weights cannot be quoted precisely without knowledge of origin of the element. Atomic weights are based on carbon-12. (*By permission of International Union of Pure and Applied Chemistry, 1975.*)

NONMETALS

METALS

TRANSITION METALS

periods across

PERIODS	I A	II A	III B	IV B	V B	VI B	VII B	VIII	VIII	VIII	I B	II B	III A	IV A	V A	VI A	VII A	0
1 (0)	1.0079 H[1] 1																	4.00260 He[2] 2
2 (2)	6.941 Li[3] 1	9.01218 Be[4] 2											10.81 B[5] 3	12.011 C[6] 4	14.0067 N[7] 5	15.9994 O[8] 6	18.99840 F[9] 7	20.179 Ne[10] 8
3 (2, 8)	22.98977 Na[11] 1	24.305 Mg[12] 2											26.98154 Al[13] 3	28.086 Si[14] 4	30.97376 P[15] 5	32.06 S[16] 6	35.453 Cl[17] 7	39.948 Ar[18] 8
4 (2, 8)	39.098 K[19] 8, 1	40.08 Ca[20] 8, 2	44.9559 Sc[21] 9, 2	47.90 Ti[22] 10, 2	50.9414 V[23] 11, 2	51.996 Cr[24] 13, 1	54.9380 Mn[25] 13, 2	55.847 Fe[26] 14, 2	58.9332 Co[27] 15, 2	58.70 Ni[28] 16, 2	63.546 Cu[29] 18, 1	65.38 Zn[30] 18, 2	69.72 Ga[31] 18, 3	72.59 Ge[32] 18, 4	74.9216 As[33] 18, 5	78.96 Se[34] 18, 6	79.904 Br[35] 18, 7	83.80 Kr[36] 18, 8
5 (2, 8, 18)	85.4678 Rb[37] 8, 1	87.62 Sr[38] 8, 2	88.9059 Y[39] 9, 2	91.22 Zr[40] 10, 2	92.9064 Nb[41] 12, 1	95.94 Mo[42] 13, 1	(97) Tc[43] 14, 1	101.07 Ru[44] 15, 1	102.9055 Rh[45] 16, 1	106.4 Pd[46] 18	107.868 Ag[47] 18, 1	112.41 Cd[48] 18, 2	114.82 In[49] 18, 3	118.69 Sn[50] 18, 4	121.75 Sb[51] 18, 5	127.60 Te[52] 18, 6	126.9045 I[53] 18, 7	131.30 Xe[54] 18, 8
6 (2, 8, 18)	132.9054 Cs[55] 18, 8, 1	137.33 Ba[56] 18, 8, 2	[57-71] *	178.49 Hf[72] 32, 10, 2	180.9479 Ta[73] 32, 11, 2	183.85 W[74] 32, 12, 2	186.207 Re[75] 32, 13, 2	190.2 Os[76] 32, 14, 2	192.22 Ir[77] 32, 15, 2	195.09 Pt[78] 32, 17, 1	196.9665 Au[79] 32, 18, 1	200.59 Hg[80] 32, 18, 2	204.37 Tl[81] 32, 18, 3	207.2 Pb[82] 32, 18, 4	208.9804 Bi[83] 32, 18, 5	(209) Po[84] 32, 18, 6	(210) At[85] 32, 18, 7	(222) Rn[86] 32, 18, 8
7 (2, 8, 18, 32)	(223) Fr[87] 18, 8, 1	226.0254 Ra[88] 18, 8, 2	[89-103] †	[257] Rf[104] 32, 10, 2	[260] Ha[105] 32, 11, 2	[106] 32, 12, 2	[107]	[108]										

*LANTHANIDE SERIES

138.9055 La[57] 18, 9, 2	140.12 Ce[58] 20, 8, 2	140.9077 Pr[59] 21, 8, 2	144.24 Nd[60] 22, 8, 2	(145) Pm[61] 23, 8, 2	150.4 Sm[62] 24, 8, 2	151.96 Eu[63] 25, 8, 2	157.25 Gd[64] 25, 9, 2	158.9254 Tb[65] 27, 8, 2	162.50 Dy[66] 28, 8, 2	164.9304 Ho[67] 29, 8, 2	167.26 Er[68] 30, 8, 2	168.9342 Tm[69] 31, 8, 2	173.04 Yb[70] 32, 8, 2	174.97 Lu[71] 32, 9, 2

†ACTINIDE SERIES

(227) Ac[89] 18, 9, 2	232.0381 Th[90] 18, 10, 2	231.0359 Pa[91] 20, 9, 2	238.029 U[92] 21, 9, 2	237.0482 Np[93] 23, 8, 2	(244) Pu[94] 24, 8, 2	(243) Am[95] 25, 8, 2	(247) Cm[96] 25, 9, 2	(247) Bk[97] 26, 9, 2	(251) Cf[98] 28, 8, 2	(254) Es[99] 29, 8, 2	(257) Fm[100] 30, 8, 2	(258) Md[101] 31, 8, 2	(255) No[102] 32, 8, 2	(260) Lr[103] 32, 9, 2

period, in distinction to the two short periods which begin with lithium and sodium, respectively. The other long periods are indicated in the Periodic Table. An explanation of the appearance of long and short periods in terms of atomic structure is made in Section 3.14.

The work of Moseley (Section 3.7) showed that atomic numbers rather than atomic weights are fundamental in determining the chemical properties of the elements. If you examine the arrangement of the Periodic Table, you will note that argon (At. Wt. 39.948) precedes potassium (At. Wt. 39.098), cobalt (At. Wt. 58.9332) precedes nickel (At. Wt. 58.70), tellurium (At. Wt. 127.60) precedes iodine (At. Wt. 126.9045), thorium (At. Wt. 232.0381) precedes protactinium (At. Wt. 231.0359), and uranium (At. Wt. 238.029) precedes neptunium (At. Wt. 237.0482). These reverses, on the basis of atomic weights, are due to unusual percentages of isotopes in these pairs of elements. Consider, for example, argon (At. No. 18) and potassium (At. No. 19). The abundance of the isotopes of these elements is such that the weighted average atomic weight of the naturally occurring mixture of isotopes is greater for argon than for potassium. Thus, argon consists almost entirely of the isotope with a mass number of 40, whereas potassium consists largely of the isotope with a mass number of 39. If the arrangement of the elements is in order of increasing atomic numbers, the ten elements mentioned previously fall into their proper positions in the table. The modern statement of the **Periodic Law** is: **The properties of the elements are periodic functions of their atomic numbers.**

The Periodic Table is printed on the front endleaf of this book and also in Table 3–4 for convenient reference and study by the reader.

3.14   Electronic Structure and the Periodic Law

As we discuss, in this section, the relationship of the electronic structures (electron arrangement) of the atoms to the positions of their elements in the Periodic Table, you will find it exceedingly helpful to refer to the Table as each idea is developed.

When arranged in order of increasing atomic numbers the elements with similar chemical properties recur at definite intervals, i.e., periodically. In regard to electronic structure, this suggests a periodicity in the number of electrons in the outer shells of the atoms of the elements. The electrons in the outermost shell of an atom (referred to as **valence electrons**) are largely responsible for its chemical behavior. If elements having the same number of valence electrons are grouped together, the elements falling within each group have similar chemical properties. Table 3–5 reviews the electron grouping in the shells for the elements of atomic numbers 1 to 18, inclusive. Note the periodicity in regard to the number of valence electrons.

It is seen that elements in any one vertical column (known as a **group,** or **family**) have the same number of electrons in their outermost shells. We can now understand the similarity in chemical properties among elements of the same group, because the number of electrons in the outermost shell of an atom is very important in determining its chemical properties.

As mentioned before, the horizontal rows of the Periodic Table are referred

TABLE 3-5   Periodicity of Valence Electrons

Element	H							He
Electrons in shell	**1**							**2**
Element	Li	Be	B	C	N	O	F	Ne
Electrons in shells	**2,1**	**2,2**	**2,3**	**2,4**	**2,5**	**2,6**	**2,7**	**2,8**
Element	Na	Mg	Al	Si	P	S	Cl	Ar
Electrons in shells	**2,8,1**	**2,8,2**	**2,8,3**	**2,8,4**	**2,8,5**	**2,8,6**	**2,8,7**	**2,8,8**

to as **periods.** Notice that the **first short period** of the table contains only the elements hydrogen and helium, a noble gas, Helium has a complete outer shell (K shell) of two electrons.

The **second short period** contains eight elements, beginning with lithium and ending with the noble gas neon. Neon has a complete outer shell (L shell) of eight electrons.

The **third short period** contains eight elements beginning with sodium and ending with the noble gas argon, which also contains eight electrons in the outer shell (M shell).

The **fourth period,** you will note, is the first of two long periods that contain eighteen elements each. This period includes elements from scandium (At. No. 21) through copper (At. No. 29), which are known as a **transition series of elements,** in which the M shell, in this case the shell next to the outer shell, is building from eight to eighteen electrons. Before the transition series begins, however, two electrons enter the outer shell (the N shell) in potassium (At. No. 19) and calcium (At. No. 20), respectively. Following the transition series, the outer shell builds on up, reaching eight electrons with the noble gas krypton (At. No. 36).

It is apparent from further inspection of the table that the **fifth period,** beginning with rubidium and ending with xenon, is similar to the fourth period. It has eighteen elements and contains a second transition series of elements, yttrium (At. No. 39) through silver (At. No. 47), in which the shell next to the outer shell (the N shell) is building from eight to eighteen electrons, analogous to the first transition series in the preceding period.

The **sixth period** contains 32 elements. A third transition series occurs in this period, made up of lanthanum (At. No. 57) and the elements hafnium (At. No. 72) through gold (At. No. 79), in which the second shell counting from the outside (the O shell) builds from eight to eighteen electrons. Notice that the third transition series is split, however, and between lanthanum and hafnium is a series of fourteen elements, cerium (At. No. 58) through lutetium (At. No. 71). In these fourteen elements, the **third** shell counting from the outside (the N shell, which can hold a maximum of 32 electrons—see Table 3-2) builds from 18 to 32 electrons. Lutetium, therefore, has the structure 2, 8, 18, 32, 9, 2. These elements constitute the first **inner transition series** and are referred to as the **lanthanide elements,** or the **rare earth elements.** Inasmuch as lanthanum has a similar outer configuration, its properties are very much like those of the elements cerium through lutetium; hence, for convenience, it is often included in the lanthanide series even though in terms of electron structure it is more properly considered as the first element of the third transition series. Following lutetium, the elements hafnium through gold complete the third transition series, building the second shell counting from the outer shell (O shell) up to eighteen electrons. The outer shell (P shell) then builds up, reaching eight electrons with the noble gas radon (At. No. 86).

The **seventh period** is incomplete. The first two members are francium (At. No. 87) and radium (At. No. 88), with K, L, M, N, O, P, and Q shells of 2, 8, 18, 32, 18, 8, 1 and 2, 8, 18, 32, 18, 8, 2 electrons, respectively. Actinium (At. No. 89), with the structure 2, 8, 18, 32, 18, 9, 2, is the first element of the fourth transition series. The next three naturally occurring elements (thorium, protactinium, and uranium) and the eleven transuranium elements, all artificially produced (neptunium, plutonium, americium, curium, berkelium, californium, einsteinium, fermium, mendelevium, nobelium, and lawrencium) constitute the second inner transition series, in which the third shell counting from the outside, the O shell, is building from 18 to 32 electrons. These elements, in which actinium is sometimes included because of its similarity in properties, are called the **actinide elements.**

Although only three elements (At. Nos. 104, 105, and 106) beyond lutetium have been reported, it is logical to assume that they and any elements immediately beyond them (elements 107, 108, etc.) that may exist or be artificially produced will be a part of fourth transition series in which the shell next to the outer shell (the P shell, in this case) builds toward 18 electrons.

It is interesting to note that a number of irregularities occur in the atomic structures of the elements. For example, chromium (At. No. 24), which might be expected to have a structure 2, 8, 12, 2, instead has the structure 2, 8, 13, 1; copper (At. No. 29) would normally be expected to have a structure 2, 8, 17, 2, but actually is 2, 8, 18, 1; palladium (At. No. 46) might logically be 2, 8, 18, 16, 2, but is instead 2, 8, 18, 18, 0. The several irregularities occur because of similarities in energies of various electron shells. Very little difference in energy, for example, is involved between the M and N shells through the first transition series. In the chromium atom the M shell actually possesses sufficiently lower energy to allow one of the electrons to move down from the N shell. For the next element, manganese (At. No. 25), the N shell is then somewhat lower in energy than the M shell, so that the regularity is restored, with the additional electron going into the N shell to give it a total of two electrons again.

The vertical columns (families, or groups) of elements in the Periodic Table are numbered IA, IIA, IIIB, IVB, VB, VIB, VIIB, VIII, IB, IIB, IIIA, IVA, VA, VIA, VIIA, and 0. The elements in a given "A" group show some resemblances to the corresponding elements in the "B" group of the same number.

Tables 3–3 and 3–4 show the electron distribution for each element.

3.15 Summary Classification of Elements in Terms of the Periodic Table

It is convenient to classify the elements in the Periodic Table into four categories, according to their atomic structures.

1. *Noble gases.* Elements in which the outer shell is complete with eight electrons

(two electrons for helium). The noble gases are helium (He), neon (Ne), argon (Ar), krypton (Kr), xenon (Xe), and radon (Rn).

2. *Representative elements.* Elements in which the added electron enters the outermost shell but in which the outermost shell is incomplete. The representative elements are those in Groups IA, IIA, IIB, IIIA, IVA, VA, VIA, and VIIA of the Periodic Table.

3. *Transition elements.* Elements in which the second shell counting from the outside is building from eight to eighteen electrons. The four transition series are

First transition series: Scandium (Sc) through copper (Cu).
Second transition series: Yttrium (Y) through silver (Ag).
Third transition series: Lanthanum (La); and hafnium (Hf) through gold (Au).
Fourth transition series (incomplete): Actinium (Ac); and 106—.

4. *Inner transition elements.* Elements in which the third shell counting from the outside is building from eighteen to thirty-two electrons. The two inner transition series are:

First inner transition series: Cerium (Ce) through lutetium (Lu).
Second inner transition series: Thorium (Th) through lawrencium (Lr).

(Lanthanum and actinium, because of their similarities to the other members of the series, are sometimes included as the first elements of the first and second inner transition series, respectively.)

3.16 Variation of Properties Within Periods and Groups

As we have already pointed out, the properties of the elements are determined largely by their atomic structures. Differences in chemical properties are caused primarily by differences in three characteristics: (1) the magnitude of the nuclear charge and the number of electrons in the shells surrounding the nucleus, both of which are equal to the atomic number; (2) the number of shells of electrons and the number of electrons in these shells, particularly in the valence shells; and (3) the distances of the electrons in the various shells from each other and from the nucleus. These topics and additional interesting facts concerning the Periodic Table will be discussed in Chapter 7.

3.17 Subshells, Orbitals, and Quantum Numbers

There are many more lines in the spectrum of an atom than can be accounted for in terms of electron transitions from the energy levels, or shells (designated by K, L, M, N, O, P, and Q or by 1, 2, 3, 4, 5, 6, and 7). Many of these additional atomic spectral lines result from the facts that the principal shell is divisible into **subshells** and electron transitions are possible between subshells. The energies involved in electron transitions from different subshells of the same principal shell differ only slightly but, of course, still give rise to radiations of different wavelengths. The first principal shell has a single subshell, designated as 1*s*; the second principal shell has two subshells, designated as 2*s* and 2*p*; the third principal shell has three subshells, designated as 3*s*, 3*p*, and 3*d*; and the fourth principal shell has four subshells designated as 4*s*, 4*p*, 4*d*, and 4*f*. The subshells are further divisible into **orbitals,** with each orbital containing a maximum of two electrons. Thus, an *s* subshell, which is made up of one orbital, can contain a maximum of two electrons; a *p* subshell has three orbitals and can contain up to six electrons; a *d* subshell has five orbitals and can contain up to ten electrons; and an *f* subshell has seven orbitals and can contain a maximum of fourteen electrons. On the average, an *s* electron (an electron is an *s* subshell) will approach the nucleus more closely than a *p* electron of the same major shell, a *p* electron more closely than a *d* electron, and a *d* electron more closely than an *f* electron.

In general, the motions of the electrons about the nucleus can be described in terms of four quantum numbers:

■ **1. Principal Quantum Number, *n*.**   The principal quantum number designates, in general, the effective volume of the space in which the electron moves. An increase in the value of *n* indicates an increase in the energy associated with the electron energy level, or major shell, (see Fig. 3–14) and an increase in the *average* distance of the electron from the nucleus (see Fig. 3–13). Values of *n* are whole numbers which theoretically can vary from 1 to ∞.

Shell	n	Shell	n
K	1	O	5
L	2	P	6
M	3	Q	7
N	4		

■ **2. Subsidiary Quantun Number, *l*.**   The subsidiary quantum number designates the shape of the region which the electron occupies. Values of *l* are whole numbers which can vary from 0 to (*n* − 1). Thus if *n* = 1, *l* can be only 0; if *n* = 2, *l* can be either 0 or 1; etc.

n	l	n	l
1	0	4	0, 1, 2, 3
2	0, 1	5	0, 1, 2, 3, 4
3	0, 1, 2	6	0, 1, 2, 3, 4, 5

The electrons are often designated by the letters *s*, *p*, *d*, *f*, as indicated earlier in this section. These letters correspond to the values of *l*, and continue alphabetically following *f*.

l	Electron Designation	l	Electron Designation
0	*s*	4	*g*
1	*p*	5	*h*
2	*d*	6	*i*
3	*f*	–	–

■ **3. Magnetic Quantum Number, *m*.** The magnetic quantum number designates in a general way the orientation of the electron path in space. Values of *m* are whole numbers and can vary from $-\ell$ through zero to $+\ell$.

n	*ℓ*	*m*
1	0	0
2	0	0
	1	$-1, 0, +1$
3	0	0
	1	$-1, 0, +1$
	2	$-2, -1, 0, +1, +2$
4	0	0
	1	$-1, 0, +1$
	2	$-2, -1, 0, +1, +2$
	3	$-3, -2, -1, 0, +1, +2, +3$

■ **4. Spin Quantum Number, *s*.** The spin quantum number specifies the direction of spin of the electron about *its own axis* as the electron moves around the nucleus. The spin can be either counterclockwise or clockwise and is designated arbitrarily by either $+\frac{1}{2}$ or $-\frac{1}{2}$. For every possible combination of *n*, *ℓ*, and *m*, there can be two electrons differing only in the direction of spin about their own axes. The two electrons which can occupy any one given orbital differ only in spin and have identical *n*, *ℓ*, and *m* values. **The electrons enter each orbital of a given type singly and with identical spins before any pairing of electrons of opposite spin occurs within those orbitals**—a rule known as **Hund's Rule.** We will see the importance of this rule in future applications (see, for example, Sections 5.7 and 32.3). (Note that the symbol *s* for spin quantum number does not have the same meaning as the symbol *s* used to designate an electron for which $\ell = 0$).

This brings us to an important principle, called the **Pauli Exclusion Principle: No two electrons in a particular atom can have the same set of four quantum numbers.**

Table 3–6 summarizes the permissible values for each quantum number and sums up the number of electrons in each subshell and in each major shell, taking into account the Pauli Exclusion Principle.

The notation commonly used in describing electronic structures of atoms consists of a number in front of the subshell letter to designate the number of the principal, or major, shell and a superscript to designate the number of electrons in that particular subshell. For example, the notation $2p^4$ indicates four electrons in the *p* subshell of the second principal shell from the nucleus; the notation $3d^8$ indicates eight electrons in the *d* subshell of the third principal shell.

In arriving at the electronic structure of atoms of the various elements in terms of subshells, it is convenient to consider the subshells which electrons would enter if these atoms were built up in order of increasing atomic number, beginning with hydrogen. With each additional electron there is a tendency for the electron

TABLE 3-6   Summary of Permissible Values For Each Quantum Number

Shell	n	ℓ	m	s	Number of Electrons in Subshell	Total Electrons in Major Shell
K	1	0 (s)	0	$+\frac{1}{2}$	$\left.\begin{array}{c}1\\1\end{array}\right\}2$	2
	1	0	0	$-\frac{1}{2}$		
L	2	0 (s)	0	$+\frac{1}{2}, -\frac{1}{2}$	2	
	2	1 (p)	-1	$+\frac{1}{2}, -\frac{1}{2}$	$\left.\begin{array}{c}2\\2\\2\end{array}\right\}6$	8
	2	1	0	$+\frac{1}{2}, -\frac{1}{2}$		
	2	1	$+1$	$+\frac{1}{2}, -\frac{1}{2}$		
M	3	0 (s)	0	$+\frac{1}{2}, -\frac{1}{2}$	2	
		1 (p)	$-1, 0, +1$	$\pm\frac{1}{2}$ for each value of m	6	18
		2 (d)	$-2, -1, 0, +1, +2$	$\pm\frac{1}{2}$ for each value of m	10	
N	4	0 (s)	0	$\pm\frac{1}{2}$ for each value of m	2	
		1 (p)	$-1, 0, +1$	$\pm\frac{1}{2}$ for each value of m	6	32
		2 (d)	$-2, -1, 0, +1, +2$	$\pm\frac{1}{2}$ for each value of m	10	
		3 (f)	$-3, -2, -1, 0, +1, +2, +3$	$\pm\frac{1}{2}$ for each value of m	14	
O	5	0 (s)	0	$\pm\frac{1}{2}$ for each value of m	2	
		1 (p)	$-1, 0, +1$	$\pm\frac{1}{2}$ for each value of m	6	
		2 (d)	$-2, -1, 0, +1, +2$	$\pm\frac{1}{2}$ for each value of m	10	50*
		3 (f)	$-3, -2, -1, 0, +1, +2, +3$	$\pm\frac{1}{2}$ for each value of m	14	
		4 (g)	$-4, -3, -2, -1, 0, +1, +2, +3, +4$	$\pm\frac{1}{2}$ for each value of m	18*	

*As noted in Table 3–2, the total number of 50 electrons for the O shell, including the 18 electrons for the g subshell, is a prediction of the number of electrons theoretically possible. No element presently known contains more than 32 electrons in the O shell.

to occupy the available orbital of lowest energy. Electrons enter higher energy orbitals only after lower energy orbitals have been filled to capacity.

The energies corresponding to the various subshells of any given principal shell increase in the order $s < p < d < f$. In some cases the d or f subshells in a shell of lower principal quantum number overlap with the s or p subshells in the next higher shell. For example, two electrons enter the $4s$ subshell in the calcium atom (At. No. 20), $1s^2$, $2s^22p^6$, $3s^23p^6$, $4s^2$, before an electron enters the $3d$ subshell in the scandium atom (At. No. 21), $1s^2$, $2s^22p^6$, $3s^23p^63d^1$, $4s^2$. In the first transition series of the Periodic Table, beginning with scandium (see Sections 3.14 and 3.15), additional electrons are added successively to the $3d$ subshell after two electrons have already occupied the $4s$ subshell. After the $3d$ subshell is filled to its capacity with ten electrons, the $4p$ subshell fills. The electronic configurations of the elements potassium through gallium are given in Table 3–7. *Note that the electronic configurations of atoms of chromium and copper do not conform with the generalization stated above. For chromium the third subshell becomes half-filled with 5 electrons and for copper the third subshell becomes completely filled with 10 electrons.*

TABLE 3-7

Atomic Number	Element	Electronic Configuration of Atoms
19	Potassium	$1s^2\ 2s^2\ 2p^6\ 3s^2\ 3p^6\ 3d^0\ \ 4s^1$
20	Calcium	$3d^0\ \ 4s^2$
21	Scandium	$3d^1\ \ 4s^2$
22	Titanium	$3d^2\ \ 4s^2$
23	Vanadium	$3d^3\ \ 4s^2$
24	Chromium	$3d^5\ \ 4s^1$
25	Manganese	$3d^5\ \ 4s^2$
26	Iron	$3d^6\ \ 4s^2$
27	Cobalt	$3d^7\ \ 4s^2$
28	Nickel	$3d^8\ \cdot 4s^2$
29	Copper	$3d^{10}\ 4s^1$
30	Zinc	$3d^{10}\ 4s^2$
31	Gallium	$3d^{10}\ 4s^2\ 4p^1$

The extra electron needed for this to occur in both chromium and copper appears to come from the $4s$ orbital, leaving it with only one electron. It has been shown by quantum mechanics that half-filled and completely filled subshells represent the conditions of greatest stability.

The subshells listed in order of increasing energy are $1s < 2s < 2p < 3s < 3p < 4s < 3d < 4p < 5s < 4d < 5p < 6s < 4f < 5d < 6p < 7s < 5f < 6d$.

Students are sometimes troubled by the apparent exceptions to the expected order of filling orbitals. **The energies of the 4f and 5d subshells, and of the 5f and 6d subshells, are of nearly the same magnitude. In certain atoms their order of occupancy is reversed, giving rise to a number of minor exceptions to the expected order of filling orbitals.** For example, one would logically expect the electronic configuration of lanthanum (At. No. 57) to be $1s^2$, $2s^2$, $2p^6$, $3s^2$, $3p^6$, $3d^{10}$, $4s^2$, $4p^6$, $4d^{10}$, $4f^1$, $5s^2$, $5p^6$, $6s^2$. Instead, as shown in Table 3–8, it has no $4f$ electron but has one $5d$ electron: $1s^2$, $2s^2$, $2p^6$, $3s^2$, $3p^6$, $3d^{10}$, $4s^2$, $4p^6$, $4d^{10}$, $5s^2$, $5p^6$, $5d^1$, $6s^2$.

This minor exception in the case of lanthanum reflects the fact that the $4f$ and $5d$ subshells have very similar energies (see Fig. 3–14) and that electrons change from one sublevel to the other easily. A close inspection of Tables 3–3 and 3–8 will disclose a considerable number of such "exceptions." The student should realize that these exceptions result from the similar energies of various shells, and remember a few of the specific instances where they occur. Beyond this, a complete understanding of the general principles of electronic configuration and an ability to write the generalized structure for each element, based upon the expected order of addition of electrons, is sufficient for most purposes.

3.18   An Energy-Level Diagram

Although it is not possible to depict the electron energy levels exactly, a diagram representing roughly the relative energy values of all electrons in all the atoms

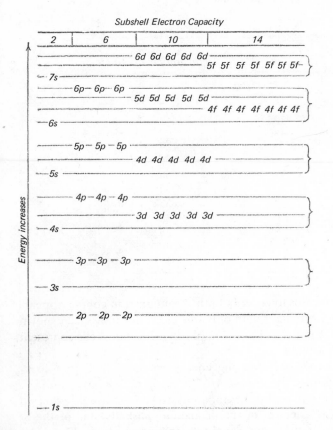

Subshell Electron Capacity

◀ **FIGURE 3–14**
Energy-level diagram for atomic orbitals.

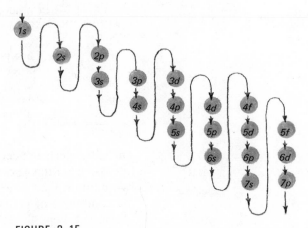

FIGURE 3–15
Order of occupancy of atomic orbitals.

is given in Fig. 3–14. The energy of an electron in an orbital is indicated by the vertical coordinate in the diagram, the orbital with electrons of lowest energy and greatest stability being the $1s$ orbital at the bottom of the diagram. Levels of nearly the same energy content are enclosed by vertical braces at the right of the figure. In general, added electrons may be expected to occupy the orbitals in order as they appear in the diagram, starting at the bottom and working up, each set of orbitals being filled before electrons enter the next set immediately above. This scheme holds strictly true only for the elements of low atomic numbers; as the atomic number increases, the relative energies of the levels change somewhat, but not all change to the same extent. This means that the elements of higher atomic number have electron arrangements slightly different from that depicted in the energy-level diagram.

The device shown in Fig. 3–15 is quite useful in arriving at the electronic configuration of an atom. Orbitals are occupied by electrons in the order indicated by the connecting lines. An exception to the diagram occurs where a single $5d$ electron is added before any $4f$ orbitals are occupied. The remaining nine $5d$ electrons enter the $5d$ orbitals after the $4f$ orbitals have been completely filled with fourteen electrons. Similarly, one or more electrons enter the $6d$ orbitals before any occupy the $5f$ orbitals.

Table 3–8 lists the electronic structures of atoms in terms of subshells for each of the known elements. Note that for each transition series of elements, ten d electrons are added to the principal shell next to the outer shell to bring the shell from eight to eighteen electrons. For each inner transition series, fourteen f electrons are added to the third shell from the outside to bring that shell from eighteen to thirty-two electrons. (See also Table 3–3 and Sections 3.14 and 3.15.)

Figure 3–16 shows in Periodic Table form the orbitals which additional electrons would enter if the atoms were built up in order of increasing atomic number ("Aufbau Principle").

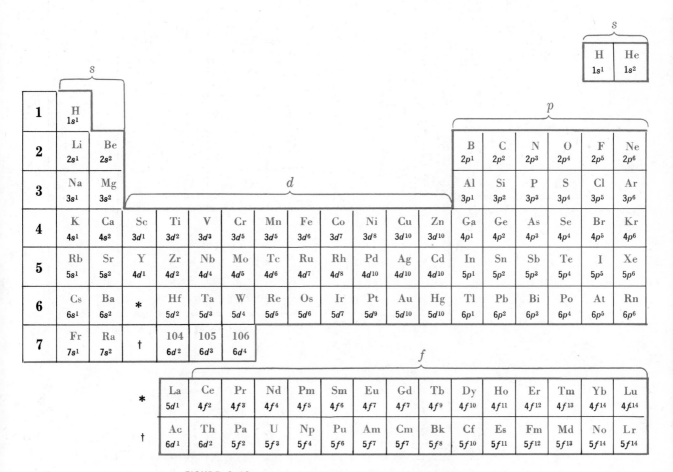

FIGURE 3-16

The order of occupancy of atomic orbitals and the Periodic Table, as the atoms are built up in order of increasing atomic number (Aufbau Principle).

TABLE 3-8 Electron Distribution, in Terms of Subshells, for the Known Elements

		K	L		M			N				O				P			Q
		1s	2s	2p	3s	3p	3d	4s	4p	4d	4f	5s	5p	5d	5f	6s	6p	6d	7s
H	1	1																	
He	2	2																	
Li	3	2	1																
Be	4	2	2																
B	5	2	2	1															
C	6	2	2	2															
N	7	2	2	3															
O	8	2	2	4															
F	9	2	2	5															
Ne	10	2	2	6															
Na	11	2	2	6	1														
Mg	12	2	2	6	2														
Al	13	2	2	6	2	1													
Si	14	2	2	6	2	2													
P	15	2	2	6	2	3													
S	16	2	2	6	2	4													
Cl	17	2	2	6	2	5													
Ar	18	2	2	6	2	6													
K	19	2	2	6	2	6		1											
Ca	20	2	2	6	2	6		2											
Sc	21	2	2	6	2	6	1	2											
Ti	22	2	2	6	2	6	2	2											
V	23	2	2	6	2	6	3	2											
Cr	24	2	2	6	2	6	5	1											
Mn	25	2	2	6	2	6	5	2											
Fe	26	2	2	6	2	6	6	2											
Co	27	2	2	6	2	6	7	2											
Ni	28	2	2	6	2	6	8	2											
Cu	29	2	2	6	2	6	10	1											
Zn	30	2	2	6	2	6	10	2											
Ga	31	2	2	6	2	6	10	2	1										
Ge	32	2	2	6	2	6	10	2	2										
As	33	2	2	6	2	6	10	2	3										
Se	34	2	2	6	2	6	10	2	4										
Br	35	2	2	6	2	6	10	2	5										
Kr	36	2	2	6	2	6	10	2	6										
Rb	37	2	2	6	2	6	10	2	6			1							
Sr	38	2	2	6	2	6	10	2	6			2							
Y	39	2	2	6	2	6	10	2	6	1		2							
Zr	40	2	2	6	2	6	10	2	6	2		2							
Nb	41	2	2	6	2	6	10	2	6	4		1							
Mo	42	2	2	6	2	6	10	2	6	5		1							
Tc	43	2	2	6	2	6	10	2	6	6		1							
Ru	44	2	2	6	2	6	10	2	6	7		1							
Rh	45	2	2	6	2	6	10	2	6	8		1							
Pd	46	2	2	6	2	6	10	2	6	10									
Ag	47	2	2	6	2	6	10	2	6	10		1							
Cd	48	2	2	6	2	6	10	2	6	10		2							
In	49	2	2	6	2	6	10	2	6	10		2	1						
Sn	50	2	2	6	2	6	10	2	6	10		2	2						
Sb	51	2	2	6	2	6	10	2	6	10		2	3						
Te	52	2	2	6	2	6	10	2	6	10		2	4						
I	53	2	2	6	2	6	10	2	6	10		2	5						
Xe	54	2	2	6	2	6	10	2	6	10		2	6						
		2	8		18			18				8							

TABLE 3-8 (Continued)

		K	L	M	N				O				P			Q
					4s	4p	4d	4f	5s	5p	5d	5f	6s	6p	6d	7s
Cs	55	2	8	18	2	6	10		2	6			1			
Ba	56	2	8	18	2	6	10		2	6			2			
La	57	2	8	18	2	6	10		2	6	1		2			
Ce	58	2	8	18	2	6	10	2	2	6			2			
Pr	59	2	8	18	2	6	10	3	2	6			2			
Nd	60	2	8	18	2	6	10	4	2	6			2			
Pm	61	2	8	18	2	6	10	5	2	6			2			
Sm	62	2	8	18	2	6	10	6	2	6			2			
Eu	63	2	8	18	2	6	10	7	2	6			2			
Gd	64	2	8	18	2	6	10	7	2	6	1		2			
Tb	65	2	8	18	2	6	10	9	2	6			2			
Dy	66	2	8	18	2	6	10	10	2	6			2			
Ho	67	2	8	18	2	6	10	11	2	6			2			
Er	68	2	8	18	2	6	10	12	2	6			2			
Tm	69	2	8	18	2	6	10	13	2	6			2			
Yb	70	2	8	18	2	6	10	14	2	6			2			
Lu	71	2	8	18	2	6	10	14	2	6	1		2			
Hf	72	2	8	18	2	6	10	14	2	6	2		2			
Ta	73	2	8	18	2	6	10	14	2	6	3		2			
W	74	2	8	18	2	6	10	14	2	6	4		2			
Re	75	2	8	18	2	6	10	14	2	6	5		2			
Os	76	2	8	18	2	6	10	14	2	6	6		2			
Ir	77	2	8	18	2	6	10	14	2	6	7		2			
Pt	78	2	8	18	2	6	10	14	2	6	9		1			
Au	79	2	8	18	2	6	10	14	2	6	10		1			
Hg	80	2	8	18	2	6	10	14	2	6	10		2			
Tl	81	2	8	18	2	6	10	14	2	6	10		2	1		
Pb	82	2	8	18	2	6	10	14	2	6	10		2	2		
Bi	83	2	8	18	2	6	10	14	2	6	10		2	3		
Po	84	2	8	18	2	6	10	14	2	6	10		2	4		
At	85	2	8	18	2	6	10	14	2	6	10		2	5		
Rn	86	2	8	18	2	6	10	14	2	6	10		2	6		
Fr	87	2	8	18	2	6	10	14	2	6	10		2	6		1
Ra	88	2	8	18	2	6	10	14	2	6	10		2	6		2
Ac	89	2	8	18	2	6	10	14	2	6	10		2	6	1	2
Th	90	2	8	18	2	6	10	14	2	6	10		2	6	2	2
Pa	91	2	8	18	2	6	10	14	2	6	10	2	2	6	1	2
U	92	2	8	18	2	6	10	14	2	6	10	3	2	6	1	2
Np	93	2	8	18	2	6	10	14	2	6	10	4	2	6	1	2
Pu	94	2	8	18	2	6	10	14	2	6	10	6	2	6		2
Am	95	2	8	18	2	6	10	14	2	6	10	7	2	6		2
Cm	96	2	8	18	2	6	10	14	2	6	10	7	2	6	1	2
Bk	97	2	8	18	2	6	10	14	2	6	10	8	2	6	1	2
Cf	98	2	8	18	2	6	10	14	2	6	10	10	2	6		2
Es	99	2	8	18	2	6	10	14	2	6	10	11	2	6		2
Fm	100	2	8	18	2	6	10	14	2	6	10	12	2	6		2
Md	101	2	8	18	2	6	10	14	2	6	10	13	2	6		2
No	102	2	8	18	2	6	10	14	2	6	10	14	2	6		2
Lr	103	2	8	18	2	6	10	14	2	6	10	14	2	6	1	2
Rf	104	2	8	18	2	6	10	14	2	6	10	14	2	6	2	2
Ha	105	2	8	18	2	6	10	14	2	6	10	14	2	6	3	2
–	106	2	8	18	2	6	10	14	2	6	10	14	2	6	4	2
		2	8	18	32				32				12			2

3.19 Elements with Atomic Number Beyond 103

Elements 104, 105, and 106 are now known, each reported recently as having been produced in the laboratory. As would be logically expected, these three elements appear to be analogs of hafnium (At. No. 72), tantalum (At. No. 73), and tungsten (At. No. 74), respectively. This places the new elements as the second, third, and fourth elements (actinium—At. No. 89— being considered as the first) of the fourth transition series, with the *d* orbitals of the principal shell next to the outer shell filling with electrons.

(104)
$1s^2, 2s^2, 2p^6, 3s^2, 3p^6, 3d^{10}, 4s^2, 4p^6, 4d^{10}, 4f^{14}, 5s^2, 5p^6, 5d^{10}, 5f^{14}, 6s^2, 6p^6, 6d^2, 7s^2$

(105)
$\ldots 6d^3, 7s^2$

(106)
$\ldots 6d^4, 7s^2$

Recall that the outer structure of actinium is $6d^17s^2$, with no $5f$ electrons present.

The next elements after 106 should complete the fourth transition series by the addition of the remaining six $6d$ electrons for a total of ten $6d$ electrons. The successive addition of six $7p$ electrons in the outer principal shell should then presumably complete the horizontal row, with the addition of the sixth $7p$ electron resulting in another "noble gas" element.

It is interesting to speculate on the structures of even more hypothetical elements, including the possibility of a series of elements in which eighteen *g* electrons successively enter the principal shell for which $n = 5$, perhaps when that shell becomes the fourth shell from the outside. Such a series could be considered a sort of "super inner transition series" and has been referred to by the American chemist Glenn Seaborg, who has had a part in the laboratory preparation of many of the elements of the second inner transition series, as a "superactinide" series. This very interesting subject of heavier elements which are not yet known and the possibility that some of them may be produced in the future is discussed in greater detail in Chapter 30, Nuclear Chemistry.

3.20 The Nature of Atomic Orbitals

Orbitals occupied by electrons in different energy levels differ from each other with regard to their size, shape, and orientation in space. The larger the number of the principal shell, the greater the volume of the corresponding orbital. For example, a 2*s* electron has an orbital of greater volume than a 1*s* electron, and a 3*s* electron has an orbital of greater volume than a 2*s* electron. It should be noted, however, that the orbitals of the higher principal shells overlap those of the lower principal shells, meaning that outer electrons penetrate the regions occupied by inner electrons.

With regard to shape, the *s* orbital is spherical (Fig. 3–17), while the *p* orbital

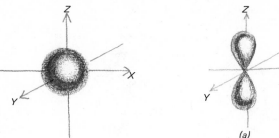

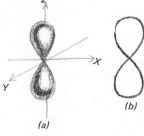

FIGURE 3-17
Atomic *s* orbitals.

FIGURE 3-18
Atomic *p* orbitals.

shape is a sort of dumbbell, represented by two lobes (Fig. 3–18a), or in cross section by a figure eight loop (b).

The number of possible orientations of atomic orbitals in space depends to an extent upon the shape of the orbital. No matter how a sphere (*s* orbital) is oriented, it still presents the same appearance to an outside observer. Orbitals of the *s* type are said, therefore, to be spherically symmetrical and without directional characteristics. The situation becomes more complex with *p* orbitals, which occur in sets of three. Although it is not possible to determine the direction of any one *p* orbital in a given set, the axes (with the nucleus of the atom at the intersection) along which the three *p* orbitals lie are mutually at right angles to each other (Fig. 3–19). The three *p* orbitals are designated, as in the diagram, p_x, p_y, and p_z to indicate their directional character.

The five *d* orbitals, each consisting of lobe-shaped regions, are arranged in space as shown in Fig. 3–20. These will be discussed in greater detail in the chapter on Coordination Compounds (Chapter 32).

There is not as good agreement on how to represent the seven *f* orbitals as there is for the *s*, *p*, and *d* orbitals. Figure 3–21 shows one way of representing the *f* orbitals. Each orbital is shown within a cube to help the student see the three-dimensional aspect of this representation.

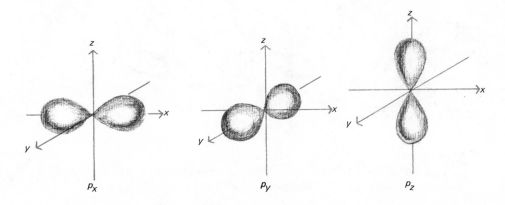

FIGURE 3-19
Directional characteristics of atomic *p* orbitals.

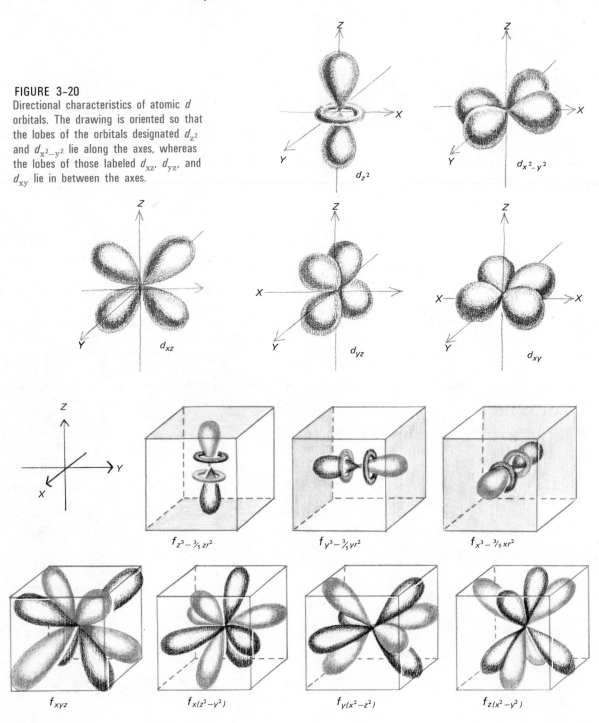

FIGURE 3-20
Directional characteristics of atomic d orbitals. The drawing is oriented so that the lobes of the orbitals designated d_{z^2} and $d_{x^2-y^2}$ lie along the axes, whereas the lobes of those labeled d_{xz}, d_{yz}, and d_{xy} lie in between the axes.

d_{z^2}

$d_{x^2-y^2}$

d_{xz}

d_{yz}

d_{xy}

$f_{z^3-\frac{3}{5}zr^2}$

$f_{y^3-\frac{3}{5}yr^2}$

$f_{x^3-\frac{3}{5}xr^2}$

f_{xyz}

$f_{x(z^2-y^2)}$

$f_{y(x^2-z^2)}$

$f_{z(x^2-y^2)}$

FIGURE 3-21
Directional characteristics of atomic f orbitals. To improve perspective, each of the seven orbitals is shown within a cube with one lobe of each pair shown in color.

QUESTIONS

1. What evidence do we have that all atoms contain electrons?
2. Relate the proton to the hydrogen atom. What is the evidence that all atoms contain protons?
3. How can atoms contain electrically charged particles and still be electrically neutral?
4. Describe the production of cathode rays in a simple vacuum discharge tube.
5. Compare the properties of cathode rays, alpha particles, beta particles, protons, neutrons, and electrons.
6. What is the experimental evidence that electrons are small negatively charged particles?
7. Explain the formation of fog tracks in a Wilson cloud chamber.
8. Define the following terms: atomic number, atomic weight, nuclear charge, isotopes.
9. In what way do the isotopes of a given element differ?
10. What logical conclusions can be drawn from the results of Rutherford's experiment concerning the structure of atoms?
11. Describe Bohr's concept of an atom.
12. What did the experiments by Rydberg demonstrate? How did they relate to Bohr's work?
13. For an electron in a hydrogen atom, what is the relationship between allowed energy values E_n and the principal quantum number, n?
14. Which one, or ones, of the four quantum numbers are important for determining the relative energies of electrons in atoms? Is your answer consistent with the Bohr model for an atom?
15. Are electrons with large values for the principal quantum number, n, lost from an atom more easily or with greater difficulty than electrons with small values for the principal quantum number? Justify your answer.
16. Does the difference in energy between successive energy levels increase or decrease as the principal quantum number, n, increases? Justify your answer.
17. What is the significance of atomic numbers? How were they determined by Moseley?
18. List the name, mass, charge, and symbol of each of the fundamental particles of matter discussed in Chapter 3, with respect to the composition of atoms.
19. How many electrons are found in the outermost energy level of each of the noble gases He, Ne, Ar, Kr, Xe, and Rn?
[S]20. (a) Why is 12.011 listed in the atomic weight table as the atomic weight of carbon, whereas the arbitrary standard for the atomic weight scale is the exact number 12 for the mass of the carbon-12 isotope?
    (b) Carbon, as it occurs in nature, contains approximately 99% $^{12}_{6}C$, 1% $^{13}_{6}C$, and an infinitesimal trace of $^{14}_{6}C$. Using the recognized standard that specifies the exact number 12 for the mass of $^{12}_{6}C$, calculate the average atomic weight for the mixture and compare this with the atomic weight listed for carbon in the atomic weight table.
21. What is the relationship between the number of electrons of a neutral atom, the number of protons in its nucleus, and its atomic number?

22. Explain why the atomic weights of many of the elements are quite different from whole numbers even though the component particles of atoms have weights which are nearly whole numbers.
23. Describe the three types of rays emitted by radioactive substances.
24. Explain the following terms: thermal emission of electrons; the photoelectric effect.
25. Determine the number of neutrons and the number of protons in each of the following nuclei: $^{14}_{7}N$, $^{52}_{24}Cr$, $^{37}_{17}Cl$, $^{141}_{59}Pr$, $^{208}_{82}Pb$, $^{249}_{98}Cf$.
26. Give the nuclear composition of the following isotopes: $^{1}_{1}H$, $^{2}_{1}H$, $^{3}_{1}H$; $^{12}_{6}C$, $^{13}_{6}C$, $^{14}_{6}C$; and $^{39}_{18}Ar$, $^{40}_{18}Ar$.
27. Make drawings to represent the complete nuclear composition and electronic structure (using principal electron shells) of the isotopes listed in Question 26.
[S] 28. Why should less energy be required in the removal of an electron from an L shell than from a K shell?
29. Write the complete symbol for the nucleus of each of the following isotopes:
    (a) The oxygen isotope possessing eight neutrons.
    (b) The cobalt isotope with a mass number of 59.
    (c) The selenium isotope with 41 neutrons.
    (d) An isotope with 80 protons and 121 neutrons.
    (e) An isotope with 91 neutrons and a mass number of 155.
    (f) An isotope with 91 neutrons and a mass number of 152.
30. Explain the use of atomic spectra in the determination of electron arrangements in atoms.
31. Compare alpha, beta, and gamma rays in terms of mass, charge, and penetrating power.
32. State the Periodic Law. How can the Periodic Law be accounted for in terms of atomic structure?
33. Give the meaning of each of the following terms as applied to the Periodic Table, and illustrate each with specific examples: period, group, short period, long period, transition series, inner transition series, noble gases, representative elements.
34. Find all of the places in the Periodic Table where the positions of the elements are not in keeping with their atomic weights. Account for the positions of these elements.
[S] 35. How many electrons would be required to balance a single proton on an analytical balance sensitive enough to weigh electrons? How many electrons would be needed to balance a neutron? How many electrons would be needed to weigh one gram?
36. By means of standard notation ($1s^2$, $2s^2$, $2p^6$, etc.), give the electron distribution in terms of s, p, d, and f subshells for each of the following, and classify each as to whether it is a representative element, a transition element, an inner transition element, or a noble gas: Al, Ti, Mg, Mo, Ar, Ce, Pa, element 106.
37. Represent the electronic structure of each of the noble gases in terms of subshell notation (s, p, d, and f).

38. (a) Give the electron distribution, in terms of s, p, d and f notation, for the following pairs of atoms: K (in Group IA of the Periodic Table) and Cu (Group IB); Sr (Group IIA) and Cd (Group IIB); Tc (Group VIIB) and I (Group VIIA).

    (b) What similarities are evident in the outer one or two subshells for the elements of each of the pairs referred to in part (a)? (The energy-level diagram of Fig. 3–14 should be considered in answering this question.)

39. Formulate definitions for the terms "noble gases," "representative elements," "transition elements," and "inner transition elements" in terms of s, p, d, and f subshells.

40. What do the following notations mean: $1s^2$; $2p^4$; $3d^04f^3$; $3s^23p^63d^2$?

41. A number of apparent exceptions occur to the expected order of filling the electron orbitals. Explain why this is so.

42. Reconcile the total maximum number of electrons in each major shell with the permitted values of the quantum numbers.

43. Distinguish between principal electron shells and subshells; subshells and orbitals.

44. Discuss the following terms: electron cloud, probable electron density, atomic orbital, radial probability density.

45. State the maximum number of s orbitals, p orbitals, and d orbitals, and relate this to the maximum number of electrons that can be present, in a given major shell. Draw diagrams to show each of the possible s, p, and d orbitals of a given major shell.

46. Name the four quantum numbers and explain, in approximate terms, the significance of each.

47. State the Pauli Exclusion Principle and explain its significance in regard to the electron structures of the elements.

48. (a) State Hund's Rule.

    (b) On the basis of the rule, calculate the number of unpaired electrons (those in singly occupied orbitals) and the number of pairs of electrons (those in doubly occupied orbitals) for the case in which a total of seven d electrons is present in the shell of principal quantum number 3 ($n = 3$).

    (c) Which element in the Periodic Table is described in part (b)?

49. Give the values for the four quantum numbers of each of the 14 electrons in the f orbitals of the fifth principal shell from the nucleus ($n = 5$).

⑤50. How does the mass of a neutron compare with the sum of the masses of a proton and an electron? If a neutron should disintegrate, producing a proton and an electron, can all the mass be accounted for?

51. (a) Make a logical prediction for the electronic configuration ($1s^2$, $2s^2$, $2p^6$, etc.) of element 114.

    (b) Into which of the four classifications (Sect. 3.15) does it fit?

52. Work out on a logical basis, by extrapolation of trends present in electronic structures of known elements, the atomic numbers and subshell electron structures ($1s^2$, $2s^2$, $2p^6$, etc.) for the next two noble gases after Rn. (NOTE: *No one answer can be "correct," as these elements have not yet been made—and*

may never be. Nevertheless, it is possible to predict logical structures, taking into account the possibility of electrons in a g orbital, as well as those in the s, p, d, and f orbitals.)

PROBLEMS

ⓢ1. There are three naturally occurring isotopes of silicon. Their atomic weights are 27.97693, 28.97650, and 29.97377. They constitute 92.210%, 4.7000%, and 3.0900% of naturally occurring silicon, respectively. Calculate the atomic weight of naturally occurring silicon (as it is given in the Periodic Table). Show your calculations.   *Ans. 28.086*

ⓢ2. Naturally occurring boron consists of two isotopes whose atomic weights are 10.01294 and 11.00931. From these data and the atomic weight of boron given in the Periodic Table, calculate the percentage of each isotope in naturally occurring boron.   *Ans. 20.00% and 80.00%, respectively*

ⓢ3. Compute the frequency of radiation emitted for the electron transitions between the following energy levels in the hydrogen atom: ⓢ(a) from $n = 2$ to $n = 1$ (i.e., $n_1 = 1$ and $n_2 = 2$); (b) from $n = 3$ to $n = 1$ ($n_1 = 1$ and $n_2 = 3$); (c) from $n = 4$ to $n = 1$.

Ans. (a) 2.467×10^{15} cycles per sec; (b) 2.924×10^{15} cycles per sec; (c) 3.083×10^{15} cycles per sec

ⓢ4. Calculate the amount of energy emitted for the electron transitions in Problem 3.

Ans. ⓢ(a) 1.634×10^{-11} erg per atom; (b) 1.937×10^{-11} erg per atom; (c) 2.043×10^{-11} erg per atom

5. (a) Calculate the energy required (absorbed) when a hydrogen atom in the ground state is ionized, i.e. when its electron is completely removed (from $n = 1$ to infinity).   *Ans. 21.79×10^{-12} erg per atom*

   (b) Calculate the energy in kilocalories per mole of hydrogen and in electron volts per atom. See Appendix C for conversion factors. (NOTE: *The energy required to remove an electron completely from an atom is called the ionization potential of the atom. Check your answer with the value for the ionization potential of hydrogen given in Table 7-7 in Chapter 7.*)

Ans. 313.6 kcal/mole; 13.60 ev/atom

6. When 18 g of water is formed from its elements, 68, 315 cal of heat energy is evolved. By the Law of Conservation of Energy, we can conclude, therefore, that conversely 68,315 cal of heat energy is needed to decompose 18 g of water into its elements. Assume that instead of using heat energy, we decompose water with the energy emitted when the electron in an excited hydrogen atom undergoes a transition from the $n = 2$ energy level to the $n = 1$ energy level. How many hydrogen atoms would have to undergo this electron transition to decompose 18 g of water into its elements? How many moles of hydrogen atoms is this?   *Ans. 1.7×10^{23} atoms; 0.29 mole*

7. How many grams of water can be decomposed into its elements by the energy needed to ionize one mole of hydrogen atoms (see Problems 5 and 6)?

Ans. 83 g

REFERENCES

"The Structure of the Proton and the Neutron," H. W. Kendall and W. Panofsky, *Sci. American,* June, 1971; p. 60.

"Atomic Weights of the Elements: Changes in Atomic Weight Values" (report of the Inorganic Chemistry Division Commission on Atomic Weights, International Union of Pure and Applied Chemistry), *Pure and Applied Chemistry,* **37** (4), 591 (1974).

"The Flash of Genius: The Bohr Atomic Model: Niels Bohr," A. B. Garrett, *J. Chem. Educ.,* **39,** 534 (1962).

"The Flash of Genius: The Neutron Identified: Sir James Chadwick," A. B. Garrett, *J. Chem. Educ.,* **39,** 638 (1962).

"The Background of Dalton's Atomic Theory," M. B. Hall, *Chem. in Britain,* **2,** 341 (1966).

"On the Discovery of the Electron," B. A. Morrow, *J. Chem. Educ.,* **46,** 584 (1969).

"Moseley and the Numbering of the Elements," W. A. Smeaton, *Chem. in Britain,* **1,** 353 (1965).

"Demonstration of the Uncertainty Principle," W. Laurita, *J. Chem. Educ.,* **45,** 461 (1968).

"The Planck Radiation Law and the Efficiency of a Light Bulb," T. A. Lehman, *J. Chem. Educ.,* **49,** 832 (1972).

"Quantum Mechanics You Can See," (Staff), *Science News,* **106,** 68 (1974).

"A Pattern of Chemistry. Hundred Years of the Periodic Table," F. Greenaway, *Chem. in Britain,* **5,** 97 (1969).

"Prospects for Further Considerable Extension of the Periodic Table," G. T. Seaborg, *J. Chem. Educ.,* **46,** 626 (1969).

"The Synthetic Elements," G. T. Seaborg, *Sci. American,* April, 1969; p. 56.

"Albert Ghiorso, Element Builder," J. F. Henehan, *Chem. & Eng. News,* Jan. 18, 1971; p. 26.

"Elements: Superheavy Ones Found," (Staff), *Chem. & Eng. News,* Feb. 22, 1971; p. 12.

"Element 105 is Long-Lived," (Staff), *Chem. & Eng. News,* May 4, 1970; p. 9.

"Element 106: Soviet and American Claims in Muted Conflict," T. H. Maugh II, *Science,* **186,** 42 (1974).

"Five Equivalent *d* Orbitals," L. Pauling and V. McClure, *J. Chem. Educ.,* **47,** 15 (1970).

"On the Shapes of *f* Orbitals," E. A. Ogryzlo, *J. Chem. Educ.,* **42,** 150 (1965).

(On *f* Orbitals), H. G. Friedman, Jr., G. R. Choppin, D. G. Feuerbacher, and C. Becker, *J. Chem. Educ.,* **42,** 151 (1965).

"Lord Ernest Rutherford," R. H. Cragg, *Chem. in Britain,* **7,** 518 (1971).

"The Texture of the Nuclear Surface," C. D. Zafiratos, *Sci. American,* Oct., 1972; p. 100.

"Ferdinand Braun and the Cathode Ray Tube," G. Shiers, *Sci. American,* March, 1974; p. 92.

"Structure of the Proton," R. F. Feynman, *Science,* **183,** 601 (1974).

"Dual-Resonance Models of Elementary Particles," J. H. Schwarz, *Sci. American,* Feb., 1975; p. 61.

Chemical Bonding, Part 1—General Concepts

4

The establishment of chemical bonds between the atoms of elements results in the union of elements to form compounds. When the atoms separate, the bonds are destroyed, and the compound no longer exists. Before the discovery of the electrical nature of the atom the character of the forces holding atoms together was a mysterious one. Now it is believed that these forces are electrical in nature and that the chemical reactions that occur between atoms involve changes in their electronic structures.

4.1 Chemical Bonding by Electron Transfer; Ionic Bonds

The electrons involved in bond formation between atoms are those in the outermost shell (sometimes in the next to the outermost shell) of the neutral atom; these electrons are called **valence electrons.** The atoms of elements which have only one or two electrons in their outermost shells (the active metals) usually lose electrons when they combine with atoms of other elements. An atom which has lost one or more valence electrons possesses a positive charge and is called a **positive ion.** The sodium atom loses its one valence electron and acquires a +1 charge when it enters into chemical combination with an atom of an element such as chlorine. The magnesium atom usually loses its two valence electrons and assumes a +2 charge.

$$\underset{\substack{\text{Sodium} \\ \text{atom}}}{\text{Na}} \longrightarrow \underset{\substack{\text{Sodium} \\ \text{ion}}}{\text{Na}^+} + e^-$$

$$\underset{\substack{\text{Magnesium} \\ \text{atom}}}{\text{Mg}} \longrightarrow \underset{\substack{\text{Magnesium} \\ \text{ion}}}{\text{Mg}^{2+}} + 2e^-$$

The smaller the number of valence electrons in the atom, the greater the tendency for the element to lose electrons and thus to form positive ions during chemical combination with atoms of other elements. The energy required to remove an electron from a neutral atom in the formation of a positive ion is called the **ionization potential** of the atom. This energy is commonly expressed in electron volts. Some metals have small ionization potentials and readily form positive ions. The nonmetals, which have more electrons in their outer shells than the metals, have large ionization potentials and show little tendency toward the formation of positive ions.

Atoms which lack one or two electrons of having an outermost shell of eight electrons tend to gain sufficient electrons from certain other atoms, such as sodium and magnesium, to make a full complement of eight electrons in the outside shell. Neutral atoms become **negative ions** by gaining electrons. The nonmetals, such as F, Cl, Br, I, O, and S, readily form negative ions.

$$Cl \quad + e^- \longrightarrow \quad Cl^-$$
Chlorine   Chloride
atom   ion

$$S \quad + 2e^- \longrightarrow \quad S^{2-}$$
Sulfur   Sulfide
atom   ion

The attraction of a neutral atom for electrons is known as its **electron affinity.** The nonmetals have high electron affinities and the metals have very low electron affinities. Thus, the nonmetals tend to form negative ions during chemical combination with metals.

When a positive ion and a negative ion are brought close together, strong electrostatic attractive forces between the charges of opposite sign are set up, and the ions are said to be held together by **ionic bonding.** The term **electrovalence** is sometimes used to designate this type of bonding.

The changes in electronic structure which take place during chemical bonding can be simply expressed with a system of notation in which the symbol of an atom represents all of the atom except the valence electrons; valence electrons are written around the symbol. Such notations are referred to as **valence electronic symbols.** Valence electronic symbols may be used to show the formulas for compounds. Such formulas are referred to as **valence electronic formulas,** or **Lewis formulas.** Valence electrons are designated by the symbols ($\cdot$), ($\times$), and ($\circ$). Different symbols for electrons are used to distinguish their sources; it must be remembered, however, that all electrons are identical, regardless of their origin.

The use of valence electronic symbols and Lewis formulas to show the formation of some ionic compounds by electron transfer is illustrated below.

$$Na\cdot \quad + \quad \overset{\times\times}{\underset{\times\times}{\times}Cl\overset{}{\times}} \quad \longrightarrow \quad Na^+ \left[\overset{\times\times}{\underset{\times\times}{\times}Cl\overset{}{\times}} \right]^-$$
Sodium atom   Chlorine atom   Sodium chloride (ionic)

$$Mg: \quad + \quad \overset{\times\times}{\underset{\times\times}{O}\overset{}{\times}} \quad \longrightarrow \quad Mg^{2+} \left[\overset{\times\times}{\underset{\times\times}{:O}\overset{}{\times}} \right]^{2-}$$
Magnesium atom   Oxygen atom   Magnesium oxide (ionic)

$$Ca: \quad + \quad 2\,\overset{\times\times}{\underset{\times\times}{\times}F\overset{}{\times}} \quad \longrightarrow \quad \left[\overset{\times\times}{\underset{\times\times}{\times}F\overset{}{\times}} \right]^- Ca^{2+} \left[\overset{\times\times}{\underset{\times\times}{\times}F\overset{}{\times}} \right]^-$$
Calcium atom   Fluorine atoms   Calcium fluoride (ionic)

FIGURE 4-1
The arrangement of sodium and chloride ions in a crystal of sodium chloride (common salt). The smaller spheres represent sodium ions and the larger ones chloride ions in this drawing.

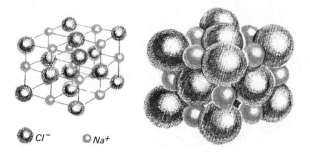

Cl^- Na^+

The atom and the ion of an element have distinctly different physical and chemical properties. Sodium is a soft, silvery-white metal that burns vigorously in air and reacts rapidly with water. Chlorine is a greenish-yellow gas which is extremely corrosive to most metals and is very poisonous to animals and plants. Sodium chloride, common table salt, is formed in a vigorous reaction between sodium atoms and chlorine atoms. This compound, which contains ions of sodium and chlorine, exists as a crystalline solid with properties entirely different from those of sodium and chlorine as elements. Chlorine is poisonous, but sodium chloride is essential to life.

A crystal of sodium chloride consists of a regular geometrical arrangement of sodium ions and chloride ions (Fig. 4–1). Each sodium ion is surrounded by six chloride ions and each chloride ion is surrounded by six neighboring sodium ions. The force which holds the ions together in the crystal is the electrostatic attraction between ions of opposite charge. Any given ion in the crystal exerts a similar force on all of its six immediate neighbors of opposite charge, and it is therefore impossible to identify any one sodium ion and chloride ion as constituting a molecule of sodium chloride. In truly ionic compounds, then, no molecules are present and a crystal of an ionic compound is an aggregation of ions. The formula of an ionic compound represents the relative number of ions necessary to give an algebraic balance of ionic charges. Since a crystal of sodium chloride is electrically neutral, it must contain the same numbers of Na^+ and Cl^- ions, and its formula is NaCl. A crystal of sodium oxide contains twice as many Na^+ as O^{2-} ions. Its formula is Na_2O. It follows that the term "molecular weight" has no significance in connection with ionic substances. The formula NaCl represents one **formula weight** of sodium chloride; but it cannot be said to represent a molecular weight, for there are no molecules of sodium chloride. The International System of Units (SI) recommends that the term **molar mass** be used instead of either "molecular weight" or "formula weight" for all compounds.

Ionic compounds are usually hard, crystalline solids with high melting points and low volatility. They conduct electricity when melted or dissolved in a suitable solvent, conditions under which the ions become mobile.

4.2   Ions with a Noble Gas Structure

When the sodium atom loses its one valence electron, it acquires an electronic structure identical with that of the noble gas neon. A magnesium atom acquires this same electronic structure by losing its two valence electrons.

	Nucleus	Electron Configurations	
Neon atom, Ne	$10p^+$ / $10n$	$2e^-$	$8e^-$
Sodium ion, Na$^+$	$11p^+$ / $12n$	$2e^-$	$8e^-$
Magnesium ion, Mg^{2+}	$12p^+$ / $12n$	$2e^-$	$8e^-$

Lithium ions (Li$^+$) and beryllium ions (Be^{2+}) have the same electron arrangement as the helium atom; Na$^+$ and Mg^{2+}, the same as neon; K$^+$ and Ca^{2+}, the same as argon; Rb$^+$ and Sr^{2+}, the same as krypton; Cs$^+$ and Ba^{2+}, the same as xenon; and Fr$^+$ and Ra^{2+}, the same as radon. The electron configurations in atoms of the noble gases are reviewed in Table 4–1, in terms first of principal shells (energy levels) and then of sublevels.

TABLE 4-1   Electron Configurations in Principal Shells of the Noble Gases

Element	Number of Electrons						Subshell Configurations					
	K	L	M	N	O	P	K	L	M	N	O	P
He	2						$1s^2$					
Ne	2	8					$1s^2$	$2s^22p^6$				
Ar	2	8	8				$1s^2$	$2s^22p^6$	$3s^23p^6$			
Kr	2	8	18	8			$1s^2$	$2s^22p^6$	$3s^23p^63d^{10}$	$4s^24p^6$		
Xe	2	8	18	18	8		$1s^2$	$2s^22p^6$	$3s^23p^63d^{10}$	$4s^24p^64d^{10}$	$5s^25p^6$	
Rn	2	8	18	32	18	8	$1s^2$	$2s^22p^6$	$3s^23p^63d^{10}$	$4s^24p^64d^{10}4f^{14}$	$5s^25p^65d^{10}$	$6s^26p^6$

The negative chloride and sulfide ions resemble the argon atom in electronic configuration.

	Nucleus	Electron Configurations		
Argon atom, Ar	$18p^+$ / $22n$	$2e^-$	$8e^-$	$8e^-$
Chloride ion, Cl$^-$	$17p^+$ / $18n$	$2e^-$	$8e^-$	$8e^-$
Sulfide ion, S^{2-}	$16p^+$ / $16n$	$2e^-$	$8e^-$	$8e^-$

Furthermore, each of the negative ions F^-, Br^-, I^-, H^-, O^{2-}, and N^{3-} has an electronic structure like that of one of the noble gases. It is a remarkable fact that every positive and negative ion mentioned above contains the same number of electrons as one of the noble gases. The noble gas arrangement of electrons is undeniably one of stability. Many positive ions do have other electron arrangements, however, that are stable in aqueous solution. Some examples of such ions are Cu^{2+}, Zn^{2+}, Ag^+, Cd^{2+}, Hg^{2+}, Au^+, Cr^{2+}, Cr^{3+}, Mn^{2+}, Fe^{2+}, Fe^{3+}, Ni^{2+}, Co^{2+}, and there are many others.

4.3   Chemical Bonding by Sharing Electrons; Covalent Bonds

In Section 4.1 we considered chemical compounds which contain ions held together by strong electrostatic forces. There are many compounds, however, which do not consist of ions. These nonionic compounds consist of atoms bonded tightly together in the form of molecules. The bonds holding the atoms together are called **shared-electron-pair bonds,** or **covalent bonds.** The simplest substance in which atoms are covalently bonded is the hydrogen molecule, H_2. Each hydrogen atom has one electron in its $1s$ shell. The two electrons (an electron pair) from two hydrogen atoms are shared by the two nuclei (Fig. 4–2). The two

FIGURE 4–2

Combination of two hydrogen atoms to form a hydrogen molecule, H_2, by covalent bonding. Notice the two dots, where the spheres overlap, which represent the shared pair of electrons.

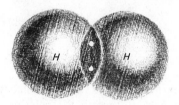

$$H\cdot\ +\ {}_\times H\ \longrightarrow\ H{}_\times^\cdot H$$

Hydrogen atoms   Hydrogen molecule

electrons are held jointly by the two nuclei and serve to bond the nuclei together. The bond is very strong, as evidenced by the large amount of energy required to break it—104 kilocalories per mole. Conversely, this same quantity of energy is evolved when a mole of molecular hydrogen is formed from hydrogen atoms.

It is evident that the bonding in the hydrogen molecule cannot be the result of electron transfer, as in ionic compounds, because the two hydrogen atoms have identical ionization potentials and identical electron affinities. Thus, no ions are formed when two atoms unite by the sharing of a pair of electrons; the product of the union is a molecule. We saw in the preceding section that there is a strong tendency for certain metals and the nonmetals to gain stability by assuming the electronic arrangements of noble gases through the transfer of electrons. This same tendency is operative when atoms unite to form covalent molecules by electron-pair sharing. When the two electrons of the covalent bond are counted for each hydrogen atom, each atom has the electronic arrangement of the stable helium atom. The $1s$ orbital of each hydrogen atom in the H_2 molecule is, in effect, occupied by both electrons of the shared pair. The electron pair occupies the whole molecule, spending an equal amount of time near each nucleus. Figure 4–3

FIGURE 4–3(a)
Separate atomic orbitals of hydrogen.

FIGURE 4–3(b)
Atomic orbitals of hydrogen approach each other closely enough to begin to act upon each other.

FIGURE 4–3(c)
Distribution of charge within the hydrogen molecules. The electron pair occupies the whole molecule, spending an equal amount of time near each nucleus.

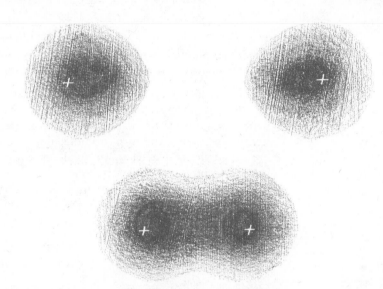

illustrates the distribution of charge in the H_2 molecule. The shading represents the intensity of negative charge, that is, the relative probability of finding the electron pair at a given location. In Chapter 5, we shall discover that when two hydrogen atoms combine to form a hydrogen molecule the electron pair goes into a molecular orbital, which is formed from the two atomic orbitals (see Section 5.3).

The bonding in a molecule of chlorine, Cl_2, furnishes a second example of covalent bonding. Each atom of chlorine has seven electrons in its outer shell and differs from the noble gas argon in its electronic configuration by one electron. The sharing of one pair of electrons between two atoms in a molecule of chlorine gives each atom the same stable electronic structure as an atom of argon.

$$:\overset{..}{\underset{..}{Cl}}\cdot \quad + \quad \overset{\times\times}{\underset{\times\times}{\times Cl}}\overset{}{\times} \quad \longrightarrow \quad :\overset{..}{\underset{..}{Cl}}\times\overset{\times\times}{\underset{\times\times}{Cl}}\overset{}{\times}$$

Chlorine   Chlorine
atoms   molecule

The bonding in the molecules F_2, Br_2, I_2, and At_2 (all in the same family with Cl_2) is like that in the chlorine molecule, but with the formation of electronic structures like those of other noble gases.

Many atoms share more than one pair of electrons, if that is necessary to give each atom a full complement of eight electrons in its valence shell. For example, the atoms in the nitrogen molecule, N_2, share three pairs of electrons. This makes a total of eight electrons in the valence shell of each nitrogen atom.

$$:\overset{\displaystyle .}{\underset{\displaystyle .}{N}}\cdot \; + \; \overset{\displaystyle \times}{\underset{\displaystyle \times}{\times}N}\overset{\times}{\underset{\times}{\,}} \longrightarrow :N \overset{\times}{\underset{\times}{:}} N\overset{\times}{\underset{\times}{\,}}$$

Nitrogen atoms   Nitrogen
                        molecule

The two atoms in the N_2 molecule are said to be held together by a triple covalent bond. We shall learn more about this as we discuss, in Chapter 5, the molecular orbitals involved in the bonding (Sections 5.9 and 5.15).

4.4   Covalent Bonds between Unlike Atoms

Unlike atoms may also combine through covalent bond formation. For example, one hydrogen atom combines covalently with one chlorine atom in the formation of a molecule of hydrogen chloride.

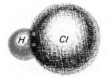

$$H\cdot \; + \; \times\overset{\times\times}{\underset{\times\times}{Cl}}\overset{\times}{\underset{\times}{\,}} \longrightarrow H \overset{\times\times}{\underset{\times\times}{:Cl}}\overset{\times}{\underset{\times}{\,}}$$

Although the electrons of the pair are shared between the hydrogen atom and chlorine atom, the electrons are not shared equally, as they are in H_2 and Cl_2. A chlorine atom attracts electrons more strongly than a hydrogen atom does, causing the electrons of the shared pair to be more closely associated with the chlorine nucleus. This results in the development of a small positive charge (often referred to as a partial positive charge) on the hydrogen atom and a partial negative charge on the chlorine atom. This does not imply, however, that the hydrogen atom has lost its electron; it means that the electrons of the pair spend more time *on the average* in the vicinity of the chlorine nucleus than they do near the hydrogen nucleus. Another way of stating this is to say that the electron density, or the density of the electron cloud, is greater around the chlorine nucleus than around the hydrogen nucleus.

The oxygen atom has six valence electrons and completes an octet of electrons (eight electrons) by sharing one electron pair with each of two hydrogen atoms in the covalent water molecule.

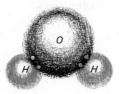

$$H\overset{\times}{\,}\overset{\times\times}{\underset{\cdot\times}{O}}\overset{\times}{\underset{\,}{\,}}$$
$$H$$

The nitrogen atom with five valence electrons shares electron pairs with three hydrogen atoms in the covalent ammonia molecule, NH_3. In a molecule of methane, CH_4, the carbon atom with four valence electrons completes its octet by forming covalent bonds with four atoms of hydrogen.

The carbon dioxide molecule contains one atom of carbon with four valence electrons and two atoms of oxygen, each with six valence electrons. The sharing of two electron pairs between the carbon atom and each of the two oxygen atoms gives each of the three atoms in the molecule an octet of electrons.

Ammonia   Methane   Carbon dioxide

4.5 Covalently Bonded Atoms without Noble Gas Structure

As has been pointed out, stable compounds do exist in which all the atoms of the molecule do not have the noble gas arrangement. For example, boron, with three valence electrons, shares electron pairs with three chlorine atoms in the stable molecule BCl_3. Although this union of atoms gives each chlorine atom an argon structure, boron with six electrons in its outer shell does not have the electron arrangement of a noble gas. Furthermore, atoms of the elements in which the outermost electron shell is the M shell or higher can participate in covalent bonding with other atoms in which more than four pairs of electrons are shared. For example, the phosphorus atom in PF_5 shares five pairs of electrons, or ten electrons in all, whereas atoms of the noble gases are restricted to a maximum of eight electrons in their outer shells. The outermost, or M, shell of phosphorus has a theoretical maximum capacity of eighteen electrons, or nine electron pairs, but no examples of molecules are known in which this condition exists. Sulfur shares six electron pairs (twelve electrons) in the SF_6 molecule; and iodine shares seven electron pairs in IF_7. There are eight shared pairs of electrons in OsF_8. In some cases, the number of electrons in the outer shell exceeds eight, even though some of the pairs are not shared. This is the case with IF_5 and XeF_4.

The formation of covalent compounds such as BCl_3, PCl_5, SF_6, and IF_7 conforms to a rule which supplements the noble gas structure or octet rule, namely, that electrons tend to occur in pairs in molecular structures. In other words, we may state that *the atoms in most covalent molecules appear to have reached a stable condition by sharing pairs of electrons with each other*. Thus, the boron atom, which can form only three bonds by sharing its electrons because it has only three valence electrons, attains a condition of stability by forming three electron-pair bonds in the molecule BCl_3.

$$\dot{B}\cdot + 3 \times \overset{\times\times}{\underset{\times\times}{Cl}}\times \longrightarrow \quad \underset{\underset{\times\times}{\overset{\times}{\underset{\times}{Cl}}\times}}{\overset{\overset{\times\times}{\overset{\times}{\underset{\times}{Cl}}\times}}{B\overset{\cdot\times}{\underset{\cdot\times}{Cl}}\overset{\times\times}{\underset{\times\times}{Cl}}\times}}$$

Boron trichloride

4.6 Electronegativity of the Elements

We have seen that the chlorine atom in the hydrogen chloride molecule attracts the electrons of the electron-pair bond more strongly than does the hydrogen atom so that the electrons are not shared equally by the two atoms. This power of attraction that an atom shows for electrons in a covalent bond is known as electronegativity. **Electronegativity** is a measure of the attraction of an atom for electrons in its outer shell. The values of the electronegativities of many of the elements are given in Table 4–2. These values are based upon an arbitrary scale, meaning that we cannot say, for example, that fluorine (4.0) is twice as electronegative as boron (2.0). The electronegativity values are not a measure of absolute electronegativity, but they do provide a measure of differences in electronegativity. For example, the difference in electronegativity between boron (2.0) and nitrogen (3.0) is the same as that between nitrogen (3.0) and fluorine (4.0).

Note that the nonmetals, in general, have higher electronegativity values than the metals. Fluorine, the most chemically active nonmetal, has the highest electronegativity (4.0), and cesium, the most chemically active metal (with the possible exception of francium), has the lowest electronegativity (0.7). Because the metals

TABLE 4–2 Electronegativity Values of Some of the Elements, According to the Periodic Table Arrangement

H 2.1																		He ...
Li 1.0	Be 1.5											B 2.0	C 2.5	N 3.0	O 3.5	F 4.0		Ne ...
Na 0.9	Mg 1.2											Al 1.5	Si 1.8	P 2.1	S 2.5	Cl 3.0		Ar ...
K 0.8	Ca 1.0	Sc 1.3	Ti 1.5	V 1.6	Cr 1.6	Mn 1.5	Fe 1.8	Co 1.8	Ni 1.8	Cu 1.9	Zn 1.6	Ga 1.6	Ge 1.8	As 2.0	Se 2.4	Br 2.8		Kr ...
Rb 0.8	Sr 1.0	Y 1.2	Zr 1.4	Nb 1.6	Mo 1.8	Tc 1.9	Ru 2.2	Rh 2.2	Pd 2.2	Ag 1.9	Cd 1.7	In 1.7	Sn 1.8	Sb 1.9	Te 2.1	I 2.5		Xe ...
Cs 0.7	Ba 0.9	La-Lu 1.1-1.2	Hf 1.3	Ta 1.5	W 1.7	Re 1.9	Os 2.2	Ir 2.2	Pt 2.2	Au 2.4	Hg 1.9	Tl 1.8	Pb 1.8	Bi 1.9	Po 2.0	At 2.2		Rn ...
Fr 0.7	Ra 0.9	Ac-Lr 1.1-																

have relatively low electronegativities and tend to assume positive charges in compounds, they are often spoken of as being **electropositive;** conversely, non-metals are said to be **electronegative.**

4.7   Polarity of Substances and the Chemical Bond

A system of bonded atoms is said to be **polar** if its center of positive charge does not coincide with its center of negative charge. An extreme case of polarity is represented by an ionic compound such as sodium chloride (Na^+Cl^-), in which the sodium ion is completely positive and the chloride ion is completely negative. When a covalent bond is formed between atoms of different electronegativities, the pair of electrons will be more closely associated with the more electronegative atom, and the resulting covalent bond will be somewhat polar. We have mentioned previously that in the hydrogen chloride molecule, the chlorine atom attracts the pair of electrons of the covalent bond more strongly than does the hydrogen atom (Section 4.4). The hydrogen-chlorine bond is polar, the chlorine atom becoming somewhat negative and the hydrogen atom becoming somewhat positive as the bond is formed. Since the centers of positive and negative charges do not coincide, the molecule of hydrogen chloride is electrically unsymmetrical. Because of the separation of centers of charge, molecules held together by polar bonds tend to turn when placed in an electric field, with the positive end of the molecule oriented toward the negative plate and the negative end toward the positive plate (Fig. 4–4).

FIGURE 4–4

Polar molecules, such as hydrogen chloride, tend to line up when in an electric field, with the positive ends oriented toward the negative plate and the negative ends toward the positive plate.

Polar molecules randomly oriented in the absence of an electric field

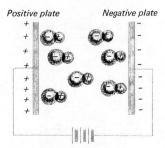

Positive plate   Negative plate

Polar molecules lined up in an electric field

The greater the difference between the electronegativities of the two atoms involved in the bond, the greater the polarity of the bond. Thus, the polarity of the bond in the hydrogen halides increases in the order HI < HBr < HCl < HF, corresponding to an increase in electronegativity of the halogens: I(2.5), Br(2.8), Cl(3.0), and F(4.0). If the difference in electronegativity between the two atoms is sufficiently large, the electron furnished by the atom of lower electronegativity will be transferred completely to the more electronegative atom, and ionic bonding, rather than covalent bonding, will result. The other extreme may be achieved when identical atoms share a pair of electrons as in *nonpolar*

the case of H×H, where the bonding is covalent with no polarity. It becomes apparent then that there is no sharp dividing line between compounds in which the bonding is covalent and those in which the bonding is ionic. In the intermediate cases the molecules will have bonds which possess some of the nature of both covalent and ionic bonds and are often referred to as **covalent bonds with partial ionic character,** or **polar covalent bonds.**

It is possible to have bonds which are of the polar covalent type but where the molecule as a whole is nonpolar. If a molecule contains several polar covalent bonds directed in such a way as to give a symmetrical molecule, then the molecule is nonpolar. This is illustrated by $HgCl_2$, in which each of the covalent bonds is polar while the molecule as a whole is nonpolar. The centers of positive and negative electricity for the molecule are identical. Each chlorine is negative with respect to positive mercury, and each mercury-chlorine bond has some polar character. However, these bond polarities counterbalance each other because the bonds are directed in such a manner as to give an electrically symmetrical molecule.

$$Cl^- \longleftarrow {}^+Hg^+ \longrightarrow {}^-Cl$$

Covalent compounds may exist as solids, liquids, or gases at ordinary temperatures. In general, they have low melting points and are volatile. Their solutions conduct electricity only when they form ions by reacting with the solvent (see Chapter 14).

4.8 Coordinate Covalence

We have noted that covalent bonding involves the sharing of electron pairs between atoms, with both atoms involved in the bond furnishing equal numbers of electrons. When only one of the two atoms involved in the linkage furnishes both electrons of the electron-pair bond, the bonding is called **coordinate covalence.** An example of coordinate covalence is provided by the ammonium ion, NH_4^+. The bonds in the ammonia molecule itself are of the covalent type. The unshared pair of electrons originally belonging to the nitrogen atom is available for use in bond formation as indicated by the readiness with which ammonia will combine with a hydrogen ion to form the ammonium ion.

$$\begin{array}{ccc}
\text{H} & & \left[\begin{array}{c}\text{H}\end{array}\right]^+ \\
\text{H}\overset{\cdot\times}{\underset{\cdot\times}{\times}}\text{N}\overset{}{\underset{}{\times}} \;+\; \text{H}^+ \longrightarrow & & \text{H}\times\overset{\cdot\times}{\underset{\cdot\times}{\text{N}}}\times\text{H} \\
\text{H} & & \text{H}
\end{array}$$

Ammonia molecule, NH_3 Ammonium ion, NH_4^+

Because NH_3 is a neutral molecule, the union with a hydrogen ion (proton) gives a unit positive charge to the resulting ammonium ion. In a similar fashion, water molecules combine with hydrogen ions to form hydronium ions.

$$H \overset{\times\times}{\underset{\overset{\bullet\times}{H}}{\colon O \colon}} + H^+ \longrightarrow \left[H \overset{\times\times}{\underset{\overset{\bullet\times}{H}}{\colon O \colon}} H \right]^+$$

Water   Hydronium
molecule   ion, H_3O^+

Even though one electron pair on the H_3O^+ is still available for bonding, a second hydrogen ion (positive) is not likely to be attracted to the positive hydronium ion. The formation of a coordinate covalent bond is possible only between an atom or ion with an unshared pair of electrons in its valence shell and an atom or ion that needs a pair of electrons to acquire a more stable electronic configuration. The chief difference between the coordinate covalent bond and the normal covalent bond is in the mode of formation. Once established, they are indistinguishable.

In our discussion of bonding, we have thus far considered the electrons as being located within atomic orbitals of the atoms undergoing bonding. Actually, the valence electrons move about over the molecule as a whole in what are called **molecular orbitals,** instead of being restricted to the atomic orbitals of specific individual atoms. In the next chapter, we shall discuss the manner in which atomic orbitals combine to form molecular orbitals and also point out some of the ways in which this concept is especially useful in explaining the properties of molecules.

4.9   Oxidation Numbers

The **oxidation number,** sometimes referred to as the **oxidation state,** is used to designate the positive and negative character of atoms. When valence electrons are removed or shifted away from an atom during a chemical reaction, the atom is assigned a **positive oxidation number** and is said to be in a **positive oxidation state.** When electrons are gained by or shifted toward an atom during a chemical reaction, the atom is given a **negative oxidation number** and is said to be in a **negative oxidation state.** The numerical value of the oxidation number depends upon the number of electrons involved per atom in the transfer or shift to or away from the atom.

For ionic materials, the oxidation number of an element is equal to the charge on the ion. In sodium chloride, NaCl, the oxidation number for sodium is $+1$ and for chlorine is -1; in magnesium oxide, MgO, the oxidation number for magnesium is $+2$ and for oxygen is -2; in calcium bromide, $CaBr_2$, the oxidation number for calcium is $+2$ and for bromine is -1. It is convenient to indicate the oxidation number of an atom by placing the proper number and sign over its symbol.

$$\overset{+1\,-1}{\text{NaCl}} \qquad \overset{+2\,-2}{\text{MgO}} \qquad \overset{+2\,-1}{\text{CaBr}_2}$$

Sodium   Magnesium   Calcium
chloride   oxide   bromide

For covalent materials, the oxidation number concept is more arbitrary but is nevertheless useful in writing formulas and, as we shall see in Chapter 16, in balancing oxidation-reduction equations. In covalent compounds containing two elements, the more electronegative element (see Table 4–2) is assigned a negative oxidation number. The more positive element is assigned a positive oxidation number. In the covalent molecule of hydrogen chloride, HCl, the hydrogen atom has an oxidation number of $+1$ due to a shift (but not a transfer) of the valence electron of the hydrogen toward the more electronegative chlorine atom. The chlorine atom in HCl has an oxidation number of -1. The electronegativity of carbon is greater than that of hydrogen; so in methane, CH_4, it is customary to speak of the oxidation number of carbon as -4 and that of hydrogen as $+1$; the electrons are shifted toward the carbon atom. In carbon dioxide, CO_2, the electronegativity of oxygen is greater than that of carbon; therefore, in this compound the oxidation number of oxygen is -2 and that of carbon is $+4$. In water, H_2O, each hydrogen atom has an oxidation number of $+1$ and oxygen an oxidation number of -2.

$\overset{+1\ -1}{HCl}$	$\overset{-4\ +1}{CH_4}$	$\overset{+4\ -2}{CO_2}$	$\overset{+4\ -1}{CCl_4}$	$\overset{+1\ -2}{H_2O}$	$\overset{-3\ +1}{NH_3}$
Hydrogen chloride	Methane	Carbon dioxide	Carbon tetrachloride	Water	Ammonia

Elements in the free state are always assigned an oxidation number of zero. The zero oxidation number for free elements is based upon the fact that all of the atoms of the same element have the same electronegativity; there is no transfer or net shift of electrons occurring during bond formation between atoms of the same element.

Many elements exhibit more than one oxidation number in their various compounds. For example, iron has an oxidation number of $+2$ in $FeCl_2$ and an oxidation number of $+3$ in $FeCl_3$. Tin exhibits oxidation numbers of $+2$ and $+4$ in $SnCl_2$ and $SnCl_4$, respectively. The oxidation number of chlorine in each of the above examples is -1. Chlorine, however, exhibits an oxidation number of $+1$ in $NaClO$, $+3$ in $NaClO_2$, $+5$ in $NaClO_3$, and $+7$ in $NaClO_4$.

If the oxidation numbers are known for all but one kind of atom in a compound, the remaining oxidation number can be calculated. **The algebraic sum of the positive oxidation numbers and the negative oxidation numbers of the atoms and ions present in a compound must always be zero.** In Na_2SO_4, for example, the oxidation number for sulfur can be calculated from the known oxidation numbers for sodium and oxygen. The two sodium atoms, each with an oxidation number of $+1$, total $+2$; the four oxygen atoms, each with an oxidation number of -2, total -8. For the sum of the oxidation numbers to be zero, sulfur must have an oxidation number of $+6$. For Na_2SO_3, a similar calculation shows the oxidation number of sulfur in that compound to be $+4$. In H_2S, the oxidation number of sulfur is -2.

Several of the more commonly observed oxidation numbers are given in Table 4–3.

It should be emphasized that, although the concept of oxidation numbers is of great convenience in writing formulas and in balancing oxidation-reduction equations, the concept is quite arbitrary. This is especially apparent in a compound in which the calculation of oxidation number results in a fractional value. For

TABLE 4-3   Examples of Common Oxidation Numbers

Element	Oxidation Number
H	+1 (except −1 in metal hydrides; for example, NaH)
Li	+1
Na	+1
K	+1
Mg	+2
Ca	+2
Zn	+2
Al	+3
Cl	−1 ⎫ (−1 in compounds with only two elements; variable oxidation num-
Br	−1 ⎬ bers in compounds containing more than two elements, such as
I	−1 ⎭ NaClO$_3$ and NaClO$_4$)
O	−2 (except −1 in peroxides—for example, H$_2$O$_2$ and Na$_2$O$_2$—and −½ in "superoxides"—for example, KO$_2$)
Hg	+1 and +2
Fe	+2 and +3
Sn	+2 and +4
Pb	+2 (+4 in PbO$_2$)

example, with Fe$_3$O$_4$ the four oxygen atoms, each −2, would contribute a total of −8. For the molecule to be neutral, the three iron atoms must together contribute a total of +8 units of charge, or $+\frac{8}{3}$ or $+2\frac{2}{3}$ each.

$$\overset{+2\frac{2}{3}}{\text{Fe}_3}\ \overset{-2}{\text{O}_4}$$

The oxidation number always refers to one atom of an element and hence is $+2\frac{2}{3}$ for iron in this example. It can be noted for this particular case that if one atom of iron is considered to be +2 (as in FeO) and two atoms of iron are considered to be +3 each (as in Fe$_2$O$_3$) the *average weighted* oxidation state for the iron atoms is $2\frac{2}{3}$. The Fe$_3$O$_4$ molecule can thus be considered as if it were a composite of one molecule of FeO combined with one molecule of Fe$_2$O$_3$. However, analogous reasoning cannot be applied to all molecules possessing an atom with a fractional oxidation number.

The distribution of electrons in the molecule is a more fundamental property of a molecule than oxidation state, but in many cases a close relationship exists between electron distribution and oxidation state. One should always think of the calculation of oxidation states as a useful but quite arbitrary concept.

4.10   Application of Oxidation Numbers to Writing Formulas

One can write the formulas of a great many compounds by knowing the oxidation numbers of the constituent elements of each compound. The principal oxidation number of an element is, in general, predictable from the position of the element in the Periodic Table and from a knowledge of the electronic structure of the atom. The writing of formulas by using oxidation numbers is possible because the algebraic sum of the units of positive and negative oxidation number for any molecule must be equal to zero.

Example 1. Let us write the formula for aluminum oxide by using the oxidation numbers of its constituent elements.

The oxidation number of aluminum is +3 and that of oxygen is −2. Place the element with the positive oxidation number before the one with the negative oxidation number: $\overset{+3\,-2}{\text{AlO}}$. Because +3 plus −2 does not give 0, then AlO is not the correct formula for aluminum oxide. By inspection it is readily seen that 2 atoms of aluminum would give a total of 6 units of positive oxidation number, that three atoms of oxygen would give 6 units of negative oxidation number, and that the algebraic sum of the oxidation numbers would be zero. The correct simplest formula for aluminum oxide is, therefore, Al_2O_3.

Example 2. Write the formula for magnesium chloride.

The formula cannot be $\overset{+2\,-1}{\text{MgCl}}$, because +2 and −1 do not add up to 0. For the total of the oxidation numbers to be zero for the compound, the ions must be in a ratio of one magnesium ion to two chloride ions, or $MgCl_2$.

4.11   Ions Containing More than One Atom

Many ions, referred to as **polyatomic ions,** contain more than one atom. Several examples, selected from many such ions, are given in Table 4–4.

For ions containing more than one atom, the algebraic sum of the positive and negative oxidation numbers of the constituent atoms must equal the charge on the ion. Hence, for the OH^- ion, the −2 oxidation number of oxygen and the +1 oxidation number of hydrogen add to give the −1 charge for the ion.

It is customary in writing formulas of compounds which include more than one unit of a given polyatomic ion to enclose the formula of the ion in parentheses and to indicate with a subscript the number of such ions in the compound. Examples are $(NH_4)_2CO_3$ and $Al_2(SO_4)_3$. In $(NH_4)_2CO_3$, two ammonium ions, each with a +1 ionic charge, are necessary to balance the −2 ionic charge of the carbonate ion. In $Al_2(SO_4)_3$, two aluminum ions, each with a charge of +3,

TABLE 4-4   Some Common Polyatomic Ions

Ammonium	NH_4^+	Carbonate	CO_3^{2-}	Phosphate	PO_4^{3-}
Acetate	CH_3COO^-	Sulfate	SO_4^{2-}	Pyrophosphate	$P_2O_7^{4-}$
Nitrate	NO_3^-	Sulfite	SO_3^{2-}	Arsenate	AsO_4^{3-}
Nitrite	NO_2^-	Thiosulfate	$S_2O_3^{2-}$	Arsenite	AsO_3^{3-}
Hydroxide	OH^-	Peroxide	O_2^{2-}		
Hypochlorite	ClO^-	Chromate	CrO_4^{2-}		
Chlorite	ClO_2^-	Dichromate	$Cr_2O_7^{2-}$		
Chlorate	ClO_3^-	Silicate	SiO_3^{2-}		
Perchlorate	ClO_4^-				
Permanganate	MnO_4^-				

and three sulfate ions, each with a charge of -2, are required to balance the charges. It should be noted that the sum of the total positive and negative oxidation numbers for the various atoms, as well as the sum of the total charges on the ions, equals zero for each compound.

The following examples illustrate the process of writing formulas for compounds containing polyatomic ions.

Example 1.   Write the formula for iron perchlorate, given that the oxidation number for iron in the compound is $+3$.

Consideration of the charge of $+3$ for the iron ion and the charge of -1 for the perchlorate ion (see Table 4–4) shows the formula $\overset{+3}{Fe}\overset{-1}{(ClO_4)}$ to be incorrect. By using three perchlorate ions and one iron ion, the sum of the positive and negative charges becomes zero; the correct formula is $Fe(ClO_4)_3$.

Example 2.   Write the formula for calcium phosphate.

Inasmuch as the charge for the calcium ion is $+2$ and the charge on the phosphate ion is -3, the formula cannot be $\overset{+2}{Ca}\overset{-3}{(PO_4)}$, because the algebraic sum of the charges on the ions must be zero for the compound. By using three calcium ions and two phosphate ions the algebraic sum becomes zero, $3(+2) + 2(-3) = 0$. Thus, $Ca_3(PO_4)_2$ is the correct formula for calcium phosphate.

4.12   The Names of Compounds

■ **1. Binary Compounds.**   Binary compounds are those containing two different elements. The name of a binary compound consists of the name of the more electropositive element followed by the name of the more electronegative element with its ending replaced by the suffix "ide." Some examples are

NaCl, sodium chloride	CdS, cadmium sulfide
KBr, potassium bromide	Mg_3N_2, magnesium nitride
CaI_2, calcium iodide	Ca_3P_2, calcium phosphide
AgF, silver fluoride	Al_4C_3, aluminum carbide
HCl, hydrogen chloride	LiH, lithium hydride
Na_2O, sodium oxide	Mg_2Si, magnesium silicide

A few polyatomic ions have special names and are treated as if they were single atoms in naming their compounds; thus NaOH is called sodium hydroxide; HCN, hydrogen cyanide; and NH_4Cl, ammonium chloride.

If a binary hydrogen compound is an acid when it is dissolved in water, the prefix "hydro" is used, and the suffix "ic" replaces the suffix "ide" when one is referring to the solution.

HCl, hydrochloric acid	H_2S, hydrosulfuric acid
HBr, hydrobromic acid	HCN, hydrocyanic acid

When an element of variable valence unites with another element to form more than one compound, the compounds may be distinguished from each other by the Greek prefixes *mono-* (meaning one), *di-* (two), *tri-* (three), *tetra-* (four), *penta-* (five), *hexa-* (six), *hepta-* (seven), and *octa-* (eight). The prefixes precede the name of the constituent to which they refer.

CO, carbon monoxide	PbO, lead monoxide
CO_2, carbon dioxide	PbO_2, lead dioxide
NO_2, nitrogen dioxide	SO_2, sulfur dioxide
N_2O_4, dinitrogen tetraoxide	SO_3, sulfur trioxide
N_2O_5, dinitrogen pentaoxide	BCl_3, boron trichloride

A second method of naming different binary compounds containing the same elements involves the use of Roman numerals placed in parentheses to indicate the oxidation number of the more positive element, and following the names of the elements to which they refer. This method of naming binary compounds is usually applied to those in which the electropositive element is a metal, but it is occasionally applied to other compounds as well.

$FeCl_2$, iron(II) chloride	SO_2, sulfur(IV) oxide
$FeCl_3$, iron(III) chloride	SO_3, sulfur(VI) oxide
Hg_2O, mercury(I) oxide	NO, nitrogen(II) oxide
HgO, mercury(II) oxide	NO_2, nitrogen(IV) oxide

Although the system of nomenclature used in this book is, for the most part, an improved system formulated by a committee of the International Union of Pure and Applied Chemistry, it is essential that the student become familiar also with the "old" system, because it will be encountered frequently.

According to the "old" system, when two elements form more than one compound with each other, and when both elements are nonmetals, the distinction is made by indicating only the number of atoms of the more electronegative element by Greek prefixes. NO_2, N_2O_3, N_2O_4, and N_2O_5 are nitrogen dioxide, trioxide, tetraoxide, and pentaoxide, respectively. When the more electropositive element is a metal, the lower oxidation number of the metal is indicated by using the suffix *-ous* on the name of the metal. The higher oxidation number is designated by the suffix *-ic*. Thus $FeCl_2$ is ferrous chloride, and $FeCl_3$ is ferric chloride, Hg_2O is mercurous oxide, and HgO is mercuric oxide.

■ **2. Ternary Compounds.** Ternary compounds are those containing three different elements. It has already been noted that a few ternary compounds, such as NH_4Cl, KOH, and HCN, are named as if they were binary compounds. Chlorine, nitrogen, sulfur, phosphorus, and several other elements each form oxyacids (ternary compounds with hydrogen and oxygen) which usually differ from each other in their oxygen content. Normally, the most common acid of a series bears the name of the acid-forming element ending with the suffix *-ic*. This may be noted in the names chloric acid ($HClO_3$), sulfuric acid (H_2SO_4), nitric acid (HNO_3), and phosphoric acid (H_3PO_4). If the "central" element (Cl, S, etc.) of an acid has a higher oxidation number than the *-ic* acid the suffix *-ic* is retained and the prefix *per-* is added. The name perchloric acid for $HClO_4$ illustrates this rule. An acid which has a lower oxidation number than the *-ic* acid for the central element is named with the suffix *-ous*. Examples are chlorous acid ($HClO_2$), sulfurous acid (H_2SO_3), nitrous acid (HNO_2), and phosphorous acid (H_3PO_3). If two acids should have lower oxidation numbers than the *-ic* acid for the central element, the lower of the two is named by adding the prefix *hypo-* and retaining the ending *-ous*. Thus, HClO is hypochlorous acid and H_2SO_2 is hyposulfurous acid.

Metal salts of the oxyacids (compounds in which a metal replaces the hydrogen of the acid) are named by identifying the metal and then the negative acid ion. The ending *-ic* of the oxyacid name is changed to *-ate* and the ending *-ous* of the acid is changed to *-ite* for the salt. The salts of perchloric acid are perchlorates, those of sulfuric acid are sulfates, those of nitrous acid are nitrites, and those

TABLE 4-5   Names of Oxyacids of Chlorine and the Corresponding Sodium Salts

Acids	*Salts*
HClO, hypochlorous acid	NaClO, sodium hypochlorite
$HClO_2$, chlorous acid	$NaClO_2$, sodium chlorite
$HClO_3$, chloric acid	$NaClO_3$, sodium chlorate
$HClO_4$, perchloric acid	$NaClO_4$, sodium perchlorate

of hyposulfurous acid are hyposulfites. This system of naming applies to all inorganic oxyacids and their salts. The names of the oxyacids of chlorine and their corresponding sodium salts are given in Table 4-5.

The system of nomenclature for a class of compounds known as **coordination compounds,** or **complex compounds,** will be described in Chapter 32 (see Section 32.2).

QUESTIONS

1. How are positive and negative ions formed from neutral atoms?
2. What kind of elements tend to form positive ions? negative ions?
3. From the following list of atoms, select the ones which you would expect to form positive ions and the ones which you would expect to form negative ions: Ca, Br, Li, N, Zr, Ar, Sr, S.
4. Relate the electronic structures of positive and negative ions to those of the noble gases. Cite several examples of ions which have a noble gas electronic structure and several ions which do not.
5. How is the number of valence electrons in an atom of an element related to its tendency to gain or lose electrons during compound formation?
6. Why is it logical that a sample of sodium chloride should contain equal numbers of sodium and chloride ions? Would this also be the case for calcium (Ca) chloride?
7. Why may we not look upon sodium chloride as being a molecular compound?
8. What is the nature of the bonding force which holds the sodium and chloride ions together in a crystal of sodium chloride?
9. Explain the meaning of the term "ionization potential." How is the tendency

of an element to form positive (or negative) ions related to its ionization potential?

10. Compare metals and nonmetals with regard to electron affinity.

11. Cite evidence that shows there is a chemical difference between fluorine molecules and fluoride ions.

12. Define the term "covalence." Give examples of covalent bonding between like atoms and between unlike atoms.

13. Contrast the general physical properties typical of covalent compounds with those typical of ionic compounds.

14. Relate oxidation number to the number of electrons in the outermost shell of a neutral atom.

15. Define the following in terms of electrons: single bond, double bond, and triple bond.

16. When is the formation of a coordinate covalent bond possible? Give an example.

17. Under what conditions do ionic materials conduct an electric current?

18. What is meant by the electronegativity of an element?

19. From the electronegativity scale, which of the following compounds would you expect to be covalent and which ones ionic?

Magnesium fluoride, MgF_2	Bromine monochloride, BrCl
Sulfur dioxide, SO_2	Chloroform, $HCCl_3$
Copper(II) sulfide, CuS	Calcium oxide, CaO
Carbon tetrabromide, CBr_4	Arsine, AsH_3
Acetylene, C_2H_2	Silane, SiH_4
Cyanogen, C_2N_2	Carbon dioxide, CO_2

20. How do the electronegativities of the metals compare with those of the nonmetals?

21. Relate bond polarity to electronegativity differences between the atoms involved in the bond.

22. Give an example of a nonpolar covalent bond, a polar covalent bond, and a bond of extreme polarity.

23. Many molecules which contain polar bonds are polar molecules. However, some molecules which have polar bonds are nonpolar molecules. Explain.

24. Which of the following molecules contain polar bonds: Br_2, S_8, CO_2, H_2O, N_2, H_2S, HBr?

25. Which of the following are polar species: Br_2, CO_2 (known to be a straight line linear molecule), H_2O (known to be an angular molecule), HBr, CH_4 (known to be a tetrahedral structure with carbon in the center and the four hydrogens symmetrically distributed around the carbon at the corners of the tetrahedron)?

26. Do the suffixes -*ous* and -*ic* correspond to any particular oxidation state? What is their significance?

27. Name the following compounds: KBr, NaCl, MgF_2, AgBr, HI, RaO, ZnS, AlN, $Ba_3(PO_4)_2$, NaH, Ca_2Si, LiCN, PCl_5, PCl_3, $FeBr_2$, $Al(OH)_3$, NH_4NO_3, $HBrO_4$, KNO_2, KNO_3, Na_2SO_3, $Ca(ClO)_2$, KIO_4.

28. Write valence electronic structures for the following species: F_2, HF, F^-, Cs^+, NF_3, SCl_2.
29. Write valence electronic structures for each of the following: LiF, $MgCl_2$, NH_3, NH_4^+, H_2O, H_3O^+, $SiCl_4$, HI, Br_2, SF_6, BF_3, and ICl.
30. Calculate the oxidation number for each element in each of the substances listed in Questions 27, 28, and 29.
31. Calculate the oxidation number of nitrogen in each of the following: N_2O, NO, N_2O_3, N_2O_4, HNO_3, HNO_2, NH_3, NH_4Cl, N_2H_4, NH_2OH, AlN.
32. Calculate the oxidation number for each element in CH_3OH, C_2H_5Cl, Na_2O, Na_2O_2, KO_2, MnO_4^- (permanganate ion), MnO_4^{2-} (manganate ion), PbO, PbO_2, Pb_3O_4, $CaCO_3$, CH_2O, $PbSiF_6$, and $C_3H_6Cl_2$.
33. With the aid of oxidation numbers, write formulas for the following compounds: sodium fluoride, calcium hydride, aluminum oxide, tin(II) fluoride, uranium hexafluoride, calcium phosphate, sodium nitrite, potassium carbonate, potassium hypobromite, sodium chlorate, iron(II) sulfate, tin(IV) chloride, sulfurous acid, mercury(II) chloride, ammonium perchlorate, sodium oxide, and calcium chlorite.
34. Write formulas for the following compounds: calcium acetate, potassium nitrate, iron(II) hydroxide, iron(III) hydroxide, barium hypochlorite, potassium chlorite, hydrogen perchlorate, potassium permanganate, zinc oxide, phosphorus(III) chloride, iron(II) phosphate, strontium carbonate, sodium sulfite, lead(II) sulfate, tin(IV) sulfate, cobalt(III) fluoride, sodium arsenate, potassium silicate, sodium peroxide, sodium thiosulfate, mercury(II) sulfide, cadmium chloride, silver carbonate, magnesium sulfate, chromium(III) oxide, ammonium phosphate, gold(III) chloride, and hydrogen phosphate.
35. Name each of the following and calculate the oxidation number of manganese in each (MnO_4^- is the permanganate ion): $MnCl_2$, $MnCl_3$, $CaMnO_3$, $KMnO_4$, Mn_2O_3, MnO_2.

REFERENCES

"Early Views on Forces Between Atoms," L. Holliday, *Sci. American*, May, 1970; p. 116.

"Principles of Chemical Bonding," R. T. Sanderson, *J. Chem. Educ.*, **38**, 382 (1961).

"Revised Inorganic (Stock) Nomenclature for the General Chemistry Student," R. C. Brasted, *J. Chem. Educ.*, **35**, 136 (1958).

"Oxidation States in Inorganic Compounds," D. B. Sowerby and M. F. A. Dove, *Educ. in Chemistry*, **1**, 83 (1964).

"The Electron Repulsion Theory of the Chemical Bond," W. F. Luder, *J. Chem. Educ.*, **44**, 206, 269 (1967).

"Anticipating 'Valences' from Electron Configurations," J. W. Eichinger, Jr., *J. Chem. Educ.*, **44**, 689 (1967).

"The Chemical Bond and the Geochemical Distribution of the Elements," L. H. Ahrens, *Chem. in Britain*, **2**, 14 (1966).

"Hydrogen-Like Wave Functions," B. Perlmutter-Hayman, *J. Chem. Educ.*, **46**, 428 (1969).

"Werner, Kekulé, and the Demise of the Doctrine of Constant Valency," G. B. Kaufman, *J. Chem. Educ.*, **49**, 813 (1972).

"Chemistry and the Spinning Electron," (Staff), *New Scientist*, **60** (867), 128 (1973).

"Strengths of Chemical Bonds," J. D. Christian, *J. Chem. Educ.*, **50**, 176 (1973).

Chemical Bonding, Part 2—Molecular Orbitals

5

In the previous chapter, we considered bonding from the viewpoint that electrons participating in bonding are principally located in the atomic orbitals of the bonded atoms. In the Molecular Orbital Theory, the atomic orbitals of the bonding atoms are postulated as combining to form **molecular orbitals,** in which electrons move about over the molecule as a whole.

5.1   Molecular Orbital Theory

The Molecular Orbital Theory treats the distribution of electrons in molecules in much the same way as the distribution of electrons in atomic orbitals is treated for atoms.

Molecular orbitals are postulated as being around the various nuclei present in a molecule. These molecular orbitals are the regions in which a given electron is most likely to be found in the molecule. The valence electrons are not thought of as being located within individual atoms but rather as moving about over the molecule as a whole.

The molecular orbitals bear a resemblance to the atomic orbitals of the atoms that make up the molecule. The shapes of the molecular orbitals, in fact, can be approximated from appropriate combinations of the atomic orbitals.

Two molecular orbitals result from the combination of two atomic orbitals. One of the molecular orbitals results from the *subtraction* of the parts of the atomic orbitals that overlap and is known as an **antibonding orbital.** The other results from the *addition* of the parts of the atomic orbitals which overlap and is known as a **bonding orbital.**

Of the two molecular orbitals formed by the combination of two atomic orbitals,

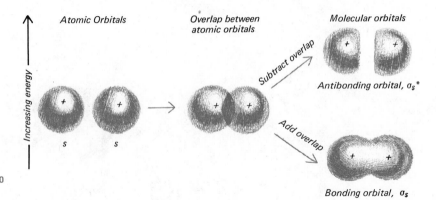

Atomic Orbitals

Overlap between
atomic orbitals

Molecular orbitals

Subtract overlap

Antibonding orbital, σ_s*

Increasing energy

s s

Add overlap

Bonding orbital, σ_s

FIGURE 5-1

The combination of two s atomic orbitals to provide two molecular orbitals.

the bonding orbital is of lower energy than the antibonding orbital. Electrons tend, therefore, to fill the lower energy *bonding* molecular orbital in preference to the higher energy *antibonding* molecular orbital, just as electrons tend to fill atomic orbitals of lower energy before filling those of higher energy.

As we have seen in Chapter 3 (Section 3.20), the s atomic orbitals are thought to be spherical in shape. An s orbital of one atom is postulated as combining with an s orbital of another atom to provide two molecular orbitals (Fig. 5–1). Such combinations of s atomic orbitals are said to give **sigma (σ)** molecular orbitals. It should be noted that adding the overlap provides a bonding orbital which is of lower energy than either of the atomic orbitals that combined to produce it. The antibonding orbital produced by subtraction of overlap is of higher energy than either of the two atomic orbitals. As might be expected, the electrons favor the lower energy bonding orbital. This is because a given electron is attracted by both nuclei in the bonding orbital but largely by only one nucleus either in the antibonding orbital or in the two atomic orbitals. Notice in Fig. 5–1 that the antibonding orbital does not include some of the space between the nuclei.

The symbol for an antibonding orbital often includes an asterisk to distinguish it from its corresponding bonding orbital. A subscript letter is frequently used to designate the type of atomic orbitals giving rise to the molecular orbitals. For example, the bonding and antibonding orbitals resulting from the combination of two s orbitals are designated σ_s and σ_s*, respectively, as indicated in the diagram (Fig. 5–1). They are commonly referred to in words as "sigma-s" and "sigma-s star."

Two p atomic orbitals, each of which, as we have noted, possesses two lobes (see Section 3.20) combine in either of two ways—end-to-end or side-by-side—to produce molecular orbitals (Fig. 5–2). When two p atomic orbitals, of two atoms, combine end-to-end, the resulting two molecular orbitals are referred to as **sigma (σ)** orbitals, as in the case of s atomic orbitals. However, when two p orbitals combine side-by-side, the resulting molecular orbitals are called **pi (π)** orbitals.

Each atom contains three p atomic orbitals—p_x, p_y, and p_z (see Section 3.20). One of these (e.g., p_x) combines end-to-end with a corresponding p atomic orbital of another atom to form two σ molecular orbitals, σ_{p_x} and σ_{p_x}*. The other two p atomic orbitals, p_y and p_z, each combine side-by-side with a corresponding p

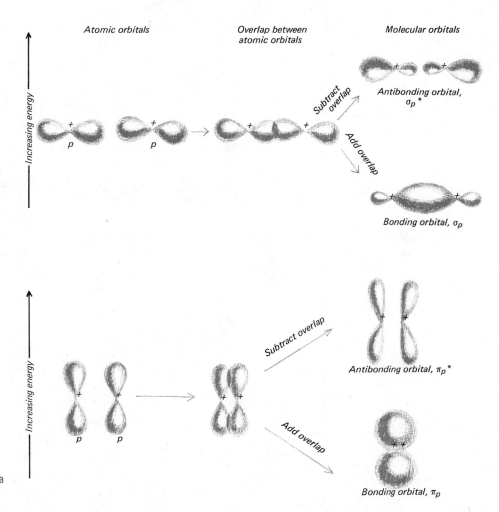

FIGURE 5-2
The combination of two p atomic orbitals to form sigma (σ) and pi (π) orbitals.

atomic orbital of another atom giving rise to two sets of π molecular orbitals. The two sets are oriented at approximately right angles to each other as if one set were oriented along the y-axis of a set of coordinates and one along the z-axis. The notations π_{p_y} and π_{p_z} (commonly referred to as "pi-p-y" and "pi-p-z") are thus applied to the bonding orbitals, and $\pi_{p_y}^*$ and $\pi_{p_z}^*$ ("pi-p-y star" and "pi-p-z star") to the antibonding orbitals. Except for their orientation, the π_{p_y} and π_{p_z} orbitals are identical and have the same energy. The $\pi_{p_y}^*$ and $\pi_{p_z}^*$ antibonding orbitals likewise are the same except for their orientation, and they possess equal energy.

Thus, a total of six molecular orbitals result from the combination of the six atomic p orbitals in two atoms. If the σ_p and σ_p^* orbitals are thought of as being oriented along the x-axis, and the two sets of π_p and π_p^* orbitals as being oriented along the y-axis and z-axis, respectively, the six molecular orbitals are σ_{p_x} and $\sigma_{p_x}^*$, π_{p_y} and $\pi_{p_y}^*$, and π_{p_z} and $\pi_{p_z}^*$.

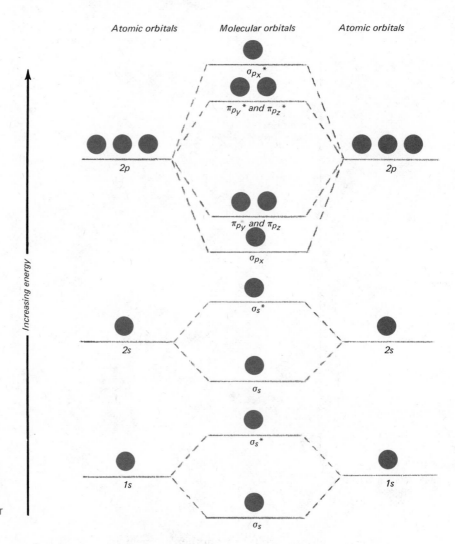

FIGURE 5-3

Molecular orbital energy diagram. A colored disc represents an atomic or molecular orbital which can hold one or two electrons.

5.2 Molecular Orbital Energy Diagram

The relative energy levels of the lower energy atomic and molecular orbitals are typically as shown in Fig. 5–3. Each colored disc represents one atomic or molecular orbital which can hold one or two electrons but no more than two.

It should be mentioned that in some specific cases, the energy levels of the σ_p and the two π_p bonding orbitals are reversed with the π_{p_y} and π_{p_z} bonding orbitals being slightly lower in energy than the σ_{p_x} orbital. There is a logical reason for this. We have assumed up to now in our discussion that s orbitals interact only with s orbitals to form σ_s molecular orbitals and that p orbitals interact only with p orbitals to form σ_p or π_p molecular orbitals. The main interactions do indeed occur between orbitals identical in energy. It is, however, also possible for an s orbital to interact with a p orbital, if the s and p orbitals are similar in energy.

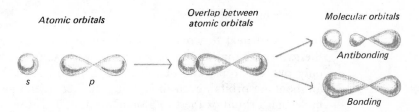

This results in the σ_s molecular orbital no longer having pure s character and the σ_p molecular orbital no longer having pure p character. The mixing of s and p character shifts the energies of the molecular orbitals and, in some cases, even changes the relative positions of the energy levels of the orbitals. Hence, the π_{p_y} and π_{p_z} molecular orbitals, normally at a higher energy than the σ_{p_x} orbital, sometimes are slightly lower than the σ_{p_x} orbital. Such energy shifts will be the greatest when the energy difference between the s and p orbitals is low so that they may interact easily.

5.3 The Hydrogen Molecule, H_2

The hydrogen molecule (H_2) is made up of two hydrogen atoms, each of which possesses one electron in a $1s$ atomic orbital. When the atomic orbitals for the two hydrogen atoms combine, the two electrons seek the molecular orbital of lowest energy, which is the σ_s bonding orbital. Each molecular orbital can hold two electrons. Hence, both electrons are in the σ_s bonding orbital for the hydrogen molecule. This can be represented by a molecular orbital energy diagram (Fig. 5-4) in which each electron within an orbital is indicated by an arrow, (↑). Two electrons of opposite spin within an orbital are designated (↑↓).

The σ_s orbital, in which both electrons of the hydrogen molecule are most likely to be found, is lower in energy than either of the two $1s$ orbitals. Hence, the hydrogen molecule, H_2, is readily formed from two hydrogen atoms.

In general, it can be said that the difference in energy of the atomic orbitals and the bonding molecular orbital will depend upon the extent of overlap of the two atomic orbitals. The greater the amount of overlap, the greater the energy difference will be, and the stronger the resulting bond will be.

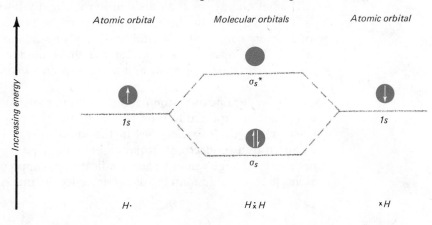

FIGURE 5-4

Molecular orbital energy diagram for the hydrogen molecule.

5.4   Helium Diatomic Species

It is logical next to wonder what happens when three electrons are present in the two combining atoms or ions. Inasmuch as the σ_s bonding orbital can hold only two electrons, one of the three electrons in such a case goes to the σ_s* antibonding orbital, even though it is higher in energy than either the bonding molecular orbital or the two atomic orbitals. This use of the antibonding orbital is illustrated by the combination of a helium atom (He) and a helium ion (He$^+$) to form the dihelium ion (He$_2$$^+$). See Fig. 5–5.

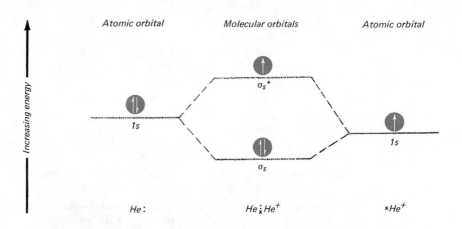

FIGURE 5-5

Molecular orbital energy diagram for the He$_2$$^+$ ion.

The helium atom, as we have seen, has two electrons, both of which are in the 1s orbital (Table 3–8). The helium ion (He$^+$) is a helium atom after it has lost one of its two 1s electrons. The charge of plus one occurs as a result of the loss of the negatively charged electron. When the helium atom and the helium ion combine, two of the electrons go to the lower energy σ_s bonding orbital. The third electron goes to the molecular orbital which is next lowest in terms of energy, the σ_s* antibonding orbital. These positions are illustrated in the molecular orbital energy level diagram of Fig. 5–5.

The net energy for the dihelium ion is slightly lower than the energy of the helium atom or of the helium ion, inasmuch as two electrons are present in the lower energy molecular orbital and only one in the higher energy molecular orbital. The lower net energy for the dihelium ion indicates an appreciable tendency for a helium atom and a helium ion to combine to form a dihelium ion.

By a similar line of reasoning, the dihelium molecule (He$_2$) with four electrons would not be expected to form readily by the combination of two helium atoms, inasmuch as the two electrons in the lower energy bonding orbital would be balanced by two electrons in the higher energy antibonding orbital. Hence, the net energy change would be zero, indicating no appreciable tendency for helium atoms to combine to form the diatomic molecule. Indeed, as predicted, the helium

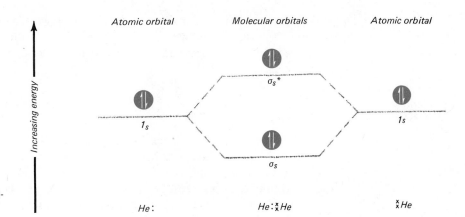

FIGURE 5-6
Molecular orbital energy diagram for He_2 (not stable).

element exists as discrete atoms, He, rather than as diatomic molecules, He_2. See Figure 5–6.

It is interesting to apply the principles we have just discussed to the possible formation of diatomic molecules of the second period of elements (from lithium to neon).

5.5 The Lithium Molecule

The combination of two lithium atoms to form a lithium molecule, Li_2, is analogous to the formation of H_2, except that the $2s$ atomic orbital is principally involved instead of the $1s$ orbital. Each of the two lithium atoms, with an electronic configuration $1s^2 2s^1$, has one valence electron. Hence, two valence electrons are available to go into the σ_{2s} bonding molecular orbital. The lower-lying $1s$ electrons in the $1s$ orbital of each atom are not appreciably involved in the bonding.

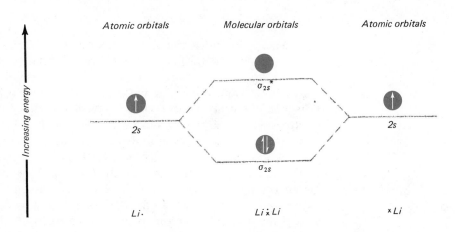

FIGURE 5-7
Molecular orbital energy diagram for Li_2.

Inasmuch as both electrons would be in the lower energy σ_{2s} bonding orbital, it should be possible to form Li_2. The molecule is, in fact, present in appreciable concentration in lithium vapor at temperatures near the boiling point of the element.

5.6   The Beryllium Molecule (Not Stable)

The diatomic molecule of beryllium, Be_2, on the other hand, would not be expected to be particularly stable, inasmuch as two of the four valence electrons would go to the σ_s bonding orbital and the other two would be forced to go to the σ_s* antibonding orbital. The net energy change, just as is the case with helium, is essentially zero, indicating no appreciable tendency for beryllium atoms to combine to form the diatomic beryllium molecule. This is in accord with the experimental finding that no stable Be_2 molecule is known.

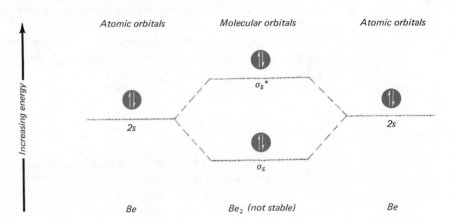

FIGURE 5-8
Molecular orbital energy diagram for Be_2 (not stable).

5.7   The Boron Molecule

The element boron has the electronic structure $1s^2 2s^2 2p^1$. Hence, beginning with boron in this period, the p orbitals make an important contribution to the bonding. We have noted that combinations of p atomic orbitals give rise to both sigma and pi molecular orbitals and that the two π_p bonding orbitals are of equal energy. **Whenever there are two or more molecular orbitals at the same energy level, electrons tend to fill the orbitals of that type singly before any pairing of electrons takes place within these orbitals.** Recall that an exactly analogous situation pertains to atomic orbitals (see Section 3.17).

We have noted earlier that usually the σ_p energy level is slightly lower than the energy level of the two π_p orbitals, but that sometimes the reverse is true. Experimental magnetic data for the B_2 molecule, which is known to exist, tell us that the molecule contains two unpaired electrons. This is an indication that

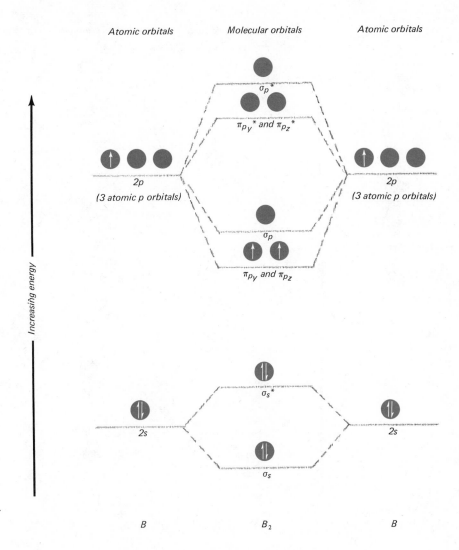

Atomic orbitals   Molecular orbitals   Atomic orbitals

FIGURE 5-9
Molecular orbital energy diagram for B_2.

in this particular case the two π_p orbitals are lower in energy than the σ_p orbital and are filled preferentially, with one electron going into each π_p orbital singly. If the σ_p orbital were lower in energy than the two π_p orbitals, the two electrons would be expected to pair within the one orbital. Figure 5–9 is the molecular orbital energy diagram for B_2 and shows the two unpaired electrons.

The four electrons in the σ_s and σ_s^* orbitals balance each other in terms of energy and do not make a significant contribution to the stability of the B_2 molecule. However, the fact that the two π_p electrons are in bonding orbitals indicates that it should be possible to form the diboron molecule by combination of boron atoms. This is verified by the fact that the molecule is known to exist.

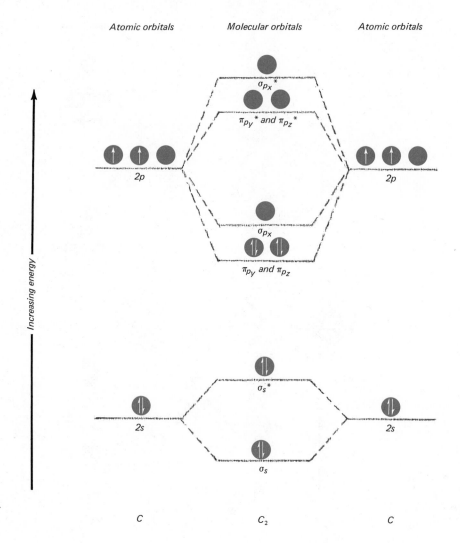

Atomic orbitals   Molecular orbitals   Atomic orbitals

FIGURE 5-10

Molecular orbital energy diagram for C_2.

5.8   The Carbon Molecule

For the formation of the carbon molecule, C_2, the situation is similar to that of B_2, except that one additional electron from each atom must be considered. For carbon, as with boron, the π_p orbitals are thought to be at a lower energy level than the σ_p orbital (Fig. 5-10). The molecular orbital energy diagram indicates that the C_2 molecule should exist, and indeed it does.

5.9   The Nitrogen Molecule

Each nitrogen atom has three $2p$ electrons. The nitrogen molecule, N_2, is well known and very stable. Experiment shows that all electrons are paired in the

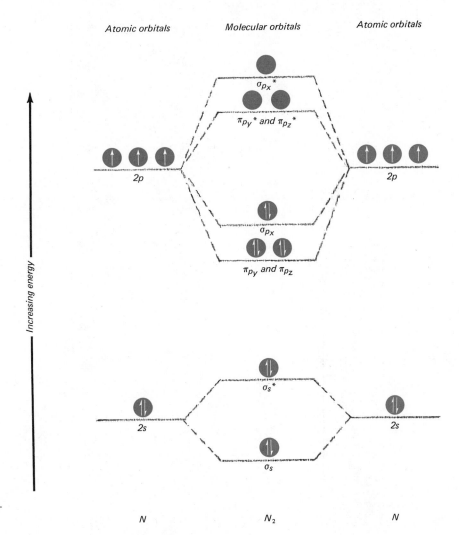

Atomic orbitals   Molecular orbitals   Atomic orbitals

FIGURE 5-11
Molecular orbital energy diagram for N_2.

molecular orbitals and that the π_p orbitals are lower in energy than the σ_p orbital. The molecular orbital energy level diagram is given in Fig. 5–11. The fact that the molecule is stable is in accord with the assumption that the six electrons arising from the $2p$ atomic orbitals are all in bonding molecular orbitals.

5.10 The Oxygen Molecule

The molecular orbital energy level diagram for the oxygen molecule, O_2, is shown in Fig. 5–12. The σ_p orbital in oxygen is usually considered to be at a lower energy level than the π_p orbitals.

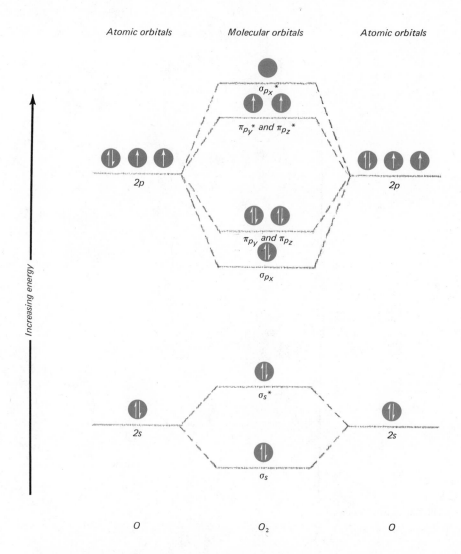

Atomic orbitals　　*Molecular orbitals*　　*Atomic orbitals*

FIGURE 5-12

Molecular orbital energy diagram for O_2.

The molecular orbital energy diagram for O_2 is in accord with the known experimental fact that the oxygen molecule has two unpaired electrons. This has proved to be difficult to explain on the basis of valence electronic formulas, but with the molecular orbital theory it is quite straightforward. In fact, the unpaired electrons of the oxygen molecule provide one of the strong pieces of support for the molecular orbital theory.

With a majority of the *p* electrons in bonding orbitals, the oxygen molecule would be expected to be relatively stable. However, the bond, with two *p* electrons in antibonding orbitals, might be expected to be less strong than in the nitrogen molecule, in which no *p* electrons are in antibonding orbitals, and this is the case as is verified by the experimental observation that much less energy is required to break the oxygen molecule bond than the nitrogen molecule bond (see Section 5.15).

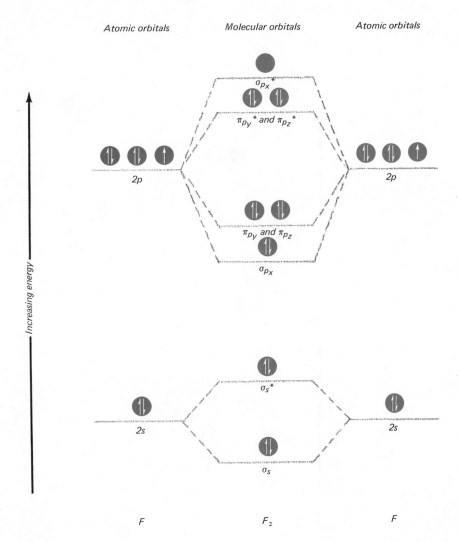

FIGURE 5-13
Molecular orbital energy diagram for F_2.

5.11 The Fluorine Molecule

The fluorine molecule has no unpaired electrons. With six of the p electrons in bonding orbitals and four in antibonding orbitals (Fig. 5-13), fluorine atoms should combine to form the fluorine molecule but with a somewhat weaker bond than in the oxygen molecule. Such is the case. In fact, the F_2 bond is one of the weakest of the covalent bonds.

5.12 The Neon Molecule (Not Stable)

The Ne_2 molecule has no appreciable tendency to form, inasmuch as it would have the same number of electrons in antibonding orbitals as in bonding orbitals. As with helium, the atoms remain as discrete atoms. See Figure 5-14.

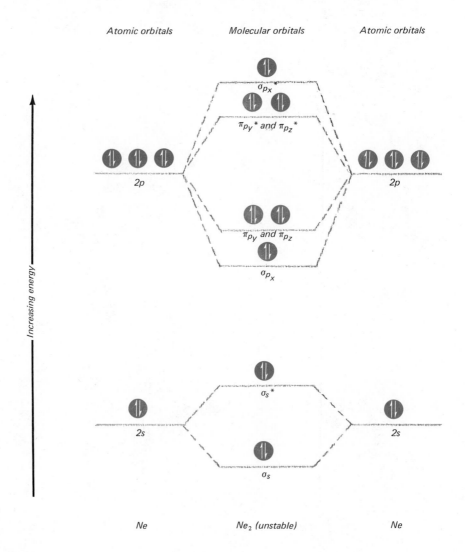

Atomic orbitals   *Molecular orbitals*   *Atomic orbitals*

FIGURE 5-14
Molecular orbital energy diagram for Ne_2 (not stable).

5.13 Diatomic Molecules with Two Different Elements

Thus far, we have considered only molecular orbitals arising from two identical atoms or from an atom and an ion of the same element. It is of interest to look briefly at some of the factors involved in an extension of the molecular orbital theory to bonds between atoms of two different elements.

Different elements usually have different electronegativities (see Table 4–2). This difference is reflected in an energy level diagram, inasmuch as the atomic orbitals are shown at different energy levels, the orbitals for the more electronegative element being at the lower level.

The molecular orbital energy level diagram shown in Fig. 5–15 is for a general molecule, QR, in which R represents the more electronegative element. We will

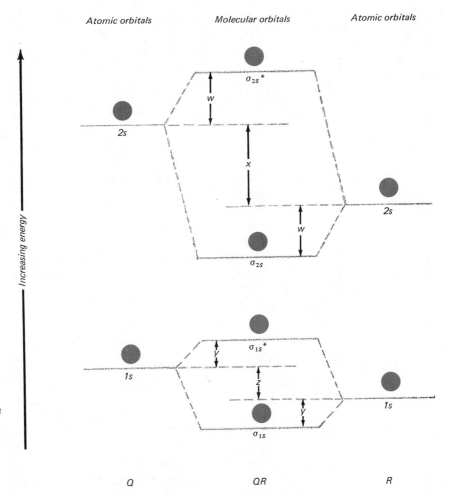

Atomic orbitals   Molecular orbitals   Atomic orbitals

FIGURE 5-15

Molecular orbital energy diagram for the general molecule QR, in which Q and R are the combining elements. Element R is more electronegative than element Q, as is shown by placing the atomic orbitals of R lower on the energy scale than the corresponding atomic orbitals of Q.

consider only the $1s$ and $2s$ atomic orbitals, although orbitals of higher energy may also be present. The electrons are not specifically shown in the diagram.

The quantities w and y in the figure provide indications of the degree of overlap between the atomic orbitals of elements Q and R as they combine to form the molecule QR. The amount of overlap is related to the degree of covalent character in the bond. In general, the greater the amount of overlap (greater magnitude of w and y) the more significant is the interaction between the two atoms; this is in the direction of a stronger covalent bond. Usually, the $2s$ orbitals, being farther from the nucleus than the $1s$ orbitals, will interact more readily. Hence, w is usually larger than y, as shown in the diagram.

The quantities x and z relate to the differences in energies of the atomic orbitals of elements Q and R and are related to the difference in electronegativities of the two atoms. In general, the greater the difference in energies of the atomic orbitals, the greater the amount of polarity or of ionic character in the bond.

This correlates with our earlier discussion relating the difference in electronegativity for two bonded atoms to the polar character in covalent bonds and to the extent of "partial ionic character" in such bonds (see Section 4.7). As mentioned before, in the hydrogen molecule the two atoms are identical, and the corresponding atomic orbitals have equal energies. Hence, there is very little polar character (or partial ionic character) in such a covalent bond.

Notice that in Fig. 5–15 for the QR molecule the σ_{1s} bonding molecular orbital is closer in energy to the $1s$ orbital of R than to the $1s$ orbital of Q. This indicates that element R has a somewhat greater attraction for a pair of electrons in the σ_{1s} molecular orbital than does element Q. This is in accord with the greater electronegativity of R and a greater electron density at the R end than at the Q end of the QR molecule.

The nitrogen(II) oxide molecule, NO, is a specific example of a diatomic molecule with two different elements. The oxygen atom is more electronegative (3.5) than nitrogen (3.0). See Table 4–2. A molecular orbital energy level diagram for the formation of the NO molecule would show one unpaired electron in a $\pi_p{}^*$ antibonding orbital. (A good test of your understanding of the molecular orbital energy level diagrams of the types we have been discussing would be to draw the energy level diagram for the formation of the NO molecule from a nitrogen atom and an oxygen atom.)

5.14 Energy Release in the Formation of a Molecule

An energy change accompanies the formation of a bond between two atoms. Usually, energy is released (exothermic); occasionally, but less frequently, energy is absorbed (endothermic). In general, a stable bond corresponds to a release of energy with the formation of the bond; the larger the amount of energy released the more stable the bond.

The energy released in bond formation can be estimated from a molecular orbital energy level diagram. For example, consider the energy level diagram for QR (Fig. 5–15) and assume the hypothetical case in which the bond involves one $1s$ electron from element Q and one $1s$ electron from element R. As can be determined from Fig. 5–15, the energy is lowered (energy release) by an amount equal to y when an electron goes from the $1s$ atomic orbital of element R to the lower energy σ_{1s} molecular orbital of QR. Similarly, the $1s$ electron of element Q loses energy (energy release) when it drops down to the lower energy level σ_{1s} molecular orbital. The energy release for the $1s$ electron of element Q is equal to $(z + y)$. Hence, the total energy release for the two electrons is $y + (z + y)$ or $2y + z$.

For another hypothetical case based upon Fig. 5–15, let us consider the bond in which two $1s$ electrons from element Q and one $1s$ electron from element R are involved in the bond formation. In this case, two of the electrons (for our purposes it does not matter which two) go into the σ_{1s} bonding molecular orbital and the third is forced into the $\sigma_{1s}{}^*$ antibonding orbital. Assuming, arbitrarily, that the two $1s$ electrons from element Q go to the σ_{1s} bonding orbital, the energy release for each would be $(y + z)$. The energy release for the two electrons would therefore be $2(y + z)$ or $(2y + 2z)$. When the electron from element R goes to the $\sigma_{1s}*$ antibonding orbital, it goes to a molecular orbital higher in energy than the atomic orbital from which it came, and hence a *gain* of energy of $(y + z)$ results, which offsets partially the release of energy of the other two electrons. The net energy change, therefore, is a release of energy equal to $(2y + 2z) - (y + z)$, or $(y + z)$. Note that this is a smaller energy release (and a weaker bond) than in the case with two electrons cited earlier (energy release of $2y + z$), the smaller energy release resulting because of the involvement of an antibonding orbital.

If two $1s$ electrons from element Q and two $1s$ electrons from element R are involved in the bonding, a consideration of Fig. 5–15 shows that the net energy change is zero (two electrons in the σ_{1s} bonding orbital balancing two electrons in the $\sigma_{1s}*$ antibonding orbital). In such a case, there should be no significant tendency for the atoms to form a molecule; they would remain as separate atoms.

One example involving $2s$ atomic orbitals is appropriate. Suppose that element Q has two $1s$ electrons and one $2s$ electron, and that element R has two $1s$ electrons and two $2s$ electrons. The $2s$ electrons in each case will be the ones expected to be significantly involved in the bond. (If the $1s$ electrons were assumed to participate, the energy change arising from the four $1s$ electrons would be zero, two in the σ_s orbital balancing the two in the σ_s* orbital. Hence, the effect of the $1s$ electrons on the bond would be insignificant.) Assuming that the two $2s$ electrons of element R go to the σ_{2s} bonding orbital, the energy release would be $2w$. The one $2s$ electron from element Q would then be forced into the molecular orbital of next lowest energy—the $\sigma_{2s}*$ antibonding orbital—with a resulting *gain* of energy equal to w, partially offsetting the energy release occurring with the electrons from R. The net energy change would be a release of energy equal to $(2w - w)$, or w.

The energy released when a bond is formed is the same as the energy required to break the bond. The energy required to break the bond is referred to as the **bond energy.** The actual bond energy for the B_2 molecule is about 70,000 calories per mole (the total bond energy for 6×10^{23} molecules). The bond energy for the C_2 molecule is considerably larger, about 83,000 calories per mole. Reference to Figs. 5–9 and 5–10 shows us that it is reasonable for the C_2 molecule to have a higher bond energy than B_2, inasmuch as it has a larger number of electrons in bonding orbitals than does B_2.

Our understanding of the energy relationships of bonds will be expanded further in Chapter 20, Chemical Thermodynamics.

5.15 Bond Order

From our study of nitrogen, oxygen, and fluorine molecules we have learned that the bond between the atoms is weaker in O_2 than in N_2 and weaker in F_2 than in O_2 (Sections 5.9, 5.10, and 5.11). This order of bond energies was predicted because of an observed increase in the ratio of the number of antibonding orbital electrons to that of the bonding orbital electrons, in the order N_2, O_2, and F_2. Values for the bond energies of the three molecules support the predictions.

Molecule	Bond energy
N_2	226,000 calories per mole
O_2	119,000 calories per mole
F_2	38,000 calories per mole

Nitrogen, with three pairs of bonding electrons in σ_p and π_p molecular orbitals, has a triple bond (see Fig. 5–11 and Section 4.3) and is said to have a *bond order of three*. In oxygen, the two additional electrons are in the $\pi_p{}^*$ antibonding orbitals (Fig. 5–12), thereby weakening the bond. The effect of six bonding electrons minus the partially offsetting effect of two antibonding electrons results in a net effect approximately equivalent to four bonding electrons (or two pairs of bonding electrons). Hence, oxygen is said to have a double bond, which is on the average weaker than a triple bond. The oxygen molecule, therefore, has a *bond order of two*. Fluorine, with six bonding and four antibonding electrons in the *p* levels (Fig. 5–13) has a net effect equivalent to two bonding electrons (or one pair of bonding electrons). The fluorine molecule, therefore, has a single bond and a *bond order of one*. A single bond is, in general, weaker than either a double bond or a triple bond.

By way of summary, **the bond order for a given bond may be determined by dividing the number of bonding electrons in the outer shell minus the number of antibonding electrons in the outer shell by two.**

$$\text{Bond order} = \frac{\text{Number of bonding electrons minus number of antibonding electrons}}{2}$$

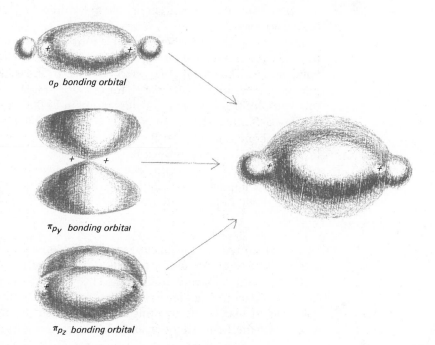

σ_p *bonding orbital*

π_{py} *bonding orbital*

π_{pz} *bonding orbital*

FIGURE 5-16

An approximation of the region in space in which the *p* electrons are most probably located in the nitrogen molecule (*right side of diagram*). This region represents a composite blend of the three molecular orbitals involved (*left side of diagram*).

5.16 Electron Distribution in Molecules

It is possible to picture in a very approximate way the region of a molecule in which the electrons are most probably located. The molecular orbitals overlap and blend together, so that the region of probable location of the electrons can be thought of as a composite blend of the molecular orbitals superimposed on each other.

Considering only the outermost orbitals, the nitrogen molecule has a pair of electrons each in the σ_{p_x}, π_{p_y}, and π_{p_z} orbitals (Fig. 5–11). A composite of these produces an electron cloud, shaped like an ellipsoid with a small lobe on each end, surrounding the two positive nuclei (Fig. 5–16). Other molecules can be pictured in a similar fashion.

QUESTIONS

1. In Section 3.17, Hund's Rule was discussed in connection with atomic orbitals. Does such a rule have application to molecular orbitals? Explain, utilizing specific examples to support your answer.
2. Compare atomic orbitals and molecular orbitals; bonding orbitals and anti-bonding orbitals; sigma orbitals and pi orbitals.
3. Show how bonding and antibonding molecular orbitals arise from combination of atomic orbitals. Draw diagrams showing the molecular orbitals that result from combination of s atomic orbitals; combination of p atomic orbitals.
4. Using molecular orbital energy level diagrams and electron occupancy in molecular orbitals, compare H_2, He_2^+, and He_2 with respect to stability. Explain your conclusions.
5. Compare the stabilities of the possible diatomic molecules for the first ten elements in the Periodic Table. Explain in terms of molecular orbital energy level diagrams and electron occupancy in molecular orbitals.
6. How does the molecular orbital energy level diagram for a diatomic molecule involving atoms of two different elements differ principally from one for a diatomic molecule made up of two atoms of the same element?
7. Paramagnetism is associated with unpaired electrons; diamagnetism, with all electrons paired. Predict whether each of the following atoms and molecules should be paramagnetic or diamagnetic. See Chapter 3 (including Section 3.17) as well as the material in this chapter before making your answer.

    (a) H   (d) N_2 (g) Cl   (j) Xe

    (b) H_2 (e) O   (h) Cl_2 (k) Bk

    (c) N   (f) O_2 (i) Co   (l) Ba
8. Draw a molecular orbital energy level diagram for chlorine (Cl_2) using only the electrons in the atomic orbitals of the outermost principal shell ($n = 3$).
9. (a) Draw a molecular orbital energy level diagram for carbon monoxide (CO).
    (b) Does the molecular orbital energy level diagram support the known fact that CO is a stable molecule? Explain.
10. Would you expect magnesium to be more likely or less likely than beryllium (which is in the same family of the Periodic Table) to form the diatomic

molecule? *It might be noted that this is a difficult question to answer, even for a professional chemist, for many complex factors are involved and all do not operate in the same direction. However, it is a good question to test your ability to think intelligently about these matters and to make reasonable predictions based upon intelligent thought. A few of the factors that you might consider are nuclear charge, atomic radius, diffuseness of the electron distribution in the outer shell, atomic orbitals, electron occupancy in the molecular orbitals, and the molecular orbital energy level diagrams.*

11. State what is meant by bond energy, bond order, heat of formation, heat of dissociation.

12. Predict and calculate the bond order for each of the following: F_2, O_2, N_2, C_2, B_2, Li_2, H_2, and CO.

13. What property do σ orbitals formed from two s atomic orbitals have in common with σ orbitals formed from two "end-to-end" p atomic orbitals?

14. What relation does the extent of covalent or ionic character of a given bond have to the molecular orbital energy level diagram?

15. Would you expect that a σ molecular orbital could be formed from an s atomic orbital and a p atomic orbital? Would you expect that an s atomic orbital and a p atomic orbital could be combined in such a way as to form a π molecular orbital? Explain your answers. Draw a diagram to show your concept of molecular orbitals which arise from the combination of an s atomic orbital with a p atomic orbital.

16. Using Fig. 5–15, calculate in terms of the energy quantities w, x, y, and z, the energy release or gain for bond formation in each of the following cases, in which the bonding electrons from element Q and element R are as follows:

    (a) Q, $1s^1$; R, $1s^2$ (c) Q, $1s^2 2s^2$; R, $1s^2$

    (b) Q, $1s^2$; R, $1s^2 2s^2$ (d) Q, $1s^2 2s^2$; R, $1s^2 2s^2$

17. How would the energy release or gain in each of the four cases in Question 16 be affected if element Q were more electronegative than element R, rather than the other way around?

18. The compound lithium hydride, LiH, is held together largely by ionic bonds. By referring to the energy level diagram in Fig. 5–15, rationalize the complete transfer of the $2s$ valence electron of the lithium atom to the $1s$ orbital of the hydrogen atom.

19. Draw an approximate picture of the region in space occupied by the valence electrons for He_2^+, B_2, and O_2, based upon the shapes of the molecular orbitals involved.

REFERENCES

"A Simple, Quantitative Molecular Orbital Theory," W. F. Cooper, G. A. Clark, and C. R. Hare, *J. Chem. Educ.*, **48**, 247 (1971).

"Chemical Bonds," N. N. Greenwood, *Educ. in Chemistry*, **4**, 164 (1967).

"Chemistry by Computer," A. C. Wahl, *Sci. American*, April, 1970; p. 54.

"Molecular Orbital Symmetry Rules," R. G. Pearson, *Chem. & Eng. News*, Sept. 28, 1970; p. 66.

"Polar Bonds," L. Melander, *J. Chem. Educ.*, **49**, 687 (1972).

"Metal Sandwiches," M. Sherwood, *New Scientist*, **60** (870), 335 (1973).

Molecular Structure

6

For the most part, the chemist is interested in the properties of molecules which contain two or more atoms. This interest includes information on the arrangement of the atoms when they are bonded together, that is, the **molecular structure.** The chemical, physical, and biological properties which depend on the structures of molecules often are related in a very sensitive manner to the exact spatial distribution of the constituent atoms. Many chemical reactions are best interpreted in terms of the detailed structures of the molecules of reactants and products. In this chapter we shall study the relationship of molecular structures to the "directed" covalent bond.

Valence Shell Electron-Pair Repulsion Theory

6.1 Prediction of Molecular Structures According to the Valence Shell Electron-Pair Repulsion Theory

We are interested here in a method for predicting the geometrical, three-dimensional arrangement of the atoms covalently bonded in a polyatomic molecule. According to the **Electron-Pair Repulsion Theory,** the electron-pairs associated with the central atom of a covalent molecule arrange themselves so as to minimize electrostatic repulsion. Minimum repulsion results when the various electron-pairs assume positions of greatest possible separation from each other. The resultant idealized geometry based on electron-pair repulsions is given in Table 6–1. The letter M denotes the central atom of the molecule.

TABLE 6-1   Idealized Geometry of Molecules Based Upon Electron-Pair Repulsions

Two Electron-pairs	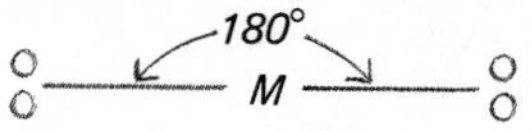	Linear, 180°
Three Electron-pairs	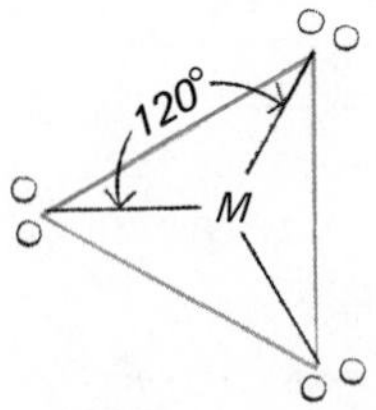	Trigonal planar, 120°
Four Electron-pairs	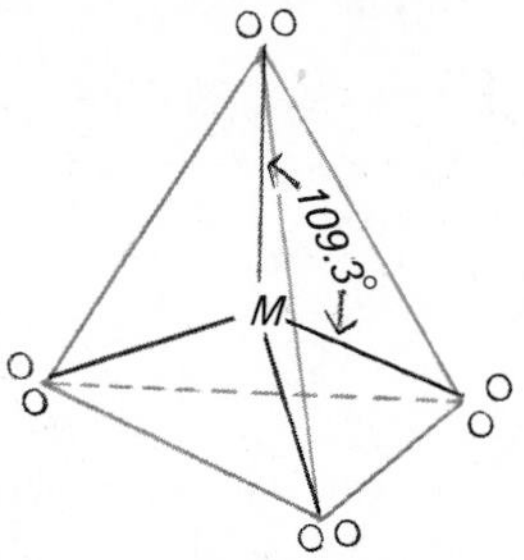	Tetrahedral, 109.3°
Five Electron-pairs	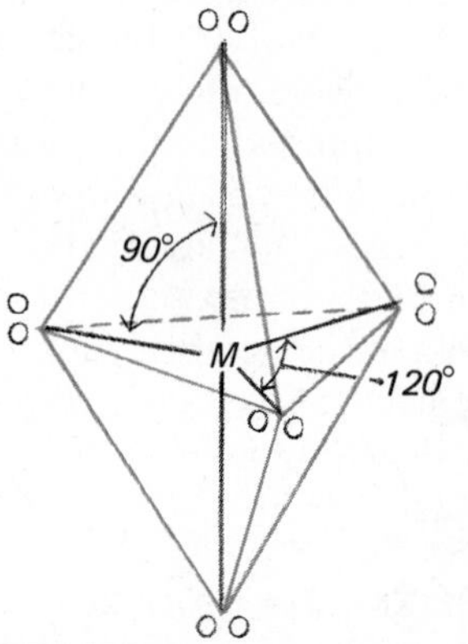	Trigonal bipyramidal, 90° or 120° (an attached atom may be equatorial, in the plane of the triangle, or axial, above or below the plane of the triangle)
Six Electron-pairs	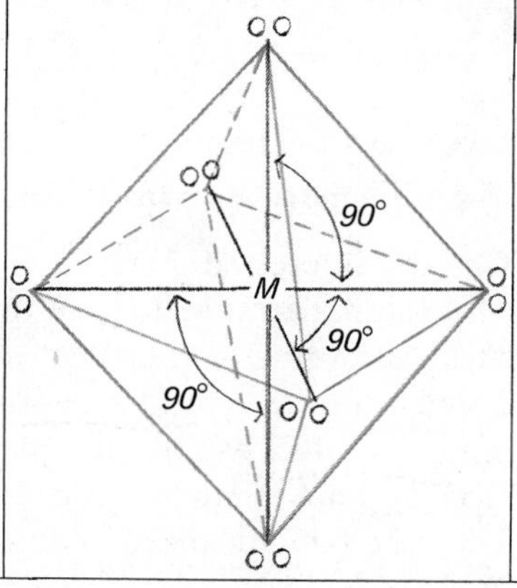	Octahedral, 90°

6.2   Rules for Predicting Molecular Structures

1. Determine the total number of electrons in the valence shell of the central atom M. This is done by adding the number of valence electrons originally belonging to the central atom to the number of electrons furnished by another atom for sharing with the central atom. In regular covalent bonds, hydrogen and the halogens (Periodic Group VIIA) donate one electron each for sharing. Elements in the oxygen group (Periodic Group VIA) are considered to donate no electrons. When acting as the central atom of the molecule, oxygen and each of the other Periodic Group VIA elements furnish all six of their valence electrons, and the halogens (Group VIIA) all seven valence electrons. If the species being considered is an ion, add or subtract the number of electrons required to account for the charge on the ion. For example, add three electrons for PO_4^{3-}; subtract one electron for NH_4^+. Finally, divide the total number of electrons surrounding the central atom M in the valence shell by two to obtain the number of electron-pairs.

2. From Table 6-1, select the proper geometry as indicated by the number of electron-pairs surrounding the central atom M. In the case of an odd number of electrons, treat the extra electron (one-half of an electron-pair) as if it were an electron-pair.

3. To draw the structure, arrange the appropriate atoms in the correct geometric shape around the central atom, using one pair of electrons to bond each atom to the central atom. Pairs of electrons left over, in excess of those necessary for bonded atoms, are also shown and are referred to as **lone electron-pairs,** or **free electron-pairs.** If more than one structure is possible, the most stable structure can often be determined using the following rules:

a. Consider only the repulsions between any two lone (free) electron-pairs, between a lone pair and a bond pair, or between two bond pairs, which are separated by the smallest angles (usually 90°). A **lone pair** is a free electron-pair which does not take part in bonding. A **bond pair** is an electron-pair which is shared between the central atom and another atom. The angles referred to are the lone pair–central atom–lone pair angle, the lone pair–central atom–bond pair angle, and the bond pair–central atom–bond pair angle.

For example, in the following one-of-three possible trigonal bypyramidal structures for chlorine trifluoride, ClF_3:

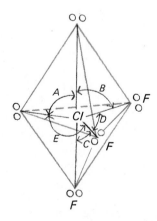

A represents a 90° lone pair–lone pair repulsion;
B represents a 90° lone pair–bond pair repulsion;
C represents a 90° bond pair–bond pair repulsion;
D represents a 120° bond pair–bond pair repulsion; and
E represents a 120° lone pair–bond pair repulsion.

b. Tabulate the total number of repulsions for the smallest angle between electron-pairs (usually 90°) for each structure as:

(1) Lone pair–lone pair repulsions; these have the greatest repulsion, causing the least stability.
(2) Lone pair–bond pair repulsions; intermediate in repulsive force.
(3) Bond pair–bond pair repulsions; smallest in repulsive force.

c. Select as the stable structure that one which for the smallest angle has the smallest number of lone pair–lone pair repulsions. If more than one structure remains, select that structure which has the fewest lone pair–bond pair repulsions. If more than one structure still remains, select the structure which has the fewest bond pair–bond pair repulsions. Occasionally, even after all of this, more than one structure remains. In such a case, the preceding rules are not sufficient to determine a difference in stability between the remaining structures. In most cases, however, use of the rules does delineate one structure as being the most stable.

6.3 Examples of Predicting Molecular Structures

Example 1. Beryllium Chloride, $BeCl_2$

The beryllium atom has two valence electrons; two chlorine atoms donate one electron each. This makes a total of four valence electrons, or two electron-pairs. The structure, therefore, is linear (Fig. 6–1).

FIGURE 6-1

The linear structure of $BeCl_2$.

Cl—Be—Cl

Example 2. Carbon Dioxide, CO_2

The carbon atom has four valence electrons; the two oxygen atoms donate no electrons according to the previously cited rules. The total of four valence electrons corresponds to two electron-pairs. Thus, the structure is linear (Fig. 6–2).

FIGURE 6-2

The linear structure of CO_2.

O—C—O

Example 3. Boron Trichloride, BCl_3

Boron has three valence electrons; three chlorine atoms donate one electron each. The total of six electrons is equivalent to three electron-pairs. Hence, the structure is trigonal planar (Fig. 6–3).

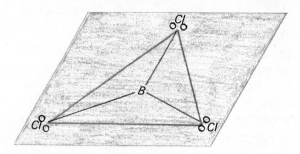

FIGURE 6-3

The trigonal planar structure of BCl_3.

Example 4. Methane, CH_4

Carbon has four valence electrons; four hydrogens donate one electron each. The total of eight electrons corresponds to four electron-pairs for the carbon. The structure is tetrahedral (Fig. 6–4).

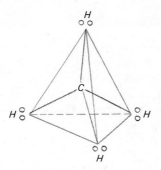

FIGURE 6-4

The tetrahedral structure of CH_4.

Example 5. Water, H_2O

Oxygen as the central atom of the molecule has six valence electrons, and two hydrogen atoms donate one electron each. The total of eight valence electrons, or four electron-pairs, indicates that the geometry is a tetrahedron with two corners occupied by the two hydrogen atoms and the other two corners by free electron-pairs. The three atoms by themselves, not taking into account the lone pairs of electrons, comprise an angular arrangement (Fig. 6–5).

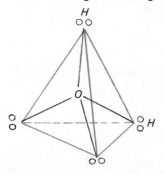

FIGURE 6-5

The tetrahedral structure of H_2O (or angular if only the three atoms are considered).

Example 6. Ammonia, NH_3

AB_3 molecule

Nitrogen has five valence electrons; three hydrogen atoms donate one electron each, thus making a total of eight valence electrons, or four electron-pairs. The structure is, therefore, a tetrahedron with one corner occupied by a free electron-pair. The four atoms, by themselves, form a trigonal pyramid (Fig. 6–6).

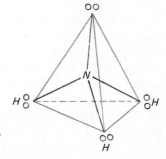

FIGURE 6-6

The tetrahedral structure of NH_3 (or trigonal pyramidal if only the four atoms are considered).

Example 7. Sulfur Tetrafluoride, SF_4

> Sulfur has six valence electrons; four fluorine atoms donate four electrons making a total of ten valence electrons, or five electron-pairs, for the sulfur atom. These five electron-pairs occupy the corners of a trigonal bipyramid. There are, however, two possible arrangements for the four fluorine atoms in the five positions, as shown in Fig. 6–7. To determine which one of the

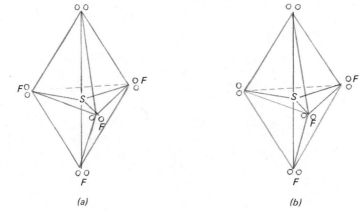

FIGURE 6-7

The possible structures of SF_4 (trigonal bipyramidal). Structure (b) is the more stable structure.

(a)   (b)

> two structures is the more probable, count the 90° repulsions (the 90° angle is the lesser of the 90° and 120° bond angles in a trigonal bipyramid structure).

Structure: (a)   (b)

(a)	(b)	
0	0	Number of 90° lone pair–lone pair repulsions
3	2	Number of 90° lone pair–bond pair repulsions
3	4	Number of 90° bond pair–bond pair repulsions

> Neither structure (a) nor structure (b) has any lone pair–lone pair repulsions. However, structure (b) has fewer 90° lone pair–bond pair repulsions than structure (a). Therefore, structure (b) is the more stable structure.

Example 8. Chlorine Trifluoride, ClF_3

> Chlorine as the central atom contributes seven valence electrons; three fluorine atoms furnish three electrons, making a total of ten valence electrons, for the chlorine, or five electron-pairs. The idealized geometry indicates a trigonal bipyramid with two positions occupied by electron-pairs. Adding the fluorine atoms around the central chlorine, we find three possible structures (Fig. 6–8).

The smallest angle between electron pairs in a trigonal bipyramid is 90°, so we catalog only 90° repulsions:

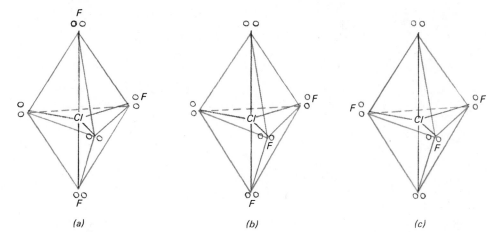

FIGURE 6-8

The possible structures of ClF_3 (trigonal bipyramidal). Structure (a) is the most stable structure.

(a)   (b)   (c)

$$F—Cl—F$$
$$|$$
$$F$$

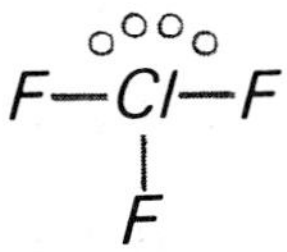

FIGURE 6-9

The T-shaped structure of ClF_3, considering only the four atoms.

Structure:	(a)	(b)	(c)	
	0	1	0	Number of 90° lone pair–lone pair repulsions
	4	3	6	Number of 90° lone pair–bond pair repulsions
	2	2	0	Number of 90° bond pair–bond pair repulsions

Structures (a) and (c) have no lone pair–lone pair repulsions. *Of these two structures*, (a) has the fewer lone pair–bond pair repulsions. Hence, structure (a) is the most stable of the three possible structures for ClF_3.

If the molecular structure is determined by the positions of the four atoms of the molecule, not taking into account the free electron-pairs, it is a T structure (Fig. 6–9).

Example 9.   Xenon Tetrafluoride, XeF_4

Xenon has eight valence electrons; the fluorine atoms each donate one electron. This makes a total of twelve electrons, or six electron-pairs. The structure is octahedral, therefore, with the four fluorine atoms occupying four of the corners and the remaining two electron-pairs occupying the other two corners. However, there are two possible arrangements, as shown in Fig. 6–10.

The smallest angle between electron-pairs in an octahedral configuration is 90°, so to determine the most stable structure we will check only the 90° repulsions:

Structure:	(a)	(b)	
	0	1	Number of 90° lone pair–lone pair repulsions
	8	6	Number of 90° lone pair–bond pair repulsions
	4	5	Number of 90° bond pair–bond pair repulsions

Structure (a) has fewer lone pair–lone pair repulsions (0) than structure (b) (1). Hence, structure (a) is the more stable of the two possible structures

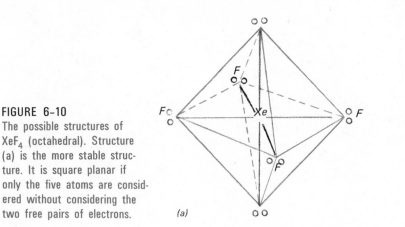

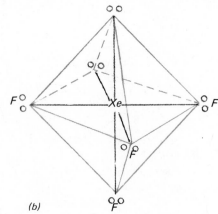

FIGURE 6-10
The possible structures of XeF_4 (octahedral). Structure (a) is the more stable structure. It is square planar if only the five atoms are considered without considering the two free pairs of electrons.

(a)   *(b)*

for XeF_4. The five atoms, themselves, are all in the same plane and form what is usually referred to as a "square planar" configuration.

Example 10.   Arsenate ion, AsO_4^{3-}

The arsenic atom has five valence electrons, and each oxygen donates none. Three more electrons are necessary to produce the charge of -3 for the ion. There is, therefore, a total of eight valence electrons, or four electron-pairs.

The structure is tetrahedral. Because there are four oxygen atoms, each electron-pair is used in bonding and only one structure is possible (Fig. 6-11).

FIGURE 6-11
The tetrahedral structure of the AsO_4^{3-} ion.

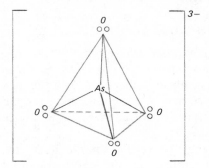

6.4   Bond Order

The foregoing method provides a means of predicting the spatial arrangement of the atoms of a molecule. However, it does not provide any indication as to the bond order of a given bond (Section 5.15), that is, whether the bond involves a single pair of shared electrons (a single bond), two pairs of shared electrons (a double bond), or three pairs of shared electrons (a triple bond). Inasmuch as many examples of each of these three types of bonds exist, the question arises as to whether one can determine if a bond is single, double, or triple. This can often be done by drawing the **valence electronic formula,** frequently referred to as a Lewis formula, using the valence electronic symbol for each atom (see Sections 4.1, 4.3, and 4.4).

For example, in the Lewis formula for methane, with a total of eight valence electrons, a complete complement of two electrons for each hydrogen atom and eight electrons for the carbon atom is provided only if each bond is a single bond, involving a single pair of electrons.

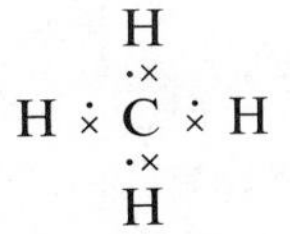

$$
\begin{array}{c}
\text{H} \\
\overset{\cdot\times}{} \\
\text{H} \overset{\cdot}{\times} \text{C} \overset{\cdot}{\times} \text{H} \\
\overset{\cdot\times}{} \\
\text{H}
\end{array}
$$

Methane

In the case of the carbon dioxide molecule, a sixteen valence-electron structure, the Lewis formula shows us that the only way to achieve an octet of valence electrons for each atom is through the use of two double bonds, each involving two pairs of electrons.

$$
\overset{\times\times}{\times}\text{O}\overset{}{\times}:\text{C}:\overset{\times\times}{\times}\text{O}\overset{}{\times} \qquad \text{O}=\text{C}=\text{O}
$$

Carbon dioxide

The ethylene molecule, C_2H_4, with twelve valence electrons, must include a double bond between the two carbon atoms and single bonds for each bond between a carbon atom and a hydrogen atom.

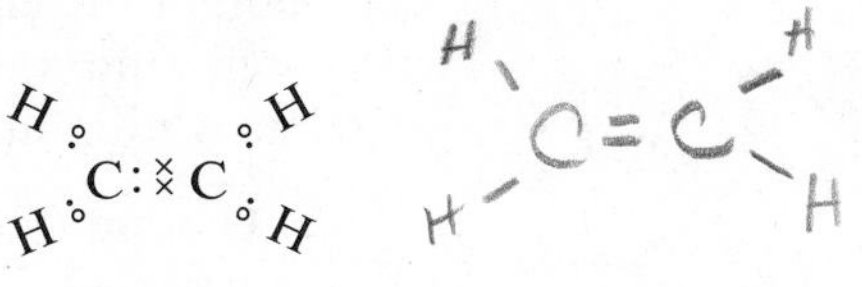

$$
\underset{\text{H}}{\overset{\text{H}}{}}\text{C}:\overset{\times}{\times}\text{C}\underset{\text{H}}{\overset{\text{H}}{}} \qquad \underset{\text{H}}{\overset{\text{H}}{}}\text{C}=\text{C}\underset{\text{H}}{\overset{\text{H}}{}}
$$

Ethylene

For acetylene, C_2H_2, a ten valence-electron structure, the carbon-carbon bond must involve three electron-pairs and each carbon-hydrogen bond one electron-pair to achieve a complete complement of two electrons for each hydrogen and eight electrons for each carbon.

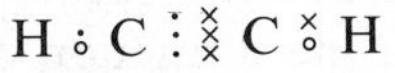

$$
\text{H}\overset{\cdot}{\circ}\text{C}\overset{\times}{\underset{\times}{\vdots}}\text{C}\overset{\times}{\circ}\text{H}
$$

Acetylene

$$
\text{H}-\text{C}=\text{C}-\text{H}
$$

Valence Bond Approach—Hybridization of Atomic Orbitals

Linus Pauling, in 1935, introduced the **Valence Bond Theory,** which postulates that nonequivalent atomic orbitals hybridize (combine) to give a set of equivalent hybrid orbitals whose characteristic orientation determines the geometry of the molecule or ion. Although the theory uses atomic orbitals rather than molecular orbitals, it is nevertheless especially useful in predicting molecular structures. Several examples for specific compounds follow:

6.5   Methane, CH_4, and Ethane, C_2H_6 (Tetrahedral Hybridization)

Each atomic orbital can accommodate two electrons of opposing spin. As we learned earlier (Section 3.17), the electrons enter each orbital of a given type singly before any pairing of electrons occurs within those orbitals (Hund's Rule). The carbon atom, therefore, has only two unpaired electrons in its normal state (ground state) with an electronic configuration $1s^2 2s^2 2p^1 2p^1$. With this structure, one would expect carbon to form only two covalent bonds, as indeed it sometimes does. However, the divalent compounds of carbon are not nearly so prevalent as are the tetravalent compounds containing four covalent bonds. We might ask, then, how the carbon atom attains four unpaired electrons with which to form four covalent bonds. The answer is, by promoting (by excitation) one of the electrons in the $2s$ orbital to the third $2p$ orbital with the resulting configuration $1s^2 2s^1 2p^1 2p^1 2p^1$. It should be noted that the energy required for promotion is more than offset by the energy given off during bond formation. Therefore, the energy state after electron promotion and bond formation is lower than before these steps; the decrease in energy corresponds to a more stable state.

The electron distribution for the carbon atom in the ground state and after promotion of the electrons can be shown as follows by representing each orbital by a circle and each electron by an arrow; two arrows pointing in opposite directions within an orbital designate two electrons of opposing spin.

C atom (ground state):

1s   2s   2p

⓵⓵ ⓵⓵ ⓵⓵〇〇

C atom (with one 2s electron promoted to the third 2p orbital):

1s   2s   2p

⓵⓵ ⓵ ⓵⓵⓵

The three *p*-orbitals are oriented at right angles (see Section 3.20); the *s* orbital is undirected because of its spherical shape.

In the formation of the methane molecule, each of the four hydrogen atoms covalently bonded to the carbon provides one electron (shown by a red arrow) to each one of the four orbitals which contain an unpaired electron.

CH$_4$ molecule:

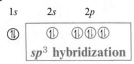

Tetrahedral structure

Experimental evidence shows that the four bonds of methane are equal in length and in strength and that they are pointed toward the corners of a regular tetrahedron. According to quantum mechanical calculations, the one 2s orbital and the three 2p orbitals of the carbon atom **hybridize** (combine) to form four new hybridized orbitals, which are exactly equivalent to one another and are so arranged in space that their lobes point to the corners of a regular tetrahedron (Fig. 6–12). They are referred to as sp^3 hybridized orbitals, to designate the hybridization of one s orbital and three p orbitals; sp^3 hybridization always results in a tetrahedral configuration for the molecule.

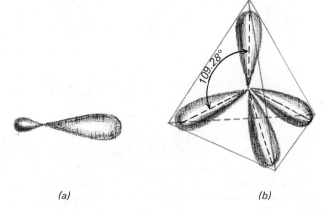

FIGURE 6–12

(a) An individual hybridized sp^3 orbital. (b) The four tetrahedral hybridized sp^3 orbitals of the carbon atom.

(a)   *(b)*

The 1s orbital of each of the four hydrogen atoms of the methane molecule can be thought of as overlapping with each one of the four sp^3 orbitals of a carbon atom to form an sp^3-s sigma bond. Notice the similarity between this and the overlap to produce molecular orbitals (see Section 5.1). This results in the formation of four very strong covalent bonds between one carbon atom and four hydrogen atoms to produce the methane molecule, CH$_4$ (Fig. 6–13).

In ethane, C$_2$H$_6$, an sp^3 orbital of one carbon atom overlaps end to end with an sp^3 orbital of a second carbon atom to form a sigma bond. Each of the other three sp^3 orbitals of each carbon atom overlaps with an s orbital of a hydrogen atom to form additional sigma bonds. The structure and overall outline of the bonding orbitals of ethane are shown in Fig. 6–14. Ethane is made up of two tetrahedra with one corner in common.

Hybridization can occur *only* when the orbitals involved have very similar energies. It is possible to hybridize 2s with 2p orbitals, or 3s with 3p orbitals, for example, but not 2s with 3s (see Chapter 3, Fig. 3–14).

FIGURE 6-13
The methane molecule. (a) Diagram showing the overlap of the four tetrahedral hybridized sp^3 orbitals with four s orbitals of four hydrogen atoms to produce the methane molecule. (b) The overall outline of the bonding orbitals in methane.

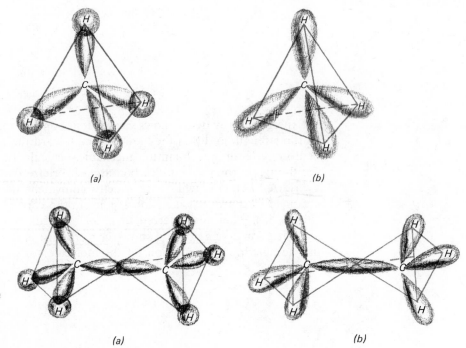

(a)   (b)

FIGURE 6-14
The ethane molecule. (a) The overlap diagram for the ethane molecule. (b) The overall outline of the bonding orbitals in ethane.

(a)   (b)

6.6   Beryllium Chloride, BeCl$_2$ (Digonal Hybridization)

The electronic configuration of Be is $1s^2 2s^2$, and it appears that in the normal, or ground, state the element should not form covalent bonds at all. However, in an excited state an electron in the $2s$ orbital can be promoted to a $2p$ orbital so that the configuration is $1s^2 2s^1 2p^1$, which means that two unpaired electrons are available for forming covalent bonds with atoms that can share electrons.

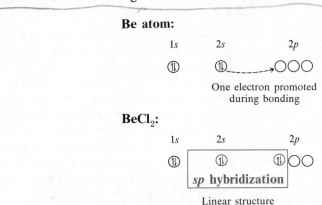

Be atom:

| $1s$ | $2s$ | $2p$ |

One electron promoted during bonding

BeCl$_2$:

| $1s$ | $2s$ | $2p$ |

sp hybridization

Linear structure

In the gaseous state, BeCl$_2$ is a linear molecule, Cl—Be—Cl (compare with Example 1, Section 6.3). All three atoms lie in a straight line and the two Be—Cl bonds have the same length and strength, so the two Be—Cl bonds are equivalent.

FIGURE 6-15
Hybridization of an *s* orbital and a *p* orbital (of the same atom) to produce two *sp* hybrid orbitals.

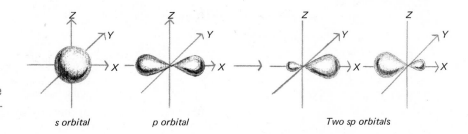

s orbital   p orbital   Two sp orbitals

It is assumed that the 2*s* and 2*p* atomic orbitals hybridize to form two *sp* hybrid orbitals. These *sp* orbitals overlap with *p* orbitals from the chlorine atoms to form sigma bonds. Because there are only two covalent bonds to the beryllium atom, and no lone pairs, the maximum angle separating the axes of the two *sp* hybrid atomic orbitals should be 180° (Figs. 6–1 and 6–15), which is the observed value. Because of the spatial orientation, the *sp* hybrid atomic orbitals are also referred to as **digonal hybrid orbitals** and always correspond to a straight-line linear structure.

6.7   Boron Trifluoride, BF₃ (Trigonal Hybridization)

Experimental evidence shows that there are three equivalent B—F bonds in boron trifluoride, BF_3. All four atoms of this molecule lie in the same plane, with the boron atom in the center and the three fluorine atoms at the corners of an equilateral triangle. The F—B—F bond angle, therefore, is 120° (see Fig. 6–16). The same is true for BCl_3 (see Fig. 6–3).

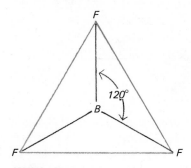

FIGURE 6-16
The trigonal planar structure of BF_3.

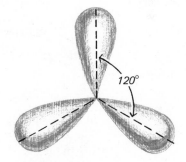

FIGURE 6-17
The shape and spatial orientation of trigonal planar, sp^2, hybrid orbitals.

The electronic structure of boron in the ground state is $1s^2 2s^2 2p^1$. During chemical reaction with fluorine (or chlorine), a 2*s* electron of boron is postulated as being promoted to a 2*p* orbital giving the structure $1s^2 2s^1 2p^1 2p^1$. The 2*s* and two of the 2*p* orbitals of boron hybridize to form three sp^2 hybrid orbitals. The sp^2 hybrid orbitals always have a triangular planar (also called trigonal planar) orientation (see Fig. 6–17 and top of next page).

B atom:

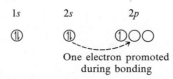

One electron promoted
during bonding

BF$_3$ molecule (also applies to BCl$_3$):

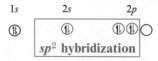

Trigonal planar structure

6.8   Ethylene, H$_2$C=CH$_2$ (Trigonal Hybridization)

As we discovered from reading Section 6.5, tetrahedral hybridization of carbon in methane and ethane is due to the mixing of a 2s orbital with three 2p orbitals of carbon. In ethylene a different situation arises, inasmuch as only two 2p orbitals are involved in the mixing with the 2s orbital, giving sp^2 hybridization as in BF$_3$ (see Section 6.7). This leaves one 2p orbital unhybridized. The sp^2 orbitals of carbon are pictured in Fig. 6–18. As in BF$_3$, the three sp^2 orbitals of carbon lie

FIGURE 6-18
Diagram illustrating three trigonal sp^2 hybridized orbitals of the carbon atom, which lie in the same plane, and the one unhybridized p orbital (shown in color), which is perpendicular to the plane.

in the same plane and are arranged so that the angles between them are 120°. The unhybridized p orbital (shown in color in Fig. 6–18) is perpendicular to the sp^2 plane.

When we bring the two sp^2 hybridized carbon atoms together in the formation of ethylene, it should be noted first (see Fig. 6–19) that the overlap of lobes of the two sp^2 orbitals end-to-end forms a strong sigma (σ) bond. Second, it should be noted that the unhybridized p orbitals are in a position to interact with one another, side-by-side. This kind of overlap results in a pi (π) bond of the type discussed in Chapter 5. The overlap to form the π bond is not as efficient as the overlap to form the σ bond and the π bond is therefore weaker than the σ bond. The two carbon atoms of ethylene are thus bound together by two kinds of bonds—one σ and one π bond. This kind of attachment is called a **double bond.** Note that in ethylene the four hydrogen atoms and two carbon atoms are all in the same plane.

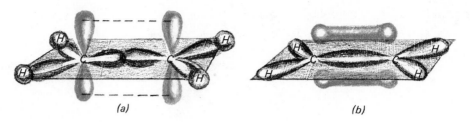

(a)   *(b)*

FIGURE 6-19

The ethylene molecule. (a) The overlap diagram of two sp^2 hybridized carbon atoms and four s orbitals from four hydrogen atoms. There are four σ C—H bonds, one σ C—C bond, and one π C—C bond (making the net carbon–carbon bond a double bond). The dashed lines, each connecting two lobes, indicate the side-by-side overlap of the two unhybridized p orbitals. The hybridized sp^2 orbitals are shown in grey, and the unhybridized p orbitals are shown in color. The hybridized sp^2 orbitals lie in a plane with the unhybridized p orbitals extending above and below the plane and perpendicular to it. (b) The overall outline of the bonding molecular orbitals in ethylene. The two portions of the π bonding orbital (shown in color), resulting from the side-by-side overlap of the unhybridized p orbitals, are above and below the plane.

6.9 Acetylene, H—C≡C—H (Digonal Hybridization)

We can now go one step further in the hybridization scheme. If just one $2p$ orbital mixes with the $2s$ orbital, we obtain two hybrid orbitals, designated as sp orbitals. This arrangement leaves two $2p$ orbitals unhybridized (Fig. 6–20).

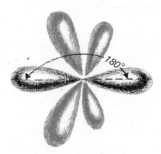

FIGURE 6-20

Diagram of the two linear, hybridized sp orbitals of the carbon atom, which lie in a straight line, and the two unhybridized p orbitals (shown in color).

When two sp hybridized orbitals of the carbon atoms react with one another, the two sp orbitals overlap end-to-end to form a strong σ bond (Fig. 6–21). In addition to this, as indicated in Fig. 6–21, the two sets of unhybridized p orbitals are in correct orientation to overlap side-by-side and hence to form two π bonds. The two carbon atoms are thus bound together by three bonds. This kind of attachment is referred to as a **triple bond.** Note that the two hydrogen atoms and the two carbon atoms are in a straight line—a linear structure.

(a)   *(b)*

FIGURE 6-21

The acetylene molecule. (a) The overlap diagram of two *sp* hybridized carbon atoms and two *s* orbitals from two hydrogen atoms. There are two σ C—H bonds, one σ C—C bond, and two π C—C bonds (making the net carbon–carbon bond a triple bond). The dashed lines, each connecting two lobes, indicate the side-by-side overlap of the four unhybridized *p* orbitals. The hybridized *sp* orbitals are shown in grey, and the unhybridized *p* orbitals are shown in color. (b) The overall outline of the bonding orbitals in acetylene. The π bonding orbitals (in color) are positioned with one above and below the line of the σ bonds and the other behind and in front of the line of the σ bonds.

ABs molecules

6.10   Phosphorus Pentachloride, PCl_5 (Trigonal Bipyramidal Hybridization)

In a molecule of phosphorus pentachloride, PCl_5, there are five P—Cl bonds and thus five pairs of valence electrons about the central phosphorus atom. To accommodate five pairs of electrons, either lone or bonding, an atom must have five atomic orbitals available. The outermost *s* and *p* orbitals of a phosphorus atom total only four. According to the Valence Bond Theory more than four orbitals can be made available by hybridization of these four orbitals with one or more *d* orbitals. The electronic structure of phosphorus in the ground state is $1s^2 2s^2 2p^6 3s^2 3p^1 3p^1 3p^1 3d^0$. A 3s electron is postulated as being promoted to the 3d orbital giving the structure $1s^2 2s^2 2p^6 3s^1 3p^1 3p^1 3p^1 3d^1$, thereby making five orbitals available for electron-pair sharing with five chlorine atoms. This is called *sp³d* hybridization.

sp³d

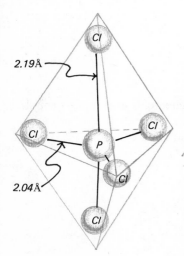

FIGURE 6-22

Structure of phosphorus pentachloride, PCl_5; sp^3d hybridization.

2.19Å

2.04Å

P atom:

1s	2s	2p	3s	3p	3d
⑪	⑪	⑪⑪⑪	⑪	①①①	○○○○○

One electron promoted
during bonding

PCl_5 molecule:

1s	2s	2p	3s	3p	3d
⑪	⑪	⑪⑪⑪	①	①①①	①○○○○

sp^3d **hybridization**

Trigonal bipyramidal structure

It is known that the PCl_5 molecule is a trigonal bipyramid (Fig. 6–22). Three of the chlorine atoms (equatorial) lie at the corners of an equilateral triangle, whereas the other two chlorine atoms (axial) lie above and below the center of the triangle. It is interesting to note that the two axial bonds are not equivalent geometrically to the equatorial bonds and also are longer than the equatorial bonds. In this respect the sp^3d hybrid orbitals differ from the hybrid orbitals discussed in Sections 6.5 through 6.9; sp, sp^2, and sp^3 orbital hybridization results in geometrically equivalent bonds. The sp^3d (or dsp^3) hybridization always corresponds to a trigonal bipyramidal structure.

6.11   Sulfur Hexafluoride, SF_6 (Octahedral Hybridization) AB_6 molecules

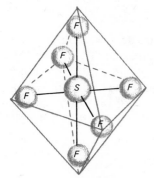

A molecule of sulfur hexafluoride has six fluorine atoms surrounding a single sulfur atom. To accommodate six pairs of electrons, in bonding with six fluorine atoms, the sulfur atom must provide six bonding orbitals. The electronic structure of sulfur in the ground state is $1s^2 2s^2 2p^6 3s^2 3p^2 3p^1 3p^1$. Promotion of an s and a p electron to d orbitals gives the structure $1s^2 2s^2 2p^6 3s^1 3p^1 3p^1 3p^1 3d^1 3d^1$. The one $3s$, three $3p$, and two $3d$ orbitals hybridize to form six hybridized sp^3d^2 orbitals. Each orbital points to the corner of an octahedron (Fig. 6–23), and hence the sp^3d^2 (or d^2sp^3) hybridization is called octahedral. The angle between any two bonds is 90°.

FIGURE 6-23
The octahedral configuration of SF_6; sp^3d^2 hybridization.

S atom:

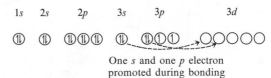

One s and one p electron promoted during bonding

SF_6 molecule:

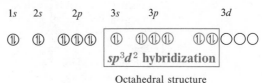

Octahedral structure

6.12   Summary

All of the examples for Valence Bond Theory given in this chapter are for molecules with regular covalent bonds in which each bonding pair of electrons consists of one electron from each of the two bonded atoms. It should be noted, however, that the geometric shapes of molecules with coordinate covalent bonds can also be explained by Valence Bond Theory. Examples are $[Ag(NH_3)_2]^+$ (linear), $[Zn(CN)_4]^{2-}$ (tetrahedral), $[Ni(CN)_4]^{2-}$ (square planar), and $[Co(NH_3)_6]^{3+}$ (octahedral). The geometry of such molecules will be discussed in the chapter on Coordination Compounds (Chapter 32). You will probably find it useful to look ahead at Table 32–1 in Chapter 32. The table gives the geometric shapes and corresponding hybridization of orbitals for several compounds and ions.

QUESTIONS

1. Predict the structure of each of the following molecules (or ions): CS_2, CH_2O, CCl_4, GeH_4, NF_2Cl, PF_5, NO_2^-, NO_3^-, OH_3^+, SO_4^{2-}, $AlCl_3$, PO_4^{3-}, AlF_6^{3-}, SO_3, SO_3^{2-}, SF_6, TeF_4, $TeCl_2$, BrF_3, ClO_2^-, BrF_4^-, $XeOF_4$, XeF_4, $XeOF_2$, XeO_3. (For the Xe compounds, consider Xe as the central atom.)

2. Define and illustrate the term "hybridization."

3. (a) Show the distribution of the valence electrons in the orbitals, just prior to bonding with other atoms by single covalent bonds, for (1) an sp^2 hybridized B atom; (2) an sp^3 hybridized Si atom; (3) an sp^3d hybridized As atom; (4) an sp^3d^2 hybridized Te atom.

   (b) Show the distribution of valence electrons in the orbitals for the central atom after covalent bonding takes place with other atoms for (1) BCl_3; (2) $SiCl_4$; (3) AsF_5; (4) $TeCl_6$.

4. Write valence electronic (Lewis) formulas for the molecules or ions listed below. Utilizing both the Valence Shell Electron-Pair Repulsion Theory and the Valence Bond Theory and remembering that single bonds are sigma bonds, double bonds are a sigma plus a pi bond, and triple bonds are a sigma plus two pi bonds, indicate the type of hybridization (sp, sp^2, sp^3) for each atom in the molecules and ions listed: BH_3, N_2, $SnCl_3^-$, SO_2, SO_3, NO_2, CO_2, NO, CO, CO_3^{2-}, SO_4^{2-}, NO_2^-, NO_3^-, HNO_3, H—O—O—H.

REFERENCES

"Hybrid Orbitals in Molecular Orbital Theory," I. Cohen and J. D. Bene, *J. Chem. Educ.*, **46**, 487 (1969).

"Simplified Molecular Orbital Approach to Inorganic Stereochemistry," R. M. Gavin, *J. Chem. Educ.*, **46**, 413 (1969).

"Size and Shape of a Molecule," M. J. Demchik and V. C. Demchik, *J. Chem. Educ.*, **48**, 770 (1971).

"Accurate Molecular Geometry," D. H. Whiffen, *Chem. in Britain*, **7**, 57 (1971).

"The Electron-Pair Repulsion Model for Molecular Geometry," R. J. Gillespie, *J. Chem. Educ.*, **47**, 18 (1970).

"A Defense of the Valence Shell Electron Pair Repulsion Theory (VSEPR) Model," R. J. Gillespie, *J. Chem. Educ.*, **51**, 367 (1974).

"A Simple Method of Generating Sets of Orthonormal Hybrid Atomic Orbitals," C.-Y. Hsu and M. Orchin, *J. Chem. Educ.*, **50**, 114 (1973).

The Relationships of the Periodic Classification to Properties of the Elements

7

At the present time, we know of 106 elements which have either been discovered in nature or made artificially (see Sections 3.19 and 30.12). In Chapter 3, it was pointed out that, though each element is different from every other element, similarities in electron structures make possible arrangements such as the Periodic Table that aid greatly in correlating the study of the properties and compounds of the elements. We shall now consider in more detail the characteristics which are important in determining the properties of the elements.

7.1 Variation of Properties within Periods and Groups of the Periodic Table

As we have mentioned previously, differences in the chemical properties of elements are caused primarily by differences in three characteristics: (a) the magnitude of the nuclear charge and the number of electrons in the shells surrounding the nucleus, both of which are equal to the atomic number, (b) the number of shells of electrons and the number of electrons in these shells, particularly in the valence shells, and (c) the distances of the electrons in the various shells from each other and from the nucleus.

■ **1. Variation in Atomic Radii.** (*See table inside back cover.*) In general, from left to right across the periods of the Periodic Table, each element has a smaller atomic radius (radius of the atom, assuming a spherical shape) than the one preceding it. The atomic radii of the noble gases at the ends of the periods, however, are larger than those of the elements of next lower atomic number; in the free state, the noble gases have completed outer shells.

Each element (except the first element, hydrogen) in the Periodic Table has a nuclear charge one higher than the preceding element. Each element also has one more electron than the preceding element, even though the number of shells is constant in each period. In general within a given period, the larger the nuclear charge and the number of electrons, the larger will be the force of electrostatic attraction between the nucleus and the electrons. This, in theory, causes the decrease in atomic radii across the period.

Atomic radii values are based upon measurements of interatomic distances in the solid state. In general, strong chemical bonds hold atoms close together in the solid state. However, with the noble gases, such as argon, only a weak electrostatic force of attraction holds the atoms together. This force, known as the **van der Waals force,** is the electrostatic attraction of the positive nucleus of one atom for the negative electron cloud of a neighboring atom (see Section 11.4). This results in abnormally large interatomic distances and atomic radii for these elements.

The gradation from highly metallic properties (of sodium, for example) to highly nonmetallic properties (of chlorine, for example) in the third period (and other comparable periods) is explainable in terms of the smaller size of the atom for each succeeding element in the period (see Table 7-1). The tendency for each

TABLE 7-1

Element:	Na	Mg	Al	Si	P	S	Cl	Ar
Atomic radius, Å	1.86	1.60	1.43	1.17	1.10	1.04	0.99	1.54
Nuclear charge	+11	+12	+13	+14	+15	+16	+17	+18
Electronic structure	2, 8, 1	2, 8, 2	2, 8, 3	2, 8, 4	2, 8, 5	2, 8, 6	2, 8, 7	2, 8, 8

succeeding element to act less like a metal results from the fact that the valence electrons are less readily lost as their distance from the positive nucleus becomes less, and the attraction of the nucleus for additional electrons from other atoms becomes greater.

Proceeding down the groups of the Periodic Table, succeeding elements have increasing atomic radii, as a result of larger numbers of electron shells (Table 7-2). These elements become more metallic (lose valence electrons more readily) as their atomic size and weight increase. This is due to the valence electrons being held less strongly with additional intervening electron shells, even though the charge on the nucleus is greater with each succeeding element down a group. Francium is the most active metal of Group IA and radium of Group IIA. Even though Groups IVA, VA, VIA, and VIIA are headed by the nonmetals carbon, nitrogen, oxygen, and fluorine respectively, this same trend toward metallicity down the groups is generally noted. Lead (Pb), bismuth (Bi), polonium (Po), and astatine (At) all exhibit metallic properties, at least to some degree.

TABLE 7-2

Element	Atomic Radius, Å	Nuclear Charge	Electronic Structure
Li	1.52	+3	2, 1
Na	1.86	+11	2, 8, 1
K	2.31	+19	2, 8, 8, 1
Rb	2.44	+37	2, 8, 18, 8, 1
Cs	2.62	+55	2, 8, 18, 18, 8, 1
Fr	2.7	+87	2, 8, 18, 32, 18, 8, 1

■ **2. Variation in Ionic Radii.** The radius of a positive ion is less than that of its parent atom. The loss of all the electrons from the outermost shell as the positive ion is formed results in a smaller radius for the system. Thus the radius of the sodium atom (2,8,1) is 1.86 Å, whereas that of the sodium ion, Na^+ (2,8,0), is 0.95 Å. Not only does the outer electron shell of the sodium atom disappear as the sodium ion is formed, but also the radii of the two remaining electron shells decrease because of an increase in *effective* nuclear charge as the valence electron is removed, giving rise to a greater average attraction of the nucleus per remaining electron. Down the groups of the Periodic Table, positive ions of succeeding elements have larger radii corresponding to larger numbers of electron shells (Table 7–3).

TABLE 7-3

Ion	Radius, Å	Nuclear Charge	Electronic Structure
Li^+	0.60	+3	2, 0
Na^+	0.95	+11	2, 8, 0
K^+	1.33	+19	2, 8, 8, 0
Rb^+	1.48	+37	2, 8, 18, 8, 0
Cs^+	1.69	+55	2, 8, 18, 18, 8, 0

A simple negative ion is formed by the addition of one or more electrons to the valence shell of an atom. This results in a greater force of repulsion among the electrons and also a decrease in the *effective* nuclear charge per electron. Both effects operate in the same direction to cause the radius of a negative ion to be greater than that of the parent atom. For example, the chlorine atom (2,8,7) has a radius of 0.99 Å, whereas that of the chloride ion (2,8,8) is 1.81 Å. For succeeding elements down the groups, negative ions have more electron shells, greater nuclear charge, and larger radii (Table 7–4).

Ions which have the same electron configuration, such as those in the series Na^+, Mg^{2+}, and Al^{3+}, and those in the series P^{3-}, S^{2-}, and Cl^-, are termed **isoelectric.** The greater the nuclear charge the smaller the ionic radius in a series of isoelectronic ions. This trend is illustrated in Table 7–5 for the ions of

TABLE 7-4

Ion	Radius, Å	Nuclear Charge	Electronic Structure
F⁻	1.36	+9	2, 8
Cl⁻	1.81	+17	2, 8, 8
Br⁻	1.95	+35	2, 8, 18, 8
I⁻	2.16	+53	2, 8, 18, 18, 8

TABLE 7-5

Ion:	Na^+	Mg^{2+}	Al^{3+}	Si^{4+}	P^{3-}	S^{2-}	Cl^-	Ar (Atom)
Radius, Å	0.95	0.65	0.50	0.41	2.12	1.84	1.81	1.54
Nuclear charge	+11	+12	+13	+14	+15	+16	+17	+18
Electronic structure	2, 8, 0	2, 8, 0	2, 8, 0	2, 8, 0	2, 8, 8	2, 8, 8	2, 8, 8	2, 8, 8

the elements of the third period. As we shall see later, many of the properties of ions can best be explained in terms of their sizes and charges.

■ **3. Variation in Ionization Potentials.** The amount of energy required to remove the most loosely bound electron from an atom is called its **ionization potential** or, more precisely, its **first ionization potential.** This change may be represented by

$$X + Energy \longrightarrow X^+ + e^-$$

In general, the greater the nuclear charge of atoms with the same number of electron shells, the greater the ionization potential (Table 7-6). This is true because the greater the nuclear charge, the greater the attraction of the nucleus for electrons. In general, therefore, succeeding elements across the periods have larger ionization potentials.

TABLE 7-6

Element:	Li	Be	B	C	N	O	F	Ne
Ionization potential (in volts)	5.39	9.32	8.30	11.26	14.54	13.61	17.42	21.60
Nuclear charge	+3	+4	+5	+6	+7	+8	+9	+10
Electronic structure	2, 1	2, 2	2, 3	2, 4	2, 5	2, 6	2, 7	2, 8

Figure 7-1 shows the relationships between first ionization potentials and atomic numbers of several elements. The values of the first ionization potentials are provided in Table 7-7. Note that the ionization potential of boron is less than that of beryllium. This is explained in terms of the relative attraction by the positive nucleus of an atom for electrons of different subshells. On the average,

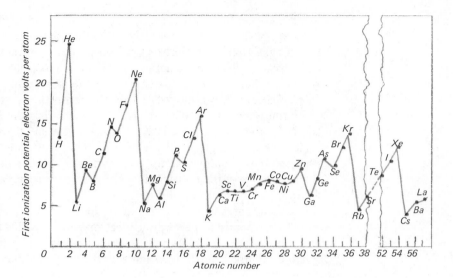

FIGURE 7-1

A graphic illustration of the periodic relationships between first ionization potentials and the atomic numbers for some of the elements.

TABLE 7-7   First Ionization Potentials of Some of the Elements
(Potentials are given in electron volts per atom)

1 H 13.6																	2 He 24.6
3 Li 5.4	4 Be 9.3											5 B 8.3	6 C 11.3	7 N 14.5	8 O 13.6	9 F 17.4	10 Ne 21.6
11 Na 5.1	12 Mg 7.6											13 Al 6.0	14 Si 8.1	15 P 11.0	16 S 10.4	17 Cl 13.0	18 Ar 15.8
19 K 4.3	20 Ca 6.1	21 Sc 6.6	22 Ti 6.8	23 V 6.7	24 Cr 6.8	25 Mn 7.4	26 Fe 7.9	27 Co 7.9	28 Ni 7.6	29 Cu 7.7	30 Zn 9.4	31 Ga 6.0	32 Ge 8.1	33 As 10	34 Se 9.8	35 Br 11.8	36 Kr 14.0
37 Rb 4.2	38 Sr 5.7	39 Y 6.6	40 Zr 7.0	41 Nb 6.8	42 Mo 7.2	43 Tc ...	44 Ru 7.5	45 Rh 7.7	46 Pd 8.3	47 Ag 7.6	48 Cd 9.0	49 In 5.8	50 Sn 7.3	51 Sb 8.6	52 Te 9.0	53 I 10.4	54 Xe 12.1
55 Cs 3.9	56 Ba 5.2	[57-71] *	72 Hf 5.5	73 Ta 6	74 W 8.0	75 Re 7.9	76 Os 8.7	77 Ir 9.2	78 Pt 9.0	79 Au 9.2	80 Hg 10.4	81 Tl 6.1	82 Pb 7.4	83 Bi 8	84 Po ...	85 At ...	86 Rn 10.7
87 Fr ...	88 Ra 5.3	[89-103] †															

* LANTHANIDE SERIES	57 La 5.6	58 Ce 6.9	59 Pr 5.8	60 Nd 6.3	61 Pm ...	62 Sm 5.6	63 Eu 5.7	64 Gd 6.2	65 Tb 6.7	66 Dy 6.8	67 Ho ...	68 Er ...	69 Tm ...	70 Yb 6.2	71 Lu 5.0
† ACTINIDE SERIES	89 Ac ...	90 Th ...	91 Pa ...	92 U 4	93 Np ...	94 Pu ...	95 Am ...	96 Cm ...	97 Bk ...	98 Cf ...	99 Es ...	100 Fm ...	101 Md ...	102 No ...	103 Lr ...

an s electron will be attracted to the nucleus more tightly than a p electron of the same principal shell, a p electron more closely than a d electron, and so on. This means that an s electron will be harder to remove from an atom than a p electron, a p electron harder to remove than a d electron, and a d electron harder to remove than an f electron, and hence the ionization potentials decrease in this order. The electron removed during the ionization of beryllium ($1s^2, 2s^2$) is an s electron whereas a p electron is removed during the ionization of boron ($1s^2, 2s^2, 2p^1$); this results in a lower ionization potential for boron even though its nuclear charge is greater by one unit. The ionization potential of nitrogen is abnormally high and that for oxygen slightly lower than for nitrogen because of the stability of the half-filled $2p$ subshell in nitrogen (Section 3.17). Analogous changes occur in succeeding periods.

The values given in Fig. 7–1 and Tables 7–6 and 7–7 are for the energy needed to remove the most loosely bound electron and are called **first ionization potentials.** The energy required to remove a second electron is called the **second ionization potential.** Third, fourth, etc., ionization potentials may be defined in a similar fashion.

Metallic elements have small ionization potentials and nonmetals large ones. This means that the metals tend to form positive ions readily by the loss of valence electrons and that the nonmetals do not tend to form positive ions during chemical changes.

The attractive force exerted by the positively charged nucleus on the valence electrons is partially counterbalanced by the repulsive forces of electrons in inner shells on each other and on the valence electrons. An electron being removed from an atom is thus shielded from the nucleus by these inner shells. This shielding, and the increasing distance of the outer electron from the nucleus, is the apparent explanation of the fact that down the groups succeeding elements have smaller ionization potentials. Members of a group as a rule have the same outer electronic configuration and the same number of valence electrons (Table 7–8).

TABLE 7-8

Element	First Ionization Potential	Nuclear Charge	Electronic Structure
Li	5.39	+3	2, 1
Na	5.14	+11	2, 8, 1
K	4.34	+19	2, 8, 8, 1
Rb	4.18	+37	2, 8, 18, 8, 1
Cs	3.89	+55	2, 8, 18, 18, 8, 1

The size of the ionization potential is to some degree a measure of the chemical activity of a metal, i.e., its tendency to form positive ions by losing valence electrons. In general, the smaller the ionization potential the more active the metal. With the possible exception of francium, cesium has the smallest ionization potential and is the most active metal.

■ **4. Variation in Electron Affinities.** Just as the ionization potential is a measure of the energy required to remove an electron from an atom to form a positive ion, another quantity, the **electron affinity,** is a measure of the energy released when an extra electron is added to an atom to form a negative ion. The change is expressed by the equation

$$X + e^- \longrightarrow X^- + \text{Energy}$$

Metal atoms have little tendency to form negative ions by gaining extra electrons; thus their electron affinities are very small (Table 7–9). Nonmetallic elements,

TABLE 7–9   Electron Affinities of Some Elements in Electron Volts Per Atom

H 0.75							He 0
Li 0.54	Be −0.6	B 0.3	C 1.1	N −0.1	O 1.47	F 3.45	Ne 0
Na 0.74	Mg −0.3	Al 0.4	Si 1.9	P 0.8	S 2.07	Cl 3.61	Ar 0
						Br 3.36	Kr 0
						I 3.06	Xe 0

on the other hand, have large electron affinities and, inasmuch as they easily add electrons, are good oxidizing agents (see Part 6 of this Section).

Generally, across the periods of the Periodic Table, succeeding elements have higher electron affinities, and nonmetallic character increases. In the second period, however, one will note exceptions for beryllium (filled 2*s* subshell), nitrogen (half-filled 2*p* subshell), and neon (all subshells filled), cases in which completely filled or half-filled subshells represent stable configurations.

■ **5. Variation in Electronegativities.** The term electronegativity was defined in Section 4.6 as a measure of the attraction of an atom for the electrons in its outer shell. Values for the electronegativities of many of the elements are provided in Table 4–2. The electronegativity of an atom is related in a general way to both its ionization potential and electron affinity. In fact, it has been suggested by R. S. Mulliken, a chemical physicist, that the average of the ionization potential and the electron affinity of an atom should be a suitable measure of its electronegativity. The most electronegative elements are found toward the end of the periods; they are the nonmetals, with large ionization potentials and large electron affinities. The most electropositive elements, those with low electronegativities (small ionization potentials and nearly zero electron affinities), are found at the beginnings of the periods. These are the alkali metals and alkaline earth metals. The elements near the middle of the periods have electronegativities which are intermediate in value. As we go down a group, succeeding elements are less electronegative (see Table 4–2).

When the difference between the electronegativities of two atoms involved in a bond is large, the bond is likely to be ionic in character. Such is the case with LiF, NaCl, K_2O, Li_3N, and CsBr, where the electronegativity difference is 2 or more. Bonds between atoms with small differences in electronegativities are primarily covalent in character. This is the case with CO_2, CCl_4, I_2O_5, NI_3, ICl, NO, and BN. It follows that bonds which are primarily ionic are formed when elements at the extreme left of the table react with elements at the extreme right of the table, whereas bonds which are primarily covalent are formed when elements close together in the table react with one another.

■ **6. Variation in Strength as Oxidizing and Reducing Agents.** Oxidation is defined as **the tendency to lose electrons.** Substances referred to as **oxidizing agents** are characterized by their tendency to gain electrons from other atoms and hence to encourage the oxidation (or loss of electrons) of the other atoms. The substance which is oxidized, by losing electrons, goes to a higher oxidation number; the oxidizing agent, by gaining electrons, goes to a lower oxidation number. The nonmetals at the extreme right of the Periodic Table, with their relatively high ionization potentials, electron affinities, and electronegativities, tend to act as oxidizing agents when they combine with other substances.

Reduction is defined as **the tendency to gain electrons; reducing agents give up electrons to other atoms** and hence cause the other atoms to undergo reduction. When an atom undergoes reduction (by gaining electrons), it goes to a lower oxidation number; the reducing agent, in losing electrons, goes to a higher oxidation number and is itself, therefore, oxidized. Reducing agents, in giving up electrons, go into higher oxidation states when they enter into chemical reactions. The reducing power is highest for the metals at the beginnings of the periods, where the ionization potentials, electron affinities, and electronegativities are low. In general, the reducing strength of the elements is progressively lower as we move across the periods and higher as we move down the groups. Thus francium (Fr) is the strongest reducing agent and fluorine the strongest oxidizing agent of all the elements.

Oxidation and reduction always occur simultaneously, for if one atom gains electrons (reduction) another atom must be present to provide those electrons by losing them (oxidation). Reactions involving oxidation and reduction are referred to as oxidation-reduction reactions and will be discussed in greater detail in Chapter 16.

■ **7. Variation in the Ionization of Hydroxyl Compounds.** Hydroxyl compounds of elements at the beginnings of the periods consist of metal ions and hydroxide ions. For example, NaOH consists of Na^+ and OH^- ions; calcium hydroxide consists of Ca^{2+} and OH^- ions. Such ionic hydroxyl compounds when dissolved in water provide hydroxide ions to the solution and are referred to as **bases;** they are said to show basic character. The basic character of hydroxyl compounds of elements which form large positive ions of low charge is particularly pronounced. On the other hand, hydroxyl compounds of elements at the nonmetal ends of the periods ionize to provide hydrogen ions to the solution and are referred to as **acids.** This is the case, for example, with hydroxyl compounds of chlorine and nitrogen.

$$HOCl \longrightarrow H^+ + OCl^-$$
$$HONO_2 \longrightarrow H^+ + ONO_2^-$$

The acid or base character of hydroxyl compounds can be postulated as arising because of the different electronegativities of the various elements. In general, an element, E, can be thought of as being associated with one or more hydroxyl groups in a hydroxyl compound (see formula following). If E has a relatively low electronegativity, its attraction for electrons is low, little tendency exists for it to form a strong covalent bond with the oxygen atom, and the bond between the element and oxygen is weaker than that between oxygen and hydrogen. Hence, the bond at a is ionic, hydroxyl ions are released to the solution, and the material behaves as a base. Large size, small nuclear charge, and low oxidation number are factors which operate in the direction of low electronegativity for E and are characteristic of the more metallic elements.

$$E \: \overset{a}{\overset{..}{\underset{..}{\text{O}}}} \: \overset{b}{\text{H}}$$

If, on the other hand, the element E has a relatively high electronegativity, it attracts the electrons which are available for sharing between it and the oxygen atom rather tightly, giving rise to a relatively strong bond between the element E and the oxygen atom. The oxygen-hydrogen bond is thereby weakened through the displacement of electrons toward E, the bond at b is ionic, releasing hydrogen ions to the solution, and the material behaves as an acid. Small size, large nuclear charge, and high oxidation number operate in the direction of high electronegativity and are characteristic of the more nonmetallic elements.

The hydroxides of intermediate elements near the heavy diagonal line separating the metals from the nonmetals in the Periodic Table are usually **amphoteric.** This means that the hydroxides act as acids toward strong bases and as bases toward strong acids.

$$Al(OH)_3 + OH^- \longrightarrow [Al(OH)_4]^-$$
$$Al(OH)_3 + 3H^+ \longrightarrow Al^{3+} + 3H_2O$$

The amphoterism of aluminum hydroxide is reflected in its solubility in both strong acids and strong bases. In strong bases, the relatively insoluble $Al(OH)_3$ is converted to the soluble $[Al(OH)_4]^-$ ion by reaction with hydroxide ion; in strong acids, the $Al(OH)_3$ is converted to the soluble Al^{3+} ion by reaction with hydrogen ion.

Acids and bases will be discussed in Chapter 15 in a much broader sense than only as compounds contributing hydrogen ions and hydroxyl ions to solutions.

7.2   Uses of the Periodic Table

The systematic arrangement of the elements found in the Periodic Table has several applications. Most important of these are (1) the classification of the elements, (2) the prediction of undiscovered elements and their properties, and (3) the stimulation of research.

■ **1. Classification of the Elements.**   The classification of elements into groups with similar properties simplifies their study. If, for example, students learn the properties of sodium of Periodic Group IA, they will also know many of the properties of the other alkali metals lithium, potassium, rubidium, cesium, and francium. Sodium is an active metal that reacts vigorously with water giving hydrogen gas and forming sodium hydroxide, a strong base. The other alkali metals react with water in a similar fashion. Chlorine, of Periodic Group VIIA, reacts vigorously with sodium, forming sodium chloride (a salt), and in fact with all the members of Group IA. Fluorine, bromine, and iodine also form salts with the Group IA metals.

The fact that elements within a group have similar properties should not be overemphasized, however, As mentioned earlier in the chapter, there is a gradation in properties of the elements down the groups so that certain properties for some elements may be quite different from those of other elements in the same group. For example, lithium reacts with water more slowly than does sodium, whereas potassium, rubidium, and cesium react much more vigorously. Chlorine forms a series of oxyacids ($HClO$, $HClO_2$, $HClO_3$, and $HClO_4$) and the corresponding sodium salts ($NaClO$, $NaClO_2$, $NaClO_3$, and $NaClO_4$). On the other hand, no oxyacids or oxysalts of fluorine have been prepared, a difference in properties which is related to the fact that fluorine is more electronegative than oxygen whereas chlorine and the other elements of Group VIIA are less electronegative than oxygen (see Table 4-2). Nitrogen, the first member of Group VA, is entirely nonmetallic in character whereas bismuth, the last member of this group, is mostly metallic in its properties.

Periodicity as regards valence electrons is reflected in periodicity of oxidation numbers (Table 7-10). Atoms of hydrogen, lithium, and sodium have one valence electron each, and each atom shows an oxidation number of $+1$ when this electron is transferred to another atom in the formation of chemical bonds. Carbon and silicon atoms with four valence electrons can form four polar covalent bonds, with the shared electron pairs being shifted away ($+4$ oxidation number) or

TABLE 7-10   Periodicity of Oxidation Numbers

$\overset{+1}{H}$							$\overset{0}{He}$
$\overset{+1}{Li}$	$\overset{+2}{Be}$	$\overset{+3}{B}$	$\overset{+4,-4}{C}$	$\overset{+5,-3}{N}$	$\overset{-2}{O}$	$\overset{-1}{F}$	$\overset{0}{Ne}$
$\overset{+1}{Na}$	$\overset{+2}{Mg}$	$\overset{+3}{Al}$	$\overset{+4,-4}{Si}$	$\overset{+5,-3}{P}$	$\overset{-2}{S}$	$\overset{-1}{Cl}$	$\overset{0}{Ar}$

towards (-4 oxidation number) the carbon or silicon atom. Fluorine and chlorine atoms, each with seven valence electrons, assume completed octets by the addition of one electron, giving them -1 oxidation numbers. It should be noted, however, that the Periodic Table emphasizes at most only the maximum and minimum oxidation numbers. Many of the elements exhibit additional intermediate oxidation numbers. For example, nitrogen of Group VA with five valence electrons (the same number as that of the group) forms N_2O_5, in which nitrogen has an oxidation number of $+5$. Because nitrogen lacks three electrons for a complete outer shell, it also forms NH_3, in which the nitrogen has an oxidation number of -3. In addition, it exhibits oxidation numbers of $+4$ in NO_2, $+3$ in N_2O_3, $+2$ in NO, $+1$ in N_2O, zero in N_2, -1 in NH_2OH, and -2 in N_2H_4. Furthermore, the number of the periodic group in which the element is located often reflects a relatively minor oxidation number of an element. Thus, copper of Group IB forms a large and important series of compounds in which it is divalent (oxidation number $+2$), whereas the compounds in which it is monovalent (oxidation number $+1$) are relatively few in number.

■ **2. Prediction of Undiscovered Elements.** A very remarkable use of the Periodic Table was made by Mendeleev. His table included only sixty-two elements, the number known at that time. In order to place related elements in the same group, he left gaps in his table and theorized that elements would be discovered later to fill these vacant places. He predicted the existence and properties of six elements corresponding to vacant places in his table. These elements have since been discovered; they are scandium (Sc), gallium (Ga), germanium (Ge), technetium (Tc), rhenium (Re), and polonium (Po), and they all have properties similar to those predicted by Mendeleev.

A comparison of the properties predicted by Mendeleev for the substance we know as germanium, which he called eka-silicon, and those determined experimentally for the element is given below.

Predicted Properties for Eka-silicon (1871)	*Observed Properties of Germanium (discovered in 1886)*
Atomic weight, 72	Atomic weight, 72.60
Specific gravity, 5.5	Specific gravity, 5.36
A gray colored metal	A grayish-white metal
Valence of 4 toward oxygen	Valence of 4 toward oxygen
EsO_2, a white solid of sp. gr. 4.7, and high m.p.	GeO_2, a white solid of sp. gr. 4.70, m.p. 1100° C
$EsCl_4$, a volatile liquid, b.p. below 100°, sp. gr. 1.9	$GeCl_4$, a volatile liquid, b.p. 83°, and sp. gr. 1.88
Es will be acted upon by acids only slightly, and not at all by alkalis, like NaOH	Ge does not react with HCl or NaOH, but dissolves in concentrated HNO_3

■ **3. Stimulation of Research.** The Periodic Table has stimulated chemists and physicists to do extensive research. Forty-four elements have been discovered since Mendeleev's original periodic classification of the elements. Before Moseley's assignment of atomic numbers to the elements (Section 3.7), much work was done to determine accurate atomic weights, which were used to place the elements in their positions in the table. For example, the element indium (In) was thought to have an oxidation number of +2, and, on the basis of this and a quantity known as the equivalent weight, which had been found to be about 38 for indium, its atomic weight was calculated as about 2×38, or 76. However, an atomic weight of 76 would place indium between arsenic (74.9) and selenium (79.0) where it obviously did not belong. There was, however, a vacant place in the table, at that time, between cadmium (112.4) and tin (118.7). The general properties of indium were such that it fitted in this Group III position much better than between arsenic and selenium, which are in Groups V and VI, respectively. Later investigations showed indium to have an oxidation number of +3 instead of +2. This meant that its atomic weight would be 3×38, or about 114, instead of 2×38, or 76. Consequently, indium was placed in its logical position in the Periodic Table.

The discovery of new compounds has often been accomplished because of predictions based on the Periodic Table relationships. Notable examples are the freon refrigerants (such as CCl_2F_2) and the silicon-containing organic compounds (such as the silicones) because it was possible to predict that such compounds would have the properties sought (see Sections 11.5, 21.6, and 29.20).

QUESTIONS

1. Give the meaning of the following terms as applied to the Periodic Table: period, group, short period, and long period.
2. State the Periodic Law. How can the Periodic Law be accounted for by use of the theory of atomic structure?
3. Why is the radius of a positive ion less than that of its parent atom?
4. Why do negative ions have larger radii than their parent atoms?
5. From its position in the Periodic Table predict the maximum positive oxidation state to be expected for each of the following: gallium, germanium, cesium, astatine, technetium, and vanadium.
6. Account for the decrease in radius in the following series of isoelectronic ions: Na^+, Mg^{2+}, and Al^{3+}.
7. Arrange the following species in order of increasing size: the potassium ion, the calcium ion, the chloride ion, the scandium ion, and the argon atom.
8. Relate metallicity to ionization potential, electron affinity, and electronegativity.
9. Account for the positions of argon, cobalt, tellurium, and thorium in the Periodic Table in view of the fact that their positions are not in keeping with their atomic weights.
10. (a) State which you would expect to have the higher ionization potential, and explain why: (1) potassium or calcium; (2) magnesium or barium; (3) beryllium or boron; and (4) phosphorus or sulfur.
    (b) Which would you expect to be more basic, $Ca(OH)_2$ or $Mg(OH)_2$? Why?

(c) Which would you expect to be more acidic, KOH or HOCl? Why?

(d) Which would you expect to be more acidic, HOCl or HOI? Why?

(e) Which would you expect to have the higher electron affinity, sulfur or selenium? Why?

(f) Which would you expect to be more electronegative, oxygen or phosphorus? Why?

11. From their positions in the Periodic Table predict which will be (a) the more nonmetallic, iodine or bromine; (b) the more acidic, the hydroxide of aluminum or that of silicon; (c) the stronger reducing agent, aluminum or iron; (d) the stronger reducing agent, lithium or boron; (e) the more electronegative, carbon or nitrogen; and (f) the larger in ionic radius, fluoride ion or chloride ion.

12. From their positions in the Periodic Table predict which will be (a) the stronger oxidizing agent, elemental oxygen or elemental fluorine; (b) the larger in atomic radius, potassium or calcium; (c) the more electropositive, magnesium or calcium; (d) the larger in ionic radius, potassium ion or rubidium ion; (e) the more metallic, germanium or arsenic; and (f) the stronger oxidizing agent, bromine or iodine.

13. A quick look at the Periodic Table shows that most elements are metals. Why do you suppose that historically the nonmetals have received a disproportionately large amount of attention from chemists?

14. How would you expect the magnitude of the second ionization potential for magnesium to compare with that of the first ionization potential for sodium? Explain your answer.

15. Summarize several important uses of the Periodic Table.

16. Decide on logical locations in the Periodic Table for elements 104, 105, 106, 107 and justify your decision.

REFERENCES

"An Electronegativity Spectrum for the Periodic Table," W. B. Guenther, *J. Chem. Educ.*, **47**, 317 (1970).

"A Pattern of Chemistry. Hundred Years of Periodic Table," F. Greenaway, *Chem. in Britain*, **5**, 97 (1969).

"Criteria of Atomic Size," J. L. Nettleship, *Educ. in Chemistry*, **2**, 241 (1965).

"American Forerunners of the Periodic Law," G. B. Kauffman, *J. Chem. Educ.*, **46**, 128 (1969).

"The Priority Conflict Between Mendeleev and Meyer," J. W. van Spronsen, *J. Chem. Educ.*, **46**, 136 (1969).

"Periodic Variation in the Abundance of Elements," R. W. Jotham, *Educ. in Chem.*, **8**, 60 (1971).

"Critical Temperature of Elements and the Periodic System," A. L. Horvath, *J. Chem. Educ.*, **50**, 335 (1973).

"Electron Affinity. The Zeroth Ionization Potential," D. W. Brooks, E. A. Meyers, F. Sicilio, and J. C. Nearing, *J. Chem. Educ.*, **50**, 487 (1973).

"Regularities and Relations Among Ionization Potentials of Nontransition Elements," J. F. Liebman, *J. Chem. Educ.*, **50**, 831 (1973).

Oxygen (O$_2$) and Ozone (O$_3$)

8

Oxygen

A study of the chemical elements appropriately begins with oxygen as it is the most abundant of all the elements (see Section 1.9). It is essential to the processes of respiration in most plants and animals and the combustion of fuels and other substances. The discovery of oxygen marked the beginning of modern chemistry.

8.1 History and Occurrence

Credit for the discovery of oxygen is usually given to Joseph Priestley, an English clergyman and scientist, who in 1774 prepared oxygen by focusing the sun's rays upon mercury(II) oxide by means of a lens. He tested the gaseous product with a burning candle and noted that the candle burned more brightly than in ordinary air. Shortly thereafter, Antoine-Laurent Lavoisier correctly interpreted the role played by oxygen in the process of combustion.

Oxygen is the most abundant and widely distributed of the terrestrial elements. It forms about 23 per cent of the air as the free element, 89 per cent of water, in which it is combined with hydrogen, and 50 per cent of the earth's crust, by weight. About 90 per cent of the volume of the earth's crust is occupied by oxygen, combined with other elements, principally silicon. In combination with carbon, hydrogen, and nitrogen, oxygen constitutes a large part of the weight of the bodies of plants and animals.

8.2   Preparation of Oxygen

Oxygen may be prepared from air or from certain oxygen-containing compounds. Because of the abundance of air and water, their availability and cheapness, and the simplicity of the processes involving their use, nearly all commercial oxygen is obtained from these two sources. Approximately 97 per cent is produced from air and 3 per cent by the electrolysis of water. Total production of oxygen in 1974 amounted to 16 million tons, third in production for all chemicals. Approximately 70% of all oxygen produced is typically used by the steel industry in processes to be described in a later chapter.

■ **1. Preparation by Fractional Evaporation of Liquid Air.**   Commercial quantities of oxygen are obtained from air by first cooling and compressing air until it liquefies and then evaporating off the lower boiling nitrogen and some other elements (see Sections 24.2 and 25.2). As a liquid, oxygen is stored and shipped in Dewar flasks (see Fig. 24–2) of various sizes. These flasks are constructed so as to be self-refrigerating by the evaporation of some of the oxygen. Much commercial oxygen, however, is stored and shipped as a compressed gas in steel cylinders. Liquid oxygen is one of the propellant components for today's rockets (see Section 25.28).

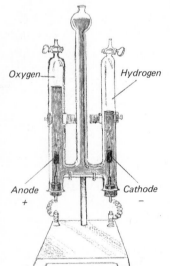

Oxygen ——

Hydrogen

Anode
+

Cathode
−

FIGURE 8-1
Laboratory apparatus for the electrolysis of water.

■ **2. Preparation from Water by Electrolysis.**   Pure water is a very poor conductor of electricity; but when a small amount of an acid, base, or salt is dissolved in water, the resulting solution readily conducts an electric current. Acids, bases, and salts are three important classes of compounds, called **electrolytes,** whose solutions carry an electric current. When a current of electricity is passed through a solution of an electrolyte, the ions of the electrolyte are the agencies that carry the current; the ions migrate toward the two electrodes, the positive ions **(cations)** moving toward the negative electrode **(cathode)** and the negative ions **(anions)** moving toward the positive electrode **(anode).** The process is called **electrolysis** (see Chapter 22). When an electric current is passed through water containing a small amount of an electrolyte such as H_2SO_4, NaOH, or Na_2SO_4, bubbles of hydrogen are formed at the cathode and oxygen is evolved at the anode (Fig. 8–1). The volume of hydrogen produced is twice that of the oxygen. The net reaction can be summarized by the equation

$$2H_2O + \text{Electrical energy} \longrightarrow 2H_2\uparrow + O_2\uparrow$$

It must be borne in mind that the equation only shows what products are formed; it does not suggest a mechanism. Actually, as we shall see in Section 22.5, the reactions taking place at the electrodes are more complex than suggested by this simple equation. The use of electrolysis for the industrial production of oxygen is limited by the high cost of electricity. Energy, in the form of electrical energy, is involved in this reaction, and it is important to note that the same amount of energy is liberated as heat and light if the hydrogen and oxygen produced by electrolysis are recombined by combustion to form water.

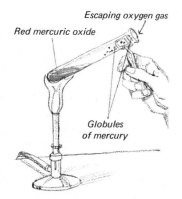

FIGURE 8-2
When mercury(II) oxide is strongly heated it decomposes, yielding gaseous oxygen and liquid mercury.

■ 3. Preparation by Heating Metal Oxides.

The oxides of mercury, silver, gold, and platinum lose oxygen when heated. For example, when red mercury(II) oxide is heated (Fig. 8-2), metallic mercury and oxygen are formed.

$$2HgO \underset{\text{Mercury(II) oxide}}{} + \triangle \longrightarrow 2Hg + O_2$$

This is the method used by Priestley and Lavoisier to produce oxygen. It is too expensive to be used commercially and is important only for its historic interest. Equations similar to that for the thermal decomposition of mercury(II) oxide can be written for silver oxide, Ag_2O; the oxides of gold, Au_2O and Au_2O_3; and the oxides of platinum, PtO and PtO_2.

When certain metal oxides are heated, only a part of their oxygen is liberated.

$$2BaO_2 \underset{\text{Barium peroxide}}{} + \triangle \longrightarrow 2BaO \underset{\text{Barium oxide}}{} + O_2$$

■ 4. Preparation by Heating Certain Salts Which Contain Oxygen.

The nitrate salts of certain metals yield oxygen upon being heated.

$$2NaNO_3 \underset{\text{Sodium nitrate}}{} + \triangle \longrightarrow 2NaNO_2 \underset{\text{Sodium nitrite}}{} + O_2$$

$$2Cu(NO_3)_2 \underset{\text{Copper(II) nitrate}}{} + \triangle \longrightarrow 2CuO \underset{\text{Copper(II) oxide}}{} + 4NO_2 \underset{\text{Nitrogen dioxide}}{} + O_2$$

Oxygen is often prepared in the laboratory on a small scale (Fig. 8-3) by heating potassium chlorate to about 50° above its melting point of 368.4°. It should be noted that when sodium nitrate is heated, only part of the oxygen is lost, but when potassium chlorate is heated, all the oxygen escapes. If manganese dioxide is mixed with the chlorate, the latter decomposes quite rapidly at about 270°, nearly 100° below its melting point.

FIGURE 8-3
In the laboratory, oxygen is usually prepared as shown. The manganese dioxide catalyst and potassium chloride remain after the potassium chlorate has decomposed.

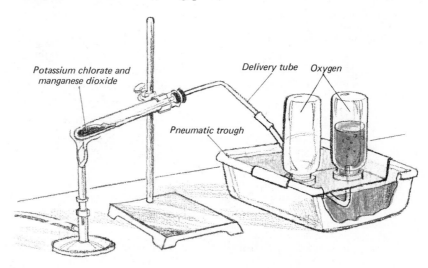

$$2KClO_3 + \triangle \xrightarrow{\text{MnO}_2} 2KCl + 3O_2$$

Potassium   Potassium
chloride   chloride

The manganese dioxide may be reclaimed chemically unchanged after the reaction is completed; it has served to "catalyze" the reaction by causing it to take place more rapidly at a lower temperature. Many other metal oxides, such as Fe_2O_3 and Cr_2O_3, also serve as catalysts for the decomposition of potassium chlorate. **A catalyst is a substance that changes the rate of a reaction without undergoing a permanent chemical change itself.** Catalysts are usually specific in their action; a substance that will catalyze one reaction is often without effect upon another. Catalysis is discussed in more detail in Chapter 17 (Section 17.6).

The preparation of oxygen by heating potassium chlorate can be very dangerous when done carelessly. Explosions can occur when combustible materials such as carbon, sulfur, rubber, or a glowing wood splint come in contact with fused (melted) potassium chlorate. The danger involved in this experiment cannot be overemphasized. Proceed with caution when preparing oxygen by this method.

■ **5. Preparation by the Action of Water upon Sodium Peroxide.** A convenient but expensive method for the preparation of oxygen involves the action of water upon sodium peroxide. Sodium peroxide is a white solid formed by burning sodium in an excess of oxygen or air.

$$2Na_2O_2 + 2H_2O \longrightarrow 4Na^+ + 4OH^- + O_2\uparrow$$

Sodium peroxide   Sodium hydroxide

The fact that the sodium hydroxide is present in the water solution as independent ions is indicated by designating it as $Na^+ + OH^-$ rather than by the formula NaOH. Evaporation of the solution to dryness would cause the ions to associate and form solid sodium hydroxide (usually written as NaOH).

8.3   Physical Properties

Oxygen is a colorless, odorless, and tasteless gas at ordinary temperatures. It is slightly more dense than air; one liter of oxygen measured at 0° and 760 mm weighs 1.429 g, whereas a liter of air under the same conditions weighs 1.292 g. Although oxygen is only slightly soluble in water (30 ml of gas in one liter of water at 20°) its solubility is very important to marine life and in the decomposition of sewage recklessly dumped into streams and lakes. Oxygen is pale blue in the liquid state and boils at −183° at atmospheric pressure. Solid oxygen, also pale blue, melts at −218.4°. The oxygen molecule is diatomic (O_2) (Fig. 8-4) and it is **paramagnetic,** i.e., attracted by a magnetic field, particularly when in the solid or liquid state. The paramagnetism is an indication that the diatomic molecule contains one or more unpaired electrons (see Section 5.10).

Recalling that the oxygen atom possesses six valence electrons, we might expect the sharing of two pairs of electrons between the two atoms in the diatomic molecule with all electrons paired. The degree of paramagnetic character of the diatomic oxygen molecule, however, indicates *two unpaired electrons* in the struc-

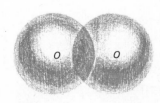

FIGURE 8-4

A representation of the oxygen molecule, O_2. Each sphere represents an oxygen atom.

ture, and, as we have seen in Section 5.10, the oxygen molecule does have two unpaired electrons, one each in the two $\pi_p{}^*$ antibonding molecular orbitals.

8.4   Chemical Properties

Oxygen is a chemically active element that combines either directly or indirectly with nearly all the other elements. The product formed when an element combines with oxygen is called an **oxide.** In combining with oxygen, the element increases in oxidation number and loses electrons, thereby undergoing oxidation (see Section 7.1, Part 6). The oxygen receives the electrons and hence is the oxidizing agent, decreases in oxidation number from zero to -2, and is itself reduced.

■ **1. Action with Metals.**   The metals down to and including copper in the activity series (see Section 9.9) combine directly with oxygen in the air to form oxides. In most cases, the action is slow at ordinary temperatures but many of the more active metals burn readily when heated in the presence of pure oxygen.

$$2Na + O_2 \longrightarrow Na_2O_2 \text{ (sodium peroxide)}$$
$$2Ca + O_2 \longrightarrow 2CaO \text{ (calcium oxide)}$$
$$2Mg + O_2 \longrightarrow 2MgO \text{ (magnesium oxide)}$$
$$4Al + 3O_2 \longrightarrow 2Al_2O_3 \text{ (aluminum oxide)}$$
$$3Fe + 2O_2 \longrightarrow Fe_3O_4 \text{ (iron(II, III) oxide)}$$

The oxide formed by burning iron in pure oxygen (Fe_3O_4) is different from the one produced when iron rusts (Fe_2O_3). In Fe_3O_4, one atom of iron can be thought of as having an oxidation number of $+2$ and the other two atoms of iron an oxidation number of $+3$ each. The two different oxidation numbers for iron in Fe_3O_4, which may also be represented as $FeO \cdot Fe_2O_3$, are indicated in the name iron(II, III) oxide. However one cannot distinguish the Fe(II) atoms from the Fe(III) atoms due to the rapidity of electron exchange between them.

Many oxides of metals dissolve in water to form aqueous solutions of hydroxide bases and hence are often called **basic oxides,** or **basic anhydrides.** The solutions turn red litmus blue, a property of aqueous solutions of bases.

$$Na_2O + H_2O \longrightarrow 2Na^+ + 2OH^- \text{ (sodium hydroxide)}$$
$$CaO + H_2O \longrightarrow Ca^{2+} + 2OH^- \text{ (calcium hydroxide)}$$

There are, however, many metal oxides, such as Fe_3O_4, which are not significantly soluble in water and thus cannot be considered to be basic oxides.

■ **2. Action with Nonmetals.**   Many of the nonmetals combine directly with oxygen to form oxides. Although oxygen does not combine directly with the halogens (fluorine, chlorine, bromine, and iodine), oxides of these elements are known. Equations for some typical reactions of nonmetals with oxygen are

$$2H_2 + O_2 \longrightarrow 2H_2O \text{ (water)}$$
$$C + O_2 \longrightarrow CO_2 \text{ (carbon dioxide)}$$
$$S + O_2 \longrightarrow SO_2 \text{ (sulfur dioxide)}$$
$$P_4 + 5O_2 \longrightarrow P_4O_{10} \text{ (phosphorus(V) oxide)}$$

Elements which exhibit more than one oxidation number may react with oxygen to form more than one oxide. For example, carbon forms CO and CO_2, sulfur forms SO_2 and SO_3, and phosphorus forms P_4O_6 and P_4O_{10}.

Most oxides of nonmetals dissolve in water to form aqueous solutions of acids and, hence, are often called **acidic oxides** or **acidic anhydrides.** The solutions turn blue litmus red, a characteristic of aqueous solutions of acids.

$$CO_2 + H_2O \longrightarrow H_2CO_3 \text{ (carbonic acid)}$$
$$SO_2 + H_2O \longrightarrow H_2SO_3 \text{ (sulfurous acid)}$$
$$P_4O_{10} + 6H_2O \longrightarrow 4H_3PO_4 \text{ (phosphoric acid)}$$

■ **3. Action with Compounds.**   If a compound is composed of elements which when free will combine with oxygen, the compound ordinarily will also react with oxygen to form oxides of the constituent elements. For example, hydrogen sulfide contains the elements hydrogen and sulfur, both of which will react directly with oxygen when they are free. Thus hydrogen sulfide burns in oxygen, forming water and sulfur dioxide.

$$2H_2S + 3O_2 \longrightarrow 2H_2O + 2SO_2$$

Additional examples of compounds that react with oxygen and the equations for the reactions follow.

$$\underset{\substack{\text{Carbon}\\\text{disulfide}}}{CS_2} + 3O_2 \longrightarrow CO_2 + 2SO_2$$

$$\underset{\text{Sugar}}{C_{12}H_{22}O_{11}} + 12O_2 \longrightarrow 12CO_2 + 11H_2O$$

$$\underset{\text{Methane}}{CH_4} + 2O_2 \longrightarrow CO_2 + 2H_2O$$

$$\underset{\text{Zinc sulfide}}{2ZnS} + 3O_2 \longrightarrow 2ZnO + 2SO_2$$

An important point that can be made now is that an equation usually represents only the *main* reaction which takes place under a given set of conditions. There may be by-products, but these are not included in writing the equation for the principal reaction. In the equation for burning sulfur in oxygen, for example, we write $S + O_2 \longrightarrow SO_2$; however, when the reaction takes place, small amounts of white smoke (SO_3) are observed. When hydrogen sulfide is burned, the chief products are water and sulfur dioxide, but we may also get small amounts of elemental sulfur or of sulfur trioxide. In the burning of carbon disulfide, possible products, in addition to carbon dioxide and sulfur dioxide, are elemental carbon, carbon monoxide, elemental sulfur, and sulfur trioxide.

Certain oxides, in which all the valence electrons of the element with which the oxygen is combined are not already involved in bonding, will unite with oxygen. Examples are

$$2CO + O_2 \longrightarrow 2CO_2$$

Carbon
monoxide

$$P_4O_6 + 2O_2 \longrightarrow P_4O_{10}$$

Phosphorus(III)
oxide

Other oxides, however, such as carbon dioxide (CO_2), silicon dioxide (SiO_2), sulfur trioxide (SO_3), and magnesium oxide (MgO) do not react with oxygen, because in these compounds all of the valence electrons of the element combined with oxygen are already involved in bonding.

8.5 Heat of Reaction

Heat is evolved in reactions involving the combination of oxygen with an element or compound. Reactions accompanied by the production of heat are said to be **exothermic.** The burning of magnesium, hydrogen, carbon, sulfur, methane, carbon disulfide, and sugar are all exothermic reactions. Once started, exothermic reactions usually proceed without addition of energy from the outside. For example, carbon will continue to burn in oxygen with the evolution of heat until either the supply of carbon or oxygen is exhausted.

Reactions accompanied by the absorption of heat are called **endothermic reactions.** Reactions of this kind require a continuous supply of energy from the outside to keep them going. For example, the decomposition of potassium chlorate into potassium chloride and oxygen will continue only so long as the compound is heated.

Compounds which are formed from their elements by highly exothermic reactions, as H_2O and CO_2, are stable toward heat; they are said to be **thermally stable.** Usually, a very high temperature is required to decompose them into their constituent elements. On the other hand, compounds resulting from endothermic reactions, such as hydrogen peroxide (H_2O_2), are thermally unstable, i.e., their internal energies tend to break the bonds holding the atoms together and the molecule may decompose even at room temperature.

The quantity of energy liberated or absorbed (usually in the form of heat) during a chemical change is referred to as the **heat of reaction.** The energy involved in the formation of compounds from their constituent elements is known as the **heat of formation.** Customarily, when energy is *liberated* during a reaction, the value for the heat of reaction is given a negative sign; conversely, when energy is *absorbed*, the value is given a positive sign. Inasmuch as 68,315 calories of heat are liberated when a mole of water is formed from hydrogen and oxygen, the heat of formation of water from hydrogen and oxygen is $-68,315$ calories per mole; that of carbon dioxide is $-94,051$ calories per mole; and of magnesium oxide is $-143,840$ calories per mole. The same amount of energy is required to decompose a compound into its constituent elements as is involved in the formation of the compound. This follows from the Law of Conservation of Energy, a very important concept in chemistry. Thus, a quantity of electrical energy equivalent to 68,315 calories is needed to decompose one mole of water.

Additional points concerning the energy relationships in chemical reactions will be brought out in the discussion of enthalpy, free energy, and entropy in Chapter 20 (Chemical Thermodynamics).

8.6   Combustion

The term **combustion** is applied to chemical reactions that are accompanied by the evolution of both light and heat. Common examples of combustion are the burning of wood, coal, or magnesium in air. Combustion, however, is not restricted to reactions involving oxygen. For example, hydrogen will burn in an atmosphere of chlorine, hydrogen chloride being formed ($H_2 + Cl_2 \longrightarrow 2HCl$).

Before combustion can take place, the substances involved must be heated to the **kindling temperature,** i.e., the temperature at which the burning is sufficiently rapid to proceed without further addition of heat from the outside. The kindling temperature for many substances, especially solids, is not definite but depends upon the extent of subdivision and other factors. For example, iron in the form of fine wire will burn readily when heated in a flame and then placed in pure oxygen. However, a rod of iron a few millimeters in diameter will not ignite under the same conditions.

Spontaneous combustion may occur when the heat evolved during a reaction is not carried away from the system and thus accumulates to raise the temperature of the reacting substances to the kindling temperature. For example, linseed oil unites with oxygen of the air at ordinary temperatures **(slow oxidation)** in an exothermic reaction. Rags soaked with linseed oil and stored in locations where there is insufficient circulation of air to carry off the heat produced by slow oxidation may ignite spontaneously; i.e., the heat of reaction accumulates and raises the temperature of the system above the kindling point. Many costly and disastrous fires have been started by spontaneous combustion of such materials as coal in large piles, uncured hay stored in unventilated barns, and waste rags containing paints or drying oils.

Whenever combustion takes place extremely rapidly, the heat of reaction is liberated almost instantly and usually a large increase in gaseous volume results. An explosion results if the gases are confined so that they cannot escape readily. The mixture of gasoline and air in the cylinder of an automobile engine explodes when ignited by a spark. The rapid burning of gunpowder confined in a small space results in an explosion. Disastrous explosions sometimes occur in flour mills and coal mines when dry finely divided particles are ignited. Any process, whether it be a chemical reaction or simply the overheating of a steam boiler, which leads to a sudden, large increase in gaseous volume in a confined space can create an explosion.

8.7   Speed of Reaction

Oxygen combines slowly with finely divided bituminous, or soft, coal at ordinary temperatures with the production of little heat and no light. Such a reaction is called **slow oxidation.** When ignited, a piece of coal may burn quietly with a flame, producing considerable heat and light. Finely divided coal and oxygen unite explosively when ignited by a spark. It is evident, then, that the speed of a given reaction may vary greatly. The rate at which substances are used up or are formed during a chemical change is known as the **speed of reaction.** Among the factors which influence the speed of a reaction are the temperature, the concentration of the reactants, the presence or absence of a catalyst (see Chapter 17), and the amount of contact between the reactants.

■ **1. Temperature.**   Heat affects the speed of most reactions. Some reactions are speeded up by heat; some are slowed down. For those that increase in speed, the reaction rates, roughly speaking, are doubled for each 10° rise in temperature.

■ **2. Concentration.**   The fact that a heated iron wire will burn in pure oxygen but not in air, which is only 21 per cent oxygen by volume, shows the effect of concentration upon the speed of oxidation of iron. The speed of a reaction increases as the concentration of the reactants is increased. The reaction of zinc with hydrochloric acid to produce zinc chloride and hydrogen proceeds more rapidly if the concentration of either zinc or hydrochloric acid is increased.

■ **3. Catalysts.**   The effect of manganese dioxide as a catalyst upon the thermal decomposition of potassium chlorate was mentioned in Section 8.2. The success of many industrial processes is due to the use of catalysts for increasing the speed of reactions that are normally too slow to be commercially feasible.

■ **4. Contact.**   Reactions between substances take place only when they are in contact, and the more intimate the contact the more rapid the reaction may become. Wood shavings burn much more rapidly than a massive wood chunk.

8.8   The Importance of Oxygen to Human Life

In animals and man the energy required for maintenance of normal body temperature is derived from the slow oxidation of materials in the body. Oxygen passes from the lungs into the blood where it combines with hemoglobin, producing oxyhemoglobin. In this form, oxygen is carried by the blood to the various tissues of the body and there consumed in reactions with oxidizable materials. Products are mainly carbon dioxide and water. The blood, which at that stage contains hemoglobin, carbon dioxide, and some oxyhemoglobin, returns through the veins to the lungs. There it gives up carbon dioxide and collects another supply of oxygen. Slow oxidation in the cells of the body involves a series of complicated reactions catalyzed by a group of substances called **enzymes.** The digestion and assimilation of food regenerates the materials which are consumed by oxidation in the body, with the same amount of energy being liberated as if the food had been burned by rapid oxidation outside the body.

8.9   Uses of Oxygen

The many applications of oxygen make use of the oxygen in the air, oxygen-enriched air, or pure oxygen. Animals and man use the oxygen in the air in the metabolic process known as respiration. Oxygen in the air is essential in combustion processes such as the burning of fuels for the production of heat; it is also essential in the decay of organic matter. Oxygen-enriched air is used in medical practice when the blood receives an inadequate supply of oxygen because of such things as shock, pneumonia, tuberculosis, and heart ailments; it is also used in high altitude flying for the same reason, not just by astronauts but also by passengers on commercial airliners. Large quantities of pure oxygen are used in the removal of carbon from iron in the production of steel. Large quantities of pure oxygen are also consumed in the cutting and welding of metals with oxyhydrogen and oxyacetylene torches. Worldwide, more than three-fourths of the pure oxygen produced is used in making other chemicals.

Liquid oxygen is commonly used in rocket engines of spacecraft as an oxidizing agent (see Sections 25.27 and 25.28). Additional quantities are carried along to provide oxygen for life support in space.

Ozone

8.10   Allotropy

When dry oxygen is passed between the two electrically charged plates of an apparatus called an **ozonizer** (Fig. 8–5), a decrease in volume of the gas occurs

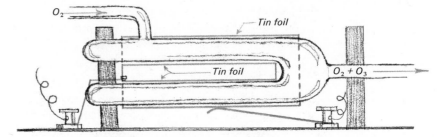

FIGURE 8–5
Laboratory apparatus for preparing ozone.

and a pale blue gaseous substance possessing a pungent (sharp, irritating) odor is formed. This substance is known as **ozone;** its molecules are each composed of three oxygen atoms; i.e., it is triatomic.

$$3O_2 + 68,200 \text{ cal} \longrightarrow 2O_3$$

Ozone and oxygen are referred to as **allotropes.** The term **allotropy** is used to designate the existence of an element in two or more forms in the same physical state. Ordinary oxygen, O_2, and ozone, O_3, are different forms of the element, both gases under ordinary conditions of temperature and pressure.

8.11   Preparation of Ozone

The formation of ozone from oxygen is an endothermic reaction in which the energy is furnished in the form of an electrical discharge, heat, or ultraviolet light.

Ozone is prepared commercially by an electrical ozonizer which is more complicated in design but more efficient than the laboratory apparatus. An alternative method of commercial production of ozone is by electrolysis of very cold, concentrated sulfuric acid. Ozone is also formed, but in small quantities, by several methods which are not suitable for commercial production; for example, it is produced by the slow oxidation of phosphorus, by a jet of burning hydrogen, by lightning, and by most electrical discharges in air. Significant amounts of ozone are apparently formed as ultraviolet light from the sun falls upon the oxygen in the upper atmosphere. Ozone introduced in the atmosphere has been found to react with engine exhaust products to form substances which are important factors contributing to smog (see Section 24.4). A great deal of the deterioration of automobile tires is due to contact with ozone in the air. Considerable discussion has arisen recently in connection with the fear that exhaust gases from supersonic airliners may react with ozone in the upper atmosphere causing a decrease in the concentration to such an extent that the ozone layer would no longer constitute a protective layer against the harmful rays of the sun. It is not yet known whether this would happen, but if it does it could have a drastic effect on the equilibria involved in the life processes on the earth.

8.12   Properties of Ozone

Ozone is a pale blue gas with a pungent odor. As would be expected from its formula, O_3, it is 1.5 times as dense as molecular oxygen, O_2.

Because energy is absorbed when ozone is formed from oxygen, it is not surprising that ozone is more active chemically than oxygen and that it decomposes readily into oxygen.

$$2O_3 \longrightarrow 3O_2 + 68,200 \text{ cal}$$

As a general rule, substances formed by endothermic reactions tend to be less stable than those formed by exothermic reactions (Section 8.5).

The presence of ozone in gas mixtures can readily be detected by passing the gas through a solution of potassium iodide containing some starch emulsion.

$$\underset{\text{Potassium iodide}}{(2K^+) + 2I^-} + O_3 + H_2O \longrightarrow (2K^+) + 2OH^- + \underset{\text{Iodine}}{I_2} + O_2\uparrow$$

The elemental iodine which is formed imparts a blue color to the starch emulsion.

The valence electronic structure of ozone, the triatomic form of oxygen, is given in Fig. 8–6. The ozone molecule is angular, as shown, and the two bonds are indistinguishable due to a phenomenon called **resonance** (Section 23.12, Part 3).

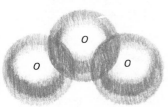

FIGURE 8-6
Representations of an ozone molecule. The Lewis formula (*above*) and the arrangement of spheres indicate the angular nature (127° angle) of the molecule.

8.13   Uses of Ozone

The uses of ozone depend upon its readiness to react with other substances. Its use as a bleaching agent for oils, waxes, fabrics, and starch involves reactions which change colored compounds in these substances to colorless compounds.

QUESTIONS

1. Why must care be exercised in the preparation of oxygen by heating $KClO_3$?
2. Which of the various methods of preparing oxygen discussed in Section 8.2 are likely to result in the production of "wet" oxygen gas?
3. Describe the preparation of oxygen by the electrolysis of water.
4. Write balanced equations for the preparation of oxygen from: $Cu(NO_3)_2$, $NaNO_3$, $KClO_3$, and H_2O.
5. Write balanced equations showing the release of oxygen gas upon heating the following oxides: Au_2O, Ag_2O, Au_2O_3, PtO_2, PtO.
6. Describe the separation of oxygen from air by fractional evaporation of liquid air.
7. Many of the familiar diatomic molecules (H_2, N_2, Cl_2, etc.) are diamagnetic, but oxygen is paramagnetic. Explain why this is so.
8. What is the function of the manganese dioxide used in the preparation of oxygen from $KClO_3$?
9. "Catalysts are specific in their action." Explain.
10. Write balanced equations for the action of oxygen upon each of the following: H_2, Ca, C, P, CS_2, CO, K, Fe, S, H_2S, CH_4, P_4O_6.
11. Which of the following compounds, when placed in contact with water, will cause the water solution to be acidic and which will cause the water solution to be basic: P_4O_6, MgO, SO_3, CO_2? Write a balanced equation for the reaction of each with water.
12. Define and illustrate each of the following: base, basic anhydride, acid, and acid anhydride.
13. Define the term "combustion." Is combustion restricted to reactions involving oxygen? Explain.
14. The heat of reaction for the combustion of carbon to produce carbon dioxide has a value which is the same as the heat of formation of carbon dioxide. Explain.
15. When may spontaneous combustion occur?
16. Why will CO_2, SiO_2, and H_2O not burn?
17. Why will magnesium ribbon burn more rapidly in pure oxygen than in air?
18. Why will wood shavings ignite and burn more rapidly than a log of wood?
19. Discuss the essentiality of oxygen to animal life.
20. Define "allotropy" and give an example.
21. Describe a chemical test for ozone.
22. Why is ozone a more active oxidizing agent than oxygen?

PROBLEMS

1. How much iron can be burned to magnetic oxide of iron, Fe_3O_4, by 850 g of 90.0% pure oxygen?   *Ans. 2.00×10^3 g*
2. What weight of CS_2 can be completely oxidized to sulfur dioxide and carbon dioxide by the oxygen liberated when 325 g of Na_2O_2 reacts with water?   *Ans. 52.9 g*

3. An oxide of iron contains 69.9% iron. Derive the formula of this compound.
   Ans. Fe_2O_3

4. What weight of potassium chlorate would be required to give enough oxygen to burn 50.0 g of methane, CH_4, to carbon dioxide and water? *Ans. 509 g*

5. How many grams of oxygen will be formed by the action of 15.6 g of Na_2O_2 upon water? From the density of oxygen (1.429 g/liter at 0° and 760 mm pressure), calculate how many liters this would be at 0° and 760 mm pressure.
   Ans. 3.20 g; 2.24 liters

6. How much electrical energy, measured in calories, would be required in the electrolytic decomposition of sufficient water to give 24.0 g of oxygen at 25°? The heat of formation of water at 25° is 68,315 calories per mole.
   Ans. 1.02×10^5 cal

7. What weight of MgO would be required to neutralize the HCl in 250 g of a 20.0% aqueous solution of the acid? *Ans. 27.6 g*

8. How many gram-atoms of oxygen are there in 3.20 moles of ozone? How many moles of oxygen will be formed by the decomposition of this quantity of ozone? *Ans. 9.60 gram-atoms; 4.80 moles*

9. What is the total weight of the products formed when 25.0 g of H_2S is oxidized by oxygen of the air to produce water and sulfur dioxide? *Ans. 60.2 g*

10. Which contains more oxygen, 100 g of $NaClO_3$ or 110 g of $KClO_3$? Show your calculations. *Ans. 100 g of $NaClO_3$*

11. How many grams of oxygen are contained in 125 g of $NaNO_3$? How many grams of oxygen will be liberated upon heating 125 g of $NaNO_3$, assuming that all the $NaNO_3$ reacts? *Ans. 70.6 g; 23.5 g*

12. What will be the approximate increase in the speed of a reaction when the temperature is raised 25°; 40°? *Ans. 5.7 times; 16 times*

13. The heat of formation of water from hydrogen and oxygen is 66,970 calories per mole at 373.15° K (100° C). How much water can be heated from 288.15° K to 373.15° K by the heat evolved during the formation of 11.11 moles of water from hydrogen and oxygen? *Ans. 8.753 kg*

14. What weight of H_2SO_4 could be produced from the SO_2 resulting from the roasting of 7,500 kg of ZnS, assuming the entire process to be 77.7% efficient?
    Ans. 5,870 kg (rounded off to three significant figures)

REFERENCES

"The Oxygen Cycle," P. Cloud and A. Gibor, *Sci. American*, Sept., 1970; p. 110.

"Oxygen in Steelmaking," J. K. Stone, *Sci. American*, April, 1968; p. 24.

"Basic Oxygen Steelmaking," H. E. McGannon, *J. Chem. Educ.*, **46**, 293 (1969).

"Recent Improvements in Explaining the Periodicity of Oxygen Chemistry," R. T. Sanderson, *J. Chem. Educ.*, **46**, 635 (1969).

"Lavoisier," D. I. Duveen, *Sci. American*, May, 1956; p. 84.

Lavoisier. The Crucial Year. The Background and Origin of His First Experiments on Combustion in 1772. H. Guerlac, Cornell University Press, Ithaca, New York, 1961.

"Combustion and Flame," R. C. Anderson, *J. Chem. Educ.*, **44**, 248 (1967).

"Why Does Methane Burn?" R. T. Sanderson, *J. Chem. Educ.*, **45**, 423 (1968).

(Fluorinated Hydrocarbons Carry Oxygen in Blood Substitute), (Staff), *Chem. & Eng. News*, May 18, 1970; p. 30.

"Medieval Uses of Air," L. White, Jr., *Sci. American*, Aug., 1970; p. 92.

"Life Support Systems for Manned Space Flights," W. H. Bowman and R. M. Lawrence, *J. Chem. Educ.*, **48**, 260 (1971).

"The Cabin Atmosphere in Manned Space Vehicles," W. H. Bowman and R. M. Lawrence, *J. Chem. Educ.*, **48**, 152 (1971).

"High Temperature Chemistry of Simple Metallic Oxides," C. B. Alcock, *Chem. in Britain*, **5**, 216 (1969).

"Catalysis," V. Haensel and R. L. Burwell, Jr., *Sci. American*, Dec., 1971; p. 46.

"A Simple Demonstration of O_2 Paramagnetism," G. H. Saban and T. F. Moran, *J. Chem. Educ.*, **50**, 217 (1973).

"Ozone: Properties, Toxicity, and Applications," F. Leh, *J. Chem. Educ.*, **50**, 404 (1973).

"Stratospheric Pollution: Multiple Threats to Earth's Ozone," A. L. Hammond and T. H. Maugh II, *Science*, **186**, 335 (1974).

"Priestley," M. Wilson, *Sci. American*, Oct., 1954; p. 68.

"The Discovery of Oxygen," J. R. Partington, *J. Chem. Educ.*, **39**, 123 (1962).

"Joseph Priestley and the Discovery of Oxygen," S. F. Mason, *Chem. in Britain*, **10**, 286 (1974).

"Steps Leading to the Discovery of Oxygen, 1774: A Bicentennial Tribute to Joseph Priestley," R. G. Neville, *J. Chem. Educ.*, **51**, 428 (1974).

"The Priestley Heritage: Prospects for Chemistry in its Third 100 Years," F. A. Long, *J. Chem. Educ.*, **52**, 12 (1975).

"Oxygen: Boon and Bane," I. Fridovich, *Amer. Scientist*, **63**, 54 (1975).

9 Hydrogen

Early in the sixteenth century the Swiss-German physician Paracelsus noted that a flammable gas was formed by the reaction of sulfuric acid with iron. However, it was not until 1766 that Cavendish, an Englishman, recognized this gas as a distinct substance and prepared it by the rather quaint method of passing steam through a red-hot gun barrel and also by the action of various acids on certain metals. Lavoisier, a French chemist, named the gas **hydrogen,** meaning "water producer," because water was formed when the gas burned in air. The hydrogen atom, which consists of one proton and one electron, has the simplest structure of any of the atoms.

9.1   Occurrence

Because it is a chemically active element, hydrogen is found in the free state in only negligible quantities. It occurs free in very small amounts in the atmosphere, in the gases of active volcanoes, in natural gas, in the gases present in some coal mines, and trapped in meteorites. The atmospheres of the sun and other stars appear to be composed largely of hydrogen.

More compounds containing hydrogen are known than those of any other element. In the combined form hydrogen comprises nearly eleven per cent of the weight of water, its most abundant compound. Hydrogen is found combined with carbon and oxygen in the tissues of all plants and animals. It is an important part of petroleum, many minerals, cellulose and starch, sugar, fats, oils, alcohols, acids and bases, and thousands of other substances.

9.2   Preparation of Hydrogen

There are many ways of liberating hydrogen from its compounds, particularly from acids, bases, and water. Where small quantities of hydrogen are required, as in the laboratory, it is usually generated from acids. When commercial quantities of hydrogen are needed, water is generally used as the raw material because it is abundant and cheap.

■ **1. From Acids.** Hydrogen is conveniently prepared in the laboratory by the reaction of an active metal with an acid. Zinc and iron are the metals most frequently used with dilute solutions of hydrochloric or sulfuric acid.

$$\text{Zn} + 2\text{H}^+ + (2\text{Cl}^-) \longrightarrow \text{Zn}^{2+} + (2\text{Cl}^-) + \text{H}_2\uparrow \qquad (1)$$
Zinc chloride

$$\text{Zn} + 2\text{H}^+ + (\text{SO}_4{}^{2-}) \longrightarrow \text{Zn}^{2+} + (\text{SO}_4{}^{2-}) + \text{H}_2\uparrow \qquad (2)$$
Zinc sulfate

$$\text{Fe} + 2\text{H}^+ + (2\text{Cl}^-) \longrightarrow \text{Fe}^{2+} + (2\text{Cl}^-) + \text{H}_2\uparrow \qquad (3)$$
Iron(II) chloride

$$\text{Fe} + 2\text{H}^+ + (\text{SO}_4{}^{2-}) \longrightarrow \text{Fe}^{2+} + (\text{SO}_4{}^{2-}) + \text{H}_2\uparrow \qquad (4)$$
Iron(II) sulfate

$$2\text{Al} + 6\text{H}^+ + (3\text{SO}_4{}^{2-}) \longrightarrow 2\text{Al}^{3+} + (3\text{SO}_4{}^{2-}) + 3\text{H}_2\uparrow \qquad (5)$$
Aluminum sulfate

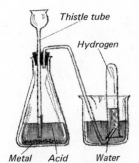

FIGURE 9–1
Apparatus used in the laboratory preparation of hydrogen.

In the laboratory preparation of hydrogen, an apparatus like that shown in Fig. 9–1 is used. The acid solution is poured through the thistle tube upon the metal, and the hydrogen is collected by the downward displacement of water. It is the hydrogen ion of the dilute aqueous acids that reacts with the active metals. The negative ion does not enter the reaction. However, on evaporation of the solution to dryness after the reaction is complete, a crystalline **salt** composed of the metal ion and the negative ion of the acid is obtained. In the reactions described by equations (1)–(5), the salts are zinc chloride, zinc sulfate, iron(II) chloride, iron(II) sulfate, and aluminum sulfate.

■ **2. By the Electrolysis of Water.** The decomposition of water into its constituent elements by means of a direct current of electricity was described in Section 8.2. Hydrogen is liberated at the cathode when water containing a small amount of an electrolyte is electrolyzed.

$$2\text{H}_2\text{O} + \text{Electrical energy} \longrightarrow 2\text{H}_2\uparrow + \text{O}_2\uparrow$$

■ **3. By the Electrolysis of Sodium Chloride Solutions.** When a concentrated aqueous solution of common table salt is electrolyzed, hydrogen is formed along with sodium hydroxide and chlorine.

$$(2\text{Na}^+) + 2\text{Cl}^- + 2\text{H}_2\text{O} + \text{Elect. energy} \longrightarrow (2\text{Na}^+) + 2\text{OH}^- + \text{Cl}_2\uparrow + \text{H}_2\uparrow$$

This is also a commercial method for producing NaOH (caustic soda) and chlorine; hydrogen is a by-product (Section 22.4). If the salt solution used is not fairly concentrated, oxygen is liberated at the anode by the reaction discussed in the preceding paragraph. This fact points up an important aspect of the process of electrolysis: The concentration of the electrolyte is a determining factor in product formation.

■ **4. By the Action of Certain Metals and Nonmetals on Water.**   Very active metals, such as sodium, potassium, and calcium, rapidly displace hydrogen from water at room temperature.

$$2Na + 2HOH \longrightarrow 2Na^+ + 2OH^- + H_2\uparrow$$
$$2K + 2HOH \longrightarrow 2K^+ + 2OH^- + H_2\uparrow$$
$$Ca + 2HOH \longrightarrow Ca^{2+} + 2OH^- + H_2\uparrow$$

The reaction of a small piece of sodium or potassium with water produces sufficient heat to ignite the hydrogen as it is produced. It is dangerous to bring large pieces of either metal into contact with water because explosions may result. An alloy of lead and sodium, or one of mercury and sodium, either of which reacts with water less vigorously than does pure sodium, is sometimes used for the preparation of hydrogen in small quantities.

Other less active metals will displace hydrogen from water at higher temperatures. Magnesium reacts only slowly with boiling water, but when steam is passed over magnesium or red-hot iron, hydrogen is liberated rapidly (Fig. 9–2).

$$Mg + H_2O \longrightarrow H_2 + \underset{\text{Magnesium oxide}}{MgO}$$

$$3Fe + 4H_2O \longrightarrow 4H_2 + \underset{\text{Iron(II, III) oxide}}{Fe_3O_4}$$

Note that all of the hydrogen of the water molecule is replaced by the metals in the reactions described by the above equations, whereas the active metals sodium and potassium replace only half of it.

Some of the hydrogen that is used in industry is produced by the reaction of iron with steam. After a mass of iron has been largely converted to Fe_3O_4 by

FIGURE 9–2

Some metals, such as magnesium and iron, react with the oxygen in steam setting hydrogen free and forming metallic oxides. The apparatus for carrying out the reaction of iron with steam to produce hydrogen and iron(II, III) oxide is illustrated.

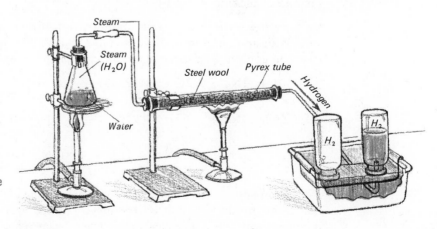

this reaction, the iron is regenerated by passing carbon monoxide, CO, over the heated iron oxide.

$$Fe_3O_4 + 4CO \longrightarrow 3Fe + 4CO_2$$

By the alternate use of steam and carbon monoxide the iron can be used over and over again.

Carbon, a nonmetal, when white hot (1,500–1,600° C) will react with steam in an endothermic reaction, producing a mixture of carbon monoxide and hydrogen. The mixture is commonly known as **water gas.**

$$C + H_2O \longrightarrow CO + H_2$$

Because both carbon monoxide and hydrogen will burn in oxygen or in air and produce much heat, water gas is a valuable industrial fuel. When water gas is mixed with steam and passed over a catalyst such as iron oxide or thorium oxide at a fairly high temperature (500° C), carbon dioxide and hydrogen are produced.

$$CO + H_2O + (H_2) \xrightarrow{\text{Catalyst}} CO_2 + H_2 + (H_2)$$

Large quantities of hydrogen produced in this way are used in the synthesis of ammonia (NH_3) and many organic compounds such as wood alcohol (CH_3OH), and in the hydrogenation of fats such as soybean oil and cottonseed oil.

■ **5. By the Action of Steam on Hydrocarbons.** When a mixture of methane and steam is heated to a high temperature in the presence of catalysts, a mixture of carbon monoxide, carbon dioxide, and hydrogen is produced.

$$CH_4 + H_2O \xrightarrow{\text{Catalyst}} CO + 3H_2$$

$$CO + H_2O \xrightarrow{\text{Catalyst}} CO_2 + H_2$$

These are typical reactions and other hydrocarbons (compounds containing only hydrogen and carbon) may be substituted for methane in this industrial method for the production of hydrogen. This is now the most important commercial method of making hydrogen.

■ **6. By the Action of Certain Elements on Active Bases.** Certain elements, such as aluminum, zinc, and silicon, react with sodium hydroxide, or other strong bases, in concentrated aqueous solution with the liberation of hydrogen.

$$2Al + (2Na^+) + 2OH^- + 6H_2O \longrightarrow (2Na^+) + 2Al(OH)_4^- + 3H_2\uparrow$$
Sodium aluminate

$$Zn + (2Na^+) + 2OH^- + 2H_2O \longrightarrow (2Na^+) + Zn(OH)_4^{2-} + H_2\uparrow$$
Sodium zincate

$$Si + (2Na^+) + 2OH^- + H_2O \longrightarrow (2Na^+) + SiO_3^{2-} + 2H_2\uparrow$$
Sodium silicate

Aluminum and zinc do not displace hydrogen from water alone, because the oxide that forms produces a film on the metal surface preventing further contact of the metal with the water. Acids and strong bases react with aluminum and zinc to produce hydrogen, because they dissolve the oxide film and allow the metal to come in direct contact with the acid or base.

■ **7. By the Action of Certain Metal Hydrides on Water.** Compounds resulting from the chemical combination of active metals and hydrogen, such as calcium hydride and sodium hydride, react vigorously with water, liberating hydrogen.

$$CaH_2 + 2H_2O \longrightarrow Ca^{2+} + 2OH^- + 2H_2\uparrow$$
$$NaH + H_2O \longrightarrow Na^+ + OH^- + H_2\uparrow$$

Metal hydrides are expensive but convenient sources of very pure hydrogen, especially where space and weight are important factors. Examples are for the inflation of life-jackets, life-rafts, military balloons, and weather balloons. Calcium hydride is so much used for these and other purposes that it is sold in commercial quantities.

■ **8. Other Methods.** Hydrogen is produced commercially from **hydrocarbons,** which are the principal components of petroleum and natural gas. Methane, the simplest hydrocarbon, is decomposed into carbon and hydrogen when heated in the presence of a suitable catalyst.

$$\underset{\text{Methane}}{CH_4} \xrightarrow[\triangle]{\text{Catalyst}} C + 2H_2\uparrow$$

Such catalyzed thermal decompositions are called **cracking reactions.** The carbon produced by the cracking of methane is an important commercial product called **carbon black.**

9.3 Physical Properties of Hydrogen

At ordinary temperatures, hydrogen is a colorless, odorless, and tasteless gas consisting of diatomic molecules, H_2. (See Figs. 4–2 and 4–3 in Chapter 4.) The bond between the two hydrogen atoms involves a pair of electrons in the sigma bonding molecular orbital (σ_s) as described in Section 5.3.

With a density of 0.08987 g/liter at 0° and 760 mm pressure, hydrogen is the lightest known substance. Because of its low density it can be collected by the downward displacement of air. The fact that hydrogen is so much lighter than air (which has a density of 1.293 g/liter) makes it useful as the lifting agent in balloons.

If sufficiently cooled and compressed, hydrogen changes to a liquid that boils at atmospheric pressure at −252.7°. The low temperature of liquid hydrogen makes it useful in cooling other materials to low temperatures. It may be used to transform all other gases into solids with the single exception of helium. Hydrogen freezes to a transparent solid at −259.14°.

Hydrogen diffuses (Section 10.10) faster than any other gas because its molecules are lighter and move faster than those of any other gas. The relative rate of diffusion of hydrogen as compared to that of air can be demonstrated using the apparatus illustrated in Fig. 9–3a. When an inverted beaker of hydrogen is lowered over the porous clay cup hydrogen diffuses through the porous wall of

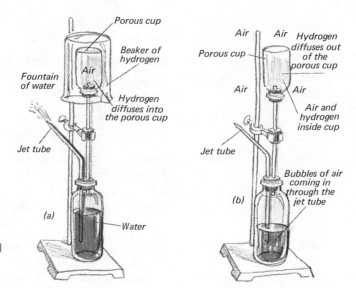

FIGURE 9-3

The "hydrogen fountain" used to demonstrate the diffusion property of hydrogen.

the cup faster than the air within the cup diffuses outward. Consequently, a pressure develops in the cup, which causes the water in the bottle to be forced out through the jet. If the beaker of hydrogen is taken away from the porous cup, which now contains both hydrogen and air (Fig. 9–3b), the hydrogen diffuses out of the porous cup faster than air diffuses inward. Thus, a partial vacuum is created within the cup which results in air entering the bottle of water through the jet tube.

9.4   The Adsorption of Hydrogen by Metals

Hydrogen is readily adsorbed by certain metals, the process being most pronounced with gold, platinum, tungsten, and especially palladium. **Adsorption is the adhesion of molecules of a gas, liquid, or dissolved substance to the surface of a solid.** The quantity of hydrogen adsorbed by a given mass of metal depends upon the condition of the metal, such as extent of subdivision, and the temperature and pressure under which the adsorption takes place. At room temperature one volume of finely divided palladium (called palladium black) adsorbs nearly 900 volumes of hydrogen. Adsorbed hydrogen is very active chemically as indicated by its rapid union with oxygen when in this state. Ordinary hydrogen and oxygen do not combine noticeably unless the mixture is ignited. There is evidence that adsorbed hydrogen is ionized or activated in some way because metals such as platinum are good catalysts for many chemical reactions involving hydrogen as a reactant. Because of the property of adsorption, hydrogen readily passes through the heated walls of containers made of metals which strongly adsorb it.

9.5   Chemical Properties of Hydrogen

At ordinary temperatures, hydrogen is relatively inactive chemically but when heated it enters into many chemical reactions.

■ **1. Reaction with Oxygen.**   When a mixture of hydrogen and oxygen is ignited, a vigorous reaction occurs, and water is formed.

$$2H_2 + O_2 \longrightarrow 2H_2O + heat$$

Because of the violence of the reaction, often resulting in explosion, great caution should be used in handling hydrogen (or any other combustible gas). Hydrogen will burn without explosion under some conditions. The fact that water is formed when hydrogen is burned in either air or oxygen can be demonstrated by allowing a jet of hydrogen to burn inside an inverted, cold, dry beaker (Fig. 9–4).

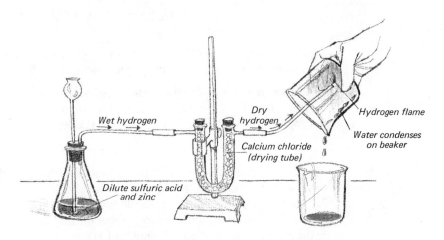

FIGURE 9-4
When hydrogen burns in air, water vapor is formed. The large beaker here serves as a condenser.

The very high heat of combustion of hydrogen in pure oxygen makes it possible to achieve temperatures up to 2,800° with the oxyhydrogen blowtorch. The hot flame of this torch can be used in "cutting" thick sheets of many metals.

When hydrogen gas, consisting of diatomic molecules, is passed through an electric arc, energy is absorbed by the hydrogen, causing the diatomic molecules to be broken up into monatomic hydrogen molecules.

$$H_2 + heat \longrightarrow 2H$$

The single atoms of hydrogen readily recombine upon collision with one another and release the heat which was absorbed when the molecule was broken up in the electric arc.

$$2H \longrightarrow H_2 + heat$$

The energy released in this reaction, added to that from the burning of the hydrogen in pure oxygen, is utilized in the atomic hydrogen torch (Fig. 9–5). This torch is useful in cutting and welding metals which melt at temperatures up to 5,000°.

■ **2. Reaction with Nonmetals Other Than Oxygen.**   Under suitable conditions, hydrogen combines directly with several nonmetals, in addition to oxygen, forming covalent compounds.

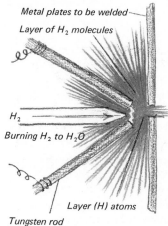

FIGURE 9-5
Atomic hydrogen torch.

Metal plates to be welded
Layer of H_2 molecules

H_2
Burning H_2 to H_2O

Layer (H) atoms

Tungsten rod

$$H_2 + F_2 \longrightarrow 2HF \text{ (Hydrogen fluoride)}$$
$$H_2 + Cl_2 \longrightarrow 2HCl \text{ (Hydrogen chloride)}$$
$$H_2 + Br_2 \longrightarrow 2HBr \text{ (Hydrogen bromide)}$$
$$H_2 + I_2 \longrightarrow 2HI \text{ (Hydrogen iodide)}$$
$$H_2 + S \longrightarrow H_2S \text{ (Hydrogen sulfide)}$$
$$3H_2 + N_2 \longrightarrow 2NH_3 \text{ (Ammonia)}$$

■**3. Reaction with Certain Metals.** Hydrogen also reacts with some of the active metals, forming crystalline ionic **hydrides.** In these compounds hydrogen becomes the negative ion, H^-.

$$2Li + H_2 \longrightarrow 2LiH \text{ (Lithium hydride)}$$
$$Ca + H_2 \longrightarrow CaH_2 \text{ (Calcium hydride)}$$

The reactions are exothermic and the resulting hydrides are relatively stable toward heat but react vigorously with water (see Section 9.2, Part 7).

■**4. Reaction with Compounds.** Hydrogen reacts with heated oxides of many metals with the formation of the free metal and water. For example, when hydrogen is passed over heated copper(II) oxide, copper and water are formed.

$$CuO + H_2 \longrightarrow Cu + H_2O$$

This is an example of an oxidation-reduction reaction. Here, hydrogen displaces oxygen from copper in the type of chemical change called reduction (see Section 7.1, Part 6). The hydrogen molecule acts as the reducing agent, each atom losing an electron, and therefore undergoes oxidation. The oxygen of the copper(II) oxide gains electrons from the hydrogen and hence causes the hydrogen to be oxidized; the copper(II) oxide thus is the oxidizing agent. As pointed out in Section 7.1, the processes of reduction and oxidation always occur simultaneously.

Hydrogen may also react with certain metal oxides reducing them to lower oxides.

$$MnO_2 + H_2 \longrightarrow MnO + H_2O$$

9.6 Reversible Reactions

When hydrogen is passed over heated magnetic oxide of iron, Fe_3O_4, the iron is reduced, and iron and water (steam) are formed.

$$Fe_3O_4 + 4H_2 \longrightarrow 3Fe + 4H_2O \tag{1}$$

This reaction is exactly the reverse of the one described in Section 9.2, i.e., the production of hydrogen by passing steam over heated iron.

$$3Fe + 4H_2O \longrightarrow Fe_3O_4 + 4H_2 \tag{2}$$

When these reactions are utilized in the laboratory, a stream of hydrogen gas is passed over a heated solid in a tube. In reaction (1), hydrogen is in excess and the steam is swept out of the reaction tube by the current of hydrogen. Thus, reaction (1) can go to completion. In reaction (2) the hydrogen is swept out of the tube by the steam and thus this reaction also can go to completion. If a mixture of iron and steam is heated in a closed tube, neither the steam nor hydrogen which forms can escape. At first, reaction (2) will be the only one to take place because only iron and steam are present, but as this reaction proceeds iron and steam will be used up while Fe_3O_4 and hydrogen are being formed. It follows that the rate of reaction (2) decreases while that of reaction (1) increases, and after a time, the two reaction rates will become equal. The two reactions will reach a state of **chemical equilibrium.** The quantity of each of the four substances in the closed tube remains unchanged at equilibrium, because each is constantly being formed at the same rate as that at which it is being consumed. Reactions of this type are said to be **reversible,** and it is customary to represent the reversibility by means of double arrows as in the equation

$$3Fe + 4H_2O \rightleftharpoons Fe_3O_4 + 4H_2$$

A more complete account of the very important phenomenon of chemical equilibrium is given in Chapter 17.

9.7 Uses of Hydrogen

Although hydrogen was once extensively used as the lifting agent in lighter-than-air craft such as blimps and dirigibles, it has largely been replaced by helium gas. Because of its inertness, helium is much safer than the highly combustible hydrogen (Fig. 9–6). Hydrogen is extensively used in a process called **hydrogenation,** in which vegetable oils are changed from liquids to solids. In the treating

FIGURE 9–6

Explosion of hydrogen on the airship *Hindenburg* at Lakehurst, New Jersey, May 6, 1937. This picture was taken immediately following the explosion of hydrogen in one compartment; explosions in the two other compartments took place almost immediately, resulting in the complete destruction of the giant airliner. *World Wide Photo*

of vegetable oils with hydrogen under pressure and in the presence of nickel as a catalyst, the hydrogen combines chemically with the oil, producing solid fats used in cooking. Crisco, Spry, and Swiftning are examples of such hydrogenated oils. Large quantities of hydrogen are united directly with nitrogen to produce ammonia and with chlorine to produce hydrogen chloride. An increasingly important use of hydrogen is in the catalytic conversion of coal dust to liquid hydrocarbons, which supplement petroleum as a source of liquid fuel. Methyl alcohol is produced synthetically by the catalyzed reaction of hydrogen with carbon monoxide

$$2H_2 + CO \xrightarrow{\text{Catalyst}} CH_3OH$$

Water gas, a mixture of hydrogen with carbon monoxide, is an important industrial fuel. Hydrogen is used in the reduction of certain metal oxides to obtain the free metal. The use of hydrogen in nuclear fusion reactions is discussed in Chapter 30 (Nuclear Chemistry).

9.8   Isotopes of Hydrogen

Using an instrument known as the mass spectrograph (see Chapter 30, Fig. 30–1) it is possible to show that hydrogen is composed of three isotopes. They are ordinary hydrogen, or **protium,** $_1^1H$; heavy hydrogen, or **deuterium,** $_1^2H$; and **tritium,** $_1^3H$. In a sample of hydrogen, there is only one atom of deuterium for every 5,000 atoms of ordinary hydrogen and one atom of tritium for every 10^7 atoms of ordinary hydrogen. The chemical properties of the different isotopic forms of hydrogen are essentially the same, as they have identical electronic structures. The differences in their atomic masses give rise to differences in physical properties, however. Deuterium and tritium have lower vapor pressures than does ordinary hydrogen. Consequently, when liquid hydrogen evaporates, the heavier isotopes are concentrated in the last portions.

Deuterium is of interest in connection with heavy water in atomic reactors (Section 12.13), and tritium with nuclear fusion reactions (Section 30.16).

9.9   The Activity, or Electromotive, Series

We noted in Section 9.2 that certain metals, such as sodium and potassium, react readily with cold water, displacing hydrogen and forming metal hydroxides. It was also pointed out that certain other metals, such as magnesium and iron, react with water only when they are heated. Sodium and potassium react much more vigorously with acids than do magnesium and iron. From experimental observations of this sort it is possible to arrange the metals in order of their chemical activities and thereby to establish an activity, or electromotive, series (Table 9–1).

Potassium is the most reactive of the common metals and so heads the series. Each succeeding metal in the series is less reactive, and gold, the least reactive of all, is found at the bottom of the series. In principle, any metal in the series will displace any other element below it in the series from dilute water solutions

TABLE 9-1 Activity, or Electromotive, Series of Common Metals

Oxides are not reduced by hydrogen	K Ba Sr Ca Na	React with cold water, liberating hydrogen
	Mg Al Mn Zn Cr	React with steam, liberating hydrogen
Oxides are reduced by hydrogen	Fe Cd	React with acids, liberating hydrogen
	Co Ni Sn Pb	
	H Sb As Bi Cu	React with oxygen, giving oxides
Oxides are reduced by heat alone	Ag Pd Hg Pt Au	Form oxides indirectly

Iron nail

Coat of copper on the nail

Solution of copper sulfate

FIGURE 9-7
An iron nail becomes coated with metallic copper when immersed in a dilute solution of copper sulfate.

of its soluble compounds, if simple ions are involved. For example, any metal above hydrogen in the series will liberate hydrogen from aqueous acids (water solutions of acids). The metals below hydrogen do not displace hydrogen from water or from aqueous acids. Another example is the displacement of copper by iron from aqueous solutions of copper(II) salts (Fig. 9-7). It should be noted that when metals are present as part of stable complex ions (Chapter 32) the principles of the activity series sometimes do not apply.

$$Fe + Cu^{2+} + (SO_4^{2-}) \longrightarrow Fe^{2+} + (SO_4^{2-}) + \underline{Cu}$$
$$\text{Copper(II) sulfate} \qquad\qquad \text{Iron(II) sulfate}$$

or
$$Fe + Cu^{2+} \longrightarrow Fe^{2+} + \underline{Cu}$$

In the equation the symbol for copper is underscored ($\underline{Cu}$), to show that copper is formed as a solid during the reaction. In a similar manner, silver is displaced by copper (Fig. 9-8), and gold is displaced by silver.

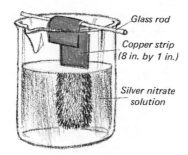

FIGURE 9-8

When a strip of copper is immersed in dilute silver nitrate solution, silver, a less active metal than copper, is displaced by the copper and deposited on the surface of the copper strip. The deposit is called a "silver tree."

Glass rod

Copper strip
(8 in. by 1 in.)

Silver nitrate
solution

The decreasing activity of the metals down the series is reflected in the decrease in tendency of the metals to form positive ions in solution. Thus, copper has less tendency to lose electrons and thereby to form positive ions in solution than zinc.

$$Zn \longrightarrow Zn^{2+} + 2e^-$$
$$Cu \longrightarrow Cu^{2+} + 2e^-$$

The reactivity of the metals toward oxygen, sulfur, and the halogens is less for each succeeding metal in progressing down the series. The heat of formation and stability of the compounds formed decreases in a similar manner. It is evident then that the activity series is very useful because it indicates the possibility of a reaction of a given metal with water, acids, salts of other metals, oxygen, sulfur, and the halogens (fluorine, chlorine, bromine, and iodine); in addition, it provides some indication of the stability of the compounds formed. It should be noted, however, that the order in which the elements are placed in the series depends somewhat on the conditions under which the activity is observed. A series determined from observation of the activity of the metals with respect to their ions in water solution will be slightly different from that showing the order of activity at high temperatures and in the absence of a solvent. The series shown in Table 9-1 is not applicable to reactions in solvents other than water (e.g., in fused salts), or when stable complex ions are formed, except in a general way.

The activity series of metals is discussed in more detail in Chapter 22 (Electrochemistry).

9.10 Hydrogen as a Secondary Energy Source

Hydrogen, like electricity, is a secondary source of energy, because it must be produced by using energy from a primary source such as coal, nuclear fusion, or solar heat. When mass-produced, hydrogen may in a few years cost no more than increasingly scarce and costly gasoline and natural gas. Unlike electricity, hydrogen can be stored conveniently and then burned to produce energy where it is needed. It is pollution-free when burned, because the only product is water.

In recent years hydrogen has been used as a fuel in the production of electricity in fuel cells (Section 22.19), a process which is much more efficient than that involving generators powered by steam. When our supply of natural gas is depleted, hydrogen conceivably can be used just as natural gas is, if safety problems can be solved. It can be transported in the nationwide network of underground pipelines, which already exists for the transport of gases such as natural gas. These lines are now linked to over sixty per cent of all U.S. homes and industries.

As an internal-combustion fuel, hydrogen is in many ways superior to gasoline and diesel oil. Engines using hydrogen start more easily, particularly in cold weather, and experience less wear. The greatest advantage is that almost no pollution is generated. A number of hydrogen-powered research cars are already on the roads. The greatest problem is the carrying of the fuel in a safe and practical manner. Because liquid hydrogen is 40 per cent lighter than conventional jet fuel, it particularly benefits aircraft performance. It is estimated that a hydrogen-powered aircraft can be designed to carry twice the current ratio of payload to plane weight.

QUESTIONS

1. What is the atomic structure of hydrogen? In what way is the hydrogen nucleus unique?
2. Account for the almost complete absence of free hydrogen on earth. Express any ideas you might have regarding the existence of free hydrogen on the moon.
3. Write ionic equations for the preparation of hydrogen by the action of Mg, Fe, Ni, Al, and Zn with hydrochloric acid; with sulfuric acid. Name the salts produced by each of these reactions.
4. Write ionic equations for the preparation of hydrogen by the reaction of each of the following metals with water: Fe, Mn, Al (in basic solution), Na, Ba.
5. Give the equation for the electrolysis of a concentrated aqueous solution of sodium chloride.
6. Contrast the products formed when each of the following elements combines with hydrogen and with oxygen, respectively: S, Li, Ca.
7. Both hydrogen and oxygen are colorless, odorless, and tasteless gases. How may one distinguish between them?
8. What happens to the hydrogen and oxygen in the water molecule during electrolysis of water?
9. What ion is common to aqueous solutions of all acids? Write equations for the ionization of hydrogen chloride and hydrogen sulfate in water.
10. Name three metals that will displace hydrogen from (a) cold water; (b) steam; (c) acids. Write a balanced equation for each reaction.
11. With the aid of the activity series, predict whether or not the following reactions will take place:
    (a) $Mg + Ni^{2+} \longrightarrow Mg^{2+} + Ni$
    (b) $MgO + H_2 \longrightarrow Mg + H_2O$
    (c) $2Ag + Cu^{2+} \longrightarrow Cu + 2Ag^+$
    (d) $Na_2O + H_2 \longrightarrow 2Na + H_2O$
    (e) $2Au_2O_3 + heat \longrightarrow 4Au + 3O_2$
    (f) $2Ag_2O + heat \longrightarrow 4Ag + O_2$
    (g) $Co + 2H^+ \longrightarrow Co^{2+} + H_2$

(h) $2Bi + 6H^+ \longrightarrow 2Bi^{3+} + 3H_2$

(i) $Pd + 2H^+ \longrightarrow Pd^{2+} + H_2$

(j) $Cd + Pb^{2+} \longrightarrow Cd^{2+} + Pb$

(k) $Ca + 2H_2O \longrightarrow Ca^{2+} + 2OH^- + H_2$

12. Under what chemical conditions may the activity series be relied upon to predict chemical behavior?

13. What is meant by a reversible reaction? What is meant by the term "chemical equilibrium"?

14. Explain how the "hydrogen fountain" (Fig. 9–3) functions.

15. Compare the rates of diffusion of helium atoms and hydrogen molecules.

16. Distinguish between exothermic and endothermic reactions.

17. Relate compound stability to heat of formation.

18. Why are higher temperatures attainable by the atomic hydrogen flame than by the oxyhydrogen flame?

19. Explain: "Metal hydrides are convenient and portable sources of hydrogen."

20. Give equations for the reaction of Al, Zn, and Si with sodium hydroxide.

21. What is the name and structure (nuclear and electronic) of each of the three isotopes of hydrogen?

22. Give the composition, mode of preparation, and one use of water gas.

23. Define and illustrate: oxidation, oxidizing agent, reduction, reducing agent.

24. Explain how hydrogen might be used as a secondary energy source.

PROBLEMS

1. How many grams of oxygen would be required to burn 100 g of hydrogen? What weight of water would be produced?   *Ans. 794 g; 894 g*

2. Determine the weight of oxygen required to burn 95.0 liters of hydrogen (density 0.08987 g/liter). What weight of water will be produced in the reaction?   *Ans. 67.8 g; 76.3 g*

3. Calculate the per cent of hydrogen in CH_3OH. What weight of CH_3OH contains 7.83 g of hydrogen?   *Ans. 12.6%; 62.2 g*

4. What weight of hydrogen can be produced from 2.40 kilograms of lithium hydride reacting with water? What volume would this mass of hydrogen occupy at 0° and 760 mm pressure? (The density of hydrogen is 0.08987 g/liter at 0° and 760 mm pressure.)   *Ans. 609 g; 6.77 × 10³ liters*

5. What weight of CaH_2 would be needed to fill a balloon of 650-liter capacity with hydrogen (density 0.08987 g/liter) assuming the reaction of the hydride with water to give a yield of 92.0%?   *Ans. 663 g*

6. Calculate the weight of carbon formed from cracking 5.00 tons of methane, CH_4.   *Ans. 3.74 tons*

7. What weight of HF can be formed from the hydrogen liberated by the action of 12.5 g of sodium upon an excess of water?   *Ans. 10.9 g*

8. What weight of tin(II) oxide can be reduced to metallic tin by 78.0 g of hydrogen gas? What weight of tin(IV) oxide can be reduced to metallic tin by 78.0 g of hydrogen gas?   *Ans. 5.21 kg; 2.92 kg*

9. Calculate the number of grams of iron needed in a reaction with aqueous copper(II) sulfate to precipitate 2.11 gram-atoms of copper metal.

Ans. 118 g

10. What is the volume of 10.0 g of hydrogen at 0° and 760 mm (density 0.08987 g/liter? *Ans. 111 liters*

11. A zinc strip weighing 36.4 g is placed in a beaker containing an excess of aqueous silver nitrate. What weight of metallic silver will eventually be deposited from the solution? *Ans. 120 g*

12. Calculate the number of gram-atoms of Na, Zn, and Al, respectively, required to liberate 11.2 liters of hydrogen (density 0.08987 g/liter) from an excess of acid. *Ans. 0.999; 0.499; 0.333*

13. How many grams of hydrogen can be displaced from an excess of acid by 3.50 g of magnesium? If the acid were hydrochloric, what weight of magnesium chloride would be produced in the reaction? *Ans. 0.290 g; 13.7 g*

14. What weight of water gas is formed by passing 175 kg of steam over white-hot carbon, assuming that 70.0% of the steam reacts? *Ans. 204 kg*

15. What weight of silicon must react with aqueous sodium hydroxide to give the same weight of hydrogen as that liberated when 45.0 g of zinc reacts with aqueous potassium hydroxide? *Ans. 9.67 g*

REFERENCES

"Significance of Hydrogen Isotopes," H. C. Urey (Willard Gibbs Medal Address), *Ind. Eng. Chem.*, **26**, 803 (1934).

"Hydrogen," M. E. Weeks, *Discovery of the Elements*, Sixth Edition, Publ. by the Journal of Chemical Education, Easton, Pa., 1956; pp. 197–205.

"The Flash of Genius: Deuterium: Harold C. Urey," A. B. Garrett, *J. Chem. Educ.*, **39**, 583 (1962).

"Principles of Hydrogen Chemistry," R. T. Sanderson, *J. Chem. Educ.*, **41**, 331 (1964).

"Heavy Hydrogen and Light Helium," J. C. Bevington, *Educ. in Chemistry*, **3**, 196 (1966).

"Thermochemical Hydrogen Generation," R. H. Wentorf, Jr., and R. E. Hanneman, *Science*, **185**, 311 (1974).

"Raising the Titanic by Electrolysis," L. C. Plumb, *J. Chem. Educ.*, **50**, 61 (1973).

"Hydrogen: Likely Fuel of the Future," (Staff), *Chem. & Eng. News*, June 26, 1972; p. 14.

"The Energy Resources of the Earth," M. K. Hubbert, *Sci. American*, Sept., 1971; p. 148.

"The Hydrogen Economy," D. P. Gregory, *Sci. American*, Jan., 1973; p. 13.

"The Hydrogen Economy," C. A. McAuliffe, *Chem. in Britain*, **9**, 559 (1973).

"The Stability of the Hydrogen Atom," F. Rioux, *J. Chem. Educ.*, **50**, 550 (1973).

"Interstellar Hydrogen in Galaxies," M. S. Roberts, *Science*, **183**, 371 (1974).

"Deuterium in the Universe," J. M. Pasachoff and W. A. Fowler, *Sci. American*, May, 1974; p. 108.

The Gaseous State and the Kinetic-Molecular Theory

10

Read through p. 194 for Exam III

10.1 Physical States of Matter

We noted in Section 1.4 that matter exists in three different physical states: **solid, liquid,** and **gas.** Water in the solid state is known as ice, in the liquid state as water, and in the gaseous state as water vapor. When water in the form of ice takes up enough heat energy, it changes to the liquid state; and when liquid water is heated to 100° at sea level, it changes to the gaseous state. These changes are reversed when heat energy is removed. Such changes in physical state are possible for a great many but not all substances. For example, we have observed in our study of the preparation of oxygen that both solid mercury(II) oxide and solid potassium chlorate decompose upon heating. Most substances which can exist in the gaseous state can be liquefied and solidified under suitable conditions of temperature and pressure.

10.2 Behavior of Matter in the Gaseous State

The volume of a gas, unlike that of a solid or a liquid, may be decreased greatly by increasing the pressure upon the gas. This property is known as **compressibility.** Increasing the amount of weight on a piston, such as that shown in Fig. 10–1, causes the volume of the gas confined in the cylinder to decrease until the pressure which the gas exerts is sufficient to support the piston and the greater weight.

To maintain a constant pressure when heated, a gas must expand to larger volumes. This fact can be understood by considering the gas contained in the cylinder illustrated in Fig. 10–1. When the gas is heated, it expands. By raising the piston we increase the volume to compensate for the expansion, thus allowing

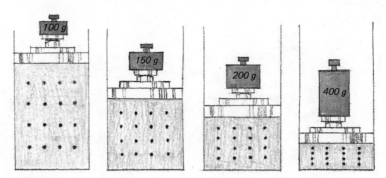

FIGURE 10-1
These figures illustrate what happens when a fixed quantity of gas at constant temperature is confined in a cylinder and subjected to more and more pressure by placing heavier and heavier weights on a movable, gas-tight piston. The number of molecules represented is but a minute fraction of the actual number in such a volume. It should be noted also that the molecules, shown lined up in the diagram for clarity, do not actually line up like this in the gaseous state.

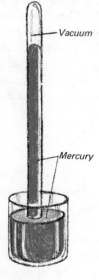

FIGURE 10-2
A mercurial barometer.

the pressure to remain constant. **Expansibility** is a characteristic of all gases. If the piston is kept from moving by adding more weight to it, then the volume remains constant. In this case the pressure of the gas increases with rise in temperature at constant volume.

A small quantity of a pungent gas released into a room in which the air is still can, in time, be detected in other parts of the room. When a sample of gas is introduced into an evacuated container, it almost instantly distributes itself and fills the container completely. These are examples of the **diffusion** of gases. Two gases introduced into the same container quickly mix by diffusion. Each gas is said to permeate the other. Thus, we say that gases possess the properties of **diffusibility** (or **diffusion**) and **permeability.**

The different gases of the atmosphere differ in density, but diffusion prevents a separation of the gases; i.e., the relative composition of the atmosphere does not vary with altitude. Mixtures of gases are homogeneous; they do not tend to separate into layers because of differences in densities.

10.3   Measurement of Gas Pressures

The pressure exerted by the atmosphere (a mixture of gases) may be measured by a simple mercury **barometer** (Fig. 10–2). Such an instrument can be made by filling a glass tube, closed at one end and about 80 cm long, with mercury and inverting it in a dish of mercury. The mercury falls in the tube to the level where the pressure exerted by the air upon the surface of the mercury in the dish is just sufficient to support the mercury in the tube. The pressure of the air is thus measured in terms of the height of the mercury column, which is the vertical distance between the surface of the mercury in the tube and that in the open

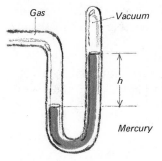

FIGURE 10-3

A simple two-arm mercurial barometer.

vessel. (Pressure means the force exerted upon a unit area of surface, e.g., pounds per square inch.) Because of the common use of the mercury barometer to measure pressure, pressure is frequently expressed in terms of **millimeters of mercury** (mm Hg, or simply mm). A pressure of one mm of Hg is alternately referred to as a **torr.**

A mercury barometer may consist of two arms, one closed and one open to the atmosphere or connected to a container filled with gas (Fig. 10-3). In this type of barometer the pressure exerted by the atmosphere or another gas is equal to that of a column of mercury whose height is the distance between the mercury levels in the two arms of the tube (h in the diagram).

The pressure of the atmosphere varies with the distance above sea level and with climatic changes. At a higher elevation, the air is less dense, and thus the pressure is less. The atmospheric pressure at 20,000 feet is only one-half of that at sea level because about half of the entire atmosphere is below this elevation. Portable **aneroid barometers** (Fig. 10-4) are made with scales graduated in feet for measuring elevations. Mountain climbers and air pilots use such barometers to determine height above sea level.

The average pressure of the atmosphere at sea level at a latitude of 45° will support a column of mercury 760 mm in height. The average sea level pressure at the latitude of 45° is thus 760 mm of mercury which is often referred to as **one atmosphere.**

The pressure unit recommended for use with the International System of Units (SI) is the **pascal,** Pa, based on the newton unit for force. A **newton,** N, is the force which when applied for one second will give to a one-kilogram mass a speed of one meter per second. Hence, a newton is expressed in the units $kg \cdot meter/sec^2$, or $kg \cdot meter \cdot sec^{-2}$. Inasmuch as pressure is force per unit area, the pressure can be expressed in terms of newtons per square meter, which are called *pascals*. Thus,

$$Pa = N/meter^2 = kg \cdot meter \cdot sec^{-2}/meter^2 = kg \cdot sec^{-2}/meter$$
$$= kg \cdot sec^{-2} \cdot meter^{-1}$$

The conversions between the various units we have considered in this section for expressing pressure are:

$$One\ atmosphere = 760\ mm\ Hg = 760\ torr = 101,325\ pascals$$

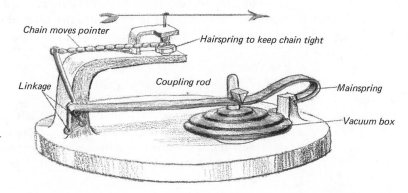

FIGURE 10-4

Diagram of an aneroid barometer. Changes in air pressure cause changes in thickness of the vacuum box to which coupling rod is attached. The pointer reacts to changes in position of the coupling rod.

Although in subsequent sections the unit we will use mainly for expressing pressures is mm Hg (equal to a torr), it is desirable to be able also to recognize and work with other units.

Example. A typical barometric pressure in a midwestern city 1,000 feet above sea level is 740 mm Hg. Express this pressure in atmospheres, torr, and pascals.

$$\frac{740 \text{ mm}}{760 \text{ mm/atm}} = 0.974 \text{ atm}$$

Inasmuch as 1 mm Hg = 1 torr,
740 mm Hg = 740 torr

$$\frac{740 \text{ mm}}{760 \text{ mm/atm}} \times 101,325 \text{ pascals/atm} = 9.87 \times 10^4 \text{ pascals}$$

10.4   Relation of Volume to Pressure at Constant Temperature; Boyle's Law

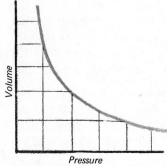

FIGURE 10-5

A graphical illustration of Boyle's Law—the inverse proportion between the volume and pressure of a gas at constant temperature.

It has been observed from experiment that, at constant temperature, the volume of a given mass of gas is reduced to half when the pressure on the gas is doubled. Conversely, the volume is doubled by a decrease in pressure to one-half. Observations of this sort were summarized by Robert Boyle in 1660 in a statement now known as **Boyle's Law: The volume of a given mass of gas held at constant temperature is inversely proportional to the pressure under which it is measured** (Fig. 10-5). The following problems will serve to illustrate Boyle's Law.

Example 1. A sample of gas has a volume of 200 ml when measured at 25° and 760 mm pressure. What volume will it occupy at 25° and 190 mm pressure?

A decrease in pressure from 760 mm to 190 mm means that the volume of the gas will increase, the temperature remaining constant. The new volume will be determined by multiplying the original volume by the ratio of the two pressures. The ratio must have the larger pressure in the numerator and the smaller pressure in the denominator so that when the original volume is multiplied by this ratio the calculation will show a change to a larger volume in accord with our prediction.

$$V_2 = \frac{P_1 V_1}{P_2} \qquad 200 \text{ ml} \times \frac{760 \text{ mm}}{190 \text{ mm}} = 800 \text{ ml}$$

Thus, the new volume is four times the original volume as the result of decreasing the pressure to one-fourth its initial value. Note that because the notation "mm" appears in both numerator and denominator of the expression, it cancels out and leaves only the notation "ml" as the unit of measurement in the answer to the problem.

$V_1 P_1 = V_2 P_2$

$P_2 = \dfrac{V_1 P_1}{V_2} = \dfrac{(400 \, ml)(750 \, mm)}{(1200 \, ml)}$

$°F = \dfrac{9}{5}(°C) + 32$

$V_2 = \dfrac{V_1 P_1}{P_2} = \dfrac{(200 \, ml)(760)}{(190 \, mm)}$

Example 2. What final pressure must be applied to a sample of gas having a volume of 400 ml at 20° and 750 mm pressure to permit the expansion of the gas to a volume of 1200 ml at 20°?

A decrease in pressure is required to permit the volume of a gas to increase at constant temperature. The new pressure will be determined by the ratio of the two volumes. By making the smaller volume the numerator and the larger volume the denominator of the ratio, and multiplying the original pressure by this ratio, the new (lower) pressure may be found.

$$750 \text{ mm} \times \frac{400 \text{ ml}}{1{,}200 \text{ ml}} = 250 \text{ mm}$$

The new pressure must be one-third the initial pressure to permit the volume to increase threefold.

10.5 The Kelvin Temperature Scale

In Section 1.21, we discussed the Kelvin temperature scale and its relationship to the Celsius and the Fahrenheit temperature scales. Now, with some knowledge of the effect of temperature on gases, we are ready to understand more about the basis for the Kelvin scale.

By experiment it has been found that when 273 ml of gas at 0° C is warmed to 1° C, its volume increases by 1 ml to 274 ml; at 20° C its volume increases to 293 ml; at 273° its volume increases to 546 ml (double that at 0°), and so on, provided that in each case the pressure remains constant. Note that the volume of the gas at 0° increases $\frac{1}{273}$ of its volume for each increase of 1° on the Celsius scale. The volume of a gas decreases in the same proportion when the temperature falls. If the temperature of 273 ml of a gas could be lowered from 0° to −273°, then the gas should have no volume at −273° because its volume should decrease at the rate of $\frac{1}{273}$ of its volume at 0° for each degree of fall in temperature. Before the temperature of −273° is reached, however, all gases become liquids or solids, to which this rate of change in volume does not apply. The temperature −273° (or more exactly −273.15° C) is called **absolute zero** and is the zero point on the **Kelvin scale** of temperature (Fig. 10–6). The freezing point of water is therefore 273° Kelvin (0° C), and the boiling point is 373° Kelvin (100° C). Temperatures on the Kelvin scale are designated as °K. As pointed out in Section 1.21, the size of the degree on the Kelvin scale is the same as on the Celsius scale, and a temperature on the Celsius scale is converted to the Kelvin scale by adding 273° to the Celsius reading. For example, 25° C = (25 + 273)° K = 298° K.

$°C = \frac{5}{9}(°F - 32)$

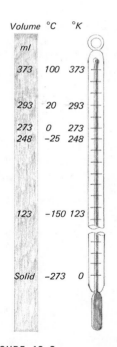

Volume	°C	°K
ml		
373	100	373
293	20	293
273	0	273
248	−25	248
123	−150	123
Solid	−273	0

FIGURE 10–6

The volume of a gas at constant pressure is directly proportional to its Kelvin temperature.

10.6 Relation of Volume to Temperature at Constant Pressure; Charles' Law

Studies of the effect of temperature upon the volume of confined gases at constant pressure by Charles in 1787 led to the generalization known as **Charles' Law,**

FIGURE 10-7

A graphical illustration of Charles' Law—the direct proportion between the volume of a gas and the Kelvin temperature at constant pressure.

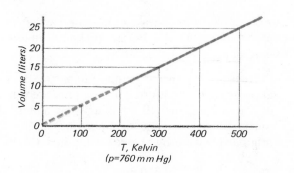

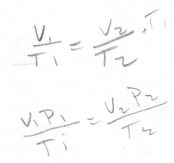

$$\frac{V_1}{T_1} = \frac{V_2}{T_2}$$

$$\frac{V_1 P_1}{T_1} = \frac{V_2 P_2}{T_2}$$

which we can state as follows: **The volume of a given mass of gas is directly proportional to its temperature on the Kelvin scale when the pressure is held constant** (see Figs. 10–6 and 10–7). Note that Charles' Law applies to the volume of a gas at its **Kelvin** temperature. The following examples are based upon this law.

$$V_2 = \frac{V_1 P_1 T_2}{T_1 P_2} = \frac{(300\,ml)(750\,mm)(30°)}{(10°)(780)}$$

Example 1. A sample of gas occupies 300 ml at 10° and 750 mm pressure. What volume will the gas have at 30° and 750 mm?

The Celsius temperatures are first converted to the Kelvin scale, $10°C + 273° = 283° K$, and $30° C + 273° = 303° K$. When the temperature of a gas is raised, the gas expands, and the ratio of temperatures used in solving for the final volume must be an improper fraction; i.e., the larger temperature must be in the numerator:

$$300 \text{ ml} \times \frac{303° \text{ K}}{283° \text{ K}} = 321 \text{ ml}$$

Example 2. A sample of gas occupies 110 ml at 27° and 740 mm pressure. What temperature will the gas have when its volume is changed to 90 ml at 740 mm?

The initial temperature of 27° C is equal to 300° K. The temperature must decrease to cause a reduction in volume at constant pressure. Thus, the ratio of volumes must be a proper fraction, i.e., the larger volume must be in the denominator:

$$300° \text{ K} \times \frac{90 \text{ ml}}{110 \text{ ml}} = 245° \text{ K}$$

Subtracting 273° from 245° K, we find the Celsius temperature to be −28°.

10.7   Standard Conditions of Temperature and Pressure

From the foregoing considerations of the variation of the volume of a given mass of gas with changes in pressure and temperature, it should be clear that the volume and the density (mass per unit volume) of a gas vary with these conditions. Thus, to be able to compare different gases with regard to their densities or to fix a definite density for any gas, it is desirable to adopt a set of **standard conditions of temperature and pressure** (S.T.P.) for all measurements of gases. Accordingly, 0° C temperature and 760 mm pressure are universally used as standard conditions.

(handwritten: 1 atmosphere — 760 tair)

10.8   Correction of the Volume of a Gas to Standard Conditions

Example.   Let us suppose that a sample of gas is found to occupy 250 ml under laboratory conditions of 27° C and 740 mm pressure. Correct the volume to standard conditions of 0° C and 760 mm.

(handwritten: $V_2 = \dfrac{V_1 P_1 T_2}{T_1 P_2} = \dfrac{(250\,ml)(740\,mm)(273°K)}{(300°K)(760\,mm)}$)

First, convert the Celsius temperatures to the Kelvin scale: 27° C = (27 + 273)° K = 300° K, and 0° C = (0 + 273)° K = 273° K. A decrease in temperature from 300° K to 273° K will cause a decrease in the volume of the gas. Therefore, we multiply the original volume by a fraction made up of the two temperatures and having a value less than unity:

$$250 \text{ ml} \times \frac{273° \text{ K}}{300° \text{ K}}$$

The increase in pressure from 740 mm to 760 mm will also decrease the volume of the gas, and this factor may be included in the same expression with the temperature factor to obtain the corrected volume:

$$250 \text{ ml} \times \frac{273° \text{ K}}{300° \text{ K}} \times \frac{740 \text{ mm}}{760 \text{ mm}} = 222 \text{ ml}$$

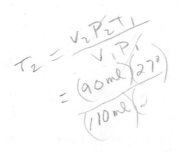

(handwritten: $T_2 = \dfrac{V_2 P_2 T_1}{V_1 P_1} = \dfrac{(90\,ml)(27°)}{(110\,ml)(\ )}$)

The volume could be corrected to any other conditions of temperature and pressure by the same type of calculation.

It is common practice to correct the volume of a gas to standard conditions, or to other conditions of temperature and pressure, in one operation. This kind of calculation can be made also by using the Ideal Gas Law Equation, which will be discussed in Sections 10.21 and 10.22.

10.9   The Pressure of a Mixture of Gases; Dalton's Law

Suppose, for example, that we have gas A at a pressure of 100 mm in a 1-liter container, and gas B also at a pressure of 100 mm and contained in a second 1-liter container. After transferring gas B to the 1-liter container containing gas A, the temperature being held constant, the total pressure of the mixture is found to be 200 mm. It is logical to assume that each gas is exerting the same pressure as before. In the absence of chemical interaction between the components of a mixture of gases, the individual gases do not interfere with the pressures of one another. The pressure exerted by each gas in a mixture is called the **partial**

pressure of that gas, and the total pressure of the mixture of gases is the sum of the partial pressures of all the gases present in the mixture. This law is known as **Dalton's Law of Partial Pressures** and may be stated as follows: **The total pressure of a mixture of gases is equal to the sum of the partial pressures of the component gases.**

A convenient method of measuring the pressure of a given volume of a gas or a mixture of gases is to collect the gas over water and make its pressure equal to the existing air pressure. This is easily done by adjusting the water level (Fig. 10–8) so that it is the same both inside and outside the container. The gas is then at the existing atmospheric pressure, which can be read on a laboratory barometer.

There is, however, another factor to consider when determining pressure by this method. When a gas is collected over water, it will soon become saturated with water vapor. The total pressure of the mixture then will be equal to the sum of the partial pressure of the gas and that of the water vapor. The pressure of the pure gas is therefore equal to the total pressure minus the water vapor pressure.

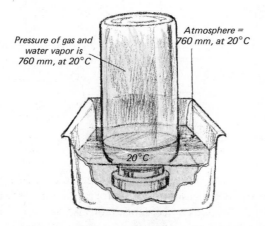

Pressure of gas and water vapor is 760 mm, at 20°C

Atmosphere = 760 mm, at 20°C

20°C

FIGURE 10-8
Method by which the pressure on a confined mixture of gases can be made equal to the atmospheric pressure.

Example 1. Suppose that 200 ml of hydrogen is collected over water at a temperature of 26° and a pressure of 750 mm. The pressure of water vapor at 26° is 25 mm. What is the pressure of the hydrogen in the dry state at 26°?

According to Dalton's Law of Partial Pressures, the pressure of the dry hydrogen is the difference between the total pressure (750 mm) and the vapor pressure of water at 26° (25 mm), or

$$P_{H_2} = P_{Total} - P_{H_2O\ vapor} = 750\ mm - 25\ mm = 725\ mm$$

The vapor pressure of water at various temperatures is given in Table 10–1. This table and Fig. 10–9 illustrate the fact that vapor pressure increases as the temperature increases.

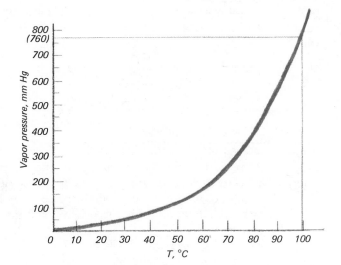

FIGURE 10-9

A graph illustrating change of water vapor pressure with change in temperature.

TABLE 10-1 Vapor Pressure of Ice and Water at Various Temperatures

Temperature, °C	Pressure, mm Hg	Temperature, °C	Pressure, mm Hg	Temperature, °C	Pressure, mm Hg
−10	2.1	18	15.5	80	355.1
−5	3.2	19	16.5	90	525.8
−2	4.0	20	17.5	95	633.9
−1	4.3	21	18.7	96	657.6
0	4.6	22	19.8	97	682.1
1	4.9	23	21.1	98	707.3
2	5.3	24	22.4	99	733.2
3	5.7	25	23.8	99.1	735.9
4	6.1	26	25.2	99.2	738.5
5	6.5	27	26.7	99.3	741.2
6	7.0	28	28.3	99.4	743.9
7	7.5	29	30.0	99.5	746.5
8	8.0	30	31.8	99.6	749.2
9	8.6	31	33.7	99.7	751.9
10	9.2	32	35.7	99.8	754.6
11	9.8	33	37.7	99.9	757.3
12	10.5	34	39.9	100.0	760.0
13	11.2	35	42.2	100.1	762.7
14	12.0	40	55.3	100.2	765.5
15	12.8	50	92.5	100.3	768.2
16	13.6	60	149.4	100.5	773.7
17	14.5	70	233.7	101.0	787.5

$\frac{P_1V_1}{T_1} = \frac{P_2V_2}{T_2}$

$\frac{23}{273}$
$\frac{273}{296}$ $800mm - 21.1m = 978.9\,mm$

$V_2 = \frac{P_1V_1T_2}{T_1P_2}$

$= \frac{(779)(500\,ml)(273°k)}{(296k)(760)}$

$= 473\,ml$

Example 2. Suppose that 500 ml of oxygen is collected over water at a temperature of 23° and a barometric pressure of 800 mm. The vapor pressure of water at 23° C is 21.1 mm. Calculate the volume of dry oxygen corrected to standard conditions.

The gas collected over water is a mixture of oxygen and water vapor. The pressure of the dry oxygen is the total pressure minus the vapor pressure of water or

$$P_{O_2} = P_{Total} - P_{H_2O\,vapor} = 800\text{ mm} - 21\text{ mm} = 779\text{ mm}$$

A decrease in temperature from 23° C (296° K) to 0° C (273° K) causes a decrease in volume. To correct for the temperature change, multiply the original volume by 273° K/296° K. A decrease in pressure from 779 mm to 760 mm causes an increase in volume. To correct for the pressure change, multiply the original volume by 779 mm/760 mm.

$$\text{Volume of dry oxygen at S.T.P.} = 500\text{ ml} \times \frac{273°\,K}{296°\,K} \times \frac{779\text{ mm}}{760\text{ mm}} = 473\text{ ml}$$

10.10 Diffusion of Gases; Graham's Law

When a sample of gas is set free in one part of a closed container, it very quickly diffuses throughout the container (see Fig. 10–10). If a mixture of gases is placed in a container the walls of which are porous to gases, diffusion of the gases through the porous walls will take place. The lighter gases will diffuse through the small openings of the porous walls more rapidly than the heavier ones. Thomas Graham, in 1832, studied the rates of diffusion of different gases and showed that **the rates of diffusion of gases are inversely proportional to the square roots of their densities (or molecular weights).**

Example. Calculate the ratio of the rate of diffusion of hydrogen to the rate of diffusion of oxygen.

Using densities:
The density of hydrogen is 0.08987 g/liter; that of oxygen is 1.429 g/liter.

$$\frac{\text{Rate of diffusion of hydrogen}}{\text{Rate of diffusion of oxygen}} = \frac{\sqrt{1.429\text{ g/liter}}}{\sqrt{0.08987\text{ g/liter}}} = \frac{1.20}{0.30} = \frac{4}{1}$$

or

Using molecular weights:

$$\frac{\text{Ratio of diffusion of hydrogen}}{\text{Ratio of diffusion of oxygen}} = \frac{\sqrt{32}}{\sqrt{2}} = \frac{\sqrt{2}\sqrt{16}}{\sqrt{2}} = \frac{\sqrt{16}}{1} = \frac{4}{1}$$

This means that hydrogen diffuses four times as rapidly as oxygen. The rates of diffusion must depend upon the speeds of the molecules; the rate of diffusion is greater for molecules of smaller mass and higher speeds than

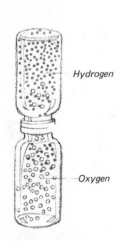

FIGURE 10-10
Despite the difference in densities of these two gases their molecules intermingle; that is, the gases diffuse. The rapid motion of the molecules and the relatively large spaces between them explain why diffusion of gases occurs.

Hydrogen

Oxygen

for molecules of larger mass and lower speeds (see Section 10.11). Note that the density of oxygen is almost exactly sixteen times that of hydrogen.

Practical application of differences in diffusion rates of gaseous substances is made in the separation of light isotopes from heavy ones. Fractional diffusion through a porous barrier from a region of higher pressure to one of lower pressure is used for the large-scale separation of gaseous $^{235}_{92}UF_6$ from $^{238}_{92}UF_6$ at the Oak Ridge atomic energy installation in Tennessee. It is said that for separation to be complete, a given volume of UF_6 must be diffused some two million times.

10.11   The Kinetic-Molecular Theory

The idea that the properties of gases such as compressibility, expansibility, and diffusibility could be accounted for by considering the gas molecules to be in continuous motion occurred to several people in the past (Bernoulli in 1738, Poule in 1851, and Kronig in 1856). During the latter half of the nineteenth century, Clausius, Maxwell, Boltzmann, and others developed this hypothesis into the detailed **kinetic-molecular theory** of gases, which is summarized as follows.

(1) Gases are composed of separate particles called molecules. The volume actually occupied by the individual molecules of a gas under ordinary conditions is insignificant compared with the total volume of the gas. Thus the molecules of a gas are relatively far apart, and they have very little attraction for one another, except when near the liquefaction point.

(2) The molecules of a gas are in continuous motion (Fig. 10–11) with varying speeds. They move in straight lines in all directions and behave as perfectly elastic bodies when they collide with the walls of the container or with each other.

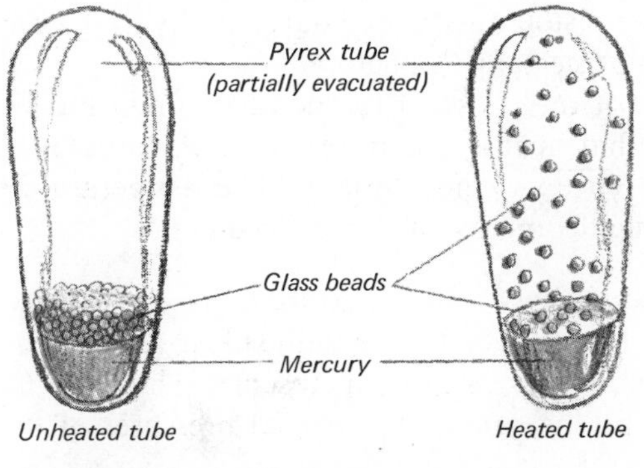

FIGURE 10–11
When the mercury in the tube is heated sufficiently to provide an appreciable vapor pressure, the glass beads move toward the top of the tube and provide visual evidence of the motion of gaseous mercury molecules.

Pyrex tube
(partially evacuated)

Glass beads

Mercury

Unheated tube   Heated tube

(3) The average kinetic energy of the molecules (energy due to their motion) of different gases is the same at the same temperature, regardless of differences in mass. The kinetic energy of gaseous molecules increases with a rise in temperature and decreases as the temperature falls. The kinetic energy of any moving body is equal to $\frac{1}{2}mu^2$, where m is its mass in grams and u is its speed in centimeters per second. It follows that molecules of small mass, like hydrogen, must move at higher speeds than molecules of larger mass, like oxygen, since both have the same average kinetic energy at the same temperature.

10.12 Relation of the Behavior of Gases to the Kinetic-Molecular Theory

The gas laws may be readily explained in light of the kinetic theory.

■ **1. Boyle's Law.** The pressure exerted by a gas upon the walls of its container is caused by the bombardment of the walls by rapidly moving molecules of the gas and varies directly with the number of molecules, confined within a given volume, hitting the walls per unit time. Reducing the volume of a given mass of gas to one-half will double the number of molecules per unit volume. It follows that the number of impacts per unit time upon the same area of wall surface will also be doubled. The doubling of the number of impacts per unit area results in twice as much pressure. These results are in accordance with Boyle's law relating volume and pressure.

$$P_1 V_1 = P_2 V_2$$

■ **2. Charles' Law.** It is a common observation that a rise in temperature increases the pressure of a gas at constant volume. The increase in pressure of a gas with increase in temperature reflects the increase in average kinetic energy of the molecules as the temperature is raised. An increase in the average speed of the molecules results in more frequent and harder impacts upon the walls of the container, i.e., greater pressure.

$$\frac{(P_1) V_1}{T_1} = \frac{V_2}{T_2}$$

If the pressure remains constant with increasing temperature, the volume must increase so that each molecule on the average travels farther before hitting the wall. The smaller number of molecules striking the wall at a given time exactly offsets the greater force with which each molecule hits, making it possible for the pressure to remain constant. This may be visualized by picturing a cylinder fitted with a piston. As the temperature increases the molecules hit the piston harder. The piston must move out, thereby increasing the volume, if the pressure is to remain the same as before.

A study of the effect of temperature upon kinetic-molecular motion leads to the belief that heat is a manifestation of the total kinetic energy that molecules possess by virtue of their motion. The temperature of a gas is a measure of the average kinetic energy of the molecules.

■ **3. Dalton's Law.** In a mixture of gases the molecules of one component will bombard the walls of the container just as frequently in the presence of other kinds of molecules as in their absence. Thus, the total pressure of a mixture of gases will be the sum of the partial pressures of the individual gases.

$$P_T = P_A + P_B + \cdots$$

■ **4. Graham's Law.** The fact that the molecules of a gas are in rapid motion and that the free space between the molecules is very great explains the phenomenon of diffusion. At the same temperature, molecules of different gases have the same average kinetic energy. This means that molecules of small mass move with higher speeds than those of large mass. Consequently, diffusion rates of gases are related inversely to their molecular weights and densities. This can be shown mathematically by the following derivation.

For two different gases, at the same temperature and pressure,

$$\frac{R(v_1)}{R(v_2)} = \frac{\sqrt{m_2}}{\sqrt{m_1}}$$

$$\text{kinetic energy for first gas} = \frac{1}{2}m_1u_1{}^2$$

$$\text{kinetic energy for second gas} = \frac{1}{2}m_2u_2{}^2$$

(where m_1 and m_2 are the masses of individual molecules of the two gases, and u_1 and u_2 are their speeds).

The kinetic energies of the two gases are equal. Hence,

$$\frac{1}{2}m_1u_1{}^2 = \frac{1}{2}m_2u_2{}^2$$

Dividing both sides of the equation by $\frac{1}{2}$, we obtain

$$m_1u_1{}^2 = m_2u_2{}^2$$

On rearranging, we have

$$\frac{u_1{}^2}{u_2{}^2} = \frac{m_2}{m_1}$$

Extracting the square roots gives

$$\frac{u_1}{u_2} = \sqrt{\frac{m_2}{m_1}}$$

We may assume that the rate at which a gas will diffuse, R, is proportional to the average speed of the gas molecules, and that the molecular weights, M_1 and M_2, of the two gases are proportional to their molecular masses. Thus,

$$\frac{R_1}{R_2} = \sqrt{\frac{M_2}{M_1}}$$

This equation states that the ratio of the rates of diffusion of two gases, held at the same temperature and pressure, equals the inverse ratio of the square roots of the molecular weights of the two gases.

As we shall see later (Section 10.17), the volume occupied by one gram-molecular weight of a gas at a particular temperature and pressure is the same for all gases and is 22.4 liters at S.T.P.

$$\text{Density} = \frac{\text{Mass}}{\text{Volume}} \quad \text{(see Sect. 1.17)}$$

Hence, the density of any gas at S.T.P. can be calculated by dividing its gram-molecular weight by its volume (22.4 liters). The ratio of the densities (d_1 and d_2) for two gases can then be shown to be equal to the ratio of their molecular weights.

$$\frac{d_1}{d_2} = \frac{\left(\dfrac{M_1}{22.4\ \text{liters}}\right)}{\left(\dfrac{M_2}{22.4\ \text{liters}}\right)} = \frac{M_1}{M_2}$$

Therefore, Graham's Law can equally well be expressed either in terms of the square roots of the molecular weights or the square roots of the densities of the gases.

$$\frac{R_1}{R_2} = \sqrt{\frac{M_2}{M_1}} = \sqrt{\frac{d_2}{d_1}}$$

Hydrogen is the lightest of all the gases and therefore diffuses the most rapidly. The average speed of its molecules at room temperature is about one mile per second and that of oxygen molecules is about one-fourth mile per second. However, diffusion rates are much lower than would be expected from such fast moving molecules due to collisions between molecules, of which there are about 11 billion per molecule per second in hydrogen gas at standard conditions. Hydrogen molecules travel about 17×10^{-6} cm between collisions and, on the average, are halted about 60 thousand times in traveling one centimeter.

10.13   The Distribution of Molecular Velocities

On the basis of the kinetic-molecular theory of gases, we must assume that the individual molecules of a particular gas travel at different speeds. Because of the enormous number of collisions, the speeds of the molecules will vary from practically zero to nearly the speed of light. Because of the collisions of molecules and the consequent exchanges of energy, the speed of a given molecule is continually changing. However, because a large number of molecules is involved, the distribution of molecular speeds of the total number of molecules is constant. The nature of the distribution of molecular speeds is described by the **Maxwell-Boltzmann Distribution Law.**

The distribution of molecular speeds is shown in Fig. 10–12. The vertical axis represents the number of molecules, and the horizontal axis represents molecular

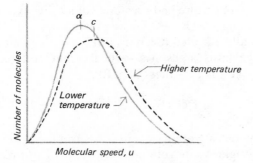

FIGURE 10-12

Distributions of molecular speed vs. number of molecules of a gas at different temperatures.

speed, u. Alpha (α) points out the most probable speed and c the average speed of molecules. The graph indicates that very few molecules move at very low or very high speeds. The number of molecules with intermediate speeds increases rapidly up to a maximum and then drops off rapidly.

At the higher temperature, the curve is flattened and shifted to higher speeds. Thus, at a higher temperature more molecules have higher speeds and fewer molecules have the lower speeds. This is in accord with expectations based upon kinetic-molecular theory.

10.14 Reactions Involving Gases; Gay-Lussac's Law

Gases combine, or react, in definite and simple proportions by volume. For example, it has been determined by experiment that one volume of nitrogen will combine with three volumes of hydrogen to give two volumes of ammonia gas, provided the volumes of the reactants and product are measured under the same conditions of temperature and pressure.

$$N_2 \quad + \quad 3H_2 \quad \longrightarrow \quad 2NH_3$$
1 Volume   3 Volumes   2 Volumes

The term "volume" is used here in a general sense, and if the volume of nitrogen is measured in liters, the volumes of hydrogen and ammonia must also be measured in liters. Experimental observations on volumes of combining gases were summarized by **Joseph Louis Gay-Lussac** (1788–1850) in the **Law of Combining Volumes: The volumes of gases involved in a reaction, at constant temperature and pressure, can be expressed as a ratio of small whole numbers.** It is important to remember that this law applies only to substances in the gaseous state and measured at the same temperature and pressure. The volumes of any solids or liquids involved in the reactions are not considered. When liquids or solids undergo chemical reaction, no generalizations can be made concerning the volumes of the reactants and products involved in the reactions. Additional examples illustrating Gay-Lussac's Law are as follows:

(1) Two volumes of hydrogen and one volume of oxygen react to give two volumes of steam.

$$2H_2 \quad + \quad O_2 \quad \longrightarrow \quad 2H_2O$$
2 Volumes   1 Volume   2 Volumes

(2) One volume of hydrogen combines with one volume of chlorine to form two volumes of hydrogen chloride.

$$H_2 \quad + \quad Cl_2 \quad \longrightarrow \quad 2HCl$$
1 Volume   1 Volume   2 Volumes

(3) Carbon (a solid) reacts with one volume of oxygen to give one volume of carbon dioxide.

$$C \quad + \quad O_2 \quad \longrightarrow \quad CO_2$$
A solid   1 Volume   1 Volume

(4) Four volumes of steam react with iron (a solid) to yield four volumes of hydrogen and Fe_3O_4 (a solid).

$$4H_2O \quad + 3Fe \quad \longrightarrow \quad 4H_2 \quad + Fe_3O_4$$
4 Volumes   Solid   4 Volumes   Solid

The simplest ratio, of course, for the gaseous volumes of water and hydrogen in this case is one to one.

10.15   An Explanation of Gay-Lussac's Law; Avogadro's Law

The law of combining volumes of gases can be satisfactorily explained in terms of the molecular nature of gases if we assume that **equal volumes of all gases, measured under the same conditions of temperature and pressure, contain the same number of molecules.** Avogadro advanced this hypothesis in 1811 to account for the behavior of gases. His hypothesis, which has since been experimentally proven, is now accepted as fact and is known as **Avogadro's Law.**

Consider the union of hydrogen and chlorine to produce hydrogen chloride, the volumes of the reactants and product being measured under the same conditions of temperature and pressure:

$$\text{1 volume of } H_2 + \text{1 volume of } Cl_2 \longrightarrow \text{2 volumes of HCl}$$

According to Avogadro's Law, equal volumes of hydrogen, chlorine, and hydrogen chloride contain the same number of molecules. During reaction, then, two molecules of hydrogen chloride are formed from one molecule each of hydrogen and chlorine. Now each molecule of hydrogen chloride must contain at least one atom of hydrogen and one atom of chlorine; hence, two molecules of hydrogen chloride will contain at least two atoms of hydrogen and two atoms of chlorine, which must have been present in one molecule of hydrogen and one molecule of chlorine.

$$\text{1 volume of } H_2 + \text{1 volume of } Cl_2 \longrightarrow \text{2 volumes of HCl}$$
$$\text{1 molecule of } H_2 + \text{1 molecule of } Cl_2 \longrightarrow \text{2 molecules of HCl}$$

No gaseous reaction has been found in which one molecule of hydrogen (or chlorine) contains enough of the element to form more than two molecules of product. We assume, therefore, that each molecule of hydrogen (and chlorine) contains two and only two atoms. If it were possible to cause all the diatomic molecules in a sample of hydrogen to dissociate to monatomic molecules, we would expect the volume to double for the same conditions of temperature and pressure. Each of the original hydrogen molecules would have dissociated into two new molecules.

$$H_2 \longrightarrow 2H$$
$$\text{1 Volume} \qquad \text{2 Volumes}$$

It can be shown similarly that elemental oxygen, nitrogen, fluorine, chlorine, bromine, and iodine also consist of diatomic molecules.

The volumetric relationship involved in the formation of hydrogen chloride from hydrogen and chlorine may be represented as in Fig. 10–13. The adherence of other gaseous reactions to Gay-Lussac's Law can be explained and pictured graphically in a manner similar to that described for the reaction of hydrogen with chlorine. Before Gay-Lussac's Law was stated, Dalton considered and then

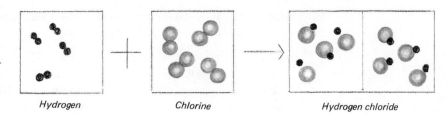

FIGURE 10-13

One volume of hydrogen combines with one volume of chlorine to yield two volumes of hydrogen chloride, HCl.

Hydrogen   *Chlorine*   *Hydrogen chloride*

rejected the idea that equal volumes of gases contain the same number of atoms; it did not occur to him that elements might exist as polyatomic molecules (e.g., H_2, O_3).

Gay-Lussac's Law can be used to determine the volumes of gases involved in a reaction.

Example. **Calculate the number of liters of hydrogen that will combine with 20 liters of nitrogen to form ammonia, the gaseous volumes being measured at the same temperature and pressure.**

First, write the balanced equation for the reaction.

$$N_2 + 3H_2 \longrightarrow 2NH_3$$

From the equation we see that one molecule of nitrogen will combine with three molecules of hydrogen. The volumes are proportional to the numbers of molecules, so 20 liters of nitrogen will combine with

$$20 \text{ liters} \times \frac{3 \text{ volumes}}{1 \text{ volume}} = 60 \text{ liters of hydrogen}$$

$$? \ell\, H_2 = 20\,\ell\, N_2 \left(\frac{3(22.4)}{1\,(22.42)N_2} \right) = 60$$

10.16   Relative Densities of Gases

Inasmuch as molecules of different substances have different masses, and equal volumes of different gases under the same conditions of temperature and pressure contain the same number of molecules, it follows that the densities of different gases will not be the same. For example, the density of oxygen at S.T.P. is 1.429 grams per liter, whereas that of hydrogen is 0.08987 gram per liter. Because equal volumes of these gases contain the same number of molecules, the ratio of the densities of the two gases is the same as the ratio of their molecular weights (see also Section 10.12, Part 4).

$$\frac{\text{Density of oxygen}}{\text{Density of hydrogen}} = \frac{1.429 \text{ g/liter}}{0.08987 \text{ g/liter}}$$

$$= 15.9 = \frac{\text{mol. wt. of oxygen}}{\text{mol. wt. of hydrogen}} = \frac{32.00}{2.016} = 15.9$$

We can take advantage of this fact in the solution of density and molecular weight problems.

Example. Suppose we wish to calculate the molecular weight of carbon dioxide, the density of which is 1.977 g/liter at S.T.P.

Noting that the density of carbon dioxide is greater than that of oxygen (1.429 g/liter), it follows that its molecular weight should be greater than that of oxygen (32.00). We may obtain the molecular weight of carbon dioxide by multiplying that of oxygen by the ratio of the two densities.

$$32 \times \frac{1.977 \text{ g/liter}}{1.429 \text{ g/liter}} = 44$$

It should be obvious to the student that if the molecular weights of two gases are known and if the density of one of them is known, the density of the other may be calculated easily.

10.17   Gram-Molecular Volume of Gases

As we shall see in Section 10.19, the volume occupied by one gram-molecular weight of a substance in the gaseous state is important in the experimental determination of the molecular weight of the substance. We can calculate the volume that one gram-molecular weight of a gas will occupy under standard conditions if we know its density at S.T.P. and its molecular weight.

Example. Oxygen has a density of 1.429 g/liter at S.T.P. and its molecular weight is 32.00. Calculate the volume occupied by one gram-molecular weight of oxygen at S.T.P.

One mole (32.00 g) of oxygen will occupy more space than 1.429 g of oxygen, which (from its density) has a volume of one liter. Therefore, set up the expression in such a manner as to give a larger volume:

$$1 \text{ liter} \times \frac{32.00 \text{ g}}{1.429 \text{ g}} = 22.4 \text{ liters}$$

The volume occupied by one gram-molecular weight of a gas is known as its gram-molecular volume (G.M.V.). A gram-molecular weight of any substance contains 6.022×10^{23} molecules of that substance (Section 2.5), and the same number of molecules of different gases occupy equal volumes under the same conditions of temperature and pressure. The highly important and interesting fact follows, then, that the gram-molecular volume at any given temperature and pressure is the same for all ideal gases; it is 22.4 liters at 0° C and 760 mm (Fig. 10–14). Inasmuch as gases do not behave exactly as *ideal* gases—that is, they do not follow perfectly the ideal conditions postulated by the kinetic-molecular theory—they show slight deviations from Avogadro's Law. Thus, the gram-molecular volume is not quite the same for different gases, and therefore not always quite equal to 22.4 liters. At ordinary pressures, however, real gases follow the gas laws quite closely, and hence 22.4 liters is a very good approximation. The subject of deviation of gases from ideal behavior is discussed in Section 10.23.

FIGURE 10-14

A gram-molecular weight of any gas occupies a volume of approximately 22.4 liters at standard conditions and is the weight in grams of 6.022×10^{23} molecules of that gas.

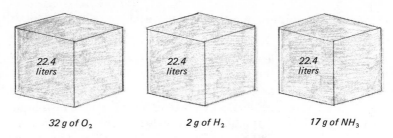

32 g of O_2 2 g of H_2 17 g of NH_3

10.18   Determination of the Volumes of Gases Involved in a Reaction

Chemical equations involving gases may be quantitatively interpreted from the fact that one gram-molecular weight of a gas occupies 22.4 liters at standard conditions.

Example.   **What volume of oxygen, measured at 0° and 760 mm, would be formed by complete decomposition of 21.66 g of mercury(II) oxide? The equation for the reaction is**

$$2HgO \longrightarrow 2Hg + O_2$$
2 moles   1 mole

A 21.66 gram quantity of HgO is equivalent to $\frac{21.66}{216.6}$ mole (gram-formula weight) or 0.1000 mole of HgO. The equation shows that two moles of HgO will liberate one mole of oxygen (volume, 22.4 liters at S.T.P.). Hence, 0.1000 mole of HgO will liberate (0.1000) $(\frac{1}{2})$ mole of O_2, which would have a volume of (0.1000) $(\frac{1}{2})$ (22.4) liters.

Summarizing:

$$\frac{21.66 \text{ g}}{216.6 \text{ g/mole}} \times \frac{1}{2} \times 22.4 \frac{\text{liters}}{\text{mole}} = 1.12 \text{ liters of oxygen}$$

Note that the unit dimensions provide the volume unit, liter, for the answer:

$$\frac{\text{g}}{\left(\frac{\text{g}}{\text{mole}}\right)} \times \frac{\text{liters}}{\text{mole}} = \cancel{\text{g}} \times \frac{\text{mole}}{\cancel{\text{g}}} \times \frac{\text{liters}}{\cancel{\text{mole}}} = \text{liters}$$

10.19   Determination of Molecular Weights of Gases or Volatile Compounds

The molecular weight of a gas can be found readily by determining experimentally the weight of a known volume of the gas under laboratory conditions of temperature and pressure and, using these data, calculating the weight of 22.4 liters of the gas at standard conditions.

Example. A sample of gas occupies 400 ml at 20.0° C and 740 mm and weighs 0.842 g. What is the molecular weight of the gas?

The volume of the gas at standard conditions can be found by multiplying the measured volume by fractions determined by the temperature change and the pressure change. The temperature change is from 293° K to 273° K, which will cause a decrease in volume; such a change will require a fraction whose value is less than unity. An increase in pressure from 740 mm to 760 mm, which will cause a decrease in volume, thus requires a fraction with a value less than unity.

$$400 \text{ ml} \times \frac{273° \text{ K}}{293° \text{ K}} \times \frac{740 \text{ mm}}{760 \text{ mm}} = 364 \text{ ml}$$

The weight of 364 ml of the gas at standard conditions is 0.842 g. Therefore, the weight of 22,400 ml (22.4 liters) of the gas at S.T.P. can be found by multiplying the weight of 364 ml of the gas by a fraction whose value is greater than unity and made up of the two volumes.

$$0.842 \text{ g} \times \frac{22,400 \text{ ml}}{364 \text{ ml}} = 52.0 \text{ g}$$

Thus, the weight of 22.4 liters of the gas at S.T.P. is 52.0 g. This weight is numerically equal to the molecular weight of the gas, 52.0. Molecular weights obtained by this method are only approximate because the gram-molecular volume at S.T.P. is not exactly 22.4 liters for all gases.

This method of determining molecular weights is applicable only to gases and to substances that can be vaporized without decomposition. Many substances decompose before they are completely changed to the vapor state, and many others vaporize at such high temperatures that it is either impossible or impractical to determine their molecular weights by this method. For substances that are soluble in suitable solvents, the molecular weights can be determined by other methods such as the depression of the freezing point, elevation of the boiling point, or change in the osmotic pressure of their solutions (Chapter 13).

10.20   Determination of Atomic Weights of Gases or Volatile Compounds

One general method of determining the atomic weight of an element is based on determining the molecular weights of different compounds of the element. Because an atom is the smallest portion of an element that can be contained in a molecule of a compound, it follows that the gram-molecular weight (mole) of any compound of an element must contain one or some whole multiple gram-atomic weight(s) of that element. Thus, the smallest weight of an element ever found in a gram-molecular weight of any of its compounds will be the gram-atomic weight of the element. Hence, it is usually possible to obtain the atomic weight of an element by experimentally determining the smallest weight of that

TABLE 10-2   Composition of Some Chlorine Compounds

Compound	Gram-Molecular Weight, g	Total Weight of Chlorine in One Gram-Molecular Weight of Compound	Formula
Hydrogen chloride	36.5	35.5	HCl
Methyl chloride	50.5	35.5	CH_3Cl
Chlorine dioxide	67.5	35.5	ClO_2
Phosgene	99.0	71.0 (2×35.5)	$COCl_2$
Phosphorus trichloride	137.4	106.4 (3×35.5)	PCl_3
Carbon tetrachloride	153.8	141.8 (4×35.5)	CCl_4
Phosphorus pentachloride	208.3	177.3 (5×35.5)	PCl_5

element in a gram-molecular volume of several of its volatile compounds. Atomic weights found by this method are only approximate because Avogadro's Law is not absolutely accurate.

To illustrate the determination of atomic weights by this method, let us consider the gram-molecular weights and analyses of several compounds of chlorine (Table 10-2).

An inspection of Table 10-2 shows that the weight of chlorine in a gram-molecular weight of these volatile compounds, and hence in a gram-molecular volume at S.T.P., is either 35.5 g or a simple whole number multiple of this number. Therefore, it is logical to conclude that the atomic weight of chlorine is very close to 35.5, if the data are sufficient.

10.21   General Gas Law Equation

The conditions relating to pressure, volume, temperature, and number of moles of a gas can be expressed in one equation, referred to as the **General Gas Law Equation,** or the **Ideal Gas Law Equation.**

Boyle's Law states that the volume, V, of a particular number of moles, n, of an ideal gas is inversely proportional to the pressure, P, at constant temperature, T. Using the symbol $\propto$ for "is proportional to," this can be written

$$V \propto \frac{1}{P} \text{ at constant } T \text{ and } n$$

Charles' Law states that the volume, V, of n moles of an ideal gas is directly proportional to the Kelvin temperature, T, at constant pressure, P.

$$V \propto T \text{ at constant } P \text{ and } n$$

According to Avogadro's Law, the volume of a gas is proportional to the number of molecules, and hence the number of moles, n, of the gas at constant pressure, P, and temperature, T.

$$V \propto n \text{ at constant } P \text{ and } T$$

In general, it can be shown that

$$V \propto \left(\frac{1}{P}\right)(T)(n)$$

or

$$V = R\left(\frac{1}{P}\right)(T)(n)$$

or

$$PV = nRT$$

where R is the proportionality constant and is often referred to as the **universal gas constant.**

The numerical value of R can be obtained by substituting actual experimental values for P, V, n, and T. At standard conditions, $P = 760$ mm or 1 atmosphere, $T = 273°$ K, and one mole of any gas occupies a volume of 22.4 liters. Substituting these values in $PV = nRT$:

$$(1 \text{ atm})(22.4 \text{ liters}) = (1 \text{ mole})(R)(273° \text{ K})$$

$$R = \frac{(1 \text{ atm})(22.4 \text{ liters})}{(1 \text{ mole})(273° \text{ K})} = 0.0821 \frac{\text{liter} \cdot \text{atm}}{\text{mole} \cdot °\text{K}}$$

When calculated from more exact data justifying more significant figures, $R = 0.08205$ liter $\cdot$ atmospheres per mole $\cdot$ degree Kelvin.

The numerical value of R for other units of P, V, n, and T will be different. It is important to remember that in using a particular numerical value for R, the values of P, V, n, and T must always be expressed in terms of the units used in evaluating R.

The following problems illustrate the use of the general gas law equation. The problems in the next examples can also be worked without use of the general equation, by methods discussed earlier in the chapter. You will find it helpful to work each of the problems both with and without use of the general gas law equation and to compare the two methods, which of course depend upon the same basic principles.

Example 1.   Calculate the volume occupied by 15 grams of nitrogen gas at 15.0° C and 735 mm pressure.

$$P = 735 \text{ mm} = \frac{735}{760} \text{ atm}$$

$$n = \frac{15}{28} \text{ moles} = \text{number of moles in 15 g nitrogen}$$

$$R = 0.0821 \frac{\text{liter} \cdot \text{atm}}{\text{mole} \cdot °\text{K}}$$

$$T = 15° \text{ C} = (273 + 15)° \text{ K} = 288° \text{ K}$$

Substituting in $PV = nRT$:

$$\left(\frac{735}{760} \text{ atm}\right)(V) = \left(\frac{15}{28} \text{ mole}\right)\left(0.0821 \frac{\text{liter} \cdot \text{atm}}{\text{mole} \cdot °\text{K}}\right)(288° \text{ K})$$

$$V = \left(\frac{15}{28} \text{ mole}\right)\left(0.0821 \frac{\text{liter} \cdot \text{atm}}{\text{mole} \cdot {}^\circ\text{K}}\right)(288^\circ \text{K}) \left(\frac{1}{\frac{735}{760} \text{ atm}}\right)$$

= 13 liters (to two significant figures, as justified by the data)

Note that units cancel out to liters, an appropriate unit for volume.

The general gas law equation, upon slight modification, provides a convenient method also for working with molecular weights and densities, as illustrated by the following examples.

Example 2. A sample of gas occupies 400 ml at 20.0° C and 740 mm and weighs 0.842 g. What is the molecular weight of the gas? (Note that this is the same problem worked in Section 10.19.)

? moles = .842g

$$PV = nRT \tag{1}$$

The number of moles of a substance, n, is equal to the number of grams of the substance, m, divided by the number of grams in each gram-molecular weight (i.e., divided by the molecular weight, M).

$$n = \frac{m}{M} \tag{2}$$

Therefore,
$$PV = \frac{m}{M}RT \tag{3}$$

$$R = 0.0821 \frac{\text{liter} \cdot \text{atm}}{\text{mole} \cdot {}^\circ\text{K}}$$

From the data given in the problem:

$$P = 740 \text{ mm} = \frac{740}{760} \text{ atm}$$

$$V = 400 \text{ ml} = 0.400 \text{ liter}$$

$$m = 0.842 \text{ g}$$

$$T = 20.0^\circ \text{ C} = 293^\circ \text{ K}$$

Substituting in (3), using units consistent with units for R:

$$\left(\frac{740}{760} \text{ atm}\right)(0.400 \text{ liter}) = \left(\frac{0.842 \text{ g}}{M}\right)\left(0.0821 \frac{\text{liter} \cdot \text{atm}}{\text{mole} \cdot {}^\circ\text{K}}\right)(293^\circ \text{ K})$$

$$M = \frac{(0.842 \text{ g})\left(0.0821 \frac{\text{liter} \cdot \text{atm}}{\text{mole} \cdot {}^\circ\text{K}}\right)(293^\circ \text{K})}{\left(\frac{740}{760} \text{ atm}\right)(0.400 \text{ liter})}$$

$$= 52.0 \frac{\text{g}}{\text{mole}}$$

Note that units cancel out to g/mole, the appropriate unit for gram-molecular weight. Hence, the molecular weight of the gas is 52.0. Compare with Section 10.19.

Example 3. Calculate the density of oxygen gas at 25.0° C and 735 mm pressure.

$$PV = nRT \tag{1}$$

$$PV = \frac{m}{M}RT \text{ (see Example 2)} \tag{2}$$

$$\text{Density} = \frac{\text{mass}}{\text{volume}}, \text{ or } d = \frac{m}{V} \tag{3}$$

Rearranging (2):

$$MP = \frac{m}{V}RT \tag{4}$$

Hence, $$\qquad\qquad\qquad MP = dRT \tag{5}$$

We know that

$$M = \text{gram-molecular weight of oxygen} = 32.0 \frac{g}{\text{mole}}$$

$$R = 0.0821 \frac{\text{liter} \cdot \text{atm}}{\text{mole} \cdot °\text{K}}$$

Given that

$$P = 735 \text{ mm} = \frac{735}{760} \text{ atm}$$

$$T = 25.0° \text{ C} = 298° \text{ K}$$

Substituting in (5):

$$\left(32.0\frac{g}{\text{mole}}\right)\left(\frac{735}{760}\text{atm}\right) = d\left(0.0821\frac{\text{liter} \cdot \text{atm}}{\text{mole} \cdot °\text{K}}\right)(298° \text{ K})$$

$$d = \frac{\left(32\frac{g}{\text{mole}}\right)\left(\frac{735}{760}\text{atm}\right)}{\left(0.0821\frac{\text{liter} \cdot \text{atm}}{\text{mole} \cdot °\text{K}}\right)(298° \text{ K})} = 1.26 \text{ g/liter}$$

Note that units cancel out to g/liter, an appropriate unit for density.

10.22 General Gas Law Equation Derived from the Kinetic-Molecular Theory

The mathematical expression of the General Gas Law Equation, $PV = nRT$, may be derived more rigorously from the kinetic-molecular theory of gases as follows: Consider **N** molecules, each having a mass m, confined within a cubical container

names for cat

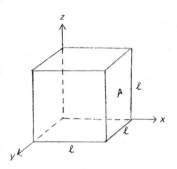

FIGURE 10–15

Drawing illustrating the concepts in the derivation of the General Gas Law Equation from the kinetic-molecular theory.

as shown in Fig. 10–15, an edge of which has a length of l cm. Although the molecules are moving in all possible directions, we may assume that one-third of the molecules ($N/3$) are moving in the direction of the X axis, one-third in the Y direction, and one-third in the Z direction. This assumption is valid because the speed of each molecule may be divided into an X component, a Y component, and a Z component.

Consider the collisions of molecules with wall A in Fig. 10–15. A molecule, on the average, travels $2l$ cm between two consecutive collisions with A as it moves back and forth across the container. If the average speed of the molecule is u cm per second, it will collide $u/2l$ times per second with wall A. Since the collisions are perfectly elastic, the molecule will rebound with a speed of $-u$, having lost no kinetic energy as a result of the collision. Because momentum is defined as the product of mass and speed, the momentum before and after collision is mu; thus the change in momentum per molecule per collision is $2mu$. There are $u/2l$ collisions per second, making the change in momentum per molecule per second $2mu \times u/2l = mu^2/l$. The total change in momentum per second for the $N/3$ molecules which can collide with wall A is $N/3 \times mu^2/l$; this represents the average force on A, because force may be defined as the time rate of change of momentum. Because pressure is defined as force per unit area, and the area of A is l^2, the pressure P on A is

$$P = \frac{\text{Force}}{\text{Area}} = \frac{1}{l^2} \times \frac{\mathbf{N}}{3} \times \frac{mu^2}{l} = \frac{\mathbf{N}}{3} \times \frac{mu^2}{l^3}$$

However, l^3 is equal to the volume of the cubical container, V; thus we have

$$P = \frac{\mathbf{N}}{3} \times \frac{mu^2}{V}$$

or

$$PV = \frac{1}{3}\mathbf{N}mu^2$$

This is the fundamental equation of the kinetic-molecular theory of gases. It must be remembered that the equation is only as exact as the gas laws themselves. It holds for real gases, however, in very good approximation at ordinary pressures.

The equation above may be written $PV = (2N/3)(mu^2/2)$. The kinetic energy, KE, of a body in motion is equal to one-half the product of its mass times the square of its velocity. Thus, $KE = mu^2/2$, and

$$PV = \left(\frac{2\mathbf{N}}{3}\right)\left(\frac{mu^2}{2}\right) = \frac{2}{3}\mathbf{N}(KE)$$

The average kinetic energy of the molecules, KE, is directly proportional to the Kelvin temperature, T.

$$KE = kT$$

The number of molecules, $\mathbf{N}$, is proportional to the number of moles of molecules, n.

$$\mathbf{N} = k'n$$

Substituting, $$PV = \frac{2}{3}(k'n)(kT) = n\left(\frac{2}{3}k'k\right)T$$

The expression $\frac{2}{3}k'k$ is a constant (being made up entirely of constants), which we can designate R and refer to as the **universal gas constant.**
    Therefore,

$$PV = nRT$$

10.23   Deviations from the Gas Laws

When gases at ordinary temperatures and pressures are compressed, the volume is reduced by crowding the molecules closer together. This reduction in volume is really a reduction in the amount of empty space between the molecules. At high pressures, the molecules are crowded so closely together that the volume which they occupy is a relatively large fraction of the total volume of the gas, including the empty space between molecules. Because the volume of the molecules themselves is not compressed, only a small fraction of the entire volume is affected by a further increase in pressure. Thus at very high pressures the whole volume is not inversely proportional to the pressure as predicted by Boyle's Law. It might be noted, nevertheless, that even at moderately high pressures, most of the volume is not occupied by molecules. If the gas behaved ideally, a pressure of about 2,000 atmospheres would be necessary to compress the volume so much that all the volume would be molecules.

    The molecules in a gas at relatively low pressures and high temperatures have practically no attraction for one another because they are far apart. However, as the molecules are crowded closer together at low temperatures and high pressures, the force of attraction between the molecules increases. This attraction has the same effect as an increase in external pressure. Consequently, when external pressure is applied to a volume of gas, especially at low temperatures, there is a slightly greater decrease in volume than would be achieved by the pressure alone. This slightly greater decrease in volume caused by intermolecular attraction is more pronounced at low temperatures because the molecules move more slowly and have less tendency to fly apart after collisions with one another.

    In 1879 van der Waals expressed the deviations of real gases from the ideal gas laws quantitatively with the following equation, which bears his name:

$$(P + n^2a/V^2)(V - nb) = nRT$$

The constant a represents the attraction between molecules, and van der Waals assumed that this force varies inversely as the square of the total volume of the gas. (See Section 11.4 for a discussion of the nature of the van der Waals force.) Since this force augments the pressure and thus tends to make the volume smaller, it is added to the term P.

    The term b in the equation represents the volume of the molecules themselves and is subtracted from V, the total volume of the gas. When V is large, both nb and n^2a/V^2 become negligible, and the van der Waals equation reduces to the simple gas equation, $PV = nRT$.

At low pressures the correction for intermolecular attraction, *a*, is more important than the correction for molecular volume, *b*. At high pressures and small volumes the correction for the volume of the molecules becomes important, because the volume of the molecules themselves is relatively incompressible and constitutes an appreciable fraction of the total volume. At some intermediate pressure the two corrections cancel one another, and the gas appears to follow the relation $PV = nRT$ over a small range of pressures.

Strictly speaking, then, the gas laws apply exactly only to gases whose molecules do not attract each other and which occupy no appreciable part of the whole volume. Because there are no gases which have these properties, we can speak of such hypothetical gases only as **ideal,** or **perfect, gases.** Under ordinary conditions, however, the deviations from the gas laws are so slight that they may be neglected.

QUESTIONS

1. State Boyle's Law. Set up a mathematical expression of this law in the form of a proportion using V_1 for the initial volume, V_2 for the final volume, P_1 for the initial pressure, and P_2 for the final pressure.
2. Give several kinds of dimensional units in which pressure may be expressed.
3. Describe a mercury barometer and explain how it functions in measuring gas pressures.
4. State Charles' Law. The fraction $\frac{1}{273}$ is sometimes called the *coefficient of volume expansion of gases*. Explain.
5. State and illustrate Dalton's Law of Partial Pressures.
6. What are the standard conditions of temperature and pressure, and why is it important to have a set of standard conditions?
7. Summarize the kinetic-molecular theory as it applies to gases.
8. Explain Boyle's Law, Charles' Law, and Dalton's Law in terms of the kinetic-molecular theory. Would all three laws apply simultaneously to a volume of gas?
9. According to the kinetic-molecular theory, the average speed of the molecules of a gas depends inversely upon a particular physical property of the gas. What physical property is this?
10. Explain the following terms as applied to the gaseous state: compressibility, expansibility, diffusibility, permeability, and homogeneity.
11. State Gay-Lussac's Law. Does this law pertain to substances in the liquid and solid states as well as the gaseous state? Explain.
12. Apply Gay-Lussac's Law to the following reactions, assuming that conditions are such that all reactants and products are gases.
    (a) $2H_2 + O_2 \longrightarrow 2H_2O$
    (b) $2NO_2 \longrightarrow N_2O_4$
    (c) $N_2 + 3H_2 \longrightarrow 2NH_3$
13. What information is given by a balanced chemical equation involving gases that can be used in determining volumetric relations?

14. Explain why the ratio of the relative densities of two gases is equal to the ratio of the molecular weights of the two gaseous substances.
15. How would you demonstrate that the molecular formula for ethane is C_2H_6 and not simply CH_3?
16. How can it be shown experimentally that molecular bromine is diatomic?
17. Relate the weight of an element present in a gram-molecular weight of any of its compounds to the atomic weight of that element.
18. State Avogadro's Law.
19. What is the term for the volume occupied by one gram-molecular weight? For gases, what is its value at S.T.P.? How is it found from the density and molecular weight of a gas?
20. Air is more dense at sea level than at higher altitudes because the gas molecules respond to the attractive force of the earth's gravity. However, the heavier molecules of oxygen do not crowd out the lighter molecules of nitrogen at the bottom of the atmosphere (the surface of the earth). Explain.
21. What properties of real gases account for "nonideal" behavior?

PROBLEMS

⑤1. The volume of a mass of gas is 4.23 liters at 747 mm and 24° C. What volume will it occupy at 700 mm and 24° C?   *Ans. 4.51 liters*

2. The volume of a mass of gas is 950 ml at 700 mm pressure. What volume will it occupy at 350 mm, the temperature remaining constant?
                                                              Ans. 1900 ml

⑤3. The volume of a given mass of gas is 410 ml at 33.0° C and 467 mm. What will its volume be if it is measured at 10.0° C and 467 mm?   *Ans. 379 ml*

4. A given mass of air occupies 900 ml at one atmosphere (760 mm) of pressure. What pressure must be applied to reduce the volume to 700 ml at constant temperature?   *Ans. 977 mm*

⑤5. The volume of a sample of a gas is 800 ml at S.T.P. What volume will it occupy at 55° C and 790 mm?   *Ans. 925 ml*

6. The volume of a mass of gas is 300 ml at 25° C and 685 mm. What volume will the gas occupy at S.T.P.?   *Ans. 248 ml*

⑤7. A sample of a gas occupies 300 ml at −10° C and 720 mm. What pressure will the gas exert in a 505 ml sealed bulb at 25° C?   *Ans. 485 mm*

8. Using the coefficient of volume expansion of gases $\frac{1}{273}$, what change in temperature (at constant pressure) would be required to double the volume of a gas measured at 0°?   *Ans. 273° change (C or K)*

⑤9. The volume of a sample of a gas collected over water at 32.0° C and 752 mm is 627 ml. What will the volume of the gas be when dried and measured at S.T.P.? (Vapor pressure of water at 32° C = 35.7 mm.)   *Ans. 529 ml*

10. The volume of a mass of gas measured at 26° C and 755 mm is 10.5 liters. What final temperature would be required to reduce the volume to 9.5 liters at constant pressure?   *Ans. 271° K or −2° C*

11. A gas occupies 275 ml at 0° C and 610 mm. What final temperature would be required to increase the pressure to 760 mm, the volume being held constant?   *Ans. 340° K or 67° C*

s 12. Calculate the relative rates of diffusion of the gases CO and Ne.
                                    Ans. Ne diffuses 1.18 times faster than CO

13. Show by calculation which of the following gases will diffuse more rapidly than oxygen: He, C_2H_2, N_2, Cl_2, CH_4, H_2S, NO_2.

14. Calculate the ratio of the rate of diffusion of HF to that of He; SO_2 to that of H_2; Cl_2 to that of N_2.   *Ans. 1:2.24; 1:5.64; 1:1.59*

15. Calculate the ratio of the rate of diffusion of $^{235}_{92}UF_6$ to that of $^{238}_{92}UF_6$.
                                                        Ans. 1.004:1

16. If the average speed of helium atoms is 0.707 mile per second at room temperature, what will be the average speed of oxygen molecules at the same temperature?   *Ans. 0.250 mile per second*

s 17. A gas of unknown composition diffuses at the rate of 10 ml/sec in a diffusion apparatus in which CH_4 gas diffuses at the rate of 30 ml/sec. Calculate the approximate molecular weight of the gas of unknown composition.
                 Ans. 140 (rounded to two significant figures in accord with data)

s 18. Calculate the volume occupied by the oxygen produced by the thermal decomposition of 75.0 g of $KClO_3$, assuming the oxygen to be measured at 20.0° C and 743 mm. (Density of oxygen at S.T.P. is 1.429 g/liter.)
                                                        Ans. 22.6 liters

19. A given mass of xenon gas has a volume of 200 ml at 25.0° C and 760 mm. To what temperature must the xenon be heated in order that it occupy 450 ml at 760 mm?   *Ans. 671° K or 398° C*

20. (a) A volume of 1.00 liter of gaseous CCl_4 is collected at 400° K and 800 mm. What will be the volume of this mass of gas measured at 700° K and standard pressure?   *Ans. 1.84 liters*

    (b) Suppose the initial volume were known only as one liter (one significant figure). What would the correct answer be? Explain your answer.
                                                        Ans. 2 liters

21. A sample of oxygen collected over water at a temperature of 29.0° C and a pressure of 764 mm has a volume of 560 ml. What volume would the dry oxygen have under the same conditions of temperature and pressure? (See Table 10–1.)   *Ans. 538 ml*

s 22. Calculate the weight of dry hydrogen in 750 ml of moist hydrogen gas collected over water at 25.0° C and 755 mm. (The density of hydrogen is 0.08987 g/liter at S.T.P.) (See Table 10–1.)   *Ans. 0.0594 g*

23. A 2.50-liter volume of hydrogen measured at the normal boiling point temperature of nitrogen, −210.0° C, is warmed to the normal boiling point of water. Calculate the new volume of the gas, assuming no change in pressure.   *Ans. 14.8 liters*

s 24. Air at S.T.P. has a density of 1.292 g/liter. What will be the weight of 4.67 liters of air at 90° C and 735 mm?   *Ans. 4.39 g*

25. What pressure will a given mass of gas measuring 230 ml at S.T.P. exert when expanded to 1,150 ml with no change in temperature?   *Ans. 152 mm*

26. The density of air at S.T.P. is 1.292 g/liter. What will be its density at 182.1° K and 506.7 mm?   *Ans. 1.292 g/liter*

27. Air contains, on the average, 0.03 per cent by volume of carbon dioxide at S.T.P. Calculate the weight of carbon dioxide in 800 liters of air measured at 97.0° F and 726 mm. (The density of CO_2 is 1.9766 g/liter at S.T.P.) *Note: Watch for significant figures.*   *Ans. 0.4 g*

28. A gas of unknown identity diffuses at the rate of 102 ml per second in a diffusion apparatus in which a second gas whose molecular weight is 66.0 diffuses at the rate of 83.3 ml per second. Calculate the molecular weight of the first gas.   *Ans. 44.0*

29. What volume is occupied by 1.00 mole of nitrogen at 27.0° C and 900 mm? (The density of nitrogen is 1.2506 g/liter at S.T.P.)   *Ans. 20.8 liters*

Ⓢ30. Calculate the partial pressures of oxygen and of nitrogen in the air at a total pressure of 760 mm if the air is 20.8 per cent oxygen and 79.2 per cent nitrogen by volume.   *Ans. 158 mm; 602 mm*

31. The volume of a mass of dry nitrogen is 250 ml at 17.0° C and 750 mm. What will be the volume of this gas maintained over water at 27.0° C and 750 mm? (See Table 10–1.)   *Ans. 268 ml*

32. In a good vacuum system, the pressure might be as low as 1.00×10^{-9} mm. If the gas remaining were oxygen, how many oxygen molecules would there be per ml at 298° K at this pressure (density of oxygen is 1.429 g/liter at S.T.P.)?   *Ans. 3.24×10^7 molecules/ml*

Ⓢ33. The time of outflow of a gas through a small opening is 32.6 minutes, while that of an equal number of moles of hydrogen is 5.50 minutes. Calculate the molecular weight of the first gas.   *Ans. 70.8*

34. Volumes of 5.0 liters of nitrogen and 15 liters of oxygen, both under the same original pressure at room temperature (25° C), are compressed into a volume of 10.0 liters in which they together exert a combined pressure of 1.5 atmospheres at room temperature. *Note: Watch for significant figures, carefully.*
    (a) What was the original pressure of the two gases?   *Ans. 0.75 atm*
    (b) What are the partial pressures of the oxygen and the nitrogen in the 10.0-liter volume?   *Ans. N_2, 0.38 atm; O_2, 1.1 atm*
    (c) What is the total pressure of the gases in the 10.0-liter container if the temperature is raised to 210° C?   *Ans. 2.4 atm*

Ⓢ35. Ⓢ(a) When two cotton plugs, one moistened with ammonia and the other with hydrochloric acid, are simultaneously inserted into opposite ends of a glass tube 97.0 cm long a white ring of NH_4Cl forms where gaseous NH_3 and gaseous HCl first come into contact [NH_3 (gas) + HCl (gas) $\longrightarrow$ NH_4Cl (solid)]. At what distance from the ammonia-moistened plug does this occur?

    (b) In an experiment, a student is trying to identify an amine which is one of the three possible compounds, CH_3NH_2, $(CH_3)_2NH$, or $(CH_3)_3N$. In this case, a white ring due to the amine hydrochloride forms at a distance of 45.9 cm from the amine-moistened plug. What is the true formula for the amine? *Ans.* (a) *57.6 cm;* (b) *$(CH_3)_2NH$*

<u>s</u>36. What volume is occupied by 40.0 g of fluorine
    <u>s</u>(a) at S.T.P.? *Ans. 23.6 liters*
    <u>s</u>(b) at 25.0° and 900 mm? *Ans. 21.7 liters*
    (c) at the normal boiling point of water and 2.50 atmospheres of pressure?
    Ans. 12.9 liters

37. Calculate the volume occupied by one mole of hydrogen at S.T.P. (density = 0.0899 g/liter); one mole of oxygen at S.T.P. (density = 1.429 g/liter); one mole of methane gas (CH_4) at S.T.P. (density = 0.715 g/liter). What is this volume called? *Ans. 22.4 liters for each gas.*

38. What volume would be occupied by each of the following gases at S.T.P.?
    (a) 0.300 mole of methane, CH_4. *Ans. 6.72 liters*
    (b) 2.40 moles of sulfur dioxide, SO_2. *Ans. 53.8 liters*
    (c) 16.0 grams of oxygen, O_2. *Ans. 11.2 liters*
    (d) 48.0 grams of ozone, O_3. *Ans. 22.4 liters*

39. Calculate the number of molecules in one cubic centimeter of a gas at 5.00° C and 850 mm. *Ans. 2.95×10^{19}*

<u>s</u>40. How many moles are represented by each of the following, measured at S.T.P.?
    <u>s</u>(a) 3.00 liters of acetylene, C_2H_2. *Ans. 0.134 mole*
    <u>s</u>(b) 50 ml of ammonia, NH_3. *Ans. 2.2×10^{-3} mole*
    (c) 12.1 liters of neon, Ne. *Ans. 0.540 mole*
    (d) 1.00 ml of diborane, B_2H_6. *Ans. 4.46×10^{-5} mole*
    (e) 7.25 ml of uranium hexafluoride, UF_6. *Ans. 3.24×10^{-4} mole*

41. How many grams of chlorine, Cl_2, are there in 106 liters of the gas at S.T.P.?
    Ans. 336 g

42. Calculate the density of the gas ethane (C_2H_6) at S.T.P. *Ans. 1.34 g/liter*

43. Calculate the specific gravity (see Section 1.18) of carbon dioxide at S.T.P. (a) referred to oxygen at S.T.P. as a standard and (b) referred to air at S.T.P. as a standard (density of air at S.T.P. = 1.29 g/liter). Are these values of specific gravity equal numerically to the density of carbon dioxide at S.T.P.? Under what conditions are the numerical values for the density and the specific gravity of a substance equal? *Ans.* (a) *1.38;* (b) *1.52*

44. (a) Calculate the weight of 20.0 liters of propane, C_3H_8, at S.T.P.
    Ans. 39.4 g
    (b) Calculate the weight of 19.2 liters of He at −20.0° C and 800 mm.
    Ans. 3.90 g
    (c) Calculate the volume of 80.0 g of N_2 at 600° C and 1,000 atmospheres pressure (1 atm = 760 mm). *Ans. 0.204 liter*

45. A volume of 1.0 liter of a gas measured at S.T.P. weighed 1.97 g. Calculate the approximate molecular weight of the gas. *Ans. 44*

<u>s</u>46. What pressure is exerted by 11.0 g of CO when contained in a 20.0-liter vessel at 27.0° C? *Ans. 0.483 atm*

47. How many grams of fluorine are there in 8.80 liters of BF_3 gas measured at S.T.P.? *Ans. 22.4 g*

s48. Calculate the density of each of the following gases at S.T.P.; at 20.0° and 720 mm: NO, SO_2, NH_3, Ne, CF_4.

Ans.	S.T.P.	20.0° and 720 mm
s NO	1.34 g/liter	1.18 g/liter
SO_2	2.86 g/liter	2.52 g/liter
NH_3	0.760 g/liter	0.671 g/liter
Ne	0.901 g/liter	0.795 g/liter
CF_4	3.93 g/liter	3.47 g/liter

49. Assume that the following reaction is carried on at a temperature and pressure such that all substances are gases. (a) How many liters of oxygen gas will be required to burn 46 liters of hydrogen gas ($2H_2 + O_2 \longrightarrow 2H_2O$)? (b) How many liters of air (21% oxygen by volume) would be required? (c) How many liters of water vapor will be formed?

Ans. (a) 23 liters; (b) 110 liters; (c) 46 liters

s50. (a) How many liters of dry hydrogen gas, measured at 27.0° and 778 mm, can be obtained by the reaction of 47.0 g of aluminum with an excess of dilute sulfuric acid? (b) What would be the volume if the hydrogen were collected over water under the same conditions?   Ans. (a) 62.8 liters; (b) 65.1 liters

s51. A mixture of 0.100 g of hydrogen, 0.500 g of nitrogen, and 0.410 g of argon is stored at S.T.P. What volume must the container have, assuming no interaction of the three gases?   Ans. 1.74 liters

52. A 15.0-ml flask contains only CO_2 at a temperature of 90° C and a pressure of 748 mm. How many molecules are in the flask?   Ans. 2.99×10^{20}

53. A quantity of a gaseous compound weighing 3.254 g had a volume of 1,029 ml at 25.0° C and 735 mm. Calculate the approximate molecular weight of the compound.   Ans. 80.0

54. A certain compound containing only carbon, chlorine, and hydrogen is found to have a vapor density of 3.168 g/liter at 100° C and 760 mm. If the atomic ratio of carbon to hydrogen to chlorine is one to one to one, what is the molecular formula of the compound?   Ans. $C_2H_2Cl_2$

55. Calculate the volume of dry hydrogen, measured at 25.0° and 750 mm, which would be produced by the reaction of 75.0 g of iron with excess steam.

Ans. 44.4 liters

s56. Calculate the weight of ethane, C_2H_6, required to produce a pressure of 1,520 mm at 20.0° when contained in a 9.00-liter vessel.   Ans. 22.5 g

57. Calculate the volume of oxygen required to burn 14.0 liters of propane gas (C_3H_8) to produce carbon dioxide and water.   Ans. 70.0 liters

58. What volume of chlorine at 25.0° C and 760 mm can be obtained by the electrolysis of 90.0 pounds of sodium chloride in a process that is 85.0% efficient? (Metallic sodium and chlorine gas, Cl_2, are the products of the electrolysis.)   Ans. 7.26×10^3 liters

59. The simplest formula of a certain compound is CH and the weight of a molar volume of the compound is 78.1 g. What must be the molecular formula of the compound?   Ans. C_6H_6

60. What volume of dry hydrogen, measured at S.T.P., would be formed by the action of 35.0 g of BaH_2 upon an excess of water? *Ans. 11.3 liters*

61. How many grams of hydrogen will be required to fill a spherical balloon 2.50 meters in diameter at 25.0° and 740 mm pressure? *Ans. 657 g*

62. The atomic weight of fluorine is 19.0, and the weight of 11.2 liters of fluorine at S.T.P. is 19.0 g. What must be the formula for molecular fluorine? *Ans. F_2*

63. The density of sulfur vapor at 1,000° C and 740 mm pressure is 0.5977 g/liter. (a) What is the molecular weight of the sulfur vapor? (b) What is the formula of the molecular sulfur vapor under these conditions? *Ans. (a) 64.1; (b) S_2*

64. What volume of fluorine (measured at S.T.P.) would be needed to react with 84.0 g of sulfur according to the reaction: $S + 3F_2 \longrightarrow SF_6$. *Ans. 176 liters*

S65. The weight of 151 ml of a certain vapor at 68.0° and 768 mm is 1.384 g. What is the molecular weight of the vapor? *Ans. 254*

66. How many grams of hydrogen would be required to reduce 125 g of mercury(II) oxide ($HgO + H_2 \longrightarrow Hg + H_2O$)? This weight of hydrogen would be equal to how many moles? How many liters would this be at S.T.P.? *Ans. 1.16 g, 0.577 mole, 12.9 liters*

67. How many moles of CF_4 are present in 33.3 liters of the gas at 100° and 1,520 mm Hg? *Ans. 2.18*

S68. What weight of MnO_2 is required for the production of 2,500 liters of Cl_2 gas at 760 mm pressure and 280° C, according to the following reaction?

$$MnO_2 + 4HCl \longrightarrow MnCl_2 + Cl_2 + 2H_2O$$

Ans. 4,790 g (to three significant figures)

S69. Calculate the pressure exerted by a mole of ethylene, C_2H_4, in a volume of 10.0 liters at 100° C (a) using the ideal gas law; (b) using the van der Waals equation. For ethylene the constant a has a value of 4.471 liter$^2 \cdot$ atm/mole^2, and the constant b a value of 0.05714 liter/mole. *Ans. (a) 3.06 atm; (b) 3.03 atm*

70. Using the information in Problem 69, calculate the per cent of the total volume taken up by the C_2H_4 molecules in a gas sample at S.T.P. assuming (a) ideal gas behavior and (b) that van der Waals equation applies. *Ans. (a) 0.255%; (b) 0.257%*

REFERENCES

"Chemical Principles Exemplified," R. C. Plumb, *J. Chem. Educ.*, **47**, 175 (1970).

"Some Early Thermometers," E. H. Brown, *J. Chem. Educ.*, **11**, 448 (1934).

"Thomas Graham's Study of the Diffusion of Gases," A. Ruckstuhl, *J. Chem. Educ.*, **28**, 594 (1951).

"Robert Boyle and His Background," D. Reilly, *J. Chem. Educ.*, **28**, 178 (1951).

"The Discovery of Boyle's Law, 1661–62," R. G. Neville, *J. Chem. Educ.*, **39**, 356 (1962).

"Robert Boyle," M. B. Hall, *Sci. American*, Aug., 1967; p. 97.

"Charles' Law: A General Chemistry Experiment," D. T. Haworth, *J. Chem. Educ.*, **44**, 353 (1967).

"Graham's Laws of Diffusion and Effusion," E. A. Mason and B. Kronstadt, *J. Chem. Educ.*, **44**, 740 (1967).

"The Range of Validity of Graham's Law," A. D. Kirk, *J. Chem. Educ.*, **44**, 745 (1967).

"Atmospheric Pressure from Barometer Readings," G. F. Kinney, *J. Chem. Educ.*, **46**, 321 (1969).

"Graham's Laws: Simple Demonstrations of Gases in Motion—Part I, Theory," E. A. Mason and R. B. Evans III, *J. Chem. Educ.*, **46**, 358 (1969). ". . . Part II, Experiments," R. B. Evans III, D. L. Love, and E. A. Mason, **46**, 423 (1969).

"The Critical Temperature: A Necessary Consequence of Gas Non-Ideality," F. L. Pilar, *J. Chem. Educ.*, **44**, 284 (1967).

"An Experimental Approach to the Ideal Gas Law," W. G. Breck and F. W. Holmes, *J. Chem. Educ.*, **44**, 293 (1967).

"Weak Intermolecular Interactions," J. E. House, Jr., *Chemistry*, **45 (4)**, 13 (1972) (April, 1972, issue).

"Determination of the Molar Volume of a Gas at Standard Temperature and Pressure," L. M. Zaborowski, *J. Chem. Educ.*, **49**, 361 (1972).

"Scuba Diving and the Gas Laws," E. D. Cooke and C. Baranowski, *J. Chem. Educ.*, **50**, 425 (1973).

"The Mole and Avogadro's Number," R. M. Hawthorne, Jr., *J. Chem. Educ.*, **50**, 282 (1973).

The Liquid and Solid States

11

The Liquid State

11.1 The Kinetic-Molecular Theory and the Liquid State

We learned in Section 10.11 that the molecules of a substance in the gaseous state are in constant and very rapid motion and that the space between the molecules is large compared to the sizes of the molecules themselves. As the molecules of a gas are brought closer together by increased pressure, the average distance between the molecules is decreased and **intermolecular attractive forces** become stronger. As a gas is cooled, the average speed of the molecules decreases and their tendency to move apart after collision decreases. If the pressure is sufficiently high and the temperature sufficiently low, the intermolecular attraction overcomes the tendency of the molecules to fly apart, and the gas changes to the liquid state. Although molecules in the liquid state cling to one another, they still retain a limited amount of motion as reflected in the capacity of liquids to flow, to take the shape of a container, to diffuse, and to evaporate.

The molecules in a liquid are held in such close contact by their mutual attractive forces that the volume of any liquid decreases very little with increased pressure; liquids are relatively incompressible compared to gases. The molecules in a liquid are able to move past one another in random fashion (diffusion) but, because of the much more limited freedom of molecular motion existing in liquids, they diffuse much more slowly than do gases.

11.2 Evaporation of Liquids

We know that water placed in an open vessel decreases in volume upon standing. We say that **evaporation** has occurred. Evaporation may be explained in terms

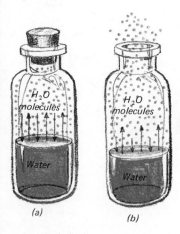

FIGURE 11-1
(a) Molecules of water do not escape from the closed bottle and a dynamic equilibrium between vapor and liquid occurs. (b) Molecules of water escape from the open bottle, evaporation occurs, and the water level is lowered.

of the motion of molecules. At any given temperature the molecules of a liquid move—some slowly, some at intermediate rates, some very fast. An average speed can be calculated. A rapidly moving molecule near the surface of the liquid may possess sufficient kinetic energy to overcome the attraction of its neighbors and escape, i.e., **evaporate,** to the space above the liquid (Fig. 11-1). Other fast moving molecules will leave the liquid phase and appear in the gaseous phase above the liquid as the process of evaporation continues. When the space above the liquid is confined, molecules cannot escape into the open but strike the walls of the container, rebound, and may strike the surface of the liquid, where they are trapped. The return of the molecules from the vapor state to the liquid state is known as **condensation.** As evaporation proceeds, the number of molecules in the vapor state increases, and in turn, the rate of condensation increases. The rate of condensation soon becomes equal to the rate of evaporation and the vapor in the closed container is in equilibrium with its liquid (Fig. 11-1). This is called a **dynamic equilibrium** because the opposing changes involved are in full operation but counterbalance each other.

$$\text{Liquid} \underset{\text{Condensation}}{\overset{\text{Evaporation}}{\rightleftharpoons}} \text{Vapor}$$

At equilibrium the space above the liquid is saturated with respect to molecules of the vapor. The pressure exerted by the vapor in equilibrium with its liquid, at a given temperature, is called the **vapor pressure** of the liquid. The area of the surface of the liquid in contact with the vapor and the size of the vessel have no effect upon the vapor pressure. Vapor pressures are commonly measured by means of a manometer (Fig. 11-2). Some vapor pressure data for water, alcohol, and ether are given in Table 11-1 and Fig. 11-3.

TABLE 11-1   Vapor Pressures (mm of Hg) of Some Common Substances at Various Temperatures

	0° C	20° C	40° C	60° C	80° C	100° C
Water	4.6	17.5	55.0	149.2	355.5	760.0
Ethyl Alcohol	12.2	43.9	135.3	352.7	812.6	1,693.3
Ethyl Ether	185.3	442.2	921.1	1,730.0	2,993.6	4,859.4

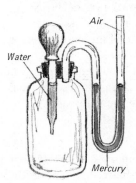

FIGURE 11-2
A manometer used to measure the vapor pressure of water. Squeezing the bulb forces water into the bottle. As the water comes to equilibrium with its vapor the vapor molecules exert a pressure on their surroundings, which include the surface of mercury in the left arm of the manometer. This surface is pushed down and the surface of the mercury in the other arm is pushed up. The difference in the levels of mercury in the two arms is a measure of the vapor pressure of the water. A third tube (*not shown*), equipped with stopcock, in the stopper allows atmospheric pressure to be maintained when the stopper is inserted in the bottle.

FIGURE 11-3

The vapor pressures of three common substances at various temperatures. The intersection of a curve with the dashed line for 760 mm pressure, referred to the temperature axis, indicates the normal boiling point of that substance. (The normal boiling point of ethyl ether is 34.6°; of ethyl alcohol, 78.4°; of water, 100°.)

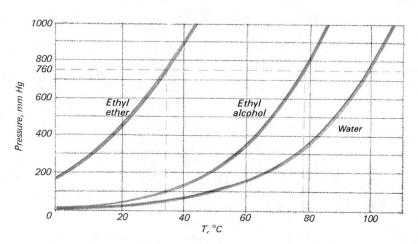

11.3   Boiling Points of Liquids

FIGURE 11-4

In a closed system water does not boil if an equilibrium exists between water vapor pressure and liquid water. Cooling the water vapor, however, lowers its pressure and boiling occurs, seemingly without heat.

It will be seen by studying Table 11-1 and Fig. 11-3 that, at a given temperature, the vapor pressure of ether is greater than that of alcohol, and that of alcohol greater than that of water. These differences in vapor pressure are related to the intermolecular attractive forces, which for the three substances are least for ether and greatest for water. Thus, the equilibrium vapor pressure of a liquid is dependent upon the particular kind of molecule composing the liquid.

Data in Table 11-1 and Fig. 11-3 show that the vapor pressures of the three liquids increase as the temperature is raised. Indeed this is true for all liquids and is due to an increase in the rate of molecular motion which accompanies an increase in temperature. This results in the escape of more molecules from the surface of the liquid per unit of time and a greater speed for each molecule which escapes. Each result contributes toward a higher equilibrium vapor pressure.

When bubbles of vapor form within a liquid and rise to the surface where they burst and release the vapor, the liquid is said to boil. A liquid exposed to the air will boil when its equilibrium vapor pressure becomes equal to the pressure of the atmosphere. The **normal boiling point** of a liquid is that temperature at which its equilibrium vapor pressure becomes exactly equal to the standard atmospheric pressure of 760 mm (Fig. 11-3). A liquid may boil at temperatures higher than normal under external pressures greater than one atmosphere; conversely, the boiling point of a liquid may be lowered below normal by decreasing the pressure on the surface of the liquid below one atmosphere (Fig. 11-4). Thus, at high altitudes where the atmospheric pressure is less than 760 mm, water boils at temperatures below its normal boiling point of 100° C. Food cooked in boiling water cooks more slowly at high altitudes, because the temperature of boiling water is lower than it would be nearer sea level. The temperature of boiling water in pressure cookers is higher than normal due to higher equilibrium vapor pressures, thus making it possible to cook foods faster than in open vessels.

11.4   Intermolecular Forces and the Boiling Point

We saw in the preceding section that differences in vapor pressures of liquids are related to differences in the forces of attraction between molecules of the liquids. All molecules exert an attraction for each other. What is the nature of these attractive forces and why do they differ in magnitude for molecules of different substances?

It was noted in Section 4.7 that molecules whose centers of positive and negative electric charge do not coincide are polar. They possess a **dipole moment.** The HCl molecule, because chlorine has a greater electronegativity than hydrogen, has a partial negative charge on the chlorine atom and a partial positive charge on the hydrogen atom; the HCl molecule has a permanent dipole moment. The electrostatic attraction of the positive end of one HCl molecule for the negative end of another constitutes an attractive force which causes HCl to have a higher boiling point than nonpolar molecules of similar molecular weight.

An additional force of attraction between molecules, which occurs for nonpolar molecules as well as polar molecules, is one known as the **van der Waals attraction,** which has its origin in the electrostatic attraction of the nucleus (positive) of one molecule for the electron cloud (negative) of a neighboring molecule (Fig. 11–5). Because of this attraction the electron distribution in Molecule A of the illustration for an instant may become unsymmetrical and be concentrated toward one side of the molecule. At this moment a temporary dipole exists in the molecule. Molecule B, in turn, is distorted at the same moment with its nucleus attracted toward the negative end of Molecule A and its negative end opposite from the negative end of Molecule A. A mutual attraction between the nucleus of one molecule and the electron cloud of the other occurs. At the same time a repulsion between the two nuclei and also a repulsion between the two centers of negative charge in the electron cloud take place. The constant motion of the electrons causes a shift so that in the next instant the concentration of negative charge in one molecule may shift to the other side. This has a corresponding effect on the distribution of charge within the other molecule, resulting in creation of two new dipoles oppositely oriented within the two molecules. These nearly instantaneous fluctuating dipoles result in an attraction between the molecules.

The van der Waals attraction is weak and is significant only when the molecules are very close together; i.e., very nearly in contact with each other. The van der Waals attraction is opposed by (1) the repulsive force of the electron clouds of the adjacent molecules and (2) the repulsion of the nuclei of neighboring atoms for one another. However, the attractive forces are somewhat stronger than the repulsive forces.

The magnitude of the van der Waals attraction increases with an increase in the number of electrons per molecule, and therefore with the molecular weight. This increase in intermolecular attraction with molecular weight is reflected in the rise in boiling point in series of related substances such as He, Ne, Ar, Kr, Xe, Rn and such as H_2, F_2, Cl_2, Br_2, I_2 (see Table 11–2).

The van der Waals force of attraction acting between molecules is of decreasing effectiveness with an increase in temperature, inasmuch as the additional thermal

Molecule A   Molecule B

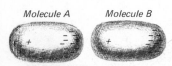

FIGURE 11–5

Diagram illustrating conditions for van der Waals attraction due to nearly instantaneous fluctuating dipoles. The plus sign indicates the nucleus and the minus signs the electron cloud in both *A* and *B*.

TABLE 11-2   Molecular Weights and Boiling Points

Substance	He	Ne	Ar	Kr	Xe	Rn
Molecular weight	4.0	20.18	39.94	83.7	131.1	222
Boiling point, °C	−268.9	−245.9	−185.7	−152.9	−107.1	−61.8

Substance	H_2	F_2	Cl_2	Br_2	I_2
Molecular weight	2.016	38.0	70.91	159.8	253.8
Boiling point, °C	−252.7	−187	−34.6	58.78	184.4

energy increases the motion of the molecules and overcomes the van der Waals force. Another way of putting it is that the kinetic energy at the higher temperature is greater than the energy associated with the van der Waals force of attraction.

It is the van der Waals attraction that causes substances such as the noble gases and the halogens to condense to liquids and to freeze into solids when the temperature is lowered sufficiently.

In general, liquids composed of discrete molecules which have no permanent dipole moments have low boiling points relative to their molecular weights, because only the weak van der Waals attraction must be overcome during vaporization. In such molecules the centers of positive and negative electric charge coincide. Examples are the molecules of the noble gases and the halogens, and other symmetrical molecules such as CH_4, SiH_4, CF_4, SiF_4, SF_6, and UF_6. Molecular substances such as H_2O, HF, and C_2H_5OH (ethyl alcohol), which have permanent dipole moments, have rather high boiling points relative to their molecular weights.

11.5   Heat of Vaporization

In Section 11.2 it was noted that evaporation involves the escape of the molecules of high kinetic energy (hotter, faster moving molecules) from the surface of a liquid. The loss of such molecules through evaporation results in a lower average kinetic energy for those remaining behind and, consequently, a lowering of the temperature of the liquid. The cooling effect due to evaporation of water is very evident to a swimmer when he comes out of the water. In this case, the skin is supplying heat energy to the water molecules, which causes the skin to feel cool.

In order that a liquid may evaporate at a constant temperature, heat must be supplied in sufficient quantities to offset the cooling effect brought about by the escape of the molecules possessing high kinetic energies. The heat energy that must be supplied to evaporate a unit mass of liquid at a constant temperature is known as the **heat of vaporization.** The heat of vaporization of water, for example, is 540 calories per gram, and that of ammonia is 327 calories per gram. The quantity of heat evolved during the condensation of a liquid is the same as that absorbed during evaporation.

The operation of mechanical refrigerators is based on the loss of heat energy to evaporating refrigerants. Heat is taken from within a refrigerator as a circulating refrigerant (usually ammonia, NH_3; or CCl_2F_2, one of the freons) absorbs the energy needed to change it to a gas. In the gaseous state, the refrigerant is then circulated through a compressor outside the refrigerator and again liquefied by combined cooling and compression. To be an effective refrigerant, a substance must be readily convertible from the gaseous to the liquid state at the working temperature and have a high heat of vaporization.

11.6   Critical Temperature and Pressure

We noted in Section 11.1 that compressing a gas and lowering its temperature favor the transition from the gaseous to the liquid state. In some cases it is possible to liquefy a gas at normal temperatures by simply compressing it. However, for each substance there exists a temperature, called the **critical temperature,** above which it cannot be liquefied no matter how much pressure is applied. The pressure required to liquefy a gas at its critical temperature is called the **critical pressure.** The critical temperatures and critical pressures of some common substances are given in Table 11-3.

TABLE 11-3   Critical Temperatures and Pressures of Some Common Substances

	Critical Temperature, °K	Critical Pressure, atm
Hydrogen	33.24	12.8
Nitrogen	126.0	33.5
Oxygen	154.3	49.7
Carbon dioxide	304.2	73.0
Ammonia	405.5	111.5
Water	647.1	217.7
Sulfur dioxide	430.3	77.7

Above the critical temperature of a substance the average kinetic energy of the gaseous molecules is sufficient to overcome their mutually attractive forces, and the molecules will not cling together closely enough to form a liquid no matter how great the pressure. If the temperature is decreased, the average kinetic energy of the molecules is decreased. At the critical temperature the intermolecular forces are sufficiently large, relative to the average kinetic energy, to liquefy the gas provided the substance is under a pressure equal to or greater than its critical pressure. The pressure aids the intermolecular forces in bringing the molecules sufficiently close together to make liquefaction possible. Below the critical temperature the pressure required for liquefaction decreases with decreasing temperature, until it reaches one atmosphere at the normal boiling temperature. Substances possessing strong intermolecular forces, such as water and ammonia, have high critical temperatures; on the other hand, substances with weak intermolecular attraction, such as hydrogen and nitrogen, have low critical temperatures.

FIGURE 11-6
Laboratory distillation apparatus. When impure water is
distilled with this apparatus
nonvolatile substances remain
in the distilling flask. The
water is vaporized, condensed,
and finally collected in the receiving flask.

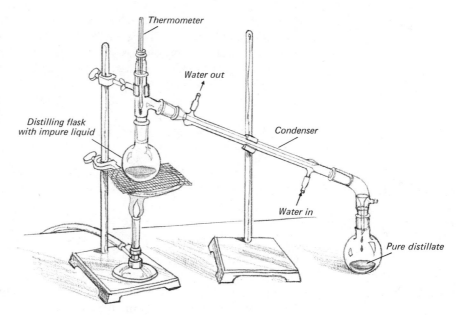

11.7 Distillation

Liquids may contain dissolved materials which make them unsuitable for a particular purpose. For example, water containing dissolved mineral matter should not be used in storage batteries (Section 22.17) because it shortens the life of the battery. Water and other liquids may be purified by a process known as **distillation.** By heating impure water in a distilling flask (Fig. 11–6), the liquid is converted to vapor which passes over into the condenser. The vapor is condensed to the liquid in the water-cooled condenser, and the liquid flows into the receiving vessel. The dissolved mineral matter, such as calcium sulfate, magnesium chloride, etc., is not volatile at the boiling point of water and remains in the distillation flask. Distillation makes use of the facts that the addition of heat to a liquid speeds up the rate of evaporation, an endothermic change, and that cooling a vapor favors condensation, an exothermic change. The separation of volatile substances by distillation is described more fully in Section 13.6.

11.8 Surface Tension

The molecules within the bulk of a liquid are attracted equally in all directions by neighboring molecules; the resultant force on any one molecule within the liquid is therefore zero. However, the molecules on the surface of a liquid are attracted only inward and sideways. This unbalanced molecular attraction pulls some of the surface molecules into the bulk of the liquid, and a condition of equilibrium is reached only when the surface area is reduced to a minimum. The surface of a liquid, therefore, behaves as if it were under a strain, or tension. This contracting force is called **surface tension.** A small drop of liquid tends to assume a spherical shape, because in a sphere the ratio of surface area to volume is at a minimum. We may define surface tension as the force which causes the surface of a liquid to contract. A liquid surface acts as if it were a stretched membrane. A steel needle carefully placed on water will float. Some insects, even though they are heavier than water, can move on its surface, being supported by the surface tension. One of the forces causing water to rise in capillary tubes (tubes with a very small bore) is its surface tension. Water is brought from the soil up through the roots and into the portion of a plant above the soil by this capillary action.

The Solid State

11.9   The Kinetic-Molecular Theory and the Solid State

When enough heat is removed from water to cause its temperature to be lowered to 0° and then more heat is removed, the molecular motion decreases and the molecules take up comparatively fixed and ordered positions, relative to each other, with only vibratory motion remaining. When this happens, the water is said to freeze, i.e., change from the liquid to the solid state, and form ice crystals which are hard and rigid. The rigidity and small compressibility of crystalline solids reflect the resistance of the molecules to change of position. However, the facts that diffusion takes place to a slight extent in the solid state and that crystalline compounds show some vapor pressure indicates that the molecules are not motionless. As the temperature of a solid is lowered, the motion of the molecules gradually decreases, and all the evidence indicates that at absolute zero (−273° C) it would cease entirely.

11.10   Crystalline Solids

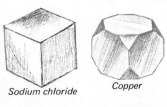

Sodium chloride   *Copper*

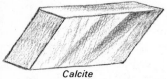

Calcite

FIGURE 11-7

The appearance of single crystals of sodium chloride, copper metal, and calcite ($CaCO_3$).

Most solid substances are crystalline in nature. Although some solids, such as sugar and table salt in the form in which we are familiar with them, are composed of single crystals, most crystalline solids with which we come in daily contact are aggregates of many interlocking small crystals. Common examples of the latter are chunks of ice and objects made of metals. **A crystal may be defined as a homogeneous body having the natural shape of a polyhedron.** Crystals are three-dimensional solids bounded by plane surfaces. The angles at which the surfaces of the crystal intersect are always the same for a given substance, and are characteristic of that substance. Figure 11–7 illustrates the appearance of single crystals of sodium chloride, copper metal, and calcite ($CaCO_3$).

It has been found by the x-ray study of crystalline forms of matter that every crystal consists of atoms, molecules, or ions arranged in a three-dimensional pattern that repeats itself regularly throughout the entire crystal. Crystals of each substance are built up according to a geometric pattern known as the **crystal lattice.** The smallest fraction of the crystal lattice which contains a representative portion of the crystal structure is referred to as the **unit cell.** The unit cell, if indefinitely repeated in three dimensions, will reproduce the crystal. The nature of the solid is determined, therefore, by the size, shape, and content of its unit

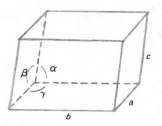

FIGURE 11-8
The unit cell.

cell. The unit cell is a parallelepiped for which the size and shape are defined by the lengths (a, b, c) of the three axes and the angles (α, β, γ) between the axes (Fig. 11-8).

The unit cells for known crystals can be grouped into seven types:

			Example
Cubic	$a = b = c$	$\alpha = \beta = \gamma = 90°$	Rock salt (NaCl)
Tetragonal	$a = b \neq c$	$\alpha = \beta = \gamma = 90°$	White tin
Orthorhombic	$a \neq b \neq c$	$\alpha = \beta = \gamma = 90°$	Mercury(II) chloride
Monoclinic	$a \neq b \neq c$	$\alpha = \gamma = 90°; \beta \neq 90°$	Potassium chlorate
Triclinic	$a \neq b \neq c$	$\alpha \neq \beta \neq \gamma \neq 90°$	Potassium dichromate
Hexagonal	$a = b \neq c$	$\alpha = \beta = 90°; \gamma = 120°$	Silica (SiO$_2$)
Rhombohedral	$a = b = c$	$\alpha = \beta = \gamma \neq 90°$	Calcite (CaCO$_3$)

Variations of the seven types of unit cell give rise to the fourteen crystal lattices shown in Fig. 11-9. The packing of the building units for the first three of these

FIGURE 11-9

The fourteen possible crystal lattices.

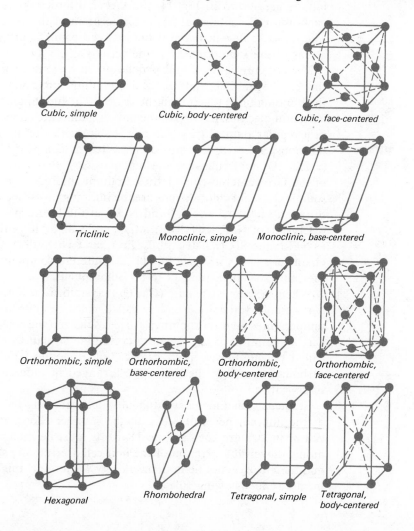

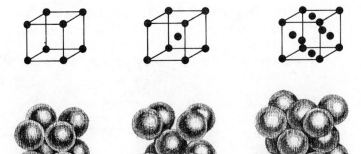

FIGURE 11-10

Unit cells showing close packing of the building units for the simple cubic, the body-centered cubic, and the face-centered cubic lattices.

Cubic, simple　　　*Cubic, body-centered*　　　*Cubic, face-centered*

lattices are shown in Fig. 11–10. A crystal lattice is said to be **body-centered** if each unit cell contains at its center an atom of the same kind as at the corners. The lattice is **face-centered** if the unit cell contains at the midpoints of its faces atoms of the same kind as at the corners.

The structure of a crystal is determined by various factors. Among these are the relative numbers and sizes of the building units and the types of bonds holding the building units together in the crystal. The building units may be either atoms, ions, or molecules. In copper, the building units are copper atoms packed together in a regular pattern (face-centered cubic) with the atoms occupying the lattice positions. Solid carbon dioxide (Dry Ice) is composed of molecular crystals in which CO_2 molecules are the unit particles; these are located at the positions of the face-centered cubic lattice. Sodium chloride forms ionic crystals in which sodium ions and chloride ions are the building units (see Fig. 11–11). Ammonium nitrate crystals have NH_4^+ and NO_3^- at the lattice positions.

Crystals of NaF, KCl, RbBr, MgO, and CaS all have the same crystal structure as NaCl; and $SrCl_2$, CdF_2, PbF_2, ZrO_2, and ThO_2 all crystallize with the structure characteristic of CaF_2 (see Fig. 11–2). Different compounds which crystallize with the same structure are said to be **isomorphous.**

When calcium carbonate ($CaCO_3$) crystallizes at low temperatures it assumes a rhombohedral lattice and is called calcite, but when it crystallizes at high temperatures an orthorhombic lattice results and the substance is called aragonite. The assumption of two or more crystalline structures by the same substance is called **polymorphism.** Both calcite and aragonite consist of calcium ions and carbonate ions, but these ions are arranged in different ways in the two kinds of crystals.

Students sometimes have difficulty in reconciling the number of atoms stated as the number per unit cell with the number shown in a diagram of the unit cell or in a figure such as Fig. 11–10. It must be remembered that some of the atoms shown in a diagram of the unit cell are actually shared by other unit cells and therefore do not lie in their entirety within one unit cell. It is helpful to keep in mind the following rules:

(1) An atom which lies completely within the unit cell belongs to that unit cell only.

(2) An atom lying on a face of a unit cell belongs equally to two unit cells and therefore counts as one-half of one atom for a particular unit cell.

(3) An atom lying on an edge is shared equally by four unit cells and thus counts as one-fourth of an atom for a particular unit cell.

(4) An atom lying at a corner is shared equally by eight unit cells, except in the hexagonal cell, and counts as one-eighth of an atom for each particular unit cell. For the hexagonal cell, an atom lying at a corner is shared equally by six unit cells. To picture this, look again at the diagram for the hexagonal crystal lattice in Fig. 11–9.

The subject of close-packing in crystals will be considered in greater detail in Chapter 31, The Metallic Elements (Section 31.5).

If the edge length of the cubic cell is known, ionic radii for the ions in the crystal lattice can be calculated, making certain assumptions.

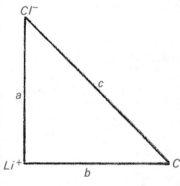

Example: (a) **The unit cell cube edge length for LiCl (NaCl structure–face-centered cubic) is 5.14 Å. Assuming anion-anion contact, calculate the ionic radius for chloride ion.**

By looking at the diagram for the NaCl-type structure in Fig. 11–11, page 229, one can visualize for a section of the unit cell of LiCl a right triangle involving two chloride ions and one lithium ion.

Inasmuch as a (the distance between the center of a chloride ion and the center of a lithium ion) is one-half the edge length of the cubic unit cell,

$$a = \frac{5.14 \text{ Å}}{2}$$

Similarly, b is the distance between the center of a chloride ion and the center of a lithium ion and hence also is one-half the edge length of the cubic unit cell.

$$b = \frac{5.14 \text{ Å}}{2}$$

By the Pythagorean Theorem, the length c (which is the distance between the centers of two chloride ions) can be calculated.

$$c^2 = a^2 + b^2$$
$$c^2 = \left(\frac{5.14}{2}\right)^2 + \left(\frac{5.14}{2}\right)^2 = 13.21$$
$$c = \sqrt{13.21} = 3.63 \text{ Å}$$

Inasmuch as the anions are assumed to touch each other, c is twice the radius of the chloride ion. Hence, the radius of the chloride ion is $\frac{1}{2}c$.

$$r_{Cl^-} = \frac{1}{2}c = \frac{1}{2}(3.63 \text{ Å}) = 1.81 \text{ Å}$$

Note that this agrees with the value given on the inside back cover for the radius of the Cl^- ion.

(b) The unit cell cube edge length for sodium chloride is 5.627 Å. Assuming anion-anion and anion-cation contact, calculate the ionic radius for Na^+.

Inspection of the diagram in Fig. 11-11 shows that the distance between the center of a sodium ion and the center of a chloride ion is one-half the cubic unit cell edge length for the NaCl unit cell, or $\frac{1}{2}(5.627) = 2.813$ Å.

Assuming anion-cation contact, 2.813 Å is the sum of the radii for Na^+ and Cl^-.

$$r_{Na^+} + r_{Cl^-} = 2.813 \text{ Å}$$

In (a), r_{Cl^-} was calculated as 1.81 Å.

Therefore,

$$r_{Na^+} = 2.813 - 1.81 = 1.00 \text{ Å}$$

Note that the calculated value (1.00 Å) is larger than the actual value given on the inside back cover (0.95 Å). This could indicate that the Na^+ is not quite large enough to touch the neighboring chloride ions, in a lattice where the larger chloride ions touch each other, and hence actually has a slightly smaller radius than if it were able to touch the chloride ions. The sodium ions could be said to "rattle around" a bit in the cation holes of the crystal lattice.

By this line of reasoning, we would expect that potassium ions or rubidium ions, being larger, might more nearly fill the cation holes of the crystal lattice. If this is true, the calculated value for the ionic radius for each of those ions should be closer to the true value.

(c) The unit cell cube edge lengths for potassium chloride and rubidium chloride, both of which exhibit the NaCl-type crystal lattice, are 6.28 Å and 6.57 Å, respectively. Assuming anion-anion and anion-cation contact, calculate the ionic radii for K^+ and Rb^+.

Following the same line of reasoning as in (b):

For K^+:

$$r_{K^+} + r_{Cl^-} = \frac{1}{2}(6.28) \text{ Å}$$

$$r_{K^+} = \frac{1}{2}(6.28) - 1.81 = 1.33 \text{ Å}$$

(Note that the actual radius for K^+, given on the inside back cover, is also 1.33 Å.)

For Rb⁺:

$$r_{Rb^+} + r_{Cl^-} = \frac{1}{2}(6.57)\ \text{Å}$$

$$r_{Rb^+} = \frac{1}{2}(6.57) - 1.81 = 1.48\ \text{Å}$$

(Actual value for radius of Rb⁺, given on the inside back cover, is also 1.48 Å.) Thus, the potassium ions and the rubidium ions would each seem to be a suitable size to fill the cation holes and hence do touch the neighboring chloride ions. However, it is important to realize that values for ionic radii calculated from crystal unit cell edge lengths depend also on numerous other assumptions, such as a perfect spherical shape for ions, which are approximations at best. Hence, such calculated values are themselves approximate, and comparisons cannot be pushed too far. Nevertheless, this is one of the methods for calculating ionic radii from experimental measurements such as x-ray crystallographic determinations, and the method has proved to be very useful.

11.11   Radius Ratio Rule for Ionic Compounds

Ionic compounds of the general formula MX usually crystallize in one of three structures: the *cesium chloride* (CsCl) structure, the *sodium chloride* (NaCl) structure, and the *zinc blende* (ZnS) structure. These structures, all of which have cubic lattices, are shown in Fig. 11-11.

Inspection of the cesium chloride model will show that a cesium ion located at the center of the cube will have as its nearest neighbors eight chloride ions. The chloride ions are in reality in contact with the central cesium ion. **This number of nearest neighbors of an atom or ion in a crystal is known as the coordination number.** [The term *coordination number* is used in a slightly different sense in connection with coordination compounds (Section 32.1).] Each chloride ion is also at the center of a cube (not shown in Fig. 11–11) made up of cesium ions at its corners so the coordination number of each chloride ion is also eight. Inasmuch as each corner ion is shared by eight cubes in the extended lattice, each unit cell of cesium chloride includes one cesium ion at its center and a net of one chloride ion at the corners ($8 \times \frac{1}{8}$). With one Cs⁺ and one Cl⁻ per unit cell, electrical neutrality is maintained.

Study of the sodium chloride structure will reveal that the central ion (sodium) has six chloride ions as nearest neighbors, situated at the corners of a regular octahedron, and thus has a coordination number of six. There are four sodium ions and four chloride ions per unit cell, resulting in a neutral net charge for the cell.

The zinc ion in the zinc blende (ZnS) structure has four sulfide ions as nearest neighbors located at the corners of a regular tetrahedron. Note that only alternate tetrahedral holes in this structure are occupied by zinc ions. There are four zinc ions and four sulfide ions per unit cell, making the unit cell neutral in net charge.

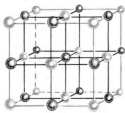

Cesium chloride
CsCl

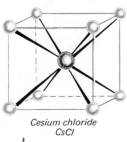

Sodium chloride
NaCl

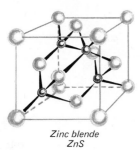

Zinc blende
ZnS

FIGURE 11-11
The crystal structures of some ionic compounds of the general formula MX. The black spheres represent positive ions, and the colored spheres represent negative ions.

The structure assumed by a given MX ionic compound, as previously described, is largely the result of simple geometric and electrostatic relationships which depend on the relative sizes of the positive ion (cation) and the negative ion (anion). A large cation can have more anions around it, while still keeping the anions apart, than can a small cation. In other words, the crystal structure and resulting coordination number depend not only upon the need for electrical neutrality but also upon the relative sizes of the ions. The limiting structure for each arrangement is expressed in terms of the radius ratio of the ions involved.

$$\text{Radius ratio} = \frac{\text{radius of cation}}{\text{radius of anion}} = \frac{r^+}{r^-}$$

There is a minimum ratio (r^+/r^-) for any coordination number, below which the anions would be touching one another. The approximate limiting values for the radius ratios for MX compounds are given in Table 11–4.

TABLE 11-4 Limiting Values for the Radius Ratio for MX Compounds (r^+ is radius of cation; r^- is radius of anion)

Coordination Number	Crystal Structure	Approximate Limiting Values of r^+/r^-
8	CsCl	above 0.732
6	NaCl	0.414 to 0.732
4	ZnS	0.225 to 0.414

Some salts with the *cesium chloride structure* and their actual radius ratios are: CsCl (0.93); CsBr (0.87); CsI (0.78); and TlCl (0.83); *NaCl structure:* NaCl (0.52); NaI (0.44); KCl (0.73); AgCl (0.70); *ZnS structure:* ZnS (0.40); ZnSe (0.37); BeO (0.22).

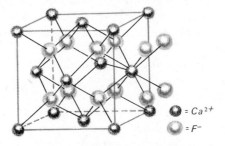

$= Ca^{2+}$
$= F^-$

FIGURE 11-12
A perspective drawing showing the arrangement of calcium ions and fluoride ions within the unit cube of calcium fluoride.

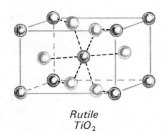

Rutile
TiO$_2$

FIGURE 11-13
Crystal structure of rutile, TiO$_2$. The black spheres represent the titanium ions, Ti^{4+}, and the colored spheres the oxygen ions, O^{2-}.

Usually ionic compounds having the general formula MX_2 crystallize in one of two structures: the fluorite (CaF$_2$) and the rutile (TiO$_2$) crystal lattices (see Figures 11–12 and 11–13). Again, one of the most important single characteristics of these structures is the coordination number. For MX_2 compounds, the coordination numbers of the cation and the anion are different because there are twice as many X$^-$ ions as M^{2+} ions. In CaF$_2$ crystals, the Ca^{2+} has a coordination number of eight while the F$^-$ has a coordination number of four. Rutile (TiO$_2$) has coordination numbers of six and three for the titanium and oxide ions, respectively.

In Table 11–5 the r^+/r^- values are given for a series of difluorides, MF_2. The experimental determination of crystal structures of all these has shown that they may be divided as indicated in the table. Note that the structure changes from one of coordination number eight to one of coordination number six when r^+/r^- falls below 0.73.

It would appear, then, that the structure of an ionic crystal depends upon the relative numbers and sizes of its ions. The radius ratio rule applies strictly only to ionic crystals. In compounds in which the covalent character of the bonds is pronounced, the rule for predicting structures does not hold.

TABLE 11–5 Structures of Difluorides (MF_2) as a Function of Radius Ratios (r^+ is radius of cation; r^- is radius of anion)

	Rutile (C.N. 6-3) Structure					Fluorite (C.N. 8-4) Structure				
	MgF_2	NiF_2	CoF_2	ZnF_2	MnF_2	CdF_2	CaF_2	SrF_2	PbF_2	BaF_2
r^+/r^-	0.48	0.51	0.53	0.54	0.59	0.71	0.73	0.83	0.88	0.99

11.12 Crystal Defects

Crystals are actually never perfect. Several types of lattice defects occur. It is quite common for some lattice points in the unit cells to be unoccupied. Less commonly, some atoms or ions may be located between lattice points. Certain distortions occur in some crystals, as for example when cations or anions are too large to fit into the lattice without distortion. Controlled minute amounts of impurities are sometimes deliberately introduced into the crystal to cause imperfections in the lattice which result in changes in the electrical conductivity of the crystal (See Section 22.18). This type of imperfection gives rise to practical applications such as the manufacture and use of semiconductors for electrical circuits.

11.13 Melting of Solids

When a crystalline solid is heated sufficiently, the vibrational energy of some of the molecules becomes great enough to overcome the intermolecular forces holding the molecules in their fixed positions in the crystal lattice, and the solid begins to melt (fuse). If heating is continued, all of the solid will pass into the liquid state even though the temperature does not rise. If, however, the heating is stopped and no heat is withdrawn, the solid and liquid phases will remain in equilibrium, the rate of melting being just balanced by the rate of freezing. The changes will continue, but the quantities of solid and liquid will remain constant. The temperature at which the solid and liquid phases of a given substance are in equilibrium is known as the **melting point** of the solid, or the **freezing point** of the liquid.

The temperature at which a solid melts reflects the strength of the forces of attraction between the building units present in the crystal. Crystals composed of small symmetrical molecules, such as H_2, N_2, O_2, and F_2, have low melting points because the intermolecular forces are of the weak electronic van der Waals type. Crystalline solids built up of unsymmetrical molecules with permanent dipole moments melt at higher temperatures; examples are ice and sugar. Diamond is an atomic crystal in which the small carbon atoms are held together in the crystal lattice by strong covalent bonds (Chapter 27, Section 27.2, and Fig. 27–1); the melting point of diamond is very high. The atoms in the crystals of metals are strongly bonded together by metallic bonds (Section 31.6), which are modified covalent bonds. In general the metals have high melting points. The electrostatic forces of attraction between the ions in ionic solids are quite strong; thus ionic crystals have high melting points.

11.14   Heat of Fusion

When heat is applied to a crystalline solid, the temperature rises until the melting point is reached. The temperature then remains constant, as additional heat is applied, until all of the solid has been changed to liquid, after which the temperature rises again. The quantity of heat that must be supplied to change a unit mass of a substance from the solid to the liquid state at constant temperature is known as the **heat of fusion** of the substance. The heat of fusion of ice is approximately 80 calories per gram. It represents the difference between the heat content of water, in which the molecules have considerable freedom of motion, and that of ice, in which the molecules vibrate about fixed positions in the crystal. The quantity of heat liberated during crystallization (freezing) is exactly the same as that absorbed during fusion. Changes in certain quantities known as enthalpy, free energy, and entropy play an important part in the process of fusion and crystallization. The concepts of enthalpy and entropy will be discussed in Chapter 20 (Sections 20.5–20.11).

11.15   Vapor Pressure of Solids

The fact that snow and ice evaporate at temperatures below their melting point and that certain solids such as naphthalene (moth balls) have characteristic odors is evidence that molecules of some solids pass directly from the solid into the vapor state. It is those solids in which the intermolecular forces are weak that exhibit measurable vapor pressures at room temperature. As one might predict, the vapor pressure of a solid increases with rise in temperature. The vapor pressure of solid iodine is 0.2 mm at 20° and 90 mm at 114°, its melting point. If iodine crystals are heated in a container to a temperature just below their melting point, evaporation of the solid proceeds rapidly and the vapor may condense to crystals in a cooler part of the container (Fig. 11–14). The combined process of a solid passing directly into the vapor state without melting and the recondensing into the solid state is called **sublimation.** Many substances, such as iodine, may be purified by sublimation if the impurities have low vapor pressures.

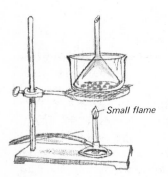

Small flame

FIGURE 11-14
A simple method of demonstrating the sublimation of iodine.

11.16   Undercooled Liquids; Amorphous Solids

The temperature of many liquids may be lowered below their freezing points before crystallization begins. A liquid existing at a temperature below its freezing point is said to be **undercooled,** or **supercooled.** An undercooled liquid is in a mestastable condition, i.e., it is not at equilibrium with its solid. Mechanical agitation such as vigorous stirring, or the introduction of a "seed" crystal of the substance, often induces crystallization by providing an ordered structure to which the slow-moving molecules can become attached.

Liquid materials such as fused glass, containing large cumbersome particles that cannot move readily into the positions of a regular crystal lattice, often show great tendencies to undercool. As the temperature is lowered, the presence of the large and irregular structural units composing the material causes the undercooled liquid to become less and less mobile, and finally to become rigid. Such materials are often spoken of as **amorphous solids,** or **glasses.** True solids are crystalline in structure with a definite internal ordered arrangement of their building units; they have sharp melting points and resist change of shape under pressure. Amorphous solids, on the other hand, are entirely lacking in a definite internal structure; they do not melt but soften and become less viscous when heated (Fig. 11–15). They are really supercooled liquids; the term "amorphous solid," though much used, is not strictly quite correct.

FIGURE 11–15
(a) A two-dimensional illustration of the ordered arrangement of atoms in a crystal of aluminum oxide. (b) An illustration showing the disorder in vitreous (amorphous) aluminum oxide.

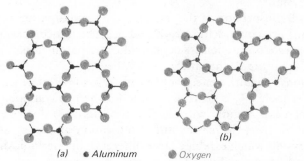

(a)   ● Aluminum   ● Oxygen   (b)

Crystalline solids such as NaCl and $NaClO_3$ melt sharply when heated, because the forces holding the ions together are all of the same strength and thus all the ions break apart at once. The gradual softening of glasses, as opposed to the sharp melting of crystalline solids, results from the structural nonequivalence of the atoms. When a glass is heated, the weakest bonds break first; as the temperature is further increased, the stronger bonds are broken. This causes a gradual decrease in the size and a corresponding increase in the mobility of the structural units as the temperature is raised.

11.17   The Lattice Energies of Ionic Crystals

The **lattice energy** of an ionic compound may be defined as the energy required to separate a mole of its ions by infinite distances. Conversely, it is also the energy released when a mole of a compound is formed by bringing together from infinite distances the necessary number of positive and negative ions. The lattice energy represents the total potential energy of interaction of the ions at their specific distances in the particular crystalline structure of the solid compound. Lattice energies can be calculated from basic principles or they can be measured experimentally.

The calculation of lattice energies is based primarily on the work of Born, Lande, and Mayer. They derived the following equation to express the lattice energy, U, of an ionic crystal:

$$U = \frac{NAZ^2e^2}{R_0}\left(1 - \frac{1}{n}\right)$$

where N is Avogadro's number, R_0 is the interionic distance (sum of the radii of positive and negative ions), Z is the largest common factor of the charges on the oppositely charged ions, and e is the unit of electrical charge (4.770 e.s.u.). A is known as the **Madelung constant** and describes the effects of the electrostatic fields of neighboring ions on the pair being considered and is determined by the geometry of the crystal, i.e., the type of crystal lattice. The symbol n is a parameter characteristic of the electron configurations of the ions; its value is either calculated theoretically or determined from the experimental compressibility of the crystal.

By inspection of the lattice energy equation one may readily deduce that the two principal factors are Z and R_0. The lattice energy increases rapidly with the charges on the ions. Keeping all other parameters constant, doubling the charge Z quadruples the lattice energy, because Z is to the second power in the expression. The lattice energy of LiF is 240 kcal per mole while that for MgO is 940 kcal per mole (R_0 is nearly the same for the two compounds). The lattice energy also increases rapidly with decreasing interionic distance in the lattice, which may result from decreasing either the cation or anion radius, or both. Examples of crystals which exhibit a large difference in interionic distances are lithium fluoride and cesium iodide.

	Interionic Distance, R_0	Lattice Energy, U
LiF	2.008 Å	240 kcal/mole
CsI	3.95 Å	136 kcal/mole

The large lattice energies of ionic compounds result from strong electrostatic forces between the ions in the crystal. These forces must be overcome in order to vaporize the crystal to separated gaseous ions, to melt the crystal, or to vaporize the liquid or solid to gaseous ion pairs. In the case of either fusion or vaporization, however, the energy requirement is less than the total lattice energy because of the interaction, respectively, of the neighboring ions in the liquid phase or the two ions of a pair in the gas phase. However, the strong interionic forces in ionic compounds are responsible for relatively large heats of fusion, sublimation, and vaporization and for the high melting and boiling temperatures, as compared to those of most covalent compounds. These same strong forces are responsible for the fact that ionic crystals are generally hard, dense, rigid, relatively incompressible, and nonvolatile. We should note, however, that while ionic compounds are characterized by such properties, certain covalent structures, such as diamond and silicon carbide, also exhibit them (see Sections 27.2 and 29.13). In such cases, the very strong forces in the crystal are due to strong covalent bonds operative in three dimensions throughout the crystal, rather than the weaker forces usually found between molecules in covalent compounds.

11.18 The Born-Haber Cycle

The direct determination of the lattice energy of an ionic crystal has been made for only a few compounds. In most cases it is not possible to measure the lattice energy directly; however, a cyclic process has been developed independently by Born and Haber; it relates the lattice energy to other thermochemical quantities and makes its calculation possible. The **Born-Haber cycle** is a thermodynamic cycle which involves the heat of formation of the compound (Sections 5.14 and 8.5), the ionization potential of the metal (Section 7.1, Part 3), the electron affinity of the nonmetal (Section 7.1, Part 4), the heat of melting and vaporization of the metal, the heat of dissociation of the nonmetal, and the lattice energy of the compound.

For example, let us consider the formation of sodium chloride from sodium metal and chlorine gas. We can assume that first the sodium metal is vaporized and the diatomic chlorine is dissociated. Then the sodium atoms are ionized, and the electrons thus obtained are transferred to the chlorine atoms to form chloride ions. Hence, sodium ions and chloride ions are the species present in the gaseous phase. The lattice energy of sodium chloride is the amount of energy evolved when one mole of solid sodium chloride is produced from one mole of gaseous sodium ions and one mole of gaseous chloride ions. The total energy evolved in the preceding hypothetical preparation of sodium chloride is equal to the experimentally determined heat of formation of the compound from its elements in their standard states (25° C and one atmosphere of pressure).

The calculation of the lattice energy of sodium chloride may be carried out using experimental data as follows.

(1) Heat of formation of sodium chloride:

$$Na(s) + \tfrac{1}{2}Cl_2(g) \longrightarrow NaCl(s) \qquad -Q = -98.2 \text{ kcal}$$

(2) Melting and vaporization energy of sodium metal:

$$Na(s) \longrightarrow Na(g) \qquad S = 26.0 \text{ kcal}$$

(3) Dissociation energy of molecular chlorine:

$$\tfrac{1}{2}Cl_2(g) \longrightarrow Cl(g) \qquad \tfrac{1}{2}D = 28.6 \text{ kcal}$$

(4) Ionization energy of sodium atom:

$$Na(g) \longrightarrow Na^+(g) + e^- \qquad I = 117.9 \text{ kcal}$$

(5) Electron affinity of chlorine atom:

$$Cl(g) + e^- \longrightarrow Cl^-(g) \qquad -E = -86.5 \text{ kcal}$$

(6) Lattice energy of sodium chloride:

$$Na^+(g) + Cl^-(g) \longrightarrow Na^+Cl^-(s) \qquad -U = ? \text{ (to be calculated)}$$

The Born-Haber cycle may be expressed diagrammatically as follows:

$$NaCl(s) \xleftarrow{\quad -U \quad} Na^+(g) + Cl^-(g)$$
$$\uparrow_{-Q} \qquad\qquad\qquad \uparrow_{+I} \quad \uparrow_{-E}$$
$$Na(s) + \tfrac{1}{2}Cl_2(g) \xrightarrow{\ +S+\frac{1}{2}D\ } Na(g) \ + Cl(g)$$

The total energy of formation of the crystal from its elements is given by the equation:

$$-Q = S + \tfrac{1}{2}D + I - E - U$$

The value of the heat of formation, Q, is accurately known for many substances. If the other thermochemical values are available, we can solve for the lattice energy (U) by rearranging the preceding equation as follows:

$$U = Q + S + \tfrac{1}{2}D + I - E$$

For sodium chloride, using the above data, the lattice energy is

$$U = (98.2 + 26.0 + 28.6 + 117.9 - 86.5) \text{ kcal} = 184.2 \text{ kcal}$$
or $\qquad -U = -184.2 \text{ kcal}$

The Born-Haber cycle may be used to calculate any one of the quantities in the equation for lattice energy, provided all of the others are known. Usually, Q, S, I, and D are known. The direct measurement of electron affinities is rather difficult, and only for the halogens have really accurate values of E been determined. For halogen compounds, then, the Born-Haber cycle can be used to calculate lattice energies, the values of which are found to agree well with those obtained by other means. The electron affinity of oxygen, an important quantity, cannot be measured directly. This value has been obtained by applying the Born-Haber cycle to the formation of various ionic oxides.

11.19 X-Ray Diffraction and Crystals

Much information concerning the arrangement of atoms, molecules, and ions within crystals has been obtained from measurements of the transmission and scattering **(diffraction)** of x rays by crystals and also from measurements of the **reflection** of x rays by crystals. X rays are electromagnetic radiations of very short wavelengths (Section 3.7) of the same order of magnitude as the distance between neighboring atoms in crystals (around 1 Å).

In 1912, the German physicist Laue pointed out that the regularly arranged

FIGURE 11-16

A transmission Laue photograph of a NaCl crystal (face-centered cubic structure). The x-ray beam was focused perpendicular to the (100) planes. A copper target x-ray tube was used, with an exposure time of 10 hours and a crystal-to-film distance of 2–3 cm. Photographed by *J. D. Wilson* and *D. C. Eldridge.*

atoms of a crystal should act as a three-dimensional diffraction grating toward x rays. He demonstrated mathematically that, on passing through a thin section of a crystal, a beam of x rays should give rise to a diffraction pattern arranged symmetrically around the primary beam as a center. A photographic plate placed perpendicular to the path of the x rays revealed, on development, a central spot due to the action of the primary beam and a set of symmetrically grouped spots due to deflected (diffracted) rays (Fig. 11–16). Photographic plates obtained by using different crystals showed a variety of geometrical patterns corresponding to different crystalline structures.

The reflection method of x-ray analysis of crystals developed by Bragg gives results which are easier to interpret than those obtained with the Laue diffraction method. The mathematical analysis of the Laue diffraction patterns is quite complex. Bragg showed that the diffraction of x rays by the atoms or ions in a crystal can be interpreted simply by considering the crystal as a reflection grating. The reflection of x rays from the face of a crystal may be compared with the reflection of a beam of light from a stack of thin glass plates of equal thickness. It is known that monochromatic (single-wavelength) light is reflected from a stack of glass plates only at definite angles, the values of which depend upon the wavelength of the incident light and upon the thickness of the plates. For crystals the planes of atoms or ions within the crystal correspond to the plates of glass. A simple mathematical relation exists between the wavelength, λ, of the x rays, the distance between the atomic planes in the crystal, and the angle of reflection.

As is shown in Fig. 11–17, when a beam of monochromatic x rays impinges at an angle θ upon two planes of a crystal it is reflected, the rays from the lower plane having traveled farther than those from the upper plane by a distance equal to $BC + CD$. If this distance is equal to one, two, or three or any other integral number of wavelengths ($n\lambda = BC + CD$), the two beams emerging from the crystal will be in phase and will reinforce each other. From the right triangle ABC of Fig. 11–17,

$$\sin BAC = \frac{BC}{AC}$$

Rearranging gives

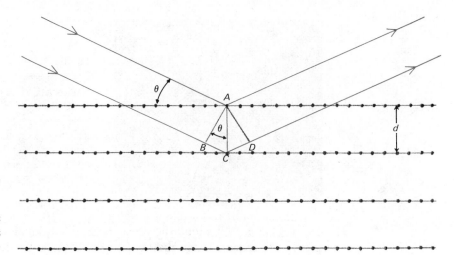

FIGURE 11-17
Reflection of a monochromatic beam of x rays by two planes of a crystal.

$$BC = AC \sin BAC$$

The distance AC is the distance between the crystal planes. We shall call this distance d. It can be shown that angle BAC is equal to θ, which can be called the angle of "grazing incidence".

Then,

$$BC = d \sin \theta$$

But the difference in length of path of the two beams is $BC + CD$. Because $BC = CD$, the total difference in length of path is $2BC$. X rays will be reflected, provided that, as indicated previously,

$$n\lambda = BC + CD = 2BC$$

Hence, $n\lambda = 2d \sin \theta$ (the **Bragg equation**)

If the x rays strike the crystal at angles other than θ the extra distance $BC + CD$ will not be an integral number of wavelengths and there will be interference of the rays reflected from the different crystal planes rather than reinforcement. This will result in weakening the intensity of the reflected rays.

The reflection corresponding to $n = 1$ is called the first order reflection, that corresponding to $n = 2$ is the second order reflection, and so forth. Each successive order gives a wider angle and weaker intensity.

The maximum intensities are found by rotating the crystal so as to change the angle θ at which the x-ray beam strikes the surface. If the wavelength of the incident radiation is known, it becomes possible to determine the spacing, d, of the planes of atoms or ions within the crystal. On the other hand, if the x rays are incident upon the face of a crystal for which d is known, the value of λ for the radiation may be determined by measuring the angles at which the reflections of maximum intensity occur.

QUESTIONS

1. What is meant by the expression "intermolecular forces"?
2. Compare the structure of solids, liquids, and gases with regard to freedom of motion of their molecules.
3. What evidence is there that atoms and molecules in solid substances are not rigidly fixed in position?
4. What feature characterizes the equilibrium between a vapor and its liquid as being dynamic?
5. Explain the cooling effect of evaporation in terms of the kinetic-molecular theory.
6. Is it possible to liquefy nitrogen at room temperature? Is it possible to liquefy ammonia at room temperature? Explain your answers.
7. Define the term "boiling point." Why is the boiling point of a liquid lowered by reducing the external pressure?
8. "Boiling point" is often distinguished from "normal boiling point." Explain what is meant by each of the two terms. A similar distinction is possible for melting and freezing points but is not utilized as often. Give a possible reason for this.
9. Will it require more heat to vaporize one mole of water or two moles of ammonia?
10. How are vapor pressures and intermolecular forces in liquids related?
11. Account for the fact that steam produces much more severe burns than does boiling water.
12. By referring to Fig. 11–3, determine the temperature at which ether boils at atmospheric pressure. Determine the vapor pressures of ethyl ether, ethyl alcohol, and water at 40° C. At what temperature does the vapor pressure of water equal that of the alcohol at 80° C?
13. What properties must a substance have if it is to be effective as a refrigerant in a mechanical refrigerator?
14. How can a needle, which is almost eight times as dense as water, float on water?
15. What are the origins of the van der Waals forces of attraction?
16. Explain the separation of a volatile substance from a nonvolatile one by distillation.
17. Methane (CH_4), ammonia (NH_3), and water (H_2O) have similar molecular weights. Yet methane has a boiling point of $-161.5°$ C, ammonia $-33.4°$ C, and water $+100°$ C. What does this suggest with respect to polar character and intermolecular attractions for each of the three compounds?
18. Crystals of copper metal have a face-centered cubic unit cell. Is there a copper atom at the center of the cell?
19. Explain in terms of internal structure the high melting point (and hardness) of diamond.
20. Relate the temperature of the melting point of a solid to the mutual attractive forces between its building units.

21. What is meant by "purification by sublimation"?
22. Determine the number of atoms per unit cell for each of the crystal lattices shown in Fig. 11-9.
23. How do crystalline solids and "amorphous solids" (supercooled liquids) differ with regard to internal structure?
24. Distinguish between isomorphism and polymorphism.
25. Contrast crystalline and "amorphous" solids (supercooled liquids) with regard to melting points.
26. How do we know that certain solids such as ice and naphthalene have vapor pressures?
27. The specific heat of ice is about 0.5 calorie per gram per degree, that of liquid water is about one calorie per gram per degree, and that of steam is about 0.5 calorie per gram per degree. Relate these values to the structures of solid, liquid, and gaseous water.
28. Demonstrate that a proposed "base-centered cubic" crystal lattice is really a simple tetragonal lattice.
29. What is the difference in heat content between a gram of water at 0° C and a gram of ice at 0° C?
30. What types of molecules are present in liquid materials which tend to undercool and form glasses?
31. From solid geometry, calculate the per cent of free space in each of the three cubic lattices if all atoms in each lattice are of equal radius and touch their nearest neighbors. Which of these lattice structures represents the most efficient packing of spheres into a volume?
32. Explain the relationship of the radius ratio concept to ionic crystalline structure.
33. Using values for ionic radii from the inside back cover, check the radius ratio value given for each MX compound listed in Section 11.11 and each MF_2 compound listed in Table 11-5.
34. Using values for ionic radii from the inside back cover, calculate the radius ratio for each of the following ionic compounds and then predict the type of crystal structure for each from the radius ratio: LiF, CsBr, MgO, ZnTe, CaTe, KI, InAs, AlP, CsI, HgF_2, FeS_2, SnF_2 (Sn^{2+} radius is 0.93 Å).

PROBLEMS

Note: In Problems 1–11, the following data for water will be useful.

Heat of vaporization of water (100° C):	540 cal/g
Heat of vaporization of water (25° C):	584 cal/g
Heat of fusion of water (0° C):	79.71 cal/g
Specific heat of water (solid):	0.50 cal/g · °C
Specific heat of water (liquid):	1.00 cal/g · °C
Specific heat of water (gas):	0.48 cal/g · °C
Heat of formation of water at 25° C:	68,315 cal/mole

⑤1. If 270 g of steam at 100° C and 950 g of ice at 0° C are combined, what is the temperature of the resultant water? (Assume no heat loss.)

Ans. 79.6° C

⑤2. How much heat is required to convert 150 g of ice at −40.0° C to steam at 135° C?

Ans. 110 kcal

3. How much heat is required to (a) melt one mole of ice? (b) change one mole of water at 100° C to steam at 100° C?   *Ans. (a) 1.44 kcal; (b) 9.73 kcal*

4. What mass of water at 0° C would have to be frozen to ice at 0° C to heat 900 g of water from 15.0° C to 47.0° C?

Ans. 361 g

⑤5. What weight of oxygen must be burned in an excess of hydrogen to furnish a quantity of heat necessary to convert 500 g of ice at 0° C to liquid water at 0° C? ·

Ans. 9.3 g

6. A sample of hydrogen weighing 88.80 g is burned in just sufficient oxygen for complete combustion. What amount of heat is evolved during the overall process? Assume that the hydrogen and oxygen are at 25° C initially and the water is cooled to 25° C at the end of the reaction.   *Ans. 3,009 kcal*

7. What weight of steam at 100° C must be condensed to supply sufficient heat to raise the temperature of 245 g of copper from 25.0° C to 27.0° C, assuming that the water formed by the condensation of the steam is kept at 100° C? (The specific heat of copper is 0.0931 cal per gram per degree.)

Ans. 0.0845 g

⑤8. What amount of liquid water at 25.0° C could be frozen by the cooling action of vaporizing 50.0 g of liquid ammonia? (The heat of vaporization of NH_3 is 327 cal/g.)

Ans. 156 g

9. How much greater is the heat content of 300 kg of steam at 100° C than the same mass of ice at 0° C?   *Ans. 2.2 × 10⁵ kcal*

10. A river is 35 ft wide and has an average depth of 4.8 ft and a current of 2.2 miles per hour. A power plant dissipates 50,000 kcal of heat into the river every second. What is the temperature difference between the water upstream and downstream from the plant?   *Ans. 3.3° C*

11. If the heat dissipated by the power plant in Problem 10 is to be removed by evaporating water at 25° C, how many gallons of water per day would be needed? What percentage of the river's daily volume does this represent? The heat of vaporization of water at 25° C is 584 cal/g.

Ans. 2.0 × 10⁶ gal/day; 0.56%

⑤12. What x-ray wavelength would be diffracted at an angle of 21.16° if the interplanar spacing is 2.50 Å?   *Ans. 1.80 Å*

13. The unit cell of metallic silver contains four silver atoms. The edge of this cubic cell is 4.0862 × 10⁻⁸ cm and the density of silver metal is 10.50 g/cm³. From these data, calculate Avogadro's Number.   *Ans. 6.023 × 10²³*

14. What is the minimum value of *d* which can diffract x rays having a wavelength of 0.771 Å?   *Ans. 0.386 Å*

15. Lead crystallizes in a cubic closest-pack lattice. The principal x-ray reflection ($n = 1$) for the planes of atoms which make up the tops and bottoms of the unit cells is at $\theta = 8.95$ degrees. The wavelength of the x rays is 1.54 Å. What is the density of the metallic lead?   *Ans. 11.4 g/cm³*

16. What interplanar spacing would correspond to an angle, θ, of 15.4° for 1.54 Å x rays?

Ans. 2.90 Å

⑤17. At what angle would 0.711 Å x rays be diffracted from planes spaced at 1.77 Å?

Ans. 11.6°

18. Using the data in Table 11–1, make graphs of the log of the vapor pressure (log P_{vap}) versus the reciprocal of temperature in °K for the three substances water (H_2O), alcohol (C_2H_5OH) and ether [$(C_2H_5)_2O$]. What is the algebraic relationship between log P_{vap} and $1/T$? The heats of vaporization are 9.73 kcal/mole, 9.67 kcal/mole, and 6.95 kcal/mole for H_2O, C_2H_5OH, and $(C_2H_5)_2O$, respectively. How do the slopes of the lines you have determined divided by the molar heats of vaporization compare for these three substances? What does this mean?

Ans. It is a linear relationship; they are all equal to −220 ± 10,
which means log P_{vap} = −220 ΔH_{vap} $(1/T)$ + b, where b is a constant,
T is in °K, P_{vap} is in mm, and ΔH_{vap} is in kcal/mole

19. Assuming cation-anion contact, and using values for ionic radii in the tables in Section 7.1, Part 2, or on the inside of the back cover, calculate what you would expect to be the internuclear distance between sodium ion and chloride ion in NaCl; the internuclear distance between magnesium ion and sulfide ion in MgS; the internuclear distance between potassium ion and fluoride ion in KF; and the internuclear distance between sodium ion and sulfide ion in Na_2S.

Ans. 2.76 Å; 2.49 Å; 2.69 Å; 2.79 Å

20. Using the data in the table below and then assuming that in LiBr and LiCl there is anion-anion contact, calculate the ionic radii for Br⁻ and Cl⁻. Then using the data for the other chlorides and bromides and assuming anion-anion and anion-cation contact, calculate ionic radii (average values) for Na⁺, K⁺, Rb⁺, and Cs⁺. Then using the data for the fluorides and iodides (except Li) calculate the ionic radii (average values) for I⁻ and F⁻. Finally using the

Compound	Unit Cell Cube Edge Length, Å	Compound	Unit Cell Cube Edge Length, Å
Face-Centered Cubic Lattice (NaCl structure):		KBr	6.578
		KI	7.052
		RbF	5.63
LiF	4.01	RbCl	6.571
LiCl	5.14	RbBr	6.868
LiBr	5.49	RbI	7.325
NaF	4.62	_ _ _ _ _ _ _ _ _	_ _ _ _ _ _ _
NaCl	5.627	Body-Centered Cubic Lattice (CsCl structure):	
NaBr	5.94		
NaI	6.46	CsCl	4.110
KF	5.33	CsBr	4.287
KCl	6.28	CsI	4.562

datum for LiF calculate the ionic radius of Li$^+$. Compare your values to those given in the tables in Section 7.1, Part 2, or on inside back cover. This problem illustrates one way in which ionic radii can be evaluated from unit cell edge lengths obtained from x-ray measurements.

REFERENCES

"An Introduction to Principles of the Solid State," P. F. Weller, *J. Chem. Educ.*, **47**, 501 (1970).

"Chemical Aspects of Dislocations in Solids," J. M. Thomas, *Chem. in Britain*, **6**, 60 (1970).

"The Undercooling of Liquids," D. Turnbull, *Sci. American*, Jan., 1965; p. 38.

"The Solid State," Sir Neville Scott, *Sci. American*, Sept., 1967; p. 80.

"The Structure of Crystal Surfaces," L. H. Germer, *Sci. American*, March, 1965; p. 32.

"The Influence of Lattice Defects on the Properties of Solids," M. I. Pope, *Educ. in Chemistry*, **2**, 127 (1965).

"The Nature of Glasses," R. J. Charles, *Sci. American*, Sept., 1967; p. 127.

"X-Ray Crystallography," Sir Lawrence Bragg, *Sci. American*, July, 1968; p. 58.

"Ionic Conduction and Diffusion in Solids," J. Kummer and M. E. Milberg, *Chem. & Eng. News*, May 12, 1969; p. 90.

"Liquid Metals," N. W. Ashcroft, *Sci. American*, July, 1969; p. 72.

"Sir Lawrence Bragg," (Staff), *Chem. in Britain*, **6**, 147, 152 (1970).

"Liquid-Crystal Display Devices," G. H. Heilmeier, *Sci. American*, April, 1970; p. 100.

"Weather Modification: A Technology Coming of Age," A. L. Hammond, *Science*, **172**, 548 (1971).

"Crystal Chemistry of $d°$ Cations," M. R. Truter, *Chem. in Britain*, **7**, 203 (1971).

"Accurate Molecular Geometry," D. H. Whiffen, *Chem. in Britain*, **7**, 57 (1971).

"The Tensile Strength of Liquids," R. E. Apfel, *Sci. American*, Dec., 1972; p. 58.

"Snow Crystals," C. Knight and N. Knight, *Sci. American*, Jan., 1973: p. 100.

"Close-Packing of Atoms," H. C. Benedict, *J. Chem. Educ.*, **50**, 419 (1973).

"The Strange Behavior of Super Cooled Water," (Staff), *New Scientist*, **59**, 311 (1973).

"The Rise and Fall of Polywater," L. Allen, *New Scientist*, **59**, 376 (1973).

"Isoelectronic Principle," R. R. Heald, *Chemistry*, **46** (6), 10 (1973) (June, 1973, issue).

"Crystals as Molecular Compounds," P. Pfeiffer (translated by G. B. Kaufman), *J. Chem. Educ.*, **50**, 279 (1973).

"The Chemistry of Liquid Metals," C. C. Addison, *Chem. in Britain*, **10**, 331 (1974).

"X-ray Crystallography: A Refinement of Technique," T. H. Maugh II, *Science*, **186**, 913 (1974).

Water and the Environment; Hydrogen Peroxide

12

Water

Water covers nearly three-fourths of the earth's surface; it is present in the atmosphere and the earth's crust and composes a large part of all living plant and animal matter. Of all the hundreds of thousands of chemical substances, none is more important to plant and animal life than water.

12.1 The Composition of Water

Until the latter part of the eighteenth century, water was thought to be an element. In 1781, Henry Cavendish showed that water is formed when hydrogen burns in air, and a few years later, Lavoisier determined the composition of water by weight.

The formula usually used for water is H_2O, and the relative weights of hydrogen and oxygen in the compound have been determined with great accuracy to be 1.0079 : 8.000. One of the methods used is that of weighing the amounts of hydrogen and oxygen liberated from water by electrolysis.

12.2 Water, an Important Natural Resource

It is a remarkable fact that each person in the United States uses about 2,500 tons of water per year. Although there is enough water in the oceans to cover the entire earth to a depth of two miles, for the most part man's needs generally can be satisfied only by fresh water. And the supply of fresh water is rapidly becoming critical in many parts of the United States (Colorado, New Mexico, Arizona, California, etc.), as the population of the country increases.

About a half-gallon of water per day is required to satisfy the biological needs of a human being. Nearly 150 gallons per day per person are used in maintaining cleanliness, in the cooking of food, and in heating and air conditioning homes. These quantities are dwarfed by the 750 gallons of water used per person per day by industries in the United States. Five gallons of water are used in the production of one gallon of milk, ten gallons for one of gasoline, 80 gallons for one kilowatt of electricity, and 65,000 gallons for one ton of steel. More water is used for irrigation in the United States than for any other one purpose; the average consumption is over 750 gallons per person per day for this purpose.

Fresh water, unlike most mineral resources, is renewable in that it is a part of a gigantic cycle. Water evaporates from the earth into the atmosphere, the energy required being supplied by radiation from the sun; it is transported in the atmosphere by the winds; and it finally falls back to the earth in some form of precipitation. The maximum supply of fresh water in an area is the amount of precipitation. A large part of the water which precipitates is lost to man's daily use, however, through evaporation and transpiration by plants.

At the present time, the oceans are becoming increasingly important as potential sources of fresh water because of recent advances in devising economical and practical desalting processes. The United States government has participated in establishing several desalting plants in various parts of the country and at the U.S. Naval Base in Cuba. The city of Key West, Florida, now obtains from a large sea water desalting plant the entire water supply for the city—over 2,700,000 gallons of pure drinking water per day.

12.3 Natural Waters

All natural waters, even when not polluted by people, are impure since they contain dissolved substances. Rain water is relatively pure, the chief impurities being dust and dissolved gases. After rain falls for a while, the air will have been washed free of dust and bacteria, and any rain that falls thereafter is quite free of such impurities. Sea water contains about 3.6 per cent of dissolved solids, principally sodium chloride. Seventy-two elements are known to exist in sea water, ranging in concentration from chlorine (19,000 mg/liter), sodium (10,600 mg/liter), and magnesium (1,300 mg/liter) to radon (9×10^{-15} mg/liter). Perhaps all naturally occurring elements exist in the sea. The impurities in the water on the earth's surface depend upon the nature of the soil and rocks which the water has passed over or through. These impurities may be classified as

(1) *Suspended solids:* sand, clay, mud, silt, organic material (such as bits of leaves), and microorganisms.

(2) *Dissolved gases:* oxygen, nitrogen, carbon dioxide, oxides of nitrogen, ammonia, and hydrogen sulfide.

(3) *Dissolved salts:* chlorides; sulfates; and hydrogen carbonates of sodium, potassium, calcium, magnesium, aluminum, and iron.

(4) *Dissolved organic substances:* from the decay of vegetable and animal matter.

12.4   Water Pollution

In addition to the natural contaminants in water, many pollutants are being added to our water supplies by people. For too long, neither the magnitude nor even the existence of the problems created by these pollutants was realized. Unfortunately, very little is known of the long-term effects on human health of the many, largely unidentified chemical compounds that enter water supply sources. A whole new field of industrial chemistry is developing rapidly whereby pollutants in our *overall* environment, not just in our water supplies, are studied in connection with both their short- and long-range effects on human health and with the efficient removal of the pollutants.

Several important water supply pollutants that are now causing concern are worthy of our attention.

■ **1. Phosphates.** Phosphates have been used widely as water softeners in laundry detergents (see Section 12.7) and hence have been released in large quantities into rivers and streams in muncipal wastes. Furthermore, some fertilizers contain phosphate [for example, a mixture of $Ca(H_2PO_4)_2$ and $CaSO_4 \cdot 2H_2O$ called "superphosphate of lime"—see Section 26.10]. In addition to the phosphates from these man-made products, a number of phosphate compounds are present in certain rocks. Thus, additional phosphates reach streams through soil runoff. The quantity of phosphates reaching streams from natural sources alone may, in the opinion of some, be great enough to make phosphates from detergents an insignificant factor in comparison.

A high concentration of phosphates in our streams and lakes presents a problem in that such substances are significant nutrients for undesirable algae, bacteria, and other plants. Over half of the problem is due to the abnormal increase of algae. The enrichment of a body of water with excessive amounts of plant nutrients is referred to as **eutrophication.** This excessive fertilization results in an inordinate growth of algae, bacteria, and rooted plants. Some such growth is desirable. Fish eat plants. Microorganisms, such as bacteria, help decompose organic matter. However, if more plants grow than fish can eat, the water fills up with the plants. The bacterial process that decomposes organic matter uses oxygen, and if excess action of this kind occurs, the dissolved oxygen in the water is depleted to such a low level that most aquatic animal life cannot survive.

Development of effective long-term controls of eutrophication will require a better understanding of the balance and form of significant nutrients in streams and lakes, the natural plant and animal populations in bodies of water, and how these are affected by externally applied factors such as the added nutrients. In the short range, suitable substitutes for phosphates in detergents are being sought, and methods of removing phosphates and controlling their undesirable effects in municipal wastes are being studied. A phosphate substitute in detergents that was widely used for a short period is nitrilotriacetic acid (NTA), $N(CH_2COOH)_3$. However, detergent manufacturers voluntarily discontinued the use of NTA in 1970. Some studies have indicated that NTA, if not completely degraded in sewage treatment, can combine with certain harmful trace metals which are present in some water, and thus provide a way for these metals to pass through the sewage treatment plants and subsequently into public water supplies.

■ **2. Mercury.**   Consumption of mercury in the United States during the last 75 years has been about 10^5 metric tons. As much as one-third of this may have been introduced into the environment as a contaminant. World production of mercury is about 9,000 metric tons per year. Discharges from industrial plants and crop applications of certain pesticides introduce mercury into the air, soil, and water. Mercury vapor in the air is deposited on the earth by rainfall. Water runoff from the land washes that mercury, and also mercury compounds arising from land application of pesticides and fungicides, into the waterways. The main source of mercury in bodies of water, however, is the direct discharge of waste mercury into them from industrial plants.

In 1971, dangerously high levels of highly toxic mercury compounds were discovered in ocean fish, such as tuna and swordfish. The accumulation of mercury in the tissues of marine life is not just a recent occurrence, however. Tuna taken from both the Atlantic and Pacific Oceans between 1878 and 1909 and preserved by the Smithsonian Institution have been tested and found to be just as contaminated with mercury as present day tuna. Mercury has polluted some inland lakes and rivers to the extent that it is unsafe to eat certain varieties of fish from them. The very poisonous compound methylmercury is produced in inland waters by the action of microorganisms in the bottom mud with mercury. Metallic mercury usually attacks the liver and kidneys, but methylmercury can be even more dangerous, inasmuch as it attacks the central nervous system. Furthermore, methylmercury is retained in the human body for long periods so that, even with small rates of intake, toxic amounts can be accumulated.

Microorganisms in the intestines of some animals, including chickens, can methylate mercury. It is not known whether such a conversion can take place in fish and mammals without the microorganisms, but it is known that fish take in methylmercury both through their food and through their gills to the extent that the concentration in their flesh is several thousand times that in the surrounding water. Hence, the possibility of mercury poisoning in humans from eating fish (or chicken and other animals) is greater than might be expected from the actual concentrations of the element in water, air, or soil.

Careful attention to prior removal of mercury from industrial effluents can be partially effective, as can controls on the use of mercury pesticides and on soil conservation to reduce water runoff. Much study is being devoted to determining whether existing deposits of mercury in inland waterways can be rendered harmless. This is an exceedingly difficult problem, but if the mercury metal on the bottom of some of our large lakes, such as Lake St. Clair, Lake Superior, and Lake Erie, is not removed or rendered harmless, the water (and the fish) will be contaminated for two or three hundred years. Tons of mercury are believed to be mixed with the mud, slime, and sand on lake bottoms.

Considerable attention is now being devoted also to pollution by cadmium, an element in the same periodic family as mercury.

■ **3. Sewage Waste Materials.**   Industrial, human, and other animal wastes introduce microorganisms and viruses into bodies of water. In many instances, "raw sewage" is dumped directly into lakes which serve as public water supply sources, or into rivers which are then the sources of public water supply a short distance downstream. Increasing effort is being expended in the study of efficient methods for treating sewage before its release to streams and lakes, but the incentives for industries and cities to undertake the trouble and expense of installing such treatment procedures are too often lacking. More strict governmental controls on a state or national level are thought by some to be essential. It has even been suggested facetiously that the problem might be considerably alleviated by requiring that a town or industry discharge its effluent *upstream* from the point at which it uses the stream as a source for its own water supply.

■ **4. Pesticides and Fertilizers.** A variety of highly toxic materials such as the double salt "Paris green" [which is copper(II) arsenite-acetate, $Cu_3(AsO_3)_2 \cdot Cu(C_2H_3O_2)_2$], the arsenates of calcium and lead, various mercury compounds, and chlorinated hydrocarbons such as dichlorodiphenyltrichloroethane (DDT) are used as pesticides in lawn care and agriculture. Fertilizers, in addition to the phosphate fertilizers mentioned previously, include many substances such as urea and a variety of nitrate and sulfate compounds. These pesticides and fertilizers, through soil runoff, reach streams either directly or through city storm sewers. The chief hazard of pesticides in the environment is their concentration in the food chain of fish and wildlife, which may result in death in these species or in decreased ability to reproduce. Fertilizers have the nutrient value, referred to earlier, for algae and plants. Strict soil conservation measures, close control on the use of certain pesticides, development of better application methods, and improved formulations can do much to alleviate these problems, which extend to *air* pollution as well as water pollution (see Sections 24.3–24.5).

■ **5. Thermal Pollution.** Industrial discharges which markedly raise the temperature of a body of water may result in harm to fish and in increased growth of algae and other microorganisms. In 1970, the Department of the Interior stated a new policy for Lake Michigan: "The minimum possible waste heat shall be added to the waters of Lake Michigan. In no event will heat discharges be permitted to exceed 1° F rise over ambient at the point of discharge." This caused great consternation in Chicago, for the implications of such a standard are far-reaching. If this standard were strictly and immediately enforced under present operating circumstances, virtually all electric power plants, both nuclear and conventional, would have to be shut down. A reasonable amount of time, however, is normally allowed for the industries, acting in good faith, to make the modifications necessary for compliance.

Interestingly, it is at least conceivable that thermal pollution, under certain special circumstances, could be beneficial. For example, the Ralston Purina Company and the Florida Power Company, which has replaced a conventional power plant with a nuclear power plant, are cooperating in a study to determine whether hot water discharges from power plants can be put to profitable use. They envision the possibility that selected saltwater fish and shellfish might thrive in the heated water, thereby improving the future of the food supply from the sea. Nevertheless, even a few degrees rise in temperature of the water through thermal pollution can decrease the solubility of oxygen in the water to a level too low for many forms of marine life to survive. Achieving a proper balance is an exceedingly delicate problem.

12.5   Purification of Water

Water is purified for city water supplies by first allowing it to stand in large reservoirs where most of the mud, clay, and silt settle out, a process called **sedimentation,** and then filtering it through beds of sand and gravel. Often, prior to filtration, lime and aluminum sulfate are added to the water in the settling reservoirs or filters. These chemicals react to form aluminum hydroxide, a gelatinous precipitate, which settles slowly carrying with it much of the suspended matter, including most of the bacteria. The equation for the precipitation of aluminum hydroxide is

$$2Al^{3+} + (3SO_4{}^{2-}) + (3Ca^{2+}) + 6OH^- \longrightarrow 2Al(OH)_3 + 3Ca^{2+} + (3SO_4{}^{2-})$$

Aluminum sulfate   Calcium hydroxide   Aluminum hydroxide   Calcium sulfate

It is common practice to kill the bacteria which remain in the water after filtration by adding chlorine.

Various synthetic materials such as cationic polyamines (substances which contain several $-NH_2$ groups) may displace aluminum sulfate as water clarifiers in some installations in the near future. They possess the advantage of being easier to remove from the purified water than aluminum hydroxide.

Relatively pure water for laboratory use is commonly prepared by distillation of tap water (see Section 11.7). Because the basic constituents of glass slowly dissolve in water, glass equipment is not satisfactory for the preparation and storage of very pure water. Distillation apparatus made of fused silica or pure tin is often used in making pure water. Gases from the air, particularly carbon dioxide, often contaminate distilled water and must be removed for some laboratory operations.

12.6   Hard Water

Water containing dissolved calcium, magnesium, and iron salts is known as **hard water.** The negative ions present in hard water are usually chloride, sulfate, and hydrogen carbonate. Hardness in water is objectionable for two reasons: (1) The calcium, magnesium, and iron ions form insoluble soaps by reaction with soluble soaps such as sodium stearate, $C_{17}H_{35}COONa$.

$$2C_{17}H_{35}COO^- + (2Na^+) + Ca^{2+} \longrightarrow \underline{Ca(C_{17}H_{35}COO)_2} + (2Na^+)$$

Insoluble soaps have no cleansing power, and due to their sticky nature adhere to fabrics, giving them a dingy appearance. Insoluble soaps also make up the "ring" in the bathtub. Soap in excess of that needed to precipitate the calcium and magnesium in hard water must be added in order to obtain cleansing action. (2) Hard water is responsible for the formation of a tightly adherent scale in boilers. At high temperatures, much of the mineral matter dissolved in hard water is precipitated as scale [insoluble substances such as calcium carbonate, magnesium carbonate, iron(II) carbonate and calcium sulfate]. Much of the scale is calcium sulfate, which is *less* soluble in hot water than in cold and precipitates partly because of that fact. The scale is a poor conductor of heat and thus causes a waste of fuel. Furthermore, boiler explosions are often due to the presence of scale. Since the scale is not a good heat conductor, the metal must be very hot (often red hot) in order to bring the water to the desired temperature. If the scale cracks, the water seeps through and comes in contact with the hot metal.

$$4H_2O + 3Fe \text{ (hot)} \longrightarrow Fe_3O_4 + 4H_2$$

The hydrogen gas, as it is formed, breaks the scale loose and more water gets in producing still more hydrogen, thereby setting the stage for a violent explosion.

It is important that the substances responsible for hardness in water be removed before the water is used for washing or in boilers. The removal of the metallic ions responsible for the hardness in hard water is known as **water-softening.**

12.7   Water-softening

When hydrogen carbonate ions are present, boiling the water drives off carbon dioxide, and the metal carbonates which are then formed precipitate.

$$Ca^{2+} + 2HCO_3^- \xrightarrow{\Delta} CaCO_3 + CO_2\uparrow + H_2O$$

The carbonates of calcium and magnesium thus precipitated form the deposits found in teakettles and boilers. Such water is said to possess **carbonate, or temporary, hardness**—temporary, because most of the hardness can be removed by boiling the water. The hydrogen carbonate ion may also be converted to the carbonate ion by the addition of a basic substance. Commercially, calcium hydroxide is added in the exact quantity needed to react with the hydrogen carbonate ions.

$$Ca^{2+} + 2OH^- + Ca^{2+} + 2HCO_3^- \longrightarrow 2CaCO_3 + 2H_2O$$

Similar equations may be written where the positive ion in the hard water is either Mg^{2+} or Fe^{2+}. The precipitated carbonates are readily removed from the water by filtration. An aqueous solution of ammonia is a basic solution that is often used in the home to remove temporary hardness in water.

$$Ca^{2+} + 2HCO_3^- + 2NH_3 \longrightarrow CaCO_3 + 2NH_4^+ + CO_3^{2-}$$

Water which contains chloride or sulfate ions in addition to the calcium, magnesium, or iron ions is said to possess **noncarbonate hardness.** Chloride and sulfate ions are not destroyed by boiling as is the hydrogen carbonate ion. Noncarbonate hardness can be eliminated by the addition of washing soda (sodium carbonate) to the water.

$$Ca^{2+} + (SO_4^{2-}) + (2Na^+) + CO_3^{2-} \longrightarrow CaCO_3 + (2Na^+) + (SO_4^{2-})$$
$$Ca^{2+} + (2Cl^-) + (2Na^+) + CO_3^{2-} \longrightarrow CaCO_3 + (2Na^+) + (2Cl^-)$$

The sodium sulfate and sodium chloride which are produced during the reaction do not interfere with the cleansing action of soap, nor do they form boiler scale.

Crude sodium hydroxide (caustic soda) is often used in large installations to remove both carbonate and noncarbonate hardness.

$$Ca^{2+} + 2HCO_3^- + (2Na^+) + 2OH^- \longrightarrow \underline{CaCO_3} + (2Na^+) + CO_3^{2-} + 2H_2O$$

The Na_2CO_3 which is produced in the removal of carbonate hardness then reacts to remove any remaining noncarbonate hardness.

$$Ca^{2+} + (SO_4^{2-}) + (2Na^+) + CO_3^{2-} \longrightarrow \underline{CaCO_3} + (2Na^+) + (SO_4^{2-})$$

Trisodium phosphate and borax also act as water softeners by forming insoluble calcium and magnesium phosphates and borates. Polymetaphosphates, such as $(NaPO_3)_x$, form soluble complexes with Ca^{2+} and Mg^{2+} and thus prevent them from reacting with soap. However, the use of phosphates in detergents is declining because of undesirable nutrient effects on algae and other plant life in rivers and streams (Section 12.4).

Ion exchange provides another useful and highly important method of softening water. Naturally occurring sodium aluminosilicates, known as **zeolites,** are used as ion-exchangers. When water containing calcium and magnesium ions filters slowly through thick layers of coarse granules of the zeolites, the sodium in the compound is replaced by calcium or magnesium, and the water is softened.

$$\underset{\substack{\text{Sodium aluminosilicate} \\ \text{(insoluble)}}}{2NaAlSi_2O_6} + Ca^{2+} \longrightarrow \underset{\substack{\text{Calcium aluminosilicate} \\ \text{(insoluble)}}}{Ca(AlSi_2O_6)_2} + 2Na^+$$

The calcium aluminosilicate can be reconverted into the sodium compound by treating the zeolite with a concentrated solution of sodium chloride.

$$Ca(AlSi_2O_6)_2 + 2Na^+ \longrightarrow 2NaAlSi_2O_6 + Ca^{2+}$$

Note that this reaction is the reverse of the preceding one; the reversibility can be represented by double arrows. The direction the reaction takes is controlled by the excess of either Ca^{2+} or Na^+ ions.

$$2Na\ Zeolite + Ca^{2+} \rightleftharpoons Ca(Zeolite)_2 + 2Na^+$$

After the sodium zeolite is regenerated, it is ready for use again. The ion exchange method of water-softening is effective for the removal of both carbonate and noncarbonate hardness (Fig. 12–1).

Synthetic resin ion-exchangers have also been developed to soften hard water. These exchangers remove both cations (positive ions) and anions (negative ions) from hard water; that is, they demineralize the water completely. Examples of these resins are Amberlite and Zeo-Carb. Metal ions in the hard water displace hydrogen ions from one type of resin.

$$2RCOOH + Ca^{2+} + (SO_4^{2-}) \longrightarrow (RCOO)_2Ca + 2H^+ + (SO_4^{2-})$$

Another type of resin then removes the ions of the acid from the water.

$$2RNH_2 + 2H^+ + (SO_4^{2-}) \longrightarrow 2RNH_3^+ + (SO_4^{2-})$$

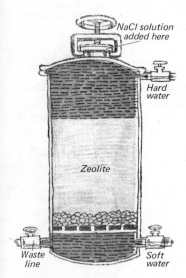

NaCl solution added here

Hard water

Zeolite

Waste line

Soft water

FIGURE 12–1

Water-softening by the ion exchange method. When regeneration is necessary, the soft water outlet valve is closed and a concentrated NaCl solution run through the zeolite and out the waste line until regeneration of the zeolite is complete.

Hence, if a hard water (or a salt solution) is passed first through an exchanger which replaces all metal ions with H^+ and then through a resin which combines with the resulting acids, it can be completely demineralized. The result is essentially the same as that achieved through distillation of water. In recent years the use of ion-exchangers in the treatment of water for industrial and domestic purposes has grown tremendously.

12.8 Physical Properties of Water

It is essential that we have a thorough knowledge of the properties of water because it is the most widely used of all substances and is a standard for the determination of a number of physical constants and units. In addition, it has many remarkable, almost unique, properties which we will study in the course of this chapter, e.g., expansion of the liquid on cooling below 4° C and abnormally high melting point, boiling point, heat of fusion and heat of vaporization for a molecule of such small molecular weight.

Pure water is an odorless, tasteless, and colorless liquid. Natural waters appear bluish-green in large bodies. The pleasant taste of drinking water is due to dissolved gases from the air and to dissolved salts from the earth.

The freezing point of water is 0° C, and its boiling point at 760 mm is 100° C. (Recall that the boiling point at 760 mm atmospheric pressure is the **normal boiling point;** see Section 11.3.) The vapor pressures of ice and water at various temperatures are listed in Table 10–1.

The unit of mass in the metric system of measurement is the mass of 1 ml of water at 4° C (the temperature of its maximum density) and is 1.00000 g. The SI unit of heat—the calorie, defined as equal to 4.1840 joules—is approximately the quantity of heat required to raise the temperature of one gram of water from 14.5° to 15.5°. Water is the reference substance also in the measurement of the specific gravity of liquids and solids (Section 1.18). The heat of fusion of water (as ice) is 80 cal/g at 0° and its heat of vaporization is 540 cal/g at 100°.

A diagram relating the pressures and temperatures at which the gaseous, liquid, and solid states (sometimes referred to as phases) can exist is called a **phase diagram.** Such a diagram is constructed by plotting the points corresponding to pressures and temperatures at which changes of state take place. The phase diagram for water is given in Fig. 12–2.

At 760 mm pressure, as heat is added, water changes from solid ice to liquid at 0°; hence, these values locate one point on the line (in color) that separates the regions for solid and liquid in the diagram. At the same pressure, a further increase of temperature causes a change of state from liquid to gas at 100°, these values corresponding to a point on the line (also in color) that separates the liquid and gas regions in the diagram. At a lower pressure, 702 mm, the change from liquid to gas takes place at 98°; these two values then locate another point. Other points are determined similarly in order to produce the complete diagram.

For each value of pressure and temperature falling within the portion of the diagram labeled *Solid,* water exists in the solid form (ice). Similarly for values of pressure and temperature within the portion labeled *Liquid,* water exists as

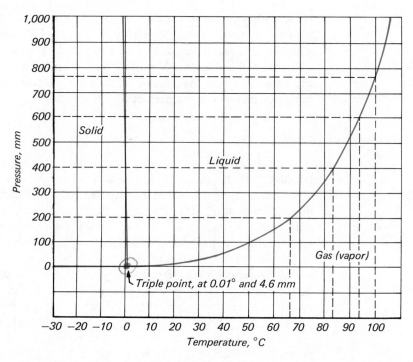

FIGURE 12-2

Phase diagram for water.

a liquid. For any combination of pressure and temperature in the region labeled *Gas,* water exists in the gaseous state. For example, at −5° and 760 mm pressure, the point on the diagram is in the region indicating that water is a solid. Moving across the diagram horizontally at a constant pressure of 760 mm, we see that as the temperature increases water becomes a liquid at 0°, which is its *normal melting temperature.* Water remains a liquid at a pressure of 760 mm for temperatures in the range 0° to 100°, changing to a gas at 100°, its *normal boiling temperature.* At temperatures above 100° at 760 mm, water is a gas.

The almost vertical colored line in Figure 12–2 shows us that as the pressure increases, the melting temperature remains almost constant, actually decreasing very slightly. On the other hand, the diagram indicates clearly that the boiling temperature increases markedly with an increase in pressure. Representative pressures and the corresponding boiling temperatures for water are given in the table at the top of the next page.

The lines which separate the various regions in Fig. 12–2 represent points at which an equilibrium exists between two states. Combinations of pressure and temperature falling on the line between the solid and liquid regions, for example, represent conditions under which the solid and liquid are in equilibrium with each other.

Note that at one point on the diagram the three lines intersect. This point is referred to as the **triple point.** At the pressure and temperature corresponding to the triple point (4.6 mm and 0.01°, respectively), all three states are in equilibrium with each other. Reference to the diagram shows that at temperatures and pressures below those at the triple point, water changes directly from solid to

Pressure, mm Hg	Boiling Temperature of Water, °C
5	1
20	22
80	44
310	77
596	93
702	98
760	(normal) 100
822	102
1,960	129
5,042	163

gas without going through the liquid state (sublimation—see Section 11.15) when the pressure on it is reduced below the triple point pressure.

Phase diagrams can also be constructed for compounds other than water, and such diagrams prove to be very useful to scientists working with the various physical states of these materials.

12.9   The Molecular Structure of Water

Water has an abnormally low vapor pressure, a high boiling point, a high heat of vaporization, and remarkable solvent properties, all of which are closely related to the structure of the water molecule.

The atoms in a molecule of water are held together by bonds that are somewhat polar in character, because oxygen is more electronegative (3.5) than hydrogen (2.1)—see Table 4-2. This indicates that the oxygen atom possesses a net negative charge and each hydrogen atom a net positive charge. If the molecule were linear ($\overset{+}{H}$—$\overset{=}{O}$—$\overset{+}{H}$), then the two bond polarities would cancel each other, and the molecule would be nonpolar. However, the two bonds make an angle less than 180° (actually 105°) with each other, and the molecule is highly polar (Fig. 12–3).

The oxygen atom has the electronic structure $1s^2 2s^2 2p^4$. The four p-subshell electrons are distributed among the three p orbitals in this manner: $2p_x^2 2p_y^1 2p_z^1$. Thus, there are two unpaired electrons in the oxygen atom. When combination with hydrogen occurs, these electrons are paired with $1s$ electrons from two hydrogen atoms (Fig. 12–4). Because p_y and p_z orbitals are at right angles to each other (see Section 3.20), a 90° angle between the two bonds in the H_2O molecule would be expected. However, the positively charged hydrogen atoms repel each other and so the angle must be somewhat larger than 90°.

If the water molecule is thought of as a tetrahedron, with oxygen in the center and the two hydrogens and two unshared pairs of electrons at the four corners of the tetrahedron (see Section 6.3), then the tetrahedron can be pictured as being within a cube with the corners of the tetrahedron at four of the eight corners of the cube (see Figs. 12–3 and 12–6). On this basis, the expected angle between the two hydrogen-oxygen bonds would be 109°.

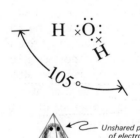

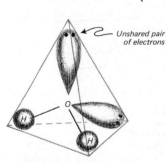

Unshared pair of electrons

FIGURE 12-3

The V-shaped character of the water molecule. The two unshared pairs of electrons can be thought of as being in p orbitals oriented toward two vertices of a tetrahedron.

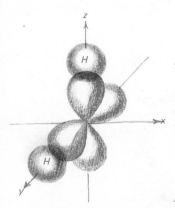

FIGURE 12-4
Orbital diagram for the water molecule showing *p* orbitals oriented at right angles to each other and superimposed on *y* and *z* axes. This orientation results in an angle of 90° between the two O—H bonds. However, the actual angle is 105°, due to repelling forces between the hydrogen atoms.

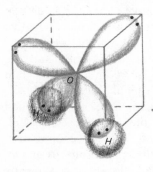

FIGURE 12-6
Orbital diagram for the water molecule showing orbitals oriented in a tetrahedral structure with the tetrahedron inscribed within a cube. This orientation would result in an angle of 109° between the two O—H bonds. As noted for Fig. 12-4, the actual angle is 105°.

Both theories of the structure of the water molecule predict an angular molecule. As we have seen, the actual angle is 105°—in between the values 90° and 109° predicted by the two theories and obviously somewhat closer to the 109° value predicted for the tetrahedral structure.

The tetrahedral structure involves two sigma bonds each of which arises from the combination of a $1s$ atomic orbital of hydrogen with a $2p$ atomic orbital of oxygen. In Chapter 5, we discussed sigma bonds arising from either the combination of two s orbitals or an end-to-end combination of two p orbitals (Section 5.1). The combination of an s orbital with a p orbital, as shown in Fig. 12-4, constitutes a third way in which a sigma bond can be formed.

When many water molecules are together, the positive hydrogen of one molecule attracts the negative oxygen of a neighboring molecule forming what is known as a **hydrogen bond** (see Section 21.11). This bond, even though weak, is effective in producing clusters of associated molecules, or **polymers** (Fig. 12-5).

$$n\mathrm{H_2O} \rightleftharpoons (\mathrm{H_2O})_n + \mathrm{Heat}$$

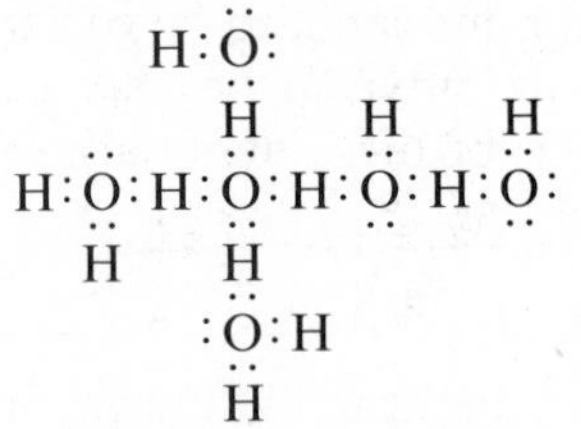

$$\mathrm{H\!:\!\overset{..}{\underset{..}{O}}\!:}$$
$$\mathrm{\overset{H}{} \quad \overset{H}{} \quad \overset{H}{}}$$
$$\mathrm{H\!:\!\overset{..}{\underset{..}{O}}\!:\!H\!:\!\overset{..}{\underset{..}{O}}\!:\!H\!:\!\overset{..}{\underset{..}{O}}\!:\!H\!:\!\overset{..}{\underset{..}{O}}\!:}$$
$$\mathrm{\overset{H}{} \quad \overset{H}{}}$$
$$\mathrm{:\!\overset{..}{\underset{..}{O}}\!:\!H}$$
$$\mathrm{\overset{H}{}}$$

FIGURE 12-5
Association of water molecules in clusters through hydrogen bonding.

The value of n in the equation depends upon such conditions as temperature and the presence of substances dissolved in the water. As the temperature is raised, some of the hydrogen bonds are disrupted, and the larger clusters are broken down into smaller ones. In the vapor state, water is composed almost entirely of single H_2O molecules. Association of the polar water molecules is responsible for the abnormally low vapor pressure, high boiling point, high heat of fusion, and high heat of vaporization of water relative to the same properties of non-associated compounds of similar formulas, such as H_2S, H_2Se, and H_2Te. These abnormal physical properties of water are due to the fact that energy is required to break the hydrogen bonds holding together clusters of associated molecules.

The abnormally large dielectric constant of water, which is responsible for the remarkable power of water to dissolve ionic substances (Chapter 14), is due to the association of the polar molecules. The product of the charge on either end of a dipole and the effective distance between the centers of the charges is the **dipole moment.** Thus, the size of the dipole moment depends upon the magnitude of the charges and the distance between them. The dipole moment of a complex made up of two water molecules is more than double that of a single molecule.

Crystal chemistry has shown that the water molecules in ice are so arranged that each oxygen atom has four hydrogen atoms as close neighbors, two attached by electron pair bonds and two by hydrogen bonds (Fig. 12-7). Such an arrangement leads to an open structure, i.e., one with relatively large holes in it. This makes ice (Fig. 12-8) a substance with a relatively low density. When ice melts, some of the hydrogen bonds are broken; the open structure is destroyed; and

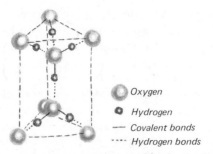

FIGURE 12-7

The arrangement of water molecules in ice, showing each oxygen atom with four hydrogen atoms as close neighbors, two attached by electron-pair bonds and two by hydrogen bonds. The oxygen atoms shown on the corners of the prism are also on the corners of adjoining prisms, as this is only a portion of a continuous lattice.

FIGURE 12-8

These are snowflakes (water crystals). Most of them are hexagonal in form. *Ewing Galloway*

the water molecules pack more closely together. This accounts for the fact that water is more dense than ice.

12.10 Chemical Properties of Water

Water is a medium in which many chemical reactions take place. The majority of the chemical changes that we study in a course in general chemistry are reactions taking place in aqueous solutions.

■ **1. The Thermal Stability of Water.** Water is the product of a highly exothermic reaction, the burning of hydrogen in oxygen; i.e., its heat of formation is high and it is accordingly a very stable compound. It decomposes to the extent of only 11.1 per cent at the very high temperature of 2,727°, the reaction being reversible.

$$2H_2O + 136,630 \text{ cal} \rightleftharpoons 2H_2 + O_2$$

As the temperature of the system is lowered, the constituent elements recombine. Thermal stability is not to be confused with nonreactivity. Although water is stable toward decomposition by heat, it is a reactive substance and enters readily into a great number and variety of reactions, even at ordinary temperatures.

■ **2. Ionization of Water.** Because aqueous solutions of acids contain hydrogen ions, H^+, and aqueous solutions of bases contain hydroxide ions, OH^-, one might ask whether or not these ions are present in pure water, which is neutral. Water dissociates to produce hydrated hydrogen ions and hydroxide ions in equal but very small concentrations. Pure water, therefore, conducts electricity, but very poorly.

It is important to note that water dissociates to form a hydrated ion. The hydrogen ion (the hydrogen atom minus its valence electron) does not exist free in solution, but combines with a water molecule to form the hydrated ion called the **hydronium ion** and commonly assigned the formula $H^+ \cdot H_2O$, or H_3O^+. It should be mentioned that the true formula for the hydrated ion is not completely agreed upon by chemists. The formula H_3O^+, however, is frequently used, and the equation for the dissociation of water is commonly written with H_3O^+, instead of H^+.

$$2H_2O \rightleftharpoons H_3O^+ + OH^-$$

For simplicity, the formula H^+ is often used even for reactions occurring in water solution, but one should always remember that in such cases the symbol H^+ actually stands for the hydrated ion, H_3O^+. The hydration of the hydrogen ion is not unique, for all ions in aqueous solution form complexes with water or at least are attracted to the water by ion-dipole forces.

The addition of an acid to water causes an increase in the hydronium ion concentration and a corresponding decrease in the hydroxide ion concentration, but not to zero. With the addition of a base to water, the hydroxide ion concentration becomes large and the hydronium ion concentration very small. Many of the chemical properties of water depend upon the hydronium ion and hydroxide ion content. A more complete account of the ionization of water is given in Chapter 18 (Ionic Equilibria of Weak Electrolytes).

■ **3. Action of Water with Metals.**   The reactivity of various metals with water was mentioned in connection with the preparation of hydrogen (Chapter 9). Only those metals which lie above cobalt in the activity series (Section 9.9) displace hydrogen from water. The more active metals react with cold water, but the less active ones must be heated to high temperatures in steam.

$$2Na + 2H_2O \text{ (cold)} \longrightarrow 2Na^+ + 2OH^- + H_2\uparrow$$
$$3Fe + 4H_2O \text{ (steam)} \longrightarrow Fe_3O_4 + 4H_2\uparrow$$

Magnesium reacts with hot water to give the hydroxide and hydrogen.

$$Mg + 2H_2O \text{ (hot)} \longrightarrow \underline{Mg(OH)_2} + H_2\uparrow$$

When magnesium burns in steam, the oxide rather than the hydroxide results.

$$Mg + H_2O \text{ (steam)} \longrightarrow MgO + H_2$$

■ **4. Action of Water with Nonmetals.**   Fluorine, chlorine, and bromine react with water at ordinary temperatures. The equation for the reaction of chlorine with water at room temperature is

$$Cl_2 + H_2O \longrightarrow \quad H^+ + Cl^- \quad + \quad HOCl$$

Hydrochloric acid   Hypochlorous acid

Hydrochloric acid is completely ionized in dilute aqueous solution, a property characteristic of **strong acids.** On the other hand, hypochlorous acid is only slightly ionized; most of it is in the form of molecules in aqueous solution, a property characteristic of **weak acids.**

Bromine reacts similarly; but the highly electronegative fluorine, with its unusually strong oxidizing power, reacts with water in a different way producing ozone, O_3.

White-hot carbon reacts with steam and produces carbon monoxide and hydrogen.

$$C + H_2O \longrightarrow CO + H_2$$

■ **5. Reaction of Water with Compounds.** The oxides of the alkali metals (Li, Na, K, Rb, and Cs) react readily with water in exothermic reactions to form the hydroxide ion.

$$Na_2O + H_2O \longrightarrow 2Na^+ + 2OH^-$$

The alkali metal hydroxides, thus produced, are too stable to be decomposed by heating, even at their boiling points, which are high. The oxides of the alkaline earth metals (Be, Mg, Ca, Sr, Ba, and Ra) react less readily with water than oxides of the alkali metals and give hydroxides, which may be decomposed into metal oxides and water by heating. Beryllium and magnesium oxides, which are highly insoluble in water, react with water only slowly and incompletely. Metal oxides which react with water to form bases are often referred to as **basic anhydrides.**

The oxides of certain nonmetals react with water to form acids. Hence, they are referred to as **acidic anhydrides.** Some examples are shown by the following equations:

$$SO_2 + H_2O \longrightarrow \underset{\text{Sulfurous acid}}{H_2SO_3}$$

$$CO_2 + H_2O \longrightarrow \underset{\text{Carbonic acid}}{H_2CO_3}$$

$$P_4O_{10} + 6H_2O \longrightarrow \underset{\text{Phosphoric acid}}{4H_3PO_4}$$

■ **6. Hydrolysis.** Water also enters into reactions with certain compounds and ions in an important way called **hydrolysis.** Reactions of this type will be considered in detail in Chapter 18.

■ **7. Hydrates.** When aqueous solutions of many of the soluble salts are evaporated, the salt separates as crystals which contain the salt and water combined in definite proportions. Such compounds are known as **hydrates,** and the water is called **water of hydration.** The formation of hydrates is not limited to salts but is common with acids and bases and even elements. Furthermore, it is not confined to crystals. Familiar examples of hydrates are blue vitriol, $CuSO_4 \cdot 5H_2O$; Epsom salts, $MgSO_4 \cdot 7H_2O$; alum, $KAl(SO_4)_2 \cdot 12H_2O$; Glauber's salt, $Na_2SO_4 \cdot 10H_2O$; $H_2SO_4 \cdot H_2O$; $Ba(OH)_2 \cdot 8H_2O$; $BaCl_2 \cdot 6H_2O$; and $Cl_2 \cdot 8H_2O$.

When blue crystals of copper(II) sulfate 5-hydrate, $CuSO_4 \cdot 5H_2O$, are heated, water is given off, the crystalline structure characteristic of the hydrated salt breaks down, and the white anhydrous salt, $CuSO_4$, remains.

$$CuSO_4 \cdot 5H_2O \rightleftharpoons CuSO_4 + 5H_2O$$

The preceding reaction is reversible in that the addition of water to the anhydrous salt produces the original hydrated salt.

Crystal chemistry has shown that in $CuSO_4 \cdot 5H_2O$ the copper(II) ion, Cu^{2+}, is combined with four water molecules, and the sulfate ion $(SO_4{}^{2-})$ with one. We may formulate the hydrate, then, as $[Cu(H_2O)_4][SO_4(H_2O)]$. All ions are undoubtedly hydrated in aqueous solution, but in many cases the water of hydration is lost when the compound separates from solution. The hydration of ions will be discussed in more detail in Section 12.12 and in the hydrolysis discussion in Chapter 18.

12.11   Deliquescence and Efflorescence

Hydrates exhibit vapor pressures, as can be demonstrated by placing a crystal of a hydrate in the vacuum above the mercury in a barometer tube. As the water vapor from the hydrate exerts pressure on the mercury, the mercury level is depressed to an extent dependent upon the magnitude of the **vapor pressure of the hydrate.** The vapor pressure of a hydrate increases with a rise in temperature and decreases as the temperature is lowered.

When a hydrate has a vapor pressure higher than the partial pressure of water in the atmosphere, the hydrate will lose part or all of its water of hydration if exposed to the air. Hydrates which lose water of hydration when exposed to the air are said to **effloresce,** and **efflorescence** (Fig. 12–9) is said to have taken place.

FIGURE 12-9

Efflorescent crystals lose water and crumble to a powder. Some deliquescent substances take up enough water from the air to dissolve.

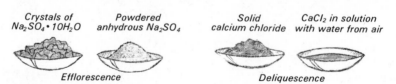

Crystals of $Na_2SO_4 \cdot 10H_2O$ Powdered anhydrous Na_2SO_4 Solid calcium chloride $CaCl_2$ in solution with water from air

Efflorescence   Deliquescence

$Na_2SO_4 \cdot 10H_2O$, which has a vapor pressure of 14 mm, effloresces rapidly when exposed to the air if the partial pressure of the water vapor in the air is less than 14 mm. $CuSO_4 \cdot 5H_2O$, with a vapor pressure of 7.8 mm, is stable in air as long as the partial pressure of the water in the air is greater than 7.8 mm. It follows that some hydrates are stable when the humidity is high but decompose when it is low. A salt such as copper(II) sulfate may form more than one hydrate, each of which possesses its own definite vapor pressure at a given temperature. The following hydrates of copper(II) sulfate are known: $CuSO_4 \cdot 5H_2O$; $CuSO_4 \cdot 3H_2O$; and $CuSO_4 \cdot H_2O$.

When certain hydrated salts of low vapor pressure, such as $CaCl_2 \cdot H_2O$, are exposed to moist air, they form higher hydrates. Such salts may be used in the removal of moisture from air or other gases. A substance that can remove moisture from the air is said to be **hygroscopic.** Concentrated sulfuric acid, a liquid, and phosphorus(V) oxide, a solid, are hygroscopic and are used as drying agents, or **desiccants.** Certain water-soluble hygroscopic solids remove sufficient water from the air to dissolve completely in this water and form solutions. Such substances are said to be **deliquescent,** and the process is termed **deliquescence** (Fig. 12–9). Very soluble salts, such as $CaCl_2 \cdot 6H_2O$, are often extremely deliquescent.

12.12 The Structure of Hydrates

It appears that water may exist in a crystal in five different forms.

■ **1. Water of Hydroxylation.** When water is present in a compound as water of **hydroxylation,** the structural unit is the OH^- ion. This is the case with many metal hydroxides such as $NaOH$, $Mg(OH)_2$, and $Al(OH)_3$. Removal of water of hydroxylation disrupts the crystal. The crystal structure characteristic of Al_2O_3, which may be formed by dehydrating $Al(OH)_3$, is entirely different from that of aluminum hydroxide.

■ **2. Water of Coordination.** Water of this type is joined to the metal ion by coordinate bonds. A crystal of $BeSO_4 \cdot 4H_2O$ has as its building units $[Be(H_2O)_4]^{2+}$ and SO_4^{2-}, and one of $NiSO_4 \cdot 6H_2O$ contains $[Ni(H_2O)_6]^{2+}$ and SO_4^{2-} as structural units. Water of coordination is essential to the stability of these crystals, and the crystal lattice collapses when even a part of the water is driven off.

■ **3. Anion Water.** This type of water is attached to the anion by hydrogen bonding (see Sections 12.9 and 20.11). A crystal of $CuSO_4 \cdot 5H_2O$ has four of the five water molecules attached to the copper ion as water of coordination; the fifth molecule of water is attached to the sulfate ion as anion water.

■ **4. Lattice Water.** In some hydrates, water molecules occupy definite positions in the crystal lattice but are not attached directly to either cation or anion. In $KAl(SO_4)_2 \cdot 12H_2O$, six water molecules are attached to the aluminum ion by coordinate bonding; the other six are lattice water, arranged in definite positions around the potassium ion but at distances too great to allow direct attachment.

■ **5. Zeolitic Water.** Such water molecules occupy relatively random positions in the crystal lattice. Zeolites, for example, are open, three-dimensional networks of SiO_4 tetrahedra, which lose or take up water without any apparent change in crystal structure.

12.13 Heavy Water

Water which is composed of deuterium, ^2_1H, and oxygen is known as **heavy water,** or deuterium oxide, D_2O. Heavy water is present in ordinary water in the ratio of one part D_2O to 6,900 parts H_2O. Heavy water was first obtained by the electrolysis of ordinary water. Deuterium oxide molecules migrate to the cathode more slowly than do ordinary water molecules during electrolysis; therefore, the heavy water tends to remain behind while the ordinary water is decomposed. Heavy water is now produced in the atomic energy program by an exchange equilibrium process, which is less costly and more efficient than the electrolytic method. In the exchange process, advantage is taken of the fact that when liquid water (which contains small quantities of D_2O) and gaseous hydrogen sulfide are mixed, the deuterium atoms exchange between sulfur and oxygen, preferentially combining with oxygen as D_2O at low temperatures and with sulfur as D_2S at elevated temperatures.

$$H_2S + D_2O \xrightleftharpoons[\text{Cold}]{\text{Hot}} D_2S + H_2O$$

If steam and hydrogen sulfide are passed through each other in opposite directions at an elevated temperature, the hydrogen sulfide becomes saturated with water and through exchange is enriched with respect to D_2S. However, when the resulting mixture of steam and hydrogen sulfide is drawn off and condensed to the liquid state by cooling, the reverse exchange process takes place, producing water which is enriched with respect to D_2O and which thereby contains a larger proportion of D_2O than before. The process typically yields a D_2O concentrate of about 15 per cent, which is then enriched to 99.8 per cent D_2O by fractional distillation and electrolysis. To produce one ton of D_2O, a plant must process 45,000 tons of water and must cycle 150,000 tons of H_2S. Several hundred tons of heavy water are now produced per year.

Heavy water resembles ordinary water in appearance but differs from it slightly in other physical properties (Table 12–1).

TABLE 12-1   Some Physical Properties of Ordinary Water and Heavy Water

Property	Ordinary Water	Heavy Water
Density, 20°	0.997	1.108
Boiling point, °C	100.00	101.41
Melting point, °C	0.00	3.79
Heat of vaporization, cal/mole	9,720	9,944
Heat of fusion, cal/mole	1,436	1,500

Heavy water is one of the "moderators" used in nuclear reactors to reduce the speed of neutrons to that required for the fission process to take place (see Sections 30.13–30.15).

Because heavy water is about 10 per cent heavier than ordinary water, it is possible to detect the presence of as little as one part of it in 100,000 parts of an aqueous solution. For this reason, heavy water and deuterium serve as valuable tracers in the study of both chemical and physiological changes. By replacing ordinary hydrogen with deuterium in the molecules of food, the processes of digestion and metabolism in the body can be studied.

Salts have slightly lower solubilities in D_2O then H_2O, and the rates of reactions of D_2O are somewhat lower than those of similar reactions involving H_2O.

Hydrogen Peroxide

Hydrogen forms another compound with oxygen, called **hydrogen peroxide**, H_2O_2, in which there is twice as much oxygen for the same weight of hydrogen as there is in water. Hydrogen peroxide was first prepared in 1818 by Thénard, who obtained it by treating barium peroxide with hydrochloric acid. Very small quantities of hydrogen peroxide are present in dew, rain, and snow, probably as a result of the action of ultraviolet light upon oxygen mixed with water vapor.

12.14   Preparation

In the nineteenth century, hydrogen peroxide was produced exclusively on a commercial scale by the reaction of an acid upon barium peroxide, which is readily formed by heating barium oxide in air ($2BaO + O_2 \longrightarrow 2BaO_2$). For the reaction of sulfuric acid with barium peroxide, the equation is

$$BaO_2 + 2H^+ + SO_4^{2-} \longrightarrow BaSO_4 + H_2O_2$$

Barium sulfate is insoluble and can be separated from the solution by filtration. The hydrogen peroxide thus produced is quite dilute.

Hydrogen peroxide can be prepared in the laboratory by adding sodium peroxide to cold water or cold dilute hydrochloric acid.

$$Na_2O_2 + 2H_2O \longrightarrow 2Na^+ + 2OH^- + H_2O_2$$

Electrochemical processes for the production of hydrogen peroxide involve the electrolysis of solutions of sulfuric acid or ammonium hydrogen sulfate to produce peroxydisulfuric acid, $H_2S_2O_8$, or its ammonium salt, at the anode and hydrogen at the cathode.

$$2H_2SO_4 + \text{Electrical energy} \longrightarrow H_2S_2O_8 + H_2\uparrow$$

$$(2NH_4^+) + 2HSO_4^- + \text{Electrical energy} \longrightarrow (2NH_4^+) + S_2O_8^{2-} + H_2\uparrow$$

The peroxydisulfuric acid is subsequently hydrolyzed (allowed to react with water) to produce hydrogen peroxide.

$$H_2S_2O_8 + 2H_2O \longrightarrow 2H_2SO_4 + H_2O_2$$

The sulfuric acid formed is recycled, i.e., it is used over again in the production of more hydrogen peroxide.

For many years, the electrochemical process accounted for 80 to 90 per cent of U.S. production of hydrogen peroxide. However, a different industrial method has gained importance rapidly in recent years and now accounts for most of the hydrogen peroxide produced. In this process, an organic compound referred to as an anthraquinone is reduced by hydrogen in a benzene solution, using palladium as a catalyst, to a compound known as a *dihydro*anthraquinone. When the dihydroanthraquinone is allowed to react with oxygen, the original anthraquinone is obtained along with hydrogen peroxide. The anthraquinone is then used over again in the production of more hydrogen peroxide.

$$\underset{\substack{\text{2-Ethyl}\\\text{anthraquinone}}}{C_{16}H_{12}O_2} + H_2 \xrightarrow{\text{Pd catalyst}} \underset{\substack{\text{2-Ethyl}\\\text{dihydroanthraquinone}}}{C_{16}H_{12}(OH)_2}$$

$$C_{16}H_{12}(OH)_2 + O_2 \longrightarrow C_{16}H_{12}O_2 + H_2O_2$$

When the hydrogen peroxide reaches a concentration, in the benzene solution, of about 5.5 grams per liter, it is extracted with water to produce an 18 per cent water solution of hydrogen peroxide, which may be further concentrated by multiple vacuum distillation. Hydrogen peroxide decomposes when heated, necessitating vacuum distillation which takes place at relatively low temperatures (lower boiling points at reduced pressures—see Section 11.3).

When very pure, H_2O_2 is quite stable. Solutions of 90 per cent concentration and higher are produced and used routinely in rockets, as a chemical reagent, and in a number of other applications. However, when impurities are present, the concentrated solutions are dangerously explosive. For storage purposes, a **stabilizer,** such as a small amount of acid or the organic compound, acetanilide, is added. Bases and heavy metal ions, such as Cu^{2+}, act as catalysts to speed up the decomposition of hydrogen peroxide into water and oxygen.

12.15   Properties

Pure hydrogen peroxide is a colorless, syrupy liquid with a sharp odor and an astringent taste. It has a density of 1.438 g/cm^3 at $20°$ C, boils at $62.8°$ at 21 mm, and freezes at $-1.70°$. It is miscible (mutually soluble) with water, alcohol, and ether in all proportions. As stated in the preceding section, hydrogen peroxide is thermally unstable and decomposes with the evolution of much heat.

$$2H_2O_2 \longrightarrow 2H_2O + O_2\uparrow - 46,870 \text{ cal}$$

Its instability may be explained by the fact that its formation is a highly endothermic process.

Hydrogen peroxide is an active oxidizing agent. It oxidizes sulfurous acid (H_2SO_3) to sulfuric acid (H_2SO_4) and sulfides to sulfates.

$$H_2SO_3 + H_2O_2 \longrightarrow H_2SO_4 + H_2O$$
$$\underset{\text{(black)}}{PbS} + 4H_2O_2 \longrightarrow \underset{\text{(white)}}{PbSO_4} + 4H_2O$$

When used as an oxidizing agent, it has an advantage over other oxidizing agents, because the only by-product of the oxidizing reactions is water.

Hydrogen peroxide not only acts as an oxidizing agent with strong reducing agents, but also as a reducing agent with strong oxidizing agents. For example, it will reduce silver oxide to the metal.

$$Ag_2O + H_2O_2 \longrightarrow 2Ag + H_2O + O_2\uparrow$$

Hydrogen peroxide reduces many other substances with the evolution of oxygen. These substances include the oxides of mercury, gold, and platinum, and the strong oxidizing agents manganese dioxide and potassium permanganate. The reaction of hydrogen peroxide with the permanganate ion in acid solution is according to

$$6H^+ + 2MnO_4^- + 5H_2O_2 \longrightarrow 5O_2\uparrow + 2Mn^{2+} + 8H_2O$$

Hydrogen peroxide is a weak acid; in water it ionizes to a very slight extent in two steps.

$$H_2O_2 \rightleftharpoons H^+ + \underset{\substack{\text{Hydroperoxide} \\ \text{ion}}}{HO_2^-}$$

$$HO_2^- \rightleftharpoons H^+ + \underset{\substack{\text{Peroxide} \\ \text{ion}}}{O_2^{2-}}$$

Its acidic character is demonstrated by the equation

$$\underset{\text{Base}}{Ba(OH)_2} + \underset{\text{Acid}}{H_2O_2} \longrightarrow \underset{\text{Salt}}{BaO_2} + \underset{\text{Water}}{2H_2O}$$

12.16 The Structure of Peroxides

The valence electronic formula for hydrogen peroxide and a diagram of its molecular structure are given in Fig. 12–10. Its structure can be likened to that of a sheet of paper folded to an angle of 94°, with the oxygen atoms located in the fold and one hydrogen atom located in each plane.

Note in the valence electronic formula in Fig. 12–10 that two atoms of oxygen

FIGURE 12-10

The electronic and molecular structure of hydrogen peroxide, H_2O_2.

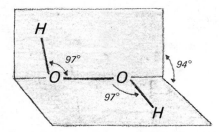

are connected through an electron-pair bond and that the charge on the peroxide group must be 2−. This is true of other peroxides as well and is illustrated by barium peroxide (BaO_2) in the following formula.

$$Ba^{2+} \left[\underset{\substack{\circ\circ \quad \times\times}}{\overset{\circ\circ \quad \times\times}{\circ O \times O \times}} \right]^{2-}$$

Barium peroxide

The usual charge for a *single* oxygen ion is two, as indicated in the valence electronic structures for barium oxide (BaO) and manganese dioxide (MnO_2).

$$Ba^{2+} \begin{bmatrix} \overset{\times\times}{\underset{\times\times}{\times \ddot{O} \ddot{}}} \end{bmatrix}^{2-} \qquad \overset{\times\times}{\times \ddot{O} \times} \overset{}{\ddot{}} Mn \overset{}{\ddot{}} \overset{\times\times}{\times \ddot{O} \times}$$

Barium oxide   Manganese dioxide

Manganese dioxide, unlike barium peroxide, does not yield hydrogen peroxide when treated with acids. This fact indicates that MnO_2 is a dioxide rather than a peroxide; i.e., the two oxygen atoms are not joined together but are each joined only to the manganese. There is no oxygen-oxygen bond in MnO_2 as there is in BaO_2.

12.17 Uses of Hydrogen Peroxide

Three per cent solutions of hydrogen peroxide by weight are used in the home as a mild antiseptic, a deodorizer, and a germicide; a somewhat more concentrated solution is a bleach for hair. Its effectiveness as a germicide is questionable, however. A 30 per cent solution of hydrogen peroxide is commonly used in the laboratory as an oxidizing agent. In commerce, the principal consumption of hydrogen peroxide is as an oxidizing agent, particularly as a bleaching agent. Substances of animal origin such as wool and hair, which are harmed by most other bleaching agents, are often bleached by hydrogen peroxide. At present, most cotton cloth is bleached with hydrogen peroxide because of the innocuous character of the decomposition products and the permanency of the whiteness produced by peroxide bleaching.

Ninety per cent hydrogen peroxide is used as an oxidizing agent in certain high explosives and in rockets. It is an ideal liquid oxidant because it has a high energy release on reaction, it is relatively stable to shock, and it is noncorrosive, nontoxic, and has a high boiling point and low freezing point (see Section 25.28).

Concentrated hydrogen peroxide may be used as a monopropellant, since it is decomposed under pressure yielding a gaseous mixture of oxygen and superheated steam.

12.18 The Law of Multiple Proportions

We have seen that hydrogen and oxygen form two distinctly different compounds, water and hydrogen peroxide. One mole of water (18.0154 g) contains 2.0160 g of hydrogen and 15.9994 g of oxygen; on the other hand, one mole of hydrogen peroxide (34.0148 g) contains the same weight of hydrogen (2.0160 g) but twice as much oxygen (31.9988 g). Many pairs of elements form more than one compound in which the proportions by weight of the constituent elements are not the same. Additional examples are Na_2O and Na_2O_2; BaO and BaO_2; FeO, Fe_2O_3, and Fe_3O_4; and N_2O, NO, NO_2, N_2O_3 and N_2O_5. This fact led Dalton (1804) to formulate the **Law of Multiple Proportions: When two elements, A and B, form more than one compound by combining with each other, the weights of element B which combine with a given weight of element A are in a ratio which can be expressed by small whole numbers.** The ratio for the weights of oxygen which combine with a given weight of hydrogen in the compounds H_2O_2 and H_2O is $2:1$. It is especially worthy of note that the law was first stated before formulas were known and, indeed, was a major factor in constructing formulas.

QUESTIONS

1. List the following physical constants for water: boiling point, freezing point, density, specific heat, heat of vaporization, and heat of fusion.
2. What metal ions are commonly responsible for the hardness of water?
3. What properties of hard water make its use objectionable?
4. What naturally occurring impurities are typically present in rain water; in sea water; in distilled water?
5. Water possessing temporary hardness may be softened by boiling. Explain.
6. How may the noncarbonate hardness of water be removed?
7. Explain the action of ion exchangers in the softening of water.
8. Distinguish between "softening" and "demineralizing" water.
9. Explain how aluminum sulfate and calcium hydroxide are used in the purification of water.
10. What are some of the substances which are of primary concern in water pollution?
11. Indicate a principal advantage of using phosphates in detergents and a principal disadvantage.
12. Explain what is meant by the term "eutrophication." Discuss its significance in terms of water.
13. How does mercury get into natural waters? How does methylmercury get into natural waters? Discuss the harmful effects of each.
14. What is "thermal pollution" of water? How does it arise? Is it bad?
15. Discuss possible procedures for helping to solve the problems of water pollution.
16. Describe the hydrogen bond.
17. Account for the polar character of the water molecule in terms of its structure.
18. Under what conditions does water exist as single H_2O molecules?
19. Account for the abnormally high boiling point of water.
20. The angle between the two O—H bonds in the water molecule is known experimentally to be about $105°$. It is sometimes said that the bond angle for water should be somewhere between $90°$ and $109°$. Explain the reasons for thinking that $90°$ and $109°$ are significant angles in connection with the structure of water and discuss why it is reasonable to assume that the angle should be (a) larger than $90°$, (b) smaller than $109°$.
21. Using information about the structure of water molecules and the interaction of water molecules with one another, discuss the fact that the heat of fusion of water is only 80 cal/g whereas the heat of vaporization of water is 540 cal/g.
22. Account for the high thermal stability of water.
23. Explain the action of drying agents.
24. How can the fact that hydrates exert vapor pressure be conveniently demonstrated?

25. In terms of relative vapor pressures, explain why $Na_2SO_4 \cdot 10H_2O$ effloresces in dry air.
26. Distinguish between the terms "hygroscopic" and "deliquescent."
27. Why is $CuSO_4 \cdot 5H_2O$ stable in moist air?
28. Relate vapor pressure of hydrates to temperature.
29. Define and illustrate the terms "acidic anhydride" and "basic anhydride."
30. What type of oxide reacts with water to form bases?
31. What is heavy water and how is it prepared?
32. Using the phase diagram for water (Fig. 12–2), determine the following:
    (a) At 600 mm, what is the approximate temperature necessary to convert water from a solid to a liquid; from a liquid to a gas?
    (b) At 37°, what is the approximate pressure at which water may be changed from a gas to a liquid? What pressure is necessary at 90°?
33. Using the phase diagram for water (Fig. 12–2), answer the following questions:
    (a) Holding the temperature constant, how may water be changed from gas to liquid and from liquid to solid?
    (b) At a temperature of −10°, can water exist as a gas; as a liquid; as a solid? Explain.
    (c) At each of the following temperatures can each of the changes listed in part (a) be accomplished: −10°; +10°? Explain.
34. Explain what is meant by the term "triple point" of a substance.
35. Determine from the phase diagram for water what physical state water would have at:
    (a) 700 mm pressure and 40°
    (b) 400 mm and −15°
    (c) 200 mm and 50°
    (d) 200 mm and 80°
36. (a) Can water be converted from a solid to a liquid at 3 mm pressure; from a gas to a liquid; from a solid to a gas; from a gas to a solid?
    (b) If the pressure is 10 mm, can the changes in state listed in part (a) be accomplished?
    (c) Explain both (a) and (b) in terms of the phase diagram for water.
37. State the Law of Multiple Proportions. Explain how NO and NO_2 illustrate this law.
38. Two sulfides of iron have the following compositions:

Fe:	63.40 per cent	46.67 per cent
S:	36.60 per cent	53.33 per cent
	100.00 per cent	100.00 per cent

Show how these compounds illustrate the Law of Multiple Proportions.
39. The heat of fusion of H_2O is 80 cal/g whereas heats of fusion of H_2S and H_2Te are 16.7 cal/g and 7.7 cal/g, respectively. (a) Explain why the heats of fusion would normally be expected to decrease in that order. (b) Explain why, even taking into account the expected trend, the heat of fusion for water is abnormally high (nearly five times as high as that for H_2S).
40. Explain the difference in vapor density of H_2O and D_2O at S.T.P.

41. How is hydrogen peroxide prepared in the laboratory? How is it prepared commercially? Write equations.
42. What is the advantage of using hydrogen peroxide as an oxidizing agent over other such agents?
43. Why is hydrogen peroxide purified by distillation under reduced pressure? What are stabilizers?
44. Distinguish between dioxides and peroxides in terms of electronic structures.
45. Under what conditions does hydrogen peroxide act as a reducing agent? Illustrate with an equation.
46. Write an equation showing that H_2O_2 is an acid.

PROBLEMS

1. What is the per cent of water by weight in each of the following compounds?
   (a) $Co(NO_3)_2 \cdot 6H_2O$ *Ans. 37.1%*
   (b) $MgSO_4 \cdot 7H_2O$ *Ans. 51.2%*
   (c) $Na_2SO_4 \cdot 10H_2O$ *Ans. 55.9%*
   (d) $Mn(C_2H_3O_2)_2 \cdot 4H_2O$ *Ans. 29.4%*
2. What weight of hydrogen is obtained when 55.5 g of magnesium reacts with steam to produce magnesium oxide and hydrogen?   *Ans. 4.60 g*
3. How many liters of hydrogen would be released at 25.0° C and one atmosphere pressure when 37.0 g of sodium reacts with water?   *Ans. 19.7 liters*
4. Calculate how many moles and how many grams of each of the following hydrates would have to be dehydrated to produce 1.00 kg of the anhydrous salt:
   (a) $Ba(OH)_2 \cdot 8H_2O$ *Ans. 5.84 moles; 1840 g*
   (b) $Na_2SO_4 \cdot 10H_2O$ *Ans. 7.04 moles; 2270 g*
   (c) $CaCl_2 \cdot 6H_2O$ *Ans. 9.01 moles; 1970 g*
   (d) $CaCl_2 \cdot 4H_2O$ *Ans. 9.01 moles; 1650 g*
5. How many moles of phosphoric acid, H_3PO_4, can be produced from 45.0 g of P_4O_{10}? What is the minimum weight of water necessary to effect this conversion?   *Ans. 0.634 mole; 17.1 g*
6. How many grams of the metal hydroxide can be produced when water is added to each of the following?
   (a) 3.00 g of Li_2O *Ans. 4.81 g*
   (b) 15.0 moles of CaO   *Ans. 1.11×10^3 g*
   (c) 15.0 g of Na_2O *Ans. 19.4 g*
   (d) 2.25 moles of MgO   *Ans. 131 g*
7. Calculate the number of moles of nitric oxide which can be prepared by the reaction of 75.0 liters of NH_3 at 600° C and 850 mm pressure, with an excess of oxygen.   *Ans. 1.17 moles*
8. (a) To break each hydrogen bond in ice requires 8.3×10^{-21} calorie. The measured heat of fusion of ice is 80 cal/g. Essentially all of the energy involved in the heat of fusion goes to break hydrogen bonds. What percentage of the hydrogen bonds are broken when ice is converted to liquid water?   *Ans. 14%*

(b) How many additional calories would be required, per gram, to break the remaining hydrogen bonds?   *Ans. 470 cal*

9. (a) What volume of oxygen is produced when 150 g of hydrogen peroxide decomposes to give oxygen and water at 20.0° C and 740 mm pressure, assuming dry oxygen? (b) If the vapor pressure from the water which is produced were taken into account, would the calculated volume of oxygen produced be higher than, lower than, or the same as the volume calculated in (a)? Explain your answer.   *Ans. (a) 54.4 liters; (b) higher*

10. What weight of a solution of hydrogen peroxide (30.0% H_2O_2 by weight) is necessary to convert 1.20 moles of sulfurous acid to sulfuric acid, assuming complete reaction of the sulfurous acid?   *Ans. 136 g*

11. Calculate the number of moles of hydrogen peroxide which can be prepared from 250 g of peroxydisulfuric acid and an excess of water, assuming complete conversion with no decomposition of the peroxides.   *Ans. 1.29 moles*

REFERENCES

"Water Pollution," R. J. Bazell, *Science*, **171**, 266 (1971).

"Mercury in the Environment: Natural and Human Factors," A. L. Hammond, *Science*, **171**, 788 (1971).

"Mercury in Swordfish," (Staff), *Chem. & Eng. News*, May 17, 1971; p. 11.

"The Biology of Heavy Water," J. J. Katz, *Sci. American*, July, 1960; p. 106.

"Thermal Pollution and Aquatic Life," J. R. Clark, *Sci. American*, Mar., 1969; p. 18.

"The Hydrogen Bond," J. N. Murrell, *Chem. in Britain*, **5**, 107 (1969).

"Chemical Reactions and the Composition of Sea Water," K. E. Chave, *J. Chem. Educ.*, **48**, 148 (1971).

"Pollution Problems, Resources, Policy, and the Scientist," A. W. Eipper, *Science*, **169**, 11 (1970).

"Weather Modification: A Technology Coming of Age," A. L. Hammond, *Science*, **172**, 548 (1971).

"Mercury in the Environment," L. J. Goldwater, *Sci. American*, May, 1971; p. 15.

"The Water Cycle," H. L. Penman, *Sci. American*, Sept., 1970; p. 98.

"The Calefaction of a River" (Thermal pollution), D. Merriman, *Sci. American*, May, 1970; p. 42.

"Mechanisms Controlling World Water Chemistry," R. J. Gibbs, *Science*, **170**, 1088 (1970).

"Pollution of the Seas," H. A. Cole, *Chem. in Britain*, **7**, 232 (1971).

"The Responsibilities of Industry" (Regarding pollution), E. Challis, *Chem. in Britain*, **7**, 235 (1971).

"The Control of the Water Cycle," J. Peixoto and M. A. Kettani, *Sci. American*, April, 1973; p. 46.

"Power, Fresh Water, and Food from Cold, Deep Sea Water," D. F. Othmer and O. A. Roels, *Science*, **182**, 121 (1973).

"The Rise and Fall of Polywater," L. Allen, *New Scientist*, **59**, 376 (1973).

"Last Word on Polywater," (Staff), *Chemistry*, **46 (8)**, 4 (1973) (April, 1973, issue).

"Mercury in British Fish," G. Grimstone, *Chem. in Britain*, **8**, 244 (1972).

"The Disposal of Waste in the Ocean," W. Bascom, *Sci. American*, August, 1974; p. 16.

"The Osmotic Pump" (Extraction of fresh water from oceans), O. Levenspiel and N. de Nevers, *Science*, **183**, 157 (1974).

"Salt Solution" (Ocean as a source of drugs, food, salt, and fresh water), K. Jacques, *Chem. in Britain*, **11**, 12 (1975).

13 Solutions

13.1 The Nature of Solutions

When crystals of sugar are stirred with a sufficient quantity of water, the sugar disappears and a clear mixture of sugar in water is formed. The sugar is said to have **dissolved** (or gone into solution) in the water. The solution consists of two components, the **solute** (the dissolved sugar) and the **solvent** (the water). In this solution the molecules of sugar are uniformly distributed among the molecules of water; i.e., the solution is a homogeneous mixture of sugar and water molecules. The molecules of sugar diffuse continuously through the water, and although each of them is heavier than a single molecule of water, the sugar does not settle out on standing.

An aqueous solution of sugar which contains only a small amount of sugar (the solute) in comparison with the amount of water is said to be **dilute;** the addition of more sugar makes the solution more **concentrated.** The solution is said to be **saturated** when the concentration is such that the dissolved solute exists **in equilibrium with the excess undissolved solute.** The **solubility** of a given solute is defined as the quantity of that solute which will dissolve in a specified quantity of solvent to produce a saturated solution. The composition of a solution may be varied between certain limits. Thus, solutions are not compounds, because the latter always contain the same elements in the same proportions by weight.

When sodium chloride, an ionic substance, dissolves in water, sodium ions and chloride ions become uniformly distributed throughout the water. This solution is a homogeneous mixture of water molecules, sodium ions, and chloride ions. As in the case of the molecular solute (for example, sugar), there is diffusion of the solute particles (ions for ionic solutes) through the water, and no settling of the solute particles takes place upon standing.

All solutions are characterized by (1) homogeneity, (2) absence of settling, and (3) the molecular or ionic state of subdivision of the components.

13.2   Kinds of Solutions

Because solutions differ in the physical state of both the solute and solvent, many kinds of solutions are possible. In view of the fact that solutions are defined as **homogeneous mixtures,** it is not surprising that almost any gas, liquid, or solid can act as a solvent for other gases, liquids, and solids. Air is a homogeneous mixture of gases and thus is a gaseous solution. Oxygen (a gas), alcohol (a liquid), and sugar (a solid) will each dissolve in water (a liquid) and form liquid solutions. Liquid solutions exhibit the general properties characteristic of liquids. The most common solutions with which we work in chemistry are those in which the solvent is a liquid. The use of water as a solvent is so general that the word "solution" has come to imply a water solution unless some other solvent is designated. Many alloys are solid solutions of one solid dissolved in another; nickel coins contain nickel dissolved in copper.

13.3   Importance of Solutions

Many reactions take place at an appreciable rate only when the reactants are in solution. For example, when dry powdered barium chloride and sodium sulfate are mixed at ordinary temperatures, there is no perceptible reaction. However, when aqueous solutions of these compounds are mixed, a reaction takes place rapidly, as indicated by the immediate formation of a precipitate of barium sulfate.

$$Ba^{2+} + (2Cl^-) + (2Na^+) + SO_4^{2-} \longrightarrow BaSO_4 + (2Na^+) + (2Cl^-)$$

When solutions of these two ionic compounds are brought together, their ions mix freely by diffusion, and when the barium ions collide with sulfate ions, they remain associated in the form of slightly soluble barium sulfate. When reactants are in the solid state, reaction is possible only for the molecules or ions in the surface layers of the particles, because collision between the molecules (or ions) is necessary before reaction can take place. In solids, the movement of the particles is greatly restricted, and the chance of collision between molecules or ions in adjacent solid particles is very slight.

Solutions are useful in the separation of substances of wide differences in solubility in a given solvent. Thus, barium sulfate, which is only slightly soluble in water, is readily separated from the quite soluble sodium chloride by filtration. The sodium and chloride ions pass through the pores of filter paper with the water, but the crystals of barium sulfate are retained by the paper. **Precipitation analysis,** a part of the field of analytical chemistry, is based upon the separation of compounds which differ greatly in their solubilities.

Solutions are extremely important in life processes. For example, oxygen and carbon dioxide are carried throughout the body in solution in the blood. Digestion is a process in which food is converted into a form that the body can use, with the useful nutrients going into solution in order that they can pass through the walls of the digestive tract and be assimilated in the blood.

The action of natural waters in dissolving substances from air and earth is one of the important processes involved in the conversion of rocks to soil, in altering the fertility of the soil, and in changing the form of the earth's surface. The deposits of many minerals in the earth's crust are the result of reactions that have taken place in solutions followed by the evaporation of the solutions.

Solutions of Gases in Liquids

13.4   Conditions Affecting the Dissolution of Gases in Liquids

The extent to which a gas dissolves in a liquid depends on the following factors: (1) the nature of the gas and the solvent, (2) the pressure, (3) the temperature.

The solubility of a gas in a given liquid is considered to be a specific property of the gas, because its solubility differs from that of other gases in the same liquid. For example, at 0° C and 760 mm, one liter of water dissolves 0.0489 liter of oxygen, or 1.713 liters of carbon dioxide, or 79.789 liters of sulfur dioxide, or 1176.0 liters of ammonia.

The solubility of a gas in a liquid can be increased by increasing the pressure on the gas. This relationship is quantitatively expressed by **Henry's Law: The weight of a gas that dissolves in a definite volume of liquid is directly proportional to the pressure on the gas.** This means that if 1 g of a gas dissolves in 1 liter of water at one atmosphere of pressure (760 mm), 5 g will dissolve at 5 atmospheres of pressure (3,800 mm). The effect of pressure does not follow Henry's Law when a chemical reaction takes place between the gas and the solvent. Thus, the solubility of ammonia in water does not increase as rapidly with increasing pressure as predicted by the law, because ammonia reacts to some extent with water to form ammonium ions and hydroxide ions.

$$NH_3 + H_2O \longrightarrow NH_4^+ + OH^-$$

The effect of increased pressure upon the solubility of a gas in a liquid is illustrated by the behavior of carbonated beverages. Carbon dioxide is forced under pressure into the beverage, and the bottle is tightly capped to maintain the pressure and prevent escape of carbon dioxide. When the cap is removed, the pressure is decreased, and some of the gas escapes. The escape of bubbles of the gas from the liquid is known as **effervescence.**

Finally, the solubility of gases in liquids decreases with an increase in temperature. For example, carbon dioxide at 760 mm pressure dissolves in 1 liter of water to the extent of 1.713 liters at 0°, 1.194 liters at 10°, 0.878 liter at 20°, and 0.359 liter at 60°. This relationship is not one of inverse proportion, however, and the solubility of a gas in a liquid at a given temperature must be determined experimentally. This decrease in solubility of gases with increasing temperature is a very important factor in thermal pollution, particularly in connection with the quantity of dissolved oxygen in water (see Section 12.4, Part 5). All gases can be expelled from solvents by boiling their solutions. Thus, the gases oxygen, nitrogen, carbon dioxide, and sulfur dioxide can be removed from water by boiling it for a few minutes. The fact that a gas is expelled from a solution by boiling is not due just to the rise in temperature of the solution. When the solution is boiled in an open container, part of the vapor of the solvent escapes, carrying with it some of the solute gas that was dissolved. This lowers the pressure of that gas above the solution, and, in accord with Henry's Law, more gas escapes.

Solutions of Liquids in Liquids

13.5   Miscibility of Liquids

Miscibility is a term used for liquids to indicate solubility of one liquid in another. Some liquids will mix with water in all proportions. Such liquids are usually either ionic in solution, so that the charged ions attract the oppositely charged ends of the polar water molecules, or they are polar substances (see Section 4.7) with polar character similar to that of water. For such polar liquids, the negative ends of the polar solute molecules attract the positive ends of the polar solvent molecules, and vice versa, with about the same degree of attraction as that by which like molecules of either substance attract each other. Hence, the two kinds of molecules mix easily in all proportions.

Liquids which mix with water in all proportions are said to be completely **miscible** with water. Ethyl alcohol, sulfuric acid, and glycerine are examples. Nonpolar covalent liquids, such as gasoline, carbon disulfide, and carbon tetrachloride, do not have a *net* charge separation in the molecule as a whole, and hence there is no effective attraction between the molecules of such liquids and the polar water molecules. Thus, the only appreciable attractions are between the water molecules, and so they effectively "squeeze out" the molecules of the nonpolar liquid making these liquids very nearly insoluble in water. Such liquids are said to be **immiscible** with water. Some other liquids, such as ether and bromine, are slightly soluble in water and are said to be partially miscible. Two layers are formed when two immiscible liquids are in contact with each other. In cases of partial miscibility, each layer is a solution of one liquid in the other.

13.6   Fractional Distillation

It will be recalled from Chapter 11 that a pure liquid exhibits a definite vapor pressure at a given temperature and that the normal boiling point of a pure liquid is the temperature at which the vapor pressure of the liquid is 760 mm. The total pressure exerted by the vapor of a solution made by dissolving one liquid in another liquid depends upon the concentration of its components. That solution vapor pressure may be (a) greater than that of either component taken separately, (b) less than that of either component, or (c) between those of the two liquids concerned. The boiling point of the solution is the temperature at which the total

vapor pressure of the mixture is equal to the atmospheric pressure. ~
boiling point of a mixture of two liquids lies between the boiling points ~
two components.

When a solution of two liquids is boiled, the vapor produced at the boiling point is richer in the lower-boiling component than was the original mixture. When this vapor is condensed (Fig. 11–6), the resulting distillate is therefore richer in the liquid of the lower-boiling component than was the original mixture. This means that the composition of the boiling mixture is constantly changing and that the boiling point rises as distillation is continued. The vapor (and distillate) contains more and more of the less volatile component and less and less of the more volatile component as distillation proceeds. By changing the receiver at intervals, successive fractions, each increasingly richer in the less volatile component, are obtained. If this process **(fractional distillation)** is repeated several times, relatively pure samples of the two liquids may be obtained.

Fractionating columns have been devised in which a single operation achieves separation of liquids that would require a great number of simple fractional distillations of the type just described. Crude oil, a complex mixture of hydrocarbons, is separated into its components by fractional distillation on an enormous industrial scale (Fig. 13–1).

FIGURE 13–1

Fractional distillation of crude oil. Oil heated to about 425° C in the furnace vaporizes when it enters the tower at the right. The vapors rise through a series of trays in the tower. As the vapors gradually cool, fractions of higher, then of lower, boiling points condense to liquids and are drawn off. The fraction of highest boiling point is drawn off at the bottom as a residue. It is heavy fuel oil. In modern refineries these fractions, which still consist of mixtures of hydrocarbons, are further processed.

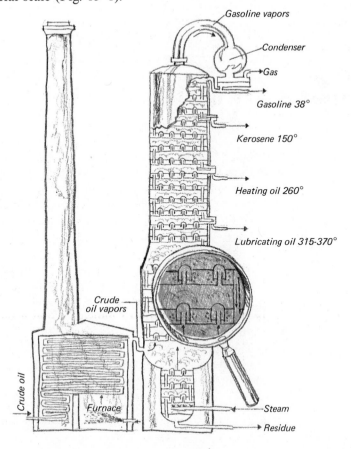

Gasoline vapors

Condenser

Gas

Gasoline 38°

Kerosene 150°

Heating oil 260°

Lubricating oil 315-370°

Crude oil vapors

Crude oil

Furnace

Steam

Residue

13.7   Constant Boiling Solutions

When a dilute aqueous solution of nitric acid is distilled, the first fraction of distillate that is formed consists mostly of water. As distillation is continued, the solution remaining in the distilling flask therefore becomes richer in nitric acid. After a time a concentration of 68 per cent HNO_3 is reached, and thereafter the solution boils at a constant temperature of 120.5° (760 mm). At this temperature the solution and the vapor have the same composition, and the solution completely distills without any further change in composition.

If a nitric acid solution more concentrated than 68 per cent is distilled, the vapor which is first formed contains a large amount of HNO_3. The solution which remains in the distilling flask contains, therefore, a greater percentage of water than at first, and the concentration of the nitric acid in the distilling flask decreases as the distillation is continued. Finally, a concentration of 68 per cent HNO_3 is reached and the solution again boils at the constant temperature of 120.5° (760 mm). The 68 per cent solution of nitric acid is referred to as a **constant boiling solution.** At this specific concentration the solution has a lower vapor pressure, and thus a higher boiling point, than at any other possible concentration. Solutions which distill without change in composition or temperature are called **azeotropic mixtures.**

Other common and important substances that form azeotropic mixtures with water are HCl (20.24 per cent, 110° at 760 mm) and H_2SO_4 (98.3 per cent, 338° at 760 mm).

Solutions of Solids in Liquids

13.8   The Effect of Temperature on Solubility

When a solid dissolves in a liquid, a change in the physical state of the solid analogous to melting takes place. Energy is absorbed in overcoming the forces which hold the molecules, atoms, or ions in their lattice positions in the crystal. This is an endothermic change that accompanies the dissolution of all crystalline solids in liquids. The physical process of solution is often accompanied by a second change, that of a chemical reaction between the solute and the solvent. This second change is commonly exothermic in character. If the heat evolved in the chemical change is greater than that absorbed in the physical change, the heat of solution is negative; i.e., the net process is exothermic. In other cases, the heat of solution is positive; i.e., the net process is endothermic (see Section 8.5). Furthermore, heat is either absorbed or evolved during the process of crystallization of a solid from solution, a process opposite to that of a solid going into solution.

The opposed processes of dissolution and crystallization under conditions of equilibrium between a solid and its saturated solution may be represented by the following systems:

Exothermic: solute + solvent $\rightleftharpoons$ solution + heat (negative heat of solution)   **(1)**
Endothermic: solute + solvent + heat $\rightleftharpoons$ solution (positive heat of solution)   **(2)**

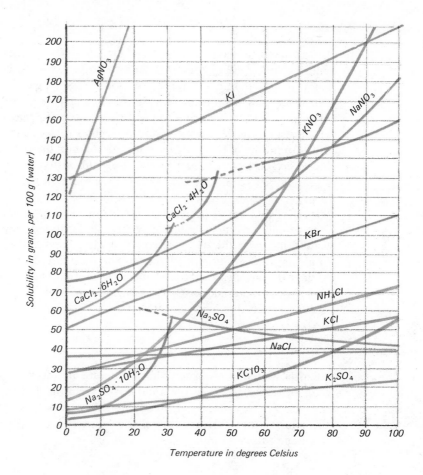

FIGURE 13-2

Graph showing the effect of temperature on the solubility of several inorganic substances.

The addition of heat (a rise in temperature) in system (2) causes more of the solute to dissolve; the added heat is absorbed by the system when the equilibrium shifts to the right. This is the case with most solid-liquid solutions. In system (1) the solubility decreases with a rise in temperature.

The dependence of solubility upon temperature for a number of inorganic substances in water is shown graphically by the solubility curves in Fig. 13–2. Note that the solubility of sodium chloride increases very slightly with a rise in temperature, whereas that of potassium nitrate increases greatly. A sharp break in a solubility curve indicates the formation of a compound whose solubility is different from that of the substance from which it was formed. For example, when $Na_2SO_4 \cdot 10H_2O$ (Glauber's salt) is heated to $32.4°$, it loses its water of hydration and forms the anhydrous salt, Na_2SO_4. The curve up to $32.4°$, the **transition point,** represents the effect of a rise in temperature upon the solubility of $Na_2SO_4 \cdot 10H_2O$ and the curve for temperatures above this point represents the effect of a rise in temperature upon the solubility of Na_2SO_4.

13.9   Saturated Solutions

It was noted in Section 13.1 that sugar dissolves in water at a given temperature until no more will dissolve, and the solution is said to be **saturated.** When more sugar is added to a saturated solution, excess sugar falls to the bottom of the container, and it appears that the dissolving process stops. Actually, molecules of sugar continue to leave the solid under the solvent action of the water and go into solution. However, as the molecules of sugar in solution move about by diffusion, some of them collide with the solid on the bottom and take up positions in the crystal lattice. This process of crystallization is the opposite of dissolution. If enough molecules return to the solid state from the saturated solution, the process of crystallization just counterbalances that of dissolution, and we have a state of dynamic equilibrium. The same reasoning holds when no crystals are formed—for example, ether in water. **A saturated solution is one which is in equilibrium with the undissolved solute.** If a crystal of imperfect shape is placed in a saturated solution of the crystalline material, it will slowly "mend" its shape by loss of particles to the solution and gain of particles from the solution. This is experimental evidence that a saturated solution in contact with its solid solute is a system in dynamic equilibrium.

13.10   Unsaturated Solutions

An unsaturated solution contains less solute dissolved in a given quantity of solvent than would be the case for a saturated solution. Excess solute will not exist in an unsaturated solution. Instead the solute dissolves, and the concentration of the solution increases until either the solution becomes saturated or all the solute has gone into solution.

13.11   Supersaturated Solutions

When saturated solutions of solid solutes are prepared at elevated temperatures and then permitted to cool, the excess solute usually separates from the solution by crystallizing. However, if a saturated solution is prepared at an elevated temperature and excess solute removed, crystallization often does not take place if the solution is allowed to cool undisturbed. **The solution contains more of the solute than it does when it is in equilibrium with the undissolved state and is called a supersaturated solution.** Such solutions are metastastable systems (unstable, but not so unstable that they cannot exist). Agitation of the solution or the addition of a "seed" crystal of the solute may start crystallization of the excess solute; after crystallization, a saturated solution in equilibrium with the crystals of solute remains.

Gases also form supersaturated solutions. For example, a bottle of a carbonated beverage may not liberate the excess carbon dioxide when opened, but if the beverage is shaken or stirred, it does. A supersaturated solution of a gas in a liquid is prepared either at low temperature, or under pressure, or in both ways.

13.12   Rate of Solution

Rate of solution pertains to the amount of solute entering the solution per unit of time. It depends upon several factors. First, more soluble substances dissolve more rapidly than less soluble ones. Second, because the dissolution of a solid solute can take place only at the surface of the particles, the more finely divided the solute is (the greater the surface area per unit mass of the substance) the greater will be its rate of solution. Third, the rate at which the dissolved molecules diffuse away from the solid solute is relatively slow, and so the solution in the immediate vicinity of the solid phase approaches the saturation concentration. Stirring or shaking the mixture brings unsaturated solution in contact with the solute and thus increases the rate of solution. Fourth, heating results in convection currents which produce the same effects as agitation of the mixture, and heating also usually increases the solubility of the solid; if the solute and solvent react during the solution process, heating causes the reaction to go faster and hence increases the rate of solution.

Expressing Concentration

13.13   Standard Solutions

The solubility of a pure solid substance in a pure solvent at a given temperature is quantitatively a definite physical property of the substance. The solubility of sodium chloride is 35.8 g per 100 g of water at 20°; that of NaF is 4.2 g; and that of silver bromide is 0.00002 g. A general idea of relative solubilities is conveyed by use of the terms **quite soluble, moderately soluble,** and **slightly soluble** or **insoluble.** Strictly speaking, no substance is absolutely insoluble, although for all practical purposes many substances appear to be so. The relative concentrations of solutions may be expressed by the terms **dilute** (containing a small proportion of solute) and **concentrated** (containing a large proportion of solute).

Chemists often work with unsaturated solutions of known concentrations, which are called **standard solutions.**

The concentrations of solutions may be expressed in a number of ways. It is important that the methods of expressing concentrations of solutions be thoroughly mastered.

13.14   Percentage Composition of Solutions

One method of expressing concentration in physical units involves the weight of solute in a given weight of solvent; for example, 1 g of NaCl in 100 g of water. A related system expresses concentration in terms of percentage composition by weight. A 10 per cent NaCl solution by weight contains 10 g of NaCl in 100 g of solution (90 g of water).

[Handwritten at top: ?g sol = 50 g NaCl (100 g sol / 10 g NaCl) = 500 g sol]

Example 1. Calculate the weight of NaCl solution that will contain 50 g of NaCl if the solution is 10 per cent NaCl by weight.

If 100 g of solution contains 10 g of NaCl (10 per cent of 100 g), then

$$100 \text{ g of solution} \times \frac{50 \text{ g of NaCl}}{10 \text{ g of NaCl}} = 500 \text{ g of solution}$$

Hence, 500 g of 10 per cent sodium chloride solution will contain 50 g of NaCl (and 450 g of water).

In order to calculate the weight of solute in a given *volume* of solution from the percentage composition by weight, it is necessary to know the specific gravity (or density) of the solution.

Example 2. Calculate the weight of hydrogen chloride (HCl) in 100 ml of concentrated hydrochloric acid of specific gravity 1.19 and containing 37.23 per cent HCl by weight.

[Handwritten at left: ?g HCl = 100 ml (1.19 g / 1 ml)(37.23 / 100 g sol)]

One ml of the solution weighs 1.19 g.
100 ml of the solution weighs 100 ml × 1.19 g/ml = 119 g.

Because the solution is 37.23 per cent HCl by weight, 119 g of the solution contains 119 g × 0.3723 = 44.3 g of HCl.

Example 3. What volume of concentrated hydrochloric acid of specific gravity 1.19 and containing 37.23 per cent HCl by weight contains 100 g of HCl?

One ml of acid contains 1.19 g × 0.3723 = 0.443 g of HCl. To obtain the volume of acid that would contain 100 g of HCl, multiply the 1 ml by an improper fraction made up of 100 g and 0.443 g.

[Handwritten at left: ?ml HCl = 100 g HCl (100 g sol / 37.23 g HCl)(1 ml / 1.19 g) = 225.7]

$$1 \text{ ml} \times \frac{100 \text{ g}}{0.443 \text{ g}} = 226 \text{ ml}$$

Thus, 226 ml of the concentrated hydrochloric acid contains 100 g of HCl.

13.15   Molar Solutions

It is frequently desirable to express concentration in terms of gram-formula weights (moles), instead of grams, of solute. **The molarity of a solution is the number of moles of solute per liter of solution.** A solution containing one gram-formula weight, or mole, of the solute in one liter of solution is a **one-molar** solution. Note that a liter of *solution* rather than a liter of solvent is specified in this definition. Because gram-formula weights of different molecular substances contain the same number of molecules, it follows that equal volumes of one-molar solutions will contain the same number of molecules of the solute. It should be obvious that by using the molar method of expressing concentration of solutions, it is easy to select a desired number of moles, molecules, or ions of the solute by measuring out the appropriate volume of solution. For example, if 1 mole of sodium hydroxide is needed for a given reaction, one can use 1 liter of a one-molar (1 M) solution or 2 liters of a 0.5 M solution of the base. If the 2 liters of 0.5 M sodium hydroxide solution were to be used in a reaction with 0.4 M hydrochloric acid, 2.5 liters of the acid would be needed, because 2.5 liters of 0.4 M acid will furnish one mole of the solute HCl.

$$\text{NaOH} \quad + \quad \text{HCl} \quad \longrightarrow \quad \text{NaCl} + \text{H}_2\text{O}$$

| 1 mole | 1 mole |
| (2 liters × 0.5 M) | (2.5 liters × 0.4 M) |

In preparing a 1 M solution of sodium bicarbonate (NaHCO$_3$), 84.0 g (1 mole) of pure sodium bicarbonate are weighed out accurately and dissolved in sufficient water to form one liter of solution. A 1 M solution of hydrochloric acid contains 36.5 g of the acid per liter of solution. A 2 M solution of sodium bicarbonate contains 168.0 g (2 moles) of NaHCO$_3$ per liter, and a 0.1 M solution of the bicarbonate contains 8.4 g of NaHCO$_3$ per liter.

Example 1. Calculate the molarity of a solution, 2.0 liters of which contains 2.6 moles of solute.

$$\text{Molarity} = \frac{\text{moles of solute}}{\text{liters of solution}} = \frac{2.6 \text{ moles}}{2.0 \text{ liters}}$$

$$= 1.3 \text{ moles/liter} = 1.3 \ M$$

The molarity of a solution is also the number of millimoles of solute per milliliter of solution. This follows because a millimole is one-thousandth of a mole and a milliliter is one-thousandth of a liter.

Example 2. Calculate the molarity of a solution, 20 ml of which contains 5.0 millimoles of solute.

$$\text{Molarity} = \frac{\text{millimoles of solute}}{\text{milliliters of solution}} = \frac{5.0 \text{ millimoles}}{20 \text{ ml}}$$

$$= 0.25 \text{ millimole/ml} = 0.25 \ M$$

Example 3. Calculate the molarity of a solution which contains 4.0 g of NaOH in 2.0 liters of solution. The gram-formula weight of NaOH is $(23.0 + 16.0 + 1.0)$ g $= 40.0$ g.

$$\text{Number of moles of NaOH} = \frac{\text{grams of NaOH}}{\text{gram-formula weight}}$$

$$= \frac{4.0 \text{ g}}{40 \text{ g/mole}} = 0.10 \text{ mole}$$

$$\text{Molarity} = \frac{\text{moles of solute}}{\text{liters of solution}} = \frac{0.10 \text{ mole}}{2.0 \text{ liters}}$$

$$= 0.05 \text{ mole/liter} = 0.05 \ M$$

$?g H_2 SO_4 = .80 \ell \left(\dfrac{.050 \text{ mole}}{1 \ell} \right) \left(\dfrac{98 g}{1 \text{ mole}} \right) =$

$?g H_2 SO \qquad \begin{matrix} 3 & 4 \\ 4 & 4 \end{matrix}$

Example 4. How many grams of H_2SO_4 are contained in 0.80 liter of 0.050 M sulfuric acid?

One liter of 0.050 M acid would contain 0.050 mole of H_2SO_4.

0.80 liter of 0.050 M acid would contain 0.80 liter $\times$ 0.050 mole/liter = 0.040 mole of H_2SO_4.

One mole of H_2SO_4 is 98 g.

0.040 mole of H_2SO_4 is 0.040 mole $\times$ 98 g/mole = 3.9 g.

Thus, 3.9 g of H_2SO_4 are contained in 0.80 liter of 0.050 M sulfuric acid.

Example 5. Calculate the molarity of a concentrated sulfuric acid solution of specific gravity 1.84 and containing 98.0 per cent H_2SO_4 by weight.

The weight of 1 ml of acid is 1.84 g.

The weight of 1 liter of acid is 1000 ml $\times$ 1.84 g/ml = 1840 g.

The weight of H_2SO_4 in one liter of the acid is 1840 g $\times$ 0.980 = 1800 g.

The gram-formula weight of H_2SO_4 is 98.0 g.

$$\text{Moles of } H_2SO_4 \text{ in 1800 g of } H_2SO_4 = \frac{1,800 \text{ g}}{98.0 \text{ g/mole}} = 18.4 \text{ moles}$$

Because the concentrated sulfuric acid contains 18.4 moles of H_2SO_4 in 1 liter of solution, its concentration is 18.4 M.

$\dfrac{? \text{ mole sol.}}{\ell} = \left(\dfrac{1840 g \text{ sol}}{\ell} \right) \left(\dfrac{98 g \, H_2SO_4}{100 g \text{ sol}} \right) \left(\dfrac{\text{mole}}{98 g \, H_2SO_4} \right) = 18.4$

13.16 Equivalent Weights of Compounds

A concept known as the **equivalent weight** is used in a method for expressing the concentration of a solution in terms of **normality.** It is necessary before discussing normality to know what is meant by the equivalent weight.

It can be shown by experiment that the sodium hydroxide in 1 liter of a 1 M solution will neutralize the hydrochloric acid in 1 liter of a 1 M solution of the acid; however, 2 liters of 1 M sodium hydroxide are required to neutralize the acid in 1 liter of 1 M sulfuric acid. This arises from the fact that one mole of sulfuric acid reacts with two moles of sodium hydroxide, whereas one mole of hydrochloric acid reacts with only one mole of sodium hydroxide.

$$HCl + NaOH \longrightarrow NaCl + H_2O$$
$$H_2SO_4 + 2NaOH \longrightarrow Na_2SO_4 + 2H_2O$$

Each mole of sulfuric acid supplies two moles of hydrogen ions, each of which can neutralize one mole of hydroxide ion: $H^+ + OH^- \longrightarrow H_2O$. In other words, one mole of sodium hydroxide is *chemically equivalent* to half a mole of sulfuric acid.

The equivalent weight of a substance is defined as the number of parts by weight of it which combine with or are otherwise chemically equivalent to 8.00 parts by weight of oxygen. A gram-equivalent weight of a substance is its equivalent weight expressed in grams.

In forming H_2O, two gram-atoms of hydrogen (2×1.008 g) combine with each gram-atom of oxygen (16.00 g); therefore, 1.008 g of hydrogen combine with (are equivalent to) 8.00 g of oxygen. Hence, a gram-equivalent weight of hydrogen is 1.008 g, which is also its gram-atomic weight. One gram-atom of magnesium (24.305 g) combines with one gram-atom of oxygen (16.00 g) in forming MgO; therefore, 12.15 g of magnesium combine with 8.00 g of oxygen. A gram-equivalent of magnesium is 12.15 g, which is one-half its gram-atomic weight. Likewise, an equivalent weight of sodium is the same as its atomic weight, because two atoms of sodium combine with one atom (two equivalent weights) of oxygen in forming Na_2O.

For determining equivalent weights, any element of known equivalent weight can be used as a standard but only because its equivalent weight has been determined in relation to the value 8.00 for oxygen. Thus, the combining capacity of hydrogen can be used in finding the equivalent weights of other elements with which it combines (or is equivalent to) because it has been established that 1.008 g of hydrogen are equivalent to 8.00 g of oxygen. It follows that the equivalent weight of chlorine in HCl is its atomic weight (35.45), that of sulfur in H_2S is one-half its atomic weight, that of nitrogen in NH_3 is one-third its atomic weight, and that of carbon in CH_4 is one-fourth its atomic weight. We observe, then, that the equivalent weight of an element is equal to its atomic weight, or it is either one-half, one-third, one-fourth, or some other simple fraction of its atomic weight. In other words, the equivalent weight of an element is numerically equal to its atomic weight divided by a small whole number—the numerical value of its oxidation number. The numerical value for the oxidation number of carbon in CH_4 is 4, that of nitrogen in NH_3 is 3, and that of sulfur in H_2S is 2. Thus, the equivalent weight of carbon in CH_4 is ($12.01/4 = 3.00$), of nitrogen in NH_3 is ($14.007 \div 3 = 4.002$), and of sulfur in H_2S is ($32.06 \div 2 = 16.03$).

The equivalent weight of an element may not always be the same, inasmuch as it depends upon the change the element undergoes in the given reaction. For example, phosphorus may combine with chlorine in one of two proportions, depending upon the conditions of the reaction, with the formation of either PCl_3 or PCl_5. The equivalent weight of chlorine in both compounds is its atomic weight (35.45). It follows that the equivalent weight of phosphorus in PCl_3 is one-third of its atomic weight ($30.97 \div 3 = 10.32$), and in PCl_5 it is one-fifth of its atomic weight ($30.97 \div 5 = 6.19$).

13.17 Determination of Equivalent Weights

The equivalent weight of an element can be determined accurately by experiment. This is done by finding the weight of the element which will combine with or displace one equivalent weight of another element of accurately known equivalent weight.

Example 1. Let us calculate the equivalent weight of aluminum from the observation that when 6.74 g of the metal are burned in oxygen, 12.74 g of aluminum oxide are formed.

The weight of oxygen which combines with 6.74 g of aluminum is the weight of the aluminum oxide minus the weight of the aluminum, or

$$(12.74 \text{ g} - 6.74 \text{ g}) = 6.00 \text{ g of oxygen}$$

The equivalent weight of aluminum is that weight of it which combines with 8.00 g of oxygen. If 6.74 g of aluminum combine with 6.00 g of oxygen, then a greater weight of aluminum will be required to combine with 8.00 g of oxygen. Hence, multiply 6.74 g of aluminum by the ratio 8.00 g/6.00 g to find the equivalent weight of aluminum.

$$6.74 \text{ g} \times \frac{8.00 \text{ g}}{6.00 \text{ g}} = 8.99 \text{ g}$$

Thus, the equivalent weight of aluminum is 8.99.

Example 2. Suppose that a 1.6345-g sample of pure zinc displaced 0.0504 g of hydrogen from a dilute solution of an acid. What is the equivalent weight of zinc?

The gram-equivalent weight of zinc is that weight of it which will displace 1.008 g of hydrogen. If 1.6345 g of zinc displace 0.0504 g of hydrogen, then a greater mass of zinc would be required in the displacement of 1.008 g of hydrogen. Multiplying 1.6345 g of zinc by the improper fraction made up of the two weights of hydrogen will give the gram-equivalent weight of zinc.

$$1.6345 \text{ g} \times \frac{1.008 \text{ g}}{0.0504 \text{ g}} = 32.7 \text{ g}$$

Thus, the equivalent weight of zinc is 32.7.

13.18   Expanded Definition of Equivalent Weights

The definition of equivalent weight as given in the preceding section does not cover all types of reactions. Consequently, the definition has been expanded to one which seems to be applicable in all ordinary cases: **Gram-equivalent weight** (often shortened to just gram-equivalent) **is the number of grams of a substance associated in a chemical reaction with the transfer of either N electrons or N protons or with the neutralization of either N negative or N positive charges, where N is Avogadro's number, 6.022×10^{23}.**

In an acid-base reaction protons, H^+, are transferred from the acid to the base, and it is upon the number of transferred protons that the calculation of the gram-equivalent weight of the acid and base depends. When hydrochloric acid reacts with sodium hydroxide, one proton is transferred from the hydronium ion (the acid) to the hydroxide ion (the base).

$$H_3O^+ + OH^- \longrightarrow 2H_2O$$

When one mole of hydrochloric acid reacts with one mole of sodium hydroxide, N protons (6.022×10^{23}) are transferred from N hydronium ions to N hydroxide ions. It follows, then, that one gram-equivalent weight of hydrochloric acid is the same as one mole (36.5 g) of HCl, and one gram-equivalent weight of sodium hydroxide is the same as one mole (40.0 g) of NaOH. When one mole of sulfuric acid reacts with two moles of sodium hydroxide $2N$ protons are transferred from the acid to the base. Hence, one gram-equivalent weight of sulfuric acid is equal to one-half mole of the acid (98.0 g $\div$ 2 = 49.0 g).

It should be emphasized that the number of protons actually transferred in a reaction determines the equivalent weight. For example, phosphoric acid may react with sodium hydroxide in one of three ways:

$$H_3PO_4 + NaOH \longrightarrow NaH_2PO_4 + H_2O \tag{1}$$

$$H_3PO_4 + 2NaOH \longrightarrow Na_2HPO_4 + 2H_2O \tag{2}$$

$$H_3PO_4 + 3NaOH \longrightarrow Na_3PO_4 + 3H_2O \tag{3}$$

The numbers of protons transferred in the three reactions are N, $2N$, and $3N$, respectively. Thus, the equivalent weight of H_3PO_4 in the three cases is equal to one mole, one-half mole, and one-third mole, respectively. *It must be emphasized again that the gram-equivalent weight to be used in determining the normality of a solution must be deduced from the reaction, not merely from the formula of the substance.*

In reaction (1) above, N protons were transferred in the formation of NaH_2PO_4; so a gram-equivalent weight of this salt is equal to one mole of it. Likewise, a gram-equivalent of Na_2HPO_4 is one-half mole, and that of Na_3PO_4 is one-third mole, because $2N$ protons and $3N$ protons were transferred during their formation, respectively.

When solutions of silver nitrate ($AgNO_3$) and sodium chloride (NaCl) are mixed, silver chloride precipitates. The equation for the reaction may be written

$$Ag^+ + Cl^- \longrightarrow \underline{Ag^+Cl^-}$$

The silver ion and the chloride ion combine to yield an ionic compound of low solubility. The ions form a crystal lattice which is electrostatically neutral. The gram-equivalent weight is calculated from the number of charges neutralized in this way. When a mole of silver nitrate and a mole of sodium chloride react, N positive charges and N negative charges are neutralized. Hence, the gram-equivalent weights of $AgNO_3$ and NaCl are each equal to one mole; i.e., to 169.87 g and 58.44 g, respectively.

Barium chloride precipitates the sulfate ion from a sodium sulfate solution according to $Ba^{2+} + SO_4^{2-} \longrightarrow \underline{Ba^{2+}SO_4^{2-}}$. When one mole of barium ion, Ba^{2+}, reacts with one mole of sulfate ion, SO_4^{2-}, $2N$ positive charges and $2N$ negative charges are neutralized; hence, the gram-equivalent weight of $BaCl_2$ is one-half the gram-formula weight, and that of Na_2SO_4 is also one-half the gram-formula weight.

13.19   Normal Solutions

Solutions containing the same number of gram-equivalent weights of different substances are chemically equivalent. **The normality of a solution is the number of gram-equivalent weights of solute per liter of solution.** A solution which contains one gram-equivalent weight of solute in one liter of solution is called a **one-normal** solution (1 N). A one-normal solution of sulfuric acid contains one gram-equivalent weight (98 g/2 = 49 g) of H_2SO_4 per liter; a 2 N solution of H_2SO_4 contains 98 g, and a 0.01 N solution contains 0.49 g. It should be evident that a 1 M solution of hydrochloric acid is also 1 N, because a gram-equivalent weight of HCl is the same as the gram-molecular weight. However, a 1 M solution of sulfuric acid is 2 N because the gram-formula weight of H_2SO_4 is equal to two gram-equivalent weights of it.

Example 1.   Calculate the normality of a solution of hydrochloric acid which contains 3.65 g of HCl in 0.50 liter of solution.

The gram-equivalent of HCl is equal to one mole, 36.5 g

$$\text{Number of gram-equivalents of HCl} = \frac{\text{grams of HCl}}{\text{gram-equivalent weight}}$$

$$= \frac{3.65 \text{ g}}{36.5 \text{ g/g-equiv}} = 0.10$$

$$\text{Normality} = \frac{\text{gram-equivalents of solute}}{\text{liters of solution}} = \frac{0.10 \text{ g-equiv}}{0.50 \text{ liter}} = 0.20 \text{ } N$$

Example 2.   How many grams of H_2SO_4 are contained in 1.2 liters of 0.50 N sulfuric acid?

One liter of 0.50 N acid would contain 0.50 gram-equivalent of H_2SO_4.

1.2 liters of 0.50 N acid would contain 1.2 liters × 0.50 g-equiv/liter = 0.60 g-equiv of H_2SO_4.

The gram-equivalent weight of H_2SO_4 is one-half the gram-formula weight (98 g/2 = 49.0 g).

0.60 g-equiv of H_2SO_4 is 0.60 g-equiv × 49.0 g/g-equiv = 29.4 g. *Inasmuch as only two significant figures are justified by the data used, the answer must be rounded off to 29 g.* Thus, we calculate that 1.2 liters of 0.50 N sulfuric acid contains 29 g of H_2SO_4.

Example 3.   What is the normality of a solution of barium hydroxide, 100.0 ml of which contain 17.14 mg of $Ba(OH)_2$?

The milligram-equivalent of $Ba(OH)_2$ is one-half its milligram-formula weight (171.4 mg/2 = 85.7 mg).

[handwritten: $\dfrac{\text{mgew}}{\text{ml}} = \overline{100\,ml}$]

Number of milligram-equivalents $= \dfrac{\text{milligrams of solute}}{\text{milligram-equivalent weight}}$

$$= \dfrac{17.14 \text{ mg}}{85.7 \text{ mg/mg-equiv wt}} = 0.200 \text{ mg-equiv}$$

Normality $= \dfrac{\text{milligram-equivalents of solute}}{\text{milliliters of solution}} = \dfrac{0.200 \text{ mg-equiv}}{100.0 \text{ ml}} = 0.0020 \ N$

13.20 Molal Solutions

The concept of molality is quite different from that of molarity and is useful in a very different way. **The molality of a solution is the number of moles of solute in 1,000 g of solvent.** A solution which contains one mole of solute in 1,000 g of solvent is called a **one-molal** solution.

Example. Calculate the molality of a solution 250 g of which contain 40.0 g of sodium chloride.

[handwritten: $\dfrac{?\,\text{moles NaCl}}{\text{kg solvent}} = \dfrac{40.0\,g\,NaCl}{.210\,\text{kg solvent}} \times \dfrac{1\,\text{mole}}{58.5\,g\,NaCl}$ $= 3.26\ m$]

The weight of water $= 250 - 40 = 210$ g of water. Hence 40.0 g of NaCl is dissolved in 210 g of water. Calculate the number of moles of NaCl dissolved in 1,000 g of water.

$$\dfrac{40.0}{58.4} \times \dfrac{1,000}{210} = 3.26 \text{ moles of sodium chloride in 1,000 g of water.}$$

Hence, the molality is 3.26 m.

13.21 Mole Fraction

The mole fraction of each component in a solution is the number of moles of the component divided by the total number of moles of all components present. The mole fractions of all components of a solution added together always equal one (1).

Example 1. Calculate the mole fraction of solute and solvent for a 3.0 molal solution of sodium chloride.

A 3.0 molal solution of sodium chloride contains 3.0 moles of NaCl dissolved in 1,000 g of water.

$$1,000 \text{ g of water} = \dfrac{1,000}{18.0} \text{ moles of water} = 55.6 \text{ moles of water.}$$

$$\text{Mole fraction of NaCl} = \dfrac{3.0}{3.0 + 55.6} = 0.051$$

$$\text{Mole fraction of } H_2O = \dfrac{55.6}{3.0 + 55.6} = 0.949$$

(Note that the sum of the two mole fractions, $0.051 + 0.949$, is 1.000.)

Example 2. Calculate the mole fraction of each component in a solution of 42 g CH_3OH, 35 g C_2H_5OH, and 50 g C_3H_7OH.

$$\frac{42}{32} = 1.31 \text{ moles of } CH_3OH$$

$$\frac{35}{46} = 0.76 \text{ mole of } C_2H_5OH$$

$$\frac{50}{60} = 0.83 \text{ mole of } C_3H_7OH$$

$$\text{Mole fraction of } CH_3OH = \frac{1.31}{1.31 + 0.76 + 0.83} = \frac{1.31}{2.90} = 0.45$$

$$\text{Mole fraction of } C_2H_5OH = \frac{0.76}{2.90} = 0.26$$

$$\text{Mole fraction of } C_3H_7OH = \frac{0.83}{2.90} = 0.29$$

(Note that the sum of the mole fractions, 0.45 + 0.26 + 0.29, is 1.00.)

13.22 Problem Combining Several Methods for Expressing Concentration

A sulfuric acid solution containing 571.6 g of H_2SO_4 per liter of solution at 20° has a density of 1.3294 g per ml. Calculate (a) the molarity, (b) the normality, (c) the molality, (d) the per cent by weight, and (e) the mole fractions for the solution.

(a) One liter of solution contains 571.6 g of H_2SO_4 of $\frac{571.6}{98.08}$ moles H_2SO_4 = 5.828 moles H_2SO_4 per liter. Hence, the solution is 5.828 M.

(b) One g-equiv wt of H_2SO_4 = $\frac{98.08}{2}$ = 49.04 g. One liter of solution contains 571.6 g H_2SO_4 = $\frac{571.6}{49.04}$ g-equiv wts = 11.66 g-equiv wts of H_2SO_4 per liter. Hence, the solution is 11.66 N.

(c) The solution contains 571.6 g of H_2SO_4 in one liter of solution. The density tells us that each ml of solution weighs 1.3294 g. One liter of solution would weigh 1,000 × 1.3294 g = 1329.4 g. Hence, 571.6 g of H_2SO_4 are present in 1329.4 g of solution.

Subtracting the weight of solute from the weight of solution in which it is contained provides the weight of solvent (water) present.

$$1329.4 \text{ g} - 571.6 \text{ g} = 757.8 \text{ g } H_2O$$

Hence 571.6 g of H_2SO_4 are present in 757.8 g of H_2O. Remember that molality is the number of moles of solute in 1,000 g of solvent. Therefore,

$$\frac{571.1}{98.08} \times \frac{1,000}{757.8} = 7.69 \text{ moles of } H_2SO_4 \text{ in } 1,000 \text{ g } H_2O = 7.691 \text{ } m$$

(d) The solution contains 571.6 g of H_2SO_4 in 1329.4 g of *solution*.

$$\frac{571.6}{1329.4} \times 100 = 43.00\% \text{ } H_2SO_4 \text{ by weight.}$$

(e) 571.6 g of $H_2SO_4 = \frac{571.6}{98.08}$ moles of $H_2SO_4 = 5.828$ moles of H_2SO_4

757.8 g $H_2O = \frac{757.8}{18.02}$ moles of $H_2O = 42.053$ moles of H_2O

Mole fraction of $H_2SO_4 = \dfrac{5.828}{5.828 + 42.053} = \dfrac{5.828}{47.881} = 0.1217$

Mole fraction of $H_2O = \dfrac{42.053}{47.881} = 0.8783$

(Note that the sum of the mole fractions, $0.1217 + 0.8783$, is 1.0000.)

13.23   Problem Based on a Precipitation Reaction

Let us calculate the volume of $0.10\,N$ silver nitrate required to precipitate the chloride ions in 200 ml of $0.05\,N$ sodium chloride solution $(Ag^+ + Cl^- \longrightarrow AgCl)$.

200 ml of $0.05\,N$ NaCl contains 200 ml $\times$ 0.05 mg-equiv/ml = 10 mg-equiv.

It is known that 10 mg-equiv of $AgNO_3$ will be required to react with 10 mg-equiv of NaCl.

Each ml of $0.10\,N$ $AgNO_3$ solution contains 0.10 mg-equiv of $AgNO_3$.

The number of ml of $0.10\,N$ silver nitrate solution that will contain 10 mg-equiv of $AgNO_3$ is therefore

$$\frac{10\ \text{mg-equiv}}{0.10\ \text{mg-equiv/ml}} = 100\ \text{ml of } AgNO_3$$

13.24   Problems Based on Dilution of Solutions

When a solution is diluted, the volume is increased by adding more solvent and the concentration is decreased, but the total amount of solute is constant.

Example 1.   A liter of $5.0\,M$ silver nitrate is diluted to a volume of 2.0 liters by adding water. What is the molarity of the diluted silver nitrate?

One liter of a $5.0\,M$ solution contains 1.0 liter $\times$ 5.0 moles/liter = 5.0 moles of solute.

After dilution, 2 liters of solution contains 5.0 moles of solute.

Molarity after dilution = $\dfrac{\text{moles of solute}}{\text{liters of solution}} = \dfrac{5.0\ \text{moles}}{2.0\ \text{liters}} = 2.5\,M$

Example 2.   What volume of water would be required to dilute 10 ml of $0.40\,N$ acid to a concentration of $0.10\,N$?

10 ml of $0.40\,N$ acid contains 10 ml $\times$ 0.40 mg-equiv/ml = 4.0 mg-equiv of pure acid.

After dilution, each ml of solution will contain 0.10 mg-equiv and the total quantity of solute will be 4.0 mg-equiv.

$$\text{Volume after dilution} = \frac{4.0 \text{ mg-equiv}}{0.10 \text{ mg-equiv/ml}} = 40 \text{ ml}$$

The volume of water required in the dilution is equal to the final volume of the solution minus the original volume: 40 ml − 10 ml = 30 ml.

Example 3.   A solution of 0.40 *M* hydrochloric acid having a volume of 100 ml is mixed with 50 ml of 0.20 *M* sodium hydroxide solution. Calculate the molarity of each solute in the resulting solution.

100 ml of 0.40 *M* hydrochloric acid contains 100 ml × 0.40 millimole/ml = 40 millimoles of HCl.

50 ml of 0.20 *M* sodium hydroxide contains 50 ml × 0.20 millimole/ml = 10 millimoles of NaOH.

10 millimoles of NaOH will neutralize 10 millimoles of HCl, forming 10 millimoles of NaCl, and leaving 40 − 10 = 30 millimoles of HCl in excess.

The total volume after mixing is 100 ml + 50 ml = 150 ml.

$$\text{Molarity of HCl} = \frac{30 \text{ millimoles of HCl}}{150 \text{ ml of solution}} = 0.20 \text{ } M$$

$$\text{Molarity of NaCl} = \frac{10 \text{ millimoles of NaCl}}{150 \text{ ml of solution}} = 0.067 \text{ } M$$

Effect of Solutes upon Properties of the Solvent

13.25   Lowering of the Vapor Pressure of the Solvent

It has been found by experiment that when a nonvolatile substance (or one with such a low vapor pressure that we can disregard it) is dissolved in a liquid the vapor pressure of the liquid is lowered. Solid solutes exhibit only negligible vapor pressures, so they may be considered to be nonvolatile. Thus, for example, the vapor pressure of an aqueous sugar solution at 20° C is less than that of pure water at 20° C. The vapor pressure of a liquid is determined by the frequency of escape of molecules from the surface of the liquid. The presence of sugar molecules in the solution lowers the frequency of escape of water molecules from the surface of the liquid. This is the reason the vapor pressure of the solution is less than that of pure water. The kind or size of solute molecule has little to do in determining the extent of this effect. The decrease in vapor pressure is proportional to the number of molecules of solute in a definite weight of the solvent. The greater the number of molecules of solute, the lower the vapor pressure exerted by the solution. These considerations lead to **Raoult's Law,** which is stated as follows: **The lowering of the vapor pressure of a solvent is directly** proportial to the number of moles of the solute dissolved in a definite weight of the solvent.

13.26 Elevation of the Boiling Point of the Solvent

The boiling point of a liquid is the temperature at which the vapor pressure of the liquid is equal to the pressure upon its surface (Section 11.3). Because the addition of a solute lowers the vapor pressure of a liquid, a higher temperature is required to bring the vapor pressure of the solution in an open vessel up to the atmospheric pressure and make the solution boil. According to Raoult's Law, the lowering of the vapor pressure of a solvent is directly proportional to the number of moles of solute which is dissolved in a definite weight of solvent. It follows, then, that the elevation of the boiling point of the solvent is also proportional to the weight of the solute which is dissolved in a definite weight of solvent. **For substances which do not dissociate into ions in solution and which either are nonvolatile or have a very low vapor pressure, the elevation of the boiling point of the solvent is the same when solutions** of equimolecular concentrations are considered. For example, one gram-molecular weight of sucrose ($C_{12}H_{22}O_{11}$) and one gram-molecular weight of glucose ($C_6H_{12}O_6$), each dissolved in 1,000 g of water, form solutions which have the same boiling points, 100.512° at 760 mm. The elevation of the boiling point of the water, then, is $100.512° - 100° = 0.512°$. It should be evident from these considerations that two gram-molecular weights of a nonelectrolyte in 1,000 g of water give a solution which boils at $100° + (2 \times 0.512°) = 101.02°$.

It should be emphasized that the extent to which the vapor pressure of a solvent is lowered and the boiling point is elevated depends upon the number of solute particles present in a given amount of solvent and not upon the mass or size of the particles. Properties of solutions which depend upon the number and not the kind of particles concerned are spoken of as **colligative** properties.

It is not surprising to find that a mole of sodium chloride, which exists as two ions in the solution, causes nearly twice as great a rise in boiling point as does a mole of a nonelectrolyte. One mole of table sugar consists of 6.022×10^{23} particles (as molecules), whereas one mole of sodium chloride consists of $2 \times 6.022 \times 10^{23}$ particles (as ions). Hence, calcium chloride ($CaCl_2$), which consists of three ions, causes nearly three times as great a rise in boiling point as does sugar. Why the elevation is not exactly twice (for NaCl) and three times (for $CaCl_2$) that of the molecular boiling point elevation is explained in Section 14.11.

13.27 Depression of the Freezing Point of the Solvent

It is a common observation that solutions freeze at lower temperatures than do pure liquids. We use aqueous solutions of various antifreezes such as ethylene glycol in place of pure water in automobile radiators, because such solutions freeze at lower temperatures. Sea water, with its large salt content, freezes at a lower temperature than fresh water. The depression of the freezing point of a solvent by an added solute is a reflection of the vapor pressure lowering caused by the solute. In Section 11.13 we learned that the freezing point of a pure liquid is the temperature at which the liquid is in equilibrium with its solid. A pure liquid and its solid have the same vapor pressure at the freezing point. Therefore, pure water and ice have the same vapor pressure at 0° C. However, the water in a solution has a lower vapor pressure than ice at this temperature. Consequently, if ice and an aqueous solution at 0° C are placed in contact, the ice melts. As the temperature is lowered below 0° C the vapor pressure of the ice decreases more rapidly than does that of the water in the solution. At a temperature somewhat below 0° C, the ice and the water have the same vapor pressure, and this is the temperature at which the solution and ice are in equilibrium; it is the freezing point of the solution. This property of solutes gives rise to the practice of using such substances as sodium chloride and calcium chloride to melt ice on streets and highways.

It has been found that one gram-molecular weight of such nonelectrolytes as sucrose, glycerine, and alcohol, when dissolved in 1,000 g of water, gives solutions which freeze at −1.86°. A gram-formula weight of sodium chloride in 1,000 g of water will show nearly twice the freezing-point depression characteristic of molecular compounds. Each ion individually produces about the same effect as a molecule upon the freezing point of a solution. In Section 14.11, we shall consider why the lowering produced by sodium chloride is not exactly twice that produced by a similar amount of a nonelectrolyte.

13.28   Phase Diagram for an Aqueous Solution of a Nonelectrolyte

In Chapter 12, we discussed the phase diagram for pure water (see Section 12.8 and Fig. 12–2). A phase diagram is given in Fig. 13–3 for an aqueous solution of a nonelectrolyte solute, such as sucrose ($C_{12}H_{22}O_{11}$). For comparison, the phase diagram for water is reproduced (broken line). The lower freezing points for the solution are shown by the solid line separating solid and liquid states being displaced to the left of the broken one (lower temperatures). Correspondingly, the higher boiling points for the solution are shown by the displacement of the solid line separating the liquid and gas states to the right of the broken one (higher temperatures). The decrease in vapor pressure for the solution, at any given temperature, is indicated by the vertical distance between the broken line and the solid line. The normal freezing-point depression and the normal boiling-point elevation may be seen on the diagram as the horizontal distance between the broken line and the solid line near 0° C and near 100° C, respectively, at a pressure of 760 mm.

13.29   Determination of Molecular Weights of Substances in Solution

We have seen in preceding sections that the lowering of the vapor pressure, the lowering of the freezing point, and the elevation of the boiling point of a solution, depend upon the number of solute particles in a definite weight of solvent, not

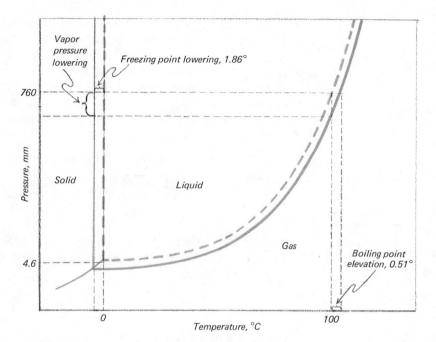

FIGURE 13-3
Phase diagram for a one-molal aqueous solution of a nonelectrolyte (solid lines) compared to that for pure water (broken lines).

upon the nature of the particles. It is thus evident that the gram-molecular weight (6.022×10^{23} molecules) of any nonelectrolyte is that weight which when dissolved in 1,000 g of water (a one-molal solution) lowers the freezing point 1.86°. Similarly, the gram-molecular weight of any nonvolatile nonelectrolyte is that weight which when dissolved in 1,000 g of water (again, a one-molal solution) raises the boiling point 0.512°. Hence, a one-molal aqueous solution of a nonelectrolyte boils at 100.512° and freezes at −1.86°. It should be emphasized that the value of 1.86° for the molal freezing-point depression and that of 0.512° for the molal boiling-point elevation are characteristic only of water. The same principles hold for other solvents, but the values of the constants are quite different from those of water. They are characteristic of the solvent. Hence, each solvent has a different **molal freezing-point depression constant** (K_f) and **molal boiling-point elevation constant** (K_b); these constants for several solvents are listed in Table 13-1.

TABLE 13-1   Boiling Points, Freezing Points, and Molal Boiling- and Freezing-Point Constants for Several Solvents

Solvent	Boiling Point (760 mm) °C	K_b	Freezing Point, °C	K_f
Water	100.0	0.512	0	1.86
Acetic acid	118.1	3.07	16.6	3.9
Benzene	80.1	2.53	5.48	5.12
Chloroform	61.26	3.63	−63.5	4.68
Nitrobenzene	210.9	5.24	5.67	8.1

$m = \dfrac{\Delta T_f}{K_f} = \dfrac{.31}{1.80°/m} = .167\,m$

$?g\ compd = 1\,kg\,H_2O\left(\dfrac{5.00g}{.1\,kg}\right) = 50g$ $\dfrac{50}{.167} = 299$

Molecular weights of nonelectrolytes are often determined by observing the effect that they have upon the freezing point or boiling point of a solvent.

Example 1. Let us suppose that 5.00 g of a nonelectrolyte dissolved in 100 g of water lowers the freezing point of the water 0.31°. We know that the molecular weight of this substance is that weight which, when dissolved in 1,000 g of water, would lower the freezing point of water 1.86°. Calculate the gram-molecular weight of the substance.

First determine the weight of the substance that would lower the freezing point of 1,000 g of water 0.31°.

$m = \dfrac{\Delta T_f}{K_f} = \dfrac{.31°C}{1.86°C/m} = .167$

$$\frac{1,000\text{ g}}{100\text{ g}} \times 5.00\text{ g} = 50.0\text{ g}$$

$?g\ compd = 1\,kg\,H_2O\left(\dfrac{5g}{.1\,kg}\right) = 50g$

Now, calculate the weight of substance required to lower the freezing point of 1,000 g of water 1.86°; this weight will be the gram-molecular weight of the substance.

$\dfrac{?g}{mole} = \dfrac{50}{.167} = 299$

$$\frac{1.86°}{0.31°} \times 50.0\text{ g} = 300\text{ g}$$

Hence, the molecular weight of the substance is 300.

$?g = 1\,kg\,H_2O\left(\dfrac{10.3}{.14\,kg}\right) = 73.6$

$\boxed{299/mole}$ $m = \dfrac{\Delta K_b}{K_b} = \dfrac{1.3}{.512} = 2.54$

Example 2. Experiment shows that 10.3 g of a certain nonelectrolyte dissolved in 140 g of water changes the boiling point of the water at atmospheric pressure (760 mm) to 101.3°. Calculate the molecular weight of the nonelectrolyte.

$m = \dfrac{1.3°}{.512°} = 2.54\,m$

The elevation of the boiling point for 10.3 g of the nonelectrolyte dissolved in 140 g of water is $(101.3 - 100.0)° = 1.3°$. The molal boiling-point elevation constant for water (K_b) is 0.512°. How much nonelectrolyte dissolved in 1,000 g of water would raise the boiling point 0.512°?

$?g = 1\,kg\,H_2O\left(\dfrac{10.3}{.14\,g\,H_2O}\right) = 73.6$

$\dfrac{?g}{mole} = \dfrac{73.6}{2.54} = 29.$

$$10.3\text{ g} \times \frac{1,000\text{ g}}{140\text{ g}} \times \frac{0.512°}{1.3°} = \frac{5274}{182} = 29\text{ g}$$

Hence, the molecular weight of the nonelectrolyte is 29.

Note that only two significant figures are justified despite the fact that all data are expressed to at least three significant figures. Why is this?

$m = \dfrac{3.16°C}{5.12°C/m} = .617$

Example 3. If 4.00 g of a particular nonelectrolyte is dissolved in 55.0 g of benzene, the resulting solution freezes at 2.32°. Calculate the molecular weight of the nonelectrolyte.

$?g\ compd = 1\,kg\,C_6H_6\left(\dfrac{4.00g}{.055\,kg\,C_6H_6}\right) = 72.7g$

According to Table 13–1, K_f for benzene is 5.12° and the freezing point of benzene is 5.48°.

$\dfrac{?g}{mole} = \dfrac{72.7}{.617} = 117.8 \approx 118$

Inasmuch as the freezing point of the solution described is 2.32°, the 4.00 g of solute have lowered the freezing point of 55.0 g of benzene from 5.48° to 2.32°, a lowering of 3.16°.

$m = \dfrac{3.16}{5.12} = .617$

$?g = 1\,kg\,C_6H_6\left(\dfrac{400}{.055\,kg}\right) = 72.7$

$\dfrac{72.7}{.617} = 118$

How much solute dissolved in 1,000 g of benzene would be required to lower the freezing point 5.12°?

$$4.00 \text{ g} \times \frac{1,000 \text{ g}}{55.0 \text{ g}} \times \frac{5.12°}{3.16°} = 118 \text{ g}$$

Hence the molecular weight of the solute is 118.

Example 4. Calculate the freezing point of a solution of 57,800 g of propylene-glycol, $C_3H_6(OH)_2$, a common antifreeze, in 10.0 gallons of water.

> 1 gallon = 3.79 liters
> 10 gallons of water = 37.9 liters = 37,900 ml = 37,900 g
> (The density of water is 1.00 g/ml.)

Hence, 57,800 g of $C_3H_6(OH)_2$ is dissolved in 37,900 g of water.

$$57,800 \text{ g of } C_3H_6(OH)_2 = \frac{57,800}{76.1} \text{ moles}$$

$$= 760 \text{ moles of } C_3H_6(OH)_2 \text{ in } 37,900 \text{ g of } H_2O$$

$$760 \times \frac{1,000}{37,900} = 20.0 \text{ moles } C_3H_6(OH)_2 \text{ in } 1,000 \text{ g of } H_2O.$$

One mole of $C_3H_6(OH)_2$ dissolved in 1,000 g of H_2O would lower the freezing point of the water 1.86° (K_f for water). Hence, 20.0 moles of $C_3H_6(OH)_2$ dissolved in 1,000 g of H_2O would lower the freezing point $20.0 \times 1.86° = 37.2°$. Therefore, the freezing point of the water would be lowered from 0° to −37.2°. It should be pointed out that actually the rules for freezing-point depression and boiling-point elevation hold very well for dilute solutions but do not apply exactly for solutions as concentrated as that in the preceding example. Therefore, the actual freezing point would probably differ from −37.2° C by a small amount.

13.30   Determination of Atomic Weights of Substances in Solution

■ **1. From Specific Heats.** The discovery by Dulong and Petit in 1819 that **the product of the specific heats** (Section 1.22) **and atomic weights of solid elements is very nearly a constant, approximately 6.4,** was a milestone in the development of the atomic theory of the structure of matter (Fig. 13–4). This law has been found to be approximately valid for all solid elements having atomic weights greater than 40 and for most metallic elements. It does not hold (at room temperature) for such elements as carbon, silicon, phosphorus, or sulfur.

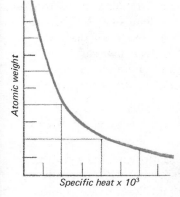

FIGURE 13–4
Variation of specific heat with atomic weight.

Example. Suppose we use the law of Dulong and Petit to calculate the approximate atomic weight of iron, which has a specific heat of 0.115 cal per gram per degree.

$$\text{Atomic weight} \times \text{specific heat} = 6.4$$

$$\text{Atomic weight} = \frac{6.4}{\text{specific heat}} = \frac{6.4}{0.115} = 56$$

The value 56 is not the *exact* atomic weight of iron, because the law of Dulong and Petit is only approximately valid.

Prior to the discovery by Dulong and Petit, atomic weights could only be conjectured from experimentally derived equivalent weights by certain unproved assumptions. With the aid of this law, however, decisions could be made as to which multiple of the equivalent weight of an element represented its atomic weight. Although proposed originally as an empirical rule, the law of Dulong and Petit has been shown in recent years to have theoretical significance as well.

■ **2. From Equivalent Weights.** The exact atomic weight of an element can be calculated after the experimental determination of (1) the exact equivalent weight, and (2) the approximate atomic weight. The calculation of the exact atomic weight of an element is illustrated by the following example.

Example. **Suppose that the least weight found for the element carbon in a gram-molecular weight of any of its compounds is 12.071 g. A very careful determination of the equivalent weight of carbon shows it to be 3.0027. Calculate the exact value for the atomic weight of carbon.**

The number 12.071 is the *approximate* atomic weight of carbon. The equivalent weight, 3.0027, can be accurately determined and it is, therefore, very nearly exact. Dividing the approximate atomic weight of carbon by the exact equivalent weight gives us the numerical value for the oxidation state of carbon.

$$\frac{12.071}{3.0027} = 4.02$$

It is evident that the exact atomic weight of carbon divided by its exact equivalent weight should give the number 4 (instead of 4.02), because oxidation states are whole numbers, and the only reason it does not is that 12.071 is not the exact atomic weight. We may now calculate the exact value of the atomic weight of carbon by multiplying the equivalent weight, 3.0027, by 4, which gives 12.011, the accepted value for the atomic weight of carbon (Section 3.10).

13.31   Osmosis and Osmotic Pressure of Solutions

When a solution and its pure solvent are separated by a semipermeable membrane (one through which the solvent but not the solute can pass), the pure solvent will diffuse through the membrane and dilute the solution. This process is known as **osmosis.** Actually, the solvent diffuses through the membrane in both directions simultaneously; however, the rate of diffusion is greater from the pure solvent to the solution than in the opposite direction. The force causing diffusion of

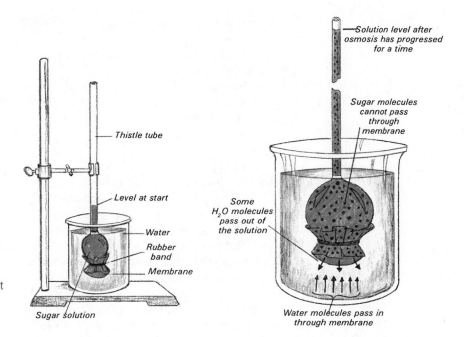

FIGURE 13-5
Apparatus for demonstrating osmosis. Drawing at the right gives some of the details of the process.

liquids produces a pressure referred to as **osmotic pressure,** which can be measured in the following way: A piece of cellophane is securely fastened over the end of a thistle tube to serve as a semipermeable membrane, and the bowl of the thistle tube is filled with a concentrated sugar solution. When the thistle tube is inverted in a beaker of water (Fig. 13–5), water will diffuse through the membrane into the solution, and a slow rise of liquid in the tube will be observed. If the membrane is sufficiently strong to withstand the pressure and if the stem of the thistle tube is long enough, the liquid will rise until its hydrostatic pressure (weight of the column of water in the tube) is equal to the osmotic pressure. The height to which the liquid rises is proportional to the concentration of the solution; in other words, the osmotic pressure is proportional to the number of solute particles in a definite volume of liquid. When gram-molecular weights of different nonelectrolytes are dissolved in 1,000 g of water, their solutions exert equal osmotic pressures, approximately 22.4 atmospheres.

QUESTIONS

1. How do solutions differ from compounds? From ordinary mixtures?
2. There being three different physical states in which matter can exist, how many different types of solutions are possible on the basis of this classification?
3. What are the principal characteristics of solutions?
4. Account for the presence of tremendous quantities of soluble substances in sea water.
5. Explain the following terms as applied to solutions: solute, solvent, dilute, concentrated, saturated, supersaturated, and unsaturated.

6. How may gases be expelled from liquid solvents?
7. Relate the solubility of gases to (a) the nature of the gas and its solvent, (b) the pressure on the gas, (c) the temperature, and (d) the reaction of the gas with the liquid.
8. Contrast the effect of heat on the solubility of gases and solids in water.
9. What is meant by miscible liquids, immiscible liquids, and partially miscible liquids? Give examples.
10. Tell how heating and agitation operate to increase the rate of solution of a soluble material.
11. Explain what is meant by the term "constant boiling solution."
12. Describe the fractional distillation of a 50 per cent mixture of isopropyl alcohol (boiling point 82.5°) and water with regard to the composition of the first and last fractions of the distillate.
13. Why are the majority of chemical reactions carried out in solutions?
14. Why can the dissolving of a solid in a liquid be considered analogous to melting? What other change often occurs when solids are dissolved in liquids?
15. What determines whether the dissolving of a solid in a liquid will be an exothermic or an endothermic process?
16. When a 50 per cent solution of nitric acid is distilled, what is the composition of the vapor compared to that of the solution? How will the boiling point change as distillation is continued? What is an azeotropic mixture?
17. What is indicated by a sharp break in a solubility curve?
18. Heat is absorbed when a certain solid is dissolved in a given liquid. How will an increase in temperature affect the solubility of the solid?
19. What factors determine the rate of solution of a solid in a liquid?
20. Why is a supersaturated solution considered to be an unstable system?
21. Explain the meaning of the statement: "A supersaturated solution is not at equilibrium."
22. How can one demonstrate that a saturated solution in contact with excess solid solute is in "dynamic" equilibrium?
23. (a) Compare the number of molecules of solute in equal volumes of 1 molar solutions of sucrose ($C_{12}H_{22}O_{11}$) and glucose ($C_6H_{12}O_6$). Would your answer be different if you were asked to compare the total number of molecules in the two solutions? Explain.
    (b) Repeat (a) but for equal volumes of 1 molal solutions of sucrose and glucose.
24. Distinguish between crystallization from a solution and crystallization from a melt.
25. Define the following terms: molar solution, equivalent weight of a compound, and normal solution.
26. Define the term "normal solution" in terms of milliequivalent weights and milliliters.
27. What is the advantage of the use of normal solutions over that of molar solutions?
28. What is the advantage of the use of molar solutions over that of normal solutions?

29. What are colligative properties? Give examples.
30. State and explain Raoult's Law.
31. Explain the lowering of the vapor pressure of a liquid by a nonvolatile solute in terms of the kinetic-molecular theory.
32. What is meant by a molal freezing-point depression constant?
33. Distinguish between molar and molal solutions.
34. Why will a mole of sodium chloride depress the freezing point of 1000 g of water about twice as much as a mole of glycerine?
35. Why will ice melt when placed in an aqueous solution which is at $0°$ and kept at that temperature?
36. Demonstrate, or take issue with, the following statement: "In all cases, the molality of a solution has a value which is larger than the molarity of the same solution."
37. Suppose you are presented with a clear solution and told what the solute and solvent are but not told whether the solution is unsaturated, saturated, or supersaturated. How could you determine which of these three conditions exist?
38. What is meant by mole fraction?
39. (a) Such substances as sodium chloride (NaCl), calcium chloride ($CaCl_2$), and urea (NH_2CONH_2) are frequently used to melt ice on streets and highways. Rank the three from high to low in terms of their effectiveness *per pound* in lowering the freezing point. Explain your ranking.
    (b) Would these substances be useful in melting ice on a day when the maximum temperature is $-20°$ F? Explain why or why not.
40. State Avogadro's Law and the Law of Dulong and Petit.
41. Define the term "equivalent weight of an element." Relate atomic weights to equivalent weights and oxidation state.
42. Why are equivalent weights more fundamental as regards quantities of elements undergoing chemical change than are atomic weights?
43. Is the equivalent weight of an element fixed at one value in the same sense that the atomic weight is?
44. How does the equivalent weight of hydrogen in the compound hydrogen peroxide (H_2O_2) compare with the equivalent weight of hydrogen in water?
45. Which equivalent weights may be used as standards in determining other equivalent weights?
46. Explain what is meant by the term "osmotic pressure." How is the osmotic pressure of a solution related to its concentration?

PROBLEMS

s1. How many grams of a 6.00% NaCl solution by weight are necessary to yield 80.0 g of NaCl?   *Ans. 1330 g*
s2. How many grams of $HC_2H_3O_2$ are contained in 60.0 ml of acetic acid of specific gravity 1.065 and containing 58.0% $HC_2H_3O_2$ by weight?
                                                                        Ans. 37.1 g

ˢ3. Calculate the weight of solute in each of the following solutions:

  ˢ(a) 2.40 liters of 0.650 M HClO$_4$.   *Ans. 157 g*

  ˢ(b) 125 ml of 0.0250 M C$_{12}$H$_{22}$O$_{11}$.   *Ans. 1.07 g*

  (c) 30.0 ml of 0.0800 M MgCl$_2$.   *Ans. 0.229 g*

  (d) 4.25 liters of 1.50 M Co(NO$_3$)$_2$·6H$_2$O.   *Ans. 1.86 kg*

ˢ4. How much sulfuric acid (98.0% by weight) is needed in the preparation of 250 g of 25.0% solution of the acid by weight?   *Ans. 63.8 g*

5. What weights of KNO$_3$ and water are contained in 184 g of a 19.0% solution of KNO$_3$ by weight?

                                            Ans. 35.0 g of KNO$_3$; 149 g of H$_2$O

ˢ6. Calculate the molarity of each of the following solutions:

  ˢ(a) 15.6 g of CsOH in 1.50 liters of aqueous solution.   *Ans. 0.0694 M*

  (b) 20 g of HNO$_3$ in 1.0 liter of aqueous solution.   *Ans. 0.32 M*

  (c) 50.0 mg of H$_2$SO$_4$ in 10.0 ml of aqueous solution.   *Ans. 0.0510 M*

  ˢ(d) 7.0 millimoles of I$_2$ in 100 ml of carbon tetrachloride solution.

                                                                   Ans. 0.070 M

  (e) 1.00 g of K$_2$Cr$_2$O$_7$ in 100 ml of aqueous solution.   *Ans. 0.0340 M*

7. It is desired to prepare 1.00 liter of a 5.00% solution of K$_2$Cr$_2$O$_7$ by weight from solid K$_2$Cr$_2$O$_7$ of 95.0% purity by weight. What weight of solid K$_2$Cr$_2$O$_7$ would be required? The specific gravity of a 5.00% solution of K$_2$Cr$_2$O$_7$ at 20° C is 1.0354.   *Ans. 54.5 g*

ˢ8. How would you prepare 90.0 g of a 1.50 per cent Ba(OH)$_2$ solution by weight starting with Ba(OH)$_2$·8H$_2$O and water?   *Ans. Dissolve 2.49 g of*

                                          Ba(OH)$_2$·8H$_2$O in 87.5 g of water

9. What volume of 0.15 M nickel sulfate would contain 47 g of nickel sulfate?

                                                                   Ans. 2.0 liters

10. How many liters of HBr measured at S.T.P. are required in the preparation of 5.00 liters of 0.500 M hydrobromic acid?   *Ans. 56.0 liters*

ˢ11. Calculate the molarity of a 500 ml volume of solution containing 3.0 g of 95.0% sulfuric acid, H$_2$SO$_4$.   *Ans. 0.058 M*

ˢ12. Calculate the molarity of a solution of D-fructose, C$_6$H$_{12}$O$_6$, for which the density is 1.06 g/ml and which contains 15.00% of the compound by weight.

                                                                   Ans. 0.883 M

13. Calculate the volume of formic acid of specific gravity 1.051 and containing 20.00% HCOOH by weight, that would contain 25.00 g of pure HCOOH.

                                                                   Ans. 118.9 ml

14. A 4.50% solution of NaHCO$_3$ by weight has a specific gravity of 1.0325. Calculate the molarity of the solution.   *Ans. 0.553 M*

ˢ15. An aqueous solution of ammonia which is 20.0% ammonia by weight has a specific gravity of 0.925. What volume of ammonia gas (S.T.P.) would be required in the preparation of 2.50 liters of this solution?   *Ans. 608 liters*

ˢ16. It is desired to produce 1.000 liter of 0.400 M sulfuric acid by diluting 12.00 M sulfuric acid. The concentrated acid is poured cautiously into the water. Calculate the volume of the concentrated acid and the volume of water required in the dilution.   *Ans. 33.3 ml of H$_2$SO$_4$; 967 ml of H$_2$O*

17. How many liters of 4.50% KCl solution by weight (specific gravity, 1.029) can be obtained by diluting 0.200 liter of 24.0% KCl by weight (specific gravity, 1.164)?   *Ans. 1.21 liters*

⑤18. Concentrated hydrochloric acid is 37.0% HCl by weight and has a specific gravity of 1.19. What volume of this acid must be diluted to 500 ml to produce a 0.300 M solution of the acid?   *Ans. 12.4 ml*

19. How many moles of sodium hydroxide would be required to react with all the hydrogen ions from:
    (a) 0.5 mole of HCl?   *Ans. 0.5 mole*
    (b) 1.5 moles of H_2SO_4?   *Ans. 3.0 moles*
    (c) 1 mole of H_3PO_4?   *Ans. 3 moles*

20. It is desired to prepare 6.00 liters of 3.00 N sulfuric acid. How many ml of 60.0% acid by weight (specific gravity, 1.50) would be required?
                                                                    Ans. 981 ml

⑤21. A 10.3-g sample of $K_2CO_3 \cdot 1.5H_2O$ is dissolved in 150 g of water. Calculate the percentage by weight of K_2CO_3 in the solution and the molality of the solution in terms of K_2CO_3.   *Ans. 5.37%; 0.411 molal*

22. A solution of K_2HPO_4 having a volume of 300 ml contains 5.369 g of $K_2HPO_4 \cdot 3H_2O$. Calculate the molarity of this solution.   *Ans. 0.0784 M*

23. What is the molarity of a solution which is prepared by dissolving 12.4 g of P_4O_{10} in sufficient water to make 500 ml of solution? Assume that the product of the reaction of P_4O_{10} with water is H_3PO_4 (orthophosphoric acid). Would it make a difference in the answer if the product of the reaction were HPO_3 (metaphosphoric acid)? Explain.   *Ans. 0.349 M*

⑤24. A standard solution of NaOH has a molarity of 0.09987. What is the molarity of a nitric acid solution if 10.0 ml of it neutralizes 24.6 ml of the NaOH solution?   *Ans. 0.246 M*

25. What weight of $K_3Fe(CN)_6$ is there in 1.00 ml of a 0.0500 M solution of the salt?   *Ans. 0.0165 g*

26. Equal volumes of 0.80 M HCl, 0.50 M KOH, and 0.60 M KCl are mixed. What is the normality of the resulting solution with regard to each of its solutes?
                                                          Ans. 0.37 M KCl; 0.10 M HCl

⑤27. Calculate the gram-equivalent weight of each of the reactants in the following:
    ⑤(a) $H_2SO_4 + 2LiOH \longrightarrow Li_2SO_4 + 2H_2O$
                                                Ans. LiOH, 23.9 g; H_2SO_4, 49.0 g
    ⑤(b) $KOH + KHCO_3 \longrightarrow K_2CO_3 + H_2O$
                                                Ans. KOH, 56.1 g; $KHCO_3$, 100.1 g
    (c) $HBr + NH_3 \longrightarrow NH_4Br$ *Ans. HBr, 80.9 g; NH_3, 17.0 g*
    ⑤(d) $Mg + 2HCl \longrightarrow MgCl_2 + H_2$ *Ans. Mg, 12.2 g; HCl, 36.5 g*
    (e) $H_2S + HgCl_2 \longrightarrow HgS + 2HCl$ *Ans. H_2S, 17.0 g; $HgCl_2$, 135.7 g*
    (f) $Sn + 2Cl_2 \longrightarrow SnCl_4$ *Ans. Sn, 29.7 g; Cl_2, 35.5 g*
    (g) $KOH + H_2SO_3 \longrightarrow KHSO_3 + H_2O$
                                                Ans. KOH, 56.1 g; H_2SO_3, 82.1 g
    (h) $Al(OH)_3 + 3HNO_3 \longrightarrow Al(NO_3)_3 + 3H_2O$
                                                Ans. $Al(OH)_3$, 26.0 g; HNO_3, 63.0 g

⑤28. Calculate the normality of each of the following solutions:

   ⑤(a) 4.0 gram-equivalent weights of LiI in 2.0 liters of solution. *Ans. 2.0 N*

   ⑤(b) 2.5 milligram-equivalent weights of $MgCl_2$ in 50 ml of solution.

                                                       Ans. 0.050 N

   (c) 0.30 gram-equivalent weights of $NiSO_4$ in 400 ml of solution.

                                                        Ans. 0.75 N

   (d) 150 milligram-equivalent weights of NaOH in 100 ml of solution.

                                                        Ans. 1.50 N

29. What is the normality of a 40.0% aqueous solution of sulfuric acid, for which the specific gravity is 1.305? *Ans. 10.6 N*

⑤30. The chloride in 50.0 ml of dilute hydrochloric acid was precipitated using an excess of silver nitrate. The weight of AgCl formed was 0.682 g. Calculate the normality of the hydrochloric acid solution. *Ans. 0.0952 N*

31. What volume of a 0.25 M solution of phosphoric acid, H_3PO_4, would be required to neutralize completely 1.00 liter of 0.35 M potassium hydroxide KOH? *Ans. 0.47 liter*

⑤32. What volume of 0.400 M HCl would be required to react completely with 2.60 g of sodium carbonate? ($Na_2CO_3 + 2HCl \longrightarrow 2NaCl + CO_2 + H_2O$)

                                                        Ans. 123 ml

33. How much water must be added to 4.50 g of 50.0% acetic acid solution (specific gravity, 1.06) to form a 0.100 M solution of the acid? *Ans. 370 ml*

⑤34. A 75.0-ml volume of gaseous ammonia measured at 25.0° and 753 mm pressure was absorbed in 50.0 ml of water. (a) What is the normality of the ammonia solution? (b) How many ml of 0.100 N sulfuric acid would be required in the neutralization of this aqueous ammonia? Assume no change in volume when the gaseous ammonia is added to the water.

                                  Ans. (a) 0.0608 N; (b) 30.4 ml

35. Calculate the weight of pure $Pb(NO_3)_2$ required in the preparation of 1,200 ml of a 1.10 M solution. *Ans. 437 g*

⑤36. What volume of 0.300 M H_2O_2 would be required to oxidize 0.634 g of Na_2SO_3 of 98.0% purity? ($Na_2SO_3 + H_2O_2 \longrightarrow Na_2SO_4 + H_2O$)

                                                  Ans. 16.4 ml

37. Calculate the volume of 0.050 M sulfuric acid necessary to precipitate the barium contained in 15.0 ml of 0.050 M $BaCl_2$. ($Ba^{2+} + SO_4^{2-} \longrightarrow BaSO_4$).

                                                     Ans. 15 ml

38. A solution of sulfuric acid contains 9.11 mg of H_2SO_4 per ml of solution. What is the normality of the solution? *Ans. 0.186 N*

⑤39. A 0.4238-gram sample of an acid was dissolved in water. It took 34.7 ml of 0.100 M NaOH solution to neutralize the acid. What is the gram equivalent weight of the acid? *Ans. 122 g*

⑤40. A mixture of NaBr and KI weighing 0.93 g was dissolved in water and then the halides were precipitated as silver salts, weighing together 1.64 g. Calculate the composition of the original mixture. *Ans. 0.79 g of NaBr; 0.14 g of KI*

41. A 1.985-g sample of an acid, HX, required 23.10 ml of KOH solution for neutralization of all the hydrogen ion. Exactly 14.2 ml of this same KOH

solution was found to neutralize 5.00 ml of 1.00 N H_2SO_4. Calculate the molecular weight of HX. *Ans. 244*

42. A sample of an organic compound (nonelectrolyte) weighing 4.173 g lowered the freezing point of 50.0 g of nitrobenzene 5.2°. The K_f of nitrobenzene is 8.1°. Calculate the approximate molecular weight of the organic compound.
Ans. 130

S43. How would you prepare a 4.21 molal aqueous solution of propyleneglycol ($C_3H_8O_2$)? What would be the freezing point of this solution?
Ans. Dissolve 320 g of propyleneglycol in 1,000 g of water; $-7.83°$ C

S44. If 35.7 g of the nonelectrolyte $C_6H_3Cl_3$ is dissolved in 220 g of chloroform, what is (a) the freezing point of the solution and S(b) the boiling point of the solution at 760 mm? For chloroform, the freezing point is $-63.5°$, the boiling point is 61.26°, K_f is 4.68°, and K_b is 3.63°.)
Ans. (a) $-67.7°$; (b) 64.5°

S45. A 4.305-gram sample of a nonelectrolyte is dissolved in 105 g of water. The solution freezes at $-1.23°$ C. Calculate the molecular weight of the substance.
Ans. 62.0

46. What would be the boiling point, at 760 mm, of a solution containing 96.7 g of sucrose ($C_{12}H_{22}O_{11}$) in 250.0 g of water? *Ans. 100.6°*

47. Which of the following compounds would serve most effectively as antifreeze agents when equal weights are employed: ethyl alcohol, C_2H_5OH; glycerine, $C_3H_8O_3$; or glucose, $C_6H_{12}O_6$? Explain. *Ans. Ethyl alcohol*

S48. How many grams of formaldehyde, CH_2O, must be added to a liter of water to prevent it from freezing at 5.00° F? *Ans. 242 g*

49. A solution which is 15.00% by weight contains 153 g of solute per liter of solution. The molecular weight of the solute is 60.0. For each mole of solute, 2N electrons are involved in the reaction in which the solution is to be utilized (where N is Avogadro's number). Calculate (a) the density, (b) the molarity, (c) the molality, and (d) the normality of the solution.
Ans. (a) 1.02 g/ml; (b) 2.55 molar; (c) 2.94 molal; (d) 5.10 normal

50. A compound is known to be a Group IVA tetrachloride (general formula, ECl_4, where E represents one of the elements in Group IVA of the Periodic Table). An approximate freezing-point determination indicates that 6 grams of the compound in 100 grams of benzene gives a solution that freezes at 4° C. What is the formula of the compound? Show your work. *Ans. $GeCl_4$*

S51. Calculate the mole fraction of solute and solvent for a 2.1 molal aqueous solution. *Ans. 0.036; 0.96*

52. Calculate the mole fraction of chloroform ($CHCl_3$), benzene (C_6H_6), and acetone (C_3H_6O) in a solution which is 25% chloroform, 35% benzene, and 40% acetone, by weight. *Ans. 0.16; 0.33; 0.51*

53. A customer wishes to add enough antifreeze to his radiator (capacity 17 quarts) to protect his car down to a temperature of $-35°$ F. Evidently the day shift has misplaced the table showing needed quantities for various radiator sizes and protection to different temperatures. However, the attendant on duty knows that his brand of antifreeze is ethyleneglycol, $C_2H_4(OH)_2$, and

its density is listed on the can as 1.09 g per ml. He also knows that the density of water is 1.00 g per ml. Recalling his freshman chemistry course, he is able to calculate the proper number of gallons of antifreeze and water, assuming that the total resulting volume is the sum of the volumes of ethyleneglycol and water used. How did he do the calculation and what are the correct quantities?   *Ans. 9.0 quarts of ethyleneglycol and 8.0 quarts of water*

54. Calculate (a) the per cent composition and (b) the molality of an aqueous solution of NH_4NO_3, if the mole fraction of NH_4NO_3 is 0.060.

Ans. (a) 22% NH_4NO_3, 78% H_2O; (b) 3.5 molal

⑤55. Concentrated hydrochloric acid is 37.0% HCl by weight and has a specific gravity of 1.19. Calculate (a) the molarity, (b) the molality, (c) the normality, and (d) the mole fraction of HCl and of H_2O.

Ans. (a) 12.1 molar; (b) 16.1 molal; (c) 12.1 normal;
(d) 0.225 mole fraction HCl; 0.775 mole fraction H_2O

REFERENCES

"Ideal Solutions," W. A. Oates, *J. Chem. Educ.*, **46**, 501 (1969).

"Gram-Equivalent Weights," W. B. Meldrum, *J. Chem. Educ.*, **32**, 48 (1955).

"The Use of the Molarity Concept in Titrimetric Analysis," J. G. Stark, *Educ. in Chemistry,* **3**, 70 (1966).

"Liquid-Liquid Extraction," P. Joseph-Nathan, *J. Chem. Educ.*, **44**, 176 (1967).

"Demonstrating Osmotic and Hydrostatic Pressures in Blood Capillaries," J. W. Ledbetter, Jr., and H. D. Jones, *J. Chem. Educ.*, **44**, 362 (1967).

"Reactions in Solutions Under Pressure," W. J. le Noble, *J. Chem. Educ.*, **44**, 729 (1967).

"Chemiluminescent Reactions in Solution," J. W. Haas, Jr., *J. Chem. Educ.*, **44**, 396 (1967).

"The Osmotic Pump," O. Levenspiel and N. de Nevers, *Science*, **183**, 157 (1974).

"Colligative Properties," F. Rioux, *J. Chem. Educ.*, **50**, 490 (1973).

Solutions of Electrolytes; Colloids

14

Electrolytes

14.1 Electrolytes and Nonelectrolytes

Substances that give solutions which conduct an electric current are known as **electrolytes.** Molten substances which conduct an electric current are electrolytes, also. The process of conducting an electric current through a solution or through a molten substance so that decomposition of either the electrolyte or the solvent results is called **electrolysis.** Most electrolytes are acids, bases, or salts. Aqueous solutions of some substances, such as sugar and alcohol, do not conduct an electric current. Substances that form nonconducting solutions are called **nonelectrolytes.**

The classification of substances as electrolytes and nonelectrolytes may be carried out experimentally by setting up a simple conductivity apparatus as shown in Fig. 14–1. The terminals of a storage battery or a 110-volt circuit (preferably direct current) are connected through an electric lamp to two electrodes in a beaker. When the beaker is filled with pure water, not enough current flows through the circuit to cause the electric lamp to emit light. When an aqueous solution of a nonelectrolyte such as sugar, alcohol, or glycerine is tested, the lamp again does not emit light. If, however, an electrolyte such as hydrochloric acid, sodium hydroxide, or sodium chloride is dissolved in the water, the lamp glows brightly.

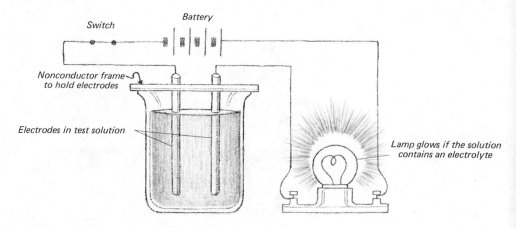

FIGURE 14-1

Apparatus for determining the conductivity of solutions.

14.2   Strong and Weak Electrolytes

If the beaker in Fig. 14–1 is filled with a 0.1 N solution of hydrochloric acid, the lamp in the circuit will glow brightly, showing that the solution is a good conductor of electricity. The same is true of 0.1 N solutions of nitric and sulfuric acids, as well as bases such as potassium, sodium, and barium hydroxides. Most salts behave in a similar fashion. Substances whose aqueous solutions are good conductors of electricity are known as **strong electrolytes.** When a 0.1 N solution of acetic acid or ammonium hydroxide is placed in the beaker, it is found that the lamp glows much less brightly than it does with acids like hydrochloric acid or bases like sodium hydroxide. Substances whose aqueous solutions are poor conductors of electricity are called **weak electrolytes.**

14.3   Theory of Electrolytic Conduction

Svante Arrhenius, a Swedish chemist, first successfully explained electrolytic conduction and presented a theory regarding certain unique properties of acids, bases, and salts. His theory explained why acids, bases, and salts (electrolytes) dissolved in water affect the vapor pressure, boiling point, freezing point, and osmotic pressure of water more than do nonelectrolytes. Although Arrhenius' theory has since been modified in the light of our present knowledge of atomic structure and chemical bonding, the modern theory of electrolytes embodies most of the principal postulates of the Arrhenius theory.

According to the modern theory, a water solution of an electrolyte contains independent positive and negative ions. We have previously defined an **ion** as an atom or group of atoms carrying an electric charge. The positive ions migrate toward the negative electrode during electrolysis and are called **cations.** The negative ions migrate toward the positive electrode during electrolysis and are called **anions.** The movement of the ions toward the electrode of opposite charge during electrolysis accounts for electrolytic conduction. The total number of positive charges on the cations in solution is just equal to the total number of

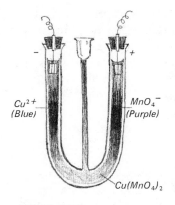

Cu^{2+}
(Blue)

MnO_4^-
(Purple)

$Cu(MnO_4)_2$

FIGURE 14-2

When an electric current flows in the system shown, blue Cu^{2+} ions and purple MnO_4^- ions travel toward the electrodes, visual evidence that the current is carried by the ions.

negative charges on the anions. The solutions of electrolytes are therefore electrically neutral. A nonelectrolytic solution does not contain ions.

The migration of ions during electrolytic conduction can be demonstrated in a striking manner. A U-tube (Fig. 14–2) is partially filled with a solution of potassium nitrate, which is colorless and which has been acidified with a few drops of sulfuric acid. Then a solution of copper(II) permanganate is carefully introduced into the bottom of the U-tube without mixing of the two solutions. When the electrical circuit is closed, the blue copper(II) ions (Cu^{2+}) begin to migrate through the colorless potassium nitrate toward the negative electrode, and the purple permanganate ions (MnO_4^-) begin to migrate toward the positive electrode. This demonstration offers visual evidence that the electric current is carried by ions, and that ions of opposite charge exist independently in the solution.

14.4 Extent of Ionization of Electrolytes

A salt is composed of ions that break away from their crystal lattice when the salt is dissolved in water. For example, sodium chloride consists of sodium ions and chloride ions held together in the crystal by strong electrostatic forces. These ions separate when sodium chloride dissolves and become distributed in the solution.

$$\overset{+}{N}a\overset{-}{Cl} \xrightarrow{H_2O} Na^+ + Cl^-$$
(crystalline)

The ions in the crystal lattice of certain metal hydroxides, such as sodium hydroxide, also separate when dissolved in water.

$$\overset{+}{N}a\overset{-}{OH} \xrightarrow{H_2O} Na^+ + OH^-$$
(crystalline)

However, certain covalent substances form ions by reacting with water; i.e., they ionize when brought into contact with water. Hydrogen chloride, which contains polar covalent molecules, ionizes as follows:

$$HCl + H_2O \longrightarrow H_3O^+ + Cl^-$$

According to the theory of electrolytes, if the solution is a good conductor of electricity the solute consists entirely or principally of ions, and if the solution is a poor conductor the solute consists principally of molecules. Strong electrolytes such as hydrochloric, nitric, and sulfuric acids are virtually completely ionized in dilute aqueous solutions. Most salts and soluble metal hydroxides are also strong electrolytes. Only a small fraction of the molecules of a weak electrolyte undergo ionization in water at any one time. The partial ionization of acetic acid, a weak electrolyte, is expressed by the equation

$$CH_3COOH + H_2O \rightleftharpoons H_3O^+ + CH_3COO^-$$

The double arrow indicates that an equilibrium exists between the ions and the un-ionized molecules of acetic acid. A 0.1 M solution of acetic acid is only 1.34

per cent ionized at 25° C. This means that 98.66 per cent of the acid is in the molecular form. As the solution is diluted the percentage of ionization increases.

Pure water is an extremely poor conductor of electricity, indicating very slight ionization. Ionization of water is effected when one molecule of water gives up a proton to another molecule of water, yielding hydronium and hydroxide ions.

$$H_2O + H_2O \rightleftharpoons H_3O^+ + OH^-$$

14.5   Properties of Ions

The chemical properties of an ion are quite different from those of the electrically neutral atom or molecule. Sodium ions and chloride ions are colorless, non-poisonous, and inert toward water (though they form unstable hydrated ions of the type described in Section 14.6). Metallic sodium, however, is a silvery-white substance which reacts violently with water, forming hydrogen gas and a solution of the strong base, sodium hydroxide; molecular chlorine, a poisonous greenish-yellow gas, reacts with water forming hypochlorous and hydrochloric acids.

The difference in properties between ions and neutral atoms or molecules comes about because ions have either more or fewer orbital electrons than the neutral atoms or molecules from which they are formed. The outer electron shell of the sodium atom contains one electron. When this electron is lost to another atom, such as chlorine, a positive sodium ion with the stable configuration of a noble gas results. Similarly, a neutral chlorine atom with seven electrons in its outside shell assumes the stable electron configuration of a noble gas by gaining an electron to become a negative ion (Section 4.2).

14.6   Mechanism of Dissolution of Ionic Compounds

Ionic compounds are ionized in the solid state as well as in solution. Since the ions in a crystal of an ionic compound are held firmly in their crystal lattice positions by strong electrostatic forces, the ions are said to be associated. Crystalline ionic compounds do not conduct an electric current even though they are 100 per cent ionized. When the ionic solid melts, however, the forces which hold the ions in their lattice positions are overcome, the ions become able to move about, and the melt conducts an electric current.

The process of dissolution of ionic compounds in water is also essentially one of separation of the ions. Water reduces the strong electrostatic forces between the ions and allows their separation. Let us consider the dissolution of potassium chloride in water. The hydrogen (positive) side of the polar water molecule is strongly attracted to the negative chloride ion, and the oxygen (negative) side of the water molecule is strongly attracted to the positive potassium ion. We may picture the water molecules as surrounding individual K^+ and Cl^- ions at the surface of the crystal and penetrating between them, thereby reducing the strong interionic forces of attraction which bind them together in the crystal and permitting them to move off into the water as hydrated ions (Fig. 14–3). Several water molecules become associated with each ion in solution as a result of this **ion-dipole**

FIGURE 14-3
The figure at the left shows
the polar nature of the water
molecule. The two drawings to
the right show how water
molecules are thought to be attracted to positive and negative ions to form hydrated
ions.

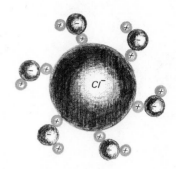

electrostatic attraction. The increase in the size of the ions brought about by hydration leads to greater effective distances between oppositely charged ions and a reduction in interionic attraction. In addition, water is a good insulator (it has a high dielectric constant) which further reduces the electrostatic attraction between the ions.

Ionic compounds usually dissolve only in polar solvents in which the polar solvent molecules can solvate and insulate the ions. In general, the higher the dielectric constant (greater polar character) of the solvent, the greater the solubility of an ionic compound put in it. This phenomenon is strikingly illustrated in the data of Table 14–1. Ionic substances in general do not dissolve appreciably in nonpolar solvents, such as benzene or carbon tetrachloride, because the nonpolar solvent molecules are not strongly attracted to ions, and nonpolar solvents have low dielectric constants.

TABLE 14-1   The Solubility of Sodium Chloride and the Dielectric Constant of the Solvent

Solvent	Solubility of NaCl (grams per 100 g of solvent, 25°)	Dielectric Constant of the Solvent
Water, H_2O	36.12	80.0
Methyl alcohol, CH_3OH	1.3	33.1
Carbon tetrachloride, CCl_4	0.00	2.2

14.7   Solubility of Salts

The solubility of an ionic compound in water is determined in large part by (1) the magnitude of the crystal forces and (2) the energy of hydration of the ions. A soluble salt is one for which the attraction of the ions for water molecules is greater than the attraction of the oppositely charged ions for each other, or, putting it another way, one for which the energy of hydration is greater than the crystal lattice energy. On the other hand, a slightly soluble salt is characterized by strong crystal forces and slight tendency of the ions to hydrate.

Now is a good time to point out that many other solvents, in addition to water, solvate ions. In this chapter, we shall be concerned mainly with solvation by water (hydration). When we discuss acids and bases in Chapter 15 we shall consider solvation by some other solvents such as liquid ammonia and liquid sulfur dioxide (low temperatures) to produce solvated ions such as $H^+ \cdot NH_3$ (the ammonium ion, NH_4^+), $[Ag(NH_3)_2]^+$, and $SO^{2+} \cdot SO_2$. Note that *solvation* is the general term; *hydration* is the specific term referring to solvation by water.

The energy quantities concerned in the solution process for an ionic crystal may be related by a cycle of the Born-Haber type (see Section 11.18). The process of solution of the ionic halide NaCl in water, for example, may be considered as the rupture of the crystal lattice to give isolated gaseous ions, followed by the hydration of the gaseous ions to produce hydrated ions in solution. We may represent this system diagrammatically as a cycle, in which L is the energy absorbed in the solution process (heat of solution), H is the hydration energy, and U the lattice energy (see Section 11.17).

$$\text{NaCl}(s) \xrightarrow[\substack{\text{mole of NaCl}}]{\substack{U = 184 \text{ kcal} \\ \text{absorbed per}}} \text{Na}^+(g) + \text{Cl}^-(g) \xrightarrow[\substack{\text{absorbed}}]{\substack{H = \text{energy}}} \text{Na}^+(aq) + \text{Cl}^-(aq)$$

$$L = 1.3 \text{ kcal absorbed per mole}$$

The energies U and H of the two steps added together should equal the total energy, L, absorbed in the solution process. In other words, the two steps, for which U and H are the energies, accomplish the same change that L accomplishes and therefore together should involve the same energy as L.

$$U + H = L$$

This relationship may be stated generally: the energy (L) absorbed in dissolving is equal to the energy (U) absorbed in disrupting the lattice plus the energy (H) absorbed in hydrating the ions.

From this relationship and the known values for U and L, we can calculate H (the energy absorbed in the hydration of the ions).

$$U + H = L$$
$$184 \text{ kcal} + H = 1.3 \text{ kcal}$$
$$H = (1.3 - 184) \text{ kcal} = -182.7 \text{ kcal}$$

Notice that the three energy quantities are expressed in terms of *absorption* of energy. Inasmuch as the sign of the calculated value for H is negative, however, heat is actually *liberated* as the ions are hydrated. In other words, minus 182.7 kcal of energy is absorbed, or 182.7 kcal of energy is liberated. Expressed still another way, if a negative 182.7 kcal of energy is absorbed, the energy present *in the system* after hydration is less than it was before hydration; hence, energy was released from the system in the process. It is usual that energy is released when ions attract water molecules and become hydrated.

A great deal of energy (184 kcal/mole) must be added to the system (absorbed) to break a crystal of sodium chloride into its gaseous ions. However, most of that energy is liberated again (182.7 kcal) as the ions hydrate. Hence, the net energy required to dissolve a mole of sodium chloride in water is quite small (1.3 kcal).

In general, because the two energies U and H are comparable in magnitude and have opposite signs, it is not surprising to find that some ionic compounds in water exhibit positive and some negative heats of solution (L). However, those with positive heats of solution (NaCl for example, $+1.3$ kcal) predominate, and hence the solubility of most ionic compounds in water increases with rising temperature. The solubility of the smaller number of ionic compounds which have negative heats of solution decreases with rising temperature. See Fig. 13–2, which shows, for example, that the solubility of anhydrous Na_2SO_4 decreases with an increase of temperature. Anhydrous Na_2SO_4 is an example of a compound that has a negative heat of solution.

In any attempt to predict the solubility of a salt, one must take into account those properties of ions which determine the magnitude of the crystal forces and the energy of hydration of the ions. Such predictions become quite involved, as can readily be seen from the following generalizations.

■ **1. Solubility and Ionic Size.** Both the crystal forces and the energy of hydration increase with decrease in ionic size, but at different rates. Thus, it is difficult to find any consistent relationship between ionic radii and solubility of ionic salts. For example, note the solubilities of the alkali metal chlorides (Table 14–2).

TABLE 14-2 Solubility of Alkali Metal Chlorides in Grams Per 100 g of Water at 0°

	LiCl	NaCl	KCl	RbCl	CsCl
Radius of M^+, Å	0.60	0.95	1.33	1.48	1.69
Solubility	67	35.7	27.6	77.0	161.4

■ **2. Solubility and Size of Ionic Charge.** With increasing ionic charge the crystal forces increase much more rapidly than the energy of hydration of the ions. Thus, solubility of ionic salts decreases very sharply as the ionic charge increases. Examples are given in Table 14–3.

TABLE 14-3 Solubility Versus Ionic Charge—Interionic Distances Relatively Constant

Ionic Charge	$+1, -1$	$+2, -1$	$+2, -2$	$+3, -2$
Compound	LiF	MgF_2	MgO	Al_2O_3
Interionic distance, Å	1.96	2.01	2.05	1.90
Solubility, moles/liter	4.6×10^{-2}	1.7×10^{-3}	1.5×10^{-4}	insoluble

■ **3. Solubility and Electronic Structure of the Cation.** The crystal forces and energy of hydration are both greater for cations with 18 electrons in the outer shell than for cations with 8 outer electrons. The sodium ion has 8 electrons and the silver ion 18 electrons, respectively, in their outer shells. The solubility of NaF in grams per 100 grams of water is 4.0, while that of AgF is 182.0. On the other hand, NaCl has a solubility of 35.7, while that of AgCl is 8.9×10^{-5} g per 100 g of water. RbF has a solubility of 130.6 g in 100 g of water and that of RbCl is 77 g. It is obvious, as indicated by these examples, that it is difficult to predict the solubility of a metallic compound on the basis of the electronic structure of its cation.

Knowledge of the solubilities of metallic compounds is nevertheless very important and useful to the student and chemist. Memorization of the solubility of individual compounds is difficult, laborious, and unnecessary. A simple and worthwhile method of acquiring knowledge of solubilities is to learn the generalizations given in the following section.

14.8   Generalizations on the Solubilities of Common Metal Compounds

It should be remembered that these generalizations are for the simple compounds of the more common metals, there being many exceptions when the less common metals and complex compounds are considered.

(1) Most nitrates and acetates are soluble in water; silver acetate, chromium(II) acetate, and mercury(I) acetate are slightly soluble; bismuth acetate hydrolyzes to bismuth oxyacetate, $BiOC_2H_3O_2$, insoluble in water.

(2) All chlorates except potassium chlorate are water soluble; potassium chlorate is slightly soluble.

(3) All chlorides are soluble except those of mercury(I), silver, lead, and copper(I) ions; lead chloride is soluble in hot water.

(4) All sulfates, except those of strontium, barium, and lead are soluble; calcium sulfate and silver sulfate are slightly soluble.

(5) Carbonates, phosphates, borates, arsenates, and arsenites—except those of ammonium and the alkali metals—are insoluble.

(6) The sulfides of ammonium and the alkali metals are soluble and other sulfides are insoluble; the alkaline earth metal sulfides are hydrolyzed in water.

(7) The hydroxides of sodium, potassium, ammonium, barium, and strontium are soluble and other hydroxides are insoluble; calcium hydroxide is slightly soluble.

14.9   Mechanism of Dissolution and Ionization of Molecular Substances

We learned in Chapter 4 that a hydrogen atom combines with a chlorine atom with the formation of a covalent bond.

$$\text{H} \cdot \; + \; \overset{\times\times}{\underset{\times\times}{\times}} \text{Cl} \overset{\times}{\underset{}{}} \; \longrightarrow \; \text{H} \overset{\times\times}{\underset{\times\times}{\times}} \text{Cl} \overset{\times}{\underset{}{}}$$

However, the electron pair involved in the covalent bond is not shared equally between the hydrogen and chlorine atoms. The chlorine atom attracts the electrons more strongly than does the hydrogen due to the greater electronegativity of the chlorine. The hydrogen chloride bond is therefore somewhat polar, with the hydrogen end positive and the chlorine end negative. Since dry hydrogen chloride in the liquid state does not conduct an electric current, we conclude that no ions are present. But an aqueous solution conducts electricity and possesses acidic properties, because hydrogen chloride reacts with water to produce hydrated hydrogen ions and chloride ions in the solution. Since no ionization of HCl occurs in such nonpolar solvents as benzene, the polar water molecules evidently play an important role in bringing about ionization.

The ionization of hydrogen chloride in water may be represented by the equation in Fig. 14–4. Electronic formulas are employed to help elucidate the process, and orbital diagrams are shown to illustrate the spatial arrangement. The polar water molecule removes a proton from the polar hydrogen chloride molecule, forming the hydronium ion (H_3O^+) and the chloride ion (Cl^-). Note that the pair of electrons that bonded the hydrogen and chlorine together in the HCl molecule remains with the chlorine, making it a chloride ion. Either of the two unshared pairs of electrons on the water molecule can be shared with the proton,

$$(1) \quad H_2O + HCl \longrightarrow H_3O^+ + Cl^-$$

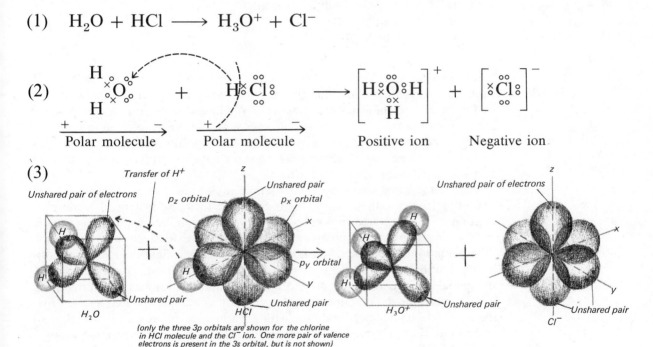

FIGURE 14–4

The ionization of hydrogen chloride in water. The reaction of water with hydrogen chloride produces the hydronium ion (H_3O^+) and the chloride ion (1). The reaction involves a transfer of a hydrogen ion from the HCl molecule to the water molecule (2). In the orbital diagrams, only the 3p valence orbitals of the chlorine are shown, oriented at right angles to each other along the x, y, and z axes. The water molecule and the hydronium ion are shown as tetrahedral structures circumscribed within a cube (3). (See also Fig. 12–6.)

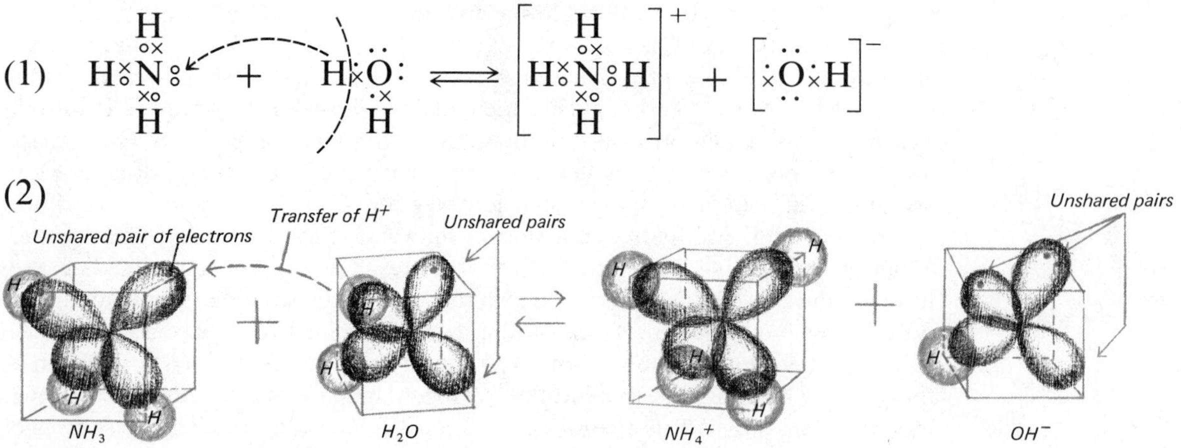

(1)

(2)

FIGURE 14–5

The ionization of ammonia in water (1). The reaction of ammonia (NH_3) with water produces the ammonium ion (NH_4^+) and the hydroxyl ion (OH^-). The reaction involves a transfer of a hydrogen ion from the H_2O molecule to the NH_3 molecule (2).

and the proton may pass quite readily from one molecule of water to another. Both the hydronium ions and chloride ions resulting from the ionization of hydrogen chloride in water are undoubtedly hydrated in solution as are all other ions. All molecular acids are polar substances and ionize in solution in the same manner as described for hydrogen chloride. It has been shown by x-ray analysis that crystalline perchloric acid monohydrate ($HClO_4 \cdot H_2O$) is actually ionic and that the two ions are H_3O^+ and ClO_4^-.

Many substances which are composed of molecules dissolve in water as hydrated molecules rather than as hydrated ions. In many instances, however, a fraction of the hydrated molecules undergo ionization. For example, ammonia (NH_3), which consists of polar molecules, dissolves extensively in water as hydrated ammonia molecules. Ionization takes place to a slight degree under ordinary conditions. Figure 14–5 shows electronic formulas for this reaction and also the spatial arrangement in terms of orbital diagrams. A proton is transferred from a water molecule to an ammonia molecule forming ammonium ions and hydroxide ions. The tendency to combine with protons is a characteristic of a class of compounds referred to as **bases.** Because the OH^- holds the proton more strongly than NH_3, the reaction proceeds only to the extent of 1.34 per cent in a 0.1 M solution of the ammonia at 25° C. This small degree of ionization classifies ammonia as a weak base.

The tendency to release protons to the solution, with subsequent hydration to form H_3O^+, is a characteristic of a class of compounds referred to as **acids.** Acids such as acetic (CH_3COOH), nitrous (HNO_2), and hydrocyanic (HCN) are also soluble in water through the hydration process, but, at any one time, only a small fraction of their hydrated polar molecules undergoes ionization. Hence, they are classified as weak acids.

$$CH_3COOH + H_2O \rightleftharpoons H_3O^+ + CH_3COO^-$$
$$HNO_2 + H_2O \rightleftharpoons H_3O^+ + NO_2^-$$
$$HCN + H_2O \rightleftharpoons H_3O^+ + CN^-$$

Certain nonpolar inorganic compounds readily dissolve in water by forming hydrated molecules. Among these compounds are the halides and cyanides of mercury, cadmium, and zinc. Molecules of these compounds have no dipole moments even though the bonds are of the polar covalent type. Their centers of positive and negative electric charge coincide because of a high degree of molecular symmetry, as in the linear (straight line) molecule, Cl—Hg—Cl. There are, however, both a net positive electric charge on the mercury atom and net negative charges on the chlorine atoms of $HgCl_2$. Water dipoles are attracted to these points of electric charge concentration and the $HgCl_2$ molecule becomes hydrated. A small fraction of the bonds between mercury and chlorine breaks, with the formation of a few hydrated mercury(II) ions and hydrated chloride ions. Compounds of this type are classed as weak electrolytes.

Ionization, then, may be defined as an interaction between a solute and solvent which results in the formation of ions. Sodium chloride is ionic in the solid state and dissolution in water merely brings about a separation of the ions. Pure hydrogen chloride is molecular, and reaction with water results in the formation of ions.

14.10   The Effect of Electrolytes on the Colligative Properties of Solutions

In Chapter 13 we learned that the effect of nonelectrolytes on the colligative properties of a solvent (vapor pressure, boiling point, freezing point, and osmotic pressure) is dependent only upon the number, and not on the kind, of particles dissolved. For example, one mole of any nonelectrolyte in solution in 1,000 g of water (a one-molal solution) produces the same lowering of the freezing point, 1.86° C, as one mole of any other nonelectrolyte because one mole of any non-electrolyte contains the same number of molecules (6.022×10^{23}). However, the molal lowering of the freezing point produced by electrolytes is much greater than for nonelectrolytes. The water in a solution which contains one mole of sodium chloride dissolved in 1,000 g of water freezes at −3.37°. This lowering of the freezing point of water is $3.37 \div 1.86 = 1.81$ times the molal depression for a nonelectrolyte. This illustrates the fact that one mole of an electrolyte produces more than 6.022×10^{23} solute particles. All acids, bases, and salts behave in this fashion when placed in solution.

14.11   Ion Activities

In 1923, Peter J. W. Debye and Erich Hückel accepted the idea that strong electrolytes completely ionize in aqueous solution, and they proposed a theory to explain the *apparent* incomplete ionization of strong electrolytes. If a strong electrolyte, such as sodium chloride, is completely dissociated in aqueous solution, it should lower the freezing point of water and raise the boiling point twice as much as an equal molal concentration of a nonelectrolyte, since it gives two ions per mole. However, a mole of sodium chloride lowers the freezing point of water only 1.81 times as much as a mole of a nonelectrolyte does, instead of twice as much. A similar discrepancy occurs in the boiling-point elevation. Debye and Hückel accounted for these discrepancies between calculated and observed values for the colligative properties of solutions of electrolytes by the following theory.

Although the forces of interionic attraction in aqueous solution are very greatly reduced by hydration of the ions and the insulating action of the polar solvent, they are not completely nullified. The **residual interionic forces of attraction** prevent the ions from behaving as totally independent particles insofar as the colligative properties of the solution or their function as carriers of the electric current are concerned (Fig. 14–6). Thus, the **activities** of the ions as independent

FIGURE 14–6
Diagrammatic representation of the various species thought to be present in a water solution of potassium chloride, which would explain unexpected values for its colligative properties.

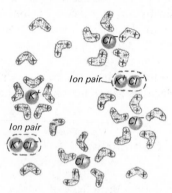

units, or their "effective concentrations," are less than indicated by their actual concentrations. The more dilute the solution, the greater the separation of oppositely charged ions and the less the residual interionic attraction. Thus, in extremely dilute solutions the effective concentrations of the ions (activities) are essentially equal to the actual concentrations.

The factor by which the actual ion concentration must be multiplied to obtain the activity of the ion is called the **activity coefficient.**

$$\text{Activity} = f \times \text{concentration}$$

The activity coefficient, f, approaches unity with increasing dilution. For a 0.100 molal solution of sodium chloride, the activity coefficient has a value of 0.778 at 25.0° C, and the activity of the ions is 0.0778 molal.

$$a = f \times \text{concentration}$$
$$a = 0.778 \times 0.100 = 0.0778 \text{ molal}$$

In solutions of weak electrolytes the concentrations of the ions are small, and the interionic forces are so slight that the activities of the ions are essentially equal to their concentrations.

14.12   Summary of the Modern Theory of Electrolytes

The modern concepts of electrolytes may be summarized as follows:

(1) Ionic compounds are completely ionized in the solid state and in solution.

(2) In aqueous solution the ions of an electrolyte are hydrated (in many other solvents, ions of an electrolyte are similarly *solvated*) and the electrostatic force of attraction between ions of opposite charge is so reduced that the charged ions move about as independent particles except for a slight residual attraction between them.

(3) The residual force of attraction between ions of opposite charge in solution is responsible for the "activity," or "effective concentration," being less than the "actual concentration" of the ions. The residual force of attraction between the ions decreases upon dilution, because the ions are on the average further apart.

(4) Acids ionize by reacting with water to form hydronium ions and acid anions of the acids.

(5) Weak electrolytes react with water to only a very limited extent and are thus only slightly ionized.

Colloid Chemistry

In Chapter 13, solutions were characterized by their homogeneity, absence of settling, and the molecular or ionic state of subdivision of their components. We shall now consider dispersions of particles somewhat larger than molecules and ions, yet not large enough to be seen under an ordinary microscope.

14.13   Colloidal Matter

When powdered starch is heated with water, a mixture is obtained which is not homogeneous, yet the particles of insoluble starch do not settle out but remain in suspension indefinitely. Such a system is called a **colloidal dispersion;** the finely divided starch is called the **dispersed phase,** and the water is called the **dispersion medium.**

The term **colloid**—from the Greek word *kolla*, meaning glue, and *eidos*, meaning like—was first used in 1861 by Thomas Graham to classify substances which usually exist in an amorphous, or gelatinous, condition such as starch and gelatin. It is now known that the properties characterized as colloidal are not peculiar to a special class of substances, but that any substance may be obtained in the colloidal form if suitable means are employed. The sizes of many single large molecules, such as synthetic polymers and proteins, place them in the category of colloids.

Colloidal matter is not limited to dispersions of solid particles in liquid media; a gas or a solid may also be the dispersion medium, and the dispersed phase may be either a gas, a liquid, or a solid. A classification of colloidal systems is given in Table 14-4. A gas dispersed in another gas is not a colloidal system because the particles are of molecular dimensions. The most important colloidal systems are those involving a solid dispersed in a liquid, and such systems are called **sols.**

TABLE 14–4   Colloidal Systems

Dispersed Phase	Dispersion Medium	Examples	Common Name
Solid	Gas	Smoke, dust	Solid aerosol
Solid	Liquid	Starch suspension, some inks, paints, milk of magnesia	Sol
Solid	Solid	Colored gems, some alloys	Solid sol
Liquid	Gas	Clouds, fogs, mists, sprays	Liquid aerosol
Liquid	Liquid	Milk, mayonnaise, butter	Emulsion
Liquid	Solid	Jellies, gels, opal (SiO_2 and H_2O), pearl ($CaCO_3$ and H_2O)	Solid emulsion
Gas	Liquid	Foams, whipped cream, beaten egg whites	Foam
Gas	Solid	Pumice, floating soaps	

14.14   Size of Colloidal Particles

Usually, from a thousand to a billion atoms are present in a colloidal particle. It is customary to express the dimensions of colloidal particles in terms of the **millimicron** ($m\mu$), which is one-millionth of a millimeter, or 10^{-7} cm. Many colloidal particles have at least one dimension between 1 $m\mu$ and 200 $m\mu$.

Colloidal particles are usually aggregates of hundreds, or even thousands, of molecules. However, some colloids consist of single, well-defined molecules, with definite molecular shape, permitting them to be packed together in a crystalline array. The viruses are giant molecules with molecular weights ranging from several hundred thousand up to billions. Tobacco mosaic virus has a molecular weight of about 40,000,000; proteins and synthetic polymers have weights in the range of a few thousand to many million. Virus molecules have the remarkable power of reproduction. Some diseases, such as the common cold, poliomyelitis, and measles, are caused by viruses. Such diseases result from the formation of a large number of virus molecules from a few, when they are in a suitable environment in the human body.

The **electron microscope** has made it possible to see and photograph virus molecules, all of which are too small to be seen with an ordinary optical microscope. Beams of electrons are used in place of beams of light in the electron microscope. Its magnifying power is about 200,000 as compared with about 2,000 for the ordinary microscope.

14.15   Preparation of Colloidal Systems

The preparation of a colloidal system is accomplished by producing particles of colloidal dimensions and distributing these particles through the dispersion medium. Particles of colloidal size are formed: (1) by **dispersion methods,** i.e., the subdivision of larger particles or masses, or (2) by **condensation methods,** i.e., growth from smaller units, such as molecules or ions.

(1) Dispersion can be accomplished by grinding large particles in special mills designed for this purpose, called colloid mills. Paint pigments are produced by this method. Colloidal systems are sometimes formed by grinding a solid substance in a liquid.

A few solid substances, when brought into contact with water, disperse spontaneously to form colloidal systems. Gelatin, glue, and starch behave in this manner, and are said to undergo **peptization.** The particles are already of colloidal size; the water simply disperses them. Some atomizers produce colloidal dispersions by a spraying process. Powdered milk of colloidal particle size is produced by dehydrating milk spray.

An **emulsion** may be prepared by shaking together two liquids which are immiscible. Agitation breaks one liquid into droplets of colloidal dimensions, which then disperse throughout the mass of the other liquid. The droplets of the dispersed phase, however, tend to coalesce, forming drops which are too large to remain in the dispersed condition, and separation of the liquids into two layers follows. Therefore, emulsions must usually be stabilized by the addition of **emulsifying agents.** These emulsifying agents may either decrease the surface tension of the two liquids and thereby reduce the tendency of the tiny droplets to coalesce and form drops, or they may form protecting layers, or films, around the droplets preventing the formation of large drops. The addition of a little soap will stabilize an emulsion of kerosene in water.

Milk is an emulsion of droplets of butterfat in water, with casein acting as the emulsifying agent. Mayonnaise is an emulsion of olive oil in vinegar, with egg yolk serving as the emulsifying agent. Oil spills in the ocean are particularly difficult to clean up because of the formation of an emulsion between the oil and the water.

(2) Condensation methods involve the formation of colloidal particles by causing smaller particles to aggregate. If the particles grow beyond the colloidal range, no colloidal system will result and larger aggregates (**precipitates**) are formed.

Condensation methods usually employ chemical reactions. A dark red colloidal suspension of iron(III) hydroxide may be prepared by mixing a concentrated solution of iron(III) chloride with hot water.

$$Fe^{3+} + (3Cl^-) + 6H_2O \longrightarrow Fe(OH)_3 + 3H_3O^+ + (3Cl^-)$$

A colloidal suspension of arsenic(III) sulfide is produced by the reaction of hydrogen sulfide with arsenic(III) oxide dissolved in water.

$$As_2O_3 + 3H_2S \longrightarrow As_2S_3 + 3H_2O$$

The formation of a colloidal gold sol is accomplished by the reduction of a very dilute solution of gold chloride by such reducing agents as formaldehyde, tin(II) chloride, or iron(II) sulfate ($Au^{3+} + 3e^- \longrightarrow Au$). Some gold sols prepared by Faraday in 1857 are still perfectly clear after 119 years.

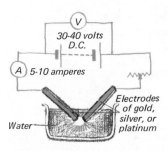

FIGURE 14-7
Diagram of the Bredig arc for preparing metallic dispersions in water.

Colloidal metals are often prepared by the condensation of metal vapors, which are produced by striking an electric arc between two electrodes of the metal. When the arc is under a liquid, the vapor condenses to particles of colloidal size (Fig. 14–7). Metals below hydrogen in the activity series, such as copper, silver, and gold, may be dispersed by striking the arc under water; metals above hydrogen are dispersed by the same process in certain liquids other than water.

The formation of synthetic polymers and protein colloids from smaller molecules is described in Chapters 27 and 28.

14.16   Detergents and Their Cleansing Action

Although the term detergent means cleansing agent and includes soap, it is now commonly used to refer to soap substitutes. Soaps are made by boiling either fats or oils with a strong base such as sodium hydroxide. (Pioneer women made soap by boiling fat with a strong basic solution made by leaching potassium carbonate (K_2CO_3) from wood ashes with hot water.) When animal fat is treated with sodium hydroxide, glycerol and sodium salts of the fatty acids (palmitic, oleic, and stearic) are formed. The sodium salt of stearic acid, sodium stearate, has the formula $C_{17}H_{35}COONa$. The cleansing action of soap depends in part upon its ability to change readily from a solid gel to a dilute colloidal suspension. The dispersed particles of soap in water adsorb fine particles of dirt and form a protective film about them. In this way the dirt particles are held in suspension and are washed away. Soap also acts as an emulsifying agent in forming stable emulsions of oils and greases in water.

Many detergents, or soap substitutes, are in common use. Each of these compounds contains a hydrocarbon chain or ring structure group which is nonpolar, such as $C_{12}H_{25}—$, or $C_{12}H_{25} \cdot C_6H_4—$; and a polar group, such as a sulfate, $—OSO_3Na$, or a sulfonate, $—SO_3Na$. Whereas soaps form insoluble calcium and magnesium compounds in hard water, these detergents form water-soluble products—a definite advantage for detergents. Numerous detergents with familiar commercial names are available in stores throughout many countries of the world.

The cleansing action of soaps and detergents can be explained in terms of the structures of the molecules involved. The molecules of both soaps and synthetic detergents consist of long hydrocarbon chains attached to a polar group.

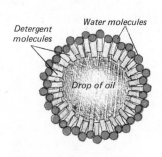

FIGURE 14-8
Diagrammatic cross-section of an emulsified drop of oil in water, with soap as the emulsifier. The negative ions of the soap are oriented at the interface between the oil particle and water. The nonpolar hydrocarbon end of the ion is in oil, and the polar carboxyl (COO—) end is in water.

$$CH_3(CH_2)_{15}CH_2\overset{-}{C}OO\overset{+}{N}a \qquad CH_3(CH_2)_{10}CH_2\overset{-}{O}SO_3\overset{+}{N}a$$
Sodium stearate (soap)   Sodium lauryl sulfate (detergent)

The hydrocarbon end of such a molecule is attracted by the dirt, oil, or grease particles and the polar group is attracted by the water (Fig. 14–8). The result is an orientation of the molecules of the cleansing agent at the interface between the dirt particles and the water in such a way that the **interfacial tension** is lowered. This enables the dirt particles to become suspended in the solution as colloidal particles, in which form they are readily washed away. Substances which lower the surface tension of liquids are sometimes called **wetting agents.**

The colloidal properties of detergents and soaps just described are, of course, useful properties. Such properties can work to a disadvantage also. As detergent use increased rapidly during the past twenty-five years, a foaming problem in streams and in sewage treatment plants developed. The foam, a colloidal system of detergent molecules dispersed in water, clogged pumps, overflowed tanks, and contaminated streams. The detergents used during that time were not **biodegradable,** that is, not decomposed by microorganisms and bacteria present in sewage treatment plants and in natural bodies of water. Inasmuch as sewage plants were unable to remove more than one-third to one-half of the detergents in the water before discharge to lakes and streams, the foam showed up in public water supplies downstream. The problem, which was first noted about 1947, became so acute by 1963 that major reformulation of detergents was required. The detergents being made at that time were largely alkyl benzene sulfonates (ABS), illustrated by the following formula in which the alkyl group (R) is a branched chain of carbon and hydrogen atoms.

$$
\begin{array}{c}
R \\
| \\
H-C-CH_3 \\
| \\
C \\
\diagup \diagdown \\
C \quad\quad C \\
| \quad\quad | \\
C \quad\quad C \\
\diagdown \diagup \\
C \\
| \\
SO_3^- \; Na^+
\end{array}
$$

An alkyl benzene
sulfonate (ABS)

Bacteria feeding on such molecules could not cope with either the branching or the ring structure, and decomposition of the molecule did not occur. Fortunately, chemists found that by using unbranched, or linear, hydrocarbon chains attached to the sulfonate (SO_3^-) group, they could produce detergents that are biodegradable. These new detergents, which are linear alkyl sulfonates (LAS), are effectively decomposed in sewage treatment plants and in natural bodies of water by microorganisms and bacteria. Hence, since about 1965, all major detergent manufacturers have changed over from ABS to LAS detergent formulations with the result that the colloidal foaming problem in water has been largely eliminated.

A current detergent problem, not related to colloids, resulting from additives such as phosphates is discussed in the section on water pollution (Section 12.4).

14.17   The Brownian Movement

When a beam of intense light is passed through a colloidal system in a darkened space and viewed with a powerful microscope against a black background and at right angles to the beam of light, the colloidal particles are seen as tiny bright flashes of light. This arrangement of microscope and light is known as the **ultramicroscope.** The colloidal particles, as seen in the ultramicroscope, appear to be in a state of irregular rapid, dancing motion (Fig. 14–9). This motion is called

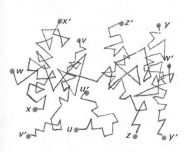

FIGURE 14–9

A diagram showing how particles in suspension may move due to unequal bombardment by molecules of the dispersion medium.

the **Brownian movement** after the botanist Robert Brown, who first observed it in 1828. He could not explain it, but now we know that the colloidal particles are small enough that bombardment by molecules of the dispersion medium gives them an irregular motion. Any particle in suspension will be bombarded on all sides by the moving molecules of the dispersion medium. For particles larger than colloidal size the bombardment force is not sufficient to impart motion, especially inasmuch as bombardment on one side of the particle is likely to be counter-balanced by an equal force on the opposite side. For particles of usual colloidal size, however, the force of bombardment is sufficient in comparison to the mass of the particle to impart momentum, and the probability of equal and simulta-neous bombardments on opposite sides of the colloidal size particle is slight. Hence, the irregular motion results. The Brownian movement explains the absence of settling of the dispersed particles in colloidal systems, even though these particles may be more dense than the medium in which they are dispersed. The kinetic-molecular theory of matter received one of its earliest confirmations as a result of studies of the Brownian movement.

14.18 Electrical Properties of Colloidal Particles

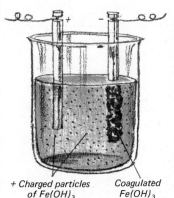

+ Charged particles   Coagulated
  of Fe(OH)₃   Fe(OH)₃

FIGURE 14–10

Colloidal iron(III) hydroxide is coagulated in an electrolytic cell.

One of the most important properties of dispersed colloidal particles is that they are usually electrically charged. When an iron(III) hydroxide sol is placed in an electrolytic cell (Fig. 14–10), the dispersed particles move to the negative electrode. Because opposite charges of electricity attract, this is good evidence that the iron(III) hydroxide particles are positively charged. At the cathode, the particles lose their charge and the colloidal particles coagulate as a precipitate. All particles in any one colloidal system have the same kind of charge with respect to sign. The similar charges on the particles help keep them dispersed, because like charges repel each other. Most hydroxides of metals have positive charges, while most sulfides of metals and the metals themselves form negatively charged colloidal dispersions.

The charges on colloidal particles result from the adsorption of ions which exist in the dispersion medium. If a colloidal particle preferentially adsorbs positive ions, it acquires a positive charge; if it preferentially adsorbs negative ions, it acquires a negative charge. Thus, iron(III) hydroxide particles become positively charged because of a preferential adsorption of iron(III) ions (Fe^{3+}), when $FeCl_3$ hydrolyzes in hot water. Arsenic(III) sulfide (As_2S_3) particles preferentially adsorb sulfide ions (S^{2-}), resulting from the ionization of H_2S, and become negatively charged.

The study of the migration of colloidal particles in an electric field, **electro-phoresis,** is a powerful tool for investigating colloidal electrolytes such as proteins.

14.19 Precipitation of Colloidal Dispersions

When a finely divided precipitate is washed, the electrolyte which has caused coagulation may be removed in the wash water, and sometimes the precipitate will again pass into colloidal solution by peptization. Peptization may be prevented by adding an electrolyte to the wash water so that when the original flocculating

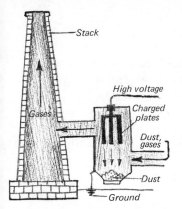

FIGURE 14-11

The principle of a Cottrell precipitator. Positively and negatively charged particles in smoke are precipitated as they pass over the electrically charged plates.

ion is dissolved, it is replaced by an ion of the electrolyte in the wash water. Because ammonium nitrate decomposes when heated and therefore can be removed readily, this salt is often added to the water used in washing sulfide precipitates.

Colloidal clay or silt particles in a river are negatively charged by the adsorption of negative hydroxide ions. When the river water reaches the salty ocean water, the negatively charged particles are neutralized by the positive sodium and magnesium ions of the sea water, and the clay and silt particles precipitate. Enough precipitate accumulates during thousands of years to form deltas at the mouths of rivers.

The carbon and dust particles in the smoke from furnace fires are often colloidally dispersed and electrically charged. A process for precipitating smoke particles was developed by Cottrell, an American chemist, to lessen the smoke nuisance in industrial centers. In this process the charged particles are attracted to highly charged electrodes, where they are neutralized and deposited as dust (Fig. 14–11). The process is also important in the recovery of many valuable products that would otherwise escape from the flues of smelters, furnaces, and kilns.

14.20   Protective Colloids

The coagulation of the dispersed phase in some colloidal systems may be prevented by the presence of another colloid. The second colloid forms a protecting film around the dispersed particles and prevents them from coalescing. Gelatin, dextrin, and gum arabic are the most familiar and widely used **protective colloids.** The gelatin on photographic plates and films maintains the colloidal dispersion of the silver halides by serving as a protective colloid. Gum arabic is used in the stabilization of certain inks.

14.21   The Adsorption Phenomenon

One of the properties of the surface of any substance is its specific ability to hold other substances to itself. This phenomenon is called **adsorption.** The atoms, molecules, or ions which compose a surface differ from those in the interior of the substance in that they are surrounded only on one side by like particles that possess equal and similar cohesive forces. The particles within the body, however, are surrounded on *all* sides by particles having equal attractive forces. The surface particles, therefore, have some attractive forces which are not satisfied. These unsatisfied forces are called **residual forces,** and they are largely responsible for adsorption.

When matter is subdivided to the extent that its particles become colloidal in dimensions, there is a tremendous increase in the surface area exposed to the surrounding medium. A cube 1 cm along each edge has a surface area of 6 cm^2. If this cube is subdivided until the particles are within the colloidal range of 10 mμ on an edge, the total surface area becomes 6,000,000 cm^2, or approximately the surface area of a square 80 feet on each side. Because colloidal particles have a very large surface area for a given volume of material, they are good adsorbers. In fact, colloidal chemistry is largely the chemistry of surface effects. Hence, the principles of colloid chemistry apply to films and to filaments, both of which have a very large surface area for a given volume of material. Dispersed solids, such as charcoal, adsorb vast quantities of gas. As pointed out in Section 14.18, colloidal dispersions of arsenic(III) sulfide and iron(III) hydroxide are stabilized by adsorption of ions from solution. Furthermore, emulsions are stabilized by adsorption of an emulsifying agent upon the surface of the droplets of the dispersed phase (Section 14.15).

14.22   Gels

Under certain conditions, the dispersed phase in colloidal systems coagulates in a manner such that the whole mass, including the liquid, sets to an extremely viscous gelatinous body known as a **gel.** For example, a hot aqueous "solution" of gelatin sets to a gel upon cooling. Because the formation of a gel is accompanied by the taking up of water or some other solvent, the gel is said to be hydrated or solvated. Apparently the fibers of the dispersed substance form a complex three-dimensional network, the interstices being filled with the liquid medium or a dilute solution of the dispersed phase.

A carbohydrate known as **pectin,** from fruit juices, is a gel-forming substance which is important in jelly making. Silica gel, a colloidal dispersion of hydrated silicon dioxide called **water glass,** is formed when dilute hydrochloric acid is added to a dilute solution of sodium silicate. Canned heat is a gel made by mixing alcohol and a saturated aqueous solution of calcium acetate. The wall of the living cell is colloidal in character, and within the cell there is a jellylike dispersion. In fact, all living tissue is colloidal, and the various life processes—nutrition, digestion, secretion—are largely those concerned with the chemistry of the colloidal state. The concentration of low grade ores by flotation processes (Section 31.4) is an important industrial application of colloid chemistry.

QUESTIONS

1. Distinguish between strong and weak electrolytes.
2. Electrolytes include what three principal classes of compounds?
3. Define the following: electrolyte, nonelectrolyte, and electrolysis.
4. Account for the electrical neutrality of electrolytic solutions.
5. Why do not crystalline ionic compounds conduct electricity, even though they are 100% ionized?
6. Why does a solution of hydrogen chloride gas in water conduct an electric current, although pure hydrogen chloride and pure water are covalent materials?
7. Why are ionic compounds that are fused (melted) good conductors, whereas in the solid state they are nonconductors?
8. How do the properties of the sodium ion differ from those of the sodium atom? How do the two differ in electronic structure?
9. Describe the chemical processes that take place when an ionic substance dissolves in water.

10. Why do not ionic compounds dissolve to an appreciable extent in such solvents as benzene or carbon tetrachloride?

11. Relate the solubility of an ionic compound in water to (a) the magnitude of the crystal forces and (b) the energy of hydration of the ions.

12. Account for the "abnormal" molal lowering of the freezing point of water produced by electrolytes.

13. If sodium chloride is completely dissociated in aqueous solution, why does it not give twice the molal depression of the freezing point of water that nonelectrolytes do?

14. What electrolyte is present in automobile storage batteries?

15. The approximate radius of a water molecule, assuming a spherical shape, is 1.40 angstrom units (Å). (1 Å $= 10^{-8}$ cm.) Assume that water molecules cluster around metal ions in solution so that the water molecules essentially touch both the metal ion and each other. On this basis, and assuming that 4, 6, 8, and 12 are the only possible coordination numbers, what is the maximum number of water molecules which can hydrate each of the following ions: Mg^{2+} (0.65 Å, radius); Al^{3+} (0.50 Å); Rb^+ (1.48 Å); and Sr^{2+} (1.13 Å)?

16. Distinguish between dispersion methods and condensation methods for preparing colloidal systems.

17. How do colloidal "solutions" differ from true solutions with regard to dispersed particle size and homogeneity?

18. Identify the dispersed phase and the dispersion medium in each of the following colloidal systems: starch dispersion, smoke, fog, pearl, whipped cream, floating soap, jelly, milk, and ruby.

19. Explain the cleansing action of soap.

20. How can it be demonstrated that colloidal particles are electrically charged? How do the particles acquire charges?

21. What is the structure of a gel?

22. Explain the phenomenon of Brownian movement. What is its relationship to the stability of the dispersed particles of a colloidal system?

23. What is the function of a protective colloid?

24. Why cannot colloidal systems be separated from electrolytes by filtration?

25. Explain the formation of river deltas.

26. Ammonium nitrate is often added to freshly precipitated metal sulfides used in analysis, and such precipitates are often later washed with a solution of an electrolyte. Explain.

PROBLEMS

[s]1. If 200 ml of 0.200 M NaCl, 300 ml of 0.100 M Na_2SO_4, 100 ml of 0.500 M K_2SO_4 and 200 ml of 0.300 M $NaNO_3$ are mixed, what is the molarity of each ion in the resulting solution? Assume complete dissociation of the substances.

Ans. 0.200 M in Na^+; 0.0500 M in Cl^-; 0.100 M in $SO_4{}^{2-}$;
0.125 M in K^+; 0.0750 M in $NO_3{}^-$

[s]2. What would be the approximate freezing point of a 0.42-molal aqueous solution of cesium iodide?   *Ans.* $-1.6°$ *C*

⒮3. A 0.500-molal aqueous solution of an acid HA shows a freezing point of −1.04° C. What is the per cent dissociation of this acid? *Ans. 11.8%*

⒮4. A solution is made by dissolving 0.498 g of potassium bromide in 100.0 g of water. The solution is observed to freeze at −0.150° C. If one assumes that the activity coefficient of the potassium ion is equal to that of the bromide ion, what is the activity coefficient of the potassium ion under these conditions? *Ans. 0.964*

5. Formic acid, HCOOH, is 2.93% ionized in a 0.2000 *M* solution of the acid and 8.98% ionized in a 0.02000 *M* solution. Calculate the molar concentration of the molecular acid and that of each of the ions in (a) 0.2000 *M* solution, and (b) 0.02000 *M* solution.
   Ans. (a) 0.194 M; 5.86 × 10⁻³ M; 5.86 × 10⁻³ M (b) 1.82 × 10⁻² M; 1.80 × 10⁻³ M; 1.80 × 10⁻³ M

⒮6. The activity coefficient for the ions in 0.100 molal HI is 0.818 at 25.0° C. Calculate the activities of these ions. *Ans. 0.0818 molal*

7. The activity of the ions in a 0.30 molal solution of RbBr at 25.0° C is 0.20 molal. Calculate the activity coefficient for the ions. *Ans. 0.67*

8. Calculate (a) the surface area of a cube for which each side is 1 cm long. Then calculate the total surface area if the cube is subdivided into cubes (b) 0.1 cm along each edge, (c) 0.01 cm along each edge, (d) 0.001 cm along each edge, and (e) 10 mμ along each edge.
   Ans. (a) 6 cm²; (b) 60 cm²; (c) 600 cm²; (d) 6,000 cm²; (e) 6,000,000 cm²

(Notice that, although the total volume remains the same, the surface area increases as the particle size becomes smaller.)

REFERENCES

"Low Energy Electrons Diffraction and Auger Electron Spectroscopy," C. S. McKee and M. W. Roberts, *Chem. in Britain*, **6**, 106 (1970).

"The History of Colloid Science," E. A. Hauser, *J. Chem. Educ.*, **32**, 2 (1955).

"Syndets and Surfactants," F. D. Snell and C. T. Snell, *J. Chem. Educ.*, **35**, 271 (1958).

"Atoms Visualized," E. W. Müller, *Sci. American,* June, 1957; p. 113.

"Poliomyelitis Virus," F. L. Schaffer, *J. Chem. Educ.*, **36**, 469 (1959).

"How Giant Molecules Are Measured," Peter Debye, *Sci. American*, Sept., 1957; p. 90.

"The Teaching of Colloid and Surface Chemistry," *J. Chem. Educ.*, **39**, 166–195 (1962).

"Brownian Motion and Potential Theory," R. Hersh and R. J. Griego, *Sci. American*, March, 1969; p. 67.

"Physical Adsorption—A Tool in the Study of the Frontiers of Matter," A. W. Adamson, *J. Chem. Educ.*, **44**, 710 (1967).

"Monomolecular Layers and Light," K. H. Drexhage, *Sci. American*, March, 1970; p. 108.

"A High-Resolution Scanning Electron Microscope," A. V. Crewe, *Sci. American*, April, 1971; p. 26.

"The Current State of Colloid Science," J. Th. G. Overbeek, *Chem. in Britain*, **8**, 370 (1972).

"Brownian Movement and Molecular Reality Prior to 1900," M. Kerker, *J. Chem. Educ.*, **51**, 764 (1974).

Acids, Bases, and Salts

15

Acids and bases are among the most important of the numerous classes of chemical compounds. Because these two classes of substances are in effect opposite each other in character and yet related through their ability to neutralize one another, it is convenient and instructive to consider them together.

The terms "acid" and "base" have been defined in a number of ways, making a brief review of the historical development of the acid-base concept of interest.

15.1 History of the Acid-Base Concept

The first significant characterization of acids and bases was made by Boyle in 1680. He noted that acids dissolve many substances, that they change the color of certain natural dyes (for example, litmus from blue to red), and that they lose these characteristic properties after coming in contact with alkalies. In the eighteenth century it was recognized that acids have a sour taste, that they react with limestone with the liberation of a gaseous substance (CO_2), and that neutral substances result from their interaction with alkalies. Lavoisier, in 1787, proposed that acids are binary compounds of oxygen, and he considered oxygen to be responsible for the acidic properties of this class of substances. The essentiality of oxygen was disproved by Davy in 1811 when he showed that hydrochloric acid contains no oxygen. Davy contributed greatly to the development of the acid-base concept by concluding that hydrogen, rather than oxygen, is the essential constituent of acids. In 1814, Gay-Lussac concluded that acids are substances which can neutralize alkalies and that these two classes of substances can be defined only in terms of each other. The ideas of Davy and Gay-Lussac provide the foundation for our modern concepts of acids and bases in water solution.

15.2 The Classical, or Arrhenius, Concept of Acids and Bases

Early in the study of elementary chemistry the student learns that acids are compounds that yield hydrogen ions (H^+) or, more properly, hydronium ions (H_3O^+) when dissolved in water, and that bases yield hydroxide ions (OH^-). The process of neutralization is then considered to be the union of hydrogen ions and hydroxyl ions to form the neutral water molecule, $H^+ + OH^- \longrightarrow H_2O$. These facts are the basis for the classical, or Arrhenius, acid-base concept. However, it is now found convenient to use the terms "acid" and "base" and "acid-base reactions" in many instances in which neither hydrogen ions, hydroxide ions, nor the union of these ions to form water is involved. Furthermore, many ions and molecules possess basic properties comparable to those of the hydroxide ion, and reactions involving salt formation occur in many solvents other than water and even in the absence of a solvent.

We shall consider first some of the properties of hydrogen acids (often referred to as protonic acids), hydroxide bases, and salts in water solution. Then we shall proceed to a consideration of some of the very useful more generalized concepts of acid-base behavior.

Protonic Acids, Hydroxide Bases, and Salts in Aqueous Solution

15.3 Properties of Protonic Acids in Aqueous Solution

The properties which are common to protonic acids in aqueous solution are those of the hydronium ion, formed when the acid dissolves in water. Although the hydronium ion is actually formed, the symbol H^+ is often used instead of H_3O^+ for the sake of convenience. It is important to remember, however, that H^+ is an abbreviated symbol and that actually the hydrogen ion is always hydrated in water solution. Even the H_3O^+ symbol is in a strict sense an abbreviation. For example, there is good recent evidence for the existence of $H_5O_2^+$, corresponding to $H^+ \cdot (H_2O)_2$.

Protonic acids in aqueous solution exhibit the following properties:

(1) They have a sour taste.

(2) They change the color of certain indicators. For example, they change litmus from blue to red, and phenolphthalein from red to colorless.

(3) They react with metals above hydrogen in the electromotive series with the liberation of hydrogen.

$$2H^+ + Zn \longrightarrow Zn^{2+} + H_2\uparrow$$

(4) They react with oxides and hydroxides of metals, forming salts and water.

$$2H^+ + (2Cl^-) + FeO \longrightarrow Fe^{2+} + (2Cl^-) + H_2O$$
$$2H^+ + (2Cl^-) + Fe(OH)_2 \longrightarrow Fe^{2+} + (2Cl^-) + 2H_2O$$

(5) They react with salts of either weaker or more volatile acids, such as carbonates or sulfides, to give a new salt and a new acid.

$$2H^+ + (2Cl^-) + CaCO_3 \longrightarrow H_2CO_3 + Ca^{2+} + (2Cl^-)$$
$$\longrightarrow H_2O + CO_2\uparrow$$

$$2H^+ + (2Cl^-) + FeS \longrightarrow H_2S\uparrow + Fe^{2+} + (2Cl^-)$$

(6) Their aqueous solutions conduct an electric current, because they contain ions; they are electrolytes.

15.4 Formation of Protonic Acids

Protonic acids may be formed by one or more of the following methods:

(1) *By the direct union of their elements.*

$$H_2 + Cl_2 \longrightarrow 2HCl$$
$$H_2 + S \longrightarrow H_2S$$

(2) *By the action of water on oxides of nonmetals.*

$$CO_2 + H_2O \longrightarrow H_2CO_3$$
$$SO_3 + H_2O \longrightarrow H_2SO_4$$
$$P_4O_{10} + 6H_2O \longrightarrow 4H_3PO_4$$

(3) *By heating salts of volatile acids with nonvolatile or slightly volatile acids.*

$$NaCl + H_2SO_4 \longrightarrow NaHSO_4 + HCl\uparrow$$
$$NaBr + H_3PO_4 \longrightarrow NaH_2PO_4 + HBr\uparrow$$

(4) *By the action of salts with other acids producing a precipitate.*

$$(H^+) + Cl^- + Ag^+ + (NO_3^-) \longrightarrow AgCl + (H^+) + (NO_3^-)$$
$$(2H^+) + SO_4^{2-} + Ba^{2+} + (2ClO_3^-) \longrightarrow BaSO_4 + (2H^+) + (2ClO_3^-)$$

(5) *By hydrolysis.*

$$PBr_3 + 3H_2O \longrightarrow H_3PO_3 + 3HBr$$
$$PCl_5 + 4H_2O \longrightarrow H_3PO_4 + 5HCl$$

(6) *By oxidation and reduction.*

$$H_2S + I_2 \longrightarrow 2H^+ + 2I^- + S$$
$$2HNO_3 + 2SO_2 + H_2O \longrightarrow 2H_2SO_4 + NO\uparrow + NO_2\uparrow$$

15.5 Properties of Hydroxide Bases in Aqueous Solution

The most important and characteristic base in aqueous solution is the hydroxide ion. In common usage, the term base often refers to hydroxide base, although, as we shall see later, many other anions are also properly classified as bases.

The hydroxides of the alkali metals (e.g., potassium and sodium) are called alkali bases, and those of the alkaline earth metals (e.g., barium, strontium, and calcium) are known as alkaline earth bases. The hydroxides of sodium, potassium, and barium are referred to as strong bases, because they give a high concentration of hydroxide ions in aqueous solution. The hydroxides of the alkali metals are very soluble in water, those of the alkaline earth metals are moderately soluble, and those of all of the other metals are only sparingly soluble.

The properties which are common to hydroxide bases in aqueous solution are due to the presence of the hydroxide ion.

(1) They have a bitter taste.

(2) They change the colors of certain indicators. For example, they change litmus from red to blue, and phenolphthalein from colorless to red.

(3) They neutralize aqueous acids, forming salts and water.

$$(M^+) + OH^- + H^+ + (A^-) \longrightarrow (M^+) + (A^-) + H_2O$$

(4) They give aqueous solutions which conduct an electric current; they are electrolytes.

15.6   Formation of Soluble Hydroxide Bases

Hydroxide bases may be prepared by one or more of the following methods:

(1) *By the action of alkali metals or alkaline earth metals with water.*

$$2K + 2H_2O \longrightarrow 2K^+ + 2OH^- + H_2\uparrow$$
$$Ca + 2H_2O \longrightarrow Ca^{2+} + 2OH^- + H_2\uparrow$$

(2) *By the action of water on the oxides of alkali metals or alkaline earth metals.*

$$Na_2O + H_2O \longrightarrow 2Na^+ + 2OH^-$$
$$CaO + H_2O \longrightarrow Ca^{2+} + 2OH^-$$

(3) *By the action of salts with other bases by which a precipitate results.*

$$(2Na^+) + CO_3{}^{2-} + Ca^{2+} + (2OH^-) \longrightarrow \underline{CaCO_3} + (2Na^+) + (2OH^-)$$

(4) *By electrolysis.*

$$(2Na^+) + 2Cl^- + 2H_2O \xrightarrow{\text{Electrolysis}} (2Na^+) + 2OH^- + H_2\uparrow + Cl_2\uparrow$$

(5) *By dissolving ammonia in water.*

$$NH_3 + H_2O \rightleftharpoons NH_4{}^+ + OH^-$$

15.7   Acid-Base Neutralization

When aqueous solutions of hydrochloric acid and sodium hydroxide are mixed in the proper proportion a reaction takes place during which the acidic and basic properties disappear. We say that **neutralization** has occurred. The hydrogen ion, which is responsible for the acidic properties, has reacted with the hydroxide ion, which is responsible for the basic properties, producing neutral water. The sodium and chloride ions have undergone no chemical change and appear in the form of crystalline sodium chloride upon evaporation of the solution. Sodium chloride is an example of the class of compounds called salts.

$$H^+ + (Cl^-) + (Na^+) + OH^- \longrightarrow H_2O + \underset{\text{A salt}}{(Na^+) + (Cl^-)}$$

Because the only change that takes place is the reaction of the hydrogen and hydroxide ions, the neutralization may be represented simply as

$$H^+ + OH^- \longrightarrow H_2O$$

This equation may be used to represent the reaction of any hydroxide base with any protonic acid. The fact that the amount of heat evolved per mole of water produced is the same for *any* completely ionized acid and metal hydroxide is evidence that acid-base neutralizations involve only the hydrogen ion of the acid and the hydroxide ion of the base; neutralizations are independent of the acid anion and the base cation.

$$H^+ + (Cl^-) + (Na^+) + OH^- \longrightarrow H_2O + (Na^+) + (Cl^-) + 13,700 \text{ cal}$$
$$H^+ + (NO_3^-) + (Na^+) + OH^- \longrightarrow H_2O + (Na^+) + (NO_3^-) + 13,700 \text{ cal}$$
$$H^+ + (Cl^-) + (K^+) + OH^- \longrightarrow H_2O + (K^+) + (Cl^-) + 13,700 \text{ cal}$$
$$2H^+ + (2Cl^-) + (Ca^{2+}) + 2OH^- \longrightarrow 2H_2O + (Ca^{2+}) + (2Cl^-) + 27,400 \text{ cal}$$

In each of the preceding equations, the metal ion and the acid anion appear on both sides of the equation, showing that they do not enter the reaction. Accordingly, these ions may be omitted, and the same equation may be written for all the reactions.

$$H^+ + OH^- \longrightarrow H_2O + 13,700 \text{ cal}$$

In the event that either the acid or base is not completely ionized, the neutralization reaction does not go to completion, and the heat of neutralization will be less than 13,700 calories. For example, acetic acid, which is weak, does not completely neutralize an equivalent of sodium hydroxide; the reaction does not proceed to completion, and the heat of neutralization is only 13,400 calories.

$$CH_3COOH + (Na^+) + OH^- \rightleftharpoons H_2O + (Na^+) + CH_3COO^- + 13,400 \text{ cal}$$

If the salt which is formed is insoluble the heat evolved is *greater* than 13,700 calories per mole of water produced because of the energy evolved in the process of crystallization.

$$2H^+ + SO_4^{2-} + Ba^{2+} + 2OH^- \rightleftharpoons 2H_2O + BaSO_4 + 31,400 \text{ cal}$$

Note that the two moles of water account for 27,400 calories. The remainder of the 31,400 calories is the energy evolved in the crystallization of barium sulfate.

Acid-base neutralization pertains to the reaction which occurs when equivalent quantities of an acid and a base are mixed. This statement does not imply, however, that the resulting solution is always neutral. The nature of the particular acid and base involved in the reaction determines whether or not the resulting solution is neutral. The following equations illustrate some reactions which take place when equal numbers of equivalents of acids and bases are brought together in aqueous solution.

(1) *A strong acid plus a strong base gives a neutral solution.*

$$H^+ + Cl^- + (Na^+) + (OH^-) \longrightarrow H_2O + (Na^+) + (Cl^-)$$

Both acid and base are completely ionized, the reaction goes to completion, and only un-ionized water and sodium ions and chloride ions remain as the final products of the reaction.

(2) *A strong acid plus a weak base gives an acidic solution.*

$$H^+ + (Cl^-) + NH_3 \rightleftharpoons NH_4^+ + (Cl^-)$$

Because ammonia is a weak base (only a small percentage is ionized) all the hydrogen ions are not taken up by the ammonia. The reaction, therefore, does not go to completion, and there is an excess of hydrogen ions in solution when equilibrium is established.

(3) *A weak acid plus a strong base gives a basic solution.*

$$HOAc + (Na^+) + OH^- \rightleftharpoons H_2O + (Na^+) + OAc^-$$

Because acetic acid is a weak acid, the proton is not completely transferred from the acid to the hydroxide ion, and there is an excess of hydroxide ions in solution when equilibrium is established.

(4) *A weak acid plus a weak base.*

$$HOAc + NH_3 \rightleftharpoons NH_4^+ + OAc^-$$

This is the most complex of the four combinations and will be discussed in detail in Chapter 18, but it is appropriate to note here that if the weak acid has the same strength, as an acid, that the weak base has, as a base, the solution will be neutral. Ammonia, for example, takes up about the same amount of protons in the formation of ammonium ion as acetic acid gives up and, although the reaction does not go to completion, the solution is neutral.

If the weak acid and the weak base are *not* of equal strengths, the solution will either be acidic or basic, depending upon the relative acid-base strengths of the two substances.

15.8   Salts

Salts are ionic compounds. Among the positive ions most frequently found in salts are the simple metal ions such as Na^+, K^+, Ca^{2+}, and Ba^{2+}; solvated metal ions such as $[Al(H_2O)_6]^{3+}$ and $[Ag(NH_3)_2]^+$ and solvated hydrogen ions such as NH_4^+ and H_3O^+ also are found. The simple negative ions of salts include F^-, Cl^-, Br^-, I^-, O^{2-}, S^{2-}, P^{3-}, N^{3-}, C_2^{2-}, and H^-. Many salts contain polyatomic negative ions such as NO_3^-, SO_4^{2-}, PO_4^{3-}, CN^-, and OH^-. Salts are to be regarded as compounds made up of positive and negative ions which may be related only indirectly to acid-base reactions. For example, the salt NaCl may be formed by the direct union of sodium metal with elemental chlorine. At this point, it must be evident that a more correct formula for sodium chloride is Na^+Cl^-. This is true also for other salts. However, it is not generally customary to use the charge signs in such formulas except in special circumstances where their use might be especially helpful.

Beginners in the study of chemistry quite often get the erroneous impression that all compounds composed of a metal and a nonmetal are salts. However, many such compounds have building units which are not ions but instead may be atoms or molecules, examples are $AlCl_3$, $SnCl_4$, $PbCl_4$, and $GeCl_4$. The bonding is either covalent or polar covalent in such compounds, and the absence of any ions is indicated by the fact that these compounds are nonconductors in the liquid state. In contrast to anhydrous aluminum chloride (a covalent compound), the hexahydrate $Al(H_2O)_6Cl_3$ is a salt composed of $[Al(H_2O)_6]^{3+}$ and Cl^- ions.

15.9   Normal Salts, Hydrogen Salts, Hydroxy- and Oxysalts

A salt which contains neither a replaceable hydrogen nor a hydroxide group is called a **normal salt**; examples are NaCl, K_2SO_4, and $Ca_3(PO_4)_2$.

When only a part of the acidic hydrogen of an acid has been replaced by a metal, the compound is known as a **hydrogen salt**; examples are sodium hydrogen carbonate, $NaHCO_3$ (generally called sodium bicarbonate, or baking soda), and sodium dihydrogen phosphate, NaH_2PO_4.

When only a part of the hydroxide groups of an ionic metal hydroxide have been replaced, the compound is known as a **hydroxysalt**; examples are barium hydroxychloride, $Ba(OH)Cl$, and bismuth dihydroxychloride, $Bi(OH)_2Cl$. Some hydroxysalts readily lose the elements of one or more molecules of water and form **oxysalts**. Thus, bismuth dihydroxychloride breaks down to give the oxychloride, BiOCl, sometimes called bismuthyl chloride.

$$Bi(OH)_2Cl \longrightarrow BiOCl + H_2O$$

15.10   Properties of Salts

Salts vary greatly in all of their properties except their ionic character. Salts may taste salty, sour, bitter, astringent, or sweet, or be tasteless. Solutions of salts may be acidic, or basic, or neutral to acid-base indicators. Fused salts and aqueous solutions of salts conduct an electric current. The reactions of salts are numerous and varied.

15.11   Preparation of Salts

Salts may be prepared by one or more of the following methods:

(1) *By the direct union of their elements.*

$$2Na + Cl_2 \longrightarrow 2NaCl$$
$$Fe + S \longrightarrow FeS$$

(2) *By the reaction of acids with metals, hydroxides of metals, or oxides of metals.*

$$Zn + 2H^+ + (SO_4^{2-}) \longrightarrow Zn^{2+} + (SO_4^{2-}) + H_2\uparrow$$
$$Fe(OH)_3 + 3H^+ + (3Cl^-) \longrightarrow Fe^{3+} + (3Cl^-) + 3H_2O$$
$$CuO + 2H^+ + (SO_4^{2-}) \longrightarrow Cu^{2+} + (SO_4^{2-}) + H_2O$$

The solid salts, often in the hydrated form, may be obtained by evaporating the solutions to dryness.

(3) *By the reaction of acid anhydrides with basic anhydrides.*

$$BaO + SO_3 \longrightarrow BaSO_4$$
$$CaO + CO_2 \longrightarrow CaCO_3$$

(4) *By the reaction of acids with salts.*

$$BaCO_3 + 2H^+ + (2Cl^-) \longrightarrow Ba^{2+} + (2Cl^-) + H_2O + CO_2\uparrow$$
$$Ba^{2+} + (2Cl^-) + (2H^+) + SO_4^{2-} \longrightarrow \underline{BaSO_4} + (2H^+) + (2Cl^-)$$

(5) *By the reaction of salts with other salts.*

$$Ag^+ + (NO_3^-) + (Na^+) + Cl^- \longrightarrow \underline{AgCl} + (Na^+) + (NO_3^-)$$
$$Zn^{2+} + (2Cl^-) + (2Na^+) + S^{2-} \longrightarrow \underline{ZnS} + (2Na^+) + (2Cl^-)$$

The Brönsted-Lowry Concept of Acids and Bases

15.12   The Protonic Concept of Acids and Bases

As a result of the work of several chemists, especially Brönsted and Lowry, the classical acid-base concept has been extended to a more general one of acids and bases. According to this view acids and bases are related as indicated by

$$\underset{\text{Acid}}{HA} \rightleftharpoons \underset{\text{Proton}}{H^+} + \underset{\text{Base}}{A^-}$$

From the protonic acid-base point of view, **an acid is any species (molecule or ion) that can give up a proton to another species. A base is any species that can combine with a proton.** Stated simply, an acid is a proton donor, and a base is a proton acceptor. These definitions do not specify a solvent, as do the classical definitions. Furthermore, species which are acids or bases according to this view do not necessarily exhibit the properties which we usually associate with aqueous acids and bases of the classical concept.

In order for an acid to act as a proton donor, a proton acceptor must be present to receive the proton. The proton acceptor in aqueous solution is water.

$$HA + H_2O \rightleftharpoons H_3O^+ + A^-$$

A specific example is the reaction of hydrogen chloride with water (see Chapter 14, Section 14.9).

$$HCl + H_2O \rightleftharpoons H_3O^+ + Cl^-$$

Thus, hydronium ions, H_3O^+, are formed when the acid HA gives up its proton to a water molecule. Because the water molecule accepts a proton, it is acting as a base. In addition, when the reversible reaction proceeds to the left, the ion A^- acts as a base by accepting a proton from the hydronium ion, an acid. The equilibrium reaction, then, involves two acids and two bases.

$$\underset{\text{Acid}_1}{HA} + \underset{\text{Base}_2}{H_2O} \rightleftharpoons \underset{\text{Acid}_2}{H_3O^+} + \underset{\text{Base}_1}{A^-}$$

Solvents other than water also can act as bases, according to the proton acid-base theory. Hydrogen chloride reacts with liquid ammonia forming ammonium ions and chloride ions. Ammonia is the proton acceptor and therefore a base.

$$\underset{\text{Acid}_1}{HCl} + \underset{\text{Base}_2}{NH_3} \rightleftharpoons \underset{\text{Acid}_2}{NH_4^+} + \underset{\text{Base}_1}{Cl^-}$$

Base$_1$ is said to be the **conjugate base** of acid$_1$ inasmuch as it is derived from acid$_1$ by the loss of a proton by acid$_1$; and acid$_1$ is the **conjugate acid** of base$_1$. Similarly, acid$_2$ is referred to as the conjugate acid of base$_2$ since in the reverse reaction it gives up a proton to form base$_2$; and base$_2$ is the conjugate base of acid$_2$.

In accordance with the definition that acids are proton donors, there may be molecular acids, such as HCl, HNO_3, H_2SO_4, CH_3COOH, HCN, H_2S, H_2CO_3, HF, and H_2O; anion acids, such as HSO_4^-, HCO_3^-, $H_2PO_4^-$, HPO_4^{2-}, and HS^-; and cation acids, such as H_3O^+, NH_4^+, $[Cu(H_2O)_4]^{2+}$, and $[Fe(H_2O)_6]^{3+}$.

As in the case of acids, there are molecular bases, such as H_2O, NH_3, and CH_3NH_2; anion bases, such as OH^-, HS^-, S^{2-}, HCO_3^-, CO_3^{2-}, HSO_4^-, SO_4^{2-}, HPO_4^{2-}, Cl^-, F^-, NO_3^-, and PO_4^{3-}; and cation bases, such as $[Fe(H_2O)_5OH]^{2+}$ and $[Cu(H_2O)_3OH]^+$. The most familiar base is the hydroxide ion, OH^-, which accepts protons from acids, forming water.

$$HA + OH^- \rightleftharpoons H_2O + A^-$$
$$H_3O^+ + OH^- \rightleftharpoons H_2O + H_2O$$

It follows from this discussion, then, that the Brönsted-Lowry concept enables us to extend the classical idea of acids and bases to include many more species. It also enables us to look upon a large number of apparently unrelated reactions as truly acid-base reactions.

15.13 Amphiprotic Species

From a study of the examples of acids and bases given in the preceding section, it is evident that certain molecules and ions are classed as both acids and bases. **A species that may either gain or lose a proton is said to be amphiprotic.** For example, water may lose a proton to a base, such as NH_3, or gain a proton from an acid, such as HCl. (See Chapter 14, Figs. 14–5 and 14–4.)

$$\underset{\text{Acid}}{H_2O} + \underset{\text{Base}}{NH_3} \rightleftharpoons NH_4^+ + OH^-$$

$$\underset{\text{Base}}{H_2O} + \underset{\text{Acid}}{HCl} \rightleftharpoons H_3O^+ + Cl^-$$

The proton-containing negative ions listed in Section 15.12 are amphiprotic, as is readily seen from the equations

$$\underset{\text{Acid}}{HS^-} + \underset{\text{Base}}{OH^-} \rightleftharpoons S^{2-} + H_2O$$

$$\underset{\text{Base}}{HS^-} + \underset{\text{Acid}}{H_3O^+} \rightleftharpoons H_2S + H_2O$$

$$\underset{\text{Acid}}{HCO_3^-} + \underset{\text{Base}}{OH^-} \rightleftharpoons CO_3^{2-} + H_2O$$

$$\underset{\text{Base}}{HCO_3^-} + \underset{\text{Acid}}{H_3O^+} \rightleftharpoons H_2CO_3 + H_2O$$

The hydroxides of certain metals are amphiprotic and so react either as acids or bases.

$$\underset{\text{Acid}}{[Al(H_2O)_3(OH)_3]} + \underset{\text{Base}}{OH^-} \rightleftharpoons [Al(H_2O)_2(OH)_4]^- + H_2O$$

$$\underset{\text{Base}}{[Al(H_2O)_3(OH)_3]} + \underset{\text{Acid}}{H_3O^+} \rightleftharpoons [Al(H_2O)_4(OH)_2]^+ + H_2O$$

In the first reaction, one of the water molecules of the hydrated aluminum hydroxide loses a proton to the hydroxide ion. In the second reaction, the aluminum hydroxide receives a proton from the hydronium ion. In both cases the hydrated aluminum hydroxide is dissolved. More will be said about this kind of reaction in the discussion on hydrolysis in Chapter 18.

15.14   Polyprotic Acids

Acids can be classified in terms of the number of protons per molecule that can be given up in a reaction. Acids, such as HCl, HNO_3, and HCN, that contain one ionizable hydrogen atom in one molecule of acid are called **monoprotic acids**. Their reaction with water is given by

$$HCl + H_2O \longrightarrow H_3O^+ + Cl^-$$

$$HNO_3 + H_2O \longrightarrow H_3O^+ + NO_3^-$$

$$HCN + H_2O \rightleftharpoons H_3O^+ + CN^-$$

Acetic acid, CH_3COOH, is monoprotic because only one of the four hydrogen atoms in a single molecule is given up as a proton in reacting with bases.

$$CH_3COOH + H_2O \rightleftharpoons CH_3COO^- + H_3O^+$$

Diprotic acids contain two ionizable hydrogen atoms in one molecule of the acid; ionization of such acids occurs in two stages. The primary ionization always takes place to a greater extent than the secondary. For example, carbonic acid ionizes as follows:

$$H_2CO_3 + H_2O \rightleftharpoons H_3O^+ + HCO_3^- \quad \text{(primary ionization)}$$
$$HCO_3^- + H_2O \rightleftharpoons H_3O^+ + CO_3^{2-} \quad \text{(secondary ionization)}$$

Triprotic acids, such as phosphoric acid, ionize in three steps:

$$H_3PO_4 + H_2O \rightleftharpoons H_3O^+ + H_2PO_4^- \quad \text{(primary ionization)}$$
$$H_2PO_4^- + H_2O \rightleftharpoons H_3O^+ + HPO_4^{2-} \quad \text{(secondary ionization)}$$
$$HPO_4^{2-} + H_2O \rightleftharpoons H_3O^+ + PO_4^{3-} \quad \text{(tertiary ionization)}$$

Often the terms monobasic, dibasic, and tribasic are used instead of monoprotic, diprotic, and triprotic.

15.15 The Strengths of Acids and Bases

In the fundamental acid-base relationship, $HA + H_2O \rightleftharpoons H_3O^+ + A^-$, the anion A^- is the conjugate base of the acid HA (Section 15.12). The strength of an acid is a measure of its tendency to liberate protons. This depends upon the ability of its conjugate base to combine with protons, as compared with that of some other base such as water. For example, the greater the extent of its ionization in water, the stronger the acid.

$$HCl + H_2O \longrightarrow H_3O^+ + Cl^- \quad \text{(complete in dil. sol.)}$$
$$HF + H_2O \rightleftharpoons H_3O^+ + F^- \quad \text{(8.5\% in 0.1 } M \text{ sol.)}$$
$$CH_3COOH + H_2O \rightleftharpoons H_3O^+ + CH_3COO^- \quad \text{(1.3\% in 0.1 } M \text{ sol.)}$$

It follows that strong bases are characterized by high proton affinities. The fluoride ion is a strong base, as indicated by the weakness of its conjugate acid, HF; i.e., the fluoride ion holds the proton firmly in the hydrogen fluoride molecule. On the other hand, the chloride ion is such a weak base that HCl readily loses its proton to the water molecule, which is a stronger base than the chloride ion. Thus, the ionization of HCl is extensive, and the acid is strong. Other weak bases are Br^-, I^-, NO_3^-, ClO_4^-, and HSO_4^-; thus, HBr, HI, HNO_3, $HClO_4$, and H_2SO_4 are strong acids; i.e., they are almost 100 per cent ionized in dilute aqueous solution. The acetate ion, CH_3COO^-, is a much stronger base than is the water molecule; it is therefore the conjugate base of a weak acid. Acetic acid is ionized only 1.3 per cent in a 0.1 M solution. Other examples of weak acids are H_2O, HS^-, HCO_3^-, NH_4^+, H_2S, HCN, and H_2SO_3; thus OH^-, S^{2-}, CO_3^{2-}, NH_3, HS^-, CN^-, and HSO_3^- are strong or moderately strong bases.

According to the Brönsted-Lowry concept, then, the strongest acids have the weakest conjugate bases and the strongest bases have the weakest conjugate acids.

15.16   The Relative Strengths of Strong Acids and Bases

The strongest acids, such as HCl, HBr, and HI, appear to have about the same strength in water solution. However, the strength of the bond between hydrogen and the negative ion in strong acids is different, even though their ionization in water does not indicate a difference. Apparently, the water molecule is such a strong base as compared to the conjugate bases Cl^-, Br^-, and I^- of the strong acids HCl, HBr, and HI, that ionization is nearly complete for all these acids in aqueous solutions. When HCl, HBr, and HI are studied in solvents less strongly basic than water, however, they are observed to differ markedly in tendency to give up a proton to the solvent. In methyl alcohol, a weaker base than water, the extent of ionization increases in the order HCl<HBr<HI. Because water tends to level off any differences in strength among strong acids, the effect is known as the **leveling effect** of water.

To a somewhat lesser extent, water exerts a leveling effect upon the base strength of very strong bases. For example the oxide ion, O^{2-}, and the amide ion, NH_2^-, are such strong bases that reaction with water by the removal of protons from the water is complete.

$$O^{2-} + H_2O \longrightarrow OH^- + OH^-$$
$$NH_2^- + H_2O \longrightarrow NH_3 + OH^-$$

Thus, O^{2-} and NH_2^- appear to have the same base strength, as indicated by their complete reaction with water. The relative strengths of weaker bases may be measured by the extent to which they yield hydroxide ions when they react with water. This point is illustrated by the following examples for 0.1 M solutions:

$$CH_3COO^- + H_2O \rightleftharpoons CH_3COOH + OH^- \qquad \text{(0.0075 per cent)}$$
$$NH_3 + H_2O \rightleftharpoons NH_4^+ + OH^- \qquad \text{(1.34 per cent)}$$
$$CO_3^{2-} + H_2O \rightleftharpoons HCO_3^- + OH^- \qquad \text{(3.9 per cent)}$$

Students are frequently puzzled as to why the acid strength *decreases* in the order HOCl>HOBr>HOI [and decreases similarly for other corresponding oxyacids of chlorine, bromine, and iodine (see Chapter 21, Section 21.15)], whereas the acid strength *increases* in the order HCl<HBr<HI. The trend in acid strength for the oxyacids is accounted for by the decrease in electronegativity with increasing size of the halogen, the subsequent increase in tendency for the E—O—H bond to break between E and O rather than between O and H, and hence a decrease in the tendency for the oxyacid to lose a proton (see Section 7.1, Part 7). The different direction in the trend of acid strengths for the binary hydrogen halide acids arises from the fact that the halogen is attached directly to a hydrogen in the binary acid (instead of to an oxygen, which is in turn attached to a hydrogen, as in the oxyacids). In either kind of compound, as we go from the larger halogen to the smaller (i.e., from I to Br to Cl), a progressive shrinking of the electron shells of the halogen takes place, including a pulling in of the adjacent atom, whether the adjacent atom be hydrogen (as in the hydrogen halides) or oxygen (as in the oxyacids). As the adjacent atom is pulled in it is brought into a

region of higher electron density. Intuitively, this can be understood by visualizing the seven valence electrons as residing within a smaller volume for the smaller halogen, giving rise to a higher density of electrons. If the adjacent atom being pulled in is a hydrogen atom (as in the case with the hydrogen halides), the hydrogen is held more tightly by the higher electron density in the smaller halogen and is pulled loose with greater difficulty. Conversely, when attached to a larger halogen the hydrogen is in a region of lower electron density (more diffuse region of electrons), is attracted less tightly and is pulled away more easily.

strength is measure of tendency to liberate H⁺

15.17 Problems Based on Acidimetry and Alkalimetry

The determination of the concentration of acidic solutions of unknown normality by means of standard basic solutions is known as **acidimetry.** The use of standard acid solutions to determine the concentration of basic solutions of unknown normality is called **alkalimetry.** In acidimetry, the quantity of acid in a solution of known volume is determined by measuring the volume of base of known normality (standard solution) required to neutralize the acid. An indicator (see Sections 18.10 and 18.12) which changes color at the point of neutralization, or **end point,** is used to indicate that equivalent amounts of acid and base have been brought together. The base is added drop by drop to the acid from a **buret** (Fig. 15–1) in a process called **titration.**

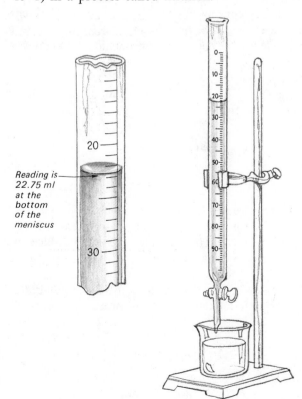

Reading is 22.75 ml at the bottom of the meniscus

FIGURE 15-1
A buret is used for accurately measuring the volume of a solution. Accurate reading (see insert) is necessary.

Example.   Let us suppose that 20 ml of 0.1 N hydrochloric acid are required in the titration of 50 ml of a basic hydroxide solution of unknown normality. Calculate the normality of the basic solution.

1 ml of 0.1 N hydrochloric acid contains 0.1 milligram-equivalent of HCl. 20 ml of 0.1 N hydrochloric acid contains 20 ml $\times$ 0.1 mg-equiv/ml = 2.0 milligram-equivalents of HCl.

For the complete neutralization of 2.0 milligram-equivalents of acid, 2.0 milligram-equivalents of base are required.

Because this quantity of base is contained in 50 ml of solution,

$$\text{Normality of base} = \frac{\text{milligram-equivalents}}{\text{milliliters of solution}} = \frac{2.0 \text{ mg-equiv}}{50 \text{ ml}} = 0.04 \ N$$

The Lewis Concept of Acids and Bases

15.18   Definitions and Examples

G. N. Lewis, in 1923, proposed a generalized theory in which acids and bases are not restricted to proton donors and proton acceptors, respectively. According to the Lewis concept, an acid is any species (molecule or ion) which can accept a pair of electrons; a base is any species (molecule or ion) which can donate a pair of electrons. An acid-base reaction then takes place with the donation of a pair of electrons by the base to the acid, with the formation of a coordinate bond between the two.

An example of an acid-base reaction that involves a proton transfer is given by the equation

$$H\overset{\times\times}{\underset{\underset{H}{\circ\times}}{O}}{}^{\times} + H\overset{\times\times}{\underset{\times\times}{Cl}}{}^{\times} \longrightarrow \left[H\overset{\times\times}{\underset{\underset{H}{\times\circ}}{O}}{}^{\times}H \right]^{+} + \left[\overset{\times\times}{\underset{\times\times}{\vdots Cl \vdots}} \right]^{-}$$

Base   Acid

In the above reaction, the oxygen atom of the water molecule donates an electron pair for the formation of a coordinate bond with the hydrogen of the hydrogen chloride molecule. Hence, water is the Lewis base and hydrogen chloride is the Lewis acid. Note that this acid-base reaction postulates the rupture of a covalent bond and the formation of a new one.

The following equations show the more general application of the acid-base theory to reactions not involving proton transfer, as proposed by Lewis:

$$2H\overset{\overset{H}{\circ\times}}{\underset{\underset{H}{\circ\times}}{N}}{}^{\times} + Ag^{+} \longrightarrow \left[H\overset{\overset{H}{\circ\times}}{\underset{\underset{H}{\circ\times}}{N}}{}^{\times}Ag{}^{\times}\overset{\overset{H}{\circ\times}}{\underset{\underset{H}{\circ\times}}{N}}{}^{\times}H \right]^{+}$$

Base   Acid

$$\left[:\!\overset{\times\times}{\underset{\times\times}{F}}\!\times\right]^{-} + \ \underset{\overset{\times}{F}}{\overset{\overset{\times}{F}}{B}}\!\times\!F\times \longrightarrow \left[\ \overset{\overset{\times\times}{F}}{\underset{\overset{\times\times}{F}}{F}}\!\times\!B\!\times\!F\times\ \right]^{-}$$

Base   Acid

$$Ca^{2+}\left[\overset{\times\times}{\underset{\times\times}{O}}\right]^{2-} + \ \underset{\overset{O}{O}}{S}\!\times\!O \longrightarrow Ca^{2+}\left[\ \overset{\overset{O}{O}}{\underset{\overset{O}{O}}{O}}\!S\!O\ \right]^{2-}$$

Base   Acid

One distinct advantage of the Lewis theory is that it explains the long-recognized basic properties of metal oxides and the acidic properties of nonmetal oxides, as illustrated in the reaction $CaO + SO_3 \longrightarrow CaSO_4$, given above.

The Solvent Systems Concept of Acids and Bases

15.19 Definitions and Examples

Work in the early twentieth century showed that reactions analogous to known acid-base reactions can occur in nonaqueous solutions. Such reactions can be classified as acid-base reactions according to the Brönsted-Lowry concept if the solvent molecules (liquid ammonia, for example) can release or accept protons. However, a considerable number of solvents, such as liquid phosgene ($COCl_2$) and liquid sulfur dioxide (SO_2), cannot release or accept protons.

To take into account the existence of acids and bases in a variety of solvents, including nonprotonic ones, the Solvent Systems concept of acids and bases was developed. It is based upon the idea of parent solvents from which acids and bases can be derived. The solvents themselves are postulated as undergoing slight auto-ionization, resulting in ions that exist in the pure solvent, as shown for typical solvents in Table 15–1.

In the Solvent Systems concept, **an acid is any material which gives a cation characteristic of the solvent.** The cation may originate either by dissociation of the solute in the solvent or by reaction of the solute with the solvent. **A base is any material which gives an anion characteristic of the solvent. Neutralization is the reaction of the solvent cation (the acid) with the solvent anion (the base) to provide the solvent.**

Several equations illustrating the Solvent Systems concept of acids and bases are given in Table 15–2.

Many reactions in solvents other than water are quite analogous to acid-base reactions in water. The following equations illustrate this by showing the amphoteric nature of $Al(OH)_3$ in water (see Section 7.1, Part 7) and the similar amphoteric nature of $Al_2(SO_3)_3$ in liquid sulfur dioxide.

TABLE 15-1

Examples of Protonic Solvents	Solvated	Unsolvated
Water	$H_2O + H_2O \rightleftharpoons (H \cdot H_2O)^+ + OH^-$	$H^+ + OH^-$
Liquid ammonia	$NH_3 + NH_3 \rightleftharpoons (H \cdot NH_3)^+ + NH_2^-$	$H^+ + NH_2^-$
Glacial acetic acid	$CH_3COOH + CH_3COOH \rightleftharpoons (H \cdot CH_3COOH)^+ + CH_3COO^-$	$H^+ + CH_3COO^-$

Examples of Nonprotonic Solvents	Solvated	Unsolvated
Liquid phosgene	$COCl_2 + COCl_2 \rightleftharpoons (COCl \cdot COCl_2)^+ + Cl^-$	$COCl^+ + Cl^-$
Liquid sulfur dioxide	$SO_2 + 2SO_2 \rightleftharpoons (SO \cdot SO_2)^{2+} + SO_3^{2-}$	$SO^{2+} + SO_3^{2-}$

TABLE 15-2

Solvent	Acid Ion	Base Ion	Typical Neutralization Reaction Acid + Base $\rightleftharpoons$ Salt + Solvent
Protonic			
Water	H^+ or H_3O^+	OH^-	$HNO_3 + NaOH \rightleftharpoons NaNO_3 + H_2O$
Liquid ammonia	H^+ or NH_4^+	NH_2^-	$NH_4NO_3 + NaNH_2 \rightleftharpoons NaNO_3 + 2NH_3$
Glacial acetic acid	H^+ or $(CH_3COOH_2)^+$	CH_3COO^-	$HClO_4 + CH_3COONa \rightleftharpoons NaClO_4 + CH_3COOH$
Nonprotonic			
Liquid phosgene	$COCl^+$	Cl^-	$[COCl][AlCl_4] + KCl \rightleftharpoons K[AlCl_4] + COCl_2$
Liquid sulfur dioxide	SO^{2+}	SO_3^{2-}	$\underset{\text{Thionyl chloride}}{SOCl_2} + Cs_2SO_3 \rightleftharpoons 2CsCl + 2SO_2$

■ **1. Aluminum Hydroxide in Water.** Aluminum ion and hydroxyl ion (the base) react to form a precipitate of aluminum hydroxide, which is insoluble in water.

$$Al^{3+} + 3OH^- \longrightarrow \underline{Al(OH)_3}$$

Aluminum hydroxide can be dissolved by adding acid (H_3O^+) again to produce aluminum ions. Aluminum hydroxide also can be dissolved by adding an excess of the base, thereby producing the soluble negative ion $[Al(OH)_4]^-$.

$$\underline{Al(OH)_3} + OH^- \longrightarrow [Al(OH)_4]^-$$

The aluminum hydroxide can be precipitated again along with production of water (the solvent) by treating the $[Al(OH)_4]^-$ ion with H_3O^+ (the acid).

$$[Al(OH)_4]^- + H_3O^+ \longrightarrow \underline{Al(OH)_3} + 2H_2O$$

Sometimes the formulas are written to show that the substances are solvated; e.g., $[Al(H_2O)_3(OH)_3]$, $[Al(H_2O)_2(OH)_4]^-$, etc. (see Chapter 18).

■ **2. Aluminum Hydroxide in Liquid Sulfur Dioxide.** Aluminum ion and sulfite ion (the base in liquid sulfur dioxide) react to form a precipitate of aluminum sulfite, $Al_2(SO_3)_3$.

$$2Al^{3+} + 3SO_3^{2-} \longrightarrow \underline{Al_2(SO_3)_3}$$

Aluminum sulfite can be dissolved by the addition of acid (SO^{2+}) to form Al^{3+}, or it can be dissolved by the addition of excess base, thereby producing the soluble negative ion $[Al(SO_3)_3]^{3-}$.

$$\underline{Al_2(SO_3)_3} + 3SO_3^{2-} \longrightarrow 2[Al(SO_3)_3]^{3-}$$

The aluminum sulfite can be precipitated again along with the production of sulfur dioxide (the solvent) by treating the $[Al(SO_3)_3]^{3-}$ ion with SO^{2+} (the acid).

$$2[Al(SO_3)_3]^{3-} + 3SO^{2+} \longrightarrow \underline{Al_2(SO_3)_3} + 6SO_2$$

Many such analogies occur in a variety of solvents. In Section 25.8, we shall consider a number of reactions in liquid ammonia.

QUESTIONS

1. Define the terms acid and base according to the classical, or Arrhenius, concept; the Brönsted-Lowry protonic concept; the Lewis concept; the Solvent Systems concept.
2. List five weak acids and their conjugate strong bases.
3. List five strong acids and their conjugate weak bases.
4. What is the conjugate acid for each of the following bases: OH^-, F^-, H_2O, HSO_4^-, HCO_3^-, NH_3, NH_2^-, H^-, N^{3-}?
5. What is the conjugate base for each of the following acids: OH^-, H_2O, HCO_3^-, HSO_4^-, NH_3, HS^-, HN_3, H_2O_2? It is stated that each substance listed is an acid. Is this correct? Explain your answer for each substance listed.
6. What is the conjugate base of each of the following acids: NH_4^+, HCl, HNO_3, $HClO_4$, $HOAc$, and HCN?
7. Write equations for the reaction of the following acids with water: HCl, HNO_3, $HClO_4$, and NH_4^+. What is the role played by water in each of these acid-base reactions?
8. Show by suitable equations that the following species are bases: OH^-, NH_3, Cl^-, NO_3^-, and H_2O.
9. Define acid-base "neutralization." Does your definition imply that the resulting solution is always neutral? Explain.
10. What is meant by titration? end point?
11. What is the nature of the resulting solution (acidic, basic, or neutral) when equivalent quantities of the following acids and bases are brought together: HNO_3 and KOH; HCl and NH_3; HNO_2 and $NaOH$; and $HOAc$ and NH_3?
12. Define the term salt and give examples of normal salts, hydrogen salts, hydroxy salts, and oxysalts.
13. What are amphiprotic substances? Illustrate with suitable equations.
14. State which of the following species are amphiprotic and write chemical equations illustrating the amphiprotic character of the species: H_2O, $H_2PO_4^-$, S^{2-}, CH_4, HSO_4^-, H_2CO_3.
15. Relate the extent of ionization of an acid in aqueous solution to the strength of the acid.

16. Write equations to illustrate three typical and characteristic reactions of aqueous acids.
17. Contrast the action of water upon oxides of metals with that of water upon oxides of nonmetals.
18. Give four methods of preparing hydroxide bases.
19. Write the equation for the essential reaction between aqueous solutions of acids and bases.
20. Write equations to show the stepwise ionization of the following polyprotic acids: H_2S, H_2CO_3, and H_3PO_4.
21. Write an equation to illustrate an acid-base reaction by the Brönsted-Lowry concept; the Lewis concept; the Solvent Systems concept.
22. What is meant by the leveling effect of water on strong acids and strong bases? How can the relative strengths of strong acids be measured?
23. State and explain the order of increasing acidity for each of the following series of compounds:
    (a) HOCl, HOBr, HOI.
    (b) HCl, HBr, HI.
    (c) HOCl, HOClO, $HOClO_2$, $HOClO_3$.
24. Describe the chemical reaction, using equations where appropriate, that you would expect from the combination of a strong acid solution with each of the following:
    (a) An active metal.
    (b) A soluble metal oxide.
    (c) A potassium hydroxide solution.
    (d) A red-colored litmus solution.
    (e) Solid aluminum hydroxide.
    (f) Solid aluminum sulfide (Al_2S_3).
    (g) Ammonia gas.
25. Write equations for four methods of preparing calcium fluoride, CaF_2.
26. Carbon dioxide dissolved in water forms carbonic acid (H_2CO_3), a weak acid. Explain why a carbonated soft drink becomes "flat" upon standing in an open container.
27. Compare the concept of amphoterism as it applies to water with its application to solvents other than water. Consider at least two solvents, one of which is nonprotonic, in your answer.
28. Predict the major products of the reaction of phosphoric acid with sodium hydroxide if the ratio of moles of the reactants are respectively, (a) 1:1, (b) 1:2, (c) 1:3, and (d) 1:4.

PROBLEMS

1. How many milliliters of 0.140 M NaOH are required to convert 2.75 g of potassium bicarbonate ($KHCO_3$) to potassium carbonate?   *Ans. 196 ml*
2. How many moles of H_3PO_3 will be prepared by treating 1.00 g of PCl_3 with an excess of water?   *Ans. 7.28×10^{-3} mole*

3. How many grams of hydrogen chloride will result from the reaction of hydrogen and chlorine in a mixture of 1.00 g of hydrogen and 10.0 liters of chlorine at S.T.P.? *Ans. 32.6 g*

4. What concentration of barium hydroxide solution will be formed by the reaction of 0.170 g of barium with 100 ml of water? What volume of hydrogen at S.T.P. will be formed? *Ans. 1.24×10^{-2} M; 2.77×10^{-2} liter*

5. How many grams of sodium carbonate will be exactly neutralized by 27.0 ml of 0.1000 M HCl? *Ans. 0.143 g*

6. Calculate the molecular weight of a diprotic acid if 253.8 mg of it is just neutralized by 47.00 ml of 0.12 M NaOH. *Ans. 90*

7. A standard acid solution has a normality of 1.60. What volume of 0.600 N base would be required to neutralize 47.0 ml of the acid? *Ans. 125 ml*

8. A 1.80-g sample of an acid, H_2X, required 14.00 ml of KOH solution for neutralization of all the hydrogen ions. Exactly 14.2 ml of this same KOH solution was found to neutralize 10.0 ml of 1.50 N H_2SO_4. Calculate the molecular weight of H_2X. *Ans. 243*

REFERENCES

"The Standard Free Energies of Solution of Anhydrous Salts in Water," D. A. Johnson, *J. Chem. Educ.*, **45**, 236 (1968).

"The Intrinsic Basicity of the Hydroxide Ion," W. L. Jolly, *J. Chem. Educ.*, **44**, 304 (1967).

"The Theory of Acids and Bases," F. M. Hall, *Educ. in Chemistry*, **1**, 91 (1964).

"The Hydrated Hydronium Ion," H. L. Clever, *J. Chem. Educ.*, **40**, 637 (1063).

"Lewis Acid-Base Titration in Fused Salts," J. M. Schlegel, *J. Chem. Educ.*, **43**, 362 (1966).

"Hard and Soft Acids and Bases," R. G. Pearson, *Chem. in Britain*, **3**, 103 (1967).

"Donor-Acceptor Interactions," R. J. Drago, *Chem. in Britain*, **3**, 516 (1967).

"Ionizing Solvents," V. Gutmann, *Chem. in Britain*, **7**, 102 (1971).

Acids, Bases, and the Chemistry of the Covalent Bond, C. A. VanderWerf, Reinhold Publishing Corp., New York, 1961 (Paperback).

Nonaqueous Solvents, R. A. Zingaro, D. C. Heath and Co., Boston, 1968 (Paperback).

Acids and Bases, R. S. Drago and N. A. Matwiyoff, D. C. Heath and Co., Boston, 1968 (Paperback).

"Ionizing Solvents," V. Gutmann, *Chem. in Britain*, **7**, 102 (1971).

"Selected Properties of Selected Solvents," G. P. Nilles and R. D. Shuetz, *J. Chem. Educ.*, **50**, 267 (1973).

"Conjugate Acid-Base and Redox Theory," R. A. Pacer, *J. Chem. Educ.*, **50**, 178 (1973).

"A Modern Approach to Acid-Base Chemistry," R. S. Drago, *J. Chem. Educ.*, **51**, 300 (1974).

Oxidation-Reduction Reactions

16

The term "oxidation" was originally applied only to reactions involving the reaction of oxygen with another element or with a compound. Likewise, the term "reduction" was used to indicate removal of oxygen from a compound. The terms oxidation and reduction are used now in a much broader sense and are applied to a great many reactions that do not involve oxygen. For example, carbon burns in fluorine in a reaction which closely resembles the union of carbon with oxygen.

$$C + 2F_2 \longrightarrow CF_4$$
$$C + O_2 \longrightarrow CO_2$$

Hydrogen burns in fluorine and chlorine as well as in oxygen.

$$H_2 + F_2 \longrightarrow 2HF$$
$$H_2 + Cl_2 \longrightarrow 2HCl$$
$$2H_2 + O_2 \longrightarrow 2H_2O$$

These reactions are all examples of the important class of reactions referred to as **oxidation-reduction** reactions.

16.1 Oxidation-Reduction and Changes in Oxidation Numbers

In Chapter 4, we discussed the concept of oxidation number. Table 16–1 lists some oxidation numbers exhibited by some of the elements. **Oxidation can be defined as an increase in the oxidation number of an atom or ion; this corresponds to a loss of electrons. Reduction is a decrease in the oxidation number of an atom**

347

TABLE 16-1   The Usual Oxidation Numbers of Some of the Common Elements*

+1	+2	+3	+4	+5	+6	+7	−1	−2	−3	−4
H	Mg	Al	C	N	S	Cl	H	O	N	C
Na	Ca	Cr	N	P	Cr	I	F	S	P	Si
K	Sr	Mn	Si	As	Mn	Mn	Cl	Se		
Cu	Ba	Fe	S	Sb		Br	Br			
Ag	Cr	Co	Mn	Bi			I			
Hg	Mn	N	Sn	Cl						
N	Fe	P	Pb	Br						
Cl	Co	As		I						
Br	Ni	Sb								
I	Cu	Bi								
	Zn	Cl								
	Cd	Br								
	Hg									
	N									
	Sn									
	Pb									

*Some of the elements listed can exhibit a variety of oxidation numbers in different compounds and hence are listed more than once in the table.

or ion; this corresponds to a gain of electrons. Hydrogen is oxidized by chlorine, which is correspondingly reduced, according to the equation

$$\overset{0}{H_2} + \overset{0}{Cl_2} \longrightarrow \overset{+1}{2}\overset{-1}{HCl}$$

The oxidation number of each hydrogen atom is increased from 0 to +1; hydrogen is oxidized. The oxidation number of each chlorine atom is reduced from 0 to −1; chlorine is reduced. Since chlorine is responsible for the increase in oxidation number of the hydrogen, chlorine is called the **oxidizing agent. Oxidizing agents are substances which increase the oxidation number of elements.** Hydrogen causes the decrease in the oxidation number of the chlorine so hydrogen is called the **reducing agent. Reducing agents are substances which decrease the oxidation number of elements.** Oxidation and reduction *always occur simultaneously* and the total increase in units of positive oxidation number exactly equals the total decrease in units of positive oxidation number in the balanced equation. Thus, two hydrogen atoms increase two units in positive oxidation number and two chlorine atoms decrease two units in positive oxidation number in the above reaction. Oxidation-reduction reactions are often conveniently referred to as **redox** reactions.

16.2   Balancing Redox Equations by the Change in Oxidation Number Method

The fact that oxidation and reduction always occur together and to the same extent makes it possible to use the changes in oxidation number in balancing redox equations. In other words, proportions of the reactants and products in the balanced redox equation must be such that the total increase in units of positive oxidation number must equal the total decrease in units of positive oxidation number.

Various types of oxidation-reduction reactions have been selected to illustrate the balancing of equations by the change in oxidation number method.

Example 1. Use the change in oxidation number method to balance the equation for the reaction of antimony and chlorine to form antimony(III) chloride.

First write the reactants and products of the reaction.

$$Sb + Cl_2 \longrightarrow SbCl_3$$

Indicate the oxidation number of each atom in the reaction (see Sections 4.9 and 4.11). Remember that elements in the free state are always assigned an oxidation number of zero.

$$\overset{0}{Sb} + \overset{0}{Cl_2} \longrightarrow \overset{+3}{Sb}\overset{-1}{Cl_3}$$

Designate the changes in oxidation number that occur during the reaction. Use an ascending arrow (↑) to denote an increase in positive oxidation number and a descending arrow (↓) to denote a decrease in positive oxidation number.

$$\left(\overset{0}{Sb} \xrightarrow{-3e^-} \overset{+3}{Sb}\right)\uparrow 3 \qquad$$ (total increase of 3 in oxidation number per formula unit, Sb)

$$\left(\overset{0}{Cl_2} \xrightarrow{+1e^-} \overset{-1}{2Cl}\right)\downarrow 2 \qquad$$ (total decrease of 2 in oxidation number per formula unit, Cl_2)

To balance the equation we must select the proper numbers of antimony and chlorine atoms so that the total increase in units of oxidation number will equal the total decrease. If we use two atoms of antimony and three molecules (six atoms) of chlorine, then there will be both an increase and a decrease of six units of positive oxidation number.

$$\overset{0}{Sb} \longrightarrow \overset{+3}{Sb}\uparrow \qquad 3 \times 2 = 6$$

$$\overset{0}{Cl_2} \longrightarrow \overset{-1}{2Cl}\downarrow \qquad 2 \times 3 = 6$$

Our final equation then becomes

$$2Sb + 3Cl_2 \longrightarrow 2SbCl_3$$

Example 2. The chloride ion is oxidized by the permanganate ion in acid solution to give the manganese(II) ion, molecular chlorine, and water. Balance the equation $MnO_4^- + Cl^- + H^+ \longrightarrow Mn^{2+} + Cl_2 + H_2O$, using the change in oxidation number method.

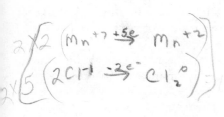

Upon assignment of oxidation numbers, it becomes evident that manganese is reduced from $+7$ to $+2$ and chlorine is oxidized from -1 to 0.

$$\overset{+7\ -2}{2MnO_4^-} + \overset{-1}{Cl^-} + \overset{+1}{H^+} \longrightarrow 2\overset{}{Mn^{2+}} + 5\overset{0}{Cl_2} + \overset{+1\ -2}{H_2O}$$

$$2Cl^- \longrightarrow \overset{0}{Cl_2}\uparrow \qquad 2 \times 5 = 10$$

$$\overset{+7}{Mn} \longrightarrow Mn^{2+}\downarrow \qquad 5 \times 2 = 10$$

Ten chloride ions will increase by ten units of oxidation number in going to five chlorine molecules (ten chlorine atoms). Two manganese atoms will decrease by ten units of oxidation number.

$$2MnO_4^- + 10Cl^- + ?H^+ \longrightarrow 2Mn^{2+} + 5Cl_2 + ?H_2O$$

To complete the balancing of this ionic equation, we may now balance the ion charges on both sides of the equation. On the right side of the equation, the only charged particles are the two Mn^{2+} with a total ionic charge of $+4$. The algebraic sum of the charges on the left must also be equal to $+4$. Excluding the H^+ for the moment, the total charge on the left is -12, $(2MnO_4^- + 10Cl^-)$. Sixteen H^+ are necessary to give an algebraic sum of $+4$, $(-2 - 10 + 16 = +4)$. The sixteen hydrogen ions are enough to produce 8 molecules of water.

$$2MnO_4^- + 10Cl^- + 16H^+ \longrightarrow 2Mn^{2+} + 5Cl_2 + 8H_2O$$

The correctness of the balancing of the equation is checked by noting that there are 8 oxygen atoms on each side of the arrow.

Example 3. Very active reducing agents such as zinc reduce dilute nitric acid to nitrous oxide (N_2O) or ammonium ion. Balance the equation for the reduction of HNO_3 with zinc, to give NH_4^+.

Write the reactants and products of the reaction producing NH_4^+, and assign oxidation numbers to the atoms that undergo a change in oxidation number.

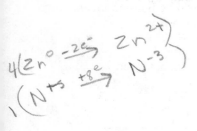

$$4\overset{0}{Zn} + \overset{+5\ -2}{NO_3^-} + H^+ \longrightarrow 4Zn^{2+} + \overset{-3\ +}{NH_4^+} + 3H_2O$$

Balance the equation with regard to the atoms that change in oxidation number.

$$\overset{0}{Zn} \longrightarrow Zn^{2+}\uparrow \quad 2 \times 4 = 8$$

$$\overset{+5}{N} \longrightarrow \overset{-3}{N}\downarrow \qquad 8 \times 1 = 8$$

$$4Zn + NO_3^- + ?H^+ \longrightarrow 4Zn^{2+} + NH_4^+ + ?H_2O$$

Balance the ion charges to find the number of H^+ required.

$$NO_3^- + ?H^+ \longrightarrow 4Zn^{2+} + NH_4^+$$

$$4Zn + \overset{-1}{NO_3^-} + \overset{+10}{10H^+} \overset{=}{\longrightarrow} \overset{+8}{4Zn^{2+}} + \overset{+1}{NH_4^+} + ?H_2O$$

Of the 10 hydrogen atoms on the left, 4 are found in the NH_4^+ on the right. Thus 6 hydrogen atoms are left to form 3 molecules of water on the right.

$$4Zn + NO_3^- + 10H^+ \longrightarrow 4Zn^{2+} + NH_4^+ + 3H_2O$$

As a check on the correctness of balancing the equation, we note there are 3 oxygen atoms on each side of the arrow.

Example 4. The chromite ion, $Cr(OH)_4^-$, is oxidized by hydrogen peroxide in basic solution to the chromate ion, CrO_4^{2-}. Balance the equation by the change in oxidation number method.

Upon assigning oxidation numbers, remember that each oxygen atom in H_2O_2 is -1.

peroxide

$$\overset{+3}{Cr}\overset{-1}{(OH)_4^-} + \overset{-1}{H_2O_2} + OH^- \longrightarrow \overset{+6}{CrO_4^{2-}} + \overset{-2}{H_2O}$$

$$\overset{+3}{Cr} \longrightarrow \overset{+6}{Cr\uparrow} \qquad 3 \times 2 = 6$$

$$\overset{-1}{O_2} \longrightarrow \overset{-2}{2O\downarrow} \qquad 2 \times 3 = 6$$

$$2Cr(OH)_4^- + 3H_2O_2 + ?OH^- \longrightarrow 2CrO_4^{2-} + ?H_2O$$

Balancing with regard to ion charges, we have

$$2Cr(OH)_4^- + ?OH^- \longrightarrow 2CrO_4^{2-}$$

$$\overset{-2}{2Cr(OH)_4^-} + 3H_2O_2 + \overset{-2}{2OH^-} \overset{=}{\longrightarrow} \overset{-4}{2CrO_4^{2-}} + ?H_2O$$

A total of 16 hydrogen atoms on the left will produce 8 molecules of water on the right.

$$2Cr(OH)_4^- + 3H_2O_2 + 2OH^- \longrightarrow 2CrO_4^{2-} + 8H_2O$$

There are 16 atoms of oxygen on each side of the balanced equation.

16.3   Balancing Redox Equations by the Ion-Electron Method

In the ion-electron method of balancing oxidation-reduction equations, the redox reaction is broken down into half-reactions, one half representing the oxidation step and the other half the reduction step.

Example 1.   Iron(II) is oxidized to iron(III) by chlorine. Use the ion-electron method and balance the following: $Fe^{2+} + Cl_2 \longrightarrow Fe^{3+} + Cl^-$.

The oxidation half-reaction is for the oxidation of Fe^{2+} to Fe^{3+}.

$$Fe^{2+} \longrightarrow Fe^{3+}$$

To balance this half-reaction in terms of ion charges and electrons, it is necessary to add one electron to the right side of the equation.

$$Fe^{2+} \longrightarrow Fe^{3+} + e^-$$
$$\phantom{Fe^{2+}}{+2} = {+3} \quad {-1}$$

To balance the reduction half-reaction, which is for the reduction of Cl_2 to Cl^-, it is necessary to add two electrons to the left side of the equation.

$$Cl_2 + 2e^- \longrightarrow 2Cl^-$$
$${-2} = {-2}$$

In the oxidation half-reaction one electron is lost, while in the reduction half-reaction two electrons are gained. To balance the electrons the oxidation half-reaction must be multiplied by two.

$$2Fe^{2+} \longrightarrow 2Fe^{3+} + 2e^-$$

On addition of the balanced half-reactions, the two electrons on either side of the equation cancel, and the balanced equation is obtained.

$$\begin{array}{r} 2Fe^{2+} \longrightarrow 2Fe^{3+} + 2e^- \\ Cl_2 + 2e^- \longrightarrow 2Cl^- \\ \hline 2Fe^{2+} + Cl_2 \longrightarrow 2Fe^{3+} + 2Cl^- \end{array}$$

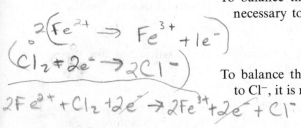

Example 2.   The dichromate ion will oxidize iron(II) to iron(III) in acid solution. Balance the equation $Cr_2O_7^{2-} + H^+ + Fe^{2+} \longrightarrow Cr^{3+} + Fe^{3+} + H_2O$ by the ion-electron method.

The oxidation half-reaction is balanced as shown in Example 1.

$$Fe^{2+} \longrightarrow Fe^{3+} + e^-$$

The reduction half-reaction is

$$Cr_2O_7^{2-} + ?H^+ \longrightarrow 2Cr^{3+} + ?H_2O$$

By inspection, the 7 atoms of oxygen from the $Cr_2O_7^{2-}$ ion require $14H^+$ ions to produce 7 molecules of water.

$$Cr_2O_7^{2-} + 14H^+ \longrightarrow 2Cr^{3+} + 7H_2O$$

To balance this reduction half-reaction in terms of ion charges and electrons, it is necessary to add 6 electrons on the left side of the equation.

$$Cr_2O_7^{2-} + 14H^+ + 6e^- \longrightarrow 2Cr^{3+} + 7H_2O$$
$${-2} \quad {+14} \quad {-6} = {+6}$$

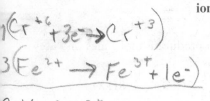

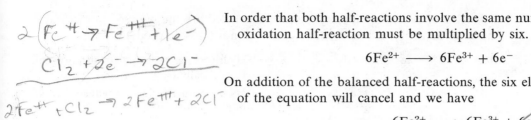

In order that both half-reactions involve the same number of electrons, the oxidation half-reaction must be multiplied by six.

$$6Fe^{2+} \longrightarrow 6Fe^{3+} + 6e^-$$

On addition of the balanced half-reactions, the six electrons on either side of the equation will cancel and we have

$$6Fe^{2+} \longrightarrow 6Fe^{3+} + 6e^-$$
$$\underline{Cr_2O_7^{2-} + 14H^+ + 6e^- \longrightarrow 2Cr^{3+} + 7H_2O}$$
$$Cr_2O_7^{2-} + 14H^+ + 6Fe^{2+} \longrightarrow 2Cr^{3+} + 6Fe^{3+} + 7H_2O$$

Example 3.   **Hydrogen peroxide in acidic solution oxidizes Fe^{2+} to Fe^{3+}. Use the ion-electron method to balance the following: $H_2O_2 + H^+ + Fe^{2+} \longrightarrow H_2O + Fe3^+$.**

The oxidation half-reaction is

$$Fe^{2+} \longrightarrow Fe^{3+} + e^-$$

The reduction half-reaction involves the reduction of H_2O_2 to H_2O in acid solution.

$$H_2O_2 + ?H^+ \longrightarrow ?H_2O$$

By inspection, it is seen that there is sufficient oxygen in one molecule of H_2O_2 to form two molecules of water. This will require two hydrogen ions.

$$H_2O_2 + 2H^+ \longrightarrow 2H_2O$$

Two electrons are required to balance the positive charges on the hydrogen ions.

$$H_2O_2 + 2H^+ + 2e^- \longrightarrow 2H_2O$$

In order that both half-reactions involve the same number of electrons, the oxidation half-reaction must be multiplied by two.

$$2Fe^{2+} \longrightarrow 2Fe^{3+} + 2e^-$$

Now let us add the two half-reactions.

$$2Fe^{2+} \longrightarrow 2Fe^{3+} + 2e^-$$
$$\underline{H_2O_2 + 2H^+ + 2e^- \longrightarrow 2H_2O}$$
$$H_2O_2 + 2H^+ + 2Fe^{2+} \longrightarrow 2H_2O + 2Fe^{3+}$$

16.4   Some Half-Reactions

When redox equations are to be balanced by the ion-electron method, it is necessary that the proper half-reactions be selected. The following half-reactions should be studied carefully, for they are very frequently encountered in the balancing of redox equations. The products listed are the *major products* under the conditions given, but in many cases are not the sole products.

$$H_2O_2 + 2H^+ + 2e^- \longrightarrow 2H_2O$$

Hydrogen peroxide as an oxidizing agent in acid solution.

$$H_2O_2 \longrightarrow O_2 + 2H^+ + 2e^-$$

Hydrogen peroxide as a reducing agent in acid solution.

$$MnO_4^- + 8H^+ + 5e^- \longrightarrow Mn^{2+} + 4H_2O$$

Permanganate ion as an oxidizing agent in acid solution.

$$MnO_4^- + 2H_2O + 3e^- \longrightarrow \underline{MnO_2} + 4OH^-$$

Permanganate ion as an oxidizing agent in neutral or basic solution.

$$Cr_2O_7^{2-} + 14H^+ + 6e^- \longrightarrow 2Cr^{3+} + 7H_2O$$

Dichromate ion as an oxidizing agent in acid solution.

$$NO_3^- + 2H^+ + e^- \longrightarrow NO_2 + H_2O$$

Concentrated nitric acid as an oxidizing agent toward less active metals, such as Cu, Ag, and Pb, and certain anions (such as Br^-).

$$NO_3^- + 4H^+ + 3e^- \longrightarrow NO + 2H_2O$$

Dilute nitric acid as an oxidizing agent toward less active metals and certain nonmetals (such as Br^-).

$$2NO_3^- + 10H^+ + 8e^- \longrightarrow N_2O + 5H_2O$$

Dilute nitric acid as an oxidizing agent toward moderately active metals such as Zn and Fe.

$$H_2SO_4 + 2H^+ + 2e^- \longrightarrow SO_2 + 2H_2O$$

Concentrated sulfuric acid as an oxidizing agent toward HBr, C, and Cu.

$$H_2SO_4 + 6H^+ + 6e^- \longrightarrow S + 4H_2O$$

Concentrated sulfuric acid as an oxidizing agent toward H_2S.

$$H_2SO_4 + 8H^+ + 8e^- \longrightarrow H_2S + 4H_2O$$

Concentrated sulfuric acid as an oxidizing agent toward HI.

$$NO_2^- + 2H^+ + e^- \longrightarrow NO + H_2O$$

Nitrous acid as an oxidizing agent.

$$NO_2^- + H_2O \longrightarrow NO_3^- + 2H^+ + 2e^-$$

Nitrous acid as a reducing agent.

$$HClO + H^+ + 2e^- \longrightarrow Cl^- + H_2O$$

Hypochlorous acid as an oxidizing agent.

$$X_2 + 2e^- \longrightarrow 2X^-$$

The halogens F_2, Cl_2, Br_2, and I_2 as oxidizing agents.

$$O_2 + 4H^+ + 4e^- \longrightarrow 2H_2O$$

Oxygen as an oxidizing agent in acid solution.

$$2H^+ + 2e^- \longrightarrow H_2$$

The hydrogen ion as an oxidizing agent.

$$H_2 \longrightarrow 2H^+ + 2e^-$$

Molecular hydrogen as a reducing agent in aqueous solution.

$$M \longrightarrow M^{n+} + ne^-$$

Metals as reducing agents.

$$M^{n+} + ne^- \longrightarrow M$$

Metal ions as oxidizing agents.

$$M^{n+} \longrightarrow M^{n+x} + xe^-$$

Oxidation of a metal ion to a higher oxidation number. For example,
$Cr^{2+} \longrightarrow Cr^{3+} + e^-$;
$Fe^{2+} \longrightarrow Fe^{3+} + e^-$;
$Sn^{2+} \longrightarrow Sn^{4+} + 2e^-$.

$$2(CrO_4^= + 4H_2O + 3e^- \rightarrow Cr(OH)_4^- + 4OH^-)$$

$$3(Sn(OH)_3^- + 3OH^- \rightarrow Sn(OH)_6^= + 2e^-)$$

$$M^{n+} + xe^- \longrightarrow M^{n-x}$$

Reduction of a metal ion to a lower oxidation number. For example,
$$Fe^{3+} + e^- \longrightarrow Fe^{2+}.$$

16.5 Gram-Equivalent Weights of Oxidizing and Reducing Agents

$$5(H_2S + 4H_2O \rightarrow SO_4^= + 10H^+ + 8e^-)$$

$$8(MnO_4^- + 8H^+ + 5e^- \rightarrow Mn^{++} + 4H_2O)$$

A redox reaction between inorganic ions always involves a transfer of electrons from the substance oxidized to the substance reduced. If the half-reactions for a redox reaction are written, the deduction of the gram-equivalent weights of the oxidizing and reducing agents is readily made. The weight of the substance yielding that quantity of the ion which gains or loses **N** electrons is the gram-equivalent weight (see Section 13.18).

For example, when potassium permanganate ($KMnO_4$) is used as an oxidizing agent in acid solution, the half-reaction is

$$MnO_4^- + 8H^+ + 5e^- \longrightarrow Mn^{2+} + 4H_2O$$

One mole of $KMnO_4$ yields one mole of permanganate ion, MnO_4^-, and, as indicated by the equation for the half-reaction, each mole of MnO_4^- requires 5**N** electrons in being reduced to Mn^{2+}. Thus, one mole of $KMnO_4$ constitutes 5 gram-equivalent weights, and the gram-equivalent weight is 158.04/5, or 31.61 g.

When iron(II) sulfate, $FeSO_4$, reacts with potassium permanganate in dilute sulfuric acid solution, Fe^{2+} is oxidized to Fe^{3+} according to the equation

$$5Fe^{2+} + MnO_4^- + 8H^+ \longrightarrow 5Fe^{3+} + Mn^{2+} + 4H_2O$$

The oxidation half-reaction is

$$Fe^{2+} \longrightarrow Fe^{3+} + e^-$$

One mole of $FeSO_4$ yields 1 mole of Fe^{2+}, which, as shown by the oxidation half-reaction, loses **N** electrons. Therefore, 1 mole (151.91 g) of $FeSO_4$ is 1 gram-equivalent weight.

This is a good place, however, to emphasize again (see Section 13.16) that the gram-equivalent weight of a substance depends upon the reaction and hence is not always the same. If, for example, the solution is very weakly acid, Fe^{2+} is oxidized by MnO_4^- to Fe^{3+}, with the formation of MnO_2, according to the equation

$$3Fe^{2+} + MnO_4^- + 4H^+ \longrightarrow 3Fe^{3+} + \underline{MnO_2} + 2H_2O$$

The half-reactions are

$$Fe^{2+} \longrightarrow Fe^{3+} + e^-$$
$$MnO_4^- + 4H^+ + 3e^- \longrightarrow \underline{MnO_2} + 2H_2O$$

One mole of Fe^{2+} again loses **N** electrons, so the equivalent weight of Fe^{2+} (and of $FeSO_4$) is the same as shown above. However, one mole of MnO_4^- requires only 3**N** electrons in being reduced to MnO_2. Hence the gram-equivalent weight of MnO_4^- (and of $KMnO_4$) is $\frac{1}{3}$ of a mole for this reaction ($\frac{158.04}{3} = 52.68$ g for $KMnO_4$) whereas it was $\frac{1}{5}$ of a mole in the example above.

16.6 Normal Solutions of Oxidizing and Reducing Agents

A one-normal solution of an oxidizing or reducing agent contains one gram-equivalent weight of the agent in a liter of solution. This is by definition of a one-normal solution (see Section 13.19). One ml of a one-normal solution of an oxidizing or reducing agent contains one milligram-equivalent weight of the agent. Because one gram-equivalent weight of an oxidizing agent reacts exactly with one gram-equivalent weight of a reducing agent, redox titrations may be conducted in the same way as described for acid and base titrations (Section 15.17) if a suitable indicator is available for determining the equivalence point.

Example 1. Suppose we wish to know the volume of 0.100 N KMnO$_4$ required to oxidize the Fe^{2+} in 200 ml of 0.050 N FeSO$_4$.

200 ml of 0.050 N FeSO$_4$ contain 200 ml $\times$ 0.050 mg-equiv wt/ml = 10.0 mg-equiv wts of Fe^{2+}. Thus, 10.0 mg-equivalent weights of KMnO$_4$ will be required to oxidize the 10.0 mg-equivalent weights of Fe^{2+}.

Each ml of 0.100 N KMnO$_4$ solution contains 0.100 mg-equivalent weight of KMnO$_4$. The number of ml of 0.100 N KMnO$_4$ solution that will contain 10.0 mg-equivalent weights of KMnO$_4$ is

$$\frac{10.0 \text{ mg-equivalent weights}}{0.100 \text{ mg-equiv wt/ml}} = 100.0 \text{ ml of } 0.100 \ N \text{ KMnO}_4$$

Example 2. Calculate the volume of 0.0500 N potassium dichromate (K$_2$Cr$_2$O$_7$) required to oxidize the chloride ion in 75.0 ml of 0.0200 N sodium chloride.

75.0 ml of 0.0200 N NaCl contain 75.0 ml $\times$ 0.0200 mg-equiv wt/ml = 1.50 mg-equiv wts of Cl$^-$. Hence, 1.50 mg-equiv wts of K$_2$Cr$_2$O$_7$ will be required to oxidize 1.50 mg-equiv wts of Cl$^-$.

Each ml of 0.0500 N K$_2$Cr$_2$O$_7$ contains 0.0500 mg-equiv wt of K$_2$Cr$_2$O$_7$. The number of ml of 0.050 N K$_2$Cr$_2$O$_7$ solution that will contain 1.50 mg-equiv wts of K$_2$Cr$_2$O$_7$ is

$$\frac{1.50 \text{ mg-equiv wts}}{0.050 \text{ mg-equiv wt/ml}} = 30.0 \text{ ml of } 0.050 \ N \text{ K}_2\text{Cr}_2\text{O}_7$$

Example 3. Calculate the normality of a K$_2$Cr$_2$O$_7$ solution if 45.0 ml of it are required to oxidize 66.0 ml of 0.150 N FeSO$_4$ solution.

66.0 ml of 0.150 N FeSO$_4$ solution contain 66.0 ml $\times$ 0.150 mg-equiv wt/ml = 9.90 mg-equiv wts of Fe^{2+}. Hence, 9.90 mg-equiv wts of K$_2$Cr$_2$O$_7$ would be required to oxidize the FeSO$_4$ and would therefore be present in the 45.0 ml of the K$_2$Cr$_2$O$_7$ solution used in the oxidation.

Normality = g-equiv wts/liter, or mg-equiv wts/ml.

9.90 mg-equiv wts in 45.0 ml = $\dfrac{9.90}{45.0}$ mg-equiv wts/ml

$$= 0.220 \text{ mg-equiv wts/ml} = 0.220 \ N \text{ K}_2\text{Cr}_2\text{O}_7$$

QUESTIONS   *answer*

1. Define oxidation and reduction in terms of change of oxidation number.
2. What are oxidizing agents? reducing agents?
3. Give the oxidation number of each of the elements contained in the following substances: FeS, $SiCl_4$, CO, Na_2O_2, K_2SO_3, H_2CO_3, $NaCl$, Li_2O, Na_2S, Na_3N, $NaHSO_4$, $Ca_3(PO_4)_2$, CaC_2, N_2H_4, CuI, Na_2CrO_4, $KMnO_4$, FeO, Fe_2O_3, Fe_3O_4.

⑤4. Balance the following redox equations by either the "change of oxidation number" method or the "ion-electron" method:

⑤(a) $H_2SO_4 + HBr \longrightarrow SO_2\uparrow + Br_2\uparrow + H_2O$
  (b) $Zn + H^+ + NO_3^- \longrightarrow Zn^{2+} + N_2\uparrow + H_2O$
⑤(c) $MnO_4^- + S^{2-} + H_2O \longrightarrow MnO_2 + S + OH^-$
  (d) $NO_3^- + I_2 + H^+ \longrightarrow IO_3^- + NO_2\uparrow + H_2O$
⑤(e) $Cu + H^+ + NO_3^- \longrightarrow Cu^{2+} + NO_2\uparrow + H_2O$
  (f) $Zn + H^+ + NO_3^- \longrightarrow Zn^{2+} + N_2O\uparrow + H_2O$
⑤(g) $Cu + H^+ + NO_3^- \longrightarrow Cu^{2+} + NO\uparrow + H_2O$
  (h) $MnO_4^- + H_2S + H^+ \longrightarrow Mn^{2+} + S + H_2O$
⑤(i) $H_2SO_4 + HI \longrightarrow H_2S + I_2 + H_2O$
  (j) $MnO_4^- + NO_2^- + H_2O \longrightarrow MnO_2 + NO_3^- + OH^-$
⑤(k) $OH^- + Cl_2 \longrightarrow ClO_3^- + Cl^- + H_2O$
  (l) $H_2O_2 + MnO_4^- + H^+ \longrightarrow Mn^{2+} + H_2O + O_2\uparrow$
⑤(m) $MnO_4^{2-} + H_2O \longrightarrow MnO_4^- + OH^- + MnO_2$
  (n) $Br_2 + SO_2 + H_2O \longrightarrow H^+ + Br^- + SO_4^{2-}$

⑤5. Balance the following redox equations by either the "change in oxidation number" method or the "ion-electron" method:

⑤(a) $MnO_4^{2-} + Cl_2 \longrightarrow MnO_4^- + Cl^-$
  (b) $OH^- + NO_2 \longrightarrow NO_3^- + NO_2^- + H_2O$
⑤(c) $HBrO \longrightarrow H^+ + Br^- + O_2\uparrow$
  (d) $Br_2 + CO_3^{2-} \longrightarrow Br^- + BrO_3^- + CO_2\uparrow$
⑤(e) $CuS + H^+ + NO_3^- \longrightarrow Cu^{2+} + S + NO\uparrow + H_2O$
  (f) $NH_3 + O_2 \longrightarrow NO + H_2O$
⑤(g) $C + HNO_3 \longrightarrow NO_2\uparrow + H_2O + CO_2\uparrow$
  (h) $ZnS + O_2 \longrightarrow ZnO + SO_2$
⑤(i) $NO_3^- + Zn + OH^- + H_2O \longrightarrow NH_3\uparrow + Zn(OH)_4^{2-}$
  (j) $HClO_3 \longrightarrow HClO_4 + ClO_2 + H_2O$
⑤(k) $H_2S + H_2O_2 \longrightarrow S + H_2O$
  (l) $ClO_3^- + H_2O + I_2 \longrightarrow IO_3^- + Cl^- + H^+$
⑤(m) $Al + H^+ \longrightarrow Al^{3+} + H_2\uparrow$
  (n) $Cr_2O_7^{2-} + HNO_2 + H^+ \longrightarrow Cr^{3+} + NO_3^- + H_2O$

⑤6. Complete and balance the following equations. When the reaction occurs in acidic solution, H^+ and/or H_2O may be added on either side of the equation, as necessary, to balance the equation properly; when the reaction occurs in basic solution, OH^- and/or H_2O may be added, as necessary, on either side of the equation. No indication of the acidity of the solution is given if neither H^+ nor OH^- is involved as a reactant or product.

⑤(a) $Ag + NO_3^- \longrightarrow Ag^+ + NO$ (acidic solution)

(b) $C_2H_4 + MnO_4^- \longrightarrow Mn^{2+} + CO_2$ (acidic solution)

⑤(c) $H_2S + I_2 \longrightarrow S + I^-$ (acidic solution)

(d) $Br_2 \longrightarrow BrO_3^- + Br^-$ (basic solution)

⑤(e) $Ag^+ + AsH_3 \longrightarrow Ag + H_3AsO_3$ (acidic solution)

(f) $PbO_2 + Cl^- \longrightarrow Pb^{2+} + Cl_2$ (acidic solution)

⑤(g) $CN^- + MnO_4^- \longrightarrow CNO^- + MnO_2$ (basic solution)

(h) $\underline{HgS} + Cl^- + NO_3^- \longrightarrow [HgCl_4]^{2-} + NO_2\uparrow + \underline{S}$ (acidic solution)

(i) $\overline{H_2O_2} + ClO_2 \longrightarrow ClO_2^- + O_2$ (basic solution)

(j) $UF_6^- + H_2O_2 \longrightarrow UO_2^{2+} + HF$ (acidic solution)

⑤7. Complete and balance the following equations (see instructions for Question 6):

⑤(a) $Fe^{2+} + MnO_4^- \longrightarrow Fe^{3+} + Mn^{2+}$ (acidic solution)

(b) $Cr_2O_7^{2-} + I^- \longrightarrow Cr^{3+} + I_2$ (acidic solution)

⑤(c) $Hg_2Cl_2 + NH_3 \longrightarrow Hg + HgNH_2Cl + NH_4^+ + Cl^-$

(d) $Fe^{3+} + I^- \longrightarrow Fe^{2+} + I_2$

⑤(e) $MnO_2 + Cl^- \longrightarrow Mn^{2+} + Cl_2$ (acidic solution)

(f) $CN^- + [Fe(CN)_6]^{3-} \longrightarrow CNO^- + [Fe(CN)_6]^{4-}$ (basic solution)

⑤(g) $Fe^{2+} + Cr_2O_7^{2-} \longrightarrow Fe^{3+} + Cr^{3+}$ (acidic solution)

(h) $I_2 + H_3AsO_3 \longrightarrow I^- + H_3AsO_4$ (acidic solution)

⑤(i) $Sn^{2+} + HgCl_2 + Cl^- \longrightarrow SnCl_6^{2-} + Hg_2Cl_2$

(j) $CrO_4^{2-} + HSnO_2^- \longrightarrow HSnO_3^- + CrO_2^-$ (basic solution)

⑤(k) $CH_2O + [Ag(NH_3)_2]^+ \longrightarrow Ag + HCO_2^- + NH_3$ (basic solution)

⑤8. Complete and balance the following equation (see instructions for Question 6). This is an exceptionally difficult equation to balance.

$$Pb(N_3)_2 + Cr(MnO_4)_2 \longrightarrow Cr_2O_3 + MnO_2 + Pb_3O_4 + NO$$

(basic solution)

PROBLEMS Work

⑤1. What weight of aluminum would be necessary to reduce all the cadmium in 240 ml of a 0.150 M solution of $CdCl_2$?

$$3Cd^{2+} + 2Al \longrightarrow 3Cd + 2Al^{3+}$$

Ans. 0.648 g

2. How many milliliters of a 0.100 M $KMnO_4$ solution would be needed to oxidize all the oxalate ($C_2O_4^{2-}$) in 556.3 milligrams of $H_2C_2O_4 \cdot 2H_2O$ (oxalic acid dihydrate)?

$$MnO_4^- + C_2O_4^{2-} + H^+ \longrightarrow Mn^{2+} + CO_2 + H_2O$$ (unbalanced)

Ans. 17.7 ml

S 3. Calculate the equivalent weight of the oxidizing and reducing agents in reactions (d), (i), (j), (l), and (n) of *Question* 4.

Ans. (*d*) NO_3^-, *62.0;* I_2, *25.4;* (*i*) H_2SO_4, *12.3;* *HI, 127.9;* (*j*) MnO_4^-, *39.6;* NO_2^-, *23.0;* (*l*) H_2O_2, *17.0;* MnO_4^-, *23.8;* (*n*) Br_2, *79.9;* SO_2, *32.0*

4. A 25.0-ml solution of 0.0500 *N* $Na_2S_2O_3$ was used in the titration of 15.0 ml of an iodine solution. Calculate the normality of the iodine solution.

$$2S_2O_3^{2-} + I_2 \longrightarrow S_4O_6^{2-} + 2I^-$$

Ans. 0.0833

5. What weight of $LiAlH_4$ is needed to reduce 10.43 grams of acetaldehyde, CH_3CHO, to ethanol, CH_3CH_2OH?

$$LiAlH_4 + 4CH_3CHO + 4H_2O \longrightarrow 4CH_3CH_2OH + LiOH + Al(OH)_3$$

Ans. 2.247 g

S 6. A solution of $(NH_4)_2SO_4 \cdot FeSO_4 \cdot 6H_2O$ containing 1.05 grams of the compound is titrated with an acidic $Na_2Cr_2O_7$ solution. If 41.6 milliliters of the $Na_2Cr_2O_7$ solution is needed, what is the normality of the $Na_2Cr_2O_7$ solution?

$$Fe^{2+} + Cr_2O_7^{2-} + H^+ \longrightarrow Cr^{3+} + Fe^{3+} + H_2O \quad \text{(unbalanced)}$$

Ans. 0.0644 N

7. How many milligrams of potassium permanganate are needed to oxidize 300 milligrams of hydrogen peroxide in acid solution?

$$MnO_4^- + H_2O_2 + H^+ \longrightarrow Mn^{2+} + H_2O + O_2 \quad \text{(unbalanced)}$$

Ans. 558 mg

S 8. The chlorine content of a solution of chlorine water is to be determined. A 50.0-ml sample (sp. gr. = 1.02) of this chlorine water is treated with an excess of potassium iodide. To titrate the liberated iodine, 34.5 ml of 0.100 *N* $Na_2S_2O_3$ is required. Calculate the percentage of chlorine by weight in the chlorine water. (Chlorine water is a solution of free chlorine in water.)

$$Cl_2 + 2I^- \longrightarrow 2Cl^- + I_2$$

Ans. 0.240%

S 9. State for each of the following whether the initial substance is an oxidizing agent or a reducing agent and make the requested conversion.

S (a) 0.3 mole of H_2S to equivalent weights of H_2S (product is S). *Ans. 0.6 eq. wt.*

(b) 6.0 equivalent weights of $Na_2Cr_2O_7$ to grams of $Na_2Cr_2O_7$ (product is Cr^{3+}). *Ans. 260 g*

S (c) 1.6 milliequivalent weights of HNO_3 to millimoles of HNO_3 (product is NH_3). *Ans. 0.20 millimole*

10. A solution of $NaMnO_4$ on the laboratory shelf is 0.0500 *M*. What would its normality be if it were used in the following reactions? (Complete and balance the equation before making the calculation.) Note: (*s*) refers to solid:

(a) $Sn^{2+} + MnO_4^- \longrightarrow Sn^{4+} + Mn^{2+}$ (acid solution). *Ans. 0.250 N*

(b) $Ga(s) + MnO_4^- \longrightarrow MnO_2(s) + Ga(OH)_3(s)$ *Ans. 0.150 N*

(c) $Fe(OH)_2(s) + MnO_4^- \longrightarrow Fe(OH)_3(s) + MnO_4^{2-}$ (basic solution). *Ans. 0.0500 N*

REFERENCES

"Redox Revisited," K. L. Lockwood, *J. Chem. Educ.*, **38,** 326 (1961).

"The Stoichiometry of an Oxidation-Reduction Reaction," W. C. Child, Jr. and R. W. Ramette, *J. Chem. Educ.*, **44,** 109 (1967).

"Mechanisms of Oxidation-Reduction Reactions," H. Taube, *J. Chem. Educ.*, **45,** 452 (1968).

"Interpretation of Oxidation-Reduction," M. P. Goodstein, *J. Chem. Educ.*, **47,** 452 (1970).

"Conjugate Acid-Base and Redox Theory," R. A. Pacer, *J. Chem. Educ.*, **50,** 178 (1973).

Chemical Equilibrium

17

The general principle of equilibrium may be stated as: **A condition of equilibrium is reached in a system when two opposing changes occur simultaneously at the same rate.** Although we have used this general principle of equilibrium often in preceding chapters, most of the examples have been systems involving physical rather than chemical changes. For example, we have defined the vapor pressure of a liquid (Section 11.2) as the pressure exerted by a vapor in equilibrium with its liquid at a given temperature. The two opposing changes in this case are evaporation and condensation. We did encounter an example of a chemical equilibrium in Section 9.6; this equilibrium involved the reduction of steam by hot iron.

$$3Fe + 4H_2O \rightleftharpoons Fe_3O_4 + 4H_2$$

At equilibrium these two opposing chemical changes occur at the same rate.

The equilibrium state of a given reversible reaction varies with the conditions under which the reaction takes place. A knowledge of the effect of a change in conditions upon an equilibrium is very important both in the laboratory and in industry. By proper selection of conditions it is possible to control the relative amounts of the substances which are present at equilibrium. For example, it was known for many years that nitrogen and hydrogen react to form ammonia ($N_2 + 3H_2 \rightleftharpoons 2NH_3$). However, the industrial manufacture of ammonia by means of this reaction was achieved only after the factors which influence the equilibrium were understood.

A clear understanding of the subject of chemical equilibrium is possible only after reaction rates and the factors which influence them have been studied.

17.1   Collision Theory of the Reaction Rate

The rate of a chemical reaction may be defined as the number of moles of a substance which disappear or are formed by the reaction per unit volume in a unit of time. The rates of reactions vary greatly. Most reactions involving ions in solution proceed very rapidly. When solutions of sodium hydroxide and hydrochloric acid are mixed, neutralization occurs almost instantaneously, and involves the union of hydronium ions and hydroxide ions with the formation of water ($H_3O^+ + OH^- \longrightarrow 2H_2O$). In general, reactions involving molecules are slower than those involving ions. The reaction between hydrogen and oxygen is extremely slow at room temperature, but when a mixture of the gases is heated to 500°, the reaction to form steam proceeds at a much faster rate. When a mixture of hydrogen and oxygen is ignited, the reaction takes place explosively.

Before two or more molecules (or ions) can react, they must collide with one another. In slow reactions only a few of the billions of collisions that occur between molecules result in reaction. The only collisions which are effective are those involving molecules possessing an energy content higher than a certain minimum value. This minimum value is called **energy of activation** (Fig. 17–1). In a slow

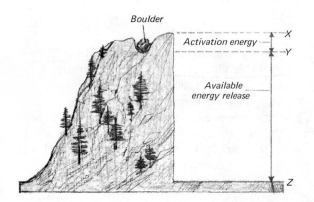

FIGURE 17–1

Illustration, by analogy, of activation energy in relation to available energy release. The boulder, potentially, can release the amount of energy created by its falling the distance from height Y to height Z. However, an amount of energy must be put into the system (the activation energy) sufficient to lift the boulder over the barrier through the distance Y to X before the boulder can fall to Z. The activation energy is regained as the boulder falls through the distance X to Y, and the additional energy is then released in the continued fall from Y to Z. The net energy released is that provided by the fall from Y to Z.

reaction, the minimum value is far above the average energy content of the molecules. The energies of a large fraction of the molecules in a system are close to an average value. However, a few of the molecules, the fast-moving ones, have relatively high energies whereas the slow-moving ones have relatively low energies. The collisions between the fast-moving molecules are those that are most apt to result in reactions. In very fast reactions the fraction of molecules in the system possessing the necessary energy of activation is large, and most collisions between the molecules result in reaction. Most reactions involving ions in solution are rapid, because the ions have a large attraction for each other and no additional energy is required to cause them to react. Practically every collision between ions which tend to combine is an effective one.

The energy relationships for the general reaction of Molecule A with Molecule B to form Molecules C and D are shown in Fig. 17–2.

$$A + B \longrightarrow C + D$$

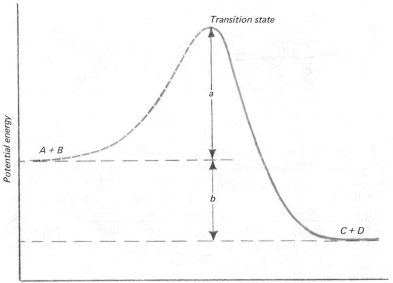

FIGURE 17-2

Potential energy relationships for the reaction $A + B \rightleftharpoons C + D$. The energy represented by the broken curve is that for the system with A and B present; the energy represented by the solid curve is that for the system with C and D present. The activation energy for the forward reaction is represented by (a), the activation energy for the reverse reaction by ($a + b$). The energy at the height of the peak corresponds to the transition state.

The diagram shows that after the activation energy, a, is exceeded, and as C and D begin to form, the system loses energy, falling to a point where it has a lower total energy than the initial mixture of A and B. The forward reaction, therefore, tends to take place readily if sufficient energy is available to exceed the activation energy. In the diagram, b represents the difference in energy between the reactants (A and B) and the products (C and D). The sum of $a + b$ represents the activation energy for the reverse reaction, $C + D \longrightarrow A + B$.

The total energy for a given reaction may involve the breaking of bonds in the reacting molecules as well as the formation of the new bonds which are present in the product molecules. Often in the process of breaking and forming bonds, a condition, referred to as a **transition state** for the reaction, arises in which an intermediate substance is formed. This intermediate substance is one which then either changes back to the original substance or forms the new products. In Fig. 17-2, the transition state occurs at the maximum energy, the height of the peak.

The dissociation of two hydrogen iodide molecules (HI) to form a hydrogen molecule and an iodine molecule is an example of a reaction involving a transition state. The process includes the net effects of breaking two H—I bonds and forming an H—H bond and an I—I bond.

$$2HI \longrightarrow H_2 + I_2$$

A possible manner in which this reaction might take place is illustrated in Fig. 17-3. According to this postulate, two hydrogen iodide molecules collide to form a transition state substance, which can then split either to form a hydrogen molecule and an iodine molecule or to revert to two hydrogen iodide molecules.

It is interesting to note that the reverse reaction, the formation of hydrogen iodide from its elements, was until rather recently thought to occur by a very

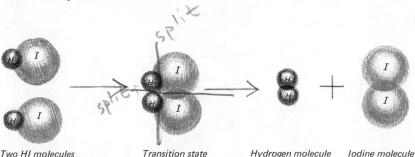

FIGURE 17-3
Probable mechanism for the dissociation of two HI molecules to produce one molecule of H_2 and one molcule of I_2.

Two HI molecules   *Transition state*   *Hydrogen molecule*   *Iodine molecule*

similar *path,* or *mechanism.* However, recent experiments have shown that the mechanism for the formation of hydrogen iodide is a much more complicated process than just a two-molecule combination (referred to as a **two-body reaction,** or **bimolecular reaction**) of a hydrogen molecule and an iodine molecule. Instead, the reaction is actually between a hydrogen molecule and two iodine atoms, and hence is a **three-body,** or **termolecular, reaction.** The iodine atoms are produced by the dissociation of an iodine molecule. The two iodine atoms then combine with one hydrogen molecule, probably with the formation of a transition state similar to that which would result from the bimolecular reaction. The transition state then splits to give two hydrogen iodide molecules (Fig. 17–4).

FIGURE 17-4
Two of the three probable steps in the mechanism for the reaction of H_2 and I_2 to produce HI. The first step (not shown) involves the dissociation of an iodine molecule into two iodine atoms ($I_2 \longrightarrow 2I$). In the second and third steps (shown here) two iodine atoms and one hydrogen molecule combine to produce two HI molecules.

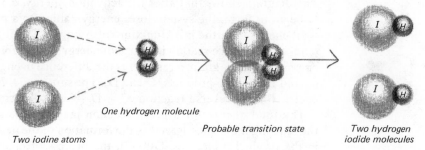

Two iodine atoms   *One hydrogen molecule*   *Probable transition state*   *Two hydrogen iodide molecules*

17.2   Reaction Mechanisms

A balanced equation for a chemical reaction indicates the materials which are reacting and the materials which are produced, but it does not tell the process by which the reaction actually takes place. The process, or pathway, by which a reaction occurs is called the **reaction mechanism,** or the **reaction path.**

A reaction often occurs in steps. Commonly, in each step two molecules (or two ions, or a molecule and an ion) react with each other—a two-body, or bimolecular, reaction. Sometimes, three molecules may react, but the probability of three molecules coming together under just the right circumstances for reaction is very much lower than for two. Hence, three-body, or termolecular, reactions are very much less common than two-body reactions. Reactions requiring reaction of four molecules or more in one step are exceedingly rare. It is reasonable to suppose, then, that a reaction such as that represented by the following equation must take place in several steps.

$$2MnO_4^- + 10Cl^- + 16H_3O^+ \longrightarrow 2Mn^{2+} + 5Cl_2 + 24H_2O$$

For the reaction to take place in one step, two permanganate ions (MnO_4^-), ten chloride ions, and 16 hydronium ions would have to collide with one another *simultaneously* in an effective collision. This, of course, is very unlikely.

Even reactions which *appear* to be simple enough to occur in one step may actually take place in more than one step. For example, we have noted in the preceding section that the reaction for the formation of hydrogen iodide from the two elements has been found to be a three-body reaction taking place in more than one step. Another example is the reaction for the formation of hydrogen bromide from hydrogen and bromine ($H_2 + Br_2 \rightleftharpoons 2HBr$).

This reaction, which might appear to be a relatively simple one-step bimolecular reaction, actually occurs in five different steps:

(1) A bromine molecule dissociates to form two bromine atoms.

$$Br_2 \rightleftharpoons 2Br$$

(2) A bromine atom combines with a hydrogen molecule to form a hydrogen bromide molecule and a hydrogen atom.

$$Br + H_2 \rightleftharpoons HBr + H$$

(3) The hydrogen atom reacts with a bromine molecule to form another hydrogen bromide molecule and a bromine atom.

$$H + Br_2 \rightleftharpoons HBr + Br$$

(4) A hydrogen atom alternatively can react with a hydrogen bromide molecule to form a hydrogen molecule and a bromine atom (an *inhibitory* step, inasmuch as some of the HBr product is used up).

$$H + HBr \rightleftharpoons H_2 + Br$$

(5) Two bromine atoms combine to form a bromine molecule.

$$2Br \rightleftharpoons Br_2$$

Note that the bromine atoms formed in Steps (3) and (4) can then react with hydrogen molecules according to Step (2), hence providing the possibility of initiating once again the series of reactions, which can repeat themselves over and over. Such a mechanism, involving repeating reactions, is known as a **chain mechanism.**

The determination of the mechanism of a reaction is important in selecting reaction conditions which provide a good yield of the desired product. A knowledge of the mechanism of a reaction sometimes makes it possible for the chemist to devise a suitable procedure for the preparation of a compound previously unknown. The study of reaction mechanisms is a very interesting but complex subject. Rather few reaction mechanisms have been completely characterized, relative to the number of chemical reactions which are known. This kind of study, referred to as a study of the **kinetics of reactions,** is one of the very active current research areas.

Some steps of a reaction are slow relative to other steps. The slowest reaction step is the one that determines the maximum rate of a reaction, inasmuch as a reaction can proceed no faster than its slowest step. It is, therefore, referred to as the **rate-determining step** of the reaction.

Reaction rates depend upon many factors, some of the more important of which we shall consider in the following sections.

17.3   Rate of Reaction and the Nature of the Reacting Substances

Similar reactions have different reaction rates under the same conditions if different reacting substances are involved. Calcium and sodium both react with water at ordinary temperatures, calcium at a moderate rate and sodium so rapidly that the reaction is of almost explosive violence. Sodium, an active metal, reacts much more rapidly with chlorine than does iron, a moderately active metal. We noted in the preceding sections that ionic substances in solution usually react nearly instantaneously, whereas molecular substances usually react much more slowly. It is evident that the rate of reaction is influenced by the nature of the substances participating in the reaction.

17.4   Rate of Reaction and the State of Subdivision

Except for substances in the gaseous state or in solution, the rate of a reaction depends to a great extent on the area of surface contact between the two phases. Finely divided solids, because of greater surface area in contact (see Problem 8, Chapter 14), react more rapidly than massive specimens of the same substances. For example, large pieces of coal burn slowly, smaller pieces more rapidly, and powdered coal may burn at an explosive rate.

17.5   Rate of Reaction and Temperature

It is a common observation that chemical reactions are accelerated by increases in temperature. The oxidation of either iron or coal is very slow at ordinary temperatures but proceeds rapidly at high temperatures. The student uses a burner or a hotplate in the laboratory to increase the speed of reactions which may proceed slowly or even at an imperceptible rate at ordinary temperatures. It is a familiar fact that foods cook faster at higher temperatures than at lower ones. In many cases, the rate of a reaction in a homogeneous system is approximately doubled by an increase in temperature of only 10° (Fig. 17–5). This rule, however, is only a rough approximation and should always be used with care.

The increase in the rate of reactions with increase in temperature is easily explained in terms of the kinetic-molecular theory. As the temperature of the system is raised, the average speed of the molecules becomes greater and more

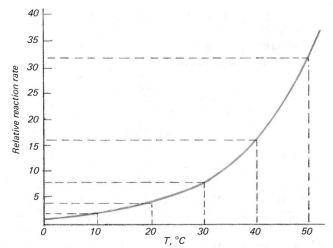

FIGURE 17-5

The effect of temperature upon reaction rate, shown graphically. The rate of reaction is often approximately doubled for each ten-degree rise in temperature.

collisions between molecules per unit time result. In addition, as the temperature rises, more molecules gain the minimum energy necessary for reaction to take place when they collide. In other words, at higher temperatures a greater fraction of the molecules acquire sufficient energy to break the bonds that hold the atoms or radicals together, thus making possible other combinations of the same particles. For many reactions, then, a rise in temperature of 10° doubles the rate of reaction, because this change in temperature doubles the number of "activated" molecules, i.e., those which possess the necessary energy of activation.

17.6   Rate of Reaction and Catalysis

Many reactions can be accelerated by the presence of small amounts of substances which are not themselves permanently used up by the reaction. Such substances, called **catalysts** (Section 8.2), may be divided into two general classes: (1) **heterogeneous catalysts,** or **contact catalysts,** and (2) those which form intermediate substances which in turn react and regenerate the catalyst, referred to as **homogeneous catalysts,** or **secondary catalysts.**

Heterogeneous catalysts act by furnishing a surface at which the reacting molecules are adsorbed and thus concentrated. Adsorption of molecules on the surface of the catalyst effectively reduces the energy necessary for activation and more collisions are therefore effective. In addition, the increase in concentration resulting from adsorption results in more collisions per unit time, thus increasing the rate of the reaction. In some cases, adsorption involves weak chemical bonding, which increases the reactivity of the adsorbed substances. Hydrogen and oxygen do not react appreciably at ordinary temperatures, but metallic platinum has the power to adsorb these gases, particularly hydrogen, so that they will react with each other at ordinary temperatures. Vanadium(V) oxide is extensively used in the contact process for the manufacture of sulfuric acid. In this process, the vanadium(V) oxide greatly accelerates the oxidation of sulfur dioxide to sulfur trioxide by oxygen of the air (Section 23.15).

Homogeneous catalysts act as "carriers" by forming an intermediate substance which then decomposes, or reacts, to regenerate the original catalyst. Nitric oxide, NO, serves as such a catalyst in the lead-chamber process for the manufacture of sulfuric acid (Section 23.15). The reaction

$$2SO_2 + O_2 \longrightarrow 2SO_3$$

is slow, while the reactions

$$2NO + O_2 \longrightarrow 2NO_2$$

$$SO_2 + NO_2 \longrightarrow SO_3 + NO$$

are fast. The NO formed in the latter reaction combines with more oxygen to produce NO_2, which with SO_2 produces more SO_3. In this way a small amount of nitric oxide suffices for the manufacture of an almost unlimited quantity of sulfuric acid. Nitric oxide in this reaction serves as a carrier for oxygen. Figure 17-6 is an energy diagram which shows the effect of a catalyst on the energy of activation.

The line between the two types of catalysts is a thin one because the distinction between adsorption and compound formation is becoming less important as we learn more about adsorption. Therefore, considerable overlap exists between heterogenous and homogeneous catalysts, and in fact recent research shows that many homogeneous catalysts can be converted into heterogeneous ones. Such conversions make possible retaining the advantage of high activity and selectivity possessed by homogeneous catalysts and gaining the advantage of ready recovery possessed by heterogeneous catalysts. The study of heterogeneous and homogeneous catalysts is one of the very active and promising areas of present-day research.

A catalyst decreases the energy of activation, resulting in a smaller required quantity of energy to initiate the reaction. Note that the catalyst decreases the

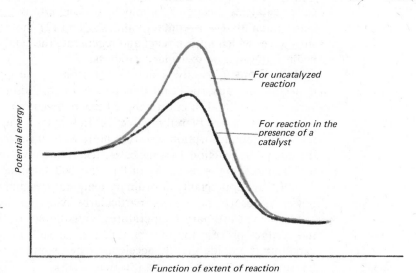

FIGURE 17-6

Potential energy diagram showing the effect of a catalyst on the activation energy of a reaction.

activation energy for both the forward and the reverse reactions and hence accelerates both the forward and the reverse reactions.

Some substances are referred to as "negative catalysts." In most cases, however, these are probably not true catalysts, but instead are substances which react with and "poison" the real catalyst and thereby prevent its action as a catalyst. For example, the presence of certain organic substances, in very small amounts, greatly decreases the rate of deterioration of rubber. However, these organic substances are not negative catalysts. They are complexing agents (see Chapter 32) that tie up such ions as Cu^{2+}, Fe^{3+}, V^{3+}, and Mn^{2+} which are present in trace amounts in rubber and which, if not complexed, catalyze its deterioration. A complexed metal ion has very different properties than the uncomplexed metal ion.

Many reactions of organic compounds are catalyzed by enzymes, which are complex substances produced by living organisms. For example, the production of alcohol from sugars by fermentation is catalyzed by the enzyme zymase, which is produced by yeast cells. Many of the chemical reactions that take place in living organisms, called **metabolic processes,** are catalyzed by enzymes. It has been estimated that there may be as many as thirty thousand different enzymes in the human body, each of which is a protein constructed in such a way as to make it effective as a catalyst for a specific chemical reaction useful to the organism. The interesting subject of enzymes will be studied in greater detail in Chapter 28, *Biochemistry.*

17.7   Rate of Reaction and Concentration

At a fixed temperature and in the absence of a catalyst (or in the presence of a catalyst if the catalyst is present in fixed amount), the rate of a given reaction in solution is largely dependent upon the concentrations of the reacting substances. Many familiar facts might be cited which illustrate this principle. For example, substances burn much more rapidly in pure oxygen than they do in air, in which only about 20 per cent of the molecules are oxygen.

The increase in reaction rate that accompanies an increase in concentration of reacting substances is readily explained in terms of the kinetic-molecular theory. With increased concentration of any or all of the reacting substances, the chances for collision between molecules are increased due to the presence of a greater number of molecules per unit volume. More collisions per unit time means a greater reaction rate. The concentration of a substance, when expressed in moles per liter, provides a direct measure of the number of molecules per unit space. For this reason concentrations are expressed in moles per liter when the effect of concentration upon reaction rate is being considered.

A quantitative relationship between reaction rate and concentration can be expressed as follows for a one-step reaction in which one molecule of each reactant appears in the equation (coefficients of 1): **The rate of a simple reaction (which might be only one step of a series of steps in a complicated mechanism), with coefficients of 1 for each reactant, is directly proportional to the concentrations of the reacting substances.**

For a reaction of the type $A \longrightarrow B + C$, such as $PCl_5 \longrightarrow PCl_3 + Cl_2$, the rate of reaction, R, is directly proportional to the concentration of A.

This relationship between reaction rate and concentration of the reactant may be expressed in the form of a **rate equation.**

$$R = k[A]$$

For PCl_5, $\qquad\qquad\qquad\qquad R = k[PCl_5]$

The reaction rate is represented by R, and the brackets mean *molar concentration of;* for example, $[PCl_5]$ means "molar concentration of PCl_5." The proportionality constant, k, is called the rate constant. The value of k remains the same as long as the temperature of the system does not change. If the temperature is raised, the reaction rate becomes greater and the value of k increases substantially. It is evident that value of k is equal to the rate of the reaction when the molar concentration of the Reactant A is unity. When the concentration of A is doubled, the rate of reaction will double, because there will be twice as many molecules to undergo decomposition per unit volume in a unit of time.

For a one-step bimolecular reaction of the type $A + B \longrightarrow C + D$, such as $NO_2 + CO \longrightarrow CO_2 + NO$ (which occurs in one step at temperatures above $225°$ C), the rate equation can be expressed as

$$R = k[A][B]$$

For NO_2 and CO, $\qquad\qquad R = k[NO_2][CO]$ $\qquad$ (temperatures above $225°$)

It must be realized that R and k each have different values for different reactions and, therefore, are not the same for the two reactions represented by $PCl_5 \longrightarrow PCl_3 + Cl_2$ and $NO_2 + CO \longrightarrow CO_2 + NO$.

In the reaction between the two reactants A and B, doubling the concentration of either one of them doubles the number of total molecular collisions and the number of effective collisions between Molecules A and B. Thus, if the concentration of A is doubled (and that of B is left the same), the rate of reaction is doubled. Doubling the concentration of both A and B quadruples the number of collisions and the rate of reaction (Fig. 17–7). If the initial concentrations of both A and B are trebled, then the reaction proceeds nine times as fast.

When more than one molecule of a reactant appears in the equation for a one-step reaction, such as $2A \longrightarrow B + C$, the reaction rate is found by experiment to be proportional to the molar concentration of the reactant raised to the power that is equal to the number of molecules of the reactant as stated in the equation. Hence, for the reaction $2A \longrightarrow B + C$, such as $2HI \longrightarrow H_2 + I_2$, the reaction rate is proportional to the square of the molar concentration of A.

$$R = k[A]^2$$

For HI, $\qquad\qquad\qquad\qquad R = k[HI]^2$

The fact that the rate of decomposition of HI is proportional to the square rather than to the first power of its concentration seems less surprising if the equation is written $HI + HI \longrightarrow H_2 + I_2$. We may suppose that this reaction can occur only when one molecule of HI collides with a second molecule of it, and hence we have $R = k[HI][HI] = k[HI]^2$.

50mm = 5cm

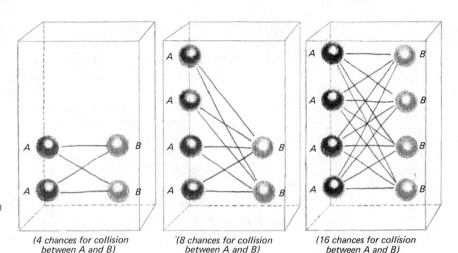

FIGURE 17-7

A schematic representation of the effect of concentration on the number of possible collisions and, hence, the rate of reaction.

(4 chances for collision between A and B)

(8 chances for collision between A and B)

(16 chances for collision between A and B)

For a one-step reaction of the type A + 2B $\longrightarrow$ C, it follows that the reaction rate is proportional to the molar concentration of A multiplied by the square of the molar concentration of B.

$$R = k[A][B]^2$$

In more general terms, for a one-step reaction of the type mA + nB $\rightleftharpoons$ C, **the reaction rate is proportional to the molar concentration of A times the molar concentration of B, with each concentration raised to the power equal to the number of molecules of that substance appearing in the equation.** This generalization is referred to as the Law of Mass Action.

$$R = k[A]^m[B]^n$$

17.8   Order of Reaction

A much used term, the **order of the reaction,** is governed by the powers to which the concentrations of the reactants are raised in the expression for the reaction rate, R. The order can be expressed either in terms of the **order with respect to each specific reactant** or in terms of the **overall order of the reaction.** The order with respect to one of the reactants is equal to the power to which the concentration of that particular reactant is raised. The overall order of the reaction is given by the sum of the exponents for the various reactants. The following example will illustrate the use of these terms.

Consider again the reaction mA + nB $\rightleftharpoons$ C (Section 17.7), for which the rate equation is

$$R = k[A]^m[B]^n$$

If the exponent m is 1, the reaction is said to be **first order** with respect to A. If m = 2, the reaction is said to be **second order** with respect to A. If n = 2, the reaction is referred to as second order with respect to B.

If m = 1 and n = 1, the reaction is first order in A and first order in B. The *overall* order of the reaction is given by m + n; therefore the overall order of the reaction in this particular case is second order.

It should be recalled that while reaction rates at a fixed temperature and in the absence of a catalyst are largely determined by the nature and concentration of the reacting substances, the reaction mechanism is also *very* important in establishing overall reaction rates (Section 17.2). It is likewise very important in establishing the overall order of the reaction. Thus, whereas we can write a valid rate equation for one step in a chemical reaction, *we cannot ordinarily write a correct rate equation or establish the order for an overall reaction which involves several steps, simply by inspection of the overall balanced equation.* It is necessary in such cases to determine the overall rate equation from experimental data. An excellent example is the reaction considered in the preceding section for the overall reaction of NO_2 and CO.

$$NO_2 + CO \longrightarrow CO_2 + NO$$

We pointed out that *for temperatures above 225° C* the rate equation is

$$R = k[NO_2][CO]$$

representing a simple one-step bimolecular reaction. For temperatures above 225° C, therefore, the order is first order with respect to NO_2 and first order with respect to CO; the overall order of the reaction is second order. It is interesting to note, however, that the reaction at *lower* temperatures proceeds in *two steps*, the first of which is slow and is therefore the rate-determining step on which the rate equation must be based.

$$NO_2 + NO_2 \longrightarrow NO_3 + NO \text{ (slow)} \qquad \textbf{(1)}$$
$$NO_3 + CO \longrightarrow NO_2 + CO_2 \text{ (fast)} \qquad \textbf{(2)}$$

(The summation of the two equations represents the net overall reaction.) Hence, for lower temperatures the rate equation, based upon the first step as the rate-determining step, is

$$R = k[NO_2]^2$$

Note that the rate equation for the lower temperatures does not even include the concentration of CO. At temperatures below 225° C, therefore, the reaction is *second* order with respect to NO_2 (and zero order with respect to CO); the overall order of the reaction still happens to be second order.

17.9   Half-life of a Reaction

■ **1. First Order Reaction.** The equation for a first order reaction relating the rate constant k to the initial concentration $[A_0]$ and to the concentration $[A]$ present after any given time t can be derived and is

$$\log \frac{[A_0]}{[A]} = \frac{kt}{2.303} \qquad \textbf{(1)}$$

An example illustrating the use of the equation follows:

Example 1.   The rate constant for a first order reaction involving Compound A was found to be 0.0820 min^{-1} when the initial concentration of A is 0.1500 mole/liter. How long will it take for the concentration of A to drop to 0.0300 mole/liter?

$$\log \frac{[A_0]}{[A]} = \frac{kt}{2.303}$$

Therefore,
$$t = \log \frac{[A_0]}{[A]} \times \frac{2.303}{k}$$

$$A_0 = 0.1500 \text{ mole/liter}$$
$$A = 0.0300 \text{ mole/liter}$$
$$k = 0.0820 \text{ min}^{-1}$$

Thus, $t = \log \dfrac{[A_0]}{[A]} \times \dfrac{2.303}{k} = \log \dfrac{0.1500 \text{ mole/liter}}{0.0300 \text{ mole/liter}} \times \dfrac{2.303}{0.0820 \text{ min}^{-1}}$

$$= 19.6 \text{ min}$$

The half-life of a reaction, $t_{\frac{1}{2}}$, is defined as the time required for half of the original concentration of the limiting reactant to be used up as the reaction takes place. In each succeeding half-life period, half of the remaining concentration of the reactant will be used up. For example, if the half-life for a reaction is ten minutes and the initial concentration of the limiting reactant is 0.20 mole/liter, the concentration will decrease to 0.10 mole/liter during the first half-life period of ten minutes, to 0.05 mole/liter during the second ten-minute period, to 0.025 mole/liter during the third ten-minute period, and so on.

The equation for determining the half-life for a first order reaction can be derived from Equation (1) as follows.

$$\log \frac{[A_0]}{[A]} = \frac{kt}{2.303}$$

$$t = \log \frac{[A_0]}{[A]} \times \frac{2.303}{k}$$

If the time t is the half-life time, $t_{\frac{1}{2}}$, the concentration of the limiting reactant at the end of that time $[A]$ is equal to one-half the initial concentration. Hence, at time $t_{\frac{1}{2}}$, $[A] = \frac{1}{2}[A_0]$.

Therefore,
$$t_{\frac{1}{2}} = \log \frac{[A_0]}{\frac{1}{2}[A_0]} \times \frac{2.303}{k}$$

$$= \log 2 \times \frac{2.303}{k}$$

$$= 0.301 \times \frac{2.303}{k}$$

Thus,
$$t_{\frac{1}{2}} = \frac{0.693}{k} \qquad\qquad (2)$$

You should note from Equation (2) that for a first order reaction the half-life is inversely proportional to the rate constant k. Hence, a fast reaction, with short half-life, has a large k value; a slow reaction, with a longer half-life, has a smaller k value.

Example 2. The reaction of Compound A to give Compounds C and D was found to be first order in A and first order overall. The rate constant for the reaction was determined to be 0.29 hr^{-1}. Calculate the half-life time for the reaction.

$$t_{\frac{1}{2}} = \frac{0.693}{k} = \frac{0.693}{0.29 \text{ hr}^{-1}} = 2.39 \text{ hr}$$

■ **2. Second Order Reaction.** For second order reactions, we will consider here (1) second order reactions which are second order with respect to one reactant and (2) second order reactions which are first order with respect to each of two reactants, but only for the special case in which the initial concentration of each reactant is the same. In such cases the equation for a second order reaction relating the rate constant k to the initial concentration $[A_0]$ for each of two reactants and to the concentration $[A]$ present after any given time t is

$$\frac{1}{[A]} - \frac{1}{[A_0]} = kt \tag{3}$$

Example 3. The reaction of an organic ester (a compound with the general formula RCOOR', where R and R' are organic groups—see Section 27.19) with a strong base is second order with a rate constant equal to 4.50 liters/mole·min. If the initial concentrations of ester and base are each 0.0200 mole/liter, what is the concentration remaining after 10.0 minutes?

$$\frac{1}{[A]} - \frac{1}{[A_0]} = kt$$

$$[A_0] = 0.0200 \text{ mole/liter}$$

$$k = 4.5 \text{ liters/mole·min}$$

$$t = 10.0 \text{ min}$$

Therefore, $\dfrac{1}{[A]} - \dfrac{1}{0.0200 \text{ mole/liter}} = 4.50 \text{ liters/mole·min} \times 10.0 \text{ min}$

or $\dfrac{1}{[A]} = 4.50 \text{ liters/mole·min} \times 10.0 \text{ min} + \dfrac{1}{0.0200 \text{ mole/liter}}$

$= 45.0 \text{ liters/mole} + \dfrac{1}{0.0200} \text{ mole}^{-1}/\text{liter}^{-1}$

$= 45.0 \text{ liters/mole} + 50.0 \text{ liters/mole}$

$= 95.0 \text{ liters/mole}$

Then $[A] = \dfrac{1}{95.0 \text{ liters/mole}} = 0.0105 \text{ mole/liter}$

Hence, 0.0105 mole each of the ester and of the strong base remains per liter at the end of ten minutes, compared to the 0.0200 mole of each originally present.

The equation used for calculating the half-life of a second order reaction in which the initial concentration of each reactant is the same can be derived from Equation (3) as follows.

$$\frac{1}{[A]} - \frac{1}{[A_0]} = kt$$

If time $t = t_{\frac{1}{2}}$, then $\quad\quad [A] = \frac{1}{2}[A_0]$

Therefore,

$$\frac{1}{\frac{1}{2}[A_0]} - \frac{1}{[A_0]} = kt_{\frac{1}{2}}$$

$$\frac{1 - \frac{1}{2}}{\frac{1}{2}[A_0]} = kt_{\frac{1}{2}}$$

$$\frac{\frac{1}{2}}{\frac{1}{2}[A_0]} = kt_{\frac{1}{2}}$$

$$\frac{1}{[A_0]} = kt_{\frac{1}{2}}$$

$$t_{\frac{1}{2}} = \frac{1}{k[A_0]} \tag{4}$$

Note that for a second order reaction, $t_{\frac{1}{2}}$ is not independent of the initial concentration, whereas for a first order reaction $t_{\frac{1}{2}}$ is independent of the initial concentration [see Equation (2)]. In this respect, the use of the half-life concept is more complex and less useful for second order reactions than for first order reactions. For second order reactions the rate constant cannot be calculated directly from the half-life, and vice versa, unless the initial concentration is known; this restriction is not the case for first order reactions.

Example 4. Calculate the half-life for the second order reaction described in Example 3 ($k = 4.50$ liters/mole·min) if the initial concentrations of each of the two reactants is 0.0200 mole/liter.

(Note before working the problem that the calculations in Example 3 show that, with an initial concentration of 0.0200 mole/liter, slightly more than half of each of the reactants (0.0105 mole/liter) remains after ten minutes. The half-life, therefore, must be a little bit longer than ten minutes. This provides us with a useful order of magnitude by which later to check our calculated answer.)

$$t_{\frac{1}{2}} = \frac{1}{k[A_0]} = \frac{1}{(4.50 \text{ liters/mole·min})(0.0200 \text{ mole/liter})} = 11.1 \text{ min}$$

In accord with our estimate, the half-life is a little longer than ten minutes, when the initial concentration is 0.0200 mole/liter.

Example 5. What is the half-life for the same reaction as in Example 4 if the initial concentration of each reactant is 0.0300 mole/liter?

$$t_{\frac{1}{2}} = \frac{1}{k[A_0]} = \frac{1}{(4.50 \text{ liters/mole} \cdot \text{min})(0.030 \text{ mole/liter})} = 7.41 \text{ min}$$

Note that the calculations in Examples 4 and 5 emphasize the dependence of half-life on the initial concentration for a second order reaction. The increase in the initial concentration from 0.0200 mole/liter to 0.0300 mole/liter shortens the half-life for this particular reaction from 11.1 minutes to 7.4 minutes.

17.10   Law of Chemical Equilibrium

Whenever the products of a chemical reaction are capable of reacting to form the reactants, two reactions are taking place simultaneously, the one tending to offset the other. As a consequence, such reactions do not go to completion and a state of equilibrium is attained. Most chemical reactions possess the quality of reversibility and do not go to completion.

Suppose we apply the law of mass action to a reversible reaction at a fixed temperature, such as $A + B \rightleftharpoons C + D$. This equation states that substances A and B react to form substances C and D, and that C and D react under the same conditions to form the original reactants A and B. Let us assume that the system at the outset contains only the original reactants A and B (see Fig. 17–8). The rate of reaction of A and B is expressed in terms of concentration in an equation referred to as the **rate equation.**

$$R_1 = k_1[A][B]$$

At first, no molecules of C and D are present, there is no reverse reaction, and $R_2 = 0$. However, as soon as some of the products C and D are formed they begin to react, and the rate equation of this reaction is

$$R_2 = k_2[C][D]$$

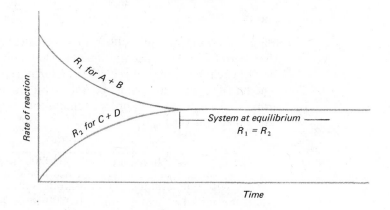

FIGURE 17–8

Rates of reaction of forward and reverse reactions for $A + B \rightleftharpoons C + D$, assuming only A and B are present initially.

Because the concentrations of C and D at first will be small, the rate of the reverse reaction, R_2, will be low. But, as the reaction between A and B proceeds, the concentrations of C and D increase and, likewise, the rate of the reverse reaction, R_2, increases. Meanwhile the concentrations of A and B are becoming less and less, so that the rate of the forward reaction, R_1, falls off. Consequently, the two reaction rates approach each other and finally become equal—a condition of *dynamic* equilibrium, which means that the opposing reactions are in full operation but at the same rate. Thus, at equilibrium $R_2 = R_1$, and we may write

$$k_2[C][D] = k_1[A][B]$$

or, by rearranging,

$$\frac{[C][D]}{[A][B]} = \frac{k_1}{k_2}$$

Because k_1 and k_2 are constants, the ratio k_1/k_2 is also constant, and the expression may be written

$$\frac{[C][D]}{[A][B]} = K_e$$

K_e is called the **equilibrium constant** for the reaction. Just as k_1 and k_2 are proportionality constants specific for each reaction at a definite temperature, K_e is likewise a constant specific to this system in equilibrium at a given temperature. The values for the molar concentrations used in the above mathematical expression must always be the concentrations present **after the reaction has reached a state of equilibrium.**

It is important to note that at equilibrium the rates of reaction, R_1 and R_2, are equal but the molar concentrations of the reactants and products in the equilibrium mixture are usually not equal. It is true, however, that the concentration of each reactant and product *at equilibrium* remains constant because the rate at which it is being used up in one reaction is equal to the rate at which it is being formed by the opposite reaction. As pointed out before, such a system is referred to as being in a state of **dynamic equilibrium.**

A general equation for chemical equilibria may be written

$$mA + nB + \ldots \rightleftharpoons xC + yD + \ldots$$

By applying the law of mass action to this reversible reaction, we may write

$$\frac{[C]^x[D]^y \ldots}{[A]^m[B]^n \ldots} = K_e$$

This is a mathematical expression of the law of chemical equilibrium which may be stated as follows: **When a reversible reaction has attained equilibrium at a given temperature, the product of the molar concentrations of the substances to the right of the arrow in the equation divided by the product of the molar concentrations of the substances to the left, each concentration raised to the power equal to the number of molecules of each substance appearing in the equation, is a constant.**

The equilibrium constant expression is valid even if the reactions take place in steps, with rates involving powers of the reactant concentrations different from the coefficients appearing in the balanced equation. In fact, the validity of the equilibrium constant expression is a consequence of the laws of thermodynamics (of physical chemistry), and a more rigorous derivation can be made at a later stage in the student's work when a background in thermodynamics and chemical kinetics has been attained.

The meaning of the mathematical expression of the law of chemical equilibrium is that, regardless of how the individual concentrations might be varied, thereby upsetting the equilibrium temporarily, the composition of the system with regard to its various components will always adjust itself to a new condition of equilibrium for which the quotient $[C]^x[D]^y \ldots /[A]^m[B]^n \ldots$ will again have the value K_e. When a mixture of A, B, C, and D is prepared in such proportions that the ratio $[C]^x[D]^y \ldots$ to $[A]^m[B]^n \ldots$ is not equal to K_e, then the system is not in a condition of equilibrium. However, its composition will change in such a direction that equilibrium will be established. If the ratio is less than K_e, the rate of reaction of A with B is greater than that of C with D, so that A and B will be used up faster than they are formed until the ratio becomes equal to K_e; conversely, if the ratio is greater than K_e, the rate of reaction of C and D will be greater than that of A with B until the ratio becomes equal to K_e.

The value of an equilibrium constant is a measure of the completeness of a reversible reaction. A large value for K_e indicates that equilibrium is attained only after the reactants A and B have been largely converted into the products C and D. When K_e is very small—much less than unity—equilibrium is attained when only a small proportion of A and B has been converted to C and D.

The following equations and equilibrium constant expressions will illustrate the law of chemical equilibrium for two reactions.

$$PCl_5 \rightleftharpoons PCl_3 + Cl_2 \quad \frac{\overset{\text{moles/liter}}{[PCl_3]}\,\overset{\text{moles/liter}}{[Cl_2]}}{\underset{\text{moles/liter}}{[PCl_5]}} = K_e \text{ moles/liter}$$

$$N_2 + 3H_2 \rightleftharpoons 2NH_3 \quad \frac{\overset{\text{(moles/liter)}}{[NH_3]^2}}{\underset{\text{moles/liter}}{[N_2]}\,\underset{\text{moles/liter}}{[H_2]^3}} = K_e \text{ moles}^{-2}/\text{liter}^{-2}$$
$$\text{(or liter}^2/\text{mole}^2)$$

17.11   Determination of Equilibrium Constants

The values of the equilibrium constants for different reactions can be determined experimentally. As one example of the determination of an equilibrium constant let us consider the reversible reaction $H_2 + I_2 \rightleftharpoons 2HI$. [As mentioned in Section 17.10, the equilibrium constant expression is valid even if the reaction takes place in steps as in the case with this particular reaction (Section 17.1)].

For a given mixture of H_2, I_2, and HI at 400°, it was found by analysis that $[H_2] = 0.221$ mole/liter, $[I_2] = 0.221$ mole/liter, and $[HI] = 1.563$ moles/liter, when the system had attained equilibrium. Substituting these values in the expression for the equilibrium constant for this system, we find

$$K_e = \frac{\underset{(\text{moles/liter})^2}{[HI]^2}}{\underset{\text{moles/liter}}{[H_2]}\underset{\text{moles/liter}}{[I_2]}} =$$

$$\frac{(1.563 \text{ moles/liter})^2}{(0.221 \text{ mole/liter})(0.221 \text{ mole/liter})} = 50.0 \text{ (units cancel out)}$$

If we start with just HI at any molar concentration, or with any mixture of H_2 and I_2, or with any mixture of H_2, I_2, and HI and hold the temperature of the system at 400° until equilibrium is established, the molar concentrations of the three substances will be such that the quotient $[HI]^2/[H_2][I_2]$ is equal to 50.0.

Example. At about 990° C, K_e for the reaction $H_2(g) + CO_2(g) \rightleftharpoons H_2O(g) + CO(g)$ is 1.6. Calculate the number of moles of each component in the final equilibrium system obtained from adding 1.00 mole of H_2, 2.0 moles of CO_2, 0.75 mole of H_2O, and 1.0 mole of CO to a 5.00-liter reactor at 990° C.

$$K_e = \frac{[H_2O][CO]}{[H_2][CO_2]} = 1.6 \qquad \frac{(.15+x)(.2+x)}{(.2-x)(.4-x)} = 1.6$$

where $[H_2O]$, $[CO]$, $[H_2]$, and $[CO_2]$ represent the *molar concentrations* (moles/liter) of H_2O, CO, H_2, and CO_2, respectively, *at equilibrium*. (Note that the units of moles/liter for each concentration cancel out, and hence the equilibrium constant in this particular case has no units.)

Let x = the number of moles of H_2 *at equilibrium* in the 5-liter volume.

Then, $\frac{x}{5}$ mole/liter represents the molar concentration of H_2 at equilibrium.

The initial quantity of H_2 which was present in the 5-liter volume before equilibrium is established, is 1.00 mole. If x moles remain after equilibrium is established, $(1.00 - x)$ mole of H_2 must have reacted with CO_2 to produce CO and H_2O in establishing equilibrium.

The balanced equation indicates that each mole of H_2 that reacts uses up one mole of CO_2 and produces one mole of H_2O and one mole of CO. Hence, when $(1.00 - x)$ mole of H_2 reacts, $(1.00 - x)$ mole of CO_2 is consumed and $(1.00 - x)$ mole each of H_2O and CO is produced.

Hence, at equilibrium, the concentration of CO_2 in the 5-liter volume is equal to the initial quantity present (2.0 moles) minus the amount which is consumed in establishing equilibrium $[(1.00 - x) \text{ mole}]$, for a net quantity of $[2.0 - (1.00 - x)]$ mole, or $(1.0 + x)$ moles, of CO_2 at equilibrium in the 5-liter volume. The molar concentration for CO_2 at equilibrium, therefore, is $\left(\dfrac{1.0 + x}{5}\right)$ mole/liter.

At the same time, $(1.00 - x)$ mole of H_2O is produced, which adds to the 0.75 mole already present for a total of $[0.75 + (1.00 - x)]$ mole, or $(1.75 - x)$ mole, of H_2O at equilibrium in the 5-liter volume. The molar concentration of H_2O at equilibrium, therefore, is $\left(\dfrac{1.75 - x}{5}\right)$ mole/liter.

Similarly, $(1.00 - x)$ mole of CO is produced, which adds to the 1.0 mole already present for a total of $[1.0 + (1.00 - x)]$ mole, or $(2.0 - x)$ mole, of CO at equilibrium in the 5-liter volume.

The molar concentration of CO at equilibrium, therefore, is $\left(\dfrac{2.0 - x}{5}\right)$ mole/liter.

Substituting the equilibrium *molar* concentrations in the expression for K_e:

$$K_e = \frac{[H_2O][CO]}{[H_2][CO_2]} = \frac{\left(\dfrac{1.75 - x}{5}\ \cancel{mole/liter}\right)\left(\dfrac{2.0 - x}{5}\ \cancel{mole/liter}\right)}{\left(\dfrac{x}{5}\ \cancel{mole/liter}\right)\left(\dfrac{1.0 + x}{5}\ \cancel{mole/liter}\right)} = 1.6$$

$$\frac{(1.75 - x)(2.0 - x)}{(x)(1.0 + x)} = 1.6$$

$$3.50 - 3.75x + x^2 = 1.6x + 1.6x^2$$

$$0.6x^2 + 5.35x - 3.50 = 0$$

By the quadratic formula (see Appendix A),

$$x = \frac{-5.35 \pm \sqrt{(5.35)^2 - 4(0.6)(-3.50)}}{2(0.6)}$$

$$= 0.61 \text{ mole of } H_2 \text{ in the 5-liter volume at equilibrium}$$

$$(1.0 + x) = 1.0 + 0.61$$
$$= 1.6 \text{ moles of } CO_2 \text{ in the 5-liter volume at equilibrium}$$

$$(1.75 - x) = 1.75 - 0.61$$
$$= 1.1 \text{ moles of } H_2O \text{ in the 5-liter volume at equilibrium}$$

$$(2.0 - x) = 2.0 - 0.61$$
$$= 1.4 \text{ moles of CO in the 5-liter volume at equilibrium}$$

17.12 Effect of Changing the Concentration upon an Equilibrium

The equilibrium existing in a chemical system may be shifted by increasing the rate of the forward or the reverse reaction. For the reversible reaction

$$A + B \rightleftharpoons C + D$$

shifts to right

when the system is in equilibrium and an additional quantity of A is added, the rate of the forward reaction is increased because the concentration of the reacting molecules is increased. This means that the rate of the forward reaction will for a time be greater than that of the reverse reaction; the system, then, is temporarily out of equilibrium. However, as the concentrations of C and D increase, the rate of the reverse reaction increases, whereas the decrease in the concentrations of A and B causes the rate of the forward reaction to decrease. The rates of the two reactions thereby will become equal again, and a second state of equilibrium attained in which the molar concentrations of A, B, C, and D have changed; however, the ratio [C][D] to [A][B] is again equal to the original value of K_e. The equilibrium is said to have been shifted to the right. In the new state of equilibrium, the substances C and D are present in greater concentration than originally, B is present in smaller concentration than originally, and A is present in greater concentration than before the addition of excess A. By increasing the concentration of B, the equilibrium can be shifted to the right in similar fashion. Increasing the concentration of either C or D, or both, will have the effect of shifting the equilibrium to the left.

shift to left

The equilibrium may also be shifted to the right by the removal of either C or D, or both. This causes a reduction in the reaction rate of the reaction to the left, and thus the reaction to the left proceeds more slowly than the reaction to the right until the reaction rates again become equal as a new condition of equilibrium is attained.

As an example of the effect of a change in concentration upon a system in equilibrium, let us consider the equilibrium $H_2 + I_2 \rightleftharpoons 2HI$, for which K_e was found to be 50.0 at 400° in the preceding section. If enough H_2 is introduced into the system to double its concentration, the rate of reaction of H_2 with I_2 to form HI increases. When equilibrium is again reached, $[H_2] = 0.374$, $[I_2] = 0.153$, and $[HI] = 1.692$. If these new values are substituted in the expression for the equilibrium constant for this system, we find

$$\frac{[HI]^2}{[H_2][I_2]} = \frac{(1.692)^2}{(0.374)(0.153)} = 50.0 = K_e$$

Hence, by doubling the concentration of H_2, we have caused the formation of more HI, used up about one-third of the I_2 present at the first equilibrium and used up some, but not all, of the excess H_2 added.

The effect of a change in concentration upon a system in equilibrium is an important application of the **principle of Le Châtelier,** which may be stated as follows: **If a stress (such as a change in concentration, pressure, or temperature) is applied to a system in equilibrium, the equilibrium is shifted in a way that tends to undo the effect of the stress.**

17.13 Effect of Change in Pressure on Equilibrium

Changes in pressure measurably affect systems in equilibrium only when gases are involved and, in such cases, only when the chemical reaction involves a change in the total number of molecules in the system. This follows from Avogadro's

Law, which states that equal volumes of gases at constant temperature and pressure contain equal numbers of molecules (see Section 2.5). As the pressure on a gaseous system is increased, the gases composing the system undergo compression, and their concentrations in moles per liter and the total number of molecules per unit volume increase. Thus, in accord with the principle of Le Châtelier, the chemical reaction that reduces the total number of molecules per unit of volume will be the one favored by an increase in pressure.

Consider the effect of an increase in pressure upon the system in which one molecule of nitrogen and three molecules of hydrogen interact to form two molecules of ammonia.

$$N_2 + 3H_2 \rightleftharpoons 2NH_3$$

The formation of ammonia decreases the total number of molecules of the system by 50 per cent, compared to the original number, thus reducing the total pressure exerted by the system. Experiment shows that an increase in pressure does drive the reaction to the right. On the other hand, lowering the pressure on the system favors decomposition of ammonia into hydrogen and nitrogen. These observations are fully in accord with Le Châtelier's principle. (Table 25–2 in the chapter on *Nitrogen and Its Compounds* shows equilibrium concentrations of ammonia at several temperatures and pressures.)

Let us now consider another reaction, one in which a molecule of nitrogen interacts with a molecule of oxygen with the formation of two molecules of nitrogen(II) oxide.

$$N_2 + O_2 \rightleftharpoons 2NO$$

Because there is no change in the total number of molecules in the system during reaction, a change in pressure does not favor either formation or decomposition of nitrogen(II) oxide at temperatures high enough for all the materials to be present in the gaseous state.

Whenever a gaseous substance is involved in a system in equilibrium, the pressure (or the partial pressure) of the gas can be substituted for the concentration in the expression for the equilibrium constant, because the concentration of a gas at constant temperature varies directly with the pressure. Thus, for the system

$$N_2(g) + 3H_2(g) \rightleftharpoons 2NH_3(g)$$

we may write

$$\frac{p_{NH_3}^2}{p_{N_2}p_{H_2}^3} = K_e$$

It is important to note that when partial pressures of gases are substituted for their concentrations the equilibrium constant, K_e, will still be a constant, but it will usually have a different numerical value and different units from those when concentrations are used.

The units for K_e for the reaction to produce ammonia, if the three partial pressures are expressed in mm, would be

$$\frac{mm^2}{mm \times mm^3} = \frac{1}{mm^2} = mm^{-2}$$

Although equilibrium constants have units (unless they happen to cancel out as they do in the reaction $N_2 + O_2 \rightleftharpoons 2NO$), the values for equilibrium constants are often given without the units.

Example 1. A 5.00-liter vessel contains CO, CO_2, H_2, and H_2O in equilibrium at 980° C. The equilibrium partial pressures are CO, 150 mm; CO_2, 200 mm; H_2, 90 mm; H_2O, 200 mm. Hydrogen is then pumped into the vessel until the equilibrium partial pressure of carbon monoxide is equal to 230 mm. Calculate the partial pressures of the other substances at the new equilibrium assuming that no temperature change occurs.

$$CO_2(g) + H_2(g) \rightleftharpoons CO(g) + H_2O(g)$$

$$K_e = \frac{p_{CO}p_{H_2O}}{p_{CO_2}p_{H_2}}$$

$$K_e = \frac{(150 \text{ mm})(200 \text{ mm})}{(200 \text{ mm})(90 \text{ mm})} = 1.67$$

(Units cancel out, and the equilibrium constant in this particular case has no units.)

At the new equilibrium $p_{CO} = 230$ mm. Therefore, p_{CO} has increased in proportion to the increased concentration, from 150 mm to 230 mm, an increase of 80 mm.

The balanced equation tells us that for each increase of one mole for CO, the H_2O concentration must also increase by one mole; hence, the pressure increase for H_2O from its initial pressure of 200 mm will be the same (80 mm) as for CO.

Thus, p_{H_2O} at the new equilibrium = 200 mm + 80 mm = 280 mm. The balanced equation also tells us that for every mole of CO produced, one mole of CO_2 must be consumed, and hence the pressure of CO_2 must decrease from its initial pressure of 200 mm by the same amount as the increase of pressure for CO (80 mm). Thus, p_{CO_2} at the new equilibrium = 200 mm − 80 mm = 120 mm.

Now, the partial pressure of the hydrogen at the new equilibrium can be calculated by substituting p_{CO}, p_{H_2O}, and p_{CO_2} for the new equilibrium into the expression for K_e and solving for p_{H_2}:

$$K_e = \frac{(230 \text{ mm})(280 \text{ mm})}{(120 \text{ mm})p_{H_2}} = 1.67$$

$$p_{H_2} = \frac{(230 \text{ mm})(280 \text{ mm})}{(120 \text{ mm})(1.67)} = 321 \text{ mm}$$

Therefore, the partial pressures for CO, H_2O, CO_2, and H_2 at the new equilibrium are 230 mm, 280 mm, 120 mm, and 321 mm, respectively.

Note that sufficient hydrogen had to be pumped into the vessel to increase the pressure of the hydrogen from 90 mm to 321 mm.

Example 2. In a 3.0-liter vessel, the following equilibrium partial pressures are measured: N_2, 380 mm; H_2, 400 mm; NH_3, 2,000 mm. Hydrogen is removed from the vessel until the partial pressure of nitrogen, at equilibrium, is equal to 450 mm. Calculate the partial pressures of the other substances under the new conditions.

$$N_2(g) + 3H_2(g) \rightleftharpoons 2NH_3(g)$$

$$K_e = \frac{p_{NH_3}^2}{p_{N_2}p_{H_2}^3}$$

$$K_e = \frac{(2000 \text{ mm})^2}{(380 \text{ mm})(400 \text{ mm})^3} = 1.644 \times 10^{-4} \text{ mm}^{-2}$$

The removal of H_2 shifts the equilibrium to the left producing additional N_2 and therefore increasing the pressure of N_2 in proportion to the increased molar concentration. At the new equilibrium $p_{N_2} = 450$ mm. Hence, the pressure of N_2 has increased by 70 mm.

The balanced equation tells us that two moles of NH_3 must be used up for each mole of N_2 that is produced. Hence, the pressure of NH_3 must decrease in proportion to the decrease in molar concentration. The pressure of NH_3 must therefore decrease by twice the amount that the pressure of N_2 increases, or a decrease of 140 mm in the pressure of NH_3.

Therefore, at the new equilibrium $p_{NH_3} = 2,000 - 140 = 1,860$ mm.

To calculate p_{H_2} at the new equilibrium, substitute p_{N_2} and p_{NH_3} in the expression for K_e:

$$K_e = \frac{(1,860 \text{ mm})^2}{(450 \text{ mm})p_{H_2}^3} = 1.644 \times 10^{-4} \text{ mm}^{-2}$$

$$p_{H_2}^3 = \frac{(1,860 \text{ mm})^2}{(450 \text{ mm})(1.644 \times 10^{-4} \text{ mm}^{-2})}$$

$$= 4.676 \times 10^7 \text{ mm}^3 = 46.76 \times 10^6 \text{ mm}^3$$

$$p_{H_2} = \sqrt[3]{46.76 \times 10^6 \text{ mm}^3} = 3.60 \times 10^2 \text{ mm} = 360 \text{ mm}$$

Hence, at the new equilibrium the partial pressures of NH_3 and H_2 are 1,860 mm and 360 mm, respectively. Note that the quantity of hydrogen that was removed is the amount which would reduce the pressure of the hydrogen from 400 mm to 360 mm.

17.14   Effect of Change in Temperature on Equilibrium

All chemical changes involve either the evolution of energy or the absorption of energy. In every system in equilibrium, an endothermic and an exothermic reaction are taking place simultaneously. The endothermic reaction is favored by an increase in temperature, providing an increase in energy, and the exothermic reaction is favored by a decrease in temperature.

In the reaction between gaseous hydrogen and gaseous iodine, heat is evolved.

$$H_2(g) + I_2(g) \rightleftharpoons 2HI(g) + heat \qquad exothermic$$

Lowering the temperature of the system favors the formation of hydrogen iodide, whereas raising the temperature favors the decomposition of hydrogen iodide. Thus, raising the temperature of the system will decrease the value of the equilibrium constant, since the molar concentration of HI will be decreased while the molar concentrations of H_2 and I_2 will be increased. The value of the equilibrium constant

$$\frac{[HI]^2}{[H_2][I_2]} = K_e$$

decreases from 67.5 at 357° to 50.0 at 400°.

The effect of temperature changes upon systems in equilibrium may be summarized by the statement: **When the temperature of a system in equilibrium is raised, the equilibrium is displaced in such a way that heat is absorbed.** This generalization is known as **van't Hoff's law,** which is a special case of Le Châtelier's principle. It is important to remember that the effect of changing the temperature is a change in the value of the equilibrium constant.

The equation for the formation of ammonia from hydrogen and nitrogen,

$$N_2 + 3H_2 \rightleftharpoons 2NH_3 + 22,040 \text{ cal}$$

indicates that the equilibrium can be shifted to the right to favor the formation of more ammonia by lowering the temperature (see Table 25–2). However, it must be remembered that, because of the rapid lowering of reaction rates with decreasing temperature, equilibrium is attained more slowly at the lower temperature. In the commercial production of ammonia from nitrogen and hydrogen, it is not feasible to use temperatures much lower than 500°, because at such temperatures, even in the presence of a catalyst, the reaction proceeds too slowly to be practical.

A **calorimeter** is used to measure the heat change which occurs during a chemical reaction. In the simplest form of calorimeter, the reaction is allowed to take place in a reaction vessel which is immersed in a known weight of a liquid within an insulated container. The change of temperature in the liquid, caused by release or absorption of heat by the reaction, is measured with a sensitive thermometer or other temperature-sensing device.

17.15 Effect of a Catalyst on Equilibrium

Iron powder is used as a catalyst in the production of ammonia from nitrogen and hydrogen to increase the rate of reaction of these two elements.

$$N_2 + 3H_2 \xrightarrow{\text{Fe}} 2NH_3$$

However, this same catalyst serves equally well to increase the rate of the reverse reaction, that is, the decomposition of ammonia into its constituent elements.

$$2NH_3 \xrightarrow{\text{Fe}} N_2 + 3H_2$$

Thus, the net effect of iron in the reversible reaction

$$N_2 + 3H_2 \rightleftharpoons 2NH_3$$

is to cause equilibrium to be reached more rapidly. *Catalysts have no effect on the value of the equilibrium constant.* They merely increase the rate of both the forward and the reverse reactions to the same extent.

17.16   Homogeneous and Heterogeneous Equilibria

A **homogeneous system,** or **a homogeneous phase,** is any homogeneous portion of matter. A phase may be solid, liquid, or gas. Furthermore, any phase may be composed of an element, a compound, or a homogeneous mixture (e.g., a solution of two or more compounds). **A homogeneous equilibrium** is an equilibrium within a single homogeneous phase. Most of the equilibria which have been considered in this chapter are homogeneous equilibria involving reversible changes in only one phase, the gas phase.

A **heterogeneous equilibrium** is an equilibrium between two or more different phases. Liquid water in equilibrium with ice, liquid water in equilibrium with water vapor, and a solid in contact with its saturated solution are examples of heterogeneous equilibria. Each of these equilibria involves some kind of boundary surface between two phases.

Suppose we consider the example of a heterogeneous equilibrium provided by the decomposition of calcium carbonate into calcium oxide and carbon dioxide according to the equation

$$CaCO_3(s) \rightleftharpoons CaO(s) + CO_2(g)$$

Let us write the expression for the equilibrium constant for this reversible reaction.

$$\frac{[CaO][CO_2]}{[CaCO_3]} = K_e$$

The concentration of a solid substance is proportional to its density and hence remains constant as long as the temperature is not changed. It follows then that the concentrations of whatever solid substances are involved in the equilibrium, since they are constant, may be included in the value of the equilibrium constant, and need not appear at all in the expression for the equilibrium constant. We may write

$$[CO_2] = \frac{[CaCO_3]}{[CaO]} \times K_e$$

Now, we have on the right-hand side of the equation only constant terms. These

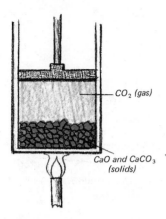

CO₂ (gas)

CaO and CaCO₃
(solids)

FIGURE 17-9

The thermal decomposition of calcium carbonate in a closed system is an example of a heterogeneous equilibrium.

can be combined into a single constant, K.

$$[CO_2] = K$$

The expression for equilibrium constant in terms of partial pressures is written

$$p_{CO_2} = K_p$$

This equation means that, at a given temperature, there is only one pressure at which carbon dioxide gas can be in equilibrium with the two solids $CaCO_3$ and CaO. Suppose a sample of $CaCO_3$ is placed in a cylinder with a movable piston, as shown in Fig. 17–9, and that the temperature of the system is raised to 900°. Under these conditions calcium carbonate will continue to decompose into calcium oxide and carbon dioxide until there is sufficient carbon dioxide present in the system to exert a pressure equal to K_p (actually 790 mm), if the piston is held stationary.

$$p_{CO_2} = K_p = 790 \text{ mm (at } 900°)$$

At this pressure, equilibrium will have been attained, and $CaCO_3$ will decompose into CaO and CO_2 at the same rate that CaO and CO_2 react to produce $CaCO_3$. If one increases the pressure by pushing the piston down and thus compressing the gas, more carbon dioxide will combine with calcium oxide to form calcium carbonate, and the pressure will drop to 790 mm again. If, on the other hand, the pressure on the system is decreased by raising the piston, just enough calcium carbonate will decompose to bring the pressure exerted by the carbon dioxide back to its original value of 790 mm.

In the commercial production of quicklime, CaO, from limestone the carbon dioxide is continuously removed as fast as it is formed by means of a stream of air; this causes the reaction to proceed essentially only to the right, and equilibrium is never established.

Tabulated values for equilibrium constants of reactions involving pure gases in equilibrium with pure liquids and/or pure solids are the K_p values based upon partial pressures.

17.17 The Distribution Law and Extraction

A good example of heterogeneous equilibria is the distribution of solid solute in two immiscible solvents which are in contact but separated by a phase boundary. Such a solute is iodine, which is soluble in both water and carbon tetrachloride. If an aqueous solution of iodine is shaken vigorously with a sufficient quantity of carbon tetrachloride, the greater part of the iodine will leave the water and go into the carbon tetrachloride. After shaking is stopped, the carbon tetrachloride, colored violet by the iodine, settles to the bottom of the container. The equilibrium between the iodine in the water and in the carbon tetrachloride may be represented by

$$I_2 \text{ (in } H_2O \text{ phase)} \rightleftharpoons I_2 \text{ (in } CCl_4 \text{ phase)}$$

Applying the law of mass action to the equilibrium system, we have

$$\frac{[I_2(CCl_4)]}{[I_2(H_2O)]} = K_e$$

The equilibrium constant for systems of this type is called the **distribution ratio,** and it has the value of 85 for this particular system at room temperature. In other words, the molar concentration of iodine in the carbon tetrachloride is 85 times that in the aqueous layer when equilibrium is established. Thus, most of the iodine in a water solution can be removed by a single treatment with carbon tetrachloride. This process is known as **extraction.** If the aqueous layer is extracted a second time with a fresh portion of carbon tetrachloride, the concentration of iodine in the water solution is reduced to a negligibly small value.

QUESTIONS

1. Characterize a reversible reaction and give an example of one.
2. Account for the increase in reaction rate brought about by catalysts.
3. List the factors that determine the rate of a reaction.
4. Relate reaction rate to activation energy.
5. What is the effect upon the rate of many reactions caused by an increase in temperature of 10°? Explain this effect in terms of the collision theory of reaction rate.
6. Using the law of mass action, derive the mathematical expression of the law of chemical equilibrium for the following reversible reactions:

$$N_2 + 3H_2 \rightleftharpoons 2NH_3$$
$$CH_4 + Cl_2 \rightleftharpoons CH_3Cl + HCl$$
$$N_2 + O_2 \rightleftharpoons 2NO$$
$$2SO_2 + O_2 \rightleftharpoons 2SO_3(g)$$
$$2H_2 + O_2 \rightleftharpoons 2H_2O(g)$$
$$CO_2 + H_2 \rightleftharpoons CO + H_2O(g)$$
$$NH_4Cl(s) \rightleftharpoons NH_3 + HCl$$
$$BaSO_3(s) \rightleftharpoons BaO(s) + SO_2$$
$$2Pb(NO_3)_2(s) \rightleftharpoons 2PbO(s) + 4NO_2 + O_2$$
$$4NH_3 + 5O_2 \rightleftharpoons 4NO + 6H_2O(g)$$
$$N_2O_4 \rightleftharpoons 2NO_2$$
$$H_2CO_3 \rightleftharpoons H_2O + CO_2$$
$$S_8 \rightleftharpoons 8S(g)$$

7. Write the mathematical expression of the law of chemical equilibrium for the reversible reaction

$$2CO + O_2 \rightleftharpoons 2CO_2 + Heat$$

What will happen to the concentration of CO_2 in the system at equilibrium, if (a) more oxygen is added? (b) CO is removed? (c) the pressure on the system is increased? (d) the temperature is decreased?

8. State Le Châtelier's principle as applied to chemical equilibria.

9. How will an increase in temperature affect the following equilibria?

$$N_2 + 3H_2 \rightleftharpoons 2NH_3 + \text{Heat}$$
$$H_2O(l) + \text{Heat} \rightleftharpoons H_2O(g)$$
$$N_2 + O_2 + \text{Heat} \rightleftharpoons 2NO$$
$$3O_2 + \text{Heat} \rightleftharpoons 2O_3$$
$$CaCO_3 + \text{Heat} \rightleftharpoons CaO + CO_2$$

10. For each of the following reactions between gases at equilibrium, determine the effect upon the equilibrium concentrations of the products when the temperature is decreased; when the external pressure on the system is decreased.

    (a) $2H_2O(g) \rightleftharpoons 2H_2 + O_2 - 115.6$ kcal
    (b) $N_2 + O_2 \rightleftharpoons 2NO - 43.2$ kcal
    (c) $N_2 + 3H_2 \rightleftharpoons 2NH_3 + 22.0$ kcal
    (d) $2O_3 \rightleftharpoons 3O_2 + 78.1$ kcal
    (e) $H_2 + F_2 \rightleftharpoons 2HF + 129.4$ kcal

11. Suggest four ways in which the equilibrium concentration of ammonia can be increased for the reaction referred to in Question 10(c).

12. For the reaction in Question 10(b), the equilibrium constant at $2000°$ C is 6.2×10^{-4}. Consider each of the two situations below and decide whether a reaction will occur and, if so, in which direction it will proceed predominantly.

    (a) A 5.0-liter box contains 0.26 mole N_2, 0.0062 mole O_2, and 0.0010 mole NO at $2,000°$ C.
    (b) A 2.5-liter box contains 0.26 mole N_2, 0.0062 mole O_2, and 0.0010 mole NO at $2,500°$ C.

13. Under what conditions do changes in pressure affect systems in equilibrium?

14. How is the speed of the reaction $A + B \longrightarrow C + D$ affected by quadrupling the initial concentrations of A and B?

15. In general, the equilibrium constant for a reaction in the gaseous phase has different units and a different numerical value when pressures rather than concentrations are used in the equilibrium constant expression. Show that such is not the case for the decomposition of HI into H_2 and I_2; show that the numerical value of the equilibrium constant would be independent of the units in which concentration is expressed for this particular reaction.

16. In the reaction between hydrogen and sulfur to produce H_2S, energy is liberated as H_2S is made. How should the pressure and temperature be adjusted in order to improve the equilibrium yield of H_2S, assuming conditions such that all reactants and products are in the gaseous state? Speculate what these conditions will do to the rate of attainment of equilibrium.

17. What is the effect of a catalyst upon the equilibrium constant of a system? Of what value is a catalyst for a system that attains equilibrium?

18. Nitric oxide gas (NO) reacts with chlorine gas according to the equation

$$2NO + Cl_2 \longrightarrow 2NOCl$$

The following initial rates of reaction have been observed for the reagent concentrations listed:

NO (moles/liter)	Cl_2 (moles/liter)	Rate (moles/liter·hour)
0.50	0.50	1.14
1.00	0.50	4.56
1.00	1.00	9.12

What is the mathematical equation (rate equation) describing the rate dependence on the concentrations of NO and Cl_2?

PROBLEMS

S1. Nitrogen reacts with hydrogen to give ammonia according to the equation $N_2 + 3H_2 \rightleftharpoons 2NH_3$. An equilibrium mixture of the above substances at 400° C was found to contain 0.45 mole of nitrogen, 0.63 mole of hydrogen, and 0.24 mole of ammonia per liter. Calculate the equilibrium constant for the system.   *Ans. 0.51*

2. Consider the equation $PCl_3(g) + Cl_2(g) \rightleftharpoons PCl_5(g)$. At equilibrium, the following quantities of the substances are present in a volume of 3.00 liters: 0.45 mole of PCl_3, 53.2 g of Cl_2, and 0.73 mole of PCl_5. Calculate K_e for the system.   *Ans. 6.5 liter/mole*

S3. The equilibrium constant for the gaseous reaction $H_2 + I_2 \rightleftharpoons 2HI$ is 50.2 at 448° C. Calculate the number of grams of HI that are in equilibrium with 1.25 moles of H_2 and 63.5 grams of iodine at this temperature. *Ans. 507 g*

S4. The equilibrium constant for the reaction represented by $CO + H_2O \rightleftharpoons CO_2 + H_2$ is 5.0 at a given temperature.

    (a) Upon analysis, an equilibrium mixture of the substances present at the given temperature was found to contain 0.20 mole of CO, 0.30 mole of water vapor, and 0.90 mole of H_2 in a liter. How many moles of CO_2 were there in the equilibrium mixture?   *Ans. 0.33*

    (b) Maintaining the same temperature, additional H_2 was added to the system, and some water vapor removed by drying. A new equilibrium mixture was thereby established which contained 0.40 mole of CO, 0.30 mole of water vapor, and 1.2 moles of H_2 in a liter. How many moles of CO_2 were in the new equilibrium mixture? Compare the value with the quantity in part (a) and discuss whether the second value is reasonable. Explain how it is possible for the water vapor concentration to be the same in the two equilibrium solutions even though some vapor was removed before the second equilibrium was established.   *Ans. 0.50*

S5. At about 990° C, K_e for the reaction represented by $H_2(g) + CO_2(g) \rightleftharpoons H_2O(g) + CO(g)$ is 1.6. Calculate the number of moles of each component in the final equilibrium system obtained from adding 1.00 mole of H_2 and 1.00 mole of CO_2 to a 5.00-liter reactor at 990° C.

Ans. 0.44 mole each of H_2 and CO_2; 0.56 mole each of H_2O and CO

[S]6. Ethanol and acetic acid interact to form ethyl acetate and water, according to the equation

$$C_2H_5OH + CH_3COOH \rightleftharpoons CH_3COOC_2H_5 + H_2O$$

When one mole each of ethanol (C_2H_5OH) and acetic acid (CH_3COOH) are allowed to react at 100° C in a sealed tube, equilibrium is established when one-third of a mole of each of the reactants remains. Calculate K_e. *Ans. 4*

7. One (1.0) mole of PCl_5 is placed in a 4.0-liter container. Dissociation takes place according to the equation $PCl_5(g) \rightleftharpoons PCl_3(g) + Cl_2(g)$. At equilibrium 0.25 mole of Cl_2 is present. Calculate the value of the equilibrium constant for this reaction under the conditions of the experiment.

Ans. 0.021 mole/liter

[S]8. If the rate of a reaction doubles for every ten degree rise in temperature, how much faster would the reaction proceed at 55° than at 25°? at 100° than at 25°? *Ans. 8 times faster; 180 times faster*

9. In an experiment, a sample of $NaClO_3$ was 90% decomposed in 48 minutes. How long would it have taken had the sample been heated 40° higher?

Ans. 3.0 min.

10. (a) Write the mathematical expression of the law of chemical equilibrium for the reversible reaction $2NO_2 \rightleftharpoons 2NO + O_2$.
    (b) At one atmosphere and 25° C, one molar NO_2 is $3.3 \times 10^{-3}\%$ decomposed into NO and O_2. Calculate the equilibrium constant.

Ans. $K_e = 1.8 \times 10^{-14}$ mole/liter

11. If 60.0 g each of acetic acid (CH_3COOH) and ethanol (C_2H_5OH) are allowed to react in a 1.00-liter sealed container at 100° C until equilibrium is established, how many moles of the ester ($CH_3COOC_2H_5$) and water are formed and how many moles of ethanol and acid remain? ($K_e = 4.00$.) The equation is $CH_3COOH + C_2H_5OH \rightleftharpoons CH_3COOC_2H_5 + H_2O$.

Ans. 0.252 mole of acid and 0.555 mole of ethanol; 0.747 mole each of ester and water

[S]12. An equilibrium mixture of N_2, H_2, and NH_3 in a 1.00-liter vessel is found to contain 0.300 mole of N_2, 0.400 mole of H_2, and 0.100 mole of NH_3. How many moles of H_2 must be introduced into the vessel in order to double the equilibrium concentration of NH_3 if the temperature remains unchanged?

Ans. 0.425 mole

[S]13. At 25° C and atmospheric pressure, the partial pressures in an equilibrium mixture of N_2O_4 and NO_2 are $p_{N_2O_4} = 0.70$ atmosphere; $p_{NO_2} = 0.30$ atmosphere. Calculate the partial pressures of these two gases when they are in equilibrium at 9.0 atmospheres and 25° C.

Ans. $p_{N_2O_4} = 8.0$ atm; $p_{NO_2} = 1.0$ atm

[S]14. Consider the gas phase reaction $N_2O_4(g) \rightleftharpoons 2NO_2(g)$. At a certain temperature K_e for this reaction is 1.1×10^{-5}. If 0.20 mole of N_2O_4 is dissolved in 400 ml of chloroform and the above reaction allowed to come to equilibrium, (a) what will be the NO_2 concentration, and (b) what will be the % dissociation of the original N_2O_4? *Ans.* (a) 2.3×10^{-3} *mole/liter;* (b) *0.23%*

15. The equilibrium pressure of carbon dioxide over calcium carbonate at various temperatures is as follows:

Temp, °C	p_{CO_2}, mm
600	10
800	180
840	320
880	580
896	760
910	1,000

(a) Plot the pressure as a function of temperature.
(b) What would be the sublimation temperature of carbon dioxide (dry ice) at sea level?   *Ans. 896°*

16. One (1.0) mole of solid ammonium chloride was placed in a box and heated, whereupon partial decomposition occurred: $NH_4Cl(s) \rightleftharpoons NH_3(g) + HCl(g)$. When the NH_3 partial pressure at equilibrium equals 1.6 atmospheres what is the equilibrium constant? What happens if the volume of the box is decreased 50% by applying pressure on a wall? What happens if a hole is punched in the box?   *Ans. $K_e = 2.6$ atm²*

17. A 5.00-liter vessel contains CO, CO_2, H_2, and H_2O in equilibrium at 980° C. The equilibrium partial pressures are CO, 300 mm; CO_2, 300 mm; H_2, 90 mm; H_2O, 150 mm. Carbon monoxide is then pumped into the vessel until the equilibrium partial pressure of hydrogen is equal to 130 mm. Calculate the partial pressures of the other substances at the new equilibrium assuming that no temperature change occurs.

Ans. CO_2, 340 mm; H_2O, 110 mm; CO, 670 mm

⑤18. In a 3.0-liter vessel, the following equilibrium partial pressures are measured: N_2, 190 mm; H_2, 317 mm; NH_3, 1000 mm. Hydrogen is removed from the vessel until the partial pressure of nitrogen, at equilibrium, is equal to 250 mm. Calculate the partial pressures of the other substances under the new conditions.   *Ans. p_{NH_3}, 880 mm; p_{H_2}, 266 mm*

⑤19. What is the minimum weight of $CaCO_3$ required to establish equilibrium at a certain temperature in a 6.50-liter container if the equilibrium constant is 0.050 mole/liter for the decomposition reaction of $CaCO_3$ at that temperature? Justify the units given in the problem for the equilibrium constant. The equation for the reaction is $CaCO_3(s) \rightleftharpoons CaO(s) + CO_2(g)$.   *Ans. 33 g*

20. What is the value of the proportionality constant for the rate equation determined previously in *Question* 18 (not Problem 18)? Reaction data provided in the question are for the reaction of NO with Cl_2 to produce NOCl.

Ans. 9.12 liters²/mole²·hour

21. A 1.00-liter vessel at 400° C contains the following equilibrium concentrations: N_2, 1.00 M; H_2, 0.50 M; and NH_3, 0.50 M. How many moles of hydrogen

must be removed from the vessel in order to increase the concentration of
nitrogen to 1.2 M?

*Ans. 0.94 (Note that more hydrogen is removed than was originally present
as elemental H_2. Is this possible?)*

§22. At high temperatures, the reaction of hydrogen and oxygen to form water
is reversible and achieves equilibrium. Under such conditions, a 1.0-liter
container holding pure hydrogen at a pressure of 0.80 atmosphere and another
1.0-liter container holding pure oxygen at a pressure of 20.0 atmospheres were
connected and opened to each other so that the gases could mix and come
to chemical equilibrium with water vapor. Then, the partial pressure of oxygen
was measured and found to be *essentially* 10.0 atmospheres. (Assume that
the quantity of oxygen which reacts is negligible compared to the quantity
originally present.) The equilibrium constant for the reaction at the temperature of the experiment has a value of 2.50 atm^{-1}. (a) Write the equilibrium
constant expression for the reaction. (b) What are the equilibrium partial
pressures of hydrogen and water? (c) Check whether the assumption made
regarding the quantity of oxygen which reacts compared to the quantity
originally present is justifiable. Could a similar assumption be justified for
hydrogen? Explain.

Ans. 0.07 atm H_2; 0.33 atm H_2O

23. For the reaction A $\longrightarrow$ B + C the following data were obtained at 30° C:

Experiment	[A] (in mole/liter)	Rate (in mole·liter^{-1}·hr^{-1})
1	0.170	0.0500
2	0.340	0.100
3	0.680	0.200

(a) What is the rate equation and order of the reaction?   *Ans. First order*
(b) Calculate k for the reaction.   *Ans. 0.294 hr^{-1}*

24. The half-life for the radioactive decay of ^{14}C (first order decay reaction) is
5730 years. What is the rate constant in units of min^{-1} and hr^{-1}?

Ans. 2.30 × 10^{-10} min^{-1}; 1.38 × 10^{-8} hr^{-1}

25. The reaction of compound A to give compounds C and D was found to be
second order in A and second order overall. The rate constant for the process
was determined to be 2.42 liter·mole^{-1}·sec^{-1}. If the initial concentration is
0.0500 mole/liter, what is the value of $t_{\frac{1}{2}}$?   *Ans. 8.26 sec*

26. The reaction of an organic ester with a strong base is second order, with
$k = 4.5$ liter·mole^{-1}·min^{-1}. The initial concentrations of ester and of base
are each 0.0200 mole/liter. Calculate the concentration each reactant will have
after (a) 15 minutes; (b) 20 minutes.   *Ans. (a) 0.0085 M; (b) 0.0071 M*

27. The half-life of a reaction involving compound A was found to be 8.50 min
when the initial concentration of A was 0.150 mole/liter. How long will it
take for the concentration to drop to 0.0300 mole/liter if the reaction is (a)
first order overall and first order with respect to A? (b) second order overall
and second order with respect to A? In your calculations, indicate the units
for each quantity, and show how the units cancel to give the proper units
in the final answer.   *Ans. (a) 19.7 min; (b) 34.0 min*

REFERENCES

"Chemical Principles Exemplified," R. C. Plumb, *J. Chem. Educ.*, **47,** 175 (1970).

"On Chemical Kinetics," C. J. Swartz, *J. Chem. Educ.*, **46,** 308 (1969).

"The Law of Mass Action," S. Berline and C. Bricker, *J. Chem. Educ.*, **46,** 499 (1969).

"Reaction Kinetics and the Law of Mass Action," P. G. Ashmore, *Educ. in Chemistry*, **2,** 160 (1965).

"Hydrogen-Iodine Reaction Not Bimolecular," (Staff report on research by Dr. J. H. Sullivan of the Los Alamos Scientific Laboratory), *Chem. & Eng. News*, Jan. 16, 1967; p. 40.

"Orbital Interactions and Reaction Paths," L. Salem, *Chem. in Britain*, **5,** 449 (1969).

"High Pressure Chemistry," H. S. Turner, *Chem. in Britain*, **4,** 245 (1968).

"Recent Developments in Theoretical Chemical Kinetics," R. A. Marcus, *J. Chem. Educ.*, **45,** 356 (1968).

"Some Aspects of Chemical Dynamics in Solution," J. Halpern, *J. Chem. Educ.*, **45,** 372 (1968).

"Some New Horizons in Chemical Kinetics," K. J. Laidler, *Chem. in Britain*, **3,** 475 (1967).

"From Stoichiometry and Rate Law to Mechanism," J. O. Edwards, E. F. Greene, and J. Ross, *J. Chem. Educ.*, **45,** 381 (1968).

"Monte Carlo Method: Plotting the Course of Complex Reactions," B. Rabinovitch, *J. Chem. Educ.*, **46,** 262 (1969).

"The Kinetics and Analysis of Very Fast Chemical Reactions," R. G. W. Norrish, *Chem. in Britain*, **1,** 289 (1965).

"Relaxation Methods in Chemistry," L. Faller, *Sci. American*, May, 1969; p. 30.

"Encounters and Slow Reactions," C. H. Langford, *J. Chem. Educ.*, **46,** 557 (1969).

"Lasers in Photochemical Kinetics," L. K. Patterson, *Chem. in Britain*, **6,** 246 (1970).

Lectures on Matter and Equilibrium, T. L. Hill, W. A. Benjamin, Inc., New York, 1966 (Paperback).

How Chemical Reactions Occur, E. L. King, W. A. Benjamin, Inc., New York, 1963 (Paperback).

Chemical Kinetics, G. M. Harris, D. C. Heath and Co., Boston, 1966 (Paperback).

Chemical Equilibrium, C. J. Nyman and R. E. Hamm, D. C. Heath and Co., Boston, 1967 (Paperback).

"Catalysis," V. Haenzel and R. L. Burwell, Jr., *Sci. American*, Dec., 1971; p. 46.

"'Heterogenizing' Homogeneous Catalysts," J. C. Bailar, Jr., *Catalysis Reviews—Sci. & Eng.*, **10** (1), 17 (1974).

Ionic Equilibria of Weak Electrolytes

18

We discovered in Section 14.4 that aqueous solutions of both strong and weak electrolytes contain ions, with the strong electrolytes being completely ionic in solution. The strong electrolytes include strong inorganic acids such as aqueous solutions of HCl, HBr, HI, HNO_3, and $HClO_4$; the hydroxides of the alkali metals and the alkaline earth metals; and salts. In solutions of weak electrolytes, ions and un-ionized molecules of the electrolyte are both present. The weak electrolytes include weak inorganic acids such as H_3PO_4, H_2SO_3, HCN, HF, and H_2CO_3; most organic acids; some inorganic hydroxides, such as ammonium hydroxide (aqueous ammonia); most divalent and trivalent hydroxides; and most of the organic bases. The polar covalent halides and cyanides of Hg, Cd, Zn, and a few other metals are also classed as weak electrolytes.

From a chemical equilibrium viewpoint the weak electrolytes are of great importance, and in this chapter we shall deal extensively with equilibria involving this class of compounds. The weak acids are the most important of all the weak electrolytes, in the interpretation not only of the many processes and reactions of a purely chemical nature, including those of qualitative analysis, but also of the processes in living systems involving the blood, the tissue and cell materials, and the glandular secretions.

18.1 Ionic Concentrations in Solutions of Strong Electrolytes

Because strong electrolytes are completely ionic in aqueous solution the ion concentrations may be found directly from the molar concentration of the solution. For example, in a 0.01 M solution of hydrochloric acid (HCl $\longrightarrow$ H$^+$ + Cl$^-$), the hydrogen ion concentration, $[H^+]$, is equal to the chloride concentration, $[Cl^-]$, and both concentrations have the value 0.01 M, or $1 \times 10^{-2} M$.* On the other hand, in a 0.01 M solution of potassium sulfate ($K_2SO_4 \longrightarrow 2K^+ + SO_4^{2-}$) the potassium ion concentration $[K^+]$ is 2×0.01 M, or 2×10^{-2} M, whereas the sulfate ion concentration $[SO_4^{2-}]$ is 0.01 M, or 1×10^{-2} M. This follows because each formula unit of potassium sulfate yields two potassium ions but only one sulfate ion. In a 0.001 M zinc chloride solution ($ZnCl_2 \longrightarrow Zn^{2+} + 2Cl^-$) the zinc ion concentration $[Zn^{2+}]$ is 1×10^{-3} M and the chloride ion concentration $[Cl^-]$ is 2×10^{-3} M. This is due to the fact that crystals (and solutions) of zinc chloride contain twice as many chloride ions as zinc ions.

18.2   Ionic Concentrations in Solutions of Weak Acids

In aqueous solution a weak acid such as acetic acid, HOAc, is only partly ionized into hydrogen ions and acid anions. The remaining portion of the acid is present in solution in the un-ionized, or molecular, form. For acetic acid, the ionization equation is

$$\text{HOAc} \rightleftharpoons \text{H}^+ + \text{OAc}^- \quad \text{(where OAc}^- \text{ is the acetate ion)}$$

or more accurately

$$\text{HOAc} + \text{H}_2\text{O} \rightleftharpoons \text{H}_3\text{O}^+ + \text{OAc}^-$$

This ionization reaction is reversible and equilibrium is attained almost instantly when the acetic acid is dissolved in the water. Therefore, in any dilute solution of acetic acid the law of chemical equilibrium (Section 17.10) may be applied.

$$\frac{[\text{H}_3\text{O}^+][\text{OAc}^-]}{[\text{HOAc}][\text{H}_2\text{O}]} = K_e \quad \text{eq constant}$$

Because the number of moles of water consumed in the formation of hydronium ions is negligibly small compared to the total number of moles of water present in dilute solution, the molar concentration of water remains practically constant and the above equation may be written

$$\frac{[\text{H}_3\text{O}^+][\text{OAc}^-]}{[\text{HOAc}]} = K_e[\text{H}_2\text{O}] = K_i$$

For this particular kind of equilibrium, the constant K_i is called an **ionization constant.** Concentrations of ions and molecules in the mathematical expression are generally given in moles per liter. The data in Table 18–1 show that K_i for acetic acid remains practically the same over a considerable range of concentrations. Note that the extent to which acetic acid ionizes increases with decreasing concentration. For example, 0.1 M acetic acid is 1.34 per cent ionized (98.66 per cent is in the molecular form), while the 0.01 M acid is 4.15 per cent ionized (95.85 per cent is in the molecular form). This results because in more dilute solution

---

*Notice the brackets here (and later) which, as explained earlier in Chapter 17, stand for "molar concentration of"; hence, $[H^+]$ reads "molar concentration of hydrogen ion."

TABLE 18-1   Ionization Constants for Acetic Acid at Different Concentrations

Molarity	Per Cent Ionized	$[H_3O^+]$ and $[OAc^-]$	[HOAc]	K_i
0.1	1.34	0.00134	0.09866	1.82×10^{-5}
0.08	1.50	0.0012	0.0788	1.83×10^{-5}
0.03	2.45	0.000735	0.02927	1.85×10^{-5}
0.01	4.15	0.000415	0.009585	1.797×10^{-5}

the ions are farther apart and tend to combine less readily to form undissociated acetic acid molecules in the equilibrium mixture.

The ionization constants of a number of weak acids are given in Table 18-2, with a more complete list given in Appendix G. For convenience, the equations in Table 18-2 are given in simplified form without showing the hydration of the hydrogen ions. Nevertheless, it should be kept in mind that the hydrogen ion is always in the hydrated form (H_3O^+) in aqueous solution.

The acids in Table 18-2 are listed in order of increasing strength as indicated by the increasing size of the ionization constant. As the fraction of the acid in the ionized form increases, the value of the constant increases. Thus, HF with its constant of 7.2×10^{-4} is a much stronger acid than HCN, which has a K_i of 4×10^{-10}.

TABLE 18-2   Ionization Constants of Some Weak Acids

Acid	Ionization	K_i at 25°
HCN	$\rightleftharpoons$ $H^+ + CN^-$	4×10^{-10}
HBrO	$\rightleftharpoons$ $H^+ + BrO^-$	2×10^{-9}
HClO	$\rightleftharpoons$ $H^+ + ClO^-$	3.5×10^{-8}
HOAc	$\rightleftharpoons$ $H^+ + OAc^-$	1.8×10^{-5}
HCOOH	$\rightleftharpoons$ $H^+ + HCOO^-$	1.8×10^{-4}
HNCO	$\rightleftharpoons$ $H^+ + NCO^-$	3.46×10^{-4}
HNO_2	$\rightleftharpoons$ $H^+ + NO_2^-$	4.5×10^{-4}
HF	$\rightleftharpoons$ $H^+ + F^-$	7.2×10^{-4}

18.3   Problems Involving the Ionization of Acetic Acid

The following exemplify the types of problems that are based upon the partial ionization of weak electrolytes.

Example 1.   In 0.100 M solution, acetic acid is 1.34 per cent ionized. Calculate $[H^+]$, $[OAc^-]$, and [HOAc] in the solution.

First write the equation for the ionization of acetic acid.

$$HOAc \rightleftharpoons H^+ + OAc^-$$

The equation shows that for each mole of HOAc that ionizes, one mole of H^+ and one mole of OAc^- are formed. Since 1.34 per cent of 0.100 mole per liter of acid ionizes, then

$$[H^+] = [OAc^-] = 0.100 \ M \times 0.0134 = 0.00134 \ M$$

and $\quad [HOAc] = (0.100 \ M - 0.00134 \ M) = 0.099 \ M$

Example 2. Using the concentrations found in Example 1, calculate the ionization constant of acetic acid.

Write down the expression for the ionization constant of acetic acid, substitute in it the values found above, and solve for K_i.*

$$\frac{[H^+][OAc^-]}{[HOAc]} = \frac{(0.00134) \times (0.00134)}{0.099} = 1.8 \times 10^{-5} = K_i$$

Example 3. Taking the K_i for acetic acid to be 1.8×10^{-5} at $25°$ C, calculate the $[H^+]$ and the per cent ionization in a $0.010 \ M$ solution of the acid.

In a 0.010 molar solution of acetic acid the total amount of acetic acid, i.e., the ionized plus the un-ionized, contained in one liter of solution is 0.010 mole. If we let x be the number of moles of acetic acid that ionize to form hydrogen ion and acetate ion to establish equilibrium, then x is the concentration of hydrogen ion $[H^+]$ and also of acetate ion $[OAc^-]$ at equilibrium. The equilibrium concentration of un-ionized acid $[HOAc]$ is $(0.010 - x)$. Then,

$$\frac{[H^+][OAc^-]}{[HOAc]} = \frac{x^2}{0.010 - x} = 1.8 \times 10^{-5} = K_i$$

The small value of K_i indicates that the ratio $x^2/(0.010 - x)$ is small and that x will be small compared with 0.010 from which it is to be subtracted. Thus, $0.010 - x$ is virtually equal to 0.010 and the foregoing expression may be simplified for an approximate solution as follows:

$$\frac{x^2}{0.010} = 1.8 \times 10^{-5}$$

$$x^2 = 1.8 \times 10^{-7} = 18.0 \times 10^{-8}$$

$$x = \sqrt{18.0 \times 10^{-8}} = 4.3 \times 10^{-4} \text{ mole per liter}$$

The concentration of H^+ and OAc^- is thus calculated to be 4.3×10^{-4} molar. It can readily be seen that we were justified in neglecting x in the expression $(0.010 - x)$ since $(0.010 - 0.00043) = 0.00957$, which is very nearly

---

*Units for equilibrium constants (moles/liter in Example 2) differ, depending upon the form of the particular equilibrium constant expression. The units are often omitted from equilibrium constant values, and this practice is usually followed in this book. The student should, however, always keep in mind the units for the constants and the fact that the units are real.

(handwritten at top:) $HClO \rightleftharpoons H^+ + ClO^-$
$.2-x \qquad x \qquad y$

$\frac{x^2}{.2} = 3.5 \times 10^{-8}$

equal to 0.010. One may use the approximate solution when x is less than about 5 per cent of the total concentration of the acid; if x is greater than 5 per cent, then the more complete quadratic equation should be solved (see Appendix A.4 for the solution of quadratic equations).

Calculation of the per cent of ionization in the 0.010 M solution of acetic acid is made by dividing the concentration of the acid in the ionic form, which is equal to the $[H^+]$, by the total concentration of the acid, and then multiplying by 100. Thus, we have

$$\frac{\text{Ionic HOAc}}{\text{Total HOAc}} \times 100 = \frac{4.3 \times 10^{-4}}{0.010} \times 100 = 4.3 \text{ per cent ionized}$$

(handwritten:) $.7 \times 10^{-8}$
$.8_{\cdot}4 \times 10^{-9}$

18.4   The Ionization of Weak Bases

Equilibria involving the ionization of weak bases may be treated mathematically in the same manner as those of weak acids. The ionization constants of several weak bases are given in Table 18–3 and in Appendix H.

TABLE 18-3   Ionization Constants of Some Weak Bases

Base	Ionization	K_i at 25°
NH_3 Ammonia	$+ H_2O \rightleftharpoons NH_4^+ + OH^-$	1.8×10^{-5}
CH_3NH_2 Methylamine	$+ H_2O \rightleftharpoons CH_3NH_3^+ + OH^-$	4.4×10^{-4}
$(CH_3)_2NH$ Dimethylamine	$+ H_2O \rightleftharpoons (CH_3)_2NH_2^+ + OH^-$	7.4×10^{-4} *strongest*
$(CH_3)_3N$ Trimethylamine	$+ H_2O \rightleftharpoons (CH_3)_3NH^+ + OH^-$	7.4×10^{-5}
$C_6H_5NH_2$ Phenylamine	$+ H_2O \rightleftharpoons C_6H_5NH_3^+ + OH^-$	4.6×10^{-10} *weakest*

The most common weak base is aqueous ammonia (ammonium hydroxide). When ammonia gas is dissolved in water the solution is distinctly basic, and the reaction is given by the equation

$$NH_3 + H_2O \rightleftharpoons NH_4^+ + OH^-$$

Application of the law of chemical equilibrium to this system gives the expression

$$\frac{[NH_4^+][OH^-]}{[NH_3][H_2O]} = K_e$$

Because only a small fraction of the water is consumed in the reaction, the concentration of water is practically constant and the above equation may be written

$$K_i = \frac{[NH_4^+][OH^-]}{[NH_3]} = K_e[H_2O] = 1.8 \times 10^{-5}$$

Aqueous ammonia as a base has about the same strength that acetic acid has as an acid; the ionization constants for the two substances are the same (to two significant figures).

It should be emphasized that the law of chemical equilibrium does not apply to aqueous solutions of strong acids such as HCl and HNO_3 and strong bases such as NaOH and KOH. These substances are completely ionic in dilute solution and there are no equilibria.

18.5   Salt Effect on Ionization

The expression for the ionization constant holds for dilute solutions of weak electrolytes when the solutions are pure. However, if the solution is not pure, the extent of ionization of a weak electrolyte may be influenced by the presence of other ions in the solution. For example, the ionization constant of acetic acid is increased from 1.8×10^{-5} to 2.2×10^{-5} when the solution is made 0.1 M with respect to sodium chloride. This means that the addition of a salt has slightly increased the degree of ionization of the weak acid, and this effect is known as the **salt effect.** The effect is due to the ions of a strong electrolyte decreasing the activities of the ions (Section 14.11) in a solution of a weak electrolyte as a result of interionic attraction. The ionization constant is, therefore, not strictly constant, but varies slightly with the ionic strength of the solution. For most purposes, and for ours, the expression for the ionization constant may be used without regard to slight errors which may arise due to the salt effect.

18.6   Common Ion Effect on Degree of Ionization

It has been found by experiment that the acidity of an aqueous solution of acetic acid is decreased by the addition of the strong electrolyte sodium acetate. This can be explained by the fact that, according to the law of mass action (Section 17.7), the addition of acetate ions to a solution of acetic acid causes the following equilibrium to shift to the left.

$$HOAc \rightleftharpoons H^+ + OAc^-$$

This decreases the concentration of H^+ and hence the acidity. The NaOAc has the acetate ion in common with HOAc, so the influence is known as the **common ion effect.** Since the equilibrium was disturbed by the increase in the concentration of OAc^-, a shift in the equilibrium will occur which will reduce the concentration of OAc^-. This is accomplished by the union of H^+ with OAc^- to form molecular acetic acid. Consequently, the $[H^+]$ of the solution is reduced.

The extent to which the concentration of the hydrogen ion is decreased by the addition of the acetate ion may be calculated from the expression for the ionization constant of acetic acid.

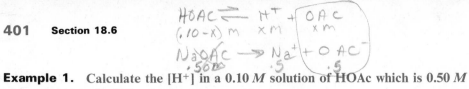

(handwritten annotations:)
$$HOAc \rightleftharpoons H^+ + OAc$$
$$(.10-x)\,m \quad xm \quad xm$$
$$NaOAc \longrightarrow Na^+ + OAc^-$$
$$.50m \qquad .5 \qquad .5$$

Example 1. Calculate the $[H^+]$ in a 0.10 M solution of HOAc which is 0.50 M with respect to NaOAc.

At equilibrium the concentration of the acetate ion is equal to the concentration of NaOAc, which is completely dissociated into Na^+ and OAc^-, plus the concentration of the acetate ion derived from the ionization of acetic acid. Let x equal the concentration of acetate ions derived from the ionization of acetic acid. Then $(0.50\ M + x)$ will equal the total concentration of the acetate ion. The concentration of the hydrogen ions will be equal to x and the concentration of un-ionized acetic acid will be equal to $(0.10\ M - x)$. Substituting in the expression for the ionization constant for acetic acid we have

$$\frac{[H^+][OAc^-]}{[HOAc]} = \frac{(x)(0.50 + x)}{0.10 - x} = 1.8 \times 10^{-5}$$

(handwritten:) $HCN \rightleftharpoons H^+ + CN$

Even in the absence of NaOAc the concentration of the acetate ion derived from the ionization of 0.10 M acetic acid is small (0.00134 mole per liter) compared to 0.50 mole per liter derived from NaOAc. Since the degree of ionization is even smaller in the presence of the high concentration of sodium acetate, it follows that the concentration of acetate ion may be taken as equal to the concentration of the sodium acetate, or that x may be neglected in the term $(0.50 + x)$. Likewise, the concentration of the un-ionized acetic acid is very nearly equal to 0.10 M, and x in the term $(0.10 - x)$ may be dropped. Thus, in an approximate solution to the problem we have

$$\frac{[H^+][OAc^-]}{[HOAc]} = \frac{(0.50)x}{0.10} = 1.8 \times 10^{-5}$$

$$x = \frac{0.10}{0.50} \times 1.8 \times 10^{-5} = 3.6 \times 10^{-6}\ M = [H^+]$$

Consequently, the concentration of the hydrogen ion is reduced from 0.00134 mole per liter to 0.0000036 mole per liter by the presence of the 0.50 M sodium acetate in the 0.10 M acetic acid solution.

(handwritten:) $NH_3 + H_2O \rightleftharpoons NH_4 + OH$ $(.10-x)M \quad x \quad 2.8 \times 10^{-6}$ $NH_4Cl \longrightarrow NH_4^+ + Cl^-$ $.10 \quad .10$

Example 2. A 0.10 M solution of aqueous ammonia, also containing ammonium chloride, has a hydroxide ion concentration of $2.8 \times 10^{-6}\ M$. What is the concentration of the ammonium ion in the solution?

First write the equation for the ionization of ammonia.

$$NH_3 + H_2O \Longrightarrow NH_4^+ + OH^-$$

The ammonium ion of the ammonium chloride present in the solution causes the ionization of ammonia to be decreased (the equilibrium to be shifted to the left) because of the common ion effect. Let x be the total ammonium ion concentration at equilibrium; this value will be the sum of the ammonium ion concentration resulting from the slight ionization of ammonia and that from the ammonium chloride. The hydroxide ion concentration is given as $2.8 \times 10^{-6} M$, and the concentration of ammonia will be $(0.10 - 2.8 \times 10^{-6}) M$, which is very nearly $0.10 M$. Substituting in the expression for the ionization constant of ammonia, we have

$$\frac{[NH_4^+][OH^-]}{[NH_3]} = \frac{x(2.8 \times 10^{-6})}{0.10} = 1.8 \times 10^{-5}$$

Solving for x, we find

$$x = \frac{(0.10)(1.8 \times 10^{-5})}{2.8 \times 10^{-6}} = \frac{1.8}{2.8} = 0.64$$

Thus, we find the concentration of the ammonium ion to be $0.64 M$ at equilibrium. Of this amount, 2.8×10^{-6} mole is formed by the ionization of ammonia and the remainder comes from the added ammonium chloride.

Example 3.   Ten ml of $4.0 M$ acetic acid is added to 20 ml of $1.0 M$ sodium hydroxide. Calculate the hydrogen ion concentration of the resulting solution.

Before mixing, the 10 ml of $4.0 M$ HOAc contains 10 ml $\times$ $4.0 M = 40$ millimoles of HOAc, and the 20 ml of $1.0 M$ NaOH contains 20 ml $\times$ $1.0 M = 20$ millimoles of NaOH. The 20 millimoles of NaOH neutralize 20 millimoles of HOAc, producing 20 millimoles of NaOAc and leaving $40 - 20 = 20$ millimoles of HOAc not neutralized. After reaction, the 20 millimoles of NaOAc and 20 millimoles of HOAc are contained in $(10 + 20)$ ml $= 30$ ml of solution, so the concentration of each is 20 millimoles/30 ml, or $0.67 M$.

The sodium acetate in the solution exhibits the common ion effect upon the ionization of the acetic acid. Let x equal the concentration of the hydrogen ion. Then the concentration of the molecular acetic acid will be $(0.67 - x) M$ and that of the acetate ion will be $(0.67 + x) M$. Neglecting the x's in the terms $(0.67 - x)$ and $(0.67 + x)$ and substituting in the expression for the ionization constant of acetic acid, we have

$$\frac{[H^+][OAc^-]}{[HOAc]} = \frac{x(0.67)}{(0.67)} = 1.8 \times 10^{-5}$$

$$x = \frac{(0.67)}{(0.67)} \times 1.8 \times 10^{-5} = 1.8 \times 10^{-5} M$$

Thus, the concentration of the hydrogen ion in the solution is $1.8 \times 10^{-5} M$.

The common ion effect is of importance in the adjustment of the hydrogen ion concentration for many of the precipitations and separations in the qualitative analysis scheme and in blood and other biological fluids.

18.7   The Ionization of Water

Water is an extremely weak electrolyte which undergoes self-ionization (see Section 12.10).

$$H_2O + H_2O \rightleftharpoons H_3O^+ + OH^-$$

or simply

$$H_2O \rightleftharpoons H^+ + OH^-$$

Application of the law of chemical equilibrium yields the expression

$$\frac{[H^+][OH^-]}{[H_2O]} = K_e$$

As long as we work with dilute aqueous solutions, the concentration of water may be considered as constant. We may therefore write

$$[H^+][OH^-] = K_e[H_2O] = K_w$$

In all problems of acidity and basicity of aqueous solutions of electrolytes, K_w is a constant of great importance. It is called the **ion-product constant of water**, and at 25° it has the value 1×10^{-14}. The value of K_w increases rapidly with increasing temperature and at 100° is approximately 1×10^{-12}, thus being 100 times as large at 100° as at 25°. This means simply that the degree of ionization of water and the concentration of the hydrogen and hydroxide ions increase with rising temperature.

The ionization of water yields the same number of hydrogen and hydroxide ions. Therefore, in pure water at 25° $[H^+] = [OH^-]$, and

$$[H^+]^2 = [OH^-]^2 = 10^{-14}$$
$$[H^+] = [OH^-] = \sqrt{10^{-14}} = 10^{-7} \, M$$

All aqueous solutions contain both hydrogen ions and hydroxide ions. If by the addition of hydrogen ions (as an acid) the $[H^+]$ is made larger than 10^{-7}, then the $[OH^-]$ will become smaller than 10^{-7}; in basic solutions the $[OH^-]$ will be greater than 10^{-7} and the $[H^+]$ will be less than 10^{-7}. The product of the $[H^+]$ and $[OH^-]$ at a given temperature always remains constant, a fact the chemist finds very useful. For example, he may wish to know the $[H^+]$ of 0.01 M NaOH at 25°.

$$[H^+][OH^-] = K_w$$
$$[H^+](10^{-2}) = 10^{-14}$$
$$[H^+] = \frac{10^{-14}}{10^{-2}} = 10^{-12} \, M$$

Or, he may wish to calculate the $[OH^-]$ in 0.001 M HCl.

$$[H^+][OH^-] = (10^{-3})[OH^-] = 10^{-14}$$
$$[OH^-] = \frac{10^{-14}}{10^{-3}} = 10^{-11} \, M$$

18.8 The pH Method of Expressing the Concentration of the Hydrogen Ion

The concentration of the hydrogen ion is a measure of the acidity or basicity of a solution. It has been found convenient to express the concentration of the hydrogen ion in terms of the negative logarithm of the hydrogen ion concentration. This is referred to as the **pH** of the solution. Expressed mathematically, we have

$$pH = -\log[H^+]$$

or

$$pH = \log\frac{1}{[H^+]}$$

The pH value is the negative power to which 10 must be raised to equal the hydrogen ion concentration.

$$[H^+] = 10^{-pH}$$

The following problems will serve to illustrate the calculation of pH values from hydrogen and hydroxide ion concentrations:

Example 1. Calculate the pH of 0.01 M HCl. Hydrochloric acid is completely ionized in dilute solution so the concentration of the hydrogen ion is 0.01 M, or 10^{-2} M.

Substituting, we have

$$pH = -\log[H^+] = -\log 10^{-2} = -(-2) = 2$$

(The use of logarithms is explained in Appendix A.3.)

Example 2. Calculate the pH of 0.0050 M HNO$_3$.

The hydrogen ion concentration of nitric acid is the same as the molar concentration of the acid, for it is a strong electrolyte. Thus, we have

$$pH = -\log(5.0 \times 10^{-3}) = -(\log 5.0 + \log 10^{-3}) = -(0.7 - 3.0)$$
$$= -(-2.3) = 2.3$$

Example 3. Calculate the pH of 0.0001 M NaOH.

Because pH is defined in terms of hydrogen ion concentration, it is first necessary to find the concentration of this ion by substituting in the ion-product expression for water. The [OH$^-$] from the sodium hydroxide is given as 10^{-4} M.

$$[H^+][OH^-] = K_w$$
$$[H^+](10^{-4}) = 10^{-14}$$
$$[H^+] = \frac{10^{-14}}{10^{-4}} = 10^{-10}$$

The hydrogen ion concentration thus found is then converted to pH as follows:

$$pH = -\log[H^+] = -\log 10^{-10} = -(-10) = 10$$

Example 4.   Water in equilibrium with the air contains 0.030 volume per cent of carbon dioxide. The resulting carbonic acid, H_2CO_3, gives to the solution a hydrogen ion concentration about twenty times larger than that of pure water, or 2.0×10^{-6} as compared to 1.0×10^{-7}. Calculate the pH of the solution.

$$pH = -\log[H^+] = -\log(2.0 \times 10^{-6}) = -(\log 2.0 + \log 10^{-6})$$
$$= -(0.3 - 6.0) = -(-5.7) = 5.7$$

Thus we see that water in contact with air is acidic, rather than neutral, due to dissolved carbon dioxide.

Example 5.   Calculate the pH of 0.100 M HOAc, which is 1.34 per cent ionized.

The hydrogen ion concentration of this weak acid is $0.100\ M \times 0.0134 = 0.00134\ M = 1.34 \times 10^{-3}\ M$ (calculated in Section 18.3).

$$pH = -\log[H^+] = -\log(1.34 \times 10^{-3}) = -(\log 1.34 + \log 10^{-3})$$
$$= -(0.13 - 3.00) = -(-2.87) = 2.87$$

The following examples illustrate the conversion of pH values to hydrogen ion concentrations.

Example 6.   Calculate the hydrogen ion concentration of a solution, the pH of which is 9.

$$pH = -\log[H^+] = 9$$
$$\log[H^+] = -9$$
$$[H^+] = \text{antilog of } (-9) = 10^{-9}$$

Example 7.   Calculate the hydrogen ion concentration of a solution whose pH is 4.4.

$$pH = -\log[H^+] = 4.4$$
$$\log[H^+] = -4.4$$

It is readily seen that -4.4 is equal to $(-5 + 0.6)$. Then we may write

$$\log[H^+] = 0.6 - 5$$
$$[H^+] = \text{antilog of } (0.6 - 5)$$
$$= (\text{antilog of } 0.6) \times (\text{antilog of } -5)$$
$$= 4 \times 10^{-5}\ M$$

18.9   The pOH Method of Expressing the Concentration of the Hydrogen Ion

We may define pOH as the negative logarithm of the hydroxide ion concentration (pOH = $-\log$ [OH$^-$]). Similarly, we may define pK_w as the negative logarithm of the ion-product constant (K_w) for water. Now we can write the ion-product expression for water in terms of pH, pOH, and pK_w.

$$[H^+][OH^-] = K_w$$
$$(-\log [H^+]) + (-\log [OH^-]) = -\log K_w$$
$$p\text{H} + p\text{OH} = pK_w$$

Because K_w has the value 10^{-14}, then

$$pK_w = -\log 10^{-14} = -(-14) = 14$$

It follows that

$$p\text{H} + p\text{OH} = 14$$
$$p\text{H} = 14 - p\text{OH}$$
$$p\text{OH} = 14 - p\text{H}$$

The relationships between the hydrogen ion and hydroxide ion concentration, and the pH and pOH of certain aqueous solutions are given in Table 18–4.

TABLE 18-4   Relationships of [H$^+$], [OH$^-$], pH, and pOH

$[H^+]$	$[OH^-]$	pH	pOH	Solution
10^1	10^{-15}	-1	15	Strongly acidic
10^0 or 1	10^{-14}	0	14	
10^{-1}	10^{-13}	1	13	
10^{-2}	10^{-12}	2	12	
10^{-3}	10^{-11}	3	11	
10^{-4}	10^{-10}	4	10	
10^{-5}	10^{-9}	5	9	
10^{-6}	10^{-8}	6	8	
10^{-7}	10^{-7}	7	7	Neutral
10^{-8}	10^{-6}	8	6	
10^{-9}	10^{-5}	9	5	
10^{-10}	10^{-4}	10	4	
10^{-11}	10^{-3}	11	3	
10^{-12}	10^{-2}	12	2	
10^{-13}	10^{-1}	13	1	Strongly basic

18.10   Acid-Base Indicators

Certain organic substances have the property of changing color in dilute solution when the hydrogen ion concentration of the solution attains a definite value. For example, phenolphthalein is a colorless substance in any aqueous solution in which the hydrogen ion concentration is greater than 10^{-9} M, or the pH is less than 9. In solutions for which the hydrogen ion concentration is less than 10^{-9} (pH greater than 9), the phenolphthalein imparts a red or pink color to the solution. Substances like phenolphthalein are called **acid-base indicators,** and they are often used to determine the pH of solutions. Acid-base indicators are either weak organic acids, HIn, or weak organic bases, InOH, in which the symbol "In" represents a complex organic group.

The equilibrium existing in a solution of a certain acid-base indicator called **methyl orange,** which is a weak acid, can be represented by the equation

$$\underset{\text{Red}}{\text{HIn}} \rightleftharpoons \text{H}^+ + \underset{\text{Yellow}}{\text{In}^-}$$

The anion of the indicator is yellow, and the un-ionized form is red. An increase in the concentration of the hydrogen ion brought about by the addition of an acid to the solution shifts the equilibrium toward the red form in accordance with the law of mass action. Application of the law of chemical equilibrium to this reversible reaction gives the expression for the ionization constant of this acid.

$$\frac{[\text{H}^+][\text{In}^-]}{[\text{HIn}]} = K_i$$

The color exhibited by the indicator is the visible result of the ratio of the concentrations of the two species HIn and In$^-$. For methyl orange,

$$\frac{[\text{In}^-]}{[\text{HIn}]} = \frac{[\text{Substance with yellow color}]}{[\text{Substance with red color}]} = \frac{K_i}{[\text{H}^+]}$$

When $[\text{H}^+]$ has the same numerical value as K_i, the ratio of $[\text{In}^-]$ to $[\text{HIn}]$ is equal to one, meaning that 50 per cent of the indicator is present in the acid (yellow form) and 50 per cent in the alkaline (red form). Under these conditions the solution appears orange in color. When the hydrogen ion concentration has been increased until the pH is 3.1, about 90 per cent of the indicator is present in the red form and 10 per cent is present in the yellow form. The eye cannot detect any change in color with further increase in the hydrogen ion concentration and corresponding increase in the concentration of the red form of the indicator.

Addition of a base to the system reduces the concentration of hydrogen ions ($\text{H}^+ + \text{OH}^- \longrightarrow \text{H}_2\text{O}$), and shifts the equilibrium toward formation of the yellow form. At a pH of 4.4 about 90 per cent of the indicator is in the form of the yellow ion, and a further decrease in the hydrogen ion concentration does not produce a color change easily detectable by the eye. The pH range between 3.1 (red) and 4.4 (yellow-orange) is the **color-change interval** of the indicator methyl orange. This means that the pronounced color change takes place between these two pH values.

A large number of acid-base indicators is known, covering a wide range of pH values. A number of indicators may be used in determining the approximate pH of any solution, and by a process of elimination the pH of a solution can be fixed within rather narrow limits. Table 18–5 lists a series of indicators, together with their colors for the corresponding pH values. The selection of indicators for specific purposes will be discussed in Section 18.12.

TABLE 18-5   Some Acid-Base Indicators

Indicator	Color in the More Acid Range	pH Range	Color in the More Basic Range
Methyl violet	yellow	0–2	violet
Thymol blue	pink	1.2–2.8	yellow
Brom-phenol blue	yellow	3.0–4.7	violet
Methyl orange	pink	3.1–4.4	yellow
Brom-cresol green	yellow	4.0–5.6	blue
Brom-cresol purple	yellow	5.2–6.8	purple
Litmus	red	4.7–8.2	blue
Phenolphthalein	colorless	8.3–10.0	pink
Thymolphthalein	colorless	9.3–10.5	blue
Alizarin yellow G	colorless	10.1–12.1	yellow
Trinitrobenzene	colorless	12.0–14.3	orange

The measurement and control of the hydrogen ion concentration are important in scientific investigations, in industry, and in agriculture. In analytical chemistry the separation and identification of many of the metallic ions depend upon the pH of the solutions containing these ions.

18.11   Buffer Solutions

Mixtures of weak acids and their salts or of weak bases and their salts are called **buffer solutions.** They resist a change in hydrogen-ion concentration upon the addition of small amounts of acids or bases. An example of a buffer solution is a mixture of $0.10\ M$ acetic acid and $0.10\ M$ sodium acetate. The pH of the solution may be found as follows:

$$HOAc \rightleftharpoons H^+ + OAc^-$$

$$\frac{[H^+][OAc^-]}{[HOAc]} = K_i$$

$$[H^+] = \frac{[HOAc]}{[OAc^-]}K_i = \frac{0.10}{0.10} \times 1.80 \times 10^{-5} = 1.80 \times 10^{-5}$$

$$pH = -\log[H^+] = -\log(1.80 \times 10^{-5}) = 4.745$$

$$= 4.74 \text{ (to justifiable limit in significant figures)}$$

Note that 4.74 represents *two* significant figures rather than three. The number to the left of the decimal in the logarithm is the "characteristic," which merely establishes the decimal in the number for which 4.74 is the logarithm. The only *significant figures* in the logarithm are those to the right of the decimal.

Now add 1.0 ml of $0.10\ M$ sodium hydroxide to 100 ml of this buffer mixture. An equivalent amount of acetic acid is neutralized by the sodium hydroxide, and sodium acetate is formed.

$$HOAc + OH^- \longrightarrow H_2O + OAc^-$$

Before reaction, 100 ml of the buffer mixture contains 100 ml $\times$ 0.10 M = 10 millimoles of HOAc, and 100 ml × 0.10 M = 10 millimoles of NaOAc. One ml of 0.10 M sodium hydroxide contains 1.0 ml × 0.10 M = 0.10 millimole of NaOH. The 0.10 millimole of NaOH neutralizes 0.10 millimole of HOAc, leaving 10 − 0.10 = 9.9 millimoles of HOAc, and producing 0.10 millimole of NaOAc; this makes a total of 10 + 0.10 = 10.1 millimoles of NaOAc. After reaction, the 9.9 millimoles of HOAc and 10.1 millimoles of NaOAc are contained in 101 ml of solution, so the concentrations are 9.9 millimoles/101 ml = 0.098 M HOAc, and 10.1 millimoles/101 ml = 0.100 M NaOAc. The final pH of the solution is then calculated as follows:

$$[H^+] = \frac{[HOAc]}{[OAc^-]} \times K_i = \frac{0.098}{0.100} \times 1.80 \times 10^{-5} = 1.76 \times 10^{-5}$$

$$pH = -\log[H^+] = -\log(1.76 \times 10^{-5}) = 4.754$$

$$= 4.75 \text{ (to justifiable limit in significant figures)}$$

Thus, the addition of the base barely changes the pH of the solution. The explanation for this buffering action lies in the fact that as the sodium hydroxide is added, the hydroxide ions unite with the few hydrogen ions present, and then more of the acetic acid ionizes, thus restoring the concentration of the hydrogen ions to near its original value. If a small amount of hydrochloric acid were to be added to the acetic acid–sodium acetate buffer solution, most of the hydrogen ions from the hydrochloric acid would unite with acetate ions from the large reserve of these ions, forming un-ionized acetic acid.

$$H^+ + OAc^- \longrightarrow HOAc$$

Thus, there would be very little increase in the concentration of the hydrogen ion, and the pH would remain practically unchanged.

Some **buffer pairs** that find extensive application are HOAc and OAc⁻, NH₃ and NH₄⁺, H₂CO₃ and HCO₃⁻, H₂PO₄⁻ and HPO₄²⁻. Blood is an important example of a buffer solution, with the principal acid and ion responsible for the buffering action being H_2CO_3 and HCO_3^-. When excess of hydrogen ion enters the blood stream, it is removed principally through the reaction

$$H^+ + HCO_3^- \longrightarrow H_2CO_3$$

and when an excess of the hydroxide ion is present, it is absorbed by the reaction

$$OH^- + H_2CO_3 \longrightarrow H_2O + HCO_3^-$$

The pH of human blood thus remains very nearly 7.35, or slightly alkaline.

18.12 Titration Curves

Plots of pH vs. volume of acid or base added in an acid-base titration (see Section 15.17) are useful in that they show, graphically, the point where equivalent quantities of acid and base are present (the equivalence point) and so aid in the choice of a proper indicator. Such plots are referred to as **titration curves.**

The simplest acid-base neutralization reactions are those for titrations involving a strong acid and a strong base. Let us consider the titration of a 25.0 ml sample of a 0.100 M hydrochloric acid solution with a 0.100 M sodium hydroxide solution. The 25.0 ml solution of 0.100 M HCl contains 0.00250 mole of HCl. Values for pH can be calculated for any given quantity of NaOH solution added. Table 18-6, column (a), provides calculated pH values for a variety of added volumes of NaOH. Figure 18-1 is a plot of the pH values (as the ordinate axis) and volume of NaOH added (as the abscissa axis).   *acidity decreases*

It is apparent from Fig. 18-1 that the pH increases very slowly at first, increases very rapidly in the middle portion of the curve, and then increases very slowly again after the middle portion. The point of inflection (the midpoint of the rapid rise in the curve) is the **equivalence point** for the titration and represents the stage of the titration where equivalent quantities of acid and base are present in the titration mixture. For the titration of a strong acid and a strong base, as illustrated in Table 18-6, column (a), and Fig. 18-1, the equivalence point occurs at a pH of 7. The solution at a pH of 7 is neutral, with hydrogen ion concentration equal to hydroxide ion concentration.

The following sample calculations of the pH values in Table 18-6, column (a), are for several points on the titration curve.

TABLE 18-6   pH Values (a) in the Titration of a Strong Acid with a Strong Base and (b) in the Titration of a Weak Acid with a Strong Base

		pH Values	
Volume of 0.100 M NaOH Added, ml	Moles of NaOH Added	(a) Titration of 25.00 ml of 0.100 M HCl	(b) Titration of 25.00 ml of 0.100 M HOAc
0.0	0.0	1.00	2.87
5.0	0.00050	1.18	4.14
10.0	0.00100	1.37	4.57
15.0	0.00150	1.60	4.92
20.0	0.00200	1.95	5.35
22.0	0.00220	2.20	5.61
24.0	0.00240	2.69	6.13
24.5	0.00245	3.00	6.44
24.9	0.00249	3.70	7.14
25.0	0.00250	7.00	8.72
25.1	0.00251	10.30	10.30
25.5	0.00255	11.00	11.00
26.0	0.00260	11.29	11.29
28.0	0.00280	11.75	11.75
30.0	0.00300	11.96	11.96
35.0	0.00350	12.22	12.22
40.0	0.00400	12.36	12.36
45.0	0.00450	12.46	12.46
50.0	0.00500	12.52	12.52

(a) Titration of 25.00 ml of 0.100 M HCl (0.00250 mole of HCl) with 0.100 M NaOH
(b) Titration of 25.00 ml of 0.100 M HOAc (0.00250 mole of HOAc) with 0.100 M NaOH

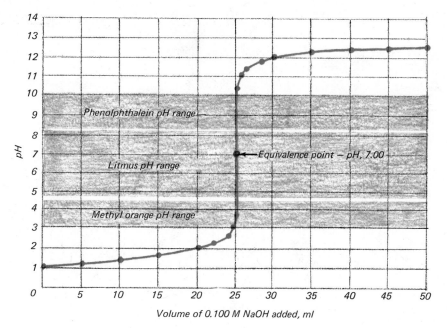

FIGURE 18–1

Titration curve for the titration of 25.00 ml of 0.100 M HCl (strong acid) with 0.100 M NaOH (strong base). The pH ranges for the color change of phenolphthalein, litmus, and methyl orange are indicated by the shaded areas.

Example 1. Calculate the pH of the initial solution before any NaOH solution is added.

The initial solution is a 0.100 M HCl solution. Inasmuch as HCl is a strong acid in water solution, $[H^+] = 0.100\ M$.

$$pH = \log\frac{1}{[H^+]} = \log\frac{1}{0.100} = \log 10.0 = 1.00$$

Example 2. Calculate the pH of the solution after 15.0 ml of 0.100 M NaOH solution has been added to the 25 ml of 0.100 M HCl solution.

The 15.0 ml of 0.100 M NaOH solution contains 0.0015 mole of NaOH, which will neutralize 0.0015 mole of HCl. Hence, the remaining HCl present is the original amount (0.0025 mole) minus the amount neutralized (0.0015 mole):

Moles of HCl present $= 0.0025 - 0.00150 = 0.00100$ mole

Therefore the amount of hydrogen ion present in solution $= 0.00100$ mole. The total volume of the solution is now the original volume (25.0 ml) plus the volume of NaOH solution added (15.0 ml):

Total volume of solution $= 25.0 + 15.0 = 40.0$ ml

Hence, the amount of H^+ is 0.00100 mole in 40.0 ml. The molarity of H^+ (moles H^+ per liter) and the pH can then be calculated:

Handwritten notes:

$HCl + NaOH \rightarrow NaCl + H_2O$

$\begin{array}{l} 2.5\ m\ moles\ HCl \\ -\ 1.5\ m\ moles\ NaOH \\ \hline 1.0\ m\ moles\ HCl\ left \end{array}$

$\dfrac{1.0}{40\ ml} = .025\ m$

$pH = -\log 2.5 \times 10^{-2}$

$= 2 - .398$

$= 1.602$

$$[\text{H}^+] = \frac{0.00100 \text{ mole}}{40.0 \text{ ml}} \times 1{,}000 \text{ ml/liter} = 0.0250 \text{ mole/liter}$$

$$= 2.50 \times 10^{-2} \, M$$

$$p\text{H} = \log\frac{1}{[\text{H}^+]} = \log\frac{1}{2.50 \times 10^{-2}} = \log 4.00 \times 10^1 = 1.60$$

Example 3. Calculate the pH of the solution when 25.0 ml of 0.100 M NaOH has been added.

At this stage, which is the equivalence point of the titration, 0.00250 mole of NaOH has been added to 0.00250 mole of HCl. Hence, 0.00250 mole of NaCl and 0.00250 mole of water have been produced, and the NaOH added has exactly neutralized the HCl originally present. A neutral solution has been produced, therefore, in which $[\text{H}^+] = [\text{OH}^-] = 10^{-7} \, M$.

$$p\text{H} = \log\frac{1}{10^{-7}} = 7.00$$

Example 4. Calculate the pH of the solution when 35.0 ml of 0.100 M NaOH solution has been added.

At all points beyond the equivalence point, an excess of NaOH is present. When 35.0 ml of NaOH has been added, the excess NaOH is $35.0 - 25.0$ ml, or 10.0 ml, and represents an excess of $0.00350 - 0.00250$ mole, or a 0.00100 mole excess.

The total volume of the solution is

$$25.0 \text{ ml} + 35.0 \text{ ml} = 60.0 \text{ ml}$$

Hence,

$$[\text{OH}^-] = \frac{0.00100 \text{ mole}}{60.0 \text{ ml}} \times 1{,}000 \text{ ml/liter}$$

$$= 0.0167 \text{ mole/liter} = 1.67 \times 10^{-2} \, M$$

$$p\text{OH} = \log\frac{1}{1.67 \times 10^{-2}} = \log 6.00 \times 10^1 = 1.78$$

$$p\text{H} = 14 - p\text{OH} = 14 - 1.78 = 12.22$$

The titration of a weak acid with a strong base or of a weak base with a strong acid is somewhat more complicated than the titrations previously discussed but follows the same basic principles. Let us consider the titration of 25.0 ml of a 0.100 M solution of acetic acid with a 0.100 M solution of sodium hydroxide and compare the titration curve with that of Fig. 18–1. Table 18–6, column (b), provides calculated pH values for several added volumes of NaOH. Figure 18–2 shows the titration curve.

The similarities in the two titration curves are readily apparent, but a close inspection also reveals several important differences. The titration curve for the titration of the weak acid begins at a higher pH value (less acidic) and maintains

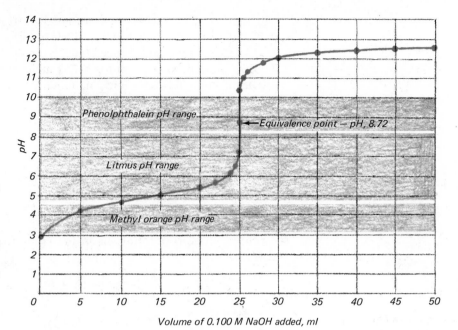

FIGURE 18-2
Titration curve for the titration of 25.00 ml of 0.100 M HOAc (weak acid) with 0.100 M NaOH (strong base). The pH ranges for the color change of phenolphthalein, litmus, and methyl orange are indicated by the shaded areas.

higher pH values for corresponding amounts of added NaOH until the equivalence point is reached. This is because the acetic acid in the solution, being a weak acid, releases only a small quantity of hydrogen ions to the solution. The pH at the equivalence point is also higher (8.72, rather than 7.00). However, following the equivalence point, the two curves are identical, as shown by the values in Table 18-6, column (b), and by the plot in Fig. 18-2. This is because after the acid is completely neutralized, whether it be a strong acid (like HCl) or a weak acid (like HOAc), it has no further significant effect upon the concentration of the hydroxide ion in the solution; the pH is dependent upon the excess of hydroxide ion in both cases.

Some sample calculations will help to clarify the points just made. The calculations are very similar to problems for weak acid equilibria discussed earlier in this chapter (see Sections 18.3, 18.6, 18.8, and 18.11).

Example 5. Calculate the pH of the initial solution before any NaOH solution is added.

The initial solution is a 0.100 M solution of acetic acid. The pH is 2.87. See Section 18.8, Example 5, for the calculation.

Example 6. Calculate the pH of the solution after 15.0 ml of 0.100 M NaOH solution has been added to the 25 ml of 0.100 M HOAc solution.

The 15.0 ml of 0.100 M NaOH solution contains 0.00150 mole of NaOH which will react with 0.00150 mole of HOAc to form 0.00150 mole of NaOAc and 0.00150 mole of water. The amount of undissociated HOAc

Handwritten top: $1 \text{ HOAC} + 1 \text{ NaOH} \rightarrow 1 \text{ NaOAC} + 1 \text{ H}_2\text{O}$
start $2.5 \quad 1.5 \quad\quad 1.5$
after 1.5

will thereby decrease from the original 0.00250 mole to 0.00100 mole (0.00250 − 0.00150 = 0.00100). The total volume of solution is 25.0 + 15.0 = 40.0 ml. Hence, 0.00150 mole of NaOAc (and hence 0.00150 mole of OAc⁻) and 0.00100 mole of undissociated HOAc are present in 40.0 ml of solution. The molarities of OAc⁻ and HOAc, therefore, are

Handwritten left:
2.5 mmoles HOAC
−1.5 mmoles NaOH
1.0 mmole HOAC
1.5 mmoles NaOH / 40 ml sol

.0375 m NaOAC
$\frac{1.0}{40} = .025$ m HOAC

$$[OAc^-] = \frac{0.00150 \text{ mole}}{40.0 \text{ ml}} \times 1{,}000 \text{ ml/liter} = 0.0375 \text{ mole/liter}$$

$$= 3.75 \times 10^{-2} M$$

$$[HOAc] = \frac{0.00100 \text{ mole}}{40.0 \text{ ml}} \times 1{,}000 \text{ ml/liter} = 0.0250 \text{ mole/liter}$$

$$= 2.50 \times 10^{-2} M$$

Handwritten left:
$\dfrac{[H^+][OAc^-] \text{ salt}}{[HOAC] \text{ acid}} = 1.8 \times 10^{-5}$

$[H^+] = \dfrac{(1.8 \times 10^{-5})(.025)}{.0375}$

$= 1.2 \times 10^{-5}$

$pH = 5 - .08$
$= 4.92$

Substituting these concentrations in the equilibrium expression for acetic acid:

$$\frac{[H^+][OAc^-]}{[HOAc]} = 1.8 \times 10^{-5}$$

$$\frac{[H^+](3.75 \times 10^{-2})}{2.50 \times 10^{-2}} = 1.8 \times 10^{-5}$$

$$[H^+] = 1.2 \times 10^{-5} M$$

$$pH = \log \frac{1}{1.2 \times 10^{-5}} = \log 8.33 \times 10^4 = 4.92$$

Example 7. Calculate the pH of the solution when 25.0 ml of 0.100 M NaOH has been added (the equivalence point).

At this stage in the titration, 0.00250 mole of NaOH has been added to 0.00250 mole of HOAc. Hence, all of the HOAc has reacted with NaOH, and in the process 0.00250 mole of NaOAc has been produced. The total volume of the solution is now 50.0 ml. The molarity of OAc⁻, therefore, is

Handwritten:
2.5 mmoles NaOAc
50 ml
.05 m NaOAC

$\dfrac{1.0 \times 10^{-14}}{1.8 \times 10^{-5}} = \dfrac{Kw}{K_{OAC}} = 5.6 \times 10^{-10}$

$\dfrac{[HOAC][OH^-]}{[OAC^-]}$

$$[OAc^-] = \frac{0.00250 \text{ mole}}{50.0 \text{ ml}} \times 1{,}000 \text{ ml} = 0.050 \text{ mole/liter} = 5.0 \times 10^{-2} M$$

The problem, therefore, becomes the calculation of the pH of a 0.050 M NaOAc solution. The calculation must take into account the hydrolysis of the OAc⁻, according to the reaction:

$$OAc^- + H_2O \rightleftharpoons HOAc + OH^-$$
Handwritten below: .05

The production of hydroxide ions makes the solution basic at the equivalence point. The actual calculation is made in Section 18.20, Example 1, later in the chapter. The pH is calculated to be 8.72.

Handwritten lower left:
$x = 5.3 \times 10^{-6}$
$= 6 - 5.72$
$pOH = 5.28$
$pH = 8.72$

Example 8. Calculate the pH of the solution when 35.0 ml of 0.100 M NaOH solution has been added.

Handwritten:
3.5 mmoles NaOH
2.5 mmoles HOAC
10 mmoles NaOH
$\dfrac{1.0}{60} = .0167$
$pOH = 1.78$
$pH = 12.22$

At all points beyond the equivalence point, the pH is controlled by the excess NaOH present. Hence, the curve is identical to that for the titration of HCl and the calculations of pH are identical. As in the calculation for HCl, the pH is 12.22.

Titration curves enable one to pick a suitable indicator which will provide a sharp color change at the equivalence point. The pH ranges for the color changes of three indicators listed in Table 18–5 are indicated by shading in Figures 18–1 and 18–2.

Phenolphthalein provides a sharp color change over a small volume of added NaOH and hence would be suitable for titration of either strong acid–strong base or weak acid–strong base. The equivalence point of 25.00 ml of NaOH is located in the steeply rising portion of the titration curve, as the curve passes through the pH range corresponding to the indicator color change.

Litmus would be a suitable indicator for the HCl titration because it would change color, as shown by the titration curve, within less than 0.10 ml addition of NaOH. However, litmus would be a poor choice of indicator for the HOAc titration because, as shown by the plot, the titration curve enters the pH range for color change of litmus when only about 12 ml of NaOH has been added and does not leave the range until 25 ml of NaOH has been added. The end of the color change would, therefore, be approximately at the equivalence point (a good feature) but the color change would be very gradual, extending over an addition of 13 ml of NaOH, making it almost useless as an indicator of the equivalence point.

Methyl orange could be used for the HCl titration (Fig. 18–1), though its use would have two disadvantages: (1) it would complete its color change slightly before the equivalence point is reached (though very close to it, so this is not particularly serious), and (2) it would change color, as the plot shows, over a range of nearly 0.5 ml of NaOH addition, which is not as sharp a color change as litmus or phenolphthalein would produce in the same system. The titration curve for HOAc shows that methyl orange would be completely useless as an indicator in that system. Its color change would not be sharp, beginning after only about one milliliter of NaOH had been added and being completed when about 8 ml had been added. More serious even is the fact that the color change would be completed long before the equivalence point (which occurs when 25.0 ml of NaOH have been added) and hence would provide no evidence of the equivalence point.

You should check Table 18–5 to see whether any of the other indicators listed would be suitable for determining the equivalence point for either of the two titrations we have discussed.

18.13 The pH Meter

The pH of a solution may be measured and the course of a neutralization titration followed by the use of an instrument called a **pH meter.** Each pH determination with the pH meter involves the measurement of the difference in electrical

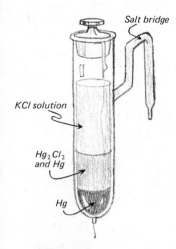

FIGURE 18-3
A calomel electrode. The most commonly used reference electrode in the pH meter.

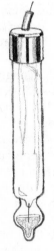

FIGURE 18-4
The glass electrode, used extensively as an indicating electrode in the pH meter.

potential between two electrodes. One electrode is called the **reference electrode,** the other the **indicating electrode.** Both are in contact with the solution whose pH is to be determined. Thus, the pH meter must consist of at least three parts: a reference electrode, an indicating electrode, and a potential-measuring meter.

The reference electrode of a pH meter is one whose potential is independent of the hydrogen ion concentration of the solution being measured. The most commonly used reference electrode is the **calomel electrode.** This electrode is composed of metallic mercury and solid mercury(I) chloride (Hg_2Cl_2), called "calomel," in contact with, and in equilibrium with, a solution of potassium chloride (Fig. 18–3). In usual practice, the potassium chloride solution is saturated.

The indicating electrode is an electrode whose potential is dependent upon the concentration of hydrogen ions in the solution being tested. The most widely used indicating electrode is the **glass electrode.** This electrode consists of a glass tube with a thin glass bulb at one end, inside of which is either a buffer solution into which a platinum wire is dipped or a $0.1\ N$ solution of hydrochloric acid in contact with a silver wire coated with a thin layer of silver chloride (Fig. 18–4). An electrical contact is made to the platinum or silver wire for connection to the potential measuring meter. When the glass electrode is in use, the bulb is immersed in the test solution. The glass membrane serves to establish a potential, the size of which is determined by the *relative* hydrogen ion concentrations on the two sides of the membrane. The hydrogen ion concentration inside the electrode is fixed. Before use, each glass electrode must be calibrated empirically by measuring the pH of a standard buffer solution. pH meters are calibrated in pH units rather than volts for direct reading of the pH of the solution being tested.

18.14   The Ionization of Weak Diprotic Acids

Polyprotic acids, those from which more than one proton may be removed, ionize in water in successive steps. An example of a **diprotic acid** is hydrogen sulfide in aqueous solution (hydrosulfuric acid). The first step in its ionization yields hydrogen and hydrosulfide ions.

$$H_2S \rightleftharpoons H^+ + HS^-$$

The expression for the ionization constant for the primary ionization of H_2S is

$$\frac{[H^+][HS^-]}{[H_2S]} = K_{H_2S} = 1.0 \times 10^{-7}$$

The hydrosulfide ion in turn ionizes and forms hydrogen and sulfide ions.

$$HS^- \rightleftharpoons H^+ + S^{2-}$$

The expression for the ionization constant for the hydrosulfide ion is

$$\frac{[H^+][S^{2-}]}{[HS^-]} = K_{HS^-} = 1.3 \times 10^{-13}$$

Note that K_{HS^-} is smaller than K_{H_2S} by almost 10^6 times. This means that very little of the HS^- formed by the ionization of H_2S ionizes to give hydrogen ions.

and sulfide ions. Thus the concentrations of H^+ and HS^- are practically equal in a pure aqueous solution of H_2S.

Example 1. The concentration of H_2S in a saturated aqueous solution of the gas at one atmosphere of pressure and room temperature is approximately 0.1 molar. Calculate the $[H^+]$, $[HS^-]$, and $[S^{2-}]$ in the solution.

If we let x equal $[H^+]$, then $[HS^-]$ will also be equal to x, and $[H_2S]$ will be equal to $0.1 - x$. Since x will be quite small compared to 0.1, it may be dropped from the term $(0.1 - x)$. Solving for x, we have

$$\frac{[H^+][HS^-]}{[H_2S]} = \frac{x^2}{0.1} = 1.0 \times 10^{-7}$$

$$x^2 = 1.0 \times 10^{-8}$$

$$x = 1 \times 10^{-4} \, M = [H^+] = [HS^-]$$

Since, in any solution, all possible equilibria must be satisfied simultaneously, the $[H^+]$ and $[HS^-]$ must be the same for the equilibria involved in the ionization of both HS^- and H_2S. Thus, substituting in the expression for the ionization constant for HS^-, we have

$$\frac{[H^+][S^{2-}]}{[HS^-]} = \frac{(1 \times 10^{-4})}{(1 \times 10^{-4})}[S^{2-}] = K_{HS^-} = 1.3 \times 10^{-13}$$

$$[S^{2-}] = 1.3 \times 10^{-13} \, M$$

(Note that in a pure aqueous solution of H_2S the $[S^{2-}]$ is equal to K_{HS^-}. In fact, for any weak diprotic acid the concentration of the divalent anion is equal to the secondary ionization constant.)

By multiplying the expressions for the ionization constants K_{H_2S} and K_{HS^-}, we obtain

$$\frac{[H^+][HS^-]}{[H_2S]} \times \frac{[H^+][S^{2-}]}{[HS^-]} = K_{H_2S} \times K_{HS^-}$$

$$\frac{[H^+]^2[S^{2-}]}{[H_2S]} = K_{H_2S} K_{HS^-} = 1.3 \times 10^{-20}$$

Since a saturated solution of hydrogen sulfide in water is 0.1 M with respect to the gas, we have

$$\frac{[H^+]^2[S^{2-}]}{0.1} = 1.3 \times 10^{-20}$$

$$[H^+]^2[S^{2-}] = 1.3 \times 10^{-21} \quad \text{(for a saturated } H_2S \text{ solution, only)}$$

Thus, in a solution saturated with hydrogen sulfide the concentration of the sulfide ion varies inversely as the square of the hydrogen ion concentration. This equation may be used to calculate the concentration of the sulfide ion when hydrogen ions in the form of a strong acid are added to a saturated hydrogen sulfide solution.

Example 2. Calculate the concentration of sulfide ion in a saturated solution of hydrogen sulfide to which sufficient hydrochloric acid has been added to make the hydrogen ion concentration of the solution 0.10 molar.

Substituting in the above equation, we have

$$[H^+]^2[S^{2-}] = (0.1)^2[S^{2-}] = 1.3 \times 10^{-21}$$

$$[S^{2-}] = \frac{1.3 \times 10^{-21}}{(0.10)^2} = 1.3 \times 10^{-19}\, M$$

(We may use the same equation in calculating the hydrogen ion concentration necessary to produce a given sulfide ion concentration.)

18.15   The Ionization of Weak Triprotic Acids

A typical example of a **triprotic acid** is phosphoric acid, which ionizes in water in three steps.

$$H_3PO_4 \rightleftharpoons H^+ + H_2PO_4^-$$
$$H_2PO_4^- \rightleftharpoons H^+ + HPO_4^{2-}$$
$$HPO_4^{2-} \rightleftharpoons H^+ + PO_4^{3-}$$

For each step in the ionization there is a corresponding ionization constant. The expressions for the ionization constants of the three stages in the ionization of phosphoric acid are

$$\frac{[H^+][H_2PO_4^-]}{[H_3PO_4]} = K_{H_3PO_4} = 7.5 \times 10^{-3}$$

$$\frac{[H^+][HPO_4^{2-}]}{[H_2PO_4^-]} = K_{H_2PO_4^-} = 6.3 \times 10^{-8}$$

$$\frac{[H^+][PO_4^{3-}]}{[HPO_4^{2-}]} = K_{HPO_4^{2-}} = 3.6 \times 10^{-13}$$

It will be noted that ionization in the successive steps becomes progressively less extensive. This is a general characteristic of polyprotic acids, successive ionization constants for most such acids differing by about 10^5.

Although this set of three dissociation reactions appears to make equilibrium calculations of equilibrium concentrations very complicated for the phosphoric acid solution, we can be reassured by noting that, as with the hydrogen sulfide calculations (Section 18.14), most equilibrium problems for polyprotic acids can be broken down into a series of calculations similar to those done for monoprotic acids.

Example. Calculate the concentrations of all species—H_3PO_4, $H_2PO_4^-$, HPO_4^{2-}, PO_4^{3-}, H^+, and OH^-—present at equilibrium in a solution containing a total phosphoric acid concentration of 0.100 M.

$$H_3PO_4 \rightleftharpoons H^+ + H_2PO_4^-$$

Let x = the number of moles of H_3PO_4 which dissociates per liter in establishing the equilibrium.

Then x moles of H^+ and x moles of $H_2PO_4^-$ would be produced. A small amount of the $H_2PO_4^-$ produced in the first step will disappear by dissociation into H^+ and HPO_4^- in the second step, and a small amount of additional H^+ will be produced in both the second and third steps. However, because K_2 and K_3 are very small compared to K_1, it is logical to assume that the decrease in $[H_2PO_4^-]$ in the second step and the increase in $[H^+]$ in the second and third steps are negligible compared to the quantity of each ion (x) produced in Step 1.

Hence, at equilibrium,

$$[H_3PO_4] = \text{original amount present minus the amount}$$
$$\text{which dissociates}$$
$$= (0.100 - x) \text{ moles/liter}$$
$$[H^+] = x \text{ moles/liter}$$
$$[H_2PO_4^-] = x \text{ moles/liter}$$
$$K_1 = \frac{[H^+][H_2PO_4^-]}{[H_3PO_4]} = 7.5 \times 10^{-3}$$
$$\frac{(x)(x)}{(0.100 - x)} = \frac{x^2}{(0.100 - x)} = 7.5 \times 10^{-3}$$

With a value of K_1 as large as 7.5×10^{-3}, it is probable that x is too large to ignore compared to 0.100. Hence, the value of x can be calculated by means of the quadratic equation (see Appendix A.4).

$$x^2 = 7.5 \times 10^{-4} - 7.5 \times 10^{-3} x$$
$$x^2 + 7.5 \times 10^{-3} x - 7.5 \times 10^{-4} = 0$$
$$x = \frac{-(7.5 \times 10^{-3}) \pm \sqrt{(7.5 \times 10^{-3})^2 - 4(-7.5 \times 10^{-4})}}{2(1)}$$
$$= 2.4 \times 10^{-2} M$$

An alternative, and very useful, method for solving the equation is by **successive approximations.**

First approximation: On a trial basis, we will neglect the x term in the denominator:

$$K_1 = \frac{x^2}{(0.100)}$$
$$= 7.5 \times 10^{-3}$$
$$x^2 = 7.5 \times 10^{-4}$$
$$x = 2.7 \times 10^{-2} M$$

Second approximation: Obviously, x as found above is *not* negligible compared to 0.10 M. A better estimate, using the approximate value of x just obtained, of the equilibrium concentration of phosphoric acid is $(0.100 - x) = (0.100 - 0.027) = 0.073 \ M$:

$$K_1 = \frac{x^2}{(0.073)} = 7.5 \times 10^{-3}$$
$$x^2 = (7.5 \times 10^{-3})(7.3 \times 10^{-2})$$
$$= 5.5 \times 10^{-4}$$
$$x = 2.3 \times 10^{-2} \ M$$

In as much as the values of x obtained from the first and second approximations differ by 15%, a third approximation is desirable.

Third approximation: A still better estimate of the equilibrium phosphoric acid concentration, using x obtained from the second approximation, is $(0.1 - x) = (0.1 - 0.023) = 0.077 \ M$.

$$K_1 = \frac{x^2}{(0.077)} = 7.5 \times 10^{-3}$$
$$x^2 = (7.5 \times 10^{-3})(0.077)$$
$$= 5.8 \times 10^{-4}$$
$$x = 2.4 \times 10^{-2} \ M$$

(Note that this value of x is the same as that calculated from the quadratic equation.)

The values of x from the last two approximations agree to within approximately 4%. Hence, we can safely conclude that

$$[H^+] = x = 0.024 \ M$$
$$[H_2PO_4^-] = x = 0.024 \ M$$
$$[H_3PO_4] = 0.100 - x = 0.100 - 0.024 = 0.076 \ M$$
$$[OH^-] = \frac{K_w}{[H^+]} = \frac{1.0 \times 10^{-14}}{2.4 \times 10^{-2}} = 4.2 \times 10^{-13} \ M$$

To calculate the concentration of HPO_4^{2-}, we must use the second step of the acid dissociation. Let y = number of moles of $H_2PO_4^-$ which dissociates in the second step to produce H^+ and HPO_4^{2-}.

	$H_2PO_4^-$	$\rightleftharpoons$	H^+	$+$	HPO_4^{2-}
Initial concentrations:	0.024 M		0.024 M		—
Equilibrium concentrations:	(0.024 − y)		(0.024 + y)		y

$$K_2 = \frac{(0.024 + y)(y)}{(0.024 - y)} = 6.3 \times 10^{-8}$$

In as much as K_2 is so small, we can neglect the additive and subtractive y terms. Then the preceding equation becomes

$$\frac{0.024 \ y}{0.024} = y = 6.3 \times 10^{-8} \ M = [HPO_4^-]$$

Note that this provides a check for our earlier assumption that the increase in H^+ and decrease in $H_2PO_4^-$ (now known to be 6.3×10^{-8} mole) is negligible compared to the quantity of each ion produced in the first step (0.024 mole).

Let us now consider the third step of dissociation, to calculate the concentration of PO_4^{3-}. Let $z =$ the number of moles of HPO_4^{2-} that dissociates to produce H^+ and PO_4^{3-}.

$$HPO_4^{2-} \rightleftharpoons H^+ + PO_4^{3-}$$

Initial concentrations: $(6.3 \times 10^{-8}\ M)$ $(0.024\ M)$ —
Equilibrium concentrations: $(6.3 \times 10^{-8} - z)$ $(0.024 + z)$ z

$$K_3 = \frac{(0.024 + z)(z)}{(6.3 \times 10^{-8} - z)} = 3.6 \times 10^{-13} \quad \text{)disregard}$$

Because K_3 is very small compared to K_2, the extent of the third dissociation is insignificant and we can neglect the additive and subtractive terms in the preceding equation; thus, we can easily solve the expression for z:

$$\frac{0.024\ z}{6.3 \times 10^{-8}} = 3.6 \times 10^{-13}$$

$$z = [PO_4^{3-}] = \frac{(3.6 \times 10^{-13})(6.3 \times 10^{-8})}{(2.4 \times 10^{-2})} = 9.5 \times 10^{-19}\ M$$

Note that the quantity of H^+ added to the solution and the decrease in HPO_4^{2-} in the third step (z) is indeed negligible compared to the amounts previously in the solution.

Hydrolysis

When a salt is dissolved in water the resulting solution may be either neutral, basic, or acidic, depending upon the nature of the salt. A salt formed from a strong acid and a strong base, such as NaCl, yields an aqueous solution which is practically neutral. A salt of a strong base and a weak acid, such as NaOAc, forms aqueous solutions which are basic. A salt of a weak base and a strong acid, such as NH_4Cl, gives an acidic reaction in water. A salt of a weak acid and a weak base may form either neutral, basic, or acidic solutions, depending upon the relative strengths or solubilities of the acid and the base. Many substances in aqueous solution react with water in such a way as to disturb the normal concentration of the hydrogen and hydroxide ions formed through the auto-ionization of water. A reaction of this type in which water is one of the reactants is called **hydrolysis.** The theory of the hydrolysis of salts of the types described above will be discussed in the sections which follow.

As has been pointed out in earlier chapters, ions known to be hydrated in water solution are often indicated for convenience in abbreviated form without showing

[handwritten: $NH_4NO_3 \rightarrow NH_4^+ + NO_3^-$]
[handwritten: $H_2O \rightarrow OH^- + H^+$]
[handwritten: $NH_4^+ + H_2O \rightarrow NH_4OH + OH^-$]

the hydration; for example, H^+ for H_3O^+, and Cu^{2+} for $[Cu(H_2O)_4]^{2+}$. However, when water is a participant in the reaction, as in hydrolysis reactions, indication of the hydration of the ions becomes important. In dealing with hydrolysis, therefore, we will use formulas which show the extent of hydration when it is important for an understanding of the reaction involved.

18.16   Salt of a Strong Acid and a Strong Base

A salt consisting of a cation from a strong hydroxide base (such as NaOH) and an anion from a strong acid (such as HCl) forms a neutral aqueous solution. This means that the ions of sodium chloride do not react with water in such a way as to alter the normal concentration of the hydrogen and hydroxyl ions which are formed through the ionization of the water ($2H_2O \rightleftharpoons H_3O^+ + OH^-$). Salts resulting from the reactions of the alkali metals and alkaline earth metals with the hydrohalic acids, nitric acid, sulfuric acid, or perchloric acid are examples of salts that give neutral aqueous solutions; i.e., these salts do not undergo appreciable hydrolysis.

18.17   Salt of a Strong Base and a Weak Acid   *[handwritten: basic]*

[handwritten: NaOAc]

Experiment shows that when sodium acetate is dissolved in water, the solution is alkaline. This result must be due to an increase in the concentration of the hydroxide ion and a decrease in the concentration of the hydrogen ion (or, more exactly, hydronium ion) in the solution. The acetate ion is the anion of a weak acid, acetic acid. The concentration of hydrogen ion supplied by the water, $10^{-7}\ M$, is too great to exist in equilibrium with the high concentration of acetate ion supplied by the salt sodium acetate. As a consequence, some undissociated acetic acid forms, according to the equation

$$OAc^- + H_3O^+ \rightleftharpoons HOAc + H_2O$$

As acetate ions combine with hydrogen ions derived from the ionization of water to form acetic acid molecules, the water ionization equilibrium shifts so that more hydrogen ions and hydroxide ions are formed. The net result is that the solution becomes slightly basic due to the accumulation of an excess of hydroxide ions in the solution. The net reaction may be expressed by

$$OAc^- + H_2O \rightleftharpoons HOAc + OH^-$$

Applying the law of chemical equilibrium to the reaction of the acetate ion with water, we obtain the expression for the equilibrium constant.

$$\frac{[HOAc][OH^-]}{[OAc^-][H_2O]} = K_e$$

Because the number of moles of water involved in the reaction is negligibly small compared with the total number of moles of water present in dilute solution, we

CN⁻ + H₂O ⇌ HCN + OH

NaOAc ⇌ Na⁺ + OAc⁻

$K_i = \dfrac{[Na][OAc]}{[NaOAc]} = \dfrac{x^2}{.10}$

may assume that the molar concentration of the water remains unchanged during the hydrolysis. Hence, the preceding equation can be written

NaCN → Na⁺ + CN⁻
H₂O → OH⁻ + H⁺

$$\frac{[HOAc][OH^-]}{[OAc^-]} = K_e[H_2O] = K_h$$

The constant K_h is called the **hydrolysis constant.** When the value for K_h is small, the degree of hydrolysis is slight; when it is large, hydrolysis is extensive.

We may obtain the value of K_h for the hydrolysis of sodium acetate from the values for the ionization constants of water and of acetic acid. This follows because the hydrolysis involves the two equilibria: (a) the ionization of water, $2H_2O \rightleftharpoons H_3O^+ + OH^-$, and (b) the ionization of acetic acid, $HOAc + H_2O \rightleftharpoons H_3O^+ + OAc^-$. For every aqueous solution

$$[H_3O^+][OH^-] = K_w \qquad \text{or} \qquad [OH^-] = \frac{K_w}{[H_3O^+]}$$

Substituting $K_w/[H_3O^+]$ for $[OH^-]$ in the expression for the hydrolysis constant, we obtain

$\dfrac{K_w}{K_{HOAc}} = \dfrac{[HOAc]}{[H^+][OAc]}$

$\dfrac{[HOAc]}{[H^+][OAc]} \times \dfrac{[H^+][OH]}{1}$

$$\frac{[H^+]}{[H^-]}\frac{[HOAc][OH^-]}{[OAc^-]} = \frac{[HOAc]K_w}{[OAc^-][H_3O^+]} = K_h$$

But the expression $\dfrac{[HOAc]}{[OAc^-][H_3O^+]}$ is the reciprocal of K_i and is equal to

$\dfrac{1}{K_i \text{ (for HOAc)}}$. Therefore, $K_h = \dfrac{K_w}{K_i} = \dfrac{1.0 \times 10^{-14}}{1.8 \times 10^{-5}} = 5.6 \times 10^{-10}$.

NaCN → Na⁺ + CN⁻

$K_h = \dfrac{1.0 \times 10^{-14}}{}$

18.18 Salt of a Weak Base and a Strong Acid

Now let us consider the hydrolysis of ammonium chloride, NH_4Cl, a salt of a weak base and a strong acid. The chloride ion is such a weak base that it does not combine appreciably with the hydrogen ion of water. However, the ammonium ion tends to react with the hydroxide ion of water according to the equation

$$NH_4^+ + OH^- \rightleftharpoons NH_3 + H_2O$$

Removal of hydroxide ions from solution causes the equilibrium

$$2H_2O \rightleftharpoons H_3O^+ + OH^-$$

to shift to the right and results in an excess of hydronium ions in solution. The hydrolysis of the ammonium ion may be represented by the net equation obtained by adding the two equations given above. Thus, we have

$$NH_4^+ + H_2O \rightleftharpoons NH_3 + H_3O^+$$

It is not sufficient in this case to write

$$NH_4^+ \longrightarrow NH_3 + H^+$$

solution is acidic

for this implies a spontaneous loss of a proton by the ammonium ion. Actually, such loss takes place only because the solvent water attracts the proton strongly enough to take protons from some of the ammonium ions.

The expression for the hydrolysis constant for the hydrolysis of the ammonium ion, based on the equation $NH_4^+ + H_2O \rightleftharpoons NH_3 + H_3O^+$, is written as

$$\frac{[NH_3][H_3O^+]}{[NH_4^+]} = K_h$$

Substituting $K_w/[OH^-]$ for $[H_3O^+]$ in the above equation, we obtain

$$\frac{[NH_3]K_w}{[NH_4^+][OH^-]} = K_h$$

Because the expression $\dfrac{[NH_3]}{[NH_4^+][OH^-]}$ is equal to $\dfrac{1}{K_i \text{ (for } NH_3)}$ we may write

$$K_h = \frac{K_w}{K_i} = \frac{1.0 \times 10^{-14}}{1.8 \times 10^{-5}} = 5.6 \times 10^{-10}$$

18.19   Salt of a Weak Base and a Weak Acid

When ammonium acetate is dissolved in water we have, in solution, ammonium ions and acetate ions, both of which undergo hydrolysis. The equations are

$$NH_4^+ + H_2O \rightleftharpoons NH_3 + H_3O^+$$
$$OAc^- + H_2O \rightleftharpoons HOAc + OH^-$$

The extent to which these reactions take place is approximately the same, inasmuch as aqueous ammonia and acetic acid have nearly equal ionization constants. Therefore, the hydrolytic products are present in nearly equal concentrations at equilibrium. As hydronium and hydroxide ions are produced by these reactions they unite to form water, because the product of their concentrations cannot exceed 1×10^{-14}. As these ions are formed and then immediately removed from solution in equal amounts, the solution remains neutral as hydrolysis proceeds.

The net reaction involved in the hydrolysis may be looked upon as the sum of the three equations

$$NH_4^+ + H_2O \rightleftharpoons NH_3 + H_3O^+$$
$$OAc^- + H_2O \rightleftharpoons HOAc + OH^-$$
$$\underline{H_3O^+ + OH^- \rightleftharpoons 2H_2O}$$
$$NH_4^+ + OAc^- \rightleftharpoons HOAc + NH_3$$

Applying the law of chemical equilibrium to this system, we obtain

$$\frac{[NH_3][HOAc]}{[NH_4^+][OAc^-]} = K_h$$

(handwritten: 6,000 − .724 = 5,276; 14,000 − 5,276 = 8,724)

To evaluate the hydrolysis constant in terms of the ion product constant for water and the ionization constants for aqueous ammonia and acetic acid, multiply the numerator and denominator of the above equation by $[H_3O^+][OH^-]$.

$$\frac{[NH_3]}{[NH_4^+][OH^-]} \times \frac{[HOAc]}{[H_3O^+][OAc^-]} \times \frac{[H_3O^+][OH^-]}{1} = K_h$$

Thus, it may readily be seen that

$$K_h = \frac{K_w}{K_i \text{ (for } NH_3) \times K_i \text{ (for HOAc)}} = \frac{1.0 \times 10^{-14}}{(1.8 \times 10^{-5})(1.8 \times 10^{-5})} = 3.1 \times 10^{-5}$$

When the hydrolytic products of a salt of a weak base and a weak acid do not have the same ionization constants, the aqueous solution of the salt is either basic or acidic, depending upon which electrolyte has the larger ionization constant. For example, ammonium cyanide hydrolyzes to give a basic solution, since K_h for NH_3 is larger than K_i for HCN. This means that the hydrolysis of the cyanide ion, $CN^- + H_2O \rightleftharpoons HCN + OH^-$, is more extensive than that of the ammonium ion, $NH_4^+ + H_2O \rightleftharpoons NH_3 + H_3O^+$, and that an excess of hydroxide ions accumulates in the solution.

(handwritten: $NaOAc \rightarrow Na^+ + OAc^-$; $H_2O \rightarrow OH^- + H^+$; $OAc^- + H_2O \rightarrow HOAc + OH^-$)

18.20   Calculations Involving the Hydrolysis of Salts

(handwritten: $K_h = \frac{1.0 \times 10^{-15}}{1.8 \times 10^{-5}}$)

Example 1.   Calculate the hydroxide ion concentration, the degree of hydrolysis, and the pH of a 0.050 M solution of sodium acetate. *(handwritten: — salt of a SB + WA)*

(handwritten: NaOAc)

We have seen that the equation for the hydrolysis of sodium acetate is

$$OAc^- + H_2O \rightleftharpoons HOAc + OH^-$$

and that the expression for the hydrolysis constant is

$$\frac{[HOAc][OH^-]}{[OAc^-]} = K_h = \frac{K_w}{K_i} = 5.6 \times 10^{-10}$$

(handwritten: $\frac{x^2}{.05} = 5.6 \times 10^{-10}$; $x^2 = 28 \times 10^{-12}$; $x = 5.3 \times 10^{-6}$ [OH])

If we let x equal the concentration of molecular acetic acid formed by the hydrolysis of the acetate ion, then the concentration of hydroxide ion will also be equal to x, and the concentration of acetate ion at equilibrium will be $0.050 - x$. Substituting in the hydrolysis constant expression above, we have

$$\frac{x^2}{0.050 - x} = 5.6 \times 10^{-10}$$

(handwritten: $pOH = 6 - .724$; $pOH = 5.276$; $pH = 8.624$)

Since x is very small compared with 0.050, we may drop it from the denominator. Then,

$$x^2 = 0.050 \times 5.6 \times 10^{-10} = 28 \times 10^{-12}$$
$$x = 5.3 \times 10^{-6} = [HOAc] = [OH^-]$$

Thus, the concentration of hydroxide ion is 5.3×10^{-6} M, and the percentage hydrolysis is equal to

$$\frac{[\text{HOAc}]}{[\text{OAc}^-]} \times 100 = \frac{5.3 \times 10^{-6}}{0.050} \times 100 = 0.011 \text{ per cent}$$

We can calculate the pH of the solution by first finding the hydronium ion concentration

$$[\text{H}_3\text{O}^+] = \frac{K_w}{[\text{OH}^-]} = \frac{1.0 \times 10^{-14}}{5.3 \times 10^{-6}} = 1.9 \times 10^{-9}$$

and then,

$$p\text{H} = -\log[\text{H}_3\text{O}^+] = -\log(1.9 \times 10^{-9}) = -(0.28 - 9) = 8.72$$

Example 2.   Calculate the sulfide ion concentration and the degree of hydrolysis in a 0.0010 M solution of sodium sulfide.

The sulfide ion hydrolyzes according to the equation

$$\text{S}^{2-} + \text{H}_2\text{O} \rightleftharpoons \text{HS}^- + \text{OH}^-$$

The hydrosulfide ion thus formed also undergoes hydrolysis, but its extent is so slight compared with that of the sulfide ion that it can be neglected. The expression for the hydrolysis constant of the sulfide ion is

$$\frac{[\text{HS}^-][\text{OH}^-]}{[\text{S}^{2-}]} = K_h = \frac{K_w}{K_i \text{ (for HS}^-)} = \frac{1.0 \times 10^{-14}}{1.3 \times 10^{-13}} = 7.7 \times 10^{-2}$$

The large value for the hydrolysis constant indicates that most of the sulfide ion undergoes hydrolysis. For this reason we shall let x equal the S^{2-} concentration when hydrolytic equilibrium has been established. Then the concentrations of HS^- and OH^-, which are equal to each other, may each be represented by $0.0010 - x$. Substituting these values in the expression, we have

$$\frac{[\text{HS}^-][\text{OH}^-]}{[\text{S}^{2-}]} = \frac{(0.0010 - x)(0.0010 - x)}{x} = 7.7 \times 10^{-2}$$

But since hydrolysis is nearly complete, x is now small as compared to 0.0010 and may be neglected in the numerator. The expression then becomes

$$\frac{(0.0010)^2}{x} = 7.7 \times 10^{-2}$$

$$x = \frac{(0.0010)^2}{7.7 \times 10^{-2}} = 1.3 \times 10^{-5} = [\text{S}^{2-}]$$

The degree of hydrolysis may be found by dividing the amount of S^{2-} hydrolyzed ($0.0010\ M - 0.000013\ M = 0.00099\ M$) by the total amount of S^{2-} originally present. Then the percentage hydrolysis is obtained by multiplying by 100.

$$\frac{0.00099}{0.0010} \times 100 = 99 \text{ per cent}$$

Example 3. Calculate the degree of hydrolysis and the pH of a solution that is 1.0 M in NH_4CN. *salt of* WB, + W.A

$HCN \to H^+ + CN^-$

$NH_4OH + H_2O \to NH_4 + H_3O^-$

When ammonium cyanide is dissolved in water, both the ammonium ion and the cyanide ion undergo hydrolysis. The equations are

$$NH_4^+ + H_2O \rightleftharpoons NH_3 + H_3O^+$$
$$CN^- + H_2O \rightleftharpoons HCN + OH^-$$
$$H_3O^+ + OH^- \rightleftharpoons 2H_2O$$

The net reaction is given by the equation

$NH_4 CN \to NH_4^+ + CN^-$

$H_2O \to OH^- + H^-$

$$NH_4^+ + CN^- \rightleftharpoons NH_3 + HCN$$

The expression for the hydrolysis constant may be written

$NH_4^+ + CN^- \to NH_4CN +$

$$\frac{[NH_3][HCN]}{[NH_4^+][CN^-]} = K_h = \frac{K_w}{K_i \text{ (for } NH_3) \times K_i \text{ (for HCN)}}$$

$$= \frac{1.0 \times 10^{-14}}{(1.8 \times 10^{-5})(4 \times 10^{-10})} = 1.4$$

It can be shown that, although the hydrolysis of cyanide ion is somewhat more extensive than that of ammonium ion, the difference in the extent of hydrolysis of the two ions is relatively small compared to the total quantity of each which undergoes hydrolysis. It is justifiable to assume that for every NH_4^+ that hydrolyzes, a CN^- also hydrolyzes, provided the NH_4CN concentration is not extremely low. Let x equal the number of moles of NH_4^+ and the number of moles of CN^- undergoing hydrolysis. At equilibrium then, the $[NH_4^+] = (1.0 - x)$ and the $[NH_3] = [HCN] = x$. Substituting, we have

$$\frac{[NH_3][HCN]}{[NH_4^+][CN^-]} = \frac{x^2}{(1.0 - x)^2} = 1.4$$

Extracting the square root of both sides of the equation, we find

$$\frac{x}{1.0 - x} = 1.18$$

$$x = 1.18 - 1.18x$$

$$2.18x = 1.18$$

$$x = 0.54 = [NH_3] = [HCN]$$

The degree of hydrolysis is the amount of NH_4CN hydrolyzed divided by the total amount of NH_4CN originally present. The per cent hydrolysis is

$$\frac{0.54}{1.0} \times 100 = 54 \text{ per cent}$$

At equilibrium, $[NH_4^+] = [CN^-] = (1.0 - x) = (1.0 - 0.54) = 0.46$, and substituting in the expression for the ionization constant for HCN, we find

$$\frac{[H^+][CN^-]}{[HCN]} = \frac{[H^+](0.46)}{0.54} = 4 \times 10^{-10}$$

$$[H^+] = \frac{0.54}{0.46} \times (4 \times 10^{-10}) = 4.7 \times 10^{-10}$$

$$pH = -\log(4.7 \times 10^{-10}) = 9$$

18.21   The Hydrolysis of Metal Ions

A large number of metal ions hydrolyze to give acidic solutions. They change the normal hydronium and hydroxide concentration of water by removing hydroxide ions from solution. For example, the aluminum ion hydrolyzes according to the equations

$$Al^{3+} + 6H_2O \rightleftharpoons [Al(H_2O)_6]^{3+}$$
$$[Al(H_2O)_6]^{3+} + H_2O \rightleftharpoons [Al(OH)(H_2O)_5]^{2+} + H_3O^+$$

leaving an excess of hydronium ions in solution. The aluminum ion hydrolyzes in stages just as polyprotic acids ionize in more than one stage, as shown by

$$Al^{3+} + 6H_2O \rightleftharpoons [Al(H_2O)_6]^{3+}$$
$$[Al(H_2O)_6]^{3+} + H_2O \rightleftharpoons [Al(OH)(H_2O)_5]^{2+} + H_3O^+$$
$$[Al(OH)(H_2O)_5]^{2+} + H_2O \rightleftharpoons [Al(OH)_2(H_2O)_4]^+ + H_3O^+$$
$$[Al(OH)_2(H_2O)_4]^+ + H_2O \rightleftharpoons [Al(OH)_3(H_2O)_3] + H_3O^+$$

Just as in the ionization of a polyprotic acid, such as H_3PO_4, the hydrolysis of cations carrying more than one charge is not extensive beyond the first stage. Additional examples of the first stage in the hydrolysis of metal ions are provided by the equations

$$[Fe(H_2O)_6]^{3+} + H_2O \rightleftharpoons [Fe(OH)(H_2O)_5]^{2+} + H_3O^+$$
$$[Cu(H_2O)_4]^{2+} + H_2O \rightleftharpoons [Cu(OH)(H_2O)_3]^+ + H_3O^+$$
$$[Zn(H_2O)_4]^{2+} + H_2O \rightleftharpoons [Zn(OH)(H_2O)_3]^+ + H_3O^+$$

Example. Let us calculate the pH of a 0.10 M solution of aluminum chloride, given that the hydrolysis constant for the reaction $[Al(H_2O)_6]^{3+} + H_2O \rightleftharpoons [Al(OH)(H_2O)_5]^{2+} + H_3O^+$ is 1.4×10^{-5}.

The expression for the hydrolysis constant is written as

$$\frac{[Al(OH)(H_2O)_5^+][H_3O^+]}{[Al(H_2O)_6^{3+}]} = K_h = 1.4 \times 10^{-5}$$

Let x equal the concentration of the hydronium ion, which is in turn equal to the concentration of $[Al(OH)(H_2O)_5]^{2+}$. Then $0.10 - x$ will equal the concentration of $[Al(H_2O)_6]^{3+}$. Substituting in the expression for the hydrolysis constant, we have

$$\frac{[Al(OH)(H_2O)_5^+][H_3O^+]}{[Al(H_2O)_6^{3+}]} = \frac{x^2}{0.10 - x} = 1.4 \times 10^{-5}$$

Because K_h is small, x will also be small, and we may drop the x in the term $0.10 - x$ in an approximate solution of the problem. Thus, we obtain

$$\frac{x^2}{0.10} = 1.4 \times 10^{-5}$$

$$x = 1.2 \times 10^{-3} = [H^+]$$

$$pH = -\log (1.2 \times 10^{-3}) = 2.9$$

Because the constants for the different stages of hydrolysis are not known for most metal ions, we cannot calculate the degree of hydrolysis for many metal ions which we know to be hydrolyzed in solution. However, if the hydroxide of a metal ion is insoluble in water, we may conclude that the metal ion will hydrolyze to give an acidic solution. In fact, practically all metal ions other than those of the alkali metals hydrolyze to give acidic solutions.

QUESTIONS

1. Why does the law of chemical equilibrium apply to solutions of weak electrolytes but not to solutions of strong electrolytes?
2. What are the ionic and molecular species present in an aqueous solution of acetic acid? a solution of hydrochloric acid? a solution of phosphoric acid?
3. Compare the extent of ionization, or dissociation, of strong electrolytes and weak electrolytes in water.
4. Classify each of the following compounds as a weak or strong electrolyte: HCl, NaOH, NaCl, HOAc, H_2SO_4, $CaCl_2$, $Hg(CN)_2$, $CdCl_2$, $Ca(OH)_2$, H_3PO_4, NH_3, $Fe(OH)_3$, and HCN.
5. What is meant by the self-ionization of water?
6. How would the extent of ionization of acetic acid in a solution be changed by the addition of some potassium acetate? by the addition of some hydrogen chloride?
7. What would be the per cent ionization of a weak base in a hypothetical infinitely dilute solution?
8. For the weak acids listed in Table 18–2, arrange the anions in order of increasing base strength.
9. The per cent dissociation of a 1.0 M methylamine solution is 2.1%. Is methylamine a stronger or weaker base than ammonia? Explain.
10. Define pH both in words and by a mathematical equation.

11. Show that the hydroxide ion concentration of a solution made by combining equal numbers of moles of a weak base and the cation of the weak base in a given volume of water is numerically equal to the value of the ionization constant.

12. Explain the change of color of acid-base indicators in terms of ionic equilibria.

13. What constitutes a buffer solution?

14. Of what practical value are buffer solutions?

15. Does the concept of buffering action have any relevance to any portion of a titration curve? Explain.

16. (a) Calculate several representative pH values and construct a titration curve for the titration of 100 ml of 0.0500 M HCN with 0.100 M NaOH. (b) Determine which indicator listed in Table 18–5 would be most suitable for determining the equivalence point. Explain your answer.

17. Explain why successive ionization constants of weak polyprotic acids become progressively smaller.

18. What determines whether a salt of a weak base and a weak acid will hydrolyze to give a solution that is either acidic, neutral, or basic?

19. Write equations to show the stepwise hydrolysis of the following hydrated metal ions: $[Al(H_2O)_6]^{3+}$, $[Cu(H_2O)_4]^{2+}$, $[Cr(H_2O)_6]^{3+}$, $[Sn(H_2O)_6]^{4+}$, and $[Zn(H_2O)_4]^{2+}$.

20. Write equations for the hydrolysis of the following ions: $CO_3{}^{2-}$, CN^-, $HPO_4{}^{2-}$, $NH_4{}^+$, $SO_3{}^{2-}$, OAc^-, $HCO_3{}^-$, F^-, $NO_2{}^-$, Zn^{2+}, Al^{3+}, and Cu^{2+}.

21. Classify solutions of the following salts as either acidic, neutral, or basic: KCl, NaNO$_3$, FeCl$_3$, NH$_4$NO$_3$, KCN, Al$_2$(SO$_4$)$_3$, CuSO$_4$, NH$_4$OAc, and Na$_2$CO$_3$.

22. The ions of which group of metals in the Periodic Table do not undergo detectable hydrolysis?

23. How does hydrolysis affect the solubility of relatively insoluble metal sulfides and carbonates?

24. Explain why aluminum sulfide cannot be recrystallized from water.

25. Calculate the hydrolysis constant for the hydroxide ion.

26. Use the hydrolysis constant for the ammonium ion (5.6×10^{-10}) and the ionization constant for aqueous ammonia to calculate the ion-product constant for water.

27. Bringing together equivalents of what kinds of acids and bases results in the formation of neutral solutions?

28. Can the numerical values of the hydrolysis constants of monobasic anions be compared directly in order to evaluate their relative tendencies to accept protons?

PROBLEMS

[S]1. Calculate the concentration of each of the ions in the following solutions: (a) 0.03 M HCl, (b) 0.002 M Ca(OH)$_2$. $Ca(OH)_2 = Ca + 2OH$

Ans. (a) $[H^+] = 0.03$ M; $[Cl^-] = 0.03$ M; $[OH^-] = 3 \times 10^{-13}$ M; (b) $[Ca^{2+}] = 0.002$ M; $[OH^-] = 0.004$ M; $[H^+] = 3 \times 10^{-12}$ M

2. The ionization constant of HClO is 3.5×10^{-8}. Calculate $[H^+]$, $[ClO^-]$, and the per cent ionization of a 0.050 M solution of the acid.

Ans. $[H^+] = [ClO^-] = 4.2 \times 10^{-5} M; \ 8.4 \times 10^{-2}\%$

3. A 0.010 M solution of HNO_2 is 19 per cent ionized. Calculate $[H^+]$, $[NO_2^-]$, and $[HNO_2]$, and the ionization constant for HNO_2.

Ans. $[H^+] = [NO_2^-] = 1.9 \times 10^{-3} M; \ [HNO_2] = 8.1 \times 10^{-3} M;$
$K_i = 4.5 \times 10^{-4}$

4. Calculate the per cent ionization of each of the following solutions. Calculate (1) using simplifying assumption and (2) without using simplifying assumption and with use of the quadratic formula. Compare the values.

   (a) 0.020 M HCOOH; $K_i = 1.8 \times 10^{-4}$ *Ans.* (1) 9.5%; (2) 9.0%
   (b) $1.0 \times 10^{-4} \ M \ H_2O_2$; $K_i = 2.4 \times 10^{-12}$ *Ans.* (1) and (2) 0.015%
   (c) 1.0 M HN_3; $K_i = 1 \times 10^{-4}$ *Ans.* (1) and (2) 1%
   (d) 0.10 M HF; $K_i = 7.2 \times 10^{-4}$ *Ans.* (1) 8.5%; (2) 8.1%

5. The ionization constant of triethylamine is 7.40×10^{-5}. $(C_2H_5)_3N + H_2O \longrightarrow [(C_2H_5)_3NH]^+ + OH^-$. Calculate the hydroxyl ion concentration in a 1.50 M triethylamine solution.   *Ans.* $1.05 \times 10^{-2} M$

6. Calculate the molarity of a solution of acetic acid which is 2.0% ionized. ($K_i = 1.8 \times 10^{-5}$.)   *Ans.* 0.044 M

7. What is the hydrogen ion concentration corresponding to each of the following pH values: 2.50; 7.20; 13.0; 4.80; 10.0; 0.500; -0.25; 14.3?

Ans. $3.2 \times 10^{-3} M; \ 6.3 \times 10^{-8} M; \ 10^{-13} M; \ 1.6 \times 10^{-5} M; \ 10^{-10} M; \ 0.32 M;$
$1.8 M; \ 5.0 \times 10^{-15} M$

8. Calculate the hydrogen ion concentration and the pH of a 0.50 M solution of propionic acid, CH_3CH_2COOH, for which $K_i = 1.34 \times 10^{-5}$.

Ans. $[H^+] = 2.6 \times 10^{-3} M; \ pH = 2.59$

9. The per cent ionization of 0.300 M chloroacetic acid is 6.60%. Calculate its ionization constant. (*Hint: The $[H^+]$ is not negligible compared with total acid concentration.*)   *Ans.* 1.40×10^{-3}

10. A dilute solution of n-butyric acid, $CH_3CH_2CH_2COOH$, containing 0.0510 mole of acid in 1,050 liters of solution, is 16.3% dissociated into ions. Calculate the ionization constant of n-butyric acid.   *Ans.* 1.54×10^{-6}

11. The ionization constant of benzoic acid, C_6H_5COOH, is 6.46×10^{-5}. What concentration of C_6H_5COOH would be required to give a solution having a hydrogen ion concentration of $2.5 \times 10^{-3} M$?   *Ans.* 0.099 M

12. Calculate the pH and $[OH^-]$ for a 0.23 M cyclohexylamine, $C_6H_{11}NH_2$, solution. (See statement in Section 18.11 concerning significant figures in pH values.) (K_i for cyclohexylamine is 4.6×10^{-4}.)

Ans. $pH = 12.01; \ [OH^-] = 0.010 M$

13. Calculate the $[H^+]$ of a solution which has a pH of 4.64. (See statement in Section 18.11 concerning significant figures in pH values.)

Ans. $2.3 \times 10^{-5} M$

14. Calculate the ionization constant for phenol, C_6H_5OH, which is $2.5 \times 10^{-2}\%$ ionized in a $2.0 \times 10^{-3} M$ solution.   *Ans.* 1.3×10^{-10}

15. A 0.350 M solution of aqueous ammonia is 0.709% ionized. What volume of this solution contains the same quantity of hydroxyl ions as 50.0 ml of $7.00 \times 10^{-3} \ M$ NaOH, which is completely dissociated?   *Ans.* 141 ml

16. The ionization constant of acetic acid is 1.8×10^{-5}. What will be the pH of a 0.10 M HOAc solution which is 0.10 M in NaOAc?   *Ans. 4.74*

ⓢ17. The ionization constant for acetic acid is 1.8×10^{-5}. How many grams of $NaC_2H_3O_2$ must be added to one liter of a 0.20 M solution of $HC_2H_3O_2$ to maintain a hydrogen ion concentration of 6.5×10^{-5} M?   *Ans. 4.5 g*

18. What is the pH of a solution which is 0.0400 M in formic acid and 0.0600 M in sodium formate? ($K_i = 1.8 \times 10^{-4}$.)   *Ans. 3.92*

19. Calculate the hydrogen and hydroxyl ion concentrations in pure water at 24° C. ($K_w = 1 \times 10^{-14}$.)   *Ans. $[H^+] = [OH^-] = 1 \times 10^{-7}$ M*

ⓢ20. Calculate the hydrogen ion concentration for a solution having a pOH of 6.34.   *Ans. 2.2×10^{-8} M*

21. Calculate the pH and pOH of water at 50°. At this temperature, $\log K_w = -13.262$.   *Ans. $pH = pOH = 6.631$*

ⓢ22. What is the pH of a solution containing aqueous ammonia and ammonium chloride in ⓢ(a) a molar ratio of 5.0 to 3.0, and (b) a molar ratio of 3.0 to 5.0?   *Ans. (a) 9.48 (b) 9.03*

23. Calculate the [H$^+$], [OH$^-$], pH, and pOH of the following solutions:
    (a) 1×10^{-3} M HBr   *Ans. 10^{-3} M; 10^{-11} M; 3.0; 11.0*
    (b) 2.0×10^{-4} M Ba(OH)$_2$ *Ans. 2.5×10^{-11} M; 4.0×10^{-4} M; 10.60; 3.40*
    (c) 0.0040 M H$_2$SO$_4$ (Remember that H$_2$SO$_4$ is a diprotic acid.)
             Ans. 8.0×10^{-3} M; 1.2×10^{-12} M; 2.10; 11.90
    (d) 0.01 M NaOH   *Ans. 1×10^{-12} M; 0.01 M; 12.0; 2.0*
    (e) 0.10 M HOAc ($K_i = 1.8 \times 10^{-5}$.)
             Ans. 1.3×10^{-3} M; 7.5×10^{-12} M; 2.87; 11.13
    (f) 0.10 M aqueous NH$_3$ ($K_i = 1.8 \times 10^{-5}$.)
             Ans. 7.5×10^{-12} M; 1.3×10^{-3} M; 11.13; 2.87

24. If abominable snowmen were to exist in the world in the same proportion to humans as do hydrogen ions to water molecules in pure water, how many abominable snowmen would there be in the world? Assume a world population of 2.2 billion people.   *Ans. 4*

ⓢ25. A 25-ml volume of 0.10 M lactic acid is added to 5.0 ml of 0.69 M sodium lactate solution. Calculate the pH of this solution. ($K_i = 1.37 \times 10^{-4}$.)
             Ans. 4.00

ⓢ26. The pH of a given HF solution is 1.57. What is the per cent ionization of the acid? ($K_i = 7.2 \times 10^{-4}$.)   *Ans. 2.6%*

ⓢ27. Forty ml of 0.20 M HCl is mixed with 50 ml of 0.30 M aqueous ammonia. Calculate the concentration of hydroxyl ions in the solution.
             Ans. 1.6×10^{-5} M

28. Calculate the pH of a solution which contains
    (a) 0.560 g of KOH in 1.0 liter of solution.   *Ans. 12.00*
    (b) 34 g of NH$_3$ in 500 ml of solution.   *Ans. 11.93*

29. Calculate the pH of a solution that is prepared from
    (a) 30 ml of 1.0 M HOAc and 10 ml of 1.0 M NaOH.   *Ans. 4.44*
    (b) 10 ml of 1.0 M aqueous ammonia and 10 ml of 0.333 M HCl. *Ans. 9.56*

ⓢ30. ⓢ(a) In what mole ratio would you combine sodium acetate and acetic acid in water solution to prepare a buffer solution of pH 7.00? 6.00? 5.08? 4.20?   *Ans. ⓢ180 to 1; 18 to 1; 2.2 to 1; 0.29 to 1*

(b) What pH would be obtained with a mole ratio of sodium acetate and acetic acid of one to one?   *Ans. 4.74*

31. Given a $0.1\ M$ solution of ammonia which is also $0.1\ M$ in ammonium chloride (a buffer mixture), and a solution of $1.8 \times 10^{-5}\ M$ NaOH, calculate (a) the initial pH in each solution, (b) the pH when a liter of each of the original solutions is treated with 0.03 mole of solid sodium hydroxide, and (c) the pH when a liter of each of the original solutions is treated with 0.03 mole of HCl. (Assume no volume change with the addition of the NaOH or the HCl.) Is the buffer solution effective in serving its intended function of holding the $[H^+]$ relatively constant compared to the unbuffered solution?
Ans. (a) 9.3 for each solution; (b) 9.5; 12.5; (c) 9.0; 1.5

⑤32. Calculate the molarity of a solution of H_2SO_4 which has a pH of 2.00. (*Hint: Be careful on this one. It is more difficult than it looks. Remember that H_2SO_4 is a diprotic acid.*)   *Ans. $6.5 \times 10^{-3}\ M$*

33. What relative number of moles of acetic acid and sodium acetate should be used in making up a buffer solution with a pH of 4.40? 6.30? Would the latter be a satisfactory buffer?
Ans. 1.0 mole of HOAc to 0.45 mole of NaOAc; 1.0 mole of HOAc to 36 moles of NaOAc (to two significant figures); No

34. It is desired to prepare a buffer solution for which the pH is 5.00. How much solid $NaC_2H_3O_2 \cdot 3H_2O$ must be added to 300 ml of $0.50\ M$ acetic acid ($K_i = 1.8 \times 10^{-5}$) to produce such a buffer solution?   *Ans. 37 g*

35. The ionization constant for lactic acid, $CH_3CHOHCOOH$, at $25°$ C, is 1.37×10^{-4}. If an 18.0-g quantity of lactic acid is diluted with water to a volume of 1.50 liters, what is the concentration of hydrogen ion in the solution?   *Ans. $4.20 \times 10^{-3}\ M$*

36. What is the fluoride ion concentration of a $0.020\ M$ solution of HF, for which K_i is 7.2×10^{-4}?   *Ans. $3.5 \times 10^{-3}\ M$*

37. The pH of a $0.150\ M$ solution of a particular monoprotic acid is 3.10. Calculate the K_i for the acid.   *Ans. 4.2×10^{-6}*

⑤38. A 50.0-ml solution of a $0.10\ M$ monoprotic acid was mixed with 20 ml of $0.10\ M$ potassium hydroxide, and the resulting mixture was diluted to 100 ml. The pH of the solution was found to be 5.25. What is the ionization constant of the acid?   *Ans. $K_i = 3.7 \times 10^{-6}$*

39. Calculate the per cent ionization of a $1.0\ M$ solution of H_3PO_4, if the ionization constant for the first ionization step is 7.5×10^{-3}. (Consider that the second and third ionization steps are negligible compared to the first.)   *Ans. 8.3%*

40. Carbonic acid (H_2CO_3) decomposes easily to form CO_2 and H_2O. What volume of CO_2 at S.T.P. will be driven off from 100 ml of H_2CO_3 which has a pH of 3.30?   *Ans. 1.3 liter*

41. A 5.36-g sample of NH_4Cl was dissolved in water, to which was added 25.0 ml of $1.00\ M$ NaOH. The resulting solution was diluted to 100 ml. Calculate the pH of the resulting solution. (K_i for ammonia is 1.8×10^{-5}.)   *Ans. 8.78*

42. A buffer solution is made up of equal volumes of $0.20\ M$ acetic acid and $0.60\ M$ sodium acetate. (a) What is the pH of this solution? (b) What is the pH of the solution which results from adding 3.00 ml of $0.034\ M$ HCl to 200 ml of the original buffer solution? (K_i for acetic acid is 1.8×10^{-5}.)

Ans. (a) 5.222; (b) 5.223 (Note that for each answer only two significant figures, or 5.22, are justified by the data provided in the problem; see note in Section 18.11 concerning the digits which can be counted as significant figures in a pH value.)

[s]43. Calculate the concentration of sulfide ion in a saturated solution of hydrogen sulfide which is 0.3 M with respect to hydrochloric acid. ($K_{H_2S} = 1 \times 10^{-7}$, and $K_{HS^-} = 1 \times 10^{-13}$.) *Ans. 1 $\times$ 10^{-20} M*

44. Calculate the pH of a solution of 6.87 liters of CO_2 (S.T.P.) in 1.0 liter of water. *Ans. 3.44*

45. What is the original concentration of an acetic acid solution which at equilibrium is (a) 50.0% ionized? (b) 5.0% ionized? (c) 0.50% ionized? ($K_i = 1.8 \times 10^{-5}$.) *Ans. (a) 3.6 $\times$ 10^{-5} M; (b) 6.8 $\times$ 10^{-3} M; (c) 0.72 M*

46. How many grams of benzoic acid, C_6H_5COOH, are there in 3.00 liters of a solution of this acid if the pH of the solution is 2.40? ($K_i = 6.5 \times 10^{-5}$.) *Ans. 91 g*

[s]47. Calculate the pH of a 0.0050 M solution of hydrogen sulfide. (Neglect the second stage in the ionization of H_2S; $K_i = 1.0 \times 10^{-7}$.) *Ans. 4.65*

48. Calculate the pH of a 0.010 M solution of carbonic acid (H_2CO_3). The ionization constant for the reaction $H_2CO_3 \rightleftharpoons H^+ + HCO_3^-$ is 4.3 $\times$ 10^{-7}. (Neglect the second stage in the ionization of H_2CO_3.) *Ans. 4.18*

[s]49. Calculate (a) the carbonate ion concentration and (b) the bicarbonate ion concentration in a 0.050 M solution of carbonic acid which is also 0.10 M in hydrochloric acid. ($K_i = 4.3 \times 10^{-7}$ and $K_2 = 7 \times 10^{-11}$, for the two steps in the dissociation of H_2CO_3.) *Ans. (a) 2 $\times$ 10^{-16} M; (b) 2.2 $\times$ 10^{-7} M*

50. How many grams of formic acid, $HCOOH$, are necessary to prepare 2.5 liters of a solution for which the pH is 3.00? (K_i for formic acid is 1.8 $\times$ 10^{-4}.) *Ans. 0.75 g*

[s]51. A 0.100-liter volume of 0.010 M aqueous ammonia solution is placed in a chamber exposed to HCl gas. Calculate the pH of the solution for each of the following situations (you may assume that dissolved HCl does not increase the volume of the solution significantly):

(a) Before any HCl dissolves in the ammonia solution. *Ans. 10.63*

(b) After 1.0 $\times$ 10^{-4} mole of HCl has dissolved. *Ans. 10.21*

(c) After 5.0 $\times$ 10^{-4} mole of HCl has dissolved. *Ans. 9.26*

(d) After 9.0 $\times$ 10^{-4} mole HCl has dissolved. *Ans. 8.30*

(e) After 1.0 $\times$ 10^{-3} mole HCl has dissolved. *Ans. 5.63*

52. Calculate the general form for the ionization constant for a weak acid or a weak base of concentration C and per cent ionization α.

$$Ans. \quad \frac{\alpha^2 C}{100(100 - \alpha)}$$

53. At $-50°$ C, the equilibrium concentrations of ammonium ion and amide ion (NH_2^-) in liquid ammonia are equal to 1 $\times$ 10^{-15} M. Calculate the ion-product constant for ammonia and the "pNH_4" of a neutral solution of ammonia at $-50°$ C. $pNH_4 = \log \dfrac{1}{[NH_4^+]}$.

Ans. 1 $\times$ 10^{-30}; 15.0

54. Verify by calculation each of the pH values listed in Table 18–6 for the titration of 25.00 ml of 0.100 M HOAc with 0.100 M NaOH.

55. Verify by calculation each of the pH values listed in Table 18–6 for the titration of 25.00 ml of 0.100 M HCl with 0.100 M NaOH.

56. Calculate the pH for each of the following solutions:

    (a) 50 ml of 0.025 M acetic acid + 5.0 ml of 0.025 M potassium hydroxide.
    Ans. 3.79

    (b) 50 ml of 0.025 M acetic acid + 20.0 ml of 0.025 M potassium hydroxide.
    Ans. 4.57

    (c) 50 ml of 0.025 M acetic acid + 40.0 ml of 0.025 M potassium hydroxide.
    Ans. 5.35

[S]57. Assuming that the hydrogen ion concentration in a solution is controlled by the reaction $H_2AsO_4^- \longrightarrow H^+ + HAsO_4^{2-}$, calculate the ratio of concentrations of $HAsO_4^{2-}$ and $H_2AsO_4^-$ necessary to make up a buffer solution of pH equal to (a) 6.00, (b) 7.20, (c) 8.00. (K_i for the reaction $H_2AsO_4^- \rightleftharpoons H^+ + HAsO_4^{2-}$ is 1.0×10^{-7}.)   *Ans. [S](a) 0.10; (b) 1.6; (c) 10*

[S]58. Calculate the pH of a 0.10 M sodium acetate solution.   *Ans. 8.87*

[S]59. Calculate the hydrolysis constant for each of the following ions:

    [S](a) $CH_3CH_2COO^-$ (K_i for $CH_3CH_2COOH = 1.3 \times 10^{-5}$.) *Ans. 7.7 $\times$ 10^{-10}*
    (b) ClO^- (K_i for $HClO = 3.5 \times 10^{-8}$.)   *Ans. 2.9 $\times$ 10^{-7}*
    (c) N_3^- (K_i for $HN_3 = 1 \times 10^{-4}$.)   *Ans. 1 $\times$ 10^{-10}*
    (d) $C_2O_4^{2-}$ (K_i for $HC_2O_4^- = 6.4 \times 10^{-5}$.)   *Ans. 1.6 $\times$ 10^{-10}*

[S]60. Calculate the pH of (a) 0.500 M solution of ammonium chloride; (b) a 0.040 M solution of NaF. (K_i for NH_3 is 1.8×10^{-5}, and K_i for HF is 7.2×10^{-4}.)   *Ans. [S](a) 4.78, (b) 7.87*

61. The ionization constant for HBrO is 2.0×10^{-9}. Calculate the pH and percentage hydrolysis in a 0.020 M KBrO solution.   *Ans. 10.50; 1.6%*

62. A 0.010 M solution of NaClO is found to be hydrolyzed to the extent of 0.53 per cent. Calculate the value of the ionization constant for HClO.
    Ans. 3.5 $\times$ 10^{-8}

[S]63. What will be the final pH of a solution made by mixing 0.20 mole of NaOH and 0.20 mole of NH_4NO_3 in enough water to make 1.0 liter of solution?
    Ans. 11.28

64. Calculate the pH of a 0.010 M solution of $Al(NO_3)_3$. (The K_h for the reaction, $[Al(H_2O)_6]^{3+} + H_2O \rightleftharpoons [Al(H_2O)_5(OH)]^{2+} + H_3O^+$, is 1.4×10^{-5}.)
    Ans. 3.44

[S]65. Calculate the pH of a 0.350 M Na_2HAsO_4 solution. Neglect the dissociation of the $HAsO_4^{2-}$ ion. (K_i for $H_2AsO_4^-$ is 1.0×10^{-7}.)   *Ans. 10.27*

66. What is the concentration of molecular HCN in a 0.50 M solution of NaCN?
    Ans. 3.5 $\times$ 10^{-3}

67. What concentration of potassium phenolate, C_6H_5OK, will give a solution in which the per cent hydrolysis is 3.0%? (K_i for phenol, C_6H_5OH, = 1.3×10^{-10}.)   *Ans. 0.083 M*

68. The hydrogen ion concentration of a 0.010 M solution of $NaNO_2$ is 2.1×10^{-8} gram-ion per liter. Calculate (a) the hydrolysis constant of $NaNO_2$, and (b) the ionization constant for HNO_2.
    Ans. (a) 2.3 $\times$ 10^{-11}; (b) 4.4 $\times$ 10^{-4}

69. Calculate the final pH of a solution which is made by adding 30.0 ml of 0.10 M HCl to 30.0 ml of 1.0 M NH$_3$. (K_i for NH$_3$ = 1.8 × 10^{-5}.) *Ans. 10.21*

[s]70. Calculate the pH of a solution made by mixing equal volumes of 0.040 M aqueous aniline, C$_6$H$_5$NH$_2$, and 0.040 M nitric acid. (K_i for C$_6$H$_5$NH$_2$ = 4.6 × 10^{-10}.) *Ans. 3.18*

71. Calculate the pH of a solution resulting from mixing 10 ml of 0.10 M HCl with 40 ml of 0.025 M ammonia. *Ans. 5.48*

[s]72. How many moles of NaC$_2$H$_3$O$_2$ must be used to prepare 2.0 liters of an aqueous solution with pH = 9.40? (K_i for HC$_2$H$_3$O$_2$ = 1.8 × 10^{-5}.)

Ans. 2.3

73. Calculate the concentration of molecular CH$_3$NH$_2$ in a 0.030 M solution of CH$_3$NH$_3$Cl. (K_i for CH$_3$NH$_2$ = 4.4 × 10^{-4}.) *Ans. 8.3 × 10^{-7}*

[s]74. Calculate the pH of each of the solutions described below:
    (a) 40 ml of 0.10 M barbituric acid (K_i = 9.8 × 10^{-5}.) *Ans. 2.50*
    (b) 40 ml of 0.10 M barbituric acid and 20 ml of 0.10 M KOH *Ans. 4.01*
    (c) 40 ml of 0.10 M barbituric acid and 39 ml of 0.10 M KOH *Ans. 5.60*
    (d) 40 ml of 0.10 M barbituric acid and 40 ml of 0.10 M KOH *Ans. 8.35*
    (e) 40 ml of 0.10 M barbituric acid and 41 ml of 0.10 M KOH *Ans. 11.09*

75. Calculate the K_i of benzoic acid, C$_6$H$_5$COOH, from the hydrolysis constant of potassium benzoate, C$_6$H$_5$COOK. The hydrolysis constant is 1.5 × 10^{-10}.

Ans. 6.7 × 10^{-5}

76. A 0.5000-gram sample of an impure monobasic amine is titrated to a good end point with 75.00 ml of 0.100 M HCl. The pH after 40.00 ml of acid is added is 10.65. The molecular weight of the pure base is known to be 59.1. (a) Calculate the acidity constant of this base and (b) its per cent purity. (c) Which indicator from the following list would be most suitable for this titration? State your reasoning.
    (a) Orange IV (red—yellow; pH 1.2–2.6)
    (b) Methyl orange (red—yellow; pH 3.1–4.4)
    (c) Methyl red (red—yellow; pH 4.8–6.0)
    (d) Bromthymol blue (yellow—blue; pH 6.0–7.6)
    (e) Phenolphthalein (colorless—pink; pH 8.3–10.0)
    (f) Thymolphthalein (colorless—blue; pH 9.3–10.5)
    Ans. (a) 5.1 × 10^{-4}; (b) 88.7%; (c) methyl red

77. Calculate the concentration of zinc chloride required to give a solution with pH = 5.00. The hydrolysis constant for the reaction [Zn(H$_2$O)$_4$]$^{2+}$ + H$_2$O $\rightleftharpoons$ [Zn(H$_2$O)$_3$(OH)]$^+$ + H$_3$O$^+$ is 2.45 × 10^{-10}.

Ans. 0.41 M

[s]78. The acids (CH$_3$)$_2$AsO$_2$H, ClCH$_2$COOH, and CH$_3$COOH have ionization constants of 6.40 × 10^{-7}, 1.40 × 10^{-3}, and 1.76 × 10^{-5}, respectively. (a) Which acid would be best for preparing a buffer solution at pH = 6.50? (b) How many grams of this acid and of NaOH would be needed to prepare 1.00 liter of buffer solution in which the sum of the concentrations of the acid and its conjugate base is equal to 1.00 M?
    Ans. (a) (CH$_3$)$_2$AsO$_2$H; (b) 138 g of the acid, 26.8 g of NaOH

$\boxed{\text{s}}$ 79. Calculate the concentration of each of the ions in a 0.400 M H_2SO_4 solution (first step, essentially 100% dissociation into ions; second step, $K_2 = 1.2 \times 10^{-2}$).

Ans. $[H^+] = 0.41\ M;\ [SO_4{}^{2-}] = 0.011\ M;\ [HSO_4{}^-] = 0.39\ M;$
$[OH^-] = 2.4 \times 10^{-14}\ M$

80. H_4EDTA is a tetraprotic acid with $K_1 = 1.0 \times 10^{-2}$, $K_2 = 2.16 \times 10^{-3}$, $K_3 = 6.92 \times 10^{-7}$, and $K_4 = 5.50 \times 10^{-11}$. Calculate what fraction of this material exists as the tetraanion at pH 12, pH 7, and pH 1.

Ans. $0.98;\ 4.8 \times 10^{-4};\ 7.5 \times 10^{-18}$

REFERENCES

"Stepwise Formation Constants of Complex Ions," W. B. Guenther, *J. Chem. Educ.*, **44**, 46 (1967).

"Component Concentrations in Solutions of Weak Acids," D. M. Goldish, *J. Chem. Educ.*, **47**, 65 (1970).

"Hydrolysis of Sodium Carbonate," F. S. Nakayama, *J. Chem. Educ.*, **47**, 67 (1970).

"Errors in Calculating Hydrogen Ion Concentration," J. E. House, Jr. and R. C. Reiter, *J. Chem. Educ.*, **45**, 679 (1968).

"pH," Duncan A. MacInnes, *Sci. American*, Jan., 1951; p. 40.

"Development of the pH Concept; A Historical Perspective," F. Szabadváry (transl. by Ralph E. Oesper), *J. Chem. Educ.*, **41**, 105 (1964).

"Indicators; A Historical Perspective," F. Szabadváry (transl. by Ralph E. Oesper), *J. Chem. Educ.*, **41**, 285 (1964).

"Equilibrium Constants from Spectrophotometric Data: Principles, Practice, and Programming," R. W. Ramette, *J. Chem. Educ.*, **44**, 647 (1967).

"Rapid Graphical Method for Determining Formation Constants," S. D. Christian, *J. Chem. Educ.*, **45**, 713 (1968).

"Acid-Base Indicators: An Experiment in Aqueous Equilibria," W. E. Brown and J. A. Campbell, *J. Chem. Educ.*, **45**, 674 (1968).

"Corrections for Simple Equations for Titration Curves of Monoprotic Acids," W. P. Cortelyou, *J. Chem. Educ.*, **45**, 677 (1968).

"Is a Weak Acid Monoprotic? A New Look at Titration Curves," P. E. Sturrock, *J. Chem. Educ.*, **45**, 258 (1968).

"Acid-Base Titration and Distribution Curves," J. Waser, *J. Chem. Educ.*, **44**, 274 (1967).

"Dissociation of Weak Acids and Bases at Infinite Dilution," D. I. Stock, *J. Chem. Educ.*, **44**, 764 (1967).

"pK Values for D_2O and H_2O," R. G. Bates, R. A. Robinson, and A. K. Covington, *J. Chem. Educ.*, **44**, 635 (1967).

"Ratio Diagrams: A Simple Graphical Representation of Complicated Equilibria," R. de Levie, *J. Chem. Educ.*, **47**, 187 (1970).

"Highly Basic Media," J. R. Jones, *Chem. in Britain*, **7**, 336 (1971).

"The Equilibrium between a Solid Solution and an Aqueous Solution of its Ions," A. F. Berndt and R. L. Stearns, *J. Chem. Educ.*, **50**, 415 (1973).

The Solubility
Product Principle

19

In chemical analyses which involve precipitations, the chemist works with slightly soluble substances which are strong electrolytes. Equilibrium systems consisting of slightly soluble strong electrolytes (ionic substances) in contact with their saturated aqueous solutions are particularly important in analytical chemistry. In this chapter we shall be concerned with the theory and application of the precipitation and dissolution of slightly soluble ionic substances.

19.1 The Solubility Product Constant

A slightly soluble electrolyte dissolves in water to the extent that a saturated solution of its ions is formed. A **saturated solution** is defined as one which is in equilibrium with undissolved solute. It can be shown by the use of radioactive tracers that at equilibrium the solid (the undissolved solute) continues to dissolve, the process of crystallization is taking place, and the opposing processes have equal rates. Let us consider a saturated solution of silver chloride in equilibrium with some crystals of silver chloride. The equilibrium can be expressed by

$$AgCl \text{ (solid)} \underset{\text{Crystallization}}{\overset{\text{Dissolution}}{\rightleftharpoons}} Ag^+ + Cl^- \text{ (saturated solution)}$$

Applying the law of chemical equilibrium to this system, we obtain the mathematical expression

$$\frac{[Ag^+][Cl^-]}{[AgCl]} = K_e$$

Because the concentration of the solid silver chloride remains essentially constant, we can write

$$[Ag^+][Cl^-] = K_e[AgCl] = K_{sp}$$

The K_{sp} is called the **solubility product constant,** or sometimes simply the **solubility product.** In a saturated solution of a slightly soluble electrolyte the product of the molar concentrations of its ions is a constant, at a given temperature, if the different kinds of ions are formed in equal numbers. If the slightly soluble electrolyte contains more than one ion of one kind in its formula, the concentration of that ion is raised to the corresponding power in the expression for the solubility product constant. For example, calcium phosphate dissociates according to the equation $Ca_3(PO_4)_2 \rightleftharpoons 3Ca^{2+} + 2PO_4^{3-}$.

The expression for its solubility product constant is

$$[Ca^{2+}]^3[PO_4^{3-}]^2 = K_{sp}$$

For a salt like magnesium ammonium phosphate, which dissociates according to the equation $MgNH_4PO_4 \rightleftharpoons Mg^{2+} + NH_4^+ + PO_4^{3-}$, we have

$$[Mg^{2+}][NH_4^+][PO_4^{3-}] = K_{sp}$$

As is the case for all equilibrium constants (see for example, Section 18.3), solubility product constants have units [for example, $mole^2/liter^2$ for K_{sp} of AgCl, $mole^5/liter^5$ for K_{sp} of $Ca_3(PO_4)_2$, and $mole^3/liter^3$ for K_{sp} of $MgNH_4PO_4$] depending upon the form of the particular equilibrium constant expression. As pointed out earlier, the units are often omitted from equilibrium constant values, and this practice is usually followed in this book. The student should, however, always keep in mind the units for the constants and the fact that the units are real.

The solubility product principle is based upon experimental studies of saturated solutions of *slightly* soluble electrolytes and does not hold for moderately or quite soluble salts, such as NaCl and $KClO_3$. Furthermore, the presence of high concentrations of ions from an added electrolyte causes an increase in the solubility of a slightly soluble substance—the salt effect (Section 18.5). The salt effect varies with the concentration of the added electrolyte and with the charges on the ions the salt supplies to the solution. Hence, the solubility product constant in some cases is *not* constant, and we must conclude that the solubility product principle is strictly valid only for saturated solutions in which the total ionic concentration is extremely small.

19.2 Calculation of Solubility Products from Molar Solubilities

When the solubility of a slightly soluble electrolyte is known, its solubility product can be calculated. In all calculations involving solubility products, the concentrations are expressed in moles per liter. Thus, when the concentrations are given in other units, such as grams per liter, they must be converted to moles per liter. Two examples of calculations of solubility product constants follow.

Example 1. The solubility of $BaSO_4$ in water is 2.42×10^{-3} gram per liter. Calculate the solubility product constant for $BaSO_4$.

First calculate the molar solubility of $BaSO_4$. The formula weight of $BaSO_4$ is 233.4.

$$\frac{2.42 \times 10^{-3} \text{ g/liter}}{233.4 \text{ g/mole}} = 1.04 \times 10^{-5} \text{ mole per liter}$$

Since the salt ($BaSO_4$) dissolves to give Ba^{2+} and SO_4^{2-} in equal amounts, $[Ba^{2+}] = [SO_4^{2-}] = 1.04 \times 10^{-5}$ M. When these values are substituted in the expression for the solubility product constant of $BaSO_4$ the numerical value of the constant may be obtained.

$$BaSO_4 \rightleftharpoons Ba^{2+} + SO_4^{2-}$$
$$[Ba^{2+}][SO_4^{2-}] = K_{sp}$$
$$(1.04 \times 10^{-5})(1.04 \times 10^{-5}) = 1.08 \times 10^{-10} = K_{sp}$$

Example 2. Calculate the solubility product constant for silver chromate, whose molar solubility is 1.3×10^{-4}.

Writing the equation for the dissolution of Ag_2CrO_4

$$Ag_2CrO_4 \rightleftharpoons 2Ag^+ + CrO_4^{2-}$$

we notice that the concentration of silver ion will be twice the molar solubility of Ag_2CrO_4, and that the concentration of chromate ion will be equal to the molar solubility of the salt, or that $[Ag^+] = 2 \times 1.3 \times 10^{-4}$ M, or 2.6×10^{-4} M, and $[CrO_4^{2-}] = 1.3 \times 10^{-4}$ M. Substituting in the expression for the solubility product, we find

$$[Ag^+]^2[CrO_4^{2-}] = K_{sp}$$
$$(2.6 \times 10^{-4})^2(1.3 \times 10^{-4}) = 8.8 \times 10^{-12} = K_{sp}$$

A table of solubility product constants of some slightly soluble electrolytes at 25° is given in Appendix E.

19.3 Calculation of the Molar Solubility from the Solubility Product

When the solubility product constant for a slightly soluble electrolyte is known, its solubility in moles per liter can be readily calculated.

Example 1. The solubility product constant for silver chloride is 1.8×10^{-10}. Calculate the molar solubility of silver chloride.

Silver chloride dissolves in water to form a saturated solution according to the equation

$$AgCl \rightleftharpoons Ag^+ + Cl^-$$

Let x be the solubility of AgCl in moles per liter. The molar concentration of Ag^+ and Cl^- will also be x since the salt is completely dissociated in aqueous solution. Substituting in the expression for the solubility product constant for AgCl, we have

$$[Ag^+][Cl^-] = K_{sp} = 1.8 \times 10^{-10}$$
$$x^2 = 1.8 \times 10^{-10}$$
$$x = \sqrt{1.8 \times 10^{-10}} = 1.3 \times 10^{-5} \text{ mole/liter}$$

Thus, the molar solubility of silver chloride is 1.3×10^{-5}.

Example 2.   The solubility product constant of $Fe(OH)_3$ is 1.1×10^{-36}. Calculate the molar solubility of iron(III) hydroxide.

Let x be the molar solubility of iron(III) hydroxide. From the equation for the dissolution of iron(III) hydroxide, $Fe(OH)_3 \rightleftharpoons Fe^{3+} + 3OH^-$, it is apparent that each mole of $Fe(OH)_3$ that dissolves yields one mole of Fe^{3+} and three moles of OH^-. Thus, if the molar solubility of $Fe(OH)_3$ is x, then $[Fe^{3+}] = x$, and $[OH^-] = 3x$. These values are substituted in the expression for the solubility product constant of ferric hydroxide, and the value of x is calculated.

$$[Fe^{3+}][OH^-]^3 = K_{sp} = 1.1 \times 10^{-36}$$
$$x(3x)^3 = 1.1 \times 10^{-36}$$
$$27x^4 = 1.1 \times 10^{-36}$$
$$x^4 = 0.0407 \times 10^{-36} = 407 \times 10^{-40}$$
$$x = 4.5 \times 10^{-10} \text{ moles/liter}$$

Thus, $4.5 \times 10^{-10} = $ molar solubility of $Fe(OH)_3 = [Fe^{3+}]$, and $[OH^-] = 3x = 13.5 \times 10^{-10} = 1.4 \times 10^{-9}$.

19.4   The Precipitation of Slightly Soluble Electrolytes

In order to precipitate a slightly soluble electrolyte, it is necessary to bring together in the solution the ions of that electrolyte in such quantities that the product of the ion concentrations, raised to the appropriate powers, exceeds the solubility product constant of the compound. Thus, when equal volumes of a $2 \times 10^{-4} M$ solution of $AgNO_3$ and a $2 \times 10^{-4} M$ solution of NaCl are mixed, we can show that silver chloride will precipitate. Because the volume of the solution is doubled by mixing equal volumes of the two solutions, each concentration is reduced to half its initial value. Consequently, $[Ag^+]$ and $[Cl^-]$ will each be $2 \times 10^{-4}/2 M = 1 \times 10^{-4} M$ immediately upon mixing. Substituting in the expression for the solubility product constant of silver chloride,

$$[Ag^+][Cl^-] = K_{sp} = 1.8 \times 10^{-10}$$
$$(1 \times 10^{-4})(1 \times 10^{-4}) = 1 \times 10^{-8} > K_{sp}$$

we see that, momentarily, the product of $[Ag^+]$ and $[Cl^-]$ is greater than ($>$) K_{sp} for AgCl. According to the principle of chemical equilibrium, some of the silver and chloride ions will unite under these conditions with the formation of silver chloride precipitate, and the precipitation will continue until the product of [Ag+] and [Cl−] still remaining in the solution attains a value equal to the solubility product constant for silver chloride.

In the above case, the Ag+ and Cl− were present in equal concentrations. Since an equal number of Ag+ and Cl− combine to form solid AgCl, the saturated solution in contact with the precipitate will contain Ag+ and Cl− in equal concentrations, their product being just equal to the solubility product constant of AgCl. If to this saturated solution of silver chloride we add more NaCl, we again shall momentarily have the product of [Ag+] and [Cl−] greater than the solubility product constant of AgCl, and then more Ag+ and Cl− will unite and form solid AgCl until the product of [Ag+] and [Cl−] again attains the value of K_{sp}. In the new equilibrium the [Ag+] will be less and the [Cl−] greater than they were in the saturated solution of AgCl in pure water. The greater we make the [Cl−], the less the [Ag+] will become. However, we can never reduce [Ag+] to zero because the product of [Ag+] and [Cl−] will always be equal to the solubility product constant. It is interesting to note that adding a *very large* excess of NaCl will cause AgCl to dissolve because of the formation of another species, $AgCl_2^-$. This introduces a new equilibrium reaction to the system.

19.5 Calculation of the Concentration of an Ion Necessary to Form a Precipitate

When the concentration of one of the ions of a slightly soluble electrolyte and the value for the solubility product constant are known, we can calculate the concentration which the other ion must exceed before precipitation can occur.

Example. If a solution contains 0.001 mole of CrO_4^{2-} per liter, what concentration of Ag+ ion must be exceeded by adding $AgNO_3$ to the solution before Ag_2CrO_4 will begin to precipitate, neglecting any increase in volume on adding the solid silver nitrate? (K_{sp} is 9×10^{-12}.)

The equilibrium involved in the problem is

$$Ag_2CrO_4 \rightleftharpoons 2Ag^+ + CrO_4^{2-}$$

The expression for the solubility product constant is $[Ag^+]^2[CrO_4^{2-}] = K_{sp}$. Substituting in the above expression the values for the $[CrO_4^{2-}]$ and K_{sp}, we have

$$[Ag^+]^2(0.001) = 9 \times 10^{-12}$$

$$[Ag^+]^2 = \frac{9 \times 10^{-12}}{10^{-3}} = 9 \times 10^{-9} = 90 \times 10^{-10}$$

$$[Ag^+] = 9.5 \times 10^{-5} \text{ or } 1 \times 10^{-4} \text{ mole/liter (rounded)}$$

A concentration of Ag+ greater than $1 \times 10^{-4} M$ is necessary to cause precipitation of Ag_2CrO_4.

19.6   Calculation of the Concentration of an Ion Left in Solution after Precipitation

It is often advantageous to the chemist to know the concentration of an ion remaining in solution after precipitation. This can also be determined using the solubility product constant.

Example.   What is the concentration of Ag^+ left in solution if AgCl is precipitated by adding sufficient hydrochloric acid to a solution of $AgNO_3$ to make the final chloride concentration 0.10 M?

$$[Ag^+][Cl^-] = K_{sp}$$
$$[Ag^+](0.10) = 1.8 \times 10^{-10}$$
$$[Ag^+] = \frac{1.8 \times 10^{-10}}{1.0 \times 10^{-1}} = 1.8 \times 10^{-9}\ M$$

The concentration of Ag^+ left in solution is 1.8×10^{-9} mole per liter.

19.7   Supersaturation of Slightly Soluble Electrolytes

When the product of the concentrations of the ions, raised to the proper powers, exceeds the value of the solubility product constant, precipitation does not always occur; sometimes a supersaturated solution is formed. For example, if magnesium is being precipitated as magnesium ammonium phosphate from a solution of low magnesium ion content, the solution may stand several hours before a visible precipitate appears. Furthermore, even though precipitation may be immediate under ordinary conditions, it does not follow that it is complete. The first crystals to precipitate are often very small and hence more soluble than the larger ones, which form as the solution stands in contact with the precipitate (see Section 19.8). Solutions should be allowed to stand before filtering, to insure maximum precipitation.

19.8   Solubility and Crystal Size

It has been shown by experiment that small crystals of any substance are more soluble than larger ones. A notable example of this is barium sulfate. The solubility of "fine" crystals of barium sulfate has been found to be about twice as great as that of "coarse" crystals of this salt. This results from the fact that ions in the interior of a crystal are bound with greater forces than those of the faces and edges. A greater fraction of the ions in a crystal occupy surface positions when the crystal is small, as opposed to a large crystal, making the tendency for ions to enter solution greater for the small crystal. Because small crystals are more soluble than large crystals, the smallest crystals will in time dissolve and the larger ones will grow still larger. True equilibrium is not reached until large, perfect crystals are formed. The solubility product constants are calculated for solutions in contact with relatively large crystals.

Because very small crystals will pass through filters, or may resist sedimentation by centrifugation, precipitates are often "digested" to increase the size of the crystals. Heat increases the rate at which large crystals will grow from the smaller ones. Allowing the suspended precipitate to stand for a time will also often result in the formation of larger crystals.

19.9   Calculations Involving Both Ionization Constants and Solubility Product Constants

In the theory and practice of precipitation analysis we often deal with systems which involve more than one equilibrium. The following examples illustrate the point.

Example 1.   Calculate the concentration of hydrogen ion required to prevent the precipitation of ZnS in a solution which is 0.10 M in $ZnCl_2$ and saturated with H_2S.

The maximum concentration of sulfide ion that can be present without causing ZnS to precipitate may be calculated from the solubility product expression for ZnS, as shown below.

$$[Zn^{2+}][S^{2-}] = K_{sp} = 1.1 \times 10^{-21}$$

$$[S^{2-}] = \frac{K_{sp}}{[Zn^{2+}]} = \frac{1.1 \times 10^{-21}}{0.10} = 1.1 \times 10^{-20}$$

The sulfide ion concentration varies inversely with the square of the hydrogen ion concentration in a saturated solution of hydrogen sulfide containing a strong acid such as HCl (see Section 18.14).

$$[H^+]^2[S^{2-}] = 1.3 \times 10^{-21}$$

Substituting in this equation the sulfide ion concentration found above, we have

$$[H^+]^2 = \frac{1.3 \times 10^{-21}}{[S^{2-}]} = \frac{1.3 \times 10^{-21}}{1.1 \times 10^{-20}} = 1.2 \times 10^{-1}$$

$$[H^+] = \sqrt{0.12} = 0.35 \ M$$

Thus, it is found that a hydrogen ion concentration of 0.35 M will prevent the precipitation of zinc sulfide in the solution described in the problem.

Example 2.   Calculate the quantity of Cd^{2+} which will remain unprecipitated in a 0.30 M HCl solution which is saturated with H_2S.

The concentration of $[Cd^{2+}]$ remaining in the solution will be determined by the final sulfide ion concentration which in turn is determined by the hydrogen ion concentration of the 0.30 M HCl solution. Solving for the sulfide ion concentration, we have

$$[H^+]^2[S^{2-}] = (0.30)^2[S^{2-}] = 1.3 \times 10^{-21}$$

$$[S^{2-}] = \frac{1.3 \times 10^{-21}}{(0.30)^2} = \frac{1.3 \times 10^{-21}}{0.090} = 1.4 \times 10^{-20} \ M$$

Substituting the value of the sulfide ion concentration in the expression for the solubility product constant of CdS, we have

$$[Cd^{2+}][S^{2-}] = [Cd^{2+}](1.4 \times 10^{-20}) = K_{sp} = 3.6 \times 10^{-29}$$
$$[Cd^{2+}] = \frac{3.6 \times 10^{-29}}{1.4 \times 10^{-20}} = 2.8 \times 10^{-9}\ M$$

Multiplying the molar concentration of Cd^{2+} by the gram-atomic weight of Cd, we find

$$2.8 \times 10^{-9}\ \text{mole/liter} \times 112.4\ \text{g/mole} =$$
$$3.1 \times 10^{-7}\ \text{g/liter, or } 3.1 \times 10^{-7}\ \text{mg/ml}$$

Hence, the precipitation of Cd^{2+} as CdS in a 0.30 M HCl solution saturated with H_2S is nearly complete.

(Calculations of the type shown in Examples 1 and 2 are important in the separation of the metal ions of Group II from those of Group III in typical qualitative analysis procedures.)

Example 3. Calculate the concentration of ammonium ion, supplied by NH_4Cl, required to prevent the precipitation of $Mg(OH)_2$ in a liter of solution containing 0.10 mole of ammonia and 0.10 mole of Mg^{2+}.

To find the concentration of NH_4^+ required to prevent precipitation of $Mg(OH)_2$ we must first find the maximum concentration of OH^- that can be present in solution without precipitating $Mg(OH)_2$. To do so we use the solubility product constant for $Mg(OH)_2$.

$$[Mg^{2+}][OH^-]^2 = (0.10)[OH^-]^2 = K_{Mg(OH)_2} = 1.5 \times 10^{-11}$$
$$[OH^-]^2 = \frac{1.5 \times 10^{-11}}{0.10} = 1.5 \times 10^{-10}$$
$$[OH^-] = 1.2 \times 10^{-5}\ M$$

Thus, $Mg(OH)_2$ will not precipitate if the $[OH^-]$ does not exceed $1.2 \times 10^{-5}\ M$.

To calculate the $[NH_4^+]$ supplied by NH_4Cl and needed to repress the ionization of NH_3 so that the $[OH^-]$ will not exceed 1.2×10^{-5}, we use the expression for the ionization constant of aqueous ammonia.

$$\frac{[NH_4^+][OH^-]}{[NH_3]} = K_{NH_3} = 1.8 \times 10^{-5}$$
$$[NH_4^+] = \frac{[NH_3]}{[OH^-]} \times K_{NH_3} = \frac{(0.1)}{(1.2 \times 10^{-5})} \times 1.8 \times 10^{-5} = 0.15\ M$$

The total concentration of ammonium ion that must be present is 0.15 M. Because the amount of ammonium ion formed by the ionization of ammonia is very small, it can be neglected.

19.10 Fractional Precipitation

When two anions form sparingly soluble compounds with the same cation or when two cations form sparingly soluble compounds with the same anion, the less soluble compound will precipitate first on the addition of a precipitant to a solution containing both. An additional quantity of the less soluble compound will precipitate along with the precipitation of the more soluble compound if addition of the precipitant is continued (**coprecipitation**). NaI   NaCl

Consider the case in which a solution containing both sodium iodide and sodium chloride is treated with silver nitrate. Silver iodide, being less soluble than silver chloride, is precipitated first. Only after most of the iodide is precipitated will the chloride begin to precipitate.

Example. A solution contains 0.010 mole of KI and 0.10 mole of KCl per liter. $AgNO_3$ is gradually added to this solution. Which will be precipitated first, AgI or AgCl?

(a) Calculate the $[Ag^+]$ necessary to start the precipitation of AgI.

$$[Ag^+][I^-] = K_{AgI} = 1.5 \times 10^{-16}$$

$$[Ag^+] = \frac{K_{AgI}}{[I^-]} = \frac{1.5 \times 10^{-16}}{0.010} = 1.5 \times 10^{-14}\ M$$

(b) Calculate the $[Ag^+]$ necessary to start the precipitation of AgCl.

$$[Ag^+][Cl^-] = K_{AgCl} = 1.8 \times 10^{-10}$$

$$[Ag^+] = \frac{K_{AgCl}}{[Cl^-]} = \frac{1.8 \times 10^{-10}}{0.10} = 1.8 \times 10^{-9}\ M$$

A greater concentration of Ag^+ is necessary to cause precipitation of AgCl than for AgI, so AgI will precipitate first.

(c) What will be the concentration of I^- in this solution when AgCl begins to precipitate as a result of the continued addition of Ag^+? The $[Ag^+]$ necessary to initiate the precipitation of AgCl is 1.8×10^{-9}. For this concentration of Ag^+, the I^- concentration will be

$$[I^-] = \frac{K_{AgI}}{[Ag^+]} = \frac{1.5 \times 10^{-16}}{1.8 \times 10^{-9}} = 8.3 \times 10^{-8}\ M$$

(d) What fraction of the amount of I^- originally present remains in solution when AgCl begins to precipitate?

$$\frac{[I^-]\ \text{when precipitation of AgCl begins}}{[I^-]\ \text{originally present}} = \frac{8.3 \times 10^{-8}}{0.010} = 8.3 \times 10^{-6}\ M$$

(e) What will be the concentration of I^- in the solution after half the Cl^- initially present is precipitated as AgCl?

$$[Ag^+] = \frac{K_{AgCl}}{[Cl^-]} = \frac{1.8 \times 10^{-10}}{0.050} = 3.6 \times 10^{-9}\ M$$

$$[I^-] = \frac{K_{AgI}}{[Ag^+]} = \frac{1.5 \times 10^{-16}}{3.6 \times 10^{-9}} = 0.417 \times 10^{-7} = 4.2 \times 10^{-8} \ M$$

Note that half of the I^- remaining in solution when AgCl begins to precipitate is coprecipitated as half of the Cl^- is precipitated.

The Dissolution of Precipitates

When the molar concentrations of the ions, raised to the appropriate powers, are such that their product is less than ($<$) the solubility product constant, the solid electrolyte either completely dissolves or dissolves until the ion concentration product is equal to the solubility product constant. For example, calcium carbonate dissolves when

$$[Ca^{2+}][CO_3^{2-}] < K_{sp}$$

To achieve this condition it is necessary to make the concentration of at least one of the ions of the electrolyte less than that in a saturated solution. Ions can be removed from saturated solutions of sparingly soluble electrolytes, and hence solid electrolytes dissolved, in the following ways: (1) by the formation of a weak electrolyte; (2) changing an ion to another species; (3) by formation of a complex ion.

19.11   Dissolution of a Precipitate by the Formation of a Weak Electrolyte

Slightly soluble electrolytes which are derived from weak acids may often be dissolved by strong acids. For example, $CaCO_3$, FeS, and $Ca_3(PO_4)_2$ are dissolved by HCl due to the fact that the acids from which they are derived (H_2CO_3, H_2S, and $H_2PO_4^-$) are weak acids.

When hydrochloric acid is added to a precipitate of calcium carbonate in equilibrium with a standard solution of its ions, the hydrogen ion from the acid combines with the carbonate ion and forms the hydrogen carbonate ion, a weak electrolyte.

$$H^+ + CO_3^{2-} \rightleftharpoons HCO_3^-$$

Then, as the concentration of the hydrogen carbonate ion increases, these ions unite with hydrogen ions from the hydrochloric acid, according to the equation

$$H^+ + HCO_3^- \rightleftharpoons H_2CO_3 \quad \text{Carbonic acid}$$

Finally, the solution becomes saturated with the slightly ionized and unstable carbonic acid, and carbon dioxide gas is evolved, as shown by the equation

$$H_2CO_3 \rightleftharpoons H_2O + CO_2$$

The preceding reactions cause the carbonate ion concentration to be reduced and maintained at such a low level that the product of the calcium and carbonate ion concentrations is less than the solubility product constant of calcium carbonate.

$$[Ca^{2+}][CO_3^{2-}] < K_{sp}$$

Consequently, the calcium carbonate dissolves. Even an acid as weak as acetic acid gives a concentration of hydrogen ion sufficiently high to bring about the dissolution of calcium carbonate because the ionization constant of acetic acid is greater than that for either the hydrogen carbonate ion or carbonic acid.

The dissolution of magnesium hydroxide in aqueous ammonium chloride can be brought about by the formation of a weak electrolyte, such as aqueous ammonia.

$$Mg(OH)_2 + 2NH_4^+ \rightleftharpoons Mg^{2+} + 2NH_3 + 2H_2O$$

When an excess of ammonium ion is added to a suspension of magnesium hydroxide, the following reaction takes place.

$$NH_4^+ + OH^- \rightleftharpoons NH_3 + H_2O$$

Consequently, the hydroxide ion concentration is lowered to the level such that

$$[Mg^{2+}][OH^-]^2 < K_{sp}$$

and the magnesium hydroxide dissolves.

Slightly soluble lead sulfate dissolves readily in solutions of ammonium acetate when the formation of slightly ionized (but soluble) lead acetate reduces the product of the lead and sulfate ion concentrations below the value of the solubility product constant for lead sulfate.

$$PbSO_4 \rightleftharpoons Pb^{2+} + SO_4^{2-}$$
$$Pb^{2+} + 2OAc^- \rightleftharpoons Pb(OAc)_2$$
$$[Pb^{2+}][SO_4^{2-}] < K_{sp}$$

The formation of slightly ionized water causes many sparingly soluble metal hydroxides such as $Mg(OH)_2$, $Al(OH)_3$, and $Fe(OH)_3$ to dissolve in solutions of acids.

$$Mg(OH)_2 + 2H^+ \longrightarrow Mg^{2+} + 2H_2O$$
$$[Mg^{2+}][OH^-]^2 < K_{sp}$$
$$Al(OH)_3 + 3H^+ \longrightarrow Al^{3+} + 3H_2O$$
$$[Al^{3+}][OH^-]^3 < K_{sp}$$

19.12 Dissolution of a Precipitate by Changing an Ion to Another Species

Many metal sulfides have solubility product constants sufficiently large that the hydrogen ion provided by strong acids will lower the sulfide ion concentration enough (by forming the weak electrolyte hydrogen sulfide) to dissolve the sulfide. For example, iron(II) sulfide is readily dissolved by hydrochloric acid according to the equation

$$FeS + 2H^+ \rightleftharpoons Fe^{2+} + H_2S\uparrow$$
$$[Fe^{2+}][S^{2-}] < K_{sp}$$

However, a number of metal sulfides, such as lead sulfide, furnish such low concentrations of sulfide ion in their saturated solutions that not even the high concentration of hydrogen ion provided by a strong acid is sufficient to exceed the ion product constant of hydrogen sulfide and bring about production of hydrogen sulfide and the subsequent dissolution of the metal sulfide precipitate. For the dissolution of such sulfides, the sulfide ion concentration may be decreased by oxidizing it to elemental sulfur with nitric acid.

$$3S^{2-} + 2NO_3^- + 8H^+ \longrightarrow 3\underline{S} + 2NO\uparrow + 4H_2O$$

Lead sulfide dissolves in nitric acid, then, because

$$[Pb^{2+}][S^{2-}] < K_{sp}$$

resulting from the fact that much of the sulfide ion is oxidized to sulfur by the nitric acid.

19.13   Dissolution of a Precipitate by the Formation of a Complex Ion

Many slightly soluble electrolytes are dissolved through the formation of complex ions. Several examples which are important to qualitative analysis follow.

$$AgCl + 2NH_3 \rightleftharpoons Ag(NH_3)_2^+ + Cl^-$$
$$CuCN + CN^- \rightleftharpoons Cu(CN)_2^-$$
$$Zn(OH)_2 + 2OH^- \rightleftharpoons Zn(OH)_4^{2-}$$
$$Sn(OH)_2 + OH^- \rightleftharpoons Sn(OH)_3^-$$
$$Al(OH)_3 + OH^- \rightleftharpoons Al(OH)_4^-$$
$$As_2S_3 + 3S^{2-} \rightleftharpoons 2AsS_3^{3-}$$
$$HgS + S^{2-} \rightleftharpoons HgS_2^{2-}$$
$$Sb_2S_3 + 3S^{2-} \rightleftharpoons 2SbS_3^{3-}$$

As an example of dissolution by complex ion formation, let us consider the case of the dissolution of silver chloride in aqueous ammonia. When silver ions and ammonia molecules are brought together in solution the complex diammine-silver ion, $Ag(NH_3)_2^+$, is formed. This complex ion dissociates reversibly in aqueous solution according to the equation

$$Ag(NH_3)_2^+ \rightleftharpoons Ag^+ + 2NH_3$$

Applying the law of chemical equilibrium to this system, we obtain

$$\frac{[Ag^+][NH_3]^2}{[Ag(NH_3)_2^+]} = K_d = 6.3 \times 10^{-8}$$

The value of the equilibrium constant K_d is a measure of the stability of the complex ion and is called the **dissociation constant;** the smaller the value of the constant, the more stable the complex. In a saturated solution of silver chloride the Ag^+ concentration is determined by the value of the solubility product, K_{sp}.

This Ag^+ concentration is greater than that which can exist in equilibrium with the ammonia which has been added.

$$\frac{[Ag^+][NH_3]^2}{[Ag(NH_3)_2^+]} > K_d$$

Therefore, silver ions and ammonia molecules combine, thus lowering the concentration of silver ions and bringing the product of the silver ion concentration and chloride ion concentration below the solubility product constant for silver chloride.

$$[Ag^+][Cl^-] < K_{sp}$$

anytime → *dissociation*

Dissolution of some silver chloride then results. If the concentration of ammonia is great enough, equilibrium will be reached only after all of the silver chloride has dissolved.

The copper(II) ion forms coordinate bonds with four ammonia molecules to produce the complex $Cu(NH_3)_4^{2+}$. The expression for the dissociation constant is

$$\frac{[Cu^{2+}][NH_3]^4}{[Cu(NH_3)_4^{2+}]} = K_d$$

When a copper(II) solution is treated with dilute aqueous ammonia, copper(II) hydroxide first precipitates since the hydroxide ion concentration of the dilute ammonia solution is sufficiently large that the solubility product constant of copper(II) hydroxide is exceeded.

$$[Cu^{2+}][OH^-]^2 > K_{sp}$$

As the concentration of ammonia is increased, the dissociation constant of the complex ion $Cu(NH_3)_4^{2+}$ is exceeded.

$$\frac{[Cu^{2+}][NH_3]^4}{[Cu(NH_3)_4^{2+}]} > K_d$$

Note that the concentration of the copper(II) ion varies inversely with the second power of the concentration of hydroxide ion and with the fourth power of the concentration of ammonia. The combination of copper(II) ion with ammonia molecules makes the concentration of Cu^{2+} so small that the condition for the dissolution of copper(II) hydroxide is attained.

$$[Cu^{2+}][OH^-]^2 < K_d$$

19.14   Calculations Involving Complex Ions

Example 1.   Calculate the concentration of the silver ion in a solution which is 0.10 M with respect to $Ag(NH_3)_2^+$.

The dissociation of the complex ion may be represented by the equilibrium

$$Ag(NH_3)_2^+ \rightleftharpoons Ag^+ + 2NH_3$$

$(.10-x) \qquad x \qquad 2x$

Let x be the number of moles of $Ag(NH_3)_2^+$ which dissociate per liter. Then at equilibrium $[Ag^+]$ will be x, and $[NH_3]$ will be $2x$. From the small value of the dissociation constant we know that the dissociation must be small, and we can assume that the complex is practically 0.10 M at equilibrium. Therefore,

$$\frac{[Ag^+][NH_3]^2}{[Ag(NH_3)_2^+]} = \frac{x(2x)^2}{0.10} = K_d = 6.3 \times 10^{-8}$$

$$4x^3 = 6.3 \times 10^{-9}$$
$$x^3 = 1.6 \times 10^{-9}$$
$$x = 1.2 \times 10^{-3} = [Ag^+]$$
$$2x = 2.4 \times 10^{-3} = [NH_3]$$

Example 2. Calculate the number of moles of ammonia that must be added to one liter of water just to dissolve 1.0×10^{-3} mole of silver chloride.

Writing the equation for the reaction,

$$AgCl + 2NH_3 \rightleftharpoons Ag(NH_3)_2^+ + Cl^-$$

we notice that when 1.0×10^{-3} mole of AgCl is dissolved, 1.0×10^{-3} mole of $Ag(NH_3)_2^+$ and 1.0×10^{-3} mole of Cl^- are formed. Two equilibria are involved when AgCl dissolves in aqueous ammonia.

$$AgCl \text{ (solid)} \rightleftharpoons Ag^+ + Cl^-$$
$$Ag(NH_3)_2^+ \rightleftharpoons Ag^+ + 2NH_3$$

The $[Ag^+]$ in solution applies to both equilibria when both solid AgCl and $Ag(NH_3)_2^+$ ions are present in the system. Since the $[Cl^-]$ is 1.0×10^{-3}, we can calculate the $[Ag^+]$ using the solubility product expression for AgCl.

$$[Ag^+][Cl^-] = [Ag^+](1.0 \times 10^{-3}) = 1.8 \times 10^{-10}$$

$$[Ag^+] = \frac{1.8 \times 10^{-10}}{1.0 \times 10^{-3}} = 1.8 \times 10^{-7}$$

By substituting in the expression for the dissociation constant of $Ag(NH_3)_2^+$ the values for $[Ag^+]$ and $[Ag(NH_3)_2^+]$, we can calculate the $[NH_3]$ required at equilibrium.

$$\frac{[Ag^+][NH_3]^2}{[Ag(NH_3)_2^+]} = \frac{(1.8 \times 10^{-7})[NH_3]^2}{1.0 \times 10^{-3}} = 6.3 \times 10^{-8}$$

$$[NH_3]^2 = \frac{(1.0 \times 10^{-3})(6.3 \times 10^{-8})}{1.8 \times 10^{-7}} = 3.5 \times 10^{-4}$$

$$[NH_3] = 1.9 \times 10^{-2}$$

In this calculation the amount of ammonia consumed in forming the complex ion is $2(1.0 \times 10^{-3})$, or 0.002 mole, which is to be added to the 1.9×10^{-2}, or 0.019, mole of ammonia required at equilibrium. Thus, the total amount of ammonia required for dissolution of 1.0×10^{-3} mole of AgCl in a liter of water is 0.021 mole.

19.15   Effect of Hydrolysis on Dissolution of Precipitates

Many salts hydrolyze when they are dissolved in water with the formation of precipitates, with the evolution of gases, or with both. For example, when aluminum sulfide, Al_2S_3, is dissolved in water, insoluble aluminum hydroxide is formed and hydrogen sulfide gas is evolved. The aluminum ions combine with the hydroxide ions of water, and the sulfide ions unite with the hydrogen ions of water according to the equations

$$2[Al(H_2O)_6]^{3+} + 6OH^- \rightleftharpoons 2[Al(OH)_3(H_2O)_3] + 6H_2O$$
$$3S^{2-} + 6H_3O^+ \rightleftharpoons 3H_2S\uparrow + 6H_2O$$
$$12H_2O \rightleftharpoons 6H_3O^+ + 6OH^-$$

The net reaction is given by

$$2[Al(H_2O)_6]^{3+} + 3S^{2-} \rightleftharpoons 2[Al(OH)_3(H_2O)_3] + 3H_2S\uparrow$$

It is evident that the aluminum sulfide cannot be recovered by crystallization after it has undergone hydrolytic decomposition. Many other salts behave similarly toward water.

When a relatively insoluble sulfide such as lead sulfide, PbS, is dissolved in water until a saturated solution is formed, an appreciable amount of the sulfide ion hydrolyzes to form the hydrosulfide ion. In some cases, as we shall see later in Example 2, a significant amount of the hydrosulfide ion hydrolyzes to form hydrogen sulfide.

$$PbS \rightleftharpoons Pb^{2+} + S^{2-}$$
$$S^{2-} + H_2O \rightleftharpoons HS^- + OH^-$$
$$HS^- + H_2O \rightleftharpoons H_2S + OH^-$$

For this reason the lead ion concentration is not the same as that of the sulfide ion under these conditions. The concentration of the lead ion is equal to the sum of the concentrations of S^{2-}, HS^-, and H_2S. This means that in calculating the solubility of a relatively insoluble sulfide from the solubility product constant, one must take into account the hydrolysis of the S^{2-} ion.

Example 1.   Let us first calculate the solubility of PbS in water, neglecting the hydrolysis of the sulfide ion. (K_{sp} for PbS is 8.4×10^{-28}.)

Let x equal the molar solubility of lead sulfide in water. Then x will also be equal to the concentration of the lead ion and of the sulfide ion. Substituting in the expression for the solubility product constant, we obtain

$$[Pb^{2+}][S^{2-}] = K_{sp}$$
$$x^2 = 8.4 \times 10^{-28}$$
$$x = 2.9 \times 10^{-14}$$

Thus, if the hydrolysis of the sulfide ion were neglected we would find the solubility of PbS in water to be 2.9×10^{-14} mole per liter.

Example 2.   Now let us solve the problem of Example 1 correctly, taking into consideration the hydrolysis of the sulfide ion.

We shall consider three steps in the dissolution of PbS in water.

$$\underline{PbS} \rightleftharpoons Pb^{2+} + S^{2-} \quad K_{sp} = 8.4 \times 10^{-28} \tag{1}$$

$$S^{2-} + H_2O \rightleftharpoons HS^- + OH^- \quad K_{h_1} = \frac{K_w}{K_i \ (\text{for } HS^-)} = \frac{1.0 \times 10^{-14}}{1.3 \times 10^{-13}} = 7.7 \times 10^{-2} \tag{2}$$

$$HS^- + H_2O \rightleftharpoons H_2S + OH^- \quad K_{h_2} = \frac{K_w}{K_i \ (\text{for } H_2S)} = \frac{1.0 \times 10^{-14}}{1.0 \times 10^{-7}} = 1.0 \times 10^{-7} \tag{3}$$

We might be tempted to say that, since K_{h_2} is quite small compared to K_{h_1}, we could consider only the first stage of the hydrolysis and neglect the second. This is often a valid way to proceed, and indeed we did this in Example 2 of Section 18.20. We found in that example that the sulfide ion in a 0.0010 M sodium sulfide solution was 99% hydrolyzed. Thus, the first stage of hydrolysis would produce (0.99×0.0010) mole, or 9.9×10^{-4} mole, of OH^-. Inasmuch as this quantity of OH^- is large compared to the 1.0×10^{-7} mole furnished by the water, the total concentration of OH^- in solution was 9.9×10^{-4} mole/liter (1.0×10^{-7} is negligible compared to 9.9×10^{-4}, to several significant figures).

We can calculate the ratio of $[H_2S]$ to $[HS^-]$, from K_{h_2}, for Example 2 of Section 18.20, then, to establish the relative amounts of HS^- and H_2S produced in the first and second stages of hydrolysis, respectively.

$$\frac{[H_2S][OH^-]}{[HS^-]} = \frac{[H_2S](9.9 \times 10^{-4})}{[HS^-]} = K_{h_2} = 1.0 \times 10^{-7}$$

$$\frac{[H_2S]}{[HS^-]} = 1.0 \times 10^{-4}$$

Hence, $[H_2S]$ is only $1/10,000 \ [HS^-]$, indicating that the second stage of the hydrolysis is negligible for Example 2 of Section 18.20, compared to the first stage of the hydrolysis, and could be neglected *in that calculation*.

However, in our present problem, we have a much smaller amount of sulfide ion. Our rough calculation for the solubility of lead sulfide, neglecting hydrolysis, indicated only 2.9×10^{-14} mole of lead sulfide going into solution per liter. The hydroxide ion concentration due to the hydrolysis of the sulfide ion, $S^{2-} + H_2O \rightleftharpoons HS^- + OH^-$, would be of the order of $3 \times 10^{-14} M$. However, this hydroxide ion concentration is so small compared with that provided by the water, $1.0 \times 10^{-7} M$, that the $[OH^-]$ of the solution will still have a value of $1.0 \times 10^{-7} M$. Even some increase in the solubility of PbS and some additional OH^- produced in the second stage of hydrolysis would not likely be enough to add sufficient OH^- to be significant compared to the 1.0×10^{-7} mole/liter furnished by the water. We can now calculate the ratio of $[H_2S]$ to $[HS^-]$ from K_{h_2}, for a solution in which $[OH^-]$ is $1 \times 10^{-7} M$ (neutral solution).

$$\frac{[H_2S][OH^-]}{[HS^-]} = \frac{[H_2S](1.0 \times 10^{-7})}{[HS^-]} = K_{h_2} = 1.0 \times 10^{-7}$$

$$\frac{[H_2S]}{[HS^-]} = 1$$

Hence, in the case of a neutral solution, $[H_2S] = [HS^-]$, indicating that the second stage *is* important in the calculation and cannot be neglected here. For $[HS^-]$ to equal $[H_2S]$, half of the HS^- ions produced in the first stage of hydrolysis must hydrolyze to produce H_2S in the second stage of hydrolysis.

Now, consider the three equilibria listed earlier for the system. As PbS dissolves, equal quantities of Pb^{2+} and S^{2-} are produced. The S^{2-} produced is essentially all hydrolyzed and hence converted to HS^- in the first stage of hydrolysis. At this point, not yet taking the second stage of hydrolysis into account, $[Pb^{2+}] = [HS^-]$. However, we have established that half of the HS^- then hydrolyzes in the second stage of hydrolysis, thereby reducing the HS^- concentration to half its former value, and thus to half the concentration of Pb^{2+}. Hence, at final equilibrium, $[HS^-] = \frac{1}{2}[Pb^{2+}]$.

Using $1.0 \times 10^{-7}\ M$ for the $[OH^-]$, we can calculate the ratio of the concentration of the hydrosulfide ion to that of the sulfide ion, by substituting in K_{h_1} and then solving for $[S^{2-}]$.

$$\frac{[HS^-][OH^-]}{[S^{2-}]} = \frac{[HS^-](1.0 \times 10^{-7})}{[S^{2-}]} = K_{h_1} = 7.7 \times 10^{-2}$$

$$\frac{[HS^-]}{[S^{2-}]} = 7.7 \times 10^5, \quad \text{or} \quad [S^{2-}] = \frac{[HS^-]}{7.7 \times 10^5}$$

Substituting the value for $[S^{2-}]$ found above in the expression for the solubility product constant for lead sulfide, we have

$$[Pb^{2+}][S^{2-}] = [Pb^{2+}]\frac{[HS^-]}{7.7 \times 10^5} = K_{sp} = 8.4 \times 10^{-28}$$

Because $[HS^-] = \frac{1}{2}[Pb^{2+}]$, we can write

$$[Pb^{2+}]\left(\frac{\frac{1}{2}[Pb^{2+}]}{7.7 \times 10^5}\right) = 8.4 \times 10^{-28}$$

$$\frac{\frac{1}{2}[Pb^{2+}]^2}{7.7 \times 10^5} = 8.4 \times 10^{-28}$$

$$[Pb^{2+}]^2 = 12.94 \times 10^{-22}$$

$$[Pb^{2+}] = 3.6 \times 10^{-11} = \text{molar solubility of PbS}$$

We have found, then, that the solubility of lead sulfide in water is 3.6×10^{-11} mole per liter when hydrolysis of the sulfide ion is considered. This value is over one thousand times greater than that calculated (2.9×10^{-14} mole per liter) when hydrolysis was neglected. If one were to take into account the fact that the lead ion hydrolyzes slightly to give $Pb(OH)^+$ and $Pb(OH)_2$, the solubility of lead sulfide would be found to be somewhat larger yet.

Finally, note that the sulfide hydrolysis would produce 3.6×10^{-11} mole OH^-/liter in the first stage of hydrolysis $+\frac{1}{2}(3.6 \times 10^{-11})$ additional mole OH^-/liter in the second stage of hydrolysis, for a total of 5.4×10^{-11} mole OH^-/liter. This amount of OH^- is indeed negligible compared to the 1.0×10^{-7} mole of OH^- furnished by the water. Hence, our earlier assumption of $[OH^-] = 1.0 \times 10^{-7} M$ for this particular case was entirely justified.

The relatively insoluble carbonates behave similarly to the sulfides. When a relatively insoluble carbonate such as barium carbonate, $BaCO_3$, is dissolved in water, its solubility is increased due to hydrolysis of the carbonate ion, $(CO_3^{2-} + H_2O \rightleftharpoons HCO_3^- + OH^-$ and $HCO_3^- + H_2O \rightleftharpoons H_2CO_3 + OH^-)$. The concentration of the barium ion is equal to the sum of the concentrations of CO_3^{2-}, HCO_3^-, and H_2CO_3.

QUESTIONS

1. Write the expression for the solubility product of each of the following slightly soluble electrolytes: $PbCl_2$, $AgBr$, $Ba_3(PO_4)_2$, Ag_2S, $MgNH_4AsO_4$.
2. State the rule regarding the precipitation of slightly soluble electrolytes.
3. What is the rule regarding the dissolution of a precipitate relative to the product of the concentrations of its ions?
4. If solid silver bromide is in equilibrium with a saturated solution of its ions, Ag^+ and Br^-, will this equilibrium be affected and in what manner if (a) more solid silver bromide is added? (b) silver nitrate is added? (c) sodium bromide is added? (d) the temperature is raised (the solubility increases with temperature)?
5. A saturated solution of a slightly soluble electrolyte in contact with some of the solid phase of the electrolyte is said to be a system in equilibrium. Explain. Why is such a system called a *heterogeneous* equilibrium?
6. Explain why the concentration of a solid is essentially constant.
7. Can an ion ever be completely removed from solution by using a large excess of the precipitant? Explain your answer.
8. What three general methods are used to dissolve slightly soluble electrolytes?
9. What is meant by fractional precipitation?
10. Look up the solubility product constants for CdS and PbS in Appendix E. Would sulfide precipitation be an effective means of separating cadmium and lead ions in water solution?
11. Refer to Appendix E for solubility product constants for sulfate salts. Determine which of the sulfates listed is least soluble in moles per liter. Determine which sulfate is least soluble in grams per liter.
12. What is meant by the dissociation constant of a complex ion? Illustrate.

PROBLEMS

ⓢ1. Calculate the solubility product constant of each of the following from the solubility given: ⓢ(a) $BaCO_3$ (9.0×10^{-5} mole/liter); (b) PbF_2 (2.1×10^{-3} mole/liter); ⓢ(c) Ag_2SO_4 (4.47 g/liter).

   Ans. (a) 8.1×10^{-9}; (b) 3.7×10^{-8}; (c) 1.18×10^{-5}

ⓢ2. Calculate the molar solubility for each of the following substances from its solubility product constant.

   ⓢ(a) $SrCO_3$, $K_{sp} = 9.42 \times 10^{-10}$ *Ans. 3.07×10^{-5} mole/liter*
   (b) $Zn(OH)_2$, $K_{sp} = 4.5 \times 10^{-17}$ *Ans. 2.2×10^{-6} mole/liter*
   ⓢ(c) Ag_3PO_4, $K_{sp} = 1.8 \times 10^{-18}$ *Ans. 1.6×10^{-5} mole/liter*
   (d) $TlCl$, $K_{sp} = 1.9 \times 10^{-4}$ *Ans. 1.4×10^{-2} mole/liter*
   (e) $CaHPO_4$, $K_{sp} = 5 \times 10^{-6}$ *Ans. 2×10^{-3} mole/liter*
   (f) $Al(OH)_3$, $K_{sp} = 1.9 \times 10^{-33}$ *Ans. 2.9×10^{-9} mole/liter*

3. Calculate the maximum concentration of Cu(I) ion in a solution in which $[Cl^-]$ is 5.50×10^{-3} M. (K_{sp} of CuCl is 1.85×10^{-7}.)   *Ans. 3.36×10^{-5} M*

ⓢ4. Calculate the $[Mg^{2+}]$ required to start the precipitation of MgF_2 from a solution that is 0.0040 M in fluoride ions ($K_{sp} = 6.4 \times 10^{-9}$).

   Ans. $4.0 = 10^{-4}$ M

5. (a) Calculate the silver ion concentration in a saturated aqueous solution of silver iodide ($K_{sp} = 1.5 \times 10^{-16}$).   *Ans. 1.2×10^{-8} M*
   (b) What will be the $[Ag^+]$ when enough KI is added to make $[I^-] =$ 0.0030 M?   *Ans. 5.0×10^{-14} M*
   (c) What will be the $[I^-]$ when enough $AgNO_3$ is added to make $[Ag^+] =$ 0.020 M?   *Ans. 7.5×10^{-15} M*

ⓢ6. The K_{sp} of HgS is 3×10^{-53}. Calculate the maximum concentration of sulfide ion that can exist in a solution that is 1.0×10^{-10} M in $Hg(NO_3)_2$.

   Ans. 3×10^{-43} M

7. What is the concentration of Ba^{2+} and SO_4^{2-} in a saturated solution of $BaSO_4$? ($K_{sp} = 1.08 \times 10^{-10}$.)   *Ans. $[Ba^{2+}] = [SO_4^{2-}] = 1.04 \times 10^{-5}$ M*

ⓢ8. Calculate solubility of AgCl in mg/100 ml. ($K_{sp} = 1.8 \times 10^{-10}$.)   *Ans. 0.19*

ⓢ9. A precipitate of AgCl was washed with 100 ml of 0.10 M HCl solution and then with 100 ml of distilled water. Calculate the amount of AgCl which dissolved with each washing, assuming that the wash liquid became saturated with AgCl. (K_{sp} of AgCl = 1.8×10^{-10}.)

   Ans. 1.8×10^{-10} mole; 1.3×10^{-6} mole

10. In one experiment, a precipitate of $BaSO_4$ was washed with 100 ml of distilled water; in another experiment, a precipitate of $BaSO_4$ was washed with 100 ml of 0.010 M H_2SO_4. Calculate the quantity of $BaSO_4$ which dissolved in each experiment, assuming that the wash liquid became saturated with $BaSO_4$. (K_{sp} of $BaSO_4$ = 1.08×10^{-10}.)   *Ans. 1.04×10^{-6} mole; 1.08×10^{-9} mole*

11. Which compound has the smaller molar solubility: $BaCrO_4$ ($K_{sp} = 2.4 \times 10^{-10}$) or Ag_2CrO_4 ($K_{sp} = 8.2 \times 10^{-12}$)? Show by calculation.

   Ans. $BaCrO_4$

ⓢ12. A 300-ml volume of 2×10^{-7} M $Hg_2(NO_3)_2$ solution and 300 ml of 3×10^{-5} M KI solution are mixed. Will Hg_2I_2 precipitate? Explain your answer. (K_{sp} of Hg_2I_2 is 4.5×10^{-29}.)   *Ans. Yes*

13. Calculate the total number of metal ions and sulfide ions in: (a) one milliliter of a saturated CuS solution ($K_{sp} = 8.7 \times 10^{-36}$); (b) one milliliter of a saturated HgS solution ($K_{sp} = 3 \times 10^{-53}$). Is the answer to part (b) reasonable? Can you justify such a low number in a practical sense?

Ans. (a) 3600; (b) 7 × 10⁻⁶

14. Calculate the molar solubility of $SrCrO_4$ in 0.010 M Na_2CrO_4 solution ($K_{sp} = 3.6 \times 10^{-5}$). *Ans. 3.6 × 10⁻³ M*

15. What is the minimum concentration of chromate ion necessary to initiate the precipitation of $BaCrO_4$ from an 8.0×10^{-4} M $BaCl_2$ solution? (K_{sp} of $BaCrO_4 = 2.4 \times 10^{-10}$.) *Ans. 3.0 × 10⁻⁷ M*

⑤16. A solution which is 0.10 M in NaI and also in Na_2SO_4 is treated with solid $Pb(NO_3)_2$. Which compound, PbI_2 or $PbSO_4$, will precipitate first? What is the concentration of the anion of the least soluble compound when the more soluble one starts to precipitate? (K_{sp} of $PbI_2 = 8.7 \times 10^{-9}$; K_{sp} of $PbSO_4 = 1.8 \times 10^{-8}$.) *Ans. PbSO₄; [SO₄²⁻] = 0.021 M*

17. A solution is 0.010 M in CrO_4^{2-} ion and also in Cl^- ion. Solid $AgNO_3$ is added to the solution. Which compound, Ag_2CrO_4 or AgCl, will precipitate first? What is the concentration of the anion of the less soluble salt when the more soluble one starts to precipitate? (K_{sp} of $AgCl = 1.8 \times 10^{-10}$; K_{sp} of $Ag_2CrO_4 = 9 \times 10^{-12}$.) *Ans. AgCl; [Cl⁻] = 6 × 10⁻⁶ M*

18. The solubility product constants of Hg_2Cl_2, Hg_2Br_2 and Hg_2I_2 are 1.1×10^{-18}, 1.26×10^{-22}, and 4.5×10^{-29}, respectively. Calculate the maximum concentration of Hg_2^{2+} ion [note that the mercury(I) ion is a dimeric ion] in separate solutions which are 1.00×10^{-6} M with respect to NaCl, NaBr, and NaI, respectively. *Ans. 1.1 × 10⁻⁶ M; 1.26 × 10⁻¹⁰ M; 4.5 × 10⁻¹⁷ M*

⑤19. A 50-ml volume of solution containing 0.95 g of $MgCl_2$ is mixed with an equal volume of 1.8 M aqueous ammonia. What weight of solid NH_4Cl must be added to the resulting solution to prevent the precipitation of $Mg(OH)_2$? (K_{sp} of $Mg(OH)_2 = 1.5 \times 10^{-11}$; K_i of NH_3 is 1.8×10^{-5}.) *Ans. 7.1 g*

20. A solution is made 0.0020 M in Mg^{2+} and 0.0025 M in OH^-. The K_{sp} of $Mg(OH)_2$ is 1.5×10^{-11}. Will $Mg(OH)_2$ precipitate under the conditions stated? Why or why not? *Ans. Yes*

⑤21. What pH is necessary for initiating the precipitation of lead(II) hydroxide from a solution of lead(II) nitrate which contains 0.662 mg of lead(II) ion per ml? (K_{sp} of $Pb(OH)_2 = 2.8 \times 10^{-16}$.) *Ans. 7.47*

22. What concentration of Al^{3+} ion must be present in a solution that has a pH of 5.40 in order to initiate precipitation of $Al(OH)_3$? (K_{sp} of $Al(OH)_3 = 1.9 \times 10^{-33}$.) *Ans. 1.2 × 10⁻⁷ M*

⑤23. What is the maximum pH that a 0.075 M $Fe(NO_3)_2$ solution can have and not have FeS precipitate when the solution is saturated with H_2S? A saturated H_2S solution is 0.10 M in H_2S. (K_{sp} of $FeS = 1 \times 10^{-19}$; K_1 and K_2 for H_2S are 1.0×10^{-7} and 1.3×10^{-13}, respectively.) *Ans. 1.5*

⑤24. A solution which is 0.10 M in Cd^{2+} and 0.30 M in HCl is saturated with H_2S. What concentration of Cd^{2+} will remain in solution? (K_{sp} of $CdS = 3.6 \times 10^{-29}$. Do not neglect the hydrogen ions produced by the reaction of Cd^{2+} with H_2S.) *Ans. 6.9 × 10⁻⁹ M*

25. How many grams of NH_4Cl must be added to 25 ml of 0.50 M aqueous

ammonia to prevent precipitation of $Mg(OH)_2$ when this solution is added to 25 ml of a 0.10 M $MgCl_2$ solution? (K_i for $NH_3 = 1.8 \times 10^{-5}$; K_{sp} for $Mg(OH)_2 = 1.5 \times 10^{-11}$.)   *Ans. 0.69 g*

26. With what volume of water must a precipitate of $NiCO_3$ be washed to dissolve 1.00 g of the compound? Assume that the wash water becomes saturated with $NiCO_3$. (K_{sp} of $NiCO_3 = 1.36 \times 10^{-7}$.)   *Ans. 22.8 liters*

27. The K_{sp} of PbI_2 is 8.7×10^{-9}. What is the maximum molar concentration of Pb^{2+} in 400 ml of solution containing 20 g of KI without resulting in precipitation of lead iodide?   *Ans. 9.6 × 10⁻⁸ M*

⑤28. The K_{sp} of $Mn(OH)_2$ is 4.5×10^{-14}. What is the pH of a saturated $Mn(OH)_2$ solution?   *Ans. 9.65*

29. Calculate the pH of a saturated solution of $Ca(OH)_2$. (K_{sp} for $Ca(OH)_2 = 7.9 \times 10^{-6}$.)   *Ans. 12.40*

30. What is the molar solubility of $Cr(OH)_3$ in a solution of pH 9.00? (K_{sp} of $Cr(OH)_3$ is 6.7×10^{-31}.)   *Ans. 6.7 × 10⁻¹⁶ M*

⑤31. What is the molar solubility of MgF_2 in a 0.30 M HCl solution? (K_{sp} of $MgF_2 = 6.4 \times 10^{-9}$; K_i of HF $= 7.2 \times 10^{-4}$.) *Hint: The method of successive approximations (see Section 18.15) will be useful in working this problem.*   *Ans. 0.050 M*

32. A solution is 0.020 M in both Sr^{2+} and Ba^{2+} ions. Calculate the percentage of strontium ion which remains in solution when 99.9% of the barium ion has been precipitated as $BaSO_4$? (K_{sp} of $BaSO_4 = 1.08 \times 10^{-10}$; K_{sp} of $SrSO_4 = 2.8 \times 10^{-7}$.)   *Ans. 100%*

33. To a solution that is 0.10 M in both Pb^{2+} and Sr^{2+} ions is added solid Na_2SO_4. (K_{sp} for $PbSO_4$ is 1.8×10^{-8}; K_{sp} for $SrSO_4$ is 2.8×10^{-7}.) Assuming no volume change:
    (a) At what $[SO_4{}^{2-}]$ will $PbSO_4$ begin to precipitate?   *Ans. 1.8 × 10⁻⁷ M*
    (b) At what $[SO_4{}^{2-}]$ will $SrSO_4$ begin to precipitate?   *Ans. 2.8 × 10⁻⁶ M*
    (c) Which sulfate salt will precipitate first, $PbSO_4$ or $SrSO_4$?   *Ans. PbSO₄*
    (d) What will be the concentration of Pb^{2+} in the solution when $SrSO_4$ begins to precipitate?   *Ans. 6.4 × 10⁻³ M*
    (e) What will be the ratio of $[Sr^{2+}]$ to $[Pb^{2+}]$ at this point?   *Ans. 16 to 1*

⑤34. A solution is 0.10 M in $ZnCl_2$. What concentration of hydrogen ion must be present so that no zinc(II) sulfide will precipitate when the solution is saturated with hydrogen sulfide? A saturated solution of H_2S is 0.10 M. (K_{sp} for ZnS $= 1.1 \times 10^{-21}$.)   *Ans. 0.34 M*

35. What concentration of silver ion remains in solution when a 0.20 M solution of $AgNO_3$ is treated with solid KI in a quantity 0.10% in excess of the amount equivalent to the silver nitrate present in the solution? (K_{sp} for AgI $= 1.5 \times 10^{-16}$.)   *Ans. 7.5 × 10⁻¹³ M*

⑤36. It is found that 1.62×10^{-9} mole of PbS will dissolve in a liter of a solution which has a pH of 1.50 and which contains H_2S at 0.040 M concentration. Using the dissociation constants for hydrogen sulfide, calculate the K_{sp} for lead sulfide. Compare calculated value with value given in Appendix E.   *Ans. 8.4 × 10⁻²⁸*

37. Into a liter of 0.10 M silver nitrate solution, 0.10 mole of hydrogen chloride is dissolved. After reaction has taken place, 0.050 mole of sodium nitrite is

added. What will be the concentrations of every ion in solution after all chemical changes have reached equilibrium? (K_{sp} for AgCl = 1.8×10^{-10}; for AgOH, 2.0×10^{-8}; for AgNO$_2$, 1.2×10^{-4}. K_i for HNO$_2$ = 4.5×10^{-4}.)

Ans. $[Ag^+] = [Cl^-] = 1.3 \times 10^{-5} M$; $[OH^-] = 2.0 \times 10^{-13} M$; $[NO_2^-] = 4.4 \times 10^{-4} M$; $[H^+] = 5.0 \times 10^{-2} M$; $[Na^+] = 5.0 \times 10^{-2} M$; $[NO_3^-] = 0.10 M$

$\boxed{s}$38. Calculate the concentration of Ni^{2+} in a 1.0 M solution of [Ni(NH$_3$)$_6$](NO$_3$)$_2$. (K_d for [Ni(NH$_3$)$_6$]$^{2+}$ = 5.7×10^{-9}.)   *Ans. 0.014 M*

39. Calculate the concentration of Zn^{2+} in a 0.30 M solution of [Zn(CN)$_4$]$^{2-}$. (K_d for [Zn(CN)$_4$]$^{2-}$ = 1×10^{-19}.)   *Ans. 4 × 10^{-5} M*

$\boxed{s}$40. Calculate the minimum number of moles of cyanide ion that must be added to 100 ml of solution to dissolve 2×10^{-2} mole of AgCN. (K_{sp} for AgCN is 1.2×10^{-16}; K_d for [Ag(CN)$_2$]$^-$ is 1×10^{-20}.)   *Ans. 2 × 10^{-2} mole*

41. Calculate the minimum concentration of ammonia needed in one liter of solution to dissolve 3.0×10^{-3} moles of AgBr. (K_{sp} for AgBr = 3.3×10^{-13}; K_d for [Ag(NH$_3$)$_2$]$^+$ = 6.3×10^{-8}.)   *Ans. 1.3 M*

42. Calculate the volume of 1.50 M HOAc required to dissolve a precipitate composed of 350 mg each of CaCO$_3$, SrCO$_3$, and BaCO$_3$.   *Ans. 10.2 ml*

$\boxed{s}$43. In a titration of cyanide ion, 28.72 ml of 0.0100 M AgNO$_3$ is added before precipitation begins. (The reaction of Ag$^+$ with CN$^-$ goes to completion producing the [Ag(CN)$_2$]$^-$ complex. Precipitation of solid AgCN takes place when excess Ag$^+$ is added to the solution, above the amount needed to complete the formation of [Ag(CN)$_2$]$^-$.) How many grams of NaCN were in the original sample? What is the insoluble material? What is the nature of the silver species before precipitation begins?

Ans. 0.0281; AgCN; [Ag(CN)$_2$]$^-$

44. A 0.010-mole sample of solid AgCN is rendered soluble in one liter of solution by adding just sufficient excess cyanide ion to form [Ag(CN)$_2$]$^-$. When all of the solid silver cyanide has just dissolved, the concentration of free cyanide ion is $1.125 \times 10^{-7} M$. Neglecting hydrolysis of the cyanide ion, determine the concentration of free, uncomplexed silver ion in the solution. If more cyanide ion is added (without changing the volume) until the equilibrium concentration of free cyanide ion is $1.0 \times 10^{-6} M$, what will be the equilibrium concentration of free silver ion?   *Ans. 8 × 10^{-9} M; 1 × 10^{-10} M*

REFERENCES

"Ratio Diagrams: A Simple Graphical Representation of Complicated Equilibria," R. de Levie, *J. Chem. Educ.,* **47,** 187 (1970).

"Potentiometric Determination of Solubility Product Constants," S. L. Tackett, *J. Chem. Educ.,* **46,** 857 (1969).

"Are Solubilities and Solubility Products Related?" L. Meites, J. S. F. Pode, and Henry C. Thomas, *J. Chem. Educ.,* **43,** 667 (1966).

"Student Experiments Involving Unknown Solubility Constants," D. E. Heinz, *J. Chem. Educ.,* **44,** 114 (1967).

"The Stoichiometry of Copper Sulfide Formed in an Introductory Laboratory Exercise," D. Dingledy and W. M. Barnard, *J. Chem. Educ.,* **44,** 242 (1967).

"The Effects of Chloride Ion and Temperature on Lead Chloride Solubility," A. C. West, *J. Chem. Educ.,* **46,** 773 (1969).

Chemical Thermodynamics

20

20.1 Introduction

Five important questions that confront the chemist are:

(1) When two or more substances are put together, will they react?

(2) If they do react, what energy changes will be associated with the reaction?

(3) If a reaction occurs, at what concentrations of the reacting substances and their products will equilibrium be established?

(4) If a reaction occurs, how fast will it go—that is, what will be its rate?

(5) What is the mechanism by which the reaction occurs?

Chemical thermodynamics is concerned with the first three of these questions. It has nothing to say about the last two. Thermodynamics is not concerned with either the rate or the specific manner in which atoms, ions, or molecules come together to form new compounds.

In Section 1.3 it was pointed out that the total quantity of energy available in the universe is constant. This of course does not mean that energy can not be transformed from one form to another and transferred from one part of the universe to another. There are many examples on both large and small scales, such as the sun, the kitchen stove, and the family automobile—that produce such energy transformations and transfers. **Chemical thermodynamics is that branch of chemistry which studies the energy transformations and transfers that accompany chemical and physical changes.** It is customary to call that part of the universe that we have under study and with whose properties we are concerned the **system.** The rest of the universe is defined as the **surroundings. Chemical thermodynamics is then a study of the energy transformations that occur in the system and any transfer of energy that may occur between the system and the surroundings.**

461

There are two fundamental laws of nature that are important in thermodynamics and which apply to all systems: (1) systems tend to attain a state of minimum potential energy, and (2) systems tend toward a state of maximum disorder. These two laws can be illustrated in the following way: If you are holding a box and release it, the box falls to the floor. In so doing it goes to a state of lower potential energy. If that box contains a jig-saw puzzle that had been assembled before the box was released, the chances are great that the puzzle will be partly disassembled after the fall, thereby becoming more disordered. The latter fact may not seem very important or fundamental, but most people would agree that no one would try to assemble a jig-saw puzzle by dropping the separated pieces on the floor, whereas one can readily accomplish the reverse process by dropping the assembled puzzle on the floor. A system tends to become less orderly because there are so many more ways to be disorderly than to be orderly; the probability for a system to become more disorderly or more random is greater than for it to become more orderly. We will consider more fully the interplay between these two fundamental laws of nature and their consequences later in this chapter.

In a practical sense, thermodynamics is concerned with the amount of energy in the form of work (see Section 1.2) which can be obtained from a system when some of its energy is transformed into heat energy and transferred to the surroundings. In nonscientific language the general principle that governs such a process can be summed up by the seemingly quite obvious statement: You can never get more of any commodity than there is to be gotten, and ordinarily you can not even get quite all of that. In more scientific language and in terms of energy, one can say that there is never more energy available to do work than the amount released by the system, and in fact *all* the energy released can never be recovered with 100 per cent efficiency in the form of *useful* work energy. In order to see how thermodynamics is connected to chemistry we shall examine the three Laws of Thermodynamics and their consequences and usefulness in chemistry.

20.2   The First Law of Thermodynamics

The **First Law of Thermodynamics** is actually just the Law of Conservation of Energy (see Section 1.3) and can be stated: **The total amount of energy in the universe is constant.** The First Law is often considered in a rather special form. If we have a system in a certain internal energy state, E_1, and we add an amount of heat energy, q, to the system which in turn does some work, w, on the surroundings, the system ends up in a new internal energy state, E_2. The Law of Conservation of Energy requires that for the system

$$E_2 = E_1 + (q - w)$$

This means that the final energy state is equal to the initial energy state plus a quantity composed of the heat added minus the work done. The quantity $(q - w)$ may be either positive or negative, depending upon the relative values of q and w.

Then, $$E_2 - E_1 = q - w$$

where $(E_2 - E_1)$ is equal to the change in internal energy for the system, ΔE_{sys}. The symbol Δ stands for the difference between two quantities—in this case, energies.

Therefore, $$\Delta E_{sys} = E_2 - E_1 = q - w$$

In other words, we can say that the change in the internal energy of any given system is equal to the heat energy tranferred to the system from the surroundings minus the work energy transferred from the system to the surroundings. It is of course true that if ΔE_{sys} is the energy change for the system and ΔE_{sur} is the energy change for the surroundings, $\Delta E_{sys} + \Delta E_{sur} = 0$. This emphasizes the fact, expressed by the First Law, that the total amount of energy in the universe is constant. The following example will help clarify the First Law.

Example 1.   If 600 calories of heat energy are added to a system in energy state E_1 and the system does 450 calories of work on the surroundings, (a) What is the energy change of the system?

$$\Delta E_{sys} = q - w = 600 \text{ cal} - 450 \text{ cal} = 150 \text{ cal}$$

(b) What is the energy change of the surroundings?

$$\Delta E_{sys} + \Delta E_{sur} = 0$$
$$150 \text{ cal} + \Delta E_{sur} = 0$$
$$\Delta E_{sur} = -150 \text{ cal}$$

(c) What is the energy of the system in the new energy state, E_2?

$$E_2 - E_1 = \Delta E_{sys} = 150 \text{ cal}$$
$$E_2 = E_1 + 150 \text{ cal}$$

In the preceding example the unit used for energy is the calorie (see Section 1.22). This is the unit commonly used in thermodynamics because of the emphasis on heat energy.

There is a consequence of the First Law which may seem relatively obvious but which is very important. If a system is in a state characterized by a certain energy E_1 and the system undergoes a change to a state characterized by E_2, the energy difference is independent of how the system gets from state 1 to state 2. An example will illustrate.

Example 2.   If we start out with the same system as in our previous example in energy state E_1, add 1,000 calories of heat energy, and do 850 calories of work on the surroundings, what is the energy of the system in the new energy state, E_2?

$$\Delta E = q - w = 1,000 \text{ cal} - 850 \text{ cal} = 150 \text{ cal}$$
$$E_2 - E_1 = \Delta E = 150$$
$$E_2 = E_1 + 150 \text{ cal}$$

Thus, the state E_2 must be the same state with which we ended in the previous example. Note that q and w changed but ΔE was constant, and therefore the final state is identical in the two examples.

Example 3.   Starting again with the same system as in the previous examples, in energy state E_1, suppose we add 100 calories of heat energy to the system and do 50 calories of work on the *system*. What is the energy of the system in the new energy state, E_2?

Note that this time the work is done on the *system* instead of on the surroundings and therefore will have the opposite (negative) sign. The value of w, thus, is -50 calories.

$$\Delta E = E_2 - E_1 = q - w = 100 \text{ cal} - (-50 \text{ cal}) = 150 \text{ cal}$$
$$E_2 = E_1 + 150 \text{ cal}$$

Again, the final state is the same as in the first two examples, but the path by which it is reached is different.

20.3   State Functions

When a property of a system (its temperature, for example) is not dependent on how the system gets from state to state and is only dependent on the state itself, it is said to be a **state function.** A change in the value of a state function is independent of the path used to carry out the process. The change in the value of the state function is equal to the value of the function at the final state minus its value at the initial state. How the system goes from the initial state to the final state is irrelevant with respect to the final value. For example, if one wishes to heat water, at 1 atmosphere pressure, from 10° C to 20° C, the change in temperature (a state function) is the same if the H_2O is heated gradually to 20° or if it is heated first to 90° and then cooled to 20°. In both cases the initial and final temperatures of the water differ by only 10°, being independent of all but the initial and final states. A second illustration can be based upon a topographical map of an area consisting of a valley and a high hill. Point A, the starting point, is in the valley while the final point B is at the top of the hill. The change in elevation which one experiences in hiking from A to B is independent of the path taken. To put it another way, state functions have no memory—they forget the path used for the transformation.

As opposed to state functions, whose values depend solely upon the initial and final states, there are those functions whose values do depend on the paths followed. The distinction between the two types of functions can be illustrated by the expression in Section 20.2.

$$\Delta E = E_2 - E_1 = q - w$$

The internal energy, E, as evidenced by the examples in the preceding section, is a state function—the ΔE of a reaction is the same regardless of how that reaction is carried out. On the other hand, the values of q and w are not constant; they vary with the process used in much the same way as the total energy expended in climbing a hill is dependent upon the path followed. The change in energy, ΔE, associated with the transformation of liquid water to water vapor, for example, is always the same and is a state function. However, q and w associated with the change will depend upon the way in which the transformation is executed, and thus they are not state functions. Quantities which are state functions are designated by capital letters, as in the now familiar expression $E_2 - E_1 = q - w$.

At this point, we can conclude that a state function is characteristic of a state of matter regardless of how that state is reached.

20.4   Energy Changes in Chemical Reactions

We are now ready to apply the concept of the internal energy, a state function, to chemical reactions. Let us take as our system some chemical elements in a particular combination (state 1) and these same elements in identical quantities but in a different combination (state 2), considering all other conditions (temperature and pressure, for example) to remain constant. Then, regardless of how we get from state 1 to state 2, the internal energy change, ΔE, must be the same.

In developing the usefulness of this principle in chemistry it is necessary to consider a particular kind of work energy that accompanies many chemical reactions carried out in open beakers at atmospheric pressure on the bench top.

It should be apparent from consideration of systems such as the internal combustion engine that gases expanding against a restraining pressure are capable of doing work. This means that if we have a system which is initially in state E_1 with volume V_1 and which goes to a new energy state E_2 with a larger volume V_2, all at a constant pressure, the system does a certain amount of expansion work on the surroundings. The amount of work done is equal to $P(V_2 - V_1)$, provided the pressure, P, remains constant. Very commonly, a reaction vessel is open to the atmosphere; the pressure in such cases is the atmospheric pressure and is essentially constant.

It should be noted that a system might contract instead of expand. In this case, work is being done on the system by the surroundings, instead of being done on the surroundings by the system. Either way, however, the change in volume is often referred to in thermodynamics as "expansion," in the sense that an increase in volume is a positive expansion, and a decrease in volume is a negative expansion.

The work term $P(V_2 - V_1)$ is an energy term with energy units. This is readily apparent if it is recalled that pressure is defined as force divided by area and therefore has units of force/(distance)^2; volume has units of (distance)^3. Thus, the units for the product are

$$P(V_2 - V_1) = \frac{\text{force}}{(\text{distance})^2} \times (\text{distance})^3$$

$$= \text{force} \times \text{distance}$$

Units of force $\times$ distance are units of work, or energy, since work can be defined as a force operating over a distance.

20.5 Enthalpy Change

Expansion work at constant pressure is an important concept in chemistry, because many chemical reactions when conducted in open containers at atmospheric pressure undergo significant changes in volume. In order to take into account the PV type of work which is done in many chemical reactions at constant pressure, we add the term $P(V_2 - V_1)$, or $P\,\Delta V$, to the internal energy change, ΔE. The sum $\Delta E + P\,\Delta V$ is a value for the total amount of heat energy that a chemical reaction can provide to the surroundings. We designate this quantity as ΔH, where H is the **heat content,** or the **enthalpy,** of the system; ΔH, therefore, is the **change in heat content,** or the **change in enthalpy.** Thus, $\Delta H = \Delta E + P\,\Delta V$ for constant pressure processes. Since E, P, and V are state functions, H also is a state function. Note that if only expansion work is involved in going from state 1 to state 2 at constant pressure,

$$w = P\,\Delta V$$

Since $\Delta E = q - w$, then

$$\Delta E = q - P\,\Delta V$$

We know that $\Delta H = \Delta E + P\,\Delta V$; therefore,

$$\Delta H = (q - P\,\Delta V) + P\,\Delta V = q$$

An enthalpy change thus tells us how much heat energy is absorbed or released by a chemical system undergoing a change at constant pressure. **The change in enthalpy, ΔH, is defined as the quantity of heat absorbed when a reaction takes place at constant pressure.** If volume changes are negligible, then

$$P\,\Delta V = 0 \quad \text{and} \quad \Delta H = \Delta E$$

If ΔH for a chemical reaction conducted at constant pressure (state 1 = reactants, state 2 = products) is negative, the system *evolves* heat energy to the surroundings, and the reaction is said to be **exothermic.** If ΔH for a chemical reaction conducted at constant pressure is positive, heat energy is *absorbed* by the system from the surroundings, and the reaction is said to be **endothermic.**

Enthalpy changes for a large number of chemical reactions have been measured and tabulated to simplify the comparison of various reactions. The National Bureau of Standards has been active in the compilation of accurate thermochemical data of this type. To facilitate the tabulation of such data, values have been measured and tabulated for many substances under a specific set of conditions referred to as the **standard state.** The standard state refers to each pure substance as being at 25° C (298.15° K) and at 1 atmosphere pressure. The state of the substance (i.e., gas, liquid, or solid) must be specified. It is customary to use the symbol ΔH°_{298} to indicate the enthalpy changes which have been measured under standard state conditions. For a particular type of chemical reaction in which *one mole* of a pure substance is formed from the free elements in their most stable states under standard state conditions, the enthalpy change is referred to as the **standard molar enthalpy of formation** of the substance formed and is designated by $\Delta H°_{f_{298}}$. In the following discussion, 298° K will be implied unless indicated otherwise. Standard molar enthalpies of formation of some common substances are given in Table 20–1. A more extensive compilation is given in Appendix J. By making use of these tabulated values and the fact that the enthalpy is a state function, we may obtain the enthalpy changes for a large number of chemical reactions. Several other properties of chemical interest to chemists can be deduced as well from the enthalpy changes.

TABLE 20-1   Standard Molar Enthalpies of Formation, Standard Molar Free Energies of Formation, and Absolute Standard Molar Entropies [298.15° K, 1 atm]. (See also Appendix J for additional values.)

Substance	$\Delta H°_{f_{298}}$, kcal/mole	$\Delta G°_{f_{298}}$, kcal/mole	$S°_{298}$, cal/mole · °K
Carbon			
C(s) (graphite)	0	0	1.372
C(g)	171.29	160.44	37.760
CO(g)	−26.42	−32.78	47.219
$CO_2(g)$	−94.05	−94.25	51.06
$CH_4(g)$	−17.88	−12.13	44.492
Chlorine			
$Cl_2(g)$	0	0	53.288
Cl(g)	29.08	25.26	39.457
Copper			
Cu(s)	0	0	7.923
CuS(s)	−12.7	−12.8	15.9
Hydrogen			
$H_2(g)$	0	0	31.208
H(g)	52.10	48.58	27.391
$H_2O(g)$	−57.80	−54.63	45.104
$H_2O(l)$	−68.32	−56.69	16.71
HCl(g)	−22.06	−22.78	44.646
$H_2S(g)$	−4.93	−8.02	49.16
Oxygen			
$O_2(g)$	0	0	49.003
O(g)	59.55	55.39	38.467
Silver			
$Ag_2O(s)$	−7.42	−2.68	29.0
$Ag_2S(s)$	−7.79	−9.72	34.42

20.6   Calculations Involving Enthalpy Change

Example 1.   What would be the standard enthalpy change for the reaction of one mole of $H_2(g)$ with one mole of $Cl_2(g)$ to produce two moles of $HCl(g)$ at standard state conditions?

$$H_2(g) + Cl_2(g) \longrightarrow 2HCl(g) \qquad \Delta H° = ?$$

Since $H_2(g)$ and $Cl_2(g)$ are the most stable states of the elements hydrogen and chlorine at standard state conditions and a pure substance is being formed at standard state conditions, $\Delta H°$ must be related to the standard molar enthalpy of formation of $HCl(g)$. In this example, two moles of $HCl(g)$ are being formed; therefore

$$\Delta H° = 2\,\Delta H°_{f_{HCl(g)}}$$

From Table 20-1, $\quad \Delta H°_{f_{HCl(g)}} = -22.06$ kcal/mole

Therefore, $\qquad \Delta H° = 2(-22.06) = -44.12$ kcal

Thus, the reaction evolves heat:

$$H_2(g) + Cl_2(g) \longrightarrow 2HCl(g) + 44.12 \text{ kcal}$$

The standard molar enthalpy of formation for any free element in its most stable form is zero. The reason for this can be seen by considering the following example:

Example 2.   What is the standard molar enthalpy of formation of $H_2(g)$?

We want to know the enthalpy change for the chemical reaction in which one mole of $H_2(g)$ is formed from elemental substances in their most stable forms at standard state conditions. The most stable form of hydrogen at standard state conditions is $H_2(g)$. Therefore we want to know the enthalpy change for the chemical reaction

$$H_2(g) \longrightarrow H_2(g)$$

Since we start with one mole of $H_2(g)$ under standard state conditions and end up with the same thing, states 1 and 2 are the same, and since the enthalpy is a state function, $\Delta H° = 0$ and $\Delta H°_{f_{H_2(g)}} = 0$. In other words, we are basically determining the enthalpy change in going from state 1 to state 2; but the states are identical and hence the enthalpy change has to be zero.

Additional examples of the use of enthalpy as a state function follow:

Example 3.   What is the $\Delta H°_{298}$ for the reaction

$$CH_4(g) + 2O_2(g) \longrightarrow CO_2(g) + 2H_2O(l) \qquad \Delta H°_{298} = ?$$

This reaction can be viewed as occurring in several steps:

Step 1. $CH_4(g) \longrightarrow C(s) + 2H_2(g) \qquad\qquad \Delta H°_1 = -\Delta H°_{f_{CH_4(g)}}$

Step 2. $2O_2(g) \longrightarrow 2O_2(g) \qquad\qquad\qquad \Delta H°_2 = 2\,\Delta H°_{f_{O_2(g)}} = 0$

Step 3. $2H_2(g) + O_2(g) \longrightarrow 2H_2O(l)$ $\quad \Delta H_3^\circ = 2\,\Delta H^\circ_{f_{H_2O(l)}}$

Step 4. $C(s) + O_2(g) \longrightarrow CO_2(g)$ $\quad \Delta H_4^\circ = \Delta H^\circ_{f_{CO_2(g)}}$

Adding steps 1, 2, 3, and 4 gives

$$CH_4(g) + \overset{2}{\cancel{4}O_2(g)} + \cancel{2H_2(g)} + \cancel{C(s)} \longrightarrow \cancel{C(s)} + 2\cancel{O_2(g)}$$
$$+ \cancel{2H_2(g)} + 2H_2O(l) + CO_2(g)$$

The net result is $CH_4(g) + 2O_2(g) \longrightarrow 2H_2O(l) + CO_2(g)$ $\Delta H^\circ = ?$ In either case, i.e., by direct reaction or by the combination of reactions in Steps 1 through 4, we end up with the same result. The enthalpy change is independent of how we get from state 1 to state 2, verifying again that H is a state function. Therefore,

$$\Delta H^\circ_{298} = \Delta H_1^\circ + \Delta H_2^\circ + \Delta H_3^\circ + \Delta H_4^\circ$$

or $\quad \Delta H^\circ_{298} = -\Delta H^\circ_{f_{CH_4(g)}} + 2\,\Delta H^\circ_{f_{O_2(g)}} + 2\,\Delta H^\circ_{f_{H_2O(l)}} + \Delta H^\circ_{f_{CO_2(g)}}$

$$= -(-17.9) + 2(0) + 2(-68.3) + (-94.1)$$
$$= -212.8 \text{ kcal}$$

Hence, 212.8 kcal of heat are evolved during the combustion of one mole of methane:

$$CH_4(g) + 2O_2(g) \longrightarrow CO_2(g) + 2H_2O(l) + 212.8 \text{ kcal}$$

The additivity process utilized in the preceding solution is an example of **Hess's law: For any process which can be looked upon as being the sum of several step-wise processes, the enthalpy change for the total process must be equal to the sum of the enthalpy changes for the various steps.** A more useful statement pertinent to the above type of calculation is the following: *For any chemical reaction occurring at standard state conditions, the standard enthalpy change is equal to the sum of the standard molar enthalpies of formation of all the products, each multiplied by their coefficients in the balanced chemical reaction, minus the corresponding sum for the reactants.*

$-230.7 + 17.9$
-212.8

Example 4. Calculate ΔH°_{298} for the reaction

$$2Ag_2S(s) + 2H_2O(l) \longrightarrow 4\overset{0}{Ag}(s) + 2H_2S(g) + \overset{0}{O_2}(g)$$

The standard molar enthalpies of formation, $\Delta H^\circ_{f_{298}}$, of the compounds involved are as follows: $Ag_2S(s)$, -7.8 kcal/mole; $H_2O(l)$, -68.3 kcal/mole; $H_2S(g)$, -4.9 kcal/mole.

$$\Delta H^\circ_{298} = \overset{product}{2(-4.9 \text{ kcal})} - \overset{reactants}{[2(-7.8 \text{ kcal}) + 2(-68.3 \text{ kcal})]}$$
$$= (-9.8 \text{ kcal}) - (-152.2 \text{ kcal}) = +142.4 \text{ kcal}$$

Hence, the reaction *absorbs* heat:

$$2Ag_2S(s) + 2H_2O(l) + 142.4 \text{ kcal} \longrightarrow 4Ag(s) + 2H_2S(g) + O_2(g)$$

Example 5.   Calculate ΔH_{298}° for the reaction

$$2Na(s) + 2H_2O(l) \longrightarrow 2NaOH(s) + H_2(g)$$

The standard molar enthalpies of formation for the compounds involved are as follows: $H_2O(l)$, -68.3 kcal/mole; $NaOH(s)$, -102.0 kcal/mole.

$$\Delta H_{298}^{\circ} = 2(\overset{\text{product}}{-102.0 \text{ kcal}}) - 2(\overset{\text{reactant}}{-68.3 \text{ kcal}})$$
$$= -204.0 \text{ kcal} + 136.6 \text{ kcal} = -67.4 \text{ kcal}$$

Hence, the reaction *evolves* heat: exothermic

$$2Na(s) + 2H_2O(l) \longrightarrow 2NaOH(s) + H_2(g) + 67.4 \text{ kcal}$$

A positive value of ΔH means that the sum of the enthalpies of the products, taking into account sign as well as numerical value, is greater than the sum of the enthalpies of the reactants; a negative value of ΔH means that the sum of the enthalpies of the products is less than that of the reactants.

In terms of bond strengths within the molecules, a positive value of ΔH indicates that the bonds are stronger in the reactants than in the products, and hence heat must be absorbed for the reaction to take place. A negative value of ΔH, on the other hand, indicates that the bonds are stronger in the products than in the reactants; hence, less energy is required to break the bonds in the reactants than is evolved in the formation of the products. The net effect is evolution of heat.

20.7   Bond Energies

We have already used the concept of bond energy several times (see Sections 5.14 and 18.6 in particular). The strength of a chemical bond is measured by the energy required to break the bond, that is, to separate the atoms from their molecular grouping and leave them as distinct isolated gaseous atoms. For a diatomic molecule the bond energy, D, is essentially equal to the change in enthalpy for the reaction

$$XY(g) \longrightarrow X(g) + Y(g) \qquad \Delta H = D$$

When only one bond between Atoms X and Y is involved, the bond energy is merely the dissociation energy of the molecule, which may be determined by a variety of spectroscopic or thermochemical methods. For diatomic molecules containing atoms of oxidation number greater than one (for example, nitrogen or oxygen), the dissociation energy, and hence the bond energy, may equally well be determined, but it corresponds to the rupture of a multiple bond between the atoms. Bond energies for commonly occurring diatomic molecules range from 226 kcal per mole for N_2 (triple bond) to 36 kcal per mole for I_2 (single bond), and from 136 kcal per mole for HF to 71 kcal per mole for HI.

Molecules of three or more atoms necessarily have two or more bonds. The heats of formation of such molecules from the isolated gaseous atoms is equal to the sum of all the bond energies; thus, the heat of formation of the S_8 ring from eight sulfur atoms is eight times the energy of formation of a single S—S bond.

Single bond energies for some common bonds appear in Table 20–2.

TABLE 20-2   Some Representative Single Bond Energies (kcal per mole of bonds)

H	C	N	O	F	Si	P	S	Cl	Br	I	
104	99	93	111	136	70	76	81	103	88	71	H
	83	70	84	105	69	63	62	79	66	57	C
		38	48	65	—	50	—	48	58 (?)	—	N
			33	44	88	84	—	49	—	48	O
				38	129	117	68	61	47 (?)	—	F
					42	51	54	86	69	51	Si
						51	55	79	65	51	P
							51	60	51	—	S
								58	52	50	Cl
									46	43	Br
										36	I

The energies for double or triple bonds between two atoms are generally higher than those for single bonds, as we would expect; but a double bond is not, in general, quite twice as strong as a comparable single bond nor a triple bond quite three times as strong as a comparable single bond. Compare the following bond energies.

C=C 146 kcal   C=N 147 kcal   C=O 177 kcal   N=N 100 kcal
C≡C 200 kcal   C≡N 213 kcal   C≡O 256 kcal   N≡N 226 kcal

Enthalpy changes can be used to obtain chemical bond energies. Returning to the example in Section 18.6 for HCl, we can write two series of steps by which two moles of HCl result from one mole of H_2 and one mole of Cl_2.

$$H_2(g) + Cl_2(g) \xrightarrow{2\,\Delta H^\circ_{f_{HCl(g)}}} 2HCl(g)$$

$$\downarrow 2\,\Delta H^\circ_{f_{H(g)}} \qquad \downarrow 2\,\Delta H^\circ_{f_{Cl(g)}} \qquad \Delta H^\circ$$

$$2H(g) + 2Cl(g)$$

$$2\,\Delta H^\circ_{f_{HCl(g)}} = 2\,\Delta H^\circ_{f_{H(g)}} + 2\,\Delta H^\circ_{f_{Cl(g)}} + \Delta H^\circ$$

·44.2 -[

ΔH° is the heat energy evolved in the formation of *two* moles of hydrogen chloride starting from *atomic* hydrogen and *atomic* chlorine at standard state conditions. The corresponding heat energy evolved in the formation of *one* mole of hydrogen chloride from atomic hydrogen and atomic chlorine, therefore, equals $\frac{1}{2}\Delta H^\circ$.

This is a quantity of energy very closely related to the bond energy of hydrogen chloride. The bond energy of hydrogen chloride is the amount of energy *input* required for the reverse process—that is, to break the bond in HCl to produce atomic hydrogen and atomic chlorine. It is apparent then that the bond energy in the present example is $-\frac{1}{2}\Delta H^\circ$. Calculating the value from the data in Table 18-1, we obtain:

$$2\,\Delta H^\circ_{f_{HCl(g)}} = 2\,\Delta H^\circ_{f_{H(g)}} + 2\,\Delta H^\circ_{f_{Cl(g)}} + \Delta H^\circ$$
$$2(-22.06) = 2(52.10) + 2(29.08) + \Delta H^\circ$$
$$\Delta H^\circ = -44.12 - 104.20 - 58.16 = -206.48 \text{ kcal}$$
$$\tfrac{1}{2}\Delta H^\circ = -103.24 \text{ kcal/mole}$$

$$HCl \longrightarrow H^o_{(g)} + Cl^o_{(g)}$$

Thus, the bond energy for HCl (which is $-\frac{1}{2}\Delta H°) = -(-103.2)$ kcal/mole $= +103.2$ kcal/mole.

Note that this is the value, rounded to three significant figures, which is given in Table 20–2 for the H—Cl bond.

In addition, from the preceding example, we can determine the bond energies for H_2 and Cl_2. Referring to Table 20–1, we find that $\Delta H°_{f_{H(g)}} = 52.10$ kcal for one mole of H. Recall that a $\Delta H°_f$ value signifies the enthalpy change for the chemical reaction in which one mole of the pure substance (in this case, atomic hydrogen, H) is formed from the free elements *in their most stable states* under standard state conditions (in this case, the one element, the gaseous diatomic molecule H_2). To form atomic hydrogen, H, from H_2 requires an input of energy to *break* the H—H bond. Note that the $\Delta H°_{f_{H(g)}}$ value of 52.10 kcal is for breaking the bonds in one-half mole of H_2 to produce one mole of H, and no sign change is required in determining the bond energy. To form two moles of H from one mole of H_2 would then require $2(52.10)$ kcal $= 104.20$ kcal. This, therefore, is the bond energy for $H_2(g)$, and it checks with the rounded-off value listed in Table 20–2. Similarly, $\Delta H°_{f_{Cl(g)}} = 29.08$ kcal for one mole of Cl. To form two moles of Cl from one mole of Cl_2 would require $2(29.08)$ kcal $= 58.16$ kcal. Thus, the bond energy for $Cl_2(g)$ is 58.16 kcal, which is also the rounded-off value listed in Table 20–2.

20.8 Bond Energies and Electronegativity

If we can assume that the energy of a given single bond arises from two contributions, one from each atom, then we might expect that the bond energies would be additive. For example, the bond energy of the H—F bond should be equal to one-half the sum of the H—H and F—F bond energies. Now it is true that the H—F bond energy is intermediate between those for the H—H and F—F bonds, but it is greater than would be predicted if the bond energies were additive. In general, bond energies are not additive; the energy of a single bond between unlike atoms is usually greater than the average of the bond energies for the separate pairs of like atoms. The excess bond energy increases as the difference in electronegativity (see Section 4.6 and Table 4–2) between the bonded atoms increases. In the bonds between hydrogen and the halogen atoms the excess bonding energy increases in the order HI<HBr<HCl<HF, with increasing difference in electronegativity; the excess energy values are 1, 13, 22, and 65 kcal/mole, respectively. Thus, the rather qualitative concept of electronegative character of the elements is related to the quantitative concept of bond energies, and the latter therefore provides one method of determining electronegativity quantitatively.

Electronegativity values may be assigned to the elements such that the square of their difference for a given pair of atoms is directly proportional to the excess bond energy for the single bonds between the atoms. The scale of electronegativities is chosen so that the energy is given in electron-volts, and, because only differences in electronegativities are involved, the arbitrary zero point is selected to make the electronegativities of the second period elements carbon to fluorine have values 2.5 to 4.0. Thus, the dissociation energy, or bond energy, $[D(A—B)]$ of the single bond between Atoms A and B may be expressed in kcal/mole in the form

$$D(A—B) = \tfrac{1}{2}[D(A—A) + D(B—B)] + 23.06(X_A - X_B)^2$$

where X represents the electronegativity value and 23.06 is the conversion factor from electron-volts to kilocalories. The relation holds well except when the electronegativity difference is large, in which case more complex mathematical relationships are required.

The method just described is one of several utilized for determining electronegativities.

20.9   Enthalpy Change and Reaction Spontaneity

In many cases (though not all, for reasons that will be pointed out in Sections 20.10 and 20.11), the enthalpy change provides a means of predicting whether a given reaction will take place spontaneously under a particular set of conditions. In general, if for 25° C and 1 atm $\Delta H°$ is negative (exothermic reaction), the reaction often will occur spontaneously at constant temperature and pressure; if $\Delta H°$ is positive (endothermic reaction), the reaction usually will not take place spontaneously. Thus, for the reaction used in Example 4 of Section 20.6, $2Ag_2S(s) + 2H_2O(l) + 142.4\,kcal \longrightarrow 4Ag(s) + 2H_2S(g) + O_2(g)$, the forward reaction, on the basis of $\Delta H°$, would be expected *not* to occur spontaneously (i.e., without external energy introduced) at 25° and 1 atm; however, the reverse reaction *would* be expected to occur spontaneously at 25° and 1 atm. For the reaction of Example 5 of Section 20.6, $2Na(s) + 2H_2O(l) \longrightarrow 2NaOH(s) + H_2(g) + 67.4\,kcal$, the forward reaction would be expected, on the basis of $\Delta H°$ alone, to occur spontaneously at 25° and 1 atm, but the reverse reaction should occur only if external energy is introduced. The predictions for these reactions are in accord with experimental evidence. However, such is not always true, and predictions based upon $\Delta H°$ alone are not always valid.

A reaction which proceeds spontaneously at one temperature and pressure very often does not proceed spontaneously at another temperature and pressure. For example, at 25° and 1 atmosphere pressure, the reaction $2Ag(s) + \tfrac{1}{2}O_2(g) \longrightarrow Ag_2O(s)$ proceeds spontaneously. The $\Delta H°$ value is negative, as we would expect for spontaneous reaction. However, although the $\Delta H°$ value for the reaction changes very little with the change in temperature, at temperatures above about 200° the reaction is not spontaneous. Another example is in the physical change which occurs when a solid crystal melts. At temperatures above the melting temperature the solid melts spontaneously; below the melting temperature it does not. Yet the ΔH value changes very little. The tendency for chemical reactions and physical changes to proceed spontaneously is usually quite dependent upon temperature and pressure changes. Yet temperature and pressure changes often have virtually no effect upon the value of ΔH for a reaction.

It is evident, then, that factors in addition to the energy change occurring when a reaction takes place must be known in determining whether a given reaction will proceed spontaneously. The change in a quantity referred to as the **free energy** provides a better basis for prediction.

20.10 Free Energy Change

Much of the energy involved in a reaction can be made to do useful work, but ordinarily some of the energy is not available for useful work. J. Willard Gibbs, a famous American professor of mathematics, proposed that a reaction will take place spontaneously at constant temperature and pressure only if it is capable of doing useful work; conversely, the reaction will not take place spontaneously at constant temperature and pressure if it is not capable of doing useful work. **The maximum amount of useful work that can be accomplished by a reaction at constant temperature and pressure is referred to as the Gibbs free energy change, ΔG,** for that reaction. If ΔG is negative, the reaction occurs spontaneously at constant temperature and pressure. If ΔG is positive, the reaction does not occur spontaneously at constant temperature and pressure; in such a case the reverse reaction is spontaneous. **For the system to be at equilibrium, ΔG must equal zero.**

It should be pointed out that G is a state function. Hence, values of $\Delta G°$ can be calculated from standard molar free energies of formation, $\Delta G_f°$, in a manner analogous to that used for calculating $\Delta H°$ values from standard molar enthalpies of formation. Several standard molar free energy of formation values are given in Table 20–1 and a more extensive set of values in Appendix J. The standard molar free energy of formation of any free element is zero for the same reasons as for $\Delta H_f°$. As with ΔH values, many ΔG values have been tabulated, and to facilitate comparison the tabulations are made for standard state conditions using the symbol $\Delta G°$.

Example. **Calculate the free energy change, $\Delta G°$, for the reaction**

$$2Ag_2S(s) + 2H_2O(l) \longrightarrow 4Ag(s) + 2H_2S(g) + O_2(g)$$

The standard molar free energies of formation of the compounds involved are as follows at 25° and 1 atm: $Ag_2S(s)$, -9.7 kcal/mole; $H_2O(l)$, -56.7 kcal/mole; $H_2S(g)$, -8.0 kcal/mole.

$$\Delta G° = [2(-8.0 \text{ kcal})] - [2(-9.7 \text{ kcal}) + 2(-56.7 \text{ kcal})]$$
$$= (-16.0 \text{ kcal}) - (-132.8 \text{ kcal}) = +116.8 \text{ kcal}$$

The positive value of $\Delta G°$ indicates that the forward reaction should not occur spontaneously, but that the reverse reaction should, at 25° and 1 atmosphere.

Whereas ΔH for a reaction usually does not change appreciably with changes in temperature and pressure, ΔG changes quite a lot with changes in temperature and pressure. Neither ΔH nor ΔG, however, is affected by the path by which the reaction takes place.

The relationship of ΔG to standard reduction potentials is discussed later, in Section 22.15.

Intuitively, one might expect that the overall energy change, ΔH, during a reaction at constant temperature and pressure, must be related in some way to the maximum useful work, ΔG, that can be done by a reaction at constant temperature and pressure. Such a relationship does exist and is considered in the following section.

20.11 Entropy Change and the Relationship between Enthalpy Change and Free Energy Change

For the reaction $2Ag_2S(s) + 2H_2O(l) \longrightarrow 4Ag(s) + 2H_2S(g) + O_2(g)$, we have calculated that $\Delta H° = +142.4$ kcal, whereas $\Delta G° = +116.8$ kcal (Sections 20.6 and 20.10). The difference, $\Delta G° - \Delta H°$, is 116.8 kcal $- 142.4$ kcal $= -25.6$ kcal; or $\Delta G° = \Delta H° - 25.6$ kcal.

The quantitative difference between $\Delta G°$ and $\Delta H°$, which is negative for the above reaction, can in other cases be either negative or positive, depending upon the reaction involved. It can be shown that the difference is almost proportional to the absolute temperature and related to the "disorder," or "randomness," of the systems involved. **The randomness, or the amount of disorder, of a system can be measured quantitatively and is referred to as the entropy, S, of the system.** S, like E, H, and G, is a state function. The greater the randomness, or disorder, in a system the higher its entropy. Entropy is one of the most important concepts in science. As we shall learn in more detail later in the Section, an entropy increase, corresponding to an increase in disorder, is the major driving force in many chemical and physical processes.

Each substance has an entropy as one of its characteristic properties, just as it has color, hardness, volume, melting point, density, enthalpy, and free energy as characteristic properties. Entropy values for several substances are listed in Table 20–1 and many more in Appendix J. Note that solids (well ordered) tend to have lower entropies than liquids (less well ordered), and liquids tend to have lower entropies than gases (still less ordered). In the same physical state, substances with simple molecules tend to have lower entropies than substances with more complicated molecules. The latter have larger numbers of atoms to move about, can hence create greater randomness or disorder, and have higher entropies. Hard substances, such as diamond, tend to be more ordered and have lower entropies than softer materials, such as graphite or sodium.

The entropy change for a chemical reaction, ΔS, represents the sum of the entropies of the products of that reaction minus the sum of the entropies of the reactants. A positive value for ΔS corresponds to an increase in randomness, or disorder; a negative value for ΔS corresponds to a decrease in randomness, or a more ordered structure. For example, when a solid melts the molecules become more mobile and freer to move about in a random manner than in the solid state. This change to a more random and less ordered system when a solid melts corresponds to an increase in entropy. Similarly, evaporation is also accompanied by an increase in entropy as the molecules go to the still more disordered and random state characteristic of gases.

The amount of change in the entropy value can be thought of as a measure of the tendency of a chemical reaction or a physical change to increase the disorder, or randomness, of a system. **In nature, chemical and physical changes tend to proceed in such a way as to increase the disorder (increase in entropy)** (see Section 20.1). This is related to the fact that random molecular motion tends to produce disorder.

Putting the idea of randomness together with that of energy discussed in connection with enthalpy, we can say that **reactions tend to proceed in such a way as to obtain a state of minimum energy (negative ΔH) and maximum disorder (positive ΔS).** Thus, we come back to the two fundamental laws discussed in a more general way in Section 20.1.

The general relationship between ΔG and ΔH at constant temperature can be shown to be

$$\Delta G = \Delta H - T \, \Delta S$$

where T is the Kelvin temperature. This relationship is highly useful in predicting the tendency of a reaction to proceed spontaneously. It is one of the most important relationships in chemical thermodynamics. For the Ag_2S reaction used in previous examples we noted that

$$\Delta G^\circ = \Delta H^\circ - 25.6 \text{ kcal}$$

Hence, for this reaction, $T \Delta S^\circ = 25.6$ kcal/mole. Inasmuch as $T = 298^\circ$ K,

$$(298^\circ \text{ K}) \, \Delta S^\circ = 25.6 \text{ kcal/mole}$$

$$\Delta S^\circ = \frac{25.6}{298} \text{ kcal/mole} \cdot {}^\circ \text{K}$$

$$= 0.0859 \text{ kcal/mole} \cdot {}^\circ \text{K}$$

Again considering the equation $\Delta G = \Delta H - T \Delta S$, the term $T \Delta S$ can be thought of in one sense as a correction term for ΔH to determine ΔG. If $T \Delta S$ is small compared to ΔH, then ΔG and ΔH have nearly the same value and serve equally well as a basis for prediction of the tendency of a given reaction to proceed spontaneously. However, when the value of $T \Delta S$ is appreciable compared to that of ΔH, the ΔH value is not an accurate indicator of spontaneity and the ΔG value must be used.

Another and perhaps more fundamental way of thinking of the prediction of spontaneity of a reaction is to think in terms of the relative importance of ΔH and $T \Delta S$ as components of ΔG.

It is apparent from the foregoing discussion that in order to determine whether a given reaction will proceed spontaneously under a particular set of conditions, we must consider both the ΔH value and the ΔS value. The ΔH portion of the free energy equation $\Delta G = \Delta H - T \Delta S$ represents the energy, as heat, required for a reaction to take place at constant pressure; hence, it corresponds to a difference in energy between the initial and final states in a process and provides information concerning the relative strengths of bonds in the products and in the reactants. The ΔS portion of the free energy equation (and hence $T \Delta S$), on the other hand, corresponds to the difference in the probabilities for the initial and final states to occur and hence provides information concerning the ordering of the atoms in the products and reactants.

For processes in which the initial and final states do not differ in enthalpy (ΔH is zero) and ΔS is not zero, the probability term $T\Delta S$ controls whether the reaction occurs. For processes, on the other hand, where the probabilities of the initial and final states are the same (ΔS, and hence $T\Delta S$, is zero) and ΔH is not zero, the energy term ΔH is considered in predicting whether the reaction will occur. If ΔH and $T\Delta S$ each differ from zero, the algebraic sum of the two (that is, the algebraic sum of the energy and the probability terms) must be considered in predicting whether the reaction will occur. Inasmuch as ΔG is the algebraic sum of ΔH and $-T\Delta S$, it is therefore the key quantity in any of the above cases for predicting whether a process will occur spontaneously.

The equation $\Delta G = \Delta H - T\Delta S$ tells us that a reaction which is exothermic (negative ΔH) and produces simpler and more numerous molecules (more disorder, positive ΔS) will proceed spontaneously (both ΔH and ΔS work toward a more negative value for ΔG). A reaction which is endothermic (positive ΔH) and produces a more ordered system (negative ΔS) will tend not to proceed spontaneously (both ΔH and ΔS work toward a more positive value for ΔG).

If the reaction is exothermic (negative ΔH) but produces a more ordered system (negative ΔS), or if the reaction is endothermic (positive ΔH) but produces a less ordered system (positive ΔS), then ΔH and ΔS are working in opposite directions and the relative sizes of ΔH and $T\Delta S$ will determine the prediction.

It should be noted that T must always be positive. Hence, ΔS (which can be either positive or negative) determines the sign of the correction term $T\Delta S$. Also, it should be noted that as T increases, T plays an increasingly important role in the value of ΔG.

As pointed out in Section 20.10, equilibrium occurs only if $\Delta G = 0$. Thus, at equilibrium the energy term (ΔH) and the probability term ($T\Delta S$) balance, and there is no net tendency for either the forward or the reverse process to predominate.

Finally, it should be emphasized that a prediction as to whether a given reaction will proceed spontaneously says *nothing whatever about the rate at which it will occur*. A reaction with a negative ΔG value should proceed spontaneously, but it may do so at a very rapid rate or at an infinitesimally slow rate.

20.12   The Second Law of Thermodynamics

The Second Law of Thermodynamics states that in any spontaneous change the entropy of the universe increases.

It is difficult to use the Second Law in its pure form. However, a useful modification can be developed from the free energy equation studied earlier.

$$\Delta G = \Delta H - T\Delta S \quad \text{(all values pertaining to the system)}$$

Dividing by T gives

$$\frac{\Delta G}{T} = \frac{\Delta H}{T} - \Delta S \quad \text{(for the system)}$$

It can be shown that

$$\Delta S_{\text{sur}} = -\frac{\Delta H_{\text{sys}}}{T} \quad \text{(at constant temperature and pressure)}$$

where ΔS_{sur} is the change in entropy for the surroundings, and ΔH_{sys} is the change in enthalpy for the system.

Thus,

$$\frac{\Delta G_{\text{sys}}}{T} = -\Delta S_{\text{sur}} - \Delta S_{\text{sys}}$$

or

$$-\frac{\Delta G_{\text{sys}}}{T} = \Delta S_{\text{sur}} + \Delta S_{\text{sys}}$$

$$\Delta S_{\text{universe}} = \Delta S_{\text{sur}} + \Delta S_{\text{sys}}$$

Hence,

$$-\frac{\Delta G_{\text{sys}}}{T} = \Delta S_{\text{universe}}$$

This derived relationship is valid only at constant temperature and pressure, because the relationships from which it is derived are valid only at constant temperature and pressure.

From the derived relationship it can be seen that as the entropy of the universe increases (that is, $\Delta S_{\text{universe}}$ is positive), ΔG_{sys} must be negative, corresponding to a spontaneous change. Thus, it can be said that **for a spontaneous change at constant temperature and pressure, the entropy of the universe must increase and the free energy change of the system must be negative; that is, the system must be capable of doing useful work on the surroundings.**

20.13 Relationship between Free Energy Change and Equilibrium Constants

Since changes in free energy at constant temperature and pressure tell us something about reaction spontaneity and, in fact, tell us when a state of equilibrium has been reached, it should not be surprising that a relationship exists between free energy changes and equilibrium constants. It can be shown that the following equation is valid:

$$\Delta G^\circ = -RT \ln K_{\text{e}} \tag{1}$$

where R is the universal gas constant (Sections 10.21 and 10.22), T is the Kelvin temperature, K_{e} is the equilibrium constant, and ln is the *natural* logarithm (to the base e). A natural logarithm is equal to 2.303 times the corresponding *common* logarithm (to the base *10*).

Hence, Equation (1) in terms of common logarithms, where log signifies the common logarithm, becomes

$$\Delta G^\circ = -2.303 \, RT \log K_{\text{e}} \tag{2}$$

If the system is at equilibrium, K_{e} is equal to 1. A calculation of ΔG° when $K_{\text{e}} = 1$ shows that $\Delta G^\circ = 0$, in accord with our statement in Section 20.10 that for a system to be at equilibrium ΔG must equal zero. It is also useful to note that if K_e has a value larger than 1, the calculated value of $\Delta G°$ is negative (spontaneous reaction at constant temperature and pressure); and if K_e has a value less than 1, the calculated value of $\Delta G°$ is positive (reaction not spontaneous at constant temperature and pressure).

Note that for equation (2) to be valid, K_e must be a dimensionless number. This is because K_e is actually the ratio of the concentration quotient at equilibrium to the concentration quotient at standard state concentrations. The concentration quotient at standard state conditions is always equal to 1 (gases at 1 atm; solids and liquids at unit activity) but has the same units as the concentration quotient at equilibrium. Therefore, K_e is dimensionless (units cancel out) and numerically equal to the equilibrium constant.

The concentration quotient has the general form of the equilibrium constant as defined in Section 17.10 for concentrations, Section 17.13 for gas pressures, and 17.16 for heterogeneous equilibria consisting of gases in equilibrium with solids and/or liquids.

Example 1.   **Using data in Appendix J, calculate the value of K_e at 298° K for the reaction**

$$H_2(g) + \tfrac{1}{2}O_2(g) \longrightarrow H_2O(g)$$
$$\Delta G° = -2.303 \, RT \log K_e$$

The value of $\Delta G°_{f_{298.15}}$ given in Appendix J for $H_2O(g)$ is -54.634 kcal/mole. Therefore,

$$-54{,}634 \text{ cal/mole} = (-2.303)(1.987 \text{ cal/mole} \cdot °K)(298° \text{ K}) \log K_e$$

$$\log K_e = \frac{-54{,}634 \text{ cal/mole}}{(-2.303)(1.987 \text{ cal/mole} \cdot °K)(298°K)} = 40.064$$

$$K_e = 1.16 \times 10^{40}$$

If we combine equation (2) with the equation relating $\Delta G°$, $\Delta H°$, and $\Delta S°$ we have

$$\Delta G° = \Delta H° - T\,\Delta S° = -2.303 \, RT \log K_e \qquad (3)$$

Relationship (3) is developed for conditions of constant temperature and pressure. However, $\Delta H°$ and $\Delta S°$ are *approximately* independent of temperature. If it is assumed that they are indeed independent of temperature, we can obtain a simple equation relating the equilibrium constant and temperature.

Consider an equilibrium reaction at two different temperatures, T_1 and T_2, with equilibrium constants $K_{e_{T_1}}$ and $K_{e_{T_2}}$:

$$\Delta H° - T_1 \, \Delta S° = -2.303 \, RT_1 \log K_{e_{T_1}} \qquad (4)$$

and
$$\Delta H° - T_2 \, \Delta S° = -2.303 \, RT_2 \log K_{e_{T_2}} \qquad (5)$$

Equations (4) and (5) may be rearranged to

$$\frac{\Delta H°}{T_1} - \Delta S° = -2.303 \, R \log K_{e_{T_1}} \qquad (6)$$

and

$$\frac{\Delta H^\circ}{T_2} - \Delta S^\circ = -2.303\, R \log K_{e_{T_2}} \tag{7}$$

by dividing Equation (4) through by T_1, and Equation (5) through by T_2. Subtracting Equation (7) from Equation (6) gives

$$\frac{\Delta H^\circ}{T_1} - \frac{\Delta H^\circ}{T_2} - \Delta S^\circ - (-\Delta S^\circ) = -2.303\, R \log K_{e_{T_1}} - (-2.303\, R \log K_{e_{T_2}})$$

or

$$\Delta H^\circ \left(\frac{1}{T_1} - \frac{1}{T_2}\right) = 2.303\, R \log \frac{K_{e_{T_2}}}{K_{e_{T_1}}}$$

or, by multiplying the left side by $T_1 T_2 / T_1 T_2$, and dividing both sides by 2.303 R,

$$\frac{\Delta H^\circ (T_2 - T_1)}{2.303\, RT_1 T_2} = \log \frac{K_{e_{T_2}}}{K_{e_{T_1}}} \tag{8}$$

Equations (2) through (8) prove to be quite useful, for if we know ΔG° for 298° K, we can by means of Equation (2) obtain the equilibrium constant, K_e, at 298° K. Then, if we know ΔH° for 298° K (which is essentially independent of temperature), we can obtain K_e by means of Equation (8) for any other temperature (within limitations of both ΔH° and ΔS° being independent of T).

It should be noted that K_e for temperatures other than 298° could also be calculated from Equation (2), *provided we have the ΔG° value for the corresponding temperature.* (Remember that ΔG° values change significantly with temperature, whereas ΔH° and ΔS° values do not.) Values for ΔG° at temperatures other than 298° K are less frequently available than at 298° K, providing a severe restriction on calculating K_e values for temperatures other than 298° K by use of ΔG° alone.

Example 2.   **For the reaction**

$$CuS(s) + H_2(g) \longrightarrow Cu(s) + H_2S(g)$$

(a) Calculate ΔG° and ΔH° at 298° K and 1 atm pressure.

$\Delta G^\circ = (-8.02) - (-12.8) = +4.78$ kcal (reaction is not spontaneous)
$\Delta H^\circ = (-4.93) - (-12.7) = 12.7 - 4.93 = 7.77$ kcal (reaction is endothermic)

(b) Calculate the value for the equilibrium constant, K_e, at 298° K and 1 atm pressure.

$$K_e = \frac{p_{H_2S}}{p_{H_2}} \text{ (where } p \text{ stands for partial pressure of a gas)}$$

$$\Delta G^\circ = -RT \ln K_e = -2.303\, RT \log K_e \ (R = 1.987 \text{ cal/mole} \cdot {}^\circ K)$$

Then,
$$\frac{4{,}780 \text{ cal/mole}}{-(2.303)(1.987 \text{ cal/mole} \cdot {}^\circ K)(298.15\, {}^\circ K)} = \log K_e$$

(The units cancel out. K_e for this reaction has no units.)

$$\log K_e = -3.50349$$
$$K_e = 3.14 \times 10^{-4} \qquad \text{(at 298° K)}$$

Note that at 298° K the equilibrium constant has a value less than 1, indicating that the elements in the reaction are present in larger quantity as reactants than as products, and that the equilibrium therefore lies far to the left.

(c) **Calculate the value for K_e at 798° K and 1 atm pressure.**

$$\frac{\Delta H°(T_2 - T_1)}{2.303 \, RT_1 T_2} = \log \frac{K_{e_{T_2}}}{K_{e_{T_1}}}$$

Then,
$$\frac{7{,}770 \text{ cal/mole } (798° \text{K} - 298° \text{K})}{(2.303)(1.987 \text{ cal/mole} \cdot °\text{K})(798° \text{K})(298° \text{K})} = \log \frac{K_{e_{798}}}{K_{e_{298}}}$$

$$3.57010 = \log \frac{K_{e_{798}}}{K_{e_{298}}}$$

$$\log K_{e_{798}} - \log K_{e_{298}} = 3.57010$$
$$\log K_{e_{298}} = -3.50349 \qquad \text{[calculated in Part (b)]}$$
$$\log K_{e_{798}} = 3.57010 + (-3.50349) = 0.06661$$
$$K_e = 1.17 \qquad \text{(at 798° K)}$$

Note that the equilibrium constant at 798° K is greater than 1, indicating the elements are present as products in greater quantity than as reactants and the equilibrium has been displaced to the right.

(d) **Calculate $\Delta S°$ at 298° K and 1 atm pressure.**

$$\Delta G° = \Delta H° - T \Delta S°$$

$$\Delta S° = \frac{\Delta H° - \Delta G°}{T} = \frac{7{,}770 \text{ cal/mole} - 4{,}780 \text{ cal/mole}}{298° \text{ K}}$$

$$= 10.03 \text{ cal/mole} \cdot °\text{K}$$

Note that this value for $\Delta S°$ is positive and indicates a spontaneous reaction, whereas the positive value of $\Delta H°$ indicates a nonspontaneous reaction. This is, therefore, a case where $\Delta S°$ is sufficiently large that $\Delta H°$ is not a valid indicator of spontaneity and $\Delta G°$ must be considered. In particular, as the temperature changes, the reaction will be expected to change with respect to spontaneity.

We have found in Part (a) that $\Delta G° = +4.78$ kcal at 298° K, indicating that at a temperature of 298° K the reaction is not spontaneous. Let us now calculate $\Delta G°$ at a higher temperature.

(e) **Calculate $\Delta G°$ at 798° K and 1 atm pressure.**

$$\Delta G° = \Delta H° - T \Delta S°$$
$$= 7{,}770 - (798)(10.03)$$
$$= -234 \text{ cal} = -0.234 \text{ kcal}$$

Hence, at the higher temperature $\Delta G°$ is negative, showing that at $798°$ K the reaction *is* spontaneous. $\Delta H°$, being relatively independent of temperature, still has a value of about $+7.70$ kcal at $798°$ K and hence does not indicate the change to spontaneity.

(f) **Calculate the temperature at which the standard $\Delta G°$ is zero at 1 atm pressure.**

$$\Delta G° = \Delta H° - T\,\Delta S° \qquad [\Delta S° = 10.03, \text{ as calculated in Part (d)}]$$
$$0 = 7,770 - T(10.03)$$
$$T = \frac{7,770}{10.03} = 775° \text{ K}$$

Hence, $\Delta G° = 0$ at $775°$ K. At temperatures below $775°$ K the values of $\Delta G°$ are positive; at temperatures above $775°$ K they are negative. Therefore, above $775°$ K the reaction is spontaneous; below $775°$ K the reaction as written is not spontaneous—indeed, it is spontaneous in the opposite direction.

Although not directly related to thermodynamics, the calculation of equilibrium partial pressures of the gases present, H_2S and H_2, under a specified set of conditions, is useful.

(g) **Calculate the partial pressures of $H_2S(g)$ and $H_2(g)$ at $298°$ K and $798°$ K if the initial pressures of the gases are zero and 2.00 atmospheres, respectively.**

At $298°$ K:

The total initial pressure is $0 + 2.00$ atm $= 2.00$ atm.

Let $x =$ equilibrium partial pressure of H_2S, p_{H_2S}. Then $(2.00 - x)$ $= p_{H_2(g)}$

$$K_{e_{298}} = 3.14 \times 10^{-4} \quad [\text{calculated in Part (b)}]$$
$$K_{e_{298}} = \frac{p_{H_2S}}{p_{H_2}} = 3.14 \times 10^{-4}$$
$$\frac{x}{2.00 - x} = 3.14 \times 10^{-4}$$
$$x = (2.00 - x)(3.14 \times 10^{-4})$$
$$x = (6.28 \times 10^{-4}) - (3.14 \times 10^{-4})x$$
$$x + (3.14 \times 10^{-4})x = 6.28 \times 10^{-4}$$
$$1.000314x = 6.28 \times 10^{-4}$$
$$x = 6.28 \times 10^{-4} \text{ atm} = p_{H_2S}$$
$$2.00 - x = 1.9994 \text{ atm (or 2.00 atm, to three significant figures)}$$
$$= p_{H_2}$$

At 798° K:

$$K_{e_{798}} = 1.17 \quad \text{[calculated in Part (c)]}$$

$$\frac{x}{2.00 - x} = 1.17$$

$$x = 2.34 - 1.17x$$

$$x + 1.17x = 2.34$$

$$2.17x = 2.34$$

$$x = 1.08 \text{ atm} = p_{H_2S}$$

$$2.00 - x = 0.92 \text{ atm} = p_{H_2}$$

Hence, the calculations in Part (g) verify the statements in Parts (b) and (c), based on the numerical values of K_e, that the greater quantity of elements present are in the form of reactants at 298° K and in the form of products at 798° and, also, that the equilibrium is shifted much more to the right at 798° than at 298°. This is in accord with van't Hoff's law and Le Châtelier's principle (see Sections 17.12 and 17.14).

Summing up the considerable amount of information we have gained from the calculations for this particular reaction (the reduction of CuS with H_2), we have found that:

(a) At 298° K, the reaction is endothermic (positive $\Delta H°$) and not spontaneous (positive $\Delta G°$).

(b) At 798° K, the reaction is still endothermic (positive $\Delta H°$) but spontaneous (negative $\Delta G°$).

(c) The change of entropy is positive at 298° K, and being relatively independent of temperature is positive also at 798° K, a condition favorable to a spontaneous reaction.

(d) At 298° K, the equilibrium constant, K_e, has a value of 3.14×10^{-4} (less than 1). This indicates that a larger quantity of the elements in the reaction are present as reactants than as products, and that the equilibrium therefore lies far to the left.

(e) At 798° K, $K_e = 1.17$ (greater than 1), indicating that the elements are present as products in larger quantity than as reactants and that, when the temperature was raised, the equilibrium was displaced to the right.

(f) If the initial partial pressures of $H_2S(g)$ and $H_2(g)$ are zero and 2.00 atm, respectively, then the equilibrium partial pressures are 6.28×10^{-4} atm and 2.00 atm at 298° K (indicating that the equilibrium lies far to the left). At 798° K, the equilibrium partial pressures are 1.078 atm and 0.92 atm, respectively (indicating a large shift in equilibrium toward the right with the increase in temperature, in accord with van't Hoff's law).

From Equation (8), it can be seen that if $\Delta H°$ is negative, K_e decreases as T increases. If $\Delta H°$ is positive, K_e increases as T increases. In other words, for an exothermic reaction (negative $\Delta H°$) the equilibrium constant decreases as the temperature increases; for an endothermic reaction (positive $\Delta H°$) the equilibrium constant increases as the temperature increases. This verifies what we learned in Section 17.14, where it was pointed out that for the exothermic reaction

$$N_2 + 3H_2 \rightleftharpoons 2NH_3 + 22,040 \text{ cal}$$

the equilibrium can be shifted to the right (higher equilibrium constant) to produce a better yield of ammonia by lowering the temperature. Conversely, the equilibrium would be shifted to the left (lower equilibrium constant) by raising the temperature. It should be recalled, however, that lowering the temperature to increase the yield of ammonia rapidly decreases the *rate* at which equilibrium is attained, an important factor that must be considered in deciding on the best temperature for the commercial production of ammonia but about which $\Delta H°$ tells us nothing whatever. We will consider the optimum conditions for the commercial production of ammonia in more detail in Chapter 25.

20.14 The Third Law of Thermodynamics

In all of the previous discussion we considered changes of state and the changes in the thermodynamic state functions that accompany such changes. So far, in this chapter, we have not tried to obtain the actual value for any of the state functions E, H, G, or S in a particular state. It is possible to obtain absolute values for the entropy of pure substances at any given temperature. The reason for this is embodied in the **Third Law of Thermodynamics, which states that the entropy of any pure, perfect crystalline substance at absolute zero ($0°$ K) is equal to zero.** This is not true, however, for E, H, or G. The zero value for S means that at absolute zero all molecular motion has stopped and for a pure crystalline substance there is no disorder. For an impure substance, all molecular motion has also stopped, but the impurity can be distributed in different ways, giving rise to disorder.

If we measure the entropy change for the process in which we take a pure crystalline substance from absolute zero to any temperature T, then we have

$$S_T - S_0 = \Delta S = \text{a measured quantity}$$

But since $S_0 = 0$,

$$\Delta S = S_T = \text{a measured quantity}$$

Hence, we obtain *absolute entropies* of the pure substances, in contrast to free energy and enthalpy for which differences between two values can be determined but normally not the absolute values. Absolute entropies allow us to compare the relative amounts of disorder present in different pure substances, and they can be used to determine entropy changes. Caution must be exercised when using absolute entropies, however, since the absolute entropy of pure elemental substances at *standard state conditions* will *not* be equal to zero. Table 20-1 and Appendix J contain absolute standard molar entropies, $S°$, of some common substances at standard state conditions. To illustrate their use consider the following examples.

Example 1. (a) Determine the entropy change for the change of liquid water to gaseous water at 298° K and 1 atmosphere pressure, and, on the basis of the value obtained, discuss the relative amount of disorder in the two states.

$$H_2O(l) \longrightarrow H_2O(g)$$

$S^{\circ}_{H_2O(l)_{298}}$ = absolute entropy for $H_2O(l)$ at 298° =
$$S^{\circ}_{H_2O(l)_{298}} - S^{\circ}_{H_2O(s)_0} = 16.71 - 0 = 16.71 \text{ cal/mole } °K$$

$S^{\circ}_{H_2O(g)_{298}}$ = absolute entropy for $H_2O(g)$ at 298° =
$$S^{\circ}_{H_2O(g)_{298}} - S^{\circ}_{H_2O(s)_0} = 45.10 - 0 = 45.10 \text{ cal/mole } °K$$

ΔS° = final state − initial state = $S^{\circ}_{H_2O(g)_{298}} - S^{\circ}_{H_2O(l)_{298}} =$
$$45.10 - 16.71 = 28.39 \text{ cal/mole } °K$$

The value for ΔS° is positive, indicating greater disorder in gaseous H_2O than in liquid H_2O. This is in accord with the fact that the molecules are in much more rapid and random motion in the gaseous state than in the liquid state (see Sections 10.11 and 11.1).

(b) Determine the temperature at which liquid water and gaseous water are in equilibrium with each other at 1 atm pressure.

Since the two states are in equilibrium, the free energy change, ΔG°, in going from the liquid to the gaseous water is zero. Assuming as before that ΔH° and ΔS° are independent of temperature,

$\Delta H^{\circ} = \Delta H^{\circ}_{f_{H_2O(g)}} - \Delta H^{\circ}_{f_{H_2O(l)}} = -57.80 - (-68.32)$
$\quad = +10.52 \text{ kcal} = 10,520 \text{ cal}$
$\Delta S^{\circ} = 28.39 \text{ cal/mole } °K$ [just calculated in Example 1(a)]
$\Delta G^{\circ} = \Delta H^{\circ} - T\Delta S^{\circ} = 0$ ($\Delta G^{\circ} = 0$ at equilibrium)

$$T = \frac{\Delta H^{\circ}}{\Delta S^{\circ}} = \frac{10,520 \text{ cal/mole}}{28.39 \text{ cal/mole } °K}$$

$\quad = 371° \text{ K} = 98° \text{ C}$

The correct answer for Part (b) of the preceding example, of course, is 373° K (100° C), the boiling point of water at 1 atm pressure, but the calculation of the value 371° was based upon the assumptions that ΔH° and ΔS° are independent of temperature. This is a good place to point out that these assumptions are only approximations. They are sufficiently true to be highly useful, as we have seen, in using standard molar enthalpy and free energy values to calculate values at other temperatures; but for really exact calculations, ΔH° and ΔS° values at the temperature in question must be used, as in the following:

(c) Recalculate the temperature at which liquid water and gaseous water are in equilibrium with each other at 1 atm pressure, this time using the true tabulated values of ΔH° and ΔS° at the approximate temperature, calculated in (b), of 98° C. ($\Delta H^{\circ}_{H_2O}$ at 98° C = 9,726.5 cal/mole; $\Delta S^{\circ}_{H_2O}$ at 98° C = 26.063 cal/mole °K.)

(Note that the $\Delta H°$ value at 25° C of 10,520 cal and the $\Delta S°$ value at 25° C of 28.39 cal are close enough, as stated earlier, to the values at 98° for an approximate calculation but are not sufficiently close for an exact calculation.)

Again using $\Delta G° = \Delta H° - T\Delta S° = 0$ ($\Delta G° = 0$ at equilibrium)

$$T = \frac{\Delta H°_{371}}{\Delta S°_{371}} = \frac{9,726.5 \text{ cal/mole}}{26.063 \text{ cal/mole °K}} = 373.2° \text{ K}$$

or $373.2 - 273.2 = 100.0° \text{ C}$

In this part of the example we obtained the true value of 100° C. If the calculation is repeated with the values of $\Delta H°$ and $\Delta S°$ for 100° C (9,717.1 cal/mole and 26.0400 cal/mole °K), a value of $T = 100.0°$ is again obtained indicating that the values for $\Delta H°$ and $\Delta S°$ at the initially calculated approximate temperature are close enough to give the correct temperature of 100.0° (to four significant figures).

If we were to apply temperatures below 373° K to the relationship $\Delta G° = \Delta H° - T\Delta S°$, we would see that the value for the $T\Delta S°$ becomes smaller, and hence a smaller term is subtracted from $\Delta H°$ to give $\Delta G°$. Thus, $\Delta G°$ becomes positive (it was zero for the equilibrium state at 373°). *Hence, at temperatures below 373° K (100° C) at one atmosphere pressure, the spontaneous change is from gaseous H_2O to liquid H_2O.* If we go to higher temperatures than 373°, $T\Delta S°$ becomes larger and a larger term is subtracted from $\Delta H°$ than at 373°. Thus, $\Delta G°$ is negative, indicating that *at temperatures above 373° K (100° C) at one atmosphere pressure the spontaneous change is from liquid H_2O to gaseous H_2O.*

Notice in this example that the heat effect ($\Delta H°$) and entropy effect ($\Delta S°$) work at cross purposes, inasmuch as $\Delta S°$ is positive (maximum disorder, indicating favorable conditions for the change) and $\Delta H°$ is positive (endothermic, indicating unfavorable conditions for the change). We could then suppose that one of these factors (either $\Delta H°$ or $\Delta S°$) is of predominant importance. But we now find ourselves in a quandary, stemming from the fact that the spontaneity changes from unfavorable to favorable at 373° K; however, $\Delta H°$ and $\Delta S°$ are both *relatively* independent of temperature and do not change sufficiently with temperature to reflect this difference in spontaneity that occurs at 373° K. This verifies the earlier statement that $\Delta H°$ alone is not sufficient to predict accurately the spontaneity of a reaction. It is now apparent also that $\Delta S°$ alone is not sufficient to predict spontaneity. However, $\Delta G°$ (which combines $\Delta H°$, $\Delta S°$, and T) is a valid quantity on which to base such predictions. As shown above, $\Delta G°$ is positive at temperatures below 373° K (100° C) and negative above 373° K.

Example 2.   What is the standard entropy change for the following reaction at 298° K at 1 atm pressure?

$$H_2(g) + \tfrac{1}{2}O_2(g) \longrightarrow H_2O(g)$$

We can consider this reaction as proceeding either through the direct route or through a series of steps: (1) and (2) the $H_2(g)$ and $O_2(g)$ are cooled to $H_2(s)$ and $O_2(s)$ at absolute zero temperature ($0°$ K); (3) the two are allowed to react at absolute zero to form $H_2O(s)$; (4) the $H_2O(s)$ is heated back to $H_2O(g)$ at the original temperature $298°$ K.

$$H_2(g)_{298} + \tfrac{1}{2}O_2(g)_{298} \xrightarrow{\Delta S°} H_2O(g)_{298}$$

$$(1)\Big\downarrow \Delta S_1° \qquad (2)\Big\downarrow \Delta S_2° \qquad (4)\Big\uparrow \Delta S_4°$$

$$H_2(s)_0 + \tfrac{1}{2}O_2(s)_0 \xrightarrow[(3)]{\Delta S_3°} H_2O(s)_0$$

Since $S°$ is a state function,

$$\Delta S° = \Delta S_1° + \Delta S_2° + \Delta S_3° + \Delta S_4°$$

Inasmuch as the absolute entropy for H_2 at $0°$ K is zero,

$$\Delta S_1° = S_{H_2(s)_0}° - S_{H_2(g)_{298}}° = 0 - S_{H_2(g)_{298}}°$$

Therefore, $\quad \Delta S_1° = -S_{H_2(g)_{298}}°$

Similarly, $\quad \Delta S_2° = \tfrac{1}{2}S_{O_2(s)_0}° - \tfrac{1}{2}S_{O_2(g)_{298}}° = 0 - \tfrac{1}{2}S_{O_2(g)_{298}}°$

Therefore, $\quad \Delta S_2° = -\tfrac{1}{2}S_{O_2(g)_{298}}°$

Similarly, $\quad \Delta S_4° = S_{H_2O(g)_{298}}° - S_{H_2O(s)_0}° = S_{H_2O(g)_{298}}° - 0 = S_{H_2O(g)_{298}}°$

Also, $\Delta S_3 = 0$, inasmuch as the absolute entropies of $H_2(s)$, $O_2(s)$, and $H_2O(s)$ (all pure substances) are zero at absolute zero temperature.

Therefore, $\Delta S° = \Delta S_1° + \Delta S_2° + \Delta S_3° + \Delta S_4°$

$$= -S_{H_2(g)_{298}}° + [-\tfrac{1}{2}S_{O_2(g)_{298}}°] + 0 + S_{H_2O(g)_{298}}°$$

$$= -(31.208) - \tfrac{1}{2}(49.003) + 0 + (45.104)$$

$$= -10.606 \text{ cal/mole } °K$$

The relationship of $\Delta G°$ to standard electrode reduction potentials and to equilibrium constants through electrode potentials is discussed in the chapter on Electrochemistry (see Section 22.15).

QUESTIONS

1. How would you define *chemical thermodynamics*?
2. What is the meaning of the term *system*?
3. What is the meaning of the term *surroundings*?
4. Explain what is meant by the term *expansion work*.
5. Why is a consideration of expansion work important when considering energy changes associated with chemical reactions?
6. In a thermodynamic sense, when is a chemical reaction said to be exothermic? When is a chemical reaction said to be endothermic?
7. (a) What effect do increases in temperature have on exothermic reactions?
   (b) What effect do increases in temperature have on endothermic reactions?
8. What is the meaning of the "standard molar enthalpy of formation" of a pure substance?

9. What is meant by the term *state function*?

10. State the First Law of Thermodynamics in two forms.

11. State the Second Law of Thermodynamics in terms of entropy changes for the universe. State the Second Law of Thermodynamics in terms of free energy changes.

12. State the Third Law of Thermodynamics.

13. What is the relationship between the useful work a system can do and the free energy change of the system, when the system undergoes a change at constant temperature and pressure?

14. What is the distinction between ΔE and ΔH for systems undergoing a change at constant pressure?

15. What is the distinction between ΔH and ΔG for systems undergoing a change at constant temperature and pressure?

16. What is meant by the term *spontaneous reaction*?

17. What is the connection between entropy and disorder?

18. (a) What is the "standard absolute entropy" of a pure substance?

    (b) How can standard absolute entropies of pure substances be used?

    (c) Can standard absolute enthalpy be calculated and used in the same way as standard absolute entropy? Why or why not? What about standard absolute free energy?

19. In each of the following, give the sign of the entropy change for the indicated system going from state 1 to state 2, and explain.

State 1	State 2
(a) Egg in shell	Scrambled egg
(b) $H_2O(g)$, 298° K, 1 atm	$H_2O(l)$, 298° K, 1 atm
(c) 1 mole of ideal gas at a volume of 30 liters at 298° K	1 mole of ideal gas at a volume of 60 liters at 298° K
(d) Stretched rubber band	Slack rubber band
(e) Slack rubber band	Rubber band dissolved in appropriate solvent

20. Liquid ammonia is a less associated liquid than water (see Section 12.9). Which substance should have the higher entropy of vaporization? Explain.

21. What is the relationship between ΔG and equilibrium?

22. How is $\Delta H_{\text{sublimation}}$ of a substance related to its ΔH_{fusion} and $\Delta H_{\text{vaporization}}$?

23. There are two forms of sulfur, rhombic and monoclinic, described in Section 23.3. Read that section and then decide whether you would expect a difference in ΔH for the oxidation of one gram atom of each. Explain your answer.

PROBLEMS

⒮1. Calculate the internal energy change, ΔE, for a system when

   ⒮(a) $q = -300$ cal; $w = -750$ cal.

   (b) $q = 190$ cal; $w = 190$ cal.

   ⒮(c) One kcal of heat energy is absorbed by the system and the system does 540 cal of work on the surroundings.

(d) 250 cal of heat energy are absorbed by the system and 635 cal of work are done on the system by the surroundings.

Ans. (a) 450 cal; (b) 0; (c) 460 cal; (d) 885 cal

$\boxed{s}$2. If one liter $\cdot$ atm = 24.2173 cal and the gas constant R = 1.987 cal/mole $\cdot$ °K, what is the value of R in units of liter $\cdot$ atm/mole $\cdot$ °K?

Ans. R = 0.08205 liter $\cdot$ atm/mole $\cdot$ °K

$\boxed{s}$3. Find the work done in kilocalories when 3.00 moles of helium expand from a volume of 1.00 liter to a volume of 45.0 liters against a pressure of 2.50 atm.

Ans. 7.99 kcal

$\boxed{s}$4. How many kcal of heat energy will be liberated when 49.7 grams of manganese are burned to form $Mn_3O_4(s)$ at standard state conditions? $\Delta H^{\circ}_{f_{298}}$ of Mn_3O_4 is equal to -331.7 kcal/mole.

Ans. 100 kcal

$\boxed{s}$5. (a) Using the enthalpy of formation data in Appendix J, calculate the enthalpy change for the following reactions:

$\boxed{s}$(1) $CaO(s) + SO_3(g) + 2H_2O(l) \longrightarrow CaSO_4 \cdot 2H_2O(s)$

Ans. -99.9 kcal

(2) $PCl_5(g) + 4H_2O(l) \longrightarrow H_3PO_4(l) + 5HCl(g)$ *Ans. -50.3 kcal*

$\boxed{s}$(3) $CaSO_3 \cdot 2H_2O(s) + CO_2(g) \longrightarrow CaCO_3(s) + SO_2(g) + 2H_2O(l)$

Ans. 19.2 kcal

(4) $2B_3N_3H_6(l) + 15O_2(g) \longrightarrow 3N_2O_5(g) + 3B_2O_3(s) + 6H_2O(g)$

Ans. -992.7 kcal

(5) $PbS(s) + H_2(g) \longrightarrow Pb(s) + H_2S(g)$ *Ans. 19.1 kcal*

(b) Which of the reactions above are exothermic?   *Ans. (1), (2), and (4)*

$\boxed{s}$6. (a) Calculate the heat (enthalpy) of formation for calcium sulfide, using the enthalpy of formation data in Appendix J, for each of the following reactions:

$\boxed{s}$(1) $CaS(s) + 2O_2(g) + 2H_2O(l) \longrightarrow CaSO_4 \cdot 2H_2O(s) + 231.3$ kcal

Ans. -115.1 kcal/mole

(2) $2CaS(s) + 3O_2(g) + 4H_2O(l) \longrightarrow 2CaSO_3 \cdot 2H_2O(s) + 338.3$ kcal

Ans. -115.4 kcal/mole

(b) Using the answer to part (a), data from Appendix J, and the information from the following equation, calculate the heat (enthalpy) of formation of ozone (O_3) to two significant figures. $CaS(s) + O_3(g) + 2H_2O(l) \longrightarrow CaSO_3 \cdot 2H_2O(s) + 203.3$ kcal.   *Ans. 34 kcal/mole*

7. Calculate, using the data in Appendix J, the bond energies of F_2, Cl_2, and FCl. All are gases in their most stable form at standard state conditions.

Ans. F—F = 37.76 kcal/mole of bonds; Cl—Cl = 58.164 kcal/mole of bonds; F—Cl = 60.98 kcal/mole of bonds

$\boxed{s}$8. Calculate, using the data in Appendix J, the bond energies of N_2, O_2, and NO. All are gases in their most stable form at standard state conditions.

Ans. N—N = 226.0 kcal/mole of bonds; O—O = 119.1 kcal/mole of bonds; N—O = 151.0 kcal/mole of bonds

$\boxed{s}$9. For each of the reactions below, calculate the Gibbs free energy change, the enthalpy change, and the entropy change. Which of the reactions are spontaneous? For which are the entropy changes favorable for the reaction to proceed?

⑤(a) $Fe_2O_3(s) + 13CO(g) \longrightarrow 2Fe(CO)_5(g) + 3CO_2(g)$
Ans. $\Delta G = -12.5\ kcal;\ \Delta H = -92.5\ kcal;\ \Delta S = -268.8\ cal/°K$

(b) $2Li(OH)(s) + CO_2(g) \longrightarrow Li_2CO_3(g) + H_2O(g)$
Ans. $\Delta G = -18.8\ kcal;\ \Delta H = -21.39\ kcal;\ \Delta S = -8.4\ cal/°K$

(c) $CH_4(g) + N_2(g) \longrightarrow HCN(g) + NH_3(g)$
Ans. $\Delta G = 38.0\ kcal;\ \Delta H = 39.2\ kcal;\ \Delta S = 3.91\ cal/°K$

(d) $CS_2(g) + 3Cl_2(g) \longrightarrow CCl_4(g) + S_2Cl_2(g)$
Ans. $\Delta G = -38.1\ kcal;\ \Delta H = -57.1\ kcal;\ \Delta S = -63.5\ cal/°K$

(e) $N_2(g) + O_2(g) \longrightarrow 2NO(g)$
Ans. $\Delta G = 41.38\ kcal;\ \Delta H = 43.14\ kcal;\ \Delta S = 5.92\ cal/°K$

10. Consider the following reactions as possible means of preparing $Cu_2O(s)$:
    (a) $Cu(s) + CuO(s) \longrightarrow Cu_2O(s)$
    (b) $2CuO(s) \longrightarrow Cu_2O(s) + \frac{1}{2}O_2(g)$
    Using the data in Appendix J, calculate ΔH°_{298} and ΔG°_{298} for these reactions. Would either of these reactions be feasible for preparing $Cu_2O(s)$ at 298°? at higher temperatures?
    Ans. (a) $\Delta H^\circ_{298} = -2.7\ kcal;\ \Delta G^\circ_{298} = -3.9\ kcal$ (b) $\Delta H^\circ_{298} = +34.9\ kcal;$
    $\Delta G^\circ_{298} = +27.1\ kcal$

⑤11. For a certain process at 300° K, $\Delta G = -18.4$ kcal and $\Delta H = -13.6$ kcal. Find the entropy change for this process at this temperature.
    Ans. $\Delta S = 16.0\ cal/°K$

⑤12. (a) For the vaporization of bromine liquid to bromine gas, calculate the change in enthalpy and the change in entropy at standard state conditions.
    Ans. $\Delta H^\circ_{298} = 7.387\ kcal/mole;\ \Delta S^\circ_{298} = 22.257\ cal/mole \cdot °K$

    (b) From the calculations in (a) discuss relative disorder in bromine liquid compared to bromine gas. On the basis of the enthalpy change, state what you can about the spontaneity of the vaporization.

    (c) Calculate the value of ΔG°_{298} for the vaporization of bromine from the data in Appendix J. *Ans.* $0.751\ kcal/mole$

    (d) State what you can about the spontaneity of the process from the value you obtained for ΔG°_{298} in (c).

    (e) Calculate the temperature at which liquid and gaseous Br_2 are in equilibrium with each other at 1 atm (assume ΔH° and ΔS° are independent of temperature). *Ans.* $331.9°\ K\ or\ 58.7°\ C$

    (f) From this temperature value (Part e) state in which direction the process would be spontaneous.

    (g) Compare ΔH°, ΔS°, and ΔG° in terms of their usefulness in predicting spontaneity of the vaporization of Br_2.

13. The enthalpies of formation of $NO(g)$, $NO_2(g)$, and $N_2O_3(g)$ are 21.57 kcal/mole, 7.93 kcal/mole, and 20.01 kcal/mole, respectively. Their standard entropies are 50.347, 57.35, and 74.61 cal/mole °K, respectively.
    (a) Use the data above to calculate the Gibbs free energy change for the following reaction at 25.0° C.

    $$N_2O_3(g) \longrightarrow NO(g) + NO_2(g) \qquad Ans.\ -0.37\ kcal$$

    (b) Repeat the above calculation for 0.00° C, 50.0° C, and 100.0° C assuming that the enthalpy and entropy changes do not change with a change in temperature.   *Ans. 0.45 kcal; −1.20 kcal; −2.86 kcal*

14. Consider the reaction:

$$I_2(g) + Cl_2(g) \longrightarrow 2ICl(g)$$

   (a) For this reaction $\Delta H^\circ_{298.15} = -6.42$ kcal, and $\Delta S^\circ_{298.15} = 2.71$ cal/°K. Calculate $\Delta G^\circ_{298.15}$ for the reaction.   *Ans. −7.23 kcal*

   (b) Calculate the equilibrium constant for this reaction at 25.0° C.
   Ans. 2.00 × 10⁵

15. If the enthalpy of vaporization of CH_2Cl_2 is 6.93 kcal/mole at 25.0° C and the entropy of vaporization is 22.1 cal/mole °K, calculate a value for the normal boiling point temperature of CH_2Cl_2.   *Ans. 40.4° C*

[s]16. (a) The value of K_e is 9.23×10^{-22} at 25° C for the reaction

$$CO_2(g) + C(s) \rightleftharpoons 2CO(g)$$

   What is ΔG°_{298} for this reaction, calculated from K_e?   *Ans. 28.7 kcal*

   (b) Calculate ΔG°_{298} again, this time using the data in Appendix J. Compare with your answer in (a).   *Ans. 28.7 kcal*

17. If the entropy of vaporization of H_2O is equal to 26.0 cal/mole °K and the enthalpy of vaporization is 9,709 cal/mole, calculate the normal boiling point temperature of water in °C.   *Ans. 100° C*

[s]18. (a) If you wished to decompose $CaCO_3(s)$ into $CaO(s)$ and $CO_2(g)$ at atmospheric pressure, what would be the minimum temperature at which you would conduct the reaction?   *Ans. 835° C*

   (b) Is your answer in (a) compatible with Section 17.16 of the text?

   (c) Calculate the equilibrium vapor pressure of $CO_2(g)$ above $CaCO_3(s)$ in a closed container at 298° K and 1.00 atm.
   Ans. $p_{CO_2} = 1.14 \times 10^{-20}$ mm Hg

19. If ΔG°_{298} for a certain reaction equals −1.00 kcal, what is K_e at 298° K for this reaction?   *Ans. 5.41*

20. Using data in Appendix J, calculate the value at K_e at 25.00° C for the reaction

$$H_2(g) + I_2(g) \longrightarrow 2HI(g) \qquad Ans.\ 6.18 \times 10^2$$

21. (a) Using data in Appendix J, calculate the enthalpy and entropy of vaporization of ethanol (C_2H_5OH) at 298° K and 1 atm.
   Ans. $\Delta H^\circ_{vap_{298}} = 10.18$ kcal/mole; $\Delta S^\circ_{vap_{298}} = 29.1$ cal/mole · °K

   (b) Using your answers to (a), calculate a value for the normal boiling point temperature of ethanol in °C.   *Ans. T = 76° C*

   (c) Compare your answer in (b) to the value given in Section 27.13 for the boiling point of ethanol. Why does your answer differ from the value given in Section 27.13?

   (d) What can you say about the dependency of $\Delta H^\circ_{vap_{298}}$ and $\Delta S^\circ_{vap_{298}}$ on temperature? Can you explain why they are independent of temperature or dependent on temperature, as the case may be?

[s]22. (a) What is the equilibrium vapor pressure, in mm Hg, of $H_2O(g)$ above

pure $H_2O(l)$ at 298° K and 1 atm (use Appendix J)? Compare your answer to the value given in Section 10.9, Table 10–1).

Ans. 23.7 mm of Hg

(b) At what temperature, °C, would water boil if under an external pressure of 23.7 mm of Hg? *Ans. 25° C*

23. (a) What is the equilibrium vapor pressure, in mm Hg, of $CHCl_3(g)$ above pure liquid $CHCl_3(l)$ at 298° K and 1 atm (use Appendix J)?

Ans. 197 mm

(b) At what temperature, °C, would chloroform ($CHCl_3$) boil if under an external pressure of 197 mm of Hg? *Ans. 25° C*

24. Given the reaction:

$$SbCl_5(g) \longrightarrow SbCl_3(g) + Cl_2(g)$$

(a) Calculate (using data in Appendix J) ΔH°_{298}, ΔG°_{298}, and ΔS°_{298}.

Ans. $\Delta H^\circ_{298} = 19.3$ kcal; $\Delta G^\circ_{298} = 7.9$ kcal; $\Delta S^\circ_{298} = 37.96$ cal/° K

(b) From your answers in (a) calculate K_e at 298° K for this reaction.

Ans. 1.58×10^{-6}

(c) What is ΔG°_{800} for the given reaction? (Assume ΔH°_{298} and ΔS°_{298} independent of T.) What is K_e at 800° K? *Ans. -11.1 kcal; 1.09×10^3*

(d) If you wished to make $SbCl_5$ from $SbCl_3$ and Cl_2 would you run the reaction at high temperatures? Why or why not?

(e) What other factors might influence your choice of temperature to produce $SbCl_5$?

(f) Knowing that $K_e = 1.58 \times 10^{-6}$ for the reaction to produce $SbCl_3$ and Cl_2 by the decomposition of $SbCl_5$ at 298° K, and using the ΔH°_{298} value found in (a), calculate K_e at 800° K. To how many significant figures do your answers to this part and part (c) agree?

Ans. 1.14×10^3; agree to two significant figures

REFERENCES

"Chemical Equilibrium as a State of Maximal Entropy," L. K. Nash, *J. Chem. Educ.*, **47**, 353 (1970).

"On Squid Axons, Frog Skins, and the Amazing Uses of Thermodynamics," W. H. Cropper, *J. Chem. Educ.*, **48**, 182 (1971).

"The Scope and Limitations of Thermodynamics," K. G. Denbigh, *Chem. in Britain*, **4**, 338 (1968).

"Perpetual Motion Machines" (ingenious devices but all doomed by the laws of thermodynamics), S. W. Augrist, *Sci. American*, Jan., 1968; p. 114.

"Bond Energies in the Interpretation of Descriptive Chemistry," R. A. Howald, *J. Chem. Educ.*, **45**, 163 (1968).

"Thermodynamics and Rocket Propulsion," F. H. Verhoek, *J. Chem. Educ.*, **46**, 140 (1969).

"Energy States of Molecules," J. L. Hollenberg, *J. Chem. Educ.*, **47**, 2 (1970).

"Ion Pairs and Complexes: Free Energies, Enthalpies, and Entropies," J. E. Prue, *J. Chem. Educ.*, **46**, 12 (1969).

"Quantities of Work in Thermodynamics Equations," P. G. Wright, *J. Chem. Educ.*, **46**, 380 (1969).

"The Energy Cycle of the Earth," A. H. Oort, *Sci. American*, Sept., 1970; p. 54.

"Evolution of the Second Law of Thermodynamics," V. V. Raman, *J. Chem. Educ.*, **47,** 331 (1970).

"The Second Law—How Much, How Soon, to How Many," H. A. Bent, *J. Chem. Educ.*, **47,** 337 (1970).

"Our Freshmen Like the Second Law," N. C. Craig, *J. Chem. Educ.*, **47,** 342 (1970).

"Human Energy Production as a Process in the Biosphere," S. F. Singer, *Sci. American*, Sept., 1970; p. 175.

"Analogies Between Chemical and Mechanical Equilibria," S. G. Canagaratna and M. Selvaratnam, *J. Chem. Educ.*, **47,** 759 (1970).

"Linear Free Energy Relationships," J. Shorter, *Chem. in Britain*, **5,** 269 (1969).

Rudimentary Chemical Thermodynamics, H. F. Franzen and B. C. Gerstein, D. C. Heath and Co., Lexington, Mass., 1971 (Paperback).

"Bibliography for Thermodynamics," W. Dannhauser, *J. Chem. Educ.*, **50,** 493 (1973).

"The First Law (of Thermodynamics) for Scientists, Citizens, Poets, and Philosophers," H. Bent, *J. Chem. Educ.*, **50,** 323 (1973).

"Thermochemical Hydrogen Generation," R. H. Wentorf, Jr. and R. E. Hanneman, *Science,* **185,** 311 (1974).

"Simple Free Energy—Oxidation State Diagrams," R. T. Myers, *J. Chem. Educ.*, **51,** 444 (1974).

"Enthalpy Cycles in Inorganic Chemistry," J. L. Holm, *J. Chem. Educ.*, **51,** 461 (1974).

"Thermodynamics, Folk Culture, and Poetry," W. L. Smith, *J. Chem. Educ.*, **52,** 97 (1975).

"Illustration of Free Energy Changes in Chemical Reactions," E. Hamori, *J. Chem. Educ.*, **52,** 370 (1975).

"Entropy: A Modern Discussion," O. Redlich, *J. Chem. Educ.*, **52,** 374 (1975).

The Halogens and Some of Their Compounds

21

The Halogens

The elements of Group VIIA of the Periodic Table are known as the **halogens,** which means *salt formers.* Their binary compounds as a group are called **halides.** Fluorine, chlorine, bromine, and iodine were discovered during the period 1774 to 1886, but astatine, the fifth halogen, was first obtained in 1940. Salts of these elements (excepting astatine) are common in nature.

21.1 Occurrence of the Halogens

The halogens never occur free in nature, because of their great chemical activity. Chlorine is the most abundant element of the group, and although fluorine, bromine, and iodine are less common, they are reasonably available. The principal occurrences of the halogens are given in Table 21-1.

21.2 Preparation and Production of the Halogens

The salts in which the halogens (except iodine) exist in an oxidation state of -1 constitute the best sources of these elements. Various methods of oxidizing the halide ions to the elemental halogens (zero oxidation state) may be employed. The general oxidation half-reaction may be written

$$2X^- \rightleftharpoons X_2 + 2e^- \quad (X = halogen)$$

The ease of oxidation of the halide ions increases in the order as follows: $F^- < Cl^- < Br^- < I^- < At^-$. As you would expect, the electron removed during

TABLE 21-1   Occurrences of the Halogens

Fluorine: **Fluorite** or **fluorspar,** CaF_2 **Fluorapatite,** $Ca_{10}F_2(PO_4)_6$ **Cryolite,** Na_3AlF_6 Sea water (small amounts) Teeth, bones, blood (small amounts)	*Bromine:* Sea water (NaBr, KBr, $MgBr_2$, $CaBr_2$) Underground brines Salt deposits
Chlorine: Sea water (2.8 per cent NaCl; other chlorides 0.8 per cent) Great Salt Lake in Utah (23 per cent NaCl) Salt beds (NaCl, $MgCl_2$, $CaCl_2$) Gastric juice (0.2 to 0.4 per cent HCl)	*Iodine:* Sea water (very small amounts) $NaIO_3$ in Chilean nitrate deposits Oil well brines of California Thyroid gland in human body

oxidation is closest to the nucleus in F^- and farthest from it in At^-. There are enough differences among the methods applicable to the preparation of the free halogens to warrant separate discussions of these methods.

■ **1. Fluorine.** It is difficult to prepare fluorine because it is the most highly electronegative element known. It is necessary to resort to electrolytic oxidation of a fluoride compound in a nonaqueous electrolyte to oxidize the fluoride ion to fluorine. The electrolyte commonly used is a mixture of three parts potassium hydrogen fluoride (KHF_2) and two parts anhydrous hydrogen fluoride. When electrolysis of the fused electrolyte (melting point 72°) begins, HF is decomposed to form fluorine gas at the anode and hydrogen at the cathode.

$$2HF + \text{electrical energy} \longrightarrow H_2\uparrow + F_2\uparrow$$

The two gases are kept separate by a barrier around the cathode and are drawn from the cell continuously. Anhydrous hydrogen fluoride is added to the cell, either continuously or intermittently, to regenerate the electrolyte. After purification, fluorine gas is compressed in special steel cylinders at 400 pounds pressure.

■ **2. Chlorine.** The 10.6 million tons of chlorine produced in 1974 made it the seventh highest chemical in amount of commercial production. The bulk of the commercial chlorine produced in the world is by electrolytic oxidation of the chloride ion in aqueous sodium chloride solutions. The half cell reactions are given by the equations

(Anodic oxidation) $\qquad 2Cl^- \longrightarrow Cl_2\uparrow + 2e^-$

(Cathodic reduction) $\qquad \underline{2H_2O + 2e^- \longrightarrow 2OH^- + H_2\uparrow}$

Net reaction $\qquad 2Cl^- + 2H_2O \longrightarrow Cl_2\uparrow + 2OH^- + H_2\uparrow$

The sodium hydroxide and hydrogen, which are by-products of the process, must be kept separate from the chlorine so as to prevent them from entering into unwanted secondary reactions. Diaphragm separation of the electrodes has commonly been used in the past in such equipment as the **Nelson, Vorce,** and **Hooker cells.** When sodium chloride solutions are electrolyzed in cells such as these using graphite electrodes, the sodium hydroxide produced is mixed with undecomposed sodium chloride, from which it must be separated by crystallization if pure sodium hydroxide is to be obtained. This troublesome step is avoided in the **DeNora cell,** which is now the most commonly used cell, by employing mercury cathodes rather than graphite cathodes. In the DeNora cell, the cathode is a thin layer of mercury, spread over the bottom of the cell. Graphite anodes hang from the lid of the cell. As the brine is electrolyzed, chlorine escapes at the anode and sodium is liberated at the mercury cathode instead of hydrogen; the sodium dissolves in the mercury, forming an amalgam (an alloy with mercury).

$$\text{(Cathodic reduction) } Na^+ + e^- \xrightarrow{\text{Hg}} \quad Na \text{ (amalgam)}$$

The sodium amalgam flows out of the cell and is brought in contact with water, with which the sodium reacts forming hydrogen and a pure solution of sodium hydroxide.

$$2Na \text{ (amalgam)} + 2H_2O \longrightarrow H_2\uparrow + 2Na^+ + 2OH^-$$

Chlorine is a by-product in the production of metals such as sodium, calcium, and magnesium by the electrolytic decomposition of their fused chlorides. For example, when fused magnesium chloride is electrolyzed magnesium and chlorine are formed.

$$MgCl_2 \text{ (fused)} + \text{electrical energy} \longrightarrow Mg + Cl_2\uparrow$$

Many attempts have been made—and are still being made—to produce chlorine commercially by nonelectrolytic processes and to obtain more valuable by-products than sodium hydroxide, but so far the attempts have been without particular success.

In the laboratory the chloride ion is oxidized to free chlorine in acid solution by manganese dioxide (MnO_2), potassium permanganate ($KMnO_4$), or sodium dichromate ($Na_2Cr_2O_7$).

$$MnO_2 + 2Cl^- + 4H^+ \longrightarrow Mn^{2+} + Cl_2\uparrow + 2H_2O$$
$$2MnO_4^- + 10Cl^- + 16H^+ \longrightarrow 2Mn^{2+} + 5Cl_2\uparrow + 8H_2O$$
$$Cr_2O_7^{2-} + 6Cl^- + 14H^+ \longrightarrow 2Cr^{3+} + 3Cl_2\uparrow + 7H_2O$$

Sodium chloride and sulfuric acid are usually used as sources of the chloride ion and the hydrogen ion, respectively, for the above reactions.

■ **3. Bromine.** The methods for the laboratory scale preparation of bromine are similar to those used for chlorine. In addition, bromine may be prepared by the oxidation of the bromide ion by chlorine. Chlorine, being more electronegative than bromine, attracts an electron from bromide ion, thereby oxidizing the bromide ion to free bromine. The chlorine, through the addition of one electron per chlorine atom, is reduced to chloride ion.

$$2Br^- + Cl_2 \rightleftharpoons Br_2 + 2Cl^-$$

Oxidation of bromide ions with elemental chlorine is used in the production of bromine from sea water. The bromine thus liberated is blown out of the solution by a stream of air. The bromine is then stripped from the air by adsorption in aqueous sodium carbonate, by which a reasonably concentrated solution of sodium bromide, NaBr, and sodium bromate, $NaBrO_3$, is formed.

$$3CO_3^{2-} + 3Br_2 \longrightarrow 5Br^- + BrO_3^- + 3CO_2\uparrow$$

Acidification of this solution with sulfuric acid liberates the bromine.

$$5Br^- + BrO_3^- + 6H^+ \rightleftharpoons 3Br_2 + 3H_2O$$

In this redox reaction, the bromate ion is the oxidizing agent and the bromide ion is the reducing agent. One million pounds of sea water must be processed in order to obtain 70 pounds of elemental bromine.

One method for the production of bromine from underground brines is accomplished in a manner similar to that described for the production of bromine from sea water. Another method of producing bromine from natural brines involves electrolysis of the mother liquor remaining after most of the sodium chloride has been removed.

■ **4. Iodine.** Elemental chlorine can be used to liberate iodine from iodides. The method is similar to that described for the preparation of bromine.

$$2I^- + Cl_2 \longrightarrow \underline{I_2} + 2Cl^-$$

Most of the iodine appears as a solid precipitate and can be separated from the solution by filtration. Care must be taken to avoid the use of an excess of chlorine in this process, otherwise unwanted secondary reactions occur, with the formation of iodine chloride, ICl, and iodic acid, HIO_3. Considerable iodine is obtained from iodides concentrated in kelp and other sea plants, and from oil field brines.

Iodine occurs in the form of sodium iodate ($NaIO_3$) as an impurity in Chile saltpeter ($NaNO_3$) deposits. The iodine is liberated by reducing the iodate with sodium hydrogen sulfite:

$$2IO_3^- + 5HSO_3^- \longrightarrow 3HSO_4^- + 2SO_4^{2-} + H_2O + I_2$$

21.3 General Properties of the Halogens

It is a remarkable fact that for most of the properties of the halogens there is a gradation with increasing atomic number. A study of Table 21–2, which lists some of their more important properties, will illustrate this point.

The valence shells of the atoms of the halogens each contain seven electrons. The atoms each tend to gain one more electron and become stable univalent negative ions. The large values of their electronegativities indicate the ease with which halide ions are formed from halogen atoms. The tendency to form halide ions and to complete an octet by sharing electrons in covalent linkages decreases as the size of the halogen atom increases, because of a resultant reduction in attraction for electrons. The halogens are oxidizing agents, fluorine being the strongest and astatine the weakest in this respect. Each of the halogens is diatomic (Fig. 21–1).

In the fluorine molecule, all electrons are paired and distributed in the

TABLE 21-2 Properties of the Halogens*

	Fluorine	Chlorine	Bromine	Iodine	Astatine
Atomic number	9	17	35	53	85
Atomic weight	18.998	35.453	79.909	126.904	(210)
Electronic structure	2,7	2,8,7	2,8,18,7	2,8,18,18,7	2,8,18,32,18,7
Radius of X^-, Å	1.36	1.81	1.95	2.16	...
Covalent bond radius, Å	0.64	0.99	1.14	1.33	1.40
Physical state	gas	gas	liquid	solid	solid
Melting point, °C	-218	-101	-7.3	114	...
Boiling point, °C	-188	-34.1	58.78	184	...
Density, g/cm³	1.108 (liq.)	1.557 (liq.)	3.119 (liq.)	4.93 (sol.)	...
Color	Pale yellow	Greenish-yellow	Reddish-brown	Black (s); violet (g)	...
Electronegativity	4.0	3.0	2.8	2.5	...
Heat of vaporiza- tion, cal/mole	1,510	4,878	7,340	11,140	...

*The halogens are represented by the symbol X.

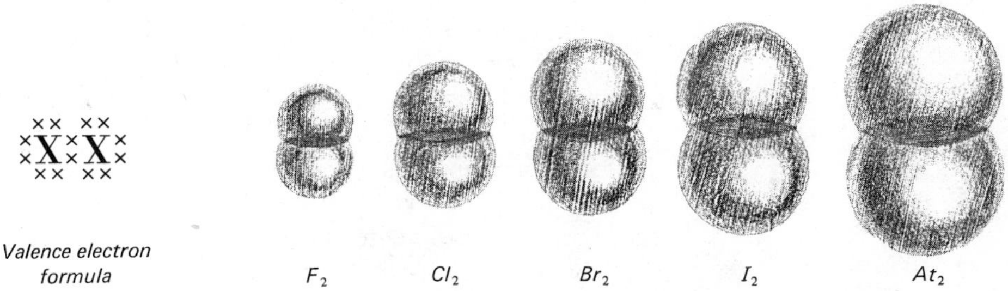

FIGURE 21-1

The halogens form diatomic molecules.

Valence electron formula F_2 Cl_2 Br_2 I_2 At_2

molecular orbitals as follows (see the molecular orbital energy diagram Fig. 5–13): one electron pair each in the σ_{1s} (bonding), σ_{1s}^* (antibonding), σ_{2s} (bonding), σ_{2s}^* (antibonding), σ_{p_x} (bonding), π_{p_y} and π_{p_z} (bonding), and $\pi_{p_y}^*$ and $\pi_{p_z}^*$ (antibonding) orbitals.

The fluorine molecule has six of its p electrons, therefore, in bonding orbitals and four in antibonding orbitals. Thus it has a somewhat weaker bond than the oxygen molecule, which has only two electrons in antibonding orbitals. In fact, as we noted in Section 5.11, the F_2 bond is one of the weakest of the covalent bonds.

The valence electrons in the major energy levels $n = 3, 4, 5,$ and 6 for chlorine, bromine, iodine, and astatine, respectively, have analogous molecular orbital electron distributions to those in the $n = 2$ level for fluorine.

For the most part, the chemical properties of the halogens differ in degree rather than in kind. For example, each reacts with hydrogen according to the equation $H_2 + X_2 \longrightarrow 2HX$; but the heat of formation (enthalpy), $\Delta H_{f_{298}}^\circ$, of HX (Table

21-4) increases in the order HF<HCl<HBr<HI, indicating a *corresponding decrease in the energy evolved* as the hydrogen halides are formed from the free elements. Remember that the enthalpy of formation is negative if heat is liberated (see Sections 8.5 and 20.5). The halogens have such high ionization potentials that the formation of positive ions would be highly unlikely, except possibly for iodine and astatine. On the other hand, positive oxidation states resulting from the sharing of electrons with elements more electronegative in character are quite common for the halogens except fluorine, the most electronegative of all the elements. Oxidation states of +1, +3, +5, and +7, and sometimes other oxidation states, are shown when the halogens share electrons with oxygen in their oxides, oxyacids, and oxysalts. See Sections 21.14 through 21.19.

The halogens oxidize a variety of metals, nonmetals, and ions. A free halogen will oxidize those halide ions which are formed from less electronegative halogens. Thus, fluorine will oxidize chloride, bromide, iodide, and astatide ions, whereas chlorine will oxidize only bromide, iodide, and astatide ions.

21.4 Physical Properties of the Halogens

As the atomic structures of the halogens become more complex with increasing atomic weight, there is a gradation in each physical property (Table 21-2). For example, fluorine is a pale yellow gas of low density; chlorine is a greenish-yellow gas 1.892 times as dense as fluorine gas; bromine is a deep reddish-brown liquid which is three times as dense as water; iodine is a grayish-black crystalline solid with a metallic appearance; and astatine is a solid with properties which indicate that it is somewhat metallic in character.

Liquid bromine has a high vapor pressure, and the reddish vapor can easily be seen in a bottle partly filled with the liquid. Iodine crystals have a high vapor pressure, and when heated gently these crystals change into a beautiful deep-violet vapor without melting; the vapor condenses readily upon cooling. This sublimation process (Section 11.15) is used in the purification of iodine.

Bromine dissolves in water, alcohol, ether, chloroform, carbon tetrachloride, and carbon disulfide, forming solutions that vary in color from yellow to reddish-brown, depending upon the concentration.

Iodine dissolves only slightly in water, giving brown solutions. However, it is quite soluble in alcohol, in ether, and in aqueous solutions of iodides, with which it forms brown solutions. The solvents in which iodine combines with the solvent molecules (**solvation**) are the ones which give brown solutions. In chloroform, carbon disulfide, and many hydrocarbons, iodine forms violet solutions. In such solutions, the iodine is present in the molecular state (I_2), and the violet color is like that of iodine in the vapor state, where it is also molecular. Solutions of hydrogen iodide, potassium iodide, or other iodides dissolve iodine very readily.

The iodine and iodide ion combine reversibly, forming the complex ion I_3^-.

$$I^- + I_2 \rightleftharpoons I_3^-$$

The negative iodide ion, when close to a large iodine molecule, whose outer electrons are far from the positive nucleus, disturbs these electronic arrangements enough to induce a dipole within the molecule. An attraction then exists between the negative iodide ion and the positive end of the polar iodine molecule. Compounds containing the complexes I_5^-, Br_3^-, Cl_3^-, ICl_2^- are known.

21.5   Chemical Properties of the Halogens

■ **1. Fluorine.**   The extreme reactivity of fluorine gas as an oxidizing agent is demonstrated by the fact that immediately upon contact with many substances, it ignites them. When a stream of the gas flows onto the surface of water it actually causes the water to burn. Wood and asbestos rapidly ignite and burn when held in a stream of fluorine. Heated glass will burn in fluorine, giving off smoke which looks much like that from wood. Most hot metals burn vigorously in fluorine. However, fluorine can be handled at ordinary or moderately elevated temperatures in containers made of copper, iron, magnesium, nickel, or Monel (an alloy of nickel, copper, and a little iron). Apparently, an adherent film of the metal fluoride protects the metal surfaces from further attack.

Fluorine readily displaces chlorine and other halogens from the solid metal halides. It reacts immediately with water in several simultaneous reactions which involve the formation of O_2, OF_2, H_2O_2, O_3, and HF. Fluorine and hydrogen react explosively, even at temperatures as low as that of liquid air.

Fluorine, as a result of its high electronegativity, reacts with the noble gases to form a variety of compounds such as XeF_2, XeF_4, $XeOF_4$ (Fig. 21–2). Additional compounds of the noble gases are discussed in Chapter 24.

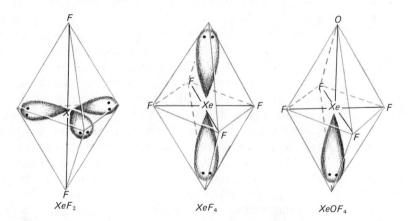

FIGURE 21–2

Structures of the molecules of XeF_2 (linear), XeF_4 (square planar), and $XeOF_4$ (square base pyramidal). Taking into account the lobes for unshared pairs of electrons, XeF_2 becomes trigonal bipyramidal, and XeF_4 and $XeOF_4$ become octahedral, as shown. Experimental evidence indicates some distortion of $XeOF_4$, with the xenon atom slightly above the plane of the fluorine atoms. Note the close similarity to ICl_2^-, ICl_4^-, and IF_5, in Fig. 21–4.

■ **2. Chlorine.** Chlorine is less active as an oxidizing agent than fluorine because of its lower electronegativity; this is a reflection of the larger atomic size of the chlorine atom.

(a) *Action with Hydrogen.* Chlorine may be mixed safely with hydrogen in the dark, as reaction between the two then is imperceptibly slow. However, when the mixture is exposed to light, the reaction is explosive.

$$H_2(g) + Cl_2(g) \longrightarrow 2HCl(g) + 44{,}124 \text{ cal} \qquad (\Delta H° = -44{,}124 \text{ cal})$$

Chemical reactions of this type, which are caused to proceed more rapidly by the effect of light, are called **photochemical reactions.** The absorption of a quantum of light energy by a chlorine molecule causes it to break up into activated (designated by an asterisk) chlorine *atoms.*

$$Cl_2 + \text{light energy} \longrightarrow 2Cl^*$$

The energy-rich, or activated, chlorine atoms react with hydrogen molecules to form hydrogen chloride molecules and activated hydrogen atoms.

$$Cl^* + H_2 \longrightarrow HCl + H^*$$

These activated hydrogen atoms then react with chlorine molecules to form hydrogen chloride molecules and activated chlorine atoms.

$$H^* + Cl_2 \longrightarrow HCl + Cl^*$$

The newly activated chlorine atoms then react in the same fashion as those originally produced by light, and thus a chain of reactions producing hydrogen chloride is set up. The chain reaction may continue until thousands of hydrogen chloride molecules have been formed. It may stop, however, when two chlorine atoms combine to form a chlorine molecule after becoming deactivated in some way, such as colliding with the walls of the container.

(b) *Action with Metals.* Chlorine is less active toward metals than fluorine, and higher temperatures are generally required in its oxidation of metals. However, powdered antimony ignites at room temperature in chlorine to form antimony(III) chloride, $SbCl_3$ (Fig. 21–3). Moist chlorine attacks such inactive metals as gold and platinum when they are heated. Apparently, perfectly dry chlorine does not react appreciably with gold or platinum, or with such metals as iron and lead. For this reason, dry liquid chlorine may be stored and shipped in steel containers. A trace of moisture, however, causes chlorine to corrode iron containers.

(c) *Action with Nonmetals.* Chlorine reacts with many nonmetals, forming covalent molecular compounds. For example, it combines with sulfur to give "sulfur monochloride," a liquid used in the vulcanization of rubber.

$$2S(s) + Cl_2(g) \longrightarrow S_2Cl_2(l) + 14.2 \text{ kcal} \qquad (\Delta H° = -14.2 \text{ kcal})$$

When the chlorine is in excess, the sulfur is oxidized to the dichloride.

$$S(s) + Cl_2(g) \longrightarrow SCl_2(l) + 12 \text{ kcal} \qquad (\Delta H° = -12 \text{ kcal})$$

Chlorine oxidizes phosphorus to the trichloride, PCl_3, when a limited supply of

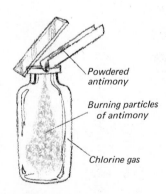

Powdered antimony

Burning particles of antimony

Chlorine gas

FIGURE 21-3
Powdered antimony burns brightly in chlorine gas, producing antimony(III) chloride.

the halogen is employed. When an excess of chlorine is used, the pentachloride, PCl_5, is formed.

$$2P(\text{excess}) + 3Cl_2 \longrightarrow 2PCl_3$$
$$2P + 5Cl_2(\text{excess}) \longrightarrow 2PCl_5$$

(d) *Action with Compounds.* The tendency of chlorine to react with hydrogen is so great that it will remove hydrogen from certain hydrogen-containing compounds. For example, turpentine, $C_{10}H_{16}$, bursts into flame when it is placed in chlorine gas.

$$C_{10}H_{16} + 8Cl_2 \longrightarrow 16HCl + 10C$$

As you can see from the equation, the carbon is displaced by the chlorine. A stepwise substitution of hydrogen by chlorine takes place when methane, CH_4, reacts with chlorine.

$$CH_4(g) + Cl_2(g) \longrightarrow CH_3Cl(g) + HCl(g) + 22.50 \text{ kcal}$$
$$(\Delta H° = -22.50 \text{ kcal})$$

$$CH_3Cl(g) + Cl_2(g) \longrightarrow CH_2Cl_2(l) + HCl(g) + 31.77 \text{ kcal}$$
$$(\Delta H° = -31.77 \text{ kcal})$$

$$CH_2Cl_2(l) + Cl_2(g) \longrightarrow CHCl_3(l) + HCl(g) + 25.17 \text{ kcal}$$
$$(\Delta H° = -25.17 \text{ kcal})$$

$$CHCl_3(l) + Cl_2(g) \longrightarrow CCl_4(l) + HCl(g) + 22.29 \text{ kcal}$$
$$(\Delta H° = -22.29 \text{ kcal})$$

When chlorine is dissolved in water, an **auto-oxidation-reduction reaction** takes place.

$$Cl_2 + H_2O \rightleftharpoons H^+ + Cl^- + \underset{\text{Hypochlorous acid}}{HClO}$$

One of the chlorine atoms of Cl_2 is oxidized to the $+1$ state in hypochlorous acid, while the other atom of chlorine is reduced to the -1 state in the chloride ion. The reaction between chlorine and water is incomplete, as is indicated by the equilibrium equation. Thus, chlorine water is a mixture of water, chlorine, the ions of hydrochloric acid, and molecules of hypochlorous acid. The solution gives off oxygen in sunlight, as the light-sensitive hypochlorous acid decomposes.

$$2HClO \xrightarrow{\text{Sunlight}} 2H^+ + 2Cl^- + O_2\uparrow$$

This reaction removes HClO from the equilibrium mixture, causing the reaction to the right to go slowly to completion. For this reason chlorine water is usually stored in colored bottles which do not permit the passage of light.

■ **3. Bromine.** The chemical properties of bromine are very similar to those of chlorine, as one would expect, but bromine atoms are larger, have a lower electronegativity, and gain electrons less readily. Thus, bromine is a weaker oxidizing agent than chlorine. The reactivity of bromine toward hydrogen, the nonmetals, the metals, and methane is less vigorous than that of chlorine. This lower reactivity is reflected in the evolution of less heat in the formation of bromides as compared to chlorides. Bromine reacts with water to form hydrobromic acid and hypobromous acid; the reaction takes place less readily than that between chlorine and water.

$$Br_2 + H_2O \rightleftharpoons H^+ + Br^- + \underset{\substack{\text{Hypobromous} \\ \text{acid}}}{HBrO}$$

Hypobromous acid decomposes in sunlight according to the equation

$$2HBrO \longrightarrow 2H^+ + 2Br^- + O_2\uparrow$$

■ **4. Iodine.** Of the four naturally occurring halogens, the iodine atom has the least attraction for an additional electron. This is due to its large radius and great number of electron shells. Thus, iodine is the weakest oxidizing agent of the four naturally occurring halogens, and the iodide ion is the most easily oxidized of the four halides.

$$2I^- \longrightarrow I_2 + 2e^-$$

Iodine combines with many of the metals but with the liberation of much less heat than the other halogens; with some metals, heating is required before reaction occurs. Although iodine does not oxidize the other halide ions (except astatide), it will remove electrons from certain other nonmetal ions such as sulfide ions, S^{2-}.

$$S^{2-} + I_2 \longrightarrow S + 2I^-$$

Iodine shows slight chemical activity with water, compared with the other halogens. A very sensitive qualitative test for elemental iodine depends upon the formation of a deep blue color when even trace quantities react with starch.

The trend down each group in the Periodic Table is from nonmetallic to metallic character. In line with this trend, iodine is the most nearly metallic of the naturally occurring halogens. Accordingly, chemists have tried to prepare compounds of iodine in which it plays the role of a "true" metal. The preparation of several compounds of tripositive iodine, such as the acetate, $I(C_2H_3O_2)_3$, the phosphate, IPO_4, the nitrate, $I(NO_3)_3$, and the perchlorate, $I(ClO_4)_3 \cdot 2H_2O$, has been achieved. In these compounds iodine is acting in the capacity of a trivalent metal. As might be expected, these compounds are readily hydrolyzed by water.

■ **5. Astatine.** Most of our knowledge of the element astatine, the fifth halogen, has come from a study of one of its isotopes (mass number 211). The isotope was first obtained in 1940 by Corson, Mackenzie, and Segre, who prepared it by bombarding bismuth with alpha particles in the 60-inch cyclotron at the University of California, Berkeley.

$$^{209}_{83}Bi + ^4_2He \longrightarrow ^{211}_{85}At + 2^1_0n$$

Astatine may exist in nature as a short-lived intermediate in a nuclear decay chain, but too little of the element is present from this source at a given time to permit its study. The name **astatine** comes from the Greek word for "unstable."

As would be expected, astatine is a solid and is more metallic than iodine. It is volatile at room temperature and is soluble in solvents such as carbon disulfide and carbon tetrachloride. Elemental astatine is the weakest oxidizing agent of the halogens, and the astatide ion, At^-, is the strongest reducing agent of the family. In addition to the negative ion, At^-, astatine forms at least two positive ions. Extensive research on the chemical properties of astatine is being conducted.

21.6   Uses of the Halogens

Fluorine gas has been used in the fluorination of organic compounds (replacing hydrogen with fluorine) since the early work of Moissan in 1886. Such reactions are usually difficult to control, however. More commonly now, the organic compounds are dissolved in anhydrous hydrogen fluoride, and the solution is electrolyzed. The **fluorocarbon** compounds thus produced are quite stable and nonflammable. They are used as lubricants, refrigerants, coolants, hydraulic liquids, plastics, and insecticides. Freon-12, which is CCl_2F_2, is widely used as a refrigerant; CCl_3F is used as an insecticide; and Teflon is a plastic composed of $-CF_2CF_2-$ units. One of the most important uses of fluorine gas is in the production of uranium hexafluoride, UF_6, which is used in separating the isotopes of uranium by the gaseous diffusion process in connection with the production of atomic energy. Another use of fluorine is in the making of sulfur(VI) fluoride, SF_6, a stable gas with high dielectric and insulating capacities for high voltage. Many community water supplies are treated with fluoride ion for the purpose of decreasing the incidence of dental cavities.

Large quantities of the chlorine produced commercially are used in bleaching wood pulp and cotton cloth. The chlorine reacts with water to form hypochlorous acid, which bleaches by oxidizing colored substances to colorless compounds. Most community water supplies are treated with small amounts of chlorine to kill bacteria. Large quantities of chlorine are used in chlorinating hydrocarbons (replacing hydrogen with chlorine) to produce such compounds as carbon tetrachloride (CCl_4), chloroform ($CHCl_3$), and paradichlorobenzene (Dichlorocide), and in the production of polyvinyl chloride and other polymers.

A principal use of bromine has been in the manufacture of ethylene dibromide, $C_2H_4Br_2$, a constituent of antiknock gasoline along with tetraethyl lead, $Pb(C_2H_5)_4$. When the gaseous fuel in an internal combustion engine explodes with extreme rapidity, "knocking" occurs with a great loss in efficiency. Tetraethyl lead retards the ignition, with a corresponding increase in the efficiency of the engine. The ethylene dibromide is added to convert the lead from tetraethyl lead to lead bromide, which escapes from the engine as a vapor. This largely prevents the accumulation of lead in the engine, but it increases the lead concentration in the air and is a cause of great concern to many persons. "Lead-free" gasolines, containing little or no tetraethyl lead, must be used in automobiles with the catalytic converters, introduced in 1975 on many models to clean up the exhaust (see Section 24.5). The tetraethyl lead quickly "poisons" the platinum catalyst in the converters.

Other applications of bromine include the production of certain organic dyes, the preparation of light-sensitive silver bromide for use in making photographic film, and the manufacture of the bromides of sodium and potassium which are used in medicine as sedatives and soporifics.

Iodine in alcohol solution with potassium iodide (**tincture of iodine**) is used as an antiseptic. In the form of iodide salts, iodine is essential in small amounts in the diet for the proper functioning of the thyroid gland. Iodine deficiency may lead to the development of goiter. Iodized table salt contains about 0.023 per cent potassium iodide. Silver iodide is used in the manufacture of photographic films and in the seeding of clouds to induce rain. Iodoform, CHI_3, is used as an antiseptic in the dressing of wounds.

21.7 Interhalogen Compounds

The compounds formed by the union of two different halogens are called **interhalogen compounds.** Molecules of these compounds consist of an atom of the heavier halogen bonded to an odd number of atoms of the lighter halogen. Examples of interhalogen compounds are given in Table 21–3.

TABLE 21–3   Interhalogen Compounds

IBr	BrCl	ClF
ICl	BrF	ClF_3
ICl_3	BrF_3	
IF_5	BrF_5	
IF_7		

Because the smaller halogen atoms are grouped about the larger, the number of smaller atoms per molecule increases as the radius ratio, $r_{larger}/r_{smaller}$, increases (see Section 11.11). Thus, iodine forms a heptafluoride, bromine merely the pentafluoride, and chlorine the trifluoride. Most of these compounds are unstable and extremely reactive chemically. The reactions of the interhalogen compounds are similar to those of the component halogens.

The **polyhalides** of the alkali metals, such as KI_3, $KICl_2$, $KICl_4$, $CsIBr_2$, and $CsClBr_2$, are closely related to the interhalogen compounds.

Typical structures of the polyhalides and interhalogens are shown in Fig. 21–4.

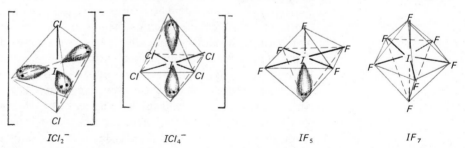

| ICl_2^- | ICl_4^- | IF_5 | IF_7 |

FIGURE 21–4

Structures of the ions ICl_2^- (linear) and ICl_4^- (square planar), and of the molecules IF_5 (square base pyramidal) and IF_7 (pentagonal bipyramidal). Taking into account the lobes for unshared pairs of electrons, the structure for ICl_2^- becomes trigonal bipyramidal, and the structures for ICl_4^- and IF_5 become octahedral. Experimental evidence indicates some distortion in IF_5, with the iodine atom slightly above the plane of the fluorine atoms. (See also Fig. 21–2.)

The Hydrogen Halides

The binary compounds which contain only hydrogen and one of the halogens are called **hydrogen halides.** These compounds have the general formula HX, in which X represents the halogen. At ordinary temperatures, molecules of hydrogen fluoride polymerize through hydrogen bonding into complex units which may be represented by the formula $(HF)_x$. Figure 21–5 shows the relative sizes of the hydrogen halide molecules.

FIGURE 21-5

Approximate relative sizes of the hydrogen halide molecules. The internuclear distance (distance between the center of the hydrogen atom and the center of the halogen atom) is indicated for each molecule.

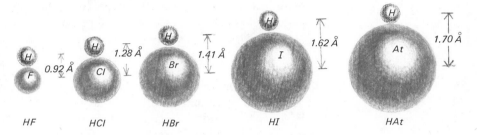

HF   HCl   HBr   HI   HAt

21.8   Preparation of the Hydrogen Halides

Various methods are used for the preparation of the hydrogen halides.

■ **1. By Direct Union.**   One method of preparing the hydrogen halides is by the direct union of their constituent elements according to the equation

$$H_2 + X_2 \longrightarrow 2HX \qquad (X = \text{symbol for halogens})$$

The enthalpies of formation of four of the hydrogen halides are given in Table 21–4. Note that the rapid increase in the enthalpies of formation of these compounds with increasing atomic weight of the halogen indicates a **corresponding decrease in the quantities of energy released** during the formation of the compounds. The difficulty of producing elemental fluorine and its relatively high cost make the direct union method of preparing hydrogen fluoride impractical. On the other hand, both hydrogen and chlorine are readily available from the electrolysis of brine (Section 21.2), and one commercial method of producing hydrogen chloride involves the combustion of hydrogen with chlorine in burners specially designed for this purpose. The reaction of bromine with hydrogen is much less vigorous than that with either chlorine or fluorine. If the reaction between hydrogen and bromine is to proceed appreciably, the mixture must be heated to about 200° C in contact with a catalyst such as platinum or carbon. The reaction is a reversible one that quickly reaches equilibrium, but it may be forced nearly to completion by using an excess of hydrogen. The excess hydrogen is recycled by liquefying the HBr and pumping away the gaseous hydrogen.

The direct union of hydrogen and iodine is unsatisfactory for the preparation of hydrogen iodide, because the reaction is slow and never complete. Heat decomposes hydrogen iodide; this means that the reaction is readily reversible and the equilibrium yield of HI is low. This reaction does not become self-sustaining as do the similar reactions of the other halogens.

The mechanisms for the preparations of hydrogen bromide and hydrogen iodide from the elements were discussed in Sections 17.1 and 17.2.

■ 2. By the Action of Concentrated Sulfuric Acid Upon a Metallic Halide.

The most convenient method of producing hydrogen fluoride and hydrogen chloride is based upon the reaction of concentrated sulfuric acid upon a halide of a metal.

$$2MX + H_2SO_4 \longrightarrow M_2SO_4 + 2HX\uparrow$$

Hydrogen fluoride is usually prepared by heating a mixture of the mineral fluorite, CaF_2, and concentrated sulfuric acid in a lead or platinum retort.

$$CaF_2 + H_2SO_4 \longrightarrow CaSO_4 + 2HF\uparrow$$

The hydrogen fluoride is evolved as a gas and if absorbed in water forms hydrofluoric acid. This acid acts upon glass, and for this reason is stored in bottles made of lead, wax, or certain plastics such as polyethylene.

Hydrogen chloride is prepared, both in the laboratory and on a commercial scale, by the action of concentrated sulfuric acid upon sodium chloride, the most plentiful and least costly of the chlorides.

$$NaCl + H_2SO_4 \longrightarrow NaHSO_4 + HCl\uparrow$$

At slightly elevated temperatures the reaction goes to completion, because hydrogen chloride is insoluble in the reaction mixture. The by-product, $NaHSO_4$, is an example of a **hydrogen salt** and is known as either sodium hydrogen sulfate, sodium acid sulfate, or sodium bisulfate. It still contains an acidic hydrogen and can be used in preparing more hydrogen chloride. By adding more sodium chloride and heating the mixture to a higher temperature, the following reaction takes place:

$$NaCl + NaHSO_4 \longrightarrow Na_2SO_4 + HCl\uparrow$$

The hydrogen chloride thus produced is usually absorbed in water and marketed as hydrochloric acid.

In the preparation of hydrogen bromide and hydrogen iodide, the use of concentrated sulfuric acid gives rise to undesirable by-products. Hydrogen bromide prepared in this way is impure with bromine and sulfur dioxide, because of the oxidation of a portion of the hydrogen bromide by the hot, concentrated sulfuric acid.

$$NaBr + H_2SO_4 \longrightarrow NaHSO_4 + HBr\uparrow$$
$$H_2SO_4(l) + 2HBr(g) + 11.76 \text{ kcal} \longrightarrow 2H_2O(l) + SO_2(g) + Br_2(g)$$
$$(\Delta H° = +11.76 \text{ kcal})$$

The relative ease of oxidation of the halide ions was discussed in Section 21.2. Concentrated sulfuric acid does not oxidize either the fluoride or chloride ion. However, a bromide ion, which is larger than either a fluoride or a chloride ion and more easily oxidized, loses an electron by reacting with sulfuric acid and becomes free bromine. This means that hydrogen bromide is a stronger reducing agent than either hydrogen fluoride or hydrogen chloride. If phosphoric acid, which is not a good oxidizing agent, is used as the nonvolatile acid in place of sulfuric to produce hydrogen bromide, the secondary reaction is avoided and nearly pure hydrogen bromide is obtained.

As would be predicted, hydrogen iodide is a still stronger reducing agent than hydrogen bromide, and the hydrogen iodide produced by the reaction of concentrated sulfuric acid upon sodium iodide reduces sulfuric acid to the hydrogen sulfide stage.

$$NaI + H_2SO_4 \longrightarrow NaHSO_4 + HI\uparrow$$
$$8HI(g) + H_2SO_4(l) \longrightarrow H_2S(g) + 4I_2(s) + 4H_2O(l) + 134.29 \text{ kcal}$$
$$(\Delta H° = -134.29 \text{ kcal})$$

Gaseous hydrogen iodide produced by this method is colored violet by iodine vapor.

■ **3. By the Hydrolysis of Nonmetallic Halides.** Another method of producing hydrogen halides is by the hydrolysis (reaction with water) of the covalent halides of the nonmetals. Typical of this class of compounds are PCl_3, PBr_3, PI_3, and SCl_4. Halides of this type generally react with water to form two acids—the hydrogen halide and an oxyacid of the other nonmetal; quite often the reaction is a vigorous one.

$$PBr_3 + 3H_2O \longrightarrow H_3PO_3 + 3HBr\uparrow$$
$$SCl_4 + 3H_2O \longrightarrow H_2SO_3 + 4HCl\uparrow$$

This method is rarely used for the commercial preparation of hydrogen chloride. However, the preparation of hydrogen bromide and hydrogen iodide by the hydrolysis of phosphorus(III) bromide and phosphorus(III) iodide, respectively, is quite common. The PBr_3 and PI_3 are prepared by the direct union of phosphorus with bromine and iodine, respectively.

■ **4. By the Halogenation of Hydrocarbons.** When fluorine, chlorine, or bromine (but not iodine) is allowed to react with a saturated or aromatic hydrocarbon, one product of the reaction is the corresponding hydrogen halide (Section 27.8). A catalyst is usually required, and the hydrogen halide is often a by-product of a reaction used to produce a desired halogenated hydrocarbon. For example, hydrogen chloride is a by-product of the manufacture of ethyl chloride (C_2H_5Cl) from ethane (C_2H_6) and chlorine. The production of chlorinated organic compounds used about half of the 10.6 million tons of chlorine produced in 1974.

$$C_2H_6(g) + Cl_2(g) \longrightarrow C_2H_5Cl(l) + HCl(g) + 34.45 \text{ kcal}$$
$$(\Delta H° = -34.45 \text{ kcal})$$

Large quantities of hydrogen chloride are produced commercially by such reactions, because the demand for halogenated hydrocarbons is large.

■ **5. Other Methods.** Hydrogen bromide and hydrogen iodide in aqueous solution (as the corresponding acids) may be produced through the reduction of the elemental halogen by means of hydrogen sulfide or sulfurous acid. These methods are illustrated by the equations

$$Br_2 + H_2S \longrightarrow 2H^+ + 2Br^- + S$$

$$Br_2 + H_2SO_3 + H_2O \longrightarrow \underset{\text{Hydrobromic acid}}{2H^+ + 2Br^-} + \underset{\text{Sulfuric acid}}{2H^+ + SO_4^{2-}}$$

$$I_2 + H_2S \longrightarrow 2H^+ + 2I^- + S$$

$$I_2 + H_2SO_3 + H_2O \longrightarrow 2H^+ + 2I^- + 2H^+ + SO_4^{2-}$$

21.9 General Properties of the Hydrogen Halides

HF, HCl, HBr, and HI are colorless gases with sharp, penetrating odors. Their constituent atoms are linked together by polar covalent bonds. The polarity of these molecules is a function of the electronegativity of the halogen atom; it is greatest for HF and least for HI. Some of the more important physical properties of the halides are given in Table 21–4.

All of the hydrogen halides decompose into their constituent elements when heated sufficiently. Of the four hydrogen halides shown in Table 21–4, hydrogen iodide is the most easily decomposed, while hydrogen chloride and hydrogen fluoride show only slight dissociation at 1,000° C. The stability of these gases decreases, then, as the formula weight increases.

The hydrogen halides fume in moist air, and all are very soluble in water. With the exception of hydrogen fluoride, the hydrogen halides are strong electrolytes; they ionize completely in dilute aqueous solution. The ionization of hydrogen chloride is given by the equation $HCl \longrightarrow H^+ + Cl^-$. If the concentration of these acids in water is greatly increased, the molecular form (HX) appears and escapes from solution.

TABLE 21-4   Some Properties of the Hydrogen Halides

	HF	*HCl*	*HBr*	*HI*
Molecular weight	20.006	36.461	80.917	127.91
Melting point, °C	−83.1	−114.2	−86.8	−50.8
Boiling point, °C	19.9	−85.1	−66.7	−35.36
Solubility in water, g/100 g of water	∞ (0°)	82.3 (0°)	221 (0°)	234 (10°)
Enthalpy of formation, $\Delta H^\circ_{f\,298}$, cal/mole	−64,800	−22,062	−8,700	+6,330
Constant-boiling solutions:				
Composition, per cent	35.35	20.24	47	57
Boiling point, °C	120	110	126	127

The anhydrous hydrogen halides are rather inactive chemically and do not attack dry metals at ordinary temperatures. Neither do they conduct electricity in the liquid state; this fact indicates that they are molecular rather than ionic compounds.

21.10   The Hydrohalic Acids

Aqueous solutions of the hydrogen halides are called **hydrofluoric acid, hydrochloric acid, hydrobromic acid,** and **hydriodic acid.**

■ **1. Properties Dependent upon Hydrogen Ion.**   Hydrofluoric acid is a weak acid, but the other hydrohalic acids are very strong. All of them react with the metals above hydrogen in the activity series, halide salts and hydrogen being formed. Their reaction with zinc is given by

$$Zn + 2H^+ + (2X^-) \longrightarrow Zn^{2+} + (2X^-) + H_2\uparrow$$

Notice in the equation that the hydrogen ion acts as an oxidizing agent. The hydrohalic acids neutralize aqueous bases, forming salts and water. Their reaction with sodium hydroxide is given by the equation

$$\underset{\text{Base}}{(Na^+) + OH^-} + \underset{\text{Acid}}{H^+ + (X^-)} \longrightarrow \underset{\text{Salt}}{(Na^+) + (X^-)} + \underset{\text{Water}}{H_2O}$$

■ **2. Properties Dependent upon Halide Ion.**   The halide ion of the hydrohalic acids can serve either as a reducing agent or as a precipitating agent, as illustrated by the following equations.

As reducing agent:

$$(2H^+) + 2Br^- + Cl_2 \rightleftharpoons (2H^+) + 2Cl^- + Br_2$$
$$14H^+ + 6Cl^- + Cr_2O_7^{2-} \rightleftharpoons 2Cr^{3+} + 3Cl_2 + 7H_2O$$

As precipitating agent:

$$Ag^+ + (NO_3^-) + (H^+) + Cl^- \rightleftharpoons AgCl + (H^+) + (NO_3^-)$$

21.11   Unique Properties of Hydrogen Fluoride and Hydrofluoric Acid

Hydrogen fluoride is unique among the hydrogen halides in that its molecules are associated. Furthermore, hydrofluoric acid is weak while the other hydrohalic acids are strong, and it attacks glass whereas the others do not.

■ **1. Hydrogen Bonding.**   Pure hydrogen fluoride differs from the other hydrogen halides because of the tendency of its molecules to associate through hydrogen bonding. In the solid state hydrogen fluoride exists as zigzag chains of molecules which extend the entire length of the crystal (Fig. 21–6).

Vapor density measurements show hydrogen fluoride gas to be a mixture of $(HF)_2$ and $(HF)_3$ at 26° C while at 88° C the vapor density corresponds to the formula HF. *The hydrogen atom is attracted simultaneously to two atoms of high*

FIGURE 21–6

Illustration showing the associative nature of HF, resulting from hydrogen bonding.

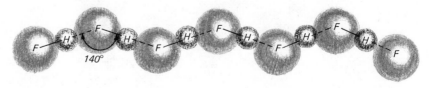

140°

electronegativity, instead of only one, and thereby acts as a bond between the two. The **hydrogen bond** is primarily electrostatic in character; it cannot be covalent because hydrogen has only a single electron in a $1s$ orbital and is therefore incapable of forming more than one covalent bond. Two covalent bonds (one from the hydrogen to each of the two highly electronegative atoms) would involve use of $2s$ or $2p$ orbitals of hydrogen. In hydrogen, these orbitals have energies too high to be useful for bonding.

Hydrogen bonds exist between highly electronegative atoms such as fluorine, oxygen, chlorine, and nitrogen. Examples are FHF, OHF, NHF, OHO, NHO, NHN, and OHCl. The strength of hydrogen bonds is about 5 to 10 per cent of that of ordinary covalent bonds. The boiling point of hydrogen fluoride is abnormally high compared to that of the other hydrogen halides, and results from the fact that hydrogen bonds must be broken before the hydrogen fluoride vaporizes, and that associated molecules of the vapor have high molecular weights compared to the undissociated molecule, HF.

Figure 21–7(a) shows a plot of boiling points and Fig. 21–7(b) heats of vaporization against the number of the period (horizontal rows) of the Periodic Table, for the binary hydrogen compounds of elements in families IVA, VA, VIA, and VIIA. Included for comparison are four of the noble gases, Ne, Ar, Kr, and Xe,

FIGURE 21-7

Graphic illustrations of (*a*) the abnormally high boiling points and (*b, next page*) the heats of vaporization caused by hydrogen bonding. H_2O, HF, and NH_3 have hydrogen bonding and exhibit abnormally high boiling points and heats of vaporization. CH_4 does not have hydrogen bonding. Plots for four of the noble gases are included to indicate typical behavior in the absence of hydrogen bonding.

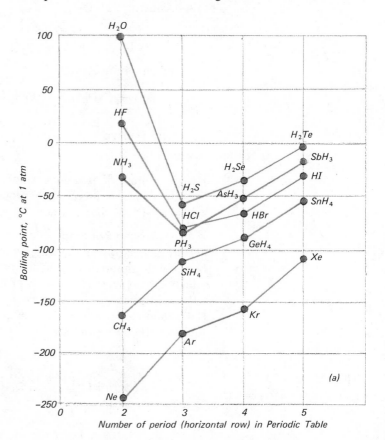

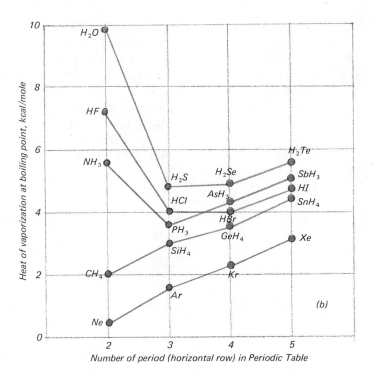

FIGURE 21–7(b)
Caption on preceding page.

which of course do not possess hydrogen bonding. Note that, in general, boiling points and heats of vaporization increase with increasing molecular or atomic weight. However, the compounds H_2O, HF, and NH_3 possess appreciable hydrogen bonding, providing clusters, and thus act as if they have larger molecular weights than they actually have. More energy is required to break the additional hydrogen bonds in order to bring the material to the gaseous (vapor) state. Hence, those substances possessing hydrogen bonds have abnormally high boiling points and heats of vaporization, as shown in the diagrams. Methane (CH_4), however, does not have hydrogen bonding and hence has a lower boiling point and heat of vaporization than the corresponding compounds of other elements in the family. Similarly, the noble gas Ne has lower values than do those noble gases that are larger than neon.

The tendency for fluorine to form hydrogen bonds with hydrogen makes it possible for many metal fluorides to form stable hydrogen difluoride salts such as $NaHF_2$ and KHF_2. The hydrogen difluoride ion has the structure

$$\left[\overset{\times\times}{\underset{\times\times}{\times}} F \overset{\times}{\underset{\times}{}} H \overset{\times}{\underset{\times}{}} \overset{\times\times}{\underset{\times\times}{}} F \times \right]^{-}$$

Hydrogen di-
fluoride ion

Hydrogen bonds do not occur in hydrogen chloride, hydrogen bromide, or hydrogen iodide, because the electronegativities of chlorine, bromine, and iodine are too low. However, it has been shown that hydrogen bonding can occur between a highly polar molecule containing hydrogen and an atom or ion in which polar character can be induced easily. The valence electrons of the larger halogen atoms are sufficiently far from their nuclei to be fairly easily attracted by a nearby polar molecule. Hence, a highly positive hydrogen atom in a compound can induce polar character in nearby bromine or iodine atoms by attracting their electrons and hence shifting the center of negative charge toward one end of the halogen atom. The O—H hydrogen in C_6H_5OH (phenol) is sufficiently positive, for example, to induce such a dipole in bromine atoms and to create reasonably strong O—H $\cdots$ X hydrogen bonds.

C_6H_5O—H

Phenol

Bromine
atom

C_6H_5O—H $\cdots$

Hydrogen bonding

■ 2. The Ionization of Hydrofluoric Acid.

Hydrofluoric acid is weak (slightly ionized) whereas the other hydrohalic acids are strong (highly ionized). In order for ionization to occur, the bond in H—X must be broken, and then hydration of the resultant ions must take place. The H—F bond is about twice as difficult to break as is the H—I bond, and the strengths of the H—Br and H—Cl bonds lie between those of the other two (see Table 20–2). The greater strength of the H—F bond as compared to that of H—I is considered to result from the smaller interatomic distance in HF (HF, 1.0 Å; HI, 1.7 Å). The ionization of hydrogen fluoride in dilute solution may be written as

$$HF \rightleftharpoons H^+ + F^-$$

However, a second ion is formed in this process due to hydrogen bonding; it is the hydrogen difluoride ion.

$$F^- + HF \rightleftharpoons HF_2^-$$

In very concentrated solutions of hydrofluoric acid, the degree of ionization rises sharply. This behavior is opposite to that observed for all other weak electrolytes and results from an increase in the concentration of the dimer, H_2F_2, which is a stronger acid than the monomer, HF.

$$H_2F_2 \rightleftharpoons H^+ + HF_2^-$$

■ 3. The Action of Hydrofluoric Acid on Glass.

The most characteristic behavior of hydrofluoric acid is its action on sand, which is silicon dioxide, and glass, which is a mixture of silicates (mostly calcium silicate). The reaction of hydrofluoric acid with sand (SiO_2) and with calcium silicate ($CaSiO_3$) is given by

$$SiO_2(s) + 4HF(g) \rightleftharpoons SiF_4(g) + 2H_2O(l) + 45.7 \text{ kcal} \qquad (\Delta H° = -45.7 \text{ kcal})$$
$$CaSiO_3 + 6HF \longrightarrow CaF_2 + SiF_4\uparrow + 3H_2O$$

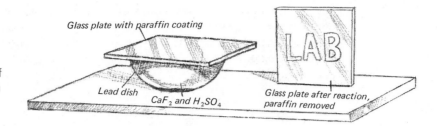

FIGURE 21-8

Laboratory demonstration of etching glass with hydrogen fluoride.

The silicon escapes in these reactions as a volatile compound, silicon tetrafluoride. When one mole of SiF_4 is formed from the elements, the amount of energy released (361 kcal) is exceedingly large. This large amount of energy provides the driving force for the conversion of silicon-oxygen compounds to silicon tetrafluoride by the action of hydrofluoric acid. For example, in the reaction with SiO_2, 80 kcal per mole of SiO_2 are released, and the escape of the gaseous SIF_4 helps drive the reaction to completion. Glass objects such as electric light bulbs can be frosted, or etched, by hydrofluoric acid. Thermometers, burets, and other glassware may be marked by use of this acid. In order that a restricted area may be etched, the glass is first covered with paraffin, and then, with a sharp-pointed instrument, part of the paraffin is removed from the glass (Fig. 21-8). Extreme care should be taken in handling hydrofluoric acid for it is very corrosive to the skin and causes painful, slow-healing burns. Hydrofluoric acid is also a local anesthetic, and if in contact with the skin may not be noticed until much damage has been done.

21.12   The Metal Halides

All metals form halides, but the variations in character among such halides are striking. The halides of the alkali metals and the heavier alkaline earth metals, almost all the metal fluorides, and some halides of the transition metals in their lower oxidation states (e.g., Fe^{2+}, Mn^{2+}, Cr^{2+}) are predominantly ionic, or saltlike, in character. These ionic halides tend to have high boiling points and high electrical conductivities in the fused state. Those which dissolve appreciably in water undergo little or no hydrolysis.

On the other hand, the halides of metals with large charge-to-size ratios are likely to be predominantly covalent in character, because of the attraction by the metal for the electrons in the outer shell of the halide ion. Aluminum chloride, $AlCl_3$, tin(IV) chloride, $SnCl_4$, and titanium(IV) chloride, $TiCl_4$, are typical of this group of halides. These covalent halides are characterized by volatility, by solubility in nonpolar solvents, by lack of electrical conductivity in the fused state, and by extensive hydrolysis in water.

As might be expected, there are metal halides intermediate in character between the ionic and covalent ones. For example, iron(III) chloride, $FeCl_3$, is volatile and readily hydrolyzed, but it is a good conductor of electricity in the fused state.

The anhydrous halides of metals may be prepared by direct union of their constituents. On the other hand, hydrated halo salts are obtained by the action of the hydrohalic acids on metals above hydrogen in the activity series, and by the action of hydrohalic acids on most oxides, hydroxides, and carbonates of these metals. Several slightly soluble metal halides may be prepared by ionic combination. For example, when solutions containing silver nitrate and sodium chloride are mixed, silver chloride precipitates as a curdy white solid.

$$Ag^+ + (NO_3^-) + (Na^+) + Cl^- \longrightarrow \underline{AgCl} + (Na^+) + (NO_3^-)$$

or simply
$$Ag^+ + Cl^- \longrightarrow \underline{AgCl}$$

Other slightly soluble halides include $AgBr$, AgI, $PbCl_2$, $PbBr_2$, PbI_2, $CuCl$, CuI, $CuBr$, $TlCl$, Hg_2Cl_2, CaF_2, and BaF_2. It is a striking fact that the solubilities of the fluorides differ markedly from those of the other halides. For example, silver fluoride is readily soluble in water whereas the other silver halides are only slightly soluble. On the other hand, calcium and barium fluorides are insoluble in water while the other halides of these metals are freely soluble.

21.13   Uses of the Hydrohalic Acids and Halides

The largest use for hydrogen fluoride is in the production of fluorocarbons for refrigerants, plastics, and propellants. The second largest use is in the production of cryolite, Na_3AlF_6, which is important in the metallurgy of aluminum. The acid is also used in making UF_6 in the atomic energy program, in the etching of glass, as a catalyst in the petroleum industry, and in the production of other inorganic compounds (such as BF_3) which serve as catalysts in the industrial synthesis of certain organic compounds. The mineral fluorspar, CaF_2, is used as a flux, because it reacts to form easily fusible products with various substances that cannot be melted readily. Sodium fluoride is used as an insecticide, a flux, and a fungicide.

Hydrochloric acid is the most important acid in industry other than sulfuric. It is used in the manufacture of metal chlorides, dyes, glue, glucose, and various other chemicals. A considerable amount is also used in removing oxide coatings from iron or steel which is to be galvanized, tinned, or enameled.

The amounts of hydrobromic and hydriodic acid used commercially are insignificant compared to the amount of hydrochloric acid used. Hydrobromic acid is used in analytical chemistry and in the synthesis of certain organic compounds.

Oxygen Compounds of the Halogens

Although the halogens tend to gain an eighth electron for their outer shells by taking electrons from certain other elements, thus becoming negative halide ions, X^-, they also quite frequently complete their outer shells by sharing electrons with other atoms. For example, electrons are shared in X_2, HX, PX_3, and SnX_4. The halogens also share electrons with oxygen in the formation of oxyacids and oxysalts, in which the halogens assume positive oxidation states. The oxygen compounds of the halogens are noted for their strength as oxidizing agents.

21.14   Binary Halogen-Oxygen Compounds

The halogens do not combine directly with oxygen. However, binary halogen-oxygen compounds can be prepared by indirect methods. Because oxygen is more electronegative than chlorine, bromine, and iodine, it is proper to call their oxygen derivatives oxides. On the other hand, the fluorine compounds with oxygen are called fluorides because fluorine is the more electronegative element. The known binary halogen-oxygen compounds are listed in Table 21-5. These compounds are characteristically reactive and unstable toward heat.

TABLE 21-5   Binary Halogen-Oxygen Compounds

Fluorine	Chlorine	Bromine	Iodine
OF_2	Cl_2O	Br_2O	I_2O_4
O_2F_2	ClO_2	Br_3O_8	I_4O_9
	Cl_2O_6	BrO_2	I_2O_5
	Cl_2O_7		

■ **1. Fluorides of Oxygen.**   Just two oxygen-fluorine compounds are known. They are **oxygen difluoride**, OF_2, a colorless gas, and **oxygen monofluoride**, O_2F_2, a red liquid. Only the difluoride is of sufficient importance for consideration here.

The bonding in OF_2 is covalent and the oxidation states for oxygen and fluorine are $+2$ and -1, respectively. Oxygen difluoride is prepared by the reaction of elemental fluorine with a solution of sodium hydroxide.

$$2F_2 + (2Na^+) + 2OH^- \longrightarrow (2Na^+) + 2F^- + OF_2\uparrow + H_2O$$

Although oxygen difluoride dissolves appreciably in water, it is apparently not the anhydride of an acid because the resulting solutions are not acidic.

■ **2. Oxides of Chlorine.**   **Chlorine monoxide,** Cl_2O, is a yellowish-red gas which is apt to decompose explosively. This compound is best prepared by passing chlorine over freshly precipitated, dried mercury(II) oxide.

$$2Cl_2 + 2HgO \longrightarrow HgCl_2 \cdot HgO + Cl_2O\uparrow$$

Chlorine monoxide is an active oxidizing agent and dissolves in water giving hypochlorous acid, HClO.

The structures of OF_2 and Cl_2O, both angular, are shown in Fig. 21-9.

Chlorine dioxide, ClO_2, is a yellow gas that is violently explosive when pure. It can be safely handled, however, if diluted with carbon dioxide or air. The chlorine dioxide molecule, which contains an odd number of electrons, is thought to be a resonance structure (see Sections 8.12 and 23.12, Part 3) involving a three-electron bond coupled with an electron-pair bond (Fig. 21-10).

The most convenient method available for the commercial preparation of chlorine dioxide involves the reaction of sodium chlorite with chlorine diluted with air.

$$2NaClO_2 + Cl_2 \xrightarrow{\text{Air}} 2NaCl + 2ClO_2\uparrow$$

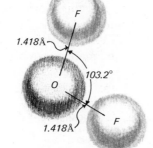

(a) OF_2

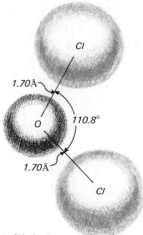

(b) Cl_2O

FIGURE 21-9

Angular structures of the OF_2 and Cl_2O molecules (drawn approximately to scale). Interatomic distances are measured from the centers of the atoms.

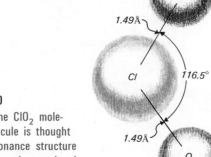

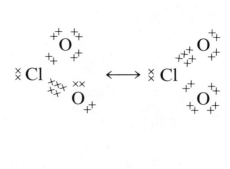

ClO_2

FIGURE 21-10

Structure of the ClO_2 molecule. The molecule is thought to have a resonance structure involving a three-electron bond coupled with an electron-pair bond (see Lewis formulas).

Chlorine heptaoxide, Cl_2O_7, is a colorless liquid, which is dangerously heat and shock sensitive. It is the acid anhydride of perchloric acid and is obtained by dehydrating perchloric acid with P_4O_{10} and then distilling the product.

$$4HClO_4(l) + P_4O_{10}(s) \longrightarrow 2Cl_2O_7(g) + 4HPO_3(s) + 89.8 \text{ kcal}$$
$$(\Delta H° = -89.8 \text{ kcal})$$

Its structure is given in Fig. 21-11.

FIGURE 21-11

The structure of the Cl_2O_7 molecule. Two tetrahedral ClO_3 groups are attached at an angle of 128° to a central oxygen atom.

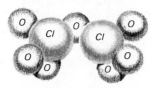

21.15   The Oxyacids of the Halogens and Their Salts

There is no definite evidence for the existence of oxyfluorine acids or oxyfluorine salts. The known oxyacids for three of the other halogens, including their formulas and the oxidation states of the halogen in them, are given in Table 21–6. The nomenclature for these acids and their salts was discussed in Section 4.12.

TABLE 21–6   Oxyacids of the Halogens

Oxidation States	Oxyacids of Chlorine	Oxyacids of Bromine	Oxyacids of Iodine
+1	$HClO$	$HBrO$	HIO
+3	$HClO_2$	$HBrO_2$	
+5	$HClO_3$	$HBrO_3$	HIO_3
+7	$HClO_4$	$HBrO_4$	HIO_4
			H_5IO_6

The oxygen compounds of chlorine, bromine, and iodine differ sufficiently to make generalizations concerning them impossible. However, trends in properties of the oxychlorine acids and salts are well established (Table 21–7), and these trends hold, in a general way, for the analogous bromine and iodine compounds.

The greater the oxygen content in members of a series of oxygen compounds

TABLE 21–7   Trends in Properties of the Oxyacids of Chlorine and their Sodium Salts

Oxidation State	Acid	Thermal Stability and Acid Strength	Oxidizing Power	Salt	Thermal Stability	Oxidizing Power and Anion Base Strength
+1	$HClO$			$NaClO$		
+3	$HClO_2$	Increase ↓	Increases ↑	$NaClO_2$	Increases ↓	Increase ↑
+5	$HClO_3$			$NaClO_3$		
+7	$HClO_4$			$NaClO_4$		

Thermal stability increases greatly →

Oxidizing power increases greatly ←

(oxyacids or their salts) of a halogen, the greater is their thermal stability. The acids decompose at much lower temperatures than do their salts. These rules apply generally to the oxygen compounds of sulfur, nitrogen, and phosphorus, as well as to those of the halogens.

The salts undergo disproportionation, or auto-oxidation-reduction, upon heating. This, too, is characteristic of the oxysalts of sulfur and phosphorus (but not of nitrogen). Examples are

$$3\,NaClO \longrightarrow NaClO_3 + 2NaCl$$

Sodium hypochlorite   Sodium chlorate

$$4\,Na_2SO_3 \longrightarrow 3\,Na_2SO_4 + Na_2S$$

Sodium sulfite   Sodium sulfate

$$4\,Na_2HPO_3 \longrightarrow 2\,Na_3PO_4 + PH_3 + Na_2HPO_4$$

Disodium hydrogen   Sodium phosphate   Disodium hydrogen
phosphite   phosphate

The same sort of reaction is characteristic of the elements.

$$2\,NaOH + Cl_2 \longrightarrow NaCl + NaClO + H_2O$$

$$6\,NaOH + 3S \longrightarrow 2\,Na_2S + Na_2SO_3 + 3\,H_2O$$

$$\downarrow 2(X-1)S \quad \downarrow S$$

$$2\,Na_2S_X \quad Na_2S_2O_3$$

$$3\,NaOH + P_4 + 3\,H_2O \longrightarrow 3\,NaH_2PO_2 + PH_3$$

The activity of the oxyacids of the halogens as oxidizing agents is related inversely to the number of oxygen atoms in the molecule, i.e., the activity is greatest for the HXO acids and least for the HXO_4 acids. Note that the oxidizing strength is greatest for the acids of lowest thermal stability.

In a series of oxyacids of any element, the acidic strength, as measured in terms of extent of ionization in aqueous solution, is greater the larger the number of oxygen atoms in the molecule. Phosphoric acid is stronger than phosphorous acid, sulfuric is stronger than sulfurous, perchloric is stronger than chloric, and so on.

The electronic (Lewis) formulas for the oxyacids of chlorine are

Hypochlorous acid   Chlorous acid   Chloric acid   Perchloric acid

In all four acids the hydrogen is bonded through an oxygen atom. The hydrogen is attached to the oxygen by a polar covalent bond and the oxygen atoms are attached to the chlorine atom by bonds that are essentially covalent in character. The oxidation state of the chlorine in these acids is dependent upon the number of its electrons that it shares with oxygen atoms. Chlorine shares one of its electrons with oxygen in HClO, three in $HClO_2$, five in $HClO_3$, and seven in $HClO_4$. Thus, the corresponding oxidation states exhibited by chlorine are $+1$, $+3$, $+5$, and $+7$, respectively. The oxidation states are positive because oxygen is more electronegative than chlorine. As the positive oxidation state of chlorine increases, the bonding electrons are shifted away from the hydrogen atom, and the tendency for the molecule to lose a proton to water and thereby form hydronium ions in solution increases (see Section 7.1, Part 7). This accounts for the correlation of the strength of the oxyacids as acids with the number of oxygen atoms in the molecule. It follows that the tendency of an oxyhalogen anion to accept protons increases as the oxygen content of the anion decreases. As we have seen in Section 15.15, this tendency of a negative ion to accept protons is a measure of its strength as a base.

The oxyacids of chlorine are more acidic than the corresponding oxyacids of bromine; and those of bromine are more acidic than the corresponding ones of iodine. These facts can be accounted for by the decrease in electronegativity with increasing size of the halogen and the resultant decreases in the tendency for the oxyacid to lose a proton to water.

21.16   Hypohalous Acids and Hypohalites

Because of their thermal instabilities, the **hypohalous acids** (HXO) have not been obtained in the pure form and are known only in aqueous solution. The hypohalous acids are all weak, with the acid strength decreasing with increasing size of the halogen atom. Thus, HClO is a stronger acid than HBrO, which in turn is stronger than HIO (Section 21.15). Because the hypohalite ions readily hydrolyze by accepting protons from water the hypohalous acids are weak. The solutions of alkali metal hypohalites are basic due to an excess of hydroxide ions.

$$\underset{\substack{\text{Hypohalite}\\\text{ion}}}{XO^-} + H_2O \rightleftharpoons \underset{\substack{\text{Hypohalous}\\\text{acid}}}{HXO} + OH^-$$

Aqueous solutions of the hypohalous acids are prepared by hydrolysis of the free halogens according to the general equation

$$X_2 + H_2O \rightleftharpoons H^+ + X^- + HOX$$

This method of preparation yields a mixture of the hypohalous acid and the hydrohalic acid. The extent to which the reaction proceeds to the right decreases in the series chlorine, bromine, iodine. The equilibrium may be shifted to the right by the addition of a base, such as NaOH, which neutralizes the acids which are formed. The net reaction may then be represented by

$$X_2 + (2Na^+) + 2OH^- \longrightarrow (Na^+) + X^- + (Na^+) + XO^- + H_2O$$

For these products to be formed, the alkali must be cold and dilute. The hypohalites are unstable at elevated temperatures and undergo auto-oxidation-reduction to form halides and halates.

$$3XO^- \xrightarrow{\Delta} 2\underset{\text{Halide}}{X^-} + \underset{\text{Halate}}{XO_3^-}$$

Sodium hypochlorite is produced commercially by the electrolysis of cold, dilute aqueous sodium chloride solutions under conditions where the sodium hydroxide and chlorine can mix. The hypochlorite ion is shown in Fig. 21–12.

$$2Cl^- + 2H_2O + \text{Electrical energy} \longrightarrow 2OH^- + Cl_2 + H_2\uparrow$$
$$Cl_2 + 2OH^- \longrightarrow Cl^- + ClO^- + H_2O$$

Net reaction: $Cl^- + H_2O + \text{Electrical energy} \xrightarrow[\text{dil.}]{\text{Cold,}} ClO^- + H_2\uparrow$

FIGURE 21–12
The structure of the hypo-chlorite ion.

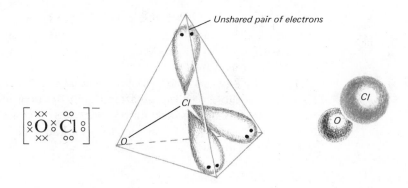

21.17 Halous Acids and Halites

None of the **halous acids,** HXO_2, have been isolated in pure form. Of the halites, only the chlorites have been adequately characterized, although the bromite ion was finally produced in small quantities in aqueous solution in 1969.

Chlorous acid, $HClO_2$, has no true anhydride, because in chlorine dioxide, ClO_2, the chlorine has an oxidation state different from that exhibited in either chlorite or chlorate. However, chlorine dioxide, ClO_2, plays the role of a **double anhydride,** in that it gives two oxyacid anions of chlorine with alkalies. The reaction is one of auto-oxidation-reduction.

$$2ClO_2 + 2OH^- \longrightarrow ClO_2^- + ClO_3^- + H_2O$$

Chlorine dioxide   Chlorite ion   Chlorate ion

An aqueous solution of chlorous acid can be prepared by the action of sulfuric acid upon a suspension of barium chlorite.

$$Ba(ClO_2)_2 + (2H^+) + SO_4^{2-} \longrightarrow \underline{BaSO_4} + (2H^+) + 2ClO_2^-$$

The insoluble barium sulfate may be removed from the solution of chlorous acid by filtering. Attempts to isolate pure chlorous acid result in its decomposition into chlorine dioxide, chloric acid, and hydrochloric acid.

The structure of the chlorite ion is shown in Fig. 21–13.

FIGURE 21–13
The structure of the chlorite ion.

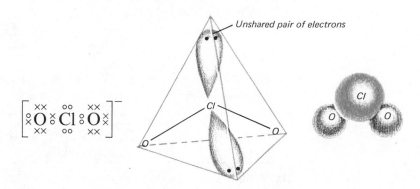

21.18 Halic Acids and Halates

Although **chloric acid,** $HClO_3$, and **bromic acid,** $HBrO_3$ decompose when attempts are made to isolate them, **iodic acid,** HIO_3, can be obtained in the form of stable colorless crystals, which melt with decomposition at 110°.

Solutions of chloric acid and bromic acid are readily obtained by the reaction of their barium salts with sulfuric acid.

$$Ba^{2+} + 2XO_3^- + (2H^+) + SO_4^{2-} \longrightarrow \underline{BaSO_4} + (2H^+) + 2XO_3^-$$

Iodic acid is easily prepared by oxidizing elemental iodine with concentrated nitric acid according to the equation

$$I_2 + 8H^+ + 10NO_3^- \longrightarrow 2IO_3^- + 10NO_2\uparrow + 4H_2O$$

All the halic acids are strong acids and very active oxidizing agents.

The water solubility of the halates decreases with increasing weight of the halogen. Most chlorates are water soluble, bromates are generally much less soluble than the chlorates, and many iodates are insoluble in water. All halates decompose when heated. Potassium chlorate yields potassium chloride and oxygen at high temperatures.

$$2KClO_3 \longrightarrow 2KCl + 3O_2\uparrow$$

At moderate temperatures potassium perchlorate and potassium chloride are formed when potassium chlorate is heated.

$$4KClO_3 \longrightarrow 3KClO_4 + KCl$$

Thermal decomposition of bromates and iodates results in a number of different products, depending on conditions under which the reactions occur. In some cases, the metal oxides and the free halogens are formed.

The structure of the chlorate ion is shown in Fig. 21–14.

FIGURE 21-14
The structure of the chlorate ion.

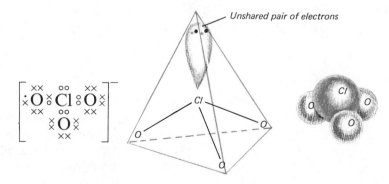

21.19 Perhalic Acids and Perhalates

■ **1. Perchloric Acid and Its Salts.** Perchloric acid ($HClO_4$) may be obtained by treating a perchlorate, such as potassium perchlorate, with sulfuric acid.

$$KClO_4 + H_2SO_4 \longrightarrow KHSO_4 + HClO_4$$

To isolate the perchloric acid the reaction mixture is distilled under reduced pressure. The acid explodes if its temperature is raised above 92°. At reduced pressure, it boils below this temperature with little decomposition and distills without exploding. Aqueous solutions of perchloric acid up to concentrations of about 60 per cent are quite stable thermally. The acid is unstable and unsafe to use and handle in concentrations above 60 per cent. It is a powerful oxidizing agent and serious explosions may occur when concentrated solutions of it are heated with substances which are easily oxidized. However, it is a very weak oxidizing agent when cold and dilute.

Perchloric acid is a valuable analytical reagent because it is the strongest of all acids, it is a good oxidizing agent, and perchlorates are quite soluble in the acid. Perchloric acid is used in the electropolishing of metals and alloys such as nickel, copper, brass, and steel, and in the quantitative determination of potassium, with which it forms slightly soluble potassium perchlorate.

The **perchlorates** of sodium and potassium are produced commercially by the prolonged electrolysis of hot solutions of their chlorides. The final step in the process is one of anodic oxidation of the chlorate ion according to the equation

$$ClO_3^- + H_2O \longrightarrow ClO_4^- + 2H^+ + 2e^-$$

The structure of the perchlorate ion is shown in Fig. 21–15.

FIGURE 21-15
The structure of the perchlorate ion.

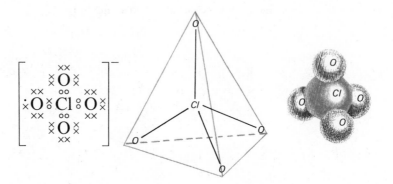

■ **2. Perbromic Acid and Its Salts.**   All attempts to prepare perbromic acid, $HBrO_4$, and its salts failed until as recently as 1968. That the acid would be so difficult to prepare was unexpected and puzzling, especially since analogous compounds of both chlorine and iodine had been known for some time and methods of preparing oxides of bromine had been discovered. Finally, in 1968, perbromate ion, BrO_4^-, was prepared by the oxidation of bromate ion with xenon difluoride, XeF_2, in aqueous solution. Later, crystalline rubidium perbromate was isolated and found to be stable at room temperature. Shortly after the successful preparation of the perbromate ion, perbromic acid in aqueous solution was prepared by acidifying rubidium perbromate with sulfuric acid. Perbromic acid is a stable compound.

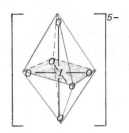

FIGURE 21-16
The octahedral structure of the paraperiodate ion, IO_6^{5-}.

■ **3. Periodic Acids and Their Salts.** Several series of periodic acids and periodates are known. Most important of these are **metaperiodic** acid, HIO_4, and **paraperiodic** acid, H_5IO_6, and their salts. The size of the iodine atom is sufficient to allow space for six oxygen atoms around it (Fig. 21–16), whereas the smaller chlorine atom can accommodate no more than four oxygen atoms. The periodic acids are strong acids and active oxidizing agents. The oxidizing properties of HIO_4 make it useful in numerous analytical procedures.

Sodium metaperiodate, $NaIO_4$, may be prepared by the oxidation of sodium iodate, $NaIO_3$, with chlorine in a hot alkaline solution. Periodates may also be prepared by the electrolytic oxidation of iodates, a method analogous to the preparation of perchlorates. The free paraperiodic acid, H_5IO_6, may be obtained as a white solid by evaporating a filtered solution resulting from the reaction of barium periodate with sulfuric acid. Salts of paraperiodic acids such as $Na_2H_3IO_6$, $Na_3H_2IO_6$, and Ag_5IO_6 have been prepared.

QUESTIONS

1. Arrange the halogens in order of increasing (a) atomic radii, (b) ionic radii, (c) electronegativity, (d) boiling points, (e) oxidizing activity, and (f) depth of color.
2. Why is it difficult to prepare elemental fluorine? How is its preparation accomplished?
3. Why must the chlorine and sodium hydroxide resulting from the electrolysis of brine be kept separate in the manufacture of chlorine? How is this done?
4. Describe the chemistry of the extraction of bromine from sea water.
5. Describe the commercial production of iodine from sodium iodate.
6. Discuss reasons for the differences in oxidizing power of the halogens.
7. A solution of iodine in carbon disulfide is violet, while an aqueous solution of this element is brown. Explain.
8. Write equations showing the action of chlorine upon a metal, hydrogen, water, sulfur, and phosphorus.
9. What product is formed when phosphorus is burned in a limited supply of chlorine? in an excess of chlorine?
10. Show by means of changes in oxidation state that the reaction of chlorine with water is one of auto-oxidation-reduction.
11. Illustrate the gradation of properties shown by the family of halogens with regard to three physical properties and two chemical properties.
12. If chlorine, iodine, a chloride, an iodide, and a starch suspension were available to you, how would you demonstrate the relative activity of chlorine and iodine?
13. Would you expect astatine to oxidize iodide ions to form elemental iodine? Why or why not?
14. How can fluorine, which is an extremely powerful oxidizing agent, be contained and stored?

15. Why is an iodide added in the preparation of the tincture of iodine?
16. What are fluorocarbons? List some general uses of these compounds.
17. Why is iodine monochloride (ICl) more polar than iodine monobromide (IBr)?
18. What products would you expect if BrCl gas were dissolved in water? Write a balanced chemical equation to describe the reaction.
19. Write chemical equations describing:
    (a) The burning of sulfur in a chlorine atmosphere.
    (b) The reaction of bromine with water.
    (c) The dissolution of iodine in potassium iodide solution.
    (d) The hydrolysis of PCl_3.
    (e) The electrolysis of fused (melted) sodium chloride.
    (f) The preparation of hydrogen fluoride by the reaction of calcium fluoride and sulfuric acid.
    (g) The thermal decomposition of $KClO_3$.
    (h) The oxidation of sulfide ion by elemental iodine.
    (i) The auto-oxidation-reduction reaction of ClO_2 with a basic solution.
    (j) The anodic oxidation of chlorate ion to perchlorate ion.
    (k) The light-induced decomposition of HClO.
    (l) The reduction of sulfate ion by bromide ion in acid solution.
    (m) The dissolution of HI in water.
20. Write equations for the interaction of hydrochloric acid and (a) MnO_2, (b) $KMnO_4$, and (c) $Na_2Cr_2O_7$.
21. Why is the direct union of constituent elements an unsatisfactory method for producing hydrogen fluoride? hydrogen iodide?
22. Why cannot pure HBr and HI be prepared by the action of concentrated sulfuric acid upon metal bromides and iodides, respectively? Write the equations involved.
23. Contrast the chemical activity of anhydrous hydrogen chloride with that of hydrochloric acid.
24. When a hydrogen halide functions as a reducing agent, which element in the molecule is oxidized? Compare HI, HBr, HCl, and HF with regard to their strengths as reducing agents.
25. List the unique properties of hydrogen fluoride and hydrofluoric acid.
26. Describe the chemistry of the etching of glass, including equations.
27. Describe the hydrogen bond. Between what kind of atoms are hydrogen bonds formed?
28. Why does hydrogen bonding have an effect on such properties as boiling point, vapor pressure, and heat of vaporization?
29. Why is a doubly charged chloride ion not observed?
30. Show by an equation that the hydrogen ion acts as an oxidizing agent toward metals above hydrogen in the activity series.
31. Contrast the properties of ionic and covalent binary metal halides. Give typical examples of each of these classes of halides.
32. Write equations for the reaction of hydrochloric acid with Ca, CaO, $Ca(OH)_2$, and $CaCO_3$.
33. How would you separate a mixture of sodium chloride and silver chloride?
34. Why is OF_2 called oxygen difluoride rather than fluorine oxide?

35. Write the formulas for five compounds in which chlorine is linked by covalent bonds; by ionic bonds.
36. Identify the oxidizing agent and reducing agent in the equation

$$I_2 + H_2S \longrightarrow 2H^+ + 2I^- + \underline{S}$$

37. Why is chlorine dioxide not considered to be a true acid anhydride?
38. Show by suitable equations that chlorine monoxide and chlorine heptaoxide are acid anhydrides.
39. What is the oxidation state of bromine in each of the oxides listed in Table 21-5?
40. Write the valence electronic structures for the oxyacids of chlorine. Calculate the oxidation state of chlorine in each of these acids, assuming hydrogen to be +1 and oxygen −2.
41. Write the formulas and names for all the oxyacids of chlorine and the corresponding calcium salts.
42. Write equations for the preparation of each of the oxyacids of chlorine and its sodium salt.
43. Relate the thermal stabilities of the series of oxyacids of chlorine and their salts to their oxygen contents. Do the same for their oxidizing power and acidic strength.
44. Explain the basic character of aqueous sodium hypochlorite.
45. Why is perchloric acid distilled only under reduced pressure?
46. Predict the products in each of the following cases:
    (a) A solution of hypochlorous acid is heated.
    (b) Potassium sulfite is treated with potassium periodate in dilute sulfuric acid solution.
    (c) Oxygen difluoride, OF_2, is brought in contact with powdered aluminum.
47. Why does not chlorine form a paraperchloric acid analogous to paraperiodic acid?
48. Why is $HClO_3$ a stronger acid than $HBrO_3$?
49. Why is $HClO_2$ a stronger acid than $HClO$?
50. Compare the structure of ICl_4^- ion with that of the IO_4^- ion.
51. List as many products as you can which can be obtained by the electrolysis of aqueous halide solutions. Explain the general procedure involved in the production of each of the substances listed.
52. Compare the photochemically induced reaction between hydrogen and chlorine with the similar reaction between hydrogen and bromine discussed in Chapter 17 (Section 17.2).

PROBLEMS

1. What volume of chlorine measured at 25.0° and 740 mm would be required to displace 400 g of bromine from a solution of sodium bromide?

   Ans. 62.9 liters

2. Calculate the densities of gaseous fluorine, chlorine, bromine, and iodine at 0.200 atm and 100°.

   Ans. F_2, 0.248 g/liter; Cl_2, 0.463 g/liter; Br_2, 1.04 g/liter; I_2, 1.66 g/liter

3. What volume of HCl gas measured at 100° and 1.00 atm can be prepared from 500 g of sodium chloride? *Ans. 262 liters*

4. An electrolytic cell consumes 90.0 kg of sodium chloride while it is producing 50.0 kg of chlorine. What is the percentage yield? *Ans. 91.6%*

5. What weight of chlorine results from the electrolysis of 3.00 tons of fused sodium chloride, assuming the process to give a 96.0% yield?
Ans. 3.49×10^3 lb

6. How many kilograms of sodium iodate are required to prepare 1.00 kg of iodine by reduction with sodium hydrogen sulfite? *Ans. 1.56 kg*

7. Calculate the molarity, normality, and molality of constant boiling hydrochloric acid, with specific gravity 1.10 and 20.24% HCl.
Ans. 6.11 M; 6.11 N; 6.96 m

8. (a) What weight of sodium dichromate ($Na_2Cr_2O_7$) would be required to react with 50 g of sodium chloride in acid solution? (b) What volume of chlorine is produced by this reaction at standard conditions of temperature and pressure? *Ans. (a) 37 g; (b) 9.6 liters*

9. Compare the rates of diffusion of Cl_2 gas, Br_2 vapor, and I_2 vapor with that of F_2 gas.
Ans. Cl_2 diffuses 0.732 times as fast as F_2; Br_2, 0.488 times as fast as F_2; I_2, 0.387 times as fast as F_2

10. What weight of calcium fluoride would be required in the production of enough HF to make 50.0 gallons of hydrogen fluoride solution which is 5.00 M? *Ans. 36.9 kg*

11. What weight of hydrogen fluoride would be required in the preparation of 500 g of constant-boiling hydrofluoric acid which is 35.35% HF by weight? What weight of methane would have to be burned in fluorine ($CH_4 + 4F_2 \longrightarrow CF_4 + 4HF$) to give this quantity of HF?
Ans. 177 g; 35.4 g

12. How much PI_3 could be produced from 250 g of iodine if the reaction gives a yield of 98.5%? *Ans. 266 g*

13. What weight of barium carbonate can be converted to barium bromide by 0.500 liter of 1.50 M hydrobromic acid? *Ans. 74.0 g*

14. What weight of HBr can be prepared from the hydrolysis of 1.00 kg of phosphorus(III) bromide? *Ans. 897 g*

15. Calculate the quantity of sodium chloride which must react with sulfuric acid to produce sufficient hydrogen chloride to make 350 ml of 40% hydrochloric acid solution (sp. gr. 1.2). *Ans. 2.7×10^2 g*

16. What volume of chlorine (STP) would be required in the combustion of 1.10 g of antimony to produce antimony(III) chloride? What weight of $SbCl_3$ would be produced? *Ans. 304 ml; 2.06 g*

17. What volume of 2.00 M iodic acid could be prepared from 150 g of iodine by reaction with nitric acid? *Ans. 591 ml*

18. What volume of hydrochloric acid (40.0% by weight; sp. gr. 1.20) would be required to dissolve 250 g of magnesium?
Ans. 1560 ml (Note that only three significant figures are justified by the data given; hence the calculated answer of 1563 ml must be rounded off to 1560 ml.)

19. How much sodium metaperiodate can be prepared by the oxidation of 50 g of sodium iodate with chlorine in hot sodium hydroxide solution?

Ans. 54 g

20. Calculate the quantity of perchloric acid which can be prepared by the treatment of 300 g of potassium perchlorate with excess sulfuric acid.

Ans. 218 g

REFERENCES

"Liquid HF Solvent Facilitates Novel Reactions," (Staff), *Chem. & Eng. News*, Feb. 23, 1970; p. 42.

(Fluorinated Hydrocarbons Carry Oxygen in Blood Substitute), (Staff), *Chem. & Eng. News,* May 18, 1970; p. 30.

"Humphry Davy and the Elementary Nature of Chlorine," R. Siegfried, *J. Chem. Educ.,* **36,** 568 (1959).

"Chemical Lasers" (Concerns use of chlorine and iodine), G. C. Pimentel, *Sci. American*, April, 1966; p. 32.

"Principles of Halogen Chemistry," R. T. Sanderson, *J. Chem. Educ.,* **41,** 361 (1964).

"The Halogen Family," M. E. Weeks, *Discovery of the Elements,* Sixth Edition, Publ. by the Journal of Chemical Education, Easton, Pa., 1956; Chapter 27, pp. 729–777.

"Fluorine Chemistry," A. K. Barbour, *Chem. in Britain,* **5,** 250 (1969).

"Organochlorine Insecticides and Birds," J. Robinson, *Chem. in Britain,* **4,** 158 (1968).

"The Photochemical Oxidation of Aqueous Iodide Solutions," P. B. Ayscough, C. E. Burchill, K. J. Ivin, and S. R. Logan, *J. Chem. Educ.,* **44,** 349 (1967).

"Chlorine Trifluoride," N. V. Steere, *J. Chem. Educ.,* **44,** A1057 (1967).

"Thermochemistry of Hypochlorite Oxidations," M. J. Bigelow, *J. Chem. Educ.,* **46,** 378 (1969).

"Applied Research in the Development of Anticaries Dentifrices," W. E. Cooley, *J. Chem. Educ.,* **47,** 177 (1970).

"Nucleophilic Reactivities of the Halide Anions," M. S. Puar, *J. Chem. Educ.,* **47,** 473 (1970).

"Endemic Goiter," R. B. Gillie, *Sci. American*, June, 1971; p. 93.

"Orbital Symmetry in Photochemical Transformations," H. Katz, *J. Chem. Educ.,* **48,** 84 (1971).

"Physiochemical Properties of Highly Fluorinated Organic Compounds," C. R. Patrick, *Chem. in Britain*, **7,** 154 (1971).

"Chlorine Shortage Shows Signs of Easing," W. F. Fallwell, *Chem. & Eng. News*, Dec. 16, 1974; p. 9.

Electrochemistry

22

Electrochemistry is that field of chemistry which deals with chemical changes produced by an electric current and with the production of electricity by chemical reactions. In **electrolytic cells,** electrical energy is used to bring about desired chemical changes. In **voltaic cells,** chemical reactions are used to produce electrical energy.

Much of our quantitative knowledge concerning the energy changes involved in chemical reactions (see Chapter 20, Chemical Thermodynamics) has come from the study of electrochemistry, because the quantity of electrical energy produced or consumed during electrochemical changes can be measured very accurately. Furthermore, an understanding of the reactions which take place at the electrodes of electrochemical cells throws much light upon the process of oxidation and reduction and chemical activity.

Many electrochemical processes are important in science and industry. The use of electrical energy in the commercial production of hydrogen, oxygen, ozone, hydrogen peroxide, chlorine, sodium hydroxide, and the oxygen compounds of the halogens has been noted in previous chapters. Additional technical applications of electrochemistry include production of many other chemicals, electrorefining of metals, electroplating of metals and alloys, the production of metal articles by electrodeposition, and the generation of electrical current by battery cells.

22.1   The Conduction of Electricity

A current of electricity is the movement of electrically charged particles through a conductor. These charged particles may be either electrons or ions (positive or negative). In a metal conductor an electric current is the movement of electrons without any obvious chemical changes in the metal and with no similar movement of the atoms composing the metal. This type of conduction of electricity is called **metallic conduction.**

Certain compounds, when either fused or in aqueous solution, conduct an electric current when placed in an electrolytic cell (Fig. 22–1). An electric generator or storage battery is connected to two electrodes by metallic wires. The electrodes, which are good conductors, dip into the liquid which conducts the current. Electrons are forced onto one of the electrodes and withdrawn from the other electrode by the battery or generator. The electrode onto which the electrons are forced becomes negatively charged, and the electrode from which electrons are withdrawn becomes positively charged. The positive ions (cations) in the electrolytic liquid are attracted by the negatively charged cathode and move toward it, and the negative ions (anions) are attracted by the positively charged anode and move toward it. This sort of motion of the ions through a liquid is known as **ionic,** or **electrolytic, conduction.** The electrolytic liquid may be a molten salt such as fused sodium chloride or an aqueous solution such as hydrochloric acid.

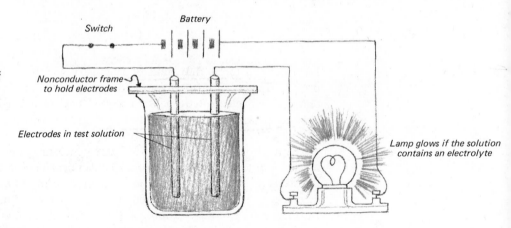

FIGURE 22–1
An electrolytic cell being used to determine the conductivity of a solution.

Electrolytic Cells

22.2   The Electrolysis of Molten Sodium Chloride

A consideration of the reactions which take place at the electrodes will help you understand the way in which the electric current passes between the electrodes and the liquid.

Fused sodium chloride consists of equal numbers of positive sodium ions and negative chloride ions, which can move about with considerable freedom. Sodium

$N\overset{+}{a} Cl^{-}$

ions are attracted to the cathode where they combine with electrons on the cathode to form sodium atoms, i.e., metallic sodium. The electrode reaction, referred to as a **half-reaction,** may be represented for the cathode by

$$Na^+ + e^- \longrightarrow Na$$

The sodium decreases in oxidation state from $+1$ to 0. This is a reduction reaction and is sometimes referred to as cathodic reduction. *The cathode is the electrode at which electrons enter the electrolytic liquid, causing reduction.* The chloride ions of the fused salt give up one electron each to the anode and become chlorine atoms, which then combine with each other to form molecules of chlorine gas. The anode half-reaction is

$$2Cl^- \longrightarrow Cl_2 + 2e^-$$

The chlorine is oxidized from -1 to 0, and the reaction is known as anodic oxidation. *The anode is the electrode through which electrons are withdrawn from the electrolytic liquid, causing oxidation.*

The net reaction for the electrolytic decomposition of the fused sodium chloride is the sum of the two electrode half-reactions. To balance the equation, the cathode reaction must be multiplied by two.

$$\begin{aligned} \textit{Cathode:} \quad & 2Na^+ + 2e^- \longrightarrow 2Na \\ \textit{Anode:} \quad & \underline{\phantom{2Na^+ +{}} 2Cl^- \longrightarrow Cl_2 + 2e^-} \\ & 2Na^+ + 2Cl^- \longrightarrow 2Na + Cl_2 \end{aligned}$$

or

$$2NaCl \longrightarrow 2Na + Cl_2$$

The production of an oxidation-reduction reaction by a direct current (D.C.) of electricity is known as electrolysis.

22.3 The Electrolysis of Aqueous Hydrochloric Acid

Let us consider what happens at the electrodes of an electrolytic cell when an aqueous solution of fairly concentrated hydrochloric acid is electrolyzed. The positive hydrogen ions of the hydrochloric acid are attracted to and reduced at the cathode by accepting electrons from it.

$$2H^+ + 2e^- \longrightarrow H_2 \quad \text{(cathodic reduction)}$$

The negative chloride ions lose electrons to the anode and thus are oxidized to chlorine gas.

$$2Cl^- \longrightarrow Cl_2 + 2e^- \quad \text{(anodic oxidation)}$$

The overall cell reaction may be obtained by adding the electrode reactions.

$$\begin{aligned} \textit{Cathode:} \quad & 2H^+ + 2e^- \longrightarrow H_2 \\ \textit{Anode:} \quad & \underline{\phantom{2H^+ +{}} 2Cl^- \longrightarrow Cl_2 + 2e^-} \\ & 2H^+ + 2Cl^- \longrightarrow H_2 + Cl_2 \end{aligned}$$

22.4 The Electrolysis of an Aqueous Sodium Chloride Solution

Suppose we consider the electrode reactions which take place during the electrolysis of an aqueous solution of sodium chloride (Fig. 22-2). If the difference in electrical potential (Section 22.8) between the anode and cathode is sufficiently great and the solution is fairly concentrated, hydrogen is evolved at the cathode and chlorine is formed at the anode. In order to account for these products it is necessary to consider all the possible electrode reactions.

FIGURE 22-2
The electrolysis of an aqueous solution of NaCl. Several different reactions occur at both the anode and cathode; the net reaction, however, produces hydrogen gas and hydroxide ions at the cathode and chlorine gas at the anode.

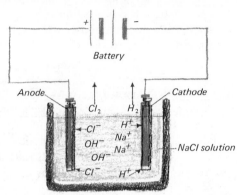

The possible cathodic (reduction) reactions are the reduction of sodium ion or the reduction of water:

$$Na^+ + e^- \longrightarrow Na \tag{1}$$
$$2H_2O + 2e^- \longrightarrow H_2 + 2OH^- \; \text{reduction} \tag{2}$$

The water is more easily reduced than the sodium ion, so reaction (2) occurs and hydrogen gas forms at the cathode. Hydroxide ions form in the vicinity of the cathode and begin to migrate toward the anode.

The possible anodic (oxidation) reactions are either the oxidation of chloride ions or the oxidation of water:

$$2Cl^- \longrightarrow Cl_2 + 2e^- \; \text{oxidation} \tag{3}$$
$$2H_2O \longrightarrow O_2 + 4H^+ + 4e^- \tag{4}$$

Chloride ion and water are oxidized with almost equal ease. Hence, concentration of the salt solution plays the dominant role in determining the product. In a concentrated sodium chloride solution, reaction (3) takes place and chlorine is formed at the anode. If the concentration of chloride ion is low, reaction (4) can take place to some extent also and some oxygen may be formed in addition to the chlorine. Little chlorine is produced when the solution of sodium chloride is very dilute.

The net reaction for the electrolysis of aqueous sodium chloride can be obtained by adding the electrode reactions.

$$
\begin{array}{ll}
\textit{Cathode:} & 2H_2O + 2e^- \longrightarrow H_2 + 2OH^- \\
\textit{Anode:} & 2Cl^- \longrightarrow Cl_2 + 2e^- \\
\hline
& 2H_2O + 2Cl^- \longrightarrow H_2\uparrow + Cl_2\uparrow + 2OH^-
\end{array}
$$

It should be noted that hydroxide ions are formed during the electrolysis and that chloride ions are removed from solution. Because the sodium ions remain unchanged, sodium hydroxide accumulates in the cell as electrolysis proceeds.

The electrolysis of aqueous sodium chloride is an important commercial process for the production of hydrogen, chlorine, and sodium hydroxide (Section 21.2).

22.5   The Electrolysis of an Aqueous Sulfuric Acid Solution

The ionization of sulfuric acid in water proceeds stepwise as follows:

$$H_2SO_4 \longrightarrow H^+ + HSO_4^- \quad \text{(essentially complete)}$$
$$HSO_4^- \rightleftharpoons H^+ + SO_4^{2-}$$

In dilute solution the anion furnished by the sulfuric acid is principally the sulfate ion. Upon electrolysis, the possible choices for the anode reaction are either oxidation of the sulfate ion or oxidation of water.

$$SO_4^{2-} \longrightarrow SO_4 + 2e^-$$
$$2H_2O \longrightarrow O_2\uparrow + 4H^+ + 4e^-$$

In practice oxygen and hydrogen ions are formed at the anode, because the water molecule is more easily oxidized than the sulfate ion; in fact, SO_4 cannot be formed by electrolytic oxidation.

The only possible cathode reaction is

$$2H^+ + 2e^- \longrightarrow H_2$$

The net reaction is the sum of the anode and cathode reactions after multiplying the cathode reaction by two.

$$
\begin{array}{ll}
\textit{Anode:} & 2H_2O \longrightarrow O_2 + 4H^+ + 4e^- \\
\textit{Cathode:} & 4H^+ + 4e^- \longrightarrow 2H_2 \\
\hline
& 2H_2O \longrightarrow 2H_2\uparrow + O_2\uparrow
\end{array}
$$

Hence, the electrolysis of an aqueous solution of sulfuric acid is the same as the electrolysis of water (see Section 8.2). Although the hydrogen ion from the sulfuric acid is consumed by cathodic reduction, this same ion is being generated at the anode at the same rate at which it disappears at the cathode. The net result is that water is used up and the sulfuric acid becomes more concentrated as electrolysis proceeds.

22.6   Electrolytic Refining of Metals

In the few examples of electrolysis we have considered thus far, the electrodes used in the electrolytic cells were made of electrically conducting materials, such as graphite or platinum, which were relatively unreactive under the conditions employed. Whenever a reactive metal is made the anode of an electrolytic cell, the anodic oxidation may involve oxidation of the metal composing the electrode. For example, in the electrolysis of a solution of copper(II) sulfate (Fig. 22–3) with

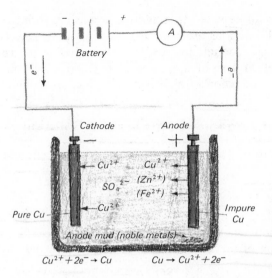

FIGURE 22-3

The electrolytic refining of copper.

$$Cu^{2+} + 2e^- \rightarrow Cu \qquad Cu \rightarrow Cu^{2+} + 2e^-$$

a strip of metallic copper as the anode, the following anodic electrode reaction takes place:

$$Cu \longrightarrow Cu^{2+} + 2e^-$$

Thus, the electrode itself goes into solution as copper(II) ions. The oxidation of the copper takes place more readily than the oxidation of either the sulfate ion or the water molecule. The copper(II) ions formed at the anode migrate toward the cathode where they are more readily reduced than the water.

$$Cu^{2+} + 2e^- \longrightarrow Cu$$

Thus, pure copper plates out on the cathode.

This process is the basis of the electrolytic refining of crude copper. When the impure copper is the anode the less active impurities, such as gold and silver, do not dissolve (are not oxidized) and fall to the bottom of the cell, forming a "mud" from which they are readily recovered. The more active metals, such as zinc and iron which may be present in the crude copper, are oxidized and go into the solution as ions. If the electrical potential between the electrodes is carefully regulated, these metallic ions are not reduced at the cathode; only the copper is reduced and deposited. Deposition of the copper in pure form is made either upon a thin sheet of pure copper, which serves as a cathode, or upon some other metal from which the deposit may be stripped.

22.7 Faraday's Law of Electrolysis

In 1832–33 Michael Faraday found that *the quantity of substances undergoing chemical change at each electrode during electrolysis is directly proportional to the quantity of electricity which passes through the electrolytic cell.* Quantity of electricity can be expressed as number of electrons. The quantity of a substance undergoing chemical change at an electrode is related to the number of electrons involved in the change and can be expressed in terms of the equivalent weight of the substance. Experimental results show that one electron reduces one silver ion, while two electrons are required to reduce one copper(II) ion.

$$Ag^+ + 1e^- \longrightarrow Ag$$
$$Cu^{2+} + 2e^- \longrightarrow Cu$$

One gram-equivalent of silver ion contains 6.022×10^{23} ions. Therefore, 6.022×10^{23} electrons are required to reduce one gram-equivalent of silver ions to silver atoms. The weight of silver produced would be one gram-equivalent, 107.868 g, which is the same as the gram-atomic weight, since the silver ion is univalent. Because two electrons are required to reduce one Cu^{2+} ion to atomic copper, 6.022×10^{23} electrons would reduce only one-half of a gram-ion weight of copper. However, Cu^{2+} being bivalent, one-half of the gram-ionic weight would be the gram-equivalent of this ion, i.e., $63.54/2 = 31.77$ g.

The term **faraday** refers to the quantity of electricity which will reduce one gram-equivalent of a substance at the cathode and oxidize one gram-equivalent at the anode, which amounts to the gain or loss of one mole of electrons, or 6.022×10^{23} electrons.

> 1 faraday = 6.022×10^{23} electrons = 96,487 coulombs
> 1 coulomb = quantity of electricity involved when a current
> of 1 ampere flows past a given point for 1 second

Faraday's Law may be stated thus: **During electrolysis, 96,487 coulombs (1 faraday) of electricity reduce and oxidize, respectively, one gram-equivalent of the oxidizing agent and the reducing agent** (Fig. 22-4).

Example.   Let us calculate the weight of copper produced by the cathodic reduction of the Cu^{2+} ion during the passage of 1.600 amperes of current through a solution of copper(II) sulfate for 1 hour.

Amperes × seconds = 1.600 amp × 60 min/hr × 60 sec/min
= 5760 coulombs

96,487 coulombs will reduce $\dfrac{63.54 \text{ g}}{2} = 31.77$ g of Cu^{2+}

5,760 coulombs will reduce 31.77 g × $\dfrac{5,760}{96,487} = 1.897$ g of copper

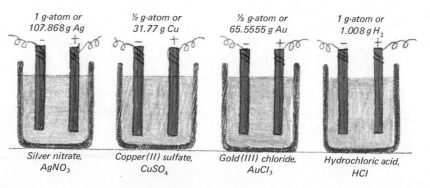

FIGURE 22-4
Amounts of various metals discharged at the cathode by one faraday (96,487 coulombs) of electricity.

1 g-atom or 107.868 g Ag	½ g-atom or 31.77 g Cu	⅓ g-atom or 65.5555 g Au	1 g-atom or 1.008 g H₂
Silver nitrate, AgNO₃	Copper(II) sulfate, CuSO₄	Gold(III) chloride, AuCl₃	Hydrochloric acid, HCl

Electrode Potentials

22.8   Single Electrodes and Electrode Potentials

When a metal such as zinc is placed in water, there is a tendency for the atoms to leave the metal and pass into solution as ions. The change may be expressed for zinc by

$$Zn \longrightarrow Zn^{2+} + 2e^-$$

The zinc ions accumulate in the solution and their valence electrons remain behind on the metal. The zinc strip thereby becomes negatively charged and the solution becomes positive. Thus, a difference in electrical potential is established between the metal and the solution. On the other hand, the negative charge which has been generated on the zinc attracts the positive zinc ions in solution and tends to reduce them to the metallic state, according to the equation

$$Zn^{2+} + 2e^- \longrightarrow Zn$$

Also, an increase in the concentration of the zinc ions in the solution, brought about by the addition of a zinc salt, tends to reverse the process. Thus the change is reversible, as indicated by double arrows, and an equilibrium can be established.

$$Zn \Longleftrightarrow Zn^{2+} + 2e^-$$

At equilibrium the potential difference between the metal and the solution depends on: (a) the tendency of the metal to form ions in aqueous solution, (b) the concentration of the metal ions in solution, and (c) the tendency of the metal ions to be reduced to the metal. The more active the metal (the more easily oxidized) the greater is its tendency to form ions and the more negative is the potential of the electrode with respect to the solution. A zinc electrode is thus more negative than a hydrogen electrode (see next section). Both the hydrogen electrode and zinc electrode are more negative than a copper electrode, because zinc is more easily oxidized than hydrogen and hydrogen is more easily oxidized than copper.

As mentioned previously, the size of the electrode potential varies with the concentration of the reactants. Thus, the potential of the zinc electrode can be varied by changing the concentration of the zinc ions in solution. With an increase in the zinc ion concentration, the equilibrium $Zn \Longleftrightarrow Zn^{2+} + 2e^-$ shifts to the left, and the difference in potential between the metal and the solution decreases. A decrease in the concentration of zinc ions shifts the equilibrium to the right and the potential difference increases. Inasmuch as the zinc electrode is a solid its concentration is constant, and the size of the electrode does not influence the electrode potential.

22.9   Standard Electrode Potentials

No satisfactory method has been devised for measuring the absolute difference in electrical potential between a metal electrode and a solution of its ions. How-

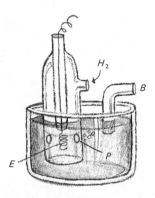

FIGURE 22-5
Diagram of a simple standard hydrogen electrode. *E*, platinized, platinum helix; *P*, port for escape of hydrogen bubbles; *B*, part of salt bridge.

ever, the difference between the potentials of two electrodes can be readily measured. To do this, we use an electrode that is taken as a standard. The hydrogen electrode, illustrated in Fig. 22–5, has been adopted as a standard electrode with which the potentials of other electrodes are compared.

Electrodes involving gases can be set up by bubbling the gas around an inert metallic conductor that will conduct electrons but will not itself enter the electrode reaction. For example, the hydrogen electrode can be constructed so that hydrogen gas is bubbled around a strip or wire of platinum covered with very finely divided platinum and immersed in a solution containing hydrogen ions. Activation of the hydrogen by the platinum results in a tendency for the gas to form hydrogen ions. There is also a tendency for the hydrogen ion to be reduced to hydrogen by acquiring electrons from the electrode. Thus, a difference in potential between the electrode and the solution is established in the same manner as described for zinc.

$$H_2 \rightleftharpoons 2H^+ + 2e^-$$

With gas electrodes, the potential increases with an increase in the partial pressure of the gas. Therefore, the **standard hydrogen electrode** is prepared by bubbling hydrogen gas at a temperature of 25° C and a pressure of 1 atm around platinized platinum immersed in a solution containing hydrogen ions at a concentration of 1 M. We arbitrarily assign the value zero to the potential for the standard hydrogen electrode.

22.10   Measurement of Electrode Potentials

To measure the electrode potential of zinc, for example, an electrochemical cell is set up consisting of a strip of zinc in contact with a 1 M solution of zinc ions as one-half of the cell and a standard hydrogen electrode as the other half (Fig. 22–6).

The two halves of the cell are connected by a **salt bridge** consisting of a saturated solution of potassium chloride in a gel of agar-agar. This bridge permits ions to

FIGURE 22-6
Measurement of the electrode potential of zinc, using the standard hydrogen electrode as reference.

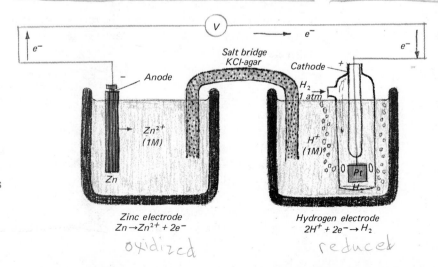

Zinc electrode
$Zn \rightarrow Zn^{2+} + 2e^-$

oxidized

Hydrogen electrode
$2H^+ + 2e^- \rightarrow H_2$

reduced

migrate through it freely from cell to cell but does not allow the two solutions to mix. The zinc strip and the platinum strip of the hydrogen electrode are connected to a voltmeter by wire. The potential difference of the cell as a whole is read in volts on the voltmeter.

Before the circuit is closed, i.e., before the zinc and platinum strips are connected by wire, we have the following electrode half-reactions, each at equilibrium.

$$H_2 \rightleftharpoons 2H^+ + 2e^-$$
$$Zn \rightleftharpoons Zn^{2+} + 2e^-$$

Because zinc has a greater tendency to go into solution as ions than does hydrogen, the zinc strip acquires a higher electron density (from the electrons released as the zinc ions are formed) than does the platinum of the hydrogen electrode. Therefore, when the electrical circuit is closed the greater electron density, or electron pressure, at the zinc electrode causes electrons to flow to the hydrogen electrode through the wire connecting the two electrodes. Then as the electron density increases at the hydrogen electrode the equilibrium is shifted to the left, causing hydrogen ions to be reduced to free hydrogen.

$$H_2 \longleftarrow 2H^+ + 2e^-$$

At the same time the electron density at the zinc electrode decreases, as the result of the flow of electrons to the hydrogen electrode, and thus the equilibrium at the zinc electrode is shifted to the right, causing more zinc to be oxidized to zinc ions.

$$Zn \longrightarrow Zn^{2+} + 2e^-$$

The difference in potential between the two half cells as measured by a voltmeter is 0.76 volt. The force operating to cause electrons to move from the zinc electrode to the hydrogen electrode is called the **electromotive force (emf)** and is 0.76 volt. The zinc is the negative electrode because its electron density is higher than that of the hydrogen electrode. If we assign the entire electromotive force of the cell to the zinc electrode (the potential of the hydrogen electrode is assumed to be zero), then we may say that the potential of the zinc electrode for the oxidation of zinc metal to zinc ion is 0.76 volt above that of the hydrogen electrode and is positive (+0.76 volt). For the reverse reaction, the reduction of zinc ion to metallic zinc,

$$Zn^{2+} + 2e^- \longrightarrow Zn$$

the electrode potential has the same absolute value but opposite sign and therefore is −0.76 volt.

It is customary, in accord with the official recommendation of the International Union of Pure and Applied Chemistry (IUPAC), to express electrode potentials in every case in terms of the reduction process and to refer to them as standard electrode potentials. The symbol $E°$ is used for standard electrode potentials. Hence, the standard electrode potential, $E°$, for zinc is −0.76 volt. The negative sign corresponds to the negative charge assumed by the zinc electrode as its electron density increases. The standard electrode potential for any metal electrode having an electron density greater than the hydrogen electrode is given a negative sign. An electrode which has less electron density than the hydrogen electrode (copper, for example) is positive with reference to the hydrogen electrode and hence is given a positive sign. It should be noted, however, that the assignment of positive or negative signs to the standard electrode potentials is arbitrary, and some scientists follow the opposite convention of defining standard electrode potentials in terms of the oxidation process and hence assigning the opposite signs to standard electrode potentials.

The overall reaction that takes place when the zinc-hydrogen cell is operating is one of oxidation and reduction.

$$
\begin{array}{ll}
\mathrm{Zn} \longrightarrow \mathrm{Zn}^{2+} + 2e^- & \text{(anodic oxidation)} \\
\underline{2\mathrm{H}^+ + 2e^- \longrightarrow \mathrm{H}_2} & \text{(cathodic reduction)} \\
\mathrm{Zn} + 2\mathrm{H}^+ \longrightarrow \mathrm{Zn}^{2+} + \mathrm{H}_2\uparrow & \text{(cell reaction)}
\end{array}
$$

This same reaction will take place when zinc metal is immersed in a solution of hydrochloric acid. However, in the zinc-hydrogen cell, the reaction takes place without contact of the reactants. The electrons involved in the oxidation-reduction reaction are transferred from the zinc to the hydrogen ions through a wire.

The standard potential of any electrode, then, **is the potential of the electrode with respect to the hydrogen electrode, measured at 25° C when the concentration of the ions in the solution is 1 M and the pressure of any gas involved is 1 atmosphere.** Compared with the standard hydrogen electrode potential of 0.00 volt, the standard electrode potential of sodium is -2.71 volts, that of zinc is -0.76 volt, and that of copper is $+0.34$ volt.

Any cell that generates an electric current by an oxidation-reduction reaction is known as a **voltaic cell.** The cell composed of the zinc and hydrogen electrodes described previously is a voltaic cell and can be diagrammed

$$
\mathrm{Zn} \,|\, \mathrm{Zn}^{2+}, M = 1 \xrightarrow[\|]{e^-} \mathrm{H}^+, M = 1 \,|\, \mathrm{H}_2(\mathrm{Pt}), 1 \text{ atm}
$$

with the double bar indicating the two halves of the cell.

The purpose of the salt bridge in the zinc-hydrogen cell is to establish an electrical connection between the solutions of the two electrodes and thereby make a completed circuit. An excess of zinc ions accumulates in the half cell in which zinc goes into solution. Similarly, an excess of anions accumulates in the half cell in which hydrogen ions are being reduced. Migration of ions through the salt bridge compensates for this situation; otherwise, no current would flow in the external circuit.

22.11   The Electromotive Series of the Elements

From the standard electrode potentials of different metals whose ion concentrations are 1 M, the relative positions of these metals in the electromotive series (Section 9.9) can be determined. It was shown that zinc has a greater tendency than hydrogen to go into solution as ions and leave electrons on the electrode. When copper metal in contact with a 1 M solution of copper(II) ions is made the other electrode of a cell containing a standard hydrogen electrode, we find that hydrogen has a greater tendency to form positive ions than does the copper. The formation of hydrogen ions from the hydrogen leaves more electrons on the platinum than are left on the copper when it forms positive ions. Consequently, electrons flow from the hydrogen electrode to the copper electrode where they reduce copper(II) ions.

$$
\begin{array}{ll}
H_2 \rightleftharpoons 2H^+ + 2e^- & \text{(anodic oxidation)} \\
\underline{Cu^{2+} + 2e^- \rightleftharpoons Cu} & \text{(cathodic reduction)} \\
H_2 + Cu^{2+} \rightleftharpoons Cu + 2H^+ & \text{(cell reaction)}
\end{array}
$$

The emf of the cell

$$
H_2(Pt), 1 \text{ atm} \,|\, H^+, M = 1 \xrightarrow{e^-} \| \;\; Cu^{2+}, M = 1 \,|\, Cu
$$

as measured by a voltmeter is 0.34 volt. Since the copper ion is reduced to the free element more easily than the hydrogen ion, the potential of the copper electrode has a value of $+0.34$ volt with respect to the hydrogen electrode.

TABLE 22-1   Electromotive Series

	Electrode Reaction		Standard Electrode
	Oxidized Form	Reduced Form	(Reduction) Potential, E° (volts)
Potassium	$K^+ + e^-$ $\rightleftharpoons$	K	-2.925
Calcium	$Ca^{2+} + 2e^-$ $\rightleftharpoons$	Ca	-2.87
Sodium	$Na^+ + e^-$ $\rightleftharpoons$	Na	-2.714
Magnesium	$Mg^{2+} + 2e^-$ $\rightleftharpoons$	Mg	-2.37
Aluminum	$Al^{3+} + 3e^-$ $\rightleftharpoons$	Al	-1.66
Manganese	$Mn^{2+} + 2e^-$ $\rightleftharpoons$	Mn	-1.18
Zinc	$Zn^{2+} + 2e^-$ $\rightleftharpoons$	Zn	-0.763
Chromium	$Cr^{3+} + 3e^-$ $\rightleftharpoons$	Cr	-0.74
Iron	$Fe^{2+} + 2e^-$ $\rightleftharpoons$	Fe	-0.440
Cadmium	$Cd^{2+} + 2e^-$ $\rightleftharpoons$	Cd	-0.40
Cobalt	$Co^{2+} + 2e^-$ $\rightleftharpoons$	Co	-0.277
Nickel	$Ni^{2+} + 2e^-$ $\rightleftharpoons$	Ni	-0.250
Tin	$Sn^{2+} + 2e^-$ $\rightleftharpoons$	Sn	-0.136
Lead	$Pb^{2+} + 2e^-$ $\rightleftharpoons$	Pb	-0.126
Hydrogen	$2H^+ + 2e^-$ $\rightleftharpoons$	H_2	0.00
Copper	$Cu^{2+} + 2e^-$ $\rightleftharpoons$	Cu	$+0.337$
Iodine	$I_2 + 2e^-$ $\rightleftharpoons$	$2I^-$	$+0.5355$
Silver	$Ag^+ + e^-$ $\rightleftharpoons$	Ag	$+0.7991$
Mercury	$Hg^{2+} + 2e^-$ $\rightleftharpoons$	Hg	$+0.854$
Bromine	$Br_2(l) + 2e^-$ $\rightleftharpoons$	$2Br^-$	$+1.0652$
Platinum	$Pt^{2+} + 2e^-$ $\rightleftharpoons$	Pt	$+1.2$
Oxygen	$O_2 + 4H^+ + 4e^-$ $\rightleftharpoons$	$2H_2O$	$+1.23$
Chlorine	$Cl_2 + 2e^-$ $\rightleftharpoons$	$2Cl^-$	$+1.3595$
Gold	$Au^+ + e^-$ $\rightleftharpoons$	Au	$+1.68$
Fluorine	$F_2 + 2e^-$ $\rightleftharpoons$	$2F^-$	$+2.87$

Standard electrode potentials of other metals can be determined in the same manner. When the standard electrode potentials are arranged in order from the most negative to the most positive, we have the electromotive force series (Table 22–1). Table 22–1 also gives the electrode potentials, often referred to as reduction potentials, of some of the nonmetals as well as the common metals. Additional electrode potential values are given in Table 22–2 and in Appendix I.

22.12 Uses of the Electromotive Series

The electromotive series of the elements has many uses; some of the more important ones are given in the following list:

(1) The metals with high negative electrode potentials at the top of the series are good reducing agents in the free state. They are the metals most easily oxidized to their ions by the removal of electrons. They have low electron affinities, low ionization potentials, and low electronegativities.

(2) The elements with positive electrode potentials at the bottom of the series are good oxidizing agents when in the oxidized form, that is, when the metals are in the form of ions and the nonmetals are in the elemental state.

(3) The reduced form of any element will reduce the oxidized form of any element below it in the series. For example, metallic zinc will reduce copper(II) ions according to the equation

$$Zn + Cu^{2+} \rightleftharpoons Cu + Zn^{2+}$$

For a voltaic cell made up of a standard zinc electrode and a standard copper electrode (Fig. 22–7), the net cell reaction is obtained by adding the two half-reactions together.

Anode half-reaction:	$Zn \rightleftharpoons Zn^{2+} + 2e^-$
Cathode half-reaction:	$Cu^{2+} + 2e^- \rightleftharpoons Cu$
Net cell reaction:	$Zn + Cu^{2+} \rightleftharpoons Zn^{2+} + Cu$

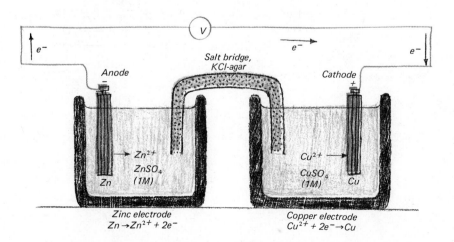

FIGURE 22-7
A zinc-copper voltaic cell.

One can measure the electromotive force (emf) of the cell by means of the voltmeter. Alternatively, the emf of the cell can be calculated by adding, algebraically, the potential of the copper electrode to that of the zinc electrode, taking into account the proper sign on each potential, based upon whether the half-reaction is taking place as oxidation or as reduction. The standard electrode potential (reduction) for zinc is -0.76 volt. For the oxidation of Zn to Zn^{2+}, therefore, the potential is $+0.76$ volt. At the copper electrode, Cu^{2+} is being reduced to Cu; hence the potential is the standard electrode potential (reduction) for copper, $+0.34$ volt. The emf for the cell = 0.76 volt + 0.34 volt = 1.10 volts.

The positive emf value for the zinc-copper cell indicates that the reaction will proceed spontaneously to the right as written and that the cell will deliver an electric current. A negative emf value for a cell would indicate that the reaction will not proceed spontaneously to the right, but would proceed spontaneously to the left.

Suppose we wish to calculate the emf of a cell composed of copper and silver electrodes in 1 M solutions of their respective ions and to determine if the cell will produce an electric current. Because copper is higher in the electromotive series than silver, copper metal will reduce the silver ion, $Cu + 2Ag^+ \rightleftharpoons Cu^{2+} + 2Ag$. To obtain this net cell reaction, we may add the half-reactions as follows:

$$
\begin{array}{ll}
Cu \rightleftharpoons Cu^{2+} + 2e^- & (-0.34 \text{ volt}) \\
2Ag^+ + 2e^- \rightleftharpoons 2Ag & +(+0.80 \text{ volt}) \\
\hline
Cu + 2Ag^+ \rightleftharpoons Cu^{2+} + 2Ag & (+0.46 \text{ volt})
\end{array}
$$

Hence, the calculated emf of the cell is 0.46 volt. Since the emf of the cell is positive, the reaction proceeds spontaneously to the right as written, and the cell delivers current.

Let us next calculate the emf of a cell made up of standard chlorine and bromine electrodes. The half-reactions and standard potentials are:

$$
\begin{array}{ll}
2Br^- \rightleftharpoons Br_2 + 2e^- & (-1.06 \text{ volts}) \\
Cl_2 + 2e^- \rightleftharpoons 2Cl^- & +(+1.36 \text{ volts}) \\
\hline
2Br^- + Cl_2 \rightleftharpoons Br_2 + 2Cl^- & (+0.30 \text{ volt})
\end{array}
$$

The emf of the cell is thus found to be 0.30 volt, meaning that the cell reaction proceeds spontaneously to the right as written. Suppose the half-reactions had been added as follows:

$$
\begin{array}{ll}
2Cl^- \rightleftharpoons Cl_2 + 2e^- & (-1.36 \text{ volts}) \\
Br_2 + 2e^- \rightleftharpoons 2Br^- & +(+1.06 \text{ volts}) \\
\hline
2Cl^- + Br_2 \rightleftharpoons Cl_2 + 2Br^- & (-0.30 \text{ volt})
\end{array}
$$

The negative emf value for the cell means that the reaction proceeds to the left spontaneously, rather than to the right. In other words, chlorine will oxidize the bromide ion to free bromine, but bromine will not oxidize the chloride ion to free chlorine (see Section 21.3).

The potentials given in the electromotive series refer only to ion concentrations of 1 M, pressures of 1 atm for any gases involved, and temperatures of 25° C. If the concentrations are not 1 M, the potentials are different and members close together in the series may even change places in the table. Furthermore, the values given do not apply to nonaqueous solutions or to fused salts. An illustration of the effect of concentration on potential is provided by the oxidation potentials for the oxidation of hydrogen to hydrogen ion at two different concentrations.

$$2H^+ (1\ M) + 2e^- \rightleftharpoons H_2 \ (1\ atm) \qquad E° + 0.00 \text{ v}$$
$$2H^+ (10^{-7}\ M) + 2e^- \rightleftharpoons H_2 \ (1\ atm) \qquad E° = -0.41 \text{ v}$$

Similar electromotive series can be constructed for solvents other than water. The electrode potentials would be quite different from those for water, and the members would fall in a different order in the series. The differences occur because of differences in energy of solvation.

22.13 Electrode Potentials for Other Half-Reactions

In addition to redox reactions between elements, there are many other redox reactions which can take place at the electrodes of electrochemical cells. The electrode consists of a relatively inactive substance such as carbon, which undergoes no change itself and merely serves as a carrier of electrons to or from the solution. The standard electrode potentials for the half-reactions of a considerable number of redox reactions are given in Table 22–2 and Appendix I.

The standard electrode potential of the Pt, Sn^{4+}, Sn^{2+} electrode (see Table 22–2) is that of the inert platinum electrode immersed in a solution which is 1 M with respect to both Sn^{4+} and Sn^{2+} ions. Similarly, the standard electrode potential of the Pt, Fe^{3+}, Fe^{2+} electrode pertains to its potential when the Fe^{3+} and Fe^{2+}

TABLE 22–2 Standard Electrode Potentials for Some Important Half-reactions

Electrode	Electrode Reaction		Standard Electrode Potential (volts)
	Oxidized Form	Reduced Form	
$Fe(OH)_2$, Fe, OH^-	$Fe(OH)_2 + 2e^-$	$\rightleftharpoons Fe + 2OH^-$	−0.877
$PbSO_4$, Pb, SO_4^{2-}	$PbSO_4 + 2e^-$	$\rightleftharpoons Pb + SO_4^{2-}$	−0.356
Pt, Sn^{4+}, Sn^{2+}	$Sn^{4+} + 2e^-$	$\rightleftharpoons Sn^{2+}$	+0.15
AgCl, Ag, Cl^-	$AgCl + e^-$	$\rightleftharpoons Ag + Cl^-$	+0.222
Hg_2Cl_2, Hg, Cl^-	$Hg_2Cl_2 + 2e^-$	$\rightleftharpoons 2Hg + 2Cl^-$	+0.27
NiO_2, $Ni(OH)_2$, OH^-	$NiO_2 + 2H_2O + 2e^-$	$\rightleftharpoons Ni(OH)_2 + 2OH^-$	+0.49
Pt, Fe^{3+}, Fe^{2+}	$Fe^{3+} + e^-$	$\rightleftharpoons Fe^{2+}$	+0.771
Pt, $Cr_2O_7^{2-}$, H^+, Cr^{3+}	$Cr_2O_7^{2-} + 14H^+ + 6e^-$	$\rightleftharpoons 2Cr^{3+} + 7H_2O$	+1.33
Pt, MnO_4^-, H^+, Mn^{2+}	$MnO_4^- + 8H^+ + 5e^-$	$\rightleftharpoons Mn^{2+} + 4H_2O$	+1.51
PbO_2, H_2SO_4, $PbSO_4$	$PbO_2 + SO_4^{2-} + 4H^+ + 2e^-$	$\rightleftharpoons PbSO_4 + 2H_2O$	+1.685

are both 1 *M*. The emf of a cell composed of the two electrodes Pt, Sn^{4+}, Sn^{2+} and Pt, Fe^{3+}, Fe^{2+} can be calculated as follows:

$$Sn^{2+} \rightleftharpoons Sn^{4+} + 2e^- \qquad (-0.15 \text{ volt})$$

Adding $\dfrac{2Fe^{3+} + 2e^- \rightleftharpoons 2Fe^{2+}}{Sn^{2+} + 2Fe^{3+} \rightleftharpoons Sn^{4+} + 2Fe^{2+}} \qquad \dfrac{+(+0.77 \text{ volt})}{(+0.62 \text{ volt})}$$

Thus, the emf of the cell is 0.62 volt when standard electrodes are used. This means that the reaction will proceed spontaneously to the right as written and will deliver a current. It also means that when solutions containing tin(II) and iron(III) are mixed, the Sn^{2+} will reduce the Fe^{3+}.

22.14   The Nernst Equation

Standard electrode potentials, abbreviated $E°$, are for what is referred to as **standard state conditions: 1 molar solutions of ions, a pressure of 1 atmosphere for gases, and a temperature of 25° C.** (For concentrations of ions, it is more accurate to use *activities* of 1, instead of molarities of 1; however, for 1 *M* solutions, concentrations can be used instead of activities, unless great accuracy is desired.) It is important to note that the terms *standard state conditions* and *standard conditions of temperature and pressure (STP)* for gases (see Section 10.7) have different meanings. Recall that STP conditions do refer to 1 atmosphere pressure but refer to a temperature of 0° C and do not include a concentration or activity term.

As we have seen in Section 22.12, the electromotive force value for a half-reaction changes when the concentration changes. An equation referred to as the **Nernst Equation** makes it possible to calculate emf values at other concentrations. The Nernst equation for 25° C is

$$E = E° - \frac{0.059}{n} \log Q$$

(For problems in which the other data justify more than two significant figures, the more accurate value for 0.059 is 0.059152.)

where E = the emf for the reaction at the new concentration
$E°$ = the *standard* electrode potential
n = the number of electrons stated in the half-reaction
Q = an expression which takes the same form as the equilibrium constant for the half-reaction without indicating the electrons.

For example, a typical half-reaction and the corresponding Nernst Equation is

$$Sn^{4+} + 2e^- \longrightarrow Sn^{2+}$$

$$E = E° - \frac{0.059}{2} \log \frac{[Sn^{2+}]}{[Sn^{4+}]}$$

or, with the standard electrode potential, $E°$, from Table 22–2 inserted,

$$E = 0.15 - \frac{0.059}{2} \log \frac{[Sn^{2+}]}{[Sn^{4+}]}$$

where brackets mean "concentration of" and "log" signifies the common logarithm.

If the concentrations of both Sn^{2+} and Sn^{4+} are 1 M (concentration of the standard state),

$$E = E^\circ - \frac{0.059}{2} \log \frac{1}{1}$$

Inasmuch as $\log 1 = 0$,

$$E = E^\circ - \frac{0.059}{2}(0) \quad \text{or} \quad E = E^\circ$$

This, of course, states that the electrode potential, E, is equal to the *standard* electrode potential, E°, when the concentrations are the standard state concentration of 1 M. (Note also that $E = E^\circ$ if $[Sn^{2+}] = [Sn^{4+}]$, even if the concentration is not 1 M.)

Example 1. Calculate E at 25° C if the concentration of Sn^{2+} is four times that of Sn^{4+}.

$$\frac{[Sn^{2+}]}{[Sn^{4+}]} = 4$$

Therefore,

$$E = 0.15 - \frac{0.059}{2} \log 4$$

$$= 0.15 - \frac{0.059}{2}(0.6021)$$

$$= 0.15 - 0.018$$
$$= 0.13 \text{ volt}$$

Example 2. Calculate E at 25° C if $[Sn^{2+}]$ is one-fourth $[Sn^{4+}]$.

$$\frac{[Sn^{2+}]}{[Sn^{4+}]} = \frac{1}{4}$$

Therefore,

$$E = 0.15 - \frac{0.059}{2} \log \frac{1}{4}$$

$$= 0.15 - \frac{0.059}{2}(\log 1 - \log 4)$$

$$= 0.15 - \frac{0.059}{2}(0 - 0.6021)$$

$$= 0.15 + \frac{0.059}{2}(0.6021)$$

$$= 0.15 + 0.018$$
$$= 0.17 \text{ volt}$$

In Section 22.12, we noted that, whereas the standard electrode potential (E°)

for the reduction of $2H^+$ to H_2 is 0.00 v, the emf value (E) at 10^{-7} M is -0.41 v. The following example shows the calculation of the -0.41 v value.

Example 3. Calculate the emf value for the reduction from H^+ at 1.0×10^{-7} M to H_2 at 1.0 atm.

$$2H^+ (1\ M) + 2e^- \rightleftharpoons H_2 (1.0\ \text{atm}) \qquad E^\circ = 0.00\ \text{v}$$

$$E = E^\circ - \frac{0.059}{n} \log \frac{p_{H_2}}{[H^+]^2}$$

$$= 0.00 - \frac{0.059}{2} \log \frac{1.0}{(1.0 \times 10^{-7})^2}$$

$$= 0.00 - \frac{0.059}{2} \log \frac{1.0}{1.0 \times 10^{-14}}$$

$$= 0.00 - \frac{0.059}{2} \log (1.0 \times 10^{14})$$

$$= 0.00 - \frac{0.059}{2} (14)$$

$$= -0.41\ \text{volt}$$

Hence, $2H^+ (1.0 \times 10^{-7}\ M) \rightleftharpoons H_2 (1.0\ \text{atm}) \qquad E = -0.41\ \text{v}$

Example 4. Calculate E for the reduction of Cl_2 at 5.0 atm to Cl^- at 0.010 M.

$$Cl_2 (1.0\ \text{atm}) + 2e^- \rightleftharpoons 2Cl^- (1.0\ M) \qquad E^\circ = 1.36\ \text{v}$$

$$E = E^\circ - \frac{0.059}{n} \log \frac{[Cl^-]^2}{p_{Cl_2}}$$

$$= 1.36 - \frac{0.059}{2} \log \frac{(0.010)^2}{(5.0)}$$

$$= 1.36 + 0.14 = 1.50\ \text{volts}$$

Hence, $Cl_2 (5.0\ \text{atm}) + 2e^- \rightleftharpoons 2Cl^- (0.010\ M) \qquad E = 1.50\ \text{v}$

Therefore, the change in concentration of Cl^- and in pressure on the Cl_2 gas results in a change in emf from 1.36 to 1.50 v.

22.15 Relationship of the Gibbs Free Energy Change, the Standard Reduction Potential, and the Equilibrium Constant

The standard Gibbs free energy change, ΔG° (see Section 20.10), is related to the standard electrode potential, E°, and to the equilibrium constant, K_e (see Section 20.13), by means of the following two equations:

$$\Delta G^\circ = -nFE^\circ \qquad \text{and} \qquad \Delta G^\circ = -RT \ln K_e$$

where n = the number of electrons stated in the half-reaction

F = a constant known as the Faraday constant, with a value of 23,061 cal/volt, or 96,487 coulombs

$E° =$ the standard electrode potential
$R =$ the universal gas constant (Section 10.21)
$T =$ the Kelvin temperature
$K_e =$ the equilibrium constant

and ln signifies the natural logarithm (to the base e)

By combining the two relationships for $\Delta G°$,

$$-nFE° = -RT \ln K_e$$

Then,

$$E° = \frac{RT}{nF} \ln K_e$$

For 25° C (298° K), after conversion to the common logarithm (to the base 10), and with substitution of the numerical value of R, the expression becomes

$$E° = \frac{0.059}{n} \log K_e \quad Nernst \;\; equation$$

As noted earlier, the more accurate value for 0.059 is 0.059152, and is for use in problems in which the other data justify more than two significant figures.

Example 1.   Calculate the standard Gibbs free energy change at 25° C for the reaction

$$Cd + Pb^{2+} \rightleftharpoons Cd^{2+} + Pb$$

The half-reactions and calculation of the emf of the cell are

	$Cd \rightleftharpoons Cd^{2+} + 2e^-$	$(+0.40 \text{ v})$
Adding	$Pb^{2+} + 2e^- \rightleftharpoons Pb$	$+(-0.13 \text{ v})$
	$Cd + Pb^{2+} \rightleftharpoons Cd^{2+} + Pb$	$(+0.27 \text{ v})$

Hence, $E°$ for the cell $= +0.27$ v.

$$\Delta G° = -nFE° = -2(23,061 \text{ cal/volt})(0.27 \text{ volt}) = -1.2 \times 10^4 \text{ cal}$$
$$= -12 \text{ kcal}$$

The negative value for $\Delta G°$ indicates that the reaction proceeds spontaneously to the right (Section 20.10).

Example 2.   Calculate the equilibrium constant at 25° C for the reaction in the preceding example.

$$E° = \frac{0.0592}{n} \log K_e$$

$$0.27 = \frac{0.0592}{2} \log K_e$$

$$\log K_e = 0.27 \times \frac{2}{0.0592} = 9.12$$

$$K_e = \text{antilog } 9.12 = 1.3 \times 10^9$$

Thus, at equilibrium the molar concentration of cadmium ion in the solution is more than a billion times that of lead ion!

Example 3.   Calculate the Gibbs free energy change and the equilibrium constant for the following reaction at 25° C:

$$Zn + Cu^{2+} (0.20\ M) \rightleftharpoons Zn^{2+} (0.50\ M) + Cu$$

Adding
$Zn \rightleftharpoons Zn^{2+} + 2e^-$	$(+0.76\ v)$
$Cu^{2+} + 2e^- \rightleftharpoons Cu$	$+(+0.34\ v)$
$Zn + Cu^{2+} \rightleftharpoons Zn^{2+} + Cu^\circ$	$(+1.10\ v)$

Hence, for the cell reaction *under standard state conditions* (Cu^{2+} and Zn^{2+} at concentrations of 1 M, or more accurately at unit activity), $E° = +1.10$ v.

The Nernst Equation enables us to calculate the potential (E) when Zn^{2+} and Cu^{2+} concentrations are 0.50 M and 0.20 M, respectively.

Zinc half-reaction:

$$E = E° - \frac{0.0592}{n} \log \frac{[Zn^{2+}]}{[Zn°]} = 0.76 - \frac{0.0592}{2} \log \frac{0.50}{1}$$

$$= 0.76 - \frac{0.0592}{2} (0.699 - 1) = 0.76 + 0.01 = 0.77\ v$$

Copper half-reaction:

$$E = E° - \frac{0.0592}{n} \log \frac{[Cu°]}{[Cu^{2+}]} = +0.34 - \frac{0.0592}{2} \log \frac{1}{0.20}$$

$$= +0.34 - \frac{0.0592}{2} (0.6990) = +0.34 - 0.02 = +0.32\ v$$

Thus, for the reaction under the specified conditions, $E = 0.77 + 0.32 = +1.09$ v (as compared to the $E°$ value of $+1.10$ v for standard state conditions).

$$\Delta G = -nFE = -2(23,060)(1.09) = -5.03 \times 10^4\ cal = -50.3\ kcal$$

The negative value for ΔG indicates that the reaction proceeds spontaneously to the right.

The value of K_e at a given temperature (25° C in this case) is calculated from $E°$ and is constant for the various concentrations.

$$E° = \frac{0.0592}{n} \log K_e$$

$$1.10 = \frac{0.0592}{2} \log K_e$$

$$\log K_e = 1.10 \times \frac{2}{0.0592} = 37.16$$

$$K_e = 1.4 \times 10^{37}$$

Voltaic Cells

22.16   Primary Voltaic Cells

Once the chemicals in some voltaic cells are consumed, further chemical action is not possible, because the electrodes and electrolytes are of such a nature that the reaction cannot be reversed to restore them to their original states by the application of an external electrical potential. Such cells are called **primary voltaic cells**, or **irreversible cells**. The Daniell cell and the "dry" cell are examples of primary cells.

■ **1. The Daniell Cell**.   The construction of a Daniell cell is shown in Fig. 22–8. Metallic copper (*A*) is placed in a saturated solution of copper(II) sulfate on the bottom of the cell with crystals of copper(II) sulfate to keep the solution saturated. A zinc electrode (*B*) is suspended near the top of the cell in a dilute solution of zinc sulfate, which floats on the more dense solution of copper(II) sulfate. When the two metals are connected by means of a wire, electrons flow through the wire from the more easily oxidized zinc to the less easily oxidized copper. The zinc is oxidized and goes into solution as zinc ions, while the copper(II) ions are reduced to metallic copper. This is the same reaction that takes place when metallic zinc becomes plated with copper upon being placed in a solution of copper(II) sulfate.

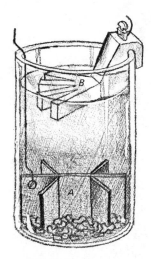

FIGURE 22–8

The Daniell, or gravity, cell. *A,* sheet copper surrounded by copper(II) sulfate solution; *B,* zinc plate surrounded by zinc sulfate solution.

$$\text{Zn} \rightleftharpoons \text{Zn}^{2+} + 2e^- \quad \text{(anodic oxidation)}$$
$$\underline{\text{Cu}^{2+} + 2e^- \rightleftharpoons \text{Cu} \qquad \text{(cathodic reduction)}}$$
$$\text{Zn} + \text{Cu}^{2+} \rightleftharpoons \text{Zn}^{2+} + \text{Cu} \quad \text{(cell reaction)}$$

If the zinc ion concentration is increased, the anode reaction is shifted to the left, resulting in a lower voltage being produced by the cell.

A cell similar to the Daniell cell in principle, but using cadmium and nickel instead of zinc and copper, is being utilized increasingly in batteries where longer cell life is desirable.

■ **2. The "Dry Cell."**   The container of the dry cell (Fig. 22–9) is made of zinc, which serves also as one of the electrodes. The container is lined with porous paper, which separates the metal from the materials within the cell. A carbon (graphite) rod in the cell acts as the other electrode. The space around the carbon rod contains a moist mixture of ammonium chloride, manganese, dioxide, zinc chloride, and some porous inactive solid such as sawdust. The cell is sealed with some sealing compound to keep the moisture in. When the cell delivers a current the zinc goes into solution as zinc ions and leaves electrons on the zinc container so that it becomes the negative electrode. This electrode, then, is called the **anode,** since this is where *oxidation* occurs. The oxidation half-reaction is

$$\textit{Anode:}\ \text{Zn} \longrightarrow \text{Zn}^{2+} + 2e^-$$

The ammonium ion is *reduced* at the carbon electrode, and this reaction is thus at the **cathode.**

$$\textit{Cathode:}\ 2\text{NH}_4^+ + 2e^- \longrightarrow 2\text{NH}_3 + 2[\text{H}]$$

FIGURE 22–9

Cross section of the Leclanché dry cell.

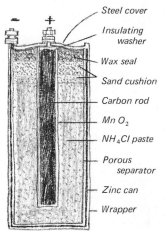

Steel cover

Insulating washer

Wax seal

Sand cushion

Carbon rod

MnO_2

NH_4Cl paste

Porous separator

Zinc can

Wrapper

The net cell reaction may be expressed by the equation

$$Zn + 2NH_4^+ \longrightarrow Zn^{2+} + 2NH_3 + 2[H]$$

The manganese dioxide in the cell oxidizes the hydrogen as it is formed. The hydrogen would otherwise collect on the cathode and stop the action of the cell, a condition called **polarization.**

$$2[H] + MnO_2 + 2H^+ \longrightarrow Mn^{2+} + 2H_2O$$

The ammonia which is formed at the cathode when the cell is working unites with part of the zinc ions, forming the complex $[Zn(NH_3)_4]^{2+}$. This reaction holds down the increase in the concentration of zinc ions, thereby keeping the potential of the zinc electrode more nearly constant. It also prevents polarization caused by the accumulation of ammonia gas on the cathode.

■ **3. Cells For Special Purposes.** Many cells are designed to meet special requirements for one specific purpose. For example, during World War II a torpedo battery was developed that had to fit into a small space and produce a lot of power, but only once over a period of just thirty seconds. A battery cell suitable for a camera light meter must produce a small amount of current but must last for a long period. A wrist-watch battery cell must produce a very small current without interruption for at least one year. A special cell used in weather balloons, which was developed to meet the special requirements of small size, light weight, and operation over a period of a few hours, is made by laying a mixture of dry magnesium powder and dry silver chloride on a strip of paper which is then rolled up; when this roll is dipped into water, the reaction begins.

The Rubin-Mallory cell, which is very small, receives widespread use in electrically operated hearing aids. It utilizes a zinc container as the anode and a carbon rod as the cathode. The electrolyte is a paste of moist mercury(II) oxide mixed with a little sodium, or potassium, hydroxide. The cell produces a potential of 1.35 volts. The half-reactions and net cell reaction are

Anode:	$Zn + 2OH^- \longrightarrow Zn(OH)_2 + 2e^-$
Cathode:	$HgO + H_2O + 2e^- \longrightarrow Hg + 2OH^-$
Net cell reaction:	$Zn + HgO + H_2O \longrightarrow Zn(OH)_2 + Hg$

22.17 Secondary Voltaic Cells

In **secondary,** or **reversible, cells,** the substances which are consumed in producing electricity may be regenerated in their original forms by causing a direct current of electricity to flow in the reverse direction of the discharge. This reversal recharges the cell. The lead storage battery and the Edison storage battery both consist of several secondary cells connected together.

■ **1. The Lead Storage Battery.** The electrodes of the lead storage battery consist of two sets of lead alloy plates in the form of grids. The openings of one set of grids are filled with lead dioxide and the openings of the other with spongy lead metal. Dilute sulfuric acid serves as the electrolyte. When the battery is delivering a current, the spongy lead is oxidized to lead ions, and these plates become negatively charged.

$$Pb \longrightarrow Pb^{2+} + 2e^- \quad \text{(oxidation)}$$

The lead ions thus formed combine with sulfate ions of the electrolyte and coat the lead electrode with lead sulfate.

$$Pb^{2+} + SO_4{}^{2-} \longrightarrow PbSO_4$$

The lead sulfate, being quite insoluble in dilute sulfuric acid, can be regenerated to spongy lead on the electrode during the recharging process.

The net lead electrode reaction when the cell is delivering a current is

$$Pb + SO_4{}^{2-} \longrightarrow PbSO_4 + 2e^-$$

The negatively charged lead electrode is the anode since it is the electrode at which oxidation occurs.

Electrons flow from the negatively charged lead electrode through the external circuit and enter the lead dioxide electrode. The lead dioxide in the presence of hydrogen ions from the electrolyte is reduced to lead(II) ions.

$$PbO_2 + 4H^+ + 2e^- \longrightarrow Pb^{2+} + 2H_2O \quad \text{(reduction)}$$

Again lead ions combine with sulfate ions of the electrolyte and the lead dioxide plate becomes coated with lead sulfate.

$$Pb^{2+} + SO_4{}^{2-} \longrightarrow PbSO_4$$

The net reaction at the positive lead dioxide electrode when the cell is delivering a current is

$$\overset{+4}{Pb}O_2 + 4H^+ + SO_4{}^{2-} + 2e^- \longrightarrow \overset{+2}{Pb}SO_4 + 2H_2O$$

The positively charged lead dioxide electrode is the cathode since it is the electrode at which reduction occurs.

The lead storage battery can be recharged by passing electrons through the cell in the reverse direction by applying an external potential. The cell is thereby made an electrolytic cell during the recharging process. The reactions are just the reverse of those which occur when the cell is operating as a voltaic cell producing a current. Electrons are forced into the lead electrode making it the cathode, at which reduction of lead ions from the lead sulfate takes place. Electrons are withdrawn from the lead dioxide electrode making it the anode, at which oxidation of lead ions from the lead sulfate takes place.

The charge and discharge at the two plates may be summarized by

$$Pb + SO_4{}^{2-} \underset{\text{Charge}}{\overset{\text{Discharge}}{\rightleftharpoons}} PbSO_4 + 2e^- \quad \text{(at the lead plate)}$$

$$PbO_2 + 4H^+ + SO_4{}^{2-} + 2e^- \underset{\text{Charge}}{\overset{\text{Discharge}}{\rightleftharpoons}} PbSO_4 + 2H_2O \quad \text{(at the lead dioxide plate)}$$

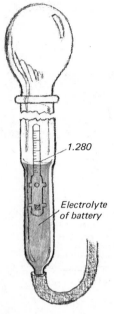

FIGURE 22-10
Hydrometers can be used to measure the density of the electrolyte of a lead storage battery and thus test the electrical charge of the cell.

1.280

Electrolyte of battery

Thus, during the charging process, the sulfate ion of the electrolyte is regenerated and lead sulfate is converted back to spongy lead at the lead electrode; hydrogen ions and sulfate ions of the electrolyte are regenerated, and lead sulfate is converted back to lead dioxide at the lead dioxide electrode.

The net cell reaction of the lead storage battery is given by

$$Pb + PbO_2 + 4H^+ + 2SO_4^{2-} \underset{\text{Charge}}{\overset{\text{Discharge}}{\rightleftharpoons}} 2PbSO_4 + 2H_2O$$

The above equation indicates that the amount of sulfuric acid decreases as the cell discharges. Conversely, charging the cell regenerates the acid. Since sulfuric acid is much heavier than water, the condition of a lead storage battery may be tested conveniently by determining the specific gravity of the electrolyte using a hydrometer (Fig. 22–10). When the specific gravity is 1.25 to 1.30, the battery is charged; when it falls much below 1.20, the battery needs charging. As can be calculated from the potentials in Table 22–2, a single, fully charged lead storage cell has a potential of $0.36 + 1.69 = 2.05$ v. The potential falls off slowly as the cell is used. A 12-volt automobile battery contains six lead storage cells, whereas a 6-volt battery contains three.

■ **2. The Edison Storage Battery.**   In the cells of the Edison battery the negative electrode consists of steel plates packed with finely divided iron and the positive electrode of steel plates with hydrated nickel dioxide. The electrolyte is a 21 per cent potassium hydroxide solution containing some lithium hydroxide. When the cell is delivering a current of electricity, oxidation takes place at the anode according to the equation

$$\textit{Anode:}\ Fe + 2OH^- \longrightarrow Fe(OH)_2 + 2e^-$$

The reduction at the cathode is expressed by

$$\textit{Cathode:}\ NiO_2 + 2H_2O + 2e^- \longrightarrow Ni(OH)_2\ \text{to}\ 2OH^-$$

These reactions are reversed while the battery is being charged. The net reaction is

$$Fe + NiO_2 + 2H_2O \underset{\text{Charge}}{\overset{\text{Discharge}}{\rightleftharpoons}} Fe(OH)_2 + Ni(OH)_2$$

It can be seen from this equation that the electrodes are restored to their original condition on recharging. The electromotive force of each cell in an Edison storage battery, as can be calculated from the potentials in Table 22–2, is about 1.4 volts. The cell must be sealed, for otherwise the potassium hydroxide and lithium hydroxide would change to carbonates by action with CO_2 from the air. The cell can be sealed because no gases are produced during reaction. This battery is lighter and more durable than the lead storage battery but more costly.

Because of environmental considerations, considerable interest has arisen recently in connection with the possibility of developing new varieties of secondary cells suitable for powering an electric automobile.

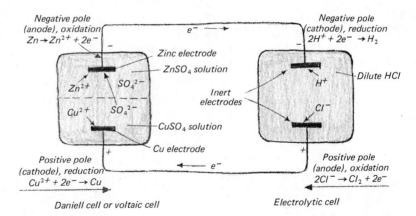

Negative pole
(anode), oxidation
$Zn \rightarrow Zn^{2+} + 2e^-$

$e^- \longrightarrow$

Zinc electrode

ZnSO₄ solution

Zn^{2+} SO_4^{2-}

Cu^{2+} SO_4^{2-}

CuSO₄ solution

Cu electrode

Positive pole
(cathode), reduction
$Cu^{2+} + 2e^- \rightarrow Cu$

$\longleftarrow e^-$

Daniell cell or voltaic cell

Negative pole
(cathode), reduction
$2H^+ + 2e^- \rightarrow H_2$

Dilute HCl

Inert
electrodes

H^+

Cl^-

Positive pole
(anode), oxidation
$2Cl^- \rightarrow Cl_2 + 2e^-$

Electrolytic cell

FIGURE 22-11

Illustration of the relationship between a voltaic cell and an electrolytic cell. The voltaic cell (*left*) provides the electric current to run the electrolytic cell (*right*). Note that the signs for the electrodes are opposite for voltaic and electrolytic cells, resulting in electrodes of like signs being connected. Note also that, regardless of whether the cell is voltaic or electrolytic, oxidation occurs at the anode and reduction at the cathode.

It should be noted that the anode is positive and the cathode negative in an electrolytic cell (see Fig. 22–3, for example), whereas the anode is negative and the cathode positive in a voltaic cell (Fig. 22–6). *For all types of cell, however, the electrode where oxidation occurs is the anode and where reduction occurs is the cathode.*

Figure 22–11 shows diagrammatically the relationships between a voltaic cell and an electrolytic cell. Note that the sign convention is such that when a voltaic cell is supplying the current for an electrolytic cell, the negative electrode of one cell is hooked to the negative electrode of the other cell, and likewise the positive electrodes of the two cells are hooked together.

22.18   The Solar Battery

It has long been known that an enormous amount of energy is given off by the sun. The earth alone receives more energy from sunlight in two days than is stored in all known reserves of fossil fuels. In order to use part of this energy, several devices are used or are under study to convert sunlight into electrical energy. The conventional photoelectric cell, used in electric-eye doors, transforms light into electrical energy (Section 3.1) but delivers as power only about one-half of one per cent of the total light energy it absorbs. A newer type of photoelectric cell is several times more efficient than the conventional types and is capable of generating electric power from sunlight at the rate of 90 watts per square yard of illuminated surface.

The basic unit of a typical solar battery is a thin wafer of very pure silicon, containing initially no more than one part of impurity per million parts of silicon. The silicon crystal lattice has a structure like that of diamond in which each silicon atom is bonded covalently through its four valence electrons to four neighboring silicon atoms at the corners of a regular tetrahedron. For use in the solar battery, a tiny amount of arsenic is added to the lattice structure of the silicon wafer. Inasmuch as arsenic has five valence electrons, compared to four for silicon, the addition of arsenic to the tetrahedral silicon crystal lattice creates an excess of free electrons within the lattice in the body of the wafer (commonly referred to as *n*-type silicon). On the surface of the wafer is placed a thin layer of silicon containing a trace of boron. Inasmuch as boron has only three valence electrons, "holes" (vacant spots for electrons) then exist in the silicon lattice at the surface of the wafer (commonly referred to as *p*-type silicon). A junction (referred to as a *p-n* junction) exists between the body of the wafer and the thin surface layer. Electrons diffuse through the junction from the wafer to "holes" in the surface layer; at the same time, "holes," or electron vacancies, move to the body of the wafer. This results in a net positive charge within the body of the wafer, which was neutral prior to the diffusion of electrons (negative charges) out of the wafer. A net negative charge, from electrons diffusing in, is correspondingly produced in the previously neutral surface layer. Thus, an electrostatic force develops between the two regions, which builds up until the negatively charged region at the surface repels any further diffusion of electrons. A counter force, opposite to the diffusion force, results as the electric field tends to attract the negative electrons back to the positively charged body of the wafer and to attract "holes" to the negatively charged surface of the wafer. When the diffusion of electrons is just balanced by the electric field, equilibrium is established between the two forces and hence also between electrons and "holes." At equilibrium, a net difference in potential exists between the two regions in the wafer.

In the solar battery, one electrical lead is attached to the body of the wafer and one lead to the surface. When the wafer is exposed to sunlight, energy from the sunlight causes electrons to be released from their positions in the lattice near the *p-n* junction, thereby upsetting the equilibrium between electrons and "holes" and causing additional electrons to move across the junction into the body of the wafer. Thus, a sufficient electron pressure (electromotive force) is built up to cause electrons to move through the electrical leads, and a current flows through the wire. The device, thus, is an electric cell, with the positive terminal at the *p*-contact and the negative terminal at the *n*-contact (Fig. 22-12).

In actual practice, a series of such wafers makes up the solar battery. The first practical application of the solar battery was in powering eight telephones on a rural telephone line in Georgia in 1955. Such cells now are used to power communication devices in spacecraft, especially those designated to remain in space for lengthy periods of time.

Two new types of solar cells, which were reported in 1975, show excellent promise for use in telephone circuits and other applications. One of these cells utilizes a layer of crystalline cadmium sulfide deposited on indium phosphide. The cell has a conversion efficiency of 12.5%, which means that 12.5% of the

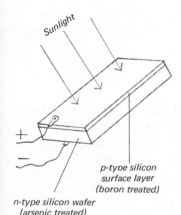

Sunlight

+

−

*p-type silicon
surface layer
(boron treated)*

*n-type silicon wafer
(arsenic treated)*

FIGURE 22-12
Solar battery cell.

sunlight which strikes the cell is converted into electrical energy. This is considered very good conversion efficiency, in terms of present technology. The other of the two new cells consists of a layer of cadmium sulfide deposited on copper indium diselenide and has a conversion efficiency of 8%.

A number of homes and a few industrial or office buildings have been heated successfully by solar energy on an experimental basis. A major problem has been to make such installations economically feasible on a commercial basis.

With the increased pollution resulting from the use of fossil fuels in electric power plants and with the inherent problems from radioactive emission and thermal pollution in nuclear power plants, it seems quite probable that solar energy will play a major part in satisfying our immense future energy needs (see Sections 12.4 and 24.3–24.5 on water and air pollution).

22.19   Fuel Cells

Fuel cells are voltaic cells in which electrode materials, usually in the form of gases, are supplied continuously to a cell and consumed to produce electricity. A typical fuel cell, currently used extensively in spacecraft, is that based upon the reaction of hydrogen (the "fuel") and oxygen (the "oxidizer") to form water. At the anode, hydrogen gas is diffused through a porous carbon electrode, in the surface of which is embedded a catalyst such as finely divided particles of platinum or palladium. At the cathode, oxygen is diffused through a porous carbon electrode impregnated with cobalt oxide, platinum, or silver as catalyst. The two electrodes are separated by an electrolyte such as a concentrated solution of sodium hydroxide or potassium hydroxide (Fig. 22–13).

As hydrogen diffuses through the anode, it is adsorbed on the electrode surface in the form of hydrogen atoms, which react with hydroxyl ions of the electrolyte to form water.

$$H_2 \xrightarrow{\text{Catalyst}} 2H$$
$$2H + 2OH^- \longrightarrow 2H_2O + 2e^-$$
Net anode reaction: $\overline{H_2 + 2OH^- \longrightarrow 2H_2O + 2e^-}$

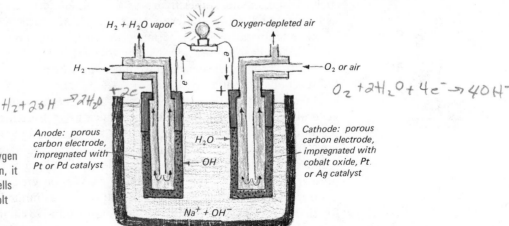

FIGURE 22–13

Diagram of a hydrogen-oxygen fuel cell. In actual operation, it would take several such cells to light an ordinary 110-volt light bulb. *From Scientific American*

FIGURE 22-14

A 1-kilowatt fuel cell designed for use in spacecraft. This particular cell is 25 inches long, weighs 68 pounds, and uses hydrogen and oxygen to produce electricity. It also produces one pint of pure, drinkable water for each kilowatt-hour of operation. *Courtesy General Electric Co.*

The electrons produced at the anode flow through the external circuit to the cathode. The oxygen, diffused through the cathode, is adsorbed on the electrode surface where it is reduced to hydroxyl ions.

$$\text{Cathode reaction: } O_2 + 2H_2O + 4e^- \longrightarrow 4OH^-$$

The hydroxyl ions complete the cycle by migrating through the electrolyte from the oxygen electrode (cathode) to the hydrogen electrode (anode). The electrical output of the cell results, as in all voltaic cells, from the flow of electrons through the external circuit from anode to cathode.

The overall cell reaction is the combination of hydrogen and oxygen to produce water (Fig. 22–14).

$$
\begin{array}{ll}
2H_2 + 4OH^- \longrightarrow 4H_2O + 4e^- & \text{(anodic oxidation)} \\
\underline{O_2 + 2H_2O + 4e^- \longrightarrow 4OH^-} & \text{(cathodic reduction)} \\
2H_2 + O_2 \longrightarrow 2H_2O & \text{(cell reaction)}
\end{array}
$$

A great deal of research effort is being expended in investigating other fuels such as methane and other hydrocarbons, other electrode systems, and cells that will operate efficiently under extremes of temperature. A fuel such as methane or other hydrocarbons can be made to react in the presence of air to produce hydrogen and carbon dioxide. The hydrogen then is fed into the cell at the anode, and air (to provide oxygen) is introduced into the cell at the cathode, to produce the reactions described earlier.

The efficiency of the fuel cell is potentially greater than that of conventional generating equipment, because of producing electric current directly from the reaction of the fuel and oxidizer without going through the inherently wasteful intermediate conversion of chemical energy to heat, which in conventional power plants can waste 60–80% of the available chemical energy. Fuel cells can, at the present stage of development, increase conversion efficiencies from about 30% in conventional power plants to over 40%. It is estimated that sufficient power for 20,000 people can be produced in a unit that is 18′ high and covers less than one-half acre of ground. An additional advantage of the fuel cell as a source of commercial power is in decreasing pollution, inasmuch as the chemical products of a cell such as those described here are essentially air, carbon dioxide, and water vapor. However, one must take into account power needs for the process, such as the energy required to decompose the fuel and to pump the cell components into the cell, in assessing the overall efficiency, space, and pollution properties of fuel cells for producing power commercially.

An interesting possibility for future application of the hydrogen-oxygen fuel cell involves capturing the power of the sun. Sunlight can be used, by a recently developed catalytic process, to decompose water into hydrogen and oxygen, which can then be used in fuel cells to produce electric current. It has been suggested, also, that fuel cells might help to cut the cost of nuclear power. Nuclear plants could be operated more nearly at peak capacity if the power generated during periods of low demand were used to electrolyze water into hydrogen and oxygen, which would then be used in fuel cells to produce electrical energy during peak demand periods.

A new potential application for fuel cells is in powering pacemakers in heart patients. In recent years, battery-powered pacemakers to maintain the regularity and proper pace of heartbeats have literally been of life-saving significance to many people. However, even with recently developed long-life batteries, a minor operation to replace the batteries has been necessary at relatively frequent intervals of one to two years. A fuel cell, with electrode materials supplied continuously over long periods of time, would make the periodic surgery unnecessary. "Biogalvanic" batteries employing a metal and the body's oxygen and fluids to generate an electric current have been used experimentally to power heart pacemakers in dogs. A metal such as aluminum or magnesium implanted subcutaneously serves as the negative electrode (anode); the body's oxygen is used for the positive electrode (cathode).

QUESTIONS

1. Explain the meaning of the terms "anode" and "cathode."
2. Name the kinds of charged particles which, when they are in motion, may constitute a current of electricity.
3. Distinguish between voltaic cells and electrolytic cells.
4. Distinguish between primary and secondary voltaic cells.
5. By means of description and chemical equations, explain the electrolytic purification of copper from impurities such as silver and zinc.
6. Give the equation for the net reaction in the electrolysis of dilute sulfuric acid.
7. Why is the hydrogen ion rather than the sodium ion reduced at the cathode during the electrolysis of aqueous sodium chloride? What is the source of the hydrogen ion?
8. Interpret the electrolysis of fused sodium chloride as an oxidation-reduction reaction.
9. Explain how the Nernst Equation is used in electrochemistry.
10. How was the electromotive series experimentally established?

11. Describe the construction of the dry cell and give its net reaction.
12. Draw a schematic diagram of a Daniell voltaic cell connected so as to supply electric current to an electrolytic cell for the electrolysis of hydrochloric acid. Label the diagram to show (a) which is anode and which is cathode in each cell; (b) direction of electron flow in the external circuit; (c) sign of each electrode of each cell (refer to Section 22.17); (d) direction of migration of each ion within the electrolyte of each cell, and (e) electrode reactions for each cell.
13. Relate the condition of charge of a lead battery to the specific gravity of its electrolyte.
14. Give equations for the net reactions at the two different kinds of plates in a lead storage battery. Relate the lead storage battery to the fact that an automobile is more difficult to start in cold weather.
15. What are the energy conversions which take place during the charging and discharging of a storage battery?
16. What is the importance of Avogadro's number to the electrolytic process?
17. State Faraday's Law of Electrolysis. What is a faraday of electricity?
18. Define the phrase "standard electrode potential."
19. List some uses of the electromotive force series.
20. What is the origin of the zero value for the standard hydrogen electrode?
21. In this and previous chapters we have noted commercial electrolytic processes for several substances, some of which are hydrogen, oxygen, chlorine, and sodium hydroxide. Write individual electrode half-reaction equations describing these commercial processes.
22. Soon after a copper metal rod is placed in a silver nitrate solution, copper ions are observed in the solution and silver metal is observed to have deposited on the rod. Can the copper rod be considered an electrode? If so, is it an anode or a cathode?
23. Show by suitable equations that the electrode reaction for the zinc electrode is a reversible one.
24. Indicate two ways in which the energy of sunlight may be converted into electrical energy.

PROBLEMS

NOTE: The value of RT/nF to five significant figures is $0.059152/n$ for common logarithms and a temperature of 298° K ($R = 1.9870$ cal/mole · °K, and $F = 23{,}061$ cal/volt). For each problem which requires the use of either the Nernst equation or other equations with the RT/nF quantity, the proper number of digits compatible with the data of the problem should be used.

[S] 1. A 1.00-liter volume of a concentrated sodium chloride solution is electrolyzed, chlorine being produced at the anode. Calculate the hydroxide ion concentration in the solution after the solution has been electrolyzed for 30 minutes at 1.0 ampere, assuming that the cell is designed so that no chlorine reacts with sodium hydroxide. *Ans.* 1.9×10^{-2} *M*

2. A current of 9.0 amperes flowed for 45 minutes through water containing a small quantity of sodium hydroxide. How many liters of gas were formed at the anode at 27.0° and 750 mm pressure? *Ans. 1.57 liters*

⒮3. How many ampere-hours of electricity are required in the electrolytic refining of 3.00 kg of copper? *Ans. 2530 ampere · hours*

4. How many grams of zinc will be deposited from a solution of zinc(II) sulfate by 3.40 faradays of electricity? *Ans. 111 g*

5. An experiment is conducted using the apparatus depicted in Fig. 22–4. How many grams of gold could be plated out of solution by the current required to plate out 4.97 g of copper? How many moles of hydrogen would simultaneously be released from the hydrochloric acid solution? How many moles of oxygen would be freed at each anode in the copper sulfate and silver nitrate solutions? *Ans. 10.3 g Au; 0.0782 mole H_2; 0.0391 mole O_2*

6. How many grams of cobalt will be deposited from a solution of cobalt(II) chloride by 65,400 coulombs of electricity? *Ans. 20.0 g*

⒮7. How many grams of platinum would be deposited from a solution of Na_2PtCl_4 by a current of 5.00 milliamperes flowing for 7.00 hours? *Ans. 0.127 g*

8. How many coulombs of electricity would be required to reduce 21.0 g of cadmium ion from a solution of $Na_2[Cd(CN)_4]$? *Ans. 36,100 (to three significant figures justified by the data)*

9. For a period of three hours and 30 minutes, an electric current of 3.50 amperes is passed successively through solutions of silver nitrate, chromium(III) sulfate, and sulfuric acid. How many grams of silver, chromium, and hydrogen are electrolytically produced during the process? *Ans. Ag, 49.3 g; Cr, 7.92 g; H_2, 0.461 g*

10. How many mercury(II) ions, thallium(I) ions, and tin(IV) ions will be reduced by 6.00×10^{22} electrons in suitable electrolytic cells? *Ans. 3.00×10^{22}; 6.00×10^{22}; 1.50×10^{22}*

⒮11. A total of 69,500 coulombs of electricity was required in the electrolytic reduction of 16.7 g of a metal from a solution of its *tripositive* ions. What is the metal? *Ans. Ga*

12. How many electrons must flow through the wires connecting a source of direct current to the electrodes to produce 0.69 g of sodium by the electrolysis of melted NaCl? *Ans. 1.8×10^{22}*

⒮13. Calculate the emf for cells made up of the pairs of standard electrodes listed below. Consult Table 22–1 and 22–2 for the standard electrode potentials. Add half-reactions to obtain cell reactions in such a way as to give positive emf values for each cell. Identify the anode and cathode in each cell. Indicate the direction of the cell reaction in each case.

⒮(a) Pb^{2+}, Pb and Mn^{2+}, Mn *Ans. 1.05 v*
(b) Ni^{2+}, Ni and Cd^{2+}, Cd *Ans. 0.15 v*
⒮(c) Pt, Sn^{4+}, Sn^{2+} and Pt, $Cr_2O_7^{2-}$, Cr^{3+}, H^+ *Ans. 1.18 v*
(d) Pt, H^+, H_2 and Ca^{2+}, Ca *Ans. 2.87 v*

14. Write equations for the electrode reactions in each of the following cases and, using potentials from Tables 22–1 and 22–2, calculate:
(a) the emf of a cadmium-nickel battery cell. *Ans. 0.15 v*

(b) the emf of an Edison storage battery cell.   *Ans. 1.37 v*

(c) the emf of a single lead storage battery cell; the emf which would be produced by three lead storage cells in series; and the emf which would be produced by six lead storage cells in series.

Ans. 2.05 v; 6.15 v; 12.30 v

15. In the production of aluminum electrolytically, oxygen is produced at the anode. Some of it is used up by reaction with the carbon electrode to produce CO_2. Calculate the volume of CO_2 produced and the volume of O_2 liberated (S.T.P.) for every pound of aluminum produced, assuming that 12% of the total oxygen initially produced reacts with the electrode.

Ans. 34 liters CO_2; 250 liters O_2

[s]16. A lead storage battery has initially 100 g of lead and 100 g of PbO_2 plus excess H_2SO_4. Theoretically, how long could this cell deliver a current of 1.00 ampere, without recharging, if it were possible to operate it so that the reaction goes to completion?   *Ans. 22.4 hours*

17. For the following cell, calculate the emf of the cell and write the equation for the cell reaction.

$$Cu\,|\,Cu^{2+},\ M = 1\,\|\,Hg^{2+},\ M = 1\,|\,Hg \qquad \textit{Ans. 0.51 v}$$

18. The international ampere is that current which under specified conditions deposits 0.001118 g of silver per second. Taking into account the atomic weight of silver and the number of significant figures justified in your answer, calculate the value of the faraday in coulombs.   *Ans. 96,480 coulombs*

[s]19. A heavy silver wire acts as the anode and a platinum wire as the cathode in 250 ml of a solution containing an unknown concentration of chloride ion. A constant current of 0.150 ampere is allowed to flow until formation of silver chloride ceases. This requires five minutes and 25 seconds. What was the concentration of chloride ion in the solution?   *Ans. 2.02×10^{-3} M*

20. Three cells, containing Bi^{3+} in the following concentrations, are connected in series. Cell no. 1 contains 0.100 M Bi^{3+}, cell no. 2 contains 0.010 M Bi^{3+}, and cell no. 3 contains 1.00 M Bi^{3+}. If 0.150 ampere is passed through the circuit for exactly two hours, how much metallic bismuth would be deposited in each cell?   *Ans. Same for each (0.780 g)*

[s]21. [s](a) A solution contains 0.0100 mole of nickel(II) iodide (NiI_2) in one liter. This solution is to be electrolyzed using inert electrodes. Nickel is plated out at the cathode; iodine is produced at the anode. Determine the standard emf for the electrolytic reaction. What is the minimum voltage which would have to be applied to cause this electrolysis to occur?

Ans. 0.786 v; 0.945 v

(b) After 0.5283 g of nickel has been plated out of the solution, what is then the minimum voltage which would have to be applied to cause continued electrolysis of the solution according to the same reaction?

Ans. 1.03 v

22. In your reading concerning halogens, you learned that when chlorine dissolves in water, it disproportionates, producing chloride ion and hypochlorous acid. At what hydrogen ion concentration does the potential (emf) for the dis-

porportionation of chlorine change from a negative value to a positive value, assuming 1.00 atm pressure and 1.00 M concentration for all species except hydrogen ion? (The standard electrode potential for the reduction of chlorine to chloride ion is 1.36 v and for hypochlorous acid to chlorine, 1.63 v.) Could chlorine be produced from hypochlorite and chloride ions in solution, through a reverse of the disproportionation reaction, by acidifying the solution with strong acid? *Ans.* $2.7 \times 10^{-5}\ M$

23. Five hundred milliliters of a 0.150 M $CuSO_4$ solution has in it two platinum electrodes. A current of one ampere is run through the solution for two hours, 40 minutes and 49 seconds. Assuming none of the solvent is lost through electrolysis, what is the final concentration of the $CuSO_4$? *Ans. 0.050 M*

[S]24. The solubility of lead(II) sulfate is $1.3 \times 10^{-4}\ M$ at 25° C. Calculate the electrode potential of a lead wire in a saturated solution of lead(II) sulfate. *Ans.* $-0.24\ v$

25. Assume that the reaction $Zn + Cu^{2+} \longrightarrow Zn^{2+} + Cu$, for which the standard potential is 1.10 volts and $K_e = 1.56 \times 10^{37}$, goes to equilibrium in a volume of solution whose volume is as large as all the oceans of the world. Assume that the equilibrium concentration of Zn^{2+} ions is $1.0 \times 10^{-3}\ M$, that the earth is a sphere with a diameter of 7920 miles, that the oceans' average depth is 2.4 miles, and that the oceans cover 71% of the earth's surface. Calculate the number of Cu^{2+} ions at equilibrium. *Ans.* 5.4×10^4

[S]26. Calculate the value for the Gibbs free energy change and the equilibrium constant under standard state conditions for the following reactions (see *Note* at beginning of Problems section):

[S](a) $Mn + Zn^{2+} \longrightarrow Zn + Mn^{2+}$ (emf = +0.42 volt)
Ans. $\Delta G° = -19$ *kcal;* $K_e = 1.6 \times 10^{14}$

(b) $Br_2 + 2Cl^- \longrightarrow Cl_2 + 2Br^-$ (emf = −0.2943 volt)
Ans. $\Delta G° = 13.57$ *kcal;* $K_e = 1.12 \times 10^{-10}$

(c) $Ca + 2H_2O \longrightarrow Ca^{2+} + 2OH^- + H_2$ (emf = 2.04 volts)
Ans. $\Delta G° = -94.1$ *kcal;* $K_e = 9.44 \times 10^{68}$

27. The change in Gibbs free energy for the following reaction has a value of 21.0 kcal. Use this information and the standard electrode potential for iodine to calculate the standard electrode potential for the $S_4O_6^{2-} + 2e^- \rightleftharpoons S_2O_3^{2-}$ couple.

$$S_4O_6^{2-} + 2I^- \rightleftharpoons I_2 + 2S_2O_3^{2-}$$

Ans. 0.08 v

28. A one-liter volume of solution contains 0.0100 mole of copper(II) bromide ($CuBr_2$). The solution is to be electrolyzed, using inert electrodes, to deposit a thin plate of copper metal on the cathode.
(a) What is the reaction which occurs at the anode?
(b) What is the standard potential for the total reaction? *Ans.* $-0.72\ v$
(c) What is the minimum voltage which would have to be applied to cause electrolysis to occur? *Ans. 0.88 v*
(d) After 0.159 g of copper has been plated on the cathode, what is the minimum voltage which would have to be applied to cause continued electrolysis according to the same reaction equation? *Ans. 0.89 v*

29. Calculate the free energy of formation of ammonia gas under standard conditions. The equilibrium constant for the following reaction is 6.78×10^5.
$N_2 + 3H_2 \rightleftharpoons 2NH_3$ *Ans. −3.98 kcal/mole*

S30. The standard reduction potentials for the reactions

$$Ag^+ + e^- \longrightarrow Ag$$
and $$AgCl + e^- \longrightarrow Ag + Cl^-$$

are 0.7991 and 0.222 volt, respectively. From these data and the Nernst equation, calculate a value for the solubility product constant (K_{sp}) for AgCl (see *Note* at beginning of Problems section). Compare your answer to the value given in Appendix E.   *Ans. 1.75 × 10⁻¹⁰*

31. Calculate the standard reduction (electrode) potential for the reaction $H_2O + e^- \longrightarrow \frac{1}{2}H_2 + OH^-$ using the Nernst equation and the fact that the standard reduction potential for the reaction $H^+ + e^- \longrightarrow \frac{1}{2}H_2$ is by definition equal to zero volts.   *Ans. −0.83 v*

32. The standard reduction potentials for the reactions

$$Ag^+ + e^- \longrightarrow Ag$$
and $$[Ag(NH_3)_2]^+ + e^- \longrightarrow Ag + 2NH_3$$

are 0.7991 and 0.373 volt, respectively. From these values and the Nernst equation, determine K_d for the $[Ag(NH_3)_2]^+$ ion (see *Note* at beginning of Problems section). Compare your answer to the value given in Appendix F.   *Ans. 6.26 × 10⁻⁸*

33. Show *by calculation* how the constant 0.059152 is obtained in converting

$$E° = \frac{RT}{nF} \ln K_e \quad \text{to} \quad E° = \frac{0.059152}{n} \log K_e$$

REFERENCES

"Faraday's Contribution to Electrolytic Solution Theory," Ollin K. Drennan, *J. Chem. Educ.*, **42**, 679 (1965).

"The Chemistry of Michael Faraday, 1791–1867," R. H. Cragg, *Chem. in Britain*, **3**, 482 (1967).

"Superconductivity at High Pressure," N. B. Brandt and N. I. Ginsburg, *Sci. American*, April, 1971; p. 83.

"The Electronics Industry—A Sizable Challenge to Chemists," L. W. Dunlap, *Chem. Eng. News*, Nov. 30, 1970; p. 42.

"Ionic Conduction and Diffusion in Solids," J. Kummer and M. E. Milberg, *Chem. Eng. News*, May 12, 1969; p. 90.

"Fuel Cells—Present and Future," D. P. Gregory, *Chem. in Britain*, **5**, 308 (1969).

"Electromotive Force of Molten Salt Concentration Cells and Association Equilibria in Solution," J. Braunstein, *J. Chem. Educ.*, **44**, 223 (1967).

"Molten Carbonate Electrolytes as Acid-Base Solvent Systems," G. J. Janz, *J. Chem. Educ.*, **44**, 581 (1967).

"Theory and Applications of the Transistor," E. A. Boucher, *J. Chem. Educ.*, **44**, A936 (1967).

"Fundamental Principles of Semiconductors," E. F. Gurnee, *J. Chem. Educ.*, **46,** 80 (1969).

"Organic Lasers," P. Sorokin, *Sci. American*, Feb., 1969; p. 30.

"Liquid Lasers," A. Lempicki and H. Samelson, *Sci. American*, June, 1967; p. 80.

"The Electrical Properties of Materials," H. Reiss, *Sci. American*, Sept., 1967; p. 210.

"Amorphous-Semiconductor Switching," H. K. Henisch, *Sci. American*, Nov., 1969; p. 30.

"Chemistry of Semiconductors," S. J. Bass, *Chem. in Britain*, **5,** 100 (1969).

"Production and Use of High-Purity Metals," J. C. Chaston, *Chem. in Britain*, **5,** 224 (1969).

"High-Purity Metals for the Electronics Industry," J. E. Wardill and D. J. Dowling, *Chem. in Britain*, **5,** 226 (1969).

"Electrochemistry '69," G. J. Hills, *Chem. in Britain*, **5,** 265 (1969).

"The Standard Electrode Potential of the Silver-Silver Bromide Electrode," R. L. Venable and D. V. Roach, *J. Chem. Educ.*, **46,** 741 (1969).

"Electrochemical Principles Involved in a Fuel Cell," A. K. Vijh, *J. Chem. Educ.*, **47,** 680 (1970).

"Large Scale Integration in Electronics," F. G. Heath, *Sci. American*, Feb., 1970; p. 22.

"The Magnetic Structure of Superconductors," U. Essmann and H. Träuble, *Sci. American*, March, 1971; p. 75.

"Solar Energy: A Feasible Source of Power?" A. L. Hammond, *Science*, **172,** 660 (1971).

"Electrochemical Cells for Space Power," R. M. Lawrence and W. H. Bowman, *J. Chem. Educ.*, **48,** 359 (1971).

"Fuel Cell Research Finally Paying Off," (Staff), *Chem. & Eng. News*, Jan. 7, 1974; p. 31.

23 Sulfur and Its Compounds

The elements sulfur, selenium, tellurium, and polonium follow oxygen in Group VIA of the Periodic Table. These elements compose what is known as the **sulfur family.** Oxygen, the first member of this periodic group, exhibits properties which set it apart from the other elements of the sulfur family, just as fluorine, the first member of the halogen family, is different in many respects from the other elements of its group. Polonium is formed only as a product of radioactive change; it is, itself, highly radioactive. Some properties are given for the elements of the sulfur family, excluding polonium, in Table 23–1. Oxygen is included for the sake of comparison.

Each atom of each element in this periodic group has six valence electrons. Oxygen, more electronegative than any other element except fluorine, usually exhibits an oxidation state of -2 in its compounds, but the larger and therefore less electronegative elements of the sulfur family exhibit both negative and positive oxidation states. As the electronegativity decreases with increasing atomic size, the strength of the elements as oxidizing agents decreases, a property which is reflected very strikingly in the heats of formation of their respective hydrides (Table 23–1).

As would be expected from the change in electronegativity, metallic character within the group increases with increasing atomic size. Oxygen is a typical non-metal, selenium evidences some metallic character, and polonium exhibits definite metallic characteristics.

TABLE 23-1   Some Properties of the Sulfur Family

Property	Oxygen	Sulfur	Selenium	Tellurium
Atomic number	8	16	34	52
Atomic weight	15.9994	32.064	78.96	127.60
Electronic structure	2,6	2,8,6	2,8,18,6	2,8,18,18,6
Radius of divalent anion, Å	1.40	1.84	1.98	2.21
Radius of covalent atom, Å	0.66	1.04	1.17	1.37
Physical state	Gas	Solid	Solid	Solid
Color	Colorless	Yellow	Red or gray	Silvery
Melting point, °C	-218.4	112.8, 119.25	217 (gray)	450
Boiling point, °C	-183	444.6	688	1390
Electronegativity	3.5	2.5	2.4	2.1
Enthalpy of formation of hydride (cal/mole)	$-68,315$	$-4,930$	$+7,100$	$+23,800$
Oxidation states	$-2, -1$	$-2, -1, +1,$ $+2, +4, +6$	$-2, +1, +4,$ $+6$	$-2, +4, +6$

23.1   Occurrence of Sulfur

Sulfur has been known from very early times because it occurs free in nature as a solid. The principal deposits that have been utilized in the United States are in Texas and Louisiana. There are also extensive deposits in Mexico and in the volcanic regions of Italy and Japan. Minerals containing sulfur in the combined form are numerous and widely distributed, and many of the metallic sulfides serve as valuable sources of the element. Sulfides of iron, zinc, lead, and copper and sulfates of sodium, calcium, barium, and magnesium are common and abundant. Sulfur is also a constituent of some proteins and therefore exists in the combined state in animal and vegetable matter.

23.2   Extraction of Sulfur

Sulfur is extracted by the **Frasch process** from enormous underground deposits in Texas and Louisiana (Fig. 23-1). In this process, superheated water (170° and 100 pounds per square inch pressure) is forced down the outermost of three concentric pipes to the deposit of sulfur located several hundred feet below the surface of the earth. When the hot water melts the sulfur, compressed air is forced down the innermost pipe. The liquid sulfur mixed with air forms an emulsion that is less dense than water, and this mixture readily flows up through the outlet pipe. The emulsified sulfur is conveyed to large settling vats where it solidifies upon cooling. Sulfur produced by this method is remarkably pure, 99.5 to 99.9 per cent, and for most of its uses requires no purification.

Increasing quantities of sulfur as a by-product from the purification of "sour" natural gas and petroleum refinery gases have caused a decline in the Frasch

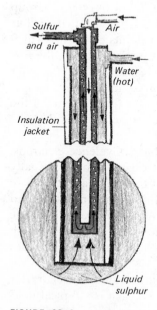

FIGURE 23-1

Diagram illustrating the Frasch method of mining sulfur.

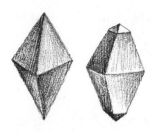

FIGURE 23-2
Crystals of orthorhombic sulfur.

FIGURE 23-3
Crystals of monoclinic sulfur.

method of obtaining sulfur. In 1970, this by-product sulfur exceeded the total production by the Frasch process in the free world for the first time. With air pollution by sulfur dioxide a major problem, increasing effort is being applied toward eliminating sulfur dioxide in power plant exhaust stack gases. While it is generally considered that a method for removing sulfur dioxide from stack gases is economically unfeasible, some method must be found if we are to have pure air to breathe (see Sections 24.3–24.5). Perhaps income from the sale of sulfur extracted from the gases can partially offset the costs involved.

23.3   Allotropic Modifications of Sulfur

Sulfur exists in several allotropic forms (Section 8.10). Native sulfur is a yellow solid, which forms crystals that belong to the orthorhombic crystal system; it is called **rhombic sulfur** (Fig. 23–2). This form of sulfur is soluble in carbon disulfide, producing a solution from which well-formed crystals separate when it is allowed to evaporate slowly. When heated to 112.8°, crystals of rhombic sulfur melt and form a straw-colored liquid known as **lambda-sulfur.** When this liquid cools and crystallizes, long transparent needles of **monoclinic sulfur** are formed (Fig. 23–3). This modification of sulfur, the stable form of sulfur above 96°, melts at 119.25° and is soluble in carbon disulfide. Upon standing at room temperature, it gradually changes to the rhombic form.

Rhombic sulfur, monoclinic sulfur, and the straw-colored liquid all contain S_8 molecules in which the atoms form eight-membered, puckered rings (Fig. 23–4a). Each atom of sulfur is linked to each of its two neighbors in the ring by single electron-pair bonds; each atom thus has a completed octet of electrons.

The straw-colored liquid form of sulfur is quite mobile, i.e., its viscosity is low, because the S_8 molecules are essentially spherical in shape and offer relatively little resistance to their motion past one another. As the temperature is raised, the S_8 rings of the yellow mobile sulfur rupture, and long chains of sulfur atoms (Fig. 23–4b) are formed. As these chains become entangled with one another and combine end to end to form still longer chains with larger numbers of sulfur atoms, an increase in the viscosity of the liquid occurs. The liquid gradually darkens in color and becomes so viscous that finally (at about 230°) it will not pour from its container. The unpaired electrons at the ends of the chains of sulfur atoms are responsible for the dark red color. This liquid form of the element is known as **μ-sulfur.** When it is cooled rapidly, a rubberlike amorphous mass, insoluble in carbon disulfide, results. This supercooled liquid is known as **plastic sulfur.**

FIGURE 23-4
Perspective drawings of (a) an S_8 molecule and (b) a chain of sulfur atoms.

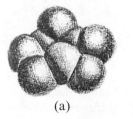

(a)

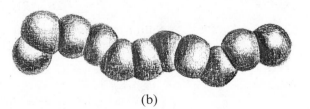

(b)

Upon standing for some time at room temperature, plastic sulfur, like all of the other allotropes, changes to the rhombic form.

Sulfur boils at 444.6° and forms a vapor consisting of S_2, S_6, and S_8 molecules; at about 1,000° the vapor density corresponds to the formula S_2.

23.4   Chemical Properties of Sulfur

Because of the change from typical nonmetallic character, with oxygen, to essentially metallic character, with tellurium, the chemical properties of the sulfur family are not readily treated systematically. The small size of the oxygen atom is responsible in part for its great reactivity as an oxidizing agent, and chemically speaking it is somewhat more like the halogens than like sulfur, selenium, and tellurium. The chemical properties of oxygen were discussed in Chapter 8.

Elemental sulfur is quite reactive, even at ordinary temperatures, although generally less so than oxygen. Many of the metals will react with solid sulfur upon coming in contact with it at room temperature. Mercury has been shown to combine with sulfur at temperatures as low as −180°. Only the noble gases and the elements iodine, nitrogen, tellurium, gold, platinum, and palladium do not combine directly with elemental sulfur.

Elemental sulfur acts as an oxidizing agent toward hydrogen, forming **hydrogen sulfide**, H_2S; toward carbon at elevated temperatures, forming **carbon disulfide,** CS_2; and toward iron when heated, forming **iron(II) sulfide.**

$$H_2 + S \longrightarrow H_2S$$
$$C + 2S \longrightarrow CS_2$$
$$Fe + S \longrightarrow FeS$$

Binary compounds in which sulfur is combined with an element more electropositive than itself are called **sulfides.** The sulfur has an oxidation state of −2 in such compounds. Sulfides are analogous to the corresponding oxides, but the properties of the two varieties of compounds are often quite different.

Sulfur acts as a reducing agent toward those nonmetals that are more electronegative than it is, such as oxygen and the halogens. When sulfur is ignited in the air, it burns with a blue flame, forming **sulfur dioxide** and a little **sulfur trioxide.**

$$S(s) + O_2(g) \longrightarrow SO_2(g) + 70{,}944 \text{ cal} \qquad (\Delta H^\circ_{298} = -70{,}944 \text{ cal})$$
$$2SO_2(g) + O_2(g) \longrightarrow 2SO_3(g) + 47{,}270 \text{ cal} \qquad (\Delta H^\circ_{298} = -47{,}270 \text{ cal})$$

As we shall see later in this chapter, the oxidation of sulfur is important in the production of **sulfurous acid,** H_2SO_3, and **sulfuric acid,** H_2SO_4. Sulfur in moist air is slowly oxidized to sulfuric acid.

$$2S + 2H_2O + 3O_2 \longrightarrow 4H^+ + 2SO_4{}^{2-}$$

Sulfur reduces chlorine in a series of reactions, the products of which depend upon the temperature and relative quantities of the reactants.

$$2S + Cl_2 \longrightarrow S_2Cl_2, \text{ "sulfur monochloride" (a misnomer)}$$
$$S_2Cl_2 + Cl_2 \rightleftharpoons 2SCl_2, \text{ sulfur dichloride}$$
$$SCl_2 + Cl_2 \rightleftharpoons SCl_4, \text{ sulfur tetrachloride}$$

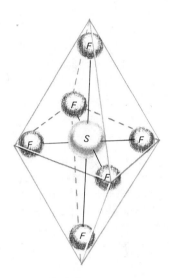

FIGURE 23-5

The octahedral configuration of the SF_6 molecule.

The valence electronic structure of S_2Cl_2 is

$$\overset{\times\times \quad \times\times \quad \times\times \quad \times\times}{\underset{\times\times \quad \times\times \quad \times\times \quad \times\times}{\times\underset{}{Cl}\times\underset{}{S}\times\underset{}{S}\times\underset{}{Cl}\times}}$$

Only one bromide of sulfur is known; it is the "monobromide," S_2Br_2, a garnet-red liquid. Sulfur forms a series of fluorides which includes the hexafluoride, SF_6, a very stable gaseous compound. In SF_6, the fluorine atoms are arranged around the sulfur atom at the corners of a regular octahedron (Fig. 23-5), and the bonding is highly covalent in character. There are twelve bonding electrons in the valence shell of sulfur in SF_6 rather than the usual eight. The existence of twelve electrons in the valence shell is not unusual; it is found also in such other materials as PF_6^-, $[Co(NH_3)_6]^{3+}$, and $[Fe(CN)_6]^{4-}$. In its binary compounds with the halogens, sulfur attains its maximum oxidation state of $+6$ only with fluorine because of the small size of the fluorine atom compared with the size of the atoms of the other halogens.

Sulfur reduces strong oxidizing agents such as concentrated nitric acid and hot, concentrated sulfuric acid. The equations are

$$S + 6H^+ + 6NO_3^- \longrightarrow 2H^+ + SO_4^{2-} + 2H_2O + 6NO_2\uparrow$$
$$S + 2H_2SO_4 \longrightarrow 3SO_2\uparrow + 2H_2O$$

23.5 Uses of Elements of the Sulfur Family

Of the elements composing the sulfur family, oxygen and sulfur are the most useful, commercially. The largest single use for sulfur is in the production of sulfuric acid, the most important acid in industry. A considerable amount of sulfur is also used in vulcanizing rubber and in producing gunpowder, sulfites, thiosulfates, fertilizers, and medicines.

Selenium is used in photoelectric cells, since it has a sufficiently low ionization potential that light can remove an electron from the outer shell. The use of a selenide in a new solar cell, announced in 1975, has been mentioned in Section 22.18. The addition of a little selenium to ordinary glass offsets the green color which such glass usually has due to the presence of iron(II) silicate. If selenium is added in larger amounts, it colors the glass red. Sodium selenide is used when selenium is to be incorporated in glass, because elemental selenium is volatile at the temperatures employed and is very toxic. In the electronics industry, selenium of high purity is used as an efficient rectifier, which is a device for changing alternating electric current to direct. Selenium is also used in the production of certain stainless steels and special copper alloys.

Tellurium is used to some extent in coloring glass blue, brown, or red. It is alloyed in small percentages with lead to increase the hardness of the metal for use in such things as battery plates and printing type. Traces of tellurium in stainless steel increase its machinability and in cast iron give a hard, wear-resistant surface.

Hydrogen Sulfide

Each member of Group VIA of the Periodic Table forms a hydride. The strength of these hydrides as acids and their reducing powers increase, whereas their thermal stabilities decrease in the order H_2O, H_2S, H_2Se, H_2Te, H_2Po.

23.6   Preparation of Hydrogen Sulfide

The gas hydrogen sulfide occurs dissolved in the water of sulfur springs and as one of the gases issuing from volcanoes. It is a product of the decay of animal matter in the absence of air. The offensive odor of spoiled eggs is responsible for the name "rotten-egg gas" that is sometimes applied to hydrogen sulfide.

Large quantities of hydrogen sulfide are produced in the refining of petroleum and thereby contribute to the air pollution problem. The production of hydrogen sulfide by the direct union of its elements is unsatisfactory because the reaction is reversible and, as it is usually carried out, not more than two per cent of the elements are combined at equilibrium.

$$H_2(g) + S(s) \rightleftharpoons H_2S(g) + 4,930 \text{ cal} \qquad (\Delta H^\circ_{298} = -4,930 \text{ cal})$$

Hydrogen sulfide can be prepared by treating iron(II) sulfide with a dilute acid.

$$FeS + 2H^+ \longrightarrow Fe^{2+} + H_2S\uparrow$$

Other sulfides may be substituted for FeS in the above reaction. Hydrogen sulfide can also be prepared by heating a mixture of paraffin (a hydrocarbon —binary compound of carbon and hydrogen—with approximately 23 to 29 carbon atoms; see Section 27.12) and sulfur. Some asbestos is added to provide porosity. The reaction stops as soon as heating stops.

$$CH_3(CH_2)_nCH_3 + \tfrac{n}{4}S \longrightarrow CH_3(CH_2CH{=}CHCH_2)_{n/4}CH_3 + \tfrac{n}{4}H_2S\uparrow$$

When an aqueous solution of hydrogen sulfide is desired, it may be prepared conveniently by the hydrolysis of the organic sulfur-containing compound **thio-acetamide,** CH_3CSNH_2. The equation for the hydrolysis is

$$CH_3CSNH_2 + 2H_2O \longrightarrow CH_3COO^- + NH_4^+ + H_2S\uparrow$$

Heating accelerates the hydrolysis considerably.

23.7   Physical Properties of Hydrogen Sulfide

Hydrogen sulfide is a colorless gas with an offensive odor. It is toxic and great care must be exercised in handling it. Hydrogen sulfide is nearly as toxic as hydrogen cyanide (prussic acid), which is used in death chambers in some states for capital punishment. Small amounts of H_2S gas in the air cause headaches, while larger amounts cause paralysis in the nerve centers of the heart and lungs which results in fainting and death. Hydrogen sulfide is particularly deceptive in that it also paralyzes the olfactory nerves, so that after a short exposure one does not smell it. Many persons have died because of this lack of warning.

It happens that small amounts of H_2S can be produced in the catalytic converters installed in automobiles beginning in 1975. At the high operating temperatures of the converter and under fuel-rich conditions, some of the hydrocarbons (binary compounds of carbon and hydrogen) of which the gasoline is composed react with sulfur compounds present in most gasolines to produce the toxic sulfide (see Sections 24.4 and 24.5). It is evident that exhaust gases from the most modern of automobiles can be perhaps just as dangerous to life as those in the past.

Liquid hydrogen sulfide freezes at $-82.9°$ and boils at $-61.8°$. The gas is slightly more dense than air and its solubility in water is 0.1 mole/liter at 18°.

The boiling points of H_2S ($-61.8°$) and H_2Se ($-42°$) are much lower than that of water (100°), indicating that the effect of molecular association through hydrogen bonding decreases greatly in the series H_2O, H_2S, H_2Se. Of the three hydrides, only water with its highly electronegative oxygen atom has pronounced hydrogen bonding (see Section 12.9). Hydrogen sulfide is much less polar than water, and hence liquid hydrogen sulfide is not a good solvent for ionic or polar compounds; it does dissolve many nonpolar compounds.

23.8   Chemical Properties of Hydrogen Sulfide

Hydrogen sulfide decomposes into hydrogen and sulfur when heated. It burns in air, forming water and sulfur dioxide. The equation for the combustion of hydrogen sulfide is

$$2H_2S(g) + 3O_2(g) \longrightarrow 2H_2O(g) + 2SO_2(g) + 247{,}620 \text{ cal}$$
$$(\Delta H^°_{298} = -247{,}620 \text{ cal})$$

When hydrogen sulfide is ignited in a limited supply of air or when a burning jet of the gas is impinged upon a cold surface, free sulfur is deposited. In this way, tons of sulfur are recovered at oil refineries each day.

$$2H_2S(g) + O_2(g) \rightarrow 2H_2O(g) + 2S(s) + 105{,}730 \text{ cal} \quad (\Delta H^°_{298} = -105{,}730 \text{ cal})$$

Most of the more reactive metals will displace hydrogen from hydrogen sulfide. Thus, lead sulfide is formed by the action of hydrogen sulfide upon metallic lead, according to the equation

$$Pb(s) + H_2S(g) \longrightarrow PbS(s) + H_2(g) + 19{,}100 \text{ cal} \quad (\Delta H^°_{298} = -19{,}100 \text{ cal})$$

Hydrogen sulfide will cause metallic silver to tarnish, black silver sulfide being formed. Hence, silver tableware tarnishes when it is used with eggs and other foods containing certain sulfur compounds.

$$4Ag(s) + 2H_2S(g) + O_2(g) \longrightarrow 2Ag_2S(s) + 2H_2O(l) + 142{,}400 \text{ cal}$$
$$(\Delta H^°_{298} = -142{,}400 \text{ cal})$$

The sulfur in hydrogen sulfide readily gives up electrons, making the hydrogen sulfide a good reducing agent. In acidic solutions, hydrogen sulfide reduces Fe^{3+} to Fe^{2+}, Br_2 to Br^-, MnO_4^- to Mn^{2+}, $Cr_2O_7^{2-}$ to Cr^{3+}, and HNO_3 to NO_2. When acting as a reducing agent the sulfur of the H_2S is usually oxidized to elemental sulfur, unless a large excess of the oxidizing agent is present. In this case the sulfide may be oxidized to SO_3^{2-} or SO_4^{2-}. In the presence of moisture, hydrogen sulfide reduces sulfur dioxide to sulfur.

$$2H_2S + SO_2 \longrightarrow 2H_2O + 3S$$

The deposits of sulfur in volcanic regions may be the result of this reaction, because both H_2S and SO_2 are constituents of volcanic gases.

The reaction of H_2S and SO_2 is particularly important as the basis of one method being investigated for the removal of sulfur dioxide, a major air pollutant, from petroleum refinery and power plant exhaust gases. For example, the reaction goes almost to completion and is very rapid in molten sulfur with ethylenediamine ($H_2NCH_2CH_2NH_2$) as a catalyst. Temperatures of 120° to 160° are utilized not only to keep the sulfur molten but also to prevent the undesirable increase in viscosity of sulfur that occurs at higher temperatures (Section 23.3). The recovery of sulfur from stack gases is becoming significant as a source of the element (Section 23.2).

23.9   Hydrogen Sulfide in Aqueous Solution

An aqueous solution of hydrogen sulfide is known as hydrosulfuric acid. The acid is weak and diprotic, i.e., it ionizes in two stages. Hydrogen sulfide yields hydrosulfide ions, HS^-, in the first stage and sulfide ions, S^{2-}, in the second.

$$H_2S \rightleftharpoons H^+ + HS^-$$
$$HS^- \rightleftharpoons H^+ + S^{2-}$$

When a solution of hydrogen sulfide is exposed to the air for a time, sulfur precipitates as the result of oxidation of the sulfide ion.

$$4H^+ + 2S^{2-} + O_2 \longrightarrow 2H_2O + 2\underline{S}$$

23.10   Sulfides

Because hydrogen sulfide is diprotic, two series of salts are possible. For example, the **normal sulfide**, Na_2S, and the **hydrosulfide**, NaHS, of sodium are known.

Sodium hydrosulfide may be made by passing an excess of hydrogen sulfide into a solution of sodium hydroxide.

$$H_2S + (Na^+) + OH^- \rightleftharpoons (Na^+) + (HS^-) + H_2O$$

By adding an equivalent amount of sodium hydroxide to a solution of sodium hydrosulfide, normal **sodium sulfide**, Na_2S, is formed.

$$(Na^+) + OH^- + (Na^+) + HS^- \rightleftharpoons (2Na^+) + S^{2-} + H_2O$$

Many normal metal sulfides are prepared by the direct union of sulfur with the metal as the following equations illustrate.

$$Fe(s) + S(s) \longrightarrow FeS(s) + 23,900 \text{ cal} \qquad (\Delta H_{298}^\circ = -23,900 \text{ cal})$$
$$2Al(s) + 3S(s) \longrightarrow Al_2S_3(s) + 173,000 \text{ cal} \qquad (\Delta H_{298}^\circ = -173,000 \text{ cal})$$

Hydrogen sulfide is useful in analytical work, because the metal sulfides show a wide variation in solubility. With such a wide variation, it is possible to divide the metallic ions into analytical groups, the members of which form sulfides of similar solubilities.

Because of the strong tendency of the sulfide ion and the hydrosulfide ion to act as proton acceptors, as indicated by the weakness of hydrogen sulfide as an acid, aqueous solutions of soluble sulfides and hydrosulfides are basic.

$$S^{2-} + H_2O \rightleftharpoons HS^- + OH^-$$
$$HS^- + H_2O \rightleftharpoons H_2S + OH^-$$

23.11   Polysulfides

When elemental sulfur is added to a solution of a soluble metal sulfide, the sulfur dissolves by combining with the sulfide ion to form complex **polysulfide** ions, formulated as S_n^{2-} (n = 2 to 5). **Disulfide ions,** S_2^{2-}, are analogous to peroxide ions in structure (Section 12.16). The sulfur atoms of these complex ions are linked together through shared electron pairs. The electronic and geometric structures of S_2^{2-}, S_5^{2-}, and H_2S_5 are shown in Fig. 23-6.

FIGURE 23-6
Lewis formulas and geometric structures of S_2^{2-}, S_5^{2-}, and H_2S_5.

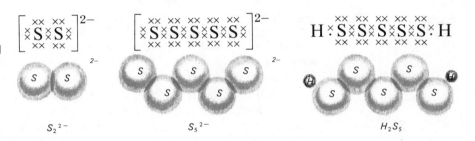

The polysulfide ions, like the peroxide ion, are oxidizing agents. For example, the disulfide ion oxidizes tin(II) sulfide to the thiostannate (IV) ion.

$$SnS + S_2^{2-} \longrightarrow SnS_3^{2-}$$

When solutions containing polysulfide ions are acidified, free sulfur, in a white, very finely divided form (milk of sulfur), and hydrogen sulfide are produced.

$$S_2^{2-} + 2H^+ \longrightarrow H_2S\uparrow + \underline{S}$$

The Oxygen Compounds of Sulfur

Some of the many oxides of sulfur are given in Table 23-2, with their names and the oxidation state of the sulfur.

In the following sections we shall consider only the more important of these oxides and their corresponding acids and other derivatives.

TABLE 23-2   Oxides and Oxidation States of Sulfur

Oxide	Compound Name	Oxidation State
SO	Sulfur monoxide	+2
S_2O_3	Sulfur sesquioxide	+3
SO_2	Sulfur dioxide	+4
SO_3	Sulfur trioxide	+6
S_2O_7	Sulfur heptaoxide	+7

23.12   Sulfur Dioxide

■ **1. Physical Properties.** The odor of burning sulfur is that of **sulfur dioxide,** SO_2. It is a colorless gas, 2.26 times as heavy as air, and very soluble in water (80 volumes of gas to 1 volume of water at S.T.P.). It readily condenses to the liquid, which boils at $-10°$ and freezes to a white solid that melts at $-75.5°$. Liquid sulfur dioxide is stored in steel cylinders and shipped in tank cars, but it must be quite dry because with moisture, even in trace amounts, enough sulfurous acid is formed to corrode the steel container.

■ **2. Occurrence and Preparation.** Sulfur dioxide occurs in volcanic gases and in the atmosphere near industrial plants that use coal or oil as energy sources. The oxide forms when sulfur compounds in oil and coal react with oxygen in the air during combustion. This is one of the primary concerns in air pollution (see Section 23.7 and Chapter 24). Sulfur dioxide is also formed in small quantities during combustion of gasoline, from sulfur compounds present in the gasoline. In the catalytic converters of some present-day automobiles the sulfur dioxide is oxidized to sulfur trioxide in the presence of the converters' catalysts at the relatively high temperatures of the converters. The sulfur trioxide then reacts with water to form a small amount of sulfuric acid mist, which can be a problem when emitted in the exhaust (see also Sections 23.7, 23.8, 23.14, 23.15, 24.4, and 24.5).

For commercial purposes, sulfur dioxide is produced by burning free sulfur and by roasting (heating in air) certain sulfide ores, such as ZnS, FeS_2, and Cu_2S. The roasting of metal sulfide ores (to oxidize the sulfur to the dioxide and to form the oxide of the metal) is the first step in the metallurgy of zinc and copper (Section 31.4).

Sulfur dioxide may be prepared conveniently in the laboratory by the action of sulfuric acid upon either sodium sulfite or sodium hydrogen sulfite. Sulfurous acid is first formed, but it quickly decomposes into sulfur dioxide and water.

$$2H^+ + SO_3^{2-} \longrightarrow H_2SO_3 \longrightarrow H_2O + SO_2\uparrow$$
$$H^+ + HSO_3^- \longrightarrow H_2SO_3 \longrightarrow H_2O + SO_2\uparrow$$

Many reducing agents react with hot concentrated sulfuric acid with the formation of sulfur dioxide. Four examples are

$$Cu(s) + 2H_2SO_4(l) \longrightarrow CuSO_4(s) + SO_2(g) + 2H_2O(l) + 2,840 \text{ cal}$$
$$(\Delta H°_{298} = -2,840 \text{ cal})$$

$$C(s) + 2H_2SO_4(l) \longrightarrow CO_2(g) + 2SO_2(g) + 2H_2O(l) - 16,527 \text{ cal}$$
$$(\Delta H°_{298} = +16,527 \text{ cal})$$

$$S(s) + 2H_2SO_4(l) \longrightarrow 3SO_2(g) + 2H_2O(l) - 39{,}634 \text{ cal}$$
$$(\Delta H^\circ_{298} = +39{,}634 \text{ cal})$$

$$2HBr(l) + H_2SO_4(l) \longrightarrow Br_2(g) + SO_2(g) + 2H_2O(l) \qquad \text{(see Section 21.8)}$$

■ 3. The Structure of the Sulfur Dioxide Molecule and Resonance.

The valence electronic formula of the sulfur dioxide molecule may be written in two ways. In any one molecule of sulfur dioxide there are believed to be two possible positions for the double bond, as shown in (a) and (b).

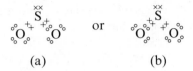

(a)   (b)

The two structures are equivalent. It has been shown experimentally that the two sulfur-oxygen bonds in SO_2 are not different. They have the same length and strength and are intermediate in character between a single and double bond. Furthermore, the SO_2 molecule is more stable than would be predicted for either structure (a) or (b). When two (or more) reasonable electronic structures can be written for a compound, we say that the "real" electronic structure is identical to neither, but is intermediate in character between the two. The concept is called **resonance,** and the structure is often called a **resonance hybrid.**

A homely analogy of resonance that has been suggested is the mule, which is the hybrid offspring of a donkey and a horse. Just as the characteristics of a mule are fixed, so the properties of a resonance hybrid are fixed, with no oscillation between the contributing electronic structures. The mule is not a donkey part of the time and a horse part of the time. It is always a mule. Correspondingly, the resonance hybrid molecule is not structure (a) part of the time and structure (b) part of the time, as would be the case with an equilibrium mixture of the two forms. Instead, the material is always in the form of the intermediate resonance hybrid. A double-ended single arrow ($\longleftrightarrow$) is utilized, therefore, to distinguish the resonance notation from the double arrow ($\rightleftharpoons$) equilibrium notation.

The geometric structure of the sulfur dioxide molecule, which is trigonal planar (see Chapter 6, Sections 6.1–6.3) with an oxygen-sulfur-oxygen bond angle of 119.5°, is shown in Fig. 23-7.

FIGURE 23-7

Structure of the sulfur dioxide molecule, SO_2.

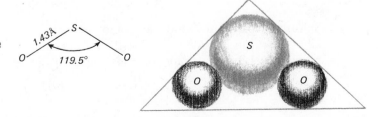

■ **4. Solvent Properties of Liquid Sulfur Dioxide.** Liquid sulfur dioxide is a good solvent for a wide variety of substances, including many salts, and it is used as a medium for carrying out certain types of reactions (see Section 15.19). Its electrical conductivity is about twice that of pure water.

■ **5. Reactions of Sulfur Dioxide.** Sulfur dioxide is slowly oxidized by oxygen of the air to sulfur trioxide, according to the equation

$$2SO_2(g) + O_2(g) \rightleftharpoons 2SO_3(g) + 47.27 \text{ kcal} \qquad (\Delta H^\circ_{298} = -47.27 \text{ kcal})$$

This oxidation is much faster in the presence of suitable catalysts, and the reaction is one step in the production of sulfuric acid (Section 23.15).

Sulfur dioxide reacts with phosphorus(V) chloride, PCl_5, to form **thionyl chloride**, $SOCl_2$ (Fig. 23-8a), and **phosphorus oxychloride, $POCl_3$.**

$$SO_2(g) + PCl_5(s) \longrightarrow SOCl_2(l) + POCl_3(l) + 24.5 \text{ kcal}$$

$$(\Delta H^\circ_{298} = -24.5 \text{ kcal})$$

FIGURE 23-8

(a) Thionyl chloride, $SOCl_2$ (trigonal pyramidal). (b) Sulfuryl chloride, SO_2Cl_2 (square pyramidal). The diagram of the SO_2Cl_2 molecule is a view looking down on the pyramid. The sulfur atom is at the apex of the pyramid; the oxygen and chlorine atoms are at the corners of the base of the pyramid.

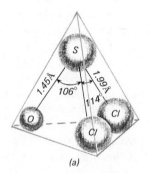

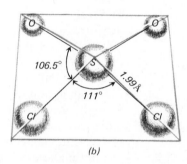

Sulfur dioxide combines with chlorine when the mixture is exposed to sunlight or, in the presence of camphor, even without sunlight; **sulfuryl chloride, SO_2Cl_2,** is the compound formed (Fig. 23-8b).

$$SO_2(g) + Cl_2(g) \longrightarrow SO_2Cl_2(l) + 23.3 \text{ kcal} \qquad (\Delta H^\circ_{298} = -23.3 \text{ kcal})$$

The reaction of sulfur dioxide with water is considered in Section 23.13.

23.13 Sulfurous Acid and Sulfites

When sulfur dioxide dissolves in water a weakly acidic solution of **sulfurous acid** results.

$$H_2O + SO_2 \rightleftharpoons H_2SO_3$$

Sulfurous acid is unstable and cannot be isolated in the anhydrous condition; all of the sulfur dioxide formed as the acid decomposes is expelled from solution by boiling. Like other diprotic acids, sulfurous acid ionizes in two steps.

$$H_2SO_3 \rightleftharpoons H^+ + HSO_3^-$$
$$HSO_3^- \rightleftharpoons H^+ + SO_3^{2-}$$

The amount of ionization is slight in both stages, but it is much less in the secondary stage than in the primary.

Both normal and hydrogen salts are formed by sulfurous acid. Sulfurous acid acts as a reducing agent toward strong oxidizing agents. Oxygen of the air oxidizes it slowly to the more stable sulfuric acid.

$$2H_2SO_3 + O_2 \longrightarrow 4H^+ + 2SO_4^{2-}$$

Solutions containing the permanganate ion (purple in color) rapidly turn colorless when sulfurous acid is added, the purple MnO_4^- ion being reduced to the colorless Mn^{2+} ion.

$$2MnO_4^- + 5H_2SO_3 \longrightarrow 2Mn^{2+} + 4H^+ + 5SO_4^{2-} + 3H_2O$$

Solid sodium hydrogen sulfite forms sodium sulfite, sulfur dioxide, and water when heated.

$$2NaHSO_3 \xrightarrow{\triangle} Na_2SO_3 + SO_2 + H_2O$$

When solid sodium sulfite is heated to high temperatures auto-oxidation-reduction occurs, with the formation of sodium sulfide and sodium sulfate.

$$\overset{+4}{4Na_2SO_3} \longrightarrow \overset{-2}{Na_2S} + \overset{+6}{3Na_2SO_4}$$

This reaction is analogous to the auto-oxidation-reduction reactions which metal hypochlorites and metal chlorates undergo (Sections 21.16 and 21.18).

Solutions of sulfites are very susceptible to air oxidation, as is sulfurous acid, and sulfates are formed. Thus, solutions of sulfites always contain sulfates after standing in contact with the air.

23.14   Sulfur Trioxide

■ 1. **Preparation.**   When sulfur burns in air small amounts of **sulfur trioxide** are formed, in addition to sulfur dioxide. When sulfur dioxide and oxygen are heated together a small amount of the trioxide is formed according to the equation

$$2SO_2(g) + O_2(g) \rightleftharpoons 2SO_3(g) + 47{,}270 \text{ cal} \qquad (\Delta H^\circ_{298} = -47{,}270 \text{ cal})$$

If the temperature of the system is raised to about 400° the equilibrium is reached more rapidly, but even at this temperature the time required to attain equilibrium is too great for the reaction to be commercially useful. The presence of a catalyst such as finely divided platinum or vanadium(V) oxide greatly decreases the time required for attainment of equilibrium. The higher the temperature, the more the equilibrium shifts to the left. See van't Hoff's law (Section 17.14).

■ 2. **Structure.**   Sulfur trioxide in the vapor state is **monomeric,** i.e., its molecules are single SO_3 units, with the sulfur atom at the center and the oxygen atoms at the corners of an equilateral triangle as shown in Fig. 23–9a (see Chapter 6, Sections 6.1–6.3). Because the sulfur-oxygen bond distance is less than that for a single bond, resonance structures involving double bonds (two pairs of electrons) and single bonds (one pair of electrons) are written for the sulfur trioxide molecule.

$$\overset{\times\overset{\times\times}{\underset{\times\times}{O}}\times}{\underset{+\overset{+}{O}++\overset{+}{O}+}{\overset{..}{\underset{..}{S}}}} \longleftrightarrow \overset{\times\overset{\times\times}{\underset{\times\times}{O}}\times}{\underset{+\overset{+}{O}++\overset{+}{O}+}{\overset{..}{\underset{.}{S}}}} \longleftrightarrow \overset{\times\overset{\times\times}{\underset{\times\times}{O}}\times}{\underset{+\overset{+}{O}++\overset{+}{O}+}{\overset{..}{\underset{.}{S}}}}$$

There are apparently three distinct solid forms of sulfur trioxide. One modification is icelike in appearance and has as building units trimeric molecular units (Fig. 23-9b). A second form is an asbestos-like solid, in which tetrahedral SO_4 units are joined to each other in long chains which extend the length of the crystal (Fig. 23-9c). In a third form (not shown) SO_3 chains (SO_4 groups joined together through common oxygen atoms at one corner of each tetrahedron) are joined together to give a layerlike arrangement.

FIGURE 23-9
Sulfur trioxide structures.

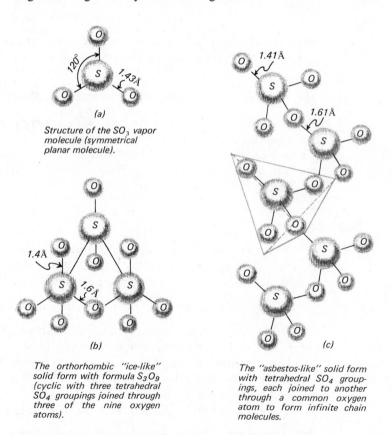

(a)

Structure of the SO_3 vapor molecule (symmetrical planar molecule).

(b)

The orthorhombic "ice-like" solid form with formula S_3O_9 (cyclic with three tetrahedral SO_4 groupings joined through three of the nine oxygen atoms).

(c)

The "asbestos-like" solid form with tetrahedral SO_4 groupings, each joined to another through a common oxygen atom to form infinite chain molecules.

■ **3. Properties.** Liquid sulfur trioxide boils at 43° and freezes at 17° to the icelike solid form. When a trace of moisture is added to liquid sulfur trioxide, it changes to the asbestos-like solid form.

Liquid sulfur trioxide fumes in moist air and dissolves in water in a highly exothermic reaction to form sulfuric acid, H_2SO_4. With a limited amount of water, the molecular acid is formed according to the equation

$$SO_3(l) + H_2O(l) \longrightarrow H_2SO_4(l) + 20,820 \text{ cal} \qquad (\Delta H^\circ_{298} = -20,820 \text{ cal})$$

Sulfur trioxide dissolves readily in concentrated sulfuric acid and forms **pyrosulfuric acid,** $H_2S_2O_7$, also known as "fuming sulfuric acid," or "oleum."

$$H_2SO_4 + SO_3 \longrightarrow H_2S_2O_7$$

At temperatures of 900° or higher, sulfur trioxide decomposes into the dioxide and oxygen. The trioxide reacts with many oxides and hydroxides in acid-base reactions (of the Lewis type) with the formation of sulfates.

$$\underset{\text{Base}}{BaO} + \underset{\text{Acid}}{SO_3} \longrightarrow \underset{\text{Salt}}{BaSO_4}$$

Sulfuric Acid

The amount of sulfuric acid used in industry exceeds that of any other manufactured compound. Its importance is so great that the amount of it produced from year to year is a fairly accurate index of industrial prosperity.

23.15   Manufacture of Sulfuric Acid

■ **1. The Contact Process.**   The chemistry of the **contact process** for the manufacture of sulfuric acid is very simple. It involves oxidizing sulfur dioxide to sulfur trioxide and combining sulfur trioxide with water. Usually, the sulfur dioxide is obtained by burning nearly pure sulfur in air.

$$S(s) + O_2(g) \longrightarrow SO_2(g) + 70,944 \text{ cal} \qquad (\Delta H^\circ_{298} = -70,944 \text{ cal})$$

The sulfur dioxide is then oxidized, by means of air in the presence of a suitable catalyst, to the trioxide.

$$2SO_2(g) + O_2(g) \longrightarrow 2SO_3(g) + 47,270 \text{ cal} \qquad (\Delta H^\circ_{298} = -47,270 \text{ cal})$$

Finely divided platinum was originally used as the "contact" catalyst, but it now has been largely replaced by vanadium(V) oxide (V_2O_5), which is more resistant to poisoning (inactivation by impurities). Although the oxidation of the dioxide to the trioxide is exothermic and thus favored by low temperatures, the reaction is too slow to be commercially feasible at low temperatures. Consequently, the oxidation step is carried out at about 400°. A yield of about 98 per cent is obtained at this temperature.

In spite of the vigorous reaction of sulfur trioxide vapor with water the sulfur trioxide if mixed with air reacts slowly with water, because the bubbles of O_2 and N_2 (which go through the solution as *big* bubbles) contain particles of liquid sulfur trioxide; the sulfur trioxide particles in the middle of these bubbles do not come in contact with the water. For this reason, the sulfur trioxide vapor is first absorbed in concentrated sulfuric acid, pyrosulfuric acid being formed.

$$H_2SO_4(l) + SO_3(g) \longrightarrow H_2S_2O_7(s) + 15,300 \text{ cal} \qquad (\Delta H^\circ_{298} = -15,300 \text{ cal})$$

The addition of water to pyrosulfuric acid gives sulfuric acid.

$$H_2S_2O_7(s) + H_2O(l) \longrightarrow 2H_2SO_4(l) + 16,400 \text{ cal} \qquad (\Delta H^\circ_{298} = -16,400 \text{ cal})$$

Since the sulfur dioxide used in this process must be quite pure (to avoid poisoning of the catalyst), the contact process gives pure acid, which may be highly concentrated. Most of the sulfuric acid used today is produced by the contact process.

■ **2. The Lead-Chamber Process.**   This process, at one time the chief method for manufacturing sulfuric acid but now a relatively minor one, received its name from the fact that the reactions involved are carried out in large lead-lined chambers. The chemistry of the chamber process differs from that of the contact process principally with regard to the method of oxidizing the sulfur dioxide to the trioxide.

The sulfur dioxide is obtained by burning sulfur or roasting sulfide ores (Section 23.12). Sulfur dioxide and a small amount of gaseous nitric acid are introduced at the bottom of the reaction chamber where a little sulfuric acid is formed by direct oxidation.

$$3SO_2(g) + 2HNO_3(g) + 2H_2O(l) \longrightarrow 3H_2SO_4(l) + 2NO(g) + 126,500 \text{ cal}$$
$$(\Delta H^\circ_{298} = -126,500 \text{ cal})$$

The remaining sulfur dioxide, nitric oxide, nitrogen dioxide, steam, and excess air enter the lead chambers. Although the reactions in the lead chambers are more complicated than the equations indicate, the net changes can be represented by the equations

$$2NO + O_2 \longrightarrow 2NO_2 \qquad \textbf{(1)}$$
$$SO_2 + NO_2 \longrightarrow SO_3 + NO \qquad \textbf{(2)}$$
$$SO_3 + H_2O \longrightarrow H_2SO_4 \qquad \textbf{(3)}$$

The nitric oxide (and some nitrogen dioxide) which is recovered at the end of the process is returned to the reaction chambers where it is again available for use as an "oxygen-carrier." The nitric oxide acts as catalyst in that it combines with oxygen of the air, reaction (1), and then is regenerated after delivering the oxygen to the sulfur dioxide, reaction (2). This type of catalyst, which actually reacts temporarily but is regenerated in a subsequent step, is referred to as a **secondary,** or **homogeneous, catalyst,** in contrast to the **contact catalyst** used in the contact process. During the process some of the oxides of nitrogen are lost mechanically, and thus a fresh supply must be furnished continually.

In contrast to the relatively pure acid produced by the contact process, sulfuric acid from the lead chamber process is impure, containing lead salts by reaction with the lead of the chamber, as well as impurities arising if a sulfide ore is used to produce the sulfur dioxide. Only about 66% H_2SO_4 can be prepared by the lead-chamber process. If more concentrated acid is produced, it attacks the lead vigorously to produce lead sulfate, $PbSO_4$, which dissolves easily as lead bisulfate, $Pb(HSO_4)_2$, in the concentrated acid.

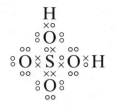

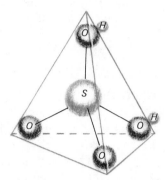

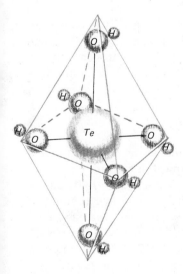

FIGURE 23-10

The electronic and tetrahedral structures of the sulfuric acid molecule. Selenic acid, H_2SeO_4, has the same kind of structure.

23.16   Physical Properties of Sulfuric Acid

Pure sulfuric acid (or **hydrogen sulfate**) is a colorless, oily liquid which freezes at 10.5°. It fumes when heated, due to decomposition of the acid to water and sulfur trioxide. More sulfur trioxide than water is lost during the heating, until a concentration of 98.33 per cent acid is reached. The acid of this concentration boils at 338° without further change in concentration and is the product sold as concentrated H_2SO_4.

Concentrated sulfuric acid dissolves in water with the evolution of a large amount of heat. The dilution may be carried out safely by pouring the concentrated acid slowly into water while the solution is stirred in order to distribute the heat of dilution. **Caution.** The addition of water to the concentrated acid may cause dangerous spattering of the acid.

Figure 23–10 shows the electronic and spatial structure (tetrahedral) of the sulfuric acid molecule.

It is interesting to note that selenic acid, H_2SeO_4, also has a tetrahedral structure, but telluric acid, with the formula H_6TeO_6, does not. Telluric acid has a more fully "hydrated" structure, attributed to the larger radius of tellurium. Its structure is octahedral, as illustrated in Fig. 23–11.

23.17   Chemical Properties of Sulfuric Acid

The sulfur atom in the sulfuric acid molecule and in sulfates is surrounded tetrahedrally by four oxygen atoms. The hydrogen atoms form hydrogen bonds between these tetrahedra, binding them together and giving the liquid a high boiling point.

The large heat of dilution of sulfuric acid is caused by hydrate and hydronium ion formation. The hydrates $H_2SO_4 \cdot H_2O$, $H_2SO_4 \cdot 2H_2O$, and $H_2SO_4 \cdot 4H_2O$ are known. The acid ionizes in two stages.

$$H_2SO_4 \longrightarrow H^+ + HSO_4^-$$
$$HSO_4^- \rightleftharpoons H^+ + SO_4^{2-}$$

In dilute solution sulfuric acid undergoes almost complete primary ionization. The secondary ionization is less complete, but even so HSO_4^- is a strong acid ($K_e = 1.2 \times 10^{-2}$).

The strong affinity of concentrated sulfuric acid for water makes it a good dehydrating agent. Gases which do not react with the acid may be dried by being passed through it. So great is the affinity of concentrated sulfuric acid for water that it will remove hydrogen and oxygen, in the form of water, from many compounds containing these elements. Organic substances containing hydrogen and oxygen in the proportion of two to one, such as cane sugar, $C_{12}H_{22}O_{11}$, and cellulose, $(C_6H_{10}O_5)_x$, are charred by concentrated sulfuric acid.

$$C_{12}H_{22}O_{11} \longrightarrow 12C + 11H_2O$$

Concentrated sulfuric acid is highly destructive to human flesh, because of its reactions with organic compounds in the flesh, and is very dangerous on that account.

FIGURE 23-11

The octahedral structure of the telluric acid molecule, H_6TeO_6.

Sulfuric acid acts as an oxidizing agent, particularly when hot and concentrated. Depending upon its concentration, the temperature, the strength of the reducing agent with which it acts, and other factors, sulfuric acid oxidizes many compounds and, in the process, undergoes reduction to either SO_2, HSO_3^-, SO_3^{2-}, S, H_2S, or S^{2-}. Its oxidizing action towards hydrogen bromide and hydrogen iodide was noted in Section 21.8, and towards metals, carbon, and sulfur in Section 23.12. The displacement of volatile acids from their salts by means of concentrated sulfuric acid has been mentioned in Section 15.4. Aqueous solutions of sulfuric acid exhibit the characteristic properties of strong acids.

23.18   Sulfates

Being a diprotic acid, sulfuric acid forms both **normal sulfates,** such as Na_2SO_4, and **hydrogen sulfates,** such as $NaHSO_4$.

The normal sulfates of barium, strontium, calcium, and lead are only slightly soluble in water. These salts occur in nature as the minerals barite, $BaSO_4$; celestite, $SrSO_4$; gypsum, $CaSO_4 \cdot 2H_2O$; and anglesite, $PbSO_4$. They can be prepared in the laboratory by ionic combination. For example, the addition of barium ion to a solution containing sulfate ions causes the precipitation of white barium sulfate.

$$Ba^{2+} + SO_4^{2-} \longrightarrow \underline{BaSO_4}$$

This reaction is the basis of a qualitative and quantitative test for the sulfate ion and the barium ion.

Although barium and its salts are very toxic, barium sulfate is sufficiently insoluble ($K_{sp} = 1.08 \times 10^{-10}$) that it can be safely put into the alimentary tract for x-ray studies of that portion of the body.

Among the important soluble normal sulfates are **Glauber's salt,** $Na_2SO_4 \cdot 10H_2O$; **Epsom salt,** $MgSO_4 \cdot 7H_2O$; **blue vitriol,** $CuSO_4 \cdot 5H_2O$; **green vitriol,** $FeSO_4 \cdot 7H_2O$; and **white vitriol,** $ZnSO_4 \cdot 7H_2O$. The hydrogen sulfates, such as $NaHSO_4$, are acids as well as salts. Sodium hydrogen sulfate is the primary ingredient in some household cleansers.

23.19   Uses of Sulfuric Acid

The 1974 production of sulfuric acid in the United States was 32,394,000 tons. For many years it has been, and it still is, the top chemical in amount produced. The major uses are in the production of ammonium sulfate (1,949,000 tons in the United States in 1974) and soluble phosphate fertilizers; in the refining of petroleum to remove impurities from such products as gasoline and kerosene; in the pickling of steel (cleaning its surface of iron rust) before coating it with tin, zinc, or enamel; in the production of dyes, drugs, and disinfectants from coal tar; in the electrometallurgy of certain metals as the electrolyte or in the production of sulfates of metals to be used as electrolytes; in the manufacture of other chemicals, such as hydrochloric and nitric acids, and the sulfates of metals; and in the production of textiles, paints, pigments, plastics, explosives, and lead storage batteries.

Other Acids of Sulfur

In addition to the acids of sulfur which we have thus far discussed, there are several others of commercial importance.

23.20 Thiosulfuric Acid

It was pointed out in Section 23.13 that sulfites are slowly oxidized to sulfates by oxygen. Sulfur plays a role similar to that of oxygen by transforming sulfites to thiosulfates. For example, when a mixture of sulfur and a solution of sodium sulfite is boiled, **sodium thiosulfate,** $Na_2S_2O_3$, is formed.

$$(2Na^+) + SO_3^{2-} + S \longrightarrow (2Na^+) + S_2O_3^{2-}$$

Crystals of the **pentahydrate,** $Na_2S_2O_3 \cdot 5H_2O$, separate when the solution is evaporated.

When a solution of sodium thiosulfate is acidified, unstable **thiosulfuric acid,** $H_2S_2O_3$, is formed.

$$(2Na^+) + S_2O_3^{2-} + 2H^+ + (2Cl^-) \longrightarrow H_2S_2O_3 + (2Na^+) + (2Cl^-)$$

The acid decomposes immediately into sulfurous acid and sulfur, the sulfur appearing either as a precipitate or in colloidal suspension.

$$H_2S_2O_3 \longrightarrow H_2SO_3 + \underline{S}$$

The electronic formula and the structure of the thiosulfate ion are compared to that of the sulfate ion in Fig. 23–12 to show that one of the oxygen atoms of the sulfate is replaced by the sulfur atom. The average oxidation state of sulfur in the thiosulfate ion is +2, but it is evident from Fig. 23–12 that the central

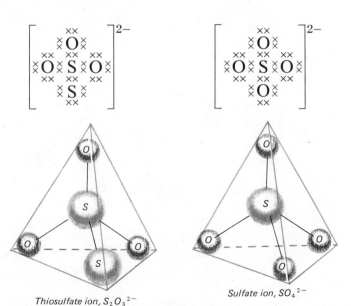

FIGURE 23–12
Electronic and tetrahedral structures of the thiosulfate ion and the sulfate ion.

Thiosulfate ion, $S_2O_3^{2-}$

Sulfate ion, SO_4^{2-}

sulfur atom is in the $+6$ oxidation state and the other sulfur atom (like oxygen in the sulfate ion) is in the -2 state.

Sodium thiosulfate, also known as "hypo," is used in the photographic process as a fixing solution to dissolve from the plate or film any silver halides which have not been reduced to metallic silver by the developer.

$$AgX + 2S_2O_3^{2-} \rightleftharpoons Ag(S_2O_3)_2^{3-} + X^- \qquad (X = \text{a halogen})$$

The thiosulfate ion is oxidized by iodine to the **tetrathionate ion,** $S_4O_6^{2-}$.

$$2S_2O_3^{2-} + I_2 \longrightarrow S_4O_6^{2-} + 2I^-$$

This reaction is used extensively in analytical chemistry. For example, in the quantitative analyses for chlorine and the copper ion, iodine is often produced by the following reactions and then titrated with a standard thiosulfate solution using a suitable indicator to determine the equivalence point (endpoint).

To determine Cl_2: $Cl_2 + 2I^- \longrightarrow I_2 + 2Cl^-$
To determine Cu^{2+}: $Cu^{2+} + 2I^- \longrightarrow I_2 + \underline{CuI}$

23.21   Peroxymonosulfuric Acid and Peroxydisulfuric Acid

Peroxymonosulfuric and peroxydisulfuric acids have the formulas H_2SO_5 and $H_2S_2O_8$, respectively. Both may be considered as derivatives of hydrogen peroxide. One of the hydrogen atoms of hydrogen peroxide has been replaced by the HSO_3 group in the case of peroxymonosulfuric acid; both hydrogen atoms of hydrogen peroxide have been replaced by HSO_3 groups in peroxydisulfuric acid. The electronic formulas and the structures of these acids are shown in Fig. 23–13.

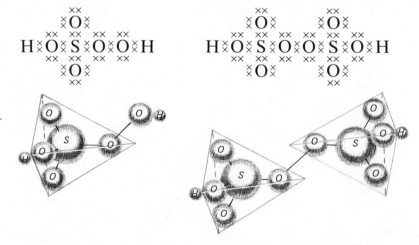

FIGURE 23–13
Structures of peroxymono-sulfuric acid (*left*) and peroxydisulfuric acid (*right*). Each acid possesses an oxygen-oxygen (peroxide) linkage; hence, the use of the prefix *peroxy-* in the name.

Peroxydisulfuric acid is produced commercially by the anodic oxidation of hydrogen sulfate ions in 45–55 per cent sulfuric acid at a low temperature.

$$2HSO_4^- \longrightarrow H_2S_2O_8 + 2e^-$$

Electrolysis of potassium hydrogen sulfate in aqueous solution gives potassium peroxydisulfate, $K_2S_2O_8$. Treatment of this salt at low temperatures with concentrated sulfuric acid produces peroxymonosulfuric acid, commonly called **Caro's acid.** This acid is also produced by the reaction of sulfur trioxide with H_2O_2.

$$H_2O_2 + SO_3 \longrightarrow H_2SO_5$$

The peroxysulfuric acids and their salts are useful as strong oxidizing agents.

23.22 Chlorosulfonic Acid and Sulfamic Acid

Chlorosulfonic acid, HSO_3Cl, and sulfamic acid, SO_3NH_3, are related structurally to sulfuric acid, as shown in Fig. 23-14. The formula for sulfamic acid, sometimes written HSO_3NH_2, is more properly designated SO_3NH_3.

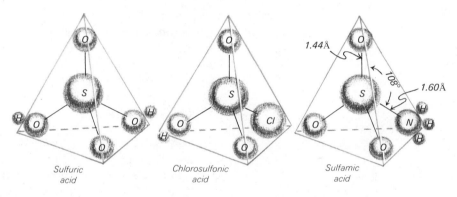

FIGURE 23-14

Structure of sulfuric acid molecule compared to the structures of two related acids. The normal bond angle in a tetrahedral structure is 109°. The O—S—N angle in sulfamic acid is 108°, indicating a slight distortion of the tetrahedron. Such a distortion arises when the structure is not symmetrical, i.e., the atoms at the corners of the tetrahedron are not all the same.

Chlorosulfonic acid is a colorless liquid which fumes in moist air, reacts vigorously with water, and is used to introduce the sulfonate group, SO_3H, into many organic compounds. Sulfamic acid is a white, crystalline, nonhygroscopic solid. It is one of the few strong monoprotic acids which can be weighed without special drying precautions being required. It is of some importance, therefore, in analytical chemistry. Sulfamates are important ingredients in some weed killers.

QUESTIONS

1. List the elements in Group VIA of the Periodic Table in order of increasing (a) atomic number, (b) melting point, (c) electronegativity, (d) strength as oxidizing agents, and (e) heat of hydride formation. Compare the directions of trends for the various properties, and, to the extent that you can, explain each order of ranking.

2. Give two properties of oxygen which set it apart from the other members of Group VIA.

3. Explain, with the aid of a diagram, the Frasch process for extracting sulfur from underground deposits.

4. What are the allotropic forms of solid sulfur and how may they be produced?

5. What is the molecular structure of sulfur in the liquid, solid, and gaseous states?

6. Account for the viscosity and color of μ-sulfur.

7. Write chemical equations describing three chemical reactions in which elemental sulfur acts as an oxidizing agent; three in which it acts as a reducing agent.

8. Write equations for the reaction of sulfur with H_2, C, Fe, O_2, Cl_2, and HNO_3.

9. What are the principal commercial uses of sulfur? of selenium? of tellurium?

10. What property of selenium makes it useful in photoelectric cells?

11. How may hydrogen sulfide be prepared conveniently in the laboratory?

12. Account for the fact that hydrogen sulfide is a gas at ordinary temperatures whereas water, which has a smaller formula weight, is a liquid.

13. Write equations for the reaction of hydrogen sulfide with Fe^{3+}, MnO_4^-, Br_2, and $Cr_2O_7^{2-}$ in acidic solution.

14. On the basis of analogy to the series of binary acids HCl, HBr, and HI (Section 15.16), predict the relative acid strengths of H_2S, H_2Se, and H_2Te. Explain your answer.

15. Why are solutions of sulfides and hydrosulfides alkaline? Write equations.

16. A convenient method for preparing hydrogen selenide involves fusing aluminum and selenium and then hydrolyzing the fused product with water. Write chemical equations describing this procedure.

17. How is sulfur dioxide produced commerically and what are its uses?

18. If you were the owner of an industrial plant using, as a fuel, coal and oil which contain sulfur, what could you do to lessen the quantity of sulfur dioxide in the exhaust gases?

19. Write three equations showing the action of sulfurous acid as a reducing agent.

20. How can sodium sulfite be made in the laboratory?

21. Determine the oxidation state of sulfur in each of the following species: HS^-, $SOCl_2$, SF_6, Na_2S, $S_2O_3^{2-}$, $S_4O_6^{2-}$, SnS_3^{2-}.

22. Why do solutions of sulfites usually contain sulfate ions?

23. Write the equations for the reactions in the contact process for the manufacture of sulfuric acid. What does the term "contact" signify in this process?

24. How did the "lead-chamber" process for sulfuric acid get its name?

25. What is the formula for pyrosulfuric acid and how is this acid formed?

26. Write chemical equations describing, respectively, the addition of water and sulfur trioxide to pure sulfuric acid.

27. The essential difference between the contact and lead-chamber processes lies in the method of oxidizing the sulfur dioxide. Explain.

28. Compare the acid ordinarily obtained by the two processes for the manufacture of sulfuric acid with regard to purity and concentration.

29. How may concentrated sulfuric acid be diluted safely?

30. What is meant by catalyst poisoning?

31. Write equations for the reactions of carbon and of copper with hot, concentrated sulfuric acid.
32. What single chemical test could be used to distinguish sulfides, polysulfides, sulfites, and sulfates from one another?
33. Interpret the following reaction in terms of an acid-base relationship according to the Lewis theory:

$$CaO + SO_3 \longrightarrow CaSO_4$$

34. Show by equations the dehydrating action of sulfuric acid upon cane sugar and upon cellulose.
35. Illustrate by equations the displacement of volatile acids from their salts by means of concentrated sulfuric acid.
36. What are the possible reduction products of sulfuric acid? What is the oxidation state of sulfur in each of these products?
37. What is "hypo" and how is it used in photography?
38. Show by writing formulas that the peroxysulfuric acids may be considered as derivatives of hydrogen peroxide.
39. Account for the formation of sulfur in a solution of sodium thiosulfate in contact with air (air contains carbon dioxide, the anhydride of carbonic acid).
40. Write the electronic formula of each of the following: S^{2-}, H_2S_2, SO_2, H_2SO_3, SO_3, Na_2SO_4, H_2SO_5, $H_2S_2O_7$, and $H_2S_2O_8$.
41. Devise the best valence electron structure you can for the SCl molecule. How satisfactory is your structure compared to that of S_2Cl_2? Comment on the relative probabilities for the existence of SCl and S_2Cl_2.
42. In Section 23.13, the equation is given for the auto-oxidation-reduction reaction of sodium sulfite, and mention is made that metal hypochlorites and metal chlorates undergo analogous auto-oxidation-reduction reactions. Write the balanced equations for the auto-oxidation-reduction reactions of sodium sulfite, sodium hypochlorite, and sodium chlorate, respectively.

PROBLEMS

1. Calculate the percentage of sulfur in the compounds S_2Cl_2, MgS, $Na_2S_2O_4$, and SF_6.   *Ans. 47.49%; 56.88%; 36.83%; 21.95%*
2. What volume of sulfur dioxide (S.T.P.) would be formed by the complete decomposition of 250.0 g of sulfurous acid? What weight of water would be formed?   *Ans. 68.23 liters; 54.88 g*
3. How much sulfuric acid can be prepared from 10 tons of sulfur by the contact process, assuming a 95% yield?   *Ans. 29 tons*
4. Calculate the volume of hydrogen sulfide at S.T.P. obtainable from the hydrolysis of 710 g of thioacetamide (CH_3CSNH_2).   *Ans. 212 liters*
5. What volume of oxygen measured at 27.0° and 750 mm would be required to convert 450 g of sulfur to sulfuric acid by the contact process?   *Ans. 525 liters*
6. What would be the density of sulfur vapor in grams per liter (measured at S.T.P.) if 65.0% of the molecules were S_8 and 35.0% were S_2?   *Ans. 8.44 g/liter*

7. A volume of 22.85 ml of a 0.1023 N standard sodium thiosulfate solution is required to titrate a 25.00-ml sample of a solution containing iodine. What is the iodine concentration in the solution?   *Ans. 0.09350 N*

8. (a) Check by calculation, using data in Appendix J, at least five of the ΔH°_{298} values given for reactions in this chapter.

   (b) For each of the same reactions, using data in Appendix J, calculate ΔG°_{298}.

   (c) On the basis of the values calculated in parts (a) and (b), state what you can as to whether or not each reaction should proceed spontaneously.

9. A fused mixture of rubidium fluoride and uranium(IV) fluoride can be oxidized with fluorine to produce a uranium compound in which the uranium is mainly but not entirely in the +5 oxidation state. The product is found to contain 54.43% uranium. A 1.0357-g sample of the product immersed in 100.0 ml of 0.1007 M acidified potassium iodide solution reacted according to the following equation:

$$2I^- + 2UF_6^- \longrightarrow 2UF_4 + I_2 + 4F^-$$

The iodine produced was titrated with 14.80 ml of 0.1494 M sodium thiosulfate solution. What per cent of the original uranium was oxidized to the +5 oxidation state?   *Ans. 93.36%*

REFERENCES

"Sulfur Isotope Distribution in Solfatoras, Yellowstone National Park," R. Schoen and R. O. Rye, *Science*, **170**, 1082 (1970).

"Variegation in Vitriol Manufacture," M. Schofield, *Chem. in Britain*, **3**, 247 (1967).

"Sulfuric Acid Plants Face Emission Limits," (Staff), *Chem. & Eng. News*, June 29, 1970; p. 43.

"Sulfur," M. E. Weeks and H. M. Leicester, *Discovery of the Elements*, Seventh Edition, Publ. by the Journal of Chemical Education, Easton, Pa., 1968; pp. 51–54.

Manufacture of Sulfuric Acid, (American Chemical Society Monograph No. 144), edited by W. W. Duecker and J. R. West, Reinhold Publ. Corp., New York, 1959.

"Chemistry of Solutions in Liquid Sulfur Dioxide," P. J. Elving and J. M. Markowitz, *J. Chem. Educ.*, **37**, 75 (1960).

"The Sulphur Cycle," J. R. Postgate, *Educ. in Chemistry*, **2**, 58 (1965).

"The Chemistry of Tetrasulfur Tetranitride," C. W. Allen, *J. Chem. Educ.*, **44**, 38 (1967).

"Sulfur," C. J. Pratt, *Sci. American*, May, 1970; p. 62.

"Modern Sulfuric Acid Technology," T. J. Browder, *Chem. Eng. Progress*, May, 1971; p. 45.

"Removing SO_2 and Acid Mist with Venturi Scrubbers," I. S. Shah, *Chem. Eng. Progress*, May, 1971; p. 51.

"SO_2 Emission Control from Acid Plants," W. G. Tucker and J. R. Burleigh, *Chem. Eng. Progress*, May, 1971; p. 57.

"An SO_2 Removal and Recovery Process," J. J. Humphries, S. B. Zdonik, and E. J. Parsi, *Chem. Eng. Progress*, May, 1971; p. 64.

"Reducing SO_2 Emission from Stationary Sources," T. H. Chilton, *Chem. Eng. Progress*, May, 1971; p. 69.

The Atmosphere, Air Pollution, and the Noble Gases

24

The Atmosphere

24.1 The Composition of the Atmosphere

The **atmosphere,** or air, is the mixture of gaseous substances which surrounds the earth. For the most part, the atmosphere consists of uncombined elements. Nitrogen, oxygen, and the noble gases are present in the atmosphere in almost constant proportion. The percentage of carbon dioxide in the air varies somewhat and that of dust and water vapor is widely variable. There are also traces of hydrogen, ammonia, hydrogen sulfide, oxides of nitrogen, sulfur dioxide, and other gases. The average composition of dry air at sea level is given in Table 24–1.

The percentage composition of dry air does not vary much with location on the earth's surface or with altitude. However, the density of air, as reflected in pressure, varies greatly with altitude. The average pressure at sea level and 45° latitude is 760 mm, at 15,000 feet it is about 400 mm, at 10 miles it is about 40 mm, and at 30 miles is only about 0.1 mm.

TABLE 24–1   Composition of the Atmosphere

Component	Per Cent by Volume	Component	Per Cent by Volume
Nitrogen	78.03	Neon	0.0012
Oxygen	20.99	Helium	0.0005
Argon	0.94	Krypton	0.0001
Carbon dioxide	0.035–0.04	Ozone	0.00006
Hydrogen	0.01	Xenon	0.000009

24.2 Liquid Air

■ **1. Preparation.** Before air can be liquefied it must be freed of moisture and carbon dioxide, because these substances change to the solid state when cooled and thus clog the pipes of the liquid-air machine (Fig. 24–1). Once free of water and carbon dioxide, the air is compressed to about 100 atmospheres and cooled to remove the heat produced by compression. It is then again compressed at about 200 atmospheres pressure (see Section 10.3) and cooled by a mixture of salt and ice. The cold compressed air is next passed into the liquefier in which it escapes through a valve and expands to a pressure of about 20 atmospheres. This expansion is accompanied by the absorption of heat from other compressed air in the small inner coil of the liquefier. As the process continues, each quantity of air that escapes from the valve is colder than that which preceded it, and finally liquefied air begins to escape.

■ **2. Properties.** Liquid air is a mobile liquid with a faint blue color. It evaporates rapidly in an open container with the absorption of a large quantity of heat. To reduce the rate of evaporation, liquid air is stored in Dewar flasks. These flasks (Fig. 24–2) have the space between the inner and outer walls evacuated. Heat

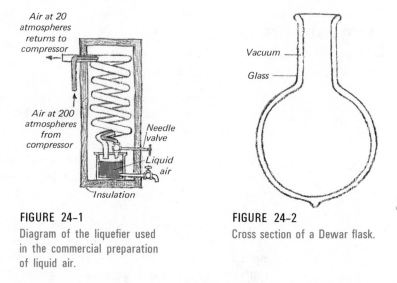

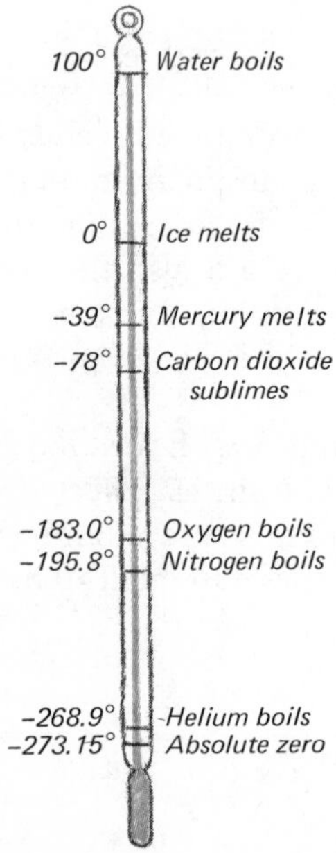

FIGURE 24–3

Comparative Celsius temperatures for changes of state of several substances.

FIGURE 24–1

Diagram of the liquefier used in the commercial preparation of liquid air.

FIGURE 24–2

Cross section of a Dewar flask.

is not transmitted through a vacuum and without heat the liquid air cannot evaporate. Dewar flasks are frequently silvered on the walls of the evacuated space to reflect heat waves which fall upon them. "Thermos" bottles are Dewar flasks.

Because liquid air is a mixture it has no definite boiling point. Oxygen boils at -183.0 and nitrogen at $-195.8°$ (Fig. 24–3). Commercial supplies of nitrogen, oxygen, and argon are obtained by the fractional distillation of liquid air. The low temperature of liquid air is of great practical value in scientific research requiring low temperatures; the study of the effects of very low temperatures on substances is called **cryogenics.** Liquid air is also a source of liquid oxygen used in rocket planes and missiles.

Air Pollution

24.3   The Problem

Air pollution is a subject of much current interest. It is a much worse problem than in years past, but it is not a new problem. William Shakespeare, for example, wrote in *Hamlet* in about 1601:

> This most excellent canopy, the air, look you, this brave o'erhanging firmament, this majestical roof fretted with golden fire, why, it appears no other thing to me than a foul and pestilent congregation of vapors.[1]

In 1272, King Edward I banned the use of a smoke-producing variety of coal in an effort to clear the smoky air of London, and the British Parliament ordered a man tortured and hanged for illegally burning the banned coal. Later, King Richard III put a high tax on the use of coal. These efforts, nevertheless, accomplished little, and London remained smoky. In 1952, in London, a combination of *smoke and fog* (commonly called smog) was responsible for the deaths of some 4,000 people during a four-day period. Again, in 1962, London smog killed 700 persons. In Donora, Pennsylvania, in 1948, twenty people died, and 5,900 were made ill during a bad period of air pollution. These are dramatic instances. Less dramatic, but no less tragic for those involved, were the years 1953, 1963, and 1966 when air pollution in New York City was a serious problem. It is estimated that during each of these years smog caused approximately 700 more deaths than the normal number from other causes. Polluted air may ultimately result in shortened lives for all of us, but it is a special problem for those suffering from respiratory diseases such as emphysema and bronchitis. Lung cancer is found to be twice as prevalent among people living in air-polluted cities as among those living in rural areas with cleaner air.

The problems resulting from air pollution extend well beyond those related to human health. Various types of air pollution can damage or ruin flowers and crops, make paint peel and discolor, damage textiles, discolor dyes, add to the expense of cleaning clothes, poison animals by contaminating feed, corrode metals, damage rubber tires, and increase lighting costs through partially blocking out the sun—to cite only some of the problems. In the United States alone, the estimate is that well over 200,000,000 tons of pollutants are emitted into the air annually. Strictly on an economic basis, it has been estimated that directly or indirectly air pollution costs the people in the United States over 12 billion dollars per year. Furthermore, over the past several years, we have had an incredible increase in the average per capita use of consumer goods and power. Truly, air pollution is a major problem. It behooves us to learn more about it and to take steps to solve the problem.

---

[1] *Hamlet, Prince of Denmark,* William Shakespeare (1601); Act II, Scene 2.

24.4 The Causes

The six major air pollutants commonly recognized are also those considered in the federal standards for air quality issued April 30, 1971. The six, with a partial list of the effects attributed to each, are:[2]

(1) *Sulfur oxides*—cause acute leaf injury and attack trees; irritate upper part of human respiratory tract; corrode metals; ruin hosiery and other textiles; disintegrate book pages and leather; destroy paint pigments; erode statues.

(2) *Particulates (solids)*—obscure vision; aggravate lung illnesses; deposit grime on buildings and personal belongings; corrode metals.

(3) *Carbon monoxide*—causes headaches, dizziness, nausea; is absorbed in the blood, reducing oxygen content, impairing mental processes, and in sufficient quantity causing death.

(4) *Nitrogen oxides*—cause leaf damage; stunt plants; irritate eyes and nose membranes; cause brown pungent irritating haze; corrode metals; damage rubber.

(5) *Hydrocarbons*—may be carcinogenic (cancer-producing); retard plant growth, cause abnormal leaf and bud development.

(6) *Photochemical oxidants:* (a) Ozone and resultant chemical products—discolor upper surface of leaves of many crops, trees, and shrubs; damage and fade textiles; reduce physical activity, including athletic performance; speed deterioration of rubber; disturb lung function; irritate eyes, nose, and throat; induce coughing. (b) Peroxyacetyl nitrate (PAN)—discolors lower leaf surface; irritates eyes; disturbs lung function.

Table 24–2 lists the average amount of daily emissions put in the air from all sources and Table 24–3 the annual amount from different sources, according to data from the Department of Health, Education, and Welfare.

The above general categories can be broken down into 119 specific polluting compounds, in itself one indication of the magnitude of the problem facing us. In addition to these compounds, many other materials can be major contaminants in specific locations under special conditions. These include, to list only a few, fluorides, lead compounds, nuclear radioactive emissions, chlorine, beryllium, mercury, and pesticides.

Chlorinated hydrocarbon insecticides, such as dichlorodiphenyltrichloroethane (DDT), are a major pollution problem both in the air and in lakes and rivers through water run-off from the soil. Chlorinated hydrocarbons are referred to as "persistent pesticides" inasmuch as their effects are long-lasting and their concentrations become magnified in the food chain. Hence, the use of most of these insecticides is now banned by law.

Most of the air pollution problem arises from combustion. Coal is largely carbon; fuel oils, gasoline, and natural gas are largely hydrocarbons (compounds

---

[2] *Air Conservation*, American Association for the Advancement of Science, No. 80; *The Effects of Air Pollution*, Health, Education, and Welfare (HEW)—National Air Pollution Control Administration (NAPCA), No. 1556; *Air Pollution Primer*, National Tuberculosis and Respiratory Disease Association; *Facts and Issues*, League of Women Voters, No. 393; *Air Pollution Injury to Vegetation*, NAPCA, No. AP-71.

TABLE 24-2   Daily Average Emissions in the United States, in Tons

Emission	All Sources	Automobiles Only
Sulfur oxides	90,000	2,000–3,000
Particulates	75,000	2,000–3,000
Carbon monoxide	270,000	115,000
Nitrogen oxides	55,000	15,000
Hydrocarbons	85,000	25,000

TABLE 24-3   Annual Emissions in the United States, in Millions of Tons

Source	Sulfur Oxides	Particulates	Carbon Monoxide	Nitrogen Oxides	Hydrocarbons	Total	Per Cent of Total
Transportation	0.8	1.2	63.8	8.1	16.6	90.5	42.3%
Fuel combustion in stationary sources (power generation, industry, space heating)	24.0	8.9	1.9	10.0	0.7	45.5	21.3
Industrial processes	7.3	7.5	9.7	0.2	4.6	29.3	13.7
Solid waste disposal	0.1	1.1	7.8	0.6	1.6	11.2	5.2
Miscellaneous (agricultural burning, forest fires, etc.)	0.6	9.6	16.9	1.7	8.5	37.3	17.5
Total	32.8	28.3	100.1	20.6	32.0	213.8	100.0%

containing only carbon and hydrogen). If combustion were complete, carbon would burn in air to produce only carbon dioxide, and a hydrocarbon would burn to produce only carbon dioxide and water, as illustrated below for carbon (C), methane (CH_4), and ethane (C_2H_6).

$$C + O_2 \longrightarrow CO_2$$
$$CH_4 + 2O_2 \longrightarrow CO_2 + 2H_2O$$
$$2C_2H_6 + 7O_2 \longrightarrow 4CO_2 + 6H_2O$$

However, combustion is usually incomplete as a result of a deficiency of oxygen and the presence of impurities such as elemental sulfur or pyrite (FeS_2) in coal. Carbon monoxide is formed when a deficiency of oxygen is present in relation to the carbon or the hydrocarbon.

$$2C + O_2 \longrightarrow 2CO$$
$$2CH_4 + 3O_2 \longrightarrow 2CO + 4H_2O$$
$$2C_2H_6 + 5O_2 \longrightarrow 4CO + 6H_2O$$

Sulfur oxides are produced when sulfur-containing fuels are burned.

$$S + O_2 \longrightarrow SO_2$$
$$2FeS_2 + 5O_2 \longrightarrow 2FeO + 4SO_2$$

The SO_2 either reacts with water to form sulfurous acid or is oxidized by nitrogen oxides, or other catalysts, to SO_3 which unites with water to form sulfuric acid (see Section 23.15). These acids are among the main substances responsible for ruining nylon hosiery and for corroding metals. Electrical generating plants using sulfur-containing coal as fuel contribute greatly to air pollution. The catalytic converters of automobile exhaust systems produce small quantities of sulfur dioxide, sulfur trioxide, sulfuric acid, and hydrogen sulfide from reactions of the sulfur compounds present in gasoline. Nitrogen oxides are produced with all combustion because of reaction of nitrogen in the air with oxygen, particularly at high temperatures (see Sections 25.14 and 25.22).

$$N_2(g) + O_2(g) + 43{,}140 \text{ cal} \longrightarrow 2NO(g) \quad (\Delta H^\circ_{298} = +43{,}140 \text{ cal})$$

$$2NO(g) + O_2(g) \longrightarrow 2NO_2(g) + 27{,}280 \text{ cal}$$
$$(\Delta H^\circ_{298} = -27{,}280 \text{ cal})$$

$$3NO_2 + H_2O \longrightarrow 2HNO_3 + NO$$

Much of the particulate matter in polluted air is noncombustible or unburned material in fuel (carbon particles, ash, etc.). Particulate matter, such as dust, carbon, and ash, may be present either as large particles that fall out of the air quickly (nevertheless creating problems where they fall) or as small particles ("aerosols") which when suspended in air as part of the colloidal system (Section 14.13) move about in air just as gases do.

Some materials, not resulting from combustion, that contribute to air pollution are from vaporization of liquids (or sometimes sublimation of solids) to gases, which then disperse in the air. Solid particles, other than those from combustion, are introduced into the air by abrasion or grinding of one material in contact with another. Examples are metal filings introduced into the air when a metal is polished on a grinding wheel, dust put into the air during a dust storm or by automobile wheels on a dusty road, and (what is said to be the most common particulate matter in human lungs) finely divided rubber added to the air by friction of automobile tires on pavements.

Secondary reactions produce additional pollution problems. One example, mentioned earlier, is the oxidation of sulfur dioxide to sulfur trioxide and the subsequent reaction with moisture in the atmosphere to produce sulfuric acid. Another example is the photochemical reaction by which unburned hydrocarbons and nitrogen oxides in automobile exhausts react in the presence of sunlight to form peroxyacetyl nitrate (PAN) and ozone. These compounds are primary constituents of smog, which is increasingly becoming a problem in many of our larger cities.

The problem of smog is intensified by the so-called air "inversion" common in Los Angeles and some other cities. Normally air near the earth's surface becomes warmer than the air above and rises, allowing cooler air to take its place. Sometimes, however, a warmer layer of air develops above, as when cool air flows down a mountain into the valleys and underneath a mass of warm air. When this happens, the cooler air is trapped and held at the earth's surface. Pollutants from auto exhausts, industrial stacks, and other sources are also trapped, sometimes for days at a time.

Tetraethyl lead has been used commonly in gasoline to provide greater antiknock capability. Its use has now been curtailed greatly, however, with the introduction of catalytic converters, because lead compounds poison the catalysts used in the converters, quickly rendering them ineffective. When used, tetraethyl lead is converted during the combustion process to lead oxide, which forms harmful deposits in the engine. Therefore, an additional compound is added, ethylene dibromide, which reacts, in the engine, with the lead oxide to form lead bromide, a volatile substance that is emitted along with other gases to the air through the exhaust. Lead compounds are known to have serious physiological effects on humans.

24.5   Some Solutions to the Air Pollution Problem

Removal of particulate (solid) pollutants from smokestack gases is easily accomplished using electrostatic Cottrell precipitators (see Section 14.19). Processes for removing sulfur oxides, however, are not well developed, though several methods are being studied and tested. But even at the present stage of development of such processes, sulfur oxides recovered from stack gases are now becoming a significant source of sulfur for the production of sulfuric acid (see Section 23.15). The "wet limestone" process for removal of sulfur oxides from stack gases is one which shows promise:

$$SO_2 + (H_2O) + CaCO_3 \longrightarrow CaSO_3 + CO_2 + (H_2O)$$
$$SO_3 + (H_2O) + CaCO_3 \longrightarrow CaSO_4 + CO_2 + (H_2O)$$

The use of low-sulfur coal, natural gas, or fuel oil, instead of high-sulfur coal, would alleviate the problem of sulfur oxide emission, but the quantity of such fuels is severely limited. It is likely that such low-sulfur fuels might most logically be used for small pollution sources such as homes and small industries. These sources, individually, do not produce a large amount of pollution but collectively are a big factor because of the large numbers, the discharge of pollutants near ground level, and the economic problem of providing pollution control devices for so many small sources.

A possible solution for the smaller industrial sources, and perhaps large ones as well, is the removal of sulfur from sulfur-containing fuels. Fuel oil can be fairly readily desulfurized, but only 12% of the sulfur oxides presently emitted by combustion sources originates in heavy fuel oil, whereas 65% originates in coal. Coal contains 2.5–3% sulfur on the average. Twenty-five per cent of the coal mined in the United States for use as a fuel is such that its sulfur content can be reduced to 1% by existing processes. For the other seventy-five per cent of coal mined, it is possible to remove only about 40% of its sulfur content.

Over 95% of nitrogen oxide emissions results from combustion; twenty-five per cent of the emissions come from combustion in steam-electric power plants. Some reduction in nitrogen oxide emissions can be made by modifying the combustion process, though with some loss of efficiency, but the technology is not well defined yet. The technology for removal of nitrogen oxides from stack gases and auto exhausts is also still in the laboratory stage.

Carbon monoxide emission can be lessened by finer control of combustion, including better tuning of auto engines, but the technology to eliminate the carbon

FIGURE 24-4

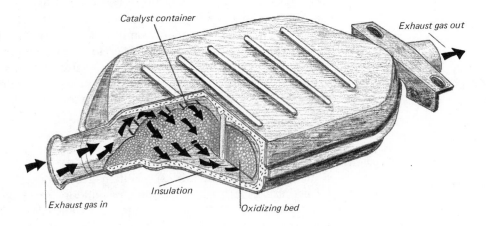

Catalyst container

Exhaust gas out

Insulation

Exhaust gas in

Oxidizing bed

monoxide problem completely, either by more efficient combustion or by removal from stacks and exhausts, is not available at present.

Catalytic converters, designed for cleaning automobile exhausts, have been installed in most automobiles now manufactured, beginning in 1975. The converter is a device added to the exhaust system (Fig. 24–4) to decrease carbon monoxide and unburned hydrocarbon pollutants to very low levels. One type of converter contains refractory beads coated with a catalytic material consisting of platinum and palladium metal in an extremely finely divided form, thereby providing a very large total surface area. The pollutants react with oxygen, from the air, on the surface of the solid catalyst.

$$2CO + O_2 \longrightarrow 2CO_2$$
$$\underset{\substack{\text{(Typical} \\ \text{hydrocarbon} \\ \text{in gasoline)}}}{C_7H_{16}} + 11O_2 \longrightarrow 7CO_2 + 8H_2O$$

In addition to decreasing carbon monoxide and hydrocarbon pollutants in the exhaust, however, the converters have been found to catalyze reactions between sulfur compounds in the gasoline and hydrocarbons, which are the principal constituents of gasoline, and between the sulfur compounds and oxygen. As mentioned previously, reaction of hydrocarbons with the sulfur compounds produces the very toxic hydrogen sulfide, but this only occurs with an abnormally rich fuel-air mixture (high ratio of fuel to air). Hence, careful tuning of the engine ordinarily eliminates that problem, though sometimes at the expense of less satisfactory performance of the engine. More serious are the catalysis of the reaction of sulfur compounds with oxygen of the air to produce sulfur dioxide and the catalysis of the oxidation of sulfur dioxide to sulfur trioxide. The sulfur trioxide then reacts with water to form small quantities of sulfuric acid mist.

Fortunately, technology is available for removing sulfur compounds from gasoline during the refining process, but there is certain to be an increase in the cost of gasoline to the consumer. The extra expense may well become necessary, however, if catalytic converters continue to be used extensively.

As pointed out in the previous section, lead compounds poison the relatively expensive catalyst in the converter after short use. Hence, low-lead or no-lead gasolines must be produced in large quantity. The undesirable emission of toxic lead compounds in the exhaust is also eliminated by the use of unleaded gasolines.

An additional pollutant, nitric oxide, results from the reaction of nitrogen and oxygen from the air during the combustion process. This reaction is promoted both by the high temperature of the converter (see Section 25.14, Part 3) and by the lean fuel-air mixture (high ratio of air to fuel) employed for more complete combustion. Some effort has been directed toward finding a suitable catalyst for the conversion of nitric oxide to harmless substances, such as elemental nitrogen and carbon dioxide. Reduction of the oxide with carbon monoxide is one possibility.

$$2NO + 2CO \longrightarrow N_2 + 2CO_2$$

Some success has been achieved recently in developing an engine which uses a stratified fuel concept to reduce undesirable emissions. The design of the engine is based on the principle that good performance requires a fuel-rich mixture only in the region of the initial spark. Each cylinder is divided into two sections, a small auxiliary combustion chamber of high temperature near the spark source and a larger main combustion chamber of lower temperature comprising the remainder of the cylinder. A fuel intake valve is provided in each chamber. The auxiliary chamber is connected to the main chamber by a small opening called a "torch nozzle" through which the initial flame can spread. The spark plug is located in the auxiliary chamber.

To operate the engine a fuel mixture with a relatively high ratio of fuel to air is introduced into the auxiliary chamber. When the mixture is ignited, even at the high temperature, very little nitric oxide is produced because of the relatively low proportion of air as a source of nitrogen and oxygen. At about the same time, an extremely lean fuel mixture that is compatible with more complete combustion is introduced into the main chamber. The lean mixture is ignited by the flame spreading into it through the torch nozzle, and the resulting explosion drives the piston. The lower temperature in the main chamber offsets the effect of the large volume of air, thereby keeping the production of nitric oxide to a low level. In addition, the more complete combustion of the leaner mixture in the main chamber reduces the amounts of carbon monoxide and unburned hydrocarbon pollutants produced.

Evaporation of gasoline from automobile gasoline tanks during filling and during use contributes surprising amounts of hydrocarbon vapors to the air. It is estimated that in one large U.S. city alone, 16 tons of hydrocarbons are introduced into the air each day by evaporation during the filling of automobile tanks at service stations. By combustion, automobile engines in the same city put over

1,700 tons of hydrocarbons and approximately 650 tons of nitrogen oxides into the air each day.

Considerable effort is being expended to study alternatives to the use of the internal combustion engine, as well as modifications of existing internal combustion engines. Alternatives include electric automobiles with rechargable batteries, steam automobiles (or ones using gases other than steam), and even wind-up automobiles in which a large spring mechanism would be wound periodically by electrical or mechanical means. It should be noted, however, that recharging of batteries or even periodic winding in turn require electrical or other forms of power and so necessitate additional generation facilities, with the subsequent increase in pollution. A change to avoid one type of pollution often introduces another type. The complete solution to the pollution dilemma involves solving exceedingly complex and intertwined problems.

Limited supplies of crude oil, low-sulfur coal, and natural gas have placed the United States, and indeed the whole world, in an energy crisis. There are, however, vast coal deposits still untouched, and a possible partial solution is gasification of coal at the mine to produce clean-burning pipeline gas. But this would introduce several technological problems, including those of cleaning the emissions in the gasification process itself at the mine.

To many persons, nuclear energy is the answer to our energy crisis. The use of nuclear energy is now increasing steadily but with it come problems of radioactive emissions (Chapter 30) and of thermal discharge, which raises the temperature of lakes and streams (Chapter 12) and of air. Thermal pollution (sometimes called "thermal loading") by both nuclear and conventional power plants is a serious problem. With nuclear power plants there is also present the possibility of an accidental explosion. Safety measures, while quite well worked out, must be handled very carefully to prevent such an explosion. Some scientists believe that thermonuclear fusion is the best hope for a clean power source. Thermonuclear fusion has a much less potential hazard from accidental explosion than fission and also a much reduced radioactive emission problem. A fission reaction, however, with its inherent problems, is often used to provide the large initial amounts of energy to initiate the fusion reaction. Nuclear fission and nuclear fusion are discussed in Chapter 30 (Sections 30.13–30.17).

The use of solar energy (energy from the sun) is in a very rudimentary state of technology but solar energy holds high promise for solving power needs in the long-term future (see Sections 22.18 and 22.19). Meanwhile, we must solve, at least on a short-term basis, the environmental problems of combustion of conventional fossil fuels.

The United States government, on April 30, 1971, announced strict air quality standards, originally specified to go into effect by July 1, 1975, but then in part postponed to a later date. These standards establish limits for the amounts of sulfur oxides, particulate matter, carbon monoxide, photochemical oxidants, nitrogen oxides, and hydrocarbons that can safely exist in the air. Full enforcement of the limits will undoubtedly cause considerable changes such as closing some sections of large cities to automobile traffic at certain hours, a greater use of public transportation and car pools, and very significant changes in the fuels used by electric generating plants and other industries.

In summary, air pollution involves the loss of natural resources, degrades the environment, and destroys health. Methods presently known to control pollution cost money, time, and effort. Development of new technology necessary for really adequate control of pollution carries a high price. It appears, nevertheless, that we really have no choice but to pay the necessary price in dollars and inconvenience.

Another equally important part of the pollution problem, that of water pollution, was discussed in Section 12.4.

The Noble Gases

24.6   Discovery of the Noble Gases

The discovery of the noble gases makes an interesting story. Helium was first discovered in the atmosphere which surrounds the sun. In 1868 the French astronomer Pierre-Jules-César Janssen went to India to study a total eclipse of the sun using a spectroscope. He observed among other bright lines one new yellow line, which caused the English chemist Professor Edward Frankland and the English astronomer Sir Norman Lockyer to conclude that the sun contained an undiscovered element. They called this new element **helium,** from the Greek word *helios*, meaning "the sun." All attempts to find this element on the earth were unsuccessful until 1895, when the Scottish chemist and physicist Sir William Ramsay showed that the gas driven off by heating the uranium mineral cleveite has a spectrum identical with that of helium.

In 1894, the British chemist Lord Rayleigh observed that a liter of nitrogen which he had prepared by removing the oxygen, carbon dioxide, and water vapor from air weighed 1.2572 grams, while a liter of nitrogen prepared from ammonia weighed only 1.2506 grams under the same conditions. This discrepancy caused Rayleigh to suspect the presence of a previously undiscovered element in the atmosphere. By passing nitrogen obtained from the air over red-hot magnesium ($3Mg + N_2 \longrightarrow Mg_3N_2$), he found a small amount of residual gas which he could not cause to combine with any other element. Lord Rayleigh and Sir William found that the residual gas showed spectral lines never before observed. In 1894, they announced the discovery of the first noble gas, which they called **argon,** meaning "the lazy one." It was later found that this residual gas was a mixture of the noble gases helium, neon, argon, krypton, and xenon.

In 1898 Sir William Ramsay and his assistant Professor Morris W. Travers isolated **neon** (meaning "the new element") by the fractional distillation of impure liquid oxygen. Shortly thereafter they showed the less volatile fractions from liquid air to contain two other new elements, **krypton** ("the hidden element") and **xenon** ("the stranger").

In 1900, the Prussian physicist Professor Friedrich E. Dorn discovered that one of the disintegration products of radium is a gas, similar in chemical properties to the noble gases. At first this gas was called radium emanation, but later its name was changed to **radon.**

24.7   Production of the Noble Gases

Helium is produced from certain natural gases, which sometimes contain as much as 2 to 5 per cent of the element by weight. Professor Hamilton P. Cady and D. F. McFarland of the University of Kansas first discovered helium in natural gas in 1905 when asked to analyze a sample of natural gas that would not ignite from a new well in southern Kansas. They found that the gas contained, by weight, 1.84 per cent helium, and so what seemed at first to be a financial fiasco turned out to be the beginning of a profitable business in the sale of helium. To isolate the helium the condensable components of the gas are liquefied in a liquid-air machine, leaving helium in the gaseous condition. The United States has most of the world's commercial supply of this element in its helium-bearing gas fields.

Argon, neon, krypton, and xenon are produced by the fractional distillation of liquid air. Radon gas is collected from radium salts.

24.8   Physical Properties of the Noble Gases

The principal properties of the noble gases are given in Table 24-4. The boiling points and melting points of the noble gases are extremely low in comparison to those of other substances of comparable atomic or molecular weights. The reason for this is that no strong chemical bonds hold the atoms together in the liquid and solid states. Only weak van der Waals forces exist between atoms in noble gases. Van der Waals forces are effective only when molecular motion is very slight, as it is at very low temperatures, so as to permit close approach of neighboring atoms.

TABLE 24-4   Properties of the Noble Gases

Property	*Helium*	*Neon*	*Argon*	*Krypton*	*Xenon*	*Radon*
Atomic number	2	10	18	36	54	86
Electrons in outer shell	2	8	8	8	8	8
Atomic weight	4.0	20.18	39.95	83.8	131.3	222
Atomic radius, Å	0.93	1.12	1.54	1.69	1.90	2.2
Melting point, °C	−269.7*	−248.6	−189.3	−157.2	−111.9	−71
Boiling point, °C	−268.9	−246.1	−186	−153.4	−108.1	−62
Critical temperature, °C	−267.9	−228.7	−122.3	−63.8	−16.6	105
Critical pressure, atm	2.26	26.9	48.3	54.3	57.6	62.4

24.9   Chemical Properties of the Noble Gases

The elements of Group 0 were formerly called the "inert gases," because until 1962 it was thought that they were chemically inert. When it was discovered that they really are not inert, the term **noble gases** was adopted. Besides being called noble gases, they are sometimes referred to as "rare gases" or "aerogens." The stability of the noble gas type of structure has been mentioned frequently in the preceding chapters. The two electrons of helium represent a completed shell, while a complement of eight electrons in the outer shell of each of the other noble gases gives them a stable configuration. Many elements form stable ions which have noble gas structures. These include the elements of high electron affinities (the halogens) which immediately precede the noble gases in the Periodic Table, and the elements of low ionization potentials (the alkali and alkaline earth metals) which immediately follow them.

Almost from the time the noble gases were first discovered, chemists have tried to force them into chemical combination by using unusual conditions. Early efforts were successful to only a limited extent. To date, helium has been forced to combine with certain other elements by using the large energies obtainable in electrical discharge tubes. Examples of species thus formed are He_2^+ and such combinations with hydrogen as HeH^+ and HeH_2^+. The molecular orbital structure of He_2^+ was discussed in Chapter 5 (Section 5.4). These combinations are not stable and have only momentary existence. On the other hand, comparatively stable helides of certain metals have been reported. These include $HgHe_{10}$, Pt_3He, and $FeHe$. "Compounds" of this type may be alloys in which the small helium atoms occupy the holes in the crystal lattice of the metal (Section 31.5).

A strong dipole, such as the water molecule, may polarize a noble gas atom so that it acts as a dipole itself and thereby attracts the original strong dipole. The larger the noble gas atom, the greater the susceptibility toward dipole induction. Thus, hydrates of argon, krypton, xenon, and radon have been prepared. These compounds have formulas such as $Ar \cdot xH_2O$, $Kr \cdot xH_2O$, and $Xe \cdot xH_2O$, where x is 1 to 6 but equals 6 only with noble gases of largest radii. These compounds are crystalline solids, which form when water and the noble gases are brought together at low temperatures and high pressures.

The fact that the noble gases have higher ionization potentials than those of any of the other elements in their respective horizontal rows of the Periodic Table (see Fig. 7–1 and Table 7–7) has encouraged the widely held concept of their inertness. However, if the ionization potentials of the larger noble gases are compared with the ionization potentials of other elements, it is apparent that these values are of the order of magnitude of those of certain other elements and indeed are even smaller than the ionization potentials for some elements. In June, 1962, Neil Bartlett, then in Canada, reported the preparation of a yellow compound of xenon to which he ascribed the formula $Xe[PtF_6]$. This discovery led a group of chemists at the Argonne National Laboratory to think that perhaps xenon might be oxidized by a highly electronegative element such as fluorine, and in September, 1962, they reported the preparation of **xenon tetrafluoride,** XeF_4. *This was the first report of a stable compound of a noble gas with another single element.* The compound was made by the surprisingly simple procedure of mixing xenon gas and fluorine gas at 400° C for one hour and cooling to −78°. The material has the form of colorless crystals (Fig. 24–5), which melt at about 90° and which are stable at room temperature, showing no evidence of reaction or decomposition after prolonged storage in glass vessels. Shortly after reporting the preparation of xenon tetrafluoride, the same group of chemists reported the preparation of

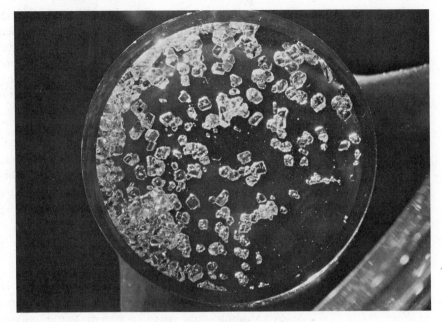

FIGURE 24-5
Crystals of xenon tetrafluoride, XeF_4, the first stable compound reported for a noble gas with another single element. The crystals shown are about four times their actual sizes.
Photograph courtesy of Argonne National Laboratory

TABLE 24-5   Formulas for Some Compounds of Three Noble Gases

Xenon			*Krypton*	*Radon*
$XePtCl_6$	XeO_3	Na_2XeF_8	KrF_2	RnF_x (where x
$XeCl_2$	XeO_4	K_2XeF_8	$KrF_2 \cdot 2SbF_5$	is not known)
XeF_2	$XeF_2 \cdot 2SbF_5$	Rb_2XeF_8	KrF_4?	
XeF_4	$XeF_2 \cdot 2TaF_5$	Cs_2XeF_8		
XeF_6	$XeF_6 \cdot 2SbF_5$	$(NO)_2{}^+[XeF_8]^{2-}$		
XeF_8?	$XeF_6 \cdot AsF_5$	$Na_4XeO_6 \cdot xH_2O$		
$XeOF_2$	$XeF_6 \cdot BF_3$	(where x = 6 and 8)		
$XeOF_4$	$RbXeF_7$	$Ba_2XeO_6 \cdot 1.5H_2O$		
XeO_2F_2	$CsXeF_7$	$K_n{}^+[XeO_3F]_n{}^-$		

XeF_2, $XeOF_4$, and XeO_3. Since that time, quite a number of additional compounds have been prepared. Table 24–5 lists the formulas for some of the noble gas compounds prepared to date.

The preparation of compounds of the noble gases not only opened the door to an exciting new area of research, but it also serves to point out that the ideas of science are subject to change. We should always keep our minds open to new possibilities and not blindly accept preconceived or entrenched ideas.

24.10   Stability of Noble Gas Compounds

The xenon fluorides are formed in exothermic reactions (loss of energy), reflected in negative values for heats of formation (Table 24–6). The energy release is sufficient to represent a substantial bond, as shown by the average bond energies listed in Table 24-6. By way of comparison, the energies of strong ionic and covalent bonds tend to be in the range of 40–120 kcal/mole. Many weak bonds, such as the hydrogen bond, have bond energies of less than 10 kcal/mole.

TABLE 24-6   Heats of Formation and Average Bond Energies for Three Noble Gas Compounds

Compound	Heat of Formation for Gaseous State (kcal/mole)	Average Bond Energy (kcal/mole)
XeF_2	−25.9	31.3
XeF_4	−51.5	31.2
XeF_6	−71.2	30.2

The oxyfluorides of xenon, in general, are formed by endothermic reactions, have positive heats of formation, and are quite unstable. $XeOF_4$ is an exception, with a heat of formation of −9 kcal/mole; it is the only oxyfluoride of a noble gas that is relatively easy to prepare and characterize.

Fluorides of krypton and radon have positive heats of formation. As would be expected, therefore, they are more difficult to make and less stable than the corresponding xenon fluorides. The heat of formation of KrF_2, as reported for the first time in 1967, is +14.4 kcal/mole.

24.11 Chemical Properties of Noble Gas Compounds

Xenon difluoride is quite soluble in water, going into solution as undissociated molecules which soon decompose to produce xenon, hydrogen fluoride, and oxygen.

$$2XeF_2 + 2H_2O \rightleftharpoons 2Xe + 4HF + O_2$$

The hydrolysis of xenon tetrafluoride is more complicated, producing $XeOF_4$ if the solutions contain the exact amount of water to react, but producing hydrated XeO_3 if an excess of water or acid is present.

$$6XeF_4 + 8H_2O \rightleftharpoons 2XeOF_4 + 4Xe + 16HF + 3O_2$$
$$6XeF_4 + 12H_2O \rightleftharpoons 2XeO_3 + 4Xe + 24HF + 3O_2$$

Dry, solid XeO_3 is an extremely sensitive explosive which must be handled with care.

In the reaction of XeF_6 with water, the +6 oxidation state for all of the xenon is retained.

$$XeF_6 + 3H_2O \longrightarrow XeO_3 + 6HF$$

Both XeF_6 and XeO_3 react with sodium hydroxide to produce Na_4XeO_6.

$$2XeF_6 + 4Na^+ + 16OH^- \longrightarrow Na_4XeO_6 + Xe + O_2 + 12F^- + 8H_2O$$
$$2XeO_3 + 4Na^+ + 4OH^- \longrightarrow Na_4XeO_6 + Xe + O_2 + 2H_2O$$

Standard reduction potentials indicate that oxidation of Xe is more difficult than oxidation of hydrogen. Hence, Xe is below hydrogen in the activity series. The oxidation reactions take place more readily in basic solution than in acidic solution. The following equations show the oxidation half-reactions and the potentials corresponding to the oxidation.

In acid solution:

$$Xe + 3H_2O \rightleftharpoons XeO_3 + 6H^+ + 6e^- \quad (E^\circ = -1.8 \text{ v})$$
$$XeO_3 + 3H_2O \rightleftharpoons H_4XeO_6 + 2H^+ + 2e^- \quad (E^\circ = -3.0 \text{ v})$$

In basic solution:

$$Xe + 7OH^- \rightleftharpoons HXeO_4^- + 3H_2O + 6e^- \quad (E^\circ = -0.9 \text{ v})$$
$$HXeO_4^- + 4OH^- \rightleftharpoons HXeO_6^{3-} + 2H_2O + 2e^- \quad (E^\circ = -0.9 \text{ v})$$

24.12 Structures of Noble Gas Compounds

The structures of XeF_2 (linear), XeF_4 (square planar), and $XeOF_4$ (slightly distorted square base pyramidal) are shown in Chapter 21 (Fig. 21–2) and are like those illustrated for polyhalides (Fig. 21–4). The structure of XeO_3 (Fig. 24–6a) is interpreted as being a trigonal pyramid with the four atoms at the corners. The XeO_4 molecule has a tetrahedral structure (Fig. 24–6b). The shape of the XeF_6 molecule has been difficult to establish experimentally. Evidence indicates that the molecule has a somewhat distorted octahedral structure.

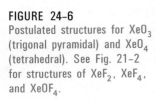

FIGURE 24–6
Postulated structures for XeO_3 (trigonal pyramidal) and XeO_4 (tetrahedral). See Fig. 21–2 for structures of XeF_2, XeF_4, and $XeOF_4$.

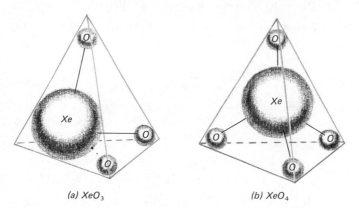

(a) XeO_3 (b) XeO_4

24.13 Uses of the Noble Gases

■ **1. Helium.** This gas is used for filling observation balloons and other lighter-than-air craft, particularly by the U.S. Navy. The fact that helium does not burn makes it safer to use than hydrogen. Although it is twice as dense as hydrogen, helium has a lifting power that is 92.6 per cent of that of hydrogen. This results from the fact that the lifting power of a gas with a density less than that of air is determined by the difference between its weight and that of an equal volume of air.

$$\begin{array}{ll}
1.293 \text{ g/liter (air)} & 1.293 \text{ g/liter (air)} \\
\underline{0.090 \text{ g/liter (hydrogen)}} & \underline{0.179 \text{ g/liter (helium)}} \\
1.203 \text{ g/liter (lifting power)} & 1.114 \text{ g/liter (lifting power)}
\end{array}$$

$$\frac{1.114 \text{ g/liter (helium)}}{1.203 \text{ g/liter (hydrogen)}} \times 100 = 92.6 \text{ per cent}$$

Because helium is less soluble than nitrogen in the blood stream, mixtures of it with oxygen are used by divers working under high pressures. The use of helium in place of nitrogen reduces the danger of a pathological condition called "bends," which is caused by the formation of bubbles of nitrogen in the blood. This results from the decrease in the solubility of nitrogen at lower pressures (true for any gas). When underwater divers breathe air under pressure for any length of time, nitrogen is forced into the blood. At the surface, the lower air pressure allows nitrogen to leave the blood too fast, causing the bubbles. The three Russian cosmonauts who lost their lives during return to earth on June 30, 1971, after a record-breaking 24 days in space, are reported to have died suddenly from a combination of a "bends-like" condition and an acute lack of oxygen, when their space cabin developed a leak and rapid depressurization occurred.

Mixtures of helium and oxygen are also beneficial in the treatment of certain respiratory diseases such as asthma. The lightness and rapid diffusion of helium decrease the muscular effort involved in breathing.

Helium is used as an inert atmosphere for the melting and welding of easily oxidizable metals and for many chemical processes which are sensitive to air. Liquid helium is of extremely great value in attaining low temperatures for cryogenic research.

■ **2. Neon.** Neon is used in the familiar neon lamps and signs. When an electric current is passed through a tube containing neon under very low pressure, a brilliant orange-red glow is emitted. The color of the light given off by a neon tube may be changed by mixing argon or mercury vapor with the neon and by utilizing tubes made of glasses of special color. Neon lamps cost less to operate than ordinary electric lamps and their light penetrates fog better.

■ **3. Argon.** This gas is used in gas-filled electric light bulbs, where its lower heat conductivity and chemical inertness make it preferable to nitrogen for inhibiting volatilization of the tungsten filament and prolonging the life of the bulb. Fluorescent tubes commonly contain a mixture of argon and mercury vapor. Many Geiger-counter tubes are also filled with argon.

■ **4. Krypton and Xenon.** A krypton-xenon photographic flash tube has been developed for taking high speed photographic exposures. An electric discharge through the tube gives a very intense light which lasts only 1/50,000 of a second.

■ **5. Radon.** This gas is collected in hospitals from disintegrating radium supplies and sealed in small tubes; the gas is then used in the radiotherapy of malignant growths.

QUESTIONS

1. What physical principles are employed in the liquefaction of air?
2. What evidence can you cite to show that air is a mixture rather than a compound?
3. Describe the construction of Dewar flasks.
4. Name some polluting components of the air in the community in which you live.
5. Discuss the uses of the noble gases in terms of their properties.
6. (a) By referring to some standard reference book, such as the *Handbook of Chemistry and Physics* (Chemical Rubber Co.) or *Lange's Handbook of Chemistry* (McGraw-Hill Book Co.), find the ionization potentials for the following species and arrange them in order of increasing tendency to lose an electron (i.e., to be oxidized): molecular hydrogen, molecular nitrogen, molecular oxygen, molecular fluorine, helium, argon, krypton, and xenon.
   (b) As pointed out in this chapter, the noble gases were thought for many years to be almost completely inert to chemical reaction. In recent years, this has proved to be incorrect. Discuss this in the light of the ionization potentials you looked up in connection with part (a).

PROBLEMS

1. Calculate the number of liters of helium at standard conditions needed to provide a lifting power of one ton. The density of air is 1.293 g/liter, that of helium is 0.179 g/liter, and that of hydrogen is 0.090 g/liter. How many liters of hydrogen would be needed to provide the same lifting power?
   Ans. 8.144×10^5 liters; 7.541×10^5 liters
2. Calculate the lifting power of 10,000 liters of helium at standard conditions. What would be the lifting power of this volume of hydrogen? (The densities of air, helium and hydrogen are given in Problem 1.)
   Ans. 11.14 kg; 12.03 kg
3. How many liters of oxygen (S.T.P.) will be produced by the action of 27.0 g of XeF_4 on an excess of H_2O?   *Ans. 1.46 liters*
4. Calculate the mass of XeF_2 which can be prepared from 3.00 liters (S.T.P.) of Xe and 10.0 liters (S.T.P.) of gaseous fluorine.   *Ans. 22.7 g*

REFERENCES

"Medieval Uses of Air," L. White, Jr., *Sci. American*, Aug., 1970; p. 92.

"How Have Sea Water and Air Got Their Present Compositions?" L. G. Sillén, *Chem. in Britain*, **3**, 291 (1967).

"The Global Circulation of Atmospheric Pollutants," R. E. Newell, *Sci. American*, Jan., 1971; p. 32.

"Pollution Problems, Resource, Policy, and the Scientist," A. W. Eipper, *Science*, **169**, 11 (1970).

"A Congregation of Vapors," *Facts and Figures*, Publication No. 393, September 1970, League of Women Voters of the United States, 1730 M Street, N.W., Washington, D.C. 20036.

"Planning Against Air Pollution," J. T. Middleton, *Amer. Scientist*, **59,** 188 (1971).

"Lead Pollution—A Growing Hazard to Public Health," D. Bryce-Smith, *Chem. in Britain*, **7,** 54 (1971).

"Lead in the Environment," A. L. Mills, *Chem. in Britain*, **7,** 160 (1971).

"Lead Poisoning," J. J. Chisolm, Jr., *Sci. American*, Feb. 1971; p. 15.

"The Energy Resources of the Earth," M. K. Hubbert, *Sci. American*, Sept., 1971; p. 60.

"Governments and the Environment," (The Stockholm Conference), *Chem. in Britain*, **8,** 236 (1972).

"Environmental Impact: Controlling the Overall Level," W. E. Westman and R. M. Gifford, *Science*, **181,** 819 (1973).

"Enforcing the Clean Air Act of 1970," N. de Nevers, *Sci. American*, June, 1973; p. 14.

"Analysis of an Important Air Pollutant: Peroxyacetyl Nitrate," E. R. Stephens and M. A. Price, *J. Chem. Educ.*, **50,** 351 (1973).

"Pollution of the Air," H. J. M. Bowen, *Educ. in Chem.*, **11,** 157 (1974).

"The Automobile and Air Pollution: A Chemical Review of the Problem," T. R. Wildeman, *J. Chem. Educ.*, **51,** 290 (1974).

"Cathedral Chemistry—Conserving the Stained Glass," R. G. Newton, *Chem. in Britain*, **10,** 89 (1974).

"Atmospheric Effects of Pollutants," P. V. Hobbs, H. Harrison, and E. Robinson, *Science*, **183,** 909 (1974).

"Coal and the Present Energy Situation," E. F. Osborn, *Science*, **183,** 477 (1974).

"The Gasification of Coal," H. Perry, *Sci. American*, Mar., 1974; p. 19.

"Energy and Life-Style," A. Mazur and E. Ross, *Science*, **186,** 607 (1974).

"Solar and Geothermal Energy: New Competition for the Atom," L. J. Carter, *Science*, **186,** 811 (1974).

"Energy Policy in the U.S.," D. J. Rose, *Sci. American*, Jan., 1974; p. 20.

"Stratospheric Ozone Depletion and Solar Ultraviolet Radiation on Earth," P. Cutchis, *Science*, **184,** 13 (1974).

"Stratospheric Pollution: Multiple Threats to Earth's Ozone," A. L. Hammond and T. H. Maugh II, *Science*, **186,** 335 (1974).

"Stratospheric Ozone Destruction by Aircraft-Induced Nitrogen Oxides," F. N. Alyea, D. M. Cunnold, and R. G. Prinn, *Science*, **188,** 117 (1975).

"Long Term Baseline Atmospheric Monitoring," M. A. Goldman, *J. Chem. Educ.*, **52,** 365, (1975).

"The Fuel Consumption of Automobiles," J. R. Pierce, *Sci. American*, Jan., 1975; p. 34.

"Energy Conservation: Better Living through Thermodynamics," W. D. Metz, *Science*, **188,** 820 (1975).

"The Correct Sizes of the Noble Gas Atoms," J. E. Huheey, *J. Chem. Educ.*, **45,** 791 (1968).

"Loyola Chemist Prepares and Isolates XeO_2F_2," (Staff report on work of J. L. Huston), *Chem. & Eng. News*, Sept. 18, 1967; p. 58.

"Discovery of the Noble Gases and Foundations of the Theory of Atomic Structure," J. E. Frey, *J. Chem. Educ.*, **43,** 371 (1966).

"'Compounds' (?) of the Noble Gases Prior to 1962," C. L. Chernick, *J. Chem. Educ.*, **41,** 185 (1964).

"The Chemistry of the Noble Gases," J. Selig, J. G. Malm, and H. H. Claassen, *Sci. American*, May, 1964; p. 66.

The Noble Gases, H. H. Claassen, D. C. Heath and Co., Boston, 1966 (Paperback).

"Introduction to the Chemistry of Compounds of the Noble Gases," H. H. Hyman and C. L. Chernick, *Inorganic Syntheses*, **8,** 249 (1966).

"Bonding in Xenon Hexafluoride," J. J. Kaufman, *J. Chem. Educ.*, **41,** 183 (1964).

"Heavy Hydrogen and Light Helium," J. C. Bevington, *Educ. in Chemistry*, **3,** 196 (1966).

"The Noble Gases and the Periodic Table," J. H. Wolfenden, *J. Chem. Educ.*, **46,** 569 (1969).

"'Second Sound' in Helium," B. Bertman and D. J. Sandiford, *Sci. American*, May, 1970; p. 92.

"Helium Conservation Program: Casting It to the Winds," W. D. Metz, *Science*, **183,** 59 (1974).

"Noble Gas Chemistry," J. H. Holloway, *Educ. in Chem.*, **10,** 140 (1973).

"A Decade of Xenon Chemistry," G. J. Moody, *J. Chem. Educ.*, **51,** 628 (1974).

Nitrogen and Its Compounds

25

The elements of Periodic Group VA make up the **nitrogen family.** Nitrogen is the first member of the family, which also includes phosphorus, arsenic, antimony, and bismuth. Nitrogen and phosphorus are among our most important and useful nonmetals. There is a regular gradation with increasing atomic weight in the nitrogen family, from the characteristics of a true nonmetal (nitrogen) to those of an almost true metal (bismuth). As has been noted for fluorine and oxygen, the lightest members of their families, some of the properties of nitrogen are anomalous, while more expected trends are found for the other members of the family (discussed in later chapters of this book).

The chemistry of nitrogen is sufficiently important and interesting to warrant its study apart from the other members of the family.

Nitrogen

25.1 History and Occurrence of Nitrogen

Nitrogen was first recognized as an element by the Scotch botanist Rutherford in 1772. He demonstrated that this gas does not support either life or combustion. Nitrogen composes 78 per cent of the atmosphere by volume and 75 per cent by weight. It has been estimated that there are more than 20 million tons of nitrogen over every square mile of the earth's surface. Natural gas also contains some free nitrogen. The most important mineral sources of nitrogen in compound form are the **saltpeter** (KNO_3) deposits in India and other countries of the Far East and **Chile saltpeter** ($NaNO_3$) deposits in South America. Nitrogen is an essential element of the proteins of all plants and animals. An important source of nitrogen compounds is the coal that is used in the production of coke and illuminating gas.

25.2 Preparation of Nitrogen

On an industrial scale, nitrogen is obtained by the fractional distillation of liquid air. Nitrogen prepared in this way usually contains small amounts of oxygen and the noble gases, particularly argon.

Pure nitrogen is prepared by heating a solution of ammonium nitrite.

$$NH_4NO_2(s) \longrightarrow 2H_2O(l) + N_2(g) + 75,300 \text{ cal} \qquad (\Delta H^{\circ}_{298} = -75,300 \text{ cal})$$

Actually, a mixture of sodium nitrite and ammonium chloride is used to furnish the ions of ammonium nitrite, because ammonium nitrite is so unstable that it cannot be stored. The equation for the reaction may be written

$$NH_4^+ + (Cl^-) + (Na^+) + NO_2^- \longrightarrow (Na^+) + (Cl^-) + 2H_2O + N_2 \uparrow$$

Nitrogen may also be prepared by the oxidation of ammonia. One such method involves passing ammonia gas over red-hot copper oxide.

$$2NH_3(g) + 3CuO(s) \longrightarrow 3H_2O(l) + N_2(g) + 3Cu(s) + 38,500 \text{ cal}$$
$$(\Delta H^{\circ}_{298} = -38,500 \text{ cal})$$

A convenient laboratory method of obtaining atmospheric nitrogen (nitrogen plus the noble gases) is to burn phosphorus in air which is confined over water. The P_4O_{10} which is formed dissolves in the water, and the residual gas is mainly nitrogen.

25.3 Properties of Nitrogen

■ 1. **Physical Properties.** Under ordinary conditions nitrogen is a colorless, odorless, and tasteless gas. It boils at $-195.8°$ under one atmosphere of pressure and freezes at $-210.0°$. Its density under standard conditions is 1.2506 g/liter; it is slightly less dense than air, for air contains the heavier molecules of oxygen as well as those of nitrogen. Under standard conditions 100 ml of water dissolves 2.4 ml of nitrogen.

The molecular orbital energy diagram for N_2 was discussed in Section 5.9 and given in Fig. 5–11. All the six electrons that come from the $2p$ atomic orbitals are in bonding molecular orbitals, accounting for the high stability of the molecule.

The triple bond between the atoms is very strong. An evidence of this is the fact that 226,000 calories of energy are absorbed when one mole of the element is decomposed into its atoms. The nitrogen molecule is the most stable diatomic molecule known.

■ 2. **Chemical Properties.** The high stability of the nitrogen molecule partially explains the lack of chemical reactivity of nitrogen under ordinary conditions. However, when heated with certain active metals, nitrogen forms **ionic nitrides,**

TABLE 25-1   Oxidation States of Nitrogen in Some of its Compounds

Oxidation State	Formula	Compound Name
−3	NH_3	Ammonia
−3	NH_4^+	Ammonium ion
−2	N_2H_4 (H_2N-NH_2)	Hydrazine
−1	NH_2OH	Hydroxylamine
0	N_2	Nitrogen
+1	N_2O	Nitrous oxide, or dinitrogen (mon)oxide
+2	NO	Nitric oxide, or nitrogen oxide
+3	N_2O_3	Nitrogen trioxide, or dinitrogen trioxide
+3	HNO_2	Nitrous acid
+4	NO_2	Nitrogen dioxide
+4	N_2O_4	Nitrogen tetraoxide, or dinitrogen tetraoxide
+5	N_2O_5	Nitrogen pentaoxide, or dinitrogen pentaoxide
+5	HNO_3	Nitric acid

which contain the nitride ion, N^{3-}. Examples are Li_3N, Ca_3N_2, and Mg_3N_2. At moderately high temperatures nitrogen combines with hydrogen to form ammonia, NH_3, and at very high temperatures with oxygen to form nitric oxide, NO.

Generally, nitrogen compounds are formed from either ammonia, nitric acid, or one of the oxides of nitrogen, rather than directly from the element. Many nitrogen compounds are unstable and revert readily to the element; this is due to the high heat of formation of the triple bond of the nitrogen molecule.

Nitrogen is a highly electronegative element, its electronegativity being exceeded only by those of oxygen and fluorine (see Section 4.6 and Table 4–2). As seen in Table 25–1, nitrogen assumes every integral oxidation state from −3 to +5.

25.4   Uses of Elemental Nitrogen

Large quantities of elemental nitrogen are used in the various processes for "fixing atmospheric nitrogen," i.e., bringing about the combination of atmospheric nitrogen with other elements. These processes are discussed later. Several uses of nitrogen are based on its inactivity. Frequently, when a chemical process requires an inert atmosphere, nitrogen is used. Nitrogen was once widely used to fill electric light bulbs to inhibit vaporization of the filament, but now it has been displaced to a large extent by argon, which is even more inert. Canned foods such as coffee and hydrogenated vegetable oils (Crisco and Spry) retain their flavor and color much better if sealed with nitrogen instead of air. Sliced luncheon meats are often packed with nitrogen gas in sealed plastic containers to retard spoilage.

Ammonia

Ammonia, NH_3, is produced in nature when any nitrogen-containing organic material is decomposed in the absence of air. The decomposition may be brought about by heat or by the bacteria which produce decay. The odor of ammonia is prevalent in stables and is frequently detected in sewage where decay is taking place.

25.5   Preparation of Ammonia

■ **1. A Laboratory Method.**   Ammonia is usually prepared in the laboratory by heating an ammonium salt with sodium hydroxide or slaked lime. For example, the reaction of NH_4Cl with NaOH produces ammonia according to the equation

$$NH_4^+ + (Cl^-) + (Na^+) + OH^- \longrightarrow NH_3\uparrow + H_2O + (Na^+) + (Cl^-)$$

Ammonia is also formed by the hydrolysis of ionic nitrides. Examples are given by the equations.

$$Mg_3N_2 + 6H_2O \longrightarrow 3Mg(OH)_2 + 2NH_3\uparrow$$
$$Li_3N + 3H_2O \longrightarrow 3Li^+ + 3OH^- + NH_3\uparrow$$

■ **2. From Coal.**   Soft coal contains about one per cent combined nitrogen by weight. About one-fifth of this nitrogen is liberated as ammonia when the coal is heated in the absence of air (destructive distillation). The ammonia gas comes off mixed with a variety of substances. The mixture is dissolved in water, and the solution treated with lime to produce pure ammonia. Usually, the ammonia is then absorbed in either sulfuric or hydrochloric acid to form ammonium sulfate or ammonium chloride, respectively. One ton of coal yields, on the average, enough ammonia to produce 20 pounds of ammonium sulfate, which is widely used as a fertilizer.

■ **3. By the Haber Process.**   Ammonia is produced in large quantities by the direct union of its elements by the Haber process.

$$N_2(g) + 3H_2(g) \xrightleftharpoons{\text{Catalyst}} 2NH_3(g) + 22{,}040 \text{ cal} \qquad (\Delta H^\circ_{298} = -22{,}040 \text{ cal})$$

The equilibrium concentrations of ammonia at several temperatures and pressures are shown in Table 25–2. Note that the reaction is exothermic and reversible, so that the yield of ammonia becomes less and less as the temperature of the system is raised. However, at low temperatures the reaction is too slow to be of practical value even when a catalyst is present. Because four volumes of the reactants (1 of N_2 and 3 of H_2) give two volumes of the product ($2NH_3$), high pressure causes an increase in the yield of ammonia at any given temperature. It would seem, then, that the process should be carried out at the lowest temperature and highest pressure practicable. Of the many catalysts which have been tried, the most efficient is a mixture of iron oxide and potassium aluminate.

In actual practice a pressure of about 200 to 600 atmospheres and a temperature of 450 to 600° C are ordinarily used. Pressures as high as 1,000 atmospheres are sometimes employed, but the added cost of equipment designed to withstand the higher pressures may more than offset the advantages of higher yield.

The hydrogen and nitrogen used in the Haber process must be very pure to

TABLE 25-2 The Equilibrium Concentrations of Ammonia (Per Cent by Volume) at Several Temperatures and Pressures for the Reaction
$$N_2 + 3H_2 \rightleftharpoons 2NH_3$$

Pressure, in Atmospheres	Temperature, °C					
	200°	300°	400°	500°	600°	800°
1	15.3	2.18	0.44	0.129	0.05	0.012
100	80.6	52.1	25.1	10.4	4.47	1.15
200	85.8	62.8	36.3	17.6	8.25	2.24
1000	98.3	92.5	80.0	57.5	31.5	—

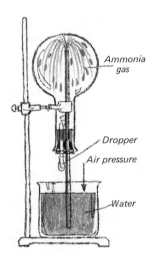

Ammonia gas

Dropper

Air pressure

Water

FIGURE 25-1
The "ammonia fountain." The fountain results when air pressure forces the water into the space left when the ammonia dissolves in water introduced from the dropper.

avoid "poisoning" the catalyst. Nitrogen for the process is obtained from liquid air, and much of the hydrogen is obtained from "water gas" (Section 9.2). The mixture of hydrogen and nitrogen is compressed and then passed over the catalyst. The ammonia is removed by liquefaction and the residual hydrogen and nitrogen are recycled through the catalyst chamber.

Fritz Haber, a German chemist, received the 1918 Nobel award in chemistry for his success in developing the direct synthesis of ammonia on a commercial scale.

25.6 Physical Properties of Ammonia

Ammonia is a colorless gas with a characteristic, irritating odor. It is a powerful heart stimulant, and people have died from inhaling it. It is lighter than air, one liter of it weighing 0.7710 g under standard conditions. Ammonia gas is readily liquefied by cooling and compressing. The liquid boils at $-33.43°$ and is colorless. The solid is white and crystalline; it melts at $-77.76°$. The heat of vaporization of liquid ammonia (328.3 calories per gram) is higher than that of any other liquid except water, so ammonia is valuable as a refrigerant. Ammonia is quite soluble in water, alcohol, and ether. One liter of water at 0° dissolves 1,185 liters of the gas. This fact makes possible the ammonia fountain (Fig. 25-1), triggered by squirting a few drops of water into the flask filled with ammonia gas. Ammonia can be completely expelled from an aqueous solution of ammonia by boiling.

In the ammonia molecule, the three hydrogen atoms are linked to the nitrogen atom by covalent bonds.

Ammonia

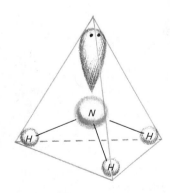

FIGURE 25-2
Structure of the ammonia molecule.

The ammonia molecule is tetrahedral in shape (see Section 6.3) with the unshared pair of electrons occupying one tetrahedral corner (Fig. 25-2).

The ammonia molecule is highly polar, a property which is partly responsible for its physical and chemical behavior.

25.7 Chemical Properties of Ammonia

Of the hydrides of the members of the nitrogen family, ammonia is the most highly associated in the liquid state, the most stable toward decomposition by heat, and the most strongly basic.

The unshared electron pair of the ammonia molecule can readily be shared by an atom or ion that has an incomplete outer shell, with the formation of a coordinate bond. The chemical properties of ammonia may be outlined as follows:

(1) Ammonia is a base since it readily accepts protons and acts as an electron pair donor. When ammonia dissolves in water some of the molecules react to form ammonium and hydroxide ions (see also Chapter 14, Fig. 14–5).

$$H\overset{ox}{\underset{ox}{\overset{H}{\underset{H}{N}}}}x + H\overset{xx}{\underset{ox}{\overset{}{O}}}x \rightleftharpoons \left[H\overset{ox}{\underset{ox}{\overset{H}{\underset{H}{N}}}}xH \right]^+ + \left[\overset{xx}{\underset{xx}{\overset{}{O}}}xH \right]^-$$

The reaction is reversible and the equilibrium lies far to the left so that ionization is only slight. **Aqueous ammonia** is, therefore, only weakly basic.

Ammonia accepts protons from acids and hydronium ions, as shown by the following equations.

$$NH_3 + HCl \rightleftharpoons NH_4^+ + Cl^-$$
$$NH_3 + H_3O^+ \rightleftharpoons NH_4^+ + H_2O$$

The ammonium ion (NH_4^+) is similar in properties to the potassium ion, and its salts are called **ammonium salts.**

Gaseous ammonia and gaseous hydrogen chloride react and form a cloud of very small crystals of ammonium chloride. This reaction is used in forming smoke screens; it also accounts largely for the white deposit so apparent on windowpanes and glassware in chemical laboratories.

Ammonia forms **ammines** by sharing electrons with certain metallic ions, just as water forms hydrates (see Chapter 32).

$$Cu^{2+} + 4\overset{x}{}NH_3 \longrightarrow [Cu(\overset{x}{}NH_3)_4]^{2+} \quad \text{[tetraamminecopper(II) ion]}$$
$$Ag^+ + 2\overset{x}{}NH_3 \longrightarrow [Ag(\overset{x}{}NH_3)_2]^+ \quad \text{(diamminesilver ion)}$$

(2) Certain compounds undergo **ammonolysis** in the presence of ammonia. This process is analogous to hydrolysis. The equation for the ammonolysis of mercury(II) chloride is

$$HgCl_2 + 2NH_3 \longrightarrow \underline{HgNH_2Cl} + NH_4^+ + Cl^-$$

$HgNH_2Cl$ is called **mercury(II) amido chloride.**

(3) Ammonia is decomposed into hydrogen and nitrogen at red heat; the action of electric sparks has the same effect. The reaction is reversible.

(4) Hot ammonia is an active reducing agent and reduces the oxides of certain metals, such as copper(II) oxide, to the free element.

$$2NH_3(g) + 3CuO(s) \longrightarrow N_2(g) + 3Cu(s) + 3H_2O(g) + 38,500 \text{ cal}$$

Ammonia also acts as a reducing agent when it burns in oxygen, forming water vapor and nitrogen.

$$4NH_3(g) + 3O_2(g) \longrightarrow 2N_2(g) + 6H_2O(g) + 302,700 \text{ cal}$$

A mixture of ammonia and air in contact with platinum at 700° forms nitric oxide.

$$4NH_3 + 5O_2 \longrightarrow 4NO + 6H_2O$$

This reaction is one step in the production of nitric acid from ammonia (Section 25.14).

(5) Ammonia forms nitrides when reacting with certain metals at high temperatures.

$$2NH_3 + 3Mg \longrightarrow Mg_3N_2 + 3H_2$$
$$2NH_3 + 6Li \longrightarrow 2Li_3N + 3H_2$$

25.8   Liquid Ammonia as a Solvent

Of the various solvents which have been studied, liquid ammonia most closely resembles water. It dissolves many electrolytes readily, and the solutions so formed are good conductors of electricity. The waterlike character of liquid ammonia is due to association of ammonia molecules through hydrogen bonding. However, because the NHN bond of associated ammonia molecules is weaker than the OHO bond of associated water, the properties which result from association are less pronounced with ammonia than with water. Ammonia is a poorer solvent for electrolytes than water, because its dielectric constant is lower than that of water. This means that ammonia does not insulate oppositely charged ions in solution as well as does water (Section 14.6).

The various types of reactions which occur in water also take place in liquid ammonia. Like water, pure ammonia is a very poor conductor of electricity, but it does ionize slightly according to the equation

$$2NH_3 \rightleftharpoons NH_4^+ + NH_2^-$$

Note the similarity of the ionization of ammonia to that of water.

$$2H_2O \rightleftharpoons H_3O^+ + OH^-$$

Reactions of ammonia and reactions in liquid ammonia as a solvent are usefully studied by means of the Concept of Solvent Systems (see Section 15.19).

Liquid ammonia possesses the ability to dissolve the more electropositive metals such as Na, K, Ba, and Ca. The free metals can be recovered by evaporation of the solvent. If the solutions are concentrated they exhibit a bronze color and conduct electricity like a pure metal. When dilute they are bright blue and conduct electricity like aqueous solutions of salts. It appears that when a metal such as sodium dissolves in liquid ammonia, it becomes a positive ion by losing its valence electron.

$$Na \rightleftharpoons Na^+ + e^-$$

The ammonia molecules then solvate the ion and electron reversibly

$$Na^+ + xNH_3 \rightleftharpoons Na(NH_3)_x^+$$
$$e^- + yNH_3 \rightleftharpoons e^-(NH_3)_y$$

The "ammoniated electron" is responsible for the blue color of dilute solutions and for the strong reducing properties shown by solutions of metals in liquid ammonia. Such solutions are abundant sources of free electrons, which may be taken up readily by reducible ions or compounds. In fact, liquid ammonia solutions of sodium are used widely in both organic and inorganic reduction reactions.

If metal-liquid ammonia solutions are allowed to stand, hydrogen is slowly liberated according to the equation (in the case of sodium)

$$\underset{\text{Sodium amide}}{2Na + 2NH_3 \longrightarrow 2Na^+ + 2NH_2^- + H_2}$$

However, if metallic iron is added as a catalyst, the reaction takes place quite rapidly.

The **amide ion,** NH_2^-, in the liquid ammonia system is analogous to the OH^- ion in the water system. The amide ion is a base in liquid ammonia, just as the hydroxide ion is a base in the water system. The ammonium ion, NH_4^+, is an acid in the liquid ammonia system, analogous to the hydronium ion, H_3O^+, in the water system. In aqueous solutions, neutralization involves the formation of the water (the solvent).

$$H_3O^+ + (Cl^-) + (Na^+) + OH^- \longrightarrow (Na^+) + (Cl^-) + 2H_2O$$
or $$H_3O^+ + OH^- \longrightarrow 2H_2O$$

Neutralization reactions in liquid ammonia result in the formation of ammonia (the solvent).

$$NH_4^+ + (Cl^-) + (Na^+) + NH_2^- \longrightarrow (Na^+) + (Cl^-) + 2NH_3$$
or $$NH_4^+ + NH_2^- \longrightarrow 2NH_3$$

Metathetical (double decomposition) reactions very often proceed in an entirely different direction in liquid ammonia than in water, because solubility relationships are not always the same in the two solvents. For example, silver chloride is soluble in ammonia but not in water, whereas barium chloride is insoluble in liquid ammonia but soluble in water. The reaction between silver nitrate and barium chloride in water is

$$2Ag^+ + (2NO_3^-) + (Ba^{2+}) + 2Cl^- \longrightarrow \underline{2AgCl} + (Ba^{2+}) + (2NO_3^-)$$

The reaction in liquid ammonia is

$$(2Ag^+) + (2NO_3^-) + Ba^{2+} + 2Cl^- \longrightarrow \underline{BaCl_2} + (2Ag^+) + (2NO_3^-)$$

Ammonia combines with the proton with the formation of a stronger bond than does water, and hence many ammonium salts are quite stable. On the other hand, most of the corresponding hydronium compounds are unstable and exist only in solution, or at low temperature. The fact that hydronium ions give up protons to ammonia molecules shows that the bond between the proton and water is weaker than the bond between the proton and ammonia. The equation for the reaction is

$$NH_3 + H_3O^+ \rightleftharpoons NH_4^+ + H_2O$$

25.9   Uses of Ammonia

About three fourths of the ammonia produced in the United States is used in fertilizers, either as the compound itself or as ammonium salts such as the sulfate and nitrate. The practice of the direct application of anhydrous ammonia and of solutions of ammonium compounds to the soil is rapidly increasing. Large quantities of ammonia are used in the production of nitric acid, urea, and other nitrogen compounds. Ammonia is the most common refrigerant used in the production of ice and for the maintenance of low temperatures in refrigerating plants. "Household ammonia" is an aqueous solution of ammonia. It is used to remove the carbonate hardness from hard water and thus save soap in cleansing and washing.

Other Compounds of Nitrogen and Hydrogen

25.10   Hydrazine

When ammonia is oxidized by sodium hypochlorite in the presence of sodium hydroxide and either gelatin or glue, **hydrazine** (N_2H_4) is produced. The reactions involved may be described by the equations

$$NH_3 + OCl^- \longrightarrow NH_2Cl + OH^-$$
$$NH_2Cl + NH_3 + OH^- \longrightarrow N_2H_4 + Cl^- + H_2O$$

A large excess of ammonia is used, and the reactants are thoroughly mixed at low temperatures. Chloramine, NH_2Cl, is an intermediate in the process.

Electronic formulas show that hydrazine may be regarded as the nitrogen analogue of hydrogen peroxide.

Hydrazine   Hydrogen peroxide

Anhydrous hydrazine is thermally stable but very reactive toward many reagents. It burns in air and reacts vigorously with the halogens. Aqueous hydrazine is weakly alkaline and forms two series of salts.

$$N_2H_4 + H_2O \rightleftharpoons N_2H_5^+ + OH^-$$
$$N_2H_4 + HCl \rightleftharpoons N_2H_5^+ + Cl^-$$
$$N_2H_4 + 2HCl \rightleftharpoons N_2H_6^{2-} + 2Cl^-$$

Anhydrous hydrazine is readily oxidized to free nitrogen and water by hydrogen peroxide.

$$N_2H_4(l) + 2H_2O_2(l) \longrightarrow N_2(g) + 4H_2O(g) + 153{,}500 \text{ cal}$$
$$(\Delta H^\circ_{298} = -153{,}500 \text{ cal})$$

Because the products of the reaction between these two liquids are both gaseous at the temperature of the reaction and the reaction is highly exothermic, there is a tremendous increase in volume of the products. Hydrazine and hydrogen peroxide have, therefore, been used as the propellant in rockets (see Section 25.28).

25.11 Hydrazoic Acid

When nitrous acid is reduced by hydrazine, **hydrazoic acid** (HN$_3$), a colorless liquid, is formed.

$$N_2H_4(l) + HNO_2(g) \longrightarrow HN_3(l) + 2H_2O(l) + 66{,}600 \text{ cal}$$
$$(\Delta H^\circ_{298} = -66{,}600 \text{ cal})$$

Hydrazoic acid detonates violently when subjected to shock. It is a weak acid and reacts with both oxidizing and reducing agents. Its salts are called **azides;** like the acid, they are unstable. Lead azide, Pb(N$_3$)$_2$, is used as a detonator for military explosives. Structures of hydrazoic acid are shown in Fig. 25–3. The three nitrogen atoms of the azide ion, N$_3^-$, lie in a straight line.

FIGURE 25-3
Structures of the hydrazoic acid molecule, HN$_3$.

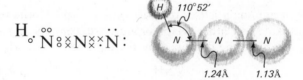

25.12 Hydroxylamine

Hydroxylamine, NH$_2$OH, may be considered to be a derivative of ammonia in which an OH group has replaced one hydrogen atom of ammonia.

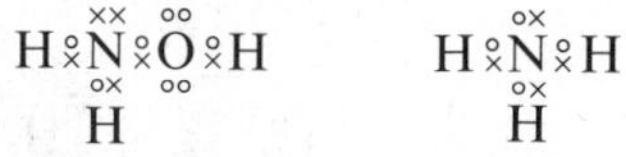

Hydroxylamine   Ammonia

As a solvent, hydroxylamine exhibits waterlike character. The pure substance is an unstable white solid that undergoes thermal decomposition at about 15° to ammonia, water, and a mixture of nitrogen and nitrous oxide. The decomposition can be explosive at elevated temperatures. Aqueous solutions of hydroxylamine are more stable. They are weakly basic, their basic strength being somewhat less than that of solutions of ammonia. Hydroxylamine is an active reducing agent. It is usually prepared and handled as a **hydroxylammonium salt,** such as the chloride, NH$_3$OH$^+$Cl$^-$.

Oxyacids of Nitrogen

25.13   Nitric Acid

Nitric acid, HNO_3, was known to the alchemists of the eighth century as *aqua fortis* (meaning "strong water"). It was prepared from KNO_3 and was used in the separation of gold and silver; it will dissolve the silver from alloys rich in silver, leaving the gold. Traces of nitric acid occur in the atmosphere after thunderstorms, and its salts are widely distributed in nature. Chile saltpeter, $NaNO_3$, is found in tremendous deposits (2 by 200 miles, and to a thickness of 5 feet) in the desert region near the boundary of Chile and Peru. Bengal saltpeter (KNO_3) is found in India and other countries of the Far East.

25.14   Preparation of Nitric Acid

■ **1. From Nitrates.** Nitric acid was formerly produced commercially by heating a mixture of sodium nitrate and concentrated sulfuric acid.

$$NaNO_3 + H_2SO_4 \rightleftharpoons NaHSO_4 + HNO_3\uparrow$$

At ordinary temperatures the reaction is reversible, and equilibrium is soon established. Because nitric acid boils at 86° while sulfuric acid boils at 338° ($NaNO_3$ and $NaHSO_4$ are nonvolatile), the nitric acid is readily removed from the reaction mixture by gentle heating. Consequently, the equilibrium is shifted to the right and the reaction goes to completion.

■ **2. From Ammonia.** The **Ostwald process** for producing nitric acid consists of the oxidation of ammonia to nitric oxide, NO, the oxidation of nitric oxide to nitrogen dioxide, NO_2, and finally the conversion of nitrogen dioxide to nitric acid. Ammonia is mixed with 10 times its volume of air. The mixture is first heated to 600°–700° and then brought into contact with platinum-rhodium alloy gauze (90% Pt, 10% Rh), which acts as a catalyst. The reaction which results is exothermic and raises the temperature of the system to about 1000°.

$$4NH_3(g) + 5O_2(g) \longrightarrow 4NO(g) + 6H_2O(g) + 216{,}420 \text{ cal}$$
$$(\Delta H^\circ_{298} = -216{,}420 \text{ cal})$$

Additional air is admitted to the reaction chamber to cool the mixture and oxidize the nitric oxide to nitrogen dioxide

$$2NO(g) + O_2(g) \longrightarrow 2NO_2(g) + 27{,}280 \text{ cal} \qquad (\Delta H^\circ_{298} = -27{,}280 \text{ cal})$$

The nitrogen dioxide, excess oxygen, and the unreacted nitrogen from the air are passed through a spray of water in an absorption tower where nitric acid and nitric oxide are formed.

$$3NO_2 + H_2O \longrightarrow 2H^+ + 2NO_3^- + NO$$

The nitric oxide is combined with more oxygen and returned to the absorption tower. The nitric acid is drawn off and concentrated. Most of our nitric acid is now produced by the Ostwald process.

■ **3. From Nitric Oxide Produced Directly from the Elements.** Nitric oxide was formerly produced by the direct union of the elements at the very high temperature of the electric arc. This is the basis of the Arc process for the production of nitric acid from the air.

$$N_2(g) + O_2(g) + 43{,}140 \text{ cal} \rightleftharpoons 2NO(g) \qquad (\Delta H^\circ_{298} = +43{,}140 \text{ cal})$$

The reaction is reversible and the union of the elements is endothermic; thus formation of NO is favored by high temperatures. However, even at 3,000° the equilibrium mixture contains only five per cent nitric oxide. The high cost of electricity and the low yield of nitric oxide caused abandonment of this process.

25.15　Physical Properties of Nitric Acid

Electronic and structural formulas for the nitric acid molecule are shown in Fig. 25–4.

FIGURE 25-4
The arrangement of atoms in the HNO_3 molecule, which in the vapor state is planar. N—O distances are (a) 1.41 Å and (b) 1.22 Å.

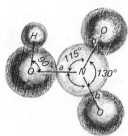

Pure nitric acid is a colorless liquid that boils at 86°, has a density of 1.50 g/ml, and freezes to a white solid at −42°. It fumes in moist air, forming a cloud of very small droplets of aqueous nitric acid. An aqueous solution containing 68 per cent of the acid has a constant boiling point of 120.5° at one atmosphere of pressure. This is the "concentrated" acid of commerce; it has a specific gravity of 1.4048 at 20°.

25.16　Chemical Properties of Nitric Acid

Pure nitric acid is unstable; it decomposes, upon being heated, to produce a mixture of nitrogen oxides, of which nitrogen dioxide is predominant.

$$4HNO_3 \xrightarrow{\triangle} 4NO_2 + 2H_2O + O_2$$

In aqueous solution nitric acid is stable and exhibits (a) the properties of a strong acid, and (b) the properties of a strong oxidizing agent. The following types of reactions illustrate the ability of nitric acid to act as an oxidizing agent.

■ **1. Action with Metals.** The products formed when nitric acid reacts with a metal depend upon (a) the concentration of the acid, (b) the activity of the metal, and (c) the temperature. A mixture of oxides and other reduction products is usually produced, but the less active metals, such as copper, silver, and lead, reduce dilute nitric acid primarily to nitric oxide. The following three equations show the oxide which is produced in largest quantity in the particular reactions.

The reaction of dilute nitric acid with copper is

$$3Cu + 8H^+ + 2NO_3^- \longrightarrow 3Cu^{2+} + 2NO\uparrow + 4H_2O$$

With concentrated nitric acid, nitrogen dioxide is formed predominantly.

$$Cu + 4H^+ + 2NO_3^- \longrightarrow Cu^{2+} + 2NO_2\uparrow + 2H_2O$$

The more active metals, such as zinc and iron, reduce dilute nitric acid to nitrous oxide, N_2O. With zinc, we have

$$4Zn + 10H^+ + 2NO_3^- \longrightarrow 4Zn^{2+} + N_2O\uparrow + 5H_2O$$

When the acid is very dilute, nitrogen or ammonium ions may be formed.

$$5Zn + 12H^+ + 2NO_3^- \longrightarrow 5Zn^{2+} + N_2\uparrow + 6H_2O$$
$$4Zn + 10H^+ + NO_3^- \longrightarrow 4Zn^{2+} + NH_4^+ + 3H_2O$$

In concentrated nitric acid, zinc, as well as copper, produces nitric oxide and nitrogen dioxide. The nitrate salts of these metals separate as crystals when the resulting solutions are evaporated. The action of nitric acid on a metal rarely produces hydrogen in more than small amounts.

■ **2. Action with Nonmetals.** Nonmetallic elements, such as sulfur, carbon, iodine, and phosphorus, are oxidized by concentrated nitric acid to their oxygen acids (or acid anhydrides), with the formation of NO_2.

$$S + 6HNO_3 \longrightarrow H_2SO_4 + 6NO_2\uparrow + 2H_2O$$
$$C + 4HNO_3 \longrightarrow CO_2 + 4NO_2\uparrow + 2H_2O$$

■ **3. Action with Inorganic Compounds.** Many compounds are oxidized by nitric acid. Hydrochloric acid is readily oxidized by concentrated nitric acid to chlorine and chlorine dioxide. A mixture of nitric and hydrochloric acids reacts vigorously with metals. This mixture is particularly useful in dissolving gold and platinum and other metals which lie below hydrogen in the Activity Series. A mixture of three parts of concentrated hydrochloric acid and one part concentrated nitric acid by volume is called **aqua regia** (meaning "royal water"). The action of aqua regia on gold may be represented, in a somewhat simplified form, by the equation

$$Au + 4HCl + 3HNO_3 \longrightarrow HAuCl_4 + 3NO_2\uparrow + 3H_2O$$

Dilute nitric acid oxidizes hydrogen sulfide to elemental sulfur and water.

$$3H_2S + 2H^+ + 2NO_3^- \longrightarrow 3S + 2NO\uparrow + 4H_2O$$

Sulfur dioxide is oxidized to sulfuric acid by nitric acid.

$$3SO_2 + 2NO_3^- + 2H_2O \longrightarrow 4H^+ + 3SO_4^{2-} + 2NO\uparrow$$

■ **4. Action with Organic Compounds.** A mixture of concentrated nitric acid and concentrated sulfuric acid acts upon glycerin with the formation of **nitroglycerin.**

$$\underset{\text{Glycerin}}{C_3H_5(OH)_3} + 3HNO_3 \longrightarrow \underset{\text{Nitroglycerin}}{C_3H_5(ONO_2)_3} + 3H_2O$$

The sulfuric acid, by combining with the water which is formed, prevents the dilution of the nitric acid. Nitroglycerin is a liquid which may be transformed quickly into gaseous products, mostly carbon dioxide, nitrogen, and steam. It is used in the manufacture of explosives such as dynamite, in which the liquid is taken up by a porous material such as infusorial earth or wood pulp. Nitroglycerin is also widely used as a drug for some heart ailments.

Guncotton is a cellulose nitrate made by nitrating cotton (cellulose) with nitric acid in the presence of sulfuric acid.

Trinitrotoluene (TNT), a valuable explosive, is produced by nitrating toluene; three NO_2 groups replace three hydrogen atoms of toluene in the reaction.

$$\underset{\text{Toluene}}{CH_3C_6H_5} + 3HNO_3 \longrightarrow \underset{\text{TNT}}{CH_3C_6H_2(NO_2)_3} + 3H_2O$$

It is interesting to note, as a side point, that nitric acid reacts with proteins, such as those in the skin, to give a yellow material by what is called the **xanthoprotein reaction.** You may have noticed in the laboratory that if you inadvertently get nitric acid on your fingers they turn a yellow color. During World War I, the explosive plants in the United States were not well ventilated, and the people who worked in them were exposed to fumes of nitric acid all day. Those persons turned a bright yellow from head to foot and consequently were referred to as "canaries." Fortunately, the bright yellow color did not show in the lamplight of that time, so their evening social life was not curtailed.

25.17   Uses of Nitric Acid

Nitric acid is used extensively in the laboratory and in the chemical industries as a strong acid and as an active oxidizing agent. It is used in the manufacture of explosives, dyes, plastics, and drugs. Salts of nitric acid (nitrates) are valuable as fertilizers. **Black gunpowder** is a mixture of potassium nitrate, sulfur, and charcoal. **Ammonal,** an explosive, is a mixture of ammonium nitrate and aluminum powder.

25.18   Nitrates

The nitrate ion has a trigonal planar structure (see Section 6.2), with each bond between nitrogen and oxygen being a single-bond, double-bond hybrid. The three resonance structures are shown.

Resonance structures of NO_3^-

The salts of nitric acid may be prepared by treating either metals or their oxides, hydroxides, or carbonates with nitric acid. All normal nitrates are soluble in water, and the oxynitrates, such as those of bismuth and mercury, are soluble in nitric acid. All nitrates decompose when heated. When sodium or potassium nitrate is heated, a nitrite is formed and oxygen is evolved.

$$2NaNO_3 \longrightarrow 2NaNO_2 + O_2$$

A nitrate of a heavy metal produces the oxide of the metal when heated. With copper(II) nitrate, the equation is

$$2Cu(NO_3)_2 \longrightarrow 2CuO + 4NO_2 + O_2$$

Ammonium nitrate produces nitrous oxide, N_2O, when heated (see Section 25.21).

25.19 Nitrous Acid and Nitrites

Nitrous acid, HNO_2, is obtained in solution (pale blue in color) when an acid, such as sulfuric, is added to a cold solution of a nitrite.

$$(2Na^+) + 2NO_2^- + 2H^+ + (SO_4^{2-}) \longrightarrow 2HNO_2 + (2Na^+) + (SO_4^{2-})$$

Nitrous acid is very unstable and is known only in solution. It changes slowly at room temperature (rapidly when heated) into nitric acid and nitric oxide.

$$3HNO_2 \longrightarrow H^+ + NO_3^- + 2NO\uparrow + H_2O$$

Nitrous acid is a weak acid. It acts as an active oxidizing agent toward strong reducing agents, but it is oxidized to nitric acid by active oxidizing agents. Figure 25–5 shows the structure of nitrous acid.

Sodium nitrite is the most important salt of nitrous acid. It is usually made by melting the nitrate with lead.

$$NaNO_3 + Pb \longrightarrow NaNO_2 + PbO$$

The nitrites are much more stable than the acid (the salts of all oxyacids are more stable than the acids themselves). Like the nitrates, they are soluble in water ($AgNO_2$ is sparingly soluble). Two resonance structures of the nitrite ion are known.

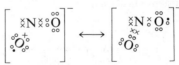

Resonance structures of NO_2^-

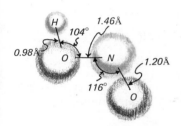

FIGURE 25–5
Structure of the nitrous acid molecule, HNO_2.

25.20 Hyponitrous Acid and Hyponitrites

Hyponitrous acid, $H_2N_2O_2$, is a white crystalline solid. The electronic formula and structure are shown in Fig. 25–6.

FIGURE 25–6
Probable electronic and geometric structures of the hyponitrous acid molecule, $H_2N_2O_2$.

Hyponitrous acid is very unstable and when it is heated, the acid decomposes explosively into nitrous oxide and water.

$$H_2N_2O_2 \longrightarrow H_2O + N_2O$$

The reverse reaction is not possible; that is, the acid cannot be formed by dissolving N_2O in water, even when high pressures are employed. The acid is weak and its salts are extensively hydrolyzed in aqueous solution. **Hyponitrites,** such as sodium hyponitrite, $Na_2N_2O_2$, may be prepared by the reduction of either nitrites or nitrates with sodium amalgam (an alloy of sodium and mercury). The free acid is prepared by allowing the silver salt, $Ag_2N_2O_2$, to react with hydrogen chloride in ether and evaporating the resulting ethereal solution.

$$Ag_2N_2O_2 + 2HCl \longrightarrow 2\underline{Ag}Cl + H_2N_2O_2$$

Oxides of Nitrogen

25.21   Nitrous Oxide

When ammonium nitrate is heated **nitrous oxide,** N_2O, is formed.

$$NH_4NO_3(s) \longrightarrow N_2O(g) + 2H_2O(g) + 8{,}612 \text{ cal} \qquad (\Delta H^{\circ}_{298} = -8{,}612 \text{ cal})$$

This is an oxidation-reduction reaction in which the nitrogen in the ammonium ion is oxidized by the nitrogen in the nitrate ion. **Caution!** If ammonium nitrate is heated too strongly an explosion will occur.

Nitrous oxide is a colorless gas possessing a mild, pleasing odor and a sweet taste. It is used as an anesthetic for minor operations, especially in dentistry, under the name "laughing gas." The structure and resonance forms of nitrous oxide are shown in Fig. 25–7.

FIGURE 25–7
Geometric structure and resonance forms of the nitrous oxide molecule, N_2O.

Nitrous oxide resembles oxygen in its behavior as an oxidizing agent with combustible substances. It decomposes when heated to form nitrogen and oxygen.

$$2N_2O(g) \xrightarrow{\triangle} 2N_2(g) + O_2(g) + 39{,}220 \text{ cal} \qquad (\Delta H^{\circ}_{298} = -39{,}000 \text{ cal})$$

Because one-third of the gas liberated is oxygen, nitrous oxide supports combustion better than air. A glowing splint will burst into flame when thrust into a bottle of this gas.

FIGURE 25-8

More "fixation" of nitrogen occurs by lightning than in any other way.

25.22 Nitric Oxide and Nitric Oxide Dimer

When copper is treated with dilute nitric acid (33 per cent, specific gravity 1.2), **nitric oxide**, NO, is formed as the chief reduction product of the nitric acid.

$$3Cu + 8H^+ + 2NO_3^- \longrightarrow 3Cu^{2+} + 2NO\uparrow + 4H_2O$$

Prepared in this way the nitric oxide is usually mixed with other oxides of nitrogen. Pure nitric oxide may be obtained by reducing nitric acid with iron(II) sulfate in dilute sulfuric acid solution.

$$3Fe^{2+} + NO_3^- + 4H^+ \longrightarrow 3Fe^{3+} + NO\uparrow + 2H_2O$$

Methods of producing nitric oxide on a commercial scale by the Ostwald process, in which ammonia is oxidized in the presence of platinum, and by the direct union of the elements at very high temperatures were discussed in Section 25.14. Nitric oxide is formed in the air by lightning during thunderstorms (Fig. 25-8). It has been estimated that as a result of this electrical phenomenon 40,000,000 tons of nitrogen are fixed annually by the series of reactions described in Section 25.26.

Nitric oxide is a colorless gas, which is only slightly soluble in water. It is the most stable thermally of the oxides of nitrogen. It is one of the air pollutants from the exhaust of the internal combustion engine, due to the reaction of nitrogen and oxygen from the air at the region of the spark in the combustion process (see Sections 24.3-24.5). Its structure is illustrated in Fig. 25-9.

Nitric oxide will act as an oxidizing agent and give up its oxygen or it will act as a reducing agent and take up more oxygen.

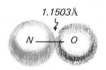

1.1503Å

FIGURE 25-9

Structure of the nitric oxide molecule, NO.

$$P_4 + 6NO \longrightarrow P_4O_6 + 3N_2$$
Oxidizing
agent

$$2NO + 2CO \longrightarrow N_2 + 2CO_2$$
Oxidizing
agent

$$2NO + O_2 \longrightarrow 2NO_2$$
Reducing
agent

Nitric oxide combines with the iron(II) ion (Fe^{2+}) to form a brown complex ion ($FeNO^{2+}$), a reaction which is used in analytical processes for the detection of nitrites and nitrates.

Two resonance forms are written for nitric oxide.

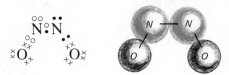

Note that these forms contain an unpaired electron which makes the substance paramagnetic. As is often, though not always, the case with molecules containing an unpaired electron, a combination of one molecule of nitric oxide with another to form a dimer can take place with some of the molecules, thereby pairing the unpaired electrons of each molecule that takes part in the dimerization. The dimer, N_2O_2, hence, is diamagnetic (Fig. 25–10).

FIGURE 25-10
Structures of the dinitrogen dioxide molecule, N_2O_2.

FIGURE 25-10
Structures of the dinitrogen dioxide molecule, N_2O_2.

25.23   Dinitrogen Trioxide

When the temperature of a mixture of one part of nitric oxide and one part of nitrogen dioxide is lowered to $-21°$ the gases become a blue liquid consisting of N_2O_3 molecules. Dinitrogen trioxide exists only in liquid and solid forms. When vaporized by heating, it forms the brown equilibrium mixture of NO and NO_2 ($N_2O_3 \rightleftharpoons NO + NO_2$).

Dinitrogen trioxide is the anhydride of nitrous acid. The structure of N_2O_3 is given in Fig. 25–11.

FIGURE 25-11
Structures of the dinitrogen trioxide molecule, N_2O_3.

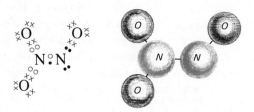

25.24   Nitrogen Dioxide and Dinitrogen Tetraoxide

Nitrogen dioxide is prepared in the laboratory (a) by heating the nitrate of a heavy metal:

$$2Pb(NO_3)_2 \xrightarrow{\triangle} 2PbO + 4NO_2 + O_2$$

or (b) by the action of concentrated nitric acid upon copper metal:

$$Cu + 4H^+ + 2NO_3^- \longrightarrow Cu^{2+} + 2NO_2 + 2H_2O$$

Nitrogen dioxide is prepared commercially by exposing nitric oxide to the air.

$$2NO + O_2 \longrightarrow 2NO_2$$

When nitrogen dioxide is dissolved in cold water, a mixture of nitric and nitrous acids is formed.

$$2NO_2 + H_2O \longrightarrow H^+ + NO_3^- + HNO_2$$

At 140° nitrogen dioxide has a deep brown color and a density corresponding to the formula NO_2. At low temperatures the color almost entirely disappears, and the density corresponds to the formula N_2O_4. At ordinary temperatures an equilibrium exists between the colored and the colorless compounds.

$$2NO_2(g) \rightleftharpoons N_2O_4(g) + 13{,}700 \text{ cal}$$
$$(\Delta H^\circ_{298} = -13{,}700 \text{ cal}; \ \Delta G^\circ_{298} = -1{,}140 \text{ cal})$$

Nitrogen dioxide contains an odd number of electrons and therefore an unpaired electron. The unpaired electron accounts for the fact that NO_2 is paramagnetic and highly colored. Two molecules of NO_2 share their unpaired electrons with each other when they combine to form N_2O_4; the tetraoxide is not paramagnetic. The resonance and geometric structures of NO_2 are shown in Fig. 25–12, along with the atomic arrangement for N_2O_4.

FIGURE 25-12

Structures of nitrogen dioxide and dinitrogen tetraoxide molecules.

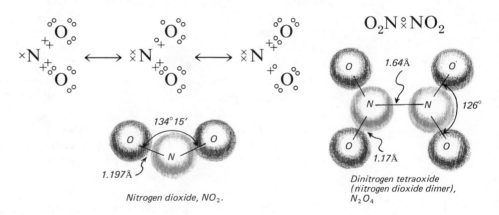

Nitrogen dioxide, NO_2.

Dinitrogen tetraoxide (nitrogen dioxide dimer), N_2O_4

25.25   Dinitrogen Pentaoxide

This compound is a white solid which is formed by the dehydrating action of phosphorus(V) oxide upon pure nitric acid.

$$P_4O_{10} + 4HNO_3 \longrightarrow 4HPO_3 + 2N_2O_5$$

Dinitrogen pentaoxide decomposes above 30°.

$$2N_2O_5 \longrightarrow 4NO_2 + O_2$$

It is the anhydride of nitric acid.

$$N_2O_5 + H_2O \longrightarrow 2HNO_3$$

Fig. 25–13 shows the electronic structure and the arrangement of atoms in N_2O_5 in the gaseous state. Crystals are made up of equal numbers of planar NO_2^+ ions (N—O bond length 1.15 Å) and NO_3^- ions (N—O bond length 1.24 Å).

FIGURE 25-13

The arrangement of atoms in the N_2O_5 molecule. The dark spheres represent nitrogen atoms, the colored spheres oxygen atoms.

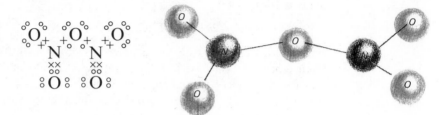

25.26   The Cycle of Nitrogen in Nature

Nitrogen is an essential constituent of all plants and animals. It is present principally in the proteins, which are complex organic materials containing carbon, hydrogen, and oxygen as well as nitrogen (Chapter 28, *Biochemistry*). Most plants obtain the nitrogen necessary for their growth through their roots as nitrogen compounds, mainly in the form of ammonium and nitrate salts. However, certain legumes, such as clover, alfalfa, peas, and beans, are able to obtain nitrogen indirectly from the air by means of nitrogen-fixing bacteria, which are found in the nodules on their roots. These bacteria convert atmospheric nitrogen into nitrites and nitrates, which are assimilated by the "host" plant in the form of proteins. Generally, the bacteria "fix" more nitrogen in the form of soluble compounds than is used by the plant itself. When legumes die, much of the combined nitrogen remains in the soil and is available for use by nonleguminous plants.

As mentioned in Section 25.22, tremendous quantities of nitrogen are fixed as nitric oxide by lightning. The nitric oxide is then oxidized to nitrogen dioxide by oxygen of the air, and the reaction of the dioxide with water forms nitrous and nitric acids (Section 25.14). These acids are carried by rain to the soil where they react with oxides and carbonates of metals to form nitrites and nitrates, respectively. Certain soil bacteria oxidize the nitrites to nitrates, a process called **nitrification.** Certain other kinds of bacteria change ammonia into nitrites. There are also denitrifying bacteria, which decompose nitrates and other nitrogen compounds, thereby returning free nitrogen to the air.

Animals obtain their nitrogenous compounds by feeding on plants and other animals. The decay of both plant and animal matter returns nitrogen to the soil in the form of nitrates, and either ammonia or free nitrogen is produced. Thus, we see that nitrogen passes through a cycle which is of fundamental importance to all plants and animals.

Rocket Propellants

25.27   Meanings of Terms

Rocket propellants contain an **oxidant** (oxidizing substance) and a **fuel** (reducing substance). Usually, the term "fuel" is used to mean the reducing component; occasionally, the term is alternatively used to indicate the propellant as a whole. If the oxidant and the fuel are two separate materials which are mixed during the burning process, the propellant is referred to as a **bipropellant,** or a **bifuel.** If the oxidant and fuel are in close association or are parts of the same molecule, the propellant is referred to as a **monopropellant,** or **monofuel.**

Propellants are classified in terms of **liquid propellants** and **solid propellants.** In rocket engines using liquid bipropellants, the oxidant and fuel are carried in separate tanks from which both are pumped and injected into the combustion chamber in fine streams. A prime problem in rocket engines is the high temperature of about 2,500–3,500° K and also the high rate of energy transfer from burning gases to the walls of the chamber. The flame temperature is high compared to the melting point of the thin-walled casing of the rocket. However, the problem exists only during the combustion process, which occurs during a very short period of time. Overheating the casing is also avoided by circulating the fuel through a honeycombed wall of the reaction chamber and around the throat of the exit nozzle. In engines using solid propellants, the combustion takes place in a cavity extending through the entire length of the charge, from the center out, and toward the thin inner wall of the casing. Ideally, the flame does not touch the casing until essentially all of the propellant has been consumed. Thus, the charge itself solves part of the insulation problem. The solid fuel engine in flight carries no mechanical pumps or valves which might stick or leak, and the propellant has less chance of getting into places not intended than a liquid would. As might be expected, however, the use of solid propellants introduces several problems (for example, uneven-burning).

25.28   Selection of Materials as Oxidants and Fuels

The commonly accepted criterion of propellant performance is **specific impulse.** Specific impulse is the number of pounds of thrust, or force, developed per second for each pound of propellant and is listed in seconds. Several typical propellants and the specific impulse of each are listed in Table 25–3.

A high energy fuel must have a high heat of combustion, low vapor pressure, good thermal stability, high density, and a high flame speed. In addition, it must have the physical properties necessary to insure safe handling during manufacture, shipping, and storage. The heat of combustion is the net energy resulting from breaking existing chemical bonds and making new ones to form different molecular arrangements.

The requirements for high energy fuels are satisfied only in molecules of low molecular weight. Hence, the choice of elements for fuel constituents is quite

TABLE 25-3   Some Typical Rocket Propellants

Oxidant	Fuel	Theoretical Specific Impulse (in Seconds)
LIQUID BIPROPELLANTS		
Hydrogen peroxide; Nitric acid (fuming); Dinitrogen tetraoxide	Alcohols, petroleum fuels, hydrazine hydrate	250–300
Oxygen (liq.)	Ethanol (92.5%)	287
	Kerosene	300
	Ammonia (liq.)	294
	Hydrazine	313
	Hydrogen (liq.)	391
Chlorine trifluoride	Kerosene	294
	Hydrazine	258
Fluorine (liq.)	Ammonia (liq.)	357
	Hydrazine	363
	Hydrogen (liq.)	410
LIQUID MONOPROPELLANTS		
	Hydrogen peroxide (90%)	147
	Ethylene oxide	180
	Hydrazine	196
	Propyl nitrate	173
	Nitromethane	220
SOLID PROPELLANTS		
Cordite (colloidally dispersed nitrocellulose in nitroglycerine)		190–235
Perchlorates	Organic polymers	180–245

limited and includes principally the elements up to about silicon in the Periodic Table. Most common rocket systems have used carbon, hydrogen, oxygen, and nitrogen as fuel constituents.

Liquid hydrogen liberates a large amount of energy for a given weight when oxidized and for this reason is an important fuel for rockets in the space program. Its low boiling point and low liquid density, however, constitute rather serious disadvantages for a high energy fuel. Materials used as fuels, therefore, are often compounds in which hydrogen is combined with another light element to yield a denser material of higher boiling point than hydrogen.

The high thermal stability of the nitrogen molecule is one of the reasons why nitrogen-containing substances are often used in fuels.

$$2N(g) = N_2(g) + 226 \text{ kcal} \qquad (\Delta H^{\circ}_{298} = -226 \text{ kcal})$$

When monopropellants are used they must be chosen carefully because of their sensitivity. Decomposition, once started through friction, electric discharge, or impact, is often self-sustaining and may change from steady burning into detonation. In general, the useful monopropellants have lower specific impulses than bifuels, because in bifuels it is possible to keep the two components physically separated until ready to mix for combustion. With monofuels, a delicate balance must be met in which the materials must have a sufficiently high burning rate to allow them to ignite without undue difficulty but not so high that they will be too easily detonated. One of the early monopropellants was hydrogen peroxide, which can also be used as the oxidant in bifuel systems. Its use as a monofuel depends upon catalytic decomposition.

$$2H_2O_2(l) = 2H_2O(l) + O_2(g) + 46.9 \text{ kcal} \qquad (\Delta H^\circ_{296} = -46.9 \text{ kcal})$$

Organic nitrates and other nitrogen compounds are also used as monofuels (Table 25–3).

The period of combustion for a rocket propellant is usually not more than two to three minutes and is often much shorter. During the brief period of burning, the rate at which power is generated is extremely high. The average rate of fuel consumption in flight may exceed a half-ton per second, which is equivalent, in the case of a propellant using oxygen as the oxidant and kerosene as the fuel, to six million B.T.U. per second, or about nine million horsepower.

Motors utilizing liquid propellants are very expensive because of the necessary equipment for pumping and mixing the liquids. Many rockets, such as the Atlas, have used liquid oxygen as the oxidant. The low boiling temperature of oxygen ($-183°$ C) is difficult to maintain in the rocket because of the undesirable weight of extra insulation.

Solid fuel boosters, by comparison, have the major advantages of ease in handling, low cost, and reliability, because of their relatively simple design. They are self-contained. They can be held for long periods of time on the launching pad without hazard. Solid fuel boosters can be reloaded and checked in a short time. There is no evaporation of fuel. Solid rocket propellants typically may contain a crystalline oxidant, such as ammonium nitrate or ammonium perchlorate, constituting about 80–90% of their weight, and a solid fuel containing a large proportion of combined hydrogen for the other 10–20% of the total weight. The mixture is cemented together in order to gain more mechanical strength and greater flexibility than what is obtained merely by compressing the materials together.

The Polaris is an example of a United States rocket using reliable solid fuel motors. Some of the recent rockets have had initial liquid stages with later stages consisting of clusters of solid fuel motors.

Nuclear power for rocket engines, a possibility for the not-too-distant future, will provide propellants which are several orders of magnitude more powerful than present propellants.

QUESTIONS

1. Relate the inactivity of molecular nitrogen, N_2, to its thermal stability.
2. How is "atmospheric nitrogen" prepared? pure nitrogen?
3. In what ways may nitrogen be "fixed"?

4. What is the oxidation state of nitrogen in each of the following: N_2, NH_4^+, $NaNO_2$, N_2H_4, NH_2OH, NO_2, N_2O_4, NH_4NO_3, N_2O, NCl_3, Li_3N, HN_3?

5. Write electronic structures for N_2, NH_3, NH_4^+, N_2H_4, HN_3, and NH_2OH.

6. What are the products of the chemical reaction occurring between an ionic nitride and water?

7. Explain the effects of temperature, pressure, and a catalyst upon the direct synthesis of ammonia.

8. Account for the solubility of electrolytes in liquid ammonia.

9. What compounds act as acids in liquid ammonia? as bases? What is the essential reaction involved in neutralizations in liquid ammonia?

10. What evidence is there that the bond between the proton and an ammonia molecule is stronger than that between the proton and water?

11. Draw a valence electron formula for the azide ion, N_3^-.

12. The salt hydroxylammonium chloride, $NH_2OH \cdot HCl$, can be formed by the reaction of hydrogen chloride and hydroxylamine. Would you expect the proton from the HCl to be found on the oxygen or on the nitrogen of the hydroxylamine?

13. Outline the chemistry of the production of nitric acid from ammonia.

14. List the possible reduction products of nitric acid. What factors determine which product will be formed when nitric acid is reduced?

15. Write equations for the reaction of concentrated HNO_3 with each of the following: Cu, S, C, $C_3H_5(OH)_3$, and SO_2.

16. Write electronic formulas for HNO_3, NO_3^-, HNO_2, $H_2N_2O_2$, N_2O, NO, NO_2, N_2O_3, N_2O_5, NCl_3, and NOCl.

17. Write equations for the preparation of nitrous acid starting with sodium nitrate.

18. What is the anhydride of nitrous acid? of nitric acid?

19. What would be the result of bubbling SO_2 through a strongly acid barium nitrate solution? Write equations for the reactions which occur.

20. Write equations for the preparation of each of the oxides of nitrogen.

21. In Section 25.24, the ΔH_{298}° and ΔG_{298}° values are given for the conversion, $2NO_2(g) \rightleftharpoons N_2O_4(g)$. The values are widely different. Discuss possible reasons for this difference, and discuss the significance of each of the two values as completely as you can. (You will probably wish to refer to Chapter 20 in answering this question.)

22. Describe the nitrogen cycle in nature.

23. Describe the fixation of nitrogen by the Arc process.

24. What are the necessary properties for a high energy rocket fuel?

25. Which is the stronger oxidizing agent, hydrazine or hydrogen peroxide? Give evidence for your answer.

26. Compare liquid rocket fuels and solid rocket fuels with respect to the advantages and disadvantages of each.

27. Given the following energy data and making use of other contributing factors, evaluate the various substances in the reactions as possible fuel constituents. Consider forward and reverse reactions in your evaluation.

$$2N = N_2 + 226 \text{ kcal}$$
$$2H_2O_2 = 2H_2O + O_2 + 46.9 \text{ kcal}$$
$$2C + O_2 = 2CO + 52.83 \text{ kcal}$$
$$2CO + O_2 = 2CO_2 + 139.3 \text{ kcal}$$
$$CO + H_2O(g) = CO_2 + H_2 + 9.84 \text{ kcal}$$
$$2H_2 + O_2 = 2H_2O + 136.6 \text{ kcal}$$
$$N_2 + 3H_2 = 2NH_3 + 22.0 \text{ kcal}$$
$$N_2H_4(l) + H_2O_2(l) = N_2(g) + 4H_2O(g) + 198.4 \text{ kcal}$$

PROBLEMS

1. What volume of nitrogen measured at 75.0° and 740 mm is formed by the decomposition of 20.0 g of ammonium nitrite? *Ans. 9.16 liters*

2. What weight of lithium nitride could be formed from the action of ammonia on 93.0 g of lithium? *Ans. 156 g*

3. What volume of ammonia measured at 1.20 atm and 25.0° is required to react with 800 ml of 0.510 M HCl solution? *Ans. 8.31 liters*

4. What is the maximum volume of ammonia which could be collected at 27° and 750 mm by the treatment of 2.00 g of ammonium chloride with 500 ml of 0.500 M sodium hydroxide? *Ans. 0.932 liter*

5. What weight of ammonia would be required to neutralize 350 ml of 0.750 M HNO₃ solution? *Ans. 4.47 g*

6. What weight of hydrazine can be obtained from 500 ml of ammonia, measured at 740 mm and 245° K, if a 65.0% yield is obtained? *Ans. 0.252 g*

7. What quantity of nitric acid could be prepared by the Ostwald process from 300 cubic feet of ammonia, measured at 4.00 atm and 250°, if a 93.0% yield is obtained based upon the original quantity of ammonia? *Ans. 46.4 kg*

8. What volume of nitrous oxide, measured at 25.0° and 730 mm, can be prepared from 190 g of ammonium nitrate? *Ans. 60.4 liters*

9. (a) Check by calculation, using values in Appendix J, at least five of the ΔH_{298}° values given for reactions in this chapter.
   (b) For each of the same reactions, using values in Appendix J, calculate ΔG_{298}°.
   (c) On the basis of your calculations in parts (a) and (b) above, state what you can as to whether or not each reaction should proceed spontaneously.

10. In Section 25.16, a mixture of three parts of concentrated hydrochloric acid and one part of concentrated nitric acid by volume (aqua regia) is described. The equation which follows that description, for the action of aqua regia on gold, indicates a ratio of four moles of hydrogen chloride to three moles of nitric acid. (a) Calculate the molar ratio of concentrated hydrochloric acid to concentrated nitric acid supplied by the aqua regia. (The specific gravities of concentrated hydrochloric acid and of concentrated nitric acid, referred to water as a standard, are 1.19 and 1.42, respectively. The percentages by weight for concentrated hydrochloric acid and concentrated nitric acid are 38% and 70%, respectively.) *Ans. 2.36 to 1*

(b) The calculation in (a) indicates that one of the acids, as supplied by aqua regia, will be in excess of that needed for the reaction with gold. Calculate which acid will be in excess and, using your calculation, explain your answer.

Ans. HCl

REFERENCES

"The Nitrogen Cycle," C. C. Delwiche, *Sci. American*, Sept., 1970; p. 136.

"Reflectors in Fishes," E. Denton, *Sci. American*, Jan., 1971; p. 64.

"Thermodynamics and Rocket Propulsion," F. H. Verhoek, *J. Chem. Educ.*, **46**, 140 (1969).

"A Computer Program for the Analysis of the N_2O_4 Dissociation Equilibrium," L. E. Erickson, *J. Chem. Educ.*, **46**, 383 (1969).

"The Physical Chemistry of Ammonia Manufacture and Nitrogenous Fertilizer Production," J. T. Gallagher and F. M. Tayler, *Educ. in Chemistry*, **4**, 30 (1967).

"Hydrazine," L. P. Lessing, *Sci. American*, July, 1953; p. 30.

"High Temperatures: Chemistry," F. Daniels, *Sci. American*, Sept., 1954; p. 109 (describes production of nitric oxide directly from air).

"Is Ammonia Like Water?" J. B. Gill, *J. Chem. Educ.*, **47**, 619 (1970).

"Chemistry in the Technology of Explosives and Propellants," I. Dunstan, *Chem. in Britain*, **7**, 62 (1971).

"Chemical Rocket Propellants," R. M. Lawrence and W. H. Bowman, *J. Chem. Educ.*, **48**, 335 (1971).

"Nitrogen Fixation: Research Efforts Intensify," J. L. Marx, *Science*, **185**, 132 (1974).

"Nitrogen Fixation," D. R. Safrany, *Sci. American*, Oct., 1974; p. 64.

"Nitrogen Fixation Research: A Key to World Food?" R. W. F. Hardy and U. D. Havelka, *Science*, **188**, 633 (1975).

"Marine Phosphorite Deposits and the Nitrogen Cycle," D. Z. Piper and L. A. Codispoti, *Science*, **188**, 15 (1975).

Phosphorus and Its Compounds

26

Phosphorus—The Element

Although phosphorus is in the same family as nitrogen and in many respects the chemistry of phosphorus closely resembles the chemistry of nitrogen, marked physical and chemical differences exist between the two elements and their compounds. Phosphorus is more metallic than nitrogen, chiefly because the phosphorus atom is larger and also less electronegative (2.1) than nitrogen (3.0). Nitrogen is quite inert except at elevated temperatures, whereas phosphorus (at least, the white modification) is very active even at low temperatures. The union of nitrogen with oxygen is an endothermic reaction, whereas phosphorus burns readily in air with the evolution of considerable heat. Nitric acid and nitrous acid are relatively strong oxidizing agents, but phosphoric acid and phosphorous acid are very weak oxidizing agents. Furthermore, phosphine, PH_3, is a much stronger reducing agent than ammonia, NH_3. On the other hand, both of these elements are nonmetals, the atoms of each have five electrons in their outer shells, they have the same oxidation states, and many of their compounds have similar formulas and structures.

Phosphorus is essential to both plants and animals. Bones, teeth, and nerve and muscle tissue contain combined phosphorus. The nucleoproteins (found in the nucleus of every cell) contain phosphorus. Phosphorus compounds are also important in the metabolism of sugar. Such foods as eggs, beans, peas, and milk furnish phosphorus for our body requirements. Plants obtain phosphorus from soluble phosphates in the soil. Soils which have become deficient in phosphorus may be enriched by the addition of soluble phosphates.

The principal commercial source of phosphorus is the so-called **phosphate rock.** It is mostly calcium phosphate, $Ca_3(PO_4)_2$, but always contains several other phosphates as well.

A statement is in order here regarding a nomenclature point, in connection with phosphorus and its compounds, that is often confusing to students. You may have noticed the difference in spelling, in the preceding introductory paragraphs, of the name of the element (phosphor*us*) and the distinguishing word for one of the acids (phosphor*ous* acid, H_3PO_3). In the latter case, the word *phosphorous* is made up of the stem of the element's name plus the *-ous* ending, which indicates a lower oxidation number for phosphorus ($+3$) than that in phosphoric acid, H_3PO_4 ($+5$). The name phosphor*ous* acid is therefore analogous to the names chlor*ous* acid, nitr*ous* acid, and sulfur*ous* acid. You will probably find it helpful to review the nomenclature discussion in Section 4.12.

26.1   Preparation of Phosphorus

Phosphorus is produced commercially by heating calcium phosphate with sand and coke in an electric furnace.

$$2Ca_3(PO_4)_2 + 6SiO_2 + 10C \longrightarrow 6CaSiO_3 + 10CO\uparrow + P_4\uparrow$$

The phosphorus distills off at the quite high furnace temperature and is condensed to the solid state, or burned to P_4O_{10}, from which other compounds of phosphorus may be manufactured. The calcium silicate, along with any calcium fluoride or ferrophosphorus, melts and collects at the bottom of the furnace where it is drawn off as a slag. The elemental phosphorus is shipped to industrial plants where it is to be made into phosphoric acid and phosphates. In this way, the cost of the freight on the oxygen and water used in the manufacture of phosphorus compounds is avoided. The profits of chemical industries often depend on such considerations.

26.2   Physical Properties of Phosphorus

Phosphorus exists in several allotropic modifications, of which only two are of general interest and importance; these are **white phosphorus** and **red phosphorus.**

As prepared by the method described in the previous section, phosphorus is a white, translucent, soft, waxlike solid. When exposed to light, white phosphorus slowly turns yellow, due to the formation of a superficial coating of red phosphorus. For this reason, white phosphorus is frequently called **yellow phosphorus.** White phosphorus is very soluble in carbon disulfide and less so in ether, chloroform, and other organic solvents. It is very nearly insoluble in water, 0.0033 g dissolving in one liter of ice water. White phosphorus melts at 44.2° and boils at 280°. Either as a solid, in solution, or as the vapor at just above the boiling point, white phosphorus exists as P_4 molecules. The four phosphorus atoms are arranged at the corners of a regular tetrahedron, and each atom is covalently bonded to the three other atoms of the molecule (Fig. 26-1). At very high temperatures phosphorus gas consists of P_2 molecules; these are assumed to have the valence electronic structure $\overset{x}{\underset{x}{:}}P\overset{x}{\underset{x}{\overset{\circ}{\underset{\circ}{:}}}}P\overset{\circ}{\underset{\circ}{:}}$, which is similar to that of the N_2 molecule.

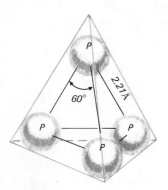

FIGURE 26-1

The white phosphorus molecule, P_4.

White phosphorus is very poisonous; a dose of about 0.15 g produces acute pains and convulsions, and may result in death. Necrosis, or death, of the bones of the jaw and nose results from continued breathing of small amounts of the fumes of phosphorus. In very small doses phosphorus stimulates the nervous system.

Red phosphorus is made by heating the white modification in a retort between 230° and 300°, with air excluded.

$$\text{P (white)} \rightleftharpoons \text{P (red)} + 4{,}400 \text{ cal}$$

This change is speeded up when catalyzed by a trace of iodine. Because the change from white to red phosphorus is exothermic, the heat of combustion of red phosphorus is less than that of the white form. Red phosphorus sublimes when heated, forming a vapor identical in molecular structure and properties to that from the white variety. The red modification is insoluble in the solvents which dissolve white phosphorus.

In red phosphorus the atoms are bonded together in an infinite polymer, the length of the polymer chain being limited only by the size of the individual piece under consideration. The exact structure of red phosphorus is not definitely known, but it is thought to be an infinite chain formed from the tetrahedral white phosphorus by the rupture of one P—P bond of each tetrahedron and a subsequent combination of the ruptured tetrahedra (Fig. 26–2). Half the phosphorus atoms (P and P″ in the figure, for example) have two bond angles of 100° and one of 60°. The other half (P′ and P‴, for example) have two bond angles of 60° and one of 100°.

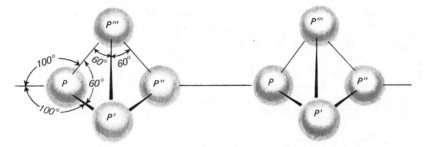

FIGURE 26–2

Probable structure of red phosphorus. Note that no bond exists between P and P″ in the first tetrahedron or between the two corresponding phosphorus atoms in the other tetrahedron; these bonds have been ruptured in the formation of red phosphorus from white phosphorus. The tetrahedra with ruptured bonds then join to form a chain structure.

26.3 Chemical Properties of Phosphorus

The most important chemical property of phosphorus is its chemical activity towards oxygen. Slow oxidation of white phosphorus at room temperature causes a rise in its temperature and spontaneous combustion results when the ignition temperature (35–45°) is reached. Because of the readiness with which white phosphorus ignites, it must be stored under water. Burns caused by phosphorus are very painful and slow to heal. *It should be handled only with forceps and put in water immediately after use.* When moist phosphorus is allowed to oxidize slowly in air, a faint light is emitted (phosphorescence) and ozone, phosphoric acid, and phosphorous acid are formed. When phosphorus is burned in either an excess of air or in oxygen, phosphorus(V) oxide, P_4O_{10}, is formed. In moist air, the cloud of solid phosphorus(V) oxide forms a fog of minute droplets of phosphoric acid.

Phosphorus combines with the halogens exothermically. When heated, it combines with sulfur and many of the metals. Concentrated nitric acid oxidizes phosphorus to orthophosphoric acid.

Red phosphorus is less active than the white form; it does not ignite in air until it is heated to about 250°. However, the products of the reactions of red phosphorus are the same as those of the white form.

26.4   Uses of Phosphorus

Formerly, white phosphorus was used in the manufacture of matches. Because it is poisonous and the workers who were exposed to the fumes suffered from necrosis of the bones, it has been replaced, in the heads of "strike anywhere" matches, by tetraphosphorus trisulfide, P_4S_3.

Large quantities of phosphorus are converted into acids and salts to be used in fertilizers, in baking powder, and in the chemical industries. Other uses are in the manufacture of special alloys such as ferrophosphorus and phosphorbronze. Considerable quantities of phosphorus are used in making fireworks, bombs, and rat poisons. Burning phosphorus has been used in the production of smoke screens during warfare.

Compounds of Phosphorus

26.5   Phosphine

Phosphorus forms a series of hydrides, the most important of which is **phosphine,** PH_3, a gaseous compound analogous to ammonia in formula and structure. Unlike ammonia, phosphine cannot be made by the direct union of the elements, but is prepared by heating white phosphorus in a concentrated solution of sodium hydroxide (Fig. 26–3).

$$(3Na^+) + 3OH^- + P_4 + 3H_2O \longrightarrow (3Na^+) + 3H_2PO_2^- + PH_3\uparrow$$
$$\text{Sodium hypophosphite}$$

Pure phosphine is not spontaneously combustible in air, but when produced by this method, it contains some diphosphorus tetrahydride, P_2H_4, which ignites on coming in contact with the air. On this account, the air in the flask is first removed

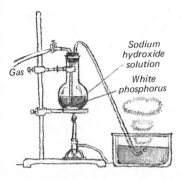

FIGURE 26-3

Apparatus used in the laboratory production of phosphine. The smoke rings consist of droplets of phosphoric acid.

by flushing the flask with nitrogen or illuminating gas. As the bubbles of phosphine and diphosphorus tetrahydride escape into the air, they ignite, forming a fog of phosphoric acid that rises in the form of smoke rings (Fig. 26-3).

$$PH_3(g) + 2O_2(g) \longrightarrow H_3PO_4(l) + 304,100 \text{ cal} \qquad (\Delta H^\circ_{298} = -304,100 \text{ cal})$$

Phosphine can also be prepared by the hydrolysis of phosphides of certain metals.

$$Ca_3P_2 + 6H_2O \longrightarrow 3Ca(OH)_2 + 2PH_3\uparrow$$

Phosphine is a colorless, very poisonous gas, which has an odor like that of decaying fish. It is easily decomposed by heat ($4PH_3 \longrightarrow P_4 + 6H_2$). Like ammonia, phosphine unites with the hydrogen halides forming the corresponding **phosphonium compounds,** PH_4Cl, PH_4Br, and PH_4I. Unlike the ammonium halides, however, the phosphonium compounds do not give the phosphonium ion, PH_4^+, in solution. Instead, phosphine escapes and the hydrogen halide remains in solution as the hydrohalic acid. Phosphine is only very slightly soluble in water and is a much weaker base than ammonia.

Phosphides of several of the metals may be prepared by reducing the corresponding phosphates with carbon.

$$Ca_3(PO_4)_2 + 8C \longrightarrow Ca_3P_2 + 8CO$$

26.6 Halides of Phosphorus

With one exception, phosphorus unites directly with all of the halogens, forming trihalides, PX_3, and pentahalides, PX_5; only the penta-iodide is not known. These halides are much more stable than the corresponding compounds of nitrogen.

It is interesting that nitrogen is the only member of the nitrogen family that cannot form pentahalides. This can be explained in terms of the availability of orbitals for bond formation. The orbitals in the valence shell of a nitrogen atom and the number of electrons in each is given by the subshell notation $2s^2 2p^1 2p^1 2p^1$. In forming NF_3, for example, each of the three $2p$ electrons is shared with a fluorine atom. The $2s$ orbital is not involved in bond formation in NF_3. However, nitrogen cannot form a pentafluoride, because five orbitals would be required and only four are available in the second principal electron shell. With phosphorus, on the other hand, the valence shell is the third principal electron shell, and $3d$ orbitals are presumably available for bond formation. Thus, the pentahalides of phosphorus are possible. In PF_5, for example, the valence electron distribution in the five bonding orbitals of phosphorus is $3s^1 3p^1 3p^1 3p^1 3d^1$. Each fluorine atom furnishes one electron to each of these five orbitals; the valence orbitals of phosphorus then contain two electrons each, $3s^2 3p^2 3p^2 3p^2 3d^2$.

The trichloride, PCl_3, and the pentachloride, PCl_5, are the most important halides of phosphorus. **Phosphorus trichloride** is prepared by passing dry chlorine over molten phosphorus ($P_4 + 6Cl_2 \longrightarrow 4PCl_3$). The trichloride is a colorless liquid which boils at 76° and freezes at −92°. The PCl_3 molecule is pyramidal, with the phosphorus atom at one corner of the pyramid (Fig. 26-4).

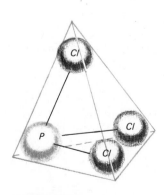

FIGURE 26-4

Structure of phosphorus trichloride, PCl_3.

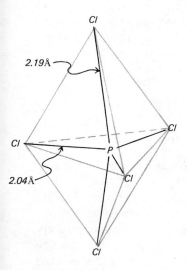

FIGURE 26-5

Structure of phosphorus pentachloride, PCl_5, in the gaseous
and liquid states.

Phosphorus trichloride is irreversibly hydrolyzed by water, with the formation of phosphorous acid and hydrogen chloride.

$$PCl_3 + 3H_2O \longrightarrow H_3PO_3 + 3HCl\uparrow$$

Because of hydrolysis, all of the halides of phosphorus fume in moist air, much like concentrated hydrochloric acid.

Phosphorus pentachloride is prepared by oxidizing the trichloride with excess chlorine ($PCl_3 + Cl_2 \longrightarrow PCl_5$). The pentachloride is a straw-colored solid, which sublimes when heated in air and decomposes reversibly into the trichloride and chlorine ($PCl_5 \rightleftharpoons PCl_3 + Cl_2$).

In both the gaseous and liquid states the PCl_5 molecule has the structure of a triangular bipyramid, with the phosphorus atom in the center and the chlorine atoms at each of the five corners (Fig. 26-5).

X-ray studies show solid phosphorus pentachloride to be an ionic compound, $[PCl_4]^+[PCl_6]^-$, in which the cation has a tetrahedral structure and the anion an octahedral one. The structure is the same when PCl_5 is in solution in solvents of high dielectric constant. Phosphorus pentachloride undergoes partial hydrolysis when treated with a limited amount of cold water, forming phosphorus(V) oxychloride, $POCl_3$, and hydrogen chloride.

$$PCl_5 + H_2O \longrightarrow POCl_3 + 2HCl\uparrow$$

When an excess of water is employed, the hydrolysis of the pentachloride is complete, and orthophosphoric acid is formed.

$$PCl_5 + 4H_2O \longrightarrow H_3PO_4 + 5HCl\uparrow$$

26.7   Phosphorus(V) Oxyhalides

FIGURE 26-6

Structure of POX_3.

Compounds of the type POX_3 are called **phosphorus(V) oxyhalides,** or **phosphoryl halides.** Their molecules have tetrahedral structures (Fig. 26-6). The chloride and bromide are readily obtained by the action of the pentahalide upon phosphorus(V) oxide.

$$P_4O_{10} + 6PX_5 \longrightarrow 10POX_3$$

Partial hydrolysis of a phosphorus pentahalide produces the corresponding oxyhalide.

$$PX_5 + H_2O \longrightarrow POX_3 + 2HX\uparrow$$

The iodide POI_3 apparently does not exist. The POX_3 compounds, especially $POCl_3$, are used in replacing OH groups in organic compounds with halogens. They hydrolyze to give orthophosphoric acid and HX.

$$POX_3 + 3H_2O \longrightarrow H_3PO_4 + 3HX\uparrow$$

26.8   Oxides of Phosphorus

Phosphorus forms several oxides, three of which are well defined. Their formulas are P_4O_6, P_8O_{16}, and P_4O_{10}. The oxidation states of phosphorus in these compounds correspond to those of nitrogen in N_2O_3, N_2O_4, and N_2O_5, respectively. The structures of P_4O_6 and P_4O_{10} are depicted in Fig. 26–7.

The combustion of phosphorus in air generally produces a mixture of P_4O_6 and P_4O_{10}. However, it is virtually impossible to get much of the P_4O_6 by burning phosphorus, even in a limited supply of air.

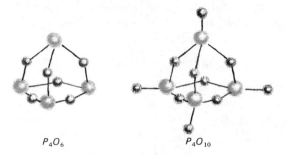

P_4O_6 P_4O_{10}

FIGURE 26–7
Structures of P_4O_6 and P_4O_{10}. The colored spheres represent phosphorus atoms and the dark spheres oxygen atoms.

■ **1. Phosphorus(III) Oxide, P_4O_6.** This compound is often formulated incorrectly as P_2O_3 and called **phosphorus trioxide,** but vapor density measurements have shown that the formula is actually P_4O_6. Phosphorus(III) oxide is a white crystalline solid which melts at 23.8° and boils at 175.3°. When the vapor of this oxide is heated to 448°, it decomposes, forming red phosphorus and P_8O_{16}. Phosphorus(III) oxide has a garliclike odor, and its vapor is very poisonous. It oxidizes slowly in air and takes fire when heated to 70°, forming P_4O_{10}. Phosphorus(III) oxide dissolves slowly in cold water to form phosphorous acid, H_3PO_3.

■ **2. Phosphorus(V) Oxide, P_4O_{10}.** This oxide of phosphorus is frequently incorrectly formulated as P_2O_5 and called **phosphorus pentaoxide.** It is a white flocculent powder which melts at 420°. Its enthalpy of formation is very high, and for this reason it is quite stable and is a very poor oxidizing agent. With a limited amount of water it forms metaphosphoric acid, $(HPO_3)_x$.

$$x\,P_4O_{10} + 2x\,H_2O \longrightarrow 4(HPO_3)_x$$

When more water is added, the metaphosphoric acid slowly changes to orthophosphoric acid, H_3PO_4. The net reaction is given by the equation

$$P_4O_{10}(s) + 6H_2O(l) \longrightarrow 4H_3PO_4(l) + 88{,}110 \text{ cal} \qquad (\Delta H^\circ_{298} = -88{,}110 \text{ cal})$$

When a piece of P_4O_{10} is dropped into water, it reacts with a hissing sound, and much heat is liberated. Because of its great affinity for water, phosphorus(V) oxide is used extensively for drying gases and removing water from many compounds.

Acids of Phosphorus and Their Salts

The important oxygen acids of phosphorus are orthophosphoric acid, H_3PO_4; pyrophosphoric (diphosphoric) acid, $H_4P_2O_7$; triphosphoric acid, $H_5P_3O_{10}$; metaphosphoric acid, $(HPO_3)_x$; phosphorous acid, H_3PO_3; and hypophosphorous acid, H_3PO_2.

26.9   Orthophosphoric Acid

The various phosphoric acids all have the same anhydride, P_4O_{10}. Each acid represents a different degree of hydration of this oxide but not a different oxidation state of phosphorus, which is $+5$ in every case. These acids are stable toward reducing agents.

One commercial method of preparing **orthophosphoric acid** (commonly called **phosphoric acid**) is the treatment of calcium rock (calcium phosphate) with concentrated sulfuric acid.

$$Ca_3(PO_4)_2 + 3H_2SO_4 \longrightarrow 2H_3PO_4 + 3CaSO_4$$

The products are diluted with water and the calcium sulfate removed by filtration. This method gives a dilute acid which is impure with calcium dihydrogen phosphate, $Ca(H_2PO_4)_2$, and certain fluorine compounds which are formed from the fluorapatite, $Ca_{10}F_2(PO_4)_6$, usually associated with the phosphate rock.

Pure orthophosphoric acid is manufactured by oxidizing phosphorus from the electric furnace (Section 26.1) to P_4O_{10} and dissolving the product in water.

The pure acid forms colorless, deliquescent crystals that melt at 42.4°. A common commercial form of the acid is an 82 per cent aqueous solution of H_3PO_4 and is known as **syrupy phosphoric acid.** Orthophosphoric acid is tetrahedral (Fig. 26-8).

FIGURE 26-8

The electronic formula and geometric structure of the orthophosphoric acid molecule.

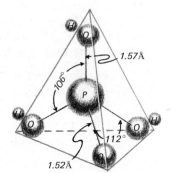

Large quantities of crude (impure) orthophosphoric acid are used in the manufacture of fertilizers. The acid is also used in medicine as an astringent, as an antipyretic, and as a stimulant.

26.10   Salts of Orthophosphoric Acid

Because orthophosphoric acid is a triprotic acid it forms three series of salts, corresponding to the three stages of ionization.

$$H_3PO_4 \rightleftharpoons H^+ + H_2PO_4^- \qquad \text{(primary ionization; } K_{e_1} = 7.5 \times 10^{-3})$$
$$H_2PO_4^- \rightleftharpoons H^+ + HPO_4^{2-} \qquad \text{(secondary ionization; } K_{e_2} = 6.2 \times 10^{-8})$$
$$HPO_4^{2-} \rightleftharpoons H^+ + PO_4^{3-} \qquad \text{(tertiary ionization; } K_{e_3} = 1 \times 10^{-12})$$

The formulas for the sodium salts of orthophosphoric acid and their names are

NaH_2PO_4 Primary sodium (ortho)phosphate, or sodium dihydrogen phosphate

Na_2HPO_4 Secondary sodium (ortho)phosphate, or disodium hydrogen phosphate

Na_3PO_4 Tertiary sodium (ortho)phosphate, or trisodium phosphate

Sodium dihydrogen phosphate forms aqueous solutions that are weakly acidic as a result of hydrolysis, as shown by

$$H_2PO_4^- + H_2O \Longrightarrow HPO_4^{2-} + H_3O^+$$

Disodium hydrogen phosphate solutions are alkaline, because the monohydrogen phosphate ion is a stronger proton acceptor than donor.

$$HPO_4^{2-} + H_2O \Longrightarrow H_2PO_4^- + OH^-$$

Trisodium phosphate solutions are strongly alkaline as a result of hydrolysis.

$$PO_4^{3-} + H_2O \Longrightarrow HPO_4^{2-} + OH^-$$

The hydrogen and ammonium phosphate salts are decomposed by heating, with volatile products such as water and ammonia being given off. Several examples are

$$x NaH_2PO_4 \longrightarrow (NaPO_3)_x + x H_2O$$
$$2 Na_2HPO_4 \longrightarrow Na_4P_2O_7 + H_2O$$
$$x NaNH_4HPO_4 \longrightarrow (NaPO_3)_x + x NH_3 + x H_2O$$
$$2 MgNH_4PO_4 \longrightarrow Mg_2P_2O_7 + 2NH_3 + H_2O$$

Sodium dihydrogen phosphate is used as a boiler cleansing compound (to prevent the formation of boiler scale) and in some baking powders; disodium hydrogen phosphate as a boiler cleansing compound and in the weighting of silk; and trisodium phosphate as a water softener, boiler cleansing compound, and detergent. **Calcium dihydrogen phosphate,** $Ca(H_2PO_4)_2$, is used as a fertilizer and as a constituent of baking powders. **Calcium monohydrogen phosphate,** $CaHPO_4$, is added to animal food for its mineral properties and is used as a polishing agent in toothpastes.

Phosphorus, in the form of soluble orthophosphates, is essential to plant growth. Tricalcium phosphate is too insoluble to furnish an adequate supply of phosphorus to plants. Calcium dihydrogen phosphate, however, is soluble in water and therefore suitable as a fertilizer. This compound is prepared commercially by treating tricalcium phosphate with sulfuric acid.

$$Ca_3(PO_4)_2 + 2H_2SO_4 + 4H_2O \longrightarrow Ca(H_2PO_4)_2 + 2(CaSO_4 \cdot 2H_2O)$$

The mixture of calcium dihydrogen phosphate and gypsum, $CaSO_4 \cdot 2H_2O$, is sold as **superphosphate of lime.** A mixture containing a higher percentage of phosphorus, **triple superphosphate of lime,** is manufactured by treating pulverized tricalcium phosphate with orthophosphoric acid.

$$Ca_3(PO_4)_2 + 4H_3PO_4 \longrightarrow 3Ca(H_2PO_4)_2$$

The direct use of $Ca_3(PO_4)_2$ and $CaHPO_4$ as phosphate fertilizers depends upon the weathering process, which changes them slowly into the soluble dihydrogen phosphate.

26.11 Pyrophosphoric Acid

Pyrophosphoric acid, $H_4P_2O_7$, may be prepared by heating the ortho acid to 250°. The equation for the reaction is

$$2H_3PO_4 \rightleftharpoons H_4P_2O_7 + H_2O$$

This reaction involves the elimination of a molecule of water from two molecules of orthophosphoric acid, as shown by the equation

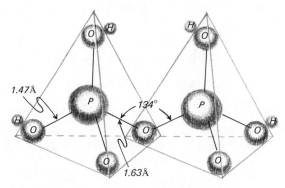

In the preceding formulas the straight line between two symbols represents a covalent bond (shared pair of electrons). The arrows to oxygen atoms designate a coordinate bond, in which the phosphorus furnishes both electrons of the shared pair. These abbreviations are often used in chemical formulas, especially when the molecules represented are rather complex. Sometimes, for convenience, the bond line between symbols is omitted, as in some of the terminal OH groups in the preceding equation. The structure of the pyrophosphoric acid molecule is shown in Fig. 26-9. The substance is a white crystalline solid which melts at 61°. When it is dissolved in water it gradually hydrolyzes to the ortho acid. Among the salts of pyrophosphoric acid that have been prepared are **tetrasodium pyrophosphate,** $Na_4P_2O_7$, **trisodium monohydrogen pyrophosphate,** $Na_3HP_2O_7$, and **disodium dihydrogen pyrophosphate,** $Na_2H_2P_2O_7$.

FIGURE 26-9

The structure of the pyrophosphoric acid molecule, $H_4P_2O_7$, consists of two tetrahedra joined at a corner.

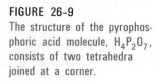

1.47Å

134°

1.63Å

26.12 Triphosphoric Acid

Triphosphoric acid (Fig. 26–10), $H_5P_3O_{10}$, is formed by the elimination of two molecules of water from three molecules of orthophosphoric acid.

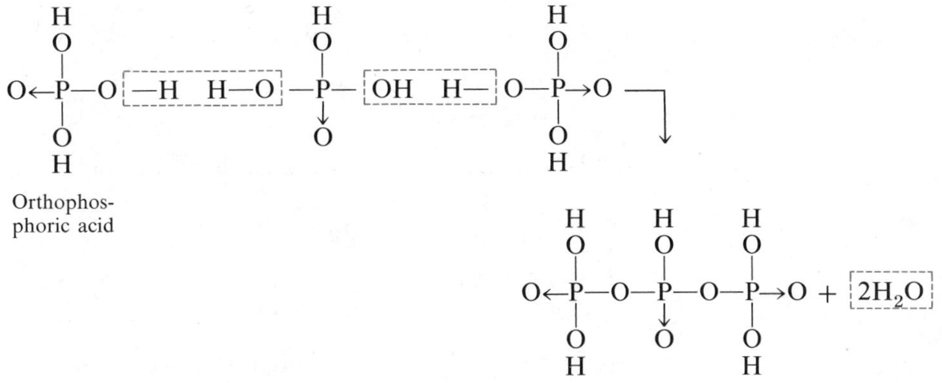

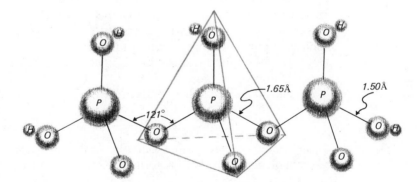

Orthophosphoric acid

Triphosphoric acid

FIGURE 26–10
Structure of triphosphoric acid, $H_5P_3O_{10}$.

The sodium salt of triphosphoric acid, $Na_5P_3O_{10}$, is used as a water softener. It may be prepared by fusing equimolar quantities of sodium pyro- and metaphosphates.

$$x\text{Na}_4\text{P}_2\text{O}_7 + (\text{NaPO}_3)_x \longrightarrow x\text{Na}_5\text{P}_3\text{O}_{10}$$

Polyphosphate linkages of the type existing in pyrophosphates and triphosphates are of great biochemical importance. The energy required for the contraction of muscles results from hydrolysis of "energy-rich" P—O—P bonds of a complex organic polyphosphate existing in muscle tissue, known as **adenosine triphosphate.** The enzyme-catalyzed hydrolysis is slow but involves the release of 7 kilocalories of energy per mole of adenosine triphosphate. The equation for the reaction is given below (R represents the complex organic portion of the molecule):

$$\underset{\text{Adenosine triphosphate}}{\text{RO}\overset{\overset{\text{O}}{\uparrow}}{\underset{\text{OH}}{\overset{|}{\text{P}}}}\text{—O—}\overset{\overset{\text{O}}{\uparrow}}{\underset{\text{OH}}{\overset{|}{\text{P}}}}\text{—O—}\overset{\overset{\text{O}}{\uparrow}}{\underset{\text{OH}}{\overset{|}{\text{P}}}}\text{—OH}} + \text{H}_2\text{O} \longrightarrow \underset{\text{Adenosine diphosphate}}{\text{RO}\overset{\overset{\text{O}}{\uparrow}}{\underset{\text{OH}}{\overset{|}{\text{P}}}}\text{—O—}\overset{\overset{\text{O}}{\uparrow}}{\underset{\text{OH}}{\overset{|}{\text{P}}}}\text{—OH}} + \underset{\text{Phosphoric acid}}{\text{HO}\overset{\overset{\text{O}}{\uparrow}}{\underset{\text{OH}}{\overset{|}{\text{P}}}}\text{—OH}}$$

This subject will be discussed in greater detail in Section 28.3.

26.13   Metaphosphoric Acid and Its Salts

The elimination of a molecule of water from a single molecule of orthophosphoric acid gives **metaphosphoric acid,** $(HPO_3)_x$.

$$x\,H_3PO_4 \longrightarrow (HPO_3)_x + x\,H_2O$$

Metaphosphoric acid is usually manufactured by heating orthophosphoric acid above 400°. It solidifies from the liquid as a glassy product called **glacial phosphoric acid.**

In metaphosphoric acid oxygen bridges are formed between adjacent phosphorus atoms to build up rings and chains, in which there are four oxygen atoms linked to every phosphorus atom. By combining with each other, the HPO_3 molecules polymerize into multiple units which may be represented by the formula $(HPO_3)_x$. Metaphosphoric acid dissolves readily in water, in which it slowly changes into orthophosphoric acid.

Sodium metaphosphate is made by heating sodium dihydrogen orthophosphate.

$$x\,NaH_2PO_4 \xrightarrow{\;\Delta\;} (NaPO_3)_x + x\,H_2O$$

If the product is heated to about 700° and then cooled rapidly, a water soluble, **glassy polymetaphosphate** is formed. This is a long-chain polymer which is best formulated as $(NaPO_3)_x$, although it has been called "sodium hexametaphosphate" because it was once thought to have the composition $(NaPO_3)_6$. The metaphosphates form soluble complexes with the calcium ion and reduce the concentration of the calcium ion in water solutions to such a low value that it cannot be precipitated by soaps. For this reason the sodium polymetaphosphates and other forms of phosphate have been used extensively as water softeners. However, their use has recently been seriously questioned, due to the fact that they enhance the growth of algae in bodies of water (see Chapter 12).

26.14   Phosphorous Acid and Its Salts

Phosphorous acid, H_3PO_3, can be prepared by the action of water upon P_4O_6, PCl_3, PBr_3, or PI_3. Pure phosphorous acid is most readily obtained by hydrolyzing phosphorus trichloride.

$$PCl_3 + 3H_2O \longrightarrow H_3PO_3 + 3HCl\uparrow$$

The resulting solution is heated to expel the hydrogen chloride and to evaporate the water to a point where white crystals of phosphorous acid appear upon cooling. The crystals are deliquescent, very soluble in water, and have an odor like that of garlic. The solid crystals melt at 70.1° and decompose at about 200° by an auto-oxidation-reduction reaction into phosphine and orthophosphoric acid.

$$4H_3PO_3 \longrightarrow PH_3 + 3H_3PO_4$$

The electronic formula and tetrahedral structure of phosphorous acid is shown in Fig. 26–11.

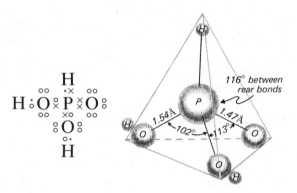

$$\text{H}$$
$$\overset{\;\;\;\circ\circ\;\;\;\cdot\times\;\;\;\circ\circ}{\text{H}\overset{\circ}{\circ}\text{O}\overset{\circ}{\circ}\text{P}\overset{\times}{\times}\text{O}\overset{\circ}{\circ}}$$
$$\overset{\circ\circ}{\text{O}}$$
$$\text{H}$$

FIGURE 26-11
Orthophosphorous acid, H_3PO_3. Note that one hydrogen atom is bonded directly to the phosphorus atom. The other two hydrogen atoms are bonded to oxygen atoms. Only those hydrogen atoms bonded through an oxygen atom are acidic.

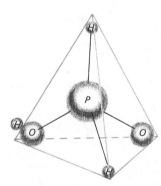

FIGURE 26-12
Hypophosphorous acid, H_3PO_2. Note that two of the three hydrogen atoms are attached directly to the phosphorus atom. Hence, only one hydrogen atom is acidic, i.e., the one bonded through an oxygen atom.

Phosphorous acid and its salts are active reducing agents, because they are readily oxidized to phosphoric acid and phosphates, respectively. Phosphorous acid reduces the silver ion to free silver, mercury(II) salts to the mercury(I) stage, and sulfurous acid to sulfur.

Phosphorous acid forms only two series of salts, such as **sodium dihydrogen phosphite**, NaH_2PO_3, and **disodium hydrogen phosphite**, Na_2HPO_3. The third atom of hydrogen cannot be replaced by a cation. The nonacidic character of the third hydrogen atom is due to its direct linkage to the phosphorus atom rather than to an oxygen atom.

26.15 Hypophosphorous Acid and Its Salts

The solution remaining from the preparation of phosphine from white phosphorus and sodium hydroxide contains sodium hypophosphite, NaH_2PO_2 (Section 26.5). The corresponding barium salt may be obtained by replacing sodium hydroxide with barium hydroxide in the preparation of phosphine. When barium hypophosphite is treated with sulfuric acid, barium sulfate precipitates and **hypophosphorous acid**, H_3PO_2, is formed in solution.

$$\text{Ba}^{2+} + 2H_2PO_2^- + 2H^+ + SO_4^{2-} \longrightarrow \underline{BaSO_4} + 2H_3PO_2$$

The acid is weak and monoprotic, forming only one series of salts. The two nonacidic hydrogen atoms are linked directly to the phosphorus atom (see Fig. 26–12). Hypophosphorous acid and its salts are strong reducing agents.

QUESTIONS

1. Compare and contrast the chemical properties of elemental phosphorus and nitrogen; of phosphoric acid and nitric acid; and of phosphine and ammonia.
2. Calculate the oxidation state of phosphorus in each of the following: PH_3, H_3PO_2, H_3PO_3, H_3PO_4, $H_4P_2O_7$, PH_4I, P_4, PCl_3.
3. Write electronic formulas for each of the compounds given in Question 2.
4. Contrast the properties of white and red phosphorus.
5. What is the reduction product containing nitrogen which results from the action of concentrated nitric acid on elemental phosphorus to produce ortho-phosphoric acid? Write a chemical equation describing the reaction.
6. Write equations showing the stepwise ionization of phosphoric acid.
7. Write equations for the preparation and hydrolysis of calcium phosphide. Compare the hydrolysis of metallic phosphides to that of metallic nitrides.
8. How may the phosphorus in insoluble tricalcium phosphate be made available for plant nutrition?
9. Write equations showing the hydrolysis of PO_4^{3-}, HPO_4^{2-}, and $H_2PO_4^-$.
10. Relate $H_5P_3O_{10}$, $H_4P_2O_7$, H_3PO_4, and HPO_3 to P_4O_{10} in terms of extent of hydration of the latter compound.
11. Write equations to show the action of heat upon NaH_2PO_4, $NaNH_4HPO_4$, and $MgNH_4PO_4$.
12. Explain the action of sodium polymetaphosphates as water softeners. Why is the use of such materials in detergents being questioned (see Chapter 12)?
13. Why does phosphorous acid form only two series of salts although its molecule contains three hydrogen atoms?
14. Write equations for the preparation of hypophosphorous acid, starting with white phosphorus.
15. Show that the decomposition of phosphorous acid by heat involves an auto-oxidation-reduction reaction.
16. Explain the difference in spelling in the name of the element, phosphorus, and the name of H_3PO_3, phosphorous acid.
17. Name each of the following compounds: K_2HPO_4, $FePO_4$, NaH_2PO_2, $(NaPO_3)_x$, $Na_4P_2O_7$, $Ca(H_2PO_4)_2$, and Na_2HPO_3.
18. Write equations for each of the following preparations: (a) P_4 from $Ca_3(PO_4)_2$; (b) P_4O_{10} from P_4; (c) H_3PO_4 from P_4O_{10}; (d) Na_2HPO_4 from H_3PO_4, and (e) $Na_4P_2O_7$ from Na_2HPO_4.
19. Draw a valence electron structure for diphosphine.
20. Compare the structures of PCl_4^+, PCl_5, PCl_6^-, and $POCl_3$.

PROBLEMS

1. How much $Ca_3(PO_4)_2$ would be needed to prepare 1.0 ton of phosphorus in the electric furnace process, if a yield of 94% is obtained? *Ans. 5.3 tons*
2. A detergent is advertised as containing "only ten per cent phosphate, expressed as phosphorus pentaoxide." Is the name misleading? If so, explain why. Calculate what this percentage would be if expressed as sodium tripolyphosphate ($Na_5P_3O_{10}$), a common form for detergent phosphate. *Ans. 26%*

3. How many pounds of phosphate ion would be added to the water system of the United States in one year if there is one load of clothes washed every week for every person in the country? Assume a population of two hundred million people and that 4 ounces of detergent which is 20% sodium orthophosphate (Na_3PO_4) is used per washing. *Ans. 300,000,000 pounds*

4. How many grams of adenosine triphosphate, ATP, must be hydrolyzed to obtain one kcal of heat energy (ATP = $C_{10}H_{16}N_5O_{13}P_3$)? (See Section 26.12. *Note that only one significant figure is justified in the answer to this problem using the information in that section.*) *Ans. 70 g*

5. How much $POCl_3$ can be produced from 50.0 g of PCl_5 and the appropriate amount of P_4O_{10}? *Ans. 61.4 g*

6. (a) In Section 26.5, the value of ΔH°_{298} is given for the reaction of phosphine gas with oxygen gas to produce liquid orthophosphoric acid. An usually large amount of energy is involved. Check the value by calculation, using data in Appendix J.

   (b) Calculate the value of ΔH_{298} for the production of *solid* orthophosphoric acid from phosphine gas. Compare this value to that for liquid orthophosphoric acid. *Ans. $\Delta H_{298} = -307,000$ cal*

   (c) Calculate ΔG°_{298} for the reaction as in (b), using data in Appendix J. *Ans. $-270,700$ cal*

   (d) Compare the values for ΔH°_{298} and ΔG°_{298}, and on the basis of them draw as specific conclusions as you can about the reaction.

REFERENCES

"Mineral Cycles," E. S. Deevey, Jr. *Sci. American*, Sept., 1970; p. 148.

"Chemical Fertilizers," C. J. Pratt, *Sci. American*, June, 1965; p. 62.

"Chemistry of the Diphosphorus Compounds," J. E. Huheey, *J. Chem. Educ.*, **40**, 153 (1963).

"Ionic and Molecular Halides of the Phosphorus Family," R. R. Holmes, *J. Chem. Educ.*, **40**, 125 (1963).

"The Structures and Reactions of Phosphorus Sulfides," A. H. Cowley, *J. Chem. Educ.*, **41**, 530 (1964).

"Some Aspects of *d*-Orbital Participation in Phosphorus and Silicon Chemistry," J. E. Bissey, *J. Chem. Educ.*, **44**, 95 (1967).

"The Physical Basis of Allotropy," W. E. Addison, *Educ. in Chemistry*, **1**, 144 (1964).

"Historical Aspects of the Tetrahedron in Chemistry," D. F. Larder, *J. Chem. Educ.*, **44**, 661 (1967).

"The Chemistry of Orthophosphoric Acid and its Sodium Salts," H. A. Neidig, T. G. Teates, and R. T. Yingling, *J. Chem. Educ.*, **45**, 57 (1968).

"Applied Research in the Development of Anticaries Dentifrices," W. E. Cooley, *J. Chem. Educ.*, **47**, 177 (1970).

"Phosphate Replacements: Problems with the Washday Miracle," A. L. Hammond, *Science*, **172**, 361 (1971).

"Inorganic Polymers," H. R. Allcock, *Sci. American,* Mar., 1974; p. 66.

"Marine Phosphorite Deposits and the Nitrogen Cycle," D. Z. Piper and L. A. Codispoti, *Science*, **188**, 15 (1975).

Carbon and Its Compounds

27

Carbon—The Element

Carbon is the first member of Group IVA of the Periodic Table; the other members of this group are silicon, germanium, tin, and lead. Carbon and silicon are predominantly nonmetallic in character whereas germanium, tin, and lead are metallic. Each of the elements in Group IVA has four valence electrons, and each exhibits a maximum oxidation state of $+4$; these elements, especially the heavier ones, also show an oxidation state of $+2$ when combining with other elements.

27.1 Occurrence of Carbon

Carbon is nineteenth among the elements in abundance; it constitutes only about 0.027 per cent of the earth's crust. It is found in the free state in the form of diamond and graphite; it occurs combined in natural gas, petroleum, coal, plants and animals, limestone, dolomite, coral, sponges, and chalk. Carbon is found in the air as carbon dioxide and in natural waters as carbon dioxide, carbonic acid, and carbonates.

27.2 Diamonds

Carbon exists in two allotropic forms, diamond and graphite. X-ray studies have shown that amorphous carbon has the same crystalline structure as graphite. Charcoal, coke, and carbon black are also microcrystalline, or amorphous, forms of carbon.

Diamonds are found in South Africa, the Congo, the Gold Coast, Brazil, India, Australia, and in the United States in Arkansas. Most of the world's supply comes from Africa. The accepted theory is that diamonds are formed when very pure carbon is subjected to a high temperature and very great pressure. This theory follows from the fact that diamonds are found in the conical pipes of extinct volcanoes, where it appears that they were formed when carbon was trapped in the molten lava and held at great heat and pressure.

Many attempts have been made during the past one hundred years to synthesize diamonds, but apparently none were successful until 1954. In that year, H. Tracy Hall and his co-workers successfully produced the first authentic man-made diamonds. They found that when graphite, in molten FeS as a solvent, was subjected to high temperature and pressure under carefully controlled conditions, small crystals of diamond were formed. This method of making diamonds has become increasingly important because man-made diamonds are superior to the natural stones for certain industrial uses. The manufacture of gem-quality diamonds has not, however, yet proved to be feasible.

Diamond is very brittle, and (with the possible exception of boron carbide, B_4C) it is the hardest substance known. The specific gravity of diamond averages about 3.5. Diamonds burn when heated in air or oxygen. When heated to 1,000° in the absence of air, diamond changes to graphite. Diamond is inert to all chemicals at ordinary temperatures.

A diamond crystal belongs to the cubic system, with each atom covalently bonded to four others located at the corners of a regular tetrahedron (Fig. 27–1).

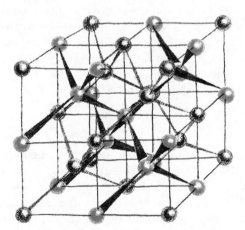

FIGURE 27–1

The crystal structure of diamond. Notice in the diagram that the atom shared by all eight cubes is bonded to the center atom of four of the cubes. Bonds in these four cubes are shown heavier and the atoms colored for distinction.

These covalent bonds bind all of the atoms in the diamond crystal into a single giant molecule. Because the carbon-carbon bonds are very strong and extend throughout the crystal in its three dimensions, the crystal is very hard and has a high melting point (probably the highest of all elements). The diamond is a nonconductor of electricity, because there are no mobile electrons in the crystal; its electrons are all used in bond formation.

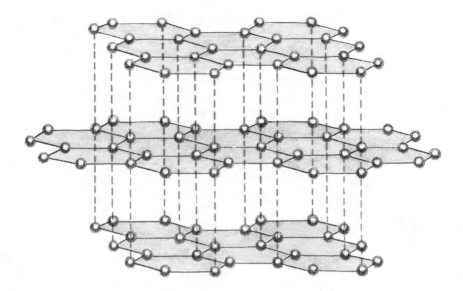

FIGURE 27-2
The crystalline structure of graphite.

27.3   Graphite

Graphite, which is also known as **plumbago,** or **black lead,** is distinctly different in crystalline form and physical properties from diamond. It is a soft, grayish-black solid which crystallizes in hexagonal plates that have a metallic luster and conduct electricity. The word *graphite* comes from the Greek meaning "to write," because the substance is used for this purpose. It has a specific gravity of 2.2, it melts at 3,527°, and it is inert to most chemical reagents. The internal energy of graphite is less than that of diamond. This means that graphite is the more stable of the two crystalline forms, in accord with the principle that substances seek the lowest energy form possible (see Section 20.11). More energy is released in the combustion of diamond, therefore, than in the combustion of graphite.

$$\text{C (diamond)} + O_2 \longrightarrow CO_2 + 94{,}504 \text{ cal} \qquad (\Delta H^\circ_{298} = -94{,}504 \text{ cal})$$
$$\text{C (graphite)} + O_2 \longrightarrow CO_2 + 94{,}051 \text{ cal} \qquad (\Delta H^\circ_{298} = -94{,}051 \text{ cal})$$
$$\text{C (diamond)} \longrightarrow \text{C (graphite)} + 453 \text{ cal} \qquad (\Delta H^\circ_{298} = -453 \text{ cal})$$

The comparative softness and electrical conductivity of graphite are in accord with its structure (Fig. 27–2). The crystal consists of layers of atoms, each atom of which has three carbon atoms as its near neighbors. Each carbon atom is bonded to two others by single covalent bonds, and to the third by a double bond. This arrangement completes the octet of each atom. The double bonds are not fixed but resonate in such a manner as to give each bond some double bond character. These covalent bonds bind the atoms very tightly together into layers of hexagons. However, the forces holding the layers to each other are weak, and the layers can be separated easily. The flaky, soft character of graphite results from this weak binding. The loosely held electrons of the mobile double bonds within the layers of carbon atoms are responsible for the electrical conductivity and the color of graphite.

27.4   Preparation and Uses of Graphite

Natural graphite is mined in Mexico, Madagascar, Ceylon, and Canada. The United States leads in the production of synthetic graphite. To synthesize graphite a mixture of amorphous carbon and a little sand and iron oxide (as catalysts) is heated in an electric furnace for 24 to 30 hours at a temperature of about 3500°. The carbon vapor that is formed condenses as graphite.

Because of its resistance to heat and chemical action, its high vaporization temperature, and its ability to conduct electricity, graphite is used in making electrodes and crucibles. It is also used in the manufacture of paints, commutator brushes, and "lead" pencils. Suspensions of colloidal graphite in water (aqua-dag) or in oil (oil-dag) provide excellent lubricants for many purposes.

27.5   Coal

Large quantities of coal have been formed in nature from vegetable matter by a process called **carbonization.** This process is essentially the slow decomposition of vegetable matter (chiefly cellulose, a compound containing carbon, hydrogen, and oxygen) at low temperatures, under high pressures, in the presence of water, and in the absence of air. This decomposition results in the loss of a large proportion of the hydrogen in the form of volatile hydrocarbons, such as methane, CH_4, and of the oxygen in the form of carbon dioxide, CO_2. The residue becomes richer in carbon and poorer in hydrogen, oxygen, nitrogen, and water. During the progressive stages of carbonization the several varieties of coal (peat, lignite, bituminous, and anthracite) are formed. The average free carbon content of peat is 11 per cent; lignite, 22 per cent; bituminous, 60 per cent; and anthracite, 88 per cent.

27.6   Coke and Charcoal

The residue remaining after coal is heated in the absence of air is known as coke. This solid product is composed chiefly of carbon and ash. The by-products of the process are ammonia, illuminating gas, and coal tar. Among the compounds obtained from the coal tar are benzene, toluene, phenol, and naphthalene. Coke is used in large quantities in the reduction of metallic ores, as a smokeless fuel, and in the production of "water gas" (see Section 27.29).

Wood charcoal is produced by the destructive distillation of wood, i.e., the heating of wood in the absence of air. Among the volatile by-products of this process are wood alcohol (methanol), acetic acid, acetone, and a tarry mixture. The very porous solid product (charcoal) retains the shape of the wood from which it was made.

Charcoal is used to a limited extent as a fuel. More important is its use as a decolorizing agent for certain liquids such as sugar solutions, alcohol, and petroleum products. This use depends upon the ability of charcoal (and other amorphous forms of carbon) to adsorb certain substances which are responsible for the discoloration of the liquids. **Adsorption** is a surface phenomenon in which forces existing on the surface of the adsorbing agent attract and hold molecules of the substance being adsorbed. Gases, as well as solids and liquids, are adsorbed on the surface of the very porous charcoal. One milliliter of finely powdered charcoal may have a total surface of 1,000 square meters. Charcoal is "activated" by heating it in steam. This treatment increases its adsorptive capacity by breaking down the granules and removing gas already adsorbed. Charcoal is used in removing odoriferous gases from petroleum products and in gas masks to protect the wearer who may be exposed to poisonous gases.

27.7 Chemical Properties of Carbon

Carbon is almost chemically inert toward most reagents at ordinary temperatures. However, graphite is slowly oxidized by a mixture of nitric acid and sodium chlorate; charcoal is rapidly oxidized by this mixture. The activity of carbon increases rapidly with rising temperatures and at elevated temperatures it is very active. At high temperatures carbon unites with oxygen, forming either carbon monoxide, CO, or carbon dioxide, CO_2, depending on the amount of oxygen present. Carbon with sulfur forms carbon disulfide, CS_2; with certain metals, the carbides, such as iron carbide, Fe_3C; and with fluorine, carbon tetrafluoride, CF_4. Hot carbon combines directly with hydrogen in the presence of a suitable catalyst, forming such products as methane, CH_4, and acetylene, C_2H_2, in very low yields.

Compounds of Carbon

The term "organic" appears to have been used for the first time about 1777 and was applied to those materials occurring in or derived from living organisms. Accordingly, such substances as starch, alcohol, and urea were classified as organic, for starch is produced by living plants, alcohol is a product of fermentation caused by microorganisms, and urea is contained in urine. In 1824, however, the German chemist Wöhler synthesized urea, and the original meaning of the term "organic" no longer applied. **Organic compounds,** in the modern sense, are the compounds of carbon. Many thousands of carbon compounds which are not found in or derived from living organisms have been produced by chemists, and well over a million organic compounds are already known.

The existence of so many organic compounds is due primarily to the ability of carbon atoms to combine with other carbon atoms, forming chains of different lengths and rings of different sizes. The elements, other than carbon, most frequently found in organic compounds are hydrogen, oxygen, nitrogen, sulfur, the halogens, phosphorus, and some of the metals.

Although the number of organic compounds is vast, the study of organic chemistry is greatly simplified by the fact that the chemistry of carbon compounds can be organized around "functional groups," each of which imparts similarities in chemical properties. In this chapter and in the chapter on Biochemistry, we shall discuss a few of these groups.

Hydrocarbons

The simplest organic compounds are those containing only carbon and hydrogen. Such compounds are known as **hydrocarbons.** Some of them are found in nature, where they were derived from plant or animal forms of life. Several types of hydrocarbons are possible. The tetrahedral, trigonal, and digonal hybridization that occurs in the bonding for the various types of hydrocarbons was discussed in Chapter 6 (see Sections 6.3, 6.5, 6.8, and 6.9).

27.8   The Alkanes

The series of compounds which have the general empirical formula C_nH_{2n+2}, where (n) is an integer, is called the **alkane,** or **paraffin, series.** A few of the alkanes and some of their properties are listed in Table 27–1.

TABLE 27-1   Some Members of the Alkane, or Paraffin, Series

	Empirical Formula	Melting Point, °C	Boiling Point, °C	Usual Form	Number of Possible Isomers
Methane	CH_4	−182.5	−161.5	Gas	0
Ethane	C_2H_6	−183.2	−88.6	Gas	0
Propane	C_3H_8	−187.7	−42.1	Gas	0
n-Butane	C_4H_{10}	−138.3	−0.5	Gas	2
n-Pentane	C_5H_{12}	−129.7	36.1	Liquid	3
n-Hexane	C_6H_{14}	−95.3	68.7	Liquid	5
n-Heptane	C_7H_{16}	−90.6	98.4	Liquid	9
n-Octane	C_8H_{18}	−56.8	125.7	Liquid	18
n-Nonane	C_9H_{20}	−53.6	150.8	Liquid	35
n-Decane	$C_{10}H_{22}$	−29.7	174.0	Liquid	75
n-Undecane	$C_{11}H_{24}$	−25.6	195.8	Liquid	159
n-Dodecane	$C_{12}H_{26}$	−9.6	216.3	Liquid	355
n-Tridecane	$C_{13}H_{28}$	−5.4	235.4	Liquid	802
n-Tetradecane	$C_{14}H_{30}$	5.9	253.5	Liquid	1,858
n-Octadecane	$C_{18}H_{38}$	28.2	316.1	Solid	60,523

The lighter members of the alkane series are gases, the members of intermediate weight are liquids, and the heavier members are solids. From the formulas of the compounds of this series it is evident that each member differs from the preceding one by the increment CH_2. A series of compounds in which each member differs from the one before it by a common increment is called a **homologous series.**

The electronic formulas for methane, ethane, and propane are easily written. Although the electronic formula representation makes the molecules seem planar, it must be remembered that in all the alkanes the C—C and C—H bonds have the geometrical arrangement set by the tetrahedral configuration of each carbon unit (see Sections 6.3 and 6.5).

$$
\begin{array}{ccc}
\text{H} & \text{H}\ \ \text{H} & \text{H}\ \ \text{H}\ \ \text{H} \\
\text{H}\!:\!\overset{\cdot\cdot}{\underset{\cdot\cdot}{\text{C}}}\!:\!\text{H} & \text{H}\!:\!\overset{\cdot\cdot}{\underset{\cdot\cdot}{\text{C}}}\!:\!\overset{\cdot\cdot}{\underset{\cdot\cdot}{\text{C}}}\!:\!\text{H} & \text{H}\!:\!\overset{\cdot\cdot}{\underset{\cdot\cdot}{\text{C}}}\!:\!\overset{\cdot\cdot}{\underset{\cdot\cdot}{\text{C}}}\!:\!\overset{\cdot\cdot}{\underset{\cdot\cdot}{\text{C}}}\!:\!\text{H} \\
\text{H} & \text{H}\ \ \text{H} & \text{H}\ \ \text{H}\ \ \text{H} \\
\text{Methane} & \text{Ethane} & \text{Propane}
\end{array}
$$

The structure of the methane molecule is shown in Figs. 6–4 and 6–13 of Chapter 6. The structure of the ethane molecule is shown in Fig. 6–14 of Chapter 6.

There are two hydrocarbons having the formula C_4H_{10}, and these are known as normal butane and isobutane. The two butanes are **isomers.** They have the same empirical formula, and hence the same composition, but different physical and chemical properties, because they differ in the arrangement of the atoms in their molecules. Normal butane, *n*-butane, is a "straight chain" molecule, and *iso*butane is a "branched chain" molecule.

$$
\begin{array}{cc}
\overset{\text{H}\ \ \ \text{H}\ \ \ \text{H}\ \ \ \text{H}}{\underset{\text{H}\ \ \ \text{H}\ \ \ \text{H}\ \ \ \text{H}}{\text{H}-\text{C}-\text{C}-\text{C}-\text{C}-\text{H}}} & \overset{\text{H}\ \ \ \text{H}\ \ \ \text{H}}{\underset{\text{H}\ \ \ |\ \ \ \text{H}}{\text{H}-\text{C}-\text{C}-\text{C}-\text{H}}} \\
\textit{n}\text{-Butane} & \underset{\text{H}}{\overset{|}{\text{H}-\text{C}-\text{H}}} \\
& |\\
& \text{H} \\
& \textit{Iso}\text{butane}
\end{array}
$$

The number of possible isomers increases with increasing molecular weight; the hydrocarbons of large molecular weight have great numbers of isomers (see Table 27–1).

Alkanes are characterized by being rather unreactive chemically; hence, the use of the name **paraffin,** which means "having little affinity." The following reaction is a typical one for the alkanes and is referred to as a substitution reaction.

$$
\underset{\text{Ethane}}{\overset{\text{H}\ \ \text{H}}{\underset{\text{H}\ \ \text{H}}{\text{H}-\text{C}-\text{C}-\text{H}}}} + \text{Cl}_2 \xrightarrow[\text{catalyst}]{\text{Light or}} \underset{\text{Ethyl chloride}}{\overset{\text{H}\ \ \text{H}}{\underset{\text{H}\ \ \text{H}}{\text{H}-\text{C}-\text{C}-\text{Cl}}}} + \text{HCl}
$$

This reaction is a chain reaction which proceeds in several steps (Section 17.2). An important point concerning this particular reaction is that it transforms an alkane molecule into one which has a more reactive "functional group" on it—a halogen in this specific case. The functional group makes it possible for the molecule to take part in many different kinds of reactions.

Another very important reaction of alkanes is based upon their ability to burn in air in a highly exothermic oxidation-reduction reaction. A typical combustion reaction is that for ethane.

$$2C_2H_6(g) + 7O_2(g) \longrightarrow 4CO_2(g) + 6H_2O(g) + 682,500 \text{ cal}$$
$$(\Delta H^\circ_{298} = -682,500 \text{ cal})$$

Methane is the principal component of cooking gas. Gasoline is a mixture of straight and branched chain alkanes containing 5–9 carbon atoms plus various additives. Kerosene, diesel oil, and fuel oil are primarily mixtures of alkanes with higher molecular weights.

27.9   The Alkenes

Hydrocarbon molecules which contain a double bond are members of a second homologous series, referred to as **alkenes.** As was pointed out in Chapter 6, the two carbon atoms linked by a double bond are bound together by two kinds of bonds, one sigma C—C bond and one pi C—C bond (see Section 6.8).

The alkenes have the general empirical formula C_nH_{2n}. Ethylene, C_2H_4, the simplest alkene, has a trigonal planar structure, as illustrated in Fig. 6–19, Section 6.8. The second member of the series is propylene and then come the butene isomers. The series builds up in a manner analogous to the alkane series. The electronic formula for ethylene is shown.

In writing formulas for members of the ethylene series, the two shared electron pairs are usually represented by two lines.

Ethylene   or   Ethylene   Propylene

1-Butene   2-Butene

Note that carbon atoms with single bonds have sp^3 hybridization, and those with double bonds have sp^2 hybridization. In the nomenclature for 1-butene and 2-butene the numbers refer to the number of the carbon atom preceding the double bond, counting each carbon beginning at the left. Actually, two isomers of 2-butene exist, because the double bond is quite rigid and hence rotation about it is very difficult. Writing the formulas for the butene molecules in a slightly different fashion makes it possible to show the three butene isomers more clearly.

$$H_2C=CH-C_2H_5 \qquad \underbrace{H_3C-CH=CH-CH_3 \quad\quad H_3C-CH=CH-CH_3}_{\text{2-Butene}}$$

1-Butene   *cis* isomer   *trans* isomer

The 2-butene isomer which has the two CH_3 groups next to each other is referred to as the *cis* isomer; the one with the two CH_3 groups opposite each other is called the *trans* isomer. The interesting subject of *cis* and *trans* isomerism will be discussed again in the chapter on Coordination Compounds (see Section 32.4).

Alkenes are much more reactive than alkanes. The π bond, being a relatively weaker bond, is disrupted much more easily than a σ bond. Thus, the reactions characteristic of alkenes are those in which the π bond is broken.

In the presence of a suitable catalyst (Pt, Pd, or Ni, for example), alkenes add hydrogen to form the corresponding alkanes.

$$H_2C=CH_2 + H_2 \xrightarrow{\text{Pt, pressure}} H_3C-CH_3$$

Ethylene   Ethane

Chlorine breaks the double bond and adds to the two carbons adjacent to the double bond in the alkenes, instead of replacing a hydrogen as in the alkanes.

$$H_2C=CH_2 + Cl_2 \longrightarrow ClH_2C-CH_2Cl$$

Ethylene   1,2-Dichloroethane

(The numbers in the name of the compound refer to the carbon atom, numbering from the left, on which a chlorine atom is located.)

Many other reagents react with alkenes by breaking the double bond. An example is the acid-catalyzed addition of water to an alkene.

$$H_2C=CH_2 + H_2O \xrightarrow[\text{Catalyst}]{\text{Acid}} H_3C-CH_2OH$$

Ethylene   Ethanol
(Ethyl alcohol)

A very important property of the alkenes is their ability to add to themselves. This reaction is a very significant one in the plastics industry.

$$n\,(H_2C=CH_2) \xrightarrow{\text{Catalyst}} {\left(\!-CH_2-CH_2-\!\right)}_n \qquad \text{(where } n \text{ is large)}$$

Polyethylene

Because ethylene and the other alkenes have the ability to *add* chlorine and other atoms, they are called **unsaturated compounds.** On the other hand, ethane and the other alkanes react with chlorine by *substitution* only, so they are called **saturated compounds.**

Ethylene is a colorless gas with a rather sweet odor. It occurs in natural gas and is formed when coal or wood is submitted to destructive distillation. Ethylene, mixed with oxygen, is used as an anesthetic in dentistry and surgery. It has the interesting property of causing green fruits to ripen and is used commercially for this purpose. The production of 11,790,000 tons of ethylene in 1974 made it the fifth highest of all chemicals produced in the United States for that year.

27.10   The Alkynes

Hydrocarbon molecules which contain the triple bond are called **alkynes;** they make up another series of unsaturated hydrocarbons. As was discussed in Chapter 6, two carbon atoms joined by a triple bond are bound together by one sigma C—C bond and two pi C—C bonds (see Section 6.9). The alkynes have the general empirical formula C_nH_{2n-2}.

The simplest and most important member of the alkyne homologous series is acetylene, C_2H_2. The electronic and line formulas for acetylene are

$$H:C::\!:C:H \qquad \text{or} \qquad H—C≡C—H$$
Acetylene

The bonding in the acetylene molecule, which has a linear structure, is illustrated in Fig. 6–21, Section 6.9.

Acetylene is liberated when calcium carbide is treated with water.

$$CaC_2 + 2HOH \longrightarrow Ca(OH)_2 + C_2H_2\uparrow$$

Chemically, the alkynes are similar to alkenes except that with two π bonds they react even more readily, adding twice as much reagent in addition reactions. The reaction of acetylene with bromine is a typical example.

$$H—C≡C—H + 2Br_2 \longrightarrow H—\overset{\displaystyle Br}{\underset{\displaystyle Br}{\overset{|}{\underset{|}{C}}}}—\overset{\displaystyle Br}{\underset{\displaystyle Br}{\overset{|}{\underset{|}{C}}}}—H$$
Tetrabromoethane

Acetylene and all the other alkynes burn very easily.

$$2H—C≡C—H(g) + 5O_2(g) \longrightarrow 4CO_2(g) + 2H_2O(g) + 600,180 \text{ cal}$$
$$(\Delta H^{\circ}_{298} = -600,180 \text{ cal})$$

The triple bond is a high energy bond, and this energy is released when the compound is transformed into carbon dioxide and water. Thus, the flame produced by burning acetylene is very hot; it is used in welding and cutting metal. When acetylene burns, some of it breaks down a second way according to the equation $C_2H_2 \longrightarrow 2C + H_2$. At the temperature of the flame, the particles of carbon become heated to incandescence; i.e., they give off a brilliant luminous white light. At one time, acetylene was used to light homes and in lamps for bicycles and automobiles.

27.11 Cyclic Hydrocarbons

The general empirical formula C_nH_{2n}, which applies for alkenes, also fits a saturated homologous series known as **cycloalkanes.** As the name implies, the molecules of these substances are cyclic, possessing ring structures. The smallest member of the series is cyclopropane;

Cyclopropane   Cyclobutane   Cyclopentane

next are cyclobutane, cyclopentane, etc. Chemically, the cycloalkanes are very similar to open chain alkanes, except for cyclopropane, which has unique properties characteristic of a three-membered ring.

It is possible to introduce double bonds into the cycloalkanes to form the unsaturated **cycloalkenes.** These in turn behave similarly to open chain alkenes.

Cyclopentene   Cyclohexene

We can take this idea one step further and introduce a second double bond into a ring to form **cyclic dialkenes,** often referred to simply as cyclic dienes.

1,3-cyclohexadiene

It is possible also to introduce a third double bond into the cyclohexane ring. If this experiment is performed, it is found that the compound which is formed has unique chemical properties, not very much like an alkene. We could call the product cyclohexatriene, but it actually is named **benzene.** The empirical formula is C_6H_6.

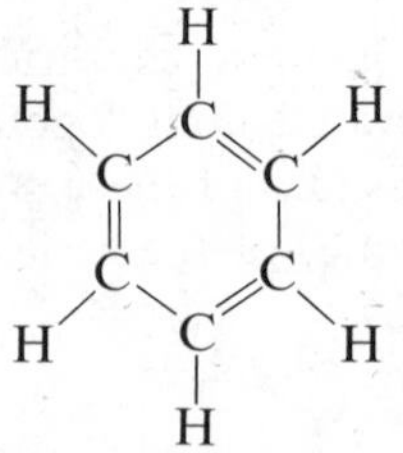

Benzene

The unique properties of benzene are related to its molecular orbital structure. As can be surmised from the structural formula above, all the carbons in benzene are sp^2 hybridized. The three sp^2 hybrid orbitals lie in a plane, and hence the benzene molecule is planar. Consider now the unhybridized p orbitals. There are six of them, with their lobes sticking above and below the plane of the six-membered ring [see (*a*) and (*c*) of Fig. 27–3]. It can be seen from the figure that the p orbitals are in the correct orientation for side-by-side overlap in either direction to produce π orbitals. Both structures, which can thereby result, are exactly equivalent and are shown in (*b*) and (*d*) of the figure. The two forms are resonance forms.

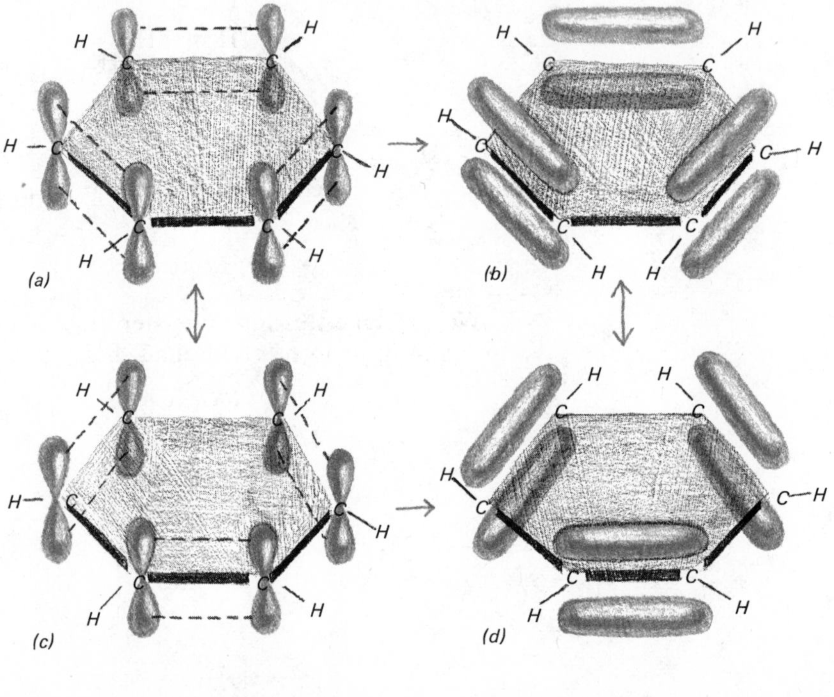

FIGURE 27–3

The benzene molecule. The six sp^2 hybridized carbon atoms lie in a plane. The six remaining unhybridized p orbitals, shown in color in (*a*) and (*c*), extend above and below the plane and are perpendicular to it. The dashed lines between lobes of p orbitals indicate side-by-side overlap. The overlap can occur in either direction between adjacent lobes as shown in (*a*) and (*c*). The resulting molecular π orbitals are shown in (*b*) and (*d*), with the orbital portion of each (in color) lying above and below the hexagonal plane. The resonance structures shown in (*a*) and (*c*) and again in (*b*) and (*d*) are exactly equivalent. The true structure is intermediate in character (hybrid) between the two (see Fig. 27–4).

FIGURE 27-4
The hybrid bonding molecular orbital structure for benzene.

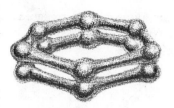

The structure of the sulfur dioxide molecule was described in terms of two equivalent resonance forms in Section 23.12, Part 3. Other examples of resonance structures have been discussed in connection with ozone and various compounds of sulfur and nitrogen. It was pointed out in the discussion of the sulfur dioxide molecule that neither of the two forms is the true representation of the structure, but rather that the true structure is intermediate in character between the two. The sulfur dioxide molecule is a **resonance hybrid** of the two contributing structures. The same situation holds true for benzene. Benzene is a resonance hybrid of the two contributing structures shown in Fig. 27-3 (*b* and *d*). The hybrid bonding molecular orbital structure is pictured in Fig. 27-4. The resonance structures and the hybrid structure are for convenience commonly written as shown.

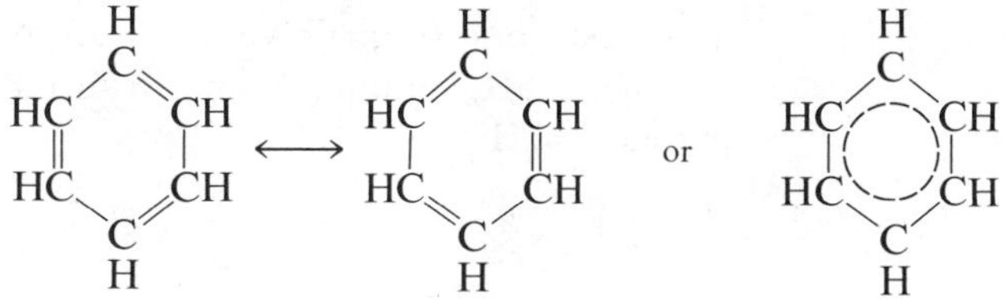

As a result of this resonance hybridization, benzene does not have the characteristics of an alkene. Each bond between two carbon atoms is actually neither a single nor a double bond; each bond is equivalent to each of the others and is intermediate in character **(hybrid)** between a single and a double bond. It is said to have a "bond order" of $1\frac{1}{2}$.

Benzene, a volatile liquid, is the first and best known member of another homologous series in which each member differs from its predecessor by the increment CH_2. Benzene and its homologs, which have one ring, have the general formula C_nH_{2n-6}. Compounds of this series are in a class called **aromatic hydrocarbons,** because many of them have pleasant odors (though most are poisonous and should be smelled gently, if at all).

If one hydrogen atom on each of two different carbon atoms of benzene is replaced by a CH_3 group, the product is xylene. Thus, xylene may be thought of as a derivative of benzene. Xylene has three isomers since substitutions to introduce the two CH_3 groups may be made in three different relative positions on the ring.

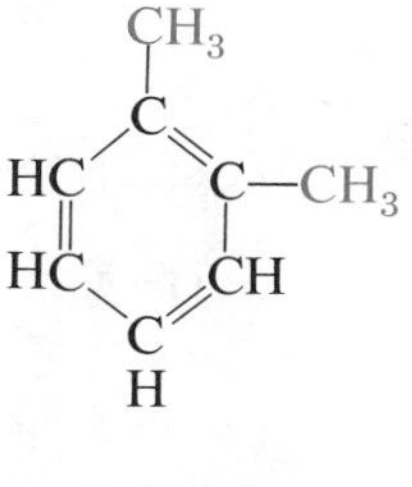

Ortho-xylene

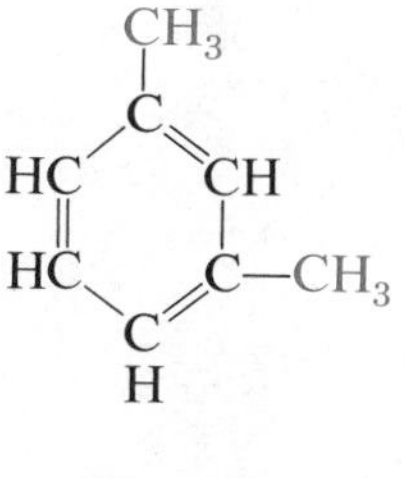

Meta-xylene

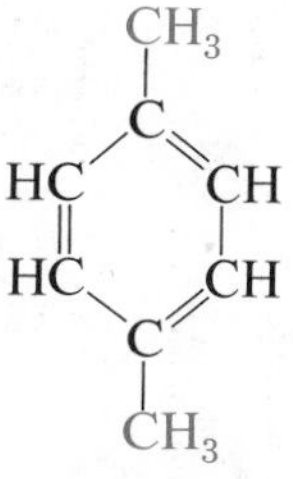

Para-xylene

There are many derivatives of benzene. The hydrogen atoms can be replaced by many different kinds of groups. The following are typical examples.

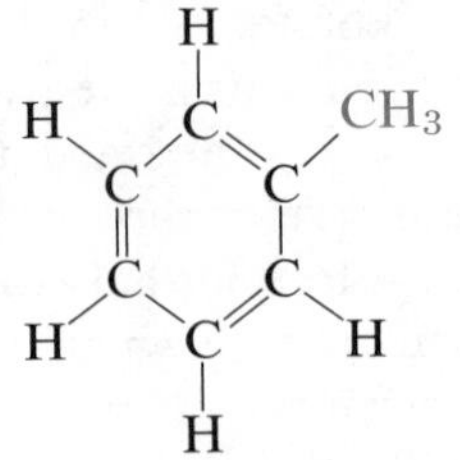

Toluene 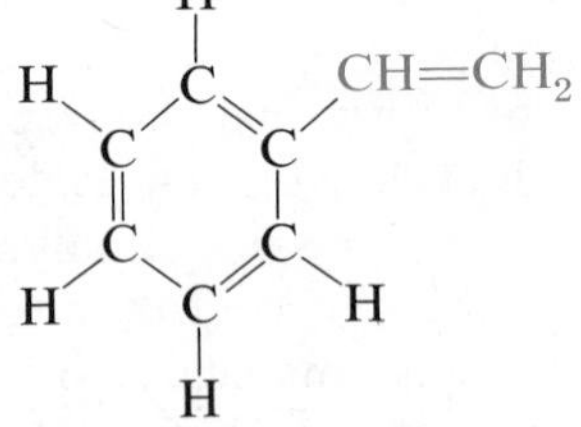   Styrene 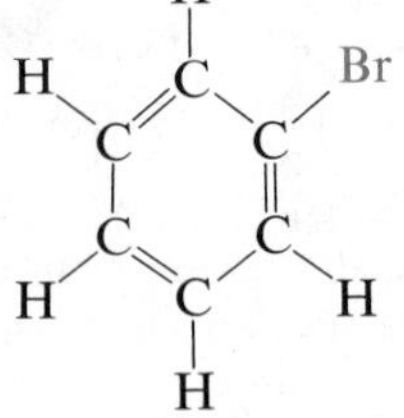   Bromobenzene

Some of the important aromatic hydrocarbons and their derivatives contain more than one ring (polynuclear aromatic hydrocarbons). Examples are naphthalene, $C_{10}H_8$, familiarly known in moth balls; anthracene, $C_{14}H_{10}$; and benzpyrene, $C_{20}H_{12}$.

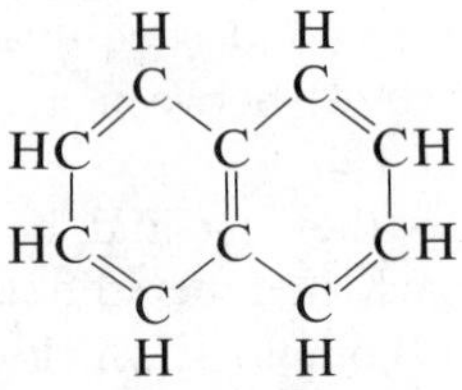

Naphthalene 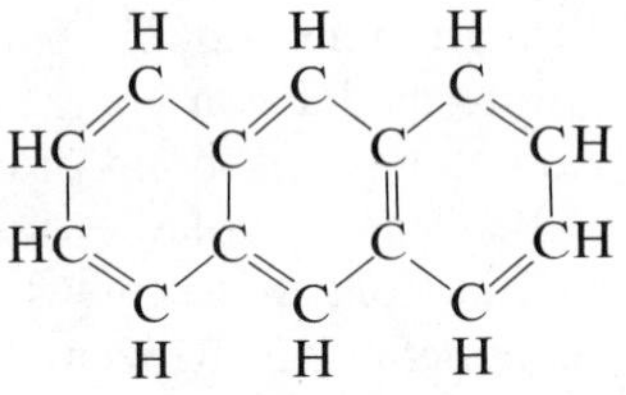   Anthracene 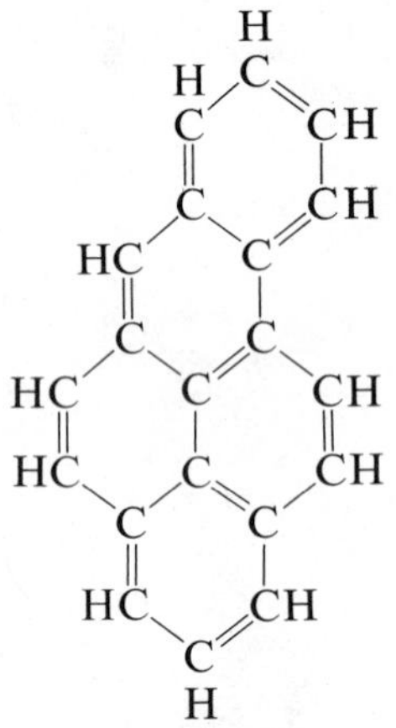   Benzpyrene

The aromatic hydrocarbons are widely used in the manufacture of dyes, synthetic drugs, explosives, plastics, and many other substances. For example, the important explosive trinitrotoluene (TNT), $C_6H_2(CH_3)(NO_2)_3$, is synthesized from toluene by replacing three hydrogen atoms with nitro groups, $-NO_2$.

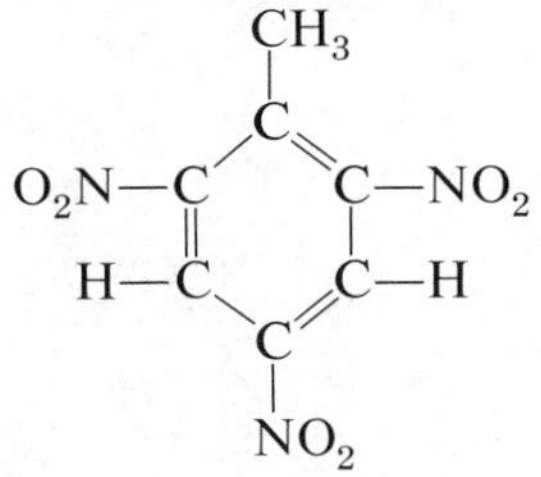

Trinitrotoluene

Graphite, one allotropic form of elemental carbon (Section 27.3), can be considered as being composed of infinite sheets of benzene ring structures (see Fig. 27–2).

27.12   Petroleum

Petroleum is composed chiefly of saturated (paraffin) hydrocarbons, but may also contain unsaturated hydrocarbons, aromatic hydrocarbons and their derivatives, nitrogen compounds, and sulfur compounds. The main step in the refining of petroleum is its separation by distillation into a number of fractions (fractional distillation, Section 13.6), each of which is a complex mixture of hydrocarbons and has properties which make it commercially valuable. The composition of each fraction depends upon the temperature range over which it is collected. Some of the more important products obtained from the refining of petroleum, classified according to number of carbon atoms, are listed in Table 27–2. Before the distillation products of petroleum are ready for use, further purification is usually necessary.

TABLE 27–2   Some Hydrocarbon Products from Petroleum

	Approximate Composition	*Boiling-Point Range*	*Uses*
Petroleum ether	C_4 to C_{10}	35–80°	Solvent
Gasoline	C_4 to C_{13}	40–225°	Motor fuel
Kerosene	C_{10} to C_{16}	175–300°	Fuel, lighting
Lubricating oils	C_{20} up	350° up	Lubrication
Paraffin	C_{23} to C_{29}	50–60° (m.p.)	Candles, waxed paper
Asphalt		Viscous liquids	Paving, roofing
Coke		Solid	Fuel

Ordinary gasoline is principally a mixture of *n*-hexane, *n*-heptane, *n*-octane, and their isomers. The gasoline obtained by the direct distillation of petroleum constitutes only about 10 per cent of that used in the United States; the bulk of the gasoline used is produced by converting larger and smaller hydrocarbon molecules to molecules of the proper weight and complexity. The conversion of the larger molecules (decomposition) is known as "cracking." This is accomplished by heating the less volatile components of petroleum to high temperatures, usually in the presence of catalysts. The simpler hydrocarbons are caused to combine, forming hydrocarbons containing 6, 7, or 8 carbon atoms. Such reactions are brought about at high temperatures and varying pressures, in the presence of catalysts and, in a sense, are the reverse of cracking.

Gasolines are rated upon an arbitrary scale in which *iso*octane, once thought to be the ideal fuel for the gasoline engine, is given the rating "100 octane." Gasolines with octane numbers less than 100 are less efficient than *iso*octane and those with octane numbers greater than 100 are more efficient. Gasoline has usually been made by blending **aliphatic** (open chain) and aromatic hydrocarbons and adding tetraethyl lead. However, manufacturers now are producing increasing quantities of gasoline without adding the tetraethyl lead, because of possible toxic effects of lead compounds in the atmosphere and the poisoning effect of tetraethyl lead on the catalysts in the catalytic converters (Sections 24.4 and 24.5). Gasoline can be produced without tetraethyl lead by blending in more highly branched hydrocarbon isomers. Studies are being made of possible nonlead additives that will permit engines to run without knocking, two such additives being methanol and ethanol. At the same time, engines are being designed to run efficiently on lower octane gasolines, and to run on other sources of power such as steam and electricity.

Derivatives of the Hydrocarbons

In previous sections of this chapter you have seen examples of derivatives of hydrocarbons—ethylene, ethanol, xylene. Derivatives of the hydrocarbons are formed, as noted previously, by replacing one or more hydrogen atoms by other groups referred to as **functional groups.**

An alcohol is closely related to a hydrocarbon in that it contains an OH group in place of a hydrogen atom. Thus, methyl alcohol, CH_3OH, is a derivative of methane, CH_4; and ethyl alcohol, C_2H_5OH, is a derivative of ethane, C_2H_6.

Methyl alcohol
(Methanol)

Ethyl alcohol
(Ethanol)

Table 27–3 lists several of the more important types of compounds derived from the hydrocarbons, the corresponding functional groups, and the empirical type-formula. In the empirical formula, R refers to the hydrocarbon "radical," or the portion of the molecule other than the functional group. For example, ROH is the empirical formula for the alcohols. For methyl alcohol (methanol), R is CH_3—; for ethyl alcohol (ethanol), R is C_2H_5—; etc.

The following sections describe the properties of some specific compounds of each type. The functional groups which are present should be noted carefully in all the examples discussed.

27.13   Alcohols, R—OH

Although all alcohols have one or more OH groups, they differ from bases such as sodium hydroxide and potassium hydroxide in that they do not furnish hydroxide ions in water, nor do they have the other usual properties of bases.

TABLE 27-3   Several Types of Derivatives of the Hydrocarbons

Derivative Type	Functional Group	Type Formula	Typical Examples	
Alcohols	—OH	R—OH	CH_3OH Methanol	C_2H_5OH Ethanol
Ethers	—O—	R—O—R	$CH_3—O—C_2H_5$ Methyl ethyl ether	$C_2H_5—O—C_2H_5$ Diethyl ether
Aldehydes	$—\overset{O}{\overset{\|}{C}}H$	$R—\overset{O}{\overset{\|}{C}}H$	$H\overset{O}{\overset{\|}{C}}H$ Formaldehyde	$CH_3\overset{O}{\overset{\|}{C}}H$ Acetaldehyde
Ketones	$—\overset{O}{\overset{\|}{C}}—$	$R—\overset{O}{\overset{\|}{C}}—R$	$CH_3—\overset{O}{\overset{\|}{C}}—CH_3$ Dimethyl ketone	$CH_3—\overset{O}{\overset{\|}{C}}—C_6H_5$ Methyl phenyl ketone
Acids	$—\overset{O}{\overset{\|}{C}}OH$	$R\overset{O}{\overset{\|}{C}}OH$	$CH_3\overset{O}{\overset{\|}{C}}OH$ Acetic acid	$C_6H_5\overset{O}{\overset{\|}{C}}OH$ Benzoic acid
Esters	$—\overset{O}{\overset{\|}{C}}OR$	$R\overset{O}{\overset{\|}{C}}—OR$	$CH_3\overset{O}{\overset{\|}{C}}OC_2H_5$ Ethyl acetate	$C_3H_7\overset{O}{\overset{\|}{C}}OC_2H_5$ Ethyl butyrate
Salts	$—\overset{O}{\overset{\|}{C}}O$ (Metal)	$R\overset{O}{\overset{\|}{C}}OM$	$CH_3\overset{O}{\overset{\|}{C}}ONa$ Sodium acetate	$C_2H_5\overset{O}{\overset{\|}{C}}OK$ Potassium propionate

Methyl alcohol (methanol) produced today is synthesized from either carbon monoxide or carbon dioxide.

$$CO + 2H_2 \longrightarrow CH_3OH$$
$$CO_2 + 3H_2 \longrightarrow CH_3OH + H_2O$$

Methyl alcohol is a colorless liquid boiling at 65°. In odor and taste it is similar to ethyl alcohol. Methyl alcohol is very poisonous, however; intoxication, blindness, and death may result when its vapors are breathed in quantities or when the liquid is taken internally.

Methyl alcohol is used in the manufacture of formaldehyde and other organic products; as a solvent for resins, gums, and shellac; and as a denaturant for ethyl alcohol (to make the ethyl alcohol unfit for human consumption).

Ethyl alcohol, C_2H_5OH, is the most important of the alcohols. It is also known as **grain alcohol, ethanol,** or simply as **alcohol.** It has long been prepared from starch, cellulose, and sugars of certain plants by the process of fermentation.

$$\underset{\text{Glucose}}{C_6H_{12}O_6} \xrightarrow[\text{yeast}]{\text{Enzymes in}} \underset{\text{Ethyl alcohol}}{2C_2H_5OH} + 2CO_2\uparrow$$

Solutions of alcohol resulting from fermentation contain from 8 to 12 per cent alcohol, but by fractional distillation 95 per cent alcohol can be obtained. Removal of the residual water by distillation over calcium oxide or barium oxide results in the production of **absolute alcohol** (100%).

Large quantities of ethyl alcohol are produced synthetically from both ethylene and acetylene. The synthesis from ethylene is summarized by the equation

$$
\begin{array}{ccc}
\text{H} & \text{H} & & & \text{H} & \text{H} \\
| & | & & & | & | \\
\text{H}-\text{C}{=}\text{C}-\text{H} + \text{HOH} & \underset{}{\overset{\text{H}^+}{\rightleftharpoons}} & \text{H}-\text{C}-\text{C}-\text{H} \\
& & & & | & | \\
& & & & \text{H} & \text{OH}
\end{array}
$$

Ethyl alcohol is a colorless liquid with a characteristic and somewhat pleasant odor. It is miscible with water in all proportions. The boiling point of the pure alcohol is 78.37°, but it forms a constant-boiling mixture with water that contains 95.57 per cent alcohol by weight and boils at 78.15°. Ethyl alcohol is the least toxic of all the alcohols and is present in all alcoholic beverages.

Of all synthetic organic compounds, ethyl alcohol ranks first in quantity produced and economic value. It is used as a solvent in the preparation of tinctures, essences, extracts, and varnishes. It is used in the preparation of iodoform, ether, medicinals, dyes, perfumes, collodion, and solvents for the lacquer industry. It is used to some extent as a motor fuel and is currently under study as an additive to gasolines containing no tetraethyl lead to increase the octane rating.

Alcohols containing two or more hydroxyl groups can be made. Ethylene glycol, $C_2H_4(OH)_2$, and glycerol, $C_3H_5(OH)_3$, are important examples.

$$
\begin{array}{cc}
& \text{H} \\
& | \\
\text{H} & \text{H}-\text{C}-\text{OH} \\
| & | \\
\text{H}-\text{C}-\text{OH} & \text{H}-\text{C}-\text{OH} \\
| & | \\
\text{H}-\text{C}-\text{OH} & \text{H}-\text{C}-\text{OH} \\
| & | \\
\text{H} & \text{H}
\end{array}
$$

Ethylene glycol   Glycerol

Ethylene glycol is used as a solvent and as an antifreeze in automobile radiators. Glycerol (also called glycerin) is produced, along with soap, when either fats or oils are heated with alkali (Section 27.20).

27.14 Phenols, ⬡—OH

Derivatives of benzene which contain one or more hydroxyl groups attached directly to carbon atoms of the benzene ring are called **phenols.** The hydroxyl group in these compounds is markedly acidic in character, and phenol itself,

C_6H_5OH, is commonly known as **carbolic acid.** It is an important constituent of coal tar but is produced synthetically from benzene. When pure, phenol is a colorless crystalline substance, melting at 42–43° and boiling at 181.4°. It has a very corrosive action on tissues, and on the skin it causes blisters to form rapidly. If taken internally, phenol causes irritation and necrosis of the mucous membranes. It may also paralyze the central nervous system, causing death. Very dilute solutions of phenol are used as disinfectants.

27.15 Ethers, R—O—R

The ethers are compounds obtained from alcohols, by the elimination of a molecule of water from two molecules of the alcohol. For example, when ethyl alcohol is treated with a limited amount of sulfuric acid and heated to 140°, **diethyl ether** (ordinary ether) is formed in a series of reactions which result also in the formation of water.

$$H-\overset{\overset{\displaystyle H}{|}}{\underset{\underset{\displaystyle H}{|}}{C}}-\overset{\overset{\displaystyle H}{|}}{\underset{\underset{\displaystyle H}{|}}{C}}-O[H + HO]\overset{\overset{\displaystyle H}{|}}{\underset{\underset{\displaystyle H}{|}}{C}}-\overset{\overset{\displaystyle H}{|}}{\underset{\underset{\displaystyle H}{|}}{C}}-H \longrightarrow H-\overset{\overset{\displaystyle H}{|}}{\underset{\underset{\displaystyle H}{|}}{C}}-\overset{\overset{\displaystyle H}{|}}{\underset{\underset{\displaystyle H}{|}}{C}}-O-\overset{\overset{\displaystyle H}{|}}{\underset{\underset{\displaystyle H}{|}}{C}}-\overset{\overset{\displaystyle H}{|}}{\underset{\underset{\displaystyle H}{|}}{C}}-H + HOH$$

Diethyl ether

In the general formula for ethers, R—O—R, the hydrocarbon radicals (R) may be the same or different. Diethyl ether is the most important compound of this class. It is a colorless volatile liquid (boiling point 35°), and highly flammable as a vapor. It has been used since 1846 as an anesthetic. Diethyl ether and other ethers are valuable solvents for gums, fats, waxes, and resins.

27.16 Aldehydes, R—$\overset{\overset{\displaystyle O}{\|}}{C}$—H

The alcohols represent the first stage of oxidation of hydrocarbons. Further oxidation leads to the production of compounds containing the group, —CHO. These compounds are known as **aldehydes.** When a mixture of methyl alcohol and air is passed through a heated tube containing either silver or a mixture of iron powder and molybdenum oxide, an aldehyde called formaldehyde is formed.

$$2H-\overset{\overset{\displaystyle H}{|}}{\underset{\underset{\displaystyle H}{|}}{C}}-OH + O_2 \xrightarrow{\text{Catalyst}} 2H-\overset{\overset{\displaystyle H}{|}}{C}=O + 2H_2O$$

Formaldehyde

Formaldehyde, HCHO, is a colorless gas with a pungent and irritating odor; it is soluble in water in all proportions. Formaldehyde is sold in an aqueous solution which contains about 37–37.5 per cent of formaldehyde by weight and is known as **formalin.** The solution also contains 7 per cent of methyl alcohol, which is added to inhibit the reaction of formaldehyde molecules with each other to form an insoluble polymer (compound of high molecular weight). The formaldehyde polymer is *Bakelite,* an important material which makes formaldehyde an industrially important compound. Formaldehyde causes coagulation of proteins, making it useful as a preservative of anatomical specimens and in embalming fluids. It is also useful as a disinfectant and as a reducing agent in the production of silvered mirrors.

Acetaldehyde, CH_3CHO, is a liquid that boils at 20.2°. It is colorless, water-soluble, and has an odor like that of freshly cut green apples. It is used in the manufacture of aniline dyes, synthetic rubber, and other organic materials.

27.17   Ketones, R—$\overset{\overset{\textstyle O}{\|}}{C}$—R

Ketones resemble aldehydes in certain respects, since both contain the **carbonyl group,** $>C=O$. In fact, a ketone may be regarded as an aldehyde in which the hydrogen in the aldehyde group is replaced by a hydrocarbon radical. In the general structural formula for ketones (see section heading) the R groups may be the same or different. **Dimethyl ketone,** CH_3COCH_3, commonly called **acetone,** is the simplest and most important ketone. It is made commercially by the fermentation of corn or molasses, or by the oxidation of petroleum gases. It is also one of the products of the destructive distillation of wood. Acetone is a colorless liquid boiling at 56.5° and possessing a characteristic pungent odor and a sweet taste. Among the many uses of acetone are: as a solvent for cellulose acetate, cellulose nitrate, acetylene, plastics, and varnishes; as a remover of paints, varnishes, and fingernail polish; and as a solvent in the manufacture of drugs, chemicals, smokeless powder, and the high explosive **cordite.**

27.18   Acids, R—$\overset{\overset{\textstyle O}{\|}}{C}$—OH

The **carboxyl group,** —COOH, is characteristic of organic acids. In general, the carboxylic acids are weak acids, yet they readily form metallic salts.

The simplest carboxylic acid is **formic acid,** HCOOH, which was first obtained in 1670 by the distillation of red ants, and its name was derived from the Latin word *formicus* for ant. It is partially responsible for the irritation of ant bites and bee stings.

Acetic acid, CH_3COOH, constitutes 3 to 6 per cent of vinegar. Cider vinegar is produced by allowing cider (apple juice) to ferment, causing the sugar present to change to ethyl alcohol. Then certain bacteria present in the juice produce an enzyme (see Section 28.27) that catalyzes the air oxidation of the alcohol to acetic acid. Pure, anhydrous acetic acid is a liquid boiling at 118.1°. It freezes at 16.6° forming a solid resembling ice in appearance; for this reason the pure acid is usually called **glacial acetic acid.** Acetic acid has a penetrating odor and a sour taste, and produces painful burns on the skin. It is an excellent solvent for many organic compounds and some inorganic compounds. In addition to its use as a solvent, acetic acid is essential in the production of cellulose acetate and has wide application in the textile and rubber industries.

Oxalic acid is a dicarboxylic acid represented by the formula $(COOH)_2$, or $H_2C_2O_4$. Its molecule consists of two carboxyl groups (dicarboxylic) bonded together.

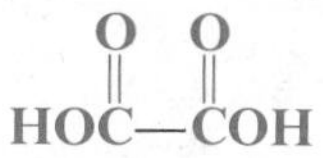

$$\underset{\text{Oxalic acid}}{\overset{\overset{\displaystyle O \quad O}{\displaystyle \| \quad \|}}{\text{HOC}-\text{COH}}}$$

Oxalic acid is a colorless crystalline solid, which is found as a metal hydrogen salt in sorrel, rhubarb, and other plants to which it imparts a sour taste. In concentrated form it is poisonous.

Lactic acid contains a hydroxyl group as well as a carboxyl group, and has the structural formula shown. This acid is produced by the fermentation of milk sugar or glucose and is the substance that imparts the sour taste to sour milk.

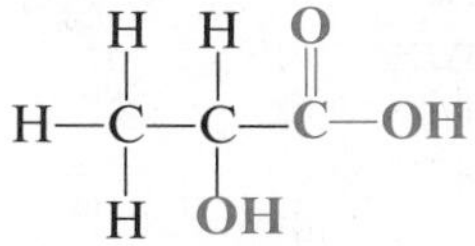

Lactic acid

Benzoic acid, C_6H_5COOH, is a colorless crystaline solid with the cyclic structural formula shown. It is a monocarboxylic acid that occurs in cranberries and coal tar. The sodium salt of benzoic acid is used in the preservation of certain foods, such as tomato ketchup and fruit juices.

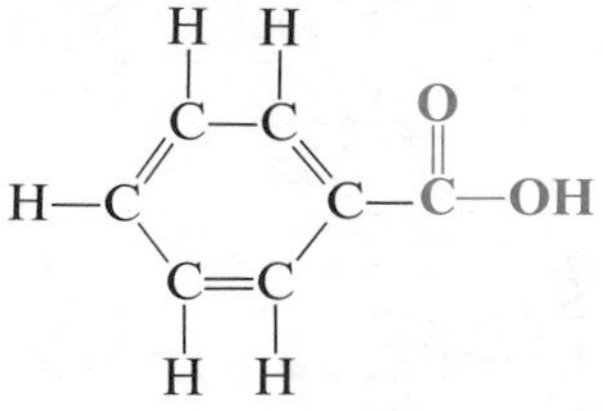

Benzoic acid

A type of organic acid commonly called a **fatty acid,** since this type of acid is commonly obtained from fats, results when a carboxyl group replaces one of the hydrogen atoms in a high molecular weight hydrocarbon (see Section 27.21). Examples are palmitic acid ($C_{15}H_{31}COOH$), stearic acid ($C_{17}H_{35}COOH$), and oleic acid ($C_{17}H_{33}COOH$).

27.19   Esters, $R\overset{\overset{\displaystyle O}{\parallel}}{-}C-OR$

Esters are the products of reaction of acids with alcohols. For example, the ester ethyl acetate, $CH_3COOC_2H_5$, is formed when acetic acid reacts with ethyl alcohol.

$$CH_3COOH + C_2H_5OH \longrightarrow CH_3COOC_2H_5 + H_2O$$

This reaction is similar to that between acetic acid and sodium hydroxide, whereby the salt sodium acetate, CH_3COONa, is formed.

$$CH_3COOH + NaOH \longrightarrow CH_3COONa + H_2O$$

However, esters differ from metallic salts of organic acids in that, generally speaking, esters are volatile liquids with a pleasing odor, they are not ionized, and they are soluble in organic solvents but not in water. The distinctive and attractive odors and flavors of many flowers and ripe fruits are due to the presence of one or more esters. Some common esters are listed in Table 27–4.

TABLE 27–4   Some Common Esters

	Formula	Odor of
Butyl acetate	$CH_3COOC_4H_9$	Bananas
Ethyl butyrate	$C_3H_7COOC_2H_5$	Pineapples
Octyl acetate	$CH_3COOC_8H_{17}$	Oranges
Methyl salicylate	$C_6H_4(OH)(COOCH_3)$	Oil of wintergreen
Methyl anthranilate	$C_6H_4(NH_2)(COOCH_3)$	Grapes

Among the most important of the natural esters are fats (such as lard, tallow, and butter) and oils (such as linseed, cottonseed, and olive). Fats and oils are esters of the trihydroxyl alcohol glycerol, $C_3H_5(OH)_3$, utilizing such high molecular weight acids as palmitic, $C_{15}H_{31}COOH$, stearic, $C_{17}H_{35}COOH$, and oleic, $C_{17}H_{33}COOH$ (see Section 28.18).

27.20   Soaps

Soaps are made by boiling natural fats and oils with strong bases such as sodium hydroxide or potassium hydroxide. When animal fat is treated with sodium hydroxide, glycerol and sodium salts of the fatty acids, palmitic, stearic, and oleic acid, are formed. The reaction in the case of glyceryl stearate (the fat) is given by

$$\underset{\text{Glyceryl stearate}}{(C_{17}H_{35}COO)_3C_3H_5} + 3NaOH \longrightarrow \underset{\text{Sodium stearate}}{3C_{17}H_{35}COONa} + \underset{\text{Glycerol}}{C_3H_5(OH)_3}$$

For the soap (the sodium salt of the fatty acid) to be free of glycerol and water, sodium chloride is added to precipitate, or "salt out," the soap. The soap, being lighter, collects as a crust on the top of the mixture, from which it is removed, partially dried, and pressed into cakes for use.

The cleansing action of soap is discussed in Section 14.16, and the action of soaps with hard water, in Section 12.6.

Polymers

Polymers, sometimes referred to as **plastics,** are compounds of very high molecular weight which are built up of a large number of simple molecules which have been caused to react with each other. Among the polymers which occur in nature are rubber, cellulose, starch, and proteins (see Chapter 28). Familiar synthetic polymers include synthetic rubber, nylon, rayon, Bakelite, and Dacron. The 1975 automobiles each contains approximately 175 pounds of polymeric materials; it is estimated that as much as 400 pounds of polymeric materials will be used per automobile in 1980 and perhaps as much as 600 pounds by 1985.

27.21 Rubber

Natural rubber is obtained mainly from the sap, called **latex,** of the rubber tree. Rubber consists of very long molecules, which are polymers formed by the union of isoprene units, C_5H_8.

$$\underset{\text{Isoprene}}{CH_2{=}\overset{\overset{\displaystyle CH_3}{|}}{C}{-}CH{=}CH_2} + \underset{\text{Isoprene}}{CH_2{=}\overset{\overset{\displaystyle CH_3}{|}}{C}{-}CH{=}CH_2} \longrightarrow$$

$$\cdots -CH_2{-}\overset{\overset{\displaystyle CH_3}{|}}{C}{=}CH{-}CH_2{\vdots}CH_2{-}\overset{\overset{\displaystyle CH_3}{|}}{C}{=}CH{-}CH_2 \cdots$$

The number of isoprene units in the rubber molecule is about 2,000, giving it a molecular weight of approximately 136,000.

Rubber has the undesirable property of becoming sticky when warmed, but the stickiness is eliminated by the process of **vulcanization.** Rubber is vulcanized by heating it with sulfur to about 140°. During the process sulfur atoms add at the double bonds in the linear polymer and form bridges which bind one rubber molecule to another. In this way a linear polymer is converted into a three-dimensional polymer. During vulcanization, fillers are also added to increase the wearing qualities of the rubber and to yield colored products. Among the substances used as fillers are carbon black, zinc oxide, antimony(V) sulfide, barium sulfate, and titanium dioxide. Most rubber also contains oil and is said to be "oil extended."

Synthetic rubbers are ordinarily not identical with natural rubber, although they resemble it and often are superior to it in several properties. For example, neoprene is a synthetic elastomer with rubberlike properties.

$$n \ CH_2=CH-\underset{\underset{Cl}{|}}{C}=CH_2 \xrightarrow{\text{Polymerization}} \left[-CH_2-CH=\underset{\underset{Cl}{|}}{C}-CH_2-\right]_n$$

Chloroprene   Neoprene

The repeating unit, chloroprene, is similar to isoprene except for a chlorine atom instead of a methyl group, $-CH_3$. Neoprene is more elastic than natural rubber, resists abrasion well, and is less affected by oil and gasoline. It is used for making gasoline and oil hose, automobile and refrigerator parts, and electrical insulation. There are many other synthetic elastomers of this type.

27.22   Synthetic Fibers

A synthetic fiber familiar to everyone is *nylon*, which is a **condensation polymer**. Two kinds of molecules take part in the condensation reaction which produces *nylon*, each of which contains two identical groups. The identical groups are $-NH_2$ and $-COOH$, respectively. One of the reactants is hexamethylenediamine and the other is adipic acid, which is a dicarboxylic acid. During condensation, molecules are formed of the type $R-NH-CO-R$, and water is eliminated. The part shown in color is called the **amide linkage**.

$$NH_2-(CH_2)_6-\overset{H}{N}\boxed{H + HO}OC-(CH_2)_4-COOH \longrightarrow$$

Hexamethylenediamine   Adipic acid

$$NH_2-(CH_2)_6-NH-CO-(CH_2)_4-COOH + H_2O$$

Because the molecule resulting from the condensation has the $-NH_2$ group at one end and the $-COOH$ group at the other, the condensation process can be repeated many times to form a linear polymer of great length.

$$(n + 1) \ H_2N(CH_2)_6NH_2 + (n + 1) \ HOOC(CH_2)_4COOH \longrightarrow$$

Hexamethylenediamine   Adipic acid

$$H_2N(CH_2)_6NH\left[\overset{O}{\overset{||}{C}}(CH_2)_4\overset{O}{\overset{||}{C}}NH(CH_2)_6NH\right]_n\overset{O}{\overset{||}{C}}(CH_2)_4COOH + 2n \ H_2O$$

Nylon

Nylon may be made into fine threads by melting and extruding through a spinneret. It is used in making hosiery and other clothing, bristles for toothbrushes, surgical sutures, strings for tennis rackets, fishing line leaders, and many other things.

Dacron is made by a similar condensation process from ethylene glycol and terephthalic acid.

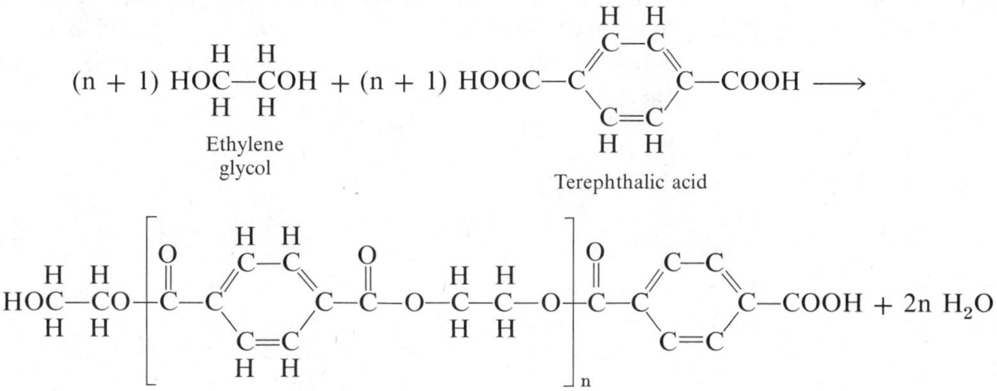

$$(n + 1)\ HOC-COH + (n + 1)\ HOOC-C \overset{\displaystyle C-C}{\underset{\displaystyle C=C}{\big\langle\big\rangle}} C-COOH \longrightarrow$$

Ethylene glycol

Terephthalic acid

Dacron

27.23 Polyethylene

The use of polyethylene in plastic bottles, bags for fruits and vegetables, and many other items has become commonplace. Polyethylene is a flexible, tough polymer which has high water resistance and excellent electrical insulating properties. Originally, it was made by polymerization of ethylene, $CH_2=CH_2$, at high temperatures and pressures using a peroxide catalyst. It is now possible, with a type of catalyst called a Ziegler catalyst, to polymerize the ethylene at atmospheric pressure and at only slightly elevated temperatures.

$$n\ CH_2=CH_2 \xrightarrow{\text{Polymerization}} [-CH_2CH_2-]_n$$

Ethylene   Polyethylene

Polypropylene is an analogous polymer, made by the polymerization of propylene.

Various derivatives of ethylene, such as vinyl chloride ($CH_2=CHCl$), vinyl acetate ($CH_2=CH-OOC-CH_3$), and tetrafluoroethylene ($CF_2=CF_2$), can be similarly polymerized. The polymerization of tetrafluoroethylene produces Teflon, a polymer which is unusually resistant to solvents and acids.

Medical Agents

27.24 Sulfa Drugs

The name **sulfa drugs** is applied to a series of sulfur-containing aromatic compounds which are exceedingly effective in the treatment of certain diseases. The parent substance in the series is sulfanilamide, which has the structural formula shown. About 1300 derivatives of sulfanilamide have been synthesized by substituting various radicals for the hydrogens of the —NH$_2$ groups of the sulfanilamide molecule. These derivatives include sulfapyridine, sulfathiazole, and sulfadiazine. They have been used in the treatment of blood poisoning, pneumonia, boils, intestinal infections, gonorrhea, meningitis, and a number of other infectious diseases.

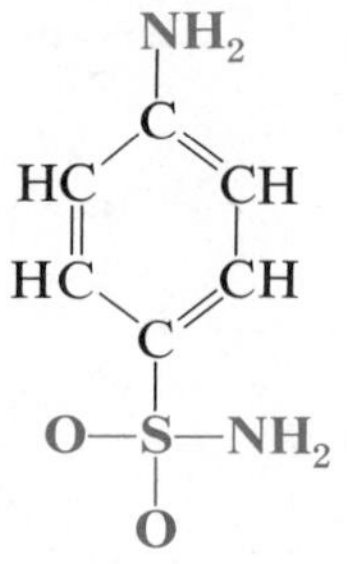

Sulfanilamide

27.25 Antibiotics

A great step forward in medical treatment was the discovery that certain soil-inhabiting bacteria and fungi produce complex organic compounds which are very active against a wide variety of pathogenic (disease producing) bacteria. The first of these so-called **antibiotics** to be produced was penicillin, which was discovered in 1928 by Dr. Alexander Fleming. It is produced by a mold, *Penicillium notatum,* which is now cultured on a commercial scale. Penicillin is particularly effective in the treatment of pneumonia, gonorrhea, syphilis, meningitis, gas gangrene, wound infections, mastoid infections, and many other infectious diseases. The chemical structure of penicillin has been determined, and the compound can now be synthesized in the laboratory. Various types of penicillin are produced by varying the radical R in the formula.

The tremendous success of penicillin as a medicinal agent led to the search for other antibiotics produced by living organisms. Several such agents have been discovered. These include streptomycin, aureomycin, neomycin, and chloromycetin. Two of these, chloromycetin and aureomycin, are capable of controlling viral infections.

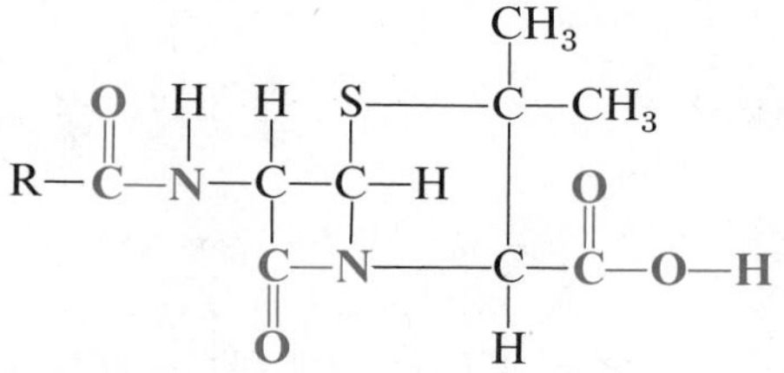

Penicillin

Other Carbon Compounds

27.26   Carbon Dioxide

Carbon dioxide is produced when any form of carbon is burned in an excess of oxygen ($C + O_2 \longrightarrow CO_2$). The same is true of almost all compounds of carbon. Some examples are

$$CH_4 + 2O_2 \longrightarrow CO_2 + 2H_2O$$
$$C_2H_5OH + 3O_2 \longrightarrow 2CO_2 + 3H_2O$$

Many carbonates liberate carbon dioxide when they are heated. An example of commercial importance is the reaction which takes place when quicklime, CaO, is produced by heating limestone.

$$CaCO_3 \longrightarrow CaO + CO_2\uparrow$$

Large quantities of carbon dioxide are obtained for commercial purposes as a by-product of the fermentation of sugar (glucose) during the preparation of alcohol and alcoholic beverages. The net reaction is given by

$$\underset{\text{Glucose}}{C_6H_{12}O_6} \xrightarrow{\text{Yeast}} \underset{\text{Ethanol}}{2C_2H_5OH} + 2CO_2\uparrow$$

In the laboratory, carbon dioxide is produced by the action of acids on carbonates.

$$CaCO_3 + 2H^+ \longrightarrow Ca^{2+} + H_2O + CO_2\uparrow$$

Carbon dioxide is a colorless and odorless gas which is 1.5 times as heavy as air. It is a component of all "carbonated" beverages. With water it forms carbonic acid, which has a mildly acid taste. One liter of water at 20° dissolves 0.9 liter of carbon dioxide. The gas is easily liquefied by compression because its critical temperature is relatively high (31.1°). When the liquid is allowed to evaporate, the vapor freezes to a snowlike solid at −56.2°. The solid vaporizes without melting (sublimes) because its vapor pressure is 1 atmosphere at −78.5°. This property makes solid carbon dioxide valuable as a refrigerant that is always free from the liquid; for this reason it is called **Dry Ice.** Temperatures as low as −77° may be obtained by mixing Dry Ice with a volatile liquid such as ether or chloroform. Such mixtures are commonly used in the laboratory for attaining low temperatures.

The valence electron formula for carbon dioxide is $\overset{\circ\circ}{\underset{\circ\circ}{O}} {\!}^{\circ}_{\circ}\! \overset{\times}{C} {\!}^{\times}_{\times}\! \overset{\circ\circ}{\underset{\circ\circ}{O}}$. The double bonds are strong, and the molecule is quite stable towards heat. At 2,000°C the dissociation into carbon monoxide and oxygen is only 1.8 per cent ($2CO_2 \rightleftharpoons 2CO + O_2$). When the heating is carried out in the presence of carbon, carbon monoxide is formed in good yield ($CO_2 + C \longrightarrow 2CO$). Burning magnesium in a carbon dioxide atmosphere reduces carbon dioxide to carbon.

$$CO_2 + 2Mg \longrightarrow 2MgO + C$$

Large quantities of carbon dioxide are used in the manufacture of **washing soda,** $Na_2CO_3 \cdot 10H_2O$, and **baking soda,** $NaHCO_3$, by the Solvay process (discussed later); in the production of **white lead,** $Pb(OH)_2 \cdot 2PbCO_3$, a paint pigment; and in the preparation of carbonated soft drinks.

Carbon dioxide is a valuable fire extinguisher because it does not support the combustion of ordinary substances, it is easily generated, and it is cheap. Air containing as little as 2.5 per cent carbon dioxide will extinguish a flame. Solid carbon dioxide which has been compressed into blocks (Dry Ice) is used as a refrigerant.

Carbon dioxide is not toxic; it can be harmful to life, however, because a large concentration in air can cause suffocation, i.e., lack of oxygen. Breathing air which contains a higher than usual percentage of carbon dioxide stimulates the respiratory centers and causes rapid breathing. Thus, carbon dioxide mixed with oxygen is used in the treatment of conditions in which rapid breathing is desirable.

27.27 Carbonic Acid and Carbonates

An aqueous solution of carbon dioxide exhibits the properties of a weak acid. **Carbonic acid,** H_2CO_3, is unstable and has never been isolated as such. The solubility of carbon dioxide in water is in accordance with Henry's Law (Section 13.4) up to a pressure of about four atmospheres. At higher pressures the solubility is greater than we should expect from Henry's Law, probably because of the reaction of the gas to form carbonic acid. When the pressure on a solution of carbon dioxide in water is decreased, effervescence takes place. This phenomenon may be observed when a bottle of a carbonated beverage is opened.

Like other diprotic acids, carbonic acid ionizes in two steps:

$$H_2CO_3 \rightleftharpoons H^+ + HCO_3^-$$
$$HCO_3^- \rightleftharpoons H^+ + CO_3^{2-}$$

When a solution of sodium hydroxide is saturated with carbon dioxide, **sodium hydrogen carbonate** is formed.

$$(Na^+) + OH^- + CO_2 \longrightarrow (Na^+) + HCO_3^-$$

The compound $NaHCO_3$ is also called sodium bicarbonate, or baking soda. Solutions of $NaHCO_3$ are weakly alkaline by hydrolysis.

$$HCO_3^- + H_2O \rightleftharpoons H_2CO_3 + OH^-$$

When an equivalent amount of sodium hydroxide is added to a solution of sodium hydrogen carbonate and crystals are allowed to form, the hydrate $Na_2CO_3 \cdot 10H_2O$ is obtained. The hydrate, washing soda, is commonly used as a water softener in the home. If this substance is heated gently, the anhydrous salt called **soda ash,** Na_2CO_3, is formed.

$$(Na^+) + HCO_3^- + (Na^+) + OH^- \rightleftharpoons (2Na^+) + CO_3^{2-} + H_2O$$

Solutions of sodium carbonate are strongly basic due to extensive hydrolysis of the carbonate ion.

$$CO_3^{2-} + H_2O \rightleftharpoons HCO_3^- + OH^-$$

27.28   The Carbon Dioxide Cycle in Nature

The atmosphere contains about 0.04 per cent by volume of carbon dioxide and thus serves as a huge reservoir of this compound. Green plants absorb carbon dioxide from the air or water, and under the influence of sunlight and chlorophyll (as a catalyst) the carbon dioxide and water become sugar and free oxygen. The reactions which take place during photosynthesis are complex, but the process may be summarized by

$$6CO_2 + 6H_2O \longrightarrow C_6H_{12}O_6 + 6O_2$$

The sugar produced during photosynthesis is glucose, $C_6H_{12}O_6$; it is the sugar that green plants use to synthesize substances such as other sugars, starches, cellulose, and fats.

Carbon dioxide is a product of respiration and is returned to the air by the green plants and by animals. The organisms (mostly nongreen plants) which produce decay of both plant and animal matter and those that cause fermentation of sugars also assist in the production of carbon dioxide, as does the combustion of carbon-containing fuels. The air from volcanoes and from certain other geological formations are other sources of carbon dioxide in the atmosphere. Because of the solubility of carbon dioxide in water, oceans and lakes are great reservoirs of this compound. The ocean reservoir is to a great extent responsible for the fact that the carbon dioxide content of the air is almost constant. Nevertheless, the carbon dioxide content of the atmosphere has increased detectably in the last few years because we burn so much fuel to generate power.

27.29   Carbon Monoxide

A second oxide of carbon, the **monoxide** (CO), is prepared in the laboratory by heating crystals of oxalic acid, $H_2C_2O_4$, with concentrated sulfuric acid; the latter compound dehydrates the oxalic acid.

$$H_2C_2O_4 \xrightarrow{\text{Concd. } H_2SO_4} H_2O + CO_2 + CO$$

The carbon monoxide may be removed from the mixture of gases by passing the mixture through solid sodium hydroxide, which absorbs the carbon dioxide.

When a limited supply of air is passed through hot coke, a mixture of carbon monoxide and nitrogen, in a ratio of about 1 to 2 by volume, is obtained. This product, called **producer gas,** is used as an inexpensive and economical gaseous fuel for certain industrial operations.

When steam is passed through a bed of red-hot coke, a mixture of hydrogen and carbon monoxide, called **water gas,** is formed.

$$C(s) + H_2O(g) + 31{,}380 \text{ cal} \longrightarrow H_2(g) + CO(g) \qquad (\Delta H^\circ_{298} = +31{,}380 \text{ cal})$$

Because the reaction is endothermic, the coke soon becomes too cool to reduce the steam. Hence, oxygen and steam are passed together through the coke to keep it heated to redheat. Water gas is used extensively as an industrial fuel and as a source of hydrogen.

Carbon monoxide is a colorless, odorless, and tasteless gas, which is only slightly soluble in water. Liquid carbon monoxide boils at $-192°$ and freezes at $-205°$.

The valence electron formula for carbon monoxide is ${}^{\circ}_{\circ}C{}^{\circ}_{\circ}{}^{\times}_{\times}O{}^{\times}_{\times}$. This arrangement of electrons gives a completed shell to each atom. The chemical behavior of carbon monoxide is in contrast with that of carbon dioxide. Carbon monoxide readily burns in oxygen forming the dioxide.

$$2CO(g) + O_2(g) \longrightarrow 2CO_2(g) + 135{,}270 \text{ cal} \qquad (\Delta H^\circ_{298} = -135{,}270 \text{ cal})$$

This reaction makes carbon monoxide valuable as a gaseous fuel.

At high temperatures carbon monoxide will reduce many metallic oxides and is important, therefore, as a reducing agent in metallurgical processes. Two examples are

$$CuO + CO \longrightarrow Cu + CO_2$$
$$FeO + CO \longrightarrow Fe + CO_2$$

Carbon monoxide combines with chlorine in the presence of sunlight and charcoal (as a catalyst) forming **carbonyl chloride,** $COCl_2$, which is also called **phosgene.**

Carbon monoxide combines directly with several metals forming **metallic carbonyls.** Examples of this type of compound are **iron carbonyl,** $Fe(CO)_5$; **nickel carbonyl,** $Ni(CO)_4$; and **cobalt carbonyl,** $Co_2(CO)_8$.

Carbon monoxide is a very dangerous poison. It is especially dangerous because it is odorless and tasteless and therefore gives no warning of its presence. It combines with the hemoglobin of the blood and forms a compound which is too stable to be broken down by the body processes. In this way carbon monoxide destroys the ability of the blood to carry oxygen. An anemic condition is produced when small amounts of the gas are breathed over a long period. One volume of carbon monoxide in 800 volumes of air will cause death in about thirty minutes, because the carbon monoxide paralyzes the respiratory organs.

The danger of carbon monoxide poisoning from the exhaust gas from automobile engines is particularly great. A small engine idling in a closed garage may produce sufficient carbon monoxide to cause death within five minutes. Other sources of carbon monoxide poisoning are faulty ventilation of stoves and furnaces, and leaky connections in the gas supply.

27.30   Carbon Disulfide

At the temperature of the electric furnace carbon is oxidized by sulfur to **carbon disulfide.**

$$C + 2S \longrightarrow CS_2$$

Air must be excluded because the volatile carbon disulfide is highly flammable and burns according to the equation

$$CS_2 + 3O_2 \longrightarrow CO_2 + 2SO_2$$

Carbon disulfide is also produced commercially by burning methane in sulfur vapor at about 700° in the presence of either silica gel or activated alumina.

Pure carbon disulfide is a colorless liquid, which boils at 46.3°. The commercial product is yellow and has a disagreeable odor. The liquid is heavy (specific gravity 1.27 to 1.29), has a high index of refraction, and is immiscible with water. The vapor is heavier than air, very poisonous, and highly flammable.

Large quantities of carbon disulfide are used in making rayon by the viscose process and in the manufacture of cellophane and carbon tetrachloride.

27.31 Carbon Tetrachloride

Carbon tetrachloride, CCl_4, is manufactured by passing chlorine into carbon disulfide containing iodine or antimony pentachloride, which act as catalysts.

$$CS_2 + 3Cl_2 \longrightarrow CCl_4 + S_2Cl_2$$

The carbon tetrachloride (b.p. 76.7°) is readily separated from the higher-boiling "sulfur monochloride" (b.p. 138°) by fractional distillation.

Carbon tetrachloride is a colorless, pleasant-smelling liquid, with a specific gravity of 1.58, and a freezing point of −23°. It is an excellent solvent for fats, oils, and greases; therefore it is used in the dry cleaning of various fabrics. Because its vapor is about five times as heavy as air and it does not burn, it makes a good fire extinguisher. The fact that it is a nonconductor of electricity makes carbon tetrachloride particularly useful in fighting fires involving electrical equipment where water might cause short circuits. It is a poison and the vapor should not be inhaled.

27.32 Calcium Carbide

Calcium carbide, CaC_2, is an important commercial product, which is made by heating quicklime with coke in an electric furnace.

$$CaO + 3C \longrightarrow CaC_2 + CO$$

The product, which is liquid at the temperature of the furnace, is drawn off and solidified by cooling. Calcium carbide is an ionic compound with the valence electron formula $Ca^{2+}[\overset{\times}{\underset{\circ}{C}}\overset{\circ}{\underset{\circ}{\circ}}\overset{\times}{\underset{\times}{C}}\overset{\times}{\underset{\circ}{}}]^{2-}$. It hydrolyzes in water forming acetylene, $HC{\equiv}CH$ (Section 25.11).

$$CaC_2 + 2H_2O \longrightarrow Ca(OH)_2 + C_2H_2\uparrow$$

Acetylene is an important compound that is used in the synthesis of many organic materials such as ethyl alcohol, synthetic fabrics, synthetic rubber, and plastics. It is used also as an illuminant, and as a fuel for cutting and welding metals.

27.33 Cyanogen

Cyanogen, C_2N_2, is a colorless, very poisonous gas that burns with a blue flame. The molecule is linear and has the structure $N\equiv C-C\equiv N$. It may be prepared by heating a solution containing copper(II) ions and cyanide ions.

$$2Cu^{2+} + 6CN^- \longrightarrow 2Cu(CN)_2^- + C_2N_2\uparrow$$

Cyanogen hydrolyzes in a manner analogous to that of molecular chlorine, as shown by the equation

$$C_2N_2 + H_2O \longrightarrow \underset{\text{Hydrogen cyanide}}{HCN} + \underset{\text{Cyanic acid}}{HOCN}$$

27.34 Hydrogen Cyanide

Hydrogen cyanide, HCN, is a gas with an odor like that of bitter almonds. It is formed when a cyanide is treated with an acid.

$$(Na^+) + CN^- + H^+ + (Cl^-) \longrightarrow (Na^+) + (Cl^-) + HCN\uparrow$$

The liquid boils at $25.7°$ and freezes at $-13.2°$. When hydrogen cyanide is dissolved in water, hydrocyanic acid is formed. This acid is so weak that it will not turn blue litmus red. Its salts, such as sodium cyanide, form solutions which are alkaline by hydrolysis.

Hydrogen cyanide is very poisonous. A dose of about 0.05 g is fatal to man; moreover, its toxic action is very rapid. It paralyzes the central nervous system and inhibits all tissue respiration by combining with the iron in the respiration enzymes which promote respiration.

Cyanide ions resemble halide ions in several ways—to an extent, in fact, that cyanide ions are sometimes even referred to as "pseudo-halide" ions. Silver cyanide, like silver chloride, is almost insoluble in water. Cyanide ions form stable complexes with many metals. For example, sodium cyanide is used extensively in the extraction of gold and silver from their ores; these processes involve the formation of complex cyanides.

$$4Au + 8NaCN + 2H_2O + O_2 \longrightarrow 4NaAu(CN)_2 + 4NaOH$$

QUESTIONS

1. Describe the crystal structure of graphite and diamond and relate the physical properties of each to their structures.
2. What experimental evidence shows that diamond contains more internal energy than graphite? How is this fact related to the manufacture of synthetic diamonds?
3. Write chemical equations describing three reactions in which elemental carbon acts as a reducing agent and three reactions in which it acts as an oxidizing agent.

4. In terms of the bonding properties of carbon, account for the fact that it forms such a great number of compounds.
5. What is meant by the "destructive distillation" of coal and what are the products of this process?
6. Explain the formation of coal by the carbonization of vegetable matter.
7. Why is charcoal such an effective adsorbent?
8. What is meant by a homologous series of compounds?
9. Write general formulas for and name the types of hydrocarbon derivatives studied in this chapter.
10. Write the equations for a substitution reaction of ethane and an addition reaction of ethylene.
11. Write the structural formulas for the five isomers of hexane.
12. Draw the structural formulas for all possible isomers of C_5H_{10}, $C_3H_8O_2$.
13. Write structural formulas for ethylbenzene, chlorobenzene, dichloromethane, and orthodibromobenzene.
14. Write the equations for the synthesis of acetylene from limestone and coke.
15. By what process are heavy molecular weight hydrocarbon molecules rendered suitable for use as a motor fuel? By what process are light molecular weight hydrocarbon molecules rendered suitable for use as a motor fuel?
16. Describe the two conversion processes for producing gasoline synthetically.
17. Draw the structure for cyclohexane, C_6H_{12}.
18. What is the difference between the structures of the saturated and unsaturated hydrocarbons?
19. Write a general empirical formula for alkenes having *two* double bonds.
20. What is the oxidation state of carbon in each of the following compounds: CH_4, CO_2, CCl_4, CH_3Cl, CH_3OH, CH_3OCH_3, and $HCOOH$?
21. How do alcohols differ from inorganic hydroxides, and esters from salts?
22. What are the products which may be formed in the stepwise oxidation of methane? of ethyl alcohol?
23. Why is C_2H_5OH soluble in water whereas C_2H_6 is not?
24. What is absolute alcohol and how is it prepared?
25. Write structural formulas for propionaldehyde, methyl ethyl ether, diethyl ketone, methyl acetate, and propionic acid.
26. Describe the preparation of the soap sodium stearate.
27. What two kinds of hydrocarbon derivatives are represented by lactic acid?
28. Which of the following pairs of compounds could be distinguished from one another by means of quantitative analysis for carbon and hydrogen: *n*-butane and *iso*butane; cyclohexane and hexane; pentadiene and pentyne; ethylene and polyethylene; cycloheptane and heptene; acetylene and ethylene; benzene and phenol; ethanol and dimethyl ether; cyclohexatriene and benzene; ethyl acetate and diethylketone; formaldehyde and formic acid; ethylene glycol and methanol; cyclohexene and propene?
29. What are polymers? List five naturally occurring polymers.
30. Describe the chemical composition of several synthetic polymers.
31. Describe the chemistry involved in the vulcanization of rubber.
32. How can Na_2CO_3 and $NaHCO_3$ be made from $NaOH$ and CO_2?

33. How is carbon dioxide produced commercially?
34. Describe the production of Dry Ice. What are the advantages of its use as a refrigerant over "water ice"?
35. Relate photosynthesis and respiration to the carbon dioxide cycle in nature.
36. Why does not carbon dioxide support the combustion of most combustible substances?
37. Write the equation describing the chemical reaction that occurs when carbon dioxide is bubbled through sodium hydroxide solution.
38. Explain the toxicity of carbon monoxide.
39. Compare the heat of combustion of carbon monoxide to that of carbon. Explain.
40. Write the electronic structures for carbon monoxide and for phosgene.
41. Write balanced equations for the complete combustion of CH_4, CH_3OH, CH_2O_2, and C_2H_2.
42. How are "producer gas" and "water gas" manufactured?
43. Write equations for the production of carbon disulfide, carbon tetrachloride, calcium carbide, acetylene, cyanogen, and hydrogen cyanide.
44. On the basis of orbital hybridization in the carbon atoms, explain why cyclohexane exists in the form of a "puckered" (bent) ring and benzene in the form of a planar ring.
45. Discuss the concept of resonance in terms of molecular orbitals, as it applies to benzene. Write structural formulas for two other carbon compounds that you would expect to undergo resonance.
46. Explain in terms of molecular orbitals the increasing reactivity of corresponding compounds in the three series, in the order: alkanes, alkenes, alkynes.

PROBLEMS

1. What volume of water gas at 600° K and 750 mm would result from the action of 100 g of water on carbon?   *Ans. 554 liters*
2. How much $CaCO_3$ would be formed upon the addition of 10.0 liters of CO_2, measured at 27° and 770 mm, to an excess of limewater?   *Ans. 41.2 g*
3. How much heat energy would be released by the combustion of one gram of C_2H_6? C_2H_2?   *Ans. 11.35 kcal/g; 11.53 kcal/g*
4. (a) What volume of $C_2H_2(g)$ (S.T.P.) would result from the hydrolysis of 10.0 g of CaC_2?   *Ans. 3.49 liters*
   (b) How many grams of magnesium would have to react with HCl to provide enough hydrogen to reduce the C_2H_2 of part (a) to ethane?

                                                                    Ans. 7.58 g
5. How many head of cattle would need to be slaughtered each year to furnish sufficient soap to take the place of the phosphate-based detergent described in Problem 3 of Chapter 26? Assume that a weight of soap (sodium stearate) equal to the weight of detergent is required per washing and that an average steer furnishes 170 pounds of rendered tallow (glyceryl tristearate).

                                                                    Ans. 14,800,000

(*Note:* By way of comparison, approximately thirty million cattle are slaughtered in the United States in a typical year.)

6. (a) From the reactions

$$C_2H_4(g) + 2H_2(g) \longrightarrow 2CH_4(g)$$
$$C(g) + 2H_2(g) \longrightarrow CH_4(g)$$

and by considering the bonds that must be broken and formed to get from reactants to products, show that the C=C bond energy, D(C=C), is given by:

$$D(C{=}C) = \Delta H^\circ_{f_{C(g)}} + \Delta H^\circ_{f_{CH_4(g)}} - \Delta H_{f_{C_2H_4(g)}}$$

Calculate, using data in Appendix J, a value for D(C=C).

Ans. 141 kcal

(b) By considerations similar to those in (a) show that the reaction

$$C_2H_4(g) + H_2(g) \longrightarrow C_2H_6(g)$$

leads to the conclusion that

$$D(C{=}C)$$
$$= D(C{-}C) + 2D(C{-}H) - D(H{-}H) + \Delta H^\circ_{f_{C_2H_6(g)}} - \Delta H^\circ_{f_{C_2H_4(g)}}$$

and calculate, using data in Table 20-2 and Appendix J, a value for D(C=C). *Ans. 144 kcal*

(c) Compare your answers in (a) and (b) to each other and to the value given in Section 20.7 of Chapter 20.

7. Assuming a value of (n) equal to six in the formula for nylon, calculate how many pounds of hexamethylenediamine would be needed to produce 1.00 ton of nylon. (See Appendix C for conversion factors.) *Ans. 1.02 × 10³ lb*

8. (a) How many grams of lactic acid would be needed to neutralize the same amount of base as is neutralized by 25.0 g of benzoic acid? (Only the hydrogen of the —COOH group of each of the acids is used in neutralizing a base.) *Ans. 18.4 g*

(b) How much oxalic acid would be needed? *Ans. 9.22 g*

9. If you wished to convert penicillin to its sodium salt (i.e., neutralize the hydrogen in the carboxyl group, —COOH, with sodium hydroxide), how much of the salt could you produce from 100 g of penicillin? (Assume the R group is —CH₃.) *Ans. 109 g*

10. How much of acetic acid, in grams, is in exactly one quart of vinegar if the vinegar contains 3.00% acetic acid by volume? The density of acetic acid is 1.049 g/ml. (See Appendix C for conversion factors.) *Ans. 29.8 g*

REFERENCES

"The Shapes of Organic Molecules," J. B. Lambert, *Sci. American*, Jan., 1970; p. 58.

"Criteria for Optical Activity in Organic Molecules," D. F. Mowery, Jr., *J. Chem. Educ.*, **46**, 269 (1969).

"Organic Lasers," P. Sorokin, *Sci. American*, Feb., 1969; p. 30.

"Donor-Acceptor Interactions in Organic Chemistry," S. G. Sunderwith, *J. Chem. Educ.*, **47**, 728 (1970).

"Physicochemical Properties of Highly-Fluorinated Organic Compounds," C. R. Patrick, *Chem. in Britain*, **7**, 154 (1971).

"Control Elements in Organic Synthesis," S. Turner, *Chem. in Britain*, **7**, 191 (1971).

"Kinetics and Mechanisms in Organic Chemistry," R. Baker, *Chem. in Britain*, **4**, 250 (1968).

"Organometallic Chemistry," M. F. Lappert, *Chem. in Britain*, **5**, 342 (1969).

"Polymer Models," C. E. Carraher, Jr., *J. Chem. Educ.*, **47**, 581 (1970).

"The Nature of Polymeric Materials," H. F. Mark, *Sci. American*, Sept., 1967; p. 149.

"Chemical Collaboration Between Polymers and Environment," B. Saville, *Chem. in Britain*, **6**, 65 (1970).

"The Isomerization of Xylenes," K. G. Harbison, *J. Chem. Educ.*, **47**, 837 (1970).

"The Carbon Cycle," B. Bolin, *Sci. American*, Sept., 1970; p. 124.

"The Stereochemical Theory of Odor," J. E. Amoore, J. W. Johnston, Jr., and M. Rubin, *Sci. American*, Feb., 1964; p. 42.

"The Chemistry of Flavor," I. Hornstein and R. Teranishi, *Chem. & Eng. News*, April 3, 1967; p. 92.

"Flower Pigments," S. Clevenger, *Sci. American*, June, 1964; p. 85.

"Clathrates: Compounds in Cages," Sister M. M. Hagan, *J. Chem. Educ.*, **40**, 643 (1963).

"Molecular Isomers in Vision," R. Hubbard and A. Kropf, *Sci. American*, June, 1967; p. 64.

"Color and Chemical Constitution," N. J. Juster, *J. Chem. Educ.*, **39**, 596 (1962).

"The Synthesis of Diamond," H. T. Hall, *J. Chem. Educ.*, **38**, 484 (1961).

"The Chemistry and Manufacturing of the Lead Pencil," F. L. Encke, *J. Chem. Educ.*, **47**, 575 (1970).

"Organotin Chemistry," A. G. Davies, *Chem. in Britain*, **4**, 403 (1968).

"Hydrocarbon Reactions on Transition Metal Catalysts," J. J. Rooney, *Chem. in Britain*, **2**, 242 (1966).

"Kekulé and Benzene," C. A. Russell, *Chem. in Britain*, **1**, 141 (1965).

"Free Radicals and Aging," W. A. Pryor, *Chem. & Eng. News*, June 7, 1971; p. 34.

"Artificial Organs," H. J. Sanders, *Chem. & Eng. News*: Part I, April 6, 1971; p. 32. Part II, April 12, 1971; p. 68.

"Textiles in the Seventies," W. S. Fedor, *Chem. & Eng. News*, April 20, 1970; p. 65.

"Rubber in the Seventies," E. L. Carpenter, *Chem. & Eng. News*, April 27, 1970; p. 31.

"Petroleum Fuels and Cleaner Air," G. S. Parkinson, *Chem. in Britain*, **7**, 239 (1971).

"Catalysis," V. Haensel and R. L. Burwell, Jr., *Sci. American*, Dec., 1971; p. 46.

"The Carbon Chemistry of the Moon," G. Eglinton, J. R. Maxwell, and C. T. Pillinger, *Sci. American*, Oct., 1972; p. 80.

"The Nature of Aromatic Molecules," R. Breslow, *Sci. American*, Aug., 1972; p. 32.

"The Gasification of Coal," H. Perry, *Sci. American*, Mar., 1974; p. 19.

"Coal and the Present Energy Situation," E. F. Osborn, *Science*, **183**, 477 (1974).

"Methanol as a Gasoline Extender: A Critique," E. E. Wigg, *Science*, **186**, 785 (1974).

"Asymmetric Synthesis," J. W. Scott and D. Valentine, Jr., *Science*, **184**, 943 (1974).

"Superhard Materials," F. P. Bundy, *Sci. American*, Aug., 1974; p. 62.

"Spatial Configuration of Macromolecular Chains," P. J. Flory, *Science*, **188**, 1268 (1975).

"Cyanate and Sickle-Cell Disease," A. Cerami and C. M. Peterson, *Sci. American*, April, 1975; p. 45.

"High Energy Reactions of Carbon," R. M. Lemmon and W. R. Erwin, *Sci. American*, Jan., 1975; p. 72.

28 Biochemistry

Biochemistry is the study of the chemical composition and structure of living organisms and the chemical reactions that take place in these organisms. During the past century, biochemistry has developed into an important branch of science. Many of the molecules found in living cells are structurally and functionally among the most complicated organic substances known to mankind. Therefore, biochemists must be broadly trained in the fundamental principles of chemistry in order to study the detailed structures of these complex molecules and the mechanisms of biochemical reactions.

Living organisms are composed principally of water, proteins, carbohydrates, lipids (fats and oils), nucleic acids, porphyrins, and inorganic substances. Vitamins, enzymes, and hormones are some of the important functioning components of living organisms. Some of these will be considered briefly in this chapter.

28.1 Water in Living Organisms

From the standpoint of weight, water is the principal constituent of living matter. Most plants and animals contain 60 to 90 per cent of water by weight, the content varying with the kind of organism and with the organ or tissue concerned. For example, bones are 12 to 15 per cent water, whereas the water content of the blood is about 80 per cent.

The principal function of water in a living organism is that of a solvent for both inorganic and organic substances which must be transported through the organism. Foods, when digested, are carried to various parts of the body in aqueous solution. Likewise, many waste products are eliminated from the body as water-soluble substances. The oxygen which we breathe dissolves in water before it passes through the tissues of the lungs and into the blood. Plants extract their supply of nutrients from the soil in aqueous solutions. Food manufactured in the leaves of a plant is transported in water solution to the storage places such as roots and seeds.

The large amount of water in the body also helps to maintain the average body temperature, because of the high heat capacity of water. Since a relatively large amount of heat is required to raise the temperature of water even a few degrees, the body is able to absorb excess heat without an appreciable change in temperature. In addition, the high heat of vaporization of water makes its evaporation from the skin effective in cooling the body.

Plants use water, along with carbon dioxide and nitrogen (or nitrogen compounds), as a raw material in the production of organic molecules. On the other hand, water and carbon dioxide are formed in the animal body as end products of the oxidation of organic matter.

The water of the body fluids, such as the blood, lymph, and urine, is called **free water,** because it is free to flow. However, much of the water in plant and animal tissues does not flow out when the tissue is cut, indicating that it is somehow bound or immobilized. This is spoken of as **bound water.** Some of the bound water is held mechanically in a lattice of fibrous molecules and membranes, while other portions of it are chemically bonded to body tissues as water of hydration.

Biochemical Energetics

28.2   Metabolism

Metabolism is the general term applied to any chemical reactions that occur in the cells. Two types of metabolism are **catabolism** and **anabolism.**

Catabolism refers to the breakdown of molecules, usually organic molecules supplied as food for the cell, with the accompanying release of chemical energy. Examples are the breakdown of sugar to carbon dioxide and of amino acids to urea. Anabolism refers to the building up (synthesis) of more complex structures from simpler ones. Anabolic processes require an input of chemical energy. Examples are the fixation of nitrogen (Section 25.26), whereby molecular nitrogen is converted first to ammonia and then to complex organic nitrogen compounds; the formation of fatty acids from acetic acid; and the formation of proteins from amino acids (Section 28.7).

28.3   Energy Required by Cells

The energy required for the growth and maintenance of the cellular structures necessary for life is provided by reactions of three general types:

(1) *Photosynthesis,* a process by which light energy is converted into chemical energy that can be "stored" in organic compounds and used when needed.

(2) *Autotropic reactions,* in which organisms grow without any organic energy source or light energy. The energy normally results from oxidation in exothermic inorganic reactions such as the oxidation of iron(II) to iron(III), nitrite ion to nitrate ion, sulfur to sulfate ion, and hydrogen sulfide to sulfate ion.

(3) *Respiration or fermentation of organic compounds,* reactions which provide

energy for a large number of different kinds of cells. These reactions, by which cells obtain energy, can conveniently be classified in two categories: (a) those which occur in the absence of molecular oxygen **(anaerobic reactions),** and (b) those which require molecular oxygen for oxidation **(aerobic reactions).** Reactions which occur anaerobically can result only by the rearrangement of bonds between carbon, hydrogen, and oxygen atoms within the organic compound and water, thereby yielding compounds which have less energy than the starting materials. The extent of these energy changes, however, is quite limited compared to the large energy changes possible with the oxidation-reduction reactions occurring aerobically.

Yeast, bacteria, and in some instances muscle tissue can anaerobically break down complicated organic molecules such as glucose into simpler molecules such as ethanol, acetic acid, or lactic acid, thereby obtaining energy from these reactions. For example, the change of a mole of glucose to two moles of lactic acid produces 58,000 calories. (More than ten times this amount of energy—688,000 calories per mole—can be obtained by the complete combustion of glucose aerobically to form carbon dioxide and water.) Some of the energy produced by these reactions, however, is dissipated in the form of heat and therefore is not available for use by the cell. A considerable portion of the energy from these reactions is stored as chemical energy in compounds such as the nucleoside triphosphates. Examples of nucleoside triphosphates are adenosine triphosphate (ATP), guanosine triphosphate (GTP), uridine triphosphate (UTP), and cytosine triphosphate (CTP). These compounds can release the chemical energy by cleavage of a phosphorus-oxygen bond, usually by reaction with water (hydrolysis). The equation for the hydrolysis of ATP is shown.

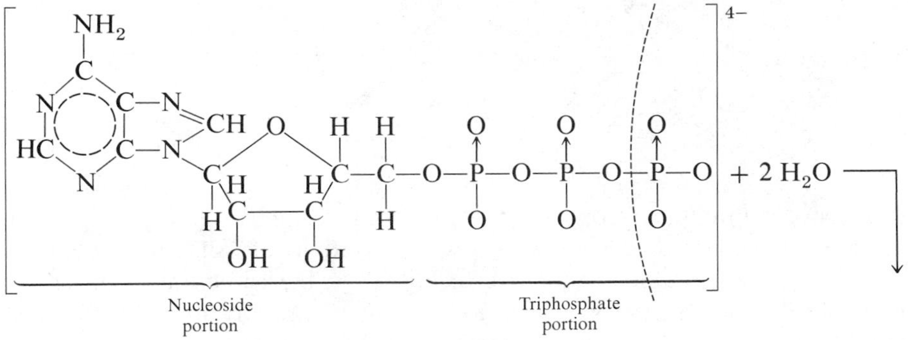

Adenosine triphosphate (ATP)

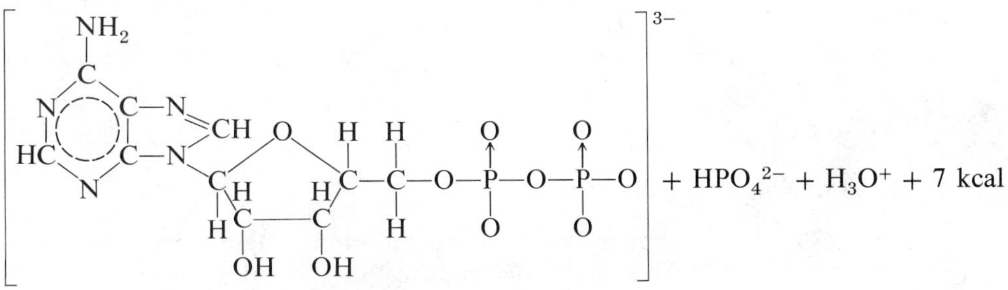

Adenosine diphosphate (ADP)

The energy released by these and similar reactions can be utilized by the cell for muscular contraction and locomotion, for initiating synthetic chemical reactions to form the structural and functional components of the cell (anabolic reactions), for the emission of light (for example, by fireflies), for the concentration of substances by the cell, for transporting substances through membranes, and for initiating specific metabolic reaction sequences.

The formation of these energy-rich phosphate compounds, which occurs in the **cytoplasm** of the cell, represents an important mechanism for extracting energy from the metabolism of foodstuffs. In aerobic cells, the phosphate compounds are formed in specific subcellular structures called **mitochondria,** which can be considered the "power plants" of the cell. The term cytoplasm refers to all parts of the cellular structure except the nucleus and the cell wall (Fig. 28–1). The nucleus is a dense, small body in the interior of the cell. The nucleus of a cell, when stained with a dye, can easily be seen with an optical microscope. The nucleus is less diffuse and heterogeneous than the cytoplasm. The mitochondria are relatively large bodies, typically about 0.002 millimeter long and 0.005 milli-

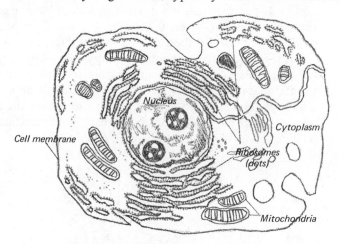

FIGURE 28–1
Simplified drawing of a living cell.

meter wide, within the cytoplasm portion of the cell. Everything within the cytoplasm of the cell that is soluble in water solutions of salts and proteins is called the **cytosol** (soluble cytoplasm). In Fig. 28–1, the white open portions of the cytoplasm constitute the cytosol. The mitochondria are surrounded and bathed by the cytosol, which is a solution, or perhaps more nearly a colloidal sol (Section 14.13), of those components of the cytoplasm that cannot be brought down by centrifuging.

Proteins

28.4   Importance of Proteins

Proteins are among the most important types of substances present in living cells. They are found in all cells of living organisms. The mitochondria each contain about one million protein molecules of typical size, many of which are identical.

The human body contains many thousands of different proteins, each with a special structure that permits it to carry on a specific function. Some proteins are specific catalysts called **enzymes** and others, insulin for example, act as **hormones** and cause specific physiological responses. Proteins also make up the contractile substance of muscle cells as well as the structural framework of animal cells.

28.5   Composition of Proteins

All proteins are nitrogenous substances, most of them containing 12–19 per cent nitrogen, 50–55 per cent carbon, 6–7 per cent hydrogen, and 20–23 per cent oxygen; some proteins contain small amounts of other elements such as sulfur, phosphorus, iron, and copper. Proteins occur as distinct molecules with highly specific structures and very large molecular weights ranging from about 10,000 to many millions.

28.6   Amino Acids

When proteins are heated with aqueous acids or bases or are treated with hydrolytic enzymes, they undergo hydrolysis (see Chapter 18), producing compounds called **amino acids.** The amino acids from proteins are carboxylic acids which contain an amino group ($-NH_2$) attached to the carbon atom adjacent to the carboxylic group ($-COOH$). This being the α-carbon, these acids are called α-amino acids. The simplest of these is glycine, H_2N-CH_2-COOH. The other natural amino acids contain other organic groups in place of one of the hydrogen atoms on the alpha carbon atom, making their general formula $H_2N-CHR-COOH$. The R in the formula represents the organic group present; for example, R is $-CH_3$ in alanine, $-CH_2OH$ in serine, and $-CH_2-C_6H_5$ in phenylalanine. Twenty-six different amino acids have been found as constituents of proteins.

28.7   Structure of Proteins

The basic structure of proteins was determined by the German chemist Emil Fischer between 1900 and 1910. He found that protein molecules consist of long chains of α-amino acids linked by acid amide bonds, $-CO-NH-$, which are referred to as **peptide links,** or **peptide bonds.** When two amino acids are joined together by a peptide link the molecule thus formed is called a **dipeptide.** The reaction is similar in principle to that for the formation of nylon (Section 27.22) and may be represented by

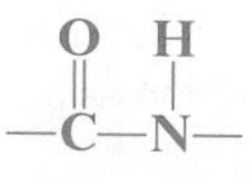

$$\begin{matrix} O & H \\ \| & | \\ -C-N- \end{matrix}$$

Peptide link

$$H_2NCHRCOH + HNCHRCOOH \rightleftharpoons H_2NCHRC-N-CHRCOOH + H_2O$$

Amino acid   Amino acid   Dipeptide

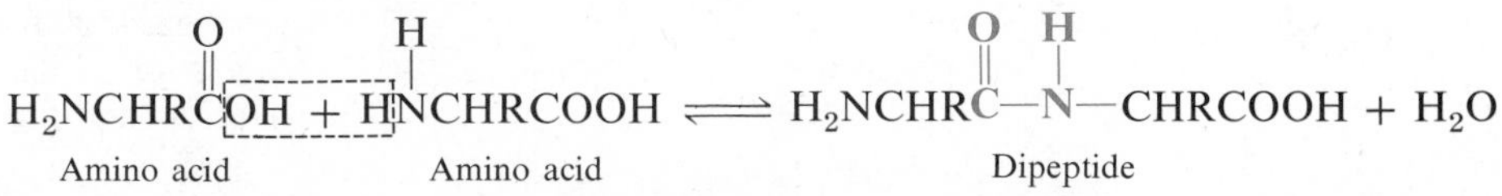

A protein of molecular weight 120,000 would be a polypeptide containing about 1,000 amino acid residues linked together by peptide bonds, inasmuch as the average molecular weight of the —NHCHRCO— group is approximately 120. Protein molecules differ from one another not only in the number and kind of amino acid residues, but also in the order in which the residues are arranged in the polypeptide chains. This means that the number of possible protein structures is extremely great. Each protein has a distinct order of arrangement of the amino acids in the peptide chain; this order of arrangement is called the **primary structure** of the protein.

It is now known that some proteins have folded sheet-like structures, whereas others are helical. The helical form, referred to as an **alpha-helical structure,** is exhibited by many fibrous proteins such as hair and muscle. These proteins consist of polypeptide chains with the configuration of a helix, the chains being arranged nearly parallel to each other and twisted in a helical shape because of hydrogen bonding between groups on the same chain. The axis of the helix lies in the direction of the fiber. The net effect in many fibrous proteins is a rope- or cable-like structure in which the polypeptide chains are twisted about one another (Figs. 28–2 and 28–3).

FIGURE 28-2

The α-helical structure of the peptide chain, referred to as the secondary structure of a protein molecule. Hydrogen bonding is indicated by the dashed lines.

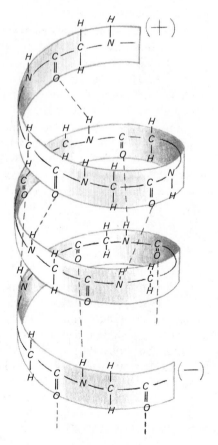

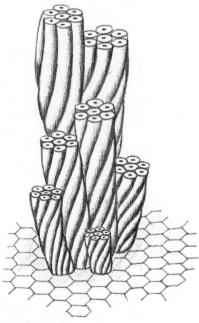

FIGURE 28-3

Diagram of the structure of a protein tissue (keratin) possessing the α-helix configuration. Fibrous proteins often consist of seven strands, arranged so as to appear cablelike. Other structures are believed to have fewer strands but each possessing the helical spatial arrangement.

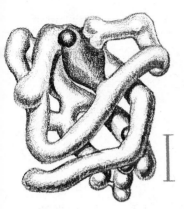

FIGURE 28-4

The tertiary structure of a protein molecule. The distance between the marks on the vertical scale at the right represents 10 Å.

Discoveries on protein structure have been made through the application of x-ray diffraction techniques; much of the credit for this very important work belongs to Linus Pauling, who received the Nobel Award in chemistry for the year 1954. The helical structure of the peptide chain is called the **secondary structure** of the protein. Detailed x-ray studies by Kendrew and Perutz, Nobel Award winners in 1962, have established that both the helical and nonhelical regions of the peptide chain are arranged in three-dimensional space in a highly specific manner, which gives rise to a specific **tertiary structure** of the molecule (Fig. 28-4). For example, an important factor in establishing and stabilizing the tertiary structure is the disulfide bond, made up of two covalently bonded sulfur atoms, which may link together two separate peptide chains or two places in the same peptide chain. Finally, the specific tertiary structures (subunits) can interact with each other to form more complex molecules. For example, hemoglobin, the oxygen-carrying protein of blood, consists of four tertiary structure subunits, as described in Section 28.27. The specific association of such subunits is called the **quaternary structure** of the protein.

The physiological function of the protein depends upon the maintenance of a precise primary, secondary, tertiary, and often quaternary structure of the protein. Thus, in characterizing the chemical structure of the protein all four of the aspects of protein structure must be considered.

The biosynthesis of protein occurs on subcellular structures called **ribosomes,** which are found in the cytoplasm of the cell (Fig. 28-1). In addition to the specific enzymes required to catalyze those chemical reactions which bring about the condensation of the amino acids into the peptide chain, adenosine triphosphate (ATP) and guanosine triphosphate (GTP) are also required in order to furnish the energy to form the peptide bond. The amino acid must be condensed onto a chemical **translating agent,** called **transfer ribonucleic acid** (t-RNA), or sometimes **soluble ribonucleic acid** (s-RNA). The translating agent orients one amino acid so that it will be brought to the proper adjacent amino acid when the protein is being formed. Information providing the exact sequence of amino acids to form the protein to be synthesized must also be available and is provided by **messenger ribonucleic acid** (m-RNA). The organized interaction of all of these components on the ribosome, which itself contains a third type of RNA (ribosomal-RNA) as a part of its structure, is necessary for the cell to synthesize a specific protein molecule. These processes will be discussed in greater detail in the sections on nucleic acids in this chapter.

28.8   Essentiality of Amino Acids

Proteins are essential to good health and therefore must either be constituents of foods or synthesized by the body. During digestion they are hydrolyzed by the digestive juices in the stomach and intestine into amino acids which are actively transported through the walls of the intestine into the blood stream. The blood carries the amino acids to the tissues, where they are used as building units for the manufacture of the specific proteins needed by the body. Only eight of the amino acids are essential in the diet of the adult human. The body is able to synthesize the remaining ones it needs. These are called the nonessential amino acids. The essential amino acids cannot be synthesized by mammals and must therefore come from plants and bacteria, which do synthesize them and thereby constantly replenish the supply. The principal sources of dietary protein are lean meat, cheese, eggs, milk, beans, and cereal grains. The proteins from these sources contain all the essential amino acids.

The Nucleic Acids

28.9   Introduction

The nucleic acids and the proteins in the cell, as well as interactions between these two types of molecules, provide the essential features which are characteristic of living matter. Whereas the proteins function as specific catalysts which bring about the chemical reactions occurring within the cell, the nucleic acids provide the necessary "information" for the formation of these specific proteins as well as other "information" required for the maintenance of the cell. A further function of the nucleic acids is to transmit all of the "information" present in the parent cell to the two succeeding (daughter) cells which form on cell division. This leads to the fascinating concept of the manner in which information can be stored in and abstracted from chemical structures.

28.10   The Concept of Information Storage and Retrieval in Chemical Structures

The idea of a molecule providing information to another molecule may be likened to a magnetic tape which contains information that it transmits to a tape recorder, which then translates the information into sound. Every time the tape is run through the recorder, exactly the same sound is obtained. In the same way that each portion of the tape has magnetic impulses that correspond to a certain sound, a long linear molecule can have chemical groups extending from it in a definite arrangement that always corresponds to a fixed response in another molecule. The nature of these groups in nucleic acids is discussed in the following section.

28.11   Structure of Nucleic Acids

Our understanding of the chemical nature of the nucleic acids forms the basis of our understanding of biochemical genetics. Two types of nucleic acids occur in cells: deoxyribonucleic acid, DNA, (molecular weight about 10^6 to 10^9) is located essentially in the nucleus of the cell; and ribonucleic acid, RNA, (molecular weight 25,000 to 10^6) located mostly in the cytoplasm.

The DNA molecule, actually a polymer chain, is made up principally of four nitrogenous bases—adenine, guanine, thymine, and cytosine—which are bonded to the first carbon of deoxyribose, a sugar. The polymer chain is linked through phosphate diesters bonded to the oxygens of the third and fifth carbons of adjacent deoxyribose residues. Thus, the monomer portion of the nucleic acid contains a base, a sugar, and a phosphate and is called a **nucleotide.** The structure of DNA, showing the bases adenine and thymine, is as follows.

Nucleotide

Deoxyribonucleic Acid (DNA)–showing two monomer units

The formulas for the bases cytosine and guanine, which are also present in nucleotides in DNA, are

Cytosine (base)   Guanine (base)

RNA is similar in its structure to DNA except that ribose replaces deoxyribose, and uracil replaces thymine.

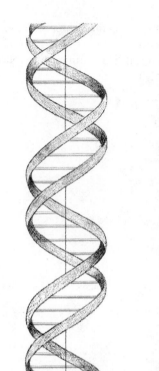

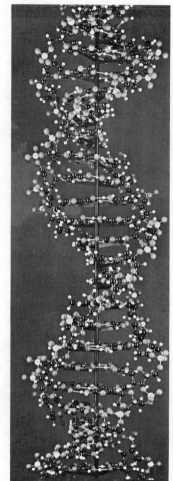

FIGURE 28-5
Deoxyribonucleic acid, DNA.
(*a*) Schematic drawing and (*b*)
model of the double-stranded
helix.

(a)   (b)

Ribose   Uracil

Watson and Crick, who received the Nobel Award in 1962, by x-ray studies
of DNA were able to show that DNA is a double-stranded helix (Fig. 28–5). The
backbone of each strand of the helix is composed of the sugar and phosphate
chain. The bases occupy the center of the helix, somewhat like the rungs in a
twisted ladder. A base from one strand backs up against a base of the other strand.
Adjacent bases in the same strand are stacked in a parallel position above and

below. The stability of the dihelical structure is due to specific hydrogen bonding between the nitrogenous bases of the two strands, illustrated below for adenine in one strand and thymine in the other.

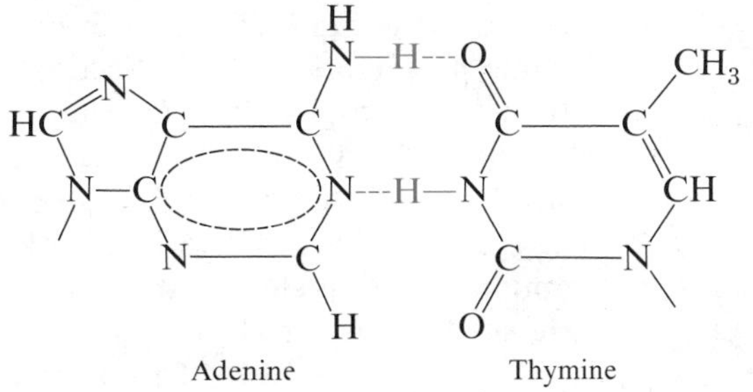

Adenine   Thymine

The bonding arrangement between cytosine and guanine is similar but with three hydrogen bonds instead of two.

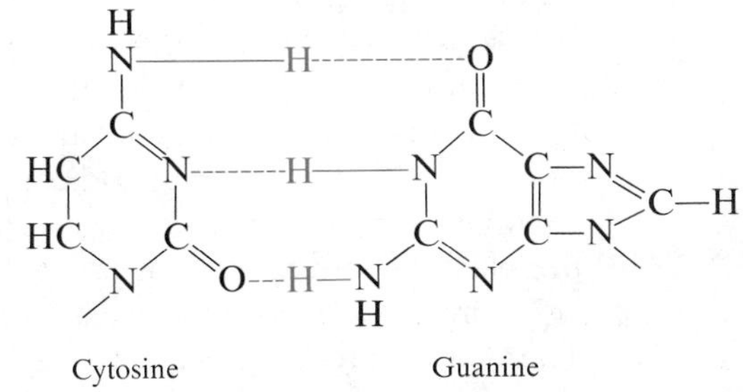

Cytosine   Guanine

The bases in the nucleic acid structure are the chemical groups which store the information (the signals) necessary for a fixed response in another molecule and which may be likened to magnetic impulses on a magnetic tape. Through the four bases, therefore, the information is stored in duplicate in a four-symbol alphabet of molecular dimensions in the DNA molecule. (It is interesting to note by comparison that most computers utilize only a two-symbol alphabet.) The sugar and phosphate portions of the chain serve as the plastic (polymeric) carrier for the signals. The stored information in DNA can be utilized by the cell for many other purposes in addition to the synthesis of specific proteins.

In the DNA molecule, the adenine is always paired with the thymine, and the guanine always with the cytosine. During cell division the two strands of the DNA molecule separate. Each daughter cell receives one of the two nucleic acid strands. Even though the two strands are not chemically identical, they each contain the information necessary to form the original dihelical structure. Therefore, each strand can serve as a pattern for the formation of a new dihelical DNA molecule which will be identical with that originally present in the parent cell. In this way, the exact sequence of bases in the DNA molecule can be preserved from parent to daughter cell through many generations. All the information required for growth and development throughout the life of the cell is stored in the DNA molecule. In higher forms of life such as man, the DNA molecule of a single cell, if unwound and measured, would probably be several feet long.

Although the information necessary to synthesize a protein is stored in DNA, the DNA does not furnish this information directly to the cell. It always works through RNA, whose function is to transcribe the information and make it available to the cell. This is accomplished through hydrogen bonding which connects the nitrogenous bases of a DNA molecule with the nucleoside triphosphate precursors of RNA (Section 28.3), originally formed in the cytoplasm and then transferred to the nucleus. Thus, in the nucleus, the hydrogen bonding produces RNA, with the bases in a specific sequence determined by one of the strands of DNA. The RNA is then transported from the nucleus to the ribosomes of the cytoplasm, where it furnishes the information for producing the proper amino acid sequence to make a protein. This RNA is referred to as **messenger ribonucleic acid,** m-RNA, one of three principal types of RNA present in the cell (see Section 28.7). The other two principal kinds of RNA are a very high molecular weight RNA associated directly with the structure and function of the ribosome (ribosomal-RNA) and a low molecular weight RNA (transfer-RNA), which functions to accept specific amino acids and translate the information present in m-RNA from nucleotide language to amino acid sequence language.

It is believed that the information from only a relatively small portion of one of the strands of the DNA molecule is used in producing each protein. (Returning for a moment to the magnetic tape analogy—one does not have to play the whole symphony if he only wants to hear one movement.) The remainder of the DNA contains information necessary to make the more than one thousand different proteins that are present in the cell.

Twenty-six amino acids exist in proteins but only four bases occur in RNA. The four different bases acting singly could designate a maximum of four amino acids. However, three consecutive nucleotides acting as a unit, with the possibility of any one of four bases in each nucleotide, can designate $4 \times 4 \times 4$, or 64, different events, which is more than enough to represent the twenty-six amino acids. The trinucleotide structure (three consecutive nucleotides) that corresponds to a specific amino acid is known as the **amino acid code** for that amino acid.

The concept that information can be stored in the chemical structure of molecules is a new and exciting concept which may be important in studying physiological processes such as memory; it may have important application in the miniaturization of information storage in computer technology.

Carbohydrates

28.12   Introduction

Carbohydrates are polyhydroxy aldehydes and polyhydroxy ketones, or their condensation products. The name is derived from the fact that many sugars have the empirical formula $C_nH_{2n}O_n$, or $C_n(H_2O)_n$, and hence the name "hydrate of carbon," or carbohydrate. This name has been retained even though it is not descriptive of the nature of these substances.

The most important carbohydrates are sugars, starches, cellulose, and substances related to them. Carbohydrates are an important source of energy for animals; they form the supporting tissues of plants and some animals; and plants use them in the manufacture of proteins and lipids. There are three important classes of carbohydrates: (a) the monosaccharides, (b) the disaccharides, and (c) the polysaccharides.

28.13   Monosaccharides

A large number of the monosaccharides that occur in nature contain five or six carbon atoms (pentoses and hexoses). They are colorless, crystalline compounds, and some have a sweet taste. The formulas of the most important monosaccharides, glucose and fructose, are shown.

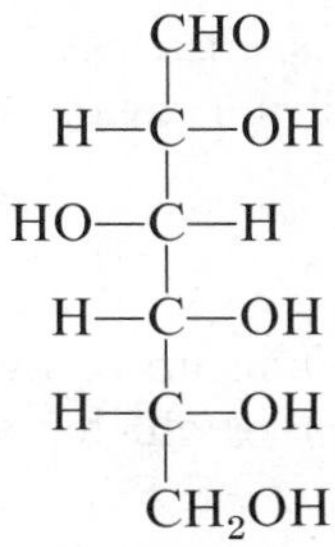

Glucose
(a polyhydroxy aldehyde)

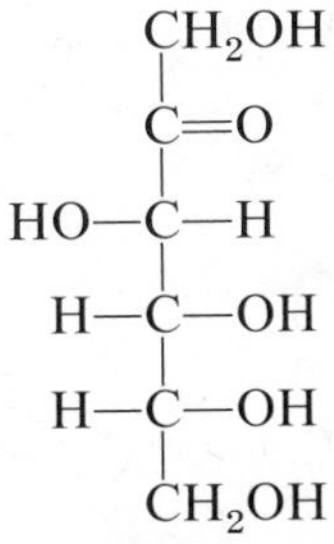

Fructose
(a polyhydroxy ketone)

Experimental evidence shows that the glucose and fructose molecules exist mainly in cyclic forms, in which two carbon atoms are linked by an oxygen atom.

Glucose

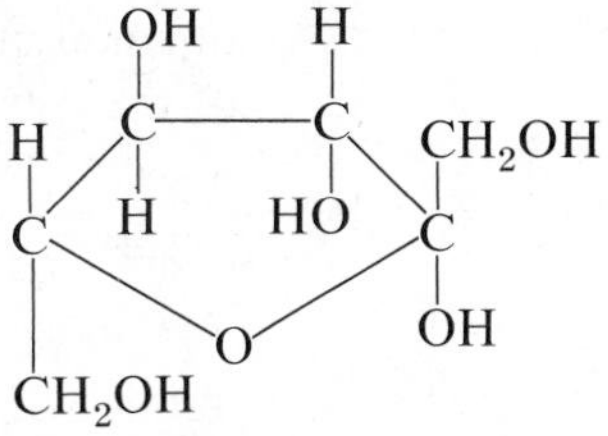

Fructose (furanose form)

Glucose, $C_6H_{12}O_6$, is also known as **dextrose** and **grape sugar.** It is widely distributed in nature. Human blood normally contains about 0.1 gram of this sugar per 100 milliliters of blood. Sometimes a solution of glucose is injected directly into the blood stream of patients seriously in need of nourishment, for this sugar is in the form that can be used immediately for this purpose. In cases of sufferers from diabetes the body is unable to assimilate glucose, and this sugar is eliminated through the kidneys. One hundred milliliters of urine may contain as much as 8 to 10 grams of glucose in such cases, and its presence there is one symptom of the disease.

Glucose is prepared commercially by the hydrolysis of starch. When corn starch, for example, is heated with water containing a small amount of acid, the changes indicated below occur.

$$(C_6H_{10}O_5)_x \xrightarrow[\text{heat}]{H^+, H_2O} \text{Dextrins} \xrightarrow[\text{heat}]{H^+, H_2O} C_{12}H_{22}O_{11} \xrightarrow[\text{heat}]{H^+, H_2O} C_6H_{12}O_6$$

Starch   Maltose   Glucose

Commercially, the hydrolysis is not carried to completion, so that the corn syrup thus produced is a mixture of glucose, maltose, and dextrins.

Fructose, $C_6H_{12}O_6$, has the same empirical formula as glucose, but its structural formula is different. The two compounds thus are isomers (Section 26.9). Fructose contains the ketone group, $-C=O$, while glucose contains the aldehyde group, $-CHO$. Fructose, also called **levulose** and **fruit sugar,** occurs in fruits and honey. Combined with glucose in the disaccharide known as sucrose, it occurs in sugar cane and sugar beets. Fructose is nearly twice as sweet as sucrose and, in fact, is the sweetest of all the sugars.

28.14   Disaccharides

A disaccharide is made up of two monosaccharide molecules, either the same or different, the union of which is accompanied by the elimination of the elements of one molecule of water. For example,

$$2C_6H_{12}O_6 \longrightarrow C_{12}H_{22}O_{11} + H_2O$$

Monosaccharide   Disaccharide
(sucrose)

The disaccharide **sucrose,** $C_{12}H_{22}O_{11}$, is the familiar table sugar, obtained largely from cane and sugar beets. Because sucrose is converted into a mixture of equal quantities of glucose and fructose by hydrolysis, it must be an anhydride of these two monosaccharides (linked together by a C—O—C bond).

(glucose unit)   (fructose unit)

Sucrose

Maltose, or **malt sugar,** $C_{12}H_{22}O_{11}$, is a disaccharide formed by the action of an enzyme called diastase upon starch, or by the partial hydrolysis of either starch or dextrins by dilute acids. Two glucose residues make up the maltose molecule, as shown by the formation of two molecules of glucose upon hydrolysis. Maltose is fermentable in the presence of yeast, since yeast produces both maltase, which catalyzes the conversion of maltose to glucose, and zymase, which catalyzes alcoholic fermentation of glucose.

Lactose, or **milk sugar,** $C_{12}H_{22}O_{11}$, occurs in the milk of mammals to the extent of 3 to 5 grams per 100 milliliters of milk. Hydrolysis of lactose in the presence of either dilute mineral acid or lactase converts it to glucose and galactose (an isomer of glucose). Certain microorganisms catalyze the fermentation of lactose to either butyric or lactic acid and are in this way responsible for the souring of milk. Lactose is an important ingredient of infant foods.

28.15 Polysaccharides

The polysaccharides are carbohydrates of high molecular weight and consist of monosaccharide units linked through oxygen atoms to form chains of different lengths. They occur in large quantities in plants and animals, either as reserve food materials, such as starches and glycogen, or as skeletal substances, such as cellulose. They are nearly tasteless and either are insoluble in water or form colloidal dispersions in it. Upon hydrolysis they yield monosaccharides.

Starch, $(C_6H_{10}O_5)_x$, is accumulated by plants in seeds, tubers, and fruits. It is often the main food supply for the young plant until it has developed a leaf system and can manufacture its own food. Starch may be hydrolyzed completely to glucose by boiling it with dilute acid, so must consist entirely of glucose units. Incomplete hydrolysis of starch gives rise to dextrin, which has the same empirical formula as starch $(C_6H_{10}O_5)_x$; the difference is in the value of x, which is smaller for dextrin than for starch. Starches are commercially important as foods, but have other uses as well.

Glycogen, or animal starch, or liver starch, $(C_6H_{10}O_5)_x$, is found principally in the liver and muscles. It is the storage food of animals, corresponding to starch in plants. The breakdown of muscle glycogen yields lactic acid, which provides energy for ATP activity in muscles and other body tissues.

Cellulose, $(C_6H_{10}O_5)_x$, has the same empirical formula as starch, and its molecules are also made up of many glucose units joined together through oxygen linkages to form long chains. The cellulose chains are arranged in bundles whose long axes are parallel to the axes of the cellulose thread or fiber. Cellulose is widely distributed in nature. It is the chief structural material of the cell walls of plants. Cotton is about 90 per cent of cellulose, coniferous woods about 60 per cent, and cereal straws about 35 per cent.

Most animals do not digest cellulose because the enzymes of the digestive tract do not hydrolyze it. However, many bacteria attack cellulose, particularly those in the intestinal tract of ruminants. For this reason, these animals are able to utilize cellulose as a food.

28.16 Photosynthesis

Green plants utilize light energy to synthesize complex organic constituents from carbon dioxide, water, and inorganic salts. On the other hand, animals lack the power of synthesizing organic materials from these simple substances, and they depend upon preformed organic materials supplied in the diet. In a very real sense, just about any biological process involves *either* the cleavage of water to form hydrogen, which may then become incorporated into organic substances, and oxygen (**photosynthesis**) *or* the combination of hydrogen atoms, usually bound up in organic molecules, with oxygen to form water (**respiration**).

Photosynthesis is a series of oxidation-reduction reactions in green plants. The energy for these reactions is supplied by the sun and absorbed by the green pigment of the plant. The green pigment in plants, called **chlorophyll,** absorbs light chiefly from the red portion of the electromagnetic spectrum (see Chapter 33). In these reactions carbohydrates are formed from carbon dioxide and water, and molecular oxygen is produced. Because one mole of oxygen is formed for each mole of carbon dioxide and each mole of water consumed, the beginning and end products can be represented by the equation

$$688,000 \text{ cal} + 6CO_2 + 6H_2O \underset{\text{Animal respiration}}{\overset{\text{Green plant photosynthesis (with light)}}{\rightleftharpoons}} (CH_2O)_6 + 6O_2$$

Carbohydrate (hexose)

Animal respiration involves the reverse process, converting oxygen and carbohydrates to carbon dioxide and water, thereby completing a cycle.

The large amount of energy absorbed during photosynthesis and released during respiration, most of which is related to the formation or cleavage of water rather than to reactions of the carbon, is a key to the necessary operation of the many closely associated biological processes.

The production of oxygen by plants and microorganisms through photosynthesis is important in replenishing the supply of oxygen in the atmosphere. Thus, conservationists consider it a serious problem when a very large proportion of ground area is covered over with concrete or artificial turf.

By providing plants with water labeled with the $^{18}_{8}O$ isotope, it has been found that the molecular oxygen released during photosynthesis is derived from the water and not from carbon dioxide. The photosynthesis series of reactions is very complicated. Some of the reactions are dependent upon light and are called **light reactions.** The light reactions involve the splitting of water to form oxygen and a powerful reducing agent, as well as the generation of ATP. **Dark reactions** of photosynthesis are not dependent upon light and result in the formation of carbohydrate by the reduction of CO_2.

Lipids

28.17   Introduction

The term **lipid** is a general name for a group of substances which can be extracted from living cells by nonpolar solvents (e.g., ether, chloroform, benzene). The lipids include such substances as fats, oils, waxes, fatty acids, phospholipids, glycolipids, and steroids.

28.18   Fats and Oils

The fats and oils are esters of the fatty acids and glycerol. Glycerol is a trihydroxy alcohol, $C_3H_5(OH)_3$, and the **fatty acids** are any organic acid that occurs in a fat. The fats and oils have the general formula shown below.

$$(RCOO)_3C_3H_5 \quad \text{or} \quad \begin{array}{c} RCOOCH_2 \\ | \\ RCOOCH \\ | \\ RCOOCH_2 \end{array} \quad \text{or} \quad \begin{array}{c} \overset{\displaystyle O}{\underset{\|}{R-C}}-O-\overset{\displaystyle H}{\underset{|}{C}}-H \\[2ex] \overset{\displaystyle O}{\underset{\|}{R-C}}-O-\overset{}{\underset{|}{C}}-H \\[2ex] \overset{\displaystyle O}{\underset{\|}{R-C}}-O-\overset{}{\underset{|}{C}}-H \\ | \\ H \end{array}$$

General formula of fats and oils

The R groups in the formula may be the same or different, and the fatty acid portion may be saturated or unsaturated. In saturated fatty acid esters of glycerol, most of the acid portions have the general formula $CH_3(CH_2)_nCO-$, where n is always an even number, most frequently 14 or 16. Compounds of this type are solid and are called **fats.** When the fatty acid portions are predominantly of the unsaturated variety, the esters are liquid and are called **oils.**

The fatty acid radicals most often found in fats and oils are those of the acids listed below.

Butyric	$CH_3CH_2CH_2COOH$
Lauric	$CH_3(CH_2)_{10}COOH$
Palmitic	$CH_3(CH_2)_{14}COOH$
Stearic	$CH_3(CH_2)_{16}COOH$
Oleic	$CH_3(CH_2)_7CH{=}CH(CH_2)_7COOH$
Linolenic	$CH_3CH_2CH{=}CHCH_2CH{=}CHCH_2CH{=}CH(CH_2)_7COOH$
Ricinoleic	$CH_3(CH_2)_5CHOHCH_2CH{=}CH(CH_2)_7COOH$

Each naturally occurring fat or oil is a mixture of many different esters. The more important fats include milk fat, lard, and tallow. Familiar oils are coconut oil, cottonseed oil, linseed oil, soybean oil, and palm oil.

Like other esters, fats and oils undergo hydrolysis; a fatty acid and glycerol are formed when fats and oils are hydrolyzed.

$$\underset{\text{Glyceryl palmitate}}{(C_{15}H_{31}COO)_3C_3H_5} + \underset{\text{(excess)}}{3H_2O} \overset{\text{Catalyst, heat}}{\rightleftharpoons} \underset{\text{Palmitic acid}}{3C_{15}H_{31}COOH} + \underset{\text{Glycerol}}{C_3H_5(OH)_3}$$

Fats and oils are hydrolyzed by certain enzymes called **lipases.** During digestion the lipases of the gastric and pancreatic juices hydrolyze fats and oils to fatty acids and glycerol. Enzymes known as **phosphatases,** present in kidney, bone, the intestines, and other animal tissues, hydrolyze the lipids known as **phospholipids** (Section 28.23).

Fats and oils play two important roles in the functions of the body. Part of these substances is used in the synthesis of certain essential constituents of the tissues, such as phospholipids, and part is used as body fuel, which may be utilized immediately or stored for further use. Fats also serve to some extent to insulate the body against loss of heat and to protect other tissues against injury.

The products of complete oxidation of fats and oils in the body are carbon dioxide and water. However, under certain pathological conditions (for example, diabetes and starvation), normal oxidation is interfered with and substances such as β-hydroxybutyric acid, β-oxybutyric acid, and acetone appear in the urine in abnormally large quantities.

28.19   Waxes

The esters of the fatty acids and alcohols other than glycerol are known as **waxes.** They are widely distributed in plants and animals, where they frequently serve as protective agents. Beeswax from the honeycomb of the bee is chiefly myricyl palmitate, $C_{15}H_{31}COOC_{31}H_{63}$.

28.20   Compound Lipids

The compound lipids are fatty acid esters of glycerol containing chemical groups in addition to the fatty acid and glycerol radicals. The phospholipids, for example, are fats containing phosphoric acid and nitrogen, while the glycolipids are fats with a carbohydrate which contains nitrogen but no phosphoric acid.

The phospholipids are present in every cell of plants and animals, and are very abundant in the tissues of the brain, heart, liver, muscle, kidney, and bone marrow. Many of the phospholipids of certain tissues serve as intermediates in the metabolism of fats and oils. The general formulas of two types of phospholipids are shown below.

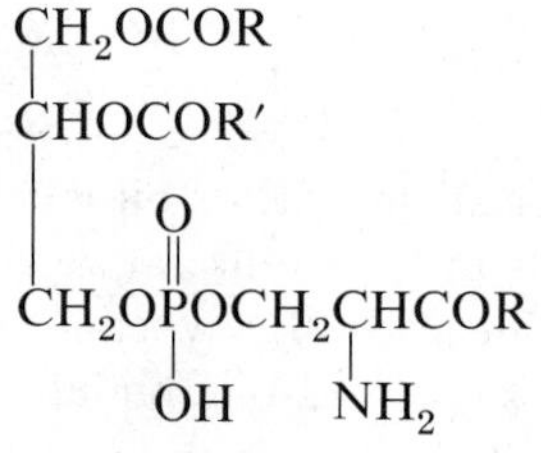

$$
\begin{array}{ll}
\text{CH}_2\text{OCOR} & \text{CH}_2\text{OCOR} \\
\text{CHOCOR}' & \text{CHOCOR}' \\
\quad\quad\ \ \overset{O}{\overset{\|}{}} & \quad\quad\ \ \overset{O}{\overset{\|}{}} \\
\text{CH}_2\text{OPOCH}_2\text{CHCOR} & \text{CH}_2\text{OPOCH}_2\text{CH}_2\text{N(CH}_3)_3 \\
\quad\ \ \text{OH}\quad\ \text{NH}_2 & \quad\ \ \text{OH}\quad\quad\ \text{OH} \\
\quad\quad\text{Cephalin} & \quad\quad\text{Lecithin}
\end{array}
$$

The **glycolipids** (also called **cerebrosides**) are associated with the phospholipids in the tissues, particularly in the brain. Upon hydrolysis they yield galactose (a sugar), sphingosine, and a fatty acid.

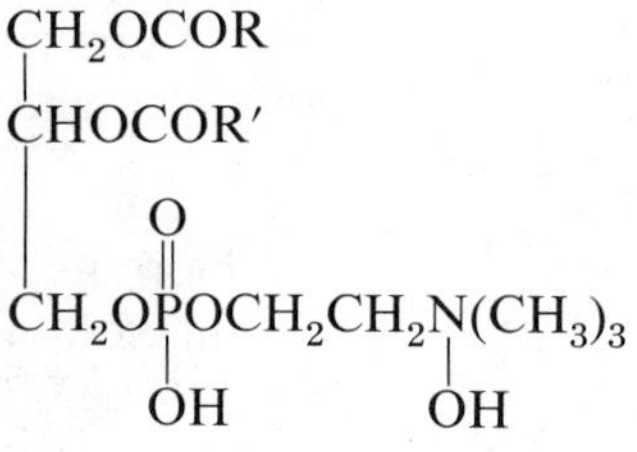

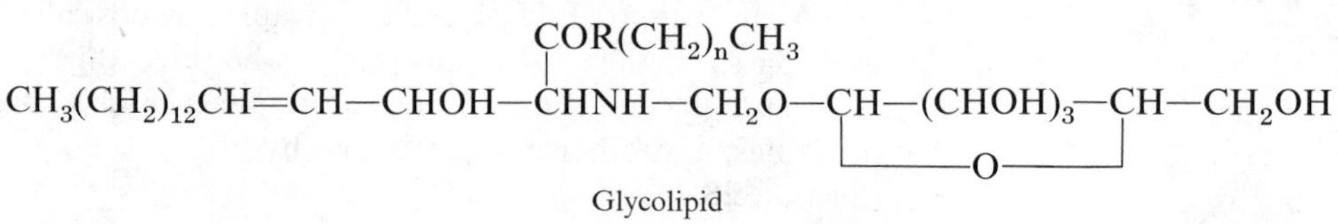

$$
\overset{\displaystyle \text{COR(CH}_2)_n\text{CH}_3}{\underset{\displaystyle |}{}}
$$
$$
\text{CH}_3(\text{CH}_2)_{12}\text{CH}=\text{CH}-\text{CHOH}-\text{CHNH}-\text{CH}_2\text{O}-\text{CH}-(\text{CHOH})_3-\text{CH}-\text{CH}_2\text{OH}
$$
$$
\underset{\displaystyle \text{O}}{\rule{3cm}{0.4pt}}
$$

Glycolipid

28.21   **Steroids**   The steroids occur widely distributed in nature and are related structurally to the complex organic compound cyclopentanoperhydrophenanthrene, the formula of which is

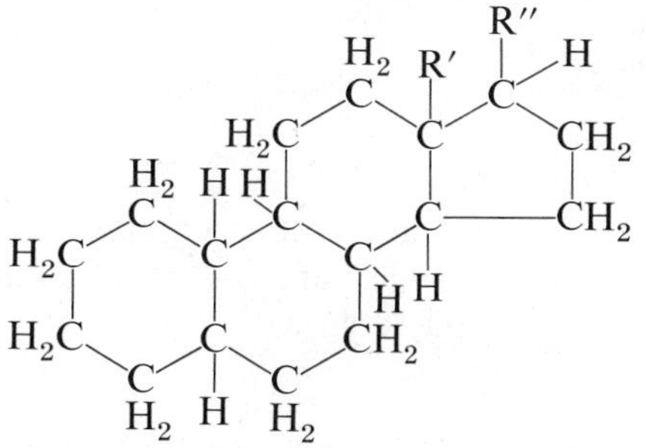

Ring system of steroids

The steroids include the sterols, bile acids, certain vitamins, and sex hormones.

The best-known sterol is **cholesterol,** $C_{27}H_{45}OH$, which is found free or as the alcohol component of a fatty acid ester in almost all tissues, but particularly in the brain, nerves, and suprarenal gland. Physiologically, cholesterol serves as a vehicle for the transportation of fatty acids; it is important to the colloidal chemistry of protoplasm; it is valuable in immunological processes; and it is a precursor of the sex hormones.

The sex hormones are responsible for the development of specific male and female sexual characteristics. The sex hormones and also the adrenal cortical hormones are related in structure to cholesterol.

Porphyrins and Their Derivatives

28.22   **Introduction**

The **porphyrins** are intensely colored compounds containing the porphyrin ring system shown below.

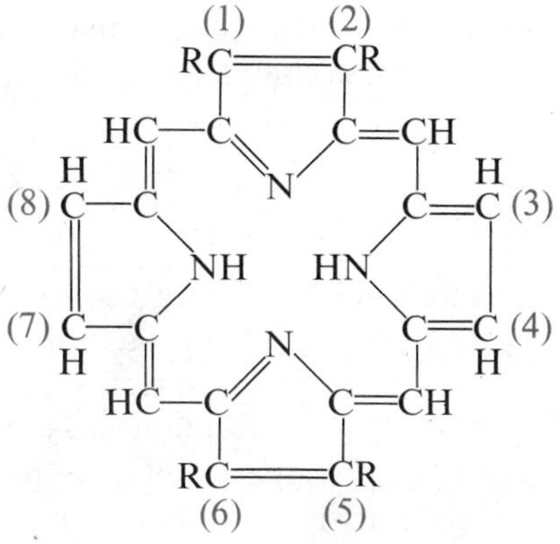

Porphyrin ring system

Most of the porphyrins found in nature are present in the form of chlorophyll, the green pigment of plants, and as hemoglobin, the red pigment of blood. In these substances it is the porphyrin portion of the molecule that is responsible for the important biological activity.

28.23 Chlorophyll

The pigment of green plants is a mixture of two closely related porphyrin derivatives which are called **chlorophyll-a** and **chlorophyll-b.** The structure of chlorophyll-a is shown.

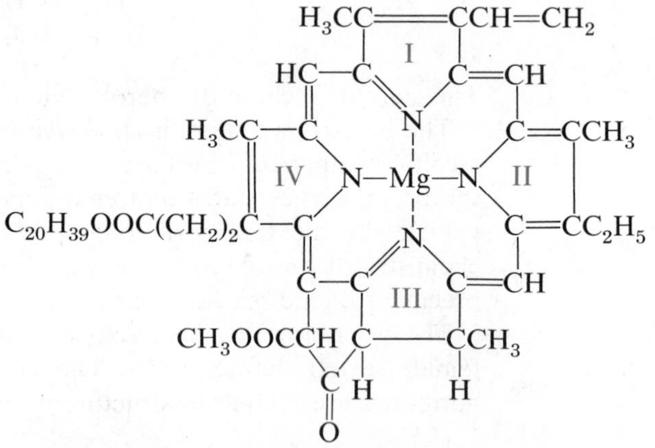

Chlorophyll-a (H. Fischer)

Chlorophyll-b differs from chlorophyll-a in that it has a —CHO group instead of the —CH_3 group on Ring II. In the plant, the pigments are chemically bound to the protein.

The function of chlorophyll is to act as an absorber of solar energy in the photosynthesis of organic material from carbon dioxide and water, but the mechanism by which the radiant energy of sunlight is converted into chemical energy is not yet completely understood.

28.24 Hemoglobin

The solid content of the red corpuscles of mammals contains, on the average, 32 per cent of the protein **hemoglobin,** which contains iron. Hemoglobin consists of a protein portion of 4 peptide chains (94%) and **heme,** the iron-containing portion (6%). Heme has the empirical formula $C_{34}H_{32}O_4N_4Fe$.

The principal function of hemoglobin is the transport of oxygen from the lungs to the tissues, where oxygen is consumed in the oxidation of such organic sub-

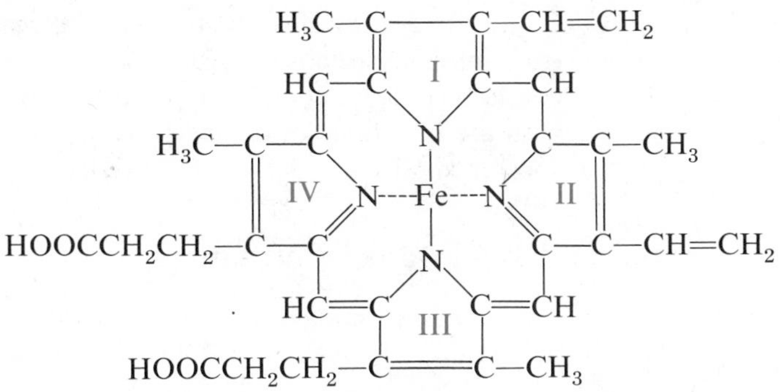

$$
\begin{array}{c}
\text{H}_3\text{C}-\text{C}\underset{\text{HC}-\text{C}}{=\!\!=}\text{C}-\text{CH}=\text{CH}_2 \\
\end{array}
$$

Heme

stances as carbohydrates, lipids, and proteins. Hemoglobin, the pigment of blood in the veins, is purplish-red. It combines with oxygen, forming bright-red **oxyhemoglobin,** the pigment of arterial blood. One hemoglobin molecule contains four heme molecules, each associated with one of the four peptide chains of the protein portion—the quaternary structure of hemoglobin referred to in Section 28.7. Since one hemoglobin molecule contains 4 heme molecules (and hence 4 atoms of iron), and the maximum amount of oxygen carried corresponds to one molecule of oxygen (O_2) per atom of iron, one hemoglobin molecule can carry 4 oxygen molecules. The union of oxygen with hemoglobin is not a reaction that can be spoken of as oxidation, for the iron remains in the +2 oxidation state when oxygen is bound to the hemoglobin molecule. The bond between hemoglobin and oxygen is a very weak one, as indicated by the fact that oxygen is liberated upon subjecting solutions of oxyhemoglobin to low pressure.

Hemoglobin readily combines with carbon monoxide, CO, forming a pinkish-red complex, called **carbon monoxide hemoglobin.** Because this complex is more stable than oxyhemoglobin, oxygen will not replace the carbon monoxide, and the blood loses its ability to combine with oxygen. This is why breathing carbon monoxide is so dangerous.

Vitamins

28.25 Introduction

Vitamins are substances which, in addition to carbohydrates, fats, proteins, and inorganic salts, are essential for the normal nutrition of animals. Most of the vitamins are synthesized in plants or bacteria, but not in the animal body. The fact that the lack of such substances can cause diseases in man was not generally recognized until about 1920. Studies of the diseases scurvy and beriberi led to the discovery of vitamins. It was found that scurvy is caused by lack of Vitamin

C and beriberi by lack of Vitamin B_1. Because Vitamin C is destroyed when food containing it is cooked, scurvy was quite common among persons whose diet consisted only of cooked or preserved foods. Beriberi appeared in the Far East with the introduction of rice-polishing machines because the indispensable food substance Vitamin B_1 was removed when the rice was polished.

28.26   Structures and Functions of Vitamins

The names, formulas, and functions of several of the important vitamins are given in Table 28-1, pages 712–713.

Enzymes

28.27   Introduction

Most of the chemical reactions occurring in living organisms are accelerated by enzymes. These reactions would proceed with extreme slowness in the absence of the enzyme. An **enzyme** is an organic substance, within a living organism, and capable of accelerating a chemical reaction. In other words, an enzyme is a biological catalyst. All the enzymes which to date have been investigated are proteins. Many enzymes contain coordinated metal ions. If the metal is removed, the enzyme loses its catalytic activity. If the metal is put back in, the catalytic activity is restored. The substances which undergo chemical reactions due to the catalytic action of enzymes are called **substrates.** Very small amounts of an enzyme suffice for the reaction of large amounts of substrate. Like all proteins, enzymes are unstable toward acids, alkalies, and heat.

Many enzymes contain a definite chemical group **(coenzyme)** which can be separated from the protein portion of the enzyme **(apoenzyme).** Neither the coenzyme nor apoenzyme shows catalytic activity after separation. Enzymes are, in general, very specific in regard to the reactions which they catalyze, and the apoenzyme is responsible for the substrate specificity and catalytic properties of the total enzyme.

It has been shown that the enzyme (E) combines with the substrate (S) forming the intermediate enzyme-substrate complex (ES), according to

$$E + S \rightleftharpoons ES$$

The intermediate enzyme-substrate complex is converted into the enzyme E and the reaction product P, or may be cleaved again into E and S. The enzyme catalyzed reaction may be expressed by

$$E + S \rightleftharpoons ES \longrightarrow E + P$$

The chief classes of enzymes are (1) **oxido-reductases** which catalyze oxidation-reduction reactions; (2) **transferases,** which catalyze the transfer of chemical groups from one substance to another; (3) **hydrolases,** which catalyze hydrolytic reactions; (4) **isomerases,** which catalyze the isomerization and racemization of optical and geometric isomers; and (5) **synthetases,** which catalyze the formation of a substance by condensation of chemical groups, usually with the utilization of energy from ATP or similar substances. Two of these classes are discussed in the following sections.

28.28   Oxido-Reductases

In 1770, Lavoisier showed that animals use oxygen in the respiration process. In this process, lipids, carbohydrates, and proteins are oxidized, liberating energy. The mechanisms by which biological oxidations and reductions are carried on have not been completely explained, but during the past 30 years a great deal has been discovered and we now have a rather good picture of the process.

Biological oxidation-reduction reactions are catalyzed by enzymes. The enzymes catalyze the transfer of electrons from the substance being oxidized to the substance being reduced. The most important electron donors in biological systems are the hydrogen atoms of organic molecules. The most important electron acceptor in aerobic organisms is the oxygen molecule. Because the oxidation of organic molecules involves, most frequently, the loss of hydrogen atoms, the reaction is designated as **dehydrogenation,** and the enzymes which catalyze reactions of this type are called **dehydrogenases.** The dehydrogenases are not able to transfer hydrogen and electrons directly to molecular oxygen. Instead, certain enzymes, called **oxidases,** catalyze the reduction of oxygen by hydrogen. Thus, two enzymes—a dehydrogenase and an oxidase—are involved in the overall oxidation-reduction reaction. The electrons and hydrogen of the organic substrate are first transferred to the dehydrogenase, which in turn transfers them to the oxidase, and finally to oxygen. In many instances of enzyme-catalyzed oxidation-reduction reactions, several electron carriers operate between the dehydrogenase and the oxidase.

28.29   Hydrolases

The hydrolases are enzymes which catalyze the hydrolysis of food substances in the digestive tract. They include the esterases, the carbohydrases, and the proteinases.

■ **1. Esterases.**   These enzymes catalyze the reversible reaction

$$\underset{\text{Ester}}{RCOOR'} + H_2O \rightleftharpoons \underset{\text{Acid}}{RCOOH} + \underset{\text{Alcohol}}{R'OH}$$

The esterases include lipase, choline esterase, phosphatase, and pyrophosphatases. Pancreatic lipase, for example, hydrolyzes fats to free fatty acids and glycerol.

$$\underset{\text{Fat}}{(RCOO)_3C_3H_5} + 3H_2O \rightleftharpoons \underset{\text{Fatty acid}}{3RCOOH} + \underset{\text{Glycerol}}{C_3H_5(OH)_3}$$

■ **2. Carbohydrases.**   These are enzymes which catalyze the hydrolytic cleavage of the oxygen bonds linking the monosaccharide residues together in di- or polysaccharides. The carbohydrases include such enzymes as maltase, lactase,

TABLE 28-1   Some of the More Important Vitamins

Name of Vitamin	Formula	Source	Function
A	(chemical structure)	Green and yellow vegetables, eggs, milk and butter.	Increases resistance to infections; helps night blindness.
B_1 (Thiamin pyrophosphate)	(chemical structure)	Rice bran, germ of other cereals such as wheat, rye, barley, and oats.	Supplies need of nerve tissue; helps to correct loss of appetite, fatigue, fear, nervousness and depression.
B_2 (Riboflavin)	(chemical structure)	Liver, yeast, cereal germ.	Stimulates growth of young animals; helps to prevent dermatitis and baldness.
B_{12} (Cobalamin)	$C_{63}H_{90}N_{14}O_{14}PCo$	Normal component of most diets.	Helps to prevent pernicious anemia.
B_6 (Pyridoxine)	(chemical structure)	Liver, milk, eggs, vegetables.	Helps to prevent anemia and nervous symptoms in animals; functions in the metabolism of amino acids.

Vitamin	Structure	Sources	Function
Niacin, or nicotinic acid	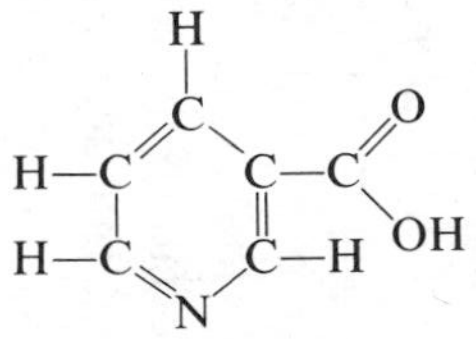 (pyridine ring structure)		Control of pellagra (a disease affecting the mucous membranes, skin and nervous system).
C (Ascorbic acid)	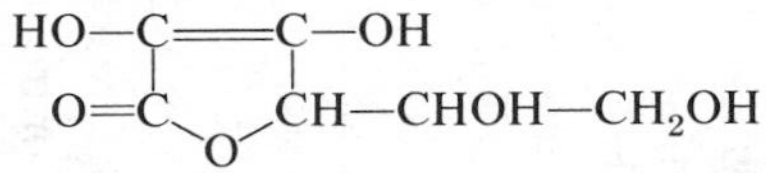 $HO-C=C-OH$, $O=C$, $CH-CHOH-CH_2OH$	Fruits (especially citrus), vegetables.	Prevents scurvy.
D (Calciferol)	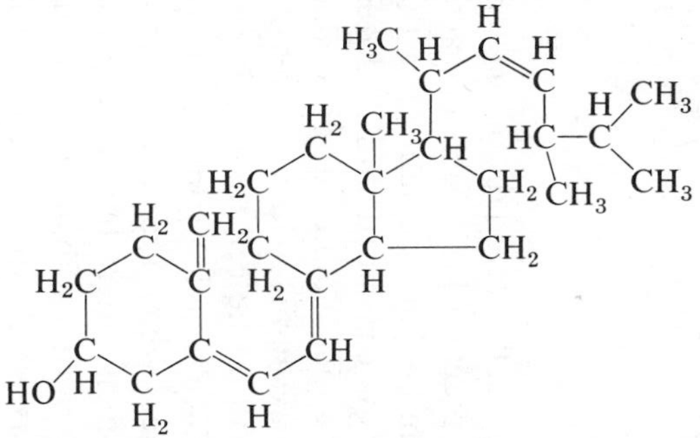 (steroid ring structure)		Prevents rickets.
E (α-Tocopherol)	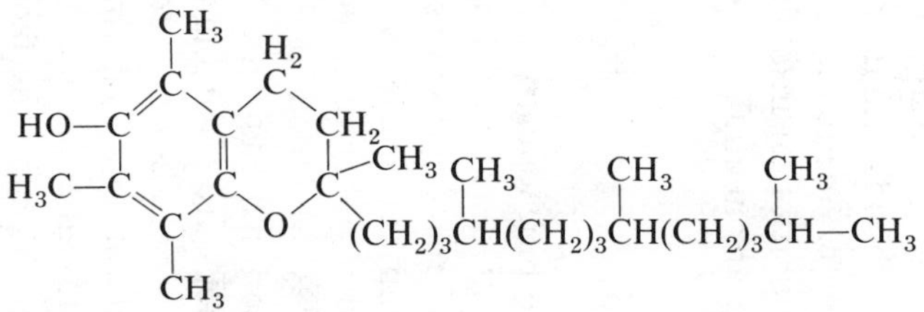 side chain $(CH_2)_3CH(CH_2)_3CH(CH_2)_3CH-CH_3$ with CH_3 branches	Wheat-germ oil, corn-germ oil, cottonseed oil; green leaves of spinach and watercress; egg yolks.	Antisterility factor.
K₁	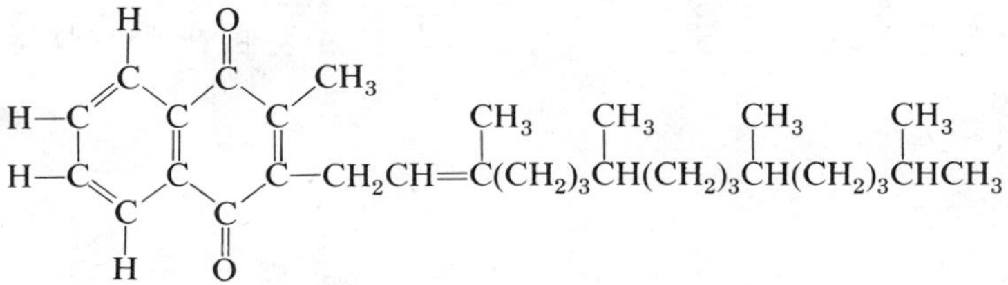 side chain $-CH_2CH=C(CH_2)_3CH(CH_2)_3CH(CH_2)_3CHCH_3$ with CH_3 branches	Leafy vegetables, fat of hog liver, fish meal, oils of grains and cereals.	Helps prevent tendency to hemorrhage by functioning in the blood-clotting mechanism.

sucrase, diastases, amylases, and cellulase. Sucrase, for example, catalyzes the hydrolysis of sucrose to glucose and fructose.

$$\underset{\text{Sucrose}}{C_{12}H_{22}O_{11}} + H_2O \rightleftharpoons \underset{\text{Glucose}}{C_6H_{12}O_6} + \underset{\text{Fructose}}{C_6H_{12}O_6}$$

■ **3. Proteinases.** The proteinases catalyze the hydrolysis of proteins. They catalyze the hydrolysis of peptide bonds inside the long peptide chains which form protein molecules. The hydrolytic cleavage of a peptide bond is shown by the equation

$$R-\overset{O}{\underset{}{C}}-\overset{H}{\underset{}{N}}-R' + H_2O \rightleftharpoons R-COOH + H_2N-R'$$

The most important proteinases involved in digestion are pepsin, trypsin, and chymotrypsin. The clotting of blood is an enzymic reaction catalyzed by the enzyme thrombin. It is a reaction in which the soluble protein fibrinogen is converted into insoluble fibrin, and it is accompanied by the splitting of several peptide bonds with the release of small polypeptides. The clotting of milk is caused by the enzyme rennin of the gastric juice of mammals. The reaction involves the formation of an insoluble compound from casein, which is soluble in solutions which are neutral.

28.30   Metabolic Pathways

In the biochemistry of the cell, a product of one enzyme-catalyzed reaction is often used as an enzyme reactant for a second reaction, and the product of the second reaction used as an enzyme reactant for a third reaction. This interrelationship between enzymes is continued until a dozen or more enzymes can be linked together in a sequential manner to give what is known as a **metabolic pathway.** Metabolic pathways indicate how a series of enzymes can work together in one functional unit. Within the cell thousands of these individual enzymatic reactions are occurring. Each reaction is one step in the sequence of one of the various metabolic sequences which are characteristic of that individual cell.

Not only is the enzymologist interested in the reactions which occur in the cell and how the enzyme exerts its catalytic function in the reaction, but he must also understand the factors which are responsible for controlling the rate of an enzymatic reaction. Uncontrolled chemical reactions can lead to a cancerous type of metabolism. The control of the metabolic pathways is brought about by feedback information, by the products of the pathway, to the enzymes which are involved in the initial stages of the pathway. Depending upon the conditions existing in the cell at a given moment, this information might call for an increase in the rate of movement of material through the pathway, a decrease, or no change. Recent studies have shown that one way the products of a metabolic pathway can alter the rates of reactions in the pathway is by increasing or decreasing the catalytic properties of an enzyme, presumably by changing its structure. The quaternary structure of an enzyme may be very important in controlling this enzymatic activity.

QUESTIONS

1. List the principal chemical constituents of living organisms. What are the specific functions of each of these constituents?
2. Explain how the high heat capacity and the high heat of vaporization of water make it especially suitable as a component of body fluids.
3. Distinguish between aerobic and anaerobic processes.
4. Into what simpler components may a protein be hydrolyzed?
5. What are amino acids?
6. Relate amino acids, with the aid of an equation, to proteins.
7. How would you attempt to determine the molecular weight of a protein?
8. What is the difference between an α-amino acid and a β-amino acid?
9. What connection is there between nucleic acids, amino acids, and proteins?
10. What was the contribution of Linus Pauling (Nobel Award in chemistry for 1954) to our knowledge of the structure of proteins?
11. Distinguish between the primary, secondary, tertiary, and quaternary structures of a protein.
12. What is the literal meaning of the term "carbohydrate," and why was it initially adopted?
13. Write the name and formula for one example each of a monosaccharide, a disaccharide, and a polysaccharide.
14. What is the chemistry involved in the souring of milk?
15. In what sense can sucrose be considered an anhydride?
16. Show that the energy we derive from foodstuffs comes initially from the sun. Do the same for energy derived from coal, petroleum, and hydro-electricity.
17. Write the general formula for fats and oils. How do fats and oils differ from each other in composition?
18. The synthesis of organic molecules in nature from carbon dioxide and water is an endothermic process. What is the source of the required energy?
19. What is the coordination number of a metal ion centered in a porphyrin ring system?
20. Why is artificial respiration often ineffective in treating carbon monoxide poisoning? What therapy might be effective?
21. Write the equation for the hydrolysis of glyceryl tristearate and name the products of the reaction.
22. Relate diabetes to the oxidation of fats and oils in the body.
23. What is the chemistry of carbon monoxide poisoning?
24. What is the composition and function of human bile?
25. Describe the function of chlorophyll; of hemoglobin.

26. What vitamin deficiency is associated with each of the following: night blindness, rickets, beriberi, scurvy, sterility, pernicious anemia, pellagra, and poor blood-clotting?
27. What are enzymes? Relate coenzyme, and substrate to enzyme.
28. Enzymes of what class are effective in the cleavage of peptide bonds?
29. What kinds of enzymes are involved in the redox reactions of living organisms?
30. What are the three main classes of hydrolytic enzymes?
31. What is the function of the enzyme thrombin?
32. Write the equation for the sucrase-catalyzed hydrolysis of sucrose.
33. Discuss the manner in which the energy requirements of cells are satisfied.
34. Discuss the subject of information storage and retrieval by chemical systems. Cite a specific example.
35. Aside from industrial pollution, why may air purity be a particular problem in heavily populated areas where paving or buildings cover practically all the ground surface?

PROBLEMS

1. What is the per cent iron in heme? magnesium in chlorophyll-a?
   Ans. 9.059%; 2.720%
2. How many grams of glucose can be prepared by the hydrolysis of one gram of corn starch?   *Ans. 1.11 g*
3. A typical vitamin capsule contains 1.0 microgram of cobalamin. How many moles of Vitamin B_{12} would a person consume in a lifetime, on the basis of one capsule daily for 70 years?   *Ans. 1.9×10^{-5} mole*

REFERENCES

"Recollections of Personalities Involved in the Early History of American Biochemistry," W. C. Rose, *J. Chem. Educ.*, **46**, 759 (1969).

"Thermodynamics of the Hydrolysis of Adenosine Triphosphate," R. A. Alberty, *J. Chem. Educ.*, **46**, 713 (1969).

"Chemical Carcinogens and their Significance for Chemists," C. E. Searle, *Chem. in Britain*, **6**, 5 (1970).

"Molecular Models of Metal Chelates to Illustrate Enzymatic Reactions," H. S. Hendrickson and P. A. Srere, *J. Chem. Educ.*, **45**, 539 (1968).

"The Chemistry of Vitamin B-12," H. A. O. Hill and R. J. P. Williams, *Chem. in Britain*, **5**, 158 (1969).

"Porphyria and King George III," I. Macalpine and R. Hunter, *Sci. American*, July, 1969; p. 38.

"Structure of Protein Molecules," L. Pauling, R. B. Corey, and R. Hayward, *Sci. American*, July, 1954; p. 51.

"Proteins," P. Doty, *Sci. American*, Sept., 1957; p. 173.

"The Genetic Code," F. H. C. Crick, *Sci. American*, Oct., 1966; p. 55.

"The Genetics of Human Populations," R. L. Cavalli-Sforza, *Sci. American*, Sept., 1974; p. 81.

"The Isolation of Genes," D. D. Brown, *Sci. American*, Aug., 1973; p. 20.

"Conformations of Cyclic and Cylindrical Peptides," C. H. Hassall and W. A. Thomas, *Chem. in Britain*, **7**, 145 (1971).

"Research on Drugs from the Sea Starts to Move Forward," R. J. Seltzer, *Chem. & Eng. News,* Dec. 16, 1974; p. 20.

"Carbon-13 as a Label in Biosynthetic Studies," U. Séquin and A. I. Scott, *Science*, **186,** 101 (1974).

"The Three-Dimensional Structure of an Enzyme Molecule," D. C. Phillips, *Sci. American*, Nov., 1966; p. 78.

"The Cell Cycle," D. Mazia, *Sci. American,* Jan., 1974; p. 54.

"The Cooperative Action of Muscle Proteins," J. M. Murray and A. Weber, *Sci. American*, Feb., 1974; p. 58.

"Nutrition and the Brain," J. D. Fernstrom and R. J. Wurtman, *Sci. American*, Feb., 1974; p. 84.

"A Dynamic Model of Cell Membranes," R. A. Capaldi, *Sci. American*, Mar., 1974; p. 27.

"High-Efficiency Photosynthesis," O. Björkman and J. Perry, *Sci. American*, Oct., 1973; p. 80.

"Improving the Efficiency of Photosynthesis," I. Zelitch, *Science*, **188**, 626, (1975).

"The Absorption of Light in Photosynthesis," R. Govindjee, *Sci. American*, Dec., 1974; p. 68.

"Adaptation of Photosynthetic Processes to Stress," J. A. Berry, *Science*, **188,** 644 (1975).

"Protein Shape and Biological Control," D. E. Koshland, Jr., *Sci. American*, Oct., 1973; p. 52.

"The Chemical Elements of Life," E. Frieden, *Sci. American*, July, 1972; p. 52.

"The Sources of Muscular Energy," R. Margaria, *Sci. American*, Mar., 1972; p. 84.

"The Structure of Cell Membranes," C. F. Fox, *Sci. American*, Feb., 1972; p. 31.

"The Nucleotide Sequence of a Nucleic Acid," R. W. Holley, *Sci. American*, Feb., 1966; p. 30.

"The Synthesis of DNA," A. Kornberg, *Sci. American*, Oct., 1968; p. 64.

"RNA-Directed DNA Synthesis," H. M. Temin, *Sci. American*, Jan., 1972, p. 24.

"DNA Ligase: Structure, Mechanism, and Function," I. R. Lehman, *Science*, **186,** 790 (1974).

"How Actinomycin Binds to DNA," H. M. Sobell, *Sci. American,* Aug., 1974; p. 82.

"RNA Processing and RNA Tumor Virus Origin and Evolution," D. Gillespie and R. C. Gallo, *Science,* **188,** 802 (1975).

"The Visualization of Genes in Action," O. L. Miller, *Sci. American,* Mar., 1973; p. 34.

"Hybrid Cells and Human Genes," F. H. Ruddle and R. A. Kucherlapati, *Sci. American*, July, 1974; p. 36.

"The Oxygen Cycle," P. Cloud and A. Gibor, *Sci. American,* Sept., 1970; p. 110.

"Oxygen: Boon and Bane," I. Fridovich, *Amer. Scientist*, **63,** 54 (1975).

"The Sex-Attractant Receptor of Moths," D. Schneider, *Sci. American*, July, 1974; p. 28.

"Hemoglobin: Model Systems Shed Light on Oxygen Binding," T. H. Maugh II, *Science*, **187,** 154 (1975).

"The Structure and History of an Ancient Protein" (Cytochrome-c), R. E. Dickerson, *Sci. American*, April, 1972; p. 58.

"A Family of Protein-Cutting Proteins," R. M. Stroud, *Sci. American*, July, 1974; p. 74.

"The Role of Wax in Oceanic Food Chains," A. A. Benson, and R. F. Lee, *Sci. American*, Mar., 1975; p. 77.

"The Most Poisonous Mushrooms," W. Litten, *Sci. American,* Mar., 1975; p. 91.

"Chromosomal Proteins and Gene Regulation," G. S. Stein, J. S. Stein, and L. J. Kleinsmith, *Sci. American*, Feb., 1975; p. 46.

"A Mechanism of Disease Resistance in Plants," G. A. Strobel, *Sci. American*, Jan., 1975; p. 81.

"The Walls of Growing Plant Cells," P. Albersheim, *Sci. American*, April, 1975; p. 81.

"The Molecular Biology of Poliovirus," D. H. Spector and D. Baltimore, *Sci. American*, May, 1975; p. 25.

"Chemical Evolution," M. Calvin, *Amer. Scientist*, **63,** 169 (1975).

29

Boron and Silicon

Although boron and silicon are members of different groups of the Periodic Table (the periodic groups IIIA and IVA, respectively), they resemble each other rather closely in chemical behavior. In addition to the similarities between elements in the same family, similarities also occur in a number of cases between elements in pairs, one element of which is in both the next family and the next group from the other element—a diagonal relationship in the Periodic Table. For example, lithium is similar in chemical properties to magnesium; beryllium is similar to aluminum; and, as we have just noted, boron is similar to silicon. Furthermore, oxygen is like chlorine, and nitrogen is somewhat similar to sulfur, although the relationship between these nonmetals is less pronounced than that with the more metallic elements. The explanation of this phenomenon lies in the fact that in each case the ratio of the atomic radius to the nuclear charge for each of the two elements is nearly the same, which means that the outer shell electrons are attracted to the nucleus with about the same force.

A comparison of some of the properties of boron with those of aluminum (the second member of Group IIIA) and with those of silicon (the second member of Group IVA) illustrates this behavior. Boron resembles aluminum in forming compounds in which these elements are trivalent—for example, B_2O_3 and Al_2O_3, $B(OH)_3$ and $Al(OH)_3$, and BCl_3 and $AlCl_3$. However, BCl_3 is more like $SiCl_4$ than $AlCl_3$ in physical and chemical properties. Boron and silicon chlorides have melting and boiling points much lower than those of aluminum chloride. Aluminum chloride hydrolyzes reversibly,

$$AlCl_3 + 3H_2O \rightleftharpoons \underline{Al(OH)_3} + 3H^+ + 3Cl^-$$

whereas boron trichloride and silicon tetrachloride hydrolyze almost completely.

$$BCl_3 + 3H_2O \longrightarrow \underline{B(OH)_3} + 3H^+ + 3Cl^-$$
$$SiCl_4 + 4H_2O \longrightarrow \underline{Si(OH)_4} + 4H^+ + 4Cl^-$$

Boron and silicon are primarily nonmetallic in their chemical properties, but both show some metallic character in the elemental state. For this reason, boron and silicon are included in the elements referred to as **semi-metallic elements,** or as **metalloids,** which are elements intermediate in nature between metals and nonmetals. Other elements commonly classified as metalloids are germanium, arsenic, antimony, tellurium, polonium, and astatine. Neither boron nor silicon occurs free in nature, but both exist as compounds with oxygen. Each forms a number of binary compounds with hydrogen.

Boron

29.1   Occurrence and Preparation of Boron

The boron-containing compound "borax" is referred to in early Latin works on chemistry. The element boron was first prepared in an impure form in 1808 by Gay-Lussac and Thénard in France and by Davy in England, by reducing boric acid with potassium.

Boron constitutes less than 0.001 per cent of the earth's crust but is widely distributed as **orthoboric acid,** H_3BO_3, in volcanic regions; and it occurs as borates, such as **borax,** $Na_2B_4O_7 \cdot 10H_2O$, **kernite,** $Na_2B_4O_7 \cdot 4H_2O$, and **colemanite,** $Ca_2B_6O_{11} \cdot 5H_2O$, in dry lake regions including the desert areas of southern California.

Impure boron is usually prepared by reducing the oxide with powdered magnesium.

$$B_2O_3 + 3Mg \longrightarrow 2B + 3MgO$$

The magnesium oxide is removed by dissolving it in hydrochloric acid. Prepared in this way, boron is an impure brown amorphous powder. Pure boron may be obtained by passing a mixture of the trichloride and hydrogen either through an electric arc or over a hot tungsten filament.

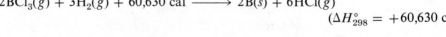

$$2BCl_3(g) + 3H_2(g) + 60{,}630 \text{ cal} \xrightarrow{1{,}500°} 2B(s) + 6HCl(g)$$
$$(\Delta H^\circ_{298} = +60{,}630 \text{ cal})$$

29.2   Properties of Boron

Crystalline boron is transparent and nearly as hard as diamond. Its brittleness and high electrical resistance cause it to be classified as a nonmetallic element.

Elemental boron crystals have a rather unusual structure, being made up of tightly packed small icosahedra. An icosahedron is a symmetrical geometric figure with 20 faces, each of which is an equilateral triangle (Fig. 29–1).

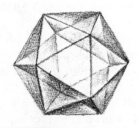

FIGURE 29–1
The icosahedron, an important geometric figure in the structure of boron and several of its compounds.

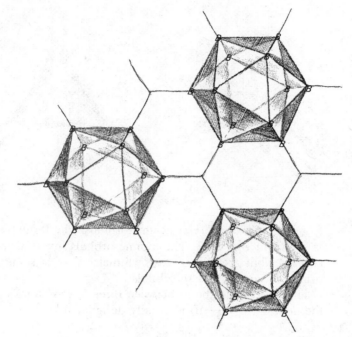

FIGURE 29-2

The structure of α-rhombo-hedral boron. Each icosahedron contains 12 boron atoms, six of which are bonded each to a boron atom in another icosahedron by two-center bonds and six of which are each bonded to two boron atoms, each in a separate icosahedron, by three-center bonds. Only the three-center bonds between icosahedra are shown here.

The faces meet at 12 corners. In the boron structure a boron atom occupies each of the 12 corners. In the most common form of boron (α-rhombohedral) the icosahedra are closely packed very nearly in a cubic structure. The length of the bonds between adjacent boron atoms in each icosahedron is 1.76 Å. However, between the icosahedra themselves, two kinds of bonds of different lengths (1.71 Å and 2.03 Å) are present (Fig. 29–2).

Half of the twelve boron atoms of an icosahedron are each joined by regular B—B two-center bonds (1.71 Å in length) to a boron atom of another icosahedron (these bonds are not shown in Fig. 29–2). In Chapters 4 and 5, we discussed the two-center bond, in which two atoms are bonded together by one pair of electrons in a molecular orbital which results from the overlap of two atomic orbitals (see Sections 5.1 and 5.2). The other half of the twelve boron atoms of each icosahedron are each bonded to *two* boron atoms, one in each of two other icosahedra, by a *three-center bond* (2.03 Å in length), as indicated in Fig. 29–2.

In a three-center bond, *three* atoms are bonded together by one pair of electrons in a molecular orbital which results from the overlap of three atomic orbitals. The three-center bond is especially likely for atoms which do not have sufficient electrons to satisfy ordinary two-center bond requirements. Of all the known elements, boron makes the greatest use of the three-center bond. It exhibits this type of bond not only in the elemental boron structure but also in several boron hydrides (Section 29.3). As was pointed out in Chapter 5, in the regular two-center bond, one bonding molecular orbital and one antibonding molecular orbital result from the overlap of two atomic orbitals (Section 5.1). In the three-center bond, one bonding molecular orbital and several antibonding molecular orbitals result

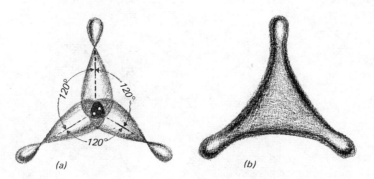

from the overlap of *three* atomic orbitals. The formation of the bonding orbital is shown in Fig. 29–3. The atomic orbitals involved for boron are largely *p* in character but are slightly less symmetric due to some hybridization with the *s* orbital to give hybrid *sp* orbitals.

The three-center bond between three boron atoms (sometimes referred to by the notation B—B—B) is usually designated

$$
\begin{array}{c}
\text{B} \\
\diagup\ \diagdown \\
\text{B}\qquad\text{B}
\end{array}
$$

and the three-center bond between two boron atoms and a hydrogen atom (sometimes referred to by the notation B—H—B) is usually designated

$$
\begin{array}{c}
\text{H} \\
\diagup\ \diagdown \\
\text{B}\qquad\text{B}
\end{array}
$$

Two other forms of elemental boron also are icosahedral in nature but have somewhat different bonding between the individual icosahedra. In one of these, the bond lengths between icosahedra are 1.68 Å and 1.81 Å. This represents rather close multiple bonding, which accounts for the hardness of this form of boron. Diamond, similar in structure to boron, has a regular structure of carbon atoms (Fig. 27–1), which are somewhat smaller than boron atoms and uniformly spaced still closer together at 1.54 Å. The closer spacing of the atoms in diamond accounts for its being harder than boron.

Boron combines with fluorine at room temperature and with chlorine, bromine, oxygen, and sulfur at elevated temperatures. It does not react with iodine directly. Boron reacts with carbon at the temperature of the electric arc and forms boron carbide, B_4C, which ranks next to diamond in hardness.

In the B_4C structure the boron atoms are at the corners of the icosahedra as in Fig. 29–2, but the icosahedra are linked together by chains of three carbon atoms. Hence, the continuing network of 12 boron atoms and three carbon atoms forms a unit with the empirical formula B_4C. The hardness of B_4C is expected, in view of the similarity of its structure to that of elemental boron.

Boron acts as a reducing agent when heated with water, sulfur dioxide, nitric oxide, carbon dioxide, or many other oxides. It is used as a deoxidizing agent in the purification of fused copper before it is cast. It also reduces both concentrated nitric and sulfuric acid, in which reactions the boron is oxidized to boric acid. In most of its chemical properties, boron acts as a nonmetal; its halides are volatile and irreversibly hydrolyzed. However, boron is intermediate in character between nonmetals and metals, as evidenced in part by $B(OH)_3$, which is a weak acid.

29.3   Boron Hydrides

Boron forms a series of volatile hydrides that somewhat resemble the hydrides of carbon and silicon. No hydride having the formula BH_3 exists. This is unexpected because boron usually exhibits a combining power of three. The simplest and most important of the boron hydrides is **diborane**, B_2H_6. However, an electronic structure cannot be written for B_2H_6 which would be in keeping with our theory of regular covalent bonds as outlined in Chapter 4 and with the properties of the compound. Fourteen valence electrons (seven electron pairs) would be required for a covalent structure of the type:

$$\begin{array}{ccc} & H & H \\ & | & | \\ H- & B-B & -H \\ & | & | \\ & H & H \end{array}$$

Actually, only twelve electrons are available for bond formation in B_2H_6. Compounds of this sort are sometimes referred to as being "electron deficient." Recent investigations have shown that two of the six hydrogen atoms of B_2H_6 are different from the other four and that rotation around the B—B linkage is hindered. The proposed structure most in keeping with the properties of diborane is one in which two "hydrogen bridges" bond the two halves of the molecule together (Fig. 29–4). Each boron atom is connected to two hydrogen atoms, all in the same flat plane, by regular two-center single covalent bonds. These four bonds use eight of the twelve available electrons. The two "bridge" hydrogens are above and below the plane, respectively. Each of the two "bridge" hydrogens is connected to the two boron atoms by a three-center B—H—B bond. The three atoms in each bond

FIGURE 29–4

The proposed structure of the diborane molecule, B_2H_6. (a) The spatial arrangement of atoms showing the "bridge" hydrogens (in color). (b) The spatial arrangement of the bonding orbitals.

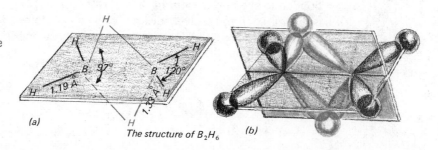

The structure of B_2H_6

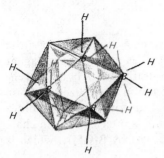

FIGURE 29-5
The B_4H_{10} molecule.

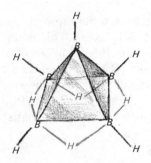

FIGURE 29-6
The B_5H_9 molecule.

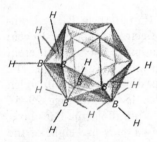

FIGURE 29-7
The B_6H_{10} molecule.

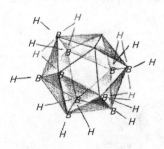

FIGURE 29-8
The $B_{10}H_{14}$ molecule.

are bonded by a molecular orbital which contains one pair of electrons, analogous to the B—B—B three-center bond in elemental boron. The two three-center bonds thereby together utilize the remaining four electrons.

Higher hydrides than B_2H_6 are also possible, and many have been made. The B_4H_{10} molecule is pictured in Fig. 29-5. Six of the ten hydrogen atoms are bonded by regular single covalent bonds; the other four by B—H—B three-center bonds. Note that the boron atoms are arranged at the corners of triangular planes, reminiscent of the icosahedral structure.

The B_5H_9 structure, shown in Fig. 29-6, is a square pyramid with boron atoms at each of the five corners. Five hydrogen atoms are bonded, one to each boron, by a regular single covalent bond. The other four hydrogen atoms are bonded through four three-center B—H—B bonds.

In the B_6H_{10} structure, diagrammed in Fig. 29-7, six hydrogen atoms are bound by regular covalent bonds and four by three-center bonds. It is interesting to note that the B_6H_{10} structure can be recognized as half of an icosahedral structure.

Several, but not all, of the known higher boron hydrides also exhibit a partial or complete icosahedral structure, with the structure of the icosahedron becoming more complete for higher members of the series. Figure 29-8 shows the $B_{10}H_{14}$ structure, in which ten of the twelve corners of the icosahedron are occupied by boron atoms. The structure has ten two-center B—H bonds and four three-center B—H—B bonds. The $B_{12}H_{12}^{2-}$ ion, in which all B—H bonds are two-center bonds, exhibits a complete icosahedron.

Diborane is readily prepared by the action of lithium aluminum hydride upon boron trifluoride in ether solution.

$$4BF_3 + 3LiAlH_4 \longrightarrow 2B_2H_6\uparrow + 3LiF + 3AlF_3$$

Diborane gas decomposes slowly at room temperature to its component elements, which then recombine to form higher boron hydrides, such as B_4H_{10}, and hydrogen. Above 300° diborane rapidly decomposes into boron and hydrogen. It ignites spontaneously in moist air, the products of combustion being boric oxide and water. Diborane reacts violently with chlorine to form the trichloride and hydrogen chloride. It is hydrolyzed by water to orthoboric acid and hydrogen.

$$B_2H_6(g) + 6H_2O(l) \longrightarrow 2H_3BO_3(s) + 6H_2(g) + 121,700 \text{ cal}$$
$$(\Delta H_{298}^\circ = -121,700 \text{ cal})$$

When an excess of sodium hydride acts upon boron trifluoride, the complex **sodium borohydride,** $NaBH_4$, is formed. The four hydrogen atoms are covalently bonded to the boron atom in the complex anion $[BH_4]^-$. The electronic formula is shown. Sodium borohydride and other similar complex hydrides, such as **lithium aluminum hydride,** $LiAlH_4$, are very useful reducing agents, particularly toward organic compounds.

$$Na^+ \left[\begin{array}{c} H \\ \overset{\circ\circ}{H \overset{\circ}{\underset{\circ\circ}{B}} H} \\ H \end{array} \right]^-$$

Sodium borohydride

29.4   The Boron Halides

As mentioned in Section 29.2, the fluoride, chloride, and bromide of boron can be prepared by direct union of the elements. The fluoride is a colorless gas that hydrolyzes in water to form boric acid and hydrofluoric acid.

$$BF_3(g) + 3H_2O(l) + 20{,}700 \text{ cal} \longrightarrow H_3BO_3(s) + 3HF(g)$$
$$(\Delta H^\circ_{298} = +20{,}700 \text{ cal})$$

The hydrofluoric acid thus formed combines with excess boron trifluoride and gives **fluoboric acid,** HBF_4.

$$BF_3 + HF \longrightarrow H^+ + BF_4^-$$

In this latter reaction the electron deficient BF_3 molecule acts as a Lewis acid (electron-pair acceptor), and hydrogen fluoride as a Lewis base (electron-pair donor), as shown by the equation

Fluoboric acid is a strong acid that so far has not been isolated in the free condition. In dilute aqueous solutions at room temperature it undergoes slow hydrolysis according to the equation

$$BF_4^- + 3H_2O \rightleftharpoons H_3BO_3 + 3H^+ + 4F^-$$

Crystal structure studies show the fluoborate ion, BF_4^-, to be tetrahedral.

The trichloride of boron is a colorless mobile liquid, the bromide is a viscous liquid, and the iodide is a white crystalline solid.

29.5   Oxide and Oxyacids of Boron

Boron takes fire at 700° in oxygen and burns to form **boric oxide,** B_2O_3. Boric oxide finds use in the production of chemical-resistant glass and certain optical glasses. The oxide dissolves in hot water to form **orthoboric acid,** H_3BO_3.

$$B_2O_3 + 3H_2O \longrightarrow 2H_3BO_3$$

The boron atom in H_3BO_3 is at the center of an equilateral triangle with oxygen atoms at the corners.

Orthoboric acid

In the solid acid these triangular units are held together by hydrogen bonding. When boric acid is heated to 100° C, a molecule of water is split out between a pair of adjacent OH groups to form **metaboric acid,** HBO_2. With further heating at 140° to 160°, B—O—B linkages occur, connecting BO_3 groups together with shared oxygen atoms, to form **tetraboric acid,** $H_2B_4O_7$. At still higher temperatures, boric oxide is formed.

$$H_3BO_3 \longrightarrow HBO_2 + H_2O$$
$$4HBO_2 \longrightarrow H_2B_4O_7 + H_2O$$
$$H_2B_4O_7 \longrightarrow 2B_2O_3 + H_2O$$

A structure which has been suggested for the $B_4O_7{}^{2-}$ portion of borax, $Na_2B_4O_7 \cdot 10H_2O$, a salt of tetraboric acid, is

$B_4O_7{}^{2-}$ unit

29.6   Borates

Commercially, the most important borate is **borax,** or **sodium tetraborate 10-hydrate,** $Na_2B_4O_7 \cdot 10H_2O$. Most of the supply of borax comes directly from dry lakes, such as Searles Lake in California, or is prepared from **kernite,** $Na_2B_4O_7 \cdot 4H_2O$. Aqueous solutions of borax are basic inasmuch as it is a salt of a strong base and weak acid. The hydrolysis of the tetraborate ion may be represented by

$$B_4O_7{}^{2-} + 7H_2O \rightleftharpoons 4H_3BO_3 + 2OH^-$$

Borax is used to soften water and to make washing compounds, uses which depend upon the alkaline character of its solutions and the insolubility of the borates of calcium and magnesium.

When heated, borax fuses to form a glass which has the property of dissolving fused metal oxides. An example is given by the equation

$$Na_2B_4O_7 + CuO \longrightarrow 2NaBO_2 + Cu(BO_2)_2$$

The use of borax to remove oxides from metal surfaces in soldering and welding depends upon this property.

Different metals give borax glasses of different colors, a fact used in detecting certain metals. Cobalt borax glass is blue, for example.

29.7 Peroxyborates

When a solution of borax is treated with sodium hydroxide and hydrogen peroxide, transparent monoclinic crystals of $NaBO_2 \cdot 3H_2O \cdot H_2O_2$ are formed. This substance is used as an antiseptic and bleaching agent under the name **sodium perborate.** It is usually considered to be peroxyborate, but erroneously so, because the salt is a **peroxyhydrate.** True **peroxyborates** such as $(KBO_3)_2 \cdot H_2O$ and $(NH_4BO_3)_2 \cdot H_2O$ are known, however, and $NaBO_3$ can be formed by the action of sodium hydroperoxide, $NaHO_2$, on boric acid. These compounds liberate hydrogen peroxide upon hydrolysis and, as a result, they are of value as oxidizing and bleaching agents.

29.8 Boron-Nitrogen Compounds

A number of compounds containing boron-nitrogen linkages have been prepared and studied. Two which are of particular interest are **boron nitride,** BN, and **triborinetriamine,** $B_3N_3H_6$ (also called **borazole** and **borazine**).

Boron nitride is the final product of the thermal decomposition of many boron-nitrogen compounds, such as $B(NH_2)_3$ and $BF_3 \cdot NH_3$. It can also be prepared by heating boron with nitrogen or ammonia, or by heating borax with ammonium chloride. Crystalline boron nitride is a white solid which sublimes somewhat below 3,000°; it melts at this temperature under pressure. It is inert to most reagents but can be decomposed by fusion with alkalies. The crystalline structure of boron nitride is analogous to that of graphite (Section 27.3 and Fig. 27–2). In fact, this form of boron nitride has been called **inorganic graphite.** The layers of atoms in boron nitride and graphite are made up of analogous hexagonal rings as shown.

Boron nitride
layer

Graphite
layer

The loosely held electrons of the resonating double bonds in graphite are responsible for its electrical conductivity and shining black color. The fact that boron nitride is a nonconductor and white in color indicates that its double bonds do not resonate. A second crystalline form of boron nitride, Borazon, has recently been synthesized. It has the diamond cubic structure (Section 27.2 and Fig. 27–1) and is second to diamond in hardness on a relative hardness scale. It is now produced commercially.

Borazine, $B_3N_3H_6$, is formed when ammoniates of the boron hydrides, such as $B_2H_6 \cdot 2NH_3$, are heated at 180° to 200°; hydrogen is the other product of the reaction. Borazine has been termed "inorganic benzene," because its structure and physical properties are similar to those of benzene. The structures of these two compounds are given.

Borazine   Benzene

Because a boron atom and a nitrogen atom bonded together have the same number of electrons as two carbon atoms and the sizes of these two groups are about the same, it is not surprising that the forms of boron nitride are similar to graphite and diamond, nor that borazine resembles benzene. Groups of this type are called **isosteres**. The concept of **isosterism** has been of particular value in accounting for similarities among the properties of compounds with apparently unrelated structures.

Silicon

The name silicon is derived from the Latin word *silex,* meaning flint. It is the second element of Periodic Group IVA, which also contains carbon, germanium, tin, and lead.

Whereas carbon, with its ability to form compounds containing carbon-carbon bonds, plays the dominant structural role in the animal and vegetable worlds, silicon, which readily forms compounds containing Si—O—Si bonds, is of prime importance in the mineral world.

29.9 Occurrence

The earth's crust is composed almost entirely of minerals in which silicon atoms are connected by oxygen atoms in complex structures involving chains, layers, and three-dimensional frameworks. The minerals constitute the bulk of most common rocks (except limestone and dolomite), and of soils, clays, and sands, which are the products of the weathering of rocks. Inorganic building materials such as granite, bricks, cement, mortar, ceramics, and glasses are composed of silicon compounds. Silicon composes nearly one-fourth of the mass of the earth's crust, being second only to oxygen in abundance. Its binary compound with

TABLE 29-1   Some Important Silicate Minerals

Asbestos	$H_4Mg_3Si_2O_9$
Natrolite (a zeolite)	$Na_2(Al_2Si_3O_{10}) \cdot 2H_2O$
Garnet	$Ca_3Al_2(SiO_4)_3$
Zircon	$ZrSiO_4$
Mica	$K_2Al_2(AlSi_3O_{10})(OH)_2$
Talc	$Mg_3(Si_4O_{10})(OH)_2$
Kaolin (a clay)	$Al_2Si_2O_5(OH)_4$
Feldspar	$KAlSi_3O_8$
Beryl	$Be_3Al_2Si_6O_{18}$

oxygen is called **silica**, SiO_2. Familiar forms of impure silica are sand and sandstone. Silica also occurs as quartz, amethyst, agate, and flint.

Silica is acidic in character and, at high temperatures, it combines with basic metal oxides forming silicates.

$$CaO + SiO_2 \longrightarrow \underset{\text{Calcium silicate}}{CaSiO_3}$$

Most silicate rocks are built up of the common metal cations and complex silicate anions. A great variety of silicates exists in nature. A few of the more important silicates are given in Table 29–1. Most of these silicates have important specific uses, but they are such stable compounds that it is not economically feasible to use them as sources of the metals which they contain.

29.10   Preparation of Silicon

Elemental silicon was first prepared in an impure, amorphous form by Berzelius in 1823 by heating silicon tetrafluoride with potassium. It can also be obtained by the action of strong reducing agents at high temperatures upon silicon dioxide. With carbon and magnesium as the reducing agents, the equations are

$$SiO_2(s) + 2C(s) + 164{,}890 \text{ cal} \longrightarrow Si(s) + 2CO(g) \quad (\Delta H^\circ_{298} = +164{,}890 \text{ cal})$$
$$SiO_2 + 2Mg \longrightarrow Si + 2MgO$$

An important alloy of iron and silicon, known as ferrosilicon, is produced by the simultaneous reduction of iron(III) oxide and silica with carbon at the high temperature (about $3{,}000°$) of the electric furnace.

29.11   Properties and Uses of Silicon

Silicon crystallizes with a diamond-like structure in which each silicon atom is covalently bonded to four neighboring silicon atoms at the corners of a regular tetrahedron. Thus, a single crystal of silicon is a three-dimensional giant molecule. Silicon is hard enough to scratch glass, melts at $1{,}410°$, and is a brittle, gray-black, metallic-appearing solid.

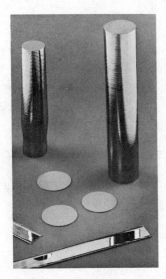

FIGURE 29-9
Single crystals of silicon. Such crystals are produced for use in solar cells. *Dow Corning Electronic Products Division*

Silicon is inactive at low temperatures and resists attack by air, water, and acids. It does react, however, with strong oxidizing agents and strong bases. It dissolves in hot sodium hydroxide or potassium hydroxide solutions forming silicates and hydrogen.

$$Si + 2OH^- + H_2O \longrightarrow SiO_3^{2-} + 2H_2\uparrow$$

Silicon is attacked by the halogens at high temperatures with the formation of volatile tetrahalides, such as SiF_4, and it burns in air to give the dioxide.

Elemental silicon is used as a deoxidizer in the production of steel, copper, and bronze, and in the manufacture of certain acid-resistant alloys such as "duriron" and "tantiron." Highly purified silicon, containing no more than one part impurity per million parts silicon, is used in semiconductor, transistor-type devices such as the solar battery (Section 22.18). See Fig. 29-9.

29.12   Silicon Hydrides

Silicon, like carbon of the same periodic group, forms a series of hydrides. This series is quite limited compared to that of carbon and includes SiH_4, Si_2H_6, Si_3H_8, Si_4H_{10}, Si_5H_{12}, and Si_6H_{14}. The chemical behavior of the hydrides of silicon is often decidedly different from that of the hydrocarbons of similar formulas. For example, the silicon hydrides are spontaneously combustible in air whereas the hydrocarbons are not. Only silicon hydrides corresponding to the alkanes are known; none are known which are analogous to the alkenes or alkynes with double or triple bonds.

Acids react with magnesium silicide to form **silane**, SiH_4, analogous in formula to methane, CH_4.

$$Mg_2Si + 4H^+ \longrightarrow 2Mg^{2+} + SiH_4\uparrow$$

Silane is a colorless gas, thermally stable at ordinary temperatures, and spontaneously combustible in air. Silicon dioxide and water are its combustion products.

$$SiH_4(g) + 2O_2(g) \longrightarrow SiO_2(s) + 2H_2O(g) + 341,500 \text{ cal}$$
$$(\Delta H_{298}^\circ = -341,500 \text{ cal})$$

Stepwise halogenation of silane is possible, though the reactions are difficult to control. A better method of obtaining partially halogenated silanes involves the use of hydrogen halides in the presence of the corresponding aluminum halide as a catalyst. With HBr and $AlBr_3$, the reactions are given by the equations

$$SiH_4 + HBr \longrightarrow SiH_3Br + H_2$$
$$SiH_3Br + HBr \longrightarrow SiH_2Br_2 + H_2$$
$$SiH_2Br_2 + HBr \longrightarrow SiHBr_3 + H_2$$
$$SiHBr_3 + HBr \longrightarrow SiBr_4 + H_2$$

Silane is extremely sensitive to alkalies, giving silicates and hydrogen.

$$SiH_4 + 2OH^- + H_2O \longrightarrow SiO_3^{2-} + 4H_2\uparrow$$

29.13 Silicon Carbide

When a mixture of sand and a large excess of coke is heated in an electric furnace, **silicon carbide (carborundum),** SiC, is produced according to the equation

$$SiO_2(s) + 3C(s) + 149,300 \text{ cal} \longrightarrow SiC(s) + 2CO(g)$$

$$(\Delta H^{\circ}_{298} = +149,300 \text{ cal})$$

The product comes from the furnace in the form of blue-black iridescent crystals, nearly as hard as diamonds and very stable at high temperatures. The crystals are crushed, the particles are screened to uniform size, mixed with a binder of clay or sodium silicate, molded into various shapes such as grinding wheels, and fired. Silicon carbide is used as an abrasive for cutting, grinding, and polishing.

Silicon carbide exists in three different crystalline forms; yet in each of these forms carbon and silicon atoms have alternate positions, and each atom is surrounded tetrahedrally by four others. One crystalline form has the diamond structure. In order to rupture a crystal of silicon carbide, one must break a number of strong covalent bonds. The high decomposition temperature (above 2,200°), the extreme hardness, the brittleness, and the chemical inactivity of silicon carbide are in accord with a diamond-like structure.

29.14 Silicon Halides

All the tetrahalides of silicon, SiX_4, have been synthesized, and several mixed halides of the type $SiCl_2F_2$ have also been prepared.

Silicon tetrachloride can be prepared by either direct chlorination at elevated temperatures or by heating silicon dioxide with chlorine and carbon. The equations are

$$Si(s) + 2Cl_2(g) \longrightarrow SiCl_4(g) + 157,030 \text{ cal} \quad (\Delta H^{\circ}_{298} = -157,030 \text{ cal})$$
$$SiO_2(s) + 2C(s) + 2Cl_2(g) + 7,860 \text{ cal} \longrightarrow SiCl_4(g) + 2CO(g)$$
$$(\Delta H^{\circ}_{298} = +7,860 \text{ cal})$$

Silicon tetrachloride is a low-boiling (57°), colorless liquid which fumes strongly in moist air to produce a dense smoke of finely divided silicic acid.

$$SiCl_4(l) + 4H_2O(l) \longrightarrow H_4SiO_4(s) + 4HCl(g) + 4,800 \text{ cal}$$
$$(\Delta H^{\circ}_{298} = -4,800 \text{ cal})$$

Elemental silicon ignites spontaneously in an atmosphere of fluorine, forming **silicon tetrafluoride,** SiF_4, which is a gas. This compound is readily prepared by the action of hydrofluoric acid upon silica or a silicate.

$$SiO_2(s) + 4HF(g) \longrightarrow SiF_4(g) + 2H_2O(l) + 45,700 \text{ cal}$$
$$(\Delta H^{\circ}_{298} = -45,700 \text{ cal})$$
$$CaSiO_3 + 6HF \longrightarrow CaF_2 + SiF_4\uparrow + 3H_2O$$

Silicon tetrafluoride hydrolyzes in water and produces **fluosilicic acid** as well as **orthosilicic acid.**

$$3SiF_4 + 4H_2O \longrightarrow \underset{\substack{\text{Orthosilicic} \\ \text{acid}}}{H_4SiO_4} + 4H^+ + \underset{\substack{\text{Fluosilicic} \\ \text{acid}}}{2SiF_6{}^{2-}}$$

Fluosilicic acid is a stronger acid than sulfuric. It is stable only in solution, however, and upon evaporation decomposes according to the equation

$$H_2SiF_6 \longrightarrow 2HF\uparrow + SiF_4\uparrow$$

29.15   Silicon Dioxide (Silica)

The usual crystalline form of silicon dioxide is quartz—a hard, brittle, refractory, colorless solid (Fig. 29–10). It is used in many ways—for architectural decorations,

FIGURE 29-10
Crystals of pure quartz. The crystals are hexagonal. Many of the crystals in this photograph have had their original surfaces eroded by washing in river gravels over a long period of time. The structure of quartz and other forms of silica may be described as consisting of SiO_4 tetrahedra bonded together by sharing the oxygen atoms at the corners between two of these tetrahedra. *August E. Miller*

FIGURE 29-11
Transparent fused quartz. The block of quartz measures $9 \times 5 \times 3$ inches. Such material is the most transparent substance known. *General Electric Co.*

semiprecious jewels, optical instruments, and frequency control in radio transmitters. The contrast in structure and physical properties between silica and its carbon analog, carbon dioxide, is interesting. The unit of structure in solid carbon dioxide (Dry Ice) is the single CO_2 molecule with very weak intermolecular forces holding the building units at the points of the crystal lattice. The low melting point and volatility of Dry Ice reflect the weak crystal forces between its building units. Each of the two oxygen atoms is attached to the central carbon atom by double covalent bonds. In contrast, a silicon atom in quartz is linked to four oxygen atoms by single bonds directed toward the corners of a regular tetrahedron, and the SiO_4 tetrahedra are bonded together by sharing the oxygen atoms at the corners between two tetrahedra. This structure extends outward in a three-dimensional continuous, giant silicon-oxygen network extending out to give a **macromolecule** of silicon dioxide, the quartz crystal. The overall ratio of silicon to oxygen atoms is one to two, and the simplest formula for the compound is SiO_2, but the formula SiO_2 does not represent a single molecule as does the formula CO_2.

At $1,600°$, quartz melts to give a viscous liquid with a random internal structure. When the liquid is cooled it does not crystallize readily, but usually undercools and forms a glass, called **silica glass** (Fig. 29–11). The SiO_4 tetrahedra in this glass have the random arrangement characteristic of undercooled liquids, and the glass has some very interesting and useful properties. Silica glass is highly transparent to both visible and ultraviolet light and so finds use in the manufacture of mercury vapor lamps, which give radiation rich in the ultraviolet region of the spectrum. It is also used in the production of certain optical instruments which operate in the ultraviolet region. The coefficient of expansion of silica glass is very low, so it is not easily fractured by sudden changes in temperature. It is also insoluble in water and inert towards acids except hydrofluoric.

$$SiO_2 + 4HF \longrightarrow SiF_4\uparrow + 2H_2O$$

This reaction is used in the quantitative separation of silica from other oxides, both products of the reaction being volatile. Hot alkali hydroxides and fused alkali carbonates convert silica into soluble silicates (SiO_4^{4-} and SiO_3^{2-}). Typical examples are

$$SiO_2 + 4OH^- \longrightarrow SiO_4^{4-} + 2H_2O$$
$$SiO_2 + Na_2CO_3 \longrightarrow Na_2SiO_3 + CO_2$$

The latter reaction is employed in the conversion of silicate rocks to soluble forms for analysis.

29.16 Silicic Acids

Orthosilicic acid, H_4SiO_4, is an extremely weak acid. It cannot be formed from its acid anhydride, SiO_2, by hydration, due to the fact that silica is very nearly insoluble in water. However, the addition of a strong mineral acid to a solution of an alkali metal silicate results in the formation of orthosilicic acid.

$$SiO_4^{4-} + 4H_3O^+ \longrightarrow \underline{H_4SiO_4} + 4H_2O$$

The orthosilicic acid first appears as a colloidal dispersion and then very shortly forms a colloidal gelatinous mass. In this form orthosilicic acid adsorbs certain ions to the extent that they cannot be removed by washing with water. It is thought that this is the mechanism by which colloidal silicic acids present in soils retain ions of soluble salts which are essential to plant growth.

The loss of water from orthosilicic acid by heating results in silica being formed as the final product of dehydration. Although there is no evidence to support the mechanism, it is thought that the dehydration is stepwise, with the formation of **metasilicic acid** (H_2SiO_3), **disilicic acid** ($H_6Si_2O_7$), **trisilicic acid** ($H_4Si_3O_8$), and perhaps other polysilicic acids as intermediates. Salts of these and other polysilicic acids are found among the naturally occurring silicates.

Silica gel is obtained when gelatinous orthosilicic acid is dehydrated until it contains only a small percentage of moisture. This gel has an open, porous structure with a large surface area per unit of mass. It has a great tendency to adsorb gases and to catalyze certain chemical reactions involving substances in the gaseous state.

29.17 Natural Silicates

As a group, the silicates are characterized by the large number of variations in the silicon-oxygen ratio which occurs from one silicate to another. In the molecules of all of the silicates, however, the silicon atoms are to be found at the centers of oxygen tetrahedra, and thus the tetravalency of silicon is maintained. The variation in the silicon-oxygen ratio is due to the fact that the silicon-oxygen tetrahedra may exist as discrete and independent building units, or they may share atoms at corners, edges, and more rarely faces, in a variety of ways. The silicon-oxygen ratio varies with the *extent* of sharing of oxygen atoms by silicon atoms in the linking together of the tetrahedra.

It is convenient to classify the silicates whose structures are known in a few groups, based upon the manner of linking of the silicon-oxygen tetrahedra.

■ **1. Individual SiO_4 Tetrahedra Existing as Independent Groups in the Crystal Lattice.** Examples are **olivine**, Mg_2SiO_4, and **zircon**, $ZrSiO_4$. The positively charged metallic ions (Mg^{2+}, Zr^{4+}) serve to bind together the negative SiO_4 radicals, which have the tetrahedral structure shown in the drawings that follow. Note that only the oxygen atoms are shown. A silicon atom is in the center of each tetrahedron and is not shown.

$SiO_4{}^{4-}$

■ **2. Two SiO_4 Tetrahedra Sharing One Oxygen Corner and thus Forming Si_2O_7 Groups which Act as Discrete Building Units.** Examples are **hardystonite,** $Ca_2ZnSi_2O_7$ and **hemimorphite,** $Zn_4(OH)_2Si_2O_7 \cdot H_2O$. The cations are to be found between the negative Si_2O_7 groups, binding them together.

$Si_2O_7{}^{6-}$

■ **3. Three Tetrahedra Sharing Corners with Each Other and Forming Closed Rings.** An example is **benitoite,** $BaTiSi_3O_9$. The Si_3O_9 rings are held together by the positive metallic ions. The ring ions are arranged in sheets with their planes parallel.

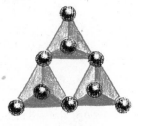

$Si_3O_9{}^{6-}$

Beryl (emerald) has the formula $Be_3Al_2Si_6O_{18}$; its structure involves six SiO_4 tetrahedra sharing corners to form a closed ring.

■ **4. Single Silicon-Oxygen Endless Chains Formed from SiO$_4$ Tetrahedra, Each Sharing Two Oxygen Atoms.** This structure gives an empirical composition of SiO$_3$, although there are no SiO$_3$ radicals present as independent groups. An example is **diopside**, CaMg(SiO$_3$)$_2$. The chains are considered to extend the full length of the crystal, and the parallel chains are held together by the attraction of the positively charged metal ions lying between them.

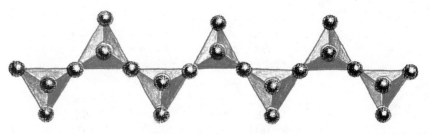

SiO$_4$ chain

■ **5. Double Silicon-Oxygen Chains.** In the single- and double-chain crystals the metallic ions link the parallel chains together. The fact that these ionic linkages are not as strong as the silicon-oxygen bonds within the chains accounts for the fibrous nature of **asbestos,** a chain-type silicate. An example is **tremoline,** Ca$_2$Mg$_5$(Si$_4$O$_{11}$)$_2$(OH)$_2$. These silicates always contain some hydroxyl groups, which are attached to the metal atoms and never to the silicon atoms. The double chains may be represented as shown.

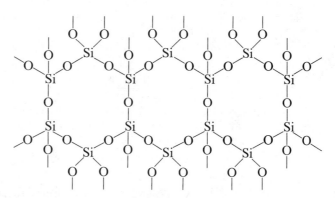

■ **6. Silicon-Oxygen Sheets Formed by the Extensions of Double Chains.** The metal ions form ionic bonds between the sheets. These ionic bonds are weaker than the silicon-oxygen bonds within the sheets. Thus, silicate minerals with this structure tend to cleave into thin layers, a property characteristic of the micas. Examples are **talc,** Mg$_3$Si$_4$O$_{10}$(OH)$_2$ and **muscovite mica,** KAl$_2$(AlSi$_3$O$_{10}$)(OH,F)$_2$.

■ **7. Three-Dimensional Silicon-Oxygen Networks in which a Portion of the Tetravalent Silicon is Replaced by Trivalent Aluminum.** The negative charge which results is neutralized by a distribution of positive ions throughout the network. Examples are **feldspar,** K(AlSi$_3$O$_8$), and the **zeolites,** such as Na$_2$(Al$_2$Si$_3$O$_{10}$)·2H$_2$O.

29.18 Glass

The common glass used for windowpanes, bottles, dishes, and the like is a mixture of sodium and calcium silicates with an excess of silica. It is made by heating together sand, sodium carbonate (or sodium sulfate), and calcium carbonate.

$$Na_2CO_3 + SiO_2 \longrightarrow Na_2SiO_3 + CO_2\uparrow$$
$$Na_2SO_4 + SiO_2 \longrightarrow Na_2SiO_3 + SO_3\uparrow$$
$$CaCO_3 + SiO_2 \longrightarrow CaSiO_3 + CO_2\uparrow$$

After the bubbles of gas have been expelled, a clear viscous melt results. This material is poured into molds or stamped with dies to produce pressed glassware. Articles such as bottles, flasks, and beakers are formed by taking a lump of the molten glass on a hollow tube, inserting it into a mold and blowing with compressed air until the outline of the mold is assumed. Ordinary window glass is "blown" in the form of a long cylinder which is then split longitudinally and opened. On heating near the melting point it flattens. The conventional method of making high quality plate glass is to draw the glass from the furnace in broad strips, roll it to the desired thickness, grind it flat, and polish it on both sides. A new technique for making plate glass involves pouring the molten glass on a layer of very pure molten tin. The tin surface is perfectly smooth, so when the glass hardens the glass surface in contact with the tin is also perfectly smooth and does not need to be ground and polished.

Glassware is "annealed" by heating it for a time just below the softening temperature and then cooling it slowly. Annealing avoids internal strains and thereby reduces the danger of breakage from shock or temperature change.

Glass is a complex mixture of silicates and is classified as an undercooled liquid. It is transparent, brittle, and entirely lacking in the ordered internal structure characteristic of crystals. When heated, it does not melt sharply. Instead it gradually softens until it reaches the liquid state.

When a colored glass is desired, an appropriate substance is added during the manufacture of the glass. Table 29–2 lists a few of the substances used in coloring glass.

If sodium is replaced by potassium in a glass melt, a higher-melting, harder, and less soluble glass is obtained. If part of the calcium is replaced by lead a glass of high density and high refractive index is formed. This variety of glass is called **flint glass** and is used in making lenses and cut-glass articles. **Pyrex glass,**

TABLE 29–2   Substances Used in the Production of Colored Glass

Substance	Color Produced	Substance	Color Produced
Iron(II) compounds	Green	Cobalt(II) oxide	Blue
Iron(III) compounds	Yellow	Manganese dioxide	Violet
Uranium compounds	Yellow, green, fluorescence	Calcium fluoride	Milky
		Tin(IV) oxide	Opaque
Colloidal selenium	Ruby	Copper(I) oxide	Red, green, blue
Colloidal gold	Red, purple, blue		

very suitable for articles subject to sudden changes in temperature and resistant to chemical action, is a borosilicate glass in which some of the silicon atoms are replaced by boron atoms.

One form of **safety glass,** used in the manufacture of automobile windshields, consists of a thin layer of plastic held between two pieces of thin plate glass. Adhesion of the glass to the flexible plastic decreases the danger from flying glass and jagged edges when the glass is broken.

Enamels on iron kitchen utensils, sinks, and bathtubs and glazes on pottery are made of easily fusible glass containing opacifiers such as titanium dioxide and tin(IV) oxide.

29.19 Cement

Portland cement is essentially powdered calcium aluminosilicate which sets to a hard mass when treated with water. It is made by pulverizing a mixture of limestone and clay, and roasting the powder in a rotary kiln heated by gas or powdered coal to a temperature of about 1,500°. This treatment yields sintered lumps called "clinker" about the size of small marbles. The clinker is then ground with a little gypsum, $CaSO_4 \cdot 2H_2O$, to a very fine powder.

Roads, walks, foundations, and floors are constructed of **concrete,** a substance made by adding water to a mixture of Portland cement, sand, and stone or gravel. Experimental mixtures incorporating such substances as rubber or plastic have been tried in efforts, at least partially successful, to improve the structural properties of concrete.

The reactions taking place as the Portland cement mixture "sets" (starts the hardening process to form concrete) are complex and not completely understood. It is known that during the process calcium aluminate hydrolyzes, forming calcium hydroxide and aluminum hydroxide. These compounds then react with the calcium silicates present, forming calcium aluminosilicate in the form of interlocking crystals. Portland cement "sets" rapidly (within 24 hours), and then "hardens" slowly, years being required for completion of the reactions.

29.20 Silicones

A modern development in the field of silicon chemistry has been the production of polymeric organosilicon compounds containing Si—O—Si linkages. These compounds are known as **silicones,** and have the general formula $(R_2SiO)_x$. They may be linear, cyclic, or cross-linked polymers of the types shown.

Linear silicone   Cyclic silicone   Cross-linked silicone

The R in the formulas represents an organic group, such as methyl (CH_3), ethyl (C_2H_5), or phenyl (C_6H_5). The linear and cyclic silicones are produced by hydrolyzing organochlorosilanes of the type R_2SiCl_2, followed by polymerization through the elimination of a molecule of water from two hydroxyl groups of adjacent $R_2Si(OH)_2$ molecules.

$$R_2SiCl_2 + 2H_2O \longrightarrow R_2Si(OH)_2 + 2HCl$$

$$\begin{matrix} & R & & & & R & & & & R & & R \\ HO-&Si&-O&\overline{\underline{H + HO}}&-&Si&-OH & \longrightarrow & HO-&Si&-O-&Si&-OH + H_2O \\ & R & & & & R & & & & R & & R \end{matrix}$$

The organosilicon polymers incorporate, to some extent, the properties of both hydrocarbons and oxysilicon compounds. Organosilicones are remarkably stable toward heat and chemical reagents and are not wetted by water. Depending upon the extent of polymerization and molecular complexity, the silicones may take the form of oils, greases, rubberlike substances, and resins. They are used as lubricants, as hydraulic fluids, for electrical insulators, and as moisture-proofing agents.

A particularly valuable property of silicone oils is their very low coefficient of viscosity, which is related to temperature in such a way that these oils can be employed as lubricants where there are extreme variations in temperature.

Such materials as paper, wool, glass, silk, and porcelain can be coated with a water-repellent film by simply exposing them for a second or two to the vapor of trimethylchlorosilane, $(CH_3)_3SiCl$. The surface becomes coated with a thin film of $(CH_3)_2Si-O$ groups, which repels water in a manner similar to that of a hydrocarbon film.

QUESTIONS

1. Compare boron and silicon with respect to atomic size, ionization potential, and electronegativity.
2. Use the ionic radii listed below (in Å) to compute values for the charge-to-volume ratio for each of the ions. Compare the values for the pairs of ions which are related by a diagonal relationship in the Periodic Table. (Li^+, 0.60; Be^{2+}, 0.31; Mg^{2+}, 0.65; B^{3+}, 0.20; Al^{3+}, 0.50; Si^{4+}, 0.41.)
3. List the physical and chemical properties of boron and silicon that would be pertinent in classifying each as a metal or as a nonmetal. How would you classify each? Explain your answer.
4. Explain how the structures of elemental boron and of diamond are related to their hardness.
5. Explain why in many respects lithium is similar to magnesium, beryllium to aluminum, and boron to silicon.
6. Write balanced equations showing the action of boron as a reducing agent toward H_2O, SO_2, NO, CO_2, and concentrated HNO_3.

7. Write balanced equations describing the reaction of boron with elemental fluorine, carbon, oxygen, and nitrogen, respectively.

8. Cite three chemical reactions in which boron exhibits nonmetallic behavior.

9. Why is B_2H_6 said to be an "electron deficient" compound?

10. Discuss the structures of the boron hydrides. What relation, if any, do these structures have to that of elemental boron?

11. Identify the oxidizing agent and the reducing agent in the chemical reaction for the preparation of diborane from boron trifluoride and lithium aluminum hydride.

12. The experimentally determined atom ratio of hydrogen to boron in a gas is 3/1. How could it be proved that the gas actually consists of B_2H_6 molecules?

13. Explain what is meant by the "irreversibility" of the hydrolysis of the halides of boron.

14. Write equations to show the formation of fluoboric acid from boron trifluoride.

15. Show by equations the relationship of the boric acids to boric oxide.

16. Why does an aqueous solution of borax turn red litmus blue?

17. Explain the chemistry of the use of borax as a flux in soldering and welding.

18. Write the structural formula for true sodium peroxyborate, showing that it is derived from hydrogen peroxide.

19. Why is boron nitride sometimes referred to as "inorganic graphite"?

20. Why should the physical properties of borazine be similar to those of benzene?

21. Evaluate the statement, "Silicon is the central element of the inorganic world."

22. By heating a mixture of sand and coke in an electric furnace, either silicon or silicon carbide may be obtained. What determines which will be formed?

23. In terms of molecular and crystal structure, account for the low melting point of Dry Ice (solid CO_2) and the high melting point of SiO_2.

24. Write balanced equations describing at least two different reactions in which solid silica exhibits acidic character.

25. How does the following reaction show the acidic character of SiO_2?

$$CaO + SiO_2 \xrightarrow{\Delta} CaSiO_3$$

26. Compare the silicon hydrides with the boron hydrides; with the hydrocarbons.

27. Relate colloidal silicic acid to soil fertility.

28. Describe the conversion of silicate rocks to soluble forms for the purpose of analysis.

29. Which would you expect to be a better Lewis base: the carbide ion, C^{4-}, or the silicide ion, Si^{4-}?

30. Account for the existence of the great variety of silicates in terms of the manner of linking of SiO_4 tetrahedra.

31. Give equations for the reactions involved in the production of common window glass.

32. How does the incorporation of the following substances affect the properties of glass: potassium oxide; cobalt(II) oxide; lead(II) oxide; boric oxide; iron(II) oxide; colloidal gold?

33. How does the internal structure of silica glass differ from that of quartz?

PROBLEMS

1. Calculate the distance between boron atoms that are next nearest neighbors in B_2H_6 (see Fig. 29-4). *Ans. 1.76 Å*

2. Calculate the weight of $LiAlH_4$ needed to produce 1.00 g of B_2H_6 by reduction of BF_3. *Ans. 2.06 g*

3. (a) Check by calculation, using data in Appendix J, the values of ΔH°_{298} for at least three of the reactions in this chapter for which values are given.

   (b) Calculate ΔG°_{298} values, using data in Appendix J, for the same reactions used in Part (a).

   (c) Compare the ΔH°_{298} and ΔG°_{298} values for each reaction, state what you can as to reasons for differences in the two quantities for each reaction, and state conclusions as to spontaneity of each reaction based upon the two quantities.

REFERENCES

"NMR of Boron Compounds," G. R. Eaton, *J. Chem. Educ.*, **46**, 547 (1969).

"Boron Chemistry in Industrial Perspective," R. Thompson, *Chem. in Britain*, **7**, 140 (1971).

"The Hydrides of Boron," A. G. Massey, *Educ. in Chemistry*, **11**, 20 (1974).

"Cage Compounds of Boron," M. S. Gaunt and A. G. Massey, *Educ. in Chemistry*, **11**, 118 (1974).

"Inorganic Polymers," H. R. Allcock, *Sci. American*, Mar., 1974; p. 66.

"The Chemical Elements of Life," E. Frieden, *Sci. American*, July, 1972; p. 52.

"Superhard Materials," F. P. Bundy, *Sci. American*, Aug., 1974; p. 62.

"Organoboranes and Organoborate Anions: New Classes of Electrophiles and Nucleophiles in Organic Synthesis," E. Negishi, *J. Chem. Educ.*, **52**, 159 (1975).

"Some Aspects of *d*-Orbital Participation in Phosphorus and Silicone Chemistry," J. E. Bissey, *J. Chem. Educ.*, **44**, 95 (1967).

"*d*-Orbitals in Main Group Elements," T. B. Brill, *J. Chem. Educ.*, **50**, 392 (1973).

"The Chemistry of Concrete," S. Brunauer and L. E. Copeland, *Sci. American*, April, 1964; p. 80.

"One Hundred Years of Organosilicon Chemistry," R. Müller (transl. by E. G. Rochow), *J. Chem. Educ.*, **42**, 41 (1965).

"The Nature of Ceramics," J. J. Gilman, *Sci. American*, Sept., 1967; p. 112.

"Hydrogen Forms Bridge from Silicon to Transition Atoms," (Staff), *Chem. & Eng. News*, June 8, 1970; p. 75.

"Artificial Organs," H. J. Sanders, *Chem. & Eng. News:* Part I, April 6, 1971; p. 32. Part II, April 12, 1971; p. 68.

"Bio-Organosilicon Chemistry," M. B. Voronkrov, *Chem. in Britain*, **9**, 411 (1973).

"Ancient Glass," R. H. Brill, *Sci. American*, Nov., 1963; p. 120.

"Chemical Strengthening of Glass," J. S. Olcott, *Science*, **140**, 1189 (1963).

"The Nature of Glasses," R. J. Charles, *Sci. American*, Sept., 1967; p. 126.

"Mikhail Lomonosov and the Manufacture of Glass and Mosaics," (historical), H. M. Leicester, *J. Chem. Educ.*, **46**, 295 (1969).

"Chemistry in Glass Technology," R. W. Douglas, *Chem. in Britain*, **5**, 349 (1969).

"Glass Formation and Crystal Structure," J. F. G. Hicks, *J. Chem. Educ.*, **51**, 28 (1974).

"Homogeneous Models Throw Light on Industrial Catalysts," D. G. H. Ballard, *Chem. in Britain*, **10**, 20 (1974).

30 Nuclear Chemistry

The chemical changes (transformations of one form of matter into another) of the types we have studied thus far have involved changes only in the electronic structures of atomic systems without alteration of their nuclear structures. Another type of transformation of matter involving changes in atomic nuclei is the basis for a branch of science which is on the borderline between physics and chemistry and which is called **nuclear chemistry.** It had its beginning with the discovery of radioactivity and has become increasingly important during the past thirty-five years.

Any atomic species, especially when it is considered in connection with its nuclear characteristics, is referred to as a **nuclide.** The term *isotope* is often used to designate a nuclide, but it must be remembered that references to isotopes are made only for forms of a given element possessing different masses but the same atomic number (Section 3.9).

Nuclear reactions, some of which occur naturally and some of which are induced artificially in the laboratory, are discussed in this chapter according to the following classification:

(1) Natural radioactivity, in which nuclei of certain atoms change spontaneously.

(2) Transmutations of elements by nuclear bombardment, whereby the nucleus of one element is changed to the nucleus of another element.

(3) Artificial radioactivity, in which radioactive elements produced artificially by transmutation reactions change spontaneously.

(4) Nuclear fission reactions.

(5) Nuclear fusion reactions.

30.1   The Structure of the Atomic Nucleus

A tremendous amount of information concerning the composition of atomic nuclei has been accumulated through research by physicists and chemists. To interpret this information they have developed a theory of the structure of the nucleus. It is thought that the heavier nuclei are built of protons and neutrons held together by intranuclear attractive forces. The forces may be designated as proton-proton forces, neutron-neutron forces, and proton-neutron forces. The three types of forces are about equal in magnitude. These intranuclear forces are different in character from either electrostatic or gravitational forces, and they are strong only when the distance between the particles is exceedingly small. For a nucleus to be stable, the nuclear attractive forces must be greater than the electrostatic repulsive forces within the nucleus. It appears that the protons and neutrons in the heavier atoms are arranged in nuclear "subdivisions," or energy levels, just as the electrons are grouped together in the energy shells of the atom. It is believed by most theoretical physicists that the nuclear particles are held together by π-*mesons*, which are particles with mass about 273 times that of an electron. The positive π-meson has a $+1$ unit charge and the negative π-meson has a -1 unit charge. According to one theory, each proton and neutron is surrounded by π-mesons.

30.2   The Determination of Atomic Weights with the Mass Spectrograph

We noted in Section 3.9 that most of the elements, as they occur in nature, are mixtures of two or more isotopes, and that different isotopes of the same element contain the same number of protons but different numbers of neutrons. Thus, ordinary oxygen is composed of 99.76 per cent $^{16}_{8}O$, 0.04 per cent $^{17}_{8}O$, and 0.2 per cent $^{18}_{8}O$; the atoms of these isotopes contain 8, 9, and 10 neutrons, respectively. The existence of isotopes was discovered by means of the mass spectrograph, an instrument designed by J. J. Thomson and improved by others, notably Aston of England and Dempster and Bainbridge of America.

The principle upon which the mass spectrograph is based is illustrated by the simple apparatus shown in Fig. 30-1. The element to be investigated is made to form positively charged gaseous ions by means of an electrical discharge through the gas at low pressure. The positive ions, travelling in straight lines,

FIGURE 30-1

A diagram showing the principles of the mass spectrograph.

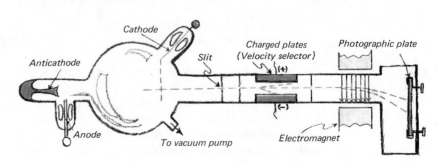

are then accelerated to high speeds by means of an electrical potential. After passing through slits the ions enter combined magnetic and electric fields where those ions having the same speed are selected. When the beam is allowed to pass through a magnetic field, the ions are deflected from a straight path, with those having the same mass and charge deflected to the same extent and striking the photographic plate in the same place. Ions of different masses are deflected to different extents and will produce lines upon the plate according to their masses. The masses of the positive ions can be calculated from the position of the lines on the plate, and the relative amounts of each ion can be obtained from the intensity of the lines.

Atomic masses obtained with the mass spectrograph are now usually reported relative to the isotope $^{12}_{6}C = 12$, which is the standard for atomic weights (see Sections 2.4 and 3.10).

If the photographic plate is replaced with separate pockets to collect ions of different masses, a separation of isotopes can be accomplished.

30.3 Packing Effect and the Binding Energy

read

Precise measurements with the mass spectrograph have shown that the mass of each isotope is very nearly a whole number. For example, the values of 19.999 and 21.998 for two of the isotopes of neon deviate from whole numbers by only slight amounts. The exact value is called the **atomic mass** and the nearest whole number is called the **mass number.** In view of the fact that isotopic masses are very nearly whole numbers, they can not be multiples of 1.0079 (the mass of the hydrogen atom) and 1.0087 (the mass of the neutron). The helium atom consists of two protons and two neutrons. If a helium nucleus were to be formed by the combination of two protons and two neutrons without change of mass, the mass of the helium atom (including the mass of the two electrons outside the nucleus) should be 4.0333 (see Section 3.5 for masses of the proton, neutron, and electron); however, it is only 4.002. This difference represents a loss in mass of 0.031 unit of mass (or atomic mass unit, amu). The union of protons and neutrons to form a nucleus involves the conversion of some of the nuclear mass to nuclear binding energy.

In summary, then, the loss in mass which accompanies the formation of a heavier atom from hydrogen atoms and neutrons is due to the fact that such reactions involve a change of mass into energy; they are thus strongly exothermic. Einstein, in his theory of relativity, related energy and mass by means of the equation

$$E = mc^2$$

in which E represents energy in ergs, m stands for mass in grams, and c is the speed of light in cm per sec. The fact that the speed of light is very high, 3.0×10^{10} cm per sec, and that this term is squared in Einstein's equation, makes it evident that a tremendous quantity of energy results from the destruction of a small quantity of matter.

In the formation of one atom of helium the calculation of the quantity of energy corresponding to the loss of 0.031 atomic mass unit is as follows.

$$E = mc^2$$
$$1 \text{ amu} = 1.66 \times 10^{-24} \text{ g}$$
Hence, m (in grams) $= (0.031 \text{ amu})(1.66 \times 10^{-24} \text{ g/amu})$
$$c = 3.0 \times 10^{10} \text{ cm/sec}$$
Therefore, $E = (0.031 \text{ amu})(1.66 \times 10^{-24} \text{ g/amu})(3.0 \times 10^{10} \text{ cm/sec})^2$
$$= 4.6 \times 10^{-5} \text{ g cm}^2/\text{sec}^2 = 4.6 \times 10^{-5} \text{ erg per atom}$$
or, since $1 \text{ erg} = 1.0 \times 10^{-7}$ joules,
$$E \text{ (in joules)} = (4.6 \times 10^{-5} \text{ ergs})(1.0 \times 10^{-7} \text{ joules/erg})$$
$$= 4.6 \times 10^{-12} \text{ joules per atom}$$
or, since $1 \text{ erg} = 6.24 \times 10^{11}$ electron volts,
$$E \text{ (in electron volts)} = (4.6 \times 10^{-5} \text{ erg})(6.2 \times 10^{11} \text{ electron volts/erg})$$
$$= 2.9 \times 10^7 \text{ electron volts per atom}$$
$$= 29 \text{ million electron volts (Mev) per atom}$$

The changes in mass in all ordinary chemical changes are negligibly small, and these changes involve the formation or breakage of chemical bonds. On the other hand, the breakdown of a nucleus into its component protons and neutrons requires very large amounts of energy and, conversely, the formation of a stable nucleus by bringing together the proper number of protons and neutrons involves the release of very large amounts of energy. The decrease in mass of the nucleus from the sum of the masses of the individual neutrons and protons from which the nucleus may be considered to be formed is called the **packing effect.** The greater the loss of mass (packing effect), the more stable the nucleus, i.e., the greater the quantity of energy liberated during the formation of the nucleus. Because this quantity of energy would have to be supplied in order to break the nucleus down completely into protons and neutrons, it is known as the **binding energy** of the nucleus.

The binding energy per nuclear particle (i.e., the total binding energy for the nucleus divided by the sum of the numbers of protons and neutrons present in the nucleus) is greatest for the nuclei of elements of mass numbers between 40 and 100, and decreases with mass numbers less than 40 or greater than 100 (Fig. 30–2).

Returning to the problem worked out previously in this section, we see that the binding energy of helium is 29 Mev per atom or $\frac{29}{4} = 7.3$ Mev *per nuclear particle*.

30.4 The Packing Fraction and Its Relationship to the Binding Energy

A quantity called the **packing fraction,** suggested originally by Aston, provides a closely related value that is useful in considering the extent of the binding forces within the nucleus.

$$\text{Packing fraction} = \frac{\text{Isotopic mass} - \text{Mass number}}{\text{Mass number}}$$

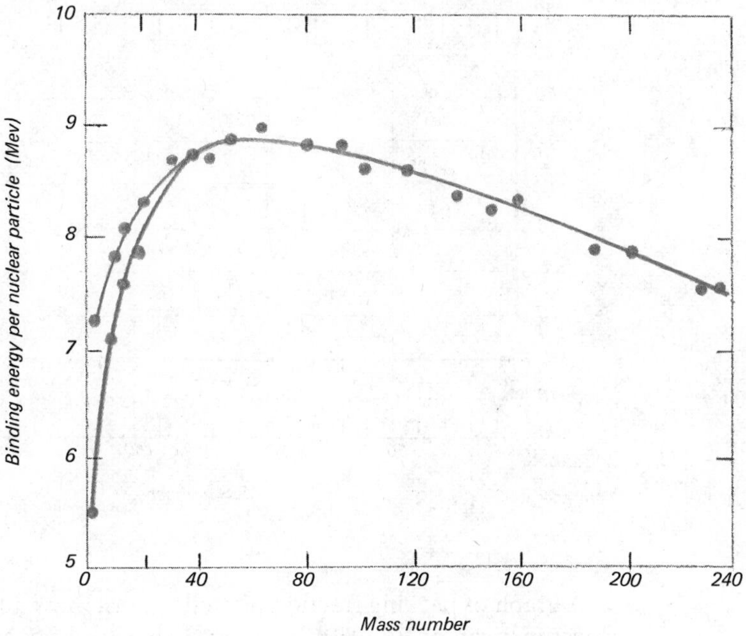

FIGURE 30-2
Binding energy curve for the elements. Note the branching in the curve for the lighter elements. Elements with even atomic numbers are on the upper branch, those with odd atomic numbers are on the lower branch.

The numerator in the packing fraction expression is sometimes called the **mass defect** and is approximately equal to, *but not exactly equal to,* the negative value of the mass lost in the formation of a nucleus from its constituent particles. The mass number, being the whole number nearest to the actual mass of the isotope, is equal to the total number of neutrons and protons within the nucleus. Hence, the packing fraction is the mass defect per *nuclear particle*.

The packing fraction for the helium isotope of mass number 4, $^4_2$He, is

$$\text{Packing fraction} = \frac{4.002 - 4}{4}$$

$$= \frac{0.002}{4} = 5 \times 10^{-4}$$

The packing fraction for the nickel isotope with mass number 58, $^{58}_{28}$Ni, which has an actual isotopic mass of 57.95, is

$$\text{Packing fraction} = \frac{57.95 - 58}{58} = -8.1 \times 10^{-4}$$

The packing fraction for the carbon isotope of mass number 12, $^{12}_6$C, which is the standard for the atomic weight scale and is assigned an isotopic mass of exactly **12** (see Section 3.10), is zero.

$$\text{Packing fraction} = \frac{12.00 - 12}{12} = 0$$

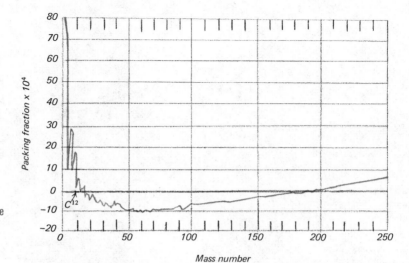

FIGURE 30-3
Packing fraction curve for the elements.

A graph of packing fraction plotted against mass number is shown in Fig. 30–3. A comparison of Figs. 30–2 and 30–3 shows that nuclei with the highest binding energies have the lowest packing fractions, and vice versa. Furthermore, a positive packing fraction indicates a lower binding energy than that of carbon-12; a negative packing fraction indicates a higher binding energy than that of carbon-12 (compare Figs. 30–2 and 30–3). The packing fraction curve shows that the lightest and heaviest atoms have positive packing fractions. This indicates that these elements have lower binding energies and less stable nuclei than carbon-12. The middle elements, with mass numbers 12 to about 200, have negative packing fractions and, in general, more stable nuclei.

When protons and neutrons are assumed to be the building blocks of all nuclei, the packing effect, m (decrease in mass of the nucleus from the sum of the masses of the individual neutrons and protons from which the nucleus may be considered to be formed), for a given nucleus of mass M can be expressed by the equation

$$m = ZM_H + (A - Z)M_n - M$$

where M_H is the mass of a proton, M_n is the mass of a neutron, Z is the atomic number, and A is the sum of the number of protons and neutrons (**nucleons**) in the nucleus.

The bonding energy, B, is expressed by

$$B = mc^2 \quad \text{(where } c \text{ is the speed of light)}$$

or
$$B = [ZM_H + (A - Z)M_n - M]c^2$$

The binding energy per nucleon then is $\dfrac{B}{A}$; and

$$\frac{B}{A} = \left[\frac{Z}{A} M_H + \frac{A - Z}{A} M_n - \frac{M}{A} \right] c^2$$

Adding and subtracting $\frac{A}{A}$ from the equation and rearranging the terms:

$$\frac{B}{A} = \left[\frac{Z}{A} M_H + \left(1 - \frac{Z}{A} \right) M_n - \frac{M}{A} + \frac{A}{A} - \frac{A}{A} \right] c^2$$

$$= \left[\frac{Z}{A} M_H + M_n - \frac{Z}{A} M_n - \frac{M}{A} + \frac{A}{A} - 1 \right] c^2$$

$$= \left[(M_n - 1) - \frac{Z}{A} (M_n - M_H) - \frac{M - A}{A} \right] c^2$$

The expression $\frac{M - A}{A}$ is the packing fraction previously described. Inasmuch as M_n and M_H are constant quantities and $\frac{Z}{A}$ varies only by about 20% from the light to the heavy elements, the above expression for binding energy per nucleon, $\frac{B}{A}$, has a close relationship to the packing fraction and can be roughly approximated as

$$\frac{B}{A} = (\text{constant} - P)c^2 \quad \text{(where } \frac{B}{A} \text{ is the binding energy per nucleon and } P \text{ is the packing fraction)}$$

or

$$P = \left(\text{constant}' - \frac{B}{A} \right) \frac{1}{c^2}$$

Inasmuch as c in this expression is a constant, the packing fraction for a nucleus, therefore, is seen to be (within the limits of constancy for Z/A) directly proportional to a constant minus the binding energy per nucleon for the nucleus. Truly, therefore, Aston's original packing fraction curve does provide a valid basis for indicating relative nuclear stabilities in terms of the tendencies for nuclei to break up into protons and neutrons. The equation shows us again that nuclei with the highest $\frac{B}{A}$ values (binding energies) should have the smallest packing fractions. Referring to Fig. 30–2, we see that the most stable nuclides are those in the vicinity of iron, cobalt, and nickel.

30.5 Factors Influencing the Change of an Unstable Nucleus to a More Stable Nucleus

A nucleus is stable if it cannot be transformed into another configuration without the addition of energy from the outside. A plot of the number of protons *vs.* the number of neutrons for stable nuclei (the curve shown in Fig. 30–4) is referred to as a **stability curve**. The straight line in Fig. 30–4 represents equal numbers of protons and neutrons and is given for comparison. The curve indicates that the lighter, stable nuclei, in general, have equal numbers of protons and neutrons. For example, nitrogen-14 has seven protons and seven neutrons. Heavier stable nuclei, however, have somewhat larger numbers of neutrons than protons, as the curve

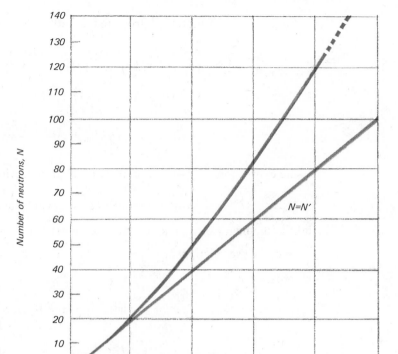

FIGURE 30-4

A plot of number of neutrons vs. number of protons for stable, naturally occurring nuclei. Broken portion of the curve represents elements with atomic number above 83. All isotopes of these elements are unstable.

shows. For example, iron-56 has 26 protons and 30 neutrons. Lead-207 has 82 protons and 125 neutrons.

The isotopes which fall to the left or right of the stability curve have unstable nuclei and are referred to as being **radioactive.** It should be noted that all isotopes with atomic number higher than 83 are radioactive; this is indicated in Fig. 30–4 by the dashed portion of the stability curve. Unstable (radioactive) nuclei tend to change spontaneously to stable configurations that would fall on or near the stability curve. The specific nature of these nuclear changes toward more stable nuclear configurations will be discussed in the following sections.

30.6 Natural Radioactivity

In 1896, the French scientist Becquerel discovered that the element uranium emitted very penetrating rays which affected a photographic plate. The subsequent work of Pierre and Marie Curie and others led to the discovery that the nuclei of certain elements, such as uranium and radium, undergo spontaneous disintegration to form nuclei of other elements. These spontaneous decompositions of atomic nuclei are called **radioactive changes** because they are always accompanied by the emission of radiations. The fact that some elements undergo these transmutations in nature has led to the use of the term **natural radioactivity** in speaking of these phenomena.

When a radioactive element disintegrates spontaneously, approaching the stability curve shown in Fig. 30–4, high speed particles are emitted from the nuclei of the radioactive atoms. It has been established that in natural radioactivity two principal kinds of particles may be given off from the nucleus: **alpha-particles,** each of which consists of two neutrons and two protons and which are identical with helium nuclei; and **beta-particles,** which are high speed electrons. In addition to alpha- and beta-particles, natural radioactivity is frequently accompanied by the emission of **gamma-rays.** These rays are electromagnetic waves, more penetrating than alpha- and beta-particles, and like x rays in character but of somewhat shorter wavelength.

The spontaneous disintegration of the radium nucleus may be represented by the following equation (see Section 3.8 for a description of the notation used):

$$^{226}_{88}\text{Ra} \longrightarrow {}^{222}_{86}\text{Rn} + {}^{4}_{2}\text{He} \quad (\alpha\text{-particle})$$

The loss of an alpha-particle, consisting of two protons and two neutrons (a mass of four units) decreases the atomic number from 88 to 86 and the mass number from 226 to 222. The loss of two protons from the nucleus is accompanied by the loss of two electrons from the electron shells of the atom, whereby the electrical balance in the atom is retained. The part of the radium atom remaining after the emission of the alpha-particle is an atom of radon. Radon atoms also decompose spontaneously and form atoms of polonium (element 84), each radon atom losing one alpha-particle.

When a beta-particle is ejected from the nucleus, a process in the nucleus takes place which is as though a neutron decomposes to give a proton and electron according to the equation

$$^{1}_{0}\text{n} \longrightarrow {}^{1}_{1}\text{p} + {}^{0}_{-1}\text{e}$$

The proton remains in the nucleus, but the electron is ejected at high speed. The formation of a proton in the nucleus increases the net positive charge on the nucleus by one unit, and thus the atomic number of the atom is increased by one unit. The ejection of an electron from the nucleus of an atom causes no appreciable change in the mass of the nucleus, because the mass of the electron is relatively very small. This is evident, for example, in the equation for the transformation of an isotope of thorium to one of protactinium, which is accompanied by the emission of an electron from the nucleus. Note that the mass numbers for both nuclides are the same.

$$^{232}_{90}\text{Th} \longrightarrow {}^{232}_{91}\text{Pa} + {}^{0}_{-1}\text{e}$$

The emission of either alpha- or beta-particles may be accompanied by gamma radiation. For example, as mentioned earlier in this section, $^{226}_{88}\text{Ra}$ disintegrates by alpha-emission to produce $^{222}_{86}\text{Rn}$. The $^{222}_{86}\text{Rn}$ then emits gamma-rays.

$$^{222}_{86}\text{Rn} \longrightarrow {}^{222}_{86}\text{Rn} + \gamma$$

The emission of gamma-rays does not alter either the atomic number or the mass number of an atom, because gamma-rays possess neither mass nor charge. The emission is associated with energy changes within the nucleus.

30.7   The Half-life

The rates at which the disintegrations of the various radioactive elements take place vary widely. However, the number of atoms of a specific element that undergo change per unit of time is a constant fraction of the total number of atoms present. The time required for one-half of the atoms in a sample to disintegrate is called its **half-life.** The half-life of $^{226}_{88}Ra$ is 1,590 years, that of $^{222}_{86}Rn$ is 3.82 days, and that of $^{218}_{84}Po$ is 3.0 minutes. Only one-half of a sample of radium will remain unchanged after 1,590 years, and at the end of another 1,590 years the sample will be reduced to one-fourth its initial mass, and so on. The half-lives of a number of nucleides are listed in Appendix L.

After the discovery of natural radioactivity, it was found that some cancerous cells are more sensitive to radiation than normal cells and are killed on exposure. Therefore, the radiations from radium and other radioactive elements can be used in the treatment of some types of cancer.

Radioactive disintegrations are first order reactions (see Section 17.8). It can be shown that the rate constant k (see Section 17.7) for a radioactive disintegration reaction (or any first order chemical reaction) can be expressed in terms of the half-life, $t_{\frac{1}{2}}$, by the following equation, the same as that which applies to all first order reactions (Section 17.9).

$$k = \frac{0.693}{t_{\frac{1}{2}}}$$

As would be expected, the relationship between the initial concentration of a radioactive isotope and the concentration remaining after a given period of time is expressed by the same logarithmic relationship that applies to all other first order reactions (Section 17.9).

$$\log \frac{c_o}{c_t} = \frac{kt}{2.303}$$

where c_o is the initial concentration and c_t is the concentration remaining at time t.

Example 1.   Calculate the rate constant for the radioactive disintegration of $^{60}_{27}Co$, an isotope used in cancer therapy. It disintegrates with a half-life of 5.2 years by β-emission to produce $^{60}_{28}Ni$.

$$k = \frac{0.693}{t_{\frac{1}{2}}}$$

$$= \frac{0.693}{5.2 \text{ years}} = 0.13 \text{ year}^{-1}$$

Example 2. Calculate the fraction and the per cent of a sample of the $^{60}_{27}$Co isotope which will remain after 15 years.

$$\log \frac{c_o}{c_t} = \frac{kt}{2.303}$$

$$= \frac{(0.13 \text{ year}^{-1})(15 \text{ years})}{2.303} = 0.847$$

$$\frac{c_o}{c_t} = \text{antilog } 0.847 = 7.031$$

The fraction remaining will be the concentration at time t divided by the initial concentration, or $\frac{c_t}{c_o}$.

$$\frac{c_t}{c_o} = \frac{1}{7.031} = 0.14$$

Hence, 0.14 of the $^{60}_{27}$Co originally present, or 14%, would still remain after 15 years.

Example 3. How long would it take for a sample of $^{60}_{27}$Co to disintegrate to the extent that only 2.0 per cent of the original concentration remains.

$$\frac{c_t}{c_o} = 0.020$$

Therefore, $\quad \frac{c_o}{c_t} = \frac{1}{0.020} = 50$

$$\log 50 = \frac{(0.13 \text{ year}^{-1})t}{2.303}$$

$$t = \frac{2.303 \log 50}{0.13 \text{ year}^{-1}} = 30 \text{ years}$$

Example 4. The half-life of $^{216}_{84}$Po is 0.16 second. How long would it take to reduce its concentration to the negligible value of $1.0 \times 10^{-5}\%$ of its original concentration (the fraction 1.0×10^{-7} times its original concentration)?

$$k = \frac{0.693}{0.16 \text{ sec}} = 4.33 \text{ sec}^{-1}$$

$$\frac{c_t}{c_o} = 1.0 \times 10^{-7}$$

Hence, $\quad \frac{c_o}{c_t} = 1.0 \times 10^7$

$$\log (1.0 \times 10^7) = \frac{(4.33 \text{ sec}^{-1})t}{2.303}$$

$$t = \frac{2.303 \log(1.0 \times 10^7)}{4.33 \text{ sec}^{-1}} = 3.7 \text{ seconds}$$

These examples indicate part of the problem of disposal of radioactive isotopes. In 3.7 seconds, any likely quantity of $^{216}_{84}Po$ would be reduced to a negligible quantity and would itself no longer be a problem, but the burst of radiation during the first second or so would be extremely intense and very hazardous. Furthermore, as indicated in Table 30–1, $^{216}_{84}Po$ is a part of a series, disintegrating by α emission into $^{212}_{82}Pb$, which is itself a β-emitter with half-life of 10.6 seconds producing $^{212}_{83}Bi$, which has a half-life of 60.5 minutes, and so on until finally the stable isotope $^{208}_{82}Pb$ is reached. Hence, in coping with radiation effects and isotope disposal all steps in the disintegration of the material must be taken into account. In contrast to the time for disintegration of $^{216}_{84}Po$, it would take 30 years for 98% of a sample of $^{60}_{27}Co$ to disintegrate, leaving 2% still present. A similar calculation for $^{226}_{88}Ra$, an isotope with a half-life of 1,590 years used commonly in cancer treatment, shows that, even after 200 years, 91.6% of the original isotope would still be present. While the radiation from $^{226}_{88}Ra$ is less intense than that of an equivalent quantity of an isotope of short half-life, the radiation is nevertheless extremely dangerous, and the level of the radiation would be diminished less than ten per cent even after two centuries of emission. Thus, it is obvious that we cannot simply leave $^{226}_{88}Ra$ or other radioactive isotopes lying around in the assumption that the radiation danger will disappear within a short period.

30.8   Radioactive Disintegration Series

All elements with atomic numbers larger than that of bismuth (83) have one or more isotopes which are radioactive. A few elements of lower atomic number, such as potassium and rubidium, have naturally occurring isotopes which also are radioactive.

The naturally occurring radioisotopes of the heavier elements belong to chains of successive disintegrations, or "decays," and all the species in one chain constitute a radioactive family, or series. Three of these series include most of the naturally radioactive elements of the Periodic Table. They are the **uranium series,** the **actinium series,** and the **thorium series.** Each series is characterized by a parent (first member) of long half-life and a series of decay processes which ultimately lead to a stable end product; i.e., an isotope on the stability curve of Fig. 30–4. In all three natural series, the end products are isotopes of lead: $^{206}_{82}Pb$ in the uranium series, $^{207}_{82}Pb$ in the actinium series, and $^{208}_{82}Pb$ in the thorium series.

Successive transformations in the natural disintegration series take place in a manner which is described by the so-called **displacement laws** originally formulated by Rutherford, Soddy, and Fajans: (1) When an atom emits an alpha-particle, the product is an isotope of an element two places to the left of the parent element in the Periodic Table. (2) When a beta-particle is emitted, the product is an isotope of an element one place to the right of the parent in the Periodic Table.

The steps in the thorium series are given in Table 30–1 as an illustration of one natural radioactive decay series.

A fourth radioactive series was discovered during World War II. This series is called the **neptunium series,** after its member of longest half-life, and was dis-

TABLE 30-1   The Thorium Series*

$$^{232}_{90}\text{Th} \xrightarrow[1.39 \times 10^{10}\ y]{\alpha} {}^{228}_{88}\text{Ra} \xrightarrow[6.7\ y]{\beta} {}^{228}_{89}\text{Ac} \xrightarrow[6.13\ h]{\beta} {}^{228}_{90}\text{Th}$$

$$\alpha \Big| 1.90\ y$$

$$^{212}_{82}\text{Pb} \xleftarrow[0.16\ sec]{\alpha} {}^{216}_{84}\text{Po} \xleftarrow[54.5\ sec]{\alpha} {}^{220}_{86}\text{Rn} \xleftarrow[3.64\ d]{\alpha} {}^{224}_{88}\text{Ra}$$

$$\beta \Big| 10.6\ h$$

$$^{212}_{83}\text{Bi} \xrightarrow[60.5\ min]{\beta} {}^{212}_{84}\text{Po}$$

$$\alpha \Big| 60.5\ min \qquad \alpha \Big| 3 \times 10^{-7}\ sec$$

$$^{208}_{81}\text{Tl} \xrightarrow[3.1\ min]{\beta} {}^{208}_{82}\text{Pb}$$

*Type of emission and half-life time is shown for each step (y means years; h, hours; min, minutes; and sec, seconds).

covered through the production of its members by artificial means (Sections 30.10 and 30.12). The end product of the series is an isotope of bismuth, $^{209}_{83}\text{Bi}$. Both the parent and end product of this series have been detected in uranium ores in recent years.

The four radioactive series are referred to as the 4 n (or thorium), 4 n + 1 (or neptunium), 4 n + 2 (or uranium), and the 4 n + 3 (or actinium) series. These numerical designations indicate whether the mass numbers of the members of the series are exactly divisible by 4, or by 4 with a remainder of 1, 2, or 3.

30.9   The Age of the Earth

One of the most interesting applications of natural radioactivity has been its use in determining the age of the earth. An estimate of the lower limit of the earth's age can be made by determining the age of various rocks and minerals, assuming that the earth must be at least as old as the rocks and minerals in its crust. Several methods involving radioactivity have been used for this purpose. One method involves the rate of disintegration in any one of the three naturally occurring radioactive series mentioned in the previous section. For example, one gram of uranium-238 would produce 0.4326 g of lead-206 and 0.0674 g of helium, and leave 0.5000 g of uranium-238 after decaying for 4.5 billion years (its half-life). Thus, by comparing the amount of lead-206 to the amount of uranium-238 in a uranium mineral, the age of the rock containing the uranium mineral can be estimated. Analyses of uranium minerals for lead-206 show the age of the earth's crust to be about 2.6 billion years. It is interesting to note that the age of the universe as calculated from the observed speed of recession of the nebulae (expanding universe) is 3.8 billion years.

30.10 Artificial Nuclear Changes (Transmutations and Artificial Radioactivity)

Atoms with stable nuclei can be converted to other atoms **(transmutations)** by bombarding their nuclei with high speed particles. The first person to produce a nucleus artificially in the laboratory was Lord Rutherford, in 1919. Using high speed alpha-particles emanating from a naturally radioactive isotope of radium

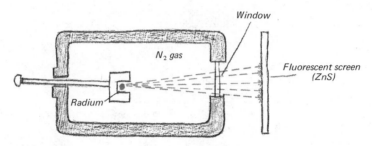

FIGURE 30-5

The type of apparatus used by Rutherford in the transmutation of nitrogen. Alpha particles from the radium, upon striking the fluorescent screen, produced flashes of light. When the screen was moved beyond the range of the alpha particles, an occasional flash of light was still noted. This was due to the screen being struck by hydrogen nuclei (protons) which were knocked from the nitrogen atom by the alpha-particle "bullets" from radium.

as projectiles, he bombarded nitrogen atoms (Fig. 30–5) and observed the nuclear reaction

$$^{14}_{7}\text{N} + ^{4}_{2}\text{He} \longrightarrow ^{17}_{8}\text{O} + ^{1}_{1}\text{H}$$

Thus, two new nuclei are formed, an $^{17}_{8}\text{O}$ nucleus and a proton. The $^{17}_{8}\text{O}$ nucleus is stable, so that this nuclear reaction does not lead to further radioactive changes. However, many elements undergo such artificially induced nuclear reactions with the formation of *unstable* nuclei, which then undergo radioactive disintegration. The disintegration in these cases is referred to as **artificial radioactivity.** An example is given below.

$$^{25}_{12}\text{Mg} + ^{4}_{2}\text{He} \longrightarrow \underset{\text{Radioactive}}{^{28}_{13}\text{Al}} + ^{1}_{1}\text{H}$$

$$\longrightarrow \underset{\text{Stable}}{^{28}_{14}\text{Si}} + ^{0}_{-1}\text{e}$$

An abbreviated notation for nuclear reactions is commonly used for convenience:

$$M\,(a, b)\,M'^{*}$$

FIGURE 30-6

A beam of deuterons (bright area) from a cyclotron at the University of California. Nuclear reactions take place when deuterons and other atomic particles strike the nuclei of atoms. *Rockefeller Foundation*

where M is the bombarded nucleus, a is the bombarding particle, b is the emitted particle, and M' is the product nucleus. An asterisk with M' indicates that the product is radioactive. The abbreviated notation for the bombardment of $^{25}_{12}\text{Mg}$ with α-particles, just discussed, is $^{25}_{12}\text{Mg}\,(\alpha, \text{p})\,^{28}_{13}\text{Al}^*$. Other examples of such abbreviated notation are

$$^{14}_{7}\text{N}\,(\alpha, \text{p})\,^{17}_{8}\text{O}$$
$$^{239}_{94}\text{Pu}\,(\alpha, \text{n})\,^{242}_{96}\text{Cm}$$

Several types of particles can be used as projectiles to bombard nuclei. Charged particles such as protons, alpha-particles, and electrons can be accelerated to very high speeds by strong magnetic and electrostatic fields. Among the instruments built to accelerate these particles are the **cyclotron** (Figs. 30–6 and 30–7), the **betatron,** and the **synchrotron.** It is also possible to use the neutral neutron as it is emitted from a nuclear reaction (Section 30.13) as a bombarding particle. Table 30–2 lists the particles most commonly used to bombard nuclei.

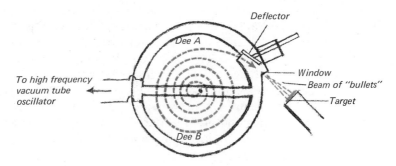

FIGURE 30-7

Diagram of the vacuum chamber of a cyclotron. By making the dees, *A* and *B*, alternately plus and minus and correctly adjusting the strength of the magnetic field, accelerating ionic "bullets" are caused to follow an ever-widening spiral path until they are finally deflected and pass through the thin metal window to the target. Some cyclotrons have only one dee.

TABLE 30-2   The More Common Particles Used for Bombarding Nuclei in Nuclear Transmutation Reactions

Bombarding Particle	Description	Source
Neutron	$^1_0 n$	Nuclear reactor
α-particle	$^4_2 He$	Cyclotron
Proton	$^1_1 H$	Cyclotron
Deuteron	$^2_1 H$	Cyclotron
Electron	e^-	Betatron
γ-rays	$h\upsilon$ (energy but no mass or charge)	Synchroton

The yield in artificial nuclear reactions is very small, because collisions between the projectiles and atomic nuclei are rare. The facts that the nucleus of an atom is very small and that positively charged projectiles such as protons and alpha-particles are repelled by positive nuclei explain the limited number of direct hits.

An example of a nuclear reaction produced by accelerated projectiles is the transmutation of calcium to scandium by the bombardment with protons.

$$^{44}_{20}Ca + {}^1_1H \longrightarrow {}^{44}_{21}Sc + {}^1_0n \qquad [^{44}_{20}Ca\,(p, n)\,^{44}_{21}Sc]$$

Deuterons, d, which are the nuclei of the heavy hydrogen isotope, $^2_1 H$, have been used to bombard aluminum atoms and cause the nuclear reaction

$$^{27}_{13}Al + {}^2_1H \longrightarrow {}^{25}_{12}Mg + {}^4_2He \qquad [^{27}_{13}Al\,(d, \alpha)\,^{25}_{12}Mg]$$

It is possible to predict on the basis of the stability curve in Figs. 30–4 and 30–8 how artificially-produced radioactive nuclides will decompose. The nuclides which fall to the left of the curve have an excessive number of neutrons and tend to decompose by β emission (Section 30.6), whereby a neutron is converted to a proton and an electron. The decrease of one neutron and increase of one proton in a nuclide decreases the neutron-proton ratio and produces a new nuclide nearer to the stability curve. The nuclides which fall to the right of the stability curve have an excessive number of protons and most often approach the stability curve by undergoing a positron emission reaction or, occasionally, by K-electron capture (E.C.). A positron, abbreviated e^+, or β^+, is a particle with the same mass as an electron but with a *positive* charge of one unit. During a positron emission reaction the positron is produced in the nucleus as if by the conversion of a proton to a neutron plus a positron.

$$^1_1 p \longrightarrow {}^1_0 n + {}^{\;\;0}_{+1}\beta^+$$

The neutron remains in the nucleus, but the positron is ejected at high speed. The conversion of a proton to a neutron increases the neutron-proton ratio and produces a nuclide nearer to the stability curve. The atomic number is decreased by one unit and the mass remains essentially the same with positron emission.

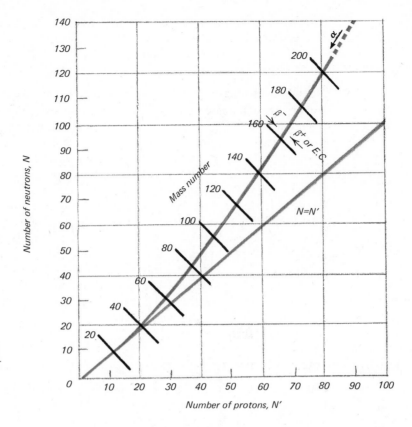

FIGURE 30-8

Stability curve for predicting how artificial, radioactive nuclides will decompose.

An example of a nuclear transmutation reaction which results in a radioactive nuclide that emits positrons is shown in the reactions

$$^{10}_{5}B + ^{4}_{2}He \longrightarrow ^{13}_{7}N^* + ^{1}_{0}n \qquad ^{10}_{5}B\ (\alpha,\ n)\ ^{13}_{7}N^*$$
$$\qquad\qquad \longrightarrow\ ^{13}_{6}C + \beta^+$$

In K-electron capture, which occurs with a relatively low neutron-proton ratio, the positive nuclear attractive forces pull an electron from the K-shell of the atom down into the nucleus, where a reaction equivalent to p + e $\longrightarrow$ n takes place. Thus, the neutron-proton ratio is increased and a nuclide is produced which is nearer to the stability curve.

Most transmutation reactions require the absorption of energy, as compared with radioactive decay, which gives up energy.

For example, for the reaction $^{14}_{7}N\ (\alpha,\ p)\ ^{17}_{8}O$:

$$^{14}_{7}N + ^{4}_{2}He + Energy \longrightarrow ^{17}_{8}O + ^{1}_{1}H$$

$^{14}_{7}N = 14.00755$ amu	$^{17}_{8}O = 17.00453$ amu
$^{4}_{2}He = \ \ 4.00387$ amu	$^{1}_{1}H = \ \ 1.00815$ amu
Total mass: $\overline{18.01142}$ amu	Total mass: $\overline{18.01268}$ amu

The increase in mass when the reactants form the products is accounted for by the conversion of energy to mass by the relationship $E = mc^2$ (Section 30.3).

30.11   Isobars

Nuclides which have the same mass number but differ in atomic number are called **isobars.** The nuclides corresponding to positions along any one of the short diagonal lines in Fig. 30-8, therefore, are isobars; hence, these lines are referred to as **isobaric lines.** Along each of the various isobaric lines, the most stable nuclei tend to be those of lowest mass. We can visualize radioactive decay by beta emission, positron emission, or K-electron capture as corresponding to a slight reduction in mass (energy) to produce a more stable nuclear configuration. When the nuclides to the right of the stability curve decompose by positron emission (or occasionally by K-electron capture) or when nuclides to the left of the stability curve decompose by beta emission, the nuclear reaction results in a movement along an isobaric line toward the stability curve.

Alpha emission occurs predominantly for nuclides with atomic number above 83, corresponding to the dashed line in Fig. 30-8. Emission of an alpha particle decreases both the number of neutrons and the number of protons and thus results in a movement down the stability curve toward the solid portion of the line which represents stable nuclides.

30.12   Production of New Elements

Not only have many of the known elements been transmuted into other elements, but a considerable number of new elements have been synthesized in the laboratory. These include, among others, element 43, technetium (Tc); element 85, astatine (At); element 87, francium (Fr); and element 61, promethium (Pm).

Prior to 1940 the heaviest known element was uranium, atomic number 92. In 1940, McMillan and Abelson were able to make element 93, neptunium (Np), by bombarding uranium with high velocity deuterons. The nuclear reaction is given by

$$^{238}_{92}\text{U} + {}^{1}_{0}\text{n} \longrightarrow {}^{239}_{92}\text{U}$$

$$^{239}_{92}\text{U} \xrightarrow{\;23 \text{ min}\;} {}^{239}_{93}\text{Np} + {}^{0}_{-1}\text{e}$$

Neptunium is also radioactive, with a half-life period of 2.3 days, and converts to plutonium (Pu), atomic number 94.

$$^{239}_{93}\text{Np} \longrightarrow {}^{239}_{94}\text{Pu} + {}^{0}_{-1}\text{e}$$

Elements 95 through 103 have likewise been prepared artificially. These elements have been named, in order, americium (Am), curium (Cm), berkelium (Bk), californium (Cf), einsteinium (Es), fermium (Fm), mendelevium (Md), nobelium (No), and lawrencium (Lr), respectively. The elements 93 through 103 are known as the **transuranium elements** and these elements, along with those of atomic numbers 89, 90, 91, and 92, make up the **actinide series.**

Typical transmutation reactions by which the transuranium elements have been produced are as follows:

$$^{239}_{94}\text{Pu} + ^{1}_{0}\text{n} \longrightarrow ^{240}_{94}\text{Pu}$$

$$^{240}_{94}\text{Pu} + ^{1}_{0}\text{n} \longrightarrow ^{241}_{94}\text{Pu}$$
$$\longrightarrow ^{241}_{95}\text{Am} + ^{0}_{-1}\text{e}$$

$$^{239}_{94}\text{Pu} + ^{4}_{2}\text{He} \longrightarrow ^{242}_{96}\text{Cm} + ^{1}_{0}\text{n}$$

$$^{241}_{95}\text{Am} + ^{4}_{2}\text{He} \longrightarrow ^{243}_{97}\text{Bk} + 2\,^{1}_{0}\text{n}$$

$$^{242}_{96}\text{Cm} + ^{4}_{2}\text{He} \longrightarrow ^{245}_{98}\text{Cf} + ^{1}_{0}\text{n}$$

$$^{238}_{92}\text{U} + 15\,^{1}_{0}\text{n} \longrightarrow ^{253}_{99}\text{Es} + 7\,^{0}_{-1}\text{e}$$

$$^{239}_{94}\text{Pu} + 15\,^{1}_{0}\text{n} \longrightarrow ^{254}_{100}\text{Fm} + 6\,^{0}_{-1}\text{e}$$

$$^{253}_{99}\text{Es} + ^{4}_{2}\text{He} \longrightarrow ^{256}_{101}\text{Md} + ^{1}_{0}\text{n}$$

$$^{246}_{96}\text{Cm} + ^{12}_{6}\text{C} \longrightarrow ^{254}_{102}\text{No} + 4\,^{1}_{0}\text{n}$$

$$^{250}_{98}\text{Cf} + ^{11}_{5}\text{B} \longrightarrow ^{257}_{103}\text{Lr} + 4\,^{1}_{0}\text{n}$$

In addition, the syntheses of elements 104, 105, and 106 have been reported recently, and element 107 may be in existence soon. Element 105 is unexpectedly long-lived, which raises the hope that these man-made elements may exist long enough for chemists to examine their chemistry in considerable detail.

The name *rutherfordium*, with the symbol *Rf*, has been suggested for element 104 by American scientists who have synthesized the element. Soviet scientists, who also have worked on the synthesis of the element, suggest the name *kurchatovium*. It has been proposed that element 105 be named *hahnium*, with the symbol *Ha*, to honor Otto Hahn, the late German scientist who won the Nobel prize for the discovery of nuclear fission.

Both a Soviet and an American group have announced the synthesis of element 106, by quite different methods. The two groups have agreed not to propose any name for the element until a decision is reached on which group will get credit for being the first to synthesize it.

The Soviet group used relatively heavy ions to bombard smaller nuclei than those used by the American group. The Soviet scientists bombarded a lead isotope with chromium-54 ions, producing an isotope reported to be either $^{257}_{106}\text{X}$ or $^{258}_{106}\text{X}$ and hence containing either 151 or 152 neutrons; it decays by spontaneous fission with the very short half-life of four to ten milliseconds.

The American group bombarded a target of californium-249 with oxygen-18 ions, resulting in emission of four neutrons and production of an isotope of element 106 having a mass number of 263 and containing 157 neutrons; it decays by α-emission with a half-life of 0.9 second, followed by two additional α-emission steps.

$$^{249}_{98}\text{Cf} + ^{18}_{8}\text{O} \longrightarrow ^{263}_{106}\text{X} + 4\,^{1}_{0}\text{n}$$
$$\longrightarrow ^{259}_{104}(\text{Rf}) + ^{4}_{2}\alpha$$
$$\longrightarrow ^{255}_{102}\text{No} + ^{4}_{2}\alpha$$
$$\longrightarrow ^{251}_{100}\text{Fm} + ^{4}_{2}\alpha$$

Because elements 104 and 105 are more stable than expected and by other reasoning based on experiment and extrapolation of known facts, it is believed that element 114 will soon be synthesized. This element, chemically similar to lead, could be so stable that tiny amounts may be found to occur naturally on earth—or, perhaps, to be present in regions of outer space where new elements seem to be forming. It is interesting to note, however, that isotopes containing more than 157 neutrons have proved to be exceedingly unstable, an experimental observation for which there is as yet no explanation.

Glenn T. Seaborg has recently suggested an extension of the Periodic Table (see Table 30–3) to include new elements whose synthesis is considered possible. (It will probably be helpful to review the material in Chapter 3 on filling energy shells and subshells by electrons.) With element 104 we enter the relatively unexplored region of the periodic system with the first member of what Seaborg has chosen to call the "trans-actinide" elements, that is, all elements beyond the inner transition or actinide series. It becomes possible to locate the position of elements 104 through 121 and, in the style of Mendeleev, to predict their chemical properties by comparing them with their analogs in the Periodic Table. Element 104 should be an analog of hafnium; 105, an analog of tantalum; and so forth until element 118, a noble gas analogous to radon, is reached.

The most striking feature of Seaborg's extension of the Periodic Table is the addition of another inner transition series of elements starting with atomic number 121 and extending through atomic number 153. He calls this grouping the "super-actinide" series. This series, except for element 121, is postulated as including those elements characterized by the progressive filling of the $5g$ electron subshell, (see Table 3–6 in Chapter 3). Element 121 can be considered as receiving the first $7d$ electron (the $n = 7$ major shell being the second from the outside), analogous to scandium, yttrium, lanthanum, and actinium. Perhaps the first of eighteen $5g$ electrons would enter at element 122, for which the $n = 5$ major shell is the fourth from the outside; the eighteenth $5g$ electron would enter at element 139. This would presumably be followed by the filling in of fourteen $6f$ electrons (third major shell from the outside) for elements 140 through 153, making this latter series an inner transition series analogous to those series for elements 58 through 71 and 90 through 103. Following the "superactinide" inner transition series, elements 154 through 168 would be analogous to elements 104 through 118, with entrance of the remainder of the ten $7d$ electrons in the second major shell from the outside and the six $8p$ electrons in the outermost major shell. Element 168 would be another noble gas—or should we say a "noble liquid"?

30.13   Nuclear Fission

The greater stability of the nuclei of elements with mass numbers of intermediate values suggests the possibility of spontaneous splitting of the less stable nuclei of the heavy elements into fragments of approximately half size, and that such nuclear fissions would be accompanied by the release of large quantities of energy. Two German scientists, Hahn and Strassman, reported in 1939 that when they bombarded uranium with slow-moving neutrons, the uranium-235 atoms split into

TABLE 30-3 Predicted Locations of New Elements in the Periodic Table (See also Fig. 3-16)

																1 H	2 He

s

1 H																	
3 Li	4 Be											5 B	6 C	7 N	8 O	9 F	10 Ne
11 Na	12 Mg											13 Al	14 Si	15 P	16 S	17 Cl	18 Ar
19 K	20 Ca	21 Sc	22 Ti	23 V	24 Cr	25 Mn	26 Fe	27 Co	28 Ni	29 Cu	30 Zn	31 Ga	32 Ge	33 As	34 Se	35 Br	36 Kr
37 Rb	38 Sr	39 Y	40 Zr	41 Nb	42 Mo	43 Tc	44 Ru	45 Rh	46 Pd	47 Ag	48 Cd	49 In	50 Sn	51 Sb	52 Te	53 I	54 Xe
55 Cs	56 Ba	[57-71] *	72 Hf	73 Ta	74 W	75 Re	76 Os	77 Ir	78 Pt	79 Au	80 Hg	81 Tl	82 Pb	83 Bi	84 Po	85 At	86 Rn
87 Fr	88 Ra	[89-103] †	104	105	106	107	108	109	110	111	112	113	114	115	116	117	118
119	120	[121-153] □	154	155	156	157	158	159	160	161	162	163	164	165	166	167	168

d | *p*

f

* LANTHANIDE SERIES	57 La	58 Ce	59 Pr	60 Nd	61 Pm	62 Sm	63 Eu	64 Gd	65 Tb	66 Dy	67 Ho	68 Er	69 Tm	70 Yb	71 Lu
† ACTINIDE SERIES	89 Ac	90 Th	91 Pa	92 U	93 Np	94 Pu	95 Am	96 Cm	97 Bk	98 Cf	99 Es	100 Fm	101 Md	102 No	103 Lr

g | *f*

□ SUPER ACTINIDES	121	122	123			139	140			153

smaller fragments, consisting of elements about in the middle of the Periodic Table, and several neutrons. The process is referred to as **fission**. Among the fission products identified were barium, krypton, lanthanum, and cerium, the nuclei of all of which are more stable than the nucleus of uranium.

$$^{235}_{92}U + ^{1}_{0}n \longrightarrow \text{Fission fragments} + 2.5 \, n + \text{energy}$$
$$\text{(Isotopes of Ba, Kr, etc.)}$$

The sum of the atomic numbers of the fission products is 92, the atomic number of the original uranium nucleus.

In this type of disintegration a loss of mass of about 0.2 amu occurs which corresponds to a fantastic quantity of energy that is released in the reaction. It is apparent from the packing fraction curve (Fig. 30–3) that $^{235}_{92}U$, with a positive packing fraction, would be expected to undergo a loss of mass in forming nuclei of medium mass number. The fission fragments produced usually have a high neutron-proton ratio (are to the left of the stability curve in Figs. 30–4 and 30–8) and hence attain stability by beta emission processes.

The fission of a pound of uranium-235 produces about 2.5 million times as much energy as the burning of a pound of coal. A combination of technological developments and the availability of reasonably large quantities of uranium and other fissionable elements in the earth's crust will undoubtedly make nuclear fission a prime source of energy in the future. It is an exceptionally clean source of energy, with respect to air pollution, except for the special problem of radioactive emissions. These emissions must be monitored and carefully controlled to avoid producing an unsafe level of radioactivity in the atmosphere.

The fission products of a U-235 nucleus include, on the average, 2.5 neutrons as well as the fission fragments. These neutrons may cause the fission of neighboring U-235 atoms, which in turn provide more neutrons for setting up a **chain reaction.** By using substances which arrest the reaction by absorbing neutrons, the chain reaction can be controlled so that the nuclear energy produced can be used as a source of industrial power.

30.14 The Atomic Bomb

An atomic bomb consists of several pounds of fissionable material, $^{235}_{92}U$ or $^{239}_{94}Pu$, and an explosive device for compressing the material quickly into a small volume. With small pieces of fissionable material the proportion of neutrons which escape at the relatively large surface area is great and a chain reaction does not take place. When the small pieces of fissionable material are brought together quickly to form a larger body with a mass larger than the "critical mass," the relative number of escaping neutrons decreases, and a chain reaction and an explosion result, with the fission of nearly all the nuclei within a very short period of time. The explosion of an atomic bomb can release more energy than that resulting from the explosion of a million tons of TNT (trinitrotoluene).

As it occurs in nature, uranium consists of a mixture of several isotopes; the most abundant of these is U-238. About one atom in every 140 atoms of uranium is the fissionable U-235 isotope. Because pure U-235 is needed for a chain reaction like that which takes place in the atomic bomb, it is necessary to separate U-235 from the nonfissionable isotopes U-238 and U-234. This proved to be a major task in the development of the atomic bomb, because there is relatively very little difference in the masses of these isotopes. Electromagnetic methods, based upon the principles of the mass spectrograph, were found to be effective, but the yields were small. The most successful of the several methods tried made use of the separation of $^{235}_{92}UF_6$ from $^{238}_{92}UF_6$ by fractional diffusion of large volumes of gaseous UF_6 at low pressure through porous diffusion barriers of very large areas. This method is based upon the fact that the lighter $^{235}_{92}UF_6$ molecules diffuse through a porous barrier faster than the heavier molecules of $^{238}_{92}UF_6$.

30.15 Nuclear Chain Reactors (Atomic Piles)

The control of chain reactions of fissionable materials is achieved in nuclear **chain reactors,** or **atomic piles.** By using efficient neutron absorbers, such as cadmium, chain propagation can be so controlled that the number of neutrons builds up to a constant level and does not increase to the explosive stage. There are three components in a nuclear chain reactor:

(1) The charge material, which includes a fissionable substance such as U-235, Pu-239, or U-233; the charge material is often uranium considerably enriched with respect to U-235.

(2) A substance to reduce the velocity of the neutrons, called a **moderator;** graphite, heavy water, and ordinary water have been used.

(3) A nonfissionable material such as cadmium or boron steel in the form of control rods to absorb neutrons.

A diagram of a nuclear chain reactor is shown in Fig. 30-9. The uranium reactor consists of cans of the fissionable metal inserted in channels in graphite blocks containing cadmium rods so arranged that they may be removed to permit the slow neutrons to strike the fissionable metal.

FIGURE 30-9

Diagram of a simple atomic pile.

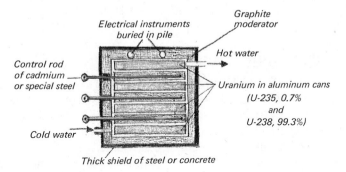

In addition to the fission reaction with U-235 (Section 30.13), the following series of reactions takes place in a uranium nuclear reactor.

$$^{238}_{92}U + ^{1}_{0}n \longrightarrow ^{239}_{92}U$$

$$^{239}_{92}U \xrightarrow{\text{23 min}} ^{239}_{93}Np + ^{0}_{-1}e$$

$$^{239}_{93}Np \xrightarrow{\text{2.3 days}} ^{239}_{94}Pu + ^{0}_{-1}e$$

Hence, the uranium pile serves as a source of the fissionable isotope $^{239}_{94}Pu$. The energy released by the pile can be utilized as heat to convert water to steam for use in conventional turbines for the production of electrical power.

Another development in the field of reactor technology is the **breeder reactor,** by means of which the large quantities of uranium and thorium may be used to produce fissionable materials. Such reactors are of much current interest, because of their potential use in producing new sources of energy. In contrast to the regular reactor, the breeder reactor is constructed with a blanket of "fertile" material (U-238 or Th-232) surrounding a core of concentrated fissionable material (U-235, Pu-239, or U-233). Extra neutrons arising from the fission of the core material are captured by the blanket of fertile material to form ("breed") more fissionable atoms. Hence, the breeder reactor is a source of energy and at the same time produces as much fuel as it consumes.

Breeder reactors are based upon one of two sets of nuclear reactions. One begins with U-238 and the other with Th-232, which are the most abundant isotopes of their respective elements and are present in trace amounts in various types of granite rock. U-238 produces the fissionable atom $^{239}_{94}Pu$ according to the reactions described above. Th-232 produces the fissionable atom $^{233}_{92}U$ by the following reactions:

$$^{232}_{90}Th + {}^{1}_{0}n \longrightarrow {}^{233}_{90}Th$$

$$^{233}_{90}Th \xrightarrow{\text{23 min}} {}^{233}_{91}Pa + {}^{0}_{-1}e$$

$$^{233}_{91}Pa \xrightarrow{\text{27 days}} {}^{233}_{92}U + {}^{0}_{-1}e$$

Several experimental breeder reactors are in operation in the United States, Great Britain, Russia, and other countries.

30.16   Nuclear Fusion and the Hydrogen Bomb

In Section 30.13 we discovered that the nuclear binding energy of heavy atoms may be increased by fission of these atoms into fragments of lower mass numbers, and that such nuclear fissions are accompanied by the liberation of extremely large amounts of energy. The process of converting very light nuclei into heavier nuclei is likewise accompanied by the conversion of mass into large amounts of energy. Such reactions are known as **nuclear fusions** and are the basis of an intensive research effort to develop a practical thermonuclear reactor. It is supposed that the principal source of energy of the sun is the fusion of four hydrogen nuclei into one helium nucleus. Four hydrogen nuclei have a greater mass (0.7 per cent) than the helium nucleus, so the fusion evidently converts the extra matter into energy.

It has been found that a deuteron, $^{2}_{1}H$, and a triton, $^{3}_{1}H$, which are the positive ions of the heavy isotopes of hydrogen, will undergo **fusion** at extremely high temperatures to form a helium nucleus and a neutron.

$$^{2}_{1}H + {}^{3}_{1}H \longrightarrow {}^{4}_{2}He + {}^{1}_{0}n$$

This change is accompanied by a conversion of a portion of the mass into energy and is the nuclear reaction of the hydrogen bomb (Fig. 30–10). In the hydrogen

FIGURE 30-10
Hydrogen bomb explosion in the Pacific Ocean test area, 1954. *Lookout Mountain Air Force Station*

bomb, a fission type of atomic bomb (uranium or plutonium) is exploded inside a charge of deuterium and tritium to provide the temperature of many millions of degrees required for the fusion of the deuterium and tritium.

If the fusion of heavy isotopes of hydrogen can be controlled, hydrogen from the water of the oceans will provide an inexhaustible supply of energy for future generations.

30.17 Applications of Nuclear Reactions

■ **1. Atomic Power.** A considerable number of atomic plants for production of power are presently in operation. Atomic power for submarines is being used extensively. As early as 1961, the United States launched the world's first nuclear-powered surface ship, the N.S. *Savannah.*

The world demand for power has been increasing about 25 per cent every ten years. If this trend continues, the world's supply of petroleum and coal may have diminished to critically small quantities by 2000 A.D. Power from atomic sources costs very little more to produce than power from conventional sources in some locations. Electricity from atomic sources is today becoming commonplace. Pound for pound, uranium has at least a million times the energy of oil. The rocks and oceans potentially provide an inexhaustible supply of energy through appropriate nuclear reactions (Section 30.15 and 30.16).

■ **2. Production of Radioactive Isotopes.** A variety of radioactive isotopes is produced by fission or by neutron bombardment of other elements in the nuclear reactor and by other suitable transmutation reactions. Large quantities of radioactive isotopes are used annually by universities, hospitals, research institutes, and industries in a variety of ways, including such diverse applications as study of the mechanisms by which chemical reactions take place, treatment of cancer, locations of flaws in metal objects intended for structural purposes, study of the digestive and milk-producing processes in the cow, determination of the effects of lubrication additives on engine wear, and study of factors involved in plant growth. Radioisotopes are being used experimentally as the source of β-radiation for the preservation of foods.

QUESTIONS

1. Describe the three types of radiation which are emitted from nuclei of naturally radioactive elements.
2. What is the change in the nucleus that gives rise to a beta-particle?
3. The loss of an alpha-particle by a nucleus causes what change in the atomic number and the mass of the nucleus? What is the change in the atomic number and mass when a beta-particle is emitted?
4. Define and illustrate the phrase "half-life period."
5. How do nuclear reactions differ from ordinary chemical changes?
6. Many nuclides of atomic number greater than 83 decay by processes such as beta emission. Rationalize the observation that the radioactive emissions resulting from these unstable nuclides normally includes alpha emission also.
7. How may charged particles be accelerated artificially for use in promoting nuclear reactions?
8. Describe the separation of the isotopes of an element by the mass spectrograph.
9. Explain the terms "packing effect" and "packing fraction." How are these related to the binding energy of a nucleus?
10. Complete the following equations:
    (a) $^{27}_{13}\text{Al} + ^{2}_{1}\text{H} \longrightarrow \quad + ^{4}_{2}\text{He}$
    (b) $^{7}_{3}\text{Li} + \quad \longrightarrow 2\,^{4}_{2}\text{He}$
    (c) $^{9}_{4}\text{Be} + ^{4}_{2}\text{He} \longrightarrow ^{12}_{6}\text{C} +$
    (d) $^{23}_{11}\text{Na} + ^{2}_{1}\text{H} \longrightarrow ^{24}_{11}\text{Na} +$
11. Fill in the atomic number of the initial nucleus and write out the complete nuclear symbol for the product of the following nuclear reactions:
    (a) $^{65}\text{Cu}\,(n, 2n)$
    (b) $^{54}\text{Fe}\,(\alpha, 2p)$
    (c) $^{33}\text{S}\,(n, p)$
    (d) $^{33}\text{S}\,(p, n)$
    (e) $^{106}\text{Rh}\,(n, p)$
    (f) $^{27}\text{Al}\,(\alpha, n)$
    (g) $^{14}\text{N}\,(p, \gamma)$

12. Complete the following notations by filling in the missing parts:
    (a) ^{2}H (d, n)
    (b)   (α, n) ^{30}P
    (c)   (d, n) ^{238}Np
    (d) ^{10}B (α,) ^{13}C
    (e) ^{232}Th (, n) ^{235}U
    (f) ^{2}H (^{12}C,) ^{10}B
    (g) ^{24}Mg (α, n)
    (h) ^{238}U ($^{12}_{6}$C, 4n)

13. Use the abbreviated notation as in the question above to describe:
    (a) the production of ^{17}O from ^{14}N by alpha bombardment.
    (b) the neutron induced fission of $^{235}_{92}$U.
    (c) the production of ^{233}Th from ^{232}Th by neutron bombardment.
    (d) the production of ^{239}U from ^{238}U by deuteron bombardment.
    (e) the production of ^{14}C from ^{14}N by neutron bombardment.

14. Below are listed several isotopes which are unstable. Predict by what mode(s) spontaneous radioactive decay might proceed in each case.
    (a) ^{34}P (n/p ratio too large)
    (b) ^{156}Eu (n/p ratio too large)
    (c) ^{235}Pa (too much mass, n/p ratio too large)
    (d) $^3_1$H
    (e) ^{18}F
    (f) ^{129}Ba
    (g) ^{237}Pu

15. Give two reasons why the atomic weights of the elements are not whole numbers.

16. Distinguish between the following terms: isotope, isobar, nuclide.

17. By reference to Fig. 30–8, predict the type of nuclear transformation, if any, that might be expected to occur for each of the following for isobaric line 100, and state how such a transformation could result in a nuclide of greater stability:
    (a) A nuclide to the left of the stability curve.
    (b) A nuclide to the right of the stability curve.
    (c) A nuclide on the stability curve.

18. Explain in terms of Fig. 30–8 how unstable heavy nuclides (atomic number greater than 83) may decompose to form nuclides of greater stability (a) if they are on the dashed portion of the stability curve, and (b) if they are to the left of the dashed portion of the stability curve.

19. Verify 18 as the maximum number of g electrons theoretically possible in a given major shell, based upon permissible values for the four quantum numbers (see Chapter 3).

20. Write out, using standard notation ($1s^2$, $2s^2$, etc.), logical predicted electron structures of elements of atomic numbers 104, 105, 106, 114, 118, 122, 130, 139, 140, 146, 153, 159, 166, 168, 170. In each case, to what known elements would the predicted element be analogous in chemical properties?

21. Distinguish between nuclear fission and nuclear fusion. Why are both processes exothermic?
22. How are atomic bombs and hydrogen bombs detonated?
23. Describe the construction and operation of a uranium nuclear reactor.
24. What is meant by the term "breeder reactor"?
25. List advantages and disadvantages of nuclear energy as a source of domestic power as compared to coal, fuel oil, natural gas, and water.
26. Discuss and compare the problems of radioactive fall-out for radioactive substances of short half-life and those of long half-life.
27. The reports describing the Soviet synthesis of an isotope of element 106 have not contained complete details of the experiment. However, on the basis of the information provided in Section 30.12, write a plausible nuclear reaction for the synthesis.

PROBLEMS

$\boxed{s}$1. The isotopic mass of $^{27}_{13}$Al is 26.98154. (a) Calculate its binding energy per atom and its packing fraction. (b) Calculate its binding energy per nucleon (nuclear particle) and compare the value with that of Fig. 30–2.

Ans. (a) 227 Mev; −6.8 × 10⁻⁴; (b) 8.41 Mev

$\boxed{s}$2. What percentage of $^{212}_{82}$Pb remains of a 1.00-gram sample, 1.0 minute after it is formed (half-life of 10.6 seconds)? ten minutes after it is formed?

Ans. 2.0%; 9.1 × 10⁻¹⁶%

$\boxed{s}$3. The isotope of ^{208}Tl undergoes beta decay with a half-life of 3.1 minutes.
  $\boxed{s}$(a) What isotope is the product of the decay?
  $\boxed{s}$(b) Is ^{208}Tl more stable or less stable than an isotope with a half-life of 54.5 seconds?
  $\boxed{s}$(c) How long will it take for 99.0% of a sample of pure ^{208}Tl to decay?

Ans. 20.6 min

  (d) What percentage of a sample of pure ^{208}Tl will remain undecayed after an hour?   *Ans. 1.5 × 10⁻⁴%*

$\boxed{s}$4. Calculate the time required for 99.999% of each of the following radioactive isotopes to decay:
  $\boxed{s}$(a) $^{226}_{88}$Ra   (half-life, 1590 years)   *Ans. 26,400 years*
  (b) $^{214}_{84}$Po   (half-life, 1.6 × 10⁻⁴ second)   *Ans. 0.0027 second*
  (c) $^{232}_{90}$Th   (half-life, 1.39 × 10¹⁰ years)   *Ans. 2.31 × 10¹¹ years*

$\boxed{s}$5. The isotope $^{90}_{38}$Sr is an extremely hazardous isotope in the fall-out from a nuclear fission explosion. A 0.500-g sample diminishes to 0.393 g in 10 years. Calculate the half-life.   *Ans. 28.8 years*

REFERENCES

"Some Recollections of Early Nuclear Age Chemistry," G. T. Seaborg, *J. Chem. Educ.*, **45**, 278 (1968).
"Discovery of Fission," O. Hahn, *Sci. American*, Febr., 1958; p. 76.
"The Size and Shape of Atomic Nuclei," M. Baranger and R. A. Sorensen, *Sci. American*, Aug., 1969; p. 59.

"The Structure of the Proton and the Neutron," H. W. Kendall and W. Panofsky, *Sci. American*, June, 1971; p. 60.

"Concepts of Nuclear Structure," A. Bohr, *Science*, **172,** 17 (1971).

"Structure of the Proton," R. F. Feynman, *Science*, **183,** 601 (1974).

"The Texture of the Nuclear Surface," C. D. Zafiratos, *Sci. American*, Oct., 1972; p. 100.

"The Periodic System of D. I. Mendeleev and Problems of Nuclear Chemistry," V. I. Gol'danskii, *J. Chem. Educ.*, **47,** 406 (1970).

"Radiation Chemistry," M. Burton, *Chem. & Eng. News,* Feb. 10, 1969; p. 87.

"Extranuclear Effects on Nuclear Decay Rates," P. K. Hopke, *J. Chem. Educ.*, **51,** 517 (1974).

"Nucleophilic Substitution—A Radioactive Study," B. A. Shaw and M. E. Shaw, *Educ. in Chemistry*, **11,** 77 (1974).

"Tin(IV) Iodide as a Radioactive Tracer," P. W. Wiggans, *Educ. in Chemistry*, **11,** 194 (1974).

"Carbon-13 as a Label in Biosynthetic Studies," U. Séquin and A. I. Scott, *Science*, **186,** 101 (1974).

"Two New Particles Found: Physicists Baffled and Delighted," W. D. Metz, *Science*, **186,** 909 (1974).

"The Age of the Elements," D. N. Schramm, *Sci. American*, Jan., 1974; p. 69.

"Deuterium in the Universe," J. M. Pasachoff and W. A. Fowler, *Sci. American*, May, 1974; p. 108.

"Experiments with Neutrino Beams," B. C. Barish, *Sci. American*, Aug., 1973; p. 30.

"Electron-Positron Collisions," A. M. Litke and R. Wilson, *Sci. American*, Oct., 1973; p. 104.

"Exotic Atoms," C. E. Wiegand, *Sci. American*, Nov., 1972; p. 102.

"Dual-Resonance Models of Elementary Particles," J. H. Schwarz, *Sci. American*, Feb., 1975; p. 61.

"Proton Interactions at High Energies," U. Amaldi, *Sci. American*, Nov., 1973; p. 36.

"Radiocarbon Dating," W. F. Libby, *Chem. in Britain*, **5,** 548 (1969).

"The Batavia Accelerator," R. R. Wilson, *Sci. American*, Feb., 1974; p. 72.

"Collective-Effect Accelerators," D. Keefe, *Sci. American*, April, 1972; p. 22.

"The First Isolations of the Transuranium Elements—A Historical Survey," J. C. Wallmann, *J. Chem. Educ.*, **36,** 340 (1959).

"The Synthetic Elements," G. T. Seaborg and J. L. Bloom, *Sci. American*, April, 1969; p. 56.

"Albert Ghiorso, Element Builder," J. F. Henehan, *Chem. & Eng. News,* Jan. 18, 1971; p. 26.

"Element 106: Soviet and American Claims in Muted Conflict," T. H. Maugh II, *Science*, **186,** 42 (1974).

"Uranium Enrichment: Laser Methods Nearing Full-Scale Test," W. D. Metz, *Science*, **185,** 602 (1974).

"Plutonium: (I) Questions of Health in a New Industry" and "—(II) Watching and Waiting for Adverse Effects," R. Gillette, *Science*, **185,** 1027 and 1140 (1974).

"Plutonium: Biomedical Research," W. J. Bair and R. C. Thompson, *Science*, **183,** 715 (1974).

"Recycling Plutonium: The National Research Council Proposes a Second Look," R. Gillette, *Science*, **188,** 818 (1975).

"Fast Breeder Reactors," G. T. Seaborg and J. L. Bloom, *Sci. American*, Nov., 1970; p. 13.

"The Energy Resources of the Earth," M. K. Hubbert, *Sci. American*, Sept., 1971; p. 60.

"Nuclear Power—the Long-Term View," G. T. Seaborg, *Chem. & Eng. News:* I, May 17, 1971; p. 3. II, May 24, 1971; p. 3.

"Fusion Power: Hallmark of the 21st Century," J. W. Landis, *J. Chem. Educ.*, **50,** 658 (1973).

"Fusion Reactors as Future Energy Sources," R. F. Post and F. L. Ribe, *Science,* **186,** 397 (1974).

"Fusion Power by Laser Implosion," J. L. Emmett, J. Nuckolls, and L. Wood, *Sci. American*, June, 1974; p. 24.

"Energy Policy in the U. S.," D. J. Rose, *Sci. American*, Jan., 1974; p. 20.

"Nuclear Strategy and Nuclear Weapons," B. E. Carter, *Sci. American*, May, 1974; p. 20.

"Nuclear Safety: Calculating the Odds of Disaster," R. Gillette, *Science*, **185,** 838 (1974).

"The Proliferation of Nuclear Weapons," W. Epstein, *Sci. American*, April, 1975; p. 18.

"Resources in Environmental Chemistry: An Annotated Bibliography of Energy and Energy-Related Topics," J. W. Moore and E. A. Moore, *J. Chem. Educ.*, **52,** 288 (1975).

"An Overview of the Energy Crisis," E. A. Walters and E. W. Wewerka, *J. Chem. Educ.*, **52,** 282 (1975).

The Metallic
Elements

31

Many times in the preceding chapters we have had occasion to classify the elements as either metallic or nonmetallic. Thus far, however, our study has been confined chiefly to the nonmetals and their compounds. Before taking up the study of the metallic elements and their compounds, a general consideration of the metals as a class of elements will be of value in correlating the study of the individual metals. Several of the chapters that follow describe the chemistry of the metals.

31.1 Metals and Nonmetals

One of the first attempts by chemists to classify the elements was based upon the observation that the oxides of some of the elements react with water to form acids, whereas others react with water to form bases. For example, sulfur dioxide and carbon dioxide react with water to form sulfurous acid and carbonic acid, respectively; sulfur and carbon are nonmetals. Sodium oxide and calcium oxide react with water to form sodium hydroxide and calcium hydroxide, respectively; sodium and calcium are metals. Metals are good conductors of heat and electricity; they are usually opaque to light and have metallic luster. These are properties of metals that are familiar to everyone. Few nonmetals have luster and nearly all are poor conductors.

The difference in chemical properties between metals and nonmetals lies chiefly in the fact that atoms of nonmetals readily fill their valence shells by sharing electrons with or using electrons transferred from other atoms. An atom of chlorine, a typical nonmetal, readily enters into chemical combination by adding one electron to an outer shell of seven electrons, either by gaining an electron from another atom or by sharing an electron pair. On the other hand, an atom of sodium, a typical metal, enters into chemical combination by the loss of its single valence electron. Metallic character of the elements decreases and non-metallic character increases with increased number of valence electrons. Furthermore, metallic character increases with the number of electron shells. Thus, the properties of succeeding elements change gradually in progressing across the periods of the Periodic Table (with increasing number of valence electrons from left to right), and down the groups (with increasing number of electron shells). This means that there is no sharp dividing line between metals and nonmetals, and most elements fall between the extremes. Elements which are definitely on the borderline and show hybrid behavior are sometimes referred to, as we learned in Chapter 29, as semi-metallic elements, or metalloids. Semi-metallic elements include boron, silicon, germanium, arsenic, antimony, tellurium, and astatine. Silicon, for example, has the physical properties characteristic of metals, but many of its chemical reactions are like those of the typical nonmetals. Of the 106 elements now known, only 17 show primarily nonmetallic character, 7 others are semi-metallic (metalloids), and 82 may be classed as metals.

Some characteristic properties of the metals and nonmetals have been summarized in Table 31–1, but it follows from the foregoing statements that there are many exceptions to these generalizations.

TABLE 31-1   Comparison of Metals and Nonmetals

Metals	Nonmetals
Physical Properties	*Physical Properties*
1. Are good conductors of heat and electricity	1. Are poor conductors
2. Are malleable and ductile in solid state	2. Are brittle, nonductile in solid state
3. Show metallic luster	3. Show no metallic luster
4. Are opaque	4. May be transparent or translucent
5. Have high density	5. Have low density
6. Are solids (except mercury)	6. Are gases, liquids, or solids
7. Have crystal structure in which each atom is surrounded by eight to twelve near neighbors; metallic bonds between atoms	7. Form molecules which consist of atoms covalently bonded; the noble gases are monatomic
Chemical Properties	*Chemical Properties*
1. Have one to four electrons in outermost shell; usually not more than three	1. Usually have four to eight electrons in outermost shell
2. Have low ionization potentials; readily form cations by losing electrons	2. Have high electron affinities; readily form anions by gaining electrons (except noble gases)
3. Are good reducing agents	3. Are good oxidizing agents (except noble gases)
4. Have hydroxides which are basic or amphoteric	4. Have hydroxides which are acidic (except noble gases)
5. Are electropositive; oxidation states are positive	5. Are electronegative; oxidation states may be either positive or negative

31.2 Occurrence of the Metals

The metals above hydrogen in the activity series (Table 9–1) are rarely found in the **free,** or **native, state.** The less active metals, which are below hydrogen in the activity series, often occur in the free state. Among the latter are copper, silver, gold, and platinum. Some metals, such as copper and silver, are found both free and combined.

It is not surprising that the metallic compounds occurring in the earth's crust are low in water solubility, while the more soluble compounds are found in sea water and in large salt beds which have been formed by the evaporation of inland seas. Water is such an excellent solvent that rains soaking into the earth dissolve many of the salts, carrying them slowly to the nearest stream, then to the nearest river, and finally down to the sea. The ocean is continually becoming richer in salts and other minerals, although man has not been observing it long enough to notice any real change. It is interesting that the age of the ocean as calculated from its salt content is between 1 and 7 billion years. This is of the same order of magnitude as that (2.6 billion years) calculated for the age of the earth by methods involving radioactive minerals (Section 30.9).

The materials in which metals or their compounds occur in the earth and from which the metals may be extracted economically are known as **ores.** Ores usually contain large percentages of rocky material called **gangue.** The important classes of ores, based on the nonmetallic element or acid radical with which the metal is combined (if any), are listed below.

(1) *Native ores:* Gold, silver, platinum, copper, mercury, arsenic, antimony, and bismuth.

(2) *Oxide ores:* Iron, aluminum, manganese, and tin.

(3) *Sulfide ores:* Zinc, cadmium, mercury, copper, lead, nickel, cobalt, silver, arsenic, and antimony.

(4) *Carbonate ores:* Iron, lead, zinc, and copper. These ores are less important as sources of the metals than are the oxides and sulfides of the same metals. The carbonates of calcium, barium, strontium, and magnesium are important as sources for the preparation of various compounds of these metals.

(5) *Halides:* The chlorides of sodium and potassium are important sources of these metals and their compounds. The halides of magnesium, calcium, and silver are also important.

(6) *Sulfates:* Calcium, strontium, barium, and lead.

(7) *Silicates:* Most silicates are unsuitable as ores because of difficulty in extracting the metals from them. However, the silicates of beryllium, zinc, and nickel are important.

31.3 The Ocean As a Source of Minerals

The oceans, which cover a large fraction of the earth's surface, have been termed "the world's greatest mine." A ton of sea water contains about 55 lb of common table salt, 2.54 lb of magnesium, 1.75 lb of sulfur, 0.8 lb of calcium, 0.75 lb of potassium, 0.125 lb of bromine, and lesser quantities of such elements as strontium, boron, fluorine, iodine, iron, copper, lead, zinc, uranium, silver, gold, and even radium. The amount of gold in the oceans is estimated to be 8.5 million tons, worth over 48,000 billion dollars at the 1975 price of $165 an ounce. Our mineral deposits in the earth's crust are being depleted at an ever increasing rate, so it is necessary to look to the oceans for future supplies of the metals.

The first mineral to be "mined" from sea water was undoubtedly salt. If sea water is trapped at high tide and the sun evaporates the water, "solar" salt with all the other solids that were in the water is obtained. This salt can be purified by fractional crystallization. Solar salt is made in large quantities on our west coast and on the coasts of Lebanon, Israel, and other Near East countries. Many chemical industries use salt as their raw material, especially the alkali and chlorine industries. Scandinavian sea salt has served as a starting point for the manufacture of soda ash (sodium carbonate) for many years.

Processes, described in the magnesium and bromine sections, are now in use by which magnesium and bromine are extracted from sea water. Extraction methods for other elements such as gold, copper, silver, potassium (for fertilizer), and especially uranium, from sea water, are as yet too expensive to be used extensively. Indirect methods of extracting these metals may be the final answer. For example, living plants have the curious ability to remove certain materials from soil or water and to store them in their tissues. The horsetail (*Equisetum*) removes silica and gold from the soil, the locoweed takes up barium, and certain seaweeds extract large amounts of potassium and iodine from sea water. When the weeds are burned, these elements are left in the ashes, from which they can be removed. The ocean is a potential source of uranium, which may well be our chief energy fuel when our deposits of coal and petroleum are exhausted. It is also the source of "manganese nodules," which occur in great quantity on the ocean floor. Much research is going into devising economically practical methods for their recovery, because they contain, in addition to manganese, other valuable metals such as nickel and copper.

31.4   Extractive Metallurgy

Extractive metallurgy pertains to the processes involved in the production of metals from their ores. Most such metallurgical processes include three principal steps: (1) preliminary treatment, (2) smelting, and (3) refining.

■ **1. Preliminary Treatment.**   Generally, ores must be subjected to certain preliminary treatments to render them suitable for chemical extraction of the metals. The first step is usually that of pulverizing the ore by crushing and grinding. This is followed by **concentration** of the metal-bearing portions by removing most of the gangue. Sometimes this is accomplished by washing away the lighter gangue particles. Slightly inclined shaking tables are used to separate the heavier ore particles from the lighter rocky material. Metal-bearing particles which are affected by a magnetic field are often separated from nonmagnetic impurities by passing the finely ground ore through a magnetic field. Certain ores are affected by an electrostatic field, and in such cases electrostatic separation may be employed. The "flotation" process is a concentration method particularly applicable to sulfide ores of lead, zinc, and copper, but it is used also for carbonates and silicates. In this process, the pulverized ore is mixed with water to which a carefully selected oil has been added. A froth is produced by blowing air into the mixture. The metal-bearing particles have little or no attraction for water (such substances are referred to as **hydrophobic**). The particles do have an attraction for the oil and are therefore preferentially coated by the oil and adhere to the air bubbles in the froth which floats on the surface (see Section 14.16). The particles of gangue (sand, rock, clay, etc.) are wetted by water (such substances are **hydrophilic**) more readily than by the oil and sink to the bottom of the flotation vat. The froth containing the metal-bearing particles is removed, and the ore is recovered in a highly concentrated form.

The preliminary treatment of ores often includes chemical changes which convert the metallic compounds into substances that may be more readily reduced to give the metals. Ores containing moisture or water chemically combined in hydrates or hydroxides are heated to expel water. Ores containing metal carbonates are heated to decompose the carbonates and drive off carbon dioxide. This treatment of heating to drive off a volatile substance is known as **calcination.** The following equations illustrate typical changes which may occur during calcination.

$$2M(OH)_3 \longrightarrow M_2O_3 + 3H_2O \qquad (M = metal)$$
$$MCO_3 \longrightarrow MO + CO_2$$

Most sulfide ores are heated in air to change the sulfides to oxides, and to expel sulfur as sulfur dioxide. This treatment is known as **roasting.** The following equation illustrates the reaction.

$$2MS + 3O_2 \longrightarrow 2MO + 2SO_2$$

■ **2. Smelting.** The next step in the metallurgical process involves the extraction of the metal in the fused state, a process called **smelting.** The chemical reaction involved in smelting is one of reduction of the metallic compound. Most ores still contain some gangue at this stage of the process; this material is removed by addition of a **flux** (from the Latin *fluere*, meaning "to flow"). The flux is a chemical which will react with the gangue to form a substance of low melting point—a **slag.** When the gangue is silica or a silicate, a basic flux such as lime or limestone is used. At the high temperature of the furnace in which the reduction is carried out, the lime reacts with the silica to produce fused calcium silicate. The molten slag is easily separated from the fused metal because of the difference in their densities and because they are insoluble in each other.

The methods commonly employed in the smelting of various ores are:

(a) *The reduction of oxide ores by carbon.*

$$ZnO + C \longrightarrow Zn + CO$$
$$SnO_2 + 2C \longrightarrow Sn + 2CO$$

When carbon is for some reason unsatisfactory, reducing agents such as aluminum, hydrogen, or iron are used.

$$Cr_2O_3 + 2Al \longrightarrow 2Cr + Al_2O_3$$
$$WO_3 + 3H_2 \longrightarrow W + 3H_2O$$
$$Sb_2S_3 + 3Fe \longrightarrow 2Sb + 3FeS$$

(b) *The electrolytic reduction of halides of such active metals as sodium, potassium, magnesium, and calcium.* Aluminum is produced by the electrolysis of aluminum oxide dissolved in fused cryolite, Na_3AlF_6.

(c) *Direct heating of some metals which occur in the native state.* The metals are smelted by heating the ore until the metals are melted. The fused metals are then drained away from the gangue.

■ **3. Refining of Metals.** Smelting usually results in the production of metals which contain greater or lesser amounts of impurities such as other metals (or nonmetals), slag, and dissolved gases. The removal of impurities is usually necessary in the preparation of a metal for at least some of its uses. The low boiling metals, such as zinc and mercury, may be purified from less volatile impurities by distillation. Other low melting metals, such as tin, when fused on an inclined table may flow away from higher melting impurities. Electrolysis is the most widely used method of refining metals. Electrolytic refining is accomplished by making the impure metal the anode and a piece of the pure metal the cathode of an electrolytic cell (see Section 22.6). Among the metals which are purified by this process are copper, gold, lead, zinc, and aluminum.

A process known as **zone refining** provides ultrahigh-purity materials essential to semiconductor and transistor devices, such as those described earlier for the solar battery (Section 22.18). When a solid bar of metal is melted at one section (zone) and then the melted zone is made to progress longitudinally through the bar, the zone carries certain impurities with it. The procedure can be repeated with increased purification each time. The process depends upon the lowering of the melting point of a substance as a result of impurities present. It is applicable to a wide range of substances and impurities.

■ **4. Hydrometallurgy.** Certain metals can be extracted from their ores by causing the metal to go into aqueous solution, as the result of chemical changes, and then precipitating the metal in the free state using a suitable reducing agent. The following equations illustrate the method.

$$4Ag + 8CN^- + O_2 + 2H_2O \longrightarrow 4Ag(CN)_2^- + 4OH^-$$

or $\quad\quad\quad\quad\quad\quad AgCl + 2CN^- \longrightarrow Ag(CN)_2^- + Cl^-$

Then, $\quad\quad\quad 2Ag(CN)_2^- + Zn \longrightarrow 2\underline{Ag} + Zn(CN)_4^{2-}$

31.5   Metallic Lattices

Metals have only a small number of valence electrons (usually 1, 2, or 3) to use in bonding. This is not enough to form either regular covalent or ionic bonds between the metal atoms. Hence, metals tend to crystallize in such a way as to form extremely compact structures in which each atom has as many nearest neighbors as possible, usually 8 or 12, to make possible an overlapping of electronic energy levels. Such structures are verified by means of x-ray diffraction

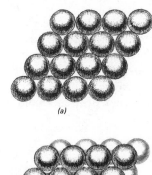

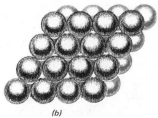

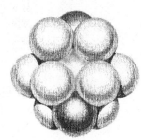

FIGURE 31-1

(*a*) Spheres in the same plane packed as closely as possible; each sphere contacts six others (view is downward). (*b*) Spheres in two planes packed as closely as possible. In an actual crystal, many more than two planes would exist. Each sphere contacts six spheres in its own layer, three spheres in the layer below, and three spheres in the layer above, making a total contact for each sphere of twelve other spheres.

studies. The three specific crystal lattices commonly found among the metals are the body-centered cubic lattice with a coordination number of 8, the close-packed, face-centered cubic lattice with a coordination number of 12, and the close-packed hexagonal lattice, also with a coordination number of 12. These lattices are shown in Fig. 11–9 of Chapter 11.

If spheres of equal size are packed together as compactly as possible in a plane, they arrange themselves as shown in Fig. 31–1(a), with their centers at the corners of equilateral triangles. Each sphere is in contact with six others. Placement of spheres in a second plane above the first layer yields an arrangement such that each sphere in the second layer is in contact with three spheres in the first layer [Fig. 31–1(b)]. Relative to the second layer, there are two different possible arrangements for construction of a third layer, this time below the first layer. One of these is to place the initial third-layer sphere directly below a sphere in the second layer two layers above. This gives rise to the **hexagonal close-packed structure** (h.c.p.), in which alternate layers are the same [Fig. 31–2(a)]. The second arrangement is obtained if the top triangular layer (Fig. 31–2) is rotated 60° so that the spheres in the trio above are no longer directly above the spheres in the trio two layers below. This gives rise to the **face-centered cubic, close-packed structure** (f.c.c.), pictured in Fig. 31–2(b). Viewed from another perspective, the face-centered cubic structure possesses atoms at each corner of a cube with other

FIGURE 31-2

The two types of crystal structure in which atoms are packed as compactly as possible. The lower diagrams show the structures expanded for clarification. Note the triangular layers are oriented in the same direction in the hexagonal structure (*left*), and in opposite directions in the face-centered cubic structure (*right*). In both structures, each atom is surrounded by twelve others in an infinite extension of the structure and is said to have a coordination number of twelve.

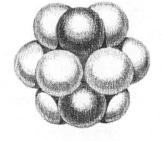

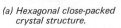

(a) Hexagonal close-packed crystal structure.

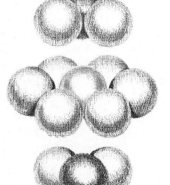

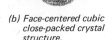

(b) Face-centered cubic close-packed crystal structure.

atoms centered in each face of the cube (see Fig. 11–9). In each of the close-packed arrangements, face-centered cubic and hexagonal, each atom has 12 equidistant neighbors, six in its own plane and three in each adjacent plane. The **body-centered cubic structure** (b.c.c.), with only eight nearest neighbors for each atom, has less effective packing and is not considered to be a close-packed structure. That arrangement has atoms occupying each corner of a cube and a central atom in the center of the cube (see Fig. 11–9).

31.6 Chemical Bonding in Metals

We have noted that the atoms of the nonmetals have a sufficient number of valence electrons to combine with each other by sharing electron pairs. For example, two chlorine atoms share an electron pair in the Cl_2 molecule, two nitrogen atoms share three electron pairs in the N_2 molecule, and each carbon atom in the diamond crystal shares electron pairs with each of four neighboring carbon atoms. However, most of the metal atoms have less than four valence electrons (many only one or two), and each atom is surrounded by eight or twelve closest neighbors in the crystal. With only one valence electron it would be impossible for an atom of sodium, for example, to be linked by electron-pair bonds to eight neighboring atoms. Two principal theories have been utilized to explain what is known as the **metallic bond.**

■ **1. Modified Covalent Bond Theory of the Metallic Bond.** The modified covalent theory assumes that in solid or liquid metals the valence electrons are free to pass from one atom to another in such a way that they may be shared by several atoms (or ions, in the sense that each atom has given up its own specific valence electrons). These shared electrons serve to hold the atoms (or ions) together and constitute what is known as a metallic bond, which may be considered as a modified covalent bond with more or less mobile electrons being shared by more than two ions. This arrangement has sometimes been rather picturesquely referred to as being made up of "metallic ions buried in a sea of electrons."

Such characteristic properties of metals as high electrical and thermal conductivity, luster, and high reflectivity are thought to be due to these relatively free, mobile valence electrons. These considerations are in accord with the fact that metallic character exists only when the material is in the massive solid or liquid state. Gaseous mercury, zinc, and sodium do not have metallic luster and the other characteristic metallic properties; their gaseous molecules are monatomic, and metallic bonding is, of course, absent.

■ **2. The Band Theory of the Metallic Bond.** The physical properties which permit us to distinguish metals from nonmetals, i.e., electrical and thermal conductivity, metallic luster, etc., may be more readily accounted for by the more recently devised **band theory** of metals than by the modified covalent bond theory.

To aid us in understanding metallic bonding from the standpoint of the band concept it will be helpful to consider the following points.

(a) Metals crystallize in structures of high coordination number, usually 8 or 12, yet have only a small number of valence electrons (usually 1, 2, or 3) to use in bonding. It is evident that no assignment of valence electrons to metallic bonding can account for such high coordination numbers, unless the valence electrons are "delocalized" (do not belong to any particular atoms).

(b) Atomic orbitals combine to yield molecular orbitals, the number of molecular orbitals depending on the number of electrons and the number of atomic orbitals. (See Section 5.1.)

(c) The number of molecular orbitals in a crystal of a metal increases as the number of atoms is increased and the energy differences between successive energy levels decrease.

(d) Close packing of atoms in the crystal causes overlapping of electronic energy levels—that is, the formation of "bands" of electronic energy levels which belong to the whole metal crystal rather than to individual atoms. Each band contains many energy levels and hence comprises a large energy range, sometimes as much as 100 kcal per mole. Only the electrons in the higher energy portion of a band are sufficiently free, or mobile, to be of significance for such properties as electrical and thermal conductivity, luster, and reflectivity.

To gain a better understanding of the band theory, consider the case of the lightest metal, lithium, whose electronic structure is $1s^2, 2s^1$. A very small crystal of lithium contains about 10^{18} atoms. For such a large number of atoms the energy difference between some successive energy levels is so small that the levels are essentially continuous. These energy levels taken together constitute an electron band. In lithium one electron band arises from the $1s$ atomic orbital which is fully occupied with two electrons. However, the second band, arising from the $2s$ orbital, is only half-filled with one electron. The differences in the energies of these two bands is so large that promotion of electrons from the lower band to the higher is prohibitive. Such energy gaps are called **forbidden zones.** Metals are characterized by having incompletely filled electron bands—for example, lithium with its half-filled band arising from the $2s$ orbital. Because there are so many electronic energy levels within a given band and because some bands are only partly filled, any electron within an unfilled band can absorb energy (for example, from an applied electric field) and be energetically promoted to other levels within the band.

In beryllium ($1s^2, 2s^2$) the $2s$ electron band is filled, and one might expect that this element should be nonmetallic, according to the band theory. It does, however, have appreciable metallic character, which can be explained by the fact, that the relative energies of successive bands depend upon distances between nuclei. In metals, which are close-packed and have relatively small internuclear distances, the adjacent bands may overlap. In the case of beryllium, therefore, the $2s$ and $2p$ bands may overlap, there is no forbidden zone, and electrons from the filled $2s$ band can be easily promoted to the unfilled $2p$ band.

The spacing of the electron bands and the filling of the bands determines whether a substance is a conductor, nonconductor (insulator), or semiconductor (metal, nonmetal, or metalloid). If the bands are completely filled or completely empty and the energy gap between bands is large, the substance will be an insulator. A substance will serve as a conductor only if it has either a partially filled band or an empty band where the energy gap is small so that adjacent bands overlap. The so-called "semiconductors" appear to be made conducting by virtue of defects in the lattice (interstitial atoms, impurities, or vacant lattice points) which introduce additional energy levels into which the electrons can move.

Let us interpret some of the physical properties of metals in the light of the electron band theory.

(a) *Metals are Good Conductors of Electricity.* When an electric field is applied to a metal, any of the many electrons at the top of the filled portion of a band may move into one of the vacant levels immediately above it in the band. The electrons thus have a net motion through the lattice in the direction of the applied field, producing conductivity.

(b) *Metals are Good Conductors of Heat.* The mobile electrons of the bands can also absorb heat energy. Transport of this energy by these electrons accounts for the high thermal conductivity of metals.

(c) *Metals are Good Reflectors of Radiant Energy.* The absorption and subsequent emission of photons of radiant energy by the mobile electrons of the bands accounts for the high reflectivity of metals.

(d) *Metals are Ductile and Malleable.* When a metallic crystal is subjected to mechanical stress, bonds are readily broken and new ones formed by the promotion of electrons from lower to higher electronic levels in the conduction bands and subsequent return of these electrons to their original levels. Thus, the crystal lattice is not altered by mechanical stress but remains essentially the same, because the bonding electrons are delocalized (do not belong to any particular atoms) and can form other bonds between any adjacent atoms.

31.7 Classification of the Metals

It is convenient to classify the metals in terms of electronic structure as representative and transition metals (see Section 3.15). The representative metals have all their valence electrons in one shell, whereas the transition metals have valence electrons in more than one shell. Those representative metals with one or two valence electrons use them in bond formation and usually exhibit only one oxidation state: e.g., sodium, +1; calcium, +2; and magnesium, +2. Some of the representative metals that have three or more valence electrons have two oxidation states: e.g., lead, +2 and +4; tin, +2 and +4; bismuth, +3 and +5. Boron and aluminum are exceptions.

The transition metals include the elements of subgroups IIIB through IB in each of the long periods of the Periodic Table. The first series of transition metals includes Sc, Ti, V, Cr, Mn, Fe, Co, Ni, and Cu. Because of the carry-over in properties, elements in subgroup IIB (Zn, Cd, and Hg) have many characteristics analogous to those of the transition elements, and for this reason they are sometimes classed with the transition elements. However, in terms of our definition of transition and representative metals, the metals of subgroup IIB are classified

TABLE 31-2   The Transition Metals of the Fourth Period

Atomic Number	Metal	Electronic Structure	Some Oxidation States
21	Scandium	$1s^2\,2s^2\,2p^6\,3s^2\,3p^6\,3d^1\ 4s^2$	$+3$
22	Titanium	$3d^2\ 4s^2$	$+2, +3, +4$
23	Vanadium	$3d^3\ 4s^2$	$+2, +3, +4, +5$
24	Chromium	$3d^5\ 4s^1$	$+2, +3, +6$
25	Manganese	$3d^5\ 4s^2$	$+1, +2, +3, +4, +6, +7$
26	Iron	$3d^6\ 4s^2$	$+2, +3, +6$
27	Cobalt	$3d^7\ 4s^2$	$+2, +3, +4$
28	Nickel	$3d^8\ 4s^2$	$+2, +3, +4$
29	Copper	$3d^{10}\,4s^1$	$+1, +2, +3$

as representative metals. The electron distribution and common oxidation states of the transition metals of the fourth period of the Periodic Table are given in Table 31–2.

The transition metals are noted for their variability in oxidation state; this is attributed to the presence of valence electrons in more than one shell. Thus, manganese has two electrons in its outside shell and five electrons in the next underlying $3d$ subshell and exhibits oxidation states of $+1$, $+2$, $+3$, $+4$, $+5$, $+6$, and $+7$. An atom of iron may lose the two electrons in its outermost shell and form an iron(II) ion, Fe^{2+}, or it may lose an additional electron from its underlying $3d$ subshell and form an iron(III) ion, Fe^{3+}. When it shares electrons with oxygen in K_2FeO_4, iron exhibits an oxidation state of $+6$.

The transition metals are further characterized by the fact that well into the series going from left to right across the Periodic Table, the properties of succeeding metals do not differ greatly from preceding ones (Table 31–3). This is attributed to the fact that, generally, succeeding elements in the series differ in electronic structure by one electron in the next to the outer valence shell rather than the outer valence shell (with two exceptions). Differences in electronic structure in

TABLE 31-3   Some Physical Properties of the Transition Metals of the Fourth Period

Property	Sc	Ti	V	Cr	Mn	Fe	Co	Ni	Cu
Atomic number	21	22	23	24	25	26	27	28	29
Density, g/cm³	2.5	4.5	5.96	7.19	7.21	7.87	8.90	8.90	8.91
Melting point, °C	1200	1725	1710	1890	1260	1535	1490	1452	1083
Atomic radius, Å	1.60	1.46	1.31	1.25	1.29	1.26	1.26	1.24	1.28
Ionization potential, electron volts	6.56	6.83	6.74	6.76	7.432	7.896	7.86	7.633	7.723

TABLE 31-4   Some Physical Properties of Three
              Representative Metals in the Third Period

Property	Na	Mg	Al
Atomic number	11	12	13
Density, g/cm^3	0.97	1.74	2.71
Melting point, °C	97.6	651	658.7
Atomic radius, Å	1.86	1.60	1.43
Ionization potential, electron volts	5.138	7.644	5.984

the outer shell have more significant effects on properties than do differences within inner shells.

In contrast to the properties of the transition metals, those of succeeding representative metals in a period differ extensively (Table 31–4). The electronic structures of succeeding elements in this series differ by one electron in the outer shell.

The transition metal ions containing incomplete underlying electron shells are usually colored, both in solid salts and in solution. The color exhibited reflects the oxidation state of the metal as indicated by the following examples: Cr^{2+} (blue), Cr^{3+} (violet); Mn^{2+} (pink), Mn^{3+} (violet); and Fe^{2+} (green), Fe^{3+} (yellow). The color is also modified by the nature of the nonmetallic element or acid radical with which the metal is combined. For example, CuO is black; $Cu(OH)_2$, pale blue; $CuCl_2$, yellow; $CuBr_2$, black; CuS, black; $CuSO_4$, white; and $CuSO_4 \cdot 5H_2O$, blue. When the electron shell underlying the outermost shell is filled, as in the case of Cu^+ and Zn^{2+}, the substance containing the ion is usually colorless. This is also usually true of compounds containing the representative metals, i.e., those with valence electrons only in the outer shells. The transition metals have two other general properties that should be mentioned—they have marked catalytic properties, and they form more stable coordination compounds (see Chapter 32) than do the representative metals.

31.8   The Transition Metals and Magnetism

Most of the transition metals and their compounds in oxidation states involving incomplete inner electron subshells are **paramagnetic**; i.e., they are substances that tend to move into a magnetic field, such as that between the poles of a magnet.

Paramagnetism is characteristic of systems containing one or more unpaired electrons. An electron in an atom spins about its own axis. Because the electron is electrically charged, the spin about its axis gives it the properties of a small magnet, with north and south poles. The two electrons of a pair spin in opposite directions. This means that their magnetic moments will cancel each other because their north and south poles are opposed. On the other hand, when an electron in an atom is unpaired the magnetic moment due to its spin confers paramagnetism upon the entire atom or ion, and in turn upon the specimen containing such atoms or ions.

The size of the magnetic moment of a system containing unpaired electrons is related directly to the number of unpaired electrons, i.e., the greater the number of such electrons, the larger the magnetic moment. Therefore, the observed magnetic moment is used to indicate the number of unpaired electrons present.

The unpaired electrons in the transition metal atoms or ions are located in the d or f subshells. As we have learned earlier, when a subshell is being filled by the stepwise addition of single electrons, the added electrons enter unoccupied orbitals first before any pairing occurs (see Section 3.17). Table 31–5 gives the distribution of electrons in the $3d$ subshell of the transition metal atoms of the fourth period. The magnetic moment of the metal ions of this series reaches a maximum with Mn^{2+} and Fe^{3+}, both of which have the d subshell configuration $3d^1 3d^1 3d^1 3d^1 3d^1$; i.e., they have the greatest number of unpaired electrons. Note that both manganese and iron lose their two $4s$ electrons and iron one $3d$ electron in forming the Mn^{2+} and Fe^{3+}, respectively.

Ferromagnetism can be thought of as an extreme form of paramagnetism. However, ferromagnetic substances may become permanently magnetized whereas paramagnetism is shown only in the presence of an applied magnetic field. Only iron, cobalt, nickel, and gadolinium (a rare earth element) exhibit ferromagnetism. In order for an element to be ferromagnetic, (a) it must have an incompletely filled d or f subshell, (b) the atoms must not be too close together in the crystal lattice, or the singly occupied orbitals of neighboring atoms overlap and the electrons become paired with those with opposed spins, and (c) the atoms must not be too far apart in the crystal, or the unpaired electrons of one atom cannot align the electron spins in the neighboring atoms.

The dependence of ferromagnetism upon a critical interatomic distance is well illustrated by the fact that some compounds of manganese (five unpaired $3d$ electrons) are ferromagnetic while the metal itself is not. The reason for this is that the atoms are close enough together in the metal to cause sharing (pairing

TABLE 31–5   Distribution of Electrons in the 3d Subshell of Fourth Period Transition Metals

Element	Order of Filling of 3d Subshell
Scandium	$3d^1 3d^0 3d^0 3d^0 3d^0 4s^2$
Titanium	$3d^1 3d^1 3d^0 3d^0 3d^0 4s^2$
Vanadium	$3d^1 3d^1 3d^1 3d^0 3d^0 4s^2$
Chromium	$3d^1 3d^1 3d^1 3d^1 3d^1 4s^1$
Manganese	$3d^1 3d^1 3d^1 3d^1 3d^1 4s^2$
Iron	$3d^2 3d^1 3d^1 3d^1 3d^1 4s^2$
Cobalt	$3d^2 3d^2 3d^1 3d^1 3d^1 4s^2$
Nickel	$3d^2 3d^2 3d^2 3d^1 3d^1 4s^2$
Copper	$3d^2 3d^2 3d^2 3d^2 3d^2 4s^1$

up) of the unpaired $3d$ electrons between neighboring atoms. In a compound such as manganese nitride the manganese atoms have been pushed apart somewhat by the nitrogen atoms, making the sharing of $3d$ electrons between atoms impossible. The same explanation may account for the ferromagnetism of the **Heusler alloys** (which consist of Cu, Al, and Mn in proportions approximating the atomic composition Cu_2AlMn) and **Alnico** (an alloy composed of Al, Ni, and Co).

31.9 Compounds of the Metals

The metals which have relatively large atomic radii and only one or two valence electrons show a marked tendency to react with nonmetals by electron transfer, with the formation of ionic compounds. For example, NaCl, KBr, CaF_2, MgO, and BaS are ionic compounds. However, those metals with relatively small atomic radii and having three or more valence electrons tend to share electrons with nonmetals and form covalently bonded molecules. For example, anhydrous aluminum chloride, anhydrous tin(IV) chloride, and titanium tetrachloride are covalent compounds.

The most metallic metals are distinguished from the nonmetals by the basic nature of their hydroxyl compounds (Section 7.1, particularly Part 7). The hydroxides of the highly electropositive metals such as sodium and potassium are ionic in the solid state and yield, therefore, large quantities of hydroxide ions in solution; they thus are strong bases. The bonding in the less soluble metal hydroxides, such as magnesium hydroxide and calcium hydroxide, is largely ionic, and they are fairly strong bases. The hydroxides of metals such as aluminum, iron, and tin are largely covalent; they are sparingly soluble in water and extremely weak as hydroxide bases.

The hydroxides of some metals such as zinc, tin, and aluminum are **amphoteric;** i.e., they can act as either acids or bases and hence are soluble in either strongly basic or strongly acidic solutions.

$$[Al(H_2O)_3(OH)_3] + OH^- \rightleftharpoons [Al(H_2O)_2(OH)_4]^- + H_2O$$
$$[Al(H_2O)_3(OH)_3] + H_3O^+ \rightleftharpoons [Al(H_2O)_4(OH)_2]^+ + H_2O$$

The hydroxyl compounds of metals which have several oxidation states, such as manganese, become less basic and more acidic as the oxidation state is increased (see Table 31–6). The larger charge, as the oxidation number becomes higher, makes the central metal ion attract electrons more readily, thereby strengthening

TABLE 31–6 Acid-Base Character of Hydroxyl Compounds of Manganese

Hydroxide	Oxidation State	Character
$Mn(OH)_2$	Mn(II)	Moderately basic
$Mn(OH)_3$	Mn(III)	Weakly basic
H_2MnO_3, $(HO)_2MnO$	Mn(IV)	Weakly acidic
H_2MnO_4, $(HO)_2MnO_2$	Mn(VI)	Definitely acidic
$HMnO_4$, $(HO)MnO_3$	Mn(VII)	Strongly acidic

the metal-oxygen bond and weakening the O—H bond. Hence, the compound releases hydrogen ions more readily.

It should be noted that manganese is typically metallic in the $+2$ oxidation state, as indicated by the basic character of $Mn(OH)_2$. In permanganic acid, $HMnO_4$, manganese has an oxidation state of $+7$ and behaves as a nonmetal; it is similar to the chlorine in perchloric acid, $HClO_4$.

An important difference between the metals and the nonmetals is demonstrated in the extent and reversibility of the hydrolysis of their chlorides. The chlorides of the more nonmetallic elements usually hydrolyze extensively—in many cases, irreversibly.

$$PCl_3 + 3H_2O \longrightarrow H_3PO_3 + 3HCl \qquad \text{(irreversible)}$$
$$AlCl_3 + 3H_2O \rightleftharpoons Al(OH)_3 + 3HCl \qquad \text{(reversible)}$$

The chlorides of very active metals (Na, K) are not hydrolyzed appreciably at ordinary temperatures.

The metal hydroxides, except those of the alkali and alkaline earth metals, yield oxides upon heating. The heating of metallic carbonates and nitrates, except those of the alkali metals, produces oxides. Metals react with the halogens and sulfur to form halides and sulfides, respectively. Mercury and the metals preceding it in the activity series (Section 9.9) react directly with oxygen. Sodium and the metals above it in the activity series displace hydrogen from cold water while the metals above hydrogen liberate hydrogen from acids and form salts. The metals below hydrogen in the series ordinarily will not liberate hydrogen from acids, but most of them react with oxidizing acids forming salts or oxides. Salts of metals may also be produced by the reaction of oxides, hydroxides, carbonates, or other salts with acids or other salts (see Chapter 15).

31.10   Metal-to-Metal Bonds

In some compounds, bonds occur between one metal and another. For example, several divalent metal acetates (monohydrated) are known to contain metal-to-metal bonds. One typical example is the copper(II) acetate monohydrate, $[Cu(OOCCH_3)_2 \cdot H_2O]_2$, which occurs as a dimer (Fig. 31–3). Each copper atom is in an octahedral environment.

Some metal carbonyl compounds (compounds of a metal with carbon monoxide) have metal-to-metal bonds also. Examples are dicobalt octacarbonyl, $Co_2(CO)_8$, and dimolybdenum decacarbonyl, $Mo_2(CO)_{10}$:

$$(OC)_4Co-Co(CO)_4$$
$$(OC)_5Mo-Mo(CO)_5$$

Note that in the copper acetate example, there are **bridging bonds** (in this case, acetate groups) between the two metal atoms in addition to the metal-to-metal bond. In the carbonyl examples shown, there are no bridging bonds. However, actually $Co_2(CO)_8$ has two isomers, one of which does have bridging (carbonyl) bonds (Fig. 31–4).

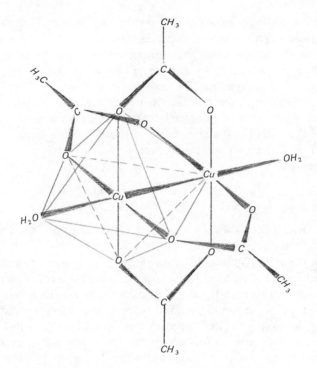

FIGURE 31–3

The structure of copper(II) acetate monohydrate dimer, which has a metal-to-metal bond. Each copper atom is in an octahedral environment. The octahedron is shown for only one of the two copper atoms.

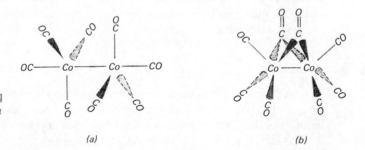

FIGURE 31–4

Two isomers of $Co_2(CO)_8$. Each has a metal-to-metal bond. Isomer (b) has bridging (carbonyl) bonds between the two metal atoms, isomer (a) does not.

31.11   Alloys

An **alloy** is usually composed of two or more metals, but it also can be made up of a metallic element and one of the less nonmetallic elements such as carbon, silicon, nitrogen, or phosphorus. Such systems are considered alloys only when they have metallic properties.

Alloys of two metals are usually prepared by fusing the metals together and allowing the melt to cool, but they are sometimes produced by simultaneous electrodeposition of two or more metals at a cathode. For example, many iron hardware fittings such as doorknobs are plated with brass (Cu 60–80 per cent, Zn 40–20 per cent) by electrodeposition from a bath containing copper and zinc in the form of soluble cyanide complexes. In general, alloys are solids, but they may be liquids, as in the case of certain amalgams, i.e., alloys of mercury. The study of alloys by x-ray methods has shown that they may be grouped into three classes, based upon their structures.

(1) *Simple mixtures*, in which the component metals are mutually insoluble and the solid alloy is composed of an intimate mixture of crystals of each metal. For example, tin and lead (in plumber's solder) are insoluble in each other in the solid state.

(2) *Solid solutions*, in which the atoms of one of the component metals take up positions in the crystal lattice of the other. The solute atoms may replace some of the atoms at the lattice points of the solvent crystal to form **substitutional solid solutions.** For example, chromium dissolves in nickel to form a solid solution in which the chromium atoms replace nickel atoms in the face-centered cubic nickel lattice. The solubility may be limited (zinc and copper; chromium and nickel) or practicaliy infinite (nickel and copper). If the solubility limit is exceeded, crystals of an alloy mixed with crystals of one of the metals are formed, as in class (1). The very small atoms of the elements H, C, B, and N may occupy the holes in the lattice of a metal, forming **interstitial solid solutions.** The solid solution of carbon in gamma-iron (austenite) is an example of this type; the iron atoms are on the face-centered cubic lattice points, and the carbon atoms occupy the interstitial positions.

(3) *Intermetallic compounds*, in which atoms of the components of the alloy appear in stoichiometric proportions which seem to be related to the ratio of the total number of electrons in the outer shell to the total number of atoms: Cu_5Zn_8 $(\frac{21}{13})$, Ag_3Al $(\frac{6}{4} = \frac{3}{2} = \frac{21}{14})$, and Cu_3Sn $(\frac{7}{4} = \frac{21}{12})$. The ratios for many intermetallic compounds (though not all) are either $\frac{21}{12}$, $\frac{21}{13}$, or $\frac{21}{14}$. Intermetallic compounds with the same ratio tend to be alike in crystal structure. In general, the formulas of such intermetallic compounds are not those which might be predicted on the basis of valence rules.

31.12   Properties of Alloys

Alloys which are solid solutions are always harder than the pure solvent metal and are apt to be less ductile; for example brass, which is a solution of zinc in copper, is much harder than pure copper and less ductile. As one metal is dissolved in another, the electrical conductivity of the solvent metal is usually lowered very sharply. Thus, copper must be quite free of impurities if it is to be used for transmission of electricity. The thermal conductivity of a solid solution alloy is less than that of the solvent metal. The melting point of a metal is usually lowered when another metal is dissolved in it. The practical application of a great many alloys depends upon their low melting, or freezing, points. For example, fuse metal, type metal, solder, and the alloys for sprinkler heads are low melting alloy systems. Alloys of iron, cobalt, and nickel with each other are strongly magnetic (ferromagnetic). The color of a solid solution alloy is impossible to predict. Copper turns gray with the addition of 23 per cent of nickel or 50 per cent of zinc. Certain gold-iron solid solutions are blue; copper-antimony is green; and gold-silver-cadmium is green. The compositions of some alloys are given in Table 31–7.

TABLE 31-7   Composition of Some Common Alloys

Trade Name	Composition, Per Cent by Weight
White cast iron	97 Fe, 3 C
Stainless steel	82.5 Fe, 16.5 Cr, 0.65 C, 0.35 Mn
Bronze	70–95 Cu, 1–25 Zn, 1–18 Sn
Type metal	70 Pb, 18 Sb, 10 Sn, 2 Cu
Wood's metal	50 Bi, 12.5 Cd, 25 Pb, 12.5 Sn
Plumber's solder	67 Pb, 33 Sn
Battery plate	94 Pb, 6 Sb
Dentist amalgam	50 Hg, 35 Ag, 13 Sn, 1.5 Cu, 0.5 Zn
Sterling silver	92.5 Ag, 7.5 Cu
Silver coin, U.S.A.	90 Ag, 10 Cu
Gold coin, U.S.A.	90 Au, 10 Cu
18 carat yellow gold	75 Au, 12.5 Ag, 12.5 Cu
18 carat white gold	75 Au, 3.5 Cu, 16.5 Ni, 5 Zn
14 carat gold	58 Au, 4–28 Ag, 14–28 Cu
Yellow brass	67 Cu, 33 Zn
Red brass	90 Cu, 10 Zn

QUESTIONS

1. Is the distinction between metals and nonmetals well defined? Explain.
2. Compare metals with nonmetals with regard to (a) number of valence electrons, (b) ionization potential, (c) electron affinity, (d) electronegativity, and (e) acidic or basic nature of hydroxides.
3. What physical properties distinguish metallic elements from nonmetallic elements?
4. What is accomplished by roasting an ore before smelting?
5. Explain the concentration of sulfide ores by flotation.
6. Which metals are likely to form ionic compounds and which ones are apt to form covalent compounds?
7. Compare the packing structure of the atoms in face-centered cubic close-packed, hexagonal close-packed, and body-centered cubic crystal lattices.
8. In what ways are metallic and covalent bonds similar and how do they differ?
9. Contrast the metallic bond with the electron-deficient bond discussed in Chapter 29 for certain boron compounds. Explain how metallic bonding holds the atoms together in a crystal lattice, and discuss how various characteristic metallic properties can be explained by means of the metallic bond.
10. Contrast the hydrolysis of the chlorides of metals and nonmetals.
11. How does the basicity of the hydroxyl compounds of manganese vary with the oxidation state of the metal? Why is this so?
12. Compare the thermal stability of the hydroxides of the alkali and alkaline earth metals with that of the other metal hydroxides.
13. Why is it that most compounds of the transition metals are colored?

14. What is the basis for the classification of metals as representative and transition?

15. Compare the electron configurations of calcium and zinc, both as atoms and as dipositive $(2+)$ ions.

16. Show that cobalt(III) and iron(II) are isoelectronic.

17. Vanadium, chromium, and manganese are elements with atomic numbers 23, 24, and 25, respectively, but have 3, 6, and 5 unpaired electrons respectively in the gaseous atoms. Explain the irregular progression in the number of unpaired electrons.

18. List five physical properties of metals that may be modified by alloying.

19. How may alloys be produced? What are alloys containing mercury called?

20. Contrast the structures of substitutional and interstitial solid solutions.

21. In Section 31.11, mention was made that many (but not all) intermetallic compounds have a ratio of total number of electrons in the outer shell to total number of atoms equal to either $\frac{21}{12}$, $\frac{21}{13}$, or $\frac{21}{14}$. Determine the ratio for each of the following known intermetallic compounds: $CuZn$, Ag_5Zn_8, Cu_3Al, $AgCd_3$, $Na_{31}Pb_8$ (note that lead has four electrons in its outer shell), Cu_9Al_4, Cu_5Sn, $CuBe_3$, Cu_5Sn, Cu_3Ge, Ag_5Al_3, $LiAg$, and $Li_{10}Pb_3$. (Two in the group do not fall into any of the three classifications.)

PROBLEMS

1. How much sulfuric acid could be formed from the sulfur dioxide released from roasting a ton of zinc ore which is 8.5% ZnS?   *Ans. 1.7×10^2 lb*

2. How many pounds of red brass could be made from one ton of copper ore which is 10.0% malachite, $Cu_2(OH)_2CO_3$?   *Ans. 128 lb*

3. Copper forms face-centered cubic crystals and has a density of 8.92 g/cm³. What is the length of the edge of the unit cell cube? What is the distance between nearest neighbor copper atoms?

Ans. 3.62×10^{-8} cm; 2.56×10^{-8} cm

REFERENCES

"Composition and Evolution of the (Earth's) Mantle and Core," D. L. Anderson, C. Sammis, and T. Jordan, *Science*, **171**, 1103 (1971).

"The Magnetic Structure of Superconductors," U. Essmann and H. Träuble, *Sci. American*, March, 1971; p. 75.

"Superplastic Metals," H. W. Hayden, R. C. Gibson, and J. H. Brophy, *Sci. American*, March, 1969; p. 28.

"The Nature of Metals," A. H. Cottrell, *Sci. American*, Sept., 1967; p. 90.

"An Analogy for Elementary Band Theory Concept in Solids," P. F. Weller, *J. Chem. Educ.*, **44**, 391 (1967).

"Chemistry of Alloy Development," J. N. Pratt and S. G. Glover, *Chem. in Britain*, **5**, 207 (1969).

"A Broader View of Close Packing to Include Body-Centered and Simple Cubic Systems," S.-M. Ho and B. E. Douglas, *J. Chem. Educ.*, **45**, 474 (1968).

"Metals: Technical and Economic," T. J. Tarring, *Chem. in Britain*, **5**, 166 (1969).

"Extraction Metallurgy," J. H. E. Jeffes, *Chem. in Britain*, **5,** 189 (1969).

"Production and Use of High-Purity Metals," J. C. Chaston, *Chem. in Britain*, **5,** 224 (1969).

"Chemical Aspects of Dislocations in Solids," J. M. Thomas, *Chem. in Britain*, **6,** (1970).

"Permanent Magnets," J. J. Becker, *Sci. American*, Dec., 1970; p. 92.

"Zone Refining," W. C. Pfann, *Sci. American*, Dec., 1967; p. 62.

"Superconductivity at High Pressure," N. B. Brandt and N. I. Ginsburg, *Sci. American*, April, 1971; p. 83.

"Chemical Reactions and the Composition of Sea Water," K. E. Chave, *J. Chem. Educ.*, **48,** 148 (1971).

"Superconductors for Power Transmission," D. P. Snowden, *Sci. American* April, 1972; p. 84.

"Conduction Electrons in Metals," M. Y. Azbel, M. I. Kaganov, and I. M. Lifshitz, *Sci. American*, Jan., 1973; p. 88.

"Metal-Oxide-Semiconductor Technology," W. C. Hittinger, *Sci. American*, Aug., 1973; p. 48.

"Electron Tunnelling and Superconductivity," I. Giaever, *Science*, **183,** 1253 (1974).

"Close-Packing of Atoms," H. C. Benedict, *J. Chem. Educ.*, **50,** 419 (1973).

"How Productive is the Sea?" I. Morris, *Chem. in Britain*, **10,** 198 (1974).

"Manganese Nodules (I): Mineral Resources on the Deep Sea Bed" and "—(II): Prospects for Deep Sea Mining," A. L. Hammond, *Science*, **183,** 502 and 644 (1974).

"Sodium Tungsten Bronzes: Oxides with Metallic Character," J.-P. Randin, *J. Chem. Educ.*, **51,** 32 (1974).

"The Chemistry of Liquid Metals," C. C. Addison, *Chem. in Britain*, **10,** 332 (1974).

"The Deformation of Metals at High Temperatures," H. J. McQueen and W. J. McG. Tegart, *Sci. American*, April, 1975; p. 116.

Coordination Compounds

32

In other chapters of this text, several examples of compounds and ions are given in which negative groups or neutral polar molecules are attached to metal ions or atoms. This type of compound, known as a **coordination,** or **complex, compound,** has many important applications in such areas as soil treatment, corrosion control, metallurgy, electroplating of metals, water softening, and catalysis. Typical examples of coordination molecules and ions are $[Ag(NH_3)_2]^+$, $[Cu(NH_3)_4]^{2+}$, $[Fe(CN)_6]^{4-}$, $[Fe(CN)_6]^{3-}$, $[Ag(CN)_2]^-$, $[Co(NH_3)_6]^{3+}$, $[Co(NH_3)_3(NO_2)_3]$, $[Pt(NH_3)_2Cl_4]$, and $[Fe(CO)_5]$.

Formation of a complex ion requires the combination of two kinds of species: (1) an ion or molecule which has at least one pair of electrons available for bonding, and (2) a metal ion or atom which has a sufficient attraction for electrons to form a coordinate covalent bond with the attaching group. As would be expected, metallic ions or atoms of small size and high nuclear charge attract electrons most readily and form the most stable complex ions. Ions of the transition metals, inner transition metals, and a few metals near these series in the Periodic Table are especially prone to react in this way.

Enthalpy values are given for several coordination compounds in Appendix J, and, for the ones for which reliable values have been established, Gibbs free energy and entropy values are also given. From these values it can be seen that, although different coordination compounds vary greatly in stability, many are exceedingly stable.

32.1 Definitions of Terms

It is important to understand the meaning of several terms which are commonly used in discussing coordination compounds.

The **coordination number** is the number of electron pairs which an acceptor metal ion attracts when a complex ion is formed. The coordination number for the silver ion in $[Ag(NH_3)_2]^+$ is *two;* that for the copper ion in $[Cu(NH_3)_4]^{2+}$ is *four;* and that for the iron(III) ion in $[Fe(CN)_6]^{3-}$ is *six*. In each of these examples, the coordination number is also equal to the number of coordinating groups attached to the metal ion, but such is not always the case. Some groups, such as ethylenediamine,

$$
\begin{array}{ccccccc}
 & \overset{H}{|} & \overset{H}{|} & \overset{H}{|} & \overset{H}{|} & & \\
H - & N - & C - & C - & N - & H \\
 & \cdot\cdot & \overset{|}{H} & \overset{|}{H} & \cdot\cdot & &
\end{array}
$$

with two donor atoms, furnish two pairs of electrons per molecule. Thus, coordination number for cobalt in $[Co(H_2NCH_2CH_2NH_2)_3]^{3+}$ is *six*. Although only three coordinating molecules are attached to the cobalt ion, *six electron pairs* are utilized in the bonding of the three groups to the cobalt. The most common coordination numbers are 2, 4, and 6, but in some complex molecules or ions coordination numbers of 3, 5, 7, or 8 occur, and sometimes (though much less commonly) 9, 10, 11, and 12. The coordination number is at times twice the oxidation state of the acceptor metal, but it is important to realize that this is not always the case. For example, $[Fe(CN)_6]^{4-}$ has a coordination number of *six*, whereas the oxidation state of iron in the complex is $+2$.

The groups attached to the metal ion, often called the **central metal ion,** are referred to as **coordinating groups,** or **ligands.** These may be either ions or neutral molecules. Within the coordinating group, the atom which is attached directly to the metal is called the **donor atom.**

The **coordination sphere,** which is usually enclosed in brackets in the formula, includes the central metal ion plus the coordinating groups (ligands) attached to it.

When one molecule or ion of a ligand attaches itself, by the use of two or more electron pairs, to more than one coordination position of the central metal ion, it is referred to as a **chelating group.** The resulting complex is referred to as a metal chelate; examples are

$$
\left[Co \left(\begin{array}{c} H_2N - CH_2 \\ | \\ H_2N - CH_2 \end{array} \right)_3 \right]^{3+}, \quad \left[Co \left(\begin{array}{c} O - C = O \\ | \\ O - C = O \end{array} \right)_3 \right]^{3-}, \text{ and } \left[Cu \left(\begin{array}{c} H_2N - CH_2 \\ | \\ O - C = O \end{array} \right)_2 \right]^{0}
$$

The **effective atomic number** of a metal in a complex is the total number of electrons in the orbitals of the central metal ion after coordination takes place. It is interesting to note that the total electronic configuration of the central ion in many complexes corresponds to that of a noble gas, a fact which may be related to the great stability of such complexes. For example, in $[Co(NH_3)_6]^{3+}$ the cobalt(III) ion, which has 24 electrons (27 − 3), obtains 12 additional electrons (two from each $:NH_3$) to make a total of 36 electrons in the atomic orbitals of the cobalt. Thus, **36** is the "effective atomic number" for the $[Co(NH_3)_6]^{3+}$ ion and is equal to the number of electrons possessed by (and hence the atomic number of) the noble gas, krypton. Other examples are as follows:

Complex Ion	*Effective Atomic Number*	*Noble Gas Structure*
$[Fe(CN)_6]^{4-}$	$26 - 2 + 12 = 36$	Krypton
$[Cd(NH_3)_4]^{2+}$	$48 - 2 + 8 = 54$	Xenon
$[PtCl_6]^{2-}$	$78 - 4 + 12 = 86$	Radon

It should be emphasized, however, that an effective atomic number equal to the atomic number of a noble gas is not required for stability, as is shown by the following examples:

Complex Ion	*Effective Atomic Number*
$[Fe(CN)_6]^{3-}$	$26 - 3 + 12 = 25$
$[Cr(NH_3)_6]^{3+}$	$24 - 3 + 12 = 33$
$[Ni(NH_3)_6]^{2+}$	$28 - 2 + 12 = 38$

The manner in which the pairs of electrons from the ligands enter the electron orbitals of the metal ion and the molecular orbital picture of complexes will be discussed later in this chapter.

32.2   The Naming of Complex Compounds

The nomenclature of coordination compounds is patterned after a system originally suggested by Alfred Werner, a Swiss chemist, whose outstanding work in the latter part of the nineteenth century and early part of the twentieth century laid the foundation for a clearer understanding of complex ions and coordination compounds. The following rules for naming complexes will be useful.

(1) The cation is named first and then the anion, in accord with usual nomenclature rules.

(2) In naming the coordination sphere, whether it be a cation, an anion, or a neutral molecule, the ligands are named first and then the central metal.

(3) The preferred order of indicating the ligands within the coordination sphere is to name the ligands alphabetically. A former, but still much used order, is to name negative ligands alphabetically first, neutral ligands alphabetically next, and positive ligands alphabetically last.

The prefixes, *di-*, *tri-*, *tetra-*, etc. (or sometimes *bis-*, *tris-*, *tetrakis-*, etc., for certain larger ligands), are used with the name of the ligand when more than one ligand of a particular kind occurs within the coordination sphere. The names used for some common ligands are: fluoro (F^-), chloro (Cl^-), bromo (Br^-), iodo (I^-), cyano (CN^-), nitro (NO_2^-), nitrito (ONO^-), nitrato (NO_3^-), hydroxo (OH^-), oxo (O^{2-}), amido (NH_2^-), oxalato ($C_2O_4^{2-}$), carbonato (CO_3^{2-}), ammine (NH_3), and aqua (H_2O).

(4) When the coordination sphere is either a cation or a neutral molecule, the name of the central metal remains intact, followed by a Roman numeral designation in parentheses to indicate the oxidation state of the metal. When the coordination sphere is an anion, the ending *-ate* is added to the stem for the name of the central metal (or sometimes to the stem of the Latin name for the metal), followed by the Roman numeral designation of the oxidation state of the metal.

Examples in which the coordination sphere is a cation:

$[Co(NH_3)_6]Cl_3$	Hexaamminecobalt(III) chloride
$[Pt(NH_3)_4Cl_2]^{2+}$	Tetraamminedichloroplatinum(IV) ion
$[Ag(NH_3)_2]^+$	Diamminesilver(I) ion
$[Cr(H_2O)_4Cl_2]Cl$	Tetraaquadichlorochromium(III) chloride
$[Co(H_2NCH_2CH_2NH_2)_3]_2(SO_4)_3$	Tris(ethylenediamine)cobalt(III) sulfate

Examples in which the coordination sphere is neutral:

$[Pt(NH_3)_2Cl_4]$	Diamminetetrachloroplatinum(IV)
$[Co(NH_3)_3(NO_2)_3]$	Triamminetrinitrocobalt(III)

Examples in which the coordination sphere is an anion:

$K_3[Co(NO_2)_6]$	Potassium hexanitrocobaltate(III)
$[PtCl_6]^{2-}$	Hexachloroplatinate(IV) ion
$Na_2[SnCl_6]$	Sodium hexachlorostannate(IV)

32.3   The Structural Chemistry of Complexes

In Chapter 6, we studied the structures of a considerable number of simple compounds and ions. We noted that many spatial configurations of the atoms are possible, some of these being linear, trigonal planar, tetrahedral, trigonal bipyramidal, and octahedral. In subsequent chapters, we have noted examples of other kinds of structures, including square planar and square pyramidal. We are now ready to extend our study to the structures of coordination molecules and ions. It will be helpful for you to reread Chapter 6 to refresh your memory of the principles and structures discussed there. You will find that many of the concepts discussed in Chapter 6 for simple compounds apply also to coordination compounds.

In 1893, Alfred Werner presented a concept of coordination compounds that laid the foundation for the further study of these substances. He suggested, as one part of his theory, that when ions or polar molecules are coordinated to a metal ion they are arranged in a definite geometrical pattern about the metal ion. With this concept he was able to account for the properties of hydrates, ammonates, and various double salts of the types $CoCl_2 \cdot 6H_2O$, $CoCl_3 \cdot 6NH_3$, and $SnCl_4 \cdot 2NaCl$, respectively. Werner assigned the formulas $[Co(H_2O)_6]Cl_2$, $[Co(NH_3)_6]Cl_3$, and $Na_2[SnCl_6]$ to these compounds and pointed out that the properties of the complexes $[Co(H_2O)_6]^{2+}$, $[Co(NH_3)_6]^{3+}$, and $[SnCl_6]^{2-}$ could be

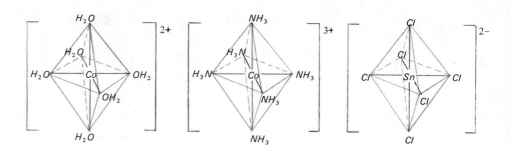

explained by postulating that the six coordinated groups are arranged about the central ion at the corners of a regular octahedron (Fig. 32–1). In such an octahedral structure, the six pairs of electrons from the six ligands may be considered as entering the orbitals of the metal ion. An isolated cobalt atom in its normal state has the electron configuration $1s^2\,2s^2\,2p^6\,3s^2\,3p^6\,3d^7\,4s^2$ (see Section 3.18). Each atomic orbital can accommodate two electrons of opposing spin. As we learned earlier (Section 3.17), the electrons enter each orbital of a given type singly before any pairing of electrons occurs within those orbitals—Hund's Rule. The orbitals for the cobalt atom, each orbital represented by a circle, can be diagrammed as follows:

Co atom:

$1s$	$2s$	$2p$	$3s$	$3p$	$3d$	$4s$
⑪	⑪	⑪⑪⑪	⑪	⑪⑪⑪	⑪⑪①①①	①

Measurement of magnetic moment for the complex indicates three unpaired electrons, in accord with the above diagram.

When the cobalt(III) ion, Co^{3+}, is formed, three electrons are lost, resulting in a structure possessing four unpaired electrons.

Co^{3+}:

$1s$	$2s$	$2p$	$3s$	$3p$	$3d$	$4s$
⑪	⑪	⑪⑪⑪	⑪	⑪⑪⑪	⑪①①①①	○

All 6-coordinate complexes, for example, $[Co(NH_3)_6]^{3+}$, $[SnCl_6]^{2-}$, $[Co(H_2O)_6]^{2+}$, $[Co(CN)_6]^{3-}$, $[Fe(CN)_6]^{4-}$, and $[Fe(CN)_6]^{3-}$, show d^2sp^3 hybridization of bonding orbitals (or sometimes sp^3d^2 hybridization, as discussed in a later chapter in connection with $[Zn(NH_3)_6]^{2+}$), and nearly all of them have the octahedral configuration.

The 6-coordinate complexes of cobalt(II) contain one more electron than those of cobalt(III). To permit d^2sp^3 bonding, this extra electron must be promoted to a $4d$ orbital (or perhaps $5s$), where it is loosely held. This electron is readily removed, as indicated by the ease with which most Co(II) complexes are oxidized to Co(III) complexes.

$[Co(NH_3)_6]^{3+}$ (no unpaired electrons):

$1s$	$2s$	$2p$	$3s$	$3p$	$3d$	$4s$	$4p$

⟨↑↓⟩ ⟨↑↓⟩ ⟨↑↓⟩⟨↑↓⟩⟨↑↓⟩ ⟨↑↓⟩ ⟨↑↓⟩⟨↑↓⟩⟨↑↓⟩ ⟨↑↓⟩⟨↑↓⟩⟨↑↓⟩ | ●● ●● ● ●●● |

d^2sp^3 hybridization

Octahedral coordination
orbitals

When six ammonia molecules bond to the cobalt(III) ion, six pairs of electrons are available for sharing with the cobalt. The postulation usually made is that the electrons which are already present in the $3d$ orbitals of the metal pair up to allow the incoming electrons from the ammonia (indicated in color) to remain paired. It should be noted, however, that once the structure is formed it is impossible to distinguish the source of the electrons, inasmuch as (so far as we know) all electrons are identical regardless of their origin. The six bonds to the central ion arise from the hybridization of two d, one s, and three p orbitals; this is spoken of as d^2sp^3 hybridization. In this notation [in contrast to the notation used for designating electron structures of atoms (see Section 3.17) in which the superscripts refer to the number of electrons] the superscripts here refer to the number of *orbitals* of each type involved in the bonding.

$[Co(CN)_6]^{4-}$ (one unpaired electron):

$1s$	$2s$	$2p$	$3s$	$3p$	$3d$	$4s$	$4p$	$4d$

⟨↑↓⟩ ⟨↑↓⟩ ⟨↑↓⟩⟨↑↓⟩⟨↑↓⟩ ⟨↑↓⟩ ⟨↑↓⟩⟨↑↓⟩⟨↑↓⟩ ⟨↑↓⟩⟨↑↓⟩⟨↑↓⟩ | ●● ●● ● ●●● | ⟨↑⟩

d^2sp^3 hybridization

Octahedral coordination
orbitals

Although good octahedral diagrams are rather difficult to draw without considerable practice, you will find it helpful in your study to be able to draw these structures readily. Sometimes, an abbreviated structure is drawn as in Fig. 32–2(a). However, a more realistic octahedron can be drawn by following the steps outlined in Fig. 32–2(b).

FIGURE 32-2

Steps in drawing an octahedron.

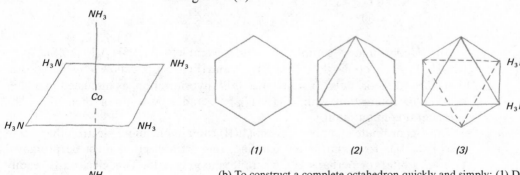

(1)   (2)   (3)   (4)

(a) An abbreviated drawing of an octahedron.

(b) To construct a complete octahedron quickly and simply: (1) Draw a regular hexagon. (2) Draw a triangle inside the hexagon. (3) Draw an upside-down triangle with dashed lines inside the hexagon and back of the triangle. (4) Add symbols for central metal and ligands.

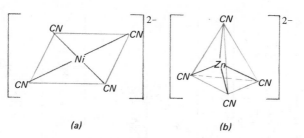

FIGURE 32-3
(a) The planar configuration of the $[Ni(CN)_4]^{2-}$ ion, and (b) the tetrahedral configuration of the $[Zn(CN)_4]^{2-}$ ion.

(a)   (b)

Complexes in which the metal shows a coordination number of four exist in two different geometric arrangements, the planar and the tetrahedral configurations. Examples of 4-coordinate complex ions with the planar configuration are $[Ni(CN)_4]^{2-}$ (Fig. 32-3a) and $[Cu(NH_3)_4]^{2+}$, and with the tetrahedral configuration are $[Zn(CN)_4]^{2-}$ (Fig. 32-3b) and $[Zn(NH_3)_4]^{2+}$.

In the case of 4-coordinate structures with a planar configuration, the four bonds of the central atom arise from dsp^2 hybridization, as illustrated for $[Ni(CN)_4]^{2-}$.

Ni atom (two unpaired electrons):

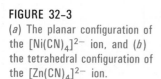

Ni^{2+} (two unpaired electrons):

$[Ni(CN)_4]^{2-}$ (no unpaired electrons):

dsp^2 hybridization

Planar coordination orbitals

For 4-coordinate complexes with a tetrahedral structure, the four bonds of the central atom arise from the hybridization of one s and three p orbitals; this is referred to as sp^3 hybridization. This is analogous to filling the carbon orbitals in methane or carbon tetrachloride to produce a tetrahedral sp^3 configuration (see Section 6.5), except that in the present instance coordinate covalent bonds are being formed with both electrons of each shared pair coming from the ligand. The tetrahedral $[Zn(CN)_4]^{2-}$ complex is represented diagrammatically as follows:

$[Zn(CN)_4]^{2-}$ (no unpaired electrons):

sp^3 hybridization

Tetrahedral coordination orbitals

Many other geometric figures are also possible. Table 32-1 shows several of the known types, with examples selected from other kinds of compounds and ions as well as coordination compounds and ions.

TABLE 32-1 Geometric Shapes and Corresponding Hybridization of Orbitals for Several Compounds and Ions

Geometric Shape	Diagram of Shape	Coordination Number	Hybridization	Examples
Straight line linear		2	sp	$[Ag(NH_3)_2]^+$, $BeCl_2$
Angular		2	p^2	H_2O, SO_2
Trigonal planar	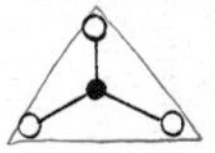	3	sp^2	BF_3, NO_3^-, CO_3^{2-}
Tetrahedral	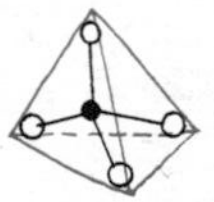	4	sp^3	$[Zn(CN)_4]^{2-}$, CH_4, alternative structure for H_2O
Square planar	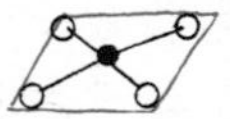	4	dsp^2	$[Ni(CN)_4]^{2-}$, $[Pt(NH_3)_4]^{2+}$
Square pyramidal	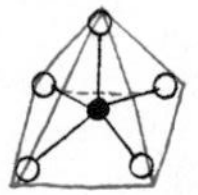	5	d^2sp^2 (or d^4s)	$[Ni(Br)_3\{(C_2H_5)_3P\}_2]$
Trigonal bipyramidal	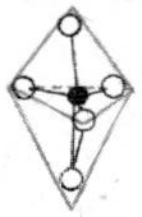	5	dsp^3 (or d^3sp)	PF_5, $[Fe(CO)_5]$
Octahedral	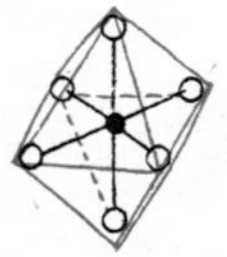	6	d^2sp^3	$[Co(NH_3)_6]^{3+}$

32.4   Isomerism in Complexes

Certain complexes, such as $[Co(NH_3)_4Cl_2]^+$—the tetraamminedichlorocobalt(III) ion, have more than one form. These different forms of the substance, possessing the same formula, are referred to as **isomers** (see also Chapter 27, *Carbon and Its Compounds*). Isomers that differ *only* in the way the atoms are oriented in space relative to each other are called **stereoisomers.** The $[Co(NH_3)_4Cl_2]^+$ ion has two isomers, one of which is violet and the other green. The violet form has been shown by crystal structure analysis, using x-ray methods and other experimental techniques, to have the *cis* configuration (the chloride ions on adjacent corners of the octahedron), and the green form to have the *trans* configuration (the chloride ions on opposite corners), as shown in Fig. 32–4.

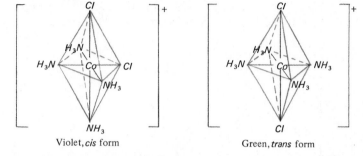

Violet, *cis* form   Green, *trans* form

FIGURE 32–4
The *cis* and *trans* isomers of $[Co(NH_3)_4Cl_2]^+$.

A complex such as $[Cr(NH_3)_2(H_2O)_2(Br)_2]^+$ has a variety of different isomeric forms. Each coordinating group can be *trans* to one like it:

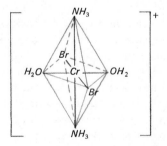

The ammonia molecules can be *trans* to each other but the water molecules and bromine atoms *cis* to groups like themselves:

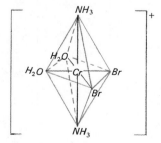

Similarly, the water molecules or the bromine atoms can be *trans* to each other with the other groups *cis* to groups like themselves:

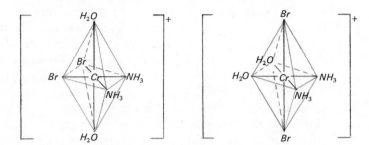

Finally, each group can be *cis* to one like itself. In this latter case, it can be shown that two different arrangements are possible:

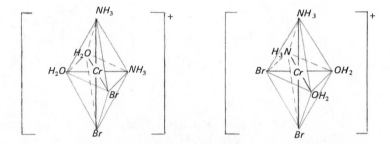

These two forms of the complex $[Cr(NH_3)_2(H_2O)_2(Br_2)]^+$ are mirror images of each other, but they are not superimposable on each other and therefore are not identical. They are called **optical isomers,** a term applied to isomers which are mirror images of each other but not identical. Diagrams for mirror images of the other isomers of $[Cr(NH_3)_2(H_2O)_2(Br)_2]^+$ can be drawn, but it can be shown in each case that the additional form can be superimposed upon the one for which it is a mirror image and is therefore identical with it. For example, the following mirror image forms are actually identical to each other, inasmuch as either may be turned 180° on an axis through the two corners occupied by the ammonia molecules to superimpose it upon the other.

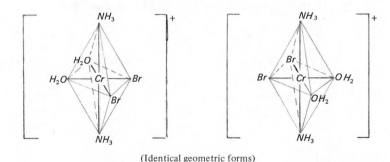

(Identical geometric forms)

The $[Cr(NH_3)_2(H_2O)_2(Br)_2]^+$ complex, thus, has a total of six stereoisomers, two of which are optical isomers of each other.

The tris(ethylenediamine)cobalt(III) ion, $[Co(H_2NCH_2CH_2NH_2)_3]^{3+}$, has two optical isomers, as shown in Fig. 32–5 (see also abbreviated form, where en = $H_2NCH_2CH_2NH_2$).

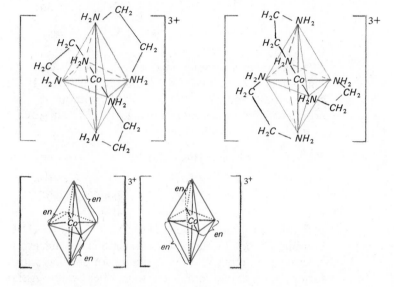

FIGURE 32–5a
Optical isomers of
$[Co(H_2NCH_2CH_2NH_2)_3]^{3+}$.

FIGURE 32–5b
Optical isomers of
$[Co(H_2NCH_2CH_2NH_2)_3]^{3+}$,
abbreviated formulas.
(en = $H_2NCH_2CH_2NH_2$)

The $[Co(en)_2Cl_2]^+$ has two *cis* isomers, which are optical isomers of each other, and one *trans* isomer. The *trans* form is symmetrical and has no possible optical isomerism. Its mirror image is superimposable on it and is therefore identical to the original *trans* form. The three isomers are shown in Fig. 32–6.

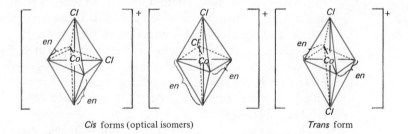

FIGURE 32–6
The three isomeric forms of
$[Co(en)_2Cl_2]^+$.

Cis forms (optical isomers)   *Trans* form

32.5   The Crystal Field Theory

The discussion of complexes, thus far, has assumed that the bonding of the ligands to the central metal ion in complex ions and compounds is primarily covalent. This assumption has been used for many years and explains a number of the properties of complexes reasonably well. Some facts, however, cannot be explained satisfactorily assuming covalent bonding. For example, measurements of magnetic moments for a group of iron(II) complexes indicate that $[Fe(CN)_6]^{4-}$ has no unpaired electrons, whereas $[Fe(H_2O)_6]^{2+}$ has four unpaired electrons. On the assumption of covalent bonding, no unpaired electrons would be predicted for either complex, and indeed $[Fe(CN)_6]^{4-}$ does not contradict the assumption.

$[Fe(CN)_6]^{4-}$:

$1s$	$2s$	$2p$	$3s$	$3p$	$3d$	$4s$	$4p$

However, a contradiction does arise in the case of the $[Fe(H_2O)_6]^{2+}$ ion, inasmuch as it does have four unpaired electrons. If $[Fe(H_2O)_6]^{2+}$ is assumed to be essentially ionically bonded, that is, without the ligand electrons entering the orbitals of the ion through sharing, the four unpaired electrons are justified, just as in the simple iron(II) ion.

Fe^{2+} and $[Fe(H_2O)_6]^{2+}$:

$1s$	$2s$	$2p$	$3s$	$3p$	$3d$	$4s$	$4p$

A differentiation between $[Fe(CN)_6]^{4-}$ and $[Fe(H_2O)_6]^{2+}$, in terms of type of bonding, is not unreasonable, inasmuch as the cyanide ion usually makes an electron pair more readily available for sharing with the central metal ion and forms more stable complexes than does water. Although the concept of ionic bonding explains some facts such as magnetic behavior more satisfactorily than does covalent bonding, it does not adequately explain certain other properties of complexes such as their structural configurations.

An alternative explanation which provides a useful modification of the concept of ionic or electrostatic bonding has proved to be of much value in explaining the various properties of complexes. This theory is referred to as the **Crystal Field Theory,** so named because some of the ideas used in the theory derive from principles formulated in connection with crystals.

In Section 3.20, the atomic orbitals were pictured for s, p, and d orbitals. It will be recalled that the orbital of an s electron is spherical, whereas that of a p electron is a dumbbell shape, and that the three p orbitals for a given major energy level are oriented at right angles to each other along the x, y, and z axes (see Fig. 3–19).

The d orbitals, which occur in sets of five, each consist of lobe-shaped regions and are arranged in space as shown in Fig. 3–20 and reproduced within an octahedral structure in Fig. 32–7.

As shown in Fig. 32–7, the lobes in two of the five orbitals point toward corners of the octahedral configuration for the metal. These orbitals are referred to as **antibonding orbitals.** Any electrons which occupy lobes of these orbitals will be repelled by electron pairs belonging to ligand groups at the corners of the octahedron. The other three orbitals, whose lobes point in between the corners of the octahedron, are called **nonbonding orbitals.** Electrons in the nonbonding

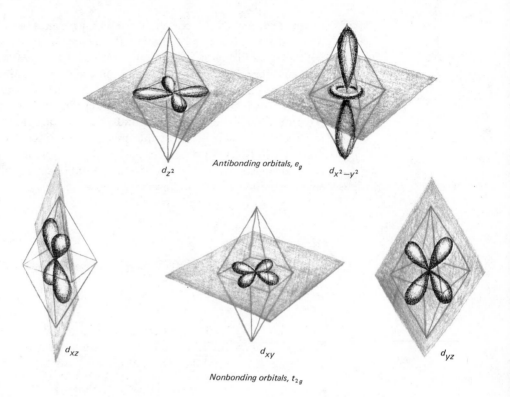

Antibonding orbitals, e_g

d_{z^2} $d_{x^2-y^2}$

d_{xz} d_{xy} d_{yz}

Nonbonding orbitals, t_{2g}

FIGURE 32-7

Diagrams showing the directional characteristics of the five *d* orbitals.

orbitals are not directly affected by the electron pairs of approaching ligands. The one or two electrons that can occupy each orbital may be any place within these regions and over a period of time will effectively occupy all the lobes of the orbital.

In a simple metal ion, the electrons will be distributed among the five 3*d* orbitals in accord with Hund's Rule, inasmuch as the electrons all have the same energy.

Ion	3d Orbitals	Ion	3d Orbitals
Ti^{3+}	⬆⚪⚪⚪⚪	Fe^{2+}	⬆⬇⬆⬆⬆
V^{3+}	⬆⬆⚪⚪⚪	Co^{3+}	⬆⬇⬆⬆⬆
Cr^{3+}	⬆⬆⬆⚪⚪	Co^{2+}	⬆⬇⬆⬆⬆
Cr^{2+}	⬆⬆⬆⬆⚪	Ni^{2+}	⬆⬇⬆⬇⬆
Mn^{2+}	⬆⬆⬆⬆⬆	Cu^{2+}	⬆⬇⬆⬇⬆
Fe^{3+}	⬆⬆⬆⬆⬆	Zn^{2+}	⬆⬇⬆⬇⬆

If a ligand with a free electron pair approaches a metal ion with that electron pair pointing toward a corner of the octahedral structure of the metal ion, two opposing forces are set up. One force tends to keep the electrons distributed within all the 3*d* orbitals according to Hund's Rule. The other force tends to concentrate the electrons in the nonbonding orbitals. This second force arises because, as the ligand approaches the corner of the octahedron, the electron pair of the ligand repels the $3d$ electrons of the metal, causing them to seek nonbonding orbitals with lobes pointing in between the corners of the octahedron. Thus, the electrons in the nonbonding orbitals become somewhat lower in energy than the electrons in the antibonding orbitals. Some ligands, such as water and the fluoride ion, produce weak fields and have very little tendency to repel the $3d$ electrons of the central ion, whereas other ligands, such as the cyanide ion, set up strong fields in which the $3d$ electrons of the metal are significantly repelled.

Hence, in $[Fe(CN)_6]^{4-}$ the strong field of the six cyanide ions concentrates the six $3d$ electrons of the iron into the three nonbonding orbitals, leaving no unpaired electrons. This result is in agreement with the experimentally measured magnetic moment (no unpaired electrons).

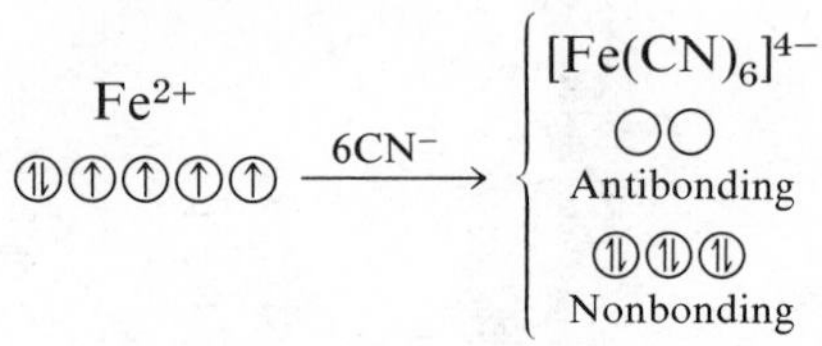

In $[Fe(H_2O)_6]^{2+}$, on the other hand, with the weak field of the water molecules, Hund's Rule predominates over the slight tendency toward repulsion by the incoming ligand, and the electrons remain distributed through all five $3d$ orbitals.

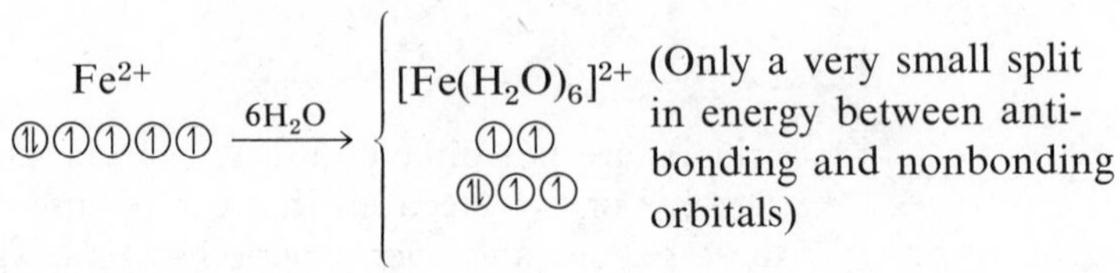

Thus, four unpaired electrons would be present, in agreement with the magnetic moment measurement, in Fe^{2+} and in $[Fe(H_2O)_6]^{2+}$. A similar line of reasoning can be followed to show why in complexes for the $+3$ oxidation state of iron the $[Fe(CN)_6]^{3-}$ has only one unpaired electron, whereas $[Fe(H_2O)_6]^{3+}$ and $[FeF_6]^{3-}$ possess five unpaired electrons each.

The difference in energy between the energy levels of the antibonding and nonbonding orbitals for a given complex is commonly designated Δ (or sometimes 10Dq). The Δ value can be estimated from spectroscopic experiments which measure the energies involved in the movement of electrons (referred to as **electronic transitions**) from one of the energy levels to the other. These spectroscopic experiments provide a measure of the relative strengths of the fields produced by various ligands. Several of the more common ligands are thereby shown to fall in the following order in terms of increasing field strength:

$$I^- < Br^- < Cl^- < F^- < H_2O < C_2O_4{}^{2-} < NH_3 < \text{Ethylenediamine} < NO_2{}^- < CN^-$$

Increasing field strength $\longrightarrow$

As might be expected, ligands that result in fields of strength between those produced by cyanide ion and those produced by water form complexes with magnetic moments intermediate between those of $[Fe(CN)_6]^{4-}$ and $[Fe(H_2O)_6]^{2+}$.

Actually, as pointed out in Section 4.7, there is no sharp dividing line between covalent and ionic bonding. The Molecular Orbital Theory, discussed in Section 32.7, seeks to introduce a covalent component into the ionic viewpoint.

A study of the structures of complexes points out that the explanations of chemical processes are never complete and are constantly open to modification as more information is obtained through research. Thus, one is constantly reminded in the study of science that there is a continuing need for research and that previous research, impressive as it is, has scarcely scratched the surface of that which is possible for the future. It is quite possible that some of the present readers of this book will, after further training, be among those contributing to a more complete knowledge of this subject or some other area of chemistry.

32.6   Crystal Field Stabilization Energy

As we have said, the difference in energy between the antibonding and the nonbonding energy levels is commonly designated either Δ or 10Dq. The two levels are frequently referred to as the e_g and t_{2g} levels, respectively.

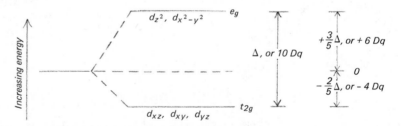

Since the splitting of the five orbitals into two energy levels does not change the total energy, the zero line on the energy axis is the weighted average energy of the d orbitals, thus making the e_g and t_{2g} energies, $+\frac{3}{5}\Delta$ and $-\frac{2}{5}\Delta$, respectively (or +6Dq and −4Dq, respectively). Hence,

$$(\tfrac{3}{5}\Delta)2 + (-\tfrac{2}{5}\Delta)3 = 0$$

or $$(\tfrac{3}{5} \times 10Dq)2 + (-\tfrac{2}{5} \times 10Dq)3 = (+6Dq)2 + (-4Dq)3 = 0$$

For each complex, it is possible to calculate a value, referred to as the **crystal field stabilization energy** (CFSE), which is related to the stability of the complex. Suppose, for example, that the metal ion is the vanadium(III) ion, which has two d electrons (referred to as d^2). The two electrons will be in two of the orbitals of the lower energy t_{2g} level, whether the vanadium is coordinated to a ligand producing a weak field or to one producing a strong field. In calculating the crystal field stabilization energy, we take account of the fact that there are two electrons of energy −4Dq relative to the zero energy line.

$$CFSE = 2(-4Dq) = -8Dq$$

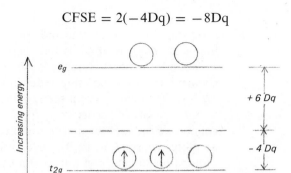

Now suppose that the metal ion is the manganese(II) ion, which has five d electrons (d^5). In this ion, these electrons will be distributed evenly throughout the five d orbitals in accord with Hund's Rule. If the ion is coordinated to a ligand producing a weak field, the five electrons will remain distributed one in each of the five orbitals; i.e., three unpaired electrons will be at the t_{2g} level and two unpaired electrons will be at the e_g level.

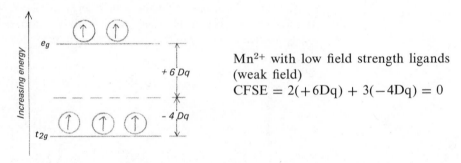

Mn^{2+} with low field strength ligands
(weak field)
CFSE = $2(+6Dq) + 3(-4Dq) = 0$

The two electrons in the e_g level will be at the higher energy level, corresponding to +6Dq, and three at the lower energy level, corresponding to −4Dq. Hence, the crystal field stabilization energy is zero in this case.

$$CFSE = 2(+6Dq) + 3(-4Dq) = 0$$

If, however, the Mn^{2+} ion is coordinated to ligands of high field strength, all five electrons will be forced into the lower energy nonbonding t_{2g} level and none will be at the higher energy e_g level.

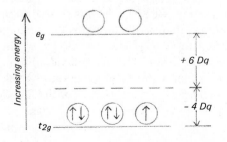

Mn^{2+} with high field strength ligands
(strong field)
CFSE = $5(-4Dq) = -20Dq$

The crystal field stabilization energy for Mn^{2+} in this case is $-20Dq$.

$$CFSE = 5(-4Dq) = -20Dq$$

In general, the more negative the value of CFSE, the more stable will be the complex. As we would expect, a complex of Mn^{2+} with ligands of high field strength is more stable than one with ligands of low field strength. Sometimes, the terms **low spin** and **spin-paired** are applied to complexes of a given metal with ligands of high field strength; conversely, the terms **high spin** and **spin-free** are sometimes used for the complexes with ligands of low field strength.

In the weak field case for Mn^{2+}, the splitting of the two energy levels (10Dq) is much smaller than for the strong field case. One can say that the possible gain in energy which would result if the fourth and fifth electrons were in the lower energy level, t_{2g}, is not sufficient to provide the energy necessary to overcome the repulsion of the electrons for each other and to pair them up. Hence, the fourth and fifth electrons go into the higher energy level, e_g, in preference to pairing.

Table 32–2 shows the crystal field stabilization energy for a variety of octahedral complexes of strong and weak fields.

TABLE 32-2   Crystal Field Stabilization Energy [CFSE] of Octahedral Complexes*

Configuration	Examples	Strong Field (Low spin)				Weak Field (High spin)			
		t_{2g}	e_g	Number of Unpaired Electrons	CFSE $(-Dq)$	t_{2g}	e_g	Number of Unpaired Electrons	CFSE $(-Dq)$
d^0	Ca^{2+}, Sc^{3+}	0	0	0	0	0	0	0	0
d^1	Ti^{3+}	1	0	1	4	1	0	1	4
d^2	V^{3+}	2	0	2	8	2	0	2	8
d^3	Cr^{3+}, V^{2+}	3	0	3	12	3	0	3	12
d^4	Cr^{2+}, Mn^{3+}	4	0	2	16	3	1	4	6
d^5	Mn^{2+}, Fe^{3+}	5	0	1	20	3	2	5	0
d^6	Fe^{2+}, Co^{3+}	6	0	0	24	4	2	4	4
d^7	Co^{2+}	6	1	1	18	5	2	3	8
d^8	Ni^{2+}	6	2	2	12	6	2	2	12
d^9	Cu^{2+}	6	3	1	6	6	3	1	6
d^{10}	Cu^+, Zn^{2+}	6	4	0	0	6	4	0	0

*Used by permission. From *Concepts and Models of Inorganic Chemistry*, B. E. Douglas and D. H. McDaniel, Ginn and Co. Blaisdell Division, New York, 1965; p. 351.

The calculated CFSE values in the table can be experimentally verified by spectroscopic methods. The values for CFSE in the table are given in $-Dq$ units; hence, $-6Dq$ is listed as 6, $-20Dq$ is listed as 20, etc. On this basis, a larger number indicates that more energy is lost in formation of the complex and that, in general, the complex is more stable for a given ligand. One must bear in mind, however, that each ligand has its own specific 10Dq value, and this must be taken into account in evaluating the stabilities of complexes with different ligands. Weak field ligands cause less splitting of the energy levels and have smaller 10Dq values, in general, than strong field ligands.

The largest value for CFSE in the table for strong field complexes is for the d^6 complexes, inasmuch as the lower energy t_{2g} orbitals are completely occupied and the higher energy e_g orbitals are empty. The largest CFSE values for weak field complexes are for the d^3 and d^8 cases. In d^3 complexes the lower energy t_{2g} orbitals are each singly occupied, and the higher energy e_g orbitals are empty. The d^8 complexes represent the lowest number of electrons for which the lower energy t_{2g} orbitals are completely filled; however, inasmuch as two electrons are present in the higher energy e_g orbitals, the highest CFSE value for weak field complexes (12) is lower than that for strong field complexes (24). Note also that the CFSE is zero, indicating very low complex stability, at two places in the table for strong field complexes, i.e., in the cases when no d electrons are present and when ten d electrons are present. In the latter case, all nonbonding and antibonding orbitals are filled, so that the stability gained in the complexing process by the electrons going to the lower energy nonbonding orbitals is exactly counterbalanced by the stability lost by electrons going to the higher energy antibonding orbitals. For weak field complexes, the d^0 and d^{10} cases have CFSE values of zero, and in addition the d^5 complexes, in which all bonding t_{2g} and antibonding e_g orbitals are singly occupied, have a CFSE value of zero.

32.7   The Molecular Orbital Theory

We have seen that a bonding theory based primarily on covalent bonding is especially useful in considering the structures of metal complexes, and that a bonding theory based primarily on ionic bonding is particularly useful in explaining magnetic properties and the energies involved in the electron transitions which give rise to the spectra of complexes. However, neither of these theories is adequate to explain certain other properties of complexes. Inasmuch as most bonds have both ionic and covalent character, an improvement in the theory of bonding in complexes can logically be made by introducing some contributions of covalent bonding into the Crystal Field Theory (ionic) by utilizing molecular orbital concepts. In Chapter 5 we discussed the Molecular Orbital Theory and in later chapters applied the theory to a variety of substances. It is an especially useful concept in understanding the nature of the bonding in coordination compounds.

The molecular orbital energy diagrams for complexes are the same in principle as those shown in Chapter 5 for diatomic species. The ones for coordination compounds are, of course, more complicated, inasmuch as several atoms are involved instead of just two, and a larger number of electrons must be considered. The molecular orbital energy diagrams for $[CoF_6]^{3-}$, possessing weak field ligands, and for $[Co(NH_3)_6]^{3+}$, possessing relatively strong field ligands, are shown in Fig. 32–8. Only the six d electrons for the Co^{3+} and the free pair of electrons for each of the six ligands (a total of 18 electrons) are indicated.

The arrangement and symbols used in Fig. 32–8 are, in general, those described in Chapter 5. In addition, the terms ($t_{1u}, a_{1g}, e_g, t_{2g}$, etc.) alongside the circles

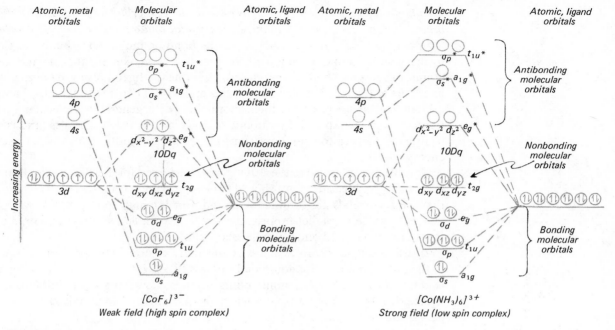

FIGURE 32-8
Molecular orbital energy diagrams for $[CoF_6]^{3-}$ and $[Co(NH_3)_6]^{3+}$.

(orbitals) are commonly used to describe the various energy levels. In some cases, the orbital description ($d_{x^2-y^2}$, d_{z^2}, d_{xy}, etc.) corresponding to the directional characteristics of the orbitals, as provided in Fig. 3–20, are also given. It can be noted that the antibonding and nonbonding energy levels discussed in Sections 32.5 and 32.6 retain the e_g and t_{2g} designations, respectively, in the molecular orbital energy level diagram. The energy difference between them is the Δ, or 10Dq, value referred to in Sections 32.5 and 32.6.

An inspection of Fig. 32–8 shows that the ligand atomic orbitals are lower in energy than are the corresponding metal atomic orbitals. Hence, as discussed in Section 5.13, this is indicative of some ionic character; that is, some electronic charge has been transferred from the metal to the ligands.

Two of the d orbitals (the e_g orbitals: $d_{x^2-y^2}$ and d_{z^2}), are directed, as indicated in Section 32.5, toward the corners of the octahedron where the ligands are located. In addition, the one $4s$ atomic orbital and the three $4p$ orbitals of the metal are also so oriented with respect to the octahedron corners. Hence, metal atomic orbital overlap with ligand atomic orbitals occurs, which results in the formation of six antibonding and six bonding molecular orbitals. The other three d orbitals (the t_{2g} orbitals: d_{xy}, d_{xz}, and d_{yz}) do not point toward the ligand orbitals and hence are not involved in σ bonding; these molecular orbitals are called nonbonding orbitals, the same term used in speaking of atomic orbitals (Sections 32.5 and 32.6). In complexes, therefore, a key observation is that the t_{2g} *d orbitals* *are unaffected by σ bonding, whereas each of the e_g d orbitals combines with ligand orbitals to give an antibonding molecular orbital and a bonding molecular orbital.*

In both $[CoF_6]^{3-}$ and $[Co(NH_3)_6]^{3+}$, the bonding molecular orbitals are closer to the ligand atomic orbitals with respect to energy content than to the metal atomic orbitals and hence most closely resemble the ligand orbitals. On the other hand, the antibonding molecular orbitals are closer in energy to the metal atomic orbitals and hence resemble them more closely. In general, a molecular orbital formed from overlap with two atomic orbitals most closely resembles (and is said to receive a larger contribution from) the atomic orbital to which it is closer in energy.

Note in Fig. 32–8 that the 10Dq value is larger for the low spin complex, $[Co(NH_3)_6]^{3+}$. Measurements show it to be approximately 69 kcal/mole. Note also that the t_{2g} orbitals are completely paired. For the high spin complex, $[CoF_6]^{3-}$, the 10Dq value is smaller (approximately 37 kcal/mole), and two unpaired electrons are in the antibonding molecular orbitals, e_g^*.

Transition metal complexes are often highly colored. This property is related to the fact that the antibonding molecular orbitals, e_g^*, are quite close in energy content to the three $3d\ t_{2g}$ nonbonding orbitals. Low-lying antibonding orbitals make possible electron transitions which involve energy that pertains to the visible part of the spectrum, giving rise to the color.

Analogous molecular orbital energy level diagrams can be constructed for coordination ions which have geometric configurations other than octahedral, such as planar and tetrahedral.

It should be noted that sometimes the term **Ligand Field Theory** is applied to the theory arising from a combination of molecular orbital concepts with the Crystal Field Theory.

32.8   Uses of Complex Compounds

Complexes have a variety of important uses. Some complexing groups when coordinated to certain metals make the metal more easily assimilable by plants; in other cases, they tie up the metal so that it cannot be effectively utilized by the plant. Hence, complexes can be an important aid to the farmer in effective soil treatment.

The rusting of iron results from the metal combining with oxygen from the air in the presence of moisture to form the reddish-brown iron(III) oxide, $Fe_2O_3 \cdot xH_2O$, which is known as iron rust. If, however, a piece of iron is treated with a suitable coordinating group to form an iron complex, the iron on the surface can be tied up in the complex so that it cannot readily react to form $Fe_2O_3 \cdot xH_2O$. Under some conditions, the iron in rust which has already formed can be complexed to form a soluble complex ion which can then be washed free from the iron. Thus, a number of commercial products are on the market which can prevent or control corrosion through complexing.

In the electroplating industry, it has been found that many metals plate out in a smoother, more uniform plate which adheres better and has better appearance if the metal is plated from a bath which contains the metal in the form of a complex ion. Thus, complexes such as $[Ag(CN)_2]^-$ and $[Au(CN)_2]^-$ are used extensively in the electroplating industry.

Complexing groups are used in medicines and sometimes in the formulation of diets to make certain metals more readily, or occasionally less readily, assimilable by the body in its metabolic processes. Vitamin B_{12} is a dark-red complex of cobalt, in which the cobalt is coordinated to a large organic complexing group.

Various chelating groups can be of aid, through their complexing action with metals, in diminishing the effects of radioactive contamination following nuclear reactions.

Chlorophyll, the green pigment in plants, is a magnesium complex (Section 28.23). The heme portion of hemoglobin, present in the red corpuscles of mammals, is an important iron complex (Section 28.24).

Complexing agents are often used in water softening, because they tie up such ions as Ca^{2+}, Mg^{2+}, and Fe^{2+} which are responsible for the hardness in water (see Section 12.6). Complexes are employed extensively in analytical laboratory work, in both qualitative and quantitative analysis.

A considerable amount of research has been done toward developing inorganic complexes which are polymeric, in the hope that such polymeric materials may be more chemically resistant and more stable at high temperatures than polymers now in use. One of the most important uses of metal complexes is as catalysts for many kinds of reactions, for example, hydrogenation reactions.

The chemistry of complexes is one of the most fascinating fields of chemistry. Many significant advances have been made during the last several years, and a rapidly expanding number of useful applications have been and continue to be developed.

QUESTIONS

1. Write formulas for each of the following, including all isomers for each case in which isomers can exist.
   (a) Tetracyanonickelate(II) ion
   (b) Triamminediaquochlorocobalt(III) bromide
   (c) Potassium amminepentachloroplatinate(IV)
   (d) Hexaamminenickel(II) ion
   (e) Chlorobis(ethylenediamine)nitrocobalt(III) ion
   (f) Tetraamminedichlorocobalt(III) chloride
   (g) Tetraamminecopper(II) tetrachloroplatinate(II)
2. Name each of the following (including isomers where applicable).
   (a) $[Co(C_2O_4)_3]^{3-}$
   (b) $[Co(H_2NCH_2CH_2NH_2)_3]Cl_3$
   (c) $[Pd(CN)_6]^{2-}$
   (d) $[Co(NH_3)_3(NO_2)_3]$
   (e) $[Cr(NH_3)_6][Co(CN)_6]$
   (f) $[Ag(NH_3)_2]^+$
   (g) $[Pt(NH_3)_2Cl_2]$
   (h) $K[Cr(NH_3)_2Cl_4]$

3. Explain what is meant by the terms (a) coordination sphere, (b) effective atomic number, (c) ligand, (d) coordination number, (e) coordination compound, and (f) chelating group.

4. Calculate the "effective atomic number" for the metal in each of the following complexes:
   (a) $[Cr(CN)_6]^{3-}$
   (b) $[PtCl_4]^{2-}$
   (c) $[IrCl_6]^{3-}$
   (d) $[Pt(NH_3)_6]^{4+}$
   (e) $[Zn(NH_3)_4]^{2+}$
   (f) $[PdCl_6]^{2-}$
   (g) $[Co(NH_3)_4Cl_2]^+$

5. Show orbital diagrams, indicate the type of hybridization you would expect, and draw a diagram showing the geometrical structure for each of the following:
   (a) $[Cu(NH_3)_4]^{2+}$
   (b) $[Zn(NH_3)_4]^{2+}$
   (c) $[Cd(CN)_4]^{2-}$
   (d) $[Co(NH_3)_6]^{3+}$
   (e) $[Fe(CN)_6]^{4-}$

6. Show by means of orbital diagrams the hybridization for each of the examples given in Table 32–1.

7. Draw diagrams for any *cis*, *trans*, and optical isomers that could exist for each of the following (en is ethylenediamine):
   (a) $[Co(en)_2(NO_2)Cl]^+$
   (b) $[Co(en)_2Cl_2]^+$
   (c) $[Cr(NH_3)_2(H_2O)_2Br_2]^+$
   (d) $[Pt(NH_3)_2Cl_4]$
   (e) $[Cr(en)_3]^{3+}$
   (f) $[Pt(NH_3)_2Cl_2]$

8. In this and other chapters we have discussed substances whose structures encompass a considerable variety of geometric shapes. (a) Name each shape; (b) draw a diagram of each; (c) give a specific example for each; (d) indicate the relation of atomic or molecular orbitals to each of the geometric shapes; (e) discuss the relationship of the geometric shape to the number of groups attached to the central element. (*Hint:* You will find it useful to make use of coordination compounds in your answer, but do not confine your answer to coordination compounds nor to the material in this chapter alone.)

9. Determine the number of unpaired electrons you would expect for $[Fe(CN)_6]^{3-}$ and for $[Fe(H_2O)_6]^{3+}$. Explain in terms of the Crystal Field Theory.

10. (a) Verify, by calculation, the crystal field stabilization energies listed in Table 32–2; (b) discuss the relationship of crystal field stabilization energies to the stabilities of complexes, and the limitations to the concept.

11. What is the significance of the energy values, relative to each other, of ligand

atomic orbitals, metal atomic orbitals, and the corresponding molecular orbitals for $[CoF_6]^{3-}$ and $[Co(NH_3)_6]^{3+}$?

12. Explain how coordination compounds may aid in (a) corrosion control; (b) soil treatment; (c) electroplating; (d) water softening.

13. It has been shown that ions of the lanthanide metals (elements number 57 through 71) can be attached to as many as nine water molecules. What geometry can you think of that might conceivably be a possibility for the $M(H_2O)_9^{3+}$ ion?

PROBLEMS

1. The bond length of Co to O in $[Co(C_5H_7O_2)_3]$ (a complex with three bidentate ligands resulting in six octahedrally coordinated oxygen atoms) is 1.898 Å. What is the shortest O—O distance? What is the longest O—O distance?

    Ans. 2.684 Å; 3.796 Å

2. Calculate the concentration of free copper ion present in equilibrium with $1.0 \times 10^{-3}\,M$ $[Cu(NH_3)_4]^{2+}$ and $1.0 \times 10^{-1}\,M$ NH_3.   *Ans. $8.5 \times 10^{-12}\,M$*

3. Propose a formula and a structure for a coordination compound with the following composition: (a) 26.36% Ni, 25.14% N, 5.39% H, 14.37% S, 28.74% O; (b) 9.91% Na, 22.42% Cr, 24.15% N, 2.59% H, 10.35% C, 30.57% Cl.

    Ans. (a) $[Ni(NH_3)_4]SO_4$; (b) Hint: Ligands are NH_3, Cl^-, and CN^-

REFERENCES

"Mechanisms of Oxidation-Reduction Reactions," H. Taube, *J. Chem. Educ.*, **45**, 452 (1968).

"Hydrocarbon Reactions on Transition Metal Catalysts," J. J. Rooney, *Chem. in Britain*, **2**, 242 (1966).

"Symmetry," C. A. Coulson, *Chem. in Britain*, **4**, 113 (1968).

"Optically Active Coordination Compounds," R. D. Gillard, *Chem. in Britain*, **3**, 205 (1967).

"Macrocyclic Tetrapyrrolic Pigments," A. W. Johnson, *Chem. in Britain*, **3**, 253 (1967).

"Magnetic Moment Measurement of a Coordination Complex," T. G. Dunne, *J. Chem. Educ.*, **44**, 142 (1967).

"Foundation of Nitrogen Stereochemistry. Alfred Werner's Inaugural Dissertation," G. B. Kauffman, *J. Chem. Educ.*, **43**, 155 (1966).

"Liquid Lasers" (describes chelate lasers which utilize rare earth chelates), A. Lempicki and H. Samelson, *Sci. American*, June, 1967; p. 81.

"Chelation in Medicine," J. Schubert, *Sci. American*, May, 1966; p. 40.

"The Stereochemistry of Complex Inorganic Compounds," D. H. Busch, *J. Chem. Educ.*, **44**, 77 (1964).

"A Simple Approach to Crystal Field Theory," R. C. Johnson, *J. Chem. Educ.*, **42**, 147 (1965).

"Ligand Field Theory," F. A. Cotton, *J. Chem. Educ.*, **41**, 467 (1964).

"Molecular Orbital Theory for Transition Metal Complexes," H. B. Gray, *J. Chem. Educ.*, **41**, 2 (1964).

Alfred Werner, Founder of Coordination Chemistry, G. B. Kauffman, Springer-Verlag, Berlin, 1966 (Paperback).

"Alfred Werner's Coordination Theory," G. B. Kauffman, *Educ. in Chemistry*, **4,** 11 (1967).

"Effect of Complexing Agents on Oxidation Potentials," J. Helsen, *J. Chem. Educ.*, **45,** 518 (1968).

"Molecular Models of Metal Chelates to Illustrate Enzymatic Reactions," H. S. Hendrickson and P. A. Srere, *J. Chem. Educ.*, **45,** 539 (1968).

"Metal Ion Control of Chemical Reactions," D. H. Busch, *Science*, **171,** 241 (1971).

"Metallation of Metallocenes," D. W. Slocum, T. R. Engelmann, C. Ernst, C. A. Jennings, W. Jones, B. Koonsvitsky, J. Lewis, and P. Chenkin, *J. Chem. Educ.*, **46,** 144 (1969).

"Computing Ligand Field Potentials and Relative Energies of *d*-Orbitals," R. Krishnamurthy and W. B. Schaap, *J. Chem. Educ.*, **46,** 799 (1969); **47,** 433 (1970).

"Simplified Molecular Orbital Approach to Inorganic Stereochemistry," R. M. Gavin, *J. Chem. Educ.*, **46,** 413 (1969).

"Crystal Field Potentials," S. F. A. Kettle, *J. Chem. Educ.*, **46,** 339 (1969).

"Organometallic Chemistry," M. F. Lappert, *Chem. in Britain*, **5,** 342 (1969).

"Mechanisms of Substitution Reactions of Metal Carbonyls," F. Basolo, *Chem. in Britain*, **5,** 505 (1969).

"Shapes of the Future," V. Klee, *Amer. Scientist*, **59,** 84 (1971).

"Metal Carbonyl Chemistry—A Simple Experiment," P. W. Wiggans, *Educ. in Chemistry*, **10,** 52 (1973).

"Hydrido Transition-Metal Cluster Complexes," H. D. Kaesz, *Chem. in Britain*, **9,** 344 (1973).

"Molecular Oxygen Adducts of Transition Metal Complexes: Structure and Mechanism," L. Klevan, J. Peone, Jr., and S. K. Madan, *J. Chem. Educ.*, **50,** 670 (1973).

"Hemoglobin: Isolation and Chemical Properties," S. F. Russo and R. B. Sorstokke, *J. Chem. Educ.*, **50,** 347 (1973).

"Homogeneous Models Throw Light on Industrial Catalysts," D. G. H. Ballard, *Chem. in Britain*, **10,** 20 (1974).

"Binuclear Cobalt(III) Complexes—A Survey of Reactions," A. G. Sykes, *Chem. in Britain*, **10,** 170 (1974).

"Rearrangements in Five- and Six-Coordinate Systems," J. I. Musher, *J. Chem. Educ.*, **51,** 94 (1974).

"Bioinorganic Drugs—1," D. R. Williams, *Educ. in Chemistry*, **11,** 124 (1974); "—2," **11,** 166 (1974).

"The Long Search for Stable Transition Metal Alkyls," G. Wilkinson, *Science*, **185,** 109 (1974).

"Thermodynamics and Kinetics of Ligand Binding to Vitamin B_{12_a}: A Laboratory Experiment," D. A. Sweigart, *J. Chem. Educ.*, **52,** 126 (1975).

"Volatile Metal Chelate Complexes," C. Kutal, *J. Chem. Educ.*, **52,** 319 (1975).

33

Spectroscopy and Chromatography

In earlier chapters we have discussed molecular structures and considered the properties of a significant number of compounds. Now is an appropriate time to look into the methods by which chemists can characterize each compound that they prepare—that is, **establish its composition, purity, and structure.** The characterization of a compound is of course especially important if the compound is a new one, never made before, but characterization is also extremely important after the preparation (often referred to as a **synthesis**) of a compound previously known and prepared. A compound, whether known previously or not, is ordinarily not of any use unless at least its composition and purity are known; for many purposes it is highly desirable, and sometimes essential, that its molecular structure be known as well.

Before World War II, a compound was usually characterized by the method of synthesis, by chemical analysis to determine the percentage composition, and (if a previously known compound) by a comparison of properties, such as melting point, color, and density, with those previously established for a pure sample of the same compound. Finally, an attempt was made to convert a sample of the compound into one or more **derivatives,** which could result only if the original compound was actually what it was thought to be. Characterization of the derivatives was, of course, essential to the method. Such procedures are still used, but in addition, during the last thirty years, several instrumental methods for characterizing compounds have been developed or improved to high levels of usefulness. Chief among these are the various **spectroscopic methods,** most of which involve **the effects of electromagnetic radiation upon matter** as detected and recorded by a spectrometer. In the following discussion, we will consider a variety of kinds of spectroscopy, often referred to also as **spectrometry:** (1) ultraviolet and visible spectroscopy, (2) infrared spectroscopy, (3) nuclear magnetic resonance spectroscopy, (4) electron paramagnetic spectroscopy, and (5) mass spectrometry. We will not, in our treatment here, consider other important kinds of spectroscopy, such as Raman spectroscopy and Mössbauer spectroscopy. The major emphasis will be placed on ultraviolet and visible, infrared, and nuclear magnetic resonance spectroscopy.

Another very useful method for characterizing compounds and also for purifying compounds and separating mixtures of compounds is chromatography. **Chromatography is a method of separating the components of a mixture by arranging a system wherein the different components of the mixture migrate through the system at different rates.** We shall consider both **gas phase chromatography** and **liquid phase chromatography** later in this chapter. As we will see, adsorption plays an important role in chromatographic procedures.

It should be mentioned that another technique, diffraction of x rays by crystals, though not a topic of this chapter, is also used for compound characterization (Section 11.19). X-ray diffraction by crystals provides one of the most direct sources of structural information, although the technique is limited to substances which form suitable crystals. X-ray diffraction involves the collection and interpretation of a large amount of data for each structural determination. Hence, the development of automated data collection systems and high-speed computers, during the past several years, has changed the method from a time-consuming and laborious process to one that is widely used and exceedingly valuable for the characterization of chemical substances.

Spectroscopy

33.1   Introduction

Most spectroscopic methods involve the absorption of energy in the form of electromagnetic radiation from a particular region of the spectrum (Sections 3.7 and 3.11). When a molecule absorbs radiation it gains an amount of energy which corresponds to some form of motion within the molecule. These motions within the molecule, which are referred to as **transitions,** can involve movement of electrons from one energy level to another (electronic transitions—see Section 32.5), vibrations and rotations of molecules, spins of electrons, or spins of nuclei. The various portions of the spectrum, in order of decreasing energy, are shown in Fig. 33–1.

How much energy the molecule gains depends upon the frequency of the radiation. As pointed out in Chapter 3, radiation can be characterized either by its wavelength, λ, or by its frequency, ν, (Section 3.11). The relationship between the two and the energy relationship, ΔE, pertaining to both, are

$$\nu_{\text{sec}^{-1}} = c_{\text{cm/sec}} / \lambda_{\text{cm}}$$
$$\Delta E_{\text{ergs/molecule}} = h\nu = hc/\lambda$$

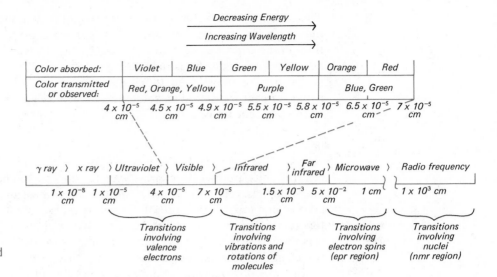

FIGURE 33-1

The various portions of the electromagnetic spectrum in order of decreasing energy and increasing wavelength.

where c is the velocity of light (2.998×10^{10} cm/sec), and h is Planck's Constant (6.626×10^{-27} erg · sec). The higher the energy of the radiation, therefore, the higher the frequency and the shorter the wavelength.

By means of a spectrophotometer, we can determine the wavelengths at which a molecule absorbs energy and also how much energy (in the form of electromagnetic radiant energy) is absorbed at that wavelength. The wavelength at which the molecule absorbs the maximum amount of energy is given the symbol λ_{max}; the quantity of radiant energy absorbed at λ_{max} is called the **molar extinction coefficient,** or **molar absorptivity,** and is given the symbol ε_{max}.

$$\varepsilon_{max} = \frac{\log I_0/I}{cl}$$

where I_0 = intensity of incident radiation (i.e., radiant energy put into the system)

I = intensity of transmitted, or emitted, radiation (i.e., the portion of incident radiation that is not absorbed by the sample and hence gets through the sample)

c = concentration of the sample solution in moles/liter

l = length of the light path for the absorption cell in centimeters (i.e., the distance the radiation travels through the sample solution)

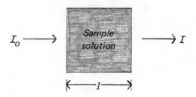

The concentration factor, c, in the expression for the molar extinction coefficient makes possible quantitative analysis by spectroscopic methods, if the other quantities in the expression are known. Log I_o/I is referred to as the **absorbance.** The ratio I/I_o (note that this is the inverse of the ratio in the log term for absorbance) is the ratio of transmitted radiation to incident radiation; hence, $(I/I_o) \times 100$ is the **per cent transmission,** sometimes referred to as the **per cent transmittance,** or the **per cent transmittancy.**

33.2   Units

Several units are used to express wavelength and frequency. **Wavelength,** as the name implies, **refers to the length of the wave** (measured, for example, from the crest of one wave to the crest of the next) and is commonly expressed in centimeters (cm). One cm $= 1 \times 10^8$ Å $= 1 \times 10^4$ microns $(\mu) = 1 \times 10^7$ millimicrons $(m\mu) = 1 \times 10^7$ nanometers (nm). One nanometer $= 1$ millimicron. Occasionally, wavelengths are expressed in meters (m). In Fig. 33–1, wavelengths are given in centimeters. Frequency is commonly expressed in terms of wave numbers. The wave number is the number of waves per centimeter and is the reciprocal of the wavelength in centimeters. Hence, wave numbers have the unit of reciprocal centimeters, cm^{-1}. Many spectroscopists prefer to use the unit Hertz (Hz) for frequency. Hertz units are the same as cycles/second (cps); thus, 1 Hz = 1 cps.

33.3   Ultraviolet and Visible Spectra

As indicated in Fig. 33–1, the ultraviolet range covers wavelengths of 1×10^{-5} cm to 4×10^{-5} cm (or 100–400 nm), and the visible range wavelengths of 4×10^{-5} cm to 7×10^{-5} cm (or 400–700 nm). Most recording spectrophotometers for ultraviolet and visible spectra operate in the range 2×10^{-5} to 8×10^{-5} cm (or 200–800 nm). Such relatively high energies are sufficient to move an electron from its ground state orbital to an orbital of higher energy. The electron can be one which is initially either in a σ orbital or in a π orbital, but in actual practice σ orbital electrons are often held so tightly that they are not moved from one orbital to another at energies corresponding to either ultraviolet or visible light. Often, the electron transitions are from a π orbital to the antibonding π^* orbital $(\pi \longrightarrow \pi^*)$ or from a nonbonding orbital (Sections 32.5–32.7) to a π^* orbital $(n \longrightarrow \pi^*)$.

Ultraviolet and visible light have limited use in characterizing organic compounds but are highly useful for compounds of metals in which the metal has an unfilled d subshell. Hence, ultraviolet and visible light are especially useful in characterizing compounds of the transition metals and the inner transition metals.

In organic compounds, electronic transitions of electrons in either single bonds or nonconjugated double bonds usually occur below 200 nm, which is below the operating range of ordinary spectrometers; but systems with conjugated double and single bonds have $\pi \longrightarrow \pi^*$ electronic transitions above 200 nm and so can be analyzed with ordinary instruments. Conjugated systems are ones in which double bonds and single bonds occur alternately along the chain, as in isoprene (Section 27.21).

$$H-\overset{\overset{\displaystyle H}{|}}{C}=\overset{\overset{\displaystyle CH_3}{|}}{C}-\overset{\overset{\displaystyle H}{|}}{C}=\overset{\overset{\displaystyle H}{|}}{C}-H$$

Isoprene

In general, the longer the conjugated system, the longer the wavelength at which the system absorbs light. For example,

$$H-\overset{\overset{\displaystyle H}{|}}{C}=\overset{\overset{\displaystyle H}{|}}{C}-H$$

Ethylene
$\lambda = 162$ nm

$$H-\overset{\overset{\displaystyle H}{|}}{C}=\overset{\overset{\displaystyle H}{|}}{C}-\overset{\overset{\displaystyle H}{|}}{C}=\overset{\overset{\displaystyle H}{|}}{C}-H$$

1,3-Butadiene
$\lambda = 215$ nm

$$H-\overset{\overset{\displaystyle H}{|}}{C}=\overset{\overset{\displaystyle H}{|}}{C}-\overset{\overset{\displaystyle H}{|}}{C}=\overset{\overset{\displaystyle H}{|}}{C}-\overset{\overset{\displaystyle H}{|}}{C}=\overset{\overset{\displaystyle H}{|}}{C}-H$$

1,3,5-Hexatriene
$\lambda = 265$ nm

Examples of other conjugated systems are aromatic rings, such as in benzene, (which have conjugated systems through resonance) and ketones with the —C=O group conjugated to carbon-carbon double bonds.

Hence, the main use of ultraviolet and visible light for characterizing organic compounds involves establishing whether or not the structure includes a conjugated system. The ultraviolet spectrum for benzaldehyde, which has a conjugated structure, is shown in Fig. 33–2.

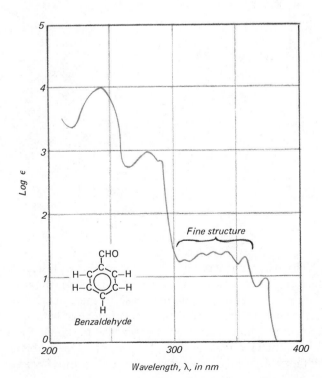

FIGURE 33–2

The ultraviolet spectrum of benzaldehyde. From *Theory and Practice in the Organic Laboratory,* J. A. Landgrebe, D. C. Heath and Co., 1973

In the use of visible light for characterization of octahedral coordination compounds of the transition metals, the splitting of the d orbitals into the three lower-energy nonbonding orbitals (t_{2g}) and the higher-energy antibonding orbitals (e_g), as described in Chapter 32, Sections 32.5–32.7, occurs at the relatively high energies of the visible absorption range. The recorded spectra that result from use of visible light energies provide a method for obtaining the 10Dq (or Δ) values for metal coordination compounds (Sections 32.5–32.7). Hence, such spectra provide considerable information concerning the bonding strengths of various ligands with different metals and the distributions of the valence electrons within the complex. From the valence electron distribution, information relative to the molecular structures of the metal complexes can be deduced.

One of the simplest cases of visible light energy absorption is that of a metal complex ion in which the metal has a d^1 configuration and an octahedral structure. An example is the Ti(III) ion in $[Ti(H_2O)_6]^{3+}$, in which the single d electron occupies a t_{2g} orbital. Absorption of light in the visible range raises the electron to the e_g orbital. The absorption band occurs in the visible light portion of the electromagnetic spectrum of $[Ti(H_2O)_6]^{3+}$ (Fig. 33–3) and accounts for the violet color of the ion.

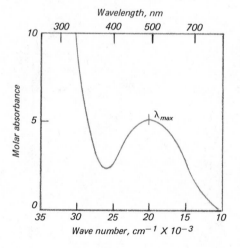

FIGURE 33-3
The visible spectrum of $[Ti(H_2O)_6]^{3+}$.

The band, or peak, in the spectrum of the $[Ti(H_2O)_6]^{3+}$ ion is broad, a characteristic common to visible spectra. The peak maximum occurs at 20,000 cm^{-1}, corresponding in this simplest case to the Δ, or 10Dq, value for the splitting of the orbital levels. In more complicated cases with larger numbers of d electrons, the Δ, or 10Dq, value can likewise be obtained from the visible spectrum but by a somewhat more involved procedure. The visible and near-ultraviolet spectra of the octahedral complexes $[Co(NH_3)_6]^{3+}$ and $[Co(en)_3]^{3+}$ are shown in Fig. 33–4.

FIGURE 33-4

The visible and near-ultraviolet spectra for $[Co(NH_3)_6]^{3+}$ (*solid line*) and $[Co(en)_3]^{3+}$ (*broken line*). Each complex has an octahedral configuration (*en* stands for ethylenediamine).

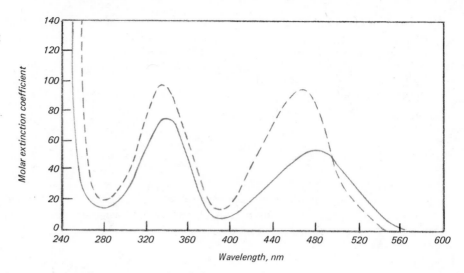

33.4   Infrared Spectra

The infrared portion of the electromagnetic spectrum covers a region of lower energy than the visible and ultraviolet portions. As shown in Fig. 33–1, the infrared region extends from 7×10^{-5} to 5×10^{-2} cm. The recorded spectra from infrared energy absorption are of perhaps the most use to the organic chemist, but they are also highly valuable to the inorganic chemist, for example in studying coordination compounds. The most common use for infrared spectra is as a "fingerprint method" for establishing the presence of certain functional groups, but other uses such as differentiating between *cis* and *trans* isomers are also important.

Many kinds of functional groups in organic molecules absorb at characteristic frequencies in the infrared region. The absorbed energy is thought to be converted within the molecule into vibrations which take the form of several stretching and bending motions in the bonds.

The stretching motions, which can be likened to those in a coiled spring, can be either unsymmetrical or symmetrical. The following diagrams for hypothetical atoms *A*, *B*, and *C* illustrate the two types of stretching vibrations:

Unsymmetrical stretching vibrations

(Continued next page)

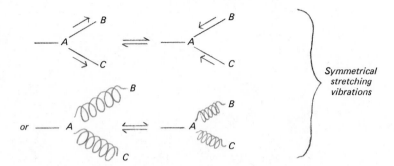

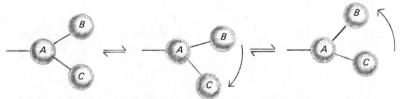

Bending vibrations can be bending in the plane (*scissoring* and *rocking*) or bending in or out of the plane (*wagging* and *twisting*), as illustrated by the following diagrams:

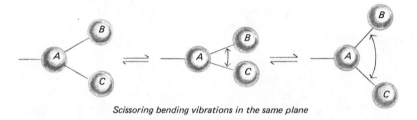

Scissoring bending vibrations in the same plane

(The B—A—C bonds alternately open up and close partially, in a motion similar to that of a pair of scissors.)

Rocking bending vibrations in the same plane

(B and C move in the same direction, alternately down and then up so that the B—A—C group rocks down and up within the plane.)

Wagging bending vibrations in and out of the plane

(Alternately both B and C bend forward and back together from the plane.)

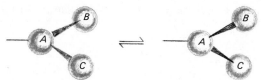

Twisting bending vibrations in and out of the plane

(Alternately, *B* bends back from the plane while *C* bends forward, then *B* bends forward while *C* bends back.)

The spectrometer detects and records the absorbed energies involved in these bond motions. Various functional groups have characteristic absorption values. Hence, by studying the absorption peaks (or bands) in the spectrum of a compound, various functional groups can be shown to be present.

It is important to note that vibrational frequencies are observable in infrared spectra, only if the vibrational stretching or bending motion results in a change in the dipole moment of the bond. The more polar the bond, the larger will be the peak observed for the stretching or bending motion.

Figure 33–5 and Table 33–1 show the wavelengths of infrared radiation ab-

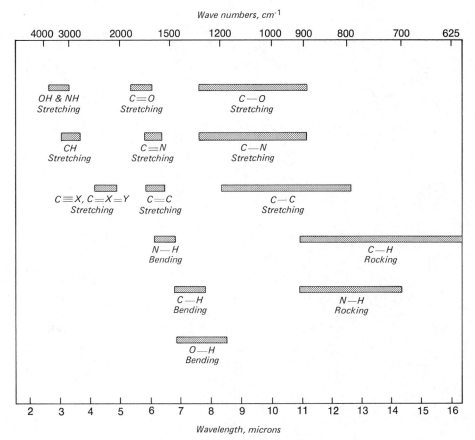

FIGURE 33-5

Regions by wavelength (bottom horizontal axis) and wave number (top horizontal axis) for some infrared stretching and bending vibrations.   From *Theory and Practice in the Organic Laboratory*, J. A. Landgrebe, D. C. Heath and Co., 1973

TABLE 33-1   Characteristic Infrared Absorption Peaks for Some Common Bonds

Type of Compound	Bond	Range of Frequency for Absorption Maximum, cm^{-1}
Alkanes	—C—H	2850–2960, 1350–1470
Alkenes	=C—H	3010–3090, 675–1000
	C=C	1600–1680
Alkynes	≡C—H	3270–3300
	C≡C	2100–2260
Aromatic rings	 ⬡	1450–1600
	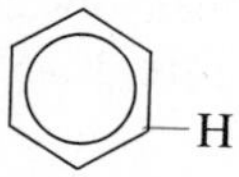 ⬡—H	3000–3100, 675–870
Alcohols and phenols		
Not hydrogen bonded	—O—H	3500–3700
Hydrogen bonded	—O—H⋯O	3100–3700
Thio alcohols	—S—H	2500
Amines	—N—H	3300–3500
Alcohols, ethers, carboxylic acids, and esters	—C—OR	1080–1300
Aldehydes, ketones, carboxylic acids, and esters	—C=O	1680–1750
Aldehydes	—C=O	1695–1740
Ketones	—C=O	1700–1730
Carboxylic acids	—C=O	1680–1730
Esters	—C=O	1720–1750
Amines	—C—NR$_2$	1180–1360
Nitriles	—C≡N	2210–2260
Nitro compounds	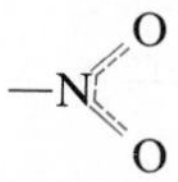 —N(=O)O	1515–1560, 1345–1385

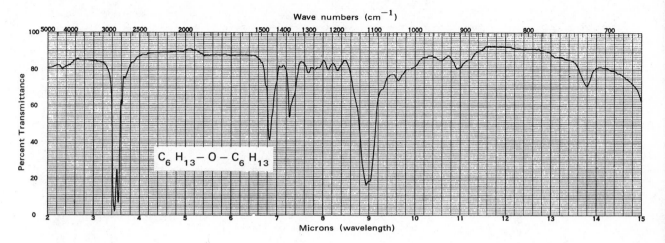

FIGURE 33-6

The infrared spectrum of di-*n*-hexyl ether.

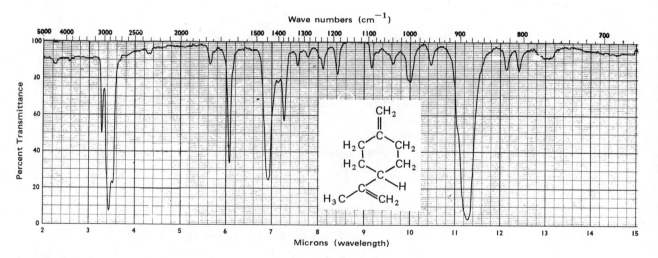

FIGURE 33-7

The infrared spectrum of 4-isopropenyl-1-methylenecyclohexane. (Note that the ring is the saturated cyclohexane ring—not the benzene ring.)

sorbed by some common bonds and the corresponding stretching, bending, and rocking vibrations.

Three examples of infrared spectra are shown in Figs. 33–6, 33–7, and 33–8. Note infrared spectrometers ordinarily plot the spectra upside down compared to ultraviolet or visible spectra. The ordinate axis is in terms of per cent transmittance in Figs. 33–6, 33–7, 33–8, a common method of plotting infrared spectra. Alternatively, infrared spectra are plotted in terms of absorbance. The abscissa axis is

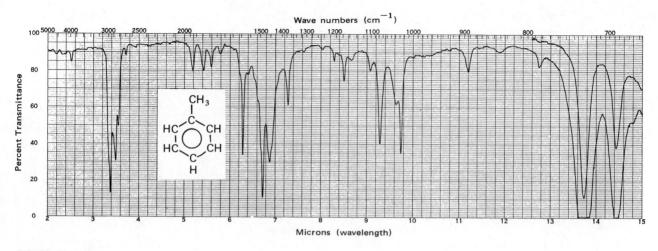

FIGURE 33-8

The infrared spectrum of toluene (toluene does contain the benzene ring, in contrast to the compound of Fig. 33-7).

plotted in terms of both wave numbers (in cm^{-1}, across the top) and wavelength (in microns, across the bottom) in Figs. 33-6, 33-7, and 33-8, and this is quite customary. It should be noted that infrared spectra are often quite complex with many peaks, or bands. Usually, no effort is made to identify all of the bands. Identification is made of a few selected bands that identify key functional groups or types of bonds.

Notice in the spectrum of di-*n*-hexyl ether (C_6H_{13}—O—C_6H_{13}), Fig. 33-6, that a strong absorption appears at about 2900 cm^{-1}, corresponding to alkane C—H stretching in the C_6H_{13} groups,

$$
\begin{array}{c}
\text{H} \quad \text{H} \quad \text{H} \quad \text{H} \quad \text{H} \quad \text{H} \\
| \quad | \quad | \quad | \quad | \quad | \\
\text{H}\!-\!\text{C}\!-\!\text{C}\!-\!\text{C}\!-\!\text{C}\!-\!\text{C}\!-\!\text{C}\!-\!\cdot \\
| \quad | \quad | \quad | \quad | \quad | \\
\text{H} \quad \text{H} \quad \text{H} \quad \text{H} \quad \text{H} \quad \text{H}
\end{array}
$$

Bands arising from alkane C—H bending occur at about 1460 cm^{-1} and 1380 cm^{-1}. At 1110 cm^{-1} there is a strong absorption for the C—O ether stretching vibration. The infrared spectrum for C_6H_{13}—O—C_6H_{13} is a relatively simple one.

The spectrum for 4-isopropenyl-1-methylenecyclohexane (Fig. 33-7) is somewhat more complex. Note that the molecule contains two olefinic double bonds. The spectrum shows a sharp olefinic =C—H stretching band, therefore, at about 3050 cm^{-1} and a broader peak at 890 cm^{-1}. It also shows a sharp olefinic C=C stretching band at 1640 cm^{-1}. The bands in the region 2850–2960 cm^{-1} can be ascribed to alkane C—H stretching in the CH_3 group. It should be pointed out that although there is a band at 1450 cm^{-1}, it cannot be ascribed to a benzene ring, inasmuch as the ring is not the benzene ring but rather the saturated cyclohexane ring. The band at 1450 cm^{-1} is instead due, at least in part, to alkane (methyl) C—H bending. These facts emphasize the need for care in making assignments for identification of unknown compounds.

The spectrum in Fig. 33–8 is for toluene, which does have a benzene ring. The benzene ring is responsible in part for the band at 1460 cm^{-1}, but alkane (methyl) C—H bending vibrations also contribute to the 1460 cm^{-1} band. In addition, strong alkane (methyl) C—H stretching bands occur in the region 2850–2960 cm^{-1}. Ar—H (hydrogen attached to an aromatic ring) stretching vibrations probably account for the band at 3010 cm^{-1} and perhaps also for the bands in the 690–730 cm^{-1} region.

Infrared spectra have been run for many hundreds, even thousands, of organic compounds. Infrared spectroscopy is one of the principal methods used by chemists for characterization of organic compounds.

Although the largest use of infrared spectroscopy is in the study of organic compounds, the method is also highly useful for a variety of other compounds, including metal coordination compounds. Many of the ligands in coordination compounds are organic in nature and hence give rise to infrared bands such as we have described. Moreover, a number of ligands other than organic ligands also give rise to infrared bands, —NO$_2$ groups being one example. Also, when a metal attaches to the carbonyl groups of a compound such as 2,4-pentanedione to form a metal chelate of the type

the carbonyl (C=O) absorption bands are shifted in frequency as compared to their absorption frequencies in the uncoordinated molecule. This provides a qualitative measure of the strength of the bonds between the metal and the donor oxygen atoms; the greater the shift, the greater is the stability of the compound.

Another interesting application of infrared spectra to metal coordination compounds is in distinguishing between *cis* and *trans* isomers of a given complex. The infrared spectra of *cis* and *trans* isomers usually differ although many similarities also exist. Very often, the spectrum of the more unsymmetric *cis* isomer is more complex, with a larger number of bands, than that of the *trans*. This is true, for example, for the *cis* and *trans* dinitrotetraamminecobalt(III) ions, [Co(NH$_3$)$_4$(NO$_2$)$_2$]$^+$, for which the spectra are shown in Fig. 33–9.

It is interesting to note that even the attachment of different atoms of a ligand to the central metal ion in a metal complex can cause a change in the infrared

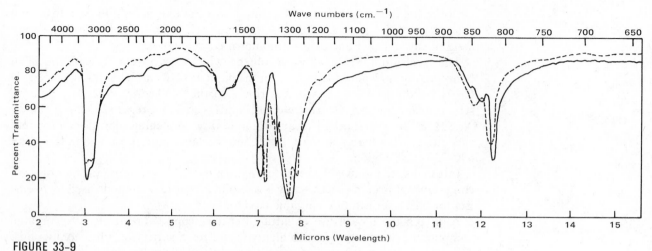

FIGURE 33-9

The infrared spectra of *cis*-[Co(NH$_3$)$_4$(NO$_2$)$_2$]$^+$ (*broken line*) and *trans*-[Co(NH$_3$)$_4$(NO$_2$)$_2$]$^+$ (*solid line*).   Reprinted with permission from the *Journal of the American Chemical Society*, Vol. 76, p. 5346: J. P. Faust and J. V. Quagliano. Copyright ⓒ 1954 by the American Chemical Society

spectrum. For example, Fig. 33-10 shows the infrared spectra of *nitro*penta-amminecobalt(III) chloride, [Co(NH$_3$)$_5$NO$_2$]Cl$_2$, (in which the nitrite ion is at-tached to the cobalt ion through the nitrogen atom) and *nitrito*pentaammineco-balt(III) chloride, [Co(NH$_3$)$_5$ONO]Cl$_2$ (in which the nitrite ion is attached through one of the oxygen atoms). Isomers of this type are known as **linkage isomers.**

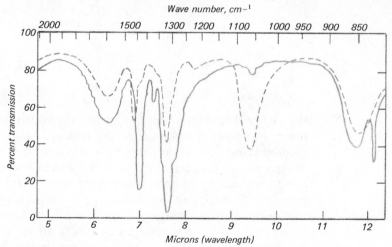

FIGURE 33-10

The infrared spectra of [Co(NH$_3$)$_5$NO$_2$]Cl$_2$, in which the NO$_2^-$ ion is attached to cobalt through the nitrogen atom (*solid line*), and [Co(NH$_3$)$_5$ONO]Cl$_2$, in which the NO$_2^-$ ion is attached to cobalt through one of the oxygen atoms (*broken line*).   Reprinted with permission from the *Journal of the American Chemical Society*, Vol. 78, p. 887: R. B. Penland, F. J. Lane, and J. V. Quagliano. Copyright ⓒ 1956 by the American Chemical Society

33.5   Nuclear Magnetic Resonance Spectra

We have learned that electrons spin about their own axes (Section 3.17), some clockwise and some counterclockwise, with spin quantum numbers of $+\frac{1}{2}$ and $-\frac{1}{2}$. Inasmuch as the spin creates a magnetic moment along the axis of spin, the electrons act as tiny bar magnets.

Nuclei of certain types also spin on their own axes, and these nuclei, like electrons, act as very small bar magnets. The hydrogen nucleus (a proton) has spin quantum numbers, like the electron, of $+\frac{1}{2}$ or $-\frac{1}{2}$, commonly referred to simply as spin number $\frac{1}{2}$. If the proton is placed in a magnetic field, it will align itself either *with* the field (lower energy condition) or *against* the field (higher energy condition). The spin quantum number $+\frac{1}{2}$ corresponds to alignment with the field and $-\frac{1}{2}$ to alignment against the field. Energy is required, and thus must be absorbed, to change the spin of the nucleus from alignment with the applied magnetic field to the higher energy, less stable, condition of alignment against the field. This energy absorption results in a signal on a nuclear magnetic resonance spectrometer and forms the basis for the **nuclear magnetic resonance** (nmr) method.

The proton is the most commonly investigated nucleus in nuclear magnetic resonance. The nmr method when applied to the proton is sometimes referred to as **proton magnetic resonance** (pmr). Certain other nuclei, such as ^{13}C, ^{19}F, and ^{31}P, each of which also has spin quantum numbers of $+\frac{1}{2}$ and $-\frac{1}{2}$, behave similarly and are sometimes used in nmr studies. However, some other nuclei have spin numbers other than $\frac{1}{2}$, such as 0, 1, $\frac{3}{2}$, and 2. The spin number of a particular nucleus depends, in a rather complex way, on the total number of protons and neutrons it contains. Nuclei with a spin number of zero, such as ^{12}C and ^{16}O, do not behave as bar magnets and an nmr signal does not result. This actually is fortunate because these very common nuclei, thus, do not interfere with the nmr signals from protons or other suitable nuclei which do cause a signal. All nuclei with spin numbers other than zero act as bar magnets, but, except for $+\frac{1}{2}$ and $-\frac{1}{2}$, only in rather special cases do useful spectra result from their signals. Examples of such nuclei that are sometimes used are ^{2}H (deuterium) and ^{14}N, each with spin number 1, and ^{11}B, with spin number $\frac{3}{2}$. In the remainder of our discussion of the nuclear magnetic resonance method we will consider only the proton.

The amount of energy necessary to change the alignment of the proton depends upon the strength of the applied magnetic field. Typically, the magnetic field in nmr instruments is arranged so that the instruments operate in either the 60 megahertz frequency range (60 million cycles per second) or the 100 megahertz range (100 million cycles per second). This field energy is in the radiofrequency range; hence, it is energy of lower frequency and longer wavelength than that for infrared radiation.

To obtain an nmr spectrum, the sample is dissolved in an appropriate solvent, such as carbon tetrachloride, and placed in a container between the poles of a magnet where it is usually spun at high speed. The sample is irradiated with sufficient radiofrequency energy to promote the flip in alignment of the nuclei. When protons go into the higher energy state, absorption of energy takes place producing an nmr signal which is detected and recorded by the instrument as a peak. To accomplish the process, either a constant magnetic field and a gradually increasing frequency of radiation can be used or, as is more common, the frequency can be held constant and the strength of the magnetic field gradually increased.

One would expect, offhand, that all protons would absorb at the same magnetic field strength. However, a proton can be "shielded" from the full force of the applied magnetic field by protons near it in its environment. In such a case, a larger *applied* field strength must be used to create the same *effective* field strength as would be required for absorption to take place if shielding did not occur. At a constant radiofrequency, all protons absorb energy at the same *effective* magnetic field strength, but at different *applied* magnetic field strengths depending upon the environment in which the proton is located. It is the *applied* field strength that is measured and plotted by the instrument against the absorption. The spectrum which results, therefore, has absorption peaks whose relative positions are dependent upon differences in environments of the various protons present in the sample. Hence, the nmr spectrum can provide considerable information about the molecular structure of the sample.

An nmr spectrum, in fact, provides even more information than the foregoing discussion suggests. Not only are the recorded positions of the nmr signals useful but also the *number* of signals, the *intensities* of the signals, and a *splitting* of signals that occurs such that a particular signal is split into several peaks. The number of signals indicates how many different kinds of proton environments are present, the intensities of the signals indicate how many protons are in each kind of environment, and the splitting of signals provides additional information about the environment of a proton with respect to other protons. We will next consider each of these factors in somewhat more detail.

■ **1. Position of Signals.** The positions of the nmr signals tell us what "kinds" of protons, in terms of their electronic environments, are present. For example, protons in aromatic environments, those in aliphatic environments, and those attached to halogens or other atoms or groups of atoms all have different electronic environments and produce nmr signals at different energies. Electrons moving about the proton itself shield the proton from the full applied magnetic field by generating a field that opposes the applied field. If the effective field, felt by the proton, is thus less than the applied field, the proton is said to be **shielded.** For a shielded proton the energy required for the applied field to reach the effective field necessary for absorption is thus larger, and the signal is said to be **shifted upfield,** as compared to that required for an unaffected proton. Electrons moving about adjacent protons can also generate a magnetic field that either opposes or reinforces the applied field. If opposition of the field takes place the proton is again shielded; if reinforcement of the field occurs the effective field felt by the proton will be larger than the applied field and the proton is said to be **deshielded.** In the latter case, the applied field required for absorption is less than for an unaffected proton, and the signal is **shifted downfield.** These shifts upfield and downfield in nmr spectra are referred to as **chemical shifts.**

The reference point for chemical shifts is usually a signal from a reference compound added to the system as an internal standard. The most common reference compound is tetramethylsilane [TMS, $(CH_3)_4Si$], which has only one kind of proton environment and hence only one signal. Considerable shielding of protons occurs in TMS, resulting in the one signal being farther upfield than the nmr signals for most other compounds. Hence, nmr signals for protons of most compounds are downfield from the TMS signal. On one common scale of measurement of chemical shifts, the delta (δ) scale, the TMS signal is arbitrarily set as zero and shifts are measured from that point on a scale of units known as ppm (parts per million). Most chemical shifts are less than 10 ppm. Sometimes, a tau (τ) scale is used, in which the TMS signal is set as 10 ppm. Thus, $\tau = 10 - \delta$.

Characteristic portions of the structure of a molecule cause particular shifts. For example, a chlorine atom attached to a carbon atom causes a downfield shift for a proton attached to the same carbon atom; two chlorine atoms cause a larger downfield shift. A chlorine atom attached to an adjacent carbon atom also causes a downfield shift but the shift is smaller than when the chlorine atom is attached to the same carbon atom. The proton in the —COOH group of an organic acid, RCOOH, has an nmr peak downfield from that in the —CHO group of an aldehyde, RCHO, with the same R group. The —CHO shift is in turn downfield from that of a proton attached to an aromatic ring. Other structural variations result in characteristic differences in the nmr spectrum.

■ **2. Number of Signals.**   Within a molecule, protons with the same environment are referred to as **equivalent protons.** Equivalent protons absorb radiofrequency energy at the same applied magnetic field strength. Protons with different environments absorb at different applied magnetic field strengths. Hence, by counting the nmr signals the number of "kinds" of protons in the molecule, in terms of their environments, can be determined.

For example, ethanol has three different kinds of protons in terms of environment.

$$
\begin{array}{c}
\quad H \quad H \\
\quad | \quad\ | \\
H-C-C-OH \\
\quad | \quad\ | \\
\quad H \quad H
\end{array}
$$

The three protons in the methyl group, CH_3—, are all equivalent, with an environment made up of the —CH_2—OH grouping; the two protons in the methylene group, —CH_2—, likewise are equivalent but are different in environment from those in the methyl group; and finally, the one proton attached to oxygen in the —OH group has a different environment from either of the other two types. Figure 33–11 shows the three peaks, corresponding to the three kinds of protons, in a low resolution nmr spectrum of ethanol. The three peaks in a high resolution spectrum are shown in Fig. 33–12.

The least shielded proton is on the electronegative oxygen atom and its signal is thus farthest downfield. The two methylene protons are somewhat more shielded than the —OH proton and their signal is upfield from that of the —OH proton. The three methyl protons are still more shielded, and hence their signal is still farther upfield.

FIGURE 33-11
The three peaks in a low
resolution nmr spectrum of
ethanol.

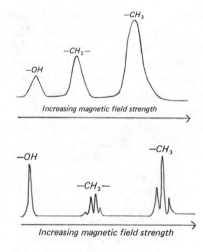

Increasing magnetic field strength

FIGURE 33-12
The three peaks in the high
resolution nmr spectrum of
ethanol.

Increasing magnetic field strength

■ **3. Intensity of Signals.** In Fig. 33–11 it is apparent that, judging roughly the sizes of the signal peaks by their heights, the sizes are approximately in proportion to the number of protons corresponding to each peak. This is reasonable, inasmuch as the greater the number of protons to be lined up against the magnetic field the larger would be the amount of energy absorbed, and hence a larger absorption peak would result. A more accurate judgment than one based on height, for the size of a peak, is based on the area under the peak. It happens that, under normal operating conditions, the area under an nmr peak is directly proportional to the number of protons responsible for the peak. Hence, the areas under the three peaks for ethanol, in either the low resolution or the high resolution spectrum, are in the ratio $1:2:3$ for the protons in the hydroxyl, methylene, and methyl groups, respectively.

The nmr spectrometer normally includes an integrator which draws a stepped curve, superimposed over the spectrum, in which the heights of the steps are proportional to the areas under the peaks. Figure 33–14 shows the stepped integration curve superimposed on the spectrum for ethyl malonic acid.

■ **4. Splitting of Signals and Spin-Spin Coupling.** A comparison of the drawing of the low resolution spectrum of ethanol in Fig. 33–11 with that of the high resolution spectrum in Fig. 33–12 shows that the signal for the —OH proton is recorded as one peak in both, but the signal for the $>$CH$_2$ protons is split into four peaks in the high resolution spectrum and the signal for the —CH$_3$ protons is split into three peaks.

The splitting of a signal into a group of peaks arises because of neighboring protons in the molecule. It is sufficient here to note that the magnetic field felt by a particular proton is either slightly increased or slightly decreased by the spin of a neighboring proton—increased if the neighboring proton at that moment is aligned with the applied magnetic field, and decreased if the neighboring proton is aligned against the applied field. The absorption by the absorbing proton in half the molecules at a given moment is thus shifted slightly downfield and for the other half slightly upfield for each neighboring proton, resulting in a splitting of the signal. The net result is that **a set of equivalent neighboring protons will split a proton signal into a number of peaks equal to one more than the number of equivalent neighboring protons.**

In the high resolution spectrum of ethanol the signal for the $-CH_3$ protons is split, because of the two equivalent protons on the neighboring methylene group, into $(2 + 1) = 3$ peaks. The signal for the methylene protons, however, is split, because of the three neighboring methyl protons, into $(3 + 1) = 4$ peaks. In a very pure sample of ethanol without acidic or basic impurities, the $-OH$ signal is split into $(2 + 1) = 3$ peaks, because of the two neighboring equivalent methylene protons (also the methylene proton signal is affected somewhat by the neighboring $-OH$ proton). However, even a small amount of acidic or basic impurity catalyzes a rapid proton exchange between the hydroxyl groups of many ethanol molecules, which makes only the average signal (one overall peak) detectable for $-OH$ in ethanol.

It might be mentioned here that even the relative heights of the individual peaks for a given signal are significant and useful in the determination of structure, but we will not go further into those details here.

Figure 33–13 shows the nmr spectrum for ethyl iodide. Note that, as in ethanol, (1) the signal for the methylene protons is downfield $(3.2\delta, 6.8\tau)$ from that for

FIGURE 33-13

The high resolution nmr spectrum for ethyl iodide, with TMS (tetramethylsilane) as internal standard. The TMS peak is the one at $\delta = 0$ $(\tau = 10)$.

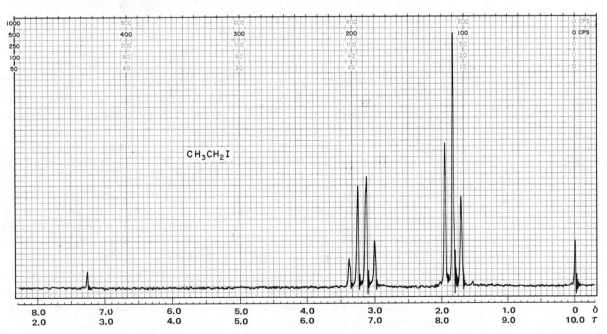

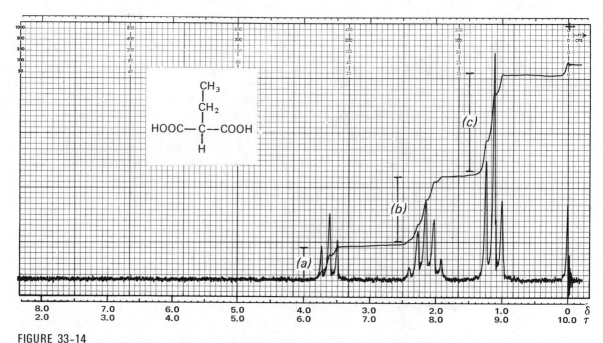

FIGURE 33-14

The high resolution nmr spectrum for ethyl malonic acid. The stepped integration curve is superimposed on the spectrum. The TMS internal standard peak is at $\delta = 0$ ($\tau = 10$).

the methyl protons (1.8δ, 8.2τ), (2) the overall signal area for the methylene protons is less than that for the methyl protons (two-thirds the area actually), in proportion to the numbers of equivalent protons for each signal, and (3) the methylene signal is split into four peaks and the methyl signal into three peaks.

Figure 33-14 is the nmr spectrum of the dicarboxylic acid ethyl malonic acid. The —COOH proton signals are off the scale in Figure 33-14 and are not shown. The signal for the —CH proton appears farthest downfield (3.6δ, 6.4τ) and is split into three peaks (a triplet) because of the two protons in the neighboring —CH$_2$ (methylene) group. The methylene proton signal, which is next upfield (2.2δ, 7.8τ), is influenced by the three protons on the neighboring methyl group *and also* by the one proton on the neighboring —CH group, and hence is split into five peaks (a quintet). Finally, the methyl proton signal is still farther upfield (1.1δ, 8.9τ) and is split into three peaks (a triplet) because of the two protons on the neighboring methylene group. The areas under the overall signals are in the ratios of 1:2:3, reflecting the numbers of equivalent protons of each type. In this spectrum, the stepped integration curve is shown, with the step for the —CH protons designated by (*a*), that for the —CH$_2$ protons designated by (*b*), and that for the —CH$_3$ protons designated by (*c*). Note that Step (*a*) has a height of 6 divisions on the recording paper, Step (*b*) a height of 12 divisions, and Step (*c*) a height of 18 divisions, reflecting the ratios 1:2:3 for the numbers of equivalent protons of each type.

The many effects, therefore, that protons and electrons within various portions of a molecule have upon the nmr spectrum make possible the determination of a very large amount of useful information concerning the structure of the molecule. Since protons are in many different kinds of inorganic and organic molecules, and also since a number of other nuclei can be used as the basis for nmr spectra, nuclear magnetic spectroscopy has come to be one of the most useful tools the chemist has for the characterization of compounds.

33.6 Electron Paramagnetic Resonance

Electron paramagnetic resonance (epr), sometimes referred to as **electron spin resonance** (esr), is analogous to nuclear magnetic resonance except that epr involves electrons rather than nuclei. When a compound contains an unpaired electron the unpaired electron spins about its own axis and creates a magnetic moment. Just as in the nmr method with nuclei, the magnetic moment of the electron can be lined up *with* an external magnetic field or *against* the field, and energy is required to change the spin of the electron from alignment with the field to the higher energy condition of alignment against the field. In the epr method, as in the nmr method, the energy is provided by electromagnetic radiation of the proper frequency. The absorption of the radiation gives rise to an epr, or esr, spectrum.

Although the epr and nmr methods are quite analogous, they differ in the amount of energy required. A much greater amount of energy is required to reverse the spin of an electron than to reverse the spin of a hydrogen nucleus (proton), because of the much larger magnetic moment created by the electron. Hence, epr absorption occurs at a much higher frequency, or lower wavelength (in the microwave region), than does nmr absorption (see Fig. 33–1).

As with nmr signals splitting and coupling also occur with epr signals, the spin-spin coupling occurring between electrons and neighboring hydrogen nuclei in much the same way and for essentially the same reasons as between two hydrogen nuclei for nmr signals. Electron paramagnetic resonance is useful in the detection of organic free radicals (organic structures with an odd, or unpaired electron) and in the study of their structures, especially with respect to where the lone electron is located within the system. The method also can be used in the study of some metal complexes and other kinds of molecules possessing unpaired electrons.

33.7 Mass Spectra

A mass spectrum can be produced by bombarding a sample of a compound in the gas phase with a beam of high energy electrons; hence, unlike the spectral methods we have studied thus far in this chapter, mass spectrometry does not utilize absorption of electromagnetic radiation.

In the mass spectrometer, bombardment of the sample in the gas phase by high energy electrons knocks one electron from a sample molecule, thereby producing a singly charged positive ion referred to as the **parent ion.** The parent ion in turn fragments to form a variety of other fast moving positive ion fragments, mainly singly charged. These ions are directed in the mass spectrometer through a magnetic field, which deflects their paths to an extent dependent upon their mass-to-charge ratios (m/e), the ions with lowest mass-to-charge ratio being deflected the most (see Section 30.2). Inasmuch as the ions are principally singly charged, the separation is primarily on the basis of mass. From the amount of deflection, taking into account the magnetic field and other characteristics of the instrument, the m/e ratio can be determined, and hence the mass, of each fragment.

The m/e ratio of the parent ion is the molecular weight of the molecule. Therefore, the mass spectrometer provides one of the best methods for obtaining very accurate molecular weights. More than that, however, the pattern of fragmentation and the masses of the parent ion and of each fragment ion provide an excellent "fingerprint" method for identifying a compound; only in a very small number of cases, such as with stereoisomers, do chemically different compounds fragment in exactly the same manner. In addition, a study of the fragmentation pattern yields considerable information about the structure of the compound and about the relative strengths of its bonds.

Chromatography

We mentioned earlier that chromatography involves the separation of a mixture of compounds by means of a system in which the components of the mixture migrate through the system at different rates. The method can be applied whether the sample is in the liquid phase or the gas phase. The method is based principally upon selective adsorption; in addition, other factors such as solubility, complexing, and hydrogen bonding play important roles in the process.

33.8   Gas Phase Chromatography

Gas phase, or vapor phase, chromatography utilizes a volatile mixture. Normally the gaseous mixture, referred to as the **moving phase,** is passed through a heated column containing a **stationary phase** consisting of an inert support material impregnated with a high-boiling liquid. The support material is typically either finely divided silicon dioxide, polymers of tetrafluoroethylene, glass beads, or porous polymeric beads of polyfluorolefins. An example of a commonly used high boiling liquid is silicone oil.

Separation of the gaseous mixture depends upon the fact that the various components will be distributed in different proportions between the moving vapor phase and the liquid part of the stationary phase. The component which has the lowest solubility in the stationary phase (or the least attraction otherwise for the stationary phase) will go through the column most rapidly, other components following in inverse order of solubility (or other attractive forces). Hence, separation is effected. The principle is analogous in some respects to that of fractional distillation (see Section 13.6). As each component is eluted from the column, a detector produces an electrical signal that is in proportion to the amount of component being eluted. The electrical signal for the concentrations of the eluted components is plotted against time on a record referred to as a **chromatogram.**

33.9   Liquid Phase Chromatography

Liquid phase chromatography is especially useful for separating mixtures of components which are not volatile or heat-sensitive and hence are not suitable for gas phase chromatography. The method can be applied either with a column (**column chromatography**) or with thin layers on a flat surface (**thin layer chromatography).**

Liquid phase chromatography can be accomplished either by **partition chromatography,** in which the stationary phase is an inert support impregnated with a liquid, or by **adsorptive chromatography,** in which the stationary inert support is a solid adsorbent material on which the components will be adsorbed but to different extents. In either partition or adsorptive chromatography the moving phase is a liquid.

A still different form of liquid phase chromatography is **ion exchange chromatography,** in which the stationary phase is a polymer containing acidic or basic groups located along its backbone. The acidic or basic groups react with ions present in the moving phase. Usually, a column is employed. The system can be set up so as to produce a quantitative exchange of ions, as in water softening by ion exchange (see Section 12.7), or so as to produce a difference in the rate at which various ionic substances in the moving phase proceed through the column. The method can be adapted to either cation or anion exchange. Acidic sites on the polymer are often carboxylic or sulfonic acids; basic sites are often substituted ammonium groups or amines.

The chromatographic process can be used to separate components of complex mixtures, such as 15 to 20 amino acids obtained from a protein, to separate components of very small samples, and to separate components in samples of very low concentrations.

A refinement and improvement in liquid chromatography is accomplished by operating the column at high pressures, making possible more efficient columns. The method, referred to as **high pressure liquid chromatography** (HPLC), is receiving a considerable amount of research effort currently and is proving to be of immense value in certain types of difficult mixture separations.

Combinations of Methods for Characterization of Compounds

The power of instrumental methods for characterization of compounds is often greatly increased by combining two or more methods together. One of several examples is the combination of gas chromatography with mass spectrometry (**gas chromatographic mass spectrometry,** GCMS). The rapidly increasing availability of small "on-line" computers suitable for use with an individual piece of equipment and the rapid development of a wide variety of automated data collection systems have in recent years made such combinations much more feasible than in the past.

QUESTIONS

1. Explain what is meant by the *characterization* of a compound.
2. List five forms of spectroscopy and distinguish between them.
3. What is the meaning of each of the following terms: molar extinction coefficient, absorbance, per cent transmittance, molar absorptivity, wavelength, wave number?
4. State which spectroscopic methods cover, respectively, the various portions of the spectrum from 100 millimicrons wavelength to above 1×10^{10} millimicrons.
5. (a) Relate the following units to each other: millimicrons, centimeters, nanometers, reciprocal centimeters, microns.
   (b) Which of the above units are used for wave number? which for wavelength?
   (c) How many microns are in a nanometer? how many centimeters in a nanometer? how many nanometers in an Angstrom unit? how many microns in an Angstrom unit?
6. How can ultraviolet spectra be used in the characterization of organic compounds?
7. How can visible spectra be used in the characterization of metal coordination compounds?
8. What is meant by the terms *stretching vibrations* and *bending vibrations*, and what is their significance to (a) spectroscopy and (b) characterization of compounds?
9. How may certain functional groups in a molecule be identified by spectroscopy?
10. Of what use is infrared spectroscopy in the characterization of metal coordination compounds?
11. Explain why nuclear magnetic resonance is helpful in determining the molecular structures of some types of compounds. What are some kinds of compounds that cannot be studied by nmr?
12. Distinguish between gas phase chromatography and liquid phase chromatography.
13. Identify as many bands (peaks) as possible in the following infrared spectrum for 2-methyl-1-penten-3-ol, $C_6H_{12}O$, considering the molecular formula given in the box on the spectrum and using the data provided in Fig. 33–5 and Table 33–1.

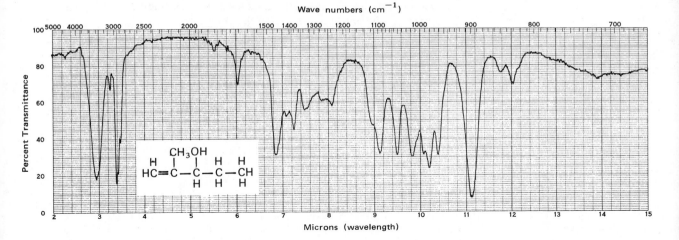

14. Identify as many bands (peaks) as possible in the following infrared spectrum for isobutyl cinnamate, $C_{13}H_{16}O_2$, considering the molecular formula given in the box on the spectrum and using the data provided in Fig. 33–5 and Table 33–1.

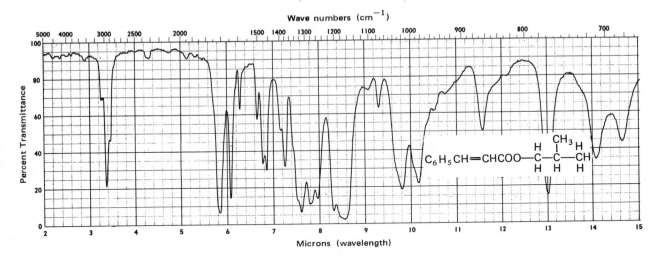

15. In the following nmr spectrum for triethylamine, $(C_2H_5)_3N$, two signals, each split into several peaks, are present; in addition there is the small tetramethylsilane peak at 0δ, or 10τ.
    (a) Explain why two signals, other than the one for TMS, would be expected.
    (b) Determine what portion of the molecule is responsible for each signal and explain the basis for your decision.
    (c) State how many peaks in theory should result from the splitting of each signal at high resolution, explain why, and determine if the spectrum agrees with the theory.

(d) Determine the pertinent heights in the stepped integration curve, by counting divisions, to determine the relative total signal areas; then check whether this agrees with the theory based upon numbers of equivalent protons of each environmental type.

(e) Explain the purpose of the tetramethylsilane peak.

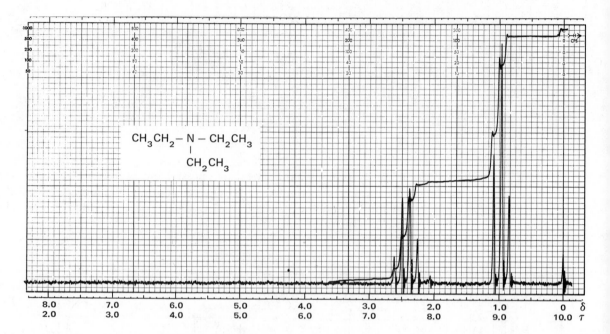

$$CH_3CH_2 - N - CH_2CH_3$$
$$|$$
$$CH_2CH_3$$

REFERENCES

"Automatic Analysis of Blood Cells," M. Ingram and K. Preston, Jr., *Sci. American*, Nov., 1970; p. 72.

"Chemical Analysis by Infrared," B. Crawford, Jr., *Sci. American*, Oct., 1953; p. 42.

"Gas Chromatography," A. Keller, *Sci. American*, Oct., 1961; p. 58.

"Infrared Determination of Stereochemistry in Metal Complexes," M. Y. Darensbourg and D. J. Darensbourg, *J. Chem. Educ.*, **47**, 33 (1970).

"Electronic Spectra of Some Transition Metal Complexes: Derivation of Dq and B," A. B. P. Lever, *J. Chem. Educ.*, **45**, 711 (1968).

"Mass Spectra of Organic Compounds," D. H. Williams, *Chem. in Britain* **4**, 5 (1968).

"NMR of Boron Compounds," G. R. Eaton, *J. Chem. Educ.*, **46**, 547 (1969).

"Modern Aspects of Mass Spectrometry," J. F. J. Todd, *Educ. in Chemistry*, **10**, 89 (1973).

"Identifying Alcohols by Thin Layer Chromatography," M. Crawford and A. K. Keenan, *Educ. in Chemistry*, **10**, 98 (1973).

"Troublesome Concepts in NMR Spectrometry," R. M. Silverstein and R. G. Silberman, *J. Chem. Educ.*, **50**, 484 (1973).

"On Thermodynamics of Solution by Gas-Liquid Chromatography," E. F. Meyer, *J. Chem. Educ.*, **50**, 191 (1973).

"Chemistry and the Spinning Electron," (Staff), *New Scientist,* **60** (867), 128 (1973).

"Spectroscopy and Structure Elucidation," C. W. Jefford, R. McCreadie, P. Muller, and J. Pfyffer, *J. Chem. Educ.*, **50,** 181 (1973).

"Test for Chemical Shift and Magnetic Equivalence in NMR," A. Ault, *J. Chem. Educ.*, **51,** 729 (1974).

"Qualitative Emission Spectroscopy in Freshman Chemistry," L. F. Druding and R. A. Lalancette, *J. Chem. Educ.*, **51,** 527 (1974).

"Calculation of Complex NMR Spectra in the Undergraduate Laboratory," R. J. Seyse, H. L. Pearce, and T. L. Rose, *J. Chem. Educ.*, **52,** 194 (1975).

"Spectroscopic Determination of Thermodynamic Quantities," M. Berger, J. A. Bell, and C. Steel, *J. Chem. Educ.*, **52,** 191 (1975).

"Lead in Blood—The Analyst's Problem," A. A. Cernik, *Chem. in Britain*, **10,** 58 (1974).

"Recent Developments in the Analysis of Toxic Elements," D. J. Lisk, *Science*, **184,** 1137 (1974).

"A Combined Infrared and Kinetic Study of Linkage Isomers: An Inorganic Experiment," W. H. Hohman, *J. Chem. Educ.*, **51,** 553 (1974).

"Infrared Determination of Stereochemistry in Metal Complexes: The Determination of Symmetry Coordinates," D. J. Darensbourg and M. Y. Darensbourg, *J. Chem. Educ.*, **51,** 787 (1974).

"Electron Spin Resonance in Transition-Metal Chemistry," J. B. Raynor, *Chem. in Britain*, **10,** 254 (1974).

"Laser Spectroscopy: Probing Biomolecular Functions," J. L. Marx, *Science*, **188,** 1002 (1975).

"Two Spectrophotometric Experiments with Alpha-Chymotrypsin," E. R. Kantrowitz and G. Eisele, *J. Chem. Educ.*, **52,** 410 (1975).

"Isomer Analysis by Spectral Methods," G. A. Poulton, *J. Chem. Educ.*, **52,** 397 (1975).

Metals of Analytical Group I

34

Mercury, Silver, and Lead

Mercury(I), silver, and lead(II) ions are precipitated as chlorides by the use of a slight excess of hydrochloric acid, whereas the chlorides of all the other common metal ions are soluble in dilute acid. Thus, Hg_2^{2+}, Ag^+, and Pb^{2+} make up Group I of our analytical scheme (Chapter 40). If the less familiar ions were also considered Analytical Group I would include Cu^+, Au^+, and Tl^+ because the chlorides of all these ions are insoluble.

The solubilities of mercury(I) chloride, Hg_2Cl_2, and silver chloride, $AgCl$, are low enough so that the mercury(I) ion and the silver ion are completely precipitated by the chloride ion in acid solution. However, lead chloride is somewhat soluble and may even fail to precipitate if the concentration of the lead ion is too low or the temperature is too high (lead chloride is readily soluble in hot water). Because of the relatively high solubility of the chloride, lead is found in Group II as well as in Group I. Some properties of the metals of Analytical Group I are listed in Table 34–1.

Mercury

34.1 Periodic Relationships of Mercury

The three elements of Periodic Group IIB are zinc, cadmium, and mercury. Cadmium is in Analytical Group II (Chapter 35), and zinc is in Group III (Chapter 36). Each of these elements has two electrons in its outer shell and eighteen in the underlying shell. They are **representative** metals and the +2 oxidation state

TABLE 34-1   Some Properties of the Metals of Analytical Group I

Property	Mercury	Silver	Lead
Atomic number	80	47	82
Atomic weight	200.59	107.868	207.2
Electronic structure	2, 8, 18, 32, 18, 2	2, 8, 18, 18, 1	2, 8, 18, 32, 18, 4
Reduction potential, volts	(Hg_2^{2+}/Hg) +0.80 (Hg^{2+}/Hg) +0.79	(Ag^+/Ag) +0.80	(Pb^{2+}/Pb) −0.13
Oxidation states	+1, +2	+1, +2, +3	+2, +4
Density, g/cm³ at 20°	13.546	10.50	11.34
Melting point, °C	−38.87	960.8	327.4
Boiling point, °C	356.57	2,193	1,750

prevails, although the +1 oxidation state of mercury (Hg_2^{2+}) is important. The melting points, boiling points, and heats of vaporization for the Group IIB members are lower than for any other group of metals except the alkali metals (Group IA). Of the three elements, zinc is the most reactive and mercury the least. The ions of all three elements are small in relation to their charges, so they show strong tendencies toward the formation of complex ions, like the transition metals, which they immediately follow in the Periodic Table. The most frequently encountered complex ions of this family are the halo-, cyano-, and ammine complexes. Coordination numbers of both 4 and 6 are quite common in the complexes of these elements. The 4-coordinate complexes are generally tetrahedral (sp^3 configuration). The 6-coordinate complexes have octahedral structures, but the bonding must be of the sp^3d^2 rather than the d^2sp^3 type, for the inner d orbitals of the simple ions are completely filled with electrons. Hybridization of the sp^3d^2 type is represented diagrammatically for $[Zn(NH_3)_6]^{2+}$ as

$$[Zn(NH_3)_6]^{2+} \quad 3d^2 \; 3d^2 \; 3d^2 \; 3d^2 \; 3d^2 \; \boxed{\begin{array}{l} 4s^2 \; 4p^2 \; 4p^2 \; 4p^2 \; 4d^2 \; 4d^2 \\ sp^3d^2 \text{ hybridization} \end{array}}$$

Octahedral coordination

34.2 History of Mercury

Mercury was one of the few metals known to the ancients. The Greek philosopher Aristotle refers to the metal as "liquid silver," or "quick silver," because of its appearance and liquid state. The alchemists named the element after *Mercury*, the swift messenger of the gods of Roman mythology. One of the early uses of mercury was for the extraction of gold from its ores, inasmuch as gold is quite soluble in mercury. The alchemists used mercury in their attempts to transmute base metals into gold, assuming that the mercury had unusual powers due to its liquid state and the property of readily forming amalgams.

34.3 Occurrence and Metallurgy of Mercury

Elemental mercury is sometimes found in rocks and as an amalgam of silver and gold. However, the most important commercial source of the metal is the dark red sulfide, HgS, known as **cinnabar.** When cinnabar is roasted, metallic mercury distills from the furnace and is condensed to the liquid state.

$$HgS(s) + O_2(g) \longrightarrow Hg(l) + SO_2(g) + 57{,}000 \text{ cal} \qquad (\Delta H^\circ_{298} = -57{,}000 \text{ cal})$$

Ordinarily, the roasting of a mineral in air results in the oxidation of the metal. However, the oxides of mercury are thermally unstable and decompose at the temperature of the furnace. Mercury may be purified by filtering it through chamois or a gold seal (the mercury "wets" the gold seal and hence passes through it; the impurities do not). It may also be purified by washing with nitric acid to oxidize the metallic impurities, or by distilling it in an atmosphere of oxygen to remove the more active metals.

34.4   Properties and Uses of Mercury

Mercury is a silvery-white metal and is the only metal which is liquid at room temperature. Due to its low freezing point ($-38.87°$ C), its high boiling point ($356.6°$ C), its uniform coefficient of expansion, and the fact that it does not wet glass, mercury is an excellent thermometric substance. Because of its chemical inactivity, mobility, high density, and electrical conductivity, it is used extensively in barometers, vacuum pumps, and liquid seals, and for electrical contacts. With the exception of iron and platinum, all metals readily dissolve in, or are wet by, mercury to form amalgams. Sodium amalgam is used as a reducing agent, because it is less active than the alkali metal alone. Amalgams of tin, silver, and gold are used in dentistry.

Mercury vapor lamps and fluorescent lamps make use of the fact that mercury vapor is a conductor of electricity and, while conducting, it emits light.

Mercury does not change chemically in air at ordinary temperatures but is slowly oxidized when heated, forming **mercury(II) oxide,** HgO, which decomposes at still higher temperatures. Being below hydrogen in the activity series, mercury is not attacked by hydrochloric acid but does dissolve in an excess of oxidizing acids such as cold nitric or hot concentrated sulfuric acid. With excess acid, mercury(II) salts are formed.

$$3Hg + 8H^+ + 2NO_3^- \longrightarrow 3Hg^{2+} + 2NO\uparrow + 4H_2O$$
$$Hg + 2H_2SO_4 \longrightarrow Hg^{2+} + SO_4^{2-} + SO_2\uparrow + 2H_2O$$

When the mercury is in excess these oxidizing acids convert the metal to the corresponding mercury(I), Hg_2^{2+}, salts rather than to the mercury(II), Hg^{2+}, salts.

The halogens attack mercury forming the halides, and sulfur combines with it to produce mercury(II) sulfide. Mercury is displaced from solutions of its ions by all metals except silver, gold, and the platinum metals (Ru, Rh, Pd, Os, Ir, and Pt). A copper wire soon becomes amalgamated when immersed in a solution of a mercury compound.

Mercury forms two series of compounds; in these series the metal shows oxidation states of $+1$ and $+2$. The mercury atom uses both its valence electrons in bonding in all of its compounds. For example, in mercury(II) chloride, $HgCl_2$, the chlorine atoms are covalently bonded to the mercury atom in a linear molecule, as shown by the electronic formula

$$\overset{\times\times}{\underset{\times\times}{\times}Cl}\overset{}{\underset{}{\,}}Hg\overset{}{\underset{}{\,}}\overset{\times\times}{\underset{\times\times}{Cl}\times}$$

With the strongly electronegative sulfate and nitrate groups, mercury(II) forms ionic compounds in which dipositive mercury(II) ions, Hg^{2+}, are present. In the series of compounds in which mercury exhibits a $+1$ oxidation state, one valence electron of each of the mercury atoms serves to form an electron-pair bond between the two mercury atoms. The Hg_2Cl_2 molecule has a linear covalent structure.

$$\overset{\times\times}{\underset{\times\times}{\times}Cl}\overset{}{\underset{}{\,}}Hg\overset{}{\underset{}{\,}}Hg\overset{}{\underset{}{\,}}\overset{\times\times}{\underset{\times\times}{Cl}\times}$$

The equal sharing of the electron pair between the two mercury atoms does not contribute to the oxidation state of either atom, but the unequal sharing of an electron pair by each mercury atom with an electronegative chlorine atom gives each mercury atom an oxidation state of $+1$. When the second electron of each mercury atom is lost completely by electron transfer, as in the formation of $Hg_2(NO_3)_2$, the double mercury(I) ion, $\overset{+}{Hg}:\overset{+}{Hg}$ is formed.

The $Hg_2{}^{2+}$ ion is the most common ionic species with a **metal-metal bond.** Evidence from various sources points to the dimeric nature of this ion. X-ray studies of crystalline mercury(I) chloride show that mercury atoms occur in pairs, and magnetic studies of solutions containing univalent mercury show that it is not paramagnetic, meaning that there are no unpaired electrons as there would be if such solutions were to contain Hg^+ ions with an odd number of electrons per ion. On the contrary, $Hg:Hg^{2+}$ contains an even number of electrons so that all the electrons present are assumed to exist in pairs. Thus, it is correct to write the formulas of mercury(I) compounds as Hg_2X_2, instead of HgX, where X is an acid radical.

34.5   Mercury(I) Compounds

A number of compounds containing univalent mercury are known. One of these, **mercury(I) chloride,** Hg_2Cl_2 (commonly called mercurous chloride), is a white, insoluble crystalline compound, which is of analytical importance because its insolubility serves to precipitate the mercury(I) ion in Group I.

$$Hg_2{}^{2+} + 2Cl^- \longrightarrow \underline{Hg_2Cl_2}$$

Mercury(I) chloride is manufactured by heating a mixture of mercury(II) sulfate, mercury, and sodium chloride.

$$HgSO_4 + Hg + 2NaCl \longrightarrow Hg_2Cl_2 + Na_2SO_4$$

Mercury(I) chloride is volatile and sublimes from the reaction mixture. It is used in the field of medicine under the name **calomel** as a cathartic and a diuretic (i.e., it stimulates the organs of secretion).

When exposed to light, mercury(I) chloride slowly decomposes ($Hg_2Cl_2 \longrightarrow HgCl_2 + Hg$); hence, it is usually stored in amber-colored bottles.

When mercury dissolves in cold, dilute nitric acid (with the mercury in excess), **mercury(I) nitrate,** $Hg_2(NO_3)_2$, is formed. This compound is important because it is the only readily soluble salt of univalent mercury. This salt hydrolyzes in aqueous solution to form the basic nitrate, $Hg_2(OH)NO_3$. The addition of nitric acid reverses the hydrolysis.

$$Hg_2(NO_3)_2 + H_2O \rightleftharpoons \underline{Hg_2(OH)NO_3} + H^+ + NO_3^-$$

On standing in contact with air, solutions of mercury(I) nitrate readily oxidize to mercury(II) nitrate. The accumulation of Hg^{2+} can be prevented by keeping a small amount of metallic mercury in the bottle in which the solution is stored ($Hg^{2+} + Hg \longrightarrow Hg_2^{2+}$).

Mercury(I) sulfide, Hg_2S, is unstable and immediately decomposes into mercury and mercury(II) sulfide when it is formed by passing hydrogen sulfide into a solution containing mercury(I) ions.

$$Hg_2^{2+} + H_2S \longrightarrow \underline{Hg_2S} + 2H^+$$
$$\quad\quad\quad\quad\quad\quad \longrightarrow \underline{Hg} + \underline{HgS}$$

Aqueous ammonia reacts with mercury(I) chloride in an oxidation-reduction reaction to form metallic mercury in a finely divided black powder form and a complex white salt, **mercury(II) amido chloride,** containing bivalent mercury.

$$Hg_2Cl_2 + 2NH_3 \longrightarrow \underline{Hg} + \underline{HgNH_2Cl} + NH_4^+ + Cl^-$$

The formation of a black material when the white mercury(I) chloride is treated with aqueous ammonia is used as the test for the mercury(I) ion in qualitative analysis.

Alkali metal hydroxides precipitate the mercury(I) ion as black Hg_2O, which is unstable and decomposes into Hg and HgO.

Reducing agents, such as tetrachlorostannate(II) ion, readily reduce mercury(I) compounds to metallic mercury. This property is used in qualitative tests for the ion.

$$\underline{Hg_2Cl_2} + SnCl_4^{2-} \longrightarrow 2\underline{Hg} + [SnCl_6]^{2-}$$

34.6 Mercury(II) Compounds

When a strong base is added to a solution of a mercury(II) compound the oxide, HgO, precipitates. When precipitated from cold solutions **mercury(II) oxide** is yellow, but when precipitated from hot solutions it is red. This difference in color has been attributed to a difference in the state of subdivision of the substance, inasmuch as the crystal structure is the same in the two cases. **Mercury(II) hydroxide** is unstable, losing the elements of a molecule of water to form the oxide.

Mercury(II) chloride, $HgCl_2$, may be formed by heating the metal with an excess of chlorine, by dissolving mercury(II) oxide in hydrochloric acid, or by the action of aqua regia upon mercury. It is prepared commercially by heating mercury(II) sulfate with sodium chloride; the $HgCl_2$ sublimes from the reaction mixture.

$$2NaCl + HgSO_4 \longrightarrow Na_2SO_4 + HgCl_2$$

The compound $HgCl_2$ is also called **bichloride of mercury** and **corrosive sublimate.** In dilute solutions it is used as an antiseptic. It is moderately soluble in water but only slightly ionized, as indicated by the low electrical conductivity of its solutions. Its solubility can be increased by adding an excess of chloride ions, which cause the formation of the complex **tetrachloromercurate(II) ion** according to the equation

$$HgCl_2 + 2Cl^- \rightleftharpoons [HgCl_4]^{2-}$$

Mercury(II) chloride hydrolyzes in water somewhat, and ammonolyzes in aqueous ammonia. These reactions are similar:

$$Cl—Hg—Cl + 2H_2O \longrightarrow Cl—Hg—OH + H_3O^+ + Cl^-$$
$$Cl—Hg—Cl + 2NH_3 \longrightarrow Cl—Hg—NH_2 + NH_4^+ + Cl^-$$

Iodide ions precipitate mercury(II) ions from solution as HgI_2, which is red in color. It is soluble in a solution containing an excess of iodide ions, tetra-iodomercurate ions being formed according to the equation

$$HgI_2 + 2I^- \longrightarrow [HgI_4]^{2-}$$

Hydrogen sulfide precipitates black **mercury(II) sulfide,** HgS, from solutions of mercury(II) salts, even in strongly acidic solutions. Because of this, bivalent mercury is a member of Analytical Group II. The precipitate formed by the interaction of $HgCl_2$ in solution and H_2S is first white, then yellow, then red, and finally black. The white compound has the formula $HgCl_2 \cdot 3HgS$. When heated, HgS becomes bright red; the red sulfide is isomeric with the black and has, in the past, been used as a pigment under the name of **vermilion.**

Mercury(II) sulfide is dissolved by aqueous solutions of sodium sulfide, Na_2S, in the presence of an excess of hydroxyl ions and forms the **thiomercurate ion.**

$$HgS + S^{2-} \rightleftharpoons [HgS_2]^{2-}$$

This reaction is used in the Group II procedures of the analytical scheme (Section 41.2).

Mercury(II) fulminate, $Hg(ONC)_2$, is an explosive compound formed by the action of nitric acid on mercury in the presence of ethyl alcohol. It is used in making detonators and percussion caps because it explodes when struck. Mercury fulminate probably has the structure shown.

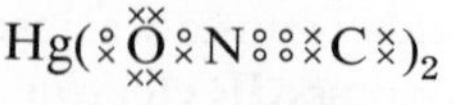

$$Hg(\overset{xx}{\underset{xx}{\ddot{O}}} \, N \, C \,)_2$$

Mercury fulminate

34.7   Physiological Action of Mercury

Metallic mercury is an accumulative poison; the breathing of air containing mercury vapor or contact of the liquid metal with the skin should be avoided. All soluble mercury compounds are poisonous, but in small doses they may be medicinal. The fatal dose of mercury(II) chloride is 0.2 to 0.4 gram. The mercury(II) ion combines with the protein tissue of the kidney and destroys the ability of this organ to remove waste products from the blood. Antidotes for mercury poisoning are egg white and milk; their proteins precipitate the mercury in the stomach.

34.8   Properties of Mercury of Analytical Importance

The mercury(I) ion is colorless and is precipitated as a white crystalline solid by hydrochloric acid, a property which makes it a member of Analytical Group I. Aqueous ammonia converts Hg_2Cl_2 to a black or gray mixture of Hg, black, and $HgNH_2Cl$, white.

The mercury(II) ion of Analytical Group II is also colorless, but $HgCl_2$ is soluble in hydrochloric acid; with hydrogen sulfide, black HgS is formed. This compound is insoluble in most acids but dissolves in aqua regia or a mixture of hydrochloric and hypochlorous acids. It is dissolved also by a mixture of NaOH and Na_2S, the complex ion $[HgS_2]^{2-}$ being formed. This property is often used in separating HgS from the sulfides of lead, cadmium, and bismuth, which do not form soluble thiocomplexes.

Both unipositive and dipositive mercury ions are readily reduced by tin(II) chloride. A black precipitate of finely divided mercury is the final product in both cases (when an excess of $SnCl_2$ is employed). Mercury is displaced from solutions of its ions by all the metals which lie above it in the activity series. A copper strip or wire soon becomes amalgamated when dipped in a solution of a mercury compound, a change which may be used as a simple test for mercury.

Silver

34.9   Periodic Relationships of Silver

Copper, silver, and gold are known as **the coinage metals** and have been used from very early times for the manufacture of ornamental objects and coins. They compose Group IB of the Periodic Table and may be classified as transition elements in that they have valence electrons in two shells. Each has one electron in its outermost shell and eighteen in the underlying shell. Because both the alkali metals of Group IA and the coinage metals of Group IB each have one electron in their outermost shell, the members of these two families might be expected to behave similarly. The elements of both families are good conductors of electricity and form a series of univalent compounds with analogous formulas (Na_2O, Ag_2O; NaCl, AgCl; and Na_2SO_4, Ag_2SO_4). But beyond this, similarities between the two families are almost entirely lacking. For example, the alkali metals are highly reactive while the coinage metals are quite unreactive. This results, in part, from the fact that the atoms of the IB family are smaller than those of the IA family, and thus their valence electrons are held more closely to the nucleus. Copper, silver, and gold, unlike the alkali metals, can use one or two electrons from the underlying shell in bond formation. Thus, copper, silver, and gold each exhibit $+1$, $+2$, and $+3$ oxidation states. However, the most common oxidation state for copper is $+2$; for silver, $+1$; and for gold, $+3$. Many of the ions and compounds of the coinage metals are colored, whereas the ions of the alkali metals are colorless and their only colored compounds are those having a color associated with the anion (e.g., $K_2Cr_2O_7$ and $KMnO_4$). The hydroxides of the alkali metals are soluble and strongly basic whereas those of copper and gold are insoluble and weakly basic. (Silver hydroxide is slightly soluble and strongly basic.) Most compounds of the alkali metals are soluble in water, while the majority of those of the coinage metals are insoluble. Finally, the alkali metals show little tendency toward the formation of complex ions, while many stable complexes, such as $[Cu(CN)_2]^-$, $[Ag(CN)_2]^-$, $[Au(CN)_2]^-$, $[Cu(NH_3)_2]^+$, $[Cu(NH_3)_4]^{2+}$, and $[Ag(NH_3)_2]^+$, are known for the coinage metals.

Among the coinage metals, trends in properties are not very regular nor readily interpreted. The chemical activity, elasticity, tensile strength, and specific heat decrease in the order copper, silver, gold; the density increases in this order. All three metals in their $+1$ oxidation states form insoluble chlorides, but only silver is included in Analytical Group I. Compounds of univalent copper are not common and are not as important as those of divalent copper, which is a member of Analytical Group II. Gold is considered a less common metal, and its chemical properties are relatively unimportant. Some physical properties of the coinage metals are given in Table 34–2.

TABLE 34–2   Some Physical Properties of the Coinage Metals

	Atomic Number	Atomic Weight	Atomic Radius, Å	Ionic (M^+) Radius, Å	Density at 20°	Melting Point, °C	Boiling Point, °C
Copper	29	63.546	1.28	0.96	8.92	1,083	2,582
Silver	47	107.868	1.44	1.26	10.50	960.8	2,193
Gold	79	196.9665	1.40	1.37	19.3	1,063	2,660

34.10   History of the Coinage Metals

These metals were the first to be used by primitive races, because they are found in the native state, they tarnish slowly, and their appearance is pleasing to the eye. Ornaments of gold have been found in Egyptian tombs which were constructed in the prehistoric Stone Age. Gold and silver coins were used as a medium of exchange in India and Egypt long before the Christian era.

It is believed that copper was the first metal to be fashioned into utensils, instruments, and weapons; its use for such purposes began sometime during the

Stone Age. The practice of alloying copper with tin and the use of the resulting bronze in place of stone introduced the Bronze Age.

The chemical symbol for gold (Au) is derived from the Latin name *aurum*, that for silver (Ag) from *argentum*, and that for copper (Cu) from *cuprum*.

34.11   Occurrence and Metallurgy of Silver

Silver is sometimes found in large nuggets but more frequently in veins and related deposits. It frequently is found alloyed with gold, copper, or mercury. It occurs combined as the chloride, AgCl **(horn silver),** and as the sulfide, Ag_2S, which is usually mixed with the sulfides of lead, copper, nickel, arsenic, and antimony. Most of the silver of commerce is a by-product of the mining of other metals such as lead and copper.

When lead is obtained from lead sulfide ore the lead is usually mixed with some silver. The silver is extracted from the fused lead by the use of zinc, in which silver is about 3000 times more soluble than it is in lead. The lead is fused and thoroughly mixed with a small quantity of zinc. Lead and zinc are immiscible. Most of the silver leaves the lead and dissolves in the zinc. When mixing is stopped, the zinc rises to the surface of the lead and solidifies. The zinc-silver alloy is removed from the lead and the more volatile zinc is separated from the silver by distillation. This extraction procedure is known as the **Parkes process.**

The extraction of silver from its ores is dependent upon the formation of the complex dicyanoargentate ion, $[Ag(CN)_2]^-$. Silver metal and all of its compounds are readily dissolved by alkali cyanides in the presence of air. Representative equations for the hydrometallurgy of silver follow.

$$4Ag + 8CN^- + O_2 + 2H_2O \longrightarrow 4[Ag(CN)_2]^- + 4OH^-$$
$$2Ag_2S + 8CN^- + O_2 + 2H_2O \longrightarrow 4[Ag(CN)_2]^- + 2S + 4OH^-$$
$$AgCl + 2CN^- \rightleftharpoons [Ag(CN)_2]^- + Cl^-$$

The silver is precipitated from the cyanide solution upon the addition of either zinc or aluminum.

$$2[Ag(CN)_2]^- + Zn \longrightarrow 2\underline{Ag} + [Zn(CN)_4]^{2-}$$

Much silver is also obtained from the anode mud formed during the electrolytic refining of copper (Section 35.8).

34.12   Properties of Silver

Silver is a white, lustrous metal and its surface is an excellent reflector of light. It is the best conductor of heat and electricity that we have other than gold, which is better but too expensive to use as a conductor. Silver is noted for its ductility and malleability.

Silver is not attacked by oxygen of the air under ordinary conditions but tarnishes quickly in the presence of hydrogen sulfide or upon contact with sulfur-containing food materials such as eggs and mustard. Its reaction with hydrogen sulfide in the presence of air is

$$4Ag(s) + 2H_2S(g) + O_2(g) \longrightarrow 2Ag_2S(s) + 2H_2O(l) + 142,350 \text{ cal}$$
$$(\Delta H^\circ_{298} = -142,350 \text{ cal})$$

Silver tarnish is a thin film of silver sulfide. The halogens react with silver forming the halides, and the metal is soluble in such oxidizing acids as nitric and hot sulfuric.

$$3Ag + 4H^+ + NO_3^- \longrightarrow 3Ag^+ + NO\uparrow + 2H_2O$$

Because silver lies below hydrogen in the activity series, it is not soluble in nonoxidizing acids such as hydrochloric. Most silver salts are sparingly soluble in water, but the nitrate is quite soluble. The common oxidation state exhibited by silver is $+1$, although the oxides AgO and Ag_2O_3 and many complex compounds containing $Ag(II)$ and $Ag(III)$ are known.

34.13   Uses of Silver

Considerable silver is used for tne making of coins, silverware, and ornaments. For most of its uses silver is too soft to wear well, and it is therefore hardened by alloying it with other metals, particularly copper. **Sterling silver** contains 7.5 per cent copper, and jewelry silver contains 20 per cent copper. Large amounts of silver are used in the preparation of dental alloys, photographic films, and mirrors.

34.14   Plating with Silver

The electroplating industry uses a large percentage of the metal produced. The object to be plated with silver is made the cathode in an electrolytic cell containing a solution of sodium dicyanoargentate, $Na[Ag(CN)_2]$, as the electrolyte. The anode is a bar of pure silver, which dissolves to replace the silver ions removed from solution as plating at the cathode proceeds. The electrode reactions are

Cathodic reduction: $[Ag(CN)_2]^- + e^- \longrightarrow Ag + 2CN^-$

Anodic oxidation: $Ag + 2CN^- \longrightarrow [Ag(CN)_2]^- + e^-$

The film of deposited silver has a flat white appearance, but it can be made to assume a brilliant luster by burnishing.

The characteristic luster can be restored to tarnished silverware by the following application of the activity series. The tarnished silver object is immersed in a solution of baking soda and salt containing a teaspoonful of each to a quart of water held in an aluminum vessel. (A piece of aluminum metal in a glass vessel will serve equally well.) The aluminum and silver must be in contact. After the tarnish is removed, the silver object should be thoroughly washed. The chemistry involved may be explained in this way: the silver sulfide of the film of tarnish dissolves somewhat in the solution, forming silver and sulfide ions. The more active aluminum displaces the silver from solution, causing it to plate out on the silver object, according to the equation

$$3Ag^+ + Al \longrightarrow 3\underline{Ag} + Al^{3+}$$

The tarnish is removed with little loss of silver; however, the "antiquing (actually silver sulfide tarnish), which often gives a desirable appearance to the crevices of the design, will be removed also. Silver polish can be used to remove tarnish but with loss of the silver in the tarnish.

Silver mirrors are formed by depositing a thin layer of silver on glass. This is accomplished by reducing an ammoniacal solution of silver nitrate with some mild reducing agent, such as glucose or formaldehyde. The equation for the reaction may be written as

$$[Ag(NH_3)_2]^+ + e^- \text{ (from reducing agent)} \longrightarrow \underline{Ag} + 2NH_3$$

The film of silver deposited on the glass is washed, dried, and varnished.

34.15   Compounds of Silver

■ **1. Silver Oxide.**   **Silver oxide** is formed when silver is exposed to ozone, or when finely divided silver is heated in oxygen under pressure. The alkalies act upon silver nitrate to give a dark-brown amorphous precipitate of silver oxide. Although the oxide is but slightly soluble in water, it dissolves enough to give a distinctly alkaline solution. The equilibrium is given by

$$Ag_2O + H_2O \rightleftharpoons 2Ag^+ + 2OH^-$$

Silver oxide is a convenient reagent, both in inorganic and organic chemistry, for preparing soluble hydroxides from the corresponding halides, because the silver halide formed at the same time may be removed conveniently by filtration. Cesium hydroxide, for example, may be prepared according to the reaction

$$2Cs^+ + 2Cl^- + Ag_2O + H_2O \longrightarrow 2Cs^+ + 2OH^- + 2\underline{AgCl}$$

This reaction proceeds to the right because AgCl is much less soluble than Ag_2O. Silver oxide dissolves readily in aqueous ammonia to form the strong base **diamminesilver hydroxide.**

$$Ag_2O + 4NH_3 + H_2O \longrightarrow 2[Ag(NH_3)_2]^+ + 2OH^-$$

When the oxide is heated at atmospheric pressure in air, it readily gives up its oxygen according to the equation

$$14{,}840 \text{ cal} + 2Ag_2O(s) \longrightarrow 4Ag(s) + O_2(g) \qquad (\Delta H^\circ_{298} = 14{,}840 \text{ cal})$$

■ **2. The Silver Halides.**   It is of interest that **silver fluoride,** AgF, is very soluble in water, whereas the chloride, bromide, and iodide are relatively insoluble—the higher the atomic weight of the halogen, the lower the solubility of the corresponding silver halide. The insoluble silver halides are formed as curdy precipitates when halide ions are added to solutions of silver salts. The chloride is white, silver bromide is pale yellow, and silver iodide is yellow. Upon exposure to light the halides of silver turn violet at first and finally black; they are decomposed by light into their elements.

$$2AgX + \text{light} \longrightarrow 2\underline{Ag} + X_2 \qquad (X = \text{halogen})$$

Silver chloride is readily soluble in an excess of dilute aqueous ammonia to form the diamminesilver complex.

$$\underline{AgCl} + 2NH_3 \longrightarrow [Ag(NH_3)_2]^+ + Cl^-$$

This reaction is used in separating silver chloride from mercury(I) chloride in the qualitative analysis scheme. Silver bromide will dissolve only if the ammonia is concentrated, and the iodide is scarcely soluble at all in this reagent. Use is made of these facts in the separation and identification of the halide ions in the anion analytical scheme (Section 45.11).

■ **3. Silver Nitrate.**   **Silver nitrate,** $AgNO_3$, is the only simple silver salt which is found to be usefully soluble. It is obtained by dissolving silver in nitric acid and evaporating the solution. Organic materials, such as the skin, readily reduce silver nitrate to give free silver, which forms a black stain. Much silver nitrate is used in the production of photographic materials, as a laboratory reagent, and in the manufacture of other silver compounds.

34.16   The Photographic Process (Black and White)

■ **1. The Photographic Film.**   Photographic films are thin sheets of cellulose acetate coated with a colloidal suspension of small silver halide crystals in gelatin. Pure solutions of silver nitrate, potassium bromide, and potassium iodide are mixed in the presence of the gelatin in the preparation of the photographic emulsion ($Ag^+ + X^- \longrightarrow AgX$). The size of the silver halide particles and the relative amounts of bromide and iodide used determine the sensitivity, or speed, of the film.

■ **2. Exposure.**   The film is exposed by focusing an image upon the light-sensitive photographic emulsion on the film. The silver halide crystals become "activated" when exposed to light, so that they are more readily reduced than before. The chemical change occurring during exposure is not well understood, but the silver halide crystals receiving the most light (from the light areas of the object) are more readily reduced, during development of the film, than those receiving little light (from the dark areas of the object). Exposure causes no visible change in the appearance of the film.

■ **3. Developing.**   After exposure the film is developed by placing it in an alkaline solution of an organic reducing agent such as either hydroquinone or pyrogallol. The reducing agent acts upon the grains of silver halide with a speed proportional to the intensity of the illumination during exposure and reduces them to metallic silver.

$$AgX + e^- \text{ (reducing agent)} \longrightarrow \underline{Ag} + X^-$$

■**4. Fixing.** After developing, the film is fixed by treating it with a solution of sodium thiosulfate (hypo) to dissolve the unreduced silver halide.

$$AgX + 2S_2O_3{}^{2-} \rightleftharpoons [Ag(S_2O_3)_2]^{3-} + X^-$$

The metallic silver remaining on the film forms the visible image and is called a **negative,** since the light portions of the original image are now dark and the dark portions of the original are now light.

■**5. Printing.** The process of printing is essentially the same as that of making a negative, but due to the fact that the sensitive printing paper is illuminated through the negative, the image is once again reversed, and now corresponds to the original image as regards light and dark areas on the print.

The print may be **toned** by replacing part of the silver of the image by gold or platinum. To do this, the print is treated with solutions of $Na[AuCl_4]$ or $K_2[PtCl_6]$. The more active silver displaces these noble metals from their salts to give a thin deposit of gold (red tone) or platinum (dark gray).

34.17 Properties of Silver of Analytical Importance

The silver ion yields a white, curdy precipitate with chloride ions in acid solution, making it a member of Analytical Group I. Silver chloride is soluble in aqueous ammonia, forming diamminesilver chloride, and is reprecipitated by acids. Silver ions are precipitated by iodide, bromide, chromate, sulfide, phosphate, and cyanide ions, yielding pale yellow AgI, yellow $AgBr$, red Ag_2CrO_4, black Ag_2S, yellow Ag_3PO_4, and white $AgCN$, respectively. Silver cyanide is soluble in an excess of soluble cyanide ($AgCN + CN^- \rightleftharpoons [Ag(CN)_2]^-$).

Lead

34.18 Periodic Relationships of Lead

The first two elements of Periodic Group IVA, carbon and silicon, have already been discussed in Chapters 27 and 29, respectively. The remaining three members of this family are germanium, tin, and lead.

Carbon and silicon are primarily nonmetallic in character, while germanium, tin, and lead, as would be expected, become increasingly metallic in properties with increasing atomic weight. In ionic compounds carbon and silicon are always found in the anion. On the other hand, germanium, tin, and lead form bivalent cations, Ge^{2+}, Sn^{2+}, and Pb^{2+}. The fact that the hydroxides of these ions are amphoteric is a reflection of some nonmetallic character. All members of this periodic group form covalent compounds or anions in which the $+4$ oxidation state is exhibited; for example, CCl_4, $SiCl_4$, $GeCl_4$, $SnCl_4$, and $PbCl_4$ are low-boiling covalent liquids; Na_2CO_3, Na_2SiO_3, Na_2GeO_3, Na_2SnO_3, and Na_2PbO_3 are ionic compounds.

Germanium and the members of Periodic Group IVB (Ti, Zr, and Hf) are

TABLE 34-3 Some Physical Properties of the Metals of Group IVA

	Atomic Number	Atomic Weight	Atomic Radius, Å	Ionic (M^{4+}) Radius, Å	Density at 20°	Melting Point, °C	Boiling Point, °C
Germanium	32	72.59	1.22	0.53	5.36	960	——
Tin	50	118.69	1.4	0.71	7.31	231.9	2,337
Lead	82	207.2	1.75	0.84	11.34	327.4	1,750

among the less familiar elements. The members of the B group usually show an oxidation state of +4 in their compounds; they are transition metals with two electrons in their outer shells and ten electrons in the next to the outer shells. Some physical properties of the Group IVA metals are given in Table 34–3.

34.19 History, Occurrence, and Metallurgy of Lead

Because lead is easily extracted from its ores, it was known to the early Egyptians and Babylonians. It is mentioned in the Bible in Job and in Numbers. Lead pipes were commonly used by the Romans for conveying water, and in the Middle Ages lead was used as a roofing material. The stained glass windows of the great cathedrals of this age were set in lead.

The principal lead ore is the sulfide, PbS, commonly called **galena.** Other common ores of lead are the carbonate, **cerrusite** ($PbCO_3$), and the sulfate, **anglesite** ($PbSO_4$), both of which may have been formed by the weathering of sulfide ores.

Lead ores are first concentrated by a series of selective flotation processes to remove the gangue materials and a large part of the zinc sulfide which is usually associated with lead ores. The concentrated ore is then roasted in air to convert most of the sulfide to the oxide.

$$2PbS(s) + 3O_2(g) \longrightarrow 2PbO(s) + 2SO_2(g) + 169,800 \text{ cal}$$
$$(\Delta H_{298}^\circ = -169,800 \text{ cal})$$

The roasted product is reduced in a blast furnace with coke and scrap iron.

$$25,520 \text{ cal} + PbO(s) + C(s) \longrightarrow Pb(s) + CO(g) \quad (\Delta H_{298}^\circ = 25,520 \text{ cal})$$
$$PbO(s) + CO(g) \longrightarrow Pb(s) + CO_2(g) + 15,700 \text{ cal}$$
$$(\Delta H_{298}^\circ = -15,700 \text{ cal})$$
$$100 \text{ cal} + PbS(s) + Fe(s) \longrightarrow Pb(s) + FeS(s) \quad (\Delta H_{298}^\circ = 100 \text{ cal})$$

The lead obtained from the blast furnace contains copper, antimony, arsenic, bismuth, gold, and silver. The crude lead is melted and stirred to bring about the oxidation of antimony, arsenic, and bismuth. The oxides of these metals rise to the surface and the molten lead is drained off for further refining. Gold and silver may be extracted from the lead by the Parkes process (Section 34.11), or by the electrolytic **Betts process.** In the Betts process thin sheets of pure lead are made the cathodes and plates of impure lead the anodes. The electrolyte is a solution containing lead hexafluosilicate, $PbSiF_6$, and hexafluosilicic acid, H_2SiF_6.

An alternative method of metallurgy for lead involves the electrolysis of lead sulfide dissolved in molten lead chloride. Lead is liberated at the cathode, and sulfur at the anode.

34.20   Properties and Uses of Lead

Lead is a soft metal having little tensile strength, and it is the heaviest of the common metals except for gold and mercury. Lead has a metallic luster when freshly cut but quickly acquires a dull gray color when exposed to moist air. In air which contains moisture and carbon dioxide lead becomes oxidized on the surface, forming a protective layer which is both compact and adherent; this film is probably the basic carbonate.

Unlike silver and mercury of Analytical Group I, lead lies above hydrogen in the activity series and therefore dissolves slowly in dilute nonoxidizing acids. Concentrated nitric acid attacks it readily. It is not dissolved by pure water in the absence of air, but in the presence of air it reacts with water to form the hydroxide.

$$2Pb + 2H_2O + O_2 \longrightarrow 2Pb(OH)_2$$

When heated in a stream of air lead burns. It also reacts with sulfur, fluorine, and chlorine.

The uses of lead depend mainly upon the ease with which it is worked, its low melting point, its great density, and its resistance to corrosion.

34.21   Compounds of Lead

Lead forms two well-defined series of compounds in which its oxidation states are $+2$ and $+4$. An atom of lead (like those of its congeners carbon, silicon, germanium, and tin) has four valence electrons; two of these are s electrons and the other two are p electrons. All four of the valence electrons are seldom, if ever, completely removed from the atom but are very often shared with electronegative elements. This latter fact accounts for the $+4$ oxidation state, which is typical of the Group IVA elements. Lead (and also tin) forms many compounds in which its two s valence electrons do not participate in the bonding but remain associated with the core of the atom as a stable electron pair. When this happens, the element assumes the $+2$ oxidation state. Because of the high oxidation potential of tetravalent lead ($Pb^{2+} \longrightarrow PbO_2$, -1.5 volts), the $+2$ oxidation state of this element is often considered to be the characteristic oxidation state.

■ **1. Oxides.** When lead is heated in air the yellow, powdery **monoxide**, PbO, is obtained.

Lead dioxide, PbO_2, is a chocolate-brown powder formed by oxidizing lead(II) compounds in alkaline solution. With sodium hypochlorite as the oxidizing agent, the equation for the oxidation of the plumbite ion is

$$Pb(OH)_3^- + ClO^- \longrightarrow \underline{PbO_2} + Cl^- + OH^- + H_2O$$

Lead dioxide is the principal constituent of the cathode of the charged lead storage battery (Section 22.17). Since lead +4 tends to revert to the stable +2 state by gaining two electrons, lead dioxide is a powerful oxidizing agent.

The oxide Pb_3O_4, called **red lead** or **trilead tetroxide,** is prepared by carefully heating the monoxide in air at temperatures between 400 and 500° C. When red lead is treated with nitric acid, two-thirds of the lead dissolves as lead nitrate (oxidation state +2), and the remaining third remains as lead dioxide (oxidation state +4). The equation for the reaction is

$$Pb_3O_4 + 4H^+ + (NO_3^-) \longrightarrow \underline{PbO_2} + 2Pb^{2+} + (NO_3^-) + 2H_2O$$

■ **2. Lead Hydroxide.** When alkali metal hydroxides are added to solutions of lead(II) compounds, white **lead hydroxide,** $Pb(OH)_2$, precipitates. An excess of alkali causes the hydroxide to dissolve according to the equation

$$\underline{Pb(OH)_2} + OH^- \longrightarrow Pb(OH)_3^-$$

which shows the amphoteric character of the compound. Its reaction with hydrogen ions is given by the equation

$$\underline{Pb(OH)_2} + 2H^+ \longrightarrow Pb^{2+} + 2H_2O$$

■ **3. Lead Chloride.** **Lead chloride,** $PbCl_2$, may be formed by the direct union of the elements, by the action of hydrochloric acid upon lead monoxide, or by precipitation from solutions containing lead(II) ions and chloride ions. It is soluble in hot water and in solutions of high chloride concentration, **tetrachloroplumbate(II) ions** being formed in the latter case.

$$\underline{PbCl_2} + 2Cl^- \rightleftharpoons PbCl_4^{2-}$$

These properties are important in the qualitative analysis of lead.

■ **4. Lead Nitrate.** When either metallic lead or lead monoxide, PbO, is dissolved in nitric acid, **lead nitrate,** $Pb(NO_3)_2$, is formed. This salt is readily soluble in water, but unless the solution is slightly acid with nitric acid, hydrolysis occurs and basic nitrates are precipitated.

$$Pb^{2+} + NO_3^- + H_2O \rightleftharpoons \underline{Pb(OH)NO_3} + H^+$$

Lead nitrate is unstable at moderately high temperatures and decomposes in the same manner as the nitrates of other heavy metals.

$$2Pb(NO_3)_2 \longrightarrow 2\underline{PbO} + 4NO_2 + O_2$$

■ **5. Carbonates.** The normal carbonate of lead, $PbCO_3$, may be prepared by the action of sodium hydrogen carbonate upon lead chloride. The basic carbonate, $Pb_3(OH)_2(CO_3)_2$, is formed when alkali metal carbonates are added to solutions containing the lead ion. This compound is important commercially as the paint pigment **white lead,** though because of the danger of lead poisoning its use is restricted (especially for objects used by children). It is prepared commercially by the action of air, carbon dioxide, and acetic acid vapor upon lead metal. The essential reactions for the process may be represented by

$$2Pb + O_2 + 2HOAc \longrightarrow 2Pb(OH)OAc \quad (\text{HOAc represents acetic acid})$$

$$6Pb(OH)OAc + 2CO_2 \longrightarrow Pb_3(OH)_2(CO_3)_2 + 3Pb(OAc)_2 + 2H_2O$$

The $Pb(OAc)_2$ which is formed in the second reaction is eventually used up in the process, with the regeneration of acetic acid.

Although white lead paints have excellent covering power, they have the disadvantage of turning dark in the presence of hydrogen sulfide due to the formation of black lead sulfide.

■ **6. Other Lead Compounds.** **Lead sulfate,** $PbSO_4$, is formed by ionic combination. It is insoluble in water but readily dissolved in solutions containing an excess of alkali metal or acetate ions. **Lead acetate,** $Pb(OAc)_2$, is one of the few soluble compounds of lead; it is a weak electrolyte, indicating that it dissolves mainly as a covalent compound rather than as an ionic one. It is extremely toxic. The **chromate,** $PbCrO_4$, is insoluble in water but dissolves readily in acids and alkali metal hydroxides. Hydrogen sulfide precipitates black **lead sulfide,** PbS, which is insoluble in dilute acids and alkali metal sulfides.

The organometallic **lead tetraethyl,** $Pb(C_2H_5)_4$, is a covalent compound which is liquid at ordinary temperatures. It has been used extensively in the production of "antiknock" gasoline, but with the present concern over air pollution and the fact that it poisons the catalysts in the catalytic converters its use in gasoline is diminishing rapidly (Section 24.4). It is prepared by the reaction of ethyl chloride, C_2H_5Cl, with a sodium-lead alloy, $Na_{31}Pb_8$, which is an intermetallic compound (Section 31.11).

34.22   Properties of Lead of Analytical Importance

Lead chloride, white, is precipitated from cold solutions which are fairly concentrated in Pb^{2+}, but not from dilute, hot solutions. Lead sulfide, black, is precipitated by hydrogen sulfide in dilute acid solutions. The sulfide is insoluble in dilute nonoxidizing acids and alkali metal sulfides but is dissolved by dilute nitric acid. The precipitation of the insoluble salts, lead sulfate (white) and lead chromate (yellow), is important in the detection of lead ions in solution.

QUESTIONS

1. The roasting of an ore of a metal usually results in the conversion of the metal to the oxide. Why does the roasting of cinnabar produce metallic mercury rather than an oxide of mercury?
2. Why is mercury not attacked by hydrochloric acid even though it dissolves readily in nitric acid or hot sulfuric acid?
3. Why are mercury(II) halides spoken of as weak electrolytes?
4. What properties make mercury valuable as a thermometric substance?
5. Write electronic formulas for $HgCl_2$ and Hg_2Cl_2.
6. Hydrogen sulfide gas is bubbled through a solution marked mercury(I) nitrate. A precipitate is observed to form. Of what substance(s) is the precipitate composed?

7. How many moles of ionic species would you predict to be present in a solution marked 1.0 M mercury(I) nitrate? How would you demonstrate the accuracy of your prediction?
8. What properties of silver have made it valuable as a coinage metal down through the ages?
9. Explain the tarnishing of silver in the presence of materials containing sulfur.
10. Compare the solubilities of the silver halides in water and in aqueous ammonia.
11. Describe the Parkes process for extracting silver from lead.
12. Why does silver dissolve in nitric acid but not in hydrochloric acid?
13. Write equations for the electrode reactions when $Na[Ag(CN)_2]$ is used as the electrolyte in silver plating.
14. Outline the chemistry of the photographic process.
15. Dilute sodium cyanide solution is slowly dripped into a slowly stirred silver nitrate solution. A white precipitate forms temporarily but dissolves as the sodium cyanide addition continues. Use chemical equations to explain these observations.
16. Account for the variable oxidation state of gold in terms of the electronic structure of its atoms.
17. Show several ways in which coordination compounds play an important part in the chemistry of silver.
18. Describe the Betts process for the refining of lead.
19. Compare the nature of the bonds in $PbCl_2$ to those in $PbCl_4$. Would you expect the existence of Pb^{4+} ions? Explain.
20. Show by suitable equations that lead(II) hydroxide is amphoteric.
21. What is the composition of the tarnish on lead?
22. Why should water to be used for human consumption not be conveyed in lead pipes?
23. Account for the solubility of $PbCl_2$ in solutions of high chloride concentration.
24. When elements in the first transition metal series form dipositive ions, they usually give up their s electrons. Does this apply to lead when it forms the $2+$ ion?
25. How is the paint pigment *white lead* prepared commercially?

PROBLEMS

1. What is the minimum number of grams of fish that one would have to consume to obtain a fatal dose of mercury, if the fish contains 5.0 parts per million mercury by weight? (Assume all the mercury from the fish ends up as mercury(II) chloride in the body and that a fatal dose is 0.20 g of $HgCl_2$.) How many pounds of fish would this be?   *Ans. 30 kg; 66 lb*
2. The dissociation constant of $[Ag(CN)_2]$ is 1.0×10^{-20}. What concentration of Ag^+ will be in equilibrium if 1.0 g of Ag is oxidized and put into one liter of solution with $1.0 \times 10^{-1} M$ CN^-?   *Ans. $1.4 \times 10^{-20} M$*

3. One hundred milliliters of a saturated solution of lead(II) iodide contains 0.41 g of solute at 100° C. Calculate the solubility product at this temperature.
Ans. 2.8 × 10⁻⁶

4. How many pounds of mercury can be obtained from ten tons of ore which is 7.4% cinnabar? *Ans. 1.3 × 10³ lb*

5. A solid 1.4820-g sample of a pure alkali metal chloride is dissolved in water and treated with excess silver nitrate. The resulting precipitate is filtered, dried, and found to weigh 2.849 g. What per cent chloride is in the original chloride compound? What is the identity of the salt? *Ans. 47.55%; KCl*

REFERENCES

"Mercury Stirs More Pollution Concern," (Staff), *Chem. & Eng. News*, June 22,1970; p. 36.

"Mercury in Swordfish," (Staff), *Chem. & Eng. News*, May 17, 1971; p. 11.

"Mercury in the Environment," L. J. Goldwater, *Sci. American*, May, 1971; p. 15.

"Mercury in the Environment: Natural and Human Factors," A. L. Hammond, *Science*, **171,** 788 (1971).

"Mercury in British Fish," G. Grimstone, *Chem. in Britain*, **8,** 244 (1972).

"Hallmarking Gold and Silver," J. S. Forbes, *Chem. in Britain*, **7,** 98 (1971).

"Resolution and Stereochemistry of Asymmetric Silicon, Germanium, Tin, and Lead Compounds," R. Belloli, *J. Chem. Educ.*, **46,** 640 (1969).

"Hydrolysis of Group IV Chlorides," C. H. Yoder, *J. Chem. Educ.*, **46,** 382 (1969).

"Electronegatives and Group IV A Chemistry," D. A. Payne, Jr. and F. H. Fink, *J. Chem. Educ.*, **43,** 654 (1966).

"Chemicals in the Manufacture of Paint," W. C. Weber, *J. Chem. Educ.*, **37,** 323 (1960).

"Lead Poisoning," J. J. Chisolm, Jr., *Sci. American*, Feb., 1971; p. 15.

"Lead Pollution—A Growing Hazard to Public Health," D. Bryce-Smith, *Chem. in Britain*, **7,** 54 (1971).

"Lead in the Environment," A. L. Mills, *Chem. in Britain*, **7,** 160 (1971).

"Unleaded Gasoline," (Staff), *Chem. & Eng. News*, March 8, 1971; p. 14.

"Our Daily Lead," T. J. Chow, *Chem. in Britain*, **9,** 258 (1973).

"Lead in Blood—the Analyst's Problem," A. A. Cernik, *Chem. in Britain*, **10,** 58 (1974).

"Lead in Petrol," F. D. Porter, *Chem. in Britain*, **10,** 61 (1974).

"Lead in Food: Are Today's Regulations Sufficient?" D. Bryce-Smith and H. A. Waldron, *Chem. in Britain*, **10,** 202 (1974).

Metals of Analytical Group II

35

Division A—Lead, Bismuth, Copper, Cadmium

Analytical Group II (see also Chapter 41) consists of the common metals whose ions form chlorides which are soluble in dilute acid but whose sulfides are precipitated by hydrogen sulfide in 0.3 molar hydrochloric acid. Lead appears in Group II as well as in Group I because its chloride is of intermediate solubility in cold water. Mercury occurs in both Groups I and II because mercury(I) chloride is insoluble whereas mercury(II) chloride is soluble in dilute hydrochloric acid.

The ions Pb^{2+}, Bi^{3+}, Cu^{2+}, and Cd^{2+}, which compose **Division A,** form sulfides insoluble in sodium sulfide solutions. This property permits the separation of these ions from those of **Division B,** Hg^{2+}, As^{3+}, Sb^{3+}, and Sn^{4+}, whose sulfides are soluble in water containing an excess of sodium sulfide.

If the ions of the less common metals were to be included in our scheme of analysis, Group II would contain ions of such metals as Au, Pt, Mo, Ru, Rh, Pd, Ir, Os, W, Re, Ge, and Te as well as the ones listed above.

The chemistry of lead was considered in Chapter 34 with that of the metals of Analytical Group I. The other metals of Division A will be considered here and those of Division B will be discussed later in this chapter. Some of the more important properties of the Division-A metals are listed in Table 35–1.

TABLE 35-1   Some Properties of Division-A Metals of Analytical Group II

Property	Lead	Bismuth	Copper	Cadmium
Atomic number	82	83	29	48
Atomic weight	207.2	208.98	63.546	112.41
Electronic structure	2, 8, 18, 32, 18, 4	2, 8, 18, 32, 18, 5	2, 8, 18, 1	2, 8, 18, 18, 2
Oxidation states	+2, +4	−3, +3, +5	+1, +2, +3	+2
Reduction potential,	(Pb^{2+}/Pb)	(Bi^{3+}/Bi)	(Cu^{2+}/Cu)	(Cd^{2+}/Cd)
volts	−0.13	+0.32	+0.34	−0.40
Density, g/cm^3, at 20°	11.34	9.78	8.92	8.65
Melting point, °C	327.4	271.0	1,083	320.9
Boiling point, °C	1,750	1,420.0	2,582	767

Bismuth

35.1   Periodic Relationships of Bismuth

Bismuth is a member of Periodic Group VA, which also includes nitrogen, phosphorus, arsenic, and antimony. The elements of this group furnish an excellent illustration of the gradation in properties characteristic of groups of the periodic system. Nitrogen, the lightest member of the group, is a typical nonmetal in that it accepts electrons from active metals and forms the nitride ion N^{3-}, as in magnesium nitride, Mg_3N_2, and lithium nitride, Li_3N. Bismuth is the heaviest member of the group, and it is almost entirely metallic in character. It gives up three of its five valence electrons to active nonmetals to form the tripositive ion, Bi^{3+}. The nonmetallic character of nitrogen and the metallic character of bismuth, each of which has five valence electrons, can be attributed, for the most part, to differences in the distance between the valence shell and the nucleus for the two atoms. Because the distance between the valence shell and the nucleus is greater for bismuth, the attraction of the positive nucleus for the valence electrons is much smaller in an atom of bismuth than in an atom of nitrogen.

The properties of phosphorus, arsenic, and antimony are intermediate between those of nitrogen and bismuth, with the nonmetallic character becoming less pronounced with increasing atomic weight. Atoms of each member of this family have five valence electrons and exhibit oxidation states of +3 and +5. Consult Chapters 25 and 26 for additional considerations of group relationships among the members of this periodic group.

35.2   Occurrence and Metallurgy of Bismuth

Bismuth is most often found in the free state in nature. It sometimes occurs in combination, as **bismuth ocher**, Bi_2O_3, and **bismuth glance**, Bi_2S_3. It is produced in the United States as a by-product of the refining of other metals, particularly lead. Ores containing the free metal are treated by heating them in inclined iron pipes, whereupon the metal melts and flows away from the gangue. The oxide and sulfide ores are roasted and then heated with charcoal. As the bismuth is set free, it melts and collects beneath the less dense material.

35.3   Properties and Uses of Bismuth

Bismuth is a lustrous, hard, and brittle metal with a reddish tint. The metal burns when heated in air, giving the oxide, Bi_2O_3, but oxidizes only superficially in moist air at ordinary temperatures to form a coating of oxide which protects it against further oxidation. Bismuth reduces the hydrogen in steam and combines directly with the halogens and sulfur. Oxidizing acids, such as hot sulfuric and nitric, dissolve it, forming the respective salts. Bismuth is dissolved slowly by hydrochloric acid in the presence of air, forming the chloride, $BiCl_3$. An unstable hydride of bismuth, BiH_3, exists; it is called **bismuthine.**

Like antimony, melted bismuth expands upon solidifying, a most unusual property. Because of this, it is used in the formation of alloys to prevent them from shrinking upon solidification. Alloys of bismuth, tin, and lead have low melting points, which makes them useful for electrical fuses, safety plugs for boilers, and automatic sprinkler systems. Some of these alloys melt even in hot water. For example, **Rose's metal** (Bi, 50 per cent; Pb, 25 per cent; Sn, 25 per cent) melts at 94° C; and **Wood's metal** (Bi, 50 per cent; Pb, 25 per cent; Sn, 12.5 per cent; Cd, 12.5 per cent) melts at 65.5° C.

35.4   Compounds of Bismuth

■ **1. Oxides and Hydroxides.**   When bismuth is burned in air or when the nitrate is strongly heated, yellow **bismuth(III) oxide**, Bi_2O_3, is formed. This oxide is basic as contrasted to the more acidic character of the lower oxide of the smaller elements in the family. Because of its basic nature, Bi_2O_3 dissolves in acids to form salts. It does not exhibit acidic properties such as the ability to react with bases. Alkali metal hydroxides or aqueous ammonia precipitate the white **hydroxide**, $Bi(OH)_3$, from solutions of trivalent bismuth salts. When a suspension of bismuth hydroxide is boiled it loses the elements of a molecule of water, forming **bismuth oxyhydroxide**, $BiO(OH)$.

Bismuth(V) oxide, Bi_2O_5, is produced by the action of very strong oxidizing agents upon the trioxide. As expected on the basis of the higher oxidation state of the bismuth, this oxide is more acidic than Bi_2O_3 and shows its acidic character by dissolving in concentrated sodium hydroxide to form **sodium bismuthate,** $NaBiO_3$. This salt reacts with nitric acid to give **bismuthic acid,** $HBiO_3$, which is a very strong oxidizing agent. It is used in analytical chemistry in the detection and estimation of manganese, which it oxidizes from Mn^{2+} to MnO_4^-.

■ **2. Bismuth(III) Sulfide.**   When hydrogen sulfide is passed into a solution of a bismuth salt, brown **bismuth(III) sulfide**, Bi_2S_3, is precipitated. The Bi_2S_3 does not dissolve in concentrated sulfide solutions to form thiosalts. This property is utilized in separating bismuth from arsenic and antimony, which do form soluble thiosalts.

■ **3. Salts of Bismuth.** Bismuth(III) oxide dissolves in acids, forming the corresponding salts such as the chloride, $BiCl_3 \cdot 2H_2O$; the nitrate, $Bi(NO_3)_3 \cdot 5H_2O$; and the sulfate, $Bi_2(SO_4)_3$. Bismuth salts hydrolyze readily when water is added, forming hydroxysalts.

$$BiCl_3 + 2H_2O \rightleftharpoons \underline{Bi(OH)_2Cl} + 2H^+ + 2Cl^-$$
$$Bi(NO_3)_3 + 2H_2O \rightleftharpoons \underline{Bi(OH)_2NO_3} + 2H^+ + 2NO_3^-$$

The dihydroxychloride loses the elements of a molecule of water to form **bismuth oxychloride.**

$$Bi(OH)_2Cl \longrightarrow \underline{BiOCl} + H_2O$$

The dihydroxynitrate, when repeatedly washed with water is converted to **bismuth hydroxide.**

$$Bi(OH)_2NO_3 + H_2O \longrightarrow \underline{Bi(OH)_3} + H^+ + NO_3^-$$

When dried, bismuth dihydroxynitrate forms the oxynitrate, $BiONO_3$.

35.5 Uses of Bismuth Compounds

The oxycarbonate, $(BiO)_2CO_3$, and the oxynitrate (under the names **bismuth subcarbonate** and **bismuth subnitrate)** are used in medicine for the treatment of stomach disorders such as gastritis and ulcers and of skin diseases, such as eczema.

35.6 Properties of Bismuth of Analytical Importance

Bismuth chloride is soluble in dilute hydrochloric acid, but hydrolyzes in water, forming the insoluble oxychloride. Bismuth(III) sulfide, brown, is precipitated when hydrogen sulfide is passed into a solution of a bismuth salt; it is insoluble in dilute hydrochloric acid and alkali metal sulfides, but is dissolved by dilute nitric acid. Bismuth hydroxide, white, is precipitated by aqueous ammonia. Upon the addition of a sodium stannite solution, bismuth(III) is reduced to metallic bismuth, black; this reaction serves as a confirmatory test for the element.

Copper

The history and periodic relationships of copper were considered in Sections 34.10 and 34.9, respectively.

35.7 Occurrence of Copper

Copper occurs in both the native and combined forms. The most important deposit of native copper in the world is that in the Michigan peninsula near Houghton. The largest single mass of native copper yet found weighs 420 tons, and it is now on display in the Smithsonian Institution in Washington. The most important copper ores are the sulfides, such as **chalcocite**, Cu_2S, and **chalcopyrite**, $CuFeS_2$. Other ores are **cuprite**, Cu_2O; **melaconite**, CuO; and **malachite**, $Cu_2(OH)_2CO_3$.

Copper is found in trace amounts in plants in regions where there are copper ores. It is also found in the brightly colored feathers of certain birds, and in the blood of certain marine animals such as lobsters, oysters, and cuttlefish, where it serves the same oxygen-carrying function as iron in the blood of higher animals (Section 28.24).

35.8   Metallurgy of Copper

In the metallurgy of copper, native copper ores are pulverized, the gangue is washed away, and the copper is melted and poured into molds to cool. Oxide and carbonate ores are often leached with sulfuric acid to produce copper(II) sulfate solutions, from which copper metal may be obtained by electrodeposition. High grade oxide or carbonate ores are reduced by heating them with coke mixed with a suitable flux.

Sulfide ores are usually low grade, containing less than 10 per cent copper. These ores are first concentrated by the flotation process (Section 31.4). The concentrate (or a high grade sulfide ore if concentration is unnecessary) is then roasted in a furnace, at a temperature below the fusion point of the ore, to drive off the moisture and remove part of the sulfur as sulfur dioxide. The remaining mixture, which is called **calcine** and consists of Cu_2S, FeS, FeO, and SiO_2, is smelted by mixing it with limestone (to serve as a flux) and heating the mixture above its melting point. The reactions taking place during the formation of the slag are given by the equations

$$CaCO_3 + SiO_2 \longrightarrow CaSiO_3 + CO_2$$
$$FeO + SiO_2 \longrightarrow FeSiO_3$$

The Cu_2S, impure with FeS, which remains after smelting is called **matte.** Reduction of the matte is accomplished in a **converter** by blowing air through the molten material. The air first oxidizes the iron(II) sulfide to iron(II) oxide and sulfur dioxide. Sand is added to form a slag of iron(II) silicate with the iron(II) oxide. After the iron has been removed, the air blast converts part of the Cu_2S to Cu_2O. As soon as copper(I) oxide is formed, it is reduced by the remaining copper(I) sulfide to metallic copper. The equations are

$$2Cu_2S + 3O_2 \longrightarrow 2Cu_2O + 2SO_2$$
$$2Cu_2O + Cu_2S \longrightarrow 6Cu + SO_2$$

The last traces of copper(II) oxide produced by the air blast are removed by reduction with H_2 and CO produced catalytically from methane or natural gas. The copper obtained in this way has a characteristic appearance due to the air blisters which it contains, and is called **blister copper.**

The impure copper is cast into large plates, which are used as anodes in the electrolytic purification of the metal (Section 22.6). Thin sheets of pure copper serve as the cathodes; copper(II) sulfate, acidified with sulfuric acid, serves as the electrolyte. The impure copper passes into solution from the anodes and pure copper plates out on the cathodes as electrolysis proceeds. Gold and silver in the anodes do not oxidize but fall to the bottom of the electrolytic cell as anode mud, along with bits of slag and Cu_2O. Silver, gold, and the platinum group metals are recovered from the anode mud as valuable by-products. The metals more active than copper, such as zinc and iron, are oxidized at the anode and their cations pass into solution, where they remain. The copper deposited on the cathode has an average purity of 99.955 per cent. The value of the precious metals recovered from the anode mud is often sufficient to pay the cost of the electrolytic refining.

35.9   Properties of Copper

Copper is a reddish-yellow metal. It is ductile and malleable, so that it is readily fashioned into wire, tubing, and sheets. Copper is the best electrical conductor of the cheaper metals, but when used for this purpose it must be quite pure, since small amounts of impurities reduce its electrical conductivity greatly.

Copper is relatively inactive chemically. In moist air it first turns brown, due to the formation of a very thin, adherent film of either copper oxide or copper sulfide. Prolonged weathering of copper causes it to become coated with a green film of the basic carbonate, $Cu_2(OH)_2CO_3$, which is similar to the mineral malachite. This compound, or the basic sulfate, is responsible for the green color of copper roofs, gutters, and downspouts that have been weathered for a considerable time. When heated in air, copper oxidizes and forms copper(II) oxide along with some copper(I) oxide. Oxidizing acids, and nonoxidizing acids in the presence of air, convert copper to the corresponding copper(II) salts. Aqueous ammonia in the presence of air will dissolve copper and give a blue solution containing $[Cu(NH_3)_4]^{2+}$. Alkali metal hydroxides do not act upon the metal. Sulfur vapor reacts with hot copper to give both Cu_2S and CuS. Hot copper burns in chlorine to give $CuCl$.

35.10   Uses of Copper

The extent and wide range of the uses of copper cause it to be regarded as second in importance to iron. The chief use of copper is in the production of electrical wiring for a wide variety of uses.

The fact that copper may be electrolytically deposited in thin sheets which are smooth and tough makes it useful in electrotyping. An impression of the type is made in wax, and this is rubbed with graphite to make its surface an electrical conductor. The wax impression is used as the cathode in an electrolytic cell containing copper sulfate as the electrolyte and an anode of copper. Copper is plated out on the graphite film until the deposit becomes the thickness of a sheet of paper. The copper sheet is then removed from the wax and strengthened by covering its back surface with lead. Plates prepared in this way were, at one time, widely used in printing books. Modern printing methods use a photographic method of producing the printing plates.

Copper is also used in the production of a great many alloys. Among the more important ones are **brass** (Cu, 60-82 per cent; Zn, 18-40 per cent), **bronze** (Cu,

70-95 per cent; Zn, 1-25 per cent; Sn, 1-18 per cent), **aluminum bronze** (Cu, 90-98 per cent; Al, 2-10 per cent), and **German silver** (Cu, 50-60 per cent; Zn, 20 per cent; Ni, 20-25 per cent). Other alloys containing copper are bell metal, gun metal, silver coin, sterling silver, gold coin, jewelry gold, and jewelry silver.

35.11   Oxidation States of Copper

Copper forms two principal series of compounds, which are based upon the oxidation states of $+1$ and $+2$, respectively. When the one electron in the outermost shell of the copper atom is involved in bond formation, the oxidation state of $+1$ is exhibited. When an additional electron from the underlying shell is removed, the copper(II) ion results. When two electrons from the underlying shell are used in bonding, copper exhibits an oxidation state of $+3$; copper(III) compounds are rare and relatively unimportant, however. Copper(I) forms a complete series of binary compounds such as the halides and the oxide, but the ternary oxygen salts are few in number and readily decomposed by water. In contrast, copper(II) forms a complete series of oxygen salts, while some of its binary compounds are unstable and break down spontaneously.

35.12   Copper(I) Compounds

Solid salts of copper(I) show a variety of colors, but their solutions are colorless.

■ **1. Copper(I) Oxide.** Cu_2O may be prepared as a reddish-brown precipitate by boiling copper(I) chloride with sodium hydroxide.

$$2CuCl + 2OH^- \longrightarrow \underline{Cu_2O} + 2Cl^- + H_2O$$

When basic solutions of copper(II) salts are heated with reducing agents, copper(I) oxide precipitates.

$$2Cu^{2+} + 2OH^- + 2e^- \text{ (reducing agent)} \longrightarrow \underline{Cu_2O} + H_2O$$

This reaction is the basis for the test for the presence of reducing sugars in the urine in the diagnosis of diabetes (Section 28.13). The reagent most often used for this purpose is Benedict's solution, which is a solution of copper(II) sulfate, sodium carbonate, and sodium citrate. The addition of a reducing sugar, such as glucose, causes the precipitation of the reddish-brown copper(I) oxide.

■ **2. Copper(I) Hydroxide.**   This compound is formed as a yellow precipitate when a cold solution of copper(I) chloride in hydrochloric acid is treated with sodium hydroxide. The hydroxide is unstable and decomposes into the oxide and water upon heating.

■ **3. Copper(I) Chloride.**   When copper metal is heated with a solution of copper(II) chloride acidified with hydrochloric acid, an oxidation-reduction reaction takes place with the precipitation of copper(I) chloride.

$$Cu^{2+} + 2Cl^- + Cu \longrightarrow 2\underline{CuCl}$$

This white crystalline compound is readily soluble in concentrated hydrochloric acid, forming the complex ion $[CuCl_2]^-$.

$$CuCl + Cl^- \rightleftharpoons [CuCl_2]^-$$

The two chlorides of the $[CuCl_2]^-$ ion are bonded to the copper by covalent bonds. Dilution with water brings about reprecipitation of the copper(I) chloride.

Among other stable and insoluble copper(I) binary salts are Cu_2S (black), CuCN (white), CuBr (white), and CuI (white).

35.13   Copper(II) Compounds

The more common compounds of copper are those in which the metal is bivalent. In the solid state the copper(II) compounds may be blue, black, green, yellow, or white. However, dilute solutions of all soluble copper(II) compounds have the blue color of the hydrated copper(II) ion $[Cu(H_2O)_4]^{2+}$.

■ **1. Copper(II) Oxide.**   **Copper(II) oxide,** CuO, is a black insoluble compound that may be obtained by heating the carbonate or nitrate, or by heating finely divided copper in oxygen.

$$CuCO_3 \longrightarrow CuO + CO_2$$
$$100{,}500 \text{ cal} + 2Cu(NO_3)_2(s) \longrightarrow 2CuO(s) + 4NO_2(g) + O_2(g)$$
$$(\Delta H^\circ_{298} = 100{,}500 \text{ cal})$$
$$2Cu(s) + O_2(g) \longrightarrow 2CuO(s) + 37{,}600 \text{ cal}$$
$$(\Delta H^\circ_{298} = -37{,}600 \text{ cal})$$

Because of the ease with which the copper can be reduced either to oxidation state +1 or to the metal, hot copper(II) oxide is a good oxidizing agent. As such, it is used in the analysis of organic compounds to determine the percentage of carbon that they contain. Copper(II) oxide oxidizes the carbon to carbon dioxide, which is absorbed by sodium hydroxide, and the gain in weight of the sodium hydroxide is taken as a measure of the quantity of carbon dioxide produced and absorbed. The copper oxide is continuously regenerated by passing a stream of air through the tube in which the organic matter and the copper oxide are reacting.

■ **2. Copper(II) Hydroxide.**   When hydroxide ions are added to cold solutions of copper(II) salts, a bluish-green gelantinous precipitate of the hydroxide, $Cu(OH)_2$, is formed. If hot solutions are employed CuO is obtained. Copper(II) hydroxide is somewhat amphoteric and dissolves slightly in solutions of concentrated alkalies, the cuprate ion, $[Cu(OH)_4]^{2-}$, being formed. The hydroxide is also soluble in aqueous ammonia, giving the **tetraamminecopper(II) hydroxide,** $[Cu(NH_3)_4](OH)_2$, which is a strong base with a deep blue color.

$$Cu(OH)_2 + 4NH_3 \longrightarrow [Cu(NH_3)_4]^{2+} + 2OH^-$$

The formation of this color is used as a qualitative test for the presence of the copper(II) ion in solution.

■ **3. Copper(II) Sulfate.**   This sulfate is the most important of the copper salts. The anhydrous salt is colorless, but when it is crystallized from aqueous solution the blue pentahydrate, $[Cu(H_2O)_4]SO_4 \cdot H_2O$, known as **blue vitriol,** is formed. As indicated by the formula, four of the water molecules are coordinated to the copper(II) ion and the fifth is attached to both the sulfate and two of the coordinated water molecules by hydrogen bonds. The structure of the pentahydrate may be represented as shown.

The dehydration of this salt proceeds in steps, yielding successively $CuSO_4 \cdot 3H_2O$ (with loss of the two nonhydrogen-bonded waters), $CuSO_4 \cdot H_2O$ [with loss of the other two water molecules coordinated to the Cu(II) ion], and finally $CuSO_4$ (with loss of the water molecule attached to the sulfate ion).

Copper(II) sulfate is produced commercially by oxidizing the sulfide either directly to the sulfate or to the oxide, which is then converted to the sulfate by reaction with sulfuric acid. In the laboratory the anhydrous sulfate may be prepared by oxidizing copper with hot concentrated sulfuric acid.

$$Cu(s) + 2H_2SO_4(l) \longrightarrow CuSO_4(s) + SO_2(g) + 2H_2O(l) + 2{,}840 \text{ cal}$$
$$(\Delta H^{\circ}_{298} = -2{,}840 \text{ cal})$$

Aqueous solutions of copper(II) sulfate are acidic by hydrolysis.

$$[Cu(H_2O)_4]^{2+} + H_2O \rightleftharpoons [Cu(H_2O)_3OH]^+ + H_3O^+$$

or simply, $Cu^{2+} + H_2O \rightleftharpoons CuOH^+ + H^+$

Copper(II) sulfate is used in the electrolytic refining of copper, in electroplating, in the Daniell cell, in the manufacture of pigments, as a mordant in the textile industry, and in the prevention of algae in reservoirs and swimming pools. Because anhydrous copper(II) sulfate is insoluble in alcohol and ether and readily takes up water, turning blue, this salt is used in detecting water in liquids such as alcohol and ether; it is also used to remove water from these liquids. Of course, copper(II) sulfate will not remove or detect water in liquids with which it reacts.

■ **4. Copper(II) Halides.**   Anhydrous **copper(II) chloride,** $CuCl_2$, may be prepared as a yellow crystalline salt by direct union of the elements. The blue-green hydrated salt $Cu(H_2O)_2Cl_2$ may be obtained by treating either the carbonate or the hydroxide of copper(II) with hydrochloric acid and evaporating the solution. Concentrated solutions of copper(II) chloride are green, because they contain both the hydrated copper(II) ion $[Cu(H_2O)_4]^{2+}$, which is blue, and the **tetrachlorocuprate(II)** ion $[CuCl_4]^{2-}$, which is yellow. (The combination of blue and yellow looks green to the eye.) Upon dilution water molecules replace the chloro groups of the $[CuCl_4]^{2-}$ ion, and the solution becomes blue.

Copper(II) bromide, $CuBr_2$, is a black solid, which may be obtained by direct union of the elements or by the action of hydrobromic acid upon the oxide or carbonate.

It is interesting that copper(II) iodide, CuI_2, does not exist. When a solution containing the copper(II) ion is treated with an excess of iodide, an oxidation-reduction reaction occurs, with precipitation of copper(I) iodide and the formation of elemental iodine.

$$2Cu^{2+} + 4I^- \longrightarrow 2\underline{Cu}I + I_2$$

This reaction is used in the quantitative determination of copper by titration of the liberated iodine with a standard solution of sodium thiosulfate. Once the amount of elemental iodine is known, the amount of copper can be calculated.

■ **5. Copper(II) Sulfide.** CuS, black, is precipitated by passing hydrogen sulfide into acid, alkaline, or neutral solutions of copper(II) salts. The sulfide readily dissolves in warm dilute nitric acid with the formation of elemental sulfur and nitric oxide.

$$3CuS + 8H^+ + 2NO_3^- \rightleftharpoons 3Cu^{2+} + 3\underline{S} + 2NO\uparrow + 4H_2O$$

Copper(II) sulfide dissolves only to a slight extent in solutions of alkali metal sulfides.

35.14 Complex Copper Salts

Copper possesses the property of forming both complex cations and anions which are quite stable. Copper(II) ions coordinate with four neutral ammonia molecules and form the deep blue **tetraamminecopper(II) ion.**

$$[Cu(H_2O)_4]^{2+} + 4NH_3 \rightleftharpoons [Cu(NH_3)_4]^{2+} + 4H_2O$$

The fact that the ammonia molecules displace the coordinated water molecules indicates the greater stability of the ammine complex. In fact, the ammine complex is so stable that most slightly soluble copper(II) compounds are dissolved by aqueous ammonia. The copper ion is located at the center of a square formed by the four attached groups, whether they be water, $[Cu(H_2O)_4]^{2+}$; or ammonia, $[Cu(NH_3)_4]^{2+}$; or the chloride ion, $[CuCl_4]^{2-}$; or the hydroxide ion, $[Cu(OH)_4]^{2-}$.

Monovalent copper forms linear complexes with ammonia, $[Cu(NH_3)_2]^+$, with the halides, $[CuX_2]^-$, and with cyanide ions, $[Cu(CN)_2]^-$. In contrast to the copper(II) complexes, most copper(I) complexes are colorless.

35.15 Properties of Copper of Analytical Importance

Chloride ions cause CuCl, white, to precipitate from solutions containing Cu(I) ions, but copper(II) chloride is soluble in water and in dilute hydrochloric acid. Sulfide ions precipitate CuS, black, from acidic, neutral, or alkaline solutions containing Cu(II) ions. This sulfide is readily dissolved by warm, dilute nitric acid, elemental sulfur and nitric oxide being formed. Alkali cyanides dissolve copper(II) sulfide with the formation of the complex dicyanocuprate(I) ion, $[Cu(CN)_2]^-$. An excess of aqueous ammonia changes light blue $[Cu(H_2O)_4]^{2+}$ ions to deep blue $[Cu(NH_3)_4]^{2+}$ ions. The precipitation of pink $Cu_2[Fe(CN)_6]$ serves as a delicate test for the copper(II) ion.

Cadmium

The periodic relationships of cadmium were discussed in Section 34.1, along with its congeners zinc and mercury.

35.16 Occurrence, Metallurgy, and Uses of Cadmium

Cadmium is found in the rare mineral **greenockite,** CdS, and in small amounts (less than 1 per cent) in several zinc ores. Most of our supply of cadmium comes from zinc smelters and from the sludge obtained from the electrolytic refining of zinc.

In the smelting of cadmium-containing zinc ores, the two metals are reduced together. Because cadmium is more volatile (b.p. 767° C) than zinc (b.p. 907° C) they may be separated by fractional distillation. Separation of the two metals is also possible by selective electrolytic deposition. Cadmium, being less active than zinc, is deposited at a lower voltage.

Cadmium is a silvery, crystalline metal resembling zinc. It is only slightly tarnished by air or water at ordinary temperatures. A large part of the cadmium produced is used in electroplating metals, such as iron and steel, to protect them from corrosion. The electrolytic bath contains tetracyanocadmate ions, $[Cd(CN)_4]^{2-}$, made by mixing cadmium cyanide and sodium cyanide. The equation for the cathodic reduction is

$$[Cd(CN)_4]^{2-} + 2e^- \longrightarrow Cd + 4CN^-$$

Cadmium-plated metals are more resistant to corrosion, more easily soldered, and more attractive in appearance than galvanized (zinc-coated) metals.

Cadmium is used in making a number of alloys. Certain ones are easily fusible; these include **Wood's metal** (12.5 per cent Cd), melting at 65.5° C, and **Lipowitz alloy** (10 per cent cadmium), melting at 70° C. Some antifriction bearing metals contain cadmium. These alloys melt at a higher temperature and have a lower friction coefficient than **Babbitt metal** (Section 35.26). Amalgamated cadmium, along with cadmium sulfate, is used in the Weston standard cell for measuring electrical potentials. Rods of cadmium are used in nuclear reactors (Section 30.15) to absorb neutrons and thus control the chain reaction.

35.17 Properties of Compounds of Cadmium

Cadmium dissolves slowly in either hot, moderately dilute hydrochloric or sulfuric acid, with the evolution of hydrogen. It is also dissolved by nitric acid, the oxides of nitrogen being evolved.

Cadmium oxide, CdO, is formed as a brown solid when cadmium burns in air.

Alkali metal hydroxides react with cadmium salts to give the white hydroxide, $Cd(OH)_2$, which is soluble in aqueous ammonia because of the formation of the soluble tetraammine complex, $[Cd(NH_3)_4]^{2+}$. Solutions of cadmium chloride are poor conductors of electricity; the compound dissolves as covalent molecules which ionize only slightly. As would be expected from the electronic configuration of the cadmium ion, the four-coordinate complexes of cadmium, unlike those of copper, are tetrahedral (see Section 32.3).

Cadmium chloride is converted almost completely into $[CdCl_4]^{2-}$ by high concentrations of the chloride ion. **Cadmium sulfide,** CdS, is formed as a bright yellow precipitate by the action of hydrogen sulfide upon solutions of cadmium salts. It is used as a paint pigment called **cadmium yellow.** The carbonate, phosphate, cyanide, and ferrocyanide are all insoluble in water. All cadmium compounds are soluble in an excess of sodium iodide, due to the formation of the complex, $[CdI_4]^{2-}$.

35.18   Properties of Cadmium of Analytical Importance

Cadmium ions are not precipitated by chloride ions. Yellow cadmium sulfide is formed when cadmium ions and sulfide ions are brought together in basic, neutral, or weakly acidic solutions. The sulfide is dissolved by strong acids, but is insoluble in alkali metal sulfide solutions. Alkali metal hydroxides precipitate cadmium hydroxide, white, which is dissolved by aqueous ammonia with the formation of $[Cd(NH_3)_4]^{2+}$. The white, insoluble salt $Cd_2[Fe(CN)_6]$ is useful in detecting cadmium in solution.

Division B—Mercury, Arsenic, Antimony, Tin

The chemistry of mercury was considered in Chapter 34 with the metals of Analytical Group I. Some properties of the Division-B metals are listed in Table 35–2.

TABLE 35–2   Some Properties of the Division-B Metals of Analytical Group II

Property	Mercury	Arsenic	Antimony	Tin
Atomic number	80	33	51	50
Atomic weight	200.59	74.9216	121.75	118.69
Electronic structure	2, 8, 18, 32, 18, 2	2, 8, 18, 5	2, 8, 18, 18, 5	2, 8, 18, 18, 4
Oxidation states	+1, +2	+3, +5	+3, +5	+2, +4
Reduction potential, volts	(Hg^{2+}/Hg) +0.85	(As^{3+}/As) +0.25	(Sb^{3+}/Sb) +0.21	(Sn^{2+}/Sn) −0.14
Density, g/cm³, at 20°	13.546	5.73	6.68	7.31
Melting point, °C	−38.87	817 (36 atm)	630.5	231.9
Boiling point, °C	356.57	610 (sublimes)	1,440	2,337

Arsenic

The periodic relationships of arsenic, a member of the nitrogen family (Periodic Group VA), have been discussed in Section 35.1.

35.19   Occurrence and Preparation of Arsenic

Arsenic is found both as the free element and combined. Its principal native compounds are **arsenopyrite**, $FeAsS$; **realgar**, As_2S_2; and **orpiment**, As_2S_3. Arsenic is widely distributed in trace amounts in the sulfide ores of many metals; consequently, the metals and the sulfuric acid obtained from these sulfides frequently contain arsenic as an impurity. A large part of the arsenic used in the United States is a by-product from the gases of copper furnaces, from which arsenic(III) oxide is collected by Cottrell precipitators (Section 14.19).

Pure arsenic is obtained by subliming native arsenic. When arsenopyrite is heated it decomposes according to the equation

$$4FeAsS(s) \longrightarrow 4FeS(s) + As_4(g) + 21{,}200 \text{ cal} \qquad (\Delta H^\circ_{298} = -21{,}200 \text{ cal})$$

When arsenic ores are roasted As_4O_6 is formed. Reduction of the oxide by heating with carbon produces the element.

$$190{,}000 \text{ cal} + As_4O_6(s) + 6C(s) \longrightarrow As_4(g) + 6CO(g)$$

$$(\Delta H^\circ_{298} = 190{,}000 \text{ cal})$$

35.20   Properties and Uses of Arsenic

Elemental arsenic exists in several allotropic modifications. Ordinary **gray arsenic** is monatomic (As). It is metallic in appearance and sublimes at $615°$ C, forming tetraatomic molecules, As_4, which are tetrahedral in structure. When arsenic vapor is cooled rapidly an unstable yellow crystalline allotrope is produced, which also consists of As_4 molecules. It is soluble in carbon disulfide. Arsenic vapor is yellow in color, has the odor of garlic, and is very poisonous. Arsenic is a member of an intermediate class of elements which possess to some degree the properties of both metals and nonmetals; hence it is a metalloid (Section 31.1). Arsenic ignites when heated, producing white clouds of As_4O_6. It is slowly attacked by hot hydrochloric acid in the presence of air, forming arsenic(III) chloride, $AsCl_3$. Hot nitric acid readily oxidizes arsenic to arsenic acid, H_3AsO_4. Arsenic combines directly with sulfur, the halogens, and many metals. The trihalides (AsX_3) are either liquids or low-melting solids with properties suggestive of covalent rather than ionic bonding, as observed in the absence of saltlike character.

Certain alloys contain arsenic as a hardening agent. Examples are bronze and lead shot.

35.21   Compounds of Arsenic

■ **1. Arsine.**   Arsine, AsH_3, a compound which has a formula analogous to that of ammonia (NH_3), is formed by reducing arsenic compounds with zinc in acid

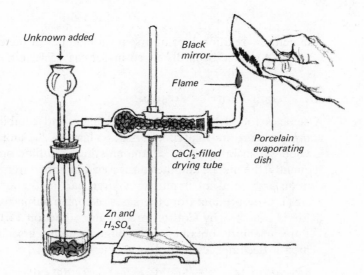

FIGURE 35-1
Laboratory apparatus used in the Marsh test for arsenic.

solution or by the action of a metallic arsenide, such as zinc arsenide, with acids.

$$As_4O_6 + 12Zn + 24H^+ \longrightarrow 4AsH_3\uparrow + 12Zn^{2+} + 6H_2O$$
$$Zn_3As_2 + 6H^+ \longrightarrow 3Zn^{2+} + 2AsH_3\uparrow$$

Arsine neither reacts with nor dissolves in water or acids to form compounds corresponding to ammonium compounds. It is unstable and decomposes into its elements when heated ($4AsH_3 \longrightarrow As_4 + 6H_2$). In the Marsh test for arsenic (Fig. 35–1), the element is deposited as a black mirror from burning arsine when a cold glazed porcelain dish is held in the flame. This test is very sensitive and is commonly employed in the detection of arsenic poisoning.

■ **2. Arsenic(III) Oxide and Arsenous Acid.**   The substance commonly referred to as "arsenic," or "white arsenic," is actually the oxide, As_4O_6. It is the product of the combustion of the element or its compounds. It has a sweet taste and is a deadly poison. Arsenic(III) oxide is slowly dissolved by water with the formation of **arsenous acid,** H_3AsO_3.

$$As_4O_6 + 6H_2O \rightleftharpoons 4H_3AsO_3$$

Arsenous acid is amphoteric as shown by the equations

$$As(OH)_3 \rightleftharpoons As(OH)_2^+ + OH^- \rightleftharpoons AsOH^{2+} + OH^- \rightleftharpoons As^{3+} + OH^-$$
$$H_3AsO_3 \rightleftharpoons H^+ + H_2AsO_3^-$$

Hence, it acts as a base in the presence of strong acids and as an acid in the presence of strong bases.

$$As(OH)_3 + 3H^+ + 3Cl^- \rightleftharpoons AsCl_3 + 3H_2O$$
$$H_3AsO_3 + (Na^+) + OH^- \longrightarrow (Na^+) + H_2AsO_3^- + H_2O$$

Arsenous acid is known only in solution; when attempts are made to isolate it as a solid, it loses water and forms the oxide. All arsenites except those of the alkali metals are insoluble in water.

■ **3. Arsenic(V) Oxide and Arsenic Acid.** **Arsenic(V) oxide,** which has the empirical formula As_2O_5 (the structure is not known), can be produced by heating **arsenic acid,** H_3AsO_4.

$$2H_3AsO_4 \longrightarrow As_2O_5 + 3H_2O$$

Arsenic acid is formed when arsenic(III) oxide is oxidized by concentrated nitric acid.

$$As_4O_6 + 8HNO_3 + 2H_2O \longrightarrow 4H_3AsO_4 + 8NO_2\uparrow$$

By careful heating, H_3AsO_4 may be converted stepwise into **pyroarsenic acid,** $H_4As_2O_7$, **meta-arsenic acid,** $HAsO_3$, and finally the oxide, As_2O_5.

Arsenic acid is a much stronger acid than arsenous acid, as would be expected from the higher oxidation state (see Section 21.15), and it does not react as a base with strong acids as does arsenous acid. Arsenic acid is rather easily reduced, so that it can be used as an oxidizing agent. The arsenates of calcium and lead are used as insecticides.

■ **4. Sulfides of Arsenic.** When hydrogen sulfide is passed into a hydrochloric acid solution of arsenic(III) chloride, the corresponding yellow sulfide is precipitated.

$$2AsCl_3 + 3H_2S \longrightarrow \underline{As_2S_3} + 6H^+ + 6Cl^-$$

Arsenic(III) sulfide dissolves readily in sodium sulfide, forming the **thioarsenite,** Na_3AsS_3.

$$As_2S_3 + 3S^{2-} \longrightarrow 2AsS_3{}^{3-}$$

Acids reprecipitate the sulfide from solutions of thioarsenites.

$$2AsS_3{}^{3-} + 6H^+ \longrightarrow \underline{As_2S_3} + 3H_2S$$

Arsenic(V) sulfide, As_2S_5, is a yellow compound obtained very slowly as a precipitate when hydrogen sulfide is passed into a solution of arsenic acid in hydrochloric acid.

$$2H_3AsO_4 + 5H_2S \longrightarrow \underline{As_2S_5} + 8H_2O$$

The arsenic acid oxidizes hydrogen sulfide in cold, weakly acidic solutions.

$$2H_3AsO_4 + 5H_2S \longrightarrow \underline{As_2S_3} + 2\underline{S} + 8H_2O$$

Iodide ions are a catalyst in the precipitation of arsenic(III) sulfide from the arsenate ion.

$$2\underline{I}^- + AsO_4{}^{3-} + 8H^+ \longrightarrow As^{3+} + 4H_2O + I_2$$
$$I_2 + H_2S \longrightarrow 2H^+ + 2I^- + \underline{S}$$
$$2As^{3+} + 3H_2S \longrightarrow \underline{As_2S_3} + 6H^+$$

Arsenic(V) sulfide dissolves in sodium sulfide with the formation of the **thioarsenate,** Na_3AsS_4. The addition of an acid reprecipitates the sulfide, As_2S_5.

35.22   Uses of Arsenic Compounds

The most important uses of arsenic compounds depend upon their poisonous character. They are used as weed killers, cattle and sheep dips, and insecticides. The compounds most widely used for these purposes are **Paris green** $[Cu_3(AsO_3)_2 \cdot Cu(C_2H_3O_2)_2]$, lead arsenate, calcium arsenate, sodium arsenite, and arsenous acid.

35.23   Properties of Arsenic of Analytical Importance

Arsenic is precipitated from solution as the yellow sulfide, As_2S_3, which is insoluble in dilute hydrochloric acid, but dissolved by alkali metal sulfides. Acids reprecipitate the sulfide from alkali metal thioarsenite solutions. Arsenic(III) sulfide is not dissolved by 6 M hydrochloric acid, but a mixture of aqueous ammonia and hydrogen peroxide dissolves it by converting it to the arsenate ion. The arsenate ion can be detected by precipitating it as white, crystalline **magnesium ammonium arsenate,** $MgNH_4AsO_4$. The Marsh test (Section 35.21) is useful for detecting traces of arsenic compounds.

Antimony

35.24   History of Antimony

Antimony has been used, both as a metal and in compounds, for at least 5,000 years. The Chaldeans used the metal in making vases and ornaments. The sulfide, Sb_2S_3, was used by the Egyptians as a pigment and for painting eyebrows at least as early as 3000 B.C. Compounds of antimony were also used in medicine. Basil Valentine, a Benedictine monk of the fifteenth century, collected all that the alchemists knew about this element in his treatise, *The Triumphal Chariot of Antimony.*

35.25   Occurrence and Metallurgy of Antimony

The principal ore of antimony is **stibnite,** Sb_2S_3, most of which is supplied to the world by China, Mexico, Argentina, Bolivia, and Chile. The sulfide is separated from earthy material of the ore by melting the Sb_2S_3; it is then reduced by heating with iron.

$$Sb_2S_3(s) + 3Fe(s) \longrightarrow 2Sb(s) + 3FeS(s) + 29{,}900 \text{ cal} \quad (\Delta H^{\circ}_{298} = -29{,}900 \text{ cal})$$

The ore may also be roasted to the oxide, from which antimony may be obtained by reduction with carbon.

$$185{,}800 \text{ cal} + Sb_4O_6(s) + 6C(s) \longrightarrow 4Sb(s) + 6CO(g) \quad (\Delta H^{\circ}_{298} = 185{,}800 \text{ cal})$$

35.26   Properties and Uses of Antimony

Antimony is a lustrous silver-white metal, brittle, and readily pulverized. It tarnishes only slightly in dry air but does oxidize slowly in moist air. The metal is readily attacked by the halogens, phosphorus, and sulfur. It is attacked, but not dissolved, by hot nitric acid, forming Sb_4O_6. It is slowly dissolved by hot concentrated sulfuric acid, forming $Sb_2(SO_4)_3$ and evolving SO_2. The fact that the metal and its oxides do not react with nitric acid to give the nitrate indicates a lack of marked base-forming properties. However, the metallic nature of the element is more pronounced than that of arsenic, which lies immediately above it in Group VA.

Lead hardened by the addition of 10 to 20 per cent of antimony is suitable for shrapnel, bullets, and bearings. Because of its resistance to corrosion by acids, it is used in making storage battery plates. Babbitt metal is an antifriction alloy of antimony with tin and copper, extensively used in machine bearings. Since antimony, like bismuth, expands on freezing it is used as a constituent of type metal, to which it confers this property. When forming a piece of type the freezing type metal expands and fills the mold, thus providing type that gives a sharp and distinct imprint.

35.27   Compounds of Antimony

■ **1. Stibine.** The compound **stibine**, SbH_3, is prepared by reducing other antimony compounds with zinc in acid solution. Stibine is poisonous and unstable towards heat. When heated in the absence of air it decomposes, and the antimony, if in contact with a cold surface, will be deposited as a black mirror, as is arsenic when arsine is burned. The two compounds may be distinguished by the fact that the antimony deposit is insoluble in sodium hypochlorite, whereas the arsenic deposit is dissolved by this reagent.

■ **2. Oxides and Acids of Antimony.** Antimony(III) oxide, Sb_4O_6, and **antimony(V) oxide**, Sb_2O_5, resemble the oxides of arsenic in their preparation and properties. Both antimony oxides form acids with water, and antimony(III) oxide exhibits some basic properties in forming salts with acids. The antimony in **antimonic acid** has a coordination number of 6, the formula being $HSb(OH)_6$. **Sodium antimonate**, $NaSb(OH)_6$, is one of the very few relatively insoluble salts of sodium. Sb_2O_4 is also known; it is formed when either Sb_4O_6 or Sb_2O_5 is heated with a free access to air.

■ **3. Antimony Halides.** Antimony forms all of the **trihalides** and **pentahalides** except the pentabromide and the pentaiodide. The trihalides and pentahalides that can be made are prepared either by direct union of the elements or by the action of the hydrohalic acids on the oxides of antimony. Hydrolysis of antimony(III) chloride forms the **oxychloride,** which is white and insoluble.

$$SbCl_3 + H_2O \rightleftharpoons \underline{SbOCl} + 2H^+ + 2Cl^-$$

An excess of hydrochloric acid changes the oxychloride to **tetrachloroantimonous acid,** $HSbCl_4$.

■ **4. Sulfides of Antimony.** Antimony(III) sulfide, Sb_2S_3, is interesting in that it occurs in two strikingly different modifications: a black form, found in nature as the mineral stibnite and also formed when antimony and sulfur are heated together; and a brilliant orange-red form, obtained by the reaction of hydrogen sulfide with slightly acidified solutions of compounds of trivalent antimony. The orange-red modification changes to the black one on being heated or on standing in contact with a dilute solution of an acid.

Antimony(III) sulfide dissolves in sodium sulfide solutions, forming **thioantimonite ions.**

$$\underline{Sb_2S_3} + 3S^{2-} \longrightarrow 2SbS_3{}^{3-}$$

Acidification of the solution causes the sulfide to reprecipitate.

$$2SbS_3{}^{3-} + 6H^+ \longrightarrow \underline{Sb_2S_3} + 3H_2S\uparrow$$

Excess hydrochloric acid dissolves antimony(III) sulfide with the formation of the **tetrachloroantimonate(III) ion,** $SbCl_4{}^-$, and hydrogen sulfide.

$$\underline{Sb_2S_3} + 6H^+ + 8Cl^- \longrightarrow 2SbCl_4{}^- + 3H_2S\uparrow$$

Antimony(V) sulfide, Sb_2S_5, is orange in color and is similar to Sb_2S_3 in many of its reactions. Treatment of antimony(V) sulfide with an excess of hydrochloric acid causes reduction of the antimony to the trivalent state.

$$\underline{Sb_2S_5} + 6H^+ + 8Cl^- \longrightarrow 2SbCl_4{}^- + 3H_2S\uparrow + 2\underline{S}$$

■ **5. Potassium Antimonyl Tartrate.** When antimony(III) oxide is heated with a solution of potassium hydrogen tartrate, $KHC_4H_4O_6$, **potassium antimonyl tartrate,** $KSbOC_4H_4O_6$, known as "tartar emetic," is formed. Like antimony oxychloride, this compound contains the univalent **antimonyl group** (SbO).

35.28   Properties of Antimony of Analytical Importance

Antimony(III) sulfide is an orange-red colored compound which precipitates when hydrogen sulfide is passed into a solution of the trichloride or weakly acidic solutions of antimonites and other compounds of trivalent antimony. When compounds of pentavalent antimony are employed **antimony(V) sulfide,** Sb_2S_5, orange-red in color, is precipitated. Both sulfides of antimony, like those of arsenic, dissolve in solutions of alkali metal sulfides. When solutions containing the thiosalts are acidified, antimony(III) sulfide is precipitated. This sulfide dissolves in 6 M hydrochloric acid with the formation of the complex ion, $SbCl_4{}^-$. Aluminum (and other active metals) reduces antimony to the metallic state. The metal can be dissolved in a mixture of nitric acid and oxalic acid, the compound $SbOHC_2O_4$ being formed. Hydrogen sulfide precipitates the orange-red sulfide, Sb_2S_3, from solutions of $SbOHC_2O_4$.

Tin

35.29   History and Occurrence of Tin

Tin found in ancient Egyptian tombs is evidence that the metal was known during the very early periods of history. Tin was obtained from deposits of cassiterite in England by the Romans and Phoenicians.

The most important and abundant ore of tin is called **cassiterite,** or **tinstone,** SnO_2. No important deposits have been found in the United States; yet more than half of the world's production of tin is used in this country. Some tin is reclaimed from the bright scrap which is left from the manufacture of tin cans. (Tin cans are actually made of steel, plated with tin.) Most detinning of tin plate is based upon the removal of the tin by alkali metal hydroxide solutions with the formation of stannites and hydrogen.

35.30 Metallurgy of Tin

The tin ore is crushed and washed with water to separate the lighter rocky material from the heavier ore. Roasting of the ore removes arsenic and sulfur as volatile oxides, and oxides of other metals are extracted by means of hydrochloric acid. The purified ore is reduced by carbon.

$$86,000 \text{ cal} + SnO_2(s) + 2C(s) \longrightarrow Sn(s) + 2CO(g) \qquad (\Delta H^{\circ}_{298} = 86,000 \text{ cal})$$

The molten tin, which collects on the bottom of the furnace, is drawn off and cast into blocks (block tin). The crude tin is remelted and permitted to flow away from the higher-melting impurities, which are chiefly compounds of iron and arsenic. Further purification of the tin is accomplished by electrolysis, using impure tin anodes, pure tin cathodes, and a bath of fluosilicic acid (H_2SiF_6) and sulfuric acid.

35.31 Properties of Tin

Tin exists in three solid allotropic forms (see Section 8.10). They are **gray tin** (cubic crystals), **malleable tin** (tetragonal crystals), and **brittle tin** (rhombic crystals). The malleable form is silvery-white with a bluish tinge. When a rod of it is bent the crystals slip over one another, producing a sound described as "tin cry." When malleable tin, or "white tin," is heated it changes to the brittle form. White tin changes slowly at low temperatures (below 13.2° C) into gray tin, a powdery form of the element. Consequently, articles made of tin are likely to disintegrate in cold weather, particularly if kept cold for a long time. The change progresses slowly from a spot of origin, with the gray tin which is formed catalyzing its own formation. An effect is thus produced which, in a way, is similar to the spread of an infection in a plant or animal body. For this reason, it is called **tin disease** or **tin pest** (Fig. 35–2).

FIGURE 35-2
A bar of tin "afflicted" with tin disease. Originally full-sized, this bar was subjected to cold temperatures during a delay in shipment in winter. One corner has crumbled and the top and bottom are distinctly grandular.

The metal dissolves slowly in hydrochloric acid when the acid is cold and dilute but rapidly when hot and concentrated, $SnCl_2$ and H_2 being produced. Very dilute cold nitric acid converts tin to tin(II) nitrate. Concentrated nitric acid rapidly converts tin into insoluble, hydrated metastannic acid, H_2SnO_3. Aklali hydroxides dissolve tin, forming stannites and hydrogen.

$$Sn + 2OH^- + 2H_2O \longrightarrow [Sn(OH)_4]^{2-} + H_2\uparrow$$

The reactions with concentrated nitric acid and sodium hydroxide show that tin is not entirely metallic in character. Dry chlorine oxidizes tin to the tetrachloride,

$SnCl_4$, and oxygen converts it to tin(IV) oxide, SnO_2. Tin burns with a white flame.

35.32   Uses of Tin

The principal use of tin is in the production of **tin plate,** i.e., sheet iron coated with tin. Most tin is now plated electrolytically. Iron is more active than tin and corrodes more rapidly when in contact with tin than otherwise. This is because iron and tin in contact with each other, and both in contact with moist air, comprise an electrolytic cell which promotes corrosion (Section 36.28). As a result, when a scratch through the tin exposes the iron, corrosion sets in rapidly. Tin is also used in making alloys, such as **bronze** (Cu, Zn, and Sn), **solder** (Sn and Pb), and **type metal** (Sn, Pb, Sb, and Cu). A plate of nickel and tin, in a ratio of about 1:2, is sometimes used as a substitute for chromium plate.

35.33   Compounds of Tin

■ **1. Oxides.** Depending upon the method of preparation, **tin(II) oxide** (stannous oxide), SnO, is a black or a green powder. It may be prepared by treating a hot solution of a tin(II) compound with an alkali carbonate or by heating tin(II) oxalate in the absence of air. The equations are

$$Sn^{2+} + CO_3^{2-} \longrightarrow \underline{SnO} + CO_2$$
$$SnC_2O_4 \longrightarrow \underline{SnO} + CO_2 + CO$$

When tin is burned in air **tin(IV) oxide** (stannic oxide), SnO_2, is formed. The product is white when cold but yellow when hot.

■ **2. Tin(II) Hydroxide.** Soluble bases precipitate tin(II) ions from solution as **tin(II) hydroxide,** $Sn(OH)_2$, which is white and gelatinous in character. An excess of alkali causes the hydroxide to dissolve, with the formation of the soluble **stannite ion.**

$$\underline{Sn(OH)_2} + OH^- \longrightarrow Sn(OH)_3^-$$

The stannite ion is an active reducing agent. It is used in qualitative analysis to test for bismuth by reducing white bismuth hydroxide to black metallic bismuth.

$$2\underline{Bi(OH)_3} + 3Sn(OH)_3^- + 3OH^- \longrightarrow 2\underline{Bi} + 3Sn(OH)_6^{2-}$$

■ **3. Tin(II) Chloride (Stannous Chloride).** **Tin(II) chloride** can be obtained as the dihydrate ($SnCl_2 \cdot 2H_2O$) by evaporating a solution formed by the reaction of tin or tin(II) oxide with hydrochloric acid. The salt is hydrolyzed by water, forming the basic hydroxychloride, Sn(OH)Cl.

$$SnCl_2 + H_2O \rightleftharpoons \underline{Sn(OH)Cl} + H^+ + Cl^-$$

Tin(II) chloride finds wide use as a reducing agent, because of the ease with which tin is oxidized to the tin(IV) condition. It follows that aqueous solutions of tin(II) must be protected against air oxidation.

■ **4. Tin(IV) Chloride (Stannic Chloride).** **Tin(IV) chloride** is formed when an excess of chlorine reacts with metallic tin. It is a colorless liquid, soluble in organic solvents such as carbon tetrachloride, and it is a nonconductor of electricity. These properties are typical of covalent compounds. When dissolved in water the tetrachloride is highly hydrolyzed (Section 35.34), whereas in hydrochloric acid, **hexachlorostannic acid** is produced.

$$SnCl_4 + 2HCl \longrightarrow H_2SnCl_6$$

■ **5. Sulfides of Tin.** Dark brown **tin(II) sulfide**, SnS, is precipitated when hydrogen sulfide is passed into a solution of a tin(II) salt. Tin(II) sulfide is not dissolved by alkali metal sulfides, whereas tin(IV) sulfide is dissolved. Tin(II) sulfide does dissolve in alkaline polysulfides.

Tin(IV) sulfide is prepared as a yellow precipitate by the interaction of hydrogen sulfide with tin(IV) ions in a moderately acid solution. It dissolves in alkali metal sulfides with the formation of **thiostannate ions**, SnS_3^{2-}. The addition of acid reprecipitates the sulfide.

$$SnS_3^{2-} + 2H^+ \longrightarrow \underline{SnS_2} + H_2S\uparrow$$

Concentrated hydrochloric acid dissolves tin(IV) sulfide according to the equation

$$\underline{SnS_2} + 4H^+ + 6Cl^- \longrightarrow SnCl_6^{2-} + 2H_2S\uparrow$$

35.34 Hydrolysis of Covalent Halides

Except for the tetrahalides of carbon, the tetrahalides of the Group IVA elements hydrolyze when dissolved in water, for the most part as indicated by the general equation

$$MX_4 + 2H_2O \longrightarrow \underline{MO_2} \text{ (or hydrate thereof)} + 4HX$$

The mechanism by which the tetrahalides of silicon, germanium, tin, and lead hydrolyze may be as follows, using silicon tetrachloride as an example. The silicon atom carries a partial net positive charge, because it is less electronegative than chlorine in the $SiCl_4$ molecule. During hydrolysis the positively charged silicon atom may attract hydroxide ions of water, forming a pentacovalent intermediate, according to the equation

$$
\begin{array}{c}
Cl \\
Cl-Si-Cl \\
Cl
\end{array}
+ \; :OH^- \longrightarrow
\left[
\begin{array}{c}
Cl \;\; OH \\
\;\;\; \overset{\circ\circ}{} \\
Cl-Si-Cl \\
Cl
\end{array}
\right]^-
$$

One of the highly negative chlorines is then easily lost from the intermediate as a chloride ion.

$$
\left[
\begin{array}{c}
Cl \;\; OH \\
\;\;\; \overset{\circ\circ}{} \\
Cl-Si-Cl \\
Cl
\end{array}
\right]^-
\longrightarrow
\begin{array}{c}
Cl \;\; OH \\
Cl-Si \\
Cl
\end{array}
+ \; Cl^-
$$

This process is repeated until all four chlorine atoms are replaced by hydroxide groups, $Si(OH)_4$ being the final hydrolytic product. The conversion of the hydroxide ions of water to coordinated hydroxo groups leaves an excess of hydrogen ions in solution, and hydrochloric acid is formed. Resistance of the carbon tetrahalides to hydrolysis is probably due to the fact that carbon is **coordinately saturated** (all its bonding orbitals are filled with electrons) in these compounds, and it cannot form intermediates of the type suggested for silicon tetrachloride. However, carbon tetrachloride, when hot, does react with water to produce phosgene ($COCl_2$), which is extremely poisonous. Hence, carbon tetrachloride fire extinguishers must be used with caution, particularly if water is also being used to fight a fire. Other halides such as sulfur hexafluoride, SF_6, do not hydrolyze when the central element is already coordinately saturated. On the other hand, tungsten hexachloride, WCl_6, hydrolyzes, because the maximum coordination number for tungsten is 8 and the central element is therefore not coordinately saturated.

The silicon halides undergo complete hydrolysis; but with halides of more ionic character, hydrolysis is often less complete and it may be suppressed by the addition of acids. Thus, aqueous solutions of tin tetrachloride can be prepared in the presence of hydrochloric acid.

35.35   Stannic Acids and Stannates

When an alkali metal hydroxide is added to a solution of a tin(IV) compound, a white precipitate forms. We would expect this compound to have the formula $Sn(OH)_4$ and to be acidic in character. However, neither the acid nor the salts derived from the acid have been found. On the other hand, alkali metal stannates are known which appear to be derived from an **orthostannic acid,** which may be formulated as $H_2[Sn(OH)_6]$. The sodium salt of the acid, for example, may be written as $Na_2[Sn(OH)_6]$. Orthostannic acid readily loses water, yielding a **metastannic acid,** H_2SnO_3, and finally the anhydride, SnO_2. When this anhydride is fused with sodium hydroxide, **sodium metastannate,** Na_2SnO_3, is formed. The metastannic acid mentioned above is sometimes called **alpha-stannic acid** to distinguish it from **beta-stannic acid,** which is formed when hot concentrated nitric acid reacts with tin. Beta-stannic acid has the same composition as the alpha form, but it is insoluble in acids and very slightly soluble in strongly alkaline solutions.

35.36   Properties of Tin of Analytical Importance

The colorless tin(II) ion forms a dark brown precipitate when hydrogen sulfide gas is passed into a solution which is dilute in hydrochloric acid. Tin(II) sulfide is insoluble in alkali metal sulfides, but is dissolved by polysulfide solutions, with the formation of thiostannate ions, SnS_3^{2-}. Solutions of tin(IV) compounds yield a yellow precipitate of tin(IV) sulfide when treated with hydrogen sulfide; this sulfide is soluble in alkali metal sulfides, forming alkali metal thiostannates, and is dissolved by concentrated hydrochloric acid, the hexachlorostannate ion, $SnCl_6^{2-}$, being formed. Active metals, such as aluminum, reduce $SnCl_6^{2-}$ to

$SnCl_3^-$ or to metallic tin. When mercury(II) chloride is added to a solution containing tin(II) ions, mercury(I) chloride (white) or metallic mercury (black) is formed.

QUESTIONS

1. What are the properties of metallic bismuth that make it commercially useful?
2. Compare the basicity and acidity of Bi_2O_3 and Bi_2O_5.
3. Write equations to show the hydrolysis of $BiCl_3$.
4. Describe the electrolytic refining process for metallic copper.
5. Copper plate makes up the surface of the Statue of Liberty in New York harbor. Why does it present a green appearance?
6. In terms of electronic structure, explain the existence of copper in three oxidation states.
7. Why must copper be quite free of impurities when used as an electrical conductor?
8. Explain the use of copper(II) sulfate in urine analysis.
9. Compare the stability of the copper(I) halides to that of copper(II) halides.
10. A white precipitate forms when copper metal is added to a solution of copper(II) chloride and hydrochloric acid. The white precipitate dissolves in excess concentrated hydrochloric acid. Dilution with water results again in a white precipitate. Write equations to justify the reactions which are involved.
11. Why are most slightly soluble copper salts readily dissolved by aqueous ammonia?
12. Write an equation to explain the acidic nature of copper(II) sulfate solutions.
13. Sketch the structure of $Cu(H_2O)_4SO_4 \cdot H_2O$.
14. A solution made by dissolving anhydrous copper sulfate in pure water has a pH which is less than 7.0. Explain.
15. Would you expect CuS to dissolve in a 1 M ammonia solution?
16. Why is cadmium chloride considered to be a weak electrolyte?
17. What property makes cadmium effective as a protective coating for other metals?
18. Predict the favored direction of the reaction shown by the equation

$$[Cd(NH_3)_4]^{2+} + Cu^{2+} \rightleftharpoons Cd^{2+} + [Cu(NH_3)_4]^{2+}$$

19. Arsenic is classed as a metalloid. Explain.
20. Compare the metallic character of arsenic with that of antimony.
21. Can the term "amphoteric" be applied properly to the sulfides of antimony and arsenic? Explain why or why not.
22. Compare and contrast arsine with ammonia.
23. Account for the oxidation states +3 and +5 of arsenic in terms of electronic structure.
24. Draw structural representations of the pyroarsenic acid and meta-arsenic acid molecules.
25. Which is the more acidic, As_4O_6 or As_2O_5?
26. Write balanced equations to show the action of Na_2S upon As_2S_3 and As_2S_5.

27. What are the composition and use of Paris green?
28. Write balanced chemical equations describing the action of hot nitric acid on elemental arsenic; the action of hot concentrated sulfuric acid on elemental antimony; the oxidation of manganous ion to permanganate ion by the action of a solution made by acidifying sodium bismuthate with nitric acid.
29. Why can it be said that iodide ions serve as a catalyst in the precipitation of arsenic(III) sulfide from arsenate ion?
30. Why is antimony used as a constituent of type metal?
31. Why does not antimony dissolve in hot nitric acid?
32. What is the coordination number of antimony in the antimonate ion?
33. Mercury(I) solutions can be protected from air oxidation to mercury(II) by placing metallic mercury in contact with the solutions. Can tin(II) solutions be similarly stabilized against oxidation to tin(IV) by keeping metallic tin in contact with the solutions? (The standard reduction potential for tin(II) to tin(IV) is $+0.15$ v.)
34. How and why is tin plate applied to the surface of iron?
35. What is meant by *tin pest,* which is also known as *tin disease?*
36. What is the action of metallic tin with HCl, with HNO_3, and with NaOH?
37. Write the equations for the reactions involved when an excess of aqueous NaOH is slowly added to a solution of tin(II) chloride.
38. Write the equation for the thermal decomposition of tin(II) oxalate.
39. Write balanced chemical equations describing:
    (a) The preparation of sodium metastannate.
    (b) The purification of tin by electrolysis.
    (c) The action of dry chlorine on tin.
    (d) The formation of thiostannate ion.
    (e) The dissolution of tin by alkali solution.
40. Why cannot $SnCl_4$ be classified as a salt?
41. Why must aqueous solutions of $SnCl_2$ be protected from the air?

PROBLEMS

1. How many grams of $CuCl_2$ contain the same mass of copper as 100 g of CuCl?
   Ans. 136 g
2. A solid 1.008-g sample of a silver-copper alloy is dissolved and treated with excess iodide ion. The liberated I_2 is titrated with a 0.1052 M $S_2O_3{}^{2-}$ solution. If 29.84 ml of this solution is required, what is the % Cu in the alloy?
   Ans. 19.79%
3. How many pounds of cadmium can be obtained from ten tons of ore which is 5.0% greenockite?   *Ans. 7.8×10^2 lb*
4. How many grams of CdS will precipitate upon treatment with an excess of sulfide ion in a solution obtained by dissolving 2.5 g of Wood's metal?
   Ans. 0.40 g
5. What would be the calculated potential of the following electrochemical cell: $Cd\,|\,Cd^{2+},\ M = 0.10\,\|\,Ni^{2+},\ M = 0.50\,|\,Ni$?   *Ans. 0.17 volt*

6. A 1.497-g type metal sample is dissolved in nitric acid, whereupon metastannic acid precipitates. This is dehydrated by heating to stannic oxide, which is found to weigh 0.4909 g. What per cent tin was the original type metal sample?

Ans. 25.83%

7. The dissociation constant of $[Cu(NH_3)_4]^{2+}$ is 8.5×10^{-13}. What concentration of Cu^{2+} will be in equilibrium if 1.0 g of Cu is oxidized and put into one liter of solution with 0.25 M NH_3?

Ans. $1.1 \times 10^{-11} M$

REFERENCES

"Factors Involved in the Stereochemistry of AX_6E Systems of the Heavy Main Group Elements," K. J. Wynne, *J. Chem. Educ.*, **50**, 328 (1973).

"*d*-Orbitals in Main Group Elements," T. B. Brill, *J. Chem. Educ.*, **50**, 392 (1973).

"Synthesis, Properties, and Hydrolysis of Antimony Trichloride," F. C. Hentz, Jr. and G. G. Long, *J. Chem. Educ.*, **52**, 189 (1975).

"The Biochemistry of Copper," E. Frieden, *Sci. American*, May, 1968; p. 102.

"Cryoscopic Measurement of the Coordination Number of Copper(II)," H. C. deLorenzo, J. A. C. Frugoni, and V. Lopez, *J. Chem. Educ.*, **46**, 113 (1969).

"Developments in Copper Smelting," E. G. West, *Chem. in Britain*, **5**, 199 (1969).

"A Potentiometric Copper Assay in Normal and Copper-Poisoned Humans," I. A. Matheson and D. R. Williams, *J. Chem. Educ.*, **50**, 345 (1973).

"Resource Utilization—Copper from Low Grade Ores," M. J. Cahalan, *Chem. in Britain*, **9**, 392 (1973).

"Organotin Chemistry," A. G. Davies, *Chem. in Britain*, **4**, 403 (1968).

"Applied Research in the Development of Anticaries Dentifrices," W. E. Cooley, *J. Chem. Educ.*, **47**, 177 (1970).

"Tin(IV) Iodide as a Radioactive Tracer," P. W. Wiggans, *Educ. in Chemistry*, **11**, 194 (1974).

Metals of Analytical Group III

36

Nickel, Cobalt, Manganese, Iron, Aluminum, Chromium, Zinc

Analytical Group III (see also Chapter 42) consists of those common cations which are precipitated as sulfides or hydroxides in solutions of aqueous ammonia saturated with hydrogen sulfide, but which are not precipitated by hydrochloric acid nor by hydrogen sulfide in solutions that are 0.3 M in hydrogen ions. The sulfides of cobalt(II), nickel(II), manganese(II), iron(II), and zinc, and the hydroxides of chromium(III) and aluminum precipitate in ammonium sulfide solutions.

Five of the metals whose ions constitute Analytical Group III (Cr, Mn, Fe, Co, and Ni) are members of the first transition series (Period 4) of the Periodic Table. This series begins with scandium and ends with copper; it is a series in which the electron shell underlying the valence shell is being increased from 9 to 18 electrons. These elements have valence electrons in two shells; associated with this property are such characteristics as variation in valence, color of the ions, and tendency toward the formation of stable complex ions. It has already been pointed out that successive members of each of the transition series show striking resemblances to each other (Section 31.7). Atomic and ionic sizes of elements with similar electronic structures are important in determining the properties of the elements. Note that the following elements of Periodic Series 4 (and Analytical Group III) have atoms of very nearly the same radius (Å): Cr, 1.25; Mn, 1.29; Fe, 1.26; Co, 1.26; Ni, 1.24. Zinc and aluminum are representative metals (valence electrons in one shell) with atomic radii of 1.33 and 1.43, respectively. Some properties of the Analytical Group III metals are given in Table 36–1.

TABLE 36-1   Some Properties of the Metals of Analytical Group III

Property	Aluminum	Chromium	Manganese	Iron	Cobalt	Nickel	Zinc
Atomic number	13	24	25	26	27	28	30
Atomic weight	26.9815	51.996	54.9280	55.847	58.9332	58.71	65.37
Electronic structure	2 8 3	2 8 13 1	2 8 13 2	2 8 14 2	2 8 15 2	2 8 16 2	2 8 18 2
Oxidation states	+3	+2 +3 +6	+2 +3 +4 +6 +7	+2 +3 +6	+2 +3 +4	+2 +3 +4	+2
Reduction potential, volts,	$\left(\dfrac{Al^{3+}}{Al}\right)$ -1.66	$\left(\dfrac{Cr^{3+}}{Cr}\right)$ -0.74	$\left(\dfrac{Mn^{2+}}{Mn}\right)$ -1.18	$\left(\dfrac{Fe^{3+}}{Fe}\right)$ -0.44	$\left(\dfrac{Co^{2+}}{Co}\right)$ -0.28	$\left(\dfrac{Ni^{2+}}{Ni}\right)$ -0.25	$\left(\dfrac{Zn^{2+}}{Zn}\right)$ -0.76
Density g/cm³ at 20°	2.70	7.1	7.20	7.86	8.71	8.9	7.14
Melting point, °C	660	1,900	1,244	1,535	1,493	1,455	419.5
Boiling point, °C	2,327	2,642	2,087	2,800	3,100	2,800	907

Nickel

Iron, cobalt, and nickel comprise the first horizontal triad of Group VIII of the Periodic Table. The second triad consists of ruthenium, rhodium, and palladium, and the third, of osmium, iridium, and platinum. In each of these horizontal series the tendency to lose electrons decreases as the nuclear charge increases. Thus iron exhibits oxidation states of +2, +3, and +6; cobalt, +2, +3, and +4; while with nickel, oxidation states higher than +2 do exist (NiO_2 in the Edison cell, described in Section 22.17, for example) but are rare.

36.1   History and Occurrence of Nickel

Because nickel ores are difficult to reduce, early attempts to produce the metal from its ores were unsuccessful. It was thought by the seventeenth century metallurgists that nickel ores were copper ores because of the similarity in appearance; but when the ores failed to yield copper, they were called *Kupfernickel* (German). *Kupfer* means copper, and *nickel* refers to a devil which was thought to prevent the extraction of copper from the ores. The metal was first obtained by Cronstedt, a German chemist, in 1751.

Nickel occurs, along with iron, as an alloy in meteorites. Metallic nickel and iron probably constitute most of the core of the earth. The most important ores of nickel are **pentlandite,** (Ni, Cu, Fe)S, and **garnierite,** which is a hydrated magnesium-nickel silicate of variable composition.

36.2 Metallurgy of Nickel

Pentlandite is roasted and then reduced with carbon. This process results in the production of an alloy containing nickel, iron, and copper. The alloy is known as **Monel metal** and, because of its resistance to corrosion, has many important industrial uses. The process used most extensively in the production of pure metallic nickel in recent years involves the separation of the sulfides of nickel, copper, and iron by selective flotation. After separation the nickel sulfide is converted to the oxide by roasting, and the oxide is reduced by carbon. The metal obtained in this way is approximately 96% pure. It is purified electrolytically to 99.98% purity. Gold, silver, and especially platinum are recovered from the anode mud.

36.3 Properties and Uses of Nickel

Nickel is a silvery-white metal that is hard, malleable, and ductile. Like iron and cobalt, it is highly magnetic. It is not oxidized by air under ordinary conditions, and it is resistant to the action of alkalies. Dilute acids slowly dissolve nickel, hydrogen being evolved. Nickel can be made passive by treatment with concentrated nitric acid; the nickel then will no longer displace hydrogen from dilute acids.

Because of its hardness, resistance to corrosion, and high reflectivity when polished, nickel is widely used in the plating of iron, steel, and copper. It is also a constituent of many important alloys such as **Monel metal** (Ni, Cu, and a little Fe), and **Permalloy** (Ni and Fe), used in instruments for the electrical transmission and reproduction of sound. **German silver** is a nickel-zinc-copper alloy. **Nichrome** and **chromel** are alloys containing nickel, iron, and chromium; they are resistant to oxidation at high temperatures and show high electrical resistance, so are used in electrical heating units such as electric stoves, pressing irons, and toasters. **Alnico** contains aluminum, nickel, iron, and cobalt; it is highly magnetic, being able to lift 4,000 times its own weight of iron. **Platinite** and **invar** are nickel alloys which have the same coefficient of expansion as glass and are therefore used for "seal-in" wires through glass, such as in electric light bulbs. Finely divided nickel is used as a catalyst in the hydrogenation of oils.

36.4 Compounds of Nickel

In combination, nickel exhibits an oxidation state of +2 almost exclusively. One notable exception is tetravalent nickel in the dioxide NiO_2, a black hydrous substance formed by the oxidation of nickel(II) salts in alkaline solution. This oxide forms one of the electrodes in the Edison storage cell (Section 22.17).

Nickel(II) oxide, NiO, is prepared by heating either the hydroxide, the carbonate, or the nitrate. When alkali metal hydroxides are added to solutions of nickel(II) salts, pale green **nickel(II) hydroxide,** $Ni(OH)_2$, precipitates. Both the oxide and the hydroxide are dissolved by aqueous ammonia with the formation of the deep blue **hexaamminenickel(II) hydroxide,** $[Ni(NH_3)_6](OH)_2$.

$$Ni(OH)_2 + 6NH_3 \longrightarrow [Ni(NH_3)_6]^{2+} + 2OH^-$$

The soluble salts of nickel are prepared from either the oxide, hydroxide, or carbonate by treatment with the proper acid. The important soluble salts of the metal are the acetate, chloride, nitrate, and sulfate, and also the hexaammine-nickel(II) sulfate. Ammoniacal solutions of the latter compound are used in nickel-plating baths. The hydrated nickel(II) ion, $[Ni(H_2O)_6]^{2+}$, imparts a pale green color to the solutions and crystallized salts of the nickel(II) ion.

Nickel(II) sulfide is produced as a black precipitate by the action of ammoniacal sulfide solutions upon nickel(II) salts.

Nickel carbonyl, $Ni(CO)_4$, is formed as a colorless volatile liquid (boiling at 43° C) when carbon monoxide is led over finely divided nickel. The carbonyl readily decomposes at higher temperatures and deposits pure nickel metal. The **Mond process** for separating nickel from other metals is based upon the formation and decomposition of nickel carbonyl. When carbon monoxide is passed over a mixture of nickel and other metals, nickel carbonyl is formed and carried along with the excess carbon monoxide, leaving the other metals behind. When heated to 200° C the nickel carbonyl deposits the metal as a fine dust.

36.5 Properties of Nickel of Analytical Importance

Aqueous solutions of nickel salts are green. The addition of aqueous ammonia first causes pale green nickel hydroxide to precipitate, but an excess of ammonia dissolves the precipitate by forming the deep blue complex, $[Ni(NH_3)_6]^{2+}$. Hydrogen sulfide causes the precipitation of brown or black nickel sulfide from neutral or alkaline solutions of nickel salts. This is readily soluble in acids, but it very quickly changes to another crystalline form that is nearly insoluble in dilute acid solutions. Nickel salts in nearly neutral solutions form a red precipitate with dimethylglyoxime.

Cobalt

36.6 History of Cobalt

The name of this metal is derived from the German word *Kobold*, which means "goblin." The reason for this was that the early metallurgists believed that goblins carefully guarded the cobalt ores and prevented man from liberating the metal. Brandt finally succeeded in isolating the metal in 1735. Even today, the metallurgy of both cobalt and nickel is difficult and complicated.

36.7 Occurrence and Metallurgy of Cobalt

The common cobalt minerals are **cobaltite,** CoAsS, and **smaltite,** $CoAs_2$, with the richest deposits of these minerals being found in Ontario, Canada. However, nearly all the world's production of cobalt is obtained as a by-product of the metallurgy of nickel, copper, iron, silver, and other metals. The metallurgy of cobalt involves its separation from nickel, copper, and iron; conversion to Co_3O_4; and the reduction of the oxide with aluminum or hydrogen.

$$3Co_3O_4(s) + 8Al(s) \longrightarrow 9Co(s) + 4Al_2O_3(s) + 963,000 \text{ cal}$$
$$(\Delta H^\circ_{298} = -963,000 \text{ cal})$$

(Notice the very large amount of energy evolved here. Very few reactions involve this much energy.) Purification of the metal is affected by electrolytic deposition on a rotating, stainless steel cathode.

36.8 Properties and Uses of Cobalt

Cobalt is similar to iron in appearance, except that it has a faint tinge of pink. Like iron and nickel, it is magnetic. It is slowly soluble in warm dilute hydrochloric or sulfuric acid and more rapidly soluble in dilute nitric acid. Like iron and nickel, cobalt is rendered passive by contact with concentrated nitric acid. It is not oxidized on exposure to the air, but at red heat cobalt reduces hydrogen in steam, with evolution of the hydrogen. The halogens, except fluorine, convert it to cobalt(II) halides. When cobalt is heated with fluorine, **cobalt(III) fluoride**, CoF_3, is formed.

Cobalt is alloyed with iron and small percentages of other metals in making high-speed cutting tools and surgical instruments. Permanent magnets are made from the alloys **Alnico** (Al, Ni, Co, Fe), **Hiperco** (Co, Fe, Cr), and **Vicalloy** (Co, Fe, V).

Finely divided cobalt metal is used as a catalyst in the hydrogenation of carbon monoxide and carbon dioxide, with the formation of hydrocarbons, and for the oxidation of ammonia.

36.9 Compounds of Cobalt

■ **1. Hydroxides and Oxides of Cobalt.** **Cobalt(II) hydroxide**, $Co(OH)_2$, is formed as a blue flocculent precipitate when an alkali metal hydroxide is added to a solution of a cobalt (II) salt. The blue color of the precipitate changes to violet and then to pink, probably as a result of hydration. The hydroxide is readily soluble in aqueous ammonia to form **hexaamminecobalt(II) hydroxide**, $[Co(NH_3)_6](OH)_2$. Solutions of the latter compound are oxidized by oxygen in air to the various cobalt(III) compounds; the oxidation is accompanied by a darkening of the solution.

When cobalt(II) hydroxide is heated in the absence of air **cobalt(II) oxide**, CoO, results. This oxide is a black substance, but when dissolved in fused glass it gives the glass a blue color; such glass is called "cobalt glass" and contains **cobalt(II) silicate**. Ignition of the hydroxide or oxide in air gives rise to **cobalt(II,III) oxide**, Co_3O_4. Cobalt(III) oxide may be produced by gently heating cobalt(II) nitrate.

$$4Co(NO_3)_2 \longrightarrow 2Co_2O_3 + 8NO_2 + O_2$$

Note that three elements in this reaction change their oxidation state. (You will find balancing this equation by the oxidation-reduction method a challenging review.)

■ **2. Cobalt(II) Chloride.** Cobalt(II) oxide and hydroxide readily dissolve in hydrochloric acid. Concentration of the solution results in crystallization of **cobalt(II) chloride 6-hydrate,** $CoCl_2 \cdot 6H_2O$, which is red in color. The hydrated cobalt(II) ion, $[Co(H_2O)_6]^{2+}$, exhibits a pink color in solution. When partially dehydrated, cobalt(II) chloride changes to a deep blue color. This is believed to result from a change in the coordination number of the cobalt(II) ion from six to four.

$$[Co(H_2O)_6]Cl_2 \longrightarrow [Co(H_2O)_4]Cl_2 + 2H_2O$$
$$\text{(Pink color)} \qquad\qquad \text{(Blue color)}$$

The same change in color is effected by dissolving $CoCl_2 \cdot 6H_2O$ in alcohol. Writing made on paper with a dilute solution of the hydrated salt is almost invisible, but when the paper is warmed, dehydration of the salt occurs and the writing becomes blue. It fades again as hydration takes place from moisture of the air. This is the chemistry of one kind of "invisible" ink.

■ **3. Cobalt(II) Sulfide.** Black **cobalt(II) sulfide,** CoS, is completely precipitated only from basic solutions. However, once precipitated this sulfide is but slightly soluble in hydrochloric acid (see Section 36.10). Aqua regia readily dissolves it.

■ **4. Complex Cobalt Compounds.** In addition to the simple salts, oxides, and hydroxides, cobalt forms a large number of complex compounds. Cobalt(II) simple salts are more stable than are those of cobalt(III); complex cobalt(III) compounds, in contrast, are much more stable than the corresponding cobalt(II) complexes. Thus, the majority of the more stable complex salts of cobalt contain the metal in the +3 oxidation state. Some of the more important cobalt(III) complexes are **hexaamminecobalt(III) chloride,** $[Co(NH_3)_6]Cl_3$, **potassium hexacyanocobaltate(III),** $K_3[Co(CN)_6]$, and **sodium hexanitrocobaltate(III),** $Na_3[Co(NO_2)_6]$. In these complex compounds the six coordinated groups, such as NH_3, CN^-, and NO_2^-, occupy the corners of a regular octahedron with the cobalt at the center (see Section 32.3).

36.10   Properties of Cobalt of Analytical Importance

Solutions of simple cobalt(II) compounds are pink. Hydrogen sulfide precipitates black cobalt(II) sulfide from basic solutions of cobalt. The sulfide changes quickly to another crystalline form which is only slightly soluble in dilute hydrochloric acid but readily soluble in aqua regia. When a concentrated solution of ammonium thiocyanate is added to a solution of a cobalt(II) salt, the complex ion $[Co(CNS)_4]^{2-}$ is formed. The characteristic blue color of the ion is accentuated by the addition of acetone to the solution. Another test for cobalt involves its oxidation by the nitrite ion from +2 to +3, and precipitation of the insoluble yellow complex salt, $K_3[Co(NO_2)_6]$.

$$Co^{2+} + 6NO_2 \longrightarrow [Co(NO_2)_6]^{4-}$$
$$[Co(NO_2)_6]^{4-} + NO_2^- + 2H^+ \longrightarrow [Co(NO_2)_6]^{3-} + NO\uparrow + H_2O$$
$$3K^+ + [Co(NO_2)_6]^{3-} \longrightarrow \underline{K_3[Co(NO_2)_6]}$$

Manganese

36.11   Periodic Relationships of Manganese

Manganese, technetium, and rhenium comprise Periodic Group VIIB. Although rhenium was not discovered until 1925, its chemistry has now been widely studied. No stable isotope of technetium is known; however, the chemistry of the element has been studied by means of radioactive isotopes prepared synthetically. Each of the three elements in Group VIIB has valence electrons in two shells, the outermost one of which contains two electrons. Contrasted with these elements are the halogens of Periodic Group VIIA, which have seven electrons in their outer and only valence shell. The halogens are active nonmetals whereas the Group VIIB elements are metals. However, in the higher oxidation states, there are striking similarities between the elements of these two groups. Thus, Mn_2O_7 and Cl_2O_7 are both volatile, explosively unstable liquids; $KMnO_4$ and $KClO_4$ are both strong oxidizing agents which form isomorphous crystals of nearly the same solubility. In the following sections, only manganese will be discussed in detail.

36.12   Occurrence and Metallurgy of Manganese

The most important ore of manganese is **pyrolusite**, MnO_2. Other manganese ores include **braunite**, Mn_2O_3, **manganite**, $MnO(OH)$ or $Mn_2O_3 \cdot H_2O$, **hausmannite**, Mn_3O_4, **franklinite**, $(Fe, Mn, Zn)O$, and **psilomelane**, MnO_2 $(BaO, K_2O, H_2O,$ etc.).

Nearly pure manganese may be produced by the high temperature reduction of the dioxide by aluminum.

$$3MnO_2(s) + 4Al(s) \longrightarrow 3Mn(s) + 2Al_2O_3(s) + 428,100 \text{ cal}$$
$$(\Delta H^\circ_{298} = -428,100 \text{ cal})$$

Since alloys of manganese and iron are extensively used in the production of steel, such alloys are usually produced instead of the pure manganese. They are prepared by reducing the mixed oxides of manganese and iron with coke in a blast furnace. The alloys high in manganese are called **ferromanganese,** and those low in manganese are called **spiegeleisen,** meaning *mirror iron* (German).

36.13   Properties of Manganese

Manganese is a gray-white metal with a slightly reddish tinge. It is brittle and has the general appearance of cast iron. It is readily oxidized by moist air and decomposes water slowly, forming manganese(II) hydroxide and hydrogen. It dissolves readily in dilute acids, forming manganese(II) salts. Manganese forms five oxides and five corresponding series of salts (Table 36–2). These compounds are to be considered in the following sections.

TABLE 36-2   Classes of Manganese Compounds

Oxidation State	Oxide	Hydroxide	Character	Derivative	Name	Color
+2	MnO	$Mn(OH)_2$	Moderately basic	$MnCl_2$	Manganese(II) chloride	Pink
+3	Mn_2O_3	$Mn(OH)_3$	Weakly basic	$MnCl_3$	Manganese(III) chloride	Violet
+4	MnO_2	H_2MnO_3	Weakly acidic	$CaMnO_3$	Calcium manganite	Brown
+6	MnO_3	H_2MnO_4	Acidic	K_2MnO_4	Potassium manganate	Green
+7	Mn_2O_7	$HMnO_4$	Strongly acidic	$KMnO_4$	Potassium permanganate	Purple

36.14   Compounds of Divalent Manganese

Although manganese forms compounds in which it exhibits oxidation states of +2, +3, +4, +6, and +7, the only stable cation is the divalent manganese ion Mn^{2+}. The common soluble manganese(II) salts are the chloride, sulfate, and nitrate. Each imparts a faint pink color to its solutions.

Alkali metal hydroxides precipitate pale pink **manganese(II) hydroxide**, $Mn(OH)_2$, which is oxidized on exposure to air to the dark brown **manganese(III) oxyhydroxide**, $MnO(OH)$. Manganese(II) hydroxide is only partially precipitated by aqueous ammonia and is dissolved by solutions of the ammonium salts of strong acids, according to the equation

$$Mn(OH)_2 + 2NH_4^+ \rightleftharpoons Mn^{2+} + 2NH_3 + 2H_2O$$

Manganese(II) hydroxide is entirely basic in character, as indicated by the fact that it dissolves in acids but not in bases such as sodium hydroxide. **Manganese(II) oxide**, MnO, may be produced by heating the corresponding hydroxide in the absence of air.

From manganese(II) solutions, alkali metal sulfides precipitate the pink **sulfide**, MnS, which is readily soluble in dilute acids.

36.15   Compounds of Trivalent Manganese

Dimanganese trioxide, Mn_2O_3, and $MnO(OH)$ occur naturally, but the manganese(III) ion is unstable in aqueous solution and is readily reduced to the manganese(II) ion. **Manganese(III) chloride** is formed in solution by the action of hydrochloric acid upon manganese dioxide at low temperature, but when the solution is warmed the manganese(III) chloride decomposes with the formation of manganese(II) chloride and chlorine.

$$MnO_2 + 4H^+ + 4Cl^- \longrightarrow MnCl_4 + 2H_2O$$
$$2MnCl_4 \longrightarrow 2MnCl_3 + Cl_2\uparrow$$
$$2MnCl_3 \longrightarrow 2Mn^{2+} + 4Cl^- + Cl_2\uparrow$$

36.16 Compounds of Tetravalent Manganese

The most important compound of tetravalent manganese is the dioxide, MnO_2. This oxide is amphoteric but is relatively inert towards acids and alkalies. Cold, concentrated hydrochloric acid acts upon the dioxide, giving a green solution of the unstable **manganese(IV) chloride.** The sulfate, $Mn(SO_4)_2$, is also unstable, but the complex salt K_2MnF_6 is not readily decomposed. When the dioxide is fused with calcium oxide, **calcium manganite,** $CaMnO_3$, is formed. This is a salt of the hypothetical **manganous acid,** H_2MnO_3.

When manganese dioxide is heated to 535° C it is transformed to **trimanganese tetraoxide,** Mn_3O_4, and oxygen, according to

$$41,200 \text{ cal} + 3MnO_2(s) \longrightarrow Mn_3O_4(s) + O_2(g) \qquad (\Delta H^\circ_{298} = 41,200 \text{ cal})$$

A note of historical interest is that Scheele, independently of Priestley and Lavoisier, discovered oxygen by the action of concentrated sulfuric acid on manganese dioxide.

$$MnO_2 + 2H_2SO_4 \longrightarrow Mn(SO_4)_2 + 2H_2O$$
$$2Mn(SO_4)_2 + 2H_2O \longrightarrow 2MnSO_4 + 2H_2SO_4 + O_2$$

36.17 Manganates, Hexavalent Manganese

When an oxide of manganese is fused with an alkali metal hydroxide or carbonate in the presence of air or some other oxidizing agent (such as potassium chlorate or potassium nitrate), a manganate is formed.

$$2MnO_2 + 4KOH + O_2 \longrightarrow 2K_2MnO_4 + 2H_2O$$

Manganates are green in color and stable only in alkaline solution. The addition of water to solutions of manganates may bring about auto-oxidation-reduction, with the precipitation of manganese dioxide and the formation of the purple permanganate ion, MnO_4^-.

$$3MnO_4^{2-} + 2H_2O \longrightarrow 2MnO_4^- + 4OH^- + \underline{MnO_2}$$

Free **manganic acid,** H_2MnO_4, is too unstable to be prepared and isolated. Solutions containing the manganate ion become active oxidizing agents when acidified. The half-reaction is given by the equation

$$MnO_4^{2-} + 8H^+ + 4e^- \longrightarrow Mn^{2+} + 4H_2O$$

36.18 Permanganates, Heptavalent Manganese

An important compound of manganese is **potassium permanganate,** $KMnO_4$. It is prepared commercially by oxidizing potassium manganate in alkaline solution with chlorine.

$$2MnO_4^{2-} + Cl_2 \longrightarrow 2MnO_4^- + 2Cl^-$$

The resultant purple solution deposits crystals when sufficiently concentrated. Potassium permanganate is a valuable laboratory reagent because it acts as a strong oxidizing agent. A solution of the free **permanganic acid,** $HMnO_4$, may be prepared by the reaction of dilute sulfuric acid and barium permanganate.

$$Ba^{2+} + 2MnO_4^- + (2H^+) + SO_4^{2-} \longrightarrow \underline{BaSO_4} + (2H^+) + 2MnO_4^-$$

Mn_2O_7, a dark brown, highly explosive liquid, is the anhydride of permanganic acid.

36.19   Uses of Manganese and Its Compounds

Manganese metal forms a number of important alloys. Two of these were mentioned in Section 36.12; they are **spiegeleisen,** which is a bright and lustrous iron alloy containing 5-20 per cent manganese, and **ferromanganese,** which contains 70-80 per cent manganese. Both of these alloys are used in making very hard steels for the manufacture of rails, safes, and heavy machinery. About 12.5 pounds of manganese metal are used in the production of every ton of steel to remove oxygen, nitrogen, and sulfur. An alloy called **manganin** (Cu, 84 per cent; Mn, 12 per cent; Ni, 4 per cent) is used in instruments for making electrical measurements, because the electrical resistance of the alloy does not change significantly with changes in temperature.

Manganese dioxide is used in glassmaking to correct the green color produced by iron(II) compounds; to color glass and enamels black; to act as an oxidizing agent, or "dryer," in black paints; and as a depolarizer in dry cells. Potassium permanganate is used as a disinfectant, a deodorant, and a germicide. Potassium permanganate is also used extensively as an oxidizing agent in analytical determinations.

36.20   Properties of Manganese of Analytical Importance

When solutions containing Mn^{2+} are treated with ammonium sulfide, flesh-colored MnS precipitates. This sulfide is soluble in acids. White $Mn(OH)_2$ is precipitated by alkalies; it soon turns brown upon exposure to the air. Manganese(II) hydroxide is soluble in solutions of ammonium salts. Bismuthate ion, BiO_3^-, in acid solution readily oxidizes Mn^{2+} to MnO_4^-, as indicated by the purple color of the resulting permanganate.

Iron

36.21   History and Importance of Iron

Iron was known at least as early as 4000 B.C. and probably was used to some extent during prehistoric times. Because free iron is not commonly found in nature, that first used in prehistoric times may well have been of meteoric origin. The early discovery and application of iron to the manufacture of tools and weapons resulted from the wide distribution of iron ores and the ease with which the iron compounds in the ores could be reduced by carbon. Charcoal was the form of carbon used for a long time in the reduction process, but the production and use of iron on a large scale did not begin until about 1620, when coal was introduced as the reducing agent.

Iron is the most widely used of all the metals. This is because its ores are abundant and widely distributed, the metal is easily and cheaply produced from the ores, and its properties can be varied over a wide range by the addition of other substances and by different methods of treatment, such as tempering and annealing. More iron is used than all the other metals combined (14 times as much).

36.22 Occurrence of Iron

Iron ranks second in abundance among the metals (aluminum is first) and fourth among all elements in the earth's crust. The central core of the earth is largely iron, as has been indicated by studies of the earth's density and the rate of transmission of earthquake shock. Metallic meteors are usually about 90 per cent iron, and the remainder is principally nickel. Practically all rocks, minerals, and soils, as well as plants, contain some iron. Iron is present in the hemoglobin of the blood, which acts as a carrier of oxygen (Section 28.24).

The important iron-bearing ores are **hematite**, Fe_2O_3, and **magnetite**, Fe_3O_4 (Fig. 36–1). The presence of Fe_2O_3 accounts largely for the red coloration of many rocks and soils. Hematite often occurs as the reddish-brown hydrate $2Fe_2O_3 \cdot 3H_2O$, which is called **limonite**. Magnetite is a black, crystalline mineral, which is strongly magnetic, sometimes possessing polarity, in which case it is called **lodestone**. Black crystalline magnetite interlocked with crystals of silica in the form of hard rock is called **taconite**. **Siderite** is a white or brownish carbonate, $FeCO_3$. **Pyrite**, FeS_2, occurs as pale yellow crystals with a metallic luster which are easily mistaken for gold; hence the name "fool's gold." Pyrite is quite abundant and useful as a source of sulfur for the manufacture of sulfuric acid; but because roasting does not remove all of the sulfur, pyrite is usually considered unsatisfactory as a source of iron, although low-grade pig iron can be made from it.

With the rather rapid depletion of domestic high-grade iron ore (over 50 per cent iron) deposits, the steel industry has turned to the conquest of the taconite ores (25-30 per cent iron). Taconite is the extremely hard rock that makes up some 95 per cent by volume of the famous Mesabi Range's vast iron-bearing deposits in upper Minnesota. Once regarded as useless for steel production, the taconite ores are now used extensively as a result of the development of satisfactory techniques for mining and concentrating them.

FIGURE 36-1
The magnetic properties of a piece of magnetite, Fe_3O_4. *Ward's Natural Science Establishment, Inc.*

36.23 Metallurgy of Iron

The first step in the metallurgy of iron is usually that of roasting the ore to remove water, decompose carbonates, and oxidize sulfides. The oxides of iron are then reduced with coke in a blast furnace 80-100 feet high and 25 feet in diameter. Near the bottom of the furnace are nozzles, or "tuyères," through which preheated air is blown into the furnace under pressure. The charge of roasted ore, coke, and flux (limestone or sand) is introduced into the top of the furnace through

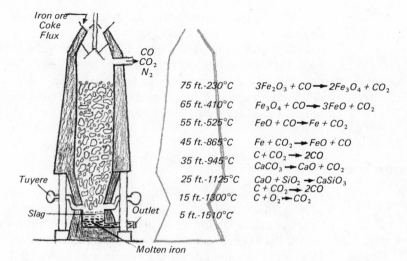

Iron ore
Coke
Flux

CO
CO₂
N₂

75 ft.-230°C	$3Fe_2O_3 + CO \rightarrow 2Fe_3O_4 + CO_2$
65 ft.-410°C	$Fe_3O_4 + CO \rightarrow 3FeO + CO_2$
55 ft.-525°C	$FeO + CO \rightarrow Fe + CO_2$
45 ft.-865°C	$Fe + CO_2 \rightarrow FeO + CO$
35 ft.-945°C	$C + CO_2 \rightarrow 2CO$ $CaCO_3 \rightarrow CaO + CO_2$
25 ft.-1125°C	$CaO + SiO_2 \rightarrow CaSiO_3$ $C + CO_2 \rightarrow 2CO$
15 ft.-1300°C	$C + O_2 \rightarrow CO_2$
5 ft.-1510°C	

Tuyere

Slag   Outlet

Molten iron

FIGURE 36–2
Reactions that occur in the blast furnace.

cone-shaped valves arranged so that the charge is deposited uniformly around the circumference of the furnace near the walls. As the charge melts, or is burned in the lower regions of the furnace, the stock column gradually settles, thus leaving room for additional charges at the top, and making the operation of the blast furnace a continuous process. The entire stock in the furnace weighs several hundred tons.

As soon as the preheated air (500° C) under pressure enters the furnace, the coke in the region of the tuyères is oxidized to carbon dioxide with the liberation of much heat, which raises the temperature to about 1,500° C. As the carbon dioxide passes upward through the overlying layer of white-hot coke, it is reduced to carbon monoxide ($CO_2 + C \longrightarrow 2CO$). The carbon monoxide thus produced serves as the reducing agent in the upper regions of the furnace. The individual reduction reactions are indicated in Fig. 36–2 and can be summarized as a single reversible reaction as follows:

$$Fe_2O_3(s) + 3CO(g) \rightleftharpoons 2Fe(s) + 3CO_2(g) + 5,900 \text{ cal}$$

$$(\Delta H_{298}^\circ = -5,900 \text{ cal})$$

The presence of an excess of carbon monoxide keeps the equilibrium shifted to the right. The iron ore is completely reduced before the formation of the slag begins in the middle region of the furnace; otherwise, part of the iron would be lost due to the reaction $FeO + SiO_2 \longrightarrow FeSiO_3$.

Just below the middle of the furnace, the temperature is high enough to melt both the iron and the slag which have been produced. The molten iron and slag collect in two layers at the bottom of the furnace, the less dense slag floating on the iron and protecting it against oxidation. Every hour or two, the liquid slag is withdrawn from the furnace. Blast furnace slag is often used in the manufacture of Portland cement. Four or five times a day, the molten iron is withdrawn through a tap hole into a large ladle lined with fire brick that holds several hundred tons of the hot metal. The ladle containing the hot metal is then trans-

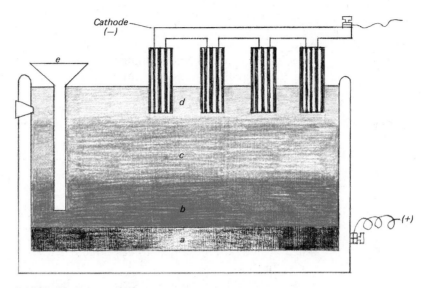

FIGURE 36-4

Electrolytic cell for the purification of aluminum. *a* is bottom portion of cell in contact with the anode layer; *b* is fused impure aluminum and copper alloy, which acts as the anode and in which the anodic oxidation takes place; *c* is an electrolyte of fused fluorides; *d* is the layer of pure aluminum deposited at the cathodes; *e* is a funnel arrangement by which impure aluminum may be added to the bottom of the cell.

aluminum hydroxide is precipitated by hydrolysis of the aluminate. The precipitated aluminum hydroxide is ignited to the oxide, which is then dissolved in fused cryolite, Na_3AlF_6. This solution is electrolyzed, whereupon aluminum metal is liberated at the cathode and oxygen, carbon monoxide, and carbon dioxide are formed at the carbon anode. A mixture of fluorides often replaces cryolite as the electrolyte. A typical mixture is $2AlF_3 \cdot 6NaF \cdot 3CaF_2$.

Aluminum prepared by electrolysis is about 99 per cent pure, the impurities being small amounts of copper, iron, silicon, and aluminum oxide. If aluminum of exceptionally high purity (99.9%) is needed, it may be obtained by the **Hoopes electrolytic process** (Fig. 36–4), which employs a fused bath consisting of three layers. The bottom layer is a fused alloy of copper and impure aluminum, the top layer is pure molten aluminum, and the middle layer (the electrolyte) consists of a fused mixture of the fluorides of barium, aluminum, and sodium, and nearly enough aluminum oxide to saturate it. The densities of the layers are such that their separation is maintained during electrolysis. The bottom layer serves as the anode and the top layer as the cathode. As electrolysis proceeds, the aluminum in the bottom layer passes into solution in the electrolyte as Al^{3+}, leaving the copper, iron, and silicon behind in the anode, for they are not oxidized under these conditions. The aluminum ion is reduced at the cathode. During electrolysis, purified aluminum is drawn off from the upper layer, and the impure molten metal is added to the lower layer through a carbon-lined funnel.

36.39   Properties and Uses of Aluminum

When freshly cut, aluminum has a silvery appearance, but it soon becomes superficially oxidized and assumes a dull white luster. The metal is very light and possesses high tensile strength. Weight for weight it is twice as good a conductor of electricity as copper. Although aluminum is a relatively reactive metal, the tenacious coating of the oxide which forms on it prevents further atmospheric corrosion. When heated in the air finely divided aluminum burns with a brilliant light. Because of the formation of a protective coat of oxide, aluminum does not decompose water. It is readily dissolved by hydrochloric acid and sulfuric acid, but is rendered passive by nitric acid. Aluminum dissolves in concentrated alkalies with the formation of aluminates and hydrogen.

$$2Al + 2OH^- + 6H_2O \longrightarrow 2Al(OH)_4^- + 3H_2\uparrow$$

This behavior is in accord with the amphoteric character of aluminum hydroxide. When heated, aluminum combines directly with the halogens, nitrogen, carbon, and sulfur.

The fact that aluminum is an excellent conductor of heat, together with its light weight and resistance to corrosion, accounts for its use in the manufacture of cooking utensils. The most important uses of aluminum are in the airplane and other transportation industries, which depend upon the lightness, toughness, and tensile strength of the metal. Aluminum is also used in the manufacture of electrical transmission wire, as a paint pigment, and, in the form of foil, as a wrapping material.

Aluminum is one of the best reflectors of heat and light, including the wavelengths in the ultraviolet. For this reason it is used as an insulating material, and as a mirror in reflecting telescopes. The 200-inch mirror in the world's largest telescope at Mount Palomar, California, is coated with aluminum.

About half of the aluminum produced in this country is converted to alloys for special uses. **Duralumin** is an alloy containing aluminum, copper, manganese, and magnesium. It is light but nearly as strong as steel, so it is useful in the construction of aircraft. **Aluminum bronzes** contain copper, and occasionally some silicon, manganese, iron, nickel, and zinc. These light-weight alloys have high tensile strength and great resistance to corrosion, so they are used extensively in the manufacture of crankcases and connecting rods for gasoline engines. **Alnico** is a magnetic alloy containing 50 per cent iron, 20 per cent aluminum, 20 per cent nickel, and 10 per cent cobalt, which will lift more than 4,000 times its own weight of iron. It is made by pressing together the constituent metals in powder form and heating the mixture just below the melting point. This is an example of **powder metallurgy**—a branch of science which is growing rapidly in importance.

When powdered aluminum and iron(III) oxide are mixed and ignited by means of a magnesium fuse a vigorous and highly exothermic reaction occurs.

$$2Al(s) + Fe_2O_3(s) \longrightarrow 2Fe(s) + Al_2O_3(s) + 203,500 \text{ cal}$$
$$(\Delta H^\circ_{298} = -203,500 \text{ cal})$$

This is known as the **thermite**, or **Goldschmidt**, reaction. The temperature of the

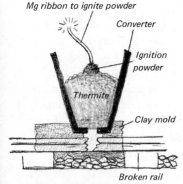

Mg ribbon to ignite powder

Converter

Ignition powder

Thermite

Clay mold

Broken rail

FIGURE 36–5
Thermite welding of a rail. Molten metal, mostly iron, flows out of the converter at the bottom and welds the broken rail without having to remove it from the roadbed.

reaction mixture rises to about 3,000° C, so the iron and aluminum oxide become liquid. The process is frequently used in welding large pieces of iron or steel (Fig. 36–5). Thermite bombs were used for incendiary purposes during World War II, because of the high temperature generated by the thermite reaction and because the reaction is not readily quenched by water. The thermite reaction is used in the reduction of metallic oxides (such as MoO_3 and WO_3) which are not readily reduced by carbon, or others (such as MnO_2 and Cr_2O_3) which do not give pure metal when reduced by carbon. The reaction is also of particular value when carbon-free alloys, such as ferrotitanium, are desired.

36.40   Aluminum Oxide

The oxide Al_2O_3 occurs in nature in pure form as the mineral **corundum,** a very hard substance which is used as an abrasive for grinding and polishing. **Emery** is aluminum oxide mixed with Fe_3O_4; it is also used as an abrasive. Several precious stones are composed of aluminum oxide and certain impurities that impart color: ruby (red, by chromium compounds); sapphire (blue, by compounds of cobalt, chromium, or titanium); oriental amethyst (violet, by manganese compounds); and oriental topaz (yellow, by iron). Artificial rubies and sapphires are now manufactured by melting aluminum oxide (m.p. 2,050° C) with small amounts of oxides to produce the desired color, and cooling the melt in such a way as to produce large crystals. These gems are indistinguishable from natural stones, except for microscopic, rounded air bubbles in the synthetic ones and flattened bubbles in the natural stones. The synthetic gems are used not only as jewelry but also as bearings ("jewels") in watches and other instruments, and as dies through which wires are drawn. Very finely divided aluminum oxide, called **activated alumina,** is used as a dehydrating agent and as a catalyst.

Various aluminum articles, such as drinking tumblers, are given a wear resistant coating of oxide by anodic oxidation in a bath of chromic, sulfuric, or oxalic acid. This surface readily adsorbs dyes and pigments which give a pleasing decorative effect to the article.

36.41   Aluminum Hydroxide

When an alkali is added to a solution of an aluminum salt, a white, gelatinous precipitate is formed.

$$Al^{3+} + 3OH^- \longrightarrow Al(OH)_3$$

or $\qquad [Al(H_2O)_6]^{3+} + 3OH^- \longrightarrow [Al(OH)_3(H_2O)_3] + 3H_2O$

Aqueous solutions of sulfides or carbonates also precipitate aluminum hydroxide, because such solutions contain hydroxide ions in considerable concentrations due to hydrolysis (see Chapter 18).

$$S^{2-} + H_2O \rightleftharpoons HS^- + OH^-$$

$$CO_3^{2-} + H_2O \rightleftharpoons HCO_3^- + OH^-$$

Aluminum hydroxide is amphoteric and is dissolved by both acids and bases.

$$[Al(OH)_4]^- \xrightleftharpoons{OH^-} Al(OH)_3 \xrightleftharpoons{3H^+} Al^{3+} + 3H_2O$$

or $$[Al(H_2O)_2(OH)_4]^- \xrightleftharpoons{OH^-} Al(H_2O)_3(OH)_3 \xrightleftharpoons{3H^+} [Al(H_2O)_6]^{3+}$$

Aluminum oxide, obtained by heating the hydroxide above 850° C, is insoluble in acids and alkalies. Products obtained below 600° C are soluble in these reagents.

Aluminum hydroxide is widely used to "fix" dyes to fabrics. Cloth which is soaked in a hot solution of aluminum acetate becomes impregnated with the aluminum hydroxide formed by hydrolysis of the salt. If the hydroxide is precipitated from a solution containing a dye, the precipitate is colored. The aluminum hydroxide adsorbs the dye and holds it fast to the cloth. When used in this way, the aluminum hydroxide is called a **mordant,** and the colored product is called a **lake.**

Aluminum hydroxide is also used in the purification of water, for its gelatinous character enables it to carry down with it any suspended material in the water, including most of the bacteria. The aluminum hydroxide for water purification is ordinarily produced by the reaction of aluminum sulfate with lime.

$$2Al^{3+} + 3SO_4^{2-} + 3Ca^{2+} + 6OH^- \longrightarrow 2\underline{Al(OH)_3} + 3\underline{CaSO_4}$$

36.42   Aluminum Chloride

Anhydrous aluminum chloride is a white crystalline solid which sublimes at 180° C and is soluble in organic solvents. Measurements of its density show that in the vapor state it is dimeric, and the formula should be written Al_2Cl_6. In these molecules each aluminum atom is tetrahedrally bonded to four chlorine atoms; two of these chlorine atoms are each bonded to two aluminum atoms.

Aluminum
chloride

During formation of the dimer, one chlorine atom of each monomeric covalent molecule donates a pair of electrons to be shared with the aluminum atom of the other molecule, and thus two "chlorine bridges" are formed; each aluminum atom then possesses an octet of electrons. X-ray studies show that there are no discrete Al_2Cl_6 molecules in solid aluminum chloride. The formula for the compound in the solid state is usually written $AlCl_3$, although the structure is actually a polymeric layer lattice that does not possess individual $AlCl_3$ molecules.

Aluminum chloride can be prepared either by direct chlorination of aluminum or by heating bauxite with carbon and chlorine.

$$2Al(s) + 3Cl_2(g) \longrightarrow 2AlCl_3(s) + 336,600 \text{ cal}$$
$$(\Delta H_{298}^\circ = -336,600 \text{ cal})$$

$$Al_2O_3(s) + 3C(s) + 3Cl_2(g) \longrightarrow 2AlCl_3(s) + 3CO(g) + 15,300 \text{ cal}$$
$$(\Delta H_{298}^\circ = -15,300 \text{ cal})$$

If aluminum or aluminum hydroxide is treated with hydrochloric acid and the solution is evaporated, crystals of the hexahydrate, $AlCl_3 \cdot 6H_2O$, are formed. When this salt is heated the oxide is produced.

$$233,600 \text{ cal} + 2[AlCl_3 \cdot 6H_2O](s) \longrightarrow Al_2O_3(s) + 6HCl(g) + 9H_2O(g)$$
$$(\Delta H^{\circ}_{298} = 233,600 \text{ cal})$$

36.43   Aluminum Sulfate

The sulfate $Al_2(SO_4)_3 \cdot 18H_2O$ is prepared by treating bauxite or clay with sulfuric acid.

$$\underset{\text{Clay}}{Al_2Si_2O_5(OH)_4} + 6H^+ + (3SO_4^{2-}) \longrightarrow 2\underline{H_2SiO_3} + 2Al^{3+} + (3SO_4^{2-}) + 3H_2O$$

The metasilicic acid (H_2SiO_3) is removed by filtration. The sulfate is the cheapest soluble salt of aluminum, so it is used as a source of aluminum hydroxide for the purification of water, the waterproofing of fabrics, the sizing of paper, and the dyeing of fabrics.

36.44   Alums

When solutions of potassium sulfate and aluminum sulfate are mixed and concentrated by evaporation, crystals of a salt called **potassium alum** are formed. The salt may best be formulated as $KAl(SO_4)_2 \cdot 12H_2O$. It is the commonest of a class of double salts known as **alums** and is frequently referred to simply as "alum." The univalent ion in an alum may be that of an alkali metal, silver, or ammonium, and the trivalent ion may be that of aluminum, chromium, iron, manganese, etc. The selenate ion, SeO_4^{2-}, may replace the sulfate ion. Thus we have **ammonium alum,** $NH_4Al(SO_4)_2 \cdot 12H_2O$; **sodium iron(III) alum,** $NaFe(SO_4)_2 \cdot 12H_2O$; and **potassium chrome alum,** $KCr(SO_4)_2 \cdot 12H_2O$. The alums are all isomorphous; i.e., they have the same crystal structure.

36.45   Aluminum Silicates

As we discovered in Section 29.17, many of the most important silicate rocks contain aluminum. Clay and sand are formed as disintegration products by the weathering of these rocks. Weathering involves the thawing and freezing of water in the rocks, and the chemical action of water and carbon dioxide upon them. For example, the chemical disintegration of feldspar may be represented by

$$\underset{\text{Feldspar}}{2KAlSi_3O_8} + 2H_2O + CO_2 \longrightarrow 2K^+ + \underset{\text{Potassium carbonate}}{CO_3^{2-}} + \underset{\text{Clay}}{\underline{Al_2Si_2O_5(OH)_4}} + \underset{\text{Sand}}{4\underline{SiO_2}}$$

The soluble potassium carbonate formed is largely removed by water, and the residue of sand and clay remains as soil. One kind of pure clay has the formula shown above; it is white and is called **kaolin.** Ordinary clay is colored by compounds of iron and other metals. **Porcelain** and **china** are made of pure kaolin, while impure clays are used in the manufacture of earthenware products such as **tile** and **brick.** The buff or red color of such products is due to the presence of iron(III) silicate.

36.46 Properties of Aluminum of Analytical Importance

Aluminum hydroxide forms as a white gelatinous precipitate when solutions containing aluminum ions are treated with aqueous ammonia or alkali hydroxides, sulfides, or carbonates. The precipitate is practically insoluble in an excess of aqueous ammonia, but dissolves readily in excess alkali, aluminate ions being formed. The dye aurin tricarboxylic acid (aluminon) forms a red lake with aluminum hydroxide; this is the basis of a sensitive test for aluminum ions.

Chromium

36.47 Periodic Relationships of Chromium

The elements which constitute Group VIB of the Periodic Table are chromium, molybdenum, and tungsten. These metals differ distinctly from the nonmetals of Periodic Group VIA, oxygen and sulfur, which we have already considered (Chapters 8 and 23). Chromium, molybdenum, and tungsten are each used in the production of commercially important alloy steels. These metals are typical transition metals; they show a variety of oxidation states, form highly colored compounds, and enter into the formation of stable complex ions. Like those of the other transition metals their lowest oxides are basic, the intermediate ones are amphoteric, and the highest ones are primarily acidic. The members of the Groups VIA and VIB show similarities in the higher oxidation states; thus the sulfates (e.g., K_2SO_4) and the chromates (K_2CrO_4) are analogous in formula and in physical and chemical properties.

36.48 Occurrence, Metallurgy, and Properties of Chromium

Chromium does not occur free in nature. Its most important ore is chromite, $FeCr_2O_4$, sometimes called **chrome iron ore.** Practically all of the chromium used in the United States is imported. Because chromium is essential to the production of special alloy steels and the supply is often short, it is classed as a strategic material by the United States Government.

Reduction of chromite by carbon in an electric furnace yields an alloy of iron and chromium, called **ferrochrome,** which is used in making chromium steels.

$$FeCr_2O_4 + 4C \longrightarrow Fe + 2Cr + 4CO$$

Pure chromium is prepared by a thermite reaction, reducing chromium(III) oxide with aluminum.

$$Cr_2O_3(s) + 2Al(s) \longrightarrow Al_2O_3(s) + 2Cr(s) + 128,100 \text{ cal}$$
$$(\Delta H^\circ_{298} = -128,100 \text{ cal})$$

Chromium can also be produced by the electrolytic reduction of its compounds from aqueous solution, a process which is used in chromium plating.

Chromium is a very hard, silvery-white, crystalline metal. It assumes a passive state by becoming coated with a thin layer of oxide, which protects it against further corrosive attack. This property, along with its metallic luster, accounts for its extensive use in the plating of iron and copper objects, such as plumbing fixtures and automobile trim. Passivity is produced by the action of concentrated nitric acid or chromic acid, or exposure of the metal to air. When not in the passive state chromium is a transition metal which shows several oxidation states, the principal ones being $+2$, $+3$, and $+6$. The compounds of chromium are highly colored, each color exhibited being dependent to some extent upon a particular oxidation state and upon the structure of the compound.

36.49 Compounds of Chromium

■ **1. Chromium(II) Compounds. Chromium(II) chloride,** $CrCl_2$, can be prepared by dissolving chromium in hydrochloric acid or by reducing a solution of chromium(III) chloride by zinc in the presence of an acid. The anhydrous salt is colorless, but its solutions have the bright blue color of the hydrated ion, $[Cr(H_2O)_6]^{2+}$. This ion is rapidly oxidized in air to the chromium(III) ion, $[Cr(H_2O)_6]^{3+}$; for this reason solutions of chromium(II) salts are frequently employed to remove the last traces of oxygen from gases. The **sulfate** $CrSO_4 \cdot 7H_2O$, is prepared by dissolving the metal in sulfuric acid. Chromium(II) **acetate,** $Cr(C_2H_3O_2)_2$, forms as a red precipitate when a saturated solution of sodium acetate is added to a solution of chromium(II) chloride. Chromium(II) acetate is one of the few exceptions to the rule that acetates are soluble. **Chromium(II) hydroxide** is basic, and chromium(II) salts are only slightly hydrolyzed.

■ **2. Chromium(III) Compounds.** Green **chromium(III) oxide,** Cr_2O_3, is formed either by heating the metal in air, by igniting ammonium dichromate, or by reducing a dichromate with sulfur or another similar reducing agent.

$$(NH_4)_2Cr_2O_7(s) \longrightarrow N_2(g) + 4H_2O(g) + Cr_2O_3(s) + 71{,}800 \text{ cal}$$
$$(\Delta H^{\circ}_{298} = -71{,}800 \text{ cal})$$
$$Na_2Cr_2O_7 + S \longrightarrow Na_2SO_4 + Cr_2O_3$$

This oxide finds use as a pigment called **chrome green.** The bluish-green gelatinous **chromium(III) hydroxide** is precipitated when either an alkali, a soluble sulfide, or a carbonate is added to a solution of a chromium(III) compound. Chromium(III) hydroxide is amphoteric, dissolving in both acids and bases.

$$Cr(OH)_3 + 3H^+ \longrightarrow Cr^{3+} + 3H_2O$$
$$Cr(OH)_3 + OH^- \longrightarrow [Cr(OH)_4]^- \text{ (chromite ion)}$$

Chromium(III) chloride 6-hydrate, $CrCl_3 \cdot 6H_2O$, and the sulfate octadecahydrate, $Cr_2(SO_4)_3 \cdot 18H_2O$, are the best known salts of trivalent chromium. Dilute solutions of the chloride are violet, the hexaaquochromium(III) ion, $[Cr(H_2O)_6]^{3+}$ being responsible for the color. In more concentrated solutions of the chloride the green $[Cr(H_2O)_4Cl_2]^+$ is formed. The hydrates of chromium(III) sulfate also exist in violet and green modifications. Trivalent chromium forms alums of the type $KCr(SO_4)_2 \cdot 12H_2O$, solutions of which are bluish-violet when cold but green when hot. **Chrome alum** is used in the tanning of leather, the printing of calico, the waterproofing of fabrics, and as a mordant. It is the most important of the soluble chromium salts.

■ **3. Chromium(VI) Compounds.**   In the $+6$ oxidation state, chromium is usually combined with oxygen in the form of the oxide **chromium trioxide, CrO_3,** or the oxyanions, chromate, CrO_4^{2-}, and dichromate, $Cr_2O_7^{2-}$. **Potassium chromate, K_2CrO_4,** is produced commercially by the atmospheric oxidation of chromite ore admixed and heated with potassium carbonate.

$$4FeCr_2O_4 + 8K_2CO_3 + 7O_2 \longrightarrow 2Fe_2O_3 + 8K_2CrO_4 + 8CO_2$$

Lead chromate, $PbCrO_4$, and **barium chromate, $BaCrO_4$,** are both insoluble in water and are used as yellow pigments.

When an acid is added to a solution containing chromate ions, the solution changes from yellow to orange-red due to the formation of the **dichromate ion, $Cr_2O_7^{2-}$.**

$$2CrO_4^{2-} + 2H^+ \Longrightarrow Cr_2O_7^{2-} + H_2O$$

The addition of a base or dilution with water reverses the reaction.

$$Cr_2O_7^{2-} + 2OH^- \Longrightarrow 2CrO_4^{2-} + H_2O$$

The conversion of chromate to dichromate involves the formation of an oxygen linkage between two chromium atoms. The electronic structure of the dichromate ion is

$$\left[\; \overset{\displaystyle :\overset{..}{\underset{..}{O}}:}{\underset{\displaystyle :\overset{..}{\underset{..}{O}}:}{:\overset{..}{O} \overset{\times}{\underset{\times\times}{Cr}} \overset{..}{\underset{..}{O}} \overset{\times}{\underset{\times\times}{Cr}} \overset{..}{O}:}} \; \right]^{2-}$$

In very strongly acid solutions, further condensation of chromate ions to **trichromate ions, $Cr_3O_{10}^{2-}$,** occurs.

A mixture of sodium dichromate and sulfuric acid is used as a **cleaning solution** for glassware in the laboratory. The cleansing action is due to the strong oxidizing power of the dichromate and the dehydrating action of the concentrated sulfuric acid. The dichromates are used as oxidizing agents in many reactions. The dichromate ion is readily reduced to the chromium(III) ion by reducing agents such as sulfurous acid.

$$Cr_2O_7^{2-} + 3H_2SO_3 + 2H^+ \longrightarrow 2Cr^{3+} + 3SO_4^{2-} + 4H_2O$$

As the reduction progresses, a change from the orange color of the dichromate to the green color of the chromium(III) ion occurs.

The trioxide CrO_3 is produced as scarlet needle-shaped crystals by the action of concentrated sulfuric acid upon a concentrated solution of potassium dichromate.

$$Cr_2O_7{}^{2-} + 2H^+ \longrightarrow 2\underline{CrO_3} + H_2O$$

The anhydride CrO_3 reacts with water in different proportions forming **chromic acid,** H_2CrO_4, **dichromic acid,** $H_2Cr_2O_7$, and **trichromic acid,** $H_2Cr_3O_{10}$.

$$CrO_3 + H_2O \longrightarrow 2H^+ + CrO_4{}^{2-}$$
$$2CrO_3 + H_2O \longrightarrow 2H^+ + Cr_2O_7{}^{2-}$$
$$3CrO_3 + H_2O \longrightarrow 2H^+ + Cr_3O_{10}{}^{2-}$$

These solutions are used as active oxidizing agents. When heated the anhydride decomposes, liberating oxygen and forming the green chromium(III) oxide.

$$118{,}800 \text{ cal} + 4CrO_3(s) \longrightarrow 2Cr_2O_3(s) + 3O_2(g) \qquad (\Delta H^\circ_{298} = 118{,}800 \text{ cal})$$

36.50   Uses of Chromium

A large portion of the chromium produced goes into steel alloys, which are very hard and strong. **Stainless steel,** which usually contains chromium and some nickel, is used in the manufacture of cutlery because of its corrosion resistance. Nonferrous chromium alloys include **nichrome** and **chromel** (Ni and Cr), which are used in various heating devices because of their electrical resistance property. Chromium is widely used as a protective and decorative coating for other metals, such as plumbing fixtures and automobile trim.

The compounds of chromium have many uses. These include uses as paint pigments and mordants and in the tanning of leather. Certain refractories which are used as lining for high-temperature furnaces are made by mixing pulverized chromite ore with clay or magnesia, MgO.

36.51   Properties of Chromium of Analytical Importance

Chromium(III) hydroxide is precipitated from solutions containing chromium (III) ions by aqueous ammonia, alkalies, and ammonium sulfide or carbonate. The bluish-green gelatinous precipitate is soluble in acids and in excess of alkali. The hydroxide is converted to soluble sodium chromate (yellow) by sodium peroxide. Acids convert the chromate to the dichromate (orange-red). Either chromate or dichromate ions precipitate lead chromate (yellow) from solutions of lead salts. Hydrogen peroxide converts dichromate ions in acid solution to a blue soluble compound whose formula has not been established but which is believed to be a peroxychromate.

Zinc

36.52   Occurrence and Metallurgy of Zinc

Sphalerite, or **zinc blende,** ZnS, is the principal ore of zinc. Less important ores include **zincite,** ZnO; **smithsonite,** $ZnCO_3$; **franklinite,** a mixture of oxides of zinc, iron, and manganese; and **willemite,** Zn_2SiO_4.

In the metallurgy of zinc, sulfide ores are first concentrated by flotation and then roasted to convert the sulfide to the oxide. From carbonate ores the oxide is obtained by simply heating.

$$2ZnS(s) + 3O_2(g) \longrightarrow 2ZnO(s) + 2SO_2(g) + 209{,}910 \text{ cal}$$
$$(\Delta H^\circ_{298} = -209{,}910 \text{ cal})$$

$$16{,}970 \text{ cal} + ZnCO_3(s) \longrightarrow ZnO(s) + CO_2(g)$$
$$(\Delta H^\circ_{298} = 16{,}970 \text{ cal})$$

The zinc oxide formed by roasting the sulfide or carbonate ores is reduced by heating it with coal in a fire-clay retort. As rapidly as the zinc is produced it distills out and is condensed. The zinc so produced is impure with cadmium, iron, lead, and arsenic. It may be purified by careful redistillation.

An electrolytic process is also employed to produce zinc. The ore is roasted in special furnaces under such conditions that most of the sulfide is converted to the sulfate and the remainder changed to the oxide. The roasted ore is leached with dilute sulfuric acid to dissolve the zinc sulfate and zinc oxide, as well as the sulfates and oxides of other metals. Silver sulfate and lead sulfate remain in the residue, for they are insoluble. The resulting solution of zinc sulfate is treated with powdered zinc to reduce the less active metals such as copper, arsenic, and antimony to the elemental condition. Any manganese present is oxidized by passing air through the solution. Manganese and iron are precipitated as hydroxides by adding the calculated amount of lime. After the solution is purified in this way, it is electrolyzed, zinc being deposited on aluminum cathodes. Sulfuric acid is regenerated as electrolysis proceeds and is used in the leaching of more roasted ore. At intervals, the cathodes are removed and the zinc scraped off, melted, and cast into ingots. Electrolytic zinc has a purity of 99.95 per cent.

36.53   Properties and Uses of Zinc

The periodic relationships of zinc were discussed in Section 34.1. Zinc is a silvery metal that quickly tarnishes to a blue-gray appearance. This color is due to an adherent coating of a basic carbonate, $Zn_2(OH)_2CO_3$, which protects the underlying metal from further corrosion. The metal is hard and brittle at ordinary temperatures but ductile and malleable at 100-150° C. When molten zinc is poured into cold water, it solidifies in irregular masses called **granulated,** or **mossy, zinc.** Zinc is a fairly active metal and will reduce steam at high temperatures.

$$Zn(s) + H_2O(g) \longrightarrow ZnO(s) + H_2(g) + 25{,}440 \text{ cal} \qquad (\Delta H^\circ_{298} = -25{,}440 \text{ cal})$$

Zinc dissolves in strong bases, liberating hydrogen and forming zincate ions.

$$Zn + 2OH^- + 2H_2O \longrightarrow [Zn(OH)_4]^{2-} + H_2\uparrow$$

A considerable amount of zinc is used in the manufacture of dry cells (Section 22.16) and in the production of alloys such as brass and bronze (Section 31.12). About half of the zinc metal produced is used to protect iron and other metals from corrosion by air and water. Its protective action depends upon the property of forming a basic carbonate film on its surface. Furthermore, it protects less active metals such as iron, because it forms an electrochemical cell with the iron, in which zinc serves as the anode and iron as the cathode (see Section 36.28). The zinc coating on iron may be applied in several ways, and the product is called **galvanized iron.**

36.54 Compounds of Zinc

■ **1. Zinc Oxide and Zinc Hydroxide.** The oxide, ZnO, is produced when zinc vapor is burned and when zinc ores are roasted in an excess of air. The oxide is yellow when hot but pure white when cold. It is used as a paint pigment, called **zinc white,** or **Chinese white;** unlike white lead it does not blacken with hydrogen sulfide, because zinc sulfide is white. It is also used in the manufacture of automobile tires and other rubber goods, and in the preparation of medical ointments.

Zinc hydroxide, $Zn(OH)_2$, is formed as a white, gelatinous precipitate when a soluble hydroxide is added to a solution of a zinc salt.

$$Zn^{2+} + 2OH^- \longrightarrow \underline{Zn(OH)_2}$$

The hydroxide is amphoteric and dissolves in an excess of alkali as well as in acids.

$$Zn(OH)_2 + 2OH^- \longrightarrow [Zn(OH)_4]^{2-} \text{ (zincate ion)}$$
$$Zn(OH)_2 + 2H^+ \longrightarrow Zn^{2+} + 2H_2O$$

Zinc hydroxide is soluble in aqueous ammonia, forming **tetraamminezinc hydroxide.**

$$Zn(OH)_2 + 4NH_3 \longrightarrow [Zn(NH_3)_4]^{2+} + 2OH^-$$

■ **2. Zinc Chloride.** When zinc and chlorine are brought together **anhydrous zinc chloride,** $ZnCl_2$, is formed as a white deliquescent solid. It is also made by dissolving either the metal, the oxide, or the carbonate in hydrochloric acid, evaporating the solution to dryness, and then fusing the residue to remove the last traces of moisture. The product, which melts at 262° C, is cast into sticks. The anhydrous salt is used as a caustic in surgery and as a dehydrating agent in certain organic reactions. Aqueous solutions of zinc chloride are acidic due to hydrolysis. Because of its acidic nature a solution of zinc chloride (along with ammonium chloride) is used to dissolve the oxides on metal surfaces before soldering. Concentrated solutions of zinc chloride dissolve cellulose, forming a gelatinous mass, which may be molded into various shapes. Fiberboard is made of this material. Zinc oxide dissolves in concentrated solutions of zinc chloride, producing a basic zinc chloride, **zinc oxychloride,** Zn_2OCl_2, which sets to a hard mass and is used as a cement.

■**3. Zinc Sulfate.** **Zinc sulfate** is produced in large quantities by roasting zinc blende at low red heat.

$$ZnS(s) + 2O_2(g) \longrightarrow ZnSO_4(s) + 185,700 \text{ cal} \qquad (\Delta H^\circ_{298} = -185,700 \text{ cal})$$

The product is extracted with water and recovered by recrystallization to form crystals of the **heptahydrate,** $ZnSO_4 \cdot 7H_2O$, which is called **white vitriol.** It is used principally in the production of the white paint pigment **lithopone,** which is a mixture of barium sulfate and zinc sulfide, formed by the reaction

$$BaS + Zn^{2+} + SO_4^{2-} \longrightarrow \underline{BaSO_4} + \underline{ZnS}$$

■**4. Zinc Sulfide.** **Zinc sulfide,** ZnS, is found in nature as zinc blende and is formed in the laboratory when ammonium sulfide is added to a solution of a zinc salt. The sulfide is insoluble in acetic acid but dissolves readily in stronger acids such as hydrochloric or sulfuric.

$$\underline{ZnS} + 2H^+ \longrightarrow Zn^{2+} + H_2S\uparrow$$

Zinc sulfide is used as a white paint pigment, either alone or mixed with zinc oxide.

36.55 Properties of Zinc of Analytical Importance

Solutions containing the zinc ion are colorless. Zinc chloride is soluble in water, and the sulfide is soluble in dilute acids but insoluble in alkaline solutions. The white hydroxide is precipitated by ammonia and by alkalies but is dissolved by an excess of these reagents.

QUESTIONS

1. Why does cobalt precede nickel in the periodic chart when cobalt has a higher atomic weight than does nickel?
2. What is meant by the term "passivity"?
3. How is nickel rendered passive?
4. What is the composition of Alnico? What use is made of this alloy?
5. Write the equation for the dissolution of nickel(II) hydroxide in aqueous ammonia.
6. What is the physical nature of nickel carbonyl?
7. Sketch the spatial configuration of $[Co(CN)_6]^{3-}$.
8. Balance the following equation by oxidation-reduction methods (you will probably find this to be a challenge):

$$Co(NO_3)_2 \longrightarrow Co_2O_3 + NO_2 + O_2$$

9. On the basis of the reactions of elemental fluorine, chlorine, and oxygen with metallic cobalt, predict which is the strongest oxidizing agent.
10. How are the ferromanganese alloys produced?
11. Outline the chemistry of the action of manganese dioxide upon hydrochloric acid.

12. Explain the dissolution of manganese(II) hydroxide in aqueous ammonium chloride in terms of an acid-base reaction.
13. Mention two uses of manganese dioxide.
14. Write the names and formulas for compounds in which manganese exhibits each of the following oxidation states: $+2$, $+3$, $+4$, $+6$, and $+7$.
15. How is potassium permanganate prepared commercially?
16. Would you expect a calcium manganite solution to have a pH greater or less than 7.0? Justify your answer.
17. How may the insolubility of barium sulfate be used to good advantage in the preparation of free permanganic acid?
18. Write the names and formulas of the principal ores of iron.
19. List the equations for the reactions that take place in the reduction of iron(III) oxide in the blast furnace.
20. What use is made of the hot exhaust gases from the blast furnace?
21. Compare the open-hearth, basic oxygen, and electric processes for the production of steel in terms of (a) speed, (b) control of composition and temperature, (c) quality of product, and (d) other factors you consider to be important.
22. The basic oxygen process for making steel has gained favor very rapidly within the last few years. Discuss the reasons why this is so.
23. Give the composition and distinguishing properties of pig iron, cast iron, and steel.
24. Identify each of the following: hematite, magnetic oxide of iron, rouge, Mohr's salt, and Prussian blue.
25. What is the composition of iron rust?
26. Describe the chemistry of an invisible ink.
27. What compound of iron is a common constituent of hard water?
28. What is the gas produced when iron(II) sulfide is treated with a nonoxidizing acid?
29. Give the general equation for the hydrolysis of iron(III) salts. Write equations showing the steps in the hydrolysis of iron(III) chloride to produce a precipitate of hydrated $Fe(OH)_3$ in solution, $[Fe(H_2O)_3(OH)_3]^0$.
30. Describe the production of metallic aluminum by electrolytic reduction.
31. What is the action of sodium hydroxide upon aluminum?
32. Illustrate the amphoteric nature of aluminum hydroxide by suitable equations.
33. How is advantage taken of the amphoteric nature of aluminum ion in the processing of bauxite?
34. Why can aluminum, which is an active metal, be used so successfully as a structural metal?
35. What is the composition of alums as a class of double salts?
36. Explain the function of aluminum hydroxide in water purification.
37. Why is it impossible to prepare anhydrous aluminum chloride by heating the hexahydrate to drive off the water?
38. What is the Goldschmidt, or thermite, process?
39. From the electron structure of thallium, what would you predict concerning the solubility of thallium(I) chloride? the basicity of thallium(I) hydroxide?

40. In what way are the Group VIB metals typical of transition metals in general?
41. Write a balanced chemical equation describing the air oxidation of aqueous chromium(II) to chromium(III).
42. Write the equation for the decomposition of ammonium dichromate by ignition.
43. Account for the extensive use of chromium in the plating of iron and copper objects, such as the trim on automobiles.
44. Reduction of chromite ore by carbon leads to an alloy of iron and chromium. How would you separate the two components of this alloy into separate solutions of their respective cations, if this were necessary?
45. What is the action of hydrogen ions upon chromate ions? This is called a condensation reaction. Explain.
46. Represent the electronic structure of trichromic acid.
47. Write an equation to illustrate the action of the dichromate ion as an oxidizing agent.
48. What is the composition and chemical action of the "cleaning solution" of the laboratory?
49. How is zinc metal separated from the impurities cadmium, iron, lead, and arsenic during the refining process?
50. Why does pure zinc react only slowly with dilute hydrochloric or sulfuric acids?
51. What is galvanized iron, how is it produced, and what properties does it have that make it useful?
52. A dilute solution is dripped into a slowly stirred solution of sodium zincate, $Na_2[Zn(OH)_4]$. A white gelatinous precipitate is formed which, upon analysis, proves to be a hydroxide. Use a chemical equation to explain these observations.
53. What is lithopone and how is it produced? In what way is lithopone superior to white lead as paint pigment?
54. Balance the following equations:

   (a) $Al + H^+ \rightleftharpoons Al^{3+} + H_2$
   (b) $NO_2^- + Al + OH^- + H_2O \rightleftharpoons NH_3 + [Al(OH)_4]^-$
   (c) $Cr_2O_7^{2-} + Fe^{2+} + H^+ \rightleftharpoons Cr^{3+} + Fe^{3+} + H_2O$
   (d) $SCN^- + MnO_4^- + H^+ \rightleftharpoons SO_4^{2-} + CN^- + Mn^{2+} + H_2O$
   (e) $Zn + OH^- \rightleftharpoons [Zn(OH)_4]^{2-} + H_2$

PROBLEMS

1. What is the per cent cobalt in sodium hexanitrocobaltate(III)?

   Ans. 14.59%

2. Iron(II) can be titrated to iron(III) by dichromate ion, which is reduced to chromic ion in acid solution. A 2.500-gram sample of iron ore is dissolved and the iron converted to the iron(II) form. Exactly 19.17 ml of 0.0100 M $Na_2Cr_2O_7$ are required in the titration. What per cent iron was the ore sample?

   Ans. 2.569%

3. How many cubic feet of air are required per ton of Fe_2O_3 to convert the oxide into iron in a blast furnace? Assume air is 19% oxygen.

Ans. 3.5 $\times$ 10^4 ft^3

REFERENCES

"From Cabul to Cobalt—A Historical View of the Mischievous Metal," F. R. Morral, *J. Chem. Educ.,* **34,** 185 (1957).

"The Production of Cobalt," J. V. Huxley, *Educ. in Chemistry,* **11,** 203 (1974).

"Manganese Nodules (I): Mineral Resources on the Deep Seabed," A. L. Hammond, *Science,* **183,** 502 (1974). "(II): Prospects for Deep Sea Mining," A. L. Hammond, *Science,* **183,** 644 (1974).

"Metallic Corrosion," S. C. Britton, *Chem. in Britain,* **5,** 228 (1969).

"Kinetics of Corrosion of Metals," F. Habashi, *J. Chem. Educ.,* **42,** 318 (1965).

"The Composition of the Earth's Interior," T. Takahashi and W. A. Bassett, *Sci. American,* June, 1965; p. 100.

"Nature's Geological Paint Pots and Pigments," R. C. Brasted, *J. Chem. Educ.,* **48,** 323 (1971).

"Turnbull's Blue and Prussian Blue: KFe(III) [Fe(II)(CN)$_6$]," L. D. Hansen, W. M. Litchman, and G. H. Daub, *J. Chem. Educ.,* **46,** 46 (1969).

"The Role of Chelation in Iron Metabolism," P. Saltman, *J. Chem. Educ.,* **42,** 683 (1965).

"The Beneficiation of Iron Ores," M. M. Fine, *Sci. American,* Jan., 1968; p. 28.

"Crystallography of the Hexagonal Ferrites," J. A. Kohn, D. W. Eckart, and C. F. Cook, Jr., *Science,* **172,** 519 (1971).

"Strong and Ductile Steels," E. H. Parker and V. F. Zackay, *Sci. American,* Nov., 1968; p. 36.

"Oxygen in Steelmaking," J. K. Stone, *Sci. American,* April, 1968; p. 24.

"Carbon Determination in Modern Steelmaking," R. V. Williams, *Chem. in Britain,* **5,** 213 (1969).

"Revolution in Iron and Steel Making," R. W. Thomas, *Educ. in Chemistry,* **2,** 167 (1965).

"Basic Oxygen Steelmaking," H. E. McGannon, *J. Chem. Educ.,* **46,** 293 (1969).

"The Steel Industry," H. C. Neely, *Chem. & Eng. News,* Sept. 21, 1970; p. 48.

"Hemoglobin: Isolation and Chemical Properties," S. F. Russo and R. B. Sorstokke, *J. Chem. Educ.,* **50,** 347 (1973).

"Iron and Susceptibility to Infectious Diseases," E. D. Weinberg, *Science,* **184,** 952 (1974).

"The Story of Hall and Aluminum," H. N. Holmes, *J. Chem. Educ.,* **7,** 232 (1930).

"Studying the Chemical Properties of Metallic Aluminum," N. Feifer, *J. Chem. Educ.,* **45,** 648 (1968).

"The Structure of Solid Aluminum Chloride," M. J. Bigelow, *J. Chem. Educ.,* **46,** 495 (1969).

"Zinc Extraction Metallurgy in the UK," A. W. Richards, *Chem. in Britain,* **5,** 203 (1969).

"Life's Essential Elements," D. R. Williams, *Educ. in Chemistry,* **10,** 56 (1973).

"Paints and Plastic Coatings," W. M. Morgans, *Educ. in Chemistry,* **10,** 12 (1973).

"Pollution and Public Health: Taconite Case Poses Major Test," L. J. Carter, *Science,* **186,** 31 (1974).

Metals of Analytical Group IV

37

Calcium, Strontium, and Barium

Analytical Group IV (see also Chapter 43) contains those common cations which are precipitated as carbonates by ammonium carbonate from an aqueous ammonia solution containing ammonium chloride, but which are not precipitated by the reagents used in the precipitation of Groups I, II, and III. All the other cations, except magnesium, which form insoluble carbonates have been separated in Groups I, II, and III. If ammonium chloride were not present, magnesium carbonate would precipitate with the ions of Group IV. However, precipitation of magnesium carbonate is not desirable because it is moderately soluble and of indefinite composition. The precipitation of the Mg^{2+} is therefore postponed. Some properties of the Analytical Group IV metals are given in Table 37-1.

TABLE 37-1   Important Properties of the Metals of Analytical Group IV

Property	Calcium	Strontium	Barium
Atomic number	20	38	56
Atomic weight	40.08	87.62	137.33
Electronic structure	2, 8, 8, 2	2, 8, 18, 8, 2	2, 8, 18, 18, 8, 2
Oxidation state	+2	+2	+2
Reduction potential, volts	$(Ca^{2+}/Ca) -2.87$	$(Sr^{2+}/Sr) -2.89$	$(Ba^{2+}/Ba) -2.90$
Density, g/cm³ at 20°	1.54	2.60	3.5
Melting point, °C	850	770	704
Boiling point, °C	1,490	1,384	1,638

37.1 Periodic Relationships of Calcium, Strontium, and Barium

The elements of Periodic Group IIA (Be, Mg, Ca, Sr, Ba, and Ra), called the **alkaline earth metals,** each have two electrons in their outer shells. All exhibit a single oxidation state, $+2$. They are all very reactive metals, the reactivity increasing with increasing atomic number. The members of the B subgroup (Zn, Cd, and Hg), on the other hand, are heavier than the corresponding A subgroup elements of the same series, and the reactivities of the B subgroup elements decrease with increasing atomic number. Beryllium, the first member of Group IIA, resembles aluminum, the second member of Group IIIA of the Periodic Table, in its chemical behavior. Beryllium and magnesium are somewhat different in chemical properties from the other elements of Group IIA. For example, the hydroxides of beryllium and magnesium are nearly insoluble in water and are easily decomposed by heat into water and metallic oxides. The hydroxide of beryllium, the smallest and most electronegative element of the group, is amphoteric, whereas the hydroxides of barium, strontium, and calcium are strong bases and their hydrated ions do not hydrolyze appreciably. Beryllium ion forms stable complex ions, but with a few exceptions the ions of other elements of Group IIA do not. In contrast to the salts of the alkali metals (Chapter 38), many of the common salts of the alkaline earth metals are insoluble in water. Most of the salts that the alkaline earth metals form with weak or moderately strong acids are sparingly soluble in water—e.g., the sulfates, phosphates, oxalates, and carbonates. Many are quite soluble in water, however, especially the cyanides, sulfides, and acetates. The solubility of the hydroxides of these metals in water increases with increasing atomic number, whereas the solubility of the carbonates and sulfates decreases in this order.

37.2 Occurrence of Calcium, Strontium, and Barium

Because of their high chemical activity, these metals never occur free in nature. Their most abundant native compounds are the carbonates and the sulfates.

■ **1. Calcium.** Calcium is the fifth most abundant element in the earth's crust and is widely distributed in nature. The common rock known as **limestone** is an impure calcium carbonate. Other natural forms of calcium carbonate include **calcite, aragonite, chalk, marl, marble, marine shells,** and **pearls. Dolomitic limestone** (dolomite) is $CaCO_3 \cdot MgCO_3$. Calcium sulfate occurs as **anhydrite,** $CaSO_4$, and **gypsum,** $CaSO_4 \cdot 2H_2O$. Other common minerals include **fluorite,** or **fluorspar,** which is CaF_2, and **apatite,** which may be formulated as $Ca_{10}(PO_4)_6(OH, Cl, F)_2$. **Hydroxyapatite** is the chief constituent of the bones and teeth of animals.

■ **2. Strontium.** The minerals containing strontium are much rarer than those of calcium. The chief ones are **celestite,** $SrSO_4$, and **strontianite,** $SrCO_3$.

■ **3. Barium.** Barium occurs more abundantly than strontium, but much less so than calcium. Its chief minerals are **barite,** or **heavy spar,** $BaSO_4$, and **witherite,** $BaCO_3$.

37.3   Preparation of Calcium, Strontium, and Barium

The most important methods for producing these metals are those in which their molten salts (generally the chlorides) are electrolyzed. Because of the high chemical activity of these metals, their salts are not readily reduced by other methods. However, in the King process for producing barium a mixture of barium oxide and barium peroxide is reduced with aluminum in a vacuum furnace. A temperature of 950° to 1,100° C is required for the reaction to take place, and at the very low pressure (10^{-3} to 10^{-4} mm) which is maintained in the furnace the metallic barium distills off and is collected on a cold surface.

37.4   Properties and Uses of Calcium, Strontium, and Barium

These metals are all silvery-white, crystalline, malleable, and ductile. Calcium is harder than lead, strontium is about as hard as lead, and barium is quite soft. Calcium, strontium, and barium are all active metals, their activity increasing in the order named. For example, calcium does not react with oxygen unless it is heated, while strontium oxidizes rapidly when exposed to air, and barium is spontaneously flammable in moist air. When heated these metals combine with hydrogen, forming hydrides; with sulfur, forming sulfides; with halogens, forming halides; with nitrogen, forming nitrides; and with phosphorus, forming phosphides. Each of the metals displaces hydrogen from water, and the corresponding hydroxide is formed.

Calcium is used as a dehydrating agent for certain organic solvents, as a reducing agent in the production of certain metals, as a scavenger (to remove gases in fused metals) in metallurgy, as a hardening agent for lead used for covering cables and making storage battery grids and bearings, in steel-making when alloyed with silicon, and for many other purposes.

Elemental strontium is not abundant and has no commercial uses. Barium is used as a degassing agent in the manufacture of vacuum tubes, and alloys of barium and nickel are used in vacuum tubes and spark plugs because of their high thermionic electron emission. It is interesting that Mg^{2+} and Ca^{2+} are not poisonous, whereas Be^{2+} and Ba^{2+} are very much so.

37.5   Oxides of Calcium, Strontium, and Barium

When the carbonates of these metals are heated sufficiently, they undergo a decomposition which yields their respective oxides and carbon dioxide ($MCO_3 \longrightarrow MO + CO_2$). The temperature required for these decompositions increases with increasing basicity of the oxide. The temperatures required to attain equilibrium pressures of carbon dioxide of 1 atm are about 900° C for $CaCO_3$, 1,250° C for $SrCO_3$, and 1,450° C for $BaCO_3$.

Pure calcium oxide is a white amorphous substance that emits an intense light, called **limelight,** when heated to a high temperature. Calcium oxide reacts vigorously and exothermally with water (Section 37.6) and therefore is often employed as a drying agent—for example, in the preparation of anhydrous alcohol and in the drying of ammonia.

In the production of barium oxide, the carbonate is heated with finely divided carbon. The purpose of the carbon is to reduce the carbon dioxide which is produced, and thus to shift the equilibrium $BaCO_3 \rightleftharpoons BaO + CO_2$ to the right. This makes possible the conversion of the carbonate to the oxide at a lower temperature than the 1,450° C otherwise necessary. Most of the barium oxide produced is used in the preparation of barium peroxide.

Calcium, strontium, and barium form peroxides of the general formula MO_2. Only barium peroxide is easily formed and important. It is produced when barium oxide is heated in air or oxygen to 500°-600° C.

$$2BaO + O_2 \rightleftharpoons 2BaO_2$$

When barium peroxide is heated to 700°-800° C the reaction is reversed, and oxygen is liberated. The equilibrium can also be shifted by varying the partial pressure of the oxygen. Barium peroxide is used for bleaching various materials, of both plant and animal origin.

37.6   Hydroxides of Calcium, Strontium, and Barium

When the oxides of calcium, strontium, and barium react with water the corresponding hydroxides are formed, and large amounts of heat are evolved.

$$CaO(s) + H_2O(l) \longrightarrow Ca(OH)_2(s) + 15{,}600 \text{ cal} \qquad (\Delta H^\circ_{298} = -15{,}600 \text{ cal})$$
$$SrO(s) + H_2O(l) \longrightarrow Sr(OH)_2(s) + 19{,}900 \text{ cal} \qquad (\Delta H^\circ_{298} = -19{,}900 \text{ cal})$$
$$BaO(s) + H_2O(l) \longrightarrow Ba(OH)_2(s) + 24{,}500 \text{ cal} \qquad (\Delta H^\circ_{298} = -24{,}500 \text{ cal})$$

The heat of hydration and the reactivity of the oxides toward water both increase with increasing cation size. The solubility of the hydroxides in water increases in the same direction: the solubilities in moles per liter at 20° C are $Ca(OH)_2$, 0.025; $Sr(OH)_2 \cdot 8H_2O$, 0.043; and $Ba(OH)_2 \cdot 8H_2O$, 0.23.

■**1. Calcium Hydroxide.** Calcium oxide is known by various names, such as **lime, quicklime,** and **unslaked lime,** while the product formed from its reaction with water, $Ca(OH)_2$, is known commercially as **slaked,** or **hydrated, lime.** Calcium hydroxide is a dry white powder that forms a pasty mass with an excess of water. Because of the high heat of reaction of quicklime with water, it should always be stored where it cannot come in contact with water, since the slaking reaction (reaction of the oxide with water to form the hydroxide) generates enough heat to ignite paper or wood. Dry hydrated lime rather than quicklime is usually sold, because it can be shipped and stored in paper bags without risk of fire.

Calcium hydroxide is classed as a strong base and, because of its activity and cheapness, it is used more extensively in commercial processes than any other base. It is sparingly soluble in water, only 1.7 grams dissolving in a liter of water at room temperature. A saturated solution of calcium hydroxide is often called **limewater.** A suspension of calcium hydroxide in water is known as **milk of lime.** Even though calcium hydroxide has a limited solubility, milk of lime furnishes a ready supply of hydroxide ions, because the solid continues to dissolve as hydroxide ions are removed from solution during chemical reaction; i.e., the equilibrium shifts to the right.

$$Ca(OH)_2 \rightleftharpoons Ca^{2+} + 2OH^-$$
$$\text{(Solid)} \qquad \text{(Dissolved)}$$

Mortar, widely used in the construction industry, is prepared by mixing slaked lime with sand and water. One part of slaked lime is used with about three or four parts of sand, and enough water is added to give the mixture the consistency of paste. It dries and becomes hard (sets) on exposure to the air by adsorbing carbon dioxide and forming crystalline calcium carbonate.

$$Ca(OH)_2 + CO_2 \longrightarrow CaCO_3 + H_2O$$

The calcium carbonate cements together the particles of sand and unchanged calcium hydroxide. Many years are required for the complete conversion of the calcium hydroxide to carbonate, especially in heavy wall construction.

Ordinary **lime plaster** for coating walls and ceilings is similar to mortar, with some binding material such as hair or fiber to help hold it in place. Cement was considered in Section 29.19.

The manufacture of at least 150 important industrial chemicals requires the use of lime. In fact, only five other raw materials are used more frequently than lime (or limestone, from which lime is made); these are salt, coal, sulfur, air, and water.

■ **2. Strontium Hydroxide.** **Strontium hydroxide** is a rather active base, which is made by treating strontium carbonate with superheated steam.

$$SrCO_3 + H_2O \longrightarrow Sr(OH)_2 + CO_2$$

Strontium hydroxide is fairly soluble in hot water and crystallizes from solution as the octahydrate, $Sr(OH)_2 \cdot 8H_2O$.

■ **3. Barium Hydroxide.** **Barium hydroxide** is made by slaking barium oxide and by treatment of the carbonate with superheated steam. Barium hydroxide is the most soluble of the alkaline earth hydroxides, and it crystallizes from solution in the form of the hydrate, $Ba(OH)_2 \cdot 8H_2O$. It is widely used in analytical chemistry because it has an advantage over hydroxides of sodium and potassium in that any carbonate which is formed due to contact with the air precipitates out as barium carbonate, leaving the solution relatively pure.

37.7   Calcium Carbonate

Calcium carbonate occurs in two distinctly different crystal forms, which are known as **calcite** and **aragonite.** The first is common while the second is comparatively rare. The lattice structure of the calcite crystal is rhombohedral (see Fig. 11–9). Calcite is the "low-temperature" form of calcium carbonate and results when precipitation occurs below 30° C. Above this temperature calcium carbonate crystallizes in rhombic prisms of aragonite, the high-temperature form.

Calcite is known in many varieties, one of which is **onyx.** This crystal form possesses the property of **birefringence,** or **double refraction;** i.e., when a beam of light enters the crystal it is broken up into two beams (Fig. 37-1).

Limestone is by far the most abundant source of calcium carbonate, being second in abundance only to silicate rocks in the earth's crust. Limestone is used more extensively than any other building stone in the United States; more than 80 per cent of it is supplied by southern Indiana. Coral reefs are composed largely of calcium carbonate of marine-animal origin. A pearl is built up of concentric layers of calcium carbonate deposited upon a foreign particle, such as a small grain of sand, which has entered the shell of an oyster.

FIGURE 37-1
Calcite, or Iceland spar, has the unusual property of birefringence, or double refraction. Notice the calcite produces two images of the word CALCITE whereas the piece of glass (*right*) produces only one image. *Bausch and Lomb Optical Co.*

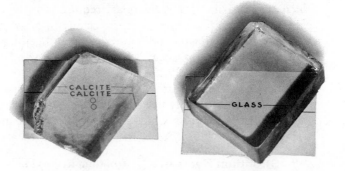

Calcium carbonate is only slightly soluble in pure water. However, it dissolves readily in water which contains dissolved carbon dioxide because of its conversion to the more soluble **calcium hydrogen carbonate,** $Ca(HCO_3)_2$.

$$CaCO_3 + H_2O + CO_2 \rightleftharpoons Ca^{2+} + 2HCO_3^-$$

This reaction is involved in the formation of limestone caves. When water containing carbon dioxide comes in contact with limestone rocks, it dissolves the rocks over which it flows. If water containing $Ca(HCO_3)_2$ finds its way into a cave where the $Ca(HCO_3)_2$ may liberate carbon dioxide, calcium carbonate may again be deposited, according to the reaction

$$Ca^{2+} + 2HCO_3^- \longrightarrow CaCO_3 + H_2O + CO_2\uparrow$$

This reaction is the basis for the formation of **stalactites,** which hang from the ceiling of limestone caves, and of **stalagmites,** which form upward from the floor of the cave where the water has dripped (Fig. 37-2).

Calcium hydrogen carbonate is readily converted to the normal carbonate when its solutions are heated. This reaction is involved in the formation of scale in kettles and boilers, and is the basis for one method of removing carbonate hardness from water (Section 12.6).

FIGURE 37-2
Stalactites and stalagmites are cave formations of calcium carbonate. These are in the Carlsbad Caverns of New Mexico. *David W. Corson from A. Devaney.*

37.8 Chlorides of Calcium, Strontium, and Barium

■ **1. Calcium Chloride.** This salt is obtained commercially as a by-product of the Solvay process for the manufacture of sodium carbonate (see Section 38.9). Calcium chloride may be produced in the laboratory by treating the oxide, hydroxide, or carbonate of calcium with hydrochloric acid. The salt crystallizes from water as the hexahydrate, $CaCl_2 \cdot 6H_2O$, which is converted upon heating to the monohydrate, $CaCl_2 \cdot H_2O$. Complete dehydration occurs upon heating to a higher temperature, but some hydrolysis always takes place during the dehydration.

$$CaCl_2 + H_2O \longrightarrow CaO + 2HCl\uparrow$$

The anhydrous salt and the monohydrate are used extensively as drying agents for gases and liquids. Calcium chloride forms compounds with ammonia, $CaCl_2 \cdot 8NH_3$, and with alcohol, $CaCl_2 \cdot 4C_2H_5OH$, and so cannot be used for drying these substances. Calcium chloride hexahydrate is very soluble in water and makes with ice an excellent freezing mixture, giving temperatures as low as $-55°$ C. Solutions of calcium chloride are quite commonly used as the cooling brine in refrigeration plants. Because it is a very deliquescent salt, $CaCl_2 \cdot 6H_2O$ is used to keep dust down on highways and in coal mines. Calcium chloride is frequently used to remove snow and ice from roads and walks.

■ **2. Strontium Chloride Hexahydrate.** This salt, $SrCl_2 \cdot 6H_2O$, is made by treating the carbonate with hydrochloric acid and evaporating the resulting solution. It is used, as are other salts of strontium, in the production of a red color in fireworks, railway fusees, and military flares.

■ **3. Barium Chloride.** This salt is prepared by heating barium sulfate with carbon and calcium chloride.

$$BaSO_4 + 4C + CaCl_2 \longrightarrow BaCl_2 + CaS + 4CO$$

Upon evaporation of solutions of the chloride, the dihydrate, $BaCl_2 \cdot 2H_2O$, separates. When a solution of the barium ion is required, this salt is usually employed. Like all other soluble barium salts, the chloride is very poisonous.

37.9 Sulfates of Calcium, Strontium, and Barium

The solubilities of the sulfates of Ca, Sr, and Ba in water decrease with increasing atomic weight of the metal; in moles per liter at 20° C they are $CaSO_4 \cdot 2H_2O$, 4.9×10^{-3}; $SrSO_4$, 5.3×10^{-4}; and $BaSO_4$, 1×10^{-5}.

■ **1. Calcium Sulfate.** **Anhydrous calcium sulfate** occurs in nature as the mineral **anhydrite,** $CaSO_4$, and as the dihydrate, **gypsum,** $CaSO_4 \cdot 2H_2O$. The latter is found in enormous deposits. Even though it is low in solubility, calcium sulfate is largely responsible for the noncarbonate hardness of ground waters (Section 12.6).

When gypsum is heated, it loses water, forming the hemihydrate $(CaSO_4)_2 \cdot H_2O$.

$$2[CaSO_4 \cdot 2H_2O] \longrightarrow (CaSO_4)_2 \cdot H_2O + 3H_2O$$

The hemihydrate is known as **plaster of Paris.** When it is ground to a fine powder and mixed with water, it sets by forming small interlocking crystals of gypsum. The setting results in an increase in volume and hence the plaster fits tightly any mold into which it is poured. Plaster of Paris is used extensively as a component of plaster for the interiors of buildings. It is also used in making statuary, stucco, wallboard, and casts of various kinds.

Gypsum is also used in making Portland cement (Section 29.19), blackboard crayon (erroneously called "chalk"), plate glass, terra cotta, pottery, and orthopedic and dental plasters. Anhydrous calcium sulfate is sold under the name "Drierite" as a drying agent for gases and organic liquids.

■ **2. Strontium Sulfate.** This compound occurs in nature in the mineral **celestite,** which is sometimes found in the form of beautiful crystals in caves where its faint tinge of blue suggested the name of the mineral.

■ **3. Barium Sulfate.** The mineral **barite,** $BaSO_4$, is the usual source of barium in the production of other barium compounds. Since the sulfate is insoluble in acids, it is first converted to the sulfide by a high temperature reduction with carbon.

$$BaSO_4 + 4C \longrightarrow BaS + 4CO$$

The sulfide can then be dissolved in acids such as hydrochloric or nitric to produce the desired salts.

A mixture of $BaSO_4$ and ZnS is known as **lithopone** and is used as a white paint pigment. It is made by the reaction of aqueous solutions of barium sulfide and zinc sulfate.

$$Ba^{2+} + S^{2-} + Zn^{2+} + SO_4{}^{2-} \longrightarrow \underline{BaSO_4} + \underline{ZnS}$$

Barium sulfate is used in taking x-ray photographs of the intestinal tract due to the fact that it is opaque to x rays. Even though Ba^{2+} is poisonous, $BaSO_4$ is so slightly soluble that it is nonpoisonous.

37.10 Properties of Calcium, Strontium, and Barium of Analytical Importance

The ions of the alkaline earth metals are colorless. Ammonium carbonate precipitates white crystalline carbonates of calcium, barium, and strontium from solutions of salts of these metals. The precipitated carbonates are dissolved by acetic acid as well as by all mineral acids. Barium can be separated from calcium and strontium by precipitation of its yellow chromate in a solution containing acetic acid buffered with ammonium acetate. The yellow strontium chromate can be precipitated from ammoniacal solutions if ethyl alcohol is added, leaving calcium ions in solution. Calcium ions are best detected by the fact that calcium oxalate is insoluble in water. Flame tests may be used to identify the salts of calcium, strontium, and barium. Calcium salts impart a brick-red color to the Bunsen flame; strontium salts, a crimson color; and barium salts, a greenish-yellow color.

QUESTIONS

1. Account for the considerable difference in size between the alkaline earth metal atoms and their dipositive ions (see table, inside back cover of text).
2. How do the metals of Group IIA differ from those of Group IA in atomic structure and general properties?
3. List several specific differences in properties between elements in the IA and IB families. Discuss the role of electron structure, with respect to these differences in properties.
4. The total of the first and second ionization potentials for beryllium atoms is larger than the corresponding sum for helium. Why does beryllium exhibit extensive chemistry and helium very little?
5. How is metallic calcium produced commercially?
6. Identify each of the following: quicklime, slaked lime, milk of lime, and soda lime.
7. Give the formula for each of the following naturally occurring calcium compounds: limestone, chalk, gypsum, fluorite, and hydroxyapatite.
8. Explain the formation of limestone caves, stalactites, and boiler scale.
9. Why cannot anhydrous calcium chloride be used for drying alcohol or ammonia?
10. Write a chemical equation describing the dehydrating action of calcium metal.
11. Give the chemistry of the preparation and setting of plaster of Paris.

12. What is the essential reaction involved in the setting of mortar?

13. Crushed limestone is used in the treatment of soils which are acid. Write equations to explain the reactions involved.

14. On the basis of atomic structure, explain why barium is more reactive than calcium.

15. Give equations for the reaction of barium with each of the following: oxygen, carbon dioxide, hydrogen, nitrogen, sulfur, and water.

16. The barium ion is poisonous. Why, then, can barium sulfate be taken internally with safety in making x-ray photographs of the intestinal tract?

17. Why is carbon added to barium carbonate when the latter is calcined to make barium oxide?

18. Explain how calcium chloride hexahydrate can be used to melt snow and ice on roads and walks and yet when mixed with ice can be an excellent "freezing mixture," much better than ice alone, for refrigeration plants and making ice cream. Are these two uses compatible?

PROBLEMS

1. Calculate the solubility of $SrSO_4$ in grams per 100 ml of water.

   Ans. 9.7 × 10⁻³ g/100 ml

2. Calculate the pH of a saturated solution of strontium hydroxide.

   Ans. 12.93

3. How much water could be heated from 25° C to 100° C with the heat given off in hydrating 1.0 kg of BaO?   *Ans. 2.1 kg*

4. What volume of calcium would have the same mass as 10.0 cm³ of copper?

   Ans. 57.9 cm³

REFERENCES

Discovery of the Elements, M. E. Weeks and H. M. Leicester, Publ. by the Journal of Chemical Education, Easton, Pa., Seventh Edition, 1968; pp. 479-502 and 774-783.

"Process Wins Beryllium from Low-Grade Ores," *Chem. & Eng. News*, April 19,1965; p. 70.

"Beryllium and Berylliosis," J. Schubert, *Sci. American*, Aug., 1958; p. 27.

"An Historical Note on Beryllium," H. Gilman, *J. Chem. Educ.*, **46**, 276 (1969).

"Calcium and Life," L. V. Heilbrunn, *Sci. American*, June, 1951; p. 60.

"How an Eggshell is Made," T. G. Taylor, *Sci. American,* March, 1970; p. 89.

"How is Muscle Turned On and Off?" G. Hoyle, *Sci. American*, April, 1970; p. 85.

"Calcitonin," H. Rasmussen and M. M. Pechet, *Sci. American*, Oct., 1970; p. 42.

"Calcium Carbonate Equilibria in Lakes," S. D. Morton and G. F. Lee, *J. Chem. Educ.*, **45**, 511 (1968).

"Calcium Carbonate Equilibria in the Oceans—Ion Pair Formation," S. D. Morton and G. F. Lee, *J. Chem. Educ.*, **45**, 513 (1968).

"Texture and Composition of Bone," D. McConnell and D. W. Foreman, Jr., *Science*, **172**, 972 (1971).

"Life's Essential Elements," D. R. Williams, *Educ. in Chemistry*, **10**, 56 (1973).

38

Metals of Analytical Group V

Sodium, Potassium, Magnesium, and Ammonium

The metals composing Analytical Group V (see also Chapter 44) are those whose cations fail to form precipitates with the reagents used in the precipitation of the first four groups. For this reason, Group V is called the soluble group, and no single reagent will precipitate all four cations. Some properties are listed in Table 38–1.

Magnesium falls in the same group of the Periodic Table as do barium, strontium, and calcium (discussed in Chapter 37) and is closely related chemically to

TABLE 38–1 Some Properties of the Metals of Analytical Group V

Property	Sodium	Potassium	Magnesium
Atomic number	11	19	12
Atomic weight	22.9898	39.0983	24.305
Electronic structure	2, 8, 1	2, 8, 8, 1	2, 8, 2
Oxidation states	+1	+1	+2
Reduction potential, volts	(Na^+/Na) −2.71	(K^+/K) −2.92	(Mg^{2+}/Mg) −2.37
Density, g/cm³ at 20°	0.97	0.86	1.74
Melting point, °C	98	63.4	650
Boiling point, °C	889	757	1,120

these metals, but it is not precipitated by ammonium carbonate in the presence of other ammonium salts.

The ammonium ion is chemically similar to ions of the alkali metals, particularly potassium. The ammonium group, NH_4, has not been isolated as a neutral molecule. However, it does form an amalgam when solutions of ammonium salts are electrolyzed with a mercury cathode, as do the alkali metals.

38.1   Periodic Relationships of the Alkali Metals

The metallic elements lithium, sodium, potassium, rubidium, cesium, and francium constitute Group IA of the Periodic Table and are known as the **alkali metals.** The heaviest of them, francium, occurs in nature in very small quantities as a short-lived radioactive isotope. The alkali metals and the metals of Group IIA (the alkaline earth metals) are often called the **light metals** because of their relatively low densities. The alkali metals have the largest atomic radii (Table 38–2) of the elements in their respective periods of the Periodic Table. There is

TABLE 38-2   Some Properties of the Alkali Metals

	Atomic Number	Atomic Weight	Atomic Radius, Å	Ionic Radius, Å	Density at 20°	Melting Point, °C	Boiling Point, °C
Lithium	3	6.941	1.52	0.60	0.534	180	1,326
Sodium	11	22.9898	1.86	0.95	0.97	98	889
Potassium	19	39.0983	2.31	1.33	0.86	63.4	757
Rubidium	37	85.4678	2.44	1.48	1.53	38.8	679
Cesium	55	132.9055	2.62	1.69	1.87	28.7	690
Francium	87	223	2.7	—	—	—	—

a single electron in the outermost shell of each of the alkali metals. Since this electron is far removed from the nucleus of the atom, it is easily lost, and the alkali metals readily form stable positive ions. The difficulty with which their ions are reduced to the metal accounts for the fact that none of the alkali metals were isolated until Sir Humphry Davy, in 1807, produced sodium and potassium by the electrolytic reduction of their respective hydroxides. As expected, the ease with which the valence electron is lost increases with increasing atomic radius. Of those members of the group which have been extensively studied, cesium is the most reactive; however, francium, with atomic radius even larger than that of cesium, should be more reactive than cesium.

The alkali metals all react vigorously with water, liberating hydrogen and giving solutions of the corresponding strong bases. They also react directly with oxygen, sulfur, nitrogen, hydrogen, and the halogens. Most of the salts of the alkali metals are readily soluble in water; however, Li_2CO_3, Li_3PO_4, and LiF are relatively insoluble. In this respect, lithium resembles magnesium, the second member of Group IIA.

Sodium and Potassium

38.2   Occurrence of Sodium and Potassium

Sodium and potassium are not found free in nature because of their high ease of oxidation. Their compounds, however, occur in abundance and are widely distributed in nature. They appear in many complex silicate rocks such as feld-spars, of which the very abundant **albite**, $NaAlSi_3O_8$, and **orthoclase**, $KAlSi_3O_8$, are typical. As these minerals are disintegrated by weathering, the sodium and potassium are converted to soluble compounds, most of which are leached out of the soil by water and are eventually carried to the sea. In passing through the soil, some of the potassium compounds are adsorbed by the colloidal particles of the soil and held there until used by plants for growth. Sodium compounds are much less retained by the soil than those of potassium, and little sodium is needed by plants. Changes in the earth's crust during past geological ages have, at various times and places, caused sections of the sea to become isolated and gradually to evaporate. Such geological changes have resulted in the forma-tion of large deposits of sodium and potassium salts.

Sodium and potassium occur also in various parts of the world in chloride, sulfate, and borate minerals.

38.3   Preparation of Sodium and Potassium

For many years metallic sodium was obtained by reducing the carbonate with carbon.

$$Na_2CO_3 + 2C \longrightarrow 2Na + 3CO$$

With the development of the dynamo for the production of cheap electricity, Davy's original electrolytic method of obtaining sodium from fused sodium hydroxide became the industrial process for producing the metal. Most of our sodium is now prepared by the electrolysis of fused sodium chloride, admixed with either sodium carbonate or calcium chloride. The energy required to melt sodium chloride and the high vapor pressure of metallic sodium at this tempera-ture prohibit the use of sodium chloride alone; the melting point of sodium chloride is 801°, while that of the mixture is about 600° C. The cell used in the United States for the electrodeposition of sodium is the Downs cell.

Sodium is shipped in the form of 12-pound bricks in air-tight steel barrels or in tank cars which are loaded by pumping in the fused metal. To unload the cars, the metal is melted and the liquid forced out by nitrogen under pressure.

Potassium is produced on a relatively small scale, since this element offers few technical advantages over the less costly sodium. Potassium may be made by the electrolytic reduction of either molten potassium chloride or potassium hydroxide but most of the production in the United States is by the reaction of sodium with fused potassium chloride ($Na + KCl \rightleftharpoons K + NaCl$). The sodium is fed into the bottom of a column of melted potassium chloride, and potassium metal escapes at the top. It should be noted that for fused salts the regular electromotive series does not apply exactly. A variety of chemical factors determine the course of the reaction. Apparently, this particular reaction proceeds, although potassium is above sodium in the electromotive series, because metallic potassium is more volatile than is sodium, thereby causing a shift in the equilibrium to the right.

38.4 Properties of Sodium and Potassium

Both sodium and potassium are soft enough to be cut by a knife; they are silvery-white in color and are excellent conductors of heat and electricity. Both metals tarnish immediately in moist air, and react vigorously with water and acids, forming hydroxides or salts of the acids, respectively, and liberating hydrogen. The heat of reaction between these metals and water may cause the evolved hydrogen to ignite. One should never touch sodium or potassium with the hands, because the heat of reaction with the moisture on the skin may cause the metal to ignite. These metals are stored under kerosene or in sealed containers.

Sodium dissolves in mercury forming an amalgam. This alloy is an active reducing agent that is more suitable for many uses than pure sodium because the mercury, being an inactive metal, retards the action of the sodium. Sodium forms alloys with many other metals such as potassium, tin, and antimony, but not with iron. Potassium and sodium, if mixed in correct proportion, give an alloy which is liquid at room temperature.

Sodium and its compounds impart a yellow color to a Bunsen flame, and potassium and its compounds color the flame violet. The color of the potassium flame is best detected by observing it through cobalt glass (blue in color), which absorbs all wavelengths of light except blue and violet.

In general, potassium is more reactive than sodium; otherwise the chemical properties of the two metals are quite similar. Potassium and sodium differ primarily in regard to the solubilities of their salts, the potassium salts being in general slightly less soluble than those of sodium; there are exceptions to this generalization, however.

38.5 Uses of Sodium and Potassium

Sodium is used as a reducing agent in the production of other metals, such as titanium and zirconium, from their chlorides or oxides. Its properties as a reducing agent make it useful in the manufacture of certain dyes, drugs, and perfumes. It is used in the preparation of sodium-lead alloys which in turn are used in the production of tetraethyl lead from ethyl chloride. It is used in sodium lights for lighting highways, because its yellow light penetrates fog well. The synthetic rubber industry consumes a considerable amount of sodium. The largest uses, however, are in the manufacture of compounds such as sodium peroxide and sodium cyanide, which cannot be made directly from sodium chloride.

Potassium has no major uses for which sodium cannot be substituted. Sodium-potassium alloys, however, are growing in importance. Their low density, low viscosity, wide liquid range, and high thermal conductivity combine to make them particularly useful as industrial heat-transfer media.

38.6   Hydrides of Sodium and Potassium

The hydrides NaH and KH are prepared by the direct union of the elements at slightly elevated temperatures and are decomposed at higher temperatures. They are saltlike compounds, with a crystal structure like that of sodium chloride. When the fused hydrides are electrolyzed, the metal is formed at the cathode and hydrogen is evolved at the anode, so the hydrides must be composed of metal ions and hydride ions, $:H^-$. The hydrides are readily decomposed by water with the evolution of hydrogen, the hydride ions combining with hydrogen ions of the water ($H^+ + :H^- \longrightarrow H_2$), and leaving an excess of hydroxide ions in solution ($NaH + H_2O \longrightarrow Na^+ + OH^- + H_2\uparrow$).

Sodium hydride is used widely in organic chemistry as a reducing agent and in the descaling of steel before it is plated or enameled.

38.7   Oxides and Peroxides of Sodium and Potassium

Oxides of most metals are produced by heating either the hydroxides, nitrates, or carbonates of the corresponding metals. The oxides of the more active alkali metals cannot be prepared in this way, however, because their hydroxides, nitrates, and carbonates are stable toward heat, decomposing only at extremely high temperatures.

Sodium forms both the **normal oxide,** Na_2O, and the **peroxide,** Na_2O_2. Sodium oxide is prepared in commercial quantities by heating sodium in a limited supply of air under carefully controlled conditions. The product is an extremely alkaline white powder. It is a very effective drying agent and is also used as a strong caustic. Sodium peroxide is the more important of the two oxides of sodium. It is manufactured by passing chips of sodium on aluminum trays through a furnace held at 300° to 400° C, while a current of air is moving in the opposite direction. This is an application of the **counter-current principle,** which is widely used in the manufacture of chemicals.

Sodium peroxide is a yellowish-white powder, used extensively as an oxidizing agent and as a bleaching agent. Solutions of the compound are alkaline by hydrolysis, as shown by the equations

$$O_2^{2-} + H_2O \rightleftharpoons HO_2^- + OH^-$$
$$HO_2^- + H_2O \rightleftharpoons H_2O_2 + OH^-$$

Potassium oxide, K_2O, may be obtained by heating potassium nitrate with elemental potassium in the absence of air. The **peroxide,** K_2O_2, is formed by burning the metal in the calculated quantity of air. When exposed to an excess of air, potassium oxide combines readily with oxygen to give **potassium superoxide,** KO_2, in which the superoxide ion has an oxidation state of -1. The superoxide of potassium is most readily formed by burning the metal in an excess of oxygen. The KO_2 is used in gas masks for mine safety work. Moisture from the air in contact with the superoxide reacts with the superoxide to give O_2 for the wearer of the mask to breathe.

$$4KO_2 + 2H_2O \longrightarrow 4KOH + 3O_2$$

When the breath is exhaled through the mask the CO_2 is absorbed by the potassium hydroxide.

$$CO_2 + KOH \longrightarrow KHCO_3$$

The gas-mask user breathes no air from outside the mask.

38.8   Hydroxides of Sodium and Potassium

Sodium hydroxide is frequently called **caustic soda** in commerce, and its solutions are sometimes referred to as **lye,** or **soda lye.** It is the sixth highest chemical in terms of production in the United States (10,865,000 tons in 1974). Its production commercially is by two processes. The older process involves the action of slaked lime, $Ca(OH)_2$, on soda ash, Na_2CO_3, and is called the **lime-soda process.**

$$(2Na^+) + CO_3{}^{2-} + Ca^{2+} + (2OH^-) \longrightarrow \underline{CaCO_3} + (2Na^+) + (2OH^-)$$

The insoluble calcium carbonate is filtered out, and the filtrate is evaporated to yield solid sodium hydroxide. The residue is fused to remove the last traces of water and is then molded into sticks or pellets.

A second and more important industrial process for the production of sodium hydroxide is the electrolysis of aqueous sodium chloride (see Sections 21.2 and 22.4).

$$(2Na^+) + 2Cl^- + 2H_2O \longrightarrow (2Na^+) + 2OH^- + H_2\uparrow + Cl_2\uparrow$$

Hydrogen is formed at the cathode and chlorine is evolved at the anode. Sodium ions and hydroxyl ions accumulate in the solution. The cell is so designed that the chlorine is not permitted to react with the hydroxide ions to form hypochlorite.

Electrolytic sodium hydroxide is impure with sodium chloride, most of which can be removed by a process called **fractional crystallization** because sodium chloride is less soluble than sodium hydroxide in water.

Pure sodium hydroxide can be obtained electrolytically from aqueous sodium chloride by using a mercury cathode cell. In this cell sodium, rather than hydrogen, is liberated at the cathode; the sodium, as it is reduced, forms an amalgam with the mercury. The amalgam is decomposed by water in a separate compartment to produce hydrogen and very pure sodium hydroxide, containing no chlorides.

$$2Na^+ + 2Cl^- + 2n Hg \xrightarrow{\text{Electrolysis}} 2Na(Hg)_n + Cl_2\uparrow$$
$$2Na(Hg)_n + 2H_2O \longrightarrow 2Na^+ + 2OH^- + 2n Hg + H_2\uparrow$$

The DeNora cell is typical of the mercury cathode type (see Section 21.2).

Sodium hydroxide is an ionic compound that melts and boils without decomposition. When fused or in aqueous solution, sodium hydroxide attacks silicate minerals and glass. The solid hydroxide and its solutions readily absorb carbon dioxide from the air, with the formation of sodium carbonate and water.

$$2OH^- + CO_2 \longrightarrow H_2O + CO_3{}^{2-}$$

Sodium hydroxide dissolves extensively in water with the evolution of a great deal of heat, and yields strongly basic solutions. It is used extensively in the production of chemicals, rayon, lye, cleaners, textiles, soap, paper and pulp, and in petroleum refining.

Potassium hydroxide is manufactured in a manner similar to that described for the production of sodium hydroxide. The method involving the treatment of aqueous potassium carbonate with slaked lime has largely given way to the electrolytic method. Potassium hydroxide is very soluble in water and, being strongly deliquescent, is used as a dehydrating agent. It forms strongly alkaline solutions with water. Because it is more expensive than sodium hydroxide, which almost always serves equally well, its use is limited.

38.9 Carbonates of Sodium and Potassium

Sodium carbonate is, after sodium hydroxide, perhaps the most important manufactured compound of sodium. The total amount used in the United States in 1974 was 7,551,000 tons, making it tenth among all industrially produced chemicals. It has the formula Na_2CO_3 and it is commonly called **soda ash,** or simply **soda.**

Practically all the sodium carbonate produced in America is made by the **Solvay process.** This process is based upon the reaction of ammonium hydrogen carbonate with a saturated solution of sodium chloride. Sodium hydrogen carbonate precipitates, since it is only slightly soluble in the reaction medium.

$$Na^+ + (Cl^-) + (NH_4^+) + HCO_3^- \longrightarrow \underline{NaHCO_3} + (NH_4^+) + (Cl^-)$$

The basic raw materials used in the process are limestone and common salt. The carbon dioxide is generated by heating limestone.

$$CaCO_3 \longrightarrow CaO + CO_2 \tag{1}$$

The ammonia is obtained by treating ammonium chloride, formed as a byproduct of the process, with calcium hydroxide.

$$2NH_4^+ + (2Cl^-) + (Ca^{2+}) + 2OH^- \longrightarrow 2NH_3\uparrow + 2H_2O + (Ca^{2+}) + (2Cl^-) \tag{2}$$

Calcium hydroxide results from slaking the lime produced in reaction (1).

$$CaO + H_2O \longrightarrow Ca^{2+} + 2OH^- \tag{3}$$

Aqueous ammonia reacts with an excess of carbon dioxide yielding ammonium hydrogen carbonate.

$$NH_3 + CO_2 + H_2O \longrightarrow NH_4^+ + HCO_3^- \tag{4}$$

Sodium hydrogen carbonate is the first desired product of the Solvay process; and, after being freed of ammonium chloride by recrystallization, it is made commercially available. The major portion of the sodium hydrogen carbonate, however, is converted to sodium carbonate by heating.

$$2NaHCO_3 \longrightarrow Na_2CO_3 + H_2O + CO_2 \tag{5}$$

The carbon dioxide from this reaction is used to produce more sodium hydrogen carbonate. The only by-product of the entire process not reused in the process is calcium chloride.

Sodium carbonate is produced to a limited extent by an electrolytic process, but here it is regarded as a by-product of the chlorine industry. We saw in Section 22.4 that the electrolysis of aqueous sodium chloride produces chlorine, hydrogen, and sodium hydroxide. By saturating the sodium hydroxide solution with carbon dioxide, sodium hydrogen carbonate is produced.

$$(Na^+) + OH^- + CO_2 \longrightarrow (Na^+) + HCO_3^-$$

The use of less carbon dioxide gives rise to the normal carbonate.

$$(2Na^+) + 2OH^- + CO_2 \longrightarrow (2Na^+) + CO_3^{2-} + H_2O$$

Sodium carbonate is used extensively in the manufacture of glass, caustic soda, other chemicals, soap, paper and pulp, cleansers, water softeners, and in petroleum refining.

When a solution of sodium carbonate is evaporated below 35.2° C, the **deca-hydrate** $Na_2CO_3 \cdot 10H_2O$, known as **washing soda**, crystallizes out; above this temperature the **monohydrate**, $Na_2CO_3 \cdot H_2O$, separates. Heating either hydrate produces the anhydrous compound. Solutions of sodium carbonate are basic, due to hydrolysis of the carbonate ions, as shown by the equation

$$CO_3^{2-} + H_2O \longrightarrow HCO_3^- + OH^-$$

For this reason sodium carbonate is generally used in commercial processes requiring an alkali which is not as strong as sodium hydroxide.

Sodium hydrogen carbonate, $NaHCO_3$, is commonly known as **bicarbonate of soda** or **baking soda**. Its solutions are weakly basic due to hydrolysis of the hydrogen carbonate:

$$HCO_3^- + H_2O \rightleftharpoons H_2CO_3 + OH^-$$

Acids act upon sodium hydrogen carbonate with the formation of carbon dioxide.

$$HCO_3^- + H_3O^+ \longrightarrow 2H_2O + CO_2\uparrow$$

This reaction is involved in the leavening, or rising, process in baking. Baking powders contain baking soda admixed with an acidic substance, such as either potassium hydrogen tartrate [$KHC_4H_4O_6$ (cream of tartar)], calcium dihydrogen phosphate [$Ca(H_2PO_4)_2$], or sodium aluminum sulfate [$NaAl(SO_4)_2 \cdot 12H_2O$]. Starch or flour is also added to keep the powder dry. When the mixture is dry no reaction takes place, but as soon as water is added carbon dioxide is given off. The acidic substance and water produce hydronium ions, which in turn react with the baking soda to form carbon dioxide.

$$HC_4H_4O_6^- + H_2O \rightleftharpoons H_3O^+ + C_4H_4O_6^{2-}$$
$$H_2PO_4^- + H_2O \rightleftharpoons H_3O^+ + HPO_4^{2-}$$
$$[Al(H_2O)_x]^{3+} + H_2O \rightleftharpoons H_3O^+ + [Al(OH)(H_2O)_{x-1}]^{2+}$$
$$HCO_3^- + H_3O^+ \longrightarrow CO_2\uparrow + 2H_2O$$

The carbon dioxide when trapped in bread dough will expand if warmed and produce spaces in the bread that give it a desirable lightness.

The lactic acid of sour milk and buttermilk or the acetic acid in vinegar will also furnish hydronium ions and thus serve the same purpose as the acidic constitutent of baking powder. Thus, it is possible to make bread rise by using either sour milk or buttermilk, with ordinary sodium bicarbonate (baking soda).

Potassium carbonate cannot be prepared by the Solvay process because of the high solubility of potassium hydrogen carbonate in solutions of aqueous ammonia. Instead, it is obtained either by treating electrolytic potassium hydroxide with carbon dioxide, or from potassium chloride. The latter method involves heating potassium chloride under pressure with magnesium carbonate, water, and carbon dioxide, whereby the slightly soluble double salt $KHCO_3 \cdot Mg(HCO_3)_2 \cdot 4H_2O$ is formed.

$$2K^+ + (2Cl^-) + 3MgCO_3 + 3CO_2 + 11H_2O \longrightarrow$$
$$2[KHCO_3 \cdot Mg(HCO_3)_2 \cdot 4H_2O] + Mg^{2+} + (2Cl^-)$$

The complex salt is removed from the solution by filtration and heated to 120°, which causes the following reaction to take place:

$$2[KHCO_3 \cdot Mg(HCO_3)_2 \cdot 4H_2O] \longrightarrow K_2CO_3 + 2MgCO_3 + 11H_2O + 3CO_2\uparrow$$

Finally, the potassium carbonate is leached from the solid mass, leaving a residue of insoluble magnesium carbonate, which is used again in the first step.

Potassium carbonate forms three hydrates, containing one, two, and three molecules of water of crystallization, respectively. Potassium carbonate is usually sold in the form of an anhydrous powder. It is deliquescent and very soluble in water, forming a strongly alkaline solution. It is used in the manufacture of soft soap, glass, pottery, and various potassium compounds.

When a solution of potassium carbonate is saturated with carbon dioxide and concentrated by evaporation, crystals of **potassium hydrogen carbonate,** $KHCO_3$, are deposited. Solutions of this salt are nearly neutral (just slightly alkaline). When heated, the hydrogen carbonate is converted to the normal carbonate.

38.10 Sodium Chloride

Sodium chloride is one of our most abundant minerals. Seawater contains 2.7 per cent sodium chloride, and the waters of the Dead Sea and the Great Salt Lake contain 23 per cent. There are vast deposits of rock salt in the Stassfurt region of Germany, and in the United States large deposits are found in New York, Michigan, West Virginia, and California, and a very extensive bed (about 400 to 500 feet thick) underlies parts of Oklahoma, Texas, and Kansas. Most of the salt consumed comes from beds which lie below the surface of the earth, rather than from seawater. The salt is removed from these underground beds by mining or by forcing water down into the deposits to form saturated brines, which are then pumped to the surface. Natural salt contains other soluble salts which make it unsuitable for many uses. Salt is purified by dissolving it in water, concentrating the solution by evaporation (often under reduced pressure), and allowing the crystals to form again. Some of the impurities are more soluble than the salt and for this reason remain in solution when the salt crystallizes. Some of the impurities which are less soluble than salt do not crystallize out because they are present in small amounts. Other of the less soluble impurities, such as calcium sulfate, do crystallize out with the salt but these are then removed by a process called **counter-current washing.** Advantage can also be taken of the decreasing solubility of calcium sulfate in sodium chloride solution as the temperature is raised above 80° C. Thus, if the brine is filtered hot, the resulting solution will not contain enough calcium sulfate to be saturated when cooled. Hence, calcium sulfate cannot precipitate with the sodium chloride, and the sodium chloride crystallizes relatively free of the impurity.

Pure sodium chloride, when the relative humidity is above 75%, or sodium chloride contaminated with chlorides of magnesium and calcium, absorbs water causing the salt to "cake." When in a shaker, the caking salt clogs the shaker. Basic magnesium carbonate or calcium aluminosilicate is usually added to serve as a drier or "anti-caking" agent.

Sodium chloride is the usual source of all other sodium and chlorine compounds. Its most extensive use is in the manufacture of chlorine, hydrochloric acid, sodium hydroxide, and sodium carbonate. Sodium chloride is an essential constituent of the foods of animals. It not only makes food more palatable, but it is the source of chlorine from which hydrochloric acid, a constituent of gastric juice, is produced. Sodium chloride is also a constituent of the blood and is essential for the life processes of the human body.

38.11   Nitrates of Sodium and Potassium

Sodium nitrate, or **Chile saltpeter,** is a very important compound that was discussed in connection with nitrogen and nitric acid (Chapter 25).

Potassium nitrate was one of the principal reagents used by the alchemists. It is formed in nature by the decay of organic matter and is prepared commercially from sodium nitrate and potassium chloride.

$$Na^+ + (NO_3^-) + (K^+) + Cl^- \longrightarrow \underline{NaCl} + (K^+) + (NO_3^-)$$

The sodium nitrate and potassium chloride are dissolved in hot water, and the solution is evaporated by boiling. Of the four compounds possible in a solution containing Na^+, NO_3^-, K^+, and Cl^-, sodium chloride is the least soluble in hot water, and hence it is the first to crystallize out when the concentration of the solution is increased by evaporation. On the other hand, potassium nitrate is very soluble in hot water, making it possible to separate the two salts by filtration. Sodium chloride has about the same solubility in cold as in hot water, so very little more of it crystallizes out when the filtrate is cooled. However, potassium nitrate is but slightly soluble in cold water, and thus it crystallizes out as the filtrate is cooled.

Potassium nitrate is a valuable fertilizer, because it furnishes both potassium and nitrogen in forms which are readily utilized by growing plants. It is also used to some extent in preserving ham and corned beef, to which it imparts a red color.

38.12   Sulfates of Sodium and Potassium

Extensive deposits of **sodium sulfate** occur in Canada, North Dakota, and the southwestern section of the United States, but most of the sodium sulfate used commercially is obtained as a by-product in the manufacture of hydrochloric acid from salt and sulfuric acid. The sodium sulfate thus formed is commonly called **salt cake** and is used in the manufacture of glass, paper, rayon, coal-tar dyes, and soap.

When sodium sulfate is crystallized from solution at temperatures below 32.28° C, the decahydrate, $Na_2SO_4 \cdot 10H_2O$, is formed. This hydrate is called **Glauber's salt,** in honor of the alchemist Glauber who used it as a medicine in the seventeenth century. The anhydrous salt, Na_2SO_4, crystallizes from solutions at temperatures above the transition point of 32.28°. This temperature is so constant that it is used as a fixed point in determining the accuracy (calibrating) of thermometer markings.

Sodium hydrogen sulfate, $NaHSO_4$, is used in the production of hydrochloric acid from sodium chloride.

$$NaCl + NaHSO_4 \longrightarrow Na_2SO_4 + HCl\uparrow$$

Sodium hydrogen sulfate is prepared by heating either sodium chloride or sodium nitrate to a moderate temperature with sulfuric acid.

$$NaCl + H_2SO_4 \longrightarrow NaHSO_4 + HCl\uparrow$$
$$NaNO_3 + H_2SO_4 \longrightarrow NaHSO_4 + HNO_3\uparrow$$

Potassium forms two sulfates, the **normal salt,** K_2SO_4, and the **hydrogen salt,** $KHSO_4$. The normal sulfate is obtained from natural salt deposits, and it is used as a fertilizer and in the preparation of alums. **Potassium hydrogen sulfate** is made by heating the normal sulfate with the proper quantity of sulfuric acid. It is used in analytical chemistry to convert metal oxides and silicates into sulfates. Heat converts potassium hydrogen sulfate to potassium pyrosulfate.

$$2KHSO_4 \longrightarrow K_2S_2O_7 + H_2O$$

When the pyrosulfate is strongly heated, it gives up sulfur trioxide, which is a very reactive acid anhydride at high temperatures.

$$K_2S_2O_7 \longrightarrow K_2SO_4 + SO_3\uparrow$$

38.13   Properties of Sodium and Potassium of Analytical Importance

■ **1. Sodium.**   Most of the salts of sodium are soluble in water. Notable exceptions are **sodium hexafluosilicate,** Na_2SiF_6, and **sodium hexahydroxoantimonate(V),** $NaSb(OH)_6$.

Of qualitative and quantitative analytical importance is the yellow complex salt, **sodium zinc uranyl acetate,** $NaZn(UO_2)_3(OAc)_9 \cdot 6H_2O$. Sodium ions may be almost completely precipitated from solution as the complex by adding a saturated acetic acid solution of zinc acetate and uranyl acetate.

The characteristic yellow color imparted to the Bunsen flame by sodium compounds serves as a qualitative test for the sodium ion.

■ 2. Potassium. With hexachloroplatinic acid, potassium ions produce **potassium hexachloroplatinate(IV)**, $K_2[PtCl_6]$, a yellow salt which is somewhat soluble in water but sparingly soluble in alcohol. Other slightly soluble salts are the yellow **hexanitrocobaltate(III)**, $K_2Na[Co(NO_2)_6]$, and the **tripotassium hexanitrocobaltate(III)**, $K_3[Co(NO_2)_6]$. Tartaric acid, $H_2C_4H_4O_6$, reacts with potassium ions in high concentrations and in neutral solutions to form white **potassium hydrogen tartrate**, $KHC_4H_4O_6$. Concentrated solutions of potassium salts react with perchloric acid, precipitating $KClO_4$, and with fluosilicic acid, precipitating K_2SiF_6.

Potassium compounds impart a violet color to a Bunsen flame, a property which is of value in detecting the element qualitatively.

Magnesium

The periodic relationships of magnesium were discussed with those of the other alkaline earth metals in Section 37.1.

38.14   Occurrence of Magnesium

Magnesium never occurs free in nature because of its reactivity, but, in combination, is abundant and widely distributed. The chloride and sulfate of magnesium are readily soluble in water and, consequently, are found in ground waters, to which they give noncarbonate hardness (Section 12.6). Seawater contains both the chloride and the sulfate, and compounds of magnesium concentrate in the mother liquor of seawater and underground brines from which sodium chloride has been crystallized. Typical magnesium minerals are **carnallite** ($KCl \cdot MgCl_2 \cdot 6H_2O$), **magnesite** ($MgCO_3$), **asbestos** ($H_4Mg_3Si_2O_9$), **dolomite** ($MgCO_3 \cdot CaCO_3$), **meerschaum** ($Mg_2Si_3O_3 \cdot 2H_2O$), **talc** or **soapstone** [$Mg_3Si_4O_{10}(OH)_2$], and **brucite** [$Mg(OH)_2$].

38.15   Preparation of Magnesium

Magnesium metal is obtained from several different sources and prepared by several different methods.

■ 1. From Underground Brines. Magnesium chloride is obtained from the underground brines of Michigan, which contain about 3 per cent magnesium chloride, 9 per cent calcium chloride, 14 per cent sodium chloride, and 0.1 per cent bromine as bromide ion. The bromine is extracted first (Section 21.2), and the brine is then treated with a suspension of magnesium hydroxide to precipitate the hydroxides of iron and certain other metals, and then the filtrate is evaporated to crystallize out the sodium chloride. The magnesium and calcium chlorides are separated by fractional crystallization. Crystalline magnesium chloride hexahydrate, $MgCl_2 \cdot 6H_2O$, is then completely dehydrated in an atmosphere of hydrogen chloride. The hydrogen chloride prevents the formation of basic magnesium salts by hydrolysis (Section 38.18). Electrolysis of the fused chloride produces magnesium metal and chlorine. Sodium chloride is usually added to lower the melting point of the electrolyte and increase the electrical conductivity. The product has a purity of 99.9 per cent.

■ **2. From Seawater.**   Seawater now serves as an important and inexhaustible source of magnesium. Nearly 6 million tons of magnesium, as the chloride and sulfate, are contained in each cubic mile of seawater. The raw materials for the process are seawater, oyster shells ($CaCO_3$) from shallow waters along the coast, salt from salt domes, fresh water, and natural gas. Nearly 800 tons of seawater must be processed to obtain one ton of magnesium metal. Oyster shells are calcined to produce lime, which is slaked by adding water. In some operations, the same seawater that was used for the recovery of bromine (Section 21.2) is treated with slaked lime to precipitate magnesium hydroxide. After filtration, the magnesium hydroxide may be treated with hydrochloric acid to produce the chloride. The magnesium chloride is crystallized from solution by evaporation. The crystallized salt is then partially dehydrated by heating. The resulting magnesium chloride, which has a composition corresponding to $MgCl_2 \cdot 1\frac{1}{2}H_2O$, is electrolyzed to produce magnesium and chlorine. The by-product chlorine is used in the burning of natural gas to produce the hydrogen chloride used in the process. In some seawater operations, magnesium carbonate and magnesium oxide are produced instead of magnesium metal.

38.16   Properties and Uses of Magnesium

Magnesium is a silvery-white metal which is malleable and ductile at high temperatures. It is the lightest of the widely used structural metals. Although it is very active, magnesium is not readily attacked by air or by water, due to the formation of a protective basic carbonate film on its surface. Magnesium is soluble in acids, including carbonic, evolving hydrogen.

$$Mg + H_2CO_3 \longrightarrow \underline{MgCO_3} + H_2 \uparrow$$
$$MgCO_3 + H_2CO_3 \longrightarrow \underline{Mg^{2+}} + 2HCO_3^-$$

It is also attacked by the alkali metal hydrogen carbonates and by various salts which give an acid reaction by hydrolysis.

Magnesium decomposes boiling water very slowly, but it rapidly reduces the hydrogen in steam, forming magnesium oxide and hydrogen. The affinity of magnesium for oxygen is so great that it will react when heated in an atmosphere of carbon dioxide, reducing the carbon of the oxide to elemental carbon.

$$2Mg + CO_2 \longrightarrow 2MgO + C$$

The great reducing power of hot magnesium is utilized in separating many metals and nonmetals, such as silicon and boron, from their oxides.

Magnesium will unite with most nonmetals. The brilliant white light emitted from burning magnesium makes it useful in flashlight powders, military flares, and incendiary bombs. When magnesium burns in air, both the oxide, MgO, and the nitride, Mg_3N_2, are formed.

Most of the magnesium produced commercially is used in making lightweight alloys, the most important of which are those of aluminum and zinc. **Magnalium** (1-15 per cent Mg, 0-1.75 per cent Cu and the remainder Al) is lighter, harder, stronger, and more easily machined than aluminum. Other important magnesium alloys include **duralumin** (0.5 per cent Mg, 0.5 per cent Mn, 3.5-5.5 per cent Cu, remainder Al), and **Dowmetal** (8.5 per cent Al, 0.15 per cent Mn, 2.0 per cent Cu, 1.0 per cent Cd, 0.5 per cent Zn, 87.85 per cent Mg).

38.17   Magnesium Oxide and Hydroxide

The oxide of magnesium, MgO, is sometimes called **magnesia.** It is formed commercially by heating **magnesite,** $MgCO_3$, to 600°-800° C, which drives off the carbon dioxide from the magnesium carbonate.

$$MgCO_3 \xrightarrow{\triangle} MgO + CO_2\uparrow$$

The product is a light, fluffy powder which still contains a small percentage of magnesium carbonate. It reacts slowly with water, forming magnesium hydroxide, and it is soluble in an aqueous solution of carbon dioxide forming magnesium hydrogen carbonate, a constituent of hard water. On the other hand, when magnesite is heated to 1,400° C or above and ignited, the product contains no magnesium carbonate and is a powder which is much more dense than the light form of the oxide. This pure form of magnesium oxide does not react with water and it does not conduct heat well. Inasmuch as it melts at 2,800° C, it is used in making fire brick and crucibles, as a lining in furnaces, and in heat insulation.

Magnesium oxide does not slake readily, so the hydroxide is best prepared by treating a solution of a magnesium salt with an alkali hydroxide. The hydroxide is slightly soluble in water, but readily soluble in solutions of ammonium salts because of the acidity of the NH_4^+.

$$Mg(OH)_2 + 2NH_4^+ \longrightarrow Mg^{2+} + 2NH_3 + 2H_2O$$

A suspension of magnesium hydroxide in water is called **milk of magnesia** and is used as a medicine to correct hyperacidity of the stomach.

38.18   Other Magnesium Compounds

The **normal carbonate,** $MgCO_3$, is found in nature as the mineral magnesite. A **basic carbonate** of approximately the composition $3MgCO_3 \cdot Mg(OH)_2 \cdot 3H_2O$ is produced by precipitation. This compound is known commercially as **magnesia alba,** and is used as a dental abrasive, as a medicine, as a cosmetic, and as a silver polish. **Magnesium hydrogen carbonate,** $Mg(HCO_3)_2$, is a constituent of many hard waters.

Magnesium chloride crystallizes from aqueous solutions as the hydrated salt, $MgCl_2 \cdot 6H_2O$. The hexahydrate is deliquescent, becoming moist in damp air.

When heated, the salt undergoes hydrolysis and forms the oxide, hydrogen chloride, and water.

$$MgCl_2 \cdot 6H_2O \longrightarrow MgO + 2HCl\uparrow + 5H_2O$$

The anhydrous salt may be obtained either by heating the hydrate in a current of hydrogen chloride, by burning magnesium in chloride, or by carefully heating the double salt $NH_4Cl \cdot MgCl_2 \cdot 6H_2O$, in which case the water of crystallization is driven off first and then the ammonium chloride is volatilized off.

Magnesium sulfate is found as the minerals **kieserite**, $MgSO_4 \cdot H_2O$, and **epsomite**, $MgSO_4 \cdot 7H_2O$. The heptahydrate in pure form is familiar as **Epsom salts.** It is used in medicine as a purgative, particularly in veterinary practice, in weighting cotton and silk, in powders for polishing, and in heat insulation.

Magnesium ammonium phosphate, $MgNH_4PO_4$, is a slightly soluble crystalline salt, which is formed whenever a soluble phosphate is added to a solution containing magnesium, ammonium, and hydroxide ions. This white substance is used in the detection and estimation of either magnesium or phosphate ions in analytical chemistry.

Anhydrous magnesium perchlorate, $Mg(ClO_4)_2$, is a highly efficient drying agent, called **Anhydrone.** It rapidly absorbs up to 35 per cent of its weight of water. The anhydrous salt is easily regenerated by heating the hydrate.

Several silicates of magnesium are of commercial importance. The mineral known as **talc,** or **soapstone,** is a hydrated magnesium silicate which feels greasy to the touch. It can be sawed and turned on a lathe, so is useful in the fabrication of tables, sinks, switchboards, and window sills. **Asbestos** is a calcium-magnesium silicate with a fibrous structure (Section 29.17), from which incombustible fabrics of considerable strength and durability can be made. Such materials are used in making automobile brake linings, paper, drop curtains for theaters, cardboard, flooring, roofing, and covering for heating pipes and boilers.

38.19   Properties of Magnesium of Analytical Importance

The alkali and alkaline earth hydroxides precipitate the magnesium ion as magnesium hydroxide, which is white and gelatinous. With aqueous ammonia alone, magnesium is partially precipitated as the hydroxide. If enough ammonium ions are present, no precipitation is effected.

The alkali metal carbonates precipitate basic magnesium carbonate of indefinite composition. Ammonium carbonate in the presence of other ammonium salts does not precipitate magnesium ions at all.

In the presence of ammonium and hydroxide ions, soluble phosphates precipitate magnesium as the white, crystalline **magnesium ammonium phosphate,** which is readily soluble in acids.

$$Mg^{2+} + NH_4^+ + HPO_4^{2-} \rightleftharpoons \underline{MgNH_4PO_4} + H^+$$

This compound serves as the basis for the qualitative and quantitative determination of magnesium.

The Ammonium Ion and Its Salts

38.20   The Ammonium Ion

The ammonium ion in general behaves chemically like the ions of the alkali metals, and in particular like the potassium ion, the two ions being of nearly the same size and having the same charge. The resemblance of ammonium and potassium salts is especially noticeable with respect to the formation of slightly soluble salts. There are two notable exceptions to the similarity between the salts of the ammonium and potassium ions: (1) The ammonium ion undergoes hydrolysis whereas the potassium ion does not,

$$NH_4^+ + H_2O \rightleftharpoons NH_3 + H_3O^+$$

and (2) ammonium salts decompose when heated whereas the potassium salts melt.

$$42{,}070 \text{ cal} + NH_4Cl(s) \longrightarrow NH_3(g) + HCl(g) \qquad (\Delta H^\circ_{298} = 42{,}070 \text{ cal})$$

$$NH_4NO_3(s) \longrightarrow N_2O(g) + 2H_2O(g) + 8{,}610 \text{ cal}$$
$$(\Delta H^\circ_{298} = -8{,}610 \text{ cal})$$

$$NH_4NO_2(s) \longrightarrow N_2(g) + 2H_2O(g) + 54{,}300 \text{ cal}$$
$$(\Delta H^\circ_{298} = -54{,}300 \text{ cal})$$

$$(NH_4)_2Cr_2O_7(s) \longrightarrow N_2(g) + 4H_2O(g) + Cr_2O_3(s) + 71{,}800 \text{ cal}$$
$$(\Delta H^\circ_{298} = -71{,}800 \text{ cal})$$

The ammonium group, NH_4, thus far has not been isolated as a neutral species; it is always found as a positively charged ion, NH_4^+, in combination with a negative ion. When attempts have been made to isolate the neutral NH_4 unit, ammonia and hydrogen have always resulted. When ammonium ions are electrolytically reduced at a mercury cathode, a voluminous, semisolid amalgam, $NH_4(Hg)_n$, is produced.

$$NH_4^+ + nHg + e^- \longrightarrow NH_4(Hg)_n \quad (\text{Cathodic reduction})$$

Ammonium amalgam can also be obtained by stirring a concentrated solution of ammonium chloride into sodium amalgam.

$$NH_4Cl + Na(Hg)_n \longrightarrow NaCl + NH_4(Hg)_n$$

Ammonium amalgam is quite unstable and soon decomposes with the formation of mercury, ammonia, and hydrogen. The fact that ammonium amalgams can be formed indicates that the neutral ammonium group is somewhat metallic in character.

38.21   Ammonium Chloride

Ammonium chloride, NH_4Cl, is known in commerce as **sal ammoniac.** It is produced by the reaction of ammonia with hydrochloric acid, and is purified by sublimation. Ammonium chloride decomposes at 350° into ammonia and hydrogen chloride.

$$42{,}070 \text{ cal} + NH_4Cl(s) \rightleftharpoons NH_3(g) + HCl(g) \quad (\Delta H^{\circ}_{298} = 42{,}070 \text{ cal})$$

For this reason it is used as a flux in soldering, because the hydrogen chloride which is formed when the salt is heated reacts with the films of metal oxides, converting them into chlorides which are either fusible or volatile, thus cleansing the metal surfaces to be joined by the solder. Ammonia and hydrogen chloride gases in the air of chemical laboratories combine to form the familiar white deposits of ammonium chloride so commonly seen on laboratory glassware and window-panes.

Much of the ammonium chloride produced is used in the manufacture of dry cells (Section 22.16). It is also used in medicine, in dyeing, in calico printing, as a laboratory reagent, and as a fertilizer.

38.22 Ammonium Nitrate

Ammonia reacts with nitric acid to form **ammonium nitrate**, NH_4NO_3, a white crystalline salt. Its production in the United States in 1974 equalled that of sodium carbonate, 7,550,000 tons, ranking them tenth among all chemicals. When heated to 166° ammonium nitrate fuses and decomposes smoothly with the formation of nitrous oxide and water. When detonated, sometimes simply by heating, it decomposes with explosive violence. The enormous explosion which destroyed a large part of Texas City and took 576 lives in April, 1947, was due to the decomposition of ammonium nitrate that was being loaded onto a ship in the harbor. Ammonium nitrate has been used for many years as an ingredient of explosives for warfare. Its principal use, however, is as a fertilizer.

38.23 Ammonium Carbonates

Ammonium hydrogen carbonate, NH_4HCO_3, is prepared by evaporating a solution made by treating an aqueous solution of ammonia with an excess of carbon dioxide. When heated the white crystalline salt decomposes rapidly.

$$NH_4HCO_3 \rightleftharpoons NH_3 + H_2O + CO_2$$

Even at ordinary temperatures, ammonium hydrogen carbonate has a faint odor of ammonia. By treating a solution of the hydrogen carbonate with an excess of ammonia, the **normal carbonate** results.

$$HCO_3^- + NH_3 \rightleftharpoons NH_4^+ + CO_3^{2-}$$

When exposed to the air, $(NH_4)_2CO_3$ gives off ammonia more readily than does NH_4HCO_3 and thus is useful in the form of "smelling salts," since ammonia is an effective heart stimulant.

38.24 Other Ammonium Salts

Large quantities of **ammonium sulfate**, $(NH_4)_2SO_4$, are produced when ammonia is absorbed in sulfuric acid. Its chief use is as a fertilizer to supply nitrogen to the soil.

When an aqueous solution of ammonia is saturated with hydrogen sulfide, **ammonium hydrogen sulfide,** NH_4HS, is formed. If the resulting solution is then treated with an equivalent amount of the aqueous ammonia, the **normal sulfide,** $(NH_4)_2S$, is formed.

38.25   Properties of the Ammonium Ion of Analytical Importance

Most ammonium salts are soluble in water. The slightly soluble salts that may be employed in the detection of the ammonium ion are **ammonium hexanitrocobaltate(III),** $(NH_4)_3[Co(NO_2)_6]$; **ammonium hydrogen tartrate,** $NH_4HC_4H_4O_6$; and **ammonium hexachloroplatinate(IV),** $(NH_4)_2[PtCl_6]$. Since potassium forms similar insoluble compounds, ammonium ions must be removed from solution prior to the precipitation of potassium salts. Unlike potassium, the ammonium ion does not give a precipitate with perchloric acid.

The alkali hydroxides and carbonates and the alkaline earth hydroxides liberate ammonia from solutions of the ammonium ion in the cold, and more rapidly upon heating.

QUESTIONS

1. Why do the alkali metals not occur in nature?
2. Correlate the positive oxidation states of the alkali metals with their atomic structures.
3. Why should metallic sodium never be handled with the fingers?
4. Why is sodium carbonate or calcium chloride added to the electrolyte in the electrolysis of fused sodium chlorides?
5. By analogy with its reaction with water, suggest a chemical formula for the product of reaction between sodium metal and ethyl alcohol.
6. Why must the chlorine and sodium hydroxide be kept separate in the manufacture of sodium hydroxide by the electrolysis of aqueous sodium chloride?
7. Outline the chemistry of the Solvay process for the production of sodium carbonate.
8. Why cannot potassium hydrogen carbonate be manufactured by the Solvay process?
9. Compare the solubilities of potassium and ammonium salts.
10. Give two examples of slightly soluble sodium salts.
11. What is the principal reaction involved when baking powder acts as a leavening agent?
12. What evidence is available to show that hydrogen is present as the negative hydride ion in sodium hydride?
13. Cite evidence that the hydride and peroxide ions are strongly basic.
14. Outline the extraction of magnesium from seawater.
15. Magnesium is an active metal; it is burned in the form of ribbons and filaments to provide flashes of brilliant light. Why is it possible to use magnesium in construction and even for fabrication of cooking grills?

16. Why cannot a magnesium fire be extinguished by either water or carbon dioxide? Suggest a method of extinguishing such a fire.
17. Write a chemical equation describing the dehydrating activity of magnesium perchlorate.
18. Explain the dissolution of magnesium hydroxide in solutions of ammonium salts.
19. Why cannot $MgCl_2 \cdot 6H_2O$ be dehydrated by heating in air? How may it be dehydrated?
20. Why is a consideration of the chemistry of the ammonium ion and its salts introduced along with a discussion of the alkali metals and their ions?
21. What evidence is there that the neutral ammonium molecule is metallic in character?
22. When potassium ion is reduced, potassium metal results. What is (are) the product(s) when ammonium ion is reduced?
23. Write equations for the thermal decomposition of the following salts: NH_4Cl, NH_4NO_3, NH_4NO_2, and $(NH_4)_2Cr_2O_7$.
24. Suggest in terms of oxidation states of nitrogen why ammonium nitrate might explode.
25. Write chemical equations describing the conversion of ammonium ion to nitride ion; of nitride ion to ammonium ion.

PROBLEMS

1. What volume of H_2 gas at 740 mm pressure and 27° C will be produced by treating with excess water an amount of sodium which produces a solution that is neutralized by 24.50 ml of 0.100 M HCl?   *Ans. 31.0 ml*
2. How much anhydrous sodium carbonate contains the same number of moles of sodium as 100 grams of $Na_2CO_3 \cdot 10H_2O$?   *Ans. 37.0 g*
3. A 10.00-ml sample of KOH solution is exactly neutralized with 14.85 ml of 0.1009 M HCl. What is the concentration of the KOH solution?
        Ans. 0.1498 M
4. Calculate the pH of a solution of 0.30 M NH_3 in water.   *Ans. 11.37*
5. What volume of lithium would have the same mass as 10.0 cm³ of gold?
        Ans. 361 cm³
6. When $MgNH_4PO_4$ is heated to 1000° C it is converted to $Mg_2P_2O_7$. A 1.203-g sample containing magnesium yielded 0.5275 g of $Mg_2P_2O_7$ after precipitation of $MgNH_4PO_4$ and heating. What per cent magnesium was present in the original sample?   *Ans. 9.577%*
7. Calculate the per cent magnesium in asbestos.   *Ans. 26.31%*

REFERENCES

Discovery of the Elements, M. E. Weeks and H. M. Leicester, Publ. by the Journal of Chemical Education, Easton, Pa., Seventh Edition, 1968; pp. 433-457 and 495-502.
"Lithium," H. Gilman and J. J. Eisch, *Sci. American*, Jan., 1963; p. 89.
"The Social Influence of Salt," M. R. Bloch, *Sci. American*, July, 1963; p. 89.

"Alkali Metal–Water Reactions," M. M. Markowitz, *J. Chem. Educ.*, **40,** 633 (1963).

"Alkali Metal Nitrides," G. L. Moody and J. D. R. Thomas, *J. Chem. Educ.*, **43,** 205 (1966).

"X-Ray Crystallography Experiment: Powder Patterns for Alkali Halides," F. P. Boer and T. H. Jordan, *J. Chem. Educ.*, **42,** 76 (1965).

"Chemical Bonds—X Rays and the Alkali Metal Chlorides," N. Booth, *Educ., in Chemistry*, **1,** 66 (1964).

"Hydrolysis of Sodium Carbonate," F. S. Nakayama, *J. Chem. Educ.*, **47,** 67 (1970).

"Industrial Emergence of Sodium and Aluminum," M. Schofield, *Chem. in Britain*, **4,** 98 (1968).

"Lithium and Mental Health," M. T. Doig(III), M. G. Heyl, and D. F. Martin, *J. Chem. Educ.*, **50,** 343 (1973).

"Life's Essential Elements," D. R. Williams, *Educ. in Chemistry*, **10,** 56 (1973).

"The Absorption Spectrum of Sodium Vapor," R. A. Ashby and H. W. Gotthard, *J. Chem. Educ.*, **51,** 408 (1974).

"How Productive is the Sea?" I. Morris, *Chem. in Britain*, **10,** 198 (1974).

"Salt Solution" (Ocean as a source of drugs, food, salt, and water), K. Jacques, *Chem. in Britain*, **11,** 12 (1975).

SEMIMICRO QUALITATIVE ANALYSIS

General Laboratory Directions

39

Your course in Qualitative Analysis has two principal objectives. One of these is to give you the reasons for the analytical procedures and results in terms of the theory of ionic equilibria, especially that relating to weak electrolytes, solubility products, complex ions, and oxidation-reduction. The other is on the practical side and is for the purpose of teaching you careful laboratory manipulation, critical observation, and logical interpretation of observed results.

39.1 Classification of the Metals into Analytical Groups

Qualitative analysis pertains to the identification of the constituents present in a sample of a substance, a mixture of substances, or a solution. In the qualitative analysis of a solution which may contain any or all of the common metal ions, the first step is that of separating the ions of the metals into several groups, each of which contains ions exhibiting a common chemical property which is the basis of the separation. The separation of the common metal ions into groups as it is usually done is outlined briefly below:

■ **The Metals of Analytical Group I.** When dilute hydrochloric acid is added to a solution containing all of the common metal ions (and NH_4^+), mercury(I) chloride, silver chloride, and lead chloride precipitate. The chlorides of all the other common metal ions are soluble in this acid solution and can be separated from those of Group I by filtration or centrifugation.

■ **The Metals of Analytical Group II.** After the Group I chlorides have been separated, the solution is made 0.3 molar in hydrochloric acid, and the Group II metals are precipitated as sulfides upon the addition of hydrogen sulfide to the solution. The precipitate formed consists of the sulfides of lead, bismuth, copper, cadmium, mercury(II), arsenic, antimony, and tin.

■ **The Metals of Analytical Group III.** After the Group II sulfides have been separated, the solution is saturated with hydrogen sulfide and then an excess of aqueous ammonia is added to the solution. Under these conditions the sulfides of cobalt, nickel, manganese, iron, and zinc, and the hydroxides of aluminum and chromium, are precipitated.

■ **The Metals of Analytical Group IV.** The Group IV metals, barium, strontium, and calcium, are precipitated as the carbonates from the filtrate or centrifugate of the Group III separation by ammonium carbonate in the presence of aqueous ammonia and ammonium chloride.

■ **The Metals of Analytical Group V.** The filtrate from the Group IV separation contains the ions of sodium, potassium, magnesium, and ammonium, which constitute Group V.

A flow sheet illustrating the schematic separations of the metal ions into the various analytical groups is given in Table 39–1. Chapters 34 through 38 are devoted to a consideration of the descriptive chemistry of the metals as they are grouped according to the analytical scheme. Chapters 40 through 44 cover the analysis procedures for the metals. Hence, Chapters 34 and 40 are both concerned with Group I of the analysis scheme; Chapters 35 and 41, Group II; Chapters 36 and 42, Group III; Chapters 37 and 43, Group IV, and Chapters 38 and 44,

TABLE 39–1   Flow Sheet of Group Separations

The flow sheet proceeds as follows.

Initial solution (with HCl):

Hg_2^{2+}, Ag^+, Pb^{2+}, Bi^{3+}, Cu^{2+}, Cd^{2+}, Hg^{2+}, As^{3+}, Sb^{3+}, Sn^{4+}, Co^{2+}, Ni^{2+}, Mn^{2+}, Fe^{3+}, Al^{3+}, Cr^{3+}, Zn^{2+}, Ba^{2+}, Sr^{2+}, Ca^{2+}, Mg^{2+}, NH_4^+, Na^+, K^+

Group I — Chlorides: $\overline{Hg_2Cl_2}$, $\overline{AgCl}$, $\overline{PbCl_2}$

After HCl, remaining (treated with $0.3\,M\ HCl$, H_2S):

Pb^{2+}, Bi^{3+}, Cu^{2+}, Cd^{2+}, Hg^{2+}, As^{3+}, Sb^{3+}, Sn^{4+}, Co^{2+}, Ni^{2+}, Mn^{2+}, Fe^{3+}, Al^{3+}, Cr^{3+}, Zn^{2+}, Ba^{2+}, Sr^{2+}, Ca^{2+}, Mg^{2+}, NH_4^+, Na^+, K^+

Group II — Sulfides: $\overline{PbS}$, $\overline{Bi_2S_3}$, $\overline{CuS}$, $\overline{CdS}$, $\overline{HgS}$, $\overline{As_2S_3}$, $\overline{Sb_2S_3}$, $\overline{SnS_2}$

Remaining (treated with NH_4Cl, $NH_3 + H_2O$, H_2S):

Co^{2+}, Ni^{2+}, Mn^{2+}, Fe^{2+}, Al^{3+}, Cr^{3+}, Zn^{2+}, Ba^{2+}, Sr^{2+}, Ca^{2+}, Mg^{2+}, NH_4^+, Na^+, K^+

Group III — Sulfides and Hydroxides: $\overline{CoS}$, $\overline{NiS}$, $\overline{MnS}$, $\overline{FeS}$, $\overline{Al(OH)_3}$, $\overline{Cr(OH)_3}$, $\overline{ZnS}$

Remaining (treated with $NH_3 + H_2O$, NH_4Cl, $(NH_4)_2CO_3$):

Ba^{2+}, Sr^{2+}, Ca^{2+}, Mg^{2+}, NH_4^+, Na^+, K^+

Group IV — Carbonates: $\overline{BaCO_3}$, $\overline{SrCO_3}$, $\overline{CaCO_3}$

Group V — Soluble Group: Mg^{2+}, NH_4^+, Na^+, K^+

Group V. The student will find it helpful to study corresponding descriptive and analytical chapters together for each analysis group.

39.2 Semimicro Qualitative Analysis

Semimicro qualitative analysis is a method of analysis employing techniques whereby the reactions and procedures used in macro work may be reliably carried out on a reduced scale. Analysis on the macro scale is made on volumes of solutions of the order of 10 to 100 ml and with ordinary test tubes, beakers, and funnels. In semimicro analysis, volumes of solutions from 1 drop to about 5 ml are employed, and small test tubes, centrifuge tubes, capillary syringes, and medicine droppers are used to carry out the separations and identification tests. This leads to a striking reduction in the consumption of reagents and usually the analyses can be carried out more rapidly than by macro methods.

Even though rather definite directions are given for the analysis to be carried out, no two analyses will be exactly alike. For this reason, directions should never be followed to the letter; but with careful thought, procedures should be adapted to the particular problem at hand.

39.3 Equipment

In Appendices M and N will be found a list of apparatus and of solutions and solid reagents which each student will need in carrying out the laboratory work of this course. After a laboratory desk has been assigned, check the apparatus in the desk according to the specific directions given by your instructor. Get from the stockroom any apparatus which may be required but is not in the desk. Wash all of the apparatus that will be employed in the course before beginning your laboratory work.

39.4 Laboratory Assignments

Below is listed a set of suggested laboratory assignments. The number of unknowns required will be specified by the instructor.

(1) Construct a wash bottle (if necessary), stirring rods, and capillary syringes.
(2) Analyze a known solution containing all the cations of Group I.
(3) Analyze an unknown solution based on the Group I cations.
(4) Analyze a known solution containing all the cations of Group II.
(5) Analyze an unknown solution based on the Group II cations.
(6) Analyze a known solution containing all the cations of Group III.
(7) Analyze an unknown solution based on the Group III cations.
(8) Analyze a known solution containing all the cations of Group IV.
(9) Analyze a known solution containing all the cations of Group V.
(10) Analyze an unknown solution based on all the cations of Groups IV and V.
(11) Analyze a general unknown solution based on all the cations of all the groups.
(12) Analyze a known salt mixture containing no oxidizing anions.

(13) Analyze an unknown salt mixture containing no oxidizing anions.

(14) Analyze a known salt mixture containing no reducing anions.

(15) Analyze an unknown salt mixture containing no reducing anions.

(16) Analyze a salt mixture for both cations and anions.

(17) Analyze an alloy.

39.5   Wash Bottle

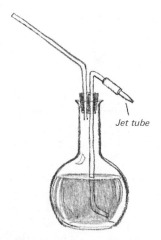

FIGURE 39-1
A laboratory constructed wash bottle.

Jet tube

Wash bottles made of plastic are supplied to students in some laboratories. Alternatively, you may use the following procedure to construct one of glass.

Using a 250-ml Florence flask and 6 mm glass tubing, construct a wash bottle of the type shown in Fig. 39–1. Make uniform bends with no constrictions, and fire-polish all ends of glass tubing. Keep the wash bottle filled with distilled water for use in the analytical procedures. Ordinary tap water contains such ions as Ca^{2+}, Mg^{2+}, Fe^{3+}, Al^{3+}, Cl^-, SO_4^{2-}, and HCO_3^-. Since these ions are among those to be tested for in the unknown solutions, all water used in the procedures and in the cleaning of glassware must be distilled.

39.6   Stirring Rods

Make at least five glass stirring rods, approximately 15 cm in length and 3 mm in diameter. Fire-polish each end of each rod in the flame.

39.7   Capillary Syringes

Standard medicine droppers of approximately 1 ml capacity are used for measuring and transferring solutions of reagents. These droppers deliver about 20 drops per milliliter. A second type of dropper, called a capillary syringe, is needed for the removal of liquids from precipitates held in small test tubes or centrifuge tubes. These may be supplied. If not available, the capillary syringes may be made from glass tubing as follows: Heat the middle portion of a 7-inch section of 8-mm glass tubing over a Bunsen flame with rotation until the glass softens. Remove the tube from the flame and slowly draw it out until the bore is about 1 mm. When the tube has cooled, cut the capillary at the mid-point and fire-polish the capillary ends. Flare the wide ends of the tubes by heating until soft and quickly pressing down against a flat metal surface. When the syringes are cold, attach medicine dropper bulbs to the flared ends. These syringes will deliver approximately 40 drops per ml.

39.8   Reagents

The solids and solutions which are called for in the analytical procedures will be stored in the 10-ml reagent bottles (Fig. 39–2) to be found in your desk. Fill these reagent bottles with the chemicals required in the analysis of each group prior to starting the analysis of that group. See Appendix N for a list of these reagents. Only a small quantity of the starred reagents will be needed during the course, so fill the bottles only about one-fourth full of these reagents.

Before filling a reagent bottle, make sure that it is perfectly clean. To avoid mistakes, it is well to label each bottle before filling it.

39.9 Precipitation

Practically all of the precipitations are carried out either in 4-ml Pyrex test tubes or 2-ml conical test tubes. Check for completeness of precipitation by adding a drop of reagent to the solution (centrifugate) obtained in the separation of the precipitate. If the addition of more reagent to the solution shows that precipitation is incomplete, separate the mixture and test the second solution for completeness of precipitation.

The precipitating agent should be added slowly, preferably from a medicine dropper, and with vigorous shaking or stirring of the reaction mixture. The formation of larger crystals of the precipitate is favored by warming the solution, and separation of the precipitate should not be attempted before the crystals become large enough to settle.

A slight excess of the precipitating agent is added to reduce the solubility of the precipitate by the common ion effect (Section 18.6). On the other hand, a very large excess of the precipitating agent should be avoided, as it may actually increase the solubility of the precipitate. For example, in precipitating silver chloride a large excess of Cl^- will bring about the formation of $AgCl_2^-$, and thereby increase the solubility of $AgCl$. Many precipitates are dissolved, at least partially, by the formation of complexes of this type.

39.10 Centrifugation of Precipitates

A precipitate may be separated from a liquid by means of a centrifuge (Fig. 39–3). By rotation of a mixture of solid and liquid at high speed in a centrifuge, the more dense precipitate is forced to the bottom of the containing tube by a centrifugal force which is many times the force of gravity. This accounts for the much shorter time required for settling of the precipitate when centrifugation is employed. Colloidal precipitates require longer centrifugation than do crystalline precipitates because of the small size of the colloidal particles.

Before centrifuging a precipitate contained in a test tube or centrifuge tube, prepare another tube to balance it in the centrifuge by filling an empty tube with water until the liquid levels in both tubes are the same. Insert the tubes in opposite positions in the centrifuge, and set the machine in motion. Allow the centrifugation to continue for 30 seconds. After the machine has come to rest, remove the tubes.

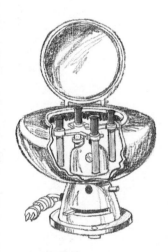

39.11 Transfer of the Centrifugate

After centrifugation, the precipitate should be found packed in the bottom of the tube. The supernatant liquid or centrifugate is separated from the precipitate by holding the tube at an angle of about 30 degrees (Fig. 39–4) and removing the supernatant liquid by slowly drawing it into a capillary syringe. The tip of the syringe is held just below the surface of the liquid and as the pressure on the bulb is slowly released, causing the liquid to rise in the syringe, the capillary is lowered into the tube until all of the liquid is removed. As the capillary approaches the bottom of the tube, the tip must not be allowed to stir up the mixture by touching the precipitate.

FIGURE 39-4
Capillary syringe is used to separate centrifugate from precipitate.

39.12   Washing of the Precipitate

The precipitate left in the tube after the removal of a supernatant liquid is still wet with a solution containing the ions of this liquid. The precipitate must be washed, usually with water, to dilute the solution adhering to the precipitate. The wash liquid is added to the precipitate, and the mixture is stirred thoroughly. The mixture is then centrifuged to cause the precipitate to settle again. After centrifugation, the washings are removed by a capillary syringe as previously described. Usually a precipitate is washed at least twice. The first wash liquid is ordinarily saved and added to the first centrifugate. If the precipitate must be transferred to another container, the reagent to be used is added, the mixture is well stirred, and then it is poured into the other container. After the precipitate has settled, the supernatant liquid may be employed to remove any precipitate remaining in the centrifuge tube.

39.13   Dissolution and Extraction of Precipitates

When all or a part of a precipitate is to be brought into solution by a reagent, the solvent is added to the precipitate which is in the centrifuge tube and the mixture is stirred. The mixture is then separated by centrifugation and the operation is repeated using fresh solvent. Oftentimes the extraction of a precipitate is more efficient at an elevated temperature.

39.14   Heating of Mixtures or Solutions

FIGURE 39-5
Individual water bath for heating reacting mixtures.

Whenever it is necessary to heat a mixture for the purpose of bringing about a precipitation or for dissolving or extracting a precipitate, the test tube or centrifuge tube is placed in a water bath (Fig. 39–5) maintained at a suitable temperature. It will be found convenient to keep the water hot in the water bath throughout the work period.

39.15   Evaporation

It is often necessary to heat solutions to boiling and to hold them at the boiling temperature in order to concentrate them, or to remove volatile acids or bases, or even to evaporate a solution to dryness. Evaporation should be carried out in a small casserole or porcelain evaporating dish. The contents of the container should be agitated constantly while the heating continues. The evaporation of solutions contained in small test tubes should be avoided because the contents of the tube may be lost due to overheating.

39.16   Cleaning Glassware

Because small amounts of contaminants may give rise to erroneous results, all glassware used in the analytical procedures should be thoroughly cleaned before it is used. The cleaning should be done with a brush and some cleansing powder such as a synthetic detergent. The apparatus should then be rinsed first with tap water and finally with distilled water. Test tube brushes and centrifuge tube brushes are available. Medicine droppers, capillary syringes, and stirring rods should be cleaned, rinsed, and stored in a beaker of distilled water.

39.17   Flame Tests

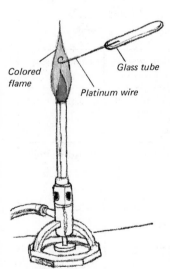

Colored flame

Glass tube

Platinum wire

Flame tests are made using a platinum wire sealed in the end of a piece of glass tubing. The wire is cleaned by first dipping the end of it in 6 M HCl and then heating it in the burner flame (Fig. 39–6). Rather than use the hottest portion of the flame, one should bring the looped end of the wire slowly up to the edge of the flame. The platinum wire should never come into the reducing part of the flame. Why? The operation of dipping in acid and heating in the flame should be repeated until the wire no longer imparts a color to the flame. The wire loop is dipped in the solution to be tested and then heated in the flame. The memory should not be relied upon in judging the color imparted to the flame; a known solution of the ion should be tested and its color compared to that given by the unknown.

FIGURE 39-6

Platinum wire mounted in glass tubing used in making the flame test for unknown ions.

39.18   Known Solutions

In order that you may know the details of the analytical procedures before attempting to analyze a solution of unknown composition, known solutions containing all the ions of a given group are provided. You should become familiar with the separations and confirmatory tests for the ions of the group by practicing on a known solution before trying to determine the ions present in an unknown solution. You should especially note the quantities of precipitates obtained and the colors of precipitates and solutions as you proceed with the analysis of a known solution. These observations and the equations for the reactions involved should be recorded in a notebook.

39.19   Unknown Solutions

When the analysis of a known solution has been satisfactorily completed, the student should prepare for an oral quiz by the instructor on the details of the analysis of the group. This quiz may include questions on the reactions involved, the theory of the separations and confirmatory tests, and the colors of precipitates, ions, and solutions. After the instructor's questions have been satisfactorily answered, a sample of an unknown may be obtained for analysis. The unknown may contain any or all of the ions of the group or groups being studied.

39.20   Unknown Report

A notebook should be obtained (the type to be designated by your instructor) in which to record the results of the analyses which are performed. All observations which are made should be recorded as soon as they have been completed. Equations should be written in the notebook to indicate the behavior of each ion with the reagents with which it comes in contact.

In reporting the results of the analysis of an unknown, fill out a report sheet in a manner similar to that shown below and present it to the instructor for grading.

SAMPLE UNKNOWN REPORT

Name: _____ Section: _____ Date: _____

Instructor's Approval: _____ Ions Found: _____ Grade: _____

CATION UNKNOWN REPORT

No.	Substance	Reagent	Result	Inference or Conclusion	Precipitate or Residue	Centrifugate or Solution
1	Group I	HCl	White ppt.	Group I present	One or more of AgCl, $PbCl_2$, Hg_2Cl_2	Possibly Pb^{2+}
2	Ppt. from 1	Hot water	No visible action	Hg_2^{2+} and/or Ag^+ present; Pb^{2+} uncertain	Hg_2Cl_2 and/or AgCl	Possibly Pb^{2+}
3	Filtrate from 2 or 1	K_2CrO_4	Yellow ppt.	Pb^{2+} present	$PbCrO_4$	
4	Residue from 2	$NH_3 + H_2O$	Residue dissolves completely	Hg_2^{2+} absent; Ag^+ probable		$[Ag(NH_3)_2]^+$
5	Filtrate from 4	HNO_3	White ppt.	Ag^+ present	AgCl	
6						

The Analysis of Group I

40

40.1 Introduction

Begin with these procedures for the analysis of a solution of a known or unknown sample containing the metal ions. Depending upon the nature of the known or unknown the solution may contain only the ions of Group I or it may contain the ions of any or all of the Groups I–V. If your sample is a general unknown, reserve a part of the original solution for the test for the ammonium ion in Group V. Review Section 39.1 for the classification of the metal ions into analytical groups before proceeding with the analysis of Group I. A schematic outline of the analysis for the Group I metals is given in the following flow sheet.

The flow sheet summarizes the chemistry of the separations and the identification tests in a form easily studied and remembered. Flow sheets similar to the one above are provided in this text to assist the students in the analysis of metals in Groups II–V.

Group I Flow Sheet

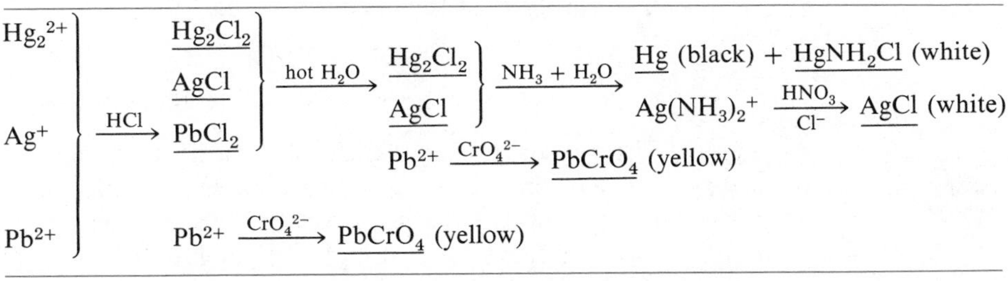

$$\left.\begin{array}{l} Hg_2^{2+} \\ \\ Ag^+ \\ \\ Pb^{2+} \end{array}\right\} \xrightarrow{HCl} \left.\begin{array}{l} Hg_2Cl_2 \\ AgCl \\ PbCl_2 \end{array}\right. \xrightarrow{hot\ H_2O} \left.\begin{array}{l} Hg_2Cl_2 \\ AgCl \end{array}\right\} \xrightarrow{NH_3+H_2O} \begin{array}{l} Hg\ (black) + HgNH_2Cl\ (white) \\ Ag(NH_3)_2^+ \xrightarrow{HNO_3 \atop Cl^-} AgCl\ (white) \end{array}$$

$$Pb^{2+} \xrightarrow{CrO_4^{2-}} PbCrO_4\ (yellow)$$

$$Pb^{2+} \xrightarrow{CrO_4^{2-}} PbCrO_4\ (yellow)$$

967

A discussion of the chemistry involved in the various steps of the analysis of each group will be given just prior to the laboratory procedure for each step. Careful study of these discussions will help the student learn the chemistry involved in the analysis. Additional details concerning the chemistry of the metals of the analytical groups are given in Chapters 34–38. Chapter 34 is concerned with the metals of Analytical Group I.

40.2   Precipitation of Group I

Hydrochloric acid is the reagent which serves as the precipitant for the metal ions of Group I. The chlorides of mercury(I), silver, and lead are the only ones of the cations under consideration in this qualitative scheme which are insoluble in acid solutions. The equations for the precipitation of the metal ions of Group I are

$$Hg_2^{2+} + 2Cl^- \rightleftharpoons Hg_2Cl_2$$
$$Ag^+ + Cl^- \rightleftharpoons AgCl$$
$$Pb^{2+} + 2Cl^- \rightleftharpoons PbCl_2$$

The solubilities and solubility product constants of the chlorides of this group are given in Table 40-1. These values show that the solubility of lead chloride

TABLE 40-1   Solubilities of the Group I Chlorides

Salt	K_{sp}	Solubility moles/liter	Solubility g/ml
Hg_2Cl_2	1.1×10^{-18}	6.5×10^{-7}	3.1×10^{-7}
$AgCl$	1.8×10^{-10}	1.3×10^{-5}	1.9×10^{-6}
$PbCl_2$	1.7×10^{-5}	1.6×10^{-2}	4.5×10^{-3}

in g/ml is about 2,800 times greater than that of silver chloride and about 14,500 times greater than that of mercury(I) chloride. Thus, it can be seen that the mercury(I) ion and the silver ion are in effect completely removed from solution when hydrochloric acid is added to a solution containing these ions. On the other hand, lead is not completely precipitated by chloride, and some of it is carried through to Group II.

A slight excess of hydrochloric acid is used in the precipitation of the group to prevent the precipitation of bismuth oxychloride and antimony oxychloride by hydrolysis of their chloride salts.

$$Bi^{3+} + Cl^- + H_2O \rightleftharpoons BiOCl + 2H^+$$
$$Sb^{3+} + Cl^- + H_2O \rightleftharpoons SbOCl + 2H^+$$

The hydrogen ions of the hydrochloric acid keep the above equilibria shifted to the left. In addition, a slight excess of chloride ions insures more complete precipitation of Hg_2Cl_2, $AgCl$, $PbCl_2$, which is due to the common ion effect (Section 18.6). For example, in a solution which is 0.30 M in chloride ions, the common ion effect will reduce the concentration of Ag^+ from 1.3×10^{-5} M in a saturated solution of AgCl to 6.0×10^{-10} M. On the other hand, a large excess of chloride ions must be avoided because silver chloride and lead chloride tend to react with chloride ions, forming soluble salts of complex ions. The equations for these reactions are

$$AgCl + Cl^- \rightleftharpoons [AgCl_2]^-$$
$$PbCl_2 + 2Cl^- \rightleftharpoons [PbCl_4]^{2-}$$

PROCEDURE 1. To 10 drops of the solution to be analyzed, add enough water to make a total volume of 1 ml. Add 2 drops of 6 M HCl to this solution (avoid using a large excess of HCl or concentrated HCl, to eliminate the possibility of formation of $[AgCl_2]^-$ or $[PbCl_4]^{2-}$.) Separate the precipitate by centrifugation. If your sample is a general unknown, reserve the solution (centrifugate) for the analysis of Groups II-V. If your sample is a Group I known or unknown, test the solution for lead as directed in Procedure (2) below. Treat the precipitate according to Procedure (2).

40.3 Separation and Identification of the Lead Ion

Lead chloride may fail to precipitate at all if the lead ion concentration is too low or if the temperature is too high. For this reason a test for lead should be made on the centrifugate from the original Group I separation.

The fact that lead chloride is about three times as soluble in hot water as in cold water enables one to separate it from silver and mercury(I) chlorides by extraction of the mixed chlorides with hot water. The presence of the lead ion may be confirmed by precipitation of the slightly soluble yellow chromate.

$$Pb^{2+} + CrO_4^{2-} \rightleftharpoons PbCrO_4$$

PROCEDURE 2. *Precipitate from (1): Hg_2Cl_2, AgCl, $PbCl_2$.* Extract the precipitate twice with 10 drops of hot water to dissolve the $PbCl_2$. Reserve the residue for Procedure (3). Add 2 drops of 1 M K_2CrO_4 to the solution (hot water extract). A yellow precipitate of $PbCrO_4$ confirms the presence of lead.

40.4 Separation and Identification of the Mercury(I) Ion

When aqueous ammonia is added to a mixture of silver chloride and mercury(I) chloride, the silver chloride dissolves by forming soluble diamminesilver chloride. The equation is

$$\underline{AgCl} + 2NH_3 \rightleftharpoons [Ag(NH_3)_2]^+ + Cl^-$$

In the presence of ammonia, mercury(I) chloride undergoes auto-oxidation-reduction with the formation of finely divided metallic mercury, which is black, and mercury(II) amido chloride, $HgNH_2Cl$, which is white. The equation is

$$\underline{Hg_2Cl_2} + 2NH_3 \longrightarrow \underline{Hg} + \underline{HgNH_2Cl} + NH_4^+ + Cl^-$$

The formation of this black (or gray) precipitate is sufficient proof of the presence of mercury(I) ions.

PROCEDURE 3. *Residue from (2): Hg₂Cl₂ and AgCl.* Extract the residue twice with 5 drops of 4 *M* aqueous ammonia. A black or gray residue confirms the presence of the mercury(I) ion. Reserve the aqueous ammonia extract for Procedure (4).

40.5   Identification of the Silver Ion

The presence of the silver ion may be confirmed by treating the aqueous ammonia extract of Hg_2Cl_2 and AgCl with nitric acid. This causes the reprecipitation of white silver chloride. The hydrogen ion supplied by the nitric acid unites with the free ammonia in equilibrium with the diamminesilver ion and causes the equilibrium, $[Ag(NH_3)_2]^+ \rightleftharpoons Ag^+ + 2NH_3$, to shift to the right. The free silver ion is then precipitated by the chloride ion which is present in the solution. The net reaction is the sum of three equilibria, as shown below.

$$[Ag(NH_3)_2]^+ \rightleftharpoons Ag^+ + 2NH_3$$
$$2NH_3 + 2H^+ \rightleftharpoons 2NH_4^+$$
$$Ag^+ + Cl^- \rightleftharpoons AgCl$$
$$\overline{[Ag(NH_3)_2]^+ + Cl^- + 2H^+ \rightleftharpoons AgCl + 2NH_4^+}$$

PROCEDURE 4. *Solution from (3): [Ag(NH₃)₂]Cl.* Add 4 *M* HNO₃ until the solution is acid to litmus. Make sure the HNO₃ is stirred into the solution to neutralize all ammonia throughout, before testing with litmus. A white precipitate (or cloudiness) of AgCl confirms the presence of silver.

QUESTIONS

1. What general statement can be made concerning the solubility of common chloride salts other than those of the Analytical Group I cations?
2. Write out the flow sheet for the Group I analysis.
3. Give the color of each of the following: AgCl, $PbCl_2$, Hg_2Cl_2, $PbCrO_4$, Hg, $HgNH_2Cl$, and $[Ag(NH_3)_2]Cl$ (in solution).
4. Why must a large excess of chloride ion be avoided in the precipitation of the Group I chlorides?
5. In the case of an unknown containing only the cations of Group I, why is it advisable to make a confirmatory test for lead on the filtrate from the Group I precipitation?
6. Select a reagent used in the analysis of Group I which will in one step separate each of the following pairs. (a) Hg_2Cl_2, $PbCl_2$, (b) Hg_2Cl_2, AgCl, (c) AgCl, $CuCl_2$, (d) Hg_2^{2+}, Hg^{2+}, (e) AgCl, $PbCl_2$.
7. In terms of ionic equilibria and solubility product theory, explain the dissolution of silver chloride in aqueous ammonia and its reprecipitation with nitric acid.

8. Show that the reaction of ammonia with mercury(I) chloride is of the oxidation-reduction type.

9. Why do not the slightly soluble oxychlorides of bismuth and antimony precipitate with the Group I chlorides?

PROBLEMS

1. A solution is 0.020 M with respect to each, Pb^{2+} and Ag^+. If Cl^- is added to this solution, what is the concentration of Ag^+ when $PbCl_2$ begins to precipitate? (K_{sp} for AgCl is 1.8×10^{-10}, and K_{sp} for $PbCl_2$ is 1.7×10^{-5}).
Ans. 6.2×10^{-9} M

2. Calculate the concentration of Ag^+ in a 0.0010 M solution of $[Ag(NH_3)_2]Cl$ which is 0.25 M in aqueous ammonia. (K_d for $[Ag(NH_3)_2]^+$ is 6.3×10^{-8}.)
Ans. 1.0×10^{-9} M

3. From the K_{sp} data given in Table 40–1, calculate the number of cations present in 1.0 ml of saturated solutions of Hg_2Cl_2, AgCl, and $PbCl_2$.
Ans. 3.9×10^{14} Hg_2^{2+} ions; 8.0×10^{15} Ag^+ ions; 9.8×10^{18} Pb^{2+} ions

4. Calculate the solubility of Hg_2Cl_2 in 0.020 M NaCl. Compare this solubility with that of Hg_2Cl_2 in water. (K_{sp} for Hg_2Cl_2 is 1.1×10^{-18}.)
Ans. 2.8×10^{-15} mole/liter

5. Calculate the solubility of AgCl in 6.0 M HCl. (The formation constant of $Ag^+ + 2Cl^- \longrightarrow [AgCl_2]^-$ is 2.5×10^5, and K_{sp} of AgCl is 1.8×10^{-10}.)
Ans. 2.7×10^{-4} mole/liter

REFERENCE

"The Effects of Chloride Ion and Temperature on Lead Chloride Solubility," A. C. West, *J. Chem. Educ.*, **46**, 773 (1969).

The Analysis of Group II

41

The solution to be analyzed may be a Group II known or unknown, or it may be the solution from the Group I separation. In either case proceed according to (1) below. See page 974 for the Group II flow sheet. You may wish also to refer to Chapter 35, which is devoted to the chemistry of the metals of analytical Group II.

41.1 Precipitation of Group II Sulfides

■ **1. Separation of Group II from Group III.** Ions of lead, bismuth, copper, cadmium, mercury(II), arsenic, antimony, and tin form sulfides which are insoluble in solutions which are 0.3 M in hydrogen ion. Cadmium sulfide is the most soluble of the sulfides of Group II and zinc sulfide is the least soluble of the sulfides of Group III. Therefore, the separation of the sulfides of Group II from those of Group III is assured if the sulfide ion concentration of the solution is controlled in such a way that the cadmium is completely precipitated as the sulfide without exceeding the solubility product constant of zinc sulfide. In a solution which is saturated with hydrogen sulfide and which is 0.3 M in hydrogen ion, the sulfide ion concentration is just right to effect the separation of cadmium from zinc and, hence, Group II from Group III. The theory for these separations is given in a quantitative manner in Examples 1 and 2 of Section 19.9.

The hydrogen sulfide required for the precipitation of the Group II cations is supplied by thioacetamide, CH_3CSNH_2, which hydrolyzes in hot acidic solutions with the formation of ammonium acetate and hydrogen sulfide.

$$CH_3CSNH_2 + 2H_2O \longrightarrow CH_3COO^- + NH_4^+ + H_2S$$

GROUP II Flow Sheet

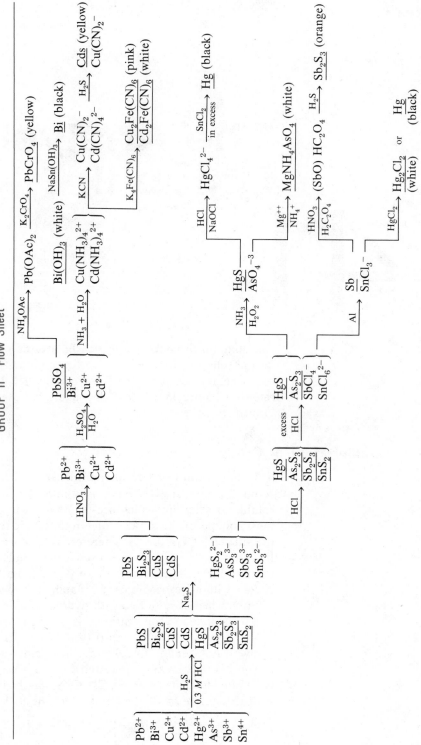

The hydrogen sulfide formed ionizes in aqueous solution according to the equation

$$H_2S \rightleftharpoons 2H^+ + S^{2-}$$

Hydrochloric acid furnishes hydrogen ions, which repress the ionization of hydrogen sulfide, and this causes a decrease in the sulfide ion concentration of the solution. (The ionization of hydrogen sulfide and its control is discussed in detail in Section 18.14.) The precipitation of the Group II sulfides may be represented by the following equations:

$$Pb^{2+} + H_2S \rightleftharpoons \underline{PbS} + 2H^+$$
$$2Bi^{3+} + 3H_2S \rightleftharpoons \underline{Bi_2S_3} + 6H^+$$
$$Cu^{2+} + H_2S \rightleftharpoons \underline{CuS} + 2H^+$$
$$Cd^{2+} + H_2S \rightleftharpoons \underline{CdS} + 2H^+$$
$$Hg^{2+} + H_2S \rightleftharpoons \underline{HgS} + 2H^+$$
$$2As^{3+} + 3H_2S \rightleftharpoons \underline{As_2S_3} + 6H^+$$
$$2Sb^{3+} + 3H_2S \rightleftharpoons \underline{Sb_2S_3} + 6H^+$$
$$Sn^{4+} + 2H_2S \rightleftharpoons \underline{SnS_2} + 4H^+$$

It should be noted that the acidity of the solution is increased by such reactions since the hydrogen ion is one of the products of the sulfide precipitation reactions.

■ **2. Mechanism of the Precipitation of the Sulfides.** The concentration of the sulfide ion in a solution saturated with hydrogen sulfide and 0.3 M in hydrogen ions is 1.4×10^{-20} mole per liter (Section 18.14). Because there are 6.022×10^{23} ions in one mole, only about 8 sulfide ions are present in each ml of solution. Yet, practically all the Pb^{2+} is precipitated immediately when hydrogen sulfide is passed into an acidic solution of a lead salt. The precipitation is so rapid that it is inconceivable that the reaction is one of simple union of lead and sulfide ions ($Pb^{2+} + S^{2-} \rightleftharpoons PbS$) when the sulfide ion is present in such a small concentration. However, the concentration of the hydrosulfide ion, HS^-, in a saturated solution of hydrogen sulfide containing 0.3 M hydrogen ions is comparatively large relative to the low concentration of the sulfide ion, as seen from the calculations:

$$\frac{[H^+][HS^-]}{[H_2S]} = \frac{(0.3)[HS^-]}{0.1} = 1 \times 10^{-7}$$

$$[HS^-] = \frac{0.1}{0.3} \times 1 \times 10^{-7} = 3 \times 10^{-8}$$

Thus, we find that the concentration of HS^- is 3×10^{-8} mole per liter, or about 10^{12} times the concentration of the sulfide ion (1.4×10^{-20}). Therefore, it is reasonable to assume that an unstable intermediate hydrosulfide, $Pb(HS)_2$, could be formed which immediately breaks down to form PbS and H_2S, as shown by

$$Pb^{2+} + 2HS^- \rightleftharpoons \underline{Pb(HS)_2} \rightleftharpoons \underline{PbS} + H_2S$$

Such processes are known for the formation of oxides through the precipitation of unstable hydroxides. A notable example is the formation of silver oxide according to the equation

$$2Ag^+ + 2OH^- \rightleftharpoons 2\underline{AgOH} \rightleftharpoons \underline{Ag_2O} + H_2O$$

Equilibrium is concerned only with the final result and not with the mechanism by which the result is obtained. This means that calculations involving the precipitation of metal sulfides are valid when concentrations of the sulfide ion are employed even though this ion may not be involved directly in the mechanism for the precipitation of the metal sulfide.

■ **3. Colors of the Sulfides of Group II.** The sulfides of lead(II), copper(II), and mercury(II) are all black. The sulfides of tin(IV), cadmium, and arsenic(III) are yellow, that of bismuth is dark brown or black, while antimony(III) sulfide is orange-red (sometimes black).

Tin(II) sulfide, SnS, forms a gelatinous precipitate which does not dissolve in sodium sulfide solutions, whereas SnS_2 is readily soluble in this reagent. For this reason tin(II) is oxidized to tin(IV) by adding nitric acid and heating the original solution before the sulfides of the Group II ions are precipitated.

PROCEDURE 1. *Precipitation of Group II: Pb^{2+}, Bi^{3+}, Cu^{2+}, Cd^{2+}, Hg^{2+}, As^{3+}, Sb^{3+}, Sn^{2+}, Sn^{4+}.* **Add 5 drops of 4 M HNO_3 to 10 drops of the known or unknown (or to the centrifugate from the Group I separation) and evaporate the solution not quite to dryness, ending the evaporation with a moist residue (Section 39.15). Cool and add 10 drops of water. Add 1 M aqueous ammonia until the solution (or mixture) is just basic to litmus. (Dip a stirring rod into the solution and then touch the moist end of the rod to a piece of litmus paper in testing for acidity or basicity of solutions.) Add 1.0 M HCl until the solution is just acidic to litmus. Now add 2 drops of 6 M HCl and dilute the solution to 1.5 ml. Add 10 drops of 5 per cent thioacetamide solution. The solution should now be 0.3 M with respect to the hydrogen ion if the directions have been followed carefully. Heat the solution contained in a test tube in a hot water bath for ten minutes. Sufficient hydrolysis of the thioacetamide to get complete precipitation of the Group II sulfides is essential. Add 1 ml of water and heat the mixture for another five minutes. Separate the precipitate and reserve the centrifugate for the analysis of Groups III-V in the case of a general unknown. Wash the precipitate with 10 drops of 0.1 M HCl twice. Add the first 10 drops of wash solution to the original centrifugate and discard the rest.**

41.2   Separation of the Group II Ions into the A and B Subdivisions

The separation of the Group II ions into subdivisions is based on the fact that HgS, As_2S_3, Sb_2S_3, and SnS_2 form soluble complex sulfides in alkaline sulfide solutions, whereas PbS, Bi_2S_3, CuS, and CdS do not form these complex ions.

The reagent used in this separation is sodium sulfide, which is made by the action of hydrogen sulfide (formed by the hydrolysis of thioacetamide) upon sodium hydroxide. The equation is

$$(2Na^+) + 2OH^- + H_2S \rightleftharpoons (2Na^+) + S^{2-} + 2H_2O$$

The presence of the hydroxide ion in high concentration, supplied by an excess of sodium hydroxide, reduces the extent of hydrolysis of the sulfide ion.

$$S^{2-} + H_2O \rightleftharpoons HS^- + OH^-$$

This provides a relatively high concentration of sulfide ion required for the dissolution of the Division-B sulfides, particularly of HgS. The equations are

$$HgS + S^{2-} \rightleftharpoons [HgS_2]^{2-} \text{ (thiomercurate)}$$
$$As_2S_3 + 3S^{2-} \rightleftharpoons 2[AsS_3]^{3-} \text{ (thioarsenite)}$$
$$Sb_2S_3 + 3S^{2-} \rightleftharpoons 2[SbS_3]^{3-} \text{ (thioantimonite)}$$
$$SnS_2 + S^{2-} \rightleftharpoons [SnS_3]^{2-} \text{ (thiostannate)}$$

The cations whose sulfides fail to dissolve in the presence of a high concentration of sulfide ions constitute Division A of Group II. They are Pb^{2+}, Bi^{3+}, Cu^{2+}, and Cd^{2+}.

PROCEDURE 2. *Precipitate from (1): PbS, Bi$_2$S$_3$, CuS, CdS, HgS, As$_2$S$_3$, Sb$_2$S$_3$, SnS$_2$. Add a mixture of 12 drops of 4 M NaOH and 4 drops of thioacetamide solution to the precipitate. Heat the mixture in a hot water bath for 5 minutes. Separate the mixture and reserve the solution (thiosalts of the Division-B ions) for Procedure (6). The residue consists of the sulfides of the ions of Division A.*

41.3   The Separation and Identification of Lead

The undissolved sulfides of Division A (PbS, Bi$_2$S$_3$, CuS, and CdS) are washed with a solution of ammonium nitrate. This is used, rather than pure water, to decrease the tendency of the sulfides to become colloidal. In the colloidal condition these sulfides would pass into the centrifugate and be lost in part.

The sulfides are then brought into solution by treatment with hot dilute nitric acid. The acid oxidizes the sulfide ion to elemental sulfur, and causes the metal sulfides to dissolve by holding the sulfide ion concentration at such a low level that the product of the concentration of the ions is less than the solubility product constant. The two equilibria involved in the system, in the case of lead sulfide, are given by

$$\underline{PbS} \rightleftharpoons Pb^{2+} + S^{2-}$$
$$3S^{2-} + 8H^+ + 2NO_3^- \rightleftharpoons 3\underline{S} + 2NO\uparrow + 4H_2O$$

and the net reaction is

$$3\underline{PbS} + 8H^+ + 2NO_3^- \rightleftharpoons 3Pb^{2+} + 3\underline{S} + 2NO\uparrow + 4H_2O$$

Lead is separated from bismuth, copper, and cadmium by precipitating the sulfate, $PbSO_4$, which is white. The sulfates of the other three metals are soluble. The nitric acid in the solution must first be removed because the sulfate ion is converted largely to the hydrogen sulfate ion by hydrogen ions in high concentration, $SO_4^{2-} + H^+ \rightleftharpoons HSO_4^-$. Under these conditions the concentration of sulfate ion is too low to allow the solubility product constant of lead sulfate to be exceeded. The removal of the nitric acid is accomplished by adding sulfuric acid and evaporating the solution to the point where sulfur trioxide fumes appear.

$$H_2SO_4 \longrightarrow H_2O + SO_3$$

The more volatile nitric acid will have been expelled before the SO_3 fumes appear. Upon dilution of the sulfuric acid solution with water, sulfate ions are formed and the lead precipitates as the sulfate. The equations are

$$H_2SO_4 \rightleftharpoons H^+ + HSO_4^-$$
$$HSO_4^- \rightleftharpoons H^+ + SO_4^{2-}$$
$$Pb^{2+} + SO_4^{2-} \rightleftharpoons \underline{PbSO_4}$$

The lead sulfate is then dissolved in a solution of ammonium acetate with the formation of slightly ionized lead acetate. The equation is

$$\underline{PbSO_4} + 2OAc^- \rightleftharpoons Pb(OAc)_2 + SO_4^{2-}$$

When chromate ions are added to this acetate solution, yellow lead chromate, $PbCrO_4$, precipitates.

$$Pb(OAc)_2 + CrO_4^{2-} \rightleftharpoons \underline{PbCrO_4} + 2OAc^-$$

PROCEDURE 3. *Residue from (2): PbS, CuS, CdS, Bi$_2$S$_3$.* **Wash the residue with 1 ml of water to which 2 drops of 1 *M* NH$_4$NO$_3$ have been added. Discard the wash solution. Add 15 drops of 6 *M* HNO$_3$ to the residue and heat the mixture in a hot water bath for several minutes to dissolve the sulfides. Separate and discard any sulfur which is formed by the oxidation of sulfide ions and transfer the solution to a casserole or evaporating dish. Add 5 drops of 4 *M* H$_2$SO$_4$ and evaporate the solution under the hood (very important) to white SO$_3$ fumes. (The SO$_3$ fumes are dense, not at all like steam. Do not prolong the heating after SO$_3$ fumes appear.) Cool and add 1 ml of water to the mixture. Warm the mixture and stir up the precipitate. (The precipitate, PbSO$_4$, may appear slowly, especially if the solution is too acidic.) Separate the mixture. Reserve the solution for Procedure (4). Wash the residue (PbSO$_4$) with a few drops of water and discard the wash solution. Extract the residue (Section 39.13) with a mixture of 5 drops of 1 *M* NH$_4$OAc and 1 drop of 1 *M* HOAc. Add 1 drop of 1 *M* K$_2$CrO$_4$ to the extract. Scratch the inside wall of the test tube with a glass rod to initiate precipitation. The formation of a yellow precipitate confirms the presence of lead as the chromate, PbCrO$_4$.**

41.4   Separation and Identification of Bismuth

Upon the addition of aqueous ammonia to the solution from the lead sulfate separation, the hydroxides of bismuth (white), copper (pale-blue), and cadmium (white) are first precipitated. The equations are

$$Bi^{3+} + 3NH_3 + 3H_2O \longrightarrow \underline{Bi(OH)_3} + 3NH_4^+$$
$$Cu^{2+} + 2NH_3 + 2H_2O \longrightarrow \underline{Cu(OH)_2} + 2NH_4^+$$
$$Cd^{2+} + 2NH_3 + 2H_2O \longrightarrow \underline{Cd(OH)_2} + 2NH_4^+$$

When an excess of aqueous ammonia is added, the hydroxides of copper and cadmium dissolve by forming tetraammine complexes, whereas the bismuth(III) hydroxide does not dissolve.

$$Bi(OH)_3 + NH_3 \longrightarrow \text{no reaction}$$
$$Cu(OH)_2 + 4NH_3 \longrightarrow [Cu(NH_3)_4]^{2+} + 2OH^-$$
$$Cd(OH)_2 + 4NH_3 \longrightarrow [Cd(NH_3)_4]^{2+} + 2OH^-$$

After it is separated from the complexes of copper and cadmium, the bismuth hydroxide is treated with sodium stannite, a strong reducing agent. The bismuth is reduced to the elemental state, in which condition it appears black. The equation is

$$2Bi(OH)_3 + 3[Sn(OH)_3]^- + 3OH^- \longrightarrow 2Bi + 3[Sn(OH)_6]^{2-}$$
$$\text{(Stannite ion)} \qquad\qquad\qquad\qquad \text{(Stannate ion)}$$

Because sodium stannite is unstable and darkens upon standing, it must be prepared just prior to use, and the test for bismuth as a black residue must be observed immediately. Sodium stannite is prepared by treating a solution of tin(II) chloride with an excess of sodium hydroxide. The equations are

$$Sn^{2+} + 2OH^- \rightleftharpoons Sn(OH)_2 \text{ (white)}$$
$$Sn(OH)_2 + \text{excess } OH^- \rightleftharpoons [Sn(OH)_3]^- \text{ (colorless)}$$

The darkening of the stannite solution upon standing is due to the formation of black tin(II) oxide. The equation is

$$[Sn(OH)_3]^- \longrightarrow SnO + H_2O + OH^-$$

PROCEDURE 4. *Solution from (3): Bi³⁺, Cu²⁺, Cd²⁺.* **Add 15 *M* aqueous ammonia dropwise to the solution until it is distinctly basic to litmus (about 5 drops). The development of a deep blue color in the solution indicates the presence of the tetraamminecopper(II) ion, [Cu(NH₃)₄]²⁺. The formation of a white precipitate indicates the presence of Bi³⁺ as Bi(OH)₃. Separate the mixture and reserve the solution for Procedure (5). Add several drops of freshly prepared sodium stannite solution to the precipitate. The immediate formation of a black residue (finely divided metallic bismuth) confirms the presence of Bi³⁺. (Preparation of sodium stannite: In a separate test tube, place 3 drops of 0.4 *M* SnCl₂ and add sufficient 4 *M* NaOH to dissolve the white precipitate which first forms. Be sure that you are not fooled by an initial precipitate which forms because of a temporary local excess of reagent and dissolves just with stirring.)**

41.5 Identification of Copper and Cadmium

The presence of copper is apparent from the deep-blue color exhibited by the complex ion, $[Cu(NH_3)_4]^{2+}$. However, if the blue color is too faint to discern, as in the case of a trace of copper, the formation of pink copper(II) hexacyanoferrate(II) serves as a delicate test for the ion. The equation for the reaction is

$$2[Cu(NH_3)_4]^{2+} + [Fe(CN)_6]^{4-} \longrightarrow Cu_2Fe(CN)_6 + 8NH_3$$

If cadmium is present, a white precipitate of cadmium hexacyanoferrate(II), $Cd_2Fe(CN)_6$, will form.

The presence of cadmium ions may be confirmed by forming yellow cadmium sulfide with hydrogen sulfide. However, if one attempts to precipitate CdS in the presence of copper, the black color of CuS obscures the yellow color of CdS. By first adding cyanide ions to a solution containing Cu^{2+} and Cd^{2+}, the very stable dicyanocopper(I) complex will be formed, which will not react with sulfide ions. The copper(II) ion is reduced to copper(I) by the cyanide ion, and the cyanide ion is oxidized to cyanogen, C_2N_2.

$$2[Cu(NH_3)_4]^{2+} + 6CN^- \longrightarrow 2[Cu(CN)_2]^- + C_2N_2\uparrow + 8NH_3$$

$$[Cu(CN)_2]^- + S^{2-} \longrightarrow \text{ no reaction}$$

The corresponding tetracyanocadmium complex is much less stable than the dicyanocopper complex, and it will react with the sulfide ion to form yellow CdS.

$$[Cd(NH_3)_4]^{2+} + 4CN^- \rightleftharpoons [Cd(CN)_4]^{2-} + 4NH_3$$

$$[Cd(CN)_4]^{2-} + S^{2-} \longrightarrow \underline{CdS} + 4CN^-$$

PROCEDURE 5. *Solution from (4):* $[Cu(NH_3)_4]^{2+}$ (*blue*), $[Cd(NH_3)_4]^{2+}$ (*colorless*). If the solution is colorless, a trace of copper may be present. Place 10 drops of the solution in a test tube, acidify with acetic acid, and add several drops of $0.1 M$ $K_4Fe(CN)_6$. The formation of a pink precipitate, $Cu_2Fe(CN)_6$, indicates the presence of a small concentration of Cu^{2+}; the formation of a white precipitate indicates the absence of Cu^{2+} and the probable presence of Cd^{2+} as $Cd_2Fe(CN)_6$. If copper(II) ions are present, add 2 drops of $1 M$ KCN and 2 drops of thioacetamide solution to the remainder of the solution from Procedure (4). Heat the mixture in a hot water bath. The formation of either a yellow or olive-green precipitate (CdS) confirms the presence of cadmium. If Cu^{2+} is found to be absent, test for Cd^{2+} as described above but leave out the KCN.

41.6   Reprecipitation of the Sulfides of the Division-B Ions

To reprecipitate the sulfides of mercury, arsenic, antimony, and tin from the solution containing their complex thiosalts, it is necessary to increase the concentration of the cations to a value such that, for example,

$$[Hg^{2+}][S^{2-}] > K_{sp}$$

This is accomplished by increasing the extent of dissociation of the complex ions through the addition of hydrochloric acid.

$$[HgS_2]^{2-} \rightleftharpoons Hg^{2+} + 2S^{2-}$$
$$2S^{2-} + 4H^+ \rightleftharpoons 2H_2S\uparrow$$
$$\overline{\text{Net reaction } [HgS_2]^{2-} + 4H^+ \rightleftharpoons Hg^{2+} + 2H_2S\uparrow}$$

As seen from the above equations, the added acid displaces the equilibria in the direction of an increased concentration of Hg^{2+} because of the formation and escape of the weak, gaseous electrolyte H_2S. This results in an increase in the concentration of the cation to such an extent that, for example, the product of the concentration of the mercury(II) ion and the sulfide ion becomes larger than the solubility product constant, and reprecipitation of the sulfide occurs. The net reactions for these reprecipitations are given by the following equations:

$$[HgS_2]^{2-} + 2H^+ \rightleftharpoons HgS + H_2S\uparrow$$
$$2[AsS_3]^{3-} + 6H^+ \rightleftharpoons As_2S_3 + 3H_2S\uparrow$$
$$2[SbS_3]^{3-} + 6H^+ \rightleftharpoons Sb_2S_3 + 3H_2S\uparrow$$
$$[SnS_3]^{2-} + 2H^+ \rightleftharpoons SnS_2 + H_2S\uparrow$$

PROCEDURE 6. *Solution from (2): Thiosalts of the Division-B Ions, $[HgS_2]^{2-}$, $[AsS_3]^{3-}$, $[SbS_3]^{3-}$, $[SnS_3]^{2-}$. Add 1 M HCl until the solution is just acidic to litmus. Heat the mixture in a hot water bath for several minutes. Separate the mixture and discard the solution. The precipitate consists of HgS, As_2S_3, Sb_2S_3, and SnS_2.*

41.7   Separation of Mercury and Arsenic from Antimony and Tin

To dissolve Sb_2S_3 and SnS_2, and leave HgS and As_2S_3 undissolved, the concentrations of the cations and sulfide ion must be made such that

$$[Sb^{3+}]^2[S^{2-}]^3 < K_{Sb_2S_3}, \quad \text{and} \quad [Hg^{2+}][S^{2-}] > K_{HgS}$$

The addition of 6 M HCl reduces the sulfide ion concentration ($S^{2-} + 2H^+ \rightleftharpoons 2H_2S$), and the concentration of antimony and tin ions ($Sb^{3+} + 4Cl^- \rightleftharpoons [SbCl_4]^-$, and $Sn^{4+} + 6Cl^- \rightleftharpoons [SnCl_6]^{2-}$) to values such that these sulfides dissolve. The net reactions are given by

$$Sb_2S_3 + 6H^+ + 8Cl^- \longrightarrow 2[SbCl_4]^- + 3H_2S\uparrow$$
$$SnS_2 + 4H^+ + 6Cl^- \longrightarrow [SnCl_6]^{2-} + 2H_2S\uparrow$$

The sulfides of mercury and arsenic remain undissolved in the presence of 6 M HCl due to the very small values of their solubility product constants.

PROCEDURE 7. *Precipitate from (6): HgS, As_2S_3, Sb_2S_3, SnS_2. Add 1 ml of 6 M HCl to the precipitate and stir the mixture. Heat the mixture in a hot water bath and then separate the residue. Add 15 drops of 6 M HCl to the residue and heat the mixture. Separate the residue and reserve the combined centrifugates for Procedure (11). Treat the residue according to Procedure (8).*

41.8   Separation of Arsenic from Mercury

The concentration of the sulfide ion in saturated solutions of the sulfides of mercury and arsenic is so small that it is not possible to dissolve these sulfides by addition of hydrogen ions. In these cases the product of $[H^+]^2$ and $[S^{2-}]$ does not exceed the ion product constant for H_2S, and equilibrium is established before these sulfides are dissolved.

In order to dissolve HgS and As_2S_3, it is necessary to reduce the concentration of the sulfide ion by oxidation, or the concentration of the cation by the formation of complex ions. A mixture of aqueous ammonia and hydrogen peroxide is used in separating arsenic from mercury. The combined reaction of the hydroxide ion and the hydrogen peroxide with As_2S_3 is given by the equation

$$As_2S_3 + 12OH^- + 14H_2O_2 \longrightarrow 2[AsO_4]^{3-} + 3SO_4^{2-} + 20H_2O$$

Thus the arsenic(III) ion is oxidized to arsenate, $[AsO_4]^{3-}$, and the sulfide ion to sulfate. Consequently, we have

$$[As^{3+}]^2[S^{2-}]^3 < K_{sp} \text{ for } As_2S_3$$

and the As_2S_3 dissolves. Mercury(II) sulfide is not dissolved by this treatment because the sulfide ion concentration in the equilibrium, $HgS \rightleftharpoons Hg^{2+} + S^{2-}$, is too small for the sulfide ion to be appreciably oxidized by hydrogen peroxide.

PROCEDURE 8. *Residue from (7): HgS and As₂S₃.* **Add 12 drops of 4 M aqueous ammonia and 6 drops of 3 per cent hydrogen peroxide to the residue. Stir the mixture and heat it in a hot water bath for 5-6 minutes. Separate the residue (HgS) and reserve the solution ($AsO_4{}^{3-}$) for Procedure (10).**

41.9 Identification of Mercury

In order to dissolve the extremely insoluble mercury(II) sulfide, it is necessary to reduce the mercury(II) and sulfide ion concentrations to the extent that $[Hg^{2+}][S^{2-}]$ will be less than the solubility product of HgS. This may be done by forming the $[HgCl_4]^{2-}$ complex and oxidizing the sulfide ion to sulfur. A mixture of hydrochloric acid and sodium hypochlorite is used to dissolve the HgS. The chloride ions from hydrochloric acid combine with mercury(II) ions and form the complex, $[HgCl_4]^{2-}$; hypochlorite ions in acid solution oxidize the sulfide ions to sulfur. The equation is

$$\underline{HgS} + 2H^+ + 3Cl^- + ClO^- \longrightarrow [HgCl_4]^{2-} + \underline{S} + H_2O$$

The solution is then boiled to decompose excess hypochlorite ions which would otherwise interfere with the confirmatory test for mercury by oxidizing tin(II) to tin(IV), thus destroying its reducing power. The decomposition of the hypochlorite ion is according to the equation

$$2H^+ + Cl^- + ClO^- \longrightarrow H_2O + Cl_2\uparrow$$

The presence of mercury(II) is confirmed by reducing the $[HgCl_4]^{2-}$ to Hg_2Cl_2 (white) or Hg (black) by means of $[SnCl_3]^-$ ions.

$$2[HgCl_4]^{2-} + [SnCl_3]^- \longrightarrow \underline{Hg_2Cl_2} + [SnCl_6]^{2-} + 3Cl^-$$

$$\underline{Hg_2Cl_2} + [SnCl_3]^- + Cl^- \longrightarrow 2\underline{Hg} + [SnCl_6]^{2-}$$

The trichlorostannate(II) ion, $[SnCl_3]^-$, is formed when tin(II) chloride is added to the hydrochloric acid solution.

$$SnCl_2 + Cl^- \longrightarrow [SnCl_3]^-$$

PROCEDURE 9. *Residue from (8): HgS.* To the black residue add 6 drops of 5 per cent NaClO and 2 drops of 6 *M* HCl. Stir the mixture, add 1 ml of water, and separate the sulfur from the solution. Heat the solution to boiling. Add 2 drops of 1 *M* SnCl₂ to the solution. The formation of a white, gray, or black precipitate confirms the presence of mercury.

41.10 Identification of Arsenic

The presence of arsenic in the form of arsenate ions is confirmed by the formation of white crystalline magnesium ammonium arsenate when magnesia mixture (MgCl₂, NH₄Cl, and aqueous ammonia) is added to a solution containing arsenate ions.

$$Mg^{2+} + NH_4^+ + AsO_4^{3-} \longrightarrow \underline{MgNH_4AsO_4}$$

PROCEDURE 10. *Solution from (8): AsO_4^{3-}.* Add 2 drops of 15 *M* aqueous ammonia and 5 drops of magnesia mixture to the solution. The formation of a white precipitate (MgNH₄AsO₄), frequently slow in forming, indicates the presence of arsenate ions.

41.11 Identification of Antimony and Tin

Antimony and tin in hydrochloric acid solutions are in the form of the complex ions, $[SbCl_4]^-$ and $[SnCl_6]^{2-}$. Aluminum metal will reduce antimony to the metallic state and tin to the divalent state in hydrochloric acid solution, thus effecting a separation of the two metals.

$$[SbCl_4]^- + Al \longrightarrow \underline{Sb} + Al^{3+} + 4Cl^-$$
$$3[SnCl_6]^{2-} + 2Al \longrightarrow 3[SnCl_3]^- + 2Al^{3+} + 9Cl^-$$

To be exact, aluminum actually reduces tin to the metallic state. When all of the aluminum is gone, the metallic tin dissolves in the hydrochloric acid, forming $[SnCl_3]^-$ ions and liberating hydrogen.

Use is made of the fact that divalent tin is a strong reducing agent in the confirmatory test for the ion. $[SnCl_3]^-$ will reduce mercury(II) to mercury(I), or metallic mercury, depending upon the amount of the tin(II) present in solution.

$$[SnCl_3]^- + 2HgCl_4^{2-} \longrightarrow [SnCl_6]^{2-} + \underline{Hg_2Cl_2} + 3Cl^-$$
$$[SnCl_3]^- + Hg_2Cl_2 + Cl^- \longrightarrow [SnCl_6]^{2-} + \underline{2Hg}$$

Antimony metal reacts with nitric acid to form the insoluble oxide, Sb_4O_6. This oxide is soluble in oxalic acid, $H_2C_2O_4$, forming antimony(III) oxyhydrogen oxalate, $(SbO)HC_2O_4$.

$$4Sb + 4H^+ + 4NO_3^- \longrightarrow \underline{Sb_4O_6} + 4NO + 2H_2O$$
$$Sb_4O_6 + 4H_2C_2O_4 \longrightarrow 4(SbO)HC_2O_4 + 2H_2O$$

The presence of antimony is then confirmed by precipitating it as the orange-red sulfide, Sb_2S_3.

$$2(SbO)HC_2O_4 + 3H_2S \longrightarrow \underline{Sb_2S_3} + 2H_2C_2O_4 + 2H_2O$$

PROCEDURE 11. *Solution from (7): [SnCl_6]^{2-} and [SbCl_4]^-*. Boil the solution until all of the H_2S has been expelled. Add a volume of water equal to that of the solution and add 2 drops of 6 M HCl. Place a piece of aluminum wire about one-eighth inch long in the solution and heat the mixture until the aluminum has completely dissolved. Add 1 drop of 6 M HCl to the mixture and heat it for a few minutes. If antimony is present, black flakes of the metal will appear. Separate the mixture. Treat the black flakes with 3 drops of 4 M HNO_3 and several drops of 1 M oxalic acid. Add 2 drops of thioacetamide to the solution and place the test tube in a hot water bath. The formation of a red-orange precipitate of Sb_2S_3 confirms the presence of antimony. To the solution obtained from the separation of metallic antimony, add a few drops of 0.2 M $HgCl_2$. The formation of a white, gray, or black precipitate confirms the presence of tin.

QUESTIONS

1. Give the color of each of the following: CuS, HgS, $[Cu(H_2O)_4]^{2+}$, $PbSO_4$, $[Cu(NH_3)_4]^{2+}$, $[Cu(CN)_2]^-$, $MgNH_4AsO_4$, PbS, $Bi(OH)_3$, Sb_2S_3, As_2S_3, SnS_2, $[Cd(NH_3)_4]^{2+}$, Bi (finely divided), Sb, and Bi_2S_3.
2. Why are the ions of lead and mercury found in both Groups I and II?
3. Explain in terms of ionic equilibria and solubility product theory why Group II can be separated from Group III by hydrogen sulfide in the presence of hydrogen ions at a concentration of 0.3 M.
4. In terms of ionic equilibria theory, discuss the effect of added hydrochloric acid on the concentration of sulfide ion in a solution of hydrogen sulfide. How would ammonia molecules and hydroxide ions influence the concentration of sulfide ions?
5. Explain in terms of ionic equilibria and solubility product theory, the dissolution of the sulfides of copper, bismuth, cadmium, and lead in nitric acid.
6. Write the equation for the hydrogen ion catalyzed hydrolysis of thioacetamide. The hydrogen sulfide, as one of the products of the hydrolysis of thioacetamide, is a diprotic acid; illustrate this property of hydrogen sulfide by suitable equations.
7. Explain the dissolution of lead sulfate in ammonium acetate and the re-precipitation of the lead as lead chromate in terms of ionic equilibria and solubility product theory.
8. Why is it necessary to remove the nitric acid present before attempting to precipitate lead as the sulfate?
9. If a yellow precipitate is obtained when an unknown solution for Group II is treated with hydrogen sulfide, what ions are probably absent?
10. The Division-A sulfides are washed with water containing ammonium nitrate. What is the function of the ammonium nitrate?
11. Why must the sodium stannite, which is used in the identification of bismuth, be prepared just prior to its use?
12. How will the separation of Groups II and III be affected if the concentration of hydronium ion in the solution saturated with hydrogen sulfide is 0.1 M? 1 M?

13. If a Group II unknown contains copper in an appreciable concentration, this fact should be evident from an inspection of the unknown solution. Why?

14. Outline the separation of the following groups of ions, leaving out all unnecessary steps: (a) Bi^{3+}, As^{3+}, Sb^{3+}; (b) Ag^+, Hg^{2+}, Co^{2+}; (c) Pb^{2+}, Cu^{2+}, Cd^{2+}.

15. Select a reagent used in the analysis of Group I or Group II which will separate each of the following pairs in one step: (a) As_2S_3, SnS_2; (b) $[SbCl_4]^-$, $[SnCl_6]^{2-}$; (c) CdS, HgS; (d) CuS, CdS; (e) Ag^+, Bi^{3+}; (f) Ag^+, Fe^{3+}; (g) Bi^{3+}, Cd^{2+}.

PROBLEMS

1. How many drops of 1 M aqueous ammonia will be required to neutralize the HCl in 2 ml of a solution to be analyzed for Group II that is 0.5 M in HCl? (1 drop is 0.05 ml.) How many drops of 6 M HCl would be required to make the resulting solution 0.3 M in HCl?   *Ans. 20 drops; 3 drops*

2. Calculate the concentration of S^{2-} in an acid solution which is saturated with H_2S and has a pH of 0.52. (Saturated H_2S is 0.1 M.)   *Ans. 1.4 × 10^{-20} M*

3. How many grams of thioacetamide are required to precipitate quantitatively the bismuth and copper from 10 ml of a solution which is 0.010 molar in each ion?   *Ans. 0.019 g*

4. Calculate the solubility of ZnS in a solution saturated with H_2S and 0.30 M in HCl.   *Ans. 0.076 M*

5. Calculate the concentration of H^+, H_2S, HS^-, and S^{2-} in a 0.052 M solution of hydrogen sulfide.

   Ans. $[H^+] = [HS^-] = 7.2 × 10^{-5}$ M;
   $[S^{2-}] = 1.3 × 10^{-13}$ M; $[H_2S] = 0.052$ M

6. Calculate the concentration of H^+ required to prevent the precipitation of CdS from 0.010 M Cd^{2+} that is saturated with H_2S. (K_{sp} for CdS is 3.6 × 10^{-29}.) Is this an attainable concentration?   *Ans. 6.0 × 10^2 M*

REFERENCE

"Some Landmarks in the History of Arsenic Testing," W. A. Campbell, *Chem. in Britain*, **1**, 198 (1965).

The Analysis of Group III

42

The solution to be analyzed may be a Group III known or unknown, or it may be the solution from the Group II separation. See the following page for the Group III flow sheet. You will probably find it helpful also to refer to Chapter 36, which discusses the chemistry of the metals of analytical Group III.

42.1 Precipitation of the Group III Ions

Analytical Group III contains the metallic ions, Ni^{2+}, Co^{2+}, Mn^{2+}, Fe^{3+}, Al^{3+}, Cr^{3+}, and Zn^{2+}. These ions are not precipitated by hydrochloric acid (the Group I precipitant), nor are they precipitated by sulfide ions in solutions which are 0.3 M in hydrogen ion (the Group II precipitant). However, an ammonium sulfide solution precipitates Ni^{2+}, Co^{2+}, Mn^{2+}, Fe^{2+}, and Zn^{2+} as sulfides and Al^{3+} and Cr^{3+} as hydroxides.

The concentration of sulfide ions in the 0.3 M hydrogen ion solution of the Group II precipitation is too small to allow the solubility products of the Group III sulfides to be exceeded. Hydrogen sulfide in an aqueous solution of ammonia has a much higher sulfide ion concentration, due to the formation of ammonium sulfide, according to the equation

$$2NH_3 + H_2S \rightleftharpoons 2NH_4^+ + S^{2-}$$

The resultant sulfide ion concentration is sufficiently large so that the solubility product constants of the sulfides of cobalt, nickel, manganese, iron, and zinc are exceeded and precipitation occurs. Likewise, the hydroxyl ion concentration of

GROUP III Flow Sheet

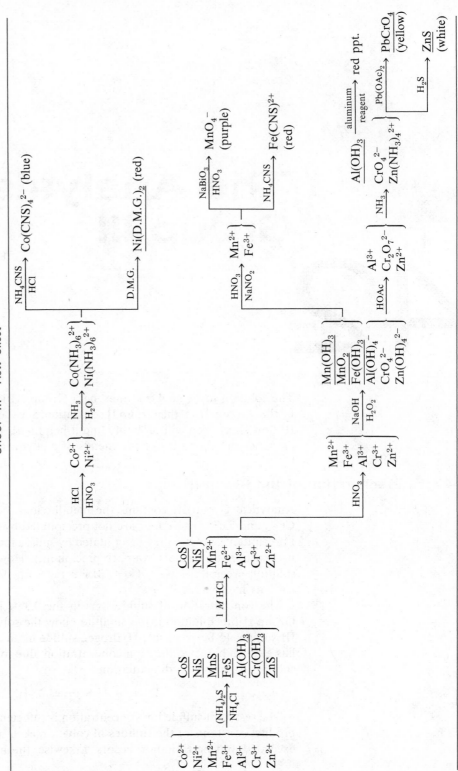

the ammonium sulfide solution is great enough to precipitate the hydroxides of aluminum and chromium. The solubility products of aluminum and chromium hydroxides are lower than those of the corresponding sulfides, thus favoring the precipitation of the former.

Hydrogen sulfide is produced for the precipitation of this group by the acidic hydrolysis of thioacetamide. Upon the addition of aqueous ammonia the following precipitations occur.

$$Co^{2+} + S^{2-} \rightleftharpoons \underline{CoS} \text{ (brown or black)}$$
$$Ni^{2+} + S^{2-} \rightleftharpoons \underline{NiS} \text{ (brown or black)}$$
$$Mn^{2+} + S^{2-} \rightleftharpoons \underline{MnS} \text{ (light-pink)}$$
$$Fe^{2+} + S^{2-} \rightleftharpoons \underline{FeS} \text{ (black)}$$
$$Zn^{2+} + S^{2-} \rightleftharpoons \underline{ZnS} \text{ (white)}$$
$$Al^{3+} + 3OH^- \rightleftharpoons \underline{Al(OH)_3} \text{ (white)}$$
$$Cr^{3+} + 3OH^- \rightleftharpoons \underline{Cr(OH)_3} \text{ (blue-green)}$$

In the event that iron is present in the original solution in the Fe^{3+} state, it will be reduced by hydrogen sulfide in the acidic solution, according to the equation

$$2Fe^{3+} + H_2S \longrightarrow 2Fe^{2+} + S + 2H^+$$

Of the cations remaining in solution, only the magnesium ion, of Group V, forms an insoluble hydroxide. The hydroxyl ion concentration is kept just low enough in the aqueous ammonia to prevent the precipitation of magnesium hydroxide by having ammonium chloride present. The ammonium ions from ammonium chloride exert the common ion effect with the ammonium ions from the ionization of ammonia, thus repressing the ionization of ammonia and reducing the hydroxyl ion concentration of the solution. This point is considered quantitatively in Example 3 of Section 19.9.

Some evidence regarding the cations present in a solution containing the Group III ions can be obtained by noting the colors of the ions: Co^{2+}, pale-red; Ni^{2+}, pale-green; Zn^{2+}, colorless; Mn^{2+}, colorless in low concentrations; Fe^{2+}, pale-green, Fe^{3+}, reddish-brown; Al^{3+}, colorless; Cr^{3+}, dark-green or blue. Due to the phenomenon of complementary colors, a solution containing certain combinations of colored ions may appear colorless.

PROCEDURE 1. *Precipitation of Group III.* **(a) If the solution to be analyzed is a known or unknown for Group III only, take 10 drops of the solution, add 1 drop of 6 *M* HCl, dilute to 1 ml, and add 5 drops of 5 per cent thioacetamide solution. Heat the solution in a hot water bath for at least 5 minutes. (b) If the solution to be analyzed is that from the Group II separation, add 5 drops of 5 per cent thioacetamide solution and heat the mixture in a hot water bath for at least 5 minutes.**

To the solution resulting from either procedure (a) or (b) above, add 5 drops of 15 *M* aqueous ammonia and stir up the precipitate. Heat the mixture for 5 minutes in the hot water bath. Separate the precipitate and wash it with a few drops of water. Reserve the solution for the analysis of Groups IV and V.

42.2   Separation of Cobalt and Nickel

Although hydrogen sulfide will precipitate the sulfides of cobalt and nickel completely only from basic solution, these sulfides are but slightly soluble in dilute HCl. It appears that these sulfides precipitate in a form which is soluble in acid but change rapidly into other crystalline modifications which have much lower solubility products. The K_{sp} of α CoS is 5.9×10^{-21}, of β CoS is 8.7×10^{-23}, of α NiS is 3×10^{-21}, and of β NiS is 1×10^{-26}. Use is made of these facts in separating CoS and NiS from MnS, FeS, Al(OH)$_3$, Cr(OH)$_3$, and ZnS, the latter five precipitates being dissolved by 1 M HCl.

$$\underline{MnS} + 2H^+ \rightleftharpoons Mn^{2+} + H_2S\uparrow \qquad \underline{Al(OH)_3} + 3H^+ \rightleftharpoons Al^{3+} + 3H_2O$$
$$\underline{FeS} + 2H^+ \rightleftharpoons Fe^{2+} + H_2S\uparrow \qquad \underline{Cr(OH)_3} + 3H^+ \rightleftharpoons Cr^{3+} + 3H_2O$$
$$\underline{ZnS} + 2H^+ \rightleftharpoons Zn^{2+} + H_2S\uparrow$$

PROCEDURE 2. *Precipitate from (1): CoS, NiS, FeS, MnS, Al(OH)$_3$, Cr(OH)$_3$, and ZnS.* **Add 10 drops of 1 M HCl to the precipitate and stir the mixture. Separate the mixture immediately because prolonged contact with the acid causes some dissolution of CoS and NiS. Wash the sulfides which remain (CoS and NiS) with 4 drops of 1 M HCl. Reserve the combined centrifugates for Procedure (4).**

42.3   Identification of Cobalt and Nickel

The sulfides of cobalt and nickel readily dissolve in a mixture of nitric and hydrochloric acids due to the higher hydrogen ion concentration and the oxidation of the sulfide ion to sulfur. The equations are

$$3CoS + 8H^+ + 2NO_3^- \rightleftharpoons 3Co^{2+} + 2NO\uparrow + 3\underline{S} + 4H_2O$$
$$3NiS + 8H^+ + 2NO_3^- \rightleftharpoons 3Ni^{2+} + 2NO\uparrow + 3\underline{S} + 4H_2O$$

After boiling the solution containing the cobalt and nickel ions to remove the oxides of nitrogen which would destroy the reagents used in the confirmation of these ions, an excess of aqueous ammonia is added. Ammonia in excess reacts with cobalt and nickel ions to form hexaamminemetal complexes.

$$Co^{2+} + 6NH_3 \rightleftharpoons [Co(NH_3)_6]^{2+} \text{ (pink)}$$
$$Ni^{2+} + 6NH_3 \rightleftharpoons [Ni(NH_3)_6]^{2+} \text{ (blue)}$$

In the ammoniacal solution, Ni^{2+} will react with dimethylglyoxime to form a very insoluble, bright red complex compound. Co^{2+} forms a brown colored soluble complex with dimethylglyoxime, which does not interfere with the test.

A concentrated solution of ammonium thiocyanate is used in the identification of the cobalt ion, with which the thiocyanate ion forms a complex ion, $[Co(CNS)_4]^{2-}$, that has a characteristic blue color.

$$[Co(NH_3)_6]^{2+} + 4CNS^- \rightleftharpoons [Co(CNS)_4]^{2-} + 6NH_3$$

When iron(III) ions are present, they interfere with this test by forming a bright-red complex ion of the formula $[Fe(CNS)]^{2+}$. This interference may be avoided by converting the iron to the colorless and very stable hexafluoroferrate(III) ion, $[FeF_6]^{3-}$. The equation is

$$Fe^{3+} + 6F^- \rightleftharpoons [FeF_6]^{3-}$$

A second confirmatory test for cobalt involves its oxidation by the nitrite ion from Co^{2+} to Co^{3+} and the precipitation of the complex yellow salt, $K_3[Co(NO_2)_6]$, tripotassium hexanitrocobaltate(III).

$$Co^{2+} + 6NO_2^- \longrightarrow [Co(NO_2)_6]^{4-}$$
$$[Co(NO_2)_6]^{4-} + NO_2^- + 2H^+ \longrightarrow [Co(NO_2)_6]^{3-} + NO\uparrow + H_2O$$
$$3K^+ + [Co(NO_2)_6]^{3-} \longrightarrow \underline{K_3[Co(NO_2)_6]}$$

PROCEDURE 3. *Residue from (2): CoS and NiS.* **Add 3 drops of 12 M HCl and 1 drop of 14 M HNO$_3$ to the residue and heat the mixture in a hot water bath. Separate any sulfur that forms and boil the solution to remove any excess nitric acid or oxides of nitrogen. Add sufficient 4 M aqueous ammonia to the solution to make it *slightly* basic to litmus. (Excess alkalinity favors the formation of the brown cobalt complex when testing for nickel with dimethylglyoxime.) Dilute the solution to 1 ml and divide it into three parts.**

3a. *Test for Nickel.* **Add one drop of dimethylglyoxime to one part of the solution. The formation of a pink or red precipitate confirms the presence of nickel.**

3b. *Test for Cobalt.* **Acidify a second portion of the solution with 1 M HCl and add several crystals of NH$_4$CNS to it. Now add an equal volume of acetone and agitate the mixture. The development of a blue color proves the presence of cobalt. If the solution becomes red upon the addition of NH$_4$CNS, iron(III) ions are present. Add 1 drop of 1 M NaF to the solution. Now, if the solution is bluish-green to green, the presence of cobalt is confirmed.**

3c. *Test for Cobalt.* **Acidify the third portion of the solution from (3) with 4 M HOAc, and add several large crystals of KNO$_2$. Warm the mixture. The formation of a yellow precipitate confirms the presence of cobalt.**

42.4 Separation of Iron and Manganese from Aluminum, Chromium, and Zinc

The solution containing the ions of the above metals is treated with nitric acid to remove the sulfide ion by oxidizing it to sulfur, and to oxidize Fe^{2+} to Fe^{3+}. The equation for the oxidation of the iron is

$$3Fe^{2+} + 4H^+ + NO_3^- \longrightarrow 3Fe^{3+} + NO\uparrow + 2H_2O$$

It is desirable to have the iron in the $+3$ condition because $Fe(OH)_3$ is less soluble and less gelatinous in character than is $Fe(OH)_2$.

In order to separate iron and manganese from aluminum, chromium, and zinc, the solution is first treated with sodium hydroxide and then with hydrogen peroxide. The precipitations which first occur, as sodium hydroxide is added to the solution, are described by the following:

$$Mn^{2+} + 2OH^- \rightleftharpoons \underline{Mn(OH)_2} \text{ (white)}$$
$$Fe^{3+} + 3OH^- \rightleftharpoons \underline{Fe(OH)_3} \text{ (reddish brown)}$$
$$Al^{3+} + 3OH^- \rightleftharpoons \underline{Al(OH)_3} \text{ (white)}$$
$$Cr^{3+} + 3OH^- \rightleftharpoons \underline{Cr(OH)_3} \text{(dark green or blue)}$$
$$Zn^{2+} + 2OH^- \rightleftharpoons \underline{Zn(OH)_2} \text{ (white)}$$

The precipitate is not separated, and the addition of an excess of NaOH dissolves the amphoteric hydroxides of aluminum, chromium, and zinc, but not the hydroxides of iron and manganese.

$$Al(OH)_3 + OH^- \rightleftharpoons [Al(OH)_4]^- \text{ (aluminate) (colorless)}$$
$$Cr(OH)_3 + OH^- \rightleftharpoons [Cr(OH)_4]^- \text{ (chromite) (green)}$$
$$Zn(OH)_2 + 2OH^- \rightleftharpoons [Zn(OH)_4]^{2-} \text{ (zincate) (colorless)}$$

Hydrogen peroxide is added to the resulting mixture to oxidize manganese(II) hydroxide to a mixture of manganese dioxide and manganese(III) hydroxide, both of which are much less soluble than $Mn(OH)_2$.

$$Mn(OH)_2 + H_2O_2 \longrightarrow MnO_2 + 2H_2O$$
$$2Mn(OH)_2 + H_2O_2 \longrightarrow 2Mn(OH)_3$$

Hydrogen peroxide oxidizes the chromite ion, $[Cr(OH)_4]^-$, to the chromate ion, CrO_4^{2-}. This is desirable because the reactions of the chromate ion permit better identification of chromium than do those of either Cr^{3+} or $[Cr(OH)_4]^-$.

$$2[Cr(OH)_4]^- + 3H_2O_2 + 2OH^- \longrightarrow 2CrO_4^{2-} + 8H_2O$$

The mixture containing MnO_2 and $Mn(OH)_2$ is brown and solutions of CrO_4^{2-} are yellow.

PROCEDURE 4. *Solution from (2): Mn^{2+}, Fe^{2+}, Al^{3+}, Cr^{3+}, and Zn^{2+}.* **Transfer the solution to a casserole, add 1 ml of 4 M HNO₃, and evaporate the solution to a moist residue. Take up the residue in 1 ml of water and transfer the solution to a test tube. Add 10 drops of 4 M NaOH beyond the amount of this reagent that is required to initiate precipitation. Now add 6 drops of 3 per cent hydrogen peroxide to the mixture and heat it in a hot water bath for 5 minutes. Separate the mixture and wash the residue with 10 drops of water to which has been added 1 drop of 4 M NaOH. Save the combined centrifugates for Procedure (6).**

42.5 Separation and Identification of Manganese and Iron

The residue containing MnO_2, $Mn(OH)_3$, and $Fe(OH)_3$ is dissolved by a mixture of HNO_3 and $NaNO_2$. The nitrite ion in acid solution reduces the manganese to the divalent condition. The iron(III) hydroxide is dissolved by nitric acid. The equations for these reactions are

$$2Mn(OH)_3 + 4H^+ + NO_2^- \longrightarrow 2Mn^{2+} + NO_3^- + 5H_2O$$
$$MnO_2 + 2H^+ + NO_2^- \longrightarrow Mn^{2+} + NO_3^- + H_2O$$
$$Fe(OH)_3 + 3H^+ \longrightarrow Fe^{3+} + 3H_2O$$

The presence of Fe^{3+} is confirmed by its reaction with the thiocyanate ion, CNS^-, to produce a blood-red complex ion, $[Fe(CNS)]^{2+}$, or one of the possible complexes containing from one to six CNS^- groups.

$$Fe^{3+} + CNS^- \rightleftharpoons [Fe(CNS)]^{2+}$$

This test for iron may be conducted on a solution containing Mn^{2+} without interference by this ion. It should be noted that there are traces of Fe^{3+} existing in many reagents. Therefore a light-pink color in this test may be due to iron as an impurity. Black tests can be run on the reagents used in the test to establish this point.

Manganese may be identified in the presence of iron by oxidizing it to the permanganate ion by means of sodium bismuthate in nitric acid. The MnO_4^- ion is purple but in dilute solutions it may appear pink.

$$2Mn^{2+} + 5BiO_3^- + 14H^+ \longrightarrow 2MnO_4^- + 5Bi^{3+} + 7H_2O$$

PROCEDURE 5. *Residue from (4): Mn(OH)$_3$, MnO$_2$, and Fe(OH)$_3$.* Treat the residue with 1 ml of 4 M HNO$_3$ and 2 drops of 1 M NaNO$_2$. Stir the mixture and heat it in a hot water bath. Separate any residue that remains. Heat the solution to boiling, then cool it, and divide it into two parts.

5a. *Test for Iron.* Dilute one part of the solution to 1 ml and add 2 or 3 crystals of NH$_4$CNS. If iron is present, a dark, blood red color will develop in the solution. (Faint or pale redness indicates a trace of iron due to impurities; a red color that does not persist indicates the presence of reducing agents but does not invalidate the test.)

5b. *Test for Manganese.* To the second part of the solution from (5) add a small quantity of solid NaBiO$_3$ and a few drops of 4 M HNO$_3$. The formation of a pink or purple color which persists confirms the presence of manganese. (Pink or purple which does not persist indicates the presence of reducing agents but does not invalidate the test.)

42.6   Separation and Identification of Aluminum

Acetic acid is added to the solution containing the $[Al(OH)_4]^-$, CrO_4^{2-}, and $[Zn(OH)_4]^{2-}$. The aluminum and zinc portray their basic character in the acidic solution and are converted to their cation forms.

$$[Al(OH)_4]^- + 4H^+ \rightleftharpoons Al^{3+} + 4H_2O$$
$$[Zn(OH)_4]^{2-} + 4H^+ \rightleftharpoons Zn^{2+} + 4H_2O$$

Acids react with yellow chromate, CrO_4^{2-}, and convert it in part to the dichromate, $Cr_2O_7^{2-}$, the extent of the conversion depending upon the concentration of the hydrogen ion.

$$2CrO_4^{2-} + 2H^+ \rightleftharpoons Cr_2O_7^{2-} + H_2O$$

The addition of excess aqueous ammonia precipitates aluminum as the hydroxide, converts $Cr_2O_7^{2-}$ to CrO_4^{2-}, and complexes the zinc as the soluble $[Zn(NH_3)_4]^{2+}$ ion.

$$Al^{3+} + 3NH_3 + 3H_2O \rightleftharpoons \underline{Al(OH)_3} + 3NH_4^+$$
$$Cr_2O_7^{2-} + 2OH^- \rightleftharpoons 2CrO_4^{2-} + H_2O$$
$$Zn^{2+} + 4NH_3 \rightleftharpoons [Zn(NH_3)_4]^{2+}$$

The presence of aluminum is confirmed by dissolving the hydroxide in acetic acid and adding the aluminum reagent and $(NH_4)_2CO_3$. Aurin tricarboxylic acid (aluminon) imparts a red color to the $Al(OH)_3$ which is formed.

PROCEDURE 6. *Solution from (4):* $[Al(OH)_4]^-$, CrO_4^{2-}, *and* $[Zn(OH)_4]^{2-}$. **Add 4 *M* HOAc to the solution until it is acid to litmus and then add two or three drops of the acid in excess. Now add 4 *M* aqueous ammonia until the solution is distinctly alkaline to litmus. If a white gelatinous precipitate forms, it is probably aluminum hydroxide. Separate the precipitate and reserve the solution for Procedure (7). Confirm the presence of aluminum by dissolving the precipitate in 4 *M* HOAc and adding 2 drops of Aluminum Reagent and enough 1 *M* $(NH_4)_2CO_3$ to make the solution basic. The formation of a reddish-colored precipitate confirms the presence of aluminum.**

42.7 Identification of Chromium and Zinc

The presence of the chromate ion is confirmed by precipitating yellow lead chromate. Zinc ions do not interfere with the chromate test.

The zinc ions are precipitated by sulfide ions from a thioacetamide solution, white zinc sulfide being formed. The chromate ion may oxidize part of the sulfide ions used in the test to elemental sulfur, but ZnS is soluble in hydrochloric acid whereas sulfur is not. A second confirmatory test for zinc may be made by boiling all of the hydrogen sulfide out of the solution and precipitating the zinc as the hexacyanoferrate(II), which is white.

$$2K^+ + Zn^{2+} + Fe(CN)_6^{4-} \longrightarrow \underline{K_2Zn[Fe(CN)_6]}$$

PROCEDURE 7. *Solution from (6):* CrO_4^{2-} *and* $[Zn(NH_3)_4]^{2+}$. **Divide the solution into two parts.**

7a. *Test for Chromium.* **To one part of the solution add 1 *M* HOAc until the solution is acid to litmus. Then add 2 drops of 0.1 *M* $Pb(OAc)_2$. The formation of a yellow precipitate, $PbCrO_4$, confirms the presence of chromium.**

7b. *Test for Zinc.* **To the other part of the solution add 5 drops of 5 per cent thioacetamide solution and heat the mixture in a hot water bath. The formation of a white precipitate, ZnS, which is soluble in 4 *M* HCl, indicates the presence of zinc. Heat the solution to expel the excess hydrogen sulfide and then neutralize it with 4 *M* aqueous ammonia. Now add 10 drops of 1 *M* HCl and 5 drops of 0.1 *M* $K_4[Fe(CN)_6]$. The formation of a white precipitate, $K_2Zn[Fe(CN)_6]$, proves the presence of zinc.**

QUESTIONS

1. A Group III unknown is colorless. What cations are probably absent? Why should one not rely definitely upon such an observation?
2. Why do aluminum and chromium precipitate as hydroxides rather than as sulfides in Group III?
3. What is the function of the ammonium chloride used in the Group III reagent?
4. Why do the sulfides of Co^{2+}, Ni^{2+}, Mn^{2+}, Fe^{2+}, and Zn^{2+} precipitate in an ammonium sulfide solution but not in a 0.3 M HCl solution of the hydrogen sulfide?
5. Why does iron precipitate as iron(II) sulfide rather than iron(III) sulfide in an ammonium sulfide solution?
6. Cite two examples of the use of complex ions in the analysis of Group III.
7. Account for the fact that CoS and NiS fail to precipitate in the 0.3 M HCl solution of Group II, yet they dissolve only very slowly in 1 M HCl.
8. Give the color of each of the following: $Fe(OH)_3$, Fe^{3+}, $Fe(OH)_2$, Fe^{2+}, $[Fe(CNS)]^{2+}$, $[FeF_6]^{3-}$.
9. Write equations showing the amphoteric nature of the hydroxides of zinc, chromium, and aluminum.
10. Outline the separation of the following groups of ions leaving out all unnecessary steps. (a) Ni^{2+}, Mn^{2+}, Zn^{2+}; (b) Hg_2^{2+}, Hg^{2+}, Cu^{2+}, Fe^{2+}; (c) Cd^{2+}, Co^{2+}, Ca^{2+}.
11. Select a reagent used in Group III that will separate each of the following pairs. (a) Al^{3+}, Zn^{2+}; (b) $[Zn(NH_3)_4]^{2+}$, CrO_4^{2-}; (c) CoS, ZnS; (d) Mn^{2+}, Mg^{2+}; (e) Fe^{3+}, Al^{3+}.
12. Why will $PbCrO_4$ precipitate when Pb^{2+} is added to a solution made up from $K_2Cr_2O_7$?
13. When and why is fluoride added in the test for cobalt using thiocyanate?
14. Show by the proper formulas that manganese can act as either a metal or a nonmetal, depending upon its oxidation state.

PROBLEMS

1. What concentration of NH_4^+ must be present in 0.20 M aqueous ammonia to prevent the precipitation of $Mg(OH)_2$ if the solution contains 1.0×10^{-3} mole of Mg^{2+} per liter? (K_{sp} for $Mg(OH)_2$ is 1.5×10^{-11}.)   *Ans. 0.029 M*
2. A solution from the Group II separation is 0.3 M in HCl and has a volume of 5.0 ml. Calculate the number of drops of 15 M aqueous ammonia required to react with the HCl (assume that 1 drop is equal to 0.05 ml).
Ans. 2 drops
3. Calculate the concentration of Pb^{2+} in a saturated solution of $PbCrO_4$. (K_{sp} for $PbCrO_4$ is 1.8×10^{-14}.)   *Ans. 1.3 $\times$ 10^{-7} M*
4. Calculate the concentration of Pb^{2+} in a saturated solution of $PbCrO_4$ in the presence of 1.0×10^{-3} M Na_2CrO_4.   *Ans. 1.8 $\times$ 10^{-11} M*
5. Using 1×10^{-19} for the solubility product of FeS, show that 0.3 M HCl should dissolve or prevent the precipitation of FeS.
Ans. These conditions will permit an iron(II) ion concentration of 7 M

The Analysis of Group IV

43

The solution to be analyzed may be a Group IV known or unknown, or it may be the solution from the Group III separation. The Group IV flow sheet follows. Note also that the chemistry of the metals of analytical Group IV is discussed in Chapter 37, to which you may find it useful to refer.

GROUP IV   Flow Sheet

$$\left.\begin{array}{c} Ba^{2+} \\ Sr^{2+} \\ Ca^{2+} \end{array}\right\} \xrightarrow[\substack{NH_4Cl \\ NH_3 + H_2O}]{(NH_4)_2CO_3} \left.\begin{array}{c} BaCO_3 \\ SrCO_3 \\ \underline{CaCO_3} \end{array}\right\} \xrightarrow{HOAc} \left.\begin{array}{c} Ba^{2+} \\ Sr^{2+} \\ Ca^{2+} \end{array}\right\} \xrightarrow[\substack{HOAc \\ NH_4OAc}]{K_2CrO_4} \left.\begin{array}{c} \underline{BaCrO_4} \\ Sr^{2+} \\ Ca^{2+} \end{array}\right\} \xrightarrow[\substack{K_2CrO_4 \\ alcohol}]{NH_3 + H_2O}$$

$$\underline{BaCrO_4} \xrightarrow{HCl} Ba^{2+} \xrightarrow{H_2SO_4} \underline{BaSO_4}\ (white)$$

$$\underline{SrCrO_4}\ (yellow)$$

$$Ca^{2+} \xrightarrow{(NH_4)_2C_2O_4} \underline{CaC_2O_4}\ (white)$$

43.1   Precipitation of the Group IV Ions

Analytical Group IV contains the metallic ions Ba^{2+}, Sr^{2+}, and Ca^{2+}. These metals form chlorides, sulfides, and hydroxides which are soluble under the conditions which prevail in the precipitation of Groups I, II, and III. The carbonates of barium, strontium, and calcium precipitate in aqueous ammonia solutions containing ammonium carbonate.

The concentration of the carbonate ion in a solution of ammonium carbonate is too low to effect a complete precipitation of barium, strontium, and calcium due to the partial hydrolysis of the carbonate ion.

$$CO_3^{2-} + H_2O \rightleftharpoons HCO_3^- + OH^-$$

This hydrolysis is repressed by increasing the hydroxide ion concentration of the solution, the additional hydroxide ions being supplied by the buffer pair—aqueous ammonia and ammonium chloride. These conditions permit a carbonate concentration high enough to precipitate $BaCO_3$, $SrCO_3$, and $CaCO_3$, but not $MgCO_3$. At the same time the hydroxide ion concentration in this buffered solution is not high enough to precipitate $Mg(OH)_2$.

In order that the solution may have the required ammonium ion concentration, the centrifugate from the Group III separation, which contains ammonium salts, is first evaporated to dryness and then heated strongly to expel the ammonium salts present. The equations are

$$NH_4Cl \longrightarrow NH_3\uparrow + HCl\uparrow$$
$$NH_4NO_3 \longrightarrow NH_3\uparrow + HNO_3\uparrow$$
$$NH_3 + HNO_3 \longrightarrow N_2O\uparrow + 2H_2O\uparrow$$

The necessary ammonium ion concentration is then obtained by adding the Group IV reagent, which consists of ammonium carbonate, ammonium chloride, and aqueous ammonia in the required concentrations. The equations for the precipitations of the Group IV carbonates are

$$Ba^{2+} + CO_3{}^{2-} \rightleftharpoons BaCO_3 \text{ (white)}$$
$$Sr^{2+} + CO_3{}^{2-} \rightleftharpoons SrCO_3 \text{ (white)}$$
$$Ca^{2+} + CO_3{}^{2-} \rightleftharpoons CaCO_3 \text{ (white)}$$

PROCEDURE 1. *Precipitation of the Group IV Ions: Ba²⁺, Sr²⁺, Ca²⁺.* **Evaporate the solution (10 drops of a Group IV known or unknown or the solution from the Group III separation) to dryness and ignite in a casserole to expel ammonium salts. Dissolve the residue in a mixture of 1 drop of 12 M HCl and 12 drops of water. Make the solution alkaline by adding 4 M aqueous ammonia. Add just one drop of aqueous ammonia in excess. Add 2 drops of 1 M (NH₄)₂CO₃, or more if necessary, to effect complete precipitation, and warm the mixture in a hot water bath. Allow the mixture to cool; separate the precipitate and reserve the solution for the Group V analysis.**

43.2   The Separation and Identification of Barium

■ **1. Dissolution of the Carbonates.**   The carbonates of barium, strontium, and calcium are dissolved readily by strong acids such as hydrochloric acid or by weak acids such as acetic acid. The latter acid is used in these procedures, and the theory involved in these dissolutions is discussed in detail in Section 19.11. The equilibria involved, as applied to barium carbonate, are given by the following equations:

$$BaCO_3 \rightleftharpoons Ba^{2+} + CO_3{}^{2-}$$
$$HOAc \rightleftharpoons H^+ + OAc^-$$
$$CO_3{}^{2-} + H^+ \rightleftharpoons HCO_3{}^-$$
$$HCO_3{}^- + H^+ \rightleftharpoons H_2CO_3$$
$$H_2CO_3 \rightleftharpoons H_2O + CO_2\uparrow$$

If the concentration of the hydrogen ion supplied by acetic acid is sufficiently high, the quantity of H_2CO_3 formed will exceed that in a saturated solution of CO_2 at atmospheric pressure and CO_2 will escape from solution. The loss of carbon dioxide from the system in this way causes all the above equilibria to be shifted to the right until the precipitate dissolves completely. The net reaction is given by the equation

$$\underline{BaCO_3} + 2HOAc \rightleftharpoons Ba^{2+} + 2OAc^- + H_2O + CO_2\uparrow$$

Similar equations may be written for the dissolution of the carbonates of strontium and calcium in acetic acid.

■ **2. Separation of Barium from Strontium and Calcium.**   By taking advantage of the fact that barium chromate is less soluble than strontium chromate and calcium chromate, barium can be separated from strontium and calcium. The solubility product constants are 2×10^{-10} for $BaCrO_4$ and 3.6×10^{-5} for $SrCrO_4$. Calcium chromate is relatively very soluble. The problem in this separation, then, is to control the concentration of the chromate ion so that the concentration of the barium ion will be reduced to at least 0.0001 M, and strontium chromate will not be precipitated from a solution 0.1 M in the strontium ion. Let us first calculate the concentration of chromate ion required to reduce the barium ion concentration to 0.0001 M.

$$[Ba^{2+}][CrO_4^{2-}] = 2 \times 10^{-10}$$

$$[CrO_4^{2-}] = \frac{2 \times 10^{-10}}{[Ba^{2+}]} = \frac{2 \times 10^{-10}}{1 \times 10^{-4}} = 2 \times 10^{-6} \, M$$

Now let us determine the maximum concentration of chromate ion that we can have without precipitating any strontium chromate when the concentration of the strontium ion is 0.1 M.

$$[Sr^{2+}][CrO_4^{2-}] = 3.6 \times 10^{-5}$$

$$[CrO_4^{2-}] = \frac{3.6 \times 10^{-5}}{[Sr^{2+}]} = \frac{3.6 \times 10^{-5}}{10^{-1}} = 3.6 \times 10^{-4} \, M$$

Thus, we find that the concentration of the chromate ion must be held between $2 \times 10^{-6} \, M$ and $3.6 \times 10^{-4} \, M$ to cause a nearly complete precipitation of $BaCrO_4$, but not the precipitation of any $SrCrO_4$.

In a solution containing chromate ions, we have the following equilibrium:

$$Cr_2O_7^{2-} + H_2O \rightleftharpoons 2CrO_4^{2-} + 2H^+$$

The concentration of chromate ions in a solution made by dissolving a dichromate in water can be controlled by adjusting the hydrogen ion concentration. Applying the law of chemical equilibrium to this system,

$$\frac{[CrO_4^{2-}]^2[H^+]^2}{[Cr_2O_7^{2-}]} = 2.4 \times 10^{-15}$$

It is readily seen from this expression that the chromate ion concentration is a function of the hydrogen ion concentration. The more acidic the solution, the lower will be the chromate ion concentration; and the more basic the solution, the higher will be the chromate ion concentration. By substituting in the above expression, it can be shown that for a solution which is about 0.01 M in dichromate ions, the hydrogen ion concentration must be adjusted to about $1 \times 10^{-5} M$ ($pH = 5$) to maintain the chromate ion concentration at about $5 \times 10^{-4} M$. As shown above, this concentration of chromate ions is approximately the value needed for the separation of Ba^{2+} from Sr^{2+}. A buffered solution composed of acetic acid and ammonium acetate is used to obtain the required hydrogen ion concentration.

As chromate ions are removed from the system through precipitation of barium chromate, the concentration of these ions does not change appreciably. This is true because the excess of dichromate serves as a reservoir for chromate ions; and as CrO_4^{2-} is used up in precipitating $BaCrO_4$, the equilibrium $Cr_2O_7^{2-} + H_2O \rightleftharpoons 2CrO_4^{2-} + 2H^+$ shifts to the right with the generation of more CrO_4^{2-} from the relatively large supply of $Cr_2O_7^{2-}$. The hydrogen ions that are formed unite with the acetate ions of the buffered solution, and a constant pH is maintained.

■ **3. Dissolution of Barium Chromate.** Even though barium chromate is precipitated from a weakly acidic solution, it is readily dissolved by strong acids. The hydrogen ion concentration of solutions of strong acids is sufficiently high to cause a reduction in the chromate ion concentration below the value required to maintain a saturated solution of barium chromate. Thus,

$$[Ba^{2+}][CrO_4^{2-}] < K_{sp}$$

and the barium chromate dissolves. A solution of 12 M HCl is used in this scheme to bring about the dissolution of $BaCrO_4$.

■ **4. Identification of Barium.** The presence of the barium ion in the solution is confirmed by precipitating it as barium sulfate, with sulfuric acid being used as the precipitant.

Additional evidence for the presence of the barium ion is obtained using the flame test. When heated in the Bunsen flame, barium salts impart to it a yellow-green color.

PROCEDURE 2. *Precipitate from (1): BaCO₃, SrCO₃, CaCO₃.* **Dissolve the precipitate in a mixture of 2 drops of 4 M HOAc and 4 drops of 1 M NH₄OAc. Add 1 drop of 1 M K₂CrO₄ to the solution. The formation of a yellow precipitate indicates the presence of barium. Separate the mixture and reserve the solution for Procedure (3). Dissolve the precipitate (BaCrO₄) in 2 drops of 12 M HCl. Make a flame test (Section 39.17) on the solution for the barium ion. The barium ion imparts a weak and fleeting greenish-yellow color to the Bunsen flame. Add 1 drop of 4 M H₂SO₄ to the remainder of the solution. A white precipitate (BaSO₄) confirms the presence of barium.**

43.3   The Separation and Identification of Strontium

■ **1. Separation of Strontium from Calcium.** Although the chromate ion concentration in an acetic acid-ammonium acetate buffered solution is too low to exceed the solubility product of strontium chromate, this compound will precipitate when the solution is made basic with aqueous ammonia. The equilibria are

$$Cr_2O_7{}^{2-} + H_2O \rightleftharpoons 2CrO_4{}^{2-} + 2H^+$$
$$2OH^- + 2H^+ \rightleftharpoons 2H_2O$$

and the net reaction is

$$Cr_2O_7{}^{2-} + 2OH^- \rightleftharpoons 2CrO_4{}^{2-} + H_2O$$

The resulting concentration of chromate ions is large enough so that the solubility product constant of strontium chromate is exceeded and precipitation occurs. The complete precipitation of strontium chromate is insured by further reducing its solubility through the addition of enough ethyl alcohol to the solution to make it 50 per cent alcohol by volume. Calcium chromate is soluble in this solution.

■ **2. Identification of Strontium.** The formation of a fine yellow crystalline precipitate of the chromate serves to confirm the presence of strontium. A crimson color imparted to the Bunsen flame is characteristic of the strontium ion.

PROCEDURE 3. *Solution from (2): Sr^{2+} and Ca^{2+}.* **Add 4 *M* aqueous ammonia to the solution until the color changes from orange to yellow. Now add a volume of ethyl alcohol equal to the volume of the solution. The formation of a fine yellow precipitate indicates $SrCrO_4$. Separate the mixture and reserve the solution for Procedure (4). Dissolve the precipitate in 2 drops of 12 *M* HCl and make a flame test on the solution for the strontium ion. A crimson color imparted to the flame is characteristic of the strontium ion.**

43.4   The Identification of Calcium

The solution from the strontium chromate separation contains the calcium ion. The addition of ammonium oxalate will precipitate calcium oxalate, CaC_2O_4, a white crystalline salt.

$$Ca^{2+} + C_2O_4{}^{2-} \rightleftharpoons \underline{CaC_2O_4}$$

Calcium ions impart a brick-red color to the Bunsen flame.

PROCEDURE 4. *Solution from (3): Ca^{2+}.* **Heat the solution to boiling and add 2 drops of 0.4 *M* $(NH_4)_2C_2O_4$. The formation of a white precipitate, which may form slowly, confirms the presence of calcium. Dissolve the precipitate in 12 *M* HCl and make a flame test on the solution. Calcium ions impart a dark-red color to the flame.**

PROCEDURE 5. *Original Solution: Ba^{2+}, Sr^{2+}, Ca^{2+}.* **Make flame tests on the original solution and compare with tests on known solutions which you make up yourself.**

QUESTIONS

1. Why are ammonium salts expelled prior to the precipitation of Group IV?
2. What reagents compose the Group IV precipitant? What is the function of each of these chemicals in the group precipitation?
3. In terms of ionic equilibria and solubility product theory, explain the dissolution of $BaCrO_4$ in HCl.
4. Barium chromate is precipitated from a solution of potassium chromate containing acetic acid. Why is the acetic acid present?
5. Write equations representing the various ionic equilibria involved in the dissolution of barium carbonate in acetic acid.
6. What are the characteristic colors imparted to the Bunsen flame by Ca^{2+}, Sr^{2+}, and Ba^{2+}?
7. Account for the color change from orange to yellow when aqueous ammonia is added to an acetic acid solution of potassium dichromate.
8. If the magnesium concentration in the unknown is quite high, or if too large an amount of carbonate is added in precipitating Group IV, some magnesium carbonate will also precipitate. What effect would such an error have on the rest of the analysis scheme for Group IV?
9. Explain why ammonium carbonate rather than sodium carbonate is used in the precipitation of Group IV.
10. What characteristic properties cause Ca^{2+}, Sr^{2+}, and Ba^{2+} to be members of the same group of the analytical scheme?
11. Outline the separation of the following groups of ions leaving out all unnecessary steps. (a) Ag^+, Cu^{2+}, Zn^{2+}, Ca^{2+}, Na^+; (b) Fe^{3+}, Ba^{2+}, Mg^{2+}; (c) Pb^{2+}, Sn^{4+}, Al^{3+}, Sr^{2+}.

PROBLEMS

1. What total weight of oxalates can be obtained from a solution containing 3.00 mg each of Ba^{2+}, Sr^{2+}, and Ca^{2+}?   *Ans. 20.5 mg*
2. Calculate the hydroxide ion concentration of a solution that is 0.40 M in NH_3 and 0.30 M in NH_4^+.   *Ans. 2.4 × 10⁻⁵ M*
3. If sodium sulfate is added to a mixture originally 0.010 M in Sr^{2+} and 0.010 M in Ba^{2+}, what per cent of the barium remains unprecipitated before any $SrSO_4$ precipitates?   *Ans. 0.039%*
4. What concentration of carbonate ion is required to initiate precipitation of (a) $CaCO_3$ from 0.010 M Ca^{2+}; (b) $SrCO_3$ from 0.010 M Sr^{2+}? (K_{sp} for $CaCO_3$ is 4.8 × 10⁻⁹; K_{sp} for $SrCO_3$ is 9.42 × 10⁻¹⁰.)
                    Ans. (a) 4.8 × 10⁻⁷ M; (b) 9.42 × 10⁻⁸ M
5. How many drops (assume 1 drop = 0.050 ml) of 1.0 M acetic acid would be required to dissolve 5 mg of $SrCO_3$?   *Ans. 2 drops*
6. Will the concentration of OH^- in Problem 2 be large enough to cause $Mg(OH)_2$ to precipitate in a 0.010 M solution of Mg^{2+}? (K_{sp} for $Mg(OH)_2$ is 1.5 × 10⁻¹¹.)   *Ans. No*

The Analysis of Group V

44

The solution for analysis may be a Group V known or unknown or it may be the solution from the Group IV separation. The flow sheet for the Group V analytical scheme is given below.

GROUP V Flow Sheet

$$
\begin{array}{l}
\left.\begin{array}{l} Na^+ \\ K^+ \\ Mg^{2+} \\ NH_4^+ \end{array}\right\}
\end{array}
$$

$\xrightarrow[OAc^-, \, H_2O]{Zn^{2+}, \, UO_2^{2+}}$ $NaZn(UO_2)_3(OAc)_9 \cdot 6H_2O$ (yellow)

$\xrightarrow[NH_3, \, H_2O]{NH_4^+, \, HPO_4^{2-}}$ $MgNH_4PO_4$ (white)

$\left.\begin{array}{l} K^+ \\ Na^+ \\ NH_4^+ \end{array}\right\}$ $\xrightarrow[heat]{HNO_3}$ $NH_3\uparrow, N_2O\uparrow$ $\left.\begin{array}{l} K^+ \\ Na^+ \end{array}\right\}$ $\xrightarrow{Na_3[Co(NO_2)_6]}$ $K_2Na[Co(NO_2)_6]$ (yellow)

$\xrightarrow[heat]{NaOH}$ $NH_3\uparrow$ $\xrightarrow{moist \, red \, litmus}$ blue litmus

The chemistry of the metals of analytical Group V is considered also in Chapter 38, to which you may wish to refer.

44.1   Removal of Traces of Calcium and Barium

The cations composing Group V (Na^+, K^+, Mg^{2+}, and NH_4^+) do not form precipitates with the reagents used to separate Groups I, II, III, and IV, and the ions of this group have no precipitant which is common to all four ions.

If ammonium salts were present in excessive quantities during the precipitation of the carbonates of the Group IV ions, then trace amounts of barium and calcium ions may have come through into the Group V solution. Because these ions will interfere with the identification of the Group V ions, they must be removed. Ammonium oxalate is added to precipitate the calcium ions and ammonium sulfate to precipitate the barium ions.

PROCEDURE 1. *Solution from the Group IV Separation: Na$^+$, K$^+$, Mg^{2+}, NH$_4^+$, (Ca^{2+}), (Ba^{2+}). Add to this solution 1 drop of 0.4 M (NH$_4$)$_2$C$_2$O$_4$ and 1 drop of 1 M (NH$_4$)$_2$SO$_4$. Separate and discard any precipitate that may form and use the solution for Procedure (2).*

44.2   Identification of Sodium

After the removal of the traces of barium and calcium, the sodium ion may be detected by the formation of the triple salt, sodium zinc uranyl acetate hexahydrate, which is a yellow crystalline compound.

$$Na^+ + Zn^{2+} + 3UO_2^{2+} + 9OAc^- + 6H_2O \rightleftharpoons \underline{NaZn(UO_2)_3(OAc)_9 \cdot 6H_2O}$$

The Sodium Reagent is a saturated solution of zinc acetate and uranyl acetate in acetic acid. Ammonium, magnesium, and potassium ions do not interfere with the test. The sodium ion imparts a yellow color to the Bunsen flame.

PROCEDURE 2. *Solution from (1): K$^+$, Na$^+$, Mg^{2+}, NH$_4^+$. To 2 drops of the solution from (1) or Group V known or unknown, add 1 M HOAc until the solution is acid to litmus. Add 1 drop of this acidified solution to 5 drops of Sodium Reagent. Shake the mixture and set it aside for an hour. The formation of a yellow crystalline precipitate indicates the presence of sodium.*

44.3   Separation and Identification of Magnesium

To a second portion of the solution for analysis is added an excess of aqueous ammonia and disodium hydrogen phosphate. The formation of a white crystalline precipitate, $MgNH_4PO_4$, is evidence of the presence of the magnesium ion. The equation is

$$Mg^{2+} + NH_4^+ + HPO_4^{2-} \rightleftharpoons \underline{MgNH_4PO_4} + H^+$$

Note that the hydrogen ion is one product of the reaction. Completeness of precipitation of the magnesium ammonium phosphate is assured by making the solution basic with aqueous ammonia, which combines with the hydrogen ion and causes a shift of the equilibrium to the right.

After the magnesium ammonium phosphate is separated from the solution containing the sodium and ammonium ions, it is dissolved in acetic acid. This treatment with acid causes the following equilibrium to be shifted to the right.

$$MgNH_4PO_4 + H^+ \rightleftharpoons Mg^{2+} + NH_4^+ + HPO_4^{2-}$$

This results because the phosphate ion in a saturated solution of $MgNH_4PO_4$ is removed from the solution by combining with hydrogen ions, making the product of the concentrations of the Mg^{2+}, NH_4^+, and PO_4^{3-} ions less than the solubility product constant. Magnesium Reagent and sodium hydroxide are added to the solution containing magnesium ions. The magnesium hydroxide which is formed adsorbs the Magnesium Reagent (a dye, *p*-nitrobenzene-azo-alpha-naphthol) and forms a blue precipitate, which confirms the presence of magnesium.

PROCEDURE 3. *Use 10 drops of the solution from (1) or 10 drops of Group V known or unknown: K⁺, Na⁺, Mg²⁺, NH₄⁺.* Add 4 M aqueous ammonia to the solution until it is alkaline to litmus and then add 1 drop in excess. Now add 2 drops of 1 M Na_2HPO_4 to the solution. The formation of a white crystalline precipitate, often slow in forming, confirms the presence of magnesium. Separate the mixture and save the solution for Procedure (4). Dissolve the precipitate in a mixture of 2 drops of 1 M HOAc and 3 drops of water. Add 1 drop of Magnesium Reagent and an excess of 4 M NaOH to the solution. The formation of a blue precipitate confirms the presence of magnesium. (Do not confuse the sky-blue color of the precipitate with the purple color of the Magnesium Reagent. Be sure a precipitate exists.)

44.4 Identification of Potassium

The identification of potassium is made from the solution obtained from the separation of magnesium ammonium phosphate. Ammonium ions must be removed prior to the test for the potassium ion because the precipitant for potassium ions will also precipitate ammonium ions. Concentrated nitric acid is added to the solution, it is evaporated to dryness, and then it is ignited to expel volatile ammonium salts.

$$NH_4Cl \longrightarrow NH_3\uparrow + HCl\uparrow$$
$$NH_4NO_3 \longrightarrow NH_3\uparrow + HNO_3\uparrow$$
$$NH_3 + HNO_3 \longrightarrow N_2O\uparrow + 2H_2O\uparrow$$

The potassium ion is precipitated from solution in the form of the yellow complex salt, $K_2Na[Co(NO_2)_6]$, upon the addition of a solution of $Na_3[Co(NO_2)_6]$.

$$2K^+ + Na^+ + [Co(NO_2)_6]^{3-} \longrightarrow \underline{K_2Na[Co(NO_2)_6]}$$

Potassium gives a characteristic violet flame test, which is masked by the intense yellow color produced by the sodium ion. By observing the flame through cobalt glass, which filters out the yellow color of the sodium ion, one may detect the potassium ion in the presence of the sodium ion.

PROCEDURE 4. *Solution from (3): K⁺, Na⁺, NH₄⁺.* Add 4 drops of concentrated HNO_3 to the solution held in a casserole, evaporate it to dryness, and then heat the dry residue for several minutes. After the casserole is cool, add 1 drop of 1 M HCl and 5 drops of 4 M HOAc to the residue and boil the resultant solution. Now add 2 drops of this solution to 5 drops of a saturated solution of $Na_3[Co(NO_2)_6]$. The formation of a yellow precipitate confirms the presence of potassium.

PROCEDURE 5. *Original Known or Unknown Solution for Group V or Solution from Procedure (1).* Make flame tests on this solution and compare with known solutions of the ions in question. When testing for potassium, observe the flame through a cobalt glass. Sodium ions impart an intense yellow color to the flame. A trace of the sodium ion as contaminant will give a yellow coloration to the flame, so do not rely entirely upon the flame test in reporting the presence of sodium ions. Potassium ions give a violet color to a flame.

44.5 Identification of Ammonium Ions

Since ammonium ions are added in the course of the analysis at various points, it becomes necessary to test the original solution of a known or unknown for this ion. The presence of the ammonium ion in a solution may be detected by the addition of a strong, nonvolatile base such as sodium hydroxide, with subsequent heating of the solution and liberation of gaseous ammonia.

$$NH_4^+ + OH^- \text{ (from NaOH)} \rightleftharpoons NH_3{\uparrow} + H_2O$$

The gaseous ammonia may be detected by moist red litmus paper and by its characteristic odor.

$$NH_3 + H_2O \rightleftharpoons NH_4^+ + OH^-$$

During the heating of the solution one must not permit the sodium hydroxide solution to come in contact with the litmus paper because this would void the test for ammonia.

PROCEDURE 6. *Original Solution of the Known or Unknown.* Add 4 *M* NaOH to the solution until it is basic. Moisten a piece of red litmus with distilled water and place it on the convex side of a watch glass. Place the watch glass on a beaker containing the solution and warm the solution gently. Avoid spattering of the solution by overheating and do not permit the litmus to come into contact with the solution. If the litmus turns blue within a short time the presence of the ammonium ion is confirmed.

QUESTIONS

1. Why must the ammonium ion be removed before making the chemical test for the potassium ion?
2. Explain why the test for the ammonium ion must be made on a portion of the original sample during the analysis of either a Group V known or unknown or a general unknown involving all of the groups.
3. Why is Group V often referred to as the soluble group?
4. Why are $(NH_4)_2SO_4$ and $(NH_4)_2C_2O_4$ added prior to the analysis of Group V?
5. If magnesium carbonate had, through an error in carrying out the procedure, precipitated with Group IV, how would you determine its presence? What error or errors might have caused its precipitation with Group IV?
6. When $MgNH_4PO_4$ is heated to a high temperature (1,000° C) it is converted to magnesium pyrophosphate ($Mg_2P_2O_7$). Write a balanced equation for this reaction.

7. Write the equation for the dissolution of $MgNH_4PO_4$ in an acid. Explain this reaction in terms of solubility product constant and ionic equilibria theory.
8. Why may red litmus turn blue upon prolonged exposure to the air of the laboratory?
9. What is the chemistry of the chemical test for potassium?
10. Explain the use of cobalt glass in the flame test for potassium.

PROBLEMS

1. What weight of $K_2Na[Co(NO_2)_6]$ can be formed from 2.5 mg of K^+?

Ans. 14 mg

2. How many ml of NH_3 gas measured at 20° C and 750 mm pressure will be evolved when a solution containing 10 mg of NH_4Cl is treated with $NaOH$ and heated?

Ans. 4.6 ml

Anion Analysis

45

45.1 Introduction

Thirteen of the more common and important negative ions are included in this scheme of anion analysis. The anions considered are carbonate, sulfide, sulfite, nitrite, sulfate, nitrate, phosphate, metaborate, oxalate, fluoride, chloride, bromide, and iodide. Among the anions that were given consideration in the cation scheme of analysis are arsenite, arsenate, stannite, stannate, permanganate, aluminate, chromate, dichromate, and zincate.

Although several schemes of anion analysis involving the separation of the ions into groups have been developed, the procedures are in general more complicated and unreliable than are those for cation analysis. Instead of using a systematic scheme of analysis on the same solution throughout the analysis, the procedures outlined below involve a series of elimination tests which prove the absence of certain anions. Tests are then made on different samples of the unknown solution for the presence of the anions whose absence was not indicated in the preliminary elimination tests. (A flow sheet of the anion elimination tests is given on the following page.)

The properties of the acids and their anions have been considered in the first part of the text in connection with the descriptive chemistry of the nonmetals. The student will find it necessary to refer to those earlier chapters in order to answer some of the questions which arise concerning the chemistry of the anions.

The anion knowns and unknowns are in general furnished in the form of mixtures of dry salts due to the fact that some of the anions react with one another in solution. Thus, certain anions may be detected in freshly prepared solutions, whereas, upon standing, these same anions may have undergone decomposition. Certain combinations of anions react almost immediately in alkaline solutions

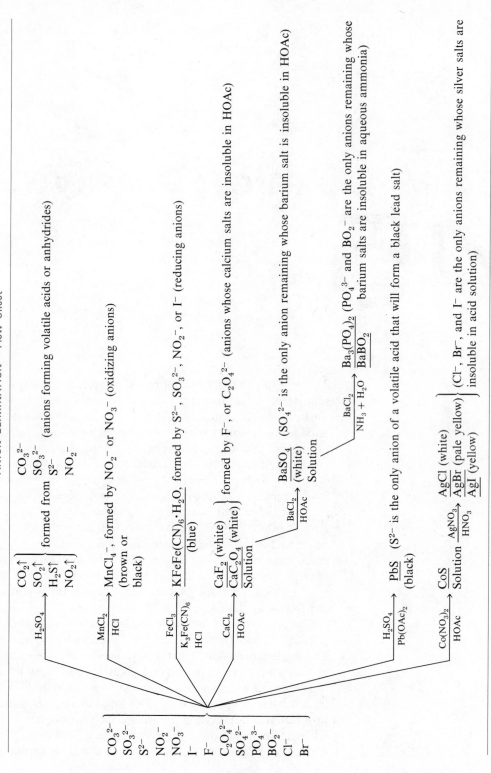

ANION ELIMINATION   Flow Sheet

while still other anions cannot exist together in acidic solutions. As an example of anion incompatibility, strong oxidizing anions usually react with strong reducing anions upon acidification.

Preliminary Elimination Tests for the Anions

45.2   Anion Elimination Chart

Prepare an Anion Elimination Chart like the one shown below. As you perform each elimination test, check off in the appropriate spaces those anions which you have found to be absent. When all the elimination tests have been made, it will be apparent which anions may be present. Tests will then be made for the anions which have not been eliminated.

Name: _____   Section: _____   Date: _____

Instructor's Approval: _____   Ions Found: _____

ANION ELIMINATION CHART

Anions	Test I	Test II	Test III	Test IV	Test V	Test VI	Test VII	Test VIII	Make Confirmatory Tests for
CO_3^{2-}									
SO_3^{2-}									
S^{2-}									
NO_2^-									
NO_3^-									
I^-									
F^-									
$C_2O_4^{2-}$									
SO_4^{2-}									
PO_4^{3-}									
BO_2^-									
Cl^-									
Br^-									

45.3 Elimination of the Anions Forming Volatile Acid Anhydrides

The ions CO_3^{2-}, SO_3^{2-}, S^{2-}, and NO_2^- are anions derived from the weak acids, H_2CO_3, H_2SO_3, H_2S, and HNO_2, respectively. As acids in solution, H_2CO_3, H_2SO_3, and HNO_2 are unstable and decompose, producing gases. Hydrogen sulfide does not decompose but has a limited solubility in water. When salts of these anions are treated with strong acids, the anions, being strong bases according to the proton theory of acids and bases, readily combine with the hydrogen ions from the strong acids. The equilibria are

$$CO_3^{2-} + H^+ \rightleftharpoons HCO_3^-$$
$$HCO_3^- + H^+ \rightleftharpoons H_2CO_3$$
$$H_2CO_3 \rightleftharpoons H_2O + CO_2\uparrow$$

$$SO_3^{2-} + H^+ \rightleftharpoons HSO_3^-$$
$$HSO_3^- + H^+ \rightleftharpoons H_2SO_3$$
$$H_2SO_3 \rightleftharpoons H_2O + SO_2\uparrow$$

$$S^{2-} + H^+ \rightleftharpoons HS^-$$
$$HS^- + H^+ \rightleftharpoons H_2S\uparrow$$

$$NO_2^- + H^+ \rightleftharpoons HNO_2$$
$$3HNO_2 \rightleftharpoons H^+ + NO_3^- + 2NO\uparrow + H_2O$$
$$2NO + O_2 \longrightarrow 2NO_2\uparrow$$

The high concentration of hydrogen ion furnished by the sulfuric acid used in this elimination test serves to shift the equilibrium to the right in each case. When the solubility of each gas is exceeded, it will escape from solution in the form of bubbles. Gentle heating decreases the solubility of the gases and aids in the test for gas-forming anions. Boiling of the solution should be avoided, because bubbles of steam may be mistaken for bubbles of gaseous reaction products.

If no gas is evolved upon acidification and gentle heating of an unknown sample, CO_3^{2-}, SO_3^{2-}, S^{2-}, and NO_2^- are absent. If a gas is evolved, then one or more of the gas-forming anions is present, and it becomes necessary to make tests for the presence of each of these ions unless they are proved absent by other elimination tests.

TEST I. *Elimination of Anions Forming Volatile Acid Anhydrides, CO_3^{2-}, SO_3^{2-}, S^{2-}, and NO_2^-.* Treat 25 mg of the solid unknown contained in a test tube with 2 drops of 1.5 M H_2SO_4. Examine the mixture for the formation of gas bubbles. If no gas is evolved, heat the mixture in the hot water bath. If no gas is evolved, then CO_3^{2-}, SO_3^{2-}, S^{2-}, and NO_2^- are proved to be absent. If these anions are absent, place a checkmark after the formula for each of these anions in the Test I column of your anion chart; leave these spaces blank if a gas is evolved.

If a gas is evolved, note its odor and color. Carbon dioxide is colorless and odorless. Sulfur dioxide is colorless and has the odor of burning sulfur. Hydrogen sulfide is colorless and exhibits a characteristic odor. If the nitrite ion is present, nitrogen dioxide, with a reddish-brown color and characteristic odor, is evolved. The solution may be pale-blue in color because of the presence of HNO_2 if the nitrite ion is present.

45.4   Preparation of the Solution for Analysis

The procedures outlined for the detection of the anions are applicable, in general, to salts containing only the alkali metals as cations. It is obvious that complications in the anion analysis will arise if certain metal ions are present to form precipitates, or possibly colored solutions, with the anions being detected.

Most of the metal ions may be removed from solution as insoluble carbonates, hydroxycarbonates, hydroxides, or oxides by treating a solution of the unknown with sodium carbonate. The ammonium ion, NH_4^+, is converted to NH_3, which is evolved from the hot solution. Since the carbonate ion is added in excess the identification of this ion must be made on the original sample. The CO_3^{2-} must be removed before certain other tests are made.

PREPARATION OF THE SOLUTION FOR ANALYSIS. *Removal of Heavy Metal Ions by Na_2CO_3.* **Place 100 mg of the powdered unknown in a test tube and add 2 ml of 1.5 M Na_2CO_3 solution. Heat the mixture for 10 minutes in the hot water bath. If ammonia is given off, continue the heating until no more gas is evolved. Separate any precipitate that forms. The solution will contain the anions in the form of sodium salts and will be referred to as the "prepared solution."**

45.5   Elimination of Oxidizing Anions

It will be noted that the nitrite ion is included in the elimination tests for both oxidizing anions and reducing anions. The oxidation potential of the nitrite ion in acid solution is intermediate in value, thus making the ion an oxidizing agent in the presence of strong reducing agents, and a reducing agent in the presence of strong oxidizing agents.

The prepared solution is tested for oxidizing anions, NO_2^- and NO_3^-, by means of a saturated solution of manganese(II) chloride in concentrated hydrochloric acid. The development of a dark-brown or black color is the result of the oxidation, in strongly acidic solution, of the manganese(II) ion, Mn^{2+}, to the manganese(III) ion, Mn^{3+}. The manganese(III) ion then unites with chloride ions and forms the dark-colored complex $MnCl_4^-$.

$$Mn^{2+} + HNO_2 + H^+ + 4Cl^- \longrightarrow MnCl_4^- + NO\uparrow + H_2O$$
$$3Mn^{2+} + 4H^+ + NO_3^- + 12Cl^- \longrightarrow 3MnCl_4^- + NO\uparrow + 2H_2O$$

TEST II. *Elimination of Oxidizing Anions: NO_3^- and NO_2^-.* **Add 6 drops of a saturated solution of $MnCl_2$ in 12 M HCl to 4 drops of the prepared solution and heat the mixture to boiling. The formation of a dark-brown or black coloration, due to $MnCl_4^-$, indicates the presence of NO_3^-, NO_2^-, or a mixture of these ions. Record the results of the test on your anion chart.**

45.6   Elimination of Reducing Anions

A sample of the prepared solution is acidified with hydrochloric acid and then iron(III) chloride and potassium hexacyanoferrate(III) are added to the solution. The prompt appearance of a blue to blue-green color or precipitate proves the presence of one or more of the reducing anions S^{2-}, SO_3^{2-}, NO_2^-, or I^-.

Iron(III), Fe^{3+}, forms a brown solution when fresh, and a green solution when old, with $Fe(CN)_6{}^{3-}$. In the presence of a strong reducing agent the Fe^{3+} may be reduced to Fe^{2+}, or the $Fe(CN)_6{}^{3-}$ reduced to $Fe(CN)_6{}^{4-}$, with the formation of $KFeFe(CN)_6 \cdot H_2O$, a blue pigment. The equations are

$$2Fe^{3+} + S^{2-} \rightleftharpoons 2Fe^{2+} + S$$
$$2Fe^{3+} + SO_3{}^{2-} + H_2O \rightleftharpoons 2Fe^{2+} + SO_4{}^{2-} + 2H^+$$
$$2Fe^{3+} + 2I^- \rightleftharpoons 2Fe^{2+} + I_2$$
$$2Fe(CN)_6{}^{3-} + NO_2{}^- + H_2O \rightleftharpoons 2Fe(CN)_6{}^{4-} + 2H^+ + NO_3{}^-$$
$$K^+ + Fe^{2+} + Fe(CN)_6{}^{3-} + H_2O \rightleftharpoons KFeFe(CN)_6 \cdot H_2O$$
$$K^+ + Fe^{3+} + Fe(CN)_6{}^{4-} + H_2O \rightleftharpoons KFeFe(CN)_6 \cdot H_2O$$

TEST III. *Elimination of Reducing Anions: S^{2-}, $SO_3{}^{2-}$, I^-, and $NO_2{}^-$.* **Mix 2 drops of a recently prepared saturated solution of $K_3Fe(CN)_6$ with 1 drop of 0.1 M $FeCl_3$, and add 2 drops of 6 M HCl. Add 2 drops of the prepared solution to this mixture and let the mixture stand for a few minutes. The development of a blue color or the formation of a blue precipitate indicates the presence of a reducing agent (S^{2-}, $SO_3{}^{2-}$, I^-, $NO_2{}^-$). Run a blank test with the same reagents but leave out the prepared solution and compare the colors and intensities of color of the unknown and blank test solutions.**

45.7 Elimination of Fluoride and Oxalate

Of the thirteen anions which may be involved in these procedures, only fluoride and oxalate form calcium salts which are insoluble in 4 M acetic acid.

$$Ca^{2+} + F^- \rightleftharpoons CaF_2 \text{ (white)}$$
$$Ca^{2+} + C_2O_4{}^{2-} \rightleftharpoons CaC_2O_4 \text{ (white)}$$

If no precipitate forms when calcium chloride is added to an acetic acid solution of the unknown, then fluoride and oxalate are absent.

TEST IV. *Elimination of Fluoride and Oxalate.* **To 20 drops of the prepared solution add 4 M HOAc (count the drops) until the solution is just acid to litmus. Now add an equal number of drops of 4 M HOAc in excess. Tap the test tube repeatedly until the excess of CO_2 from the Na_2CO_3 present is expelled. Add 8 drops of 0.1 M $CaCl_2$ and shake the contents of the tube. If a precipitate forms after a few minutes, it may be either CaF_2, CaC_2O_4, or a mixture of these compounds. If no precipitate forms F^- and $C_2O_4{}^{2-}$ are absent. Separate the precipitate and use the solution for Test V. Save the precipitate for the confirmatory test for $C_2O_4{}^{2-}$ and F^-.**

45.8 Elimination of Sulfate

After the separation of fluoride and oxalate as insoluble calcium salts, barium chloride is added to precipitate the sulfate ion as the white, crystalline barium sulfate in the solution which is still acidic with acetic acid. If the sulfite ion is present in relatively large amounts, barium sulfite may precipitate. $BaSO_4$ is insoluble in dilute hydrochloric acid, whereas $BaSO_3$ will dissolve and evolve SO_2 when treated with hydrochloric acid.

Owing to the ease with which the sulfite ion is oxidized to the sulfate ion by oxygen of the air, if sulfite is present in the original solution, you may obtain a test for sulfate although none was present at the outset.

TEST V. *Elimination of Sulfate.* **Test for completeness of precipitation in the solution from Test IV by adding 1 drop of 0.1 M $CaCl_2$. Now add 8 drops of 0.1 M $BaCl_2$. Shake the mixture and let it stand for 5 minutes. If SO_4^{2-} is present, a precipitate of $BaSO_4$ will form. Separate the mixture and reserve the solution for Test VI. If SO_3^{2-} was present in the original solution, you may obtain a test for SO_4^{2-} because sulfite is oxidized to sulfate by oxygen of the air.**

45.9   Elimination of Phosphate and Metaborate

The solution from the barium sulfate separation is acidified with hydrochloric acid and heated to expel CO_2 from CO_3^{2-} and SO_2 from SO_3^{2-}, if present. If they were not first removed, these ions would be precipitated as barium salts in the ammoniacal solution used to precipitate barium phosphate and barium metaborate.

$$3Ba^{2+} + 2PO_4^{3-} \rightleftharpoons Ba_3(PO_4)_2 \text{ (white)}$$
$$Ba^{2+} + 2BO_2^- \rightleftharpoons Ba(BO_2)_2 \text{ (white)}$$

It should be noted that phosphate and metaborate are insoluble in water, but they did not precipitate as their calcium or barium salts in the acetic acid solutions in which CaF_2, CaC_2O_4, and $BaSO_4$ were formed. Barium phosphate does not precipitate in an acetic acid solution due to the low concentration of free phosphate ions in such a solution. The phosphate ion is a strong proton acceptor

$$PO_4^{3-} + H^+ \rightleftharpoons HPO_4^{2-}$$

making its concentration so low that the solubility product constant of $Ba_3(PO_4)_2$ is not exceeded. Raising the pH of the solution by adding aqueous ammonia, on the other hand, increases the concentration of the phosphate ion

$$HPO_4^{2-} + OH^- \rightleftharpoons PO_4^{3-} + H_2O$$

and permits the solubility product constant of $Ba_3(PO_4)_2$ to be exceeded.

TEST VI. *Elimination of Phosphate and Metaborate: PO_4^{3-} and BO_2^-.* **Transfer the solution from Test V to the casserole, add 6 drops of 12 M HCl and heat to expel SO_2 and CO_2. Make the solution alkaline with aqueous ammonia and then add 5 drops in excess. A white precipitate may be $Ba_3(PO_4)_2$, $Ba(BO_2)_2$, or a mixture of these. Because barium metaborate is prone to form supersaturated solutions, it is necessary to carry out the confirmatory test for BO_2^- even though no precipitate forms in Test VI.**

45.10   Elimination of Sulfide

The addition of dilute sulfuric acid to a sample of the prepared solution causes hydrogen sulfide gas to be evolved if the sulfide ion is present. Hydrogen sulfide interacts with lead acetate and forms a black precipitate of lead sulfide.

$$Pb^{2+} + H_2S \rightleftharpoons \underline{PbS} + 2H^+$$

This elimination test also serves to confirm the sulfide ion if it is present.

TEST VII. *Elimination of Sulfide: S^{2-}.* **To 4 drops of the prepared solution contained in a small beaker add sufficient 1.5 M H_2SO_4 to make the solution acidic. Cover the beaker with a watch glass to the underside of which is attached a small piece of moist lead acetate paper. The appearance of a black stain of PbS on the paper after a few minutes indicates the presence of the S^{2-} ion.**

45.11   Elimination of Chloride, Bromide, and Iodide

If the sulfide ion is present, it is first removed as the insoluble CoS before attempting to eliminate chloride, bromide, and iodide. Silver nitrate is then added to precipitate the halides as AgCl (white), AgBr (pale-yellow), and AgI (yellow). Silver chloride is completely soluble in the mixture of aqueous ammonia and silver nitrate solutions described in Test VIII, silver bromide is partially dissolved, and silver iodide is insoluble in this reagent.

$$AgCl + 2NH_3 \rightleftharpoons [Ag(NH_3)_2]^+ + Cl^-$$
$$AgBr + 2NH_3 \rightleftharpoons [Ag(NH_3)_2]^+ + Br^- \text{ (partially soluble)}$$
$$AgI + NH_3 \rightleftharpoons \text{ no reaction}$$

These facts reflect the decreasing solubility of the silver halides with increasing atomic weight of the halogen. The solubility product constants are 1.8×10^{-10} for AgCl, 3.3×10^{-13} for AgBr, and 1.5×10^{-16} for AgI.

If a white precipitate forms upon the acidification of the aqueous ammonia extract of the silver halide precipitate, then the presence of the chloride ion is confirmed.

$$[Ag(NH_3)_2]^+ + Cl^- + 2H^+ \longrightarrow \underline{AgCl} + 2NH_4^+$$

A slight precipitate, yellow in color, may be silver bromide.

TEST VIII. *Elimination of Chloride, Bromide, and Iodide.* **If the sulfide ion is present, treat 10 drops of the prepared solution with *4 M* HOAc until the solution is acid to litmus. Now add, dropwise, with shaking, 10 drops of 1 M $(Co(NO_3)_2$. Heat in the hot water bath for five minutes and separate the precipitate of CoS. The solution may contain Cl^-, Br^-, and I^-. To the solution (or, if sulfide is absent, to 10 drops of the prepared solution acidified with 1 M HNO_3), add 5 drops of 1 M HNO_3. Now add to this solution 0.1 M $AgNO_3$ until precipitation is complete. Separate and wash the precipitate with 10 drops of water. Discard the solution. Mix 10 drops of 0.1 M $AgNO_3$, 6 drops of 4 M aqueous ammonia, and 4 ml of water. Add this mixture to the precipitate and stir thoroughly. If the precipitate**

dissolves completely, Br^- and I^- are absent, and Cl^- is indicated. Acidify the solution with 4 M HNO_3. A white precipitate confirms chloride. If the precipitate did not dissolve completely in the aqueous ammonia and silver nitrate mixture, it may have partially dissolved. Separate and acidify the solution with *4 M* HNO_3. If no precipitate forms, Cl^- is absent. The formation of a heavy white precipitate confirms the presence of Cl^-. A slight precipitate or turbidity, yellow in color, may be silver bromide.

Confirmatory Tests for the Anions

45.12   Identification of Carbonate

A sample of the original unknown is treated with hydrogen peroxide to oxidize any sulfite or nitrite which might be present. Then dilute sulfuric acid is added to decompose the carbonate and liberate carbon dioxide gas. The carbon dioxide gas which is evolved is permitted to come in contact with a drop of barium hydroxide solution which is held in a platinum wire loop and suspended above the reaction mixture. The formation of a definite turbidity in the drop of barium hydroxide indicates carbonate.

$$CO_3{}^{2-} + 2H^+ \rightleftharpoons H_2O + CO_2\uparrow$$
$$CO_2 + Ba^{2+} + 2OH^- \longrightarrow \underline{BaCO_3} + H_2O$$

If the sulfite and nitrite were not removed by oxidation to sulfate and nitrate respectively, sulfur dioxide and nitrogen dioxide would escape upon acidification, enter the drop of barium hydroxide solution, and either prevent the formation of barium carbonate or mask its presence.

PROCEDURE 1. *Identification of $CO_3{}^{2-}$*. **To 25 mg of the powdered solid unknown add 10 drops of hydrogen peroxide and heat the mixture in the hot water bath. Place a drop of $Ba(OH)_2$ solution in a loop of platinum wire. Add 3 drops of 4 M H_2SO_4 to the unknown mixture and immediately hold the drop of $Ba(OH)_2$ over the reaction mixture. The formation of a definite turbidity in the $Ba(OH)_2$ solution indicates the presence of $CO_3{}^{2-}$.**

45.13   Identification of Sulfate and Sulfite

The formation of a white precipitate of barium sulfate upon the addition of barium chloride to a hydrochloric acid solution of the unknown serves to confirm the sulfate ion. Barium salts of the other anions are soluble in 6 M hydrochloric acid.

To the solution obtained from the separation of barium sulfate is added bromine water for the purpose of oxidizing the sulfite ion to sulfate.

$$SO_3{}^{2-} + Br_2 + H_2O \rightleftharpoons SO_4{}^{2-} + 2Br^- + 2H^+$$

The appearance of a white precipitate (barium sulfate) indicates the presence of the sulfite ion in the original solution.

PROCEDURE 2. *Identification of SO_4^{2-}.* Acidify 5 drops of the prepared solution with 6 M HCl and add 2 drops in excess. Heat in the hot water bath to expel the excess of CO_2. Add 0.1 M $BaCl_2$ until precipitation is complete; then allow the mixture to stand. A white, finely divided precipitate of $BaSO_4$ confirms the presence of SO_4^{2-}. Separate the mixture and save the solution for the SO_3^{2-} test.

PROCEDURE 3. *Identification of SO_3^{2-}.* Add 5 drops of bromine water and 2 drops of 0.1 M $BaCl_2$ to the solution from (2). Heat in the hot water bath and allow to stand for 5 minutes. A white precipitate of $BaSO_4$ shows the presence of sulfite in the original solution.

PROCEDURE 4. *Identification of S^{2-}.* The sulfide ion is confirmed in elimination Test VII.

45.14   Identification of Nitrite

Use is made of the oxidizing power of the nitrite ion in acid solution in the confirmatory test for this ion. The nitrite ion oxidizes iron(II) to iron(III), and nitric oxide is formed as the reduction product of the nitrite ion.

$$NO_2^- + Fe^{2+} + 2H^+ \rightleftharpoons Fe^{3+} + NO + H_2O$$

The nitric oxide produced combines with some of the excess iron(II) ions in the reaction mixture, forming the brown complex $Fe(NO)^{2+}$.

$$Fe^{2+} + NO \longrightarrow Fe(NO)^{2+}$$

PROCEDURE 5. *Identification of NO_2^-.* To 5 drops of prepared solution add 4 M H_2SO_4 dropwise until the solution is acidic. Now add 5 drops of freshly prepared 0.1 M $FeSO_4$ solution. If NO_2^- is present the solution will assume a dark-brown color.

45.15   Identification of Nitrate

In the event that nitrite, bromide, and iodide ions are absent, one may identify the nitrate ion by the brown ring test. Iron(II) sulfate is added to a sample of the prepared solution in a test tube and then acidified with dilute sulfuric acid. The test tube containing the solution is tilted slightly and concentrated H_2SO_4 is poured down the inside wall in such a manner as to form a layer of acid in the bottom of the tube. If the nitrate ion is present, a brown ring will form at the junction of the two phases in a short while. The reduction of the nitrate ion to nitric oxide by iron(II) ions takes place rapidly only in a solution of very high hydrogen ion concentration and at relatively high temperatures, conditions which prevail at the interface of the two liquids.

$$3Fe^{2+} + NO_3^- + 4H^+ \rightleftharpoons 3Fe^{3+} + NO + 2H_2O$$
$$Fe^{2+} + NO \longrightarrow Fe(NO)^{2+}$$

If the nitrite ion is present, then it must be removed prior to the nitrate test because the entire solution will turn brown and the development of the brown ring will be obscured. The removal of the nitrite ion is accomplished by adding ammonium sulfate and evaporating nearly to dryness over a flame. The added ammonium ion reacts with the nitrite ion and forms nitrogen and water.

$$NH_4^+ + NO_2^- + heat \longrightarrow N_2\uparrow + 2H_2O$$

The residue is taken up in water and the brown ring test for nitrate is conducted on the solution.

If bromide or iodide ions are present, they will be oxidized by the concentrated sulfuric acid used in the test for nitrate, forming bromine and iodine respectively, both of which form colored layers. Therefore, these ions must be removed before making the brown ring test for nitrate. This is accomplished by the addition of solid, nitrate-free, silver sulfate to an acidified sample of the prepared solution to precipitate the bromide and iodide ions as their silver salts.

$$Ag_2SO_4 + Br^- \longrightarrow \underline{AgBr} + SO_4^{2-}$$
$$Ag_2SO_4 + I^- \longrightarrow \underline{AgI} + SO_4^{2-}$$

PROCEDURE 6. *Identification of NO_3^- in the Absence of NO_2^-, Br^-, and I^-.* To 5 drops of prepared solution in a test tube add 4 M H_2SO_4 dropwise until the solution is acidic. Now add 5 drops of freshly prepared 0.1 M $FeSO_4$ solution. Add 5 drops of 18 M H_2SO_4, holding the test tube in an inclined position so that the sulfuric acid may run down the side of the test tube and form a separate layer on the bottom of the tube. If the NO_3^- is present a brown ring will form at the junction of the two liquids, within a few minutes.

PROCEDURE 7. *Identification of NO_3^- in the Presence of NO_2^-. Removal of NO_2^-.* To 6 drops of the prepared solution add 4 M H_2SO_4 until the solution is acidic and then add 4 drops of 1 M $(NH_4)_2SO_4$ solution. Place the mixture in a casserole and slowly evaporate the solution until only a moist residue remains (do not evaporate to dryness). Add 4 drops of water and evaporate to a moist residue a second time. Dissolve the residue in 10 drops of water and transfer the mixture to a small test tube. Repeat the brown ring test for NO_3^- as described in Procedure (6) on the mixture resulting from the removal of NO_2^-.

PROCEDURE 8. *Identification of NO_3^- in the Presence of Br^- and I^-. Removal of Br^- and I^-.* To 6 drops of the prepared solution in a test tube add 10 drops of water. Acidify the solution with 4 M HOAc and then add 80 mg of powdered Ag_2SO_4 (NO_3^- free). Stir and grind the mixture in the test tube for 2–3 minutes. Separate the precipitate and transfer the solution to a test tube. Repeat the brown ring test for NO_3^- as described in Procedure (6) on the solution obtained from the removal of Br^- and I^-.

45.16   Identification of Oxalate

The precipitate formed during the elimination test for oxalate and fluoride ions is used for the identification of the oxalate ion. It consists of CaC_2O_4 and CaF_2

if both ions were present in the original solution. Treatment of the precipitate with sulfuric acid serves to dissolve the calcium oxalate and form oxalic acid.

$$CaC_2O_4 + 2H^+ \rightleftharpoons Ca^{2+} + H_2C_2O_4$$

Oxalic acid is a weak reducing agent that is readily oxidized to carbon dioxide and water by potassium permanganate in sulfuric acid solution. A dilute solution of the pink permanganate is employed, and the permanganate ion is reduced to the colorless manganese(II) ion by the oxalic acid. This bleaching of the colored permanganate solution shows the presence of the reducing agent, oxalic acid.

$$5H_2C_2O_4 + 2MnO_4^- + 6H^+ \longrightarrow 10CO_2\uparrow + 2Mn^{2+} + 8H_2O$$

PROCEDURE 9. *Identification of $C_2O_4^{2-}$.* **Wash the precipitate obtained in preliminary elimination Test IV twice. Use 10 drops of water each time and discard the washings. To the residue add 10 drops of water and 10 drops of 4 *M* H_2SO_4, and shake the mixture. Add 0.002 *M* $KMnO_4$ dropwise until 1 drop imparts a permanent pink color to the solution. If $C_2O_4^{2-}$ is present, several drops of $KMnO_4$ should be required to give a permanent pink coloration to the solution. Run a blank by repeating the test but leaving out the precipitate from elimination Test IV.**

45.17 Identification of Fluoride

In the confirmatory test for the fluoride ion, a sample of the solid unknown is treated with concentrated sulfuric acid in the presence of silica, SiO_2.

$$2NaF + H_2SO_4 \longrightarrow Na_2SO_4 + 2HF\uparrow$$

The hydrofluoric acid produced reacts with the silica present and forms gaseous silicon tetrafluoride, SiF_4.

$$SiO_2 + 4HF \longrightarrow SiF_4\uparrow + 2H_2O$$

The gaseous SiF_4 thus generated is permitted to come in contact with a drop of water contained in a loop of platinum wire suspended over the reaction mixture. The development of a white precipitate, H_4SiO_4, in the water confirms the presence of the fluoride ion.

$$3SiF_4 + 4H_2O \longrightarrow H_4SiO_4 + 4H^+ + 2SiF_6^{2-}$$

PROCEDURE 10. *Identification of F^-.* **Place 25 mg of powdered sample in a test tube and add an equal volume of powdered silica. Select a cork to fit the test tube. Cut a notch on the side of the cork and bore a hole through its center. Through the hole, insert a glass rod to one end of which has been sealed a platinum wire looped at the end. Add 2 drops of 18 *M* H_2SO_4 to the mixture in the tube. Now place the platinum wire loop with a drop of water hanging to it in the test tube directly above the reaction mixture. Warm the test tube in the hot water bath and then set it aside to cool. The formation of a white precipitate or cloudiness in the drop of water confirms the presence of the F^- ion.**

45.18   Identification of Phosphate

Magnesia mixture ($MgCl_2$, NH_4Cl, and aqueous ammonia), when added to a slightly acidic solution of the unknown, will precipitate the phosphate ion as the white crystalline magnesium ammonium phosphate. Since this compound is prone to form supersaturated solutions, the precipitate may be slow in appearing.

$$Mg^{2+} + NH_4^+ + HPO_4^{2-} \rightleftharpoons MgNH_4PO_4 + H^+$$

PROCEDURE 11. *Identification of PO_4^{3-}.* **Dilute 4 drops of prepared solution with 10 drops of water. Make the solution just acidic with 4 M HNO_3 and add 10 drops of magnesia mixture. A white precipitate, often slow in forming, confirms the presence of the PO_4^{3-} ion.**

45.19   Identification of Metaborate

A sample of the prepared solution is evaporated to a small volume and then treated with methyl alcohol and sulfuric acid. The sulfuric acid converts the metaborate ion to orthoboric acid.

$$BO_2^- + H^+ + H_2O \longrightarrow H_3BO_3$$

In the presence of the dehydrating agent, concentrated sulfuric acid, orthoboric acid reacts with methyl alcohol and forms the volatile methyl orthoborate.

$$H_3BO_3 + 3CH_3OH \rightleftharpoons B(OCH_3)_3\uparrow + 3H_2O$$

The sulfuric acid present takes up the water produced in the reaction and shifts the equilibrium to the right. Methyl orthoborate burns with a characteristic green flame when ignited.

$$2B(OCH_3)_3 + 9O_2 \longrightarrow B_2O_3 + 6CO_2 + 9H_2O$$

PROCEDURE 12. *Identification of BO_2^-.* **Evaporate 6 drops of the prepared solution to a small volume in the casserole. Add 1 ml of methyl alcohol and 5 drops of 18 M H_2SO_4. Transfer the mixture to a test tube and place the tube in the hot water bath. When the alcohol begins to boil, ignite the vapors. A green tinge to the flame confirms the presence of the metaborate ion.**

45.20   Identification of Iodide, Bromide, and Chloride

The confirmatory test for chloride was outlined in the elimination test for chloride, bromide, and iodide; no further test for this ion is necessary.

In general the methods used for the detection of the halide ions in the presence of one another depend upon the differences in the reduction potentials of the ions. Of the three ions chloride, bromide, and iodide, the iodide ion is most easily oxidized, the bromide ion is second, and the chloride ion is the most difficult to oxidize.

The iodide ion may be oxidized to iodine by iron(III) in acid solution without affecting the bromide or chloride ions

$$2I^- + 2Fe^{3+} \rightleftharpoons I_2 + 2Fe^{2+}$$

The elemental iodine is then extracted from the aqueous solution by carbon tetrachloride in which it exhibits a characteristic violet color.

After the complete removal of the iodide ion, a stronger oxidizing agent than iron(III) is used to oxidize the bromide to bromine. A dilute solution of potassium permanganate in the presence of nitric acid will oxidize bromide to bromine without interference from chloride.

$$10Br^- + 2MnO_4^- + 16H^+ \rightleftharpoons 5Br_2 + 2Mn^{2+} + 8H_2O$$

The free bromine is extracted with carbon tetrachloride, to which it imparts a yellow or orange color, depending upon the concentration of the bromine.

PROCEDURE 13. *Identification of I⁻.* **Dilute 6 drops of the prepared solution with 12 drops of water. Add 4 M HNO$_3$ until the solution is just acid to litmus and then add two drops of the acid in excess. Treat the solution with 1 ml of 0.1 M Fe(NO$_3$)$_3$ and 10 drops of CCl$_4$ and shake the test tube. The development of a violet color in the CCl$_4$ layer proves the presence of I⁻. Remove the CCl$_4$ layer by means of a capillary syringe and discard it. Add 10 more drops of CCl$_4$, shake, and remove the CCl$_4$ layer. Repeat the extraction until the CCl$_4$ layer remains colorless and use the aqueous solution for the identification of Br⁻.**

PROCEDURE 14. *Identification of Br⁻.* **Add to the solution from (13) 2 drops of 4 M HNO$_3$, then 0.1 M KMnO$_4$ solution drop by drop until the solution remains pink. Extract the solution with CCl$_4$. A yellow or orange coloration in the CCl$_4$ layer indicates the presence of Br⁻.**

QUESTIONS

1. What is meant by *incompatibility of anions?* Give an example.
2. What principle is involved in the tests for the halide ions in the presence of one another?
3. Explain why the nitrite ion will respond to both the test for oxidizing and that for reducing agents.
4. In terms of solubility product constant and ionic equilibria theory, explain the following: (a) Ca$_3$(PO$_4$)$_2$ is soluble in HCl while BaSO$_4$ is not. (b) CaCO$_3$ is soluble in HOAc while CaC$_2$O$_4$ is not.
5. Make a list of anions which are reducing agents and list their oxidation products.
6. Make a list of anions which are oxidizing agents and list their reduction products.
7. If an anion unknown contains the sulfite ion, one may obtain a test for the sulfate ion even though sulfate was not used in making up the unknown. Explain.

8. An acidic solution contains silver ions. What anions are probably absent?

9. A neutral solution contains calcium ions. What anions are probably absent?

10. Explain how the nitrite ion interferes with the test for nitrate. How is the nitrite ion removed prior to the test for nitrate?

11. A solution is slightly acid and contains sulfide ions. What cations are probably absent?

12. A solution known to contain either Na_2CO_3 or Na_2SO_4 turns red litmus blue. Which salt is present?

13. Why is it necessary to extract the iodine in the confirmatory test for the iodide ion?

14. Outline a simple test to distinguish between the ions of each pair in the following: (a) CO_3^{2-} and SO_4^{2-}, (b) Cl^- and I^-, (c) F^- and BO_2^-, (d) PO_4^{3-} and Br^-, (e) S^{2-} and SO_3^{2-}, (f) $C_2O_4^{2-}$ and Cl^-.

15. If a solution is strongly acidic, what anions cannot be present in appreciable concentrations?

16. Would you expect the test for F^- ion to work if no powdered silica were added to the reaction mixture?

PROBLEMS

1. How many ml of a solution 0.010 M in Fe^{3+} ion would be needed to oxidize 1.0 mg of the iodide ion?   *Ans. 0.79 ml*

2. How many drops of 0.020 M $KMnO_4$ solution (assume 0.05 ml per drop) would be needed to react with 1.00 mg of oxalic acid and impart a pink color to the resulting solution?   *Ans. 5 drops*

The Analysis of Solid Materials

46

The analytical procedures which have been outlined for cations and anions in the preceding chapters are usually applicable to solutions of the substances being analyzed. However, many substances are not readily soluble in water or even in acids. Thus, it becomes necessary to use special procedures for effecting the dissolution of certain solid substances before proceeding with the analyses.

46.1   The Dissolution of Nonmetallic Solids

The procedures outlined in this section are applicable to solid inorganic materials other than metals and alloys. The analysis of metals and alloys is described in Section 46.2.

PROCEDURE 1. *For Samples Soluble in Water.* **Grind the sample of solid to a fine powder. Treat 100 mg of the solid with 1 ml of water. If no dissolution is apparent, heat the mixture in the hot water bath for a few minutes. If none of the sample appears to dissolve, evaporate a few drops of the supernatant liquid on a watch glass to determine whether partial dissolution has occurred. If a solid residue remains on the watch glass, heat the sample with fresh portions of water until complete dissolution has been effected. Concentrate the combined extracts by evaporation and proceed with the analyses for cations and anions as outlined in the preceding chapters. It should be recalled that certain solid substances, such as $BiCl_3$ and $SbCl_3$, hydrolyze when treated with water to form new solid substances. In such cases proceed as with samples soluble in water.**

PROCEDURE 2. *For Samples Not Soluble in Water.* **Warm a small sample of the solid substance with 6 M HCl or, if this has no effect, with 12 M HCl. If** dissolution does not occur, or if it is incomplete, try 6 *M* HNO$_3$, 15 *M* HNO$_3$, and aqua regia (1 part 15 *M* HNO$_3$ to 3 parts 12 *M* HCl) in succession on small samples of the solid until a suitable solvent is found. If dissolution has been effected by any of these reagents, dissolve 100 mg of the sample in the solvent selected. Evaporate the solution nearly to dryness and then take up the residue in 2 ml of water. Use this solution for the analytical procedures.

PROCEDURE 3. *Treatment of the Residue with Na$_2$CO$_3$ Solution.* If HCl, HNO$_3$, or aqua regia do not completely dissolve the solid, the residue remaining after the acid treatment may be treated with a Na$_2$CO$_3$ solution. Treat the residue with 2 ml of 1.5 *M* Na$_2$CO$_3$ in a casserole and boil the mixture for ten minutes, replacing the water that is lost during evaporation. This procedure will convert many insoluble salts such as BaSO$_4$, CaSO$_4$, and PbSO$_4$, and many oxides into acid-soluble carbonates. Separate the mixture and discard the solution unless it is to be used for anion analysis. Wash the precipitate with water and then dissolve it in a few drops of 6 *M* HNO$_3$. Dilute the solution with 1 ml of water and add it to the original acid solution which is to be evaporated. If dissolution of the residue was not complete, repeat the Na$_2$CO$_3$ treatment. The halides of silver, silicate salts, some oxides, and calcined salts are not dissolved by this procedure.

PROCEDURE 4. *Reduction of Silver Halides with Zinc.* Suspend the residue containing the insoluble silver halides in 5 drops of water, add 1 ml of 1 *M* H$_2$SO$_4$ and a few granules of zinc metal. Warm and stir the mixture for several minutes. Add more zinc if the evolution of hydrogen ceases. Separate the residue of precipitated silver. Wash the residue with water, dissolve it in 6 *M* HNO$_3$, and test the solution for the Ag$^+$ ion. Conduct tests for the halide ions on the solution from the silver separation.

PROCEDURE 5. *Fusion with Sodium Carbonate.* Silicates and certain oxides and calcined salts may not be taken into solution by treatment with acid or Na$_2$CO$_3$ solution, or by reduction with zinc. Fusion with Na$_2$CO$_3$ is effective with many of these substances. Transfer the residue remaining after the zinc reduction of silver halides to a small nickel crucible. Add to the residue 100 mg of anhydrous Na$_2$CO$_3$, about half as much K$_2$CO$_3$, and a few mg of NaNO$_3$. Place the crucible in a small clay triangle and heat it in the hot flame of a Meker burner until the mixture fuses. Cool the crucible, add 1 ml of water, and warm the mixture until the solid mass has disintegrated. Separate the mixture and reserve the solution. Treat the residue with 8 drops of 6 *M* HNO$_3$ and warm in the hot water bath for a few minutes. Separate and add this nitric acid solution to the original solution. Analyze the solution for its cations.

46.2   Analysis of Metals and Alloys

All of the common metals except Al, Cr, Mn, Fe, Sb, and Sn may be taken into solution with dilute nitric acid. Al, Cr, Mn, and Fe are attacked superficially by nitric acid with the formation of an oxide film, which makes solution of these metals too slow to be practical. Metastannic acid, H$_2$SnO$_3$, and the insoluble oxides of antimony, Sb_4O_6, Sb_2O_4, and Sb_2O_5, are formed when nitric acid reacts directly with tin and antimony, respectively. Although these compounds are insoluble in nitric acid, they may be readily converted to the corresponding sulfides which are soluble in nitric acid. Hydrochloric acid is in general not suitable as a solvent for alloys of unknown composition since the volatile hydrides of sulfur, arsenic, phosphorus, and antimony may be formed; and consequently, these elements could escape detection. On the other hand, nitric acid oxidizes the first three of these elements to sulfate, arsenate, and phosphate, respectively, and antimony to oxides, all of which are nonvolatile. Aqua regia will generally dissolve alloys which resist the action of nitric acid alone.

PROCEDURE 1. *Preparation of the Sample for Dissolution.* **Convert the metal sample into a finely divided state with a large surface for interaction with the solvent. This may be accomplished by means of a steel file, a mortar and pestle for brittle metals, a hammer for malleable metals, or a knife for soft metals.**

PROCEDURE 2. *Selection of a Suitable Solvent.* **Treat a small sample of the metal with 6 M HNO_3 in a test tube and warm the mixture if necessary. If the sample reacts completely, with or without the formation of a white precipitate, proceed to (3). If the sample does not dissolve readily, try aqua regia (1 part of 15 M HNO_3 and 3 parts of 12 M HCl). If aqua regia fails to attack the sample, it may be treated with a mixture of concentrated HCl and Br_2, or fused with solid NaOH in a silver crucible.**

PROCEDURE 3. *Dissolution by HNO_3.* **Place 20 mg of the sample in a test tube, add 1 ml of 6 M HNO_3, and heat the mixture in the hot water bath. If a white residue forms, stir the mixture to remove the coating from the surface of the undissolved metallic particles. Add more HNO_3 if that is necessary to complete the reaction. After the metal has dissolved, transfer the solution or mixture to a casserole and evaporate it nearly to dryness. Add 5 drops of 15 M HNO_3 and again evaporate the solution nearly to dryness. Now add 5 drops of 6 M HNO_3 and 1 ml of water and transfer the solution to a test tube. If a clear solution is obtained, analyze it according to the procedure for the cations. If there is a residue (H_2SnO_3, Sb_2O_5, or a mixture of these), separate and analyze the solution according to the procedures for the cations, omitting the tests for tin and antimony. Add 10 drops of 4 M NaOH and 5 drops of 5 per cent thioacetamide solution to the residue. Heat the mixture in the water bath for 5 minutes. Separate any residue and analyze the solution for tin and antimony.**

PROCEDURE 4. *Dissolution by Aqua Regia.* **Place 20 mg of the sample in a test tube, and add 10 drops of 15 M HNO_3 and 30 drops of 12 M HCl. Heat the mixture in the hot water bath until the reaction is complete. Transfer the reaction mixture to a casserole and evaporate it to a small volume (not to dryness). When it is cool, add 2 ml of water and 4 drops of 6 M HCl. Separate any precipitate that forms. Analyze the centrifugate, or the clear solution if no precipitate forms, by the procedures for Groups II, III, IV, and V. The precipitate may consist of AgCl, $PbCl_2$, or SiO_2. Analyze the precipitate for Group I.**

Appendices

Appendix A

Chemical Arithmetic

In the study of general chemistry, elementary mathematics is frequently used. Of particular importance and wide application in the calculations that the student makes in chemistry are exponential arithmetic, significant figures, and logarithms.

A.1 Exponential Arithmetic

In chemistry we use the exponential method of expressing very large and very small numbers. These numbers are expressed as a product of two numbers. The first number of the product is called the digit term. This term is usually a number not less than 1 and not greater than 10. The second number of the product is called the exponential term and is written as 10 with an exponent. Some examples of the exponential method of expressing numbers are given below.

$$
\begin{aligned}
1{,}000 &= 1 \times 10^3 \\
100 &= 1 \times 10^2 \\
10 &= 1 \times 10^1 \\
1 &= 1 \times 10^0 \\
0.1 &= 1 \times 10^{-1} \\
0.01 &= 1 \times 10^{-2} \\
0.001 &= 1 \times 10^{-3} \\
2{,}386 &= 2.386 \times 1{,}000 = 2.386 \times 10^3 \\
0.123 &= 1.23 \times 0.1 = 1.23 \times 10^{-1}
\end{aligned}
$$

The power (exponent) of 10 is equal to the number of places the decimal is shifted to give the digit number. The exponential method is particularly useful as a shorthand for big numbers. For example, $1{,}230{,}000{,}000 = 1.23 \times 10^9$; and $0.000\,000\,000\,36 = 3.6 \times 10^{-10}$.

■ 1. Addition of Exponentials.   Convert all the numbers to the same power of 10 and add the digit terms of the number.

Example.   Add 5×10^{-5} and 3×10^{-3}

   Solution: $3 \times 10^{-3} = 300 \times 10^{-5}$
   $(5 \times 10^{-5}) + (300 \times 10^{-5}) = 305 \times 10^{-5} = 3.05 \times 10^{-3}$

■ 2. Subtraction of Exponentials.   Convert all the numbers to the same power of 10 and take the difference of the digit terms.

Example.   Subtract 4×10^{-7} from 5×10^{-6}

   Solution: $4 \times 10^{-7} = 0.4 \times 10^{-6}$
   $(5 \times 10^{-6}) - (0.4 \times 10^{-6}) = 4.6 \times 10^{-6}$

■ 3. Multiplication of Exponentials.   Multiply the digit terms in the usual way and add algebraically the exponents of the exponential terms.

Example.   Multiply 4.2×10^{-8} by 2×10^3

   Solution: 4.2×10^{-8}
   $\underline{2 \times 10^3}$
   8.4×10^{-5}

■ 4. Division of Exponentials.   Divide the digit term of the numerator by the digit term of the denominator and subtract algebraically the exponents of the exponential terms.

Example.   Divide 3.6×10^{-5} by 6×10^{-4}

   Solution: $\dfrac{3.6 \times 10^{-5}}{6 \times 10^{-4}} = 0.6 \times 10^{-1} = 6 \times 10^{-2}$

■ 5. Squaring of Exponentials.   Square the digit term in the usual way and multiply the exponent of the exponential term by 2.

Example.   Square the number 4×10^{-6}

   Solution: $(4 \times 10^{-6})^2 = 16 \times 10^{-12} = 1.6 \times 10^{-11}$

■ 6. Cubing of Exponentials.   Cube the digit term in the usual way and multiply the exponent of the exponential term by 3.

Example.   Cube the number 2×10^3

> *Solution:* $(2 \times 10^3)^3 = 2 \times 2 \times 2 \times 10^9 = 8 \times 10^9$

■ **7. Extraction of Square Roots of Exponentials.**   Decrease or increase the exponential term so that the power of ten is evenly divisible by 2. Extract the square root of the digit term by inspection or by logarithms and divide the exponential term by 2.

Example.   Extract the square root of 1.6×10^{-7}

> *Solution:* $1.6 \times 10^{-7} = 16 \times 10^{-8}$
> $$\sqrt{16 \times 10^{-8}} = \sqrt{16} \times \sqrt{10^{-8}} = 4 \times 10^{-4}$$

A.2   Significant Figures

A bee keeper reports that he has 525,341 bees. The last three figures of the number are obviously inaccurate, for during the time the keeper was counting the bees, some of them would have died and others would have hatched; this would have made the exact number of bees quite difficult to determine. It would have been more accurate if he had reported the number 525,000. In other words, the last three figures are not significant, except to set the position of the decimal point. Their exact values have no meaning.

In reporting any information in terms of numbers, only as many significant figures should be used as are warranted by the accuracy of the measurement. The accuracy of measurements is dependent upon the sensitivity of the measuring instruments used. For example, if the weight of an object has been reported as 2.13 g, it is assumed that the last figure (3) has been estimated and that the weight lies between 2.125 g and 2.135 g. The quantity 2.13 g represents three significant figures. The weight of this same object as determined by a more sensitive balance may have been reported as 2.134 g. In this case one would assume the correct weight to be between 2.1335 g and 2.1345 g, and the quantity 2.134 g represents 4 significant figures. Note that the last figure is estimated and is also considered as a significant figure.

A zero in a number may or may not be significant, depending upon the manner in which it is used. When one or more zeros are used in locating a decimal point, they are not significant. For example, the numbers 0.063, 0.0063, and 0.00063 have each two significant figures. When zeros appear between digits in a number they are significant. For example, 1.008 g has four significant figures. Likewise, the zero in 12.50 is significant. However, the quantity 1,370 cm has four significant figures provided the accuracy of the measurement includes the zero as a significant digit; if the digit 7 is estimated, then the number has only three significant figures.

The importance of significant figures lies in their application to fundamental computation. When adding or subtracting, the last digit that is retained in the sum or difference should correspond to the first doubtful decimal place (as indicated by underscoring).

Example.   Add 4.383 g and 0.0023 g

Solution: 4.383 g
0.0023
———————
4.385 g

When multiplying or dividing, the product or quotient should contain no more digits than the least number of significant figures in the numbers involved in the computation.

Example.   Multiply 0.6238 by 6.6

Solution: 0.6238 × 6.6 = 4.1

In rounding off numbers, increase the last digit retained by one if the following digit is five or more. Thus, 26.5 becomes 27, and 26.4 becomes 26 in the rounding-off process.

A.3   The Use of Logarithms and Exponential Numbers

The common logarithm of a number is the power to which the number 10 must be raised to equal that number. For example, the logarithm of 100 is 2 because the number 10 must be raised to the second power to be equal to 100. Additional examples follow.

Number	Number Expressed Exponentially	Logarithm
10,000	10^4	4
1,000	10^3	3
10	10^1	1
1	10^0	0
0.1	10^{-1}	-1
0.01	10^{-2}	-2
0.001	10^{-3}	-3
0.0001	10^{-4}	-4

What is the logarithm of 60? Because 60 lies between 10 and 100, which have logarithms of 1 and 2, respectively, the logarithm of 60 must lie between 1 and 2. The logarithm of 60 is 1.7782; i.e., $60 = 10^{1.7782}$.

Every logarithm is made up of two parts, called the characteristic and the mantissa. The characteristic is that part of the logarithm which lies to the left of the decimal point; thus the characteristic of the logarithm of 60 is 1. The mantissa is that part of the logarithm which lies to the right of the decimal point; thus the mantissa of the logarithm of 60 is .7782. The characteristic of the logarithm of a number greater than 1 is one less than the number of digits to the left of the decimal point in the number.

Number	Characteristic	Number	Characteristic
60	1	2.340	0
600	2	23.40	1
6,000	3	234.0	2
52,840	4	2,340.0	3

The mantissa of the logarithm of a number is found in the logarithm table (see Appendix B), and its value is independent of the position of the decimal point. Thus, 2.340, 23.40, 234.0, and 2340.0 all have the same mantissa. The logarithm of 2.340 is 0.3692, that of 23.40 is 1.3692, that of 234.0 is 2.3692, and that of 2,340.0 is 3.3692.

The meaning of the mantissa and characteristic can be better understood from a consideration of their relationship to exponential numbers. For example, 2,340 may be written 2.34×10^3. The logarithm of (2.34×10^3) = the logarithm of 2.34 + the logarithm of 10^3. The logarithm of 2.34 is 0.3692 and the logarithm of 10^3 is 3. Thus, the logarithm of 2,340 = 3 + 0.3692, or 3.3692.

The logarithm of a number less than 1 has a negative value, and a convenient method of obtaining the logarithm of such a number is given below. For example, we may obtain the logarithm of 0.00234 as follows: When expressed exponentially, $0.00234 = 2.34 \times 10^{-3}$. The logarithm of 2.34×10^{-3} = the logarithm of 2.34 + the logarithm of 10^{-3}. The logarithm of 2.34 is 0.3692 and the logarithm of 10^{-3} is -3. Thus, the logarithm of $0.00234 = 0.3692 + (-3) = 0.3692 - 3 = -2.6208$. The abbreviated form for the expression $(0.3692 - 3)$ is $\bar{3}.3692$. Note that only the characteristic has a negative value in the logarithm $\bar{3}.3692$, and that the mantissa is positive. The logarithm $\bar{3}.3692$ may also be written as $7.3692 - 10$.

To multiply two numbers we add the logarithms of the numbers. For example, suppose we multiply 412 by 353.

$$\begin{aligned} \text{Logarithm of } 412 \quad &= 2.6149 \\ \text{Logarithm of } 353 \quad &= 2.5478 \\ \text{Logarithm of product} &= \overline{5.1627} \end{aligned}$$

The number which corresponds to the logarithm 5.1627 is 145,400 or 1.454×10^5. Thus, 1.45×10^5 is the product of 412 and 353.

To divide two numbers we subtract the logarithms of the numbers. Suppose we divide 412 by 353.

$$\begin{aligned} \text{Logarithm of } 412 \quad &= 2.6149 \\ \text{Logarithm of } 353 \quad &= 2.5478 \\ \text{Logarithm of quotient} &= \overline{0.0671} \end{aligned}$$

The number which corresponds to the logarithm 0.0671 is 1.17. Thus, 412 divided by 353 is 1.17.

Suppose we wish to multiply 5,432 by 0.3124. Add the logarithm of 0.3124 to that of 5,432.

$$\begin{array}{ll}\text{Logarithm of 5,432} & = 3.7350 \\ \text{Logarithm of 0.3124} & = \overline{1}.4948 \\ \hline \text{Logarithm of the product} & = 3.2298 \end{array}$$

The number which corresponds to the logarithm 3.2298 is 1697 or 1.697×10^3.

Let us divide 5,432 by 0.3124. Subtract the logarithm of 0.3124 from that of 5,432.

$$\begin{array}{ll}\text{Logarithm of 5,432} & = 3.7350 \\ \text{Logarithm of 0.3124} & = \overline{1}.4948 \\ \hline \text{Logarithm of the quotient} & = 4.2402 \end{array}$$

The number which corresponds to the logarithm 4.2402 is 17,390 or 1.739×10^4.

The extraction of roots of numbers by means of logarithms is a simple procedure. For example, suppose we extract the cube root of 7,235. The logarithm of $\sqrt[3]{7{,}235}$ or $(7{,}235)^{1/3}$ is equal to $\frac{1}{3}$ of the logarithm of 7,235.

$$\begin{array}{l}\text{Logarithm of 7,235} = 3.8594 \\ \tfrac{1}{3} \text{ of } 3.8594 = 1.2865 \end{array}$$

The number which corresponds to the logarithm 1.2865 is 19.34. Thus, 19.34 is the cube root of 7,235.

A.4 The Solution of Quadratic Equations

Any quadratic equation can be expressed in the following form:

$$aX^2 + bX + c = 0$$

In order to solve a quadratic equation, the following formula is used.

$$X = \frac{-b \pm \sqrt{b^2 - 4ac}}{2a}$$

Example.   Solve the equation $3X^2 + 13X - 10 = 0$.

Substituting the values $a = 3$, $b = 13$, and $c = -10$ in the formula, we obtain

$$X = \frac{-13 \pm \sqrt{(13)^2 - 4 \times 3 \times (-10)}}{2 \times 3}$$

$$X = \frac{-13 \pm \sqrt{169 + 120}}{6} = \frac{-13 \pm \sqrt{289}}{6} = \frac{-13 \pm 17}{6}$$

The two roots are therefore

$$X = \frac{-13 + 17}{6} = 0.67 \quad \text{and} \quad X = \frac{-13 - 17}{6} = -5$$

Equations constructed upon physical data always have real roots, and of these real roots only those having positive values are of any significance.

A.5 Problems

In all problems, observe the principle of significant figures.

1. Carry out the following calculations:
   (a) $(4.8 \times 10^{-2}) \times (1.92 \times 10^{-4})$
   (b) $(6.1 \times 10^{-3}) \times (5.81 \times 10^{5})$
   (c) $(4.8 \times 10^{-4}) \div (2.4 \times 10^{-8})$
   (d) $(9.6 \times 10^{-4}) \div (2.00 \times 10^{3})$

2. Perform the following operations:
   (a) $\sqrt{90 \times 10^{-3}}$ (d) $(4.1 \times 10^{-2})^2$
   (b) $\sqrt[3]{2.7 \times 10^{-2}}$ (e) $(64 \times 10^{5})^{1/2}$
   (c) $(1 \times 10^{6})^4$ (f) $(5.2 \times 10^{-2})^{1/2}$

3. Add the following numbers: 2.863, 42.580, 0.02316, and 41.33.

4. Multiply 7.850 by 3.2; divide 4.6 by 2.075.

5. Carry out the following calculation:
   $(5 \times 1.008) + (2 \times 12.011) + (126.9045)$

6. Find the logarithms of the following numbers:
   (a) 51.5   (c) 512   (e) 4.80×10^3 (g) 1,000
   (b) 0.0515   (d) 51.2   (f) 4.80×10^{-3} (h) 34,200

7. Write the characteristic and mantissa of each of the following logarithms. Check the values given to verify that they are correct.
   (a) $\log 8 = 0.9031$ (d) $\log 0.8 = \bar{1}.9031$
   (b) $\log 800 = 2.9031$ (e) $\log (8 \times 10^{-2}) = -1.0969$
   (c) $\log 0.08 = \bar{2}.9031$

8. Make the following calculations using logarithms:

   (a) 35.4×8.07 (d) $\dfrac{47.2 \times 8.50}{12.1 \times 306}$

   (b) $79.2 \div 0.082$ (e) $\dfrac{1.50 \times 0.082 \times 293}{740/760}$

   (c) $\sqrt[5]{4.270}$ (f) $(9.54 \times 1.20 \times 10^2)^{1/2}$

9. Find the value of x in the following:

   (a) $x = \dfrac{-5 \pm \sqrt{5^2 - 4(0.2)(1.7)}}{2(2)}$

   (b) $x = \dfrac{11 \pm \sqrt{(-11)^2 - 4(3.5 \times 10^2)(-8.0 \times 10^{-3})}}{2(3.5 \times 10^2)}$

10. Write the quadratic equations whose solutions are shown in Problem 9.

11. Solve for x if $k = 1.8 \times 10^{-5}$ and $c = 3.8 \times 10^{-3}$:

$$x^2 + kx - kc = 0$$

12. Solve for x if $k = 5.0 \times 10^{-4}$ and $c = 2.7 \times 10^{-2}$:

$$k = \frac{x^2}{c - x}$$

dix B
Four-Place Table of Logarithms

No.	0	1	2	3	4	5	6	7	8	9	1	2	3	4	5	6	7	8	9
10	0000	0043	0086	0128	0170	0212	0253	0294	0334	0374	4	8	12	17	21	25	29	33	37
11	0414	0453	0492	0531	0569	0607	0645	0682	0719	0755	4	8	11	15	19	23	26	30	34
12	0792	0828	0864	0899	0934	0969	1004	1038	1072	1106	3	7	10	14	17	21	24	28	31
13	1139	1173	1206	1239	1271	1303	1335	1367	1399	1430	3	6	10	13	16	19	23	26	29
14	1461	1492	1523	1553	1584	1614	1644	1673	1703	1732	3	6	9	12	15	18	21	24	27
15	1761	1790	1818	1847	1875	1903	1931	1959	1987	2014	3	6	8	11	14	17	20	22	25
16	2041	2068	2095	2122	2148	2175	2201	2227	2253	2279	3	5	8	11	13	16	18	21	24
17	2304	2330	2355	2380	2405	2430	2455	2480	2504	2529	2	5	7	10	12	15	17	20	22
18	2553	2577	2601	2625	2648	2672	2695	2718	2742	2765	2	5	7	9	12	14	16	19	21
19	2788	2810	2833	2856	2878	2900	2923	2945	2967	2989	2	4	7	9	11	13	16	18	20
20	3010	3032	3054	3075	3096	3118	3139	3160	3181	3201	2	4	6	8	11	13	15	17	19
21	3222	3243	3263	3284	3304	3324	3345	3365	3385	3404	2	4	6	8	10	12	14	16	18
22	3424	3444	3464	3483	3502	3522	3541	3560	3579	3598	2	4	6	8	10	12	14	15	17
23	3617	3636	3655	3674	3692	3711	3729	3747	3766	3784	2	4	6	7	9	11	13	15	17
24	3802	3820	3838	3856	3874	3892	3909	3927	3945	3962	2	4	5	7	9	11	12	14	16
25	3979	3997	4014	4031	4048	4065	4082	4099	4116	4133	2	3	5	7	9	10	12	14	15
26	4150	4166	4183	4200	4216	4232	4249	4265	4281	4298	2	3	5	7	8	10	11	13	15
27	4314	4330	4346	4362	4378	4393	4409	4425	4440	4456	2	3	5	6	8	9	11	13	14
28	4472	4487	4502	4518	4533	4548	4564	4579	4594	4609	2	3	5	6	8	9	11	12	14
29	4624	4639	4654	4669	4683	4698	4713	4728	4742	4757	1	3	4	6	7	9	10	12	13
30	4771	4786	4800	4814	4829	4843	4857	4871	4886	4900	1	3	4	6	7	9	10	11	13
31	4914	4928	4942	4955	4969	4983	4997	5011	5024	5038	1	3	4	6	7	8	10	11	12
32	5051	5065	5079	5092	5105	5119	5132	5145	5159	5172	1	3	4	5	7	8	9	11	12
33	5185	5198	5211	5224	5237	5250	5263	5276	5289	5302	1	3	4	5	6	8	9	10	12
34	5315	5328	5340	5353	5366	5378	5391	5403	5416	5428	1	3	4	5	6	8	9	10	11
35	5441	5453	5465	5478	5490	5502	5514	5527	5539	5551	1	2	4	5	6	7	9	10	11
36	5563	5575	5587	5599	5611	5623	5635	5647	5658	5670	1	2	4	5	6	7	8	10	11
37	5682	5694	5705	5717	5729	5740	5752	5763	5775	5786	1	2	3	5	6	7	8	9	10
38	5798	5809	5821	5832	5843	5855	5866	5877	5888	5899	1	2	3	5	6	7	8	9	10
39	5911	5922	5933	5944	5955	5966	5977	5988	5999	6010	1	2	3	4	5	7	8	9	10
40	6021	6031	6042	6053	6064	6075	6085	6096	6107	6117	1	2	3	4	5	6	8	9	10
41	6128	6138	6149	6160	6170	6180	6191	6201	6212	6222	1	2	3	4	5	6	7	8	9
42	6232	6243	6253	6263	6274	6284	6294	6304	6314	6325	1	2	3	4	5	6	7	8	9
43	6335	6345	6355	6365	6375	6386	6395	6405	6415	6425	1	2	3	4	5	6	7	8	9
44	6435	6444	6454	6464	6474	6484	6493	6503	6513	6522	1	2	3	4	5	6	7	8	9
45	6532	6542	6551	6561	6571	6580	6590	6599	6609	6618	1	2	3	4	5	6	7	8	9
46	6628	6637	6646	6656	6665	6675	6684	6693	6702	6712	1	2	3	4	5	6	7	7	8
47	6721	6730	6739	6749	6758	6767	6776	6785	6794	6803	1	2	3	4	5	5	6	7	8
48	6812	6821	6830	6839	6848	6857	6866	6875	6884	6893	1	2	3	4	4	5	6	7	8
49	6902	6911	6920	6928	6937	6946	6955	6964	6972	6981	1	2	3	4	4	5	6	7	8
50	6990	6998	7007	7016	7024	7033	7042	7050	7059	7067	1	2	3	3	4	5	6	7	8
51	7076	7084	7093	7101	7110	7118	7126	7135	7143	7152	1	2	3	3	4	5	6	7	8
52	7160	7168	7177	7185	7193	7202	7210	7218	7226	7235	1	2	2	3	4	5	6	7	7
53	7243	7251	7259	7267	7275	7284	7292	7300	7308	7316	1	2	2	3	4	5	6	6	7
54	7324	7332	7340	7348	7356	7364	7372	7380	7388	7396	1	2	2	3	4	5	6	6	7
	0	1	2	3	4	5	6	7	8	9	1	2	3	4	5	6	7	8	9

Appendix B

Four-Place Table of Logarithms (continued)

No.	0	1	2	3	4	5	6	7	8	9	1	2	3	4	5	6	7	8	9
55	7404	7412	7419	7427	7435	7443	7451	7459	7466	7474	1	2	2	3	4	5	5	6	7
56	7482	7490	7497	7505	7513	7520	7528	7536	7543	7551	1	2	2	3	4	5	5	6	7
57	7559	7566	7574	7582	7589	7597	7604	7612	7619	7627	1	2	2	3	4	5	5	6	7
58	7634	7642	7649	7657	7664	7672	7679	7686	7694	7701	1	1	2	3	4	4	5	6	7
59	7709	7716	7723	7731	7738	7745	7752	7760	7767	7774	1	1	2	3	4	4	5	6	7
60	7782	7789	7796	7803	7810	7818	7825	7832	7839	7846	1	1	2	3	4	4	5	6	6
61	7853	7860	7868	7875	7882	7889	7896	7903	7910	7917	1	1	2	3	4	4	5	6	6
62	7924	7931	7938	7945	7952	7959	7966	7973	7980	7987	1	1	2	3	3	4	5	6	6
63	7992	8000	8007	8014	8021	8028	8035	8041	8048	8055	1	1	2	3	3	4	5	5	6
64	8062	8069	8075	8082	8089	8096	8102	8109	8116	8122	1	1	2	3	3	4	5	5	6
65	8129	8136	8142	8149	8156	8162	8169	8176	8182	8189	1	1	2	3	3	4	5	5	6
66	8195	8202	8209	8215	8222	8228	8235	8241	8248	8254	1	1	2	3	3	4	5	5	6
67	8261	8267	8274	8280	8287	8293	8299	8306	8312	8319	1	1	2	3	3	4	5	5	6
68	8325	8331	8338	8344	8351	8357	8363	8370	8376	8382	1	1	2	3	3	4	4	5	6
69	8388	8395	8401	8407	8414	8420	8426	8432	8439	8445	1	1	2	2	3	4	4	5	6
70	8451	8457	8463	8470	8476	8482	8488	8494	8500	8506	1	1	2	2	3	4	4	5	6
71	8513	8519	8525	8531	8537	8543	8549	8555	8561	8567	1	1	2	2	3	4	4	5	5
72	8573	8579	8585	8591	8597	8603	8609	8615	8621	8627	1	1	2	2	3	4	4	5	5
73	8633	8639	8645	8651	8657	8663	8669	8675	8681	8686	1	1	2	2	3	4	4	5	5
74	8692	8698	8704	8710	8716	8722	8727	8733	8739	8745	1	1	2	2	3	4	4	5	5
75	8751	8756	8762	8768	8774	8779	8785	8791	8797	8802	1	1	2	2	3	3	4	5	5
76	8808	8814	8820	8825	8831	8837	8842	8848	8854	8859	1	1	2	2	3	3	4	5	5
77	8865	8871	8876	8882	8887	8893	8899	8904	8910	8915	1	1	2	2	3	3	4	4	5
78	8921	8927	8932	8938	8943	8949	8954	8960	8965	8971	1	1	2	2	3	3	4	4	5
79	8976	8982	8987	8993	8998	9004	9009	9015	9020	9025	1	1	2	2	3	3	4	4	5
80	9031	9036	9042	9047	9053	9058	9063	9069	9074	9079	1	1	2	2	3	3	4	4	5
81	9085	9090	9096	9101	9106	9112	9117	9122	9128	9133	1	1	2	2	3	3	4	4	5
82	9138	9143	9149	9154	9159	9165	9170	9175	9180	9186	1	1	2	2	3	3	4	4	5
83	9191	9196	9201	9206	9212	9217	9222	9227	9232	9238	1	1	2	2	3	3	4	4	5
84	9243	9248	9253	9258	9263	9269	9274	9279	9284	9289	1	1	2	2	3	3	4	4	5
85	9294	9299	9304	9309	9315	9320	9325	9330	9335	9340	1	1	2	2	3	3	4	4	5
86	9345	9350	9355	9360	9365	9370	9375	9380	9385	9390	1	1	2	2	3	3	4	4	5
87	9395	9400	9405	9410	9415	9420	9425	9430	9435	9440	0	1	1	2	2	3	3	4	4
88	9445	9450	9455	9460	9465	9469	9474	9479	9484	9489	0	1	1	2	2	3	3	4	4
89	9494	9499	9504	9509	9513	9518	9523	9528	9533	9538	0	1	1	2	2	3	3	4	4
90	9542	9547	9552	9557	9562	9566	9571	9576	9581	9586	0	1	1	2	2	3	3	4	4
91	9590	9595	9600	9605	9609	9614	9619	9624	9628	9633	0	1	1	2	2	3	3	4	4
92	9638	9643	9647	9652	9657	9661	9666	9671	9675	9680	0	1	1	2	2	3	3	4	4
93	9685	9689	9694	9699	9703	9708	9713	9717	9722	9727	0	1	1	2	2	3	3	4	4
94	9731	9736	9741	9745	9750	9754	9759	9763	9768	9773	0	1	1	2	2	3	3	4	4
95	9777	9782	9786	9791	9795	9800	9805	9809	9814	9818	0	1	1	2	2	3	3	4	4
96	9823	9827	9832	9836	9841	9845	9850	9854	9859	9863	0	1	1	2	2	3	3	4	4
97	9868	9872	9877	9881	9886	9890	9894	9899	9903	9908	0	1	1	2	2	3	3	4	4
98	9912	9917	9921	9926	9930	9934	9939	9943	9948	9952	0	1	1	2	2	3	3	4	4
99	9956	9961	9965	9969	9974	9978	9983	9987	9991	9996	0	1	1	2	2	3	3	3	4
	0	1	2	3	4	5	6	7	8	9	1	2	3	4	5	6	7	8	9

Appendix C

Units and Conversion Factors

Base Units of International System of Units (SI)

Physical Property	Name of Unit	Symbol
Length	Meter	m
Mass	Kilogram	kg
Time	Second	s
Electric current	Ampere	A
Thermodynamic temperature	Kelvin	K
Luminous intensity	Candela	cd
Quantity of substance	Mole	mol

Units of Length

Meter (m) = 39.37 inches (in.)
Centimeter (cm) = 0.01 m
Millimeter (mm) = 0.001 m
Kilometer (km) = 1,000 m
Angstrom unit (Å) = 10^{-8} cm = 10^{-10} m

Yard = 0.9144 m
Inch = 2.54 cm

Mile (U.S.) = 1.609 km

Units of Volume

Liter (1) = volume of 1 kg of water
Milliliter (ml) = 0.001 liter

Liquid quart (U.S.) = 0.9463 liter
Cubic foot (U.S.) = 28.316 liters

Units of Weight

Gram (g) = weight of 1 ml of water
  at 4° C
Milligram (mg) = 0.001 g
Kilogram (kg) = 1,000 g
Ton (metric) = 1,000 kg = 2204.62 lb

Ounce (oz) (avoirdupois) = 28.35 g
Pound (lb) (avoirdupois) = 0.45359237 kg
Ton (short) = 2,000 lb = 907.185 kg
Ton (long) = 2,240 lb = 1.016 metric tons

Units of Energy

Thermochemical calorie (cal) = 4.1840 × 10^7 erg = 4.1840 joule (J)
Erg = 10^{-7} J
Electron volt (ev) = 1.6021 × 10^{-12} erg = 23.061 kcal/mol
Liter-atmosphere = 24.217 cal

Unit of Force

Newton (N) = 1 kg · m · s^{-2} (force which when applied for 1 second will give to a 1 kg mass a speed of 1 meter per second)

Units of Pressure

Torr = 1 mm Hg
Atmosphere (atm) = 760 mm Hg = 760 torr = 101,325 N · m^{-2}
Pascal = kg · m^{-1} · s^{-2} = N · m^{-2}

Appendix D   General Physical Constants

Avogadro's number, N_A	6.022169×10^{23} molecules/mol
Electron charge, e	$1.6021917 \times 10^{-19}$ coulomb
Electron rest mass, m_e	9.109558×10^{-28} g
Proton rest mass, m_p	1.672614×10^{-24} g
Charge-to-mass ratio for electron, e/m_e	1.7588028×10^8 coulombs/g
Faraday constant, F	9.648670×10^4 coulombs/g-equivalent weight
	23,061 cal per volt/g-equiv wt
Planck constant, h	6.626196×10^{-27} erg $\cdot$ sec
Boltzmann constant, k	1.380622×10^{-16} erg/°K
Gas constant, R	8.205×10^{-2} liter $\cdot$ atm/mol $\cdot$ °K
	1.987 cal/mol $\cdot$ °K
Speed of light (in vacuum), c	2.9979250×10^{10} cm/sec
Atomic mass unit ($= \frac{1}{12}$ the mass of an atom of the ^{12}C nuclide), u or amu	1.660531×10^{-24} g
Rydberg constant, R_∞	3.289×10^{15} cycles/sec

Appendix E   Solubility Product Constants

Substance	K_{sp} at 25°	Substance	K_{sp} at 25°
Aluminum		Cobalt	
$Al(OH)_3$	1.9×10^{-33}	$Co(OH)_2$	2×10^{-16}
Barium		$CoS(\alpha)$	5.9×10^{-21}
$BaCO_3$	8.1×10^{-9}	$CoS(\beta)$	8.7×10^{-23}
$BaC_2O_4 \cdot 2H_2O$	1.1×10^{-7}	$CoCO_3$	1.0×10^{-12}
$BaSO_4$	1.08×10^{-10}	$Co(OH)_3$	2.5×10^{-43}
$BaCrO_4$	2×10^{-10}	Copper	
BaF_2	1.7×10^{-6}	$CuCl$	1.85×10^{-7}
$Ba(OH)_2 \cdot 8H_2O$	5.0×10^{-3}	$CuBr$	5.3×10^{-9}
$Ba_3(PO_4)_2$	1.3×10^{-29}	CuI	5.1×10^{-12}
$Ba_3(AsO_4)_2$	1.1×10^{-13}	$CuCNS$	4×10^{-14}
Bismuth		Cu_2S	1.6×10^{-48}
$BiO(OH)$	1×10^{-12}	$Cu(OH)_2$	5.6×10^{-20}
$BiOCl$	7×10^{-9}	CuS	8.7×10^{-36}
Bi_2S_3	1.6×10^{-72}	$CuCO_3$	1.37×10^{-10}
Cadmium		Iron	
$Cd(OH)_2$	1.2×10^{-14}	$Fe(OH)_2$	7.9×10^{-15}
CdS	3.6×10^{-29}	$FeCO_3$	2.11×10^{-11}
$CdCO_3$	2.5×10^{-14}	FeS	1×10^{-19}
Calcium		$Fe(OH)_3$	1.1×10^{-36}
$Ca(OH)_2$	7.9×10^{-6}	Lead	
$CaCO_3$	4.8×10^{-9}	$Pb(OH)_2$	2.8×10^{-16}
$CaSO_4 \cdot 2H_2O$	2.4×10^{-5}	PbF_2	3.7×10^{-8}
$CaC_2O_4 \cdot H_2O$	2.27×10^{-9}	$PbCl_2$	1.7×10^{-5}
$Ca_3(PO_4)_2$	1×10^{-25}	$PbBr_2$	6.3×10^{-6}
$CaHPO_4$	5×10^{-6}	PbI_2	8.7×10^{-9}
CaF_2	3.9×10^{-11}	$PbCO_3$	1.5×10^{-13}
Chromium		PbS	8.4×10^{-28}
$Cr(OH)_3$	6.7×10^{-31}	$PbCrO_4$	1.8×10^{-14}

Appendix E

Solubility Product Constants (continued)

Substance	K_{sp} at 25°	Substance	K_{sp} at 25°
Lead (cont.)		Silver	
$PbSO_4$	1.8×10^{-8}	$\frac{1}{2}Ag_2O(Ag^+ + OH^-)$	2×10^{-8}
$Pb_3(PO_4)_2$	3×10^{-44}	$AgCl$	1.8×10^{-10}
Magnesium		$AgBr$	3.3×10^{-13}
$Mg(OH)_2$	1.5×10^{-11}	AgI	1.5×10^{-16}
$MgCO_3 \cdot 3H_2O$	ca. 1×10^{-5}	$AgCN$	1.2×10^{-16}
$MgNH_4PO_4$	2.5×10^{-13}	$AgCNS$	1.0×10^{-12}
MgF_2	6.4×10^{-9}	Ag_2S	1.0×10^{-51}
MgC_2O_4	8.6×10^{-5}	Ag_2CO_3	8.2×10^{-12}
Manganese		Ag_2CrO_4	9×10^{-12}
$Mn(OH)_2$	4.5×10^{-14}	$Ag_4Fe(CN)_6$	1.55×10^{-41}
$MnCO_3$	8.8×10^{-11}	Ag_2SO_4	1.18×10^{-5}
MnS	5.6×10^{-16}	Ag_3PO_4	1.8×10^{-18}
Mercury		Strontium	
$Hg_2O \cdot H_2O$	1.6×10^{-23}	$Sr(OH)_2 \cdot 8H_2O$	3.2×10^{-4}
Hg_2Cl_2	1.1×10^{-18}	$SrCO_3$	9.42×10^{-10}
Hg_2Br_2	1.26×10^{-22}	$SrCrO_4$	3.6×10^{-5}
Hg_2I_2	4.5×10^{-29}	$SrSO_4$	2.8×10^{-7}
Hg_2CO_3	9×10^{-17}	$SrC_2O_4 \cdot H_2O$	5.61×10^{-8}
Hg_2SO_4	6.2×10^{-7}	Thallium	
Hg_2S	1×10^{-45}	$TlCl$	1.9×10^{-4}
Hg_2CrO_4	2×10^{-9}	$TlCNS$	5.8×10^{-4}
HgS	3×10^{-53}	Tl_2S	1.2×10^{-24}
Nickel		$Tl(OH)_3$	1.5×10^{-44}
$Ni(OH)_2$	1.6×10^{-14}	Tin	
$NiCO_3$	1.36×10^{-7}	$Sn(OH)_2$	5×10^{-26}
$NiS(\alpha)$	3×10^{-21}	SnS	8×10^{-29}
$NiS(\beta)$	1×10^{-26}	$Sn(OH)_4$	ca. 1×10^{-56}
$NiS(\gamma)$	2×10^{-28}	Zinc	
Potassium		$ZnCO_3$	6×10^{-11}
$KClO_4$	1.07×10^{-2}	$Zn(OH)_2$	4.5×10^{-17}
K_2PtCl_6	1.1×10^{-5}	ZnS	1.1×10^{-21}
$KHC_4H_4O_6$	3×10^{-4}		

Appendix F

Dissociation Constants for Complex Ions

Equilibrium	K_d
$[AlF_6]^{3-} \rightleftharpoons Al^{3+} + 6F^-$	2×10^{-24}
$[Cd(NH_3)_4]^{2+} \rightleftharpoons Cd^{2+} + 4NH_3$	2.5×10^{-7}
$[Cd(CN)_4]^{2-} \rightleftharpoons Cd^{2+} + 4CN^-$	7.8×10^{-18}
$[Co(NH_3)_6]^{2+} \rightleftharpoons Co^{2+} + 6NH_3$	1.2×10^{-5}
$[Co(NH_3)_6]^{3+} \rightleftharpoons Co^{3+} + 6NH_3$	2.2×10^{-34}
$[Cu(CN)_2]^- \rightleftharpoons Cu^+ + 2CN^-$	1×10^{-16}
$[Cu(NH_3)_4]^{2+} \rightleftharpoons Cu^{2+} + 4NH_3$	8.5×10^{-13}
$[Fe(CN)_6]^{4-} \rightleftharpoons Fe^{2+} + 6CN^-$	1×10^{-37}
$[Fe(CN)_6]^{3-} \rightleftharpoons Fe^{3+} + 6CN^-$	1×10^{-44}

Appendix F

Dissociation Constants for Complex Ions (continued)

Equilibrium	K_d
$[Fe(CNS)_6]^{3-} \rightleftharpoons Fe^{3+} + 6CNS^-$	3.1×10^{-4}
$[HgCl_4]^{2-} \rightleftharpoons Hg^{2+} + 4Cl^-$	8.3×10^{-16}
$[Ni(NH_3)_6]^{2+} \rightleftharpoons Ni^{2+} + 6NH_3$	5.7×10^{-9}
$[AgCl_2]^- \rightleftharpoons Ag^+ + 2Cl^-$	4.0×10^{-6}
$[Ag(CN)_2]^- \rightleftharpoons Ag^+ + 2CN^-$	1×10^{-20}
$[Ag(NH_3)_2]^+ \rightleftharpoons Ag^+ + 2NH_3$	6.3×10^{-8}
$[Zn(CN)_4]^{2-} \rightleftharpoons Zn^{2+} + 4CN^-$	1×10^{-19}
$[Zn(OH)_4]^{2-} \rightleftharpoons Zn^{2+} + 4OH^-$	3.5×10^{-16}

Appendix G

Ionization Constants of Weak Acids

Acid	Formula	K_i at 25°
Acetic	$HOAc\ (CH_3COOH)$	1.8×10^{-5}
Arsenic	H_3AsO_4	4.8×10^{-3}
	$H_2AsO_4^-$	1×10^{-7}
	$HAsO_4^{2-}$	1×10^{-13}
Arsenous	H_3AsO_3	5.8×10^{-10}
Boric	H_3BO_3	5.8×10^{-10}
Carbonic	H_2CO_3	4.3×10^{-7}
	HCO_3^-	7×10^{-11}
Cyanic	$HCNO$	3.46×10^{-4}
Formic	$HCOOH$	1.8×10^{-4}
Hydrazoic	HN_3	1×10^{-4}
Hydrocyanic	HCN	4×10^{-10}
Hydrofluoric	HF	7.2×10^{-4}
Hydrogen peroxide	H_2O_2	2.4×10^{-12}
Hydrogen selenide	H_2Se	1.7×10^{-4}
	HSe^-	1×10^{-10}
Hydrogen sulfate ion	HSO_4^-	1.2×10^{-2}
Hydrogen sulfide	H_2S	1.0×10^{-7}
	HS^-	1.3×10^{-13}
Hydrogen telluride	H_2Te	2.3×10^{-3}
	HTe^-	1×10^{-5}
Hypobromous	$HBrO$	2×10^{-9}
Hypochlorous	$HClO$	3.5×10^{-8}
Nitrous	HNO_2	4.5×10^{-4}
Oxalic	$H_2C_2O_4$	5.9×10^{-2}
	$HC_2O_4^-$	6.4×10^{-5}
Phosphoric	H_3PO_4	7.5×10^{-3}
	$H_2PO_4^-$	6.2×10^{-8}
	HPO_4^{2-}	1×10^{-12}
Phosphorous	H_3PO_3	1.6×10^{-2}
	$H_2PO_3^-$	7×10^{-7}
Sulfurous	H_2SO_3	1.2×10^{-2}
	HSO_3^-	6.2×10^{-8}

Appendix H

Ionization Constants of Weak Bases

Base	Ionization Equation	K_i at 25°
Ammonia	$NH_3 + H_2O \rightleftharpoons NH_4^+ + OH^-$	1.8×10^{-5}
Dimethylamine	$(CH_3)_2NH + H_2O \rightleftharpoons (CH_3)_2NH_2^+ + OH^-$	7.4×10^{-4}
Methylamine	$CH_3NH_2 + H_2O \rightleftharpoons CH_3NH_3^+ + OH^-$	4.4×10^{-4}
Phenylamine (aniline)	$C_6H_5NH_2 + H_2O \rightleftharpoons C_6H_5NH_3^+ + OH^-$	4.6×10^{-10}
Trimethylamine	$(CH_3)_3N + H_2O \rightleftharpoons (CH_3)_3NH^+ + OH^-$	7.4×10^{-5}

Appendix I

Standard Electrode (Reduction) Potentials

Half-reactions	$E°$, volts
$Li^+ + e^- \longrightarrow Li$	-3.09
$K^+ + e^- \longrightarrow K$	-2.925
$Rb^+ + e^- \longrightarrow Rb$	-2.925
$Ra^{2+} + 2e^- \longrightarrow Ra$	-2.92
$Ba^{2+} + 2e^- \longrightarrow Ba$	-2.90
$Sr^{2+} + 2e^- \longrightarrow Sr$	-2.89
$Ca^{2+} + 2e^- \longrightarrow Ca$	-2.87
$Na^+ + e^- \longrightarrow Na$	-2.714
$La^{3+} + 3e^- \longrightarrow La$	-2.52
$Ce^{3+} + 3e^- \longrightarrow Ce$	-2.48
$Nd^{3+} + 3e^- \longrightarrow Nd$	-2.44
$Sm^{3+} + 3e^- \longrightarrow Sm$	-2.41
$Gd^{3+} + 3e^- \longrightarrow Gd$	-2.40
$Mg^{2+} + 2e^- \longrightarrow Mg$	-2.37
$Y^{3+} + 3e^- \longrightarrow Y$	-2.37
$Am^{3+} + 3e^- \longrightarrow Am$	-2.32
$Lu^{3+} + 3e^- \longrightarrow Lu$	-2.25
$\frac{1}{2}H_2 + e^- \longrightarrow H^-$	-2.25
$Sc^{3+} + 3e^- \longrightarrow Sc$	-2.08
$[AlF_6]^{3-} + 3e^- \longrightarrow Al + 6F^-$	-2.07
$Pu^{3+} + 3e^- \longrightarrow Pu$	-2.07
$Th^{4+} + 4e^- \longrightarrow Th$	-1.90
$Np^{3+} + 3e^- \longrightarrow Np$	-1.86
$Be^{2+} + 2e^- \longrightarrow Be$	-1.85
$U^{3+} + 3e^- \longrightarrow U$	-1.80
$Hf^{4+} + 4e^- \longrightarrow Hf$	-1.70
$SiO_3^{2-} + 3H_2O + 4e^- \longrightarrow Si + 6OH^-$	-1.70
$Al^{3+} + 3e^- \longrightarrow Al$	-1.66
$Ti^{2+} + 2e^- \longrightarrow Ti$	-1.63
$Zr^{4+} + 4e^- \longrightarrow Zr$	-1.53
$ZnS + 2e^- \longrightarrow Zn + S^{2-}$	-1.44
$Cr(OH)_3 + 3e^- \longrightarrow Cr + 3OH^-$	-1.3
$[Zn(CN)_4]^{2-} + 2e^- \longrightarrow Zn + 4CN^-$	-1.26
$Zn(OH)_2 + 2e^- \longrightarrow Zn + 2OH^-$	-1.245

Appendix I

Standard Electrode (Reduction) Potentials (continued)

Half-reactions	$E°$, volts
$[Zn(OH)_4]^{2-} + 2e^- \longrightarrow Zn + 4OH^-$	-1.216
$CdS + 2e^- \longrightarrow Cd + S^{2-}$	-1.21
$[Cr(OH)_4]^- + 3e^- \longrightarrow Cr + 4OH^-$	-1.2
$[SiF_6]^{2-} + 4e^- \longrightarrow Si + 6F^-$	-1.2
$V^{2+} + 2e^- \longrightarrow V$	$ca. -1.18$
$Mn^{2+} + 2e^- \longrightarrow Mn$	-1.18
$[Cd(CN)_4]^{2-} + 2e^- \longrightarrow Cd + 4CN^-$	-1.03
$[Zn(NH_3)_4]^{2+} + 2e^- \longrightarrow Zn + 4NH_3$	-1.03
$FeS + 2e^- \longrightarrow Fe + S^{2-}$	-1.01
$PbS + 2e^- \longrightarrow Pb + S^{2-}$	-0.95
$SnS + 2e^- \longrightarrow Sn + S^{2-}$	-0.94
$Cr^{2+} + e^- \longrightarrow Cr$	-0.91
$Fe(OH)_2 + 2e^- \longrightarrow Fe + 2OH^-$	-0.877
$SiO_2 + 4H^+ + 4e^- \longrightarrow Si + 2H_2O$	-0.86
$NiS + 2e^- \longrightarrow Ni + S^{2-}$	-0.83
$2H_2O + 2e^- \longrightarrow H_2 + 2OH^-$	-0.828
$Zn^{2+} + 2e^- \longrightarrow Zn$	-0.763
$Cr^{3+} + 3e^- \longrightarrow Cr$	-0.74
$HgS + 2e^- \longrightarrow Hg + S^{2-}$	-0.72
$[Cd(NH_3)_4]^{2+} + 2e^- \longrightarrow Cd + 4NH_3$	-0.597
$Ga^{3+} + 3e^- \longrightarrow Ga$	-0.53
$S + 2e^- \longrightarrow S^{2-}$	-0.48
$[Ni(NH_3)_6]^{2+} + 2e^- \longrightarrow Ni + 6NH_3\ (aq)$	-0.47
$Fe^{2+} + 2e^- \longrightarrow Fe$	-0.440
$[Cu(CN)_2]^- + e^- \longrightarrow Cu + 2CN^-$	-0.43
$Cr^{3+} + e^- \longrightarrow Cr^{2+}$	-0.41
$Cd^{2+} + 2e^- \longrightarrow Cd$	-0.403
$Se + 2H^+ + 2e^- \longrightarrow H_2Se$	-0.40
$[Hg(CN)_4]^{2-} + 2e^- \longrightarrow Hg + 4CN^-$	-0.37
$ClO_4^- + H_2O + 2e^- \longrightarrow ClO_3^- + 2OH^-$	-0.36
$PbSO_4 + 2e^- \longrightarrow Pb + SO_4^{2-}$	-0.356
$In^{3+} + 3e^- \longrightarrow In$	-0.342
$[Ag(CN)_2]^- + e^- \longrightarrow Ag + 2CN^-$	-0.31
$Co^{2+} + 2e^- \longrightarrow Co$	-0.277
$[SnF_6]^{2-} + 4e^- \longrightarrow Sn + 6F^-$	-0.25
$Ni^{2+} + 2e^- \longrightarrow Ni$	-0.250
$Sn^{2+} + 2e^- \longrightarrow Sn$	-0.136
$CrO_4^{2-} + 4H_2O + 3e^- \longrightarrow Cr(OH)_3 + 5OH^-$	-0.13
$Pb^{2+} + 2e^- \longrightarrow Pb$	-0.126
$MnO_2 + 2H_2O + 2e^- \longrightarrow Mn(OH)_2 + 2OH^-$	-0.05
$[HgI_4]^{2-} + 2e^- \longrightarrow Hg + 4I^-$	-0.04
$2H^+ + 2e^- \longrightarrow H_2$	0.00
$NO_3^- + H_2O + 2e^- \longrightarrow NO_2^- + 2OH^-$	$+0.01$
$[Ag(S_2O_3)_2]^{3-} + e^- \longrightarrow Ag^+ + 2S_2O_3^{2-}$	$+0.01$
$[Co(NH_3)_6]^{3+} + e^- \longrightarrow [Co(NH_3)_6]^{2+}$	$+0.1$

Appendix I

Standard Electrode (Reduction) Potentials (continued)

Half-reactions	$E°$, volts
$S + 2H^+ + 2e^- \longrightarrow H_2S$	$+0.141$
$Sn^{4+} + 2e^- \longrightarrow Sn^{2+}$	$+0.15$
$Co(OH)_3 + e^- \longrightarrow Co(OH)_2 + OH^-$	$+0.17$
$[HgBr_4]^{2-} + 2e^- \longrightarrow Hg + 4Br^-$	$+0.21$
$AgCl + e^- \longrightarrow Ag + Cl^-$	$+0.222$
$Hg_2Cl_2 + 2e^- \longrightarrow 2Hg + 2Cl^-$	$+0.27$
$ClO_3^- + H_2O + 2e^- \longrightarrow ClO_2^- + 2OH^-$	$+0.33$
$Cu^{2+} + 2e^- \longrightarrow Cu$	$+0.337$
$[Fe(CN)_6]^{3-} + e^- \longrightarrow [Fe(CN)_6]^{4-}$	$+0.36$
$[Ag(NH_3)_2]^+ + e^- \longrightarrow Ag + 2NH_3$	$+0.373$
$O_2 + 2H_2O + 4e^- \longrightarrow 4OH^-$	$+0.401$
$[RhCl_6]^{3-} + 3e^- \longrightarrow Rh + 6Cl^-$	$+0.44$
$Ag_2CrO_4 + 2e^- \longrightarrow 2Ag + CrO_4^{2-}$	$+0.446$
$NiO_2 + 2H_2O + 2e^- \longrightarrow Ni(OH)_2 + 2OH^-$	$+0.49$
$Cu^+ + e^- \longrightarrow Cu$	$+0.521$
$TeO_2 + 4H^+ + 4e^- \longrightarrow Te + 2H_2O$	$+0.529$
$I_2 + 2e^- \longrightarrow 2I^-$	$+0.5355$
$[PtBr_4]^{2-} + 2e^- \longrightarrow Pt + 4Br^-$	$+0.58$
$MnO_4^- + 2H_2O + 3e^- \longrightarrow MnO_2 + 4OH^-$	$+0.588$
$[PdCl_4]^{2-} + 2e^- \longrightarrow Pd + 4Cl^-$	$+0.62$
$ClO_2^- + H_2O + 2e^- \longrightarrow ClO^- + 2OH^-$	$+0.66$
$[PtCl_6]^{2-} + 2e^- \longrightarrow [PtCl_4]^{2-} + 2Cl^-$	$+0.68$
$O_2 + 2H^+ + 2e^- \longrightarrow H_2O_2$	$+0.682$
$[PtCl_4]^{2-} + 2e^- \longrightarrow Pt + 4Cl^-$	$+0.73$
$Fe^{3+} + e^- \longrightarrow Fe^{2+}$	$+0.771$
$Hg_2^{2+} + 2e^- \longrightarrow 2Hg$	$+0.789$
$Ag^+ + e^- \longrightarrow Ag$	$+0.7991$
$Hg^{2+} + 2e^- \longrightarrow Hg$	$+0.854$
$HO_2^- + H_2O + 2e^- \longrightarrow 3OH^-$	$+0.88$
$ClO^- + H_2O + 2e^- \longrightarrow Cl^- + 2OH^-$	$+0.89$
$2Hg^{2+} + 2e^- \longrightarrow Hg_2^{2+}$	$+0.920$
$NO_3^- + 3H^+ + 2e^- \longrightarrow HNO_2 + H_2O$	$+0.94$
$NO_3^- + 4H^+ + 3e^- \longrightarrow NO + H_2O$	$+0.96$
$Pd^{2+} + 2e^- \longrightarrow Pd$	$+0.987$
$Br_2(l) + 2e^- \longrightarrow 2Br^-$	$+1.0652$
$ClO_4^- + 2H^+ + 2e^- \longrightarrow ClO_3^- + H_2O$	$+1.19$
$Pt^{2+} + 2e^- \longrightarrow Pt$	$ca.\ +1.2$
$ClO_3^- + 3H^+ + 2e^- \longrightarrow HClO_2 + H_2O$	$+1.21$
$O_2 + 4H^+ + 4e^- \longrightarrow 2H_2O$	$+1.23$
$MnO_2 + 4H^+ + 2e^- \longrightarrow Mn^{2+} + 2H_2O$	$+1.23$
$Cr_2O_7^{2-} + 14H^+ + 6e^- \longrightarrow 2Cr^{3+} + 7H_2O$	$+1.33$
$Cl_2 + 2e^- \longrightarrow 2Cl^-$	$+1.3595$
$HClO + H^+ + 2e^- \longrightarrow Cl^- + H_2O$	$+1.49$
$Au^{3+} + 3e^- \longrightarrow Au$	$+1.50$
$MnO_4^- + 8H^+ + 5e^- \longrightarrow Mn^{2+} + 4H_2O$	$+1.51$

Appendix I

Standard Electrode (Reduction) Potentials (continued)

Half-reactions	$E°$, volts
$Ce^{4+} + e^- \longrightarrow Ce^{3+}$	$+1.61$
$HClO + H^+ + e^- \longrightarrow \frac{1}{2}Cl_2 + H_2O$	$+1.63$
$HClO_2 + 2H^+ + 2e^- \longrightarrow HClO + H_2O$	$+1.64$
$Au^+ + e^- \longrightarrow Au$	$ca.\ +1.68$
$NiO_2 + 4H^+ + 2e^- \longrightarrow Ni^{2+} + 2H_2O$	$+1.68$
$PbO_2 + SO_4^{2-} + 4H^+ + 2e^- \longrightarrow PbSO_4 + 2H_2O$	$+1.685$
$H_2O_2 + 2H^+ + 2e^- \longrightarrow 2H_2O$	$+1.77$
$Co^{3+} + e^- \longrightarrow Co^{2+}$	$+1.82$
$F_2 + 2e^- \longrightarrow 2F^-$	$+2.87$

Appendix J

Standard Molar Enthalpies of Formation, Standard Molar Free Energies of Formation, and Absolute Standard Entropies [298.15° K (25° C), 1 atm]

Substance	$\Delta H^°_{f_{298.15}}$, kcal/mole	$\Delta G^°_{f_{298.15}}$, kcal/mole	$S^°_{298.15}$, cal/°K·mole
Aluminum			
$Al(s)$	0	0	6.77
$Al(g)$	78.0	68.3	39.30
$Al_2O_3(s)$	−400.5	−378.2	12.17
$AlF_3(s)$	−359.5	−340.6	15.88
$AlCl_3(s)$	−168.3	−150.3	26.45
$AlCl_3 \cdot 6H_2O(s)$	−643.3	—	—
$Al_2S_3(s)$	−173.	—	—
$Al_2(SO_4)_3(s)$	−822.38	−740.95	57.2
Antimony			
$Sb(s)$	0	0	10.92
$Sb(g)$	62.7	53.1	43.06
$Sb_4O_6(s)$	−344.3	−303.1	52.8
$SbCl_3(g)$	−75.0	−72.0	80.71
$SbCl_5(g)$	−94.25	−79.91	96.04
$Sb_2S_3(s)$	−41.8	−41.5	43.5
$SbCl_3(s)$	−91.34	−77.37	44.0
$SbOCl(s)$	−89.4	—	—
Arsenic			
$As(s)$	0	0	8.4
$As(g)$	72.3	62.4	41.61
$As_4(g)$	34.4	22.1	75.0
$As_4O_6(s)$	−314.04	−275.46	51.2
$As_2O_5(s)$	−221.05	−187.0	25.2
$AsCl_3(g)$	−61.80	−58.77	78.17
$As_2S_3(s)$	−40.4	−40.3	39.1
$AsH_3(g)$	15.88	16.47	53.22
$H_3AsO_4(s)$	−216.6	—	—

Appendix J

Standard Molar Enthalpies of Formation, Standard Molar Free Energies of Formation, and Absolute Standard Entropies [298.15° K (25° C), 1 atm] (continued)

Substance	$\Delta H^\circ_{f298.15}$, kcal/mole	$\Delta G^\circ_{f298.15}$, kcal/mole	$S^\circ_{298.15}$, cal/°K · mole
Barium			
Ba(s)	0	0	16.0
Ba(g)	41.96	34.60	40.70
BaO(s)	−133.4	−126.3	16.8
BaCl$_2$(s)	−205.56	−193.8	30.0
BaSO$_4$(s)	−350.2	−323.4	31.6
Beryllium			
Be(s)	0	0	2.28
Be(g)	76.63	67.60	32.545
BeO(s)	−146.0	−139.0	3.37
Bismuth			
Bi(s)	0	0	13.56
Bi(g)	49.5	40.2	44.669
Bi$_2$O$_3$(s)	−137.16	−118.0	36.2
BiCl$_3$(s)	−90.6	−75.3	42.3
Bi$_2$S$_3$(s)	−34.2	−33.6	47.9
Boron			
B(s)	0	0	1.40
B(g)	134.5	124.0	36.65
B$_2$O$_3$(s)	−304.20	−285.30	12.90
B$_2$H$_6$(g)	8.5	20.7	55.45
B(OH)$_3$(s)	−261.55	−231.60	21.23
BF$_3$(g)	−271.75	−267.77	60.71
BCl$_3$(g)	−96.50	−92.91	69.31
B$_3$N$_3$H$_6$(l)	−129.3	−93.88	47.7
HBO$_2$(s)	−189.83	−172.9	9.
Bromine			
Br$_2$(l)	0	0	36.384
Br$_2$(g)	7.387	0.751	58.641
Br(g)	26.741	19.701	41.805
BrF$_3$(g)	−61.09	−54.84	69.89
HBr(g)	−8.70	−12.77	47.463
Cadmium			
Cd(s)	0	0	12.37
Cd(g)	26.77	18.51	40.066
CdO(s)	−61.7	−54.6	13.1
CdCl$_2$(s)	−93.57	−82.21	27.55
CdSO$_4$(s)	−223.06	−196.65	29.407
CdS(s)	−38.7	−37.4	15.5

Appendix J

Standard Molar Enthalpies of Formation, Standard Molar Free Energies of Formation, and Absolute Standard Entropies [298.15° K (25° C), 1 atm] (continued)

Substance	$\Delta H^{\circ}_{f298.15}$, kcal/mole	$\Delta G^{\circ}_{f298.15}$, kcal/mole	$S^{\circ}_{298.15}$, cal/°K · mole
Calcium			
Ca(s)	0	0	9.95
Ca(g)	46.04	37.98	36.993
CaO(s)	−151.9	−144.4	9.5
Ca(OH)$_2$(s)	−235.80	−214.33	18.2
CaSO$_4$(s)	−342.42	−315.56	25.5
CaSO$_4$·2H$_2$O(s)	−483.06	−429.19	46.36
CaCO$_3$(s) (calcite)	−288.45	−269.78	22.2
CaSO$_3$·2H$_2$O(s)	−421.2	−374.1	44.0
Carbon			
C(s) (graphite)	0	0	1.372
C(s) (diamond)	0.4533	0.6930	0.568
C(g)	171.291	160.442	37.7597
CO(g)	−26.416	−32.780	47.219
CO$_2$(g)	−94.051	−94.254	51.06
CH$_4$(g)	−17.88	−12.13	44.492
CH$_3$OH(l)	−57.04	−39.76	30.3
CH$_3$OH(g)	−47.96	−38.72	57.29
CCl$_4$(l)	−32.37	−15.60	51.72
CCl$_4$(g)	−24.6	−14.49	74.03
CHCl$_3$(l)	−32.14	−17.62	48.2
CHCl$_3$(g)	−24.65	−16.82	70.65
CS$_2$(l)	21.44	15.60	36.17
CS$_2$(g)	28.05	16.05	56.82
C$_2$H$_2$(g)	54.19	50.00	48.00
C$_2$H$_4$(g)	12.49	16.28	52.45
C$_2$H$_6$(g)	−20.24	−7.86	54.84
CH$_3$COOH(l)	−115.8	−93.2	38.2
CH$_3$COOH(g)	−103.31	−89.4	67.5
C$_2$H$_5$OH(l)	−66.37	−41.80	38.4
C$_2$H$_5$OH(g)	−56.19	−40.29	67.54
C$_3$H$_8$(g)	−24.820	−5.614	64.51
C$_6$H$_6$(g)	19.820	30.989	64.34
C$_6$H$_6$(l)	11.718	29.756	41.30
CH$_2$Cl$_2$(l)	−29.03	−16.09	42.5
CH$_2$Cl$_2$(g)	−22.10	−15.75	64.56
CH$_3$Cl(g)	−19.32	−13.72	56.04
C$_2$H$_5$Cl(l)	−32.63	−14.20	45.60
C$_2$H$_5$Cl(g)	−26.81	−14.45	65.94
C$_2$N$_2$(g)	73.84	71.07	57.79
HCN(l)	26.02	29.86	26.97
HCN(g)	32.3	29.8	48.20

Appendix J

Standard Molar Enthalpies of Formation, Standard Molar Free Energies of Formation, and Absolute Standard Entropies [298.15° K (25° C), 1 atm] (continued)

Substance	$\Delta H^\circ_{f298.15}$, kcal/mole	$\Delta G^\circ_{f298.15}$, kcal/mole	$S^\circ_{298.15}$, cal/°K · mole
Chlorine			
$Cl_2(g)$	0	0	53.288
$Cl(g)$	29.082	25.262	39.457
$ClF(g)$	−13.02	−13.37	52.05
$ClF_3(g)$	−39.0	−29.4	67.28
$Cl_2O(g)$	19.2	23.4	63.60
$Cl_2O_7(l)$	56.9	—	—
$Cl_2O_7(g)$	65.0	—	—
$HCl(g)$	−22.062	−22.777	44.646
$HClO_4(l)$	−9.70	—	—
Chromium			
$Cr(s)$	0	0	5.68
$Cr(g)$	94.8	84.1	41.68
$Cr_2O_3(s)$	−272.4	−252.9	19.4
$CrO_3(s)$	−140.9	—	—
$(NH_4)_2Cr_2O_7(s)$	−431.8	—	—
Cobalt			
$Co(s)$	0	0	7.18
$CoO(s)$	−56.87	−51.20	12.66
$Co_3O_4(s)$	−213.0	−185.0	24.5
$Co(NO_3)_2(s)$	−100.5	—	—
Copper			
$Cu(s)$	0	0	7.923
$Cu(g)$	80.86	71.34	39.74
$CuO(s)$	−37.6	−31.0	10.19
$Cu_2O(s)$	−40.3	−34.9	22.26
$CuS(s)$	−12.7	−12.8	15.9
$Cu_2S(s)$	−19.0	−20.6	28.9
$CuSO_4(s)$	−184.36	−158.2	26.
$Cu(NO_3)_2(s)$	−72.4	—	—
Fluorine			
$F_2(g)$	0	0	48.44
$F(g)$	18.88	14.80	37.917
$F_2O(g)$	−5.2	−1.1	59.11
$HF(g)$	−64.8	−65.3	41.508
Hydrogen			
$H_2(g)$	0	0	31.208
$H(g)$	52.095	48.581	27.391
$H_2O(l)$	−68.315	−56.687	16.71
$H_2O(g)$	−57.796	−54.634	45.104
$H_2O_2(l)$	−44.88	−28.78	26.2
$H_2O_2(g)$	−32.58	−25.24	55.6
$HF(g)$	−64.8	−65.3	41.508
$HCl(g)$	−22.062	−22.777	44.646

Appendix J

Standard Molar Enthalpies of Formation, Standard Molar Free Energies of Formation, and Absolute Standard Entropies [298.15° K (25° C), 1 atm] (continued)

Substance	$\Delta H^\circ_{f298.15}$, kcal/mole	$\Delta G^\circ_{f298.15}$, kcal/mole	$S^\circ_{298.15}$, cal/°K·mole
Hydrogen (cont.)			
HBr(g)	−8.70	−12.77	47.463
HI(g)	6.33	0.41	49.351
H_2S(g)	−4.93	−8.02	49.16
H_2Se(g)	7.1	3.8	52.32
Iodine			
I_2(s)	0	0	27.757
I_2(g)	14.923	4.627	62.28
I(g)	25.535	16.798	43.184
IF(g)	−22.86	−28.32	56.42
ICl(g)	4.25	−1.30	59.140
IBr(g)	9.76	0.89	61.822
IF_7(g)	−225.6	−195.6	82.8
HI(g)	6.33	0.41	49.351
Iron			
Fe(s)	0	0	6.52
Fe(g)	99.5	88.6	43.112
Fe_2O_3(s)	−197.0	−177.4	20.89
Fe_3O_4(s)	−267.3	−242.7	35.0
$Fe(CO)_5$(l)	−185.0	−168.6	80.8
$Fe(CO)_5$(g)	−175.4	−166.65	106.4
$FeSeO_3$(s)	−288.	—	—
FeO(s)	−65.0	—	—
FeAsS(s)	−10.	−12.	29.
$Fe(OH)_2$(s)	−136.0	−116.3	21.
$Fe(OH)_3$(s)	−196.7	−166.5	25.5
FeS(s)	−23.9	−24.0	14.41
Fe_3C(s)	6.0	4.8	25.0
Lead			
Pb(s)	0	0	15.49
Pb(g)	46.6	38.7	41.889
PbO(s) (yellow)	−51.94	−44.91	16.42
PbO(s) (red)	−52.34	−45.16	15.9
$Pb(OH)_2$(s)	−123.3	—	—
PbS(s)	−24.0	−23.6	21.8
$Pb(NO_3)_2$(s)	−108.0	—	—
PbO_2(s)	−66.3	−51.95	16.4
$PbCl_2$(s)	−85.90	−75.08	32.5
Lithium			
Li(s)	0	0	6.70
Li(g)	37.07	29.19	33.143
LiH(s)	−21.61	−16.72	5.9
Li(OH)(s)	−116.45	−106.1	12.0

Appendix J

Standard Molar Enthalpies of Formation, Standard Molar Free Energies of Formation, and Absolute Standard Entropies [298.15° K (25° C), 1 atm] (continued)

Substance	$\Delta H^\circ_{f298.15}$, kcal/mole	$\Delta G^\circ_{f298.15}$, kcal/mole	$S^\circ_{298.15}$, cal/°K · mole
Lithium (cont.)			
LiF(s)	−146.3	−139.6	8.57
$Li_2CO_3(s)$	−290.54	−270.66	21.60
Manganese			
Mn(s)	0	0	7.65
Mn(g)	67.1	57.0	41.49
MnO(s)	−92.07	−86.74	14.27
$MnO_2(s)$	−124.29	−111.18	12.68
$Mn_2O_3(s)$	−229.2	−210.6	26.4
$Mn_3O_4(s)$	−331.7	−306.7	37.2
Mercury			
Hg(l)	0	0	18.17
Hg(g)	14.655	7.613	41.79
HgO(s) (red)	−21.71	−13.995	16.80
HgO(s) (yellow)	−21.62	−13.964	17.0
$HgCl_2(s)$	−53.6	−42.7	34.9
$Hg_2Cl_2(s)$	−63.39	−50.377	46.0
HgS(s) (red)	−13.9	−12.1	19.7
HgS(s) (black)	−12.8	−11.4	21.1
$HgSO_4(s)$	−169.1	—	—
Nitrogen			
$N_2(g)$	0	0	45.77
N(g)	112.979	108.886	36.613
NO(g)	21.57	20.69	50.347
$NO_2(g)$	7.93	12.26	57.35
$N_2O(g)$	19.61	24.90	52.52
$N_2O_3(g)$	20.01	33.32	74.61
$N_2O_4(g)$	2.19	23.38	72.70
$N_2O_5(g)$	2.7	27.5	85.0
$NH_3(g)$	−11.02	−3.94	45.97
$N_2H_4(l)$	12.10	35.67	28.97
$N_2H_4(g)$	22.80	38.07	56.97
$NH_4NO_3(s)$	−87.37	−43.98	36.11
$NH_4Cl(s)$	−75.15	−48.15	22.6
$NH_4Br(s)$	−64.73	−41.9	27.0
$NH_4I(s)$	−48.14	−26.9	28.0
$NH_4NO_2(s)$	−61.3	—	—
$HNO_3(l)$	−41.61	−19.31	37.19
$HNO_3(g)$	−32.28	−17.87	63.63
Oxygen			
$O_2(g)$	0	0	49.003
O(g)	59.553	55.389	38.467
$O_3(g)$	34.1	39.0	57.08

Appendix J

Standard Molar Enthalpies of Formation, Standard Molar Free Energies of Formation, and Absolute Standard Entropies [298.15° K (25° C), 1 atm] (continued)

Substance	$\Delta H^\circ_{f298.15}$, kcal/mole	$\Delta G^\circ_{f298.15}$, kcal/mole	$S^\circ_{298.15}$, cal/°K · mole
Phosphorus			
P(s)	0	0	9.82
P(g)	14.08	5.85	66.89
$P_4(g)$	75.20	66.51	38.978
$PH_3(g)$	1.3	3.2	50.22
$PCl_3(g)$	−68.6	−64.0	74.49
$PCl_5(g)$	−89.6	−72.9	87.11
$P_4O_6(s)$	−392.0	—	—
$P_4O_{10}(s)$	−713.2	−644.8	54.70
$HPO_3(s)$	−226.7	—	—
$H_3PO_2(s)$	−144.5	—	—
$H_3PO_3(s)$	−230.5	—	—
$H_3PO_4(s)$	−305.7	−267.5	26.41
$H_3PO_4(l)$	−302.8	—	—
$H_4P_2O_7(s)$	−535.6	—	—
$POCl_3(l)$	−142.7	−124.5	53.17
$POCl_3(g)$	−133.48	−122.60	77.76
Potassium			
K(s)	0	0	15.2
K(g)	21.51	14.62	38.296
KF(s)	−134.46	−127.42	15.91
KCl(s)	−104.175	−97.592	19.76
Silicon			
Si(s)	0	0	4.50
Si(g)	108.9	98.3	40.12
$SiO_2(s)$	−217.72	−204.75	10.00
$SiH_4(g)$	8.2	13.6	48.88
$H_2SiO_3(s)$	−284.1	−261.1	32.
$H_4SiO_4(s)$	−354.0	−318.6	46.
$SiF_4(g)$	−385.98	−375.88	67.49
$SiCl_4(l)$	−164.2	−148.16	57.3
$SiCl_4(g)$	−157.03	−147.47	79.02
SiC(s)	−15.6	−15.0	3.97
Silver			
Ag(s)	0	0	10.17
Ag(g)	68.01	58.72	41.321
$Ag_2O(s)$	−7.42	−2.68	29.0
AgCl(s)	−30.37	−26.24	23.0
$Ag_2S(s)$	−7.79	−9.72	34.42
Sodium			
Na(s)	0	0	12.2
Na(g)	25.98	18.67	36.715
$Na_2O(s)$	−99.4	−90.0	17.4
NaCl(s)	−98.232	−91.785	17.30

Appendix J

Standard Molar Enthalpies of Formation, Standard Molar Free Energies of Formation, and Absolute Standard Entropies [298.15° K (25° C), 1 atm] (continued)

Substance	$\Delta H^\circ_{f298.15}$, kcal/mole	$\Delta G^\circ_{f298.15}$, kcal/mole	$S^\circ_{298.15}$, cal/°K · mole
Sulfur			
S(s) (rhombic)	0	0	7.60
S(g)	66.636	56.949	40.094
SO_2(g)	−70.944	−71.748	59.30
SO_3(g)	−94.58	−88.69	61.34
H_2S(g)	−4.93	−8.02	49.16
H_2SO_4(l)	−194.548	−164.938	37.501
$H_2S_2O_7$(s)	−304.4	—	—
SF_4(g)	−185.2	−174.8	69.77
SF_6(g)	−289.	−264.2	69.72
SCl_2(l)	−12.	—	—
SCl_2(g)	−4.7	—	—
S_2Cl_2(l)	−14.2	—	—
S_2Cl_2(g)	−4.4	−7.6	79.2
$SOCl_2$(l)	−58.7	—	—
$SOCl_2$(g)	−50.8	−47.4	74.01
SO_2Cl_2(l)	−94.2	—	—
SO_2Cl_2(g)	−87.0	−76.5	74.53
Tin			
Sn(s)	0	0	12.32
Sn(g)	72.2	63.9	40.243
SnO(s)	−68.3	−61.4	13.5
SnO_2(s)	−138.8	−124.2	12.5
$SnCl_4$(l)	−122.2	−105.2	61.8
$SnCl_4$(g)	−112.7	−103.3	87.4
Titanium			
Ti(s)	0	0	7.32
Ti(g)	112.3	101.6	43.066
TiO_2(s)	−225.8	−212.6	12.03
$TiCl_4$(l)	−192.2	−176.2	60.31
$TiCl_4$(g)	−182.4	−173.7	84.8
Tungsten			
W(s)	0	0	7.80
W(g)	203.0	192.9	41.549
WO_3(s)	−201.45	−182.62	18.14
Zinc			
Zn(s)	0	0	9.95
Zn(g)	31.245	22.748	38.450
ZnO(s)	−83.24	−76.08	10.43
$ZnCl_2$(s)	−99.20	−88.296	26.64
ZnS(s)	−49.23	−48.11	13.8

Appendix J

Standard Molar Enthalpies of Formation, Standard Molar Free Energies of Formation, and Absolute Standard Entropies [298.15° K (25° C), 1 atm] (continued)

Substance	$\Delta H^\circ_{f_{298.15}}$, kcal/mole	$\Delta G^\circ_{f_{298.15}}$, kcal/mole	$S^\circ_{298.15}$, cal/°K · mole
Zinc (cont.)			
$ZnSO_4(s)$	−234.9	−209.0	28.6
$ZnCO_3(s)$	−194.26	−174.85	19.7

Complexes

Substance	$\Delta H^\circ_{f_{298.15}}$	$\Delta G^\circ_{f_{298.15}}$	$S^\circ_{298.15}$
$[Co(NH_3)_4(NO_2)_2]NO_3$, cis	−214.8	—	—
$[Co(NH_3)_4(NO_2)_2]NO_3$, trans	−214.2	—	—
$NH_4[Co(NH_3)_2(NO_2)_4]$	−200.2	—	—
$[Co(NH_3)_5NO_2](NO_3)_2$	−260.2	−100.0	83
$[Co(NH_3)_6][Co(NH_3)_2(NO_2)_4]_3$	−653.2	—	—
$[Co(NH_3)_4Cl_2]Cl$, cis	−238.3	—	—
$[Co(NH_3)_4Cl_2]Cl$, trans	−238.9	—	—
$[Co(en)_2(NO_2)_2]NO_3$, cis	−164.8	—	—
$[Co(en)_2Cl_2]Cl$, cis	−162.8	—	—
$[Co(en)_2Cl_2]Cl$, trans	−161.9	—	—
$[Co(en)_3](ClO_4)_3$	−182.3	—	—
$[Co(en)_3]Br_2$	−142.4	—	—
$[Co(en)_3]I_2$	−113.6	—	—
$[Co(en)_3]I_3$	−124.1	—	—
$[Co(NH_3)_6](ClO_4)_3$	−247.3	−54.3	152
$[Co(NH_3)_5NO_2](NO_3)_2$	−260.2	−100.0	83
$[Co(NH_3)_6](NO_3)_3$	−306.4	−126.8	112
$[Co(NH_3)_5Cl]Cl_2$	−243.1	−139.3	87.5
$[Pt(NH_3)_4]Cl_2$	−174.0	—	—
$[Ni(NH_3)_6]Cl_2$	−237.6	—	—
$[Ni(NH_3)_6]Br_2$	−220.8	—	—
$[Ni(NH_3)_6]I_2$	−193.2	—	—

Appendix K

Composition of Commercial Acids and Bases

Acid or Base	Specific Gravity	Percentage by Weight	Molarity	Normality
Hydrochloric	1.19	38	12.4	12.4
Nitric	1.42	70	15.8	15.8
Sulfuric	1.84	95	17.8	35.6
Acetic	1.05	99	17.3	17.3
Aqueous ammonia	0.90	28	14.8	14.8

Appendix L

Half-Life Times for Several Radioactive Isotopes

(Symbol in parentheses indicates type of emission; *E.C.* = K-electron capture, *S.F.* = spontaneous fission; *y* = years, *d* = days, *h* = hours, *m* = minutes, s = seconds.)

$^{14}_{6}$C	5770 *y*	(β^-)	$^{226}_{88}$Ra	1590 *y*	(α)
$^{13}_{7}$N	10.0 *m*	(β^+)	$^{228}_{88}$Ra	6.7 *y*	(β^-)
$^{24}_{11}$Na	15.0 *h*	(β^-)	$^{228}_{89}$Ac	6.13 *h*	(β^-)
$^{32}_{15}$P	14.3 *d*	(β^-)	$^{228}_{90}$Th	1.90 *y*	(α)
$^{40}_{19}$K	1.3×10^9 *y*	(β^- or *E.C.*)	$^{232}_{90}$Th	1.39×10^{10} *y*	(α, β^-, or *S.F.*)
$^{60}_{27}$Co	5.2 *y*	(β^-)	$^{233}_{90}$Th	23 *m*	(β^-)
$^{87}_{37}$Rb	4.7×10^{10} *y*	(β^-)	$^{234}_{90}$Th	24.1 *d*	(β^-)
$^{90}_{38}$Sr	28 *y*	(β^-)	$^{223}_{91}$Pa	27 *d*	(β^-)
$^{115}_{49}$In	6×10^{14} *y*	(β^-)	$^{233}_{92}$U	1.62×10^5 *y*	(α)
$^{131}_{53}$I	8.05 *d*	(β^-)	$^{234}_{92}$U	2.4×10^5 *y*	(α or *S.F.*)
$^{142}_{58}$Ce	5×10^{15} *y*	(α)	$^{235}_{92}$U	7.3×10^8 *y*	(α or *S.F.*)
$^{198}_{79}$Au	64.8 *h*	(β^-)	$^{238}_{92}$U	4.5×10^9 *y*	(α or *S.F.*)
$^{208}_{81}$Tl	3.1 *m*	(β^-)	$^{239}_{92}$U	23 *m*	(β^-)
$^{210}_{82}$Pb	21 *y*	(β^-)	$^{239}_{93}$Np	2.3 *d*	(β^-)
$^{212}_{82}$Pb	10.6 *h*	(β^-)	$^{239}_{94}$Pu	24,360 *y*	(α or *S.F.*)
$^{214}_{82}$Pb	26.8 *m*	(β^-)	$^{240}_{94}$Pu	6.58×10^3 *y*	(α or *S.F.*)
$^{206}_{83}$Bi	6.3 *d*	(β^+ or *E.C.*)	$^{241}_{94}$Pu	13 *y*	(α or β^-)
$^{210}_{83}$Bi	5.0 *d*	(β^-)	$^{241}_{95}$Am	458 *y*	(α)
$^{212}_{83}$Bi	60.5 *m*	(α or β^-)	$^{242}_{96}$Cm	163 *d*	(α or *S.F.*)
$^{207}_{84}$Po	5.7 *h*	(α, β^+, or *E.C.*)	$^{243}_{97}$Bk	4.5 *h*	(α or *E.C.*)
$^{210}_{84}$Po	138.4 *d*	(α)	$^{245}_{98}$Cf	350 *d*	(α or *E.C.*)
$^{212}_{84}$Po	3×10^{-7} *s*	(α)	$^{253}_{99}$Es	20.0 *d*	(α or *S.F.*)
$^{216}_{84}$Po	0.16 *s*	(α)	$^{254}_{100}$Fm	3.24 *h*	(*S.F.*)
$^{218}_{84}$Po	3.0 *m*	(α or β^-)	$^{255}_{100}$Fm	22 *h*	(α)
$^{215}_{85}$At	10^{-4} *s*	(α)	$^{256}_{101}$Md	1.5 *h*	(*E.C.*)
$^{218}_{85}$At	1.3 *s*	(α)	$^{254}_{102}$No	3 *s*	(α)
$^{220}_{86}$Rn	54.5 *s*	(α)	$^{257}_{103}$Lr	8 *s*	(α)
$^{222}_{86}$Rn	3.82 *d*	(α)	$^{263}_{106}$(106)	0.9 *s*	(α)
$^{224}_{88}$Ra	3.64 *d*	(α)			

Appendix M

Apparatus for Qualitative Analysis (one student)

50 Reagent bottles, dropper type, 10 ml
 1 Rack for 50 reagent bottles
 1 Test tube block
 1 Flask, Florence, 250 ml
 2 Beakers, 250 ml
 2 Beakers, 50 ml
 2 Beakers, 20 ml
 1 Graduate, 10 ml
 1 Casserole, 15 ml
 6 Centrifuge tubes
 6 Test tubes, 65 × 10 mm

 6 Test tubes, 75 × 10 mm
 1 Metal rack for a water bath
 1 Micro burner
 1 Bunsen burner
 2 Watch glasses, 2.5 cm
 1 File
 1 Box of labels
 1 Bottle of litmus, red
 1 Bottle of litmus, blue
 1 Test tube holder, small
 1 Test tube brush, small, tapered

Apparatus for Qualitative Analysis (one student) (continued)

1 Test tube brush, small, not tapered
1 Wire, platinum, 2 inches
1 Wire gauze
1 Ring stand (with ring), small
1 Spatula, micro (Monel metal)
1 Forceps
1 Two-hole rubber stopper to fit
 250-ml flask
1 Wing top

1 Box matches
1 Towel
1 Box of detergent, small
6 Medicine droppers, 1 ml
6 Capillary syringes
100-cm Glass tubing, 6 mm
100-cm Glass rod, 3 mm
1 Cobalt glass

Appendix N

Reagents for Cation Analysis

GROUP I   Reagents

Hydrochloric acid, HCl, 6 M
Nitric acid, HNO_3, 4 M
Aqueous ammonia, $NH_3 + H_2O$, 4 M
* Potassium chromate, K_2CrO_4, 1 M
 (* Fill reagent bottles only about one-fourth full of starred reagents.)

GROUP II   Reagents (in addition to those listed for Group I)

Aqueous ammonia, $NH_3 + H_2O$, 6 M
Hydrochloric acid, HCl, 1.0 M
Thioacetamide, CH_3CSNH_2, 5 per cent solution
Sodium hydroxide, NaOH, 4 M
Ammonium nitrate, NH_4NO_3, 1 M
Sulfuric acid, H_2SO_4, 4 M
Acetic acid, HOAc, 1 M
Oxalic acid, $H_2C_2O_4$, 1 M
Ammonium acetate, NH_4OAc, 1 M
Aqueous ammonia, $NH_3 + H_2O$, 15 M
* Potassium hexacyanoferrate(II), $K_4Fe(CN)_6$, 0.1 M
Potassium cyanide, KCN, 1 M
Hydrogen peroxide, H_2O_2, 3 per cent solution
* Magnesia mixture: Dissolve 50 g of $MgCl_2 \cdot 6H_2O$ and 70 g of NH_4Cl in 400 ml
 of water. Add 100 ml of 15 M aqueous ammonia and dilute to 1 liter. Filter.
* Aluminum wire, Al
Mercury(II) chloride, $HgCl_2$, 0.2 M
Sodium hypochlorite, NaOCl, 5 per cent solution

Reagents for Cation Analysis (continued)

GROUP III   Reagents (in addition to those listed for Groups I and II)

Hydrochloric acid, HCl, 12 M
Hydrochloric acid, HCl, 1 M
Nitric acid, HNO_3, 14 M
*Dimethylglyoxime, 1 per cent solution, dissolve 10 g in 1 liter of alcohol
Acetic acid, HOAc, 4 M
*Ammonium thiocyanate, NH_4CNS, solid
Acetone, $(CH_3)_2CO$
*Sodium fluoride, NaF, 1 M
*Potassium nitrite, KNO_2, solid
*Sodium nitrite, $NaNO_2$, 1 M
*Sodium bismuthate, $NaBiO_3$, solid
*Aluminum Reagent: Dissolve 1 g of the ammonium salt of aurin-tricarboxylic
    acid in 1 liter of water
Lead acetate, $Pb(OAc)_2$, 0.1 M
Ammonium carbonate, $(NH_4)_2CO_3$, 1 M

GROUP IV   Reagents (in addition to those listed for Groups I-III)

Ethyl alcohol, C_2H_5OH
*Ammonium oxalate, $(NH_4)_2C_2O_4$, 0.4 M

GROUP V   Reagents (in addition to those listed for Groups I-IV)

*Ammonium sulfate, $(NH_4)_2SO_4$, 1 M
*Sodium Reagent: Mix 30 g of $UO_2(OAc)_2 \cdot 2H_2O$ with 80 g of $Zn(OAc)_2 \cdot 2H_2O$
    and 10 ml of glacial acetic acid. Dilute the solution to 250 ml, let it stand
    for several hours, and filter. Use the clear solution.
*Sodium hexanitrocobaltate(III), $Na_3Co(NO_2)_6$: Dissolve 30 g of $NaNO_2$ in 97 ml
    of water; add 3 ml of glacial HOAc and 3.3 g of $Co(NO_3)_2 \cdot 6H_2O$. Filter and
    use the clear solution.
*Disodium hydrogen phosphate, Na_2HPO_4, 1 M
*Magnesium Reagent, p-nitro-benzene-azo-alpha-naphthol: Dissolve 0.25 g of
    this reagent and 2.5 g of NaOH in sufficient water to make 250 ml of solution.

Reagents for Anion Analysis (in addition to those listed for the cation analysis)

*Sulfuric acid, H_2SO_4, 1.5 M
Sodium carbonate, Na_2CO_3, 1.5 M
*Manganese(II) chloride, $MnCl_2$: Saturate 12 M HCl with $MnCl_2$.
*Potassium hexacyanoferrate(III), $K_3Fe(CN)_6$, a freshly prepared saturated solu-
    tion

Appendix N

Reagents for Anion Analysis (continued)

* Iron(III) chloride, $FeCl_3$, 0.1 M
  Calcium chloride, $CaCl_2$, 0.1 M
  Barium chloride, $BaCl_2$, 0.1 M
* Cobalt(II) nitrate, $Co(NO_3)_2$, 1 M
* Silver nitrate, $AgNO_3$, 0.1 M
* Bromine water, $Br_2 + H_2O$, saturated solution
* Barium hydroxide, $Ba(OH)_2$, saturated solution
* Iron(II) sulfate, $FeSO_4$, 0.1 M
* Sulfuric acid, H_2SO_4, 18 M
* Silver sulfate, Ag_2SO_4, solid (nitrate free)
* Potassium permanganate, $KMnO_4$, 0.002 M
* Silica, SiO_2, powdered
* Methyl alcohol, CH_3OH
* Iron(III) nitrate, $Fe(NO_3)_3$, 0.1 M
* Carbon tetrachloride, CCl_4
* Potassium permanganate, $KMnO_4$, 0.1 M
* Lead acetate paper, filter paper moist with 0.1 M $Pb(OAc)_2$

Appendix O

Preparation of Solutions of Cations

O.1 Stock Solutions

It is recommended that all stock solutions contain the cations in question at a concentration of 50 mg per ml. These stock solutions may be prepared by grinding to a powder the weight of salt given below and adding enough water (or acid if specified) to make the volume 100 ml. Solution of the salts may be hastened by heating.

O.2 Known and Unknown Solutions

To prepare known or unknown solutions of cations, mix 20 ml of the stock solutions (40 ml of $AsCl_3$) of the cations desired and dilute the solution to 100 ml with water. This solution will contain 10 mg of cations per ml. This procedure allows for a maximum of five cations at a concentration of 10 mg of cations per ml. Dilution to 200 ml will allow a maximum of ten cations at a concentration of 5 mg per ml. Give each student about 1 ml of solution.

Stock Solutions of Cations (50 mg of Cations per ml)

Group	Ion	Formula of Salt	Grams per 100 ml of Solution
I	Ag^+	$AgNO_3$	8.0
	Pb^{2+}	$Pb(NO_3)_2$	8.0
	Hg_2^{2+}	$Hg_2(NO_3)_2$	7.0 (dissolve in 0.6 M HNO_3)

Stock Solutions of Cations (50 mg of Cations per ml) (continued)

Group	Ion	Formula of Salt	Grams per 100 ml of Solution
II	Pb^{2+}	$Pb(NO_3)_2$	8.0
	Bi^{3+}	$Bi(NO_3)_3 \cdot 5H_2O$	11.5 (dissolve in 3 M HNO_3)
	Cu^{2+}	$Cu(NO_3)_2 \cdot 3H_2O$	19.0
	Cd^{2+}	$Cd(NO_3)_2 \cdot 4H_2O$	13.8
	Hg^{2+}	$HgCl_2$	6.8
	As^{3+}	As_4O_6	3.3 (heat in 50 ml of 12 M HCl, then add 50 ml of water)
	Sb^{3+}	$SbCl_3$	9.5 (dissolve in 6 M HCl, and dilute with 2 M HCl)
	Sn^{2+}	$SnCl_2 \cdot 2H_2O$	9.5 (dissolve in 50 ml of 12 M HCl. Dilute to 100 ml with water. Add a piece of tin metal)
	Sn^{4+}	$SnCl_4 \cdot 3H_2O$	13.3 (dissolve in 6 M HCl)
III	Co^{2+}	$Co(NO_3)_2 \cdot 6H_2O$	24.7
	Ni^{2+}	$Ni(NO_3)_2 \cdot 6H_2O$	24.8
	Mn^{2+}	$Mn(NO_3)_2 \cdot 6H_2O$	26.2
	Fe^{3+}	$Fe(NO_3)_3 \cdot 9H_2O$	36.2
	Al^{3+}	$Al(NO_3)_3 \cdot 9H_2O$	69.5
	Cr^{3+}	$Cr(NO_3)_3$	23.0
	Zn^{2+}	$Zn(NO_3)_2$	14.5
IV	Ba^{2+}	$BaCl_2 \cdot 2H_2O$	8.9
	Sr^{2+}	$Sr(NO_3)_2$	12.0
	Ca^{2+}	$Ca(NO_3)_2 \cdot 4H_2O$	29.5
V	Mg^{2+}	$Mg(NO_3)_2 \cdot 6H_2O$	52.8
	NH_4^+	NH_4NO_3	22.2
	Na^+	$NaNO_3$	18.5
	K^+	KNO_3	13.0

Index

✗ cat, list of names for 201

PERIODS

IA

| 1 | ○ 0.3 H |

+1
? •   IIA

| 2 | ○ Li 1.52 | ○ Be 1.11 |

+1
0.60 ○   +2
0.31 ○

TRANSITION METALS

| 3 | ○ Na 1.86 | ○ Mg 1.60 |

VII

+1
0.95 ○   +2
0.65 ○   IIIB   IVB   VB   VIB   VIIB

| 4 | ○ K 2.31 | ○ Ca 1.97 | ○ Sc 1.60 | ○ Ti 1.46 | ○ V 1.31 | ○ Cr 1.25 | ○ Mn 1.29 | ○ Fe 1.26 | ○ 1. |

+1
1.33 ○   +2
0.99 ○   +3
0.81 ○   +4
0.64 ○   +5
0.4 ○   +6
0.52 ○   +2
0.80 ○   +2
0.75 ○   0.7

| 5 | ○ Rb 2.44 | ○ Sr 2.15 | ○ Y 1.80 | ○ Zr 1.57 | ○ Nb 1.43 | ○ Mo 1.36 | ○ Tc 1.3 | ○ Ru 1.33 | ○ 1. |

+1
1.48 ○   +2
1.13 ○   +3
0.93 ○   +4
0.87 ○   +5
0.69 ○   +6
0.62 ○

| 6 | ○ Cs 2.62 | ○ Ba 2.17 | ○ La 1.88 | ○ Hf 1.57 | ○ Ta 1.43 | ○ W 1.37 | ○ Re 1.37 | ○ Os 1.34 | ○ 1. |

+1
1.69 ○   +2
1.35 ○   +3
1.15 ○   +4
0.84 ○   +5
0.68 ○   +6
0.68 ○

| 7 | ○ Fr 2.7 | ○ Ra 2.20 | ○ Ac 2.0 |

+2
1.52 ○   +3
1.11 ○